MATRIX PARTIAL ORDERS, SHORTED OPERATORS AND APPLICATIONS

SERIES IN ALGEBRA

Series Editors: Derek J S Robinson *(University of Illinois at Urbana-Champaign, USA)*
John M Howie *(University of St. Andrews, UK)*
W Douglas Munn *(University of Glasgow, UK)*

Published

Vol. 1 Infinite Groups and Group Rings
edited by J M Corson, M R Dixon, M J Evans and F D Röhl

Vol. 2 Sylow Theory, Formations and Fitting Classes in Locally Finite Groups
by Martyn R Dixon

Vol. 3 Finite Semigroups and Universal Algebra
by Jorge Almeida

Vol. 4 Generalizations of Steinberg Groups
by T A Fournelle and K W Weston

Vol. 5 Semirings: Algebraic Theory and Applications in Computer Science
by U Hebisch and H J Weinert

Vol. 6 Semigroups of Matrices
by Jan Okninski

Vol. 7 Partially Ordered Groups
by A M W Glass

Vol. 8 Groups with Prescribed Quotient Groups and Associated Module Theory
by L Kurdachenko, J Otal and I Subbotin

Vol. 9 Ring Constructions and Applications
by Andrei V Kelarev

Vol. 10 Matrix Partial Orders, Shorted Operators and Applications
*by Sujit Kumar Mitra**, *P Bhimasankaram and Saroj B Malik*

*Deceased

SERIES IN ALGEBRA VOLUME 10

MATRIX PARTIAL ORDERS, SHORTED OPERATORS AND APPLICATIONS

Sujit Kumar Mitra
Indian Statistical Institute, India

P Bhimasankaram
University of Hyderabad, India

Saroj B Malik
Hindu College, University of Delhi, India

World Scientific

NEW JERSEY • LONDON • SINGAPORE • BEIJING • SHANGHAI • HONG KONG • TAIPEI • CHENNAI

Published by

World Scientific Publishing Co. Pte. Ltd.
5 Toh Tuck Link, Singapore 596224
USA office: 27 Warren Street, Suite 401-402, Hackensack, NJ 07601
UK office: 57 Shelton Street, Covent Garden, London WC2H 9HE

British Library Cataloguing-in-Publication Data
A catalogue record for this book is available from the British Library.

Series in Algebra — Vol. 10
MATRIX PARTIAL ORDERS, SHORTED OPERATORS AND APPLICATIONS

ISBN-13 978-981-283-844-5
ISBN-10 981-283-844-9

Printed in Singapore.

To
Asha Latha Mitra, Sunil Kumar Mitra
and
Sheila Mitra

\- Parents and Wife of Sujit Kumar Mitra

Preface

Matrix orders have fascinated mathematicians and applied scientists alike for many years. Several matrix orders have been developed during the past four decades by researchers working in Linear Algebra. Some of them are pre-orders such as the space pre-order and the Drazin pre-order, while others are partial orders like the minus order, the sharp order and the star order. These developments have grown symbiotically with advances in other areas such as Statistics and Matrix Generalized Inverses.

Two closely connected concepts - parallel sums and shorted operators for non-negative definite matrices - play a major role in the study of electrical networks. Both of these share nice relationships with matrix partial orders as also between themselves. Several extensions of these concepts have been developed in the recent past by researchers working in these and related areas.

There are many research articles in the areas of matrix orders and shorted operators that are scattered in various journals. This is the first full length monograph on these topics. The aim of this monograph is to present the developments in the fields of matrix orders and shorted operators for finite matrices in a unified way and illustrate them with suitable applications in Generalized Inverses, Statistics and Electrical Networks. In the process of this compilation, many new results have evolved. Virtually every chapter in the monograph contains results unpublished hitherto. In fact, Chapter 13 on partial orders of modified matrices comprises entirely of new material.

We believe that dissection of matrices through matrix decompositions helps in clear understanding of the anatomy of various matrix orders in a transparent manner. We employ simultaneous decompositions such as the simultaneous normal form, the simultaneous singular value decomposition and the generalized singular value decomposition extensively in developing the properties of matrix orders. Accordingly, the reader will find new and intuitive proofs to several known results in this monograph. We also pose some open problems which should be of interest to researchers in these topics. There are a number of exercises at the end of virtually every chapter. These are expected to serve the dual purpose of helping in the understanding of the topics covered in the text and of introducing other related results that have not been included in the main text.

This monograph is aimed at (i) graduate students and researchers in Matrix Theory and (ii) researchers in Statistics and Electrical Engineering who may use these concepts and results as tools in their work. The monograph can be used as a text for a one-semester graduate course in advanced topics in Matrix Theory and it can also serve as self-study text for those who have knowledge in basic Linear Algebra, perhaps at the level of [Rao and Bhimasankaram (2000)].

We deliberately avoided the study of majorization as there is an excellent book [Marshall and Olkin (1979)] on this topic. In this monograph, we consider matrix orders and shorted operators for finite matrices over a field. We do not consider matrices over more general algebraic structures or operators over more general spaces. There has been a good deal of work in these directions - see [Drazin (1958)], [Hartwig (1979)], [Hartwig (1980)], [Jain, Srivastava, Blackwood and Prasad (2009)], [Morley and William (1990a,b)] and [Mitch (1986)], to name a few. We have not included them here in order to keep the monograph focused and less bulky. We have also not gone into order preserving and order reversing transformations (for the minus, star, sharp and one-sided orders) and application of the sharp order in the analysis of Markov chains. These are exciting topics of research that may develop over the next few years.

This monograph was conceived by the first author, S. K. Mitra who was a pioneer in the development of theory of matrix partial orders and shorted operators. He introduced the sharp order, one-sided orders and a unified theory of matrix partial orders. He developed several approaches to the shorted operators along with applications in Generalized Inverses and Statistics.

The first author who is no more, has been the mentor and the main source of inspiration for the other two authors of the monograph. However, we, the second and third authors are responsible for any errors in the monograph.

July, 2009 *P. Bhimasankaram* and *Saroj B. Malik*

Acknowledgements

We are highly thankful to the SQC & OR Unit, Indian Statistical Institute, Hyderabad, the Department of Mathematics and Statistics, University of Hyderabad, Hyderabad, the Center for Analytical Finance, Indian School of Business, Hyderabad and Hindu College, Delhi University, Delhi for providing excellent facilities during different stages of the preparation of this monograph.

We are deeply indebted to Professor Debasis Sengupta for his constructive comments on parts of the manuscript in its different stages of preparation which led to significant improvement in the presentation. We are also thankful to Professors Probal Chaudhuri, Thomas Mathew and PSSNV Prasada Rao for useful discussions. We thank Professors T. Amarnath, R. Tandon, V. Suresh and Shri ALN Murthy for their encouragement during the preparation of the manuscript.

P. Bhimasankaram records his sincere thanks and appreciation for the encouragement and inspiration received from Professor Sankar De during the preparation of the manuscript. His family members, Amrita, Chandana, Shilpa, Chandu and Vijaya had to miss his company even during the weekends for years together during the preparation of the manuscript. But for the understanding, patience and cooperation received from them, he could never have completed the task satisfactorily. He expresses his deep sense of loving appreciation to them.

Saroj Malik expresses her sincere thanks to the Department of Mathematics and Statistics, University of Hyderabad for providing local hospitality during her visits to University of Hyderabad during the preparation of the manuscript (under the SAP-UGC project). She also wishes to thank the National Academy of Sciences, Delhi for partial support during one such visit. While working on the monograph, she had several useful discussions with Professor Ajeet I. Singh, Professor Emeritus, Stat.-Math. Division, Indian Statistical Institute, Delhi. She is deeply indebted to Professor Singh for all the help and encouragement. She wishes to thank her ex-Principal Dr. Kavita Sharma for her encouragement during all these years to bring this effort to a fruitful ending.

We thank Manpreet Singh and Vishal Mangla for their help in preparing the S-diagram and Naveen Reddy for his help in editing the manuscript at various stages.

We thank L. F. Kwong for her co-operation at every stage during the preparation of the manuscript. We also thank D. Rajesh Babu for the technical help. Finally, we wish to thank World Scientific Publishing Co., Singapore for providing us the necessary freedom and flexibility in completing the monograph.

Glossary of Symbols and Abbreviations

a_{ij} - the $(i,j)^{th}$ element of the matrix $\mathbf{A}$
(a_{ij}) - the matrix whose $(i,j)^{th}$ element is a_{ij}
$\mathbf{A}^t$ - transpose of the matrix $\mathbf{A}$
$\mathbf{A}^\star$ - conjugate transpose of the matrix $\mathbf{A}$
$\mathcal{C}(\mathbf{A})$ - the column space of the matrix $\mathbf{A}$
$\mathcal{C}(\mathbf{A}^t)$ - the row space of the matrix $\mathbf{A}$
$\mathcal{C}(\mathbf{A}^\perp)$ - the orthogonal complement of $\mathcal{C}(\mathbf{A})$
$\mathcal{N}(\mathbf{A})$ - the null space of the matrix $\mathbf{A}$
$tr(\mathbf{A})$ - the trace of the matrix $\mathbf{A}$
$\rho(\mathbf{A})$ - the rank of the matrix $\mathbf{A}$
$d(\mathcal{S})$ - the dimension of the subspace $\mathcal{S}$
$det(\mathbf{A})$ - the determinant of the matrix $\mathbf{A}$
$\mathbf{A}_L^{-1}$ - a left inverse of the matrix $\mathbf{A}$
$\mathbf{A}_R^{-1}$ - a right inverse of the matrix $\mathbf{A}$
$\mathbf{A}^-$ - a g-inverse of the matrix $\mathbf{A}$
$\{\mathbf{A}^-\}$ - the set of g-inverses of the matrix $\mathbf{A}$
$\mathbf{A}_r^-$ - a reflexive g-inverse of the matrix $\mathbf{A}$
$\{\mathbf{A}_r^-\}$ - the set of reflexive g-inverses of the matrix $\mathbf{A}$
$\mathbf{A}_{com}^-$ - a commuting g-inverse of $\mathbf{A}$
$\{\mathbf{A}_{com}^-\}$ - the set of commuting g-inverses of $\mathbf{A}$
$\mathbf{A}^\dagger$ - the Moore-Penrose inverse of the matrix $\mathbf{A}$
$\mathbf{A}^\sharp$ - the group inverse of the matrix $\mathbf{A}$
$\mathbf{A}^D$ - the Drazin inverse of the matrix $\mathbf{A}$
$\mathbf{A}_\ell^-$ - a least squares g-inverse of the matrix $\mathbf{A}$
$\{\mathbf{A}_\ell^-\}$ - the set of least squares g-inverses of the matrix $\mathbf{A}$
$\mathbf{A}_m^-$ - a minimum norm g-inverse of the matrix $\mathbf{A}$
$\{\mathbf{A}_m^-\}$ - the set of minimum norm g-inverses of the matrix $\mathbf{A}$
$\mathbf{A}_\rho^-$ - a commuting g-inverse of the matrix $\mathbf{A}$ with the prop-

erty that $\mathcal{C}(\mathbf{A}_{\rho}^{-t}) \subseteq \mathcal{C}(\mathbf{A}^t)$

$\{\mathbf{A}_{\rho}^{-}\}$ - the set of all ρ-inverses of the matrix $\mathbf{A}$

$\mathbf{A}_{\chi}^{-}$ - a commuting g-inverse of the matrix $\mathbf{A}$ with the property that $\mathcal{C}(\mathbf{A}_{\chi}^{-}) \subseteq \mathcal{C}(\mathbf{A})$

$\{\mathbf{A}_{\chi}^{-}\}$ - the set of all χ-inverses of the matrix $\mathbf{A}$

$\lambda_{max}(\mathbf{A})$ - the maximum eigen-value of $\mathbf{A}$

$\lambda_{min}(\mathbf{A})$ - the minimum eigen-value of $\mathbf{A}$

$\sigma(\mathbf{A})$ - the set of all singular values $\mathbf{A}$

$\mathbb{C}$ - the field of complex numbers

$\mathbb{R}$ - the field of real numbers

F - arbitrary field

$\mathbb{C}^n$ - the vector space of complex n-tuples

$\mathbb{C}^{m\times n}$ - the set of all $m \times n$ matrices over $\mathbb{C}$

F^n - the vector space of n-tuples over F

$d(V)$ - the dimension of vector space V

$\mathcal{I}_1$ - the set of all matrices of index ≤ 1

$\mathcal{I}_{1,n}$ - the set of all $n \times n$ matrices of index ≤ 1

$F^{m\times n}$ - the set of all $m \times n$ matrices over F

$\mathfrak{H}$ - the set of all hermitian matrices

$\mathfrak{H}_n$ - the set of all $n \times n$ hermitian matrices

I - the identity matrix

I_k - the $k \times k$ identity matrix

$diag(x_1, x_2, \ldots, x_n)$ - the $n \times n$ diagonal matrix with diagonal elements $x_i, i = 1, 2, \ldots, n$

$\mathcal{P}(\mathbf{A}|\mathbf{B})$ - the parallel sum of matrices $\mathbf{A}$, $\mathbf{B}$

$\mathcal{S}(\mathbf{A}|\mathbf{B})$ - the shorted matrix of $\mathbf{A}$ with respect to the matrix $\mathbf{B}$

$S(\mathbf{A}|\mathcal{S},\mathcal{T})$ - the shorted matrix of an $m\times n$ matrix $\mathbf{A}$ with respect to the subspace $\mathcal{S}$ of F^m and the subspace $\mathcal{T}$ of F^n, also read as the shorted matrix of an $m \times n$ matrix $\mathbf{A}$ indexed by subspaces $\mathcal{S}$ and $\mathcal{T}$

$\mathcal{G}(\mathbf{A})$ - a subset of $\{\mathbf{A}^-\}$

$\mathcal{G}(\mathbf{A}, \mathbf{B})$ - the set of $\{\mathbf{B}^-\mathbf{A}\mathbf{B}^- : \mathbf{B}^- \in \mathcal{G}(\mathbf{B})\}$

$\mathcal{G}_r(\mathbf{A})$ - the set $\mathcal{G}(\mathbf{A}) \cap \{\mathbf{A}_r^-\}$

$\widetilde{\mathcal{G}(\mathbf{A})}$ - completion of $\mathcal{G}(\mathbf{A})$

$\tilde{\mathcal{G}}_r(\mathbf{A})$ - the set $\widetilde{\mathcal{G}(\mathbf{A})} \cap \{\mathbf{A}_r^-\}$

$\mathfrak{P}(F^{m\times n})$ - set of all subset of $F^{m\times n}$

m-column vector - an $m \times 1$ matrix

n-row vector - a $1 \times n$ matrix

Contents

Chapter 1

Introduction

1.1 Matrix orders

Last few decades have witnessed a steady growth in the area of matrix partial orders, a central theme of this monograph. These matrix orders are developed in detail in Chapters 3-8. They play an important role in the study of shorted operators, which we treat subsequently. In this section we give a simple and intuitive interpretation of some of the matrix partial orders.

Let us begin by defining a pre-order and a partial order. A binary relation on a non-empty set is said to be a pre-order if it is reflexive and transitive. If it is also anti-symmetric, then it is called a partial order (see Appendix A).

Let f be a linear transformation from $F^n \longrightarrow F^m$, F being an arbitrary field. Then there exist bases $\mathfrak{B}$ and $\mathfrak{C}$ of F^n and F^m respectively such that f is represented by the matrix $\mathbf{diag(I,\ 0)}$ with respect to these bases (normal form). Also, if $F = \mathbb{C}$, the field of complex numbers, then there exist ortho-normal bases $\mathfrak{B}$ and $\mathfrak{C}$ of $\mathbb{C}^n$ and $\mathbb{C}^{\mathrm{m}}$ respectively such that f is represented by the matrix $\mathbf{diag(D,\ 0)}$ with respect to these bases, where $\mathbf{D}$ is a positive definite diagonal matrix (singular value decomposition). Thus, every linear transformation can be represented by a diagonal matrix by choosing the bases appropriately.

A matrix $\mathbf{G}$ is a generalized inverse (g-inverse) of a matrix $\mathbf{A}$ if $\mathbf{AGA} = \mathbf{A}$. Let $\mathbf{A}$ and $\mathbf{B}$ be matrices of the same order. Then $\mathbf{A}$ is said to be below $\mathbf{B}$ under the minus order if there exists a g-inverse $\mathbf{G}$ of $\mathbf{A}$ such that $\mathbf{AG} = \mathbf{BG}$ and $\mathbf{GA} = \mathbf{GB}$.

Again, let $\mathbf{A}$ and $\mathbf{B}$ be matrices of the same order. Let $\mathbf{G}$ be the Moore-Penrose inverse of $\mathbf{A}$, a matrix that satisfies the conditions $\mathbf{AGA} = \mathbf{A}$,

$\mathbf{GAG} = \mathbf{G}$, $\mathbf{AG}$ and $\mathbf{GA}$ are hermitian. Then $\mathbf{A}$ is said to be below $\mathbf{B}$ under the star order if the Moore-Penrose inverse $\mathbf{G}$ of $\mathbf{A}$ satisfies the conditions $\mathbf{AG} = \mathbf{BG}$ and $\mathbf{GA} = \mathbf{GB}$.

If $\mathbf{A}$ and $\mathbf{B}$ are square matrices of the same order such that $\rho(\mathbf{A}) = \rho(\mathbf{A^2})$ and $\rho(\mathbf{B}) = \rho(\mathbf{B^2})$, then $\mathbf{A}$ is below $\mathbf{B}$ under the sharp order if $\mathbf{AG} = \mathbf{BG}$ and $\mathbf{GA} = \mathbf{GB}$, the matrix $\mathbf{G}$ being the group inverse of $\mathbf{A}$. Note that the group inverse of $\mathbf{A}$ is a matrix $\mathbf{G}$ such that $\mathbf{AGA} = \mathbf{A}$, $\mathbf{GAG} = \mathbf{G}$ and $\mathbf{AG} = \mathbf{GA}$.

The minus, the star and the sharp orders are partial orders. For a pair of matrices $\mathbf{A}$ and $\mathbf{B}$, each of these orders is defined through the same type of conditions, namely: $\mathbf{AG} = \mathbf{BG}$ and $\mathbf{GA} = \mathbf{GB}$, where $\mathbf{G}$ is required to belong to a suitable subclass of g-inverses of $\mathbf{A}$.

All the above partial orders can be nicely interpreted through matrix decompositions. Let us consider two diagonal matrices $\mathbf{L} = \mathbf{diag}(\mathbf{D}, \mathbf{0}, \mathbf{0})$ and $\mathbf{M} = \mathbf{diag}(\mathbf{D}, \mathbf{E}, \mathbf{0})$, where $\mathbf{D}$ and $\mathbf{E}$ are diagonal matrices. (We say a matrix - square or rectangular - is diagonal if all the elements outside the principal diagonal are zero.) It is intuitive to say that $\mathbf{L}$ is below $\mathbf{M}$ or $\mathbf{L}$ is a section of $\mathbf{M}$. By an extension of this notion, for suitable choices of nonsingular/unitary matrices $\mathbf{P}$ and $\mathbf{Q}$, the matrix $\mathbf{PLQ}$ is also below $\mathbf{PMQ}$ in some sense. Several matrix partial orders can be expressed in this manner as we shall see in the later chapters.

For instance, the matrix $\mathbf{A}$ is below the matrix $\mathbf{B}$ under the minus order if and only if $\mathbf{A}$ and $\mathbf{B}$ have simultaneous normal form such that $\mathbf{A} = \mathbf{Pdiag}(\mathbf{I}, \mathbf{0}, \mathbf{0})\mathbf{Q}$ and $\mathbf{B} = \mathbf{Pdiag}(\mathbf{I}, \mathbf{I}, \mathbf{0})\mathbf{Q}$, where $\mathbf{P}$ and $\mathbf{Q}$ are non-singular matrices. Let $\mathbf{A}$ and $\mathbf{B}$ be matrices representing the linear transformations f and g with respect to standard bases F^n and F^m of respectively. Then $\mathbf{A}$ is below $\mathbf{B}$ under minus if and only if there exist bases $\mathfrak{B}$ and $\mathfrak{C}$ of F^n and F^m respectively such that f and g are represented by $\mathbf{diag}(\mathbf{I}, \mathbf{0}, \mathbf{0})$ and $\mathbf{diag}(\mathbf{I}, \mathbf{I}, \mathbf{0})$ respectively with respect to these bases.

Further, $\mathbf{A}$ is below $\mathbf{B}$ under the star order if and only if $\mathbf{A}$ and $\mathbf{B}$ have simultaneous singular value decomposition such that $\mathbf{A} = \mathbf{Pdiag}(\mathbf{D}, \mathbf{0}, \mathbf{0})\mathbf{Q}$ and $\mathbf{B} = \mathbf{Pdiag}(\mathbf{D}, \mathbf{E}, \mathbf{0})\mathbf{Q}$, where $\mathbf{D}$ and $\mathbf{E}$ are positive definite diagonal matrices and $\mathbf{P}$ and $\mathbf{Q}$ are unitary matrices.

Let $\mathbf{A}$ and $\mathbf{B}$ be square matrices. Then we say $\mathbf{A}$ is below $\mathbf{B}$ under the sharp order if and only if there exists a non-singular matrix $\mathbf{P}$ such that $\mathbf{A} = \mathbf{Pdiag}(\mathbf{D}, \mathbf{0}, \mathbf{0})\mathbf{P}^{-1}$ and $\mathbf{B} = \mathbf{Pdiag}(\mathbf{D}, \mathbf{E}, \mathbf{0})\mathbf{P}^{-1}$, where $\mathbf{D}$ and $\mathbf{E}$ are non-singular matrices.

Yet another partial order is the well-known Löwner order on non-negative definite matrices defined as follows. Let $\mathbf{A}$ and $\mathbf{B}$ be non-negative definite (*nnd*) matrices of the same order. Then $\mathbf{A}$ is below $\mathbf{B}$ under the Löwner order if $\mathbf{B} - \mathbf{A}$ is *nnd*. Furthermore, this happens if and only if there exists a non-singular matrix $\mathbf{P}$ such that $\mathbf{A} = \mathbf{Pdiag(D,\ 0,\ 0)P}^\star$ and $\mathbf{B} = \mathbf{Pdiag(H,\ E,\ 0)P}^\star$, where $\mathbf{D}$, $\mathbf{H}$ and $\mathbf{E}$ are positive definite diagonal matrices and each diagonal element of $\mathbf{H}$ is at least as big as the corresponding diagonal element of $\mathbf{D}$.

Thus, we see that corresponding to each of the matrix orders mentioned above $\mathbf{A}$ and $\mathbf{B}$, the matrices representing some linear transformations f and g under suitable choice of bases, are simultaneously represented by matrices $\mathbf{L}$ and $\mathbf{M}$ that have a relatively simple structure, where it is intuitively clear that $\mathbf{L}$ is below $\mathbf{M}$.

In later chapters of this book, we explore all these and other matrix orders and study their inter-relationships.

1.2 Parallel sum and shorted operator

Parallel sum and shorted operator as studied in this monograph have their origin in the study of impedance matrices of n-port electrical networks. There have been interesting extensions of these concepts to wider classes of matrices. Two matrices $\mathbf{A}$ and $\mathbf{B}$ of the same order are said to be parallel summable if the row and column spaces of $\mathbf{A}$ are contained respectively in the row and column spaces of $\mathbf{A}+\mathbf{B}$. When $\mathbf{A}$ and $\mathbf{B}$ are parallel summable, their parallel sum is defined as $\mathbf{A}(\mathbf{A}+\mathbf{B})^-\mathbf{B}$, where $(\mathbf{A}+\mathbf{B})^-$ is a g-inverse of $\mathbf{A}+\mathbf{B}$. Parallel sum has some very interesting properties. For example, its column (row) space is the intersection of the column (row) spaces of A and B. In this monograph, we study the properties of the parallel sum in detail. We also study the connection between the parallel sum and the matrix orders.

The shorted operator has been defined in more than one way. For example, the shorted operator of a matrix with respect to another matrix of the same order can be defined as the limit of a certain sequence of parallel sums. Yet another way: Let $\mathbf{A}$ be an $m \times n$ matrix, V_m and V_n be subspaces of vector spaces of m-tuples and n-tuples. The shorted operator of $\mathbf{A}$ indexed by subspaces V_m and V_n is a matrix that is the closest to $\mathbf{A}$ in the class of matrices having their column and row spaces contained in V_m and V_n respectively. These various definitions of shorted operator are based on different objectives. In this monograph, we make a comprehensive study of the shorted operators and show the equivalence of various definitions of the shorted operator.

1.3 A tour through the rest of the monograph

As indicated in Section 1.1, generalized inverses, matrix decompositions and simultaneous decompositions in particular are going to play an important role in the study of matrix partial orders, parallel sums and shorted operators. In Chapter 2, we develop the necessary background on matrix decompositions and generalized inverses.

Chapter 3 starts with the space pre-order. We study this basic matrix order in detail as we feel that this is the stepping stone to the other matrix orders that follow. We then introduce the minus order using a generalized inverse. This is the first partial order we study in this monograph. We obtain several characterizing properties of the same. We obtain the classes of matrices that lie above and below a given matrix under the minus order. We also study the minus order for certain special classes of matrices like the projectors.

In Chapter 4, we define the sharp order on the square matrices of index not exceeding 1. The sharp order is an order finer than the minus order that involves the group inverse. We make a detailed study of its characteristic and other properties. We then consider matrices which have index greater than 1. We define an order called Drazin order using the Drazin inverse (which is not a generalized inverse in the usual sense). This order turns out to be a pre-order different from the space pre-order. It does have some interesting properties. We finally extend the Drazin order to a partial order and study its properties.

The star order is perhaps more extensively studied than the minus and the sharp orders. In fact, it appeared in the literature before both the other two orders. Chapter 5 is devoted to the study of this order. We study the characterizing properties of the star order and obtain the classes of matrices above and below a given matrix under the star order. We then specialize to some interesting subclasses of matrices such as range hermitian, normal, hermitian and idempotent matrices and study the star order for such matrices. We finally consider several matrices and study when each of them is below their sum under the star order. The results are very similar to the celebrated Fisher-Cochran Theorem on distribution of quadratic forms in normal variables.

Let $\mathbf{A}$ and $\mathbf{B}$ be matrices of the same order. Then $\mathbf{A}$ is below $\mathbf{B}$ under each of the minus, the sharp and the star orders if there is a suitable g-inverse $\mathbf{G}$ of $\mathbf{A}$ such that $\mathbf{AG} = \mathbf{BG}$ and $\mathbf{GA} = \mathbf{GB}$. In Chapter 6, we consider only one of the above two conditions and show that with an additional milder condition we can get a partial order. Such an order is called a one-sided order. A one-sided order corresponding to minus order simply coincides with the minus order. However, one-sided sharp and star orders do not coincide with the sharp and the star orders respectively and lead to an interesting study. In the process of obtaining one-sided sharp order, we develop two special classes of g-inverses, which are interesting in their own way.

After a careful study of the results of Chapters 3-5, one finds several similar looking properties shared by these order relations. Is there a common thread? Is there a master characterizing property using which we can derive a number of common looking properties of these partial orders? In Chapter 7, we find an answer to these questions and give a unified theory of matrix partial orders developed via generalized inverses. Characterizations of common properties/results of all the partial orders are put under one umbrella of the unified theory. We conclude the chapter with some extensions of this unified theory.

Löwner order, usually studied for *nnd* matrices, is one of the oldest known partial orders. Some material dealing with Löwner order for hermitian matrices is also available in the literature. In Chapter 8, we bring together such results and make a comprehensive study of this order. We study the relationship of Löwner order with other partial orders studied earlier. We also study the ordering properties of generalized inverses and outer inverses under Löwner order. Finally, we consider a couple of extensions of Löwner order - one of them for rectangular matrices.

Chapter 9 is devoted to the study of the parallel sum of matrices. Besides obtaining several interesting properties of parallel sums, we also explore the relationship of parallel sum with matrix orders, particularly, the space pre-order and the Löwner order. It turns out that parallel sum of two matrices $\mathbf{A}$ and $\mathbf{B}$ is the matrix closest to $\mathbf{A}$ and $\mathbf{B}$ in the sense that any matrix which is below $\mathbf{A}$ and $\mathbf{B}$ under the space pre-order is also below their parallel sum under the space pre-order.

We study the shorted operators in Chapters 10 and 11. In Chapter 10, we provide motivation for shorted operator through Electrical Networks and Statistics. We first consider *nnd* matrices and study the shorted operator of an *nnd* matrix indexed by a subspace and develop several interesting properties. We note here that the shorted operator is a certain Schur complement. Let **A** be an *nnd* matrix of order $n \times n$ and $\mathcal{S}$ be a subspace of $\mathbb{C}^n$. Then the shorted operator of **A** with respect to $\mathcal{S}$ turns out to be the *nnd* matrix closest to **A**, both under the Löwner order and the minus order, in the class of all *nnd* matrices with column space contained in $\mathcal{S}$. We also show that the shorted operator is the limit of a sequence of certain parallel sums. We then extend the concept of the shorted operator to possibly rectangular matrices over a general field. This leads us to a concept called complementability. We examine when the shorted operator indexed by two subspaces exists and obtain an explicit expression for the same when it exists. Again, it turns out to be a Schur complement. Let **A** be an $m \times n$ matrix over a general field and let V_m and V_n be subspaces of vector spaces of m-tuples and n-tuples respectively over this field. Let the shorted operator $\mathbf{S}(\mathbf{A}|V_m, V_n)$ of **A** indexed by V_m and V_n exist. We find that the shorted operator, $\mathbf{S}(\mathbf{A}|V_m, V_n)$, whenever it exists, is the closest to **A** under the minus order in the class of all matrices the column space and row space of which are contained in V_m and V_n respectively. Further, $\mathbf{S}(\mathbf{A}|V_m, V_n)$ turns out to be a Schur complement.

In Chapter 11, we first extend the concept of shorted operator to (possibly) rectangular matrices over $\mathbb{C}$, the field of complex numbers using the approach of the limit of a sequence of parallel sums. We examine when it exists and study its properties. We then give another definition of shorted operator using the approach of closeness in the sense of rank. More precisely, for a matrix **A**, we examine when a unique matrix **B** exists such that the rank of $\mathbf{A} - \mathbf{B}$ is the least in the class of all matrices the column space and row space of which are contained in V_m and V_n respectively. When such a matrix **B** exists, we call it as the shorted operator $\mathbf{S}(\mathbf{A}|V_m, V_n)$ of **A** indexed by V_m and V_n. We show that all these approaches of defining the shorted operator lead to the same matrix. We finally make some remarks on the computational aspects of the shorted operator.

One characterizing property of the shorted operator studied in Chapters 10 and 11 is that it is a maximal element of a certain collection of matrices under a suitable partial order. This raises the following natural question. Does a set of matrices equipped with a partial order become a lattice or at least a semi-lattice under that partial order? In Chapter 12, we study this problem in some detail for three of the major partial orders studied earlier, namely, the minus, the star and the sharp orders.

Chapter 13 contain entirely new material. We make an extensive study of the matrix order relations for modified matrices. We consider here two types of modifications of matrices one: appending or deleting a row/column and the other: adding a rank 1 matrix. Let **A** be below **B** under a particular matrix order. Let **A** be modified as per one of the above modifications. We obtain the class of all modifications of **B** such that the modified matrix of **A** is below the modified matrix of **B** under the same matrix order. The matrix orders considered for this purpose are the space pre-order, the minus order, the sharp order, the star order and the Löwner order. The proofs, in general, are highly computational and lengthy. Due to space considerations, we have omitted most of the proofs. However a reader interested in the proofs of these results may write to the authors.

In Chapter 14, we give an application of the matrix partial orders in developing equivalence relations on the classes of generalized inverses and outer inverses of a matrix. This leads to the development of nice hierarchies among the various classes of inverses (both inner and outer) of a matrix. It also leads to a neat diagrammatic representation of various inverses of a matrix, which according the first author resembles a strawberry plantation. (It is often said, a picture is worth a thousand words!)

In Chapter 15, we give applications of matrix orders, parallel sum and shorted operator to Statistics and Electrical Networks. We first collect some results related to inference in linear models for the benefit of a general reader. Then we give interpretation and application of matrix orders, parallel sum and shorted operators to comparison of models and inference in linear models. We also give an application of shorted operators to the recovery of inter-block information in incomplete block designs. We give an application of shorted operators of modified matrices to obtain the modified shorted operator when a new port is included in the network.

We enlist a few open problems related to the material covered in this monograph which should be of interest to the researchers.

Finally, Appendix A contains the basic material on the algebra of relations, semi-groups, groups, partial orders and related issues for those readers who may need a little brushing up of some of these basic concepts.

Chapter 2

Matrix Decompositions and Generalized Inverses

2.1 Introduction

This chapter aims at providing some prerequisites in Linear Algebra for smooth reading of the later chapters of the monograph. We assume that the reader has the knowledge at the level of a first course in Linear Algebra. In Section 2.2, we gather several known results on matrix decompositions. Rank factorizations of matrices have been in use for a long time. However their properties are not commonly available in elementary textbooks on matrices. Therefore, we treat them here in a fairly elaborate fashion. Sections 2.3-2.6 are devoted to an exposition of generalized inverses (g-inverses), where we try to provide motivation for a gradual development of different types of g-inverses. It is perhaps a little lengthy exposition for a monograph on matrix partial orders. The justification for this is two-fold: (i) in general, a first course in Linear Algebra does not provide a proper motivation and development of g-inverse and (ii) the later chapters in the monograph require a mature understanding of g-inverses. However, in this chapter, no attempt is made to make the exposition on g-inverses exhaustive or up to date. For a detailed study of generalized inverses the reader is referred to [Rao and Mitra (1971)], [Ben-Israel and Greville (2001)] and [Campbell and Meyer (1991)]. The main purpose of this chapter is to facilitate the reader in studying the later chapters of the book comfortably. In Section 2.7, we study simultaneous diagonalization, a concept which is very important when we deal with two or more matrices together. Apart from the standard results, we include simultaneous singular value decomposition for several matrices and generalized singular value decomposition, which will prove useful in understanding some partial orders like a one-sided star order. This chapter contains a few new results. We also give new and simpler proofs to some known results. In this chapter, by and large, we work with vectors and matrices over a general field. However, if we use vectors and matrices over the complex field, we clearly mention the same before such a use.

2.2 Matrix decompositions

In this section, we enumerate several matrix decompositions (often without proof) which will be useful later in understanding the structures of generalized inverses and matrix partial orders. Proofs of most of the results mentioned here can be found in standard text books on Linear Algebra, for example [Rao and Bhimasankaram (2000)].

Theorem 2.2.1. *(Normal Form) Let* $\mathbf{A}$ *be an* $m \times n$ *matrix and let* $\rho(\mathbf{A}) = r > 0$*. Then there exist non-singular matrices* $\mathbf{P}$ *and* $\mathbf{Q}$ *of order* $m \times m$ *and* $n \times n$ *respectively such that* $\mathbf{A} = \mathbf{Pdiag}(\mathbf{I_r}, \mathbf{0})\mathbf{Q}$.

Theorem 2.2.2. *(Rank Factorization) Let* $\mathbf{A}$ *be as in Theorem 2.2.1. Then there exist matrices* $\mathbf{R}$ *and* $\mathbf{S}$ *of orders* $m \times r$ *and* $r \times n$ *respectively such that* $\rho(\mathbf{R}) = r = \rho(\mathbf{S})$ *and* $\mathbf{A} = \mathbf{RS}$*. Conversely, if* $\mathbf{A} = \mathbf{RS}$*, where* $\mathbf{R}$ *and* $\mathbf{S}$ *are full rank matrices of order* $m \times r$ *and* $r \times n$ *respectively such that* $r \leq \min\{m, n\}$*, then* $\rho(\mathbf{A}) = r$.

The first part of Theorem 2.2.2 follows easily from Theorem 2.2.1 by taking $\mathbf{R}$ as the matrix formed by taking first r columns of $\mathbf{P}$ and $\mathbf{S}$ as the matrix formed by the first r rows of $\mathbf{Q}$. The second part is easy.

Definition 2.2.3. If $\mathbf{R}$ and $\mathbf{S}$ are matrices as specified in Theorem 2.2.2, we say $(\mathbf{R}, \mathbf{S})$ is a rank factorization of $\mathbf{A}$.

Remark 2.2.4. If $(\mathbf{R}, \mathbf{S})$ is a rank factorization of a matrix $\mathbf{A}$, then
(i) the columns of $\mathbf{R}$ form a basis of $\mathcal{C}(\mathbf{A})$, the column space of $\mathbf{A}$ and
(ii) the rows of $\mathbf{S}$ form a basis of $\mathcal{C}(\mathbf{A^t})$, the column space of $\mathbf{A^t}$.

Remark 2.2.5. A matrix can have several rank factorizations. Let $(\mathbf{R}, \mathbf{S})$ be a rank factorization of $\mathbf{A}$. Then the class of matrices $\{(\mathbf{RT}, \mathbf{T}^{-1}\mathbf{S}), \mathbf{T}$ non-singular$\}$ is the class of all rank factorizations of $\mathbf{A}$.

Remark 2.2.6. Let $\mathbf{A}$ be an $m \times n$ matrix of rank r over the field of complex numbers $\mathbb{C}$. Then there exists a semi-unitary matrix $\mathbf{U}$ of order $m \times r$ (a matrix $\mathbf{U}$ such that $\mathbf{U}^{\star}\mathbf{U} = \mathbf{I_r}$) such that $(\mathbf{U}, \mathbf{V})$ is a rank factorization of $\mathbf{A}$.

Let $\mathbf{A_1}$ and $\mathbf{A_2}$ be non-null matrices of the same order. Let $(\mathbf{P_i}, \mathbf{Q_i})$ be a rank factorization of $\mathbf{A_i}, \mathbf{i} = \mathbf{1}, \mathbf{2}$. Clearly,

$$\mathbf{A_1} + \mathbf{A_2} = (\mathbf{P_1} : \mathbf{P_2}) \begin{pmatrix} \mathbf{Q_1} \\ \cdot\cdot \\ \mathbf{Q_2} \end{pmatrix}.$$

When is $\left((\mathbf{P_1} : \mathbf{P_2}), \begin{pmatrix} \mathbf{Q_1} \\ \cdot\cdot \\ \mathbf{Q_2} \end{pmatrix} \right)$ a rank factorization of $\mathbf{A_1} + \mathbf{A_2}$? Before we can answer this question we give a couple of definitions.

Definition 2.2.7. Two subspaces $\mathcal{S}$ and $\mathcal{T}$ of a vector space are said to be virtually disjoint if $\mathcal{S} \cap \mathcal{T} = \{\mathbf{0}\}$.

Definition 2.2.8. Let $\mathbf{A_1}$ and $\mathbf{A_2}$ be matrices of the same order. Then $\mathbf{A_1}$ and $\mathbf{A_2}$ are said to be disjoint if

(i) $\mathcal{C}(\mathbf{A_1}) \cap \mathcal{C}(\mathbf{A_2}) = \{\mathbf{0}\}$ and
(ii) $\mathcal{C}(\mathbf{A_1^t}) \cap \mathcal{C}(\mathbf{A_2^t}) = \{\mathbf{0}\}$

or equivalently $\rho(\mathbf{A_1} + \mathbf{A_2}) = \rho(\mathbf{A_1}) + \rho(\mathbf{A_2})$.
In such a case the sum of $\mathbf{A_1}$ and $\mathbf{A_2}$ is denoted by $\mathbf{A_1} \oplus \mathbf{A_2}$.

We now prove the following:

Theorem 2.2.9. *Let* $\mathbf{A_1}$ *and* $\mathbf{A_2}$ *be non-null matrices of the same order. Let* $(\mathbf{P_i}, \mathbf{Q_i})$ *be a rank factorization of* $\mathbf{A_i}$, $i = 1, 2$. *Then*

$$\left((\mathbf{P_1} : \mathbf{P_2}), \begin{pmatrix} \mathbf{Q_1} \\ \cdot\cdot \\ \mathbf{Q_2} \end{pmatrix} \right)$$

is a rank factorization of $\mathbf{A_1} + \mathbf{A_2}$ *if and only if* $\mathbf{A_1}$ *and* $\mathbf{A_2}$ *are disjoint.*

Proof. 'If' part

Notice that

$$\mathbf{A_1} + \mathbf{A_2} = \mathbf{P_1}\mathbf{Q_1} + \mathbf{P_2}\mathbf{Q_2} = (\mathbf{P_1} : \mathbf{P_2}) \begin{pmatrix} \mathbf{Q_1} \\ \cdot\cdot \\ \mathbf{Q_2} \end{pmatrix}.$$

Let $\mathbf{A_1}$ and $\mathbf{A_2}$ be disjoint. Then $\rho(\mathbf{A_1} + \mathbf{A_2}) = \rho(\mathbf{A_1}) + \rho(\mathbf{A_2})$.

Hence the number of columns in $(\mathbf{P_1} : \mathbf{P_2})$ = the number of rows in $\begin{pmatrix}\mathbf{Q_1}\\ \cdot\cdot \\ \mathbf{Q_2}\end{pmatrix} = \rho(\mathbf{A_1}+\mathbf{A_2})$. So, $\left((\mathbf{P_1} : \mathbf{P_2}), \begin{pmatrix}\mathbf{Q_1}\\ \cdot\cdot \\ \mathbf{Q_2}\end{pmatrix}\right)$ is a rank factorization of $\mathbf{A_1}+\mathbf{A_2}$.

'Only if' part

Since $(\mathbf{P_i}, \mathbf{Q_i})$ is a rank factorization of $\mathbf{A_i}, \mathbf{i} = \mathbf{1}, \mathbf{2}$ and $\left((\mathbf{P_1} : \mathbf{P_2}), \begin{pmatrix}\mathbf{Q_1}\\ \cdot\cdot \\ \mathbf{Q_2}\end{pmatrix}\right)$ is a rank factorization of $\mathbf{A_1}+\mathbf{A_2}$, therefore,

$$\begin{aligned}\rho(\mathbf{A_1}+\mathbf{A_2}) &= \rho(\mathbf{P_1} : \mathbf{P_2})\\ &= \text{number of columns in } (\mathbf{P_1} : \mathbf{P_2})\\ &= \rho(\mathbf{A_1}) + \rho(\mathbf{A_2}).\end{aligned}$$

□

Definition 2.2.10. A matrix $\mathbf{H} = [h_{ij}]$ of order $n \times n$ is said to be in Hermite Canonical Form (HCF) if

(i) $h_{ij} = 0$ whenever $i > j, i = 1, \ldots, n$ and $j = 1, \ldots, n$,
(ii) $h_{ii} = 1$ or 0 for $i = 1, \ldots, n$,
(iii) if $h_{ii} = 0$ for some i, then $h_{ij} = 0$ for $j = 1, \ldots, n$ and
(iv) if $h_{ii} = 1$ for some i, then $h_{ji} = 0$ for all $j \neq i$.

Thus, if $\mathbf{H}$ is in HCF, then $\mathbf{H}$ is an upper triangular matrix with each diagonal element either 0 or 1; if a diagonal element is 0, then the entire row containing this particular diagonal element is null and if a diagonal element is 1, then the remaining elements of the column containing the diagonal element are 0.

For example

$$\mathbf{H} = \begin{pmatrix}1 & 2 & 0 & 4\\ 0 & 0 & 0 & 0\\ 0 & 0 & 1 & -8\\ 0 & 0 & 0 & 0\end{pmatrix} \quad \text{is a matrix in HCF.}$$

Theorem 2.2.11. *Let* $\mathbf{A}$ *be an* $n \times n$ *matrix. Then there exists a non-singular matrix* $\mathbf{B}$ *such that* $\mathbf{H} = \mathbf{BA}$ *is in HCF.*

Remark 2.2.12. A matrix in HCF is idempotent.

Remark 2.2.13. Let $\mathbf{A}$ be an $n \times n$ matrix. Then by elementary row operations one can reduce $\mathbf{A}$ to a matrix in HCF.

Remark 2.2.14. Let $\mathbf{A}, \mathbf{B}$ and $\mathbf{H}$ be as in Theorem 2.2.11 with $\rho(\mathbf{A}) = r$ and let $i_1^{th}, \ldots, i_r^{th}$ diagonal elements of $\mathbf{H}$ be each equal to 1. Write

$$\mathbf{R} = (\mathbf{A}_{\star \mathbf{i_1}} : \quad : \mathbf{A}_{\star \mathbf{i_r}}) \text{ and } \mathbf{S} = (\mathbf{H}^{\mathbf{t}}_{\mathbf{i_1}\star} : \quad : \mathbf{H}^{\mathbf{t}}_{\mathbf{i_r}\star});$$

where $\mathbf{A}_{\star \mathbf{i_j}}$ is the i_j^{th} column of $\mathbf{A}$ and $\mathbf{H}_{\mathbf{i_k}\star}$ is the i_k^{th} row of $\mathbf{H}$. Then $(\mathbf{R}, \mathbf{S})$ is a rank factorization of $\mathbf{A}$.

Theorem 2.2.15. *(Schur Decomposition) Let $\mathbf{A}$ be an $n \times n$ matrix. then there exists a non-singular matrix $\mathbf{P}$ such that $\mathbf{A} = \mathbf{PTP^{-1}}$, where $\mathbf{T}$ is an upper triangular matrix.*

Note that in Theorem 2.2.15, the eigen-values of the matrix $\mathbf{A}$ are precisely the diagonal elements of $\mathbf{T}$. If the matrix $\mathbf{A}$ is over the field of complex numbers $\mathbb{C}$, there exists a unitary matrix $\mathbf{P}$ such that $\mathbf{A} = \mathbf{PTP^{\star}}$, where $\mathbf{T}$ is an upper triangular matrix.

Before we state our next theorem on the Jordan decomposition of a matrix, we define a Jordan block.

A Jordan block of order 1 is a 1×1 matrix of the form $[a]$.
A Jordan block $\mathbf{J} = [a_{ij}]$ of order $r > 1$ is a matrix satisfying

(i) $a_{ik} = 0$, whenever $i > k$ (upper triangular)
(ii) $a_{11} = \ldots = a_{rr}$ (identical diagonal elements)
(iii) $a_{ii+1} = 1$ for $i = 1, \ldots, r-1$ and
(iv) $a_{ik} = 0$ whenever $k > i + 1$.

For example

$$\mathbf{J} = \begin{pmatrix} 3 & 1 & 0 & 0 \\ 0 & 3 & 1 & 0 \\ 0 & 0 & 3 & 1 \\ 0 & 0 & 0 & 3 \end{pmatrix} \text{ is a Jordan block.}$$

Theorem 2.2.16. *(Jordan Decomposition) Let $\mathbf{A}$ be an $n \times n$ matrix over $\mathbb{C}$. Then there exists a non-singular matrix $\mathbf{P}$ such that $\mathbf{A} = \mathbf{P} \textbf{ diag } (\mathbf{J_1}, \ldots, \mathbf{J_r}) \ \mathbf{P}^{-1}$, where each $\mathbf{J_i}$ $i = 1, \ldots, r$ is a Jordan block.*

Remark 2.2.17. Jordan decomposition is a more specialized form of Schur decomposition.

Definition 2.2.18. Let $\mathbf{A}$ be an $n \times n$ matrix. The smallest positive integer k for which $\rho(\mathbf{A^k}) = \rho(\mathbf{A^{k+1}})$ is called the index of $\mathbf{A}$.

Remark 2.2.19. The index of a non-singular matrix $\mathbf{A}$ is 0 and the index of a null matrix is 1.

Remark 2.2.20. The matrices $\mathbf{A}$ and $\mathbf{PAP^{-1}}$ have the same index for each non-singular matrix $\mathbf{P}$.

Theorem 2.2.21. *(Core-Nilpotent Decomposition) Let $\mathbf{A}$ be an $n \times n$ matrix. Then $\mathbf{A}$ can be written as the sum of matrices $\mathbf{A_1}$ and $\mathbf{A_2}$ i.e. $\mathbf{A} = \mathbf{A_1} + \mathbf{A_2}$ where*

(i) $\rho(\mathbf{A_1}) = \rho(\mathbf{A_1^2})$ *i.e. $\mathbf{A_1}$ is of index ≤ 1*
(ii) *$\mathbf{A_2}$ is nilpotent i.e. there is a positive integer k such that $\mathbf{A_2^k} = \mathbf{0}$ and*
(iii) $\mathbf{A_1A_2} = \mathbf{A_2A_1} = \mathbf{0}$.

Here one or both of $\mathbf{A_1}$ and $\mathbf{A_2}$ can be null.
A proof of Theorem 2.2.21 will be given in Section 2.4.

Remark 2.2.22. Theorem 2.2.21 can be deduced from Theorem 2.2.16 for the matrices over complex field.

Theorem 2.2.23. *Let $\mathbf{A}$ be an $n \times n$ matrix. The following are equivalent:*

(i) $\rho(\mathbf{A}) = \rho(\mathbf{A^2})$
(ii) $\mathcal{C}(\mathbf{A}) \cap \mathcal{N}(\mathbf{A}) = \{\mathbf{0}\}$
(iii) $F^n = \mathcal{C}(\mathbf{A}) \oplus \mathcal{N}(\mathbf{A})$ *and*
(iv) *There exists a non-singular matrix $\mathbf{P}$ such that $\mathbf{A} = \mathbf{P}\mathbf{diag}(\mathbf{T}, \mathbf{0})\mathbf{P}^{-1}$, where $\mathbf{T}$ is non-singular.*

Remark 2.2.24. Theorem 2.2.21 can be restated as follows:
Let $\mathbf{A}$ be an $n \times n$ matrix. Then there exists a non-singular matrix $\mathbf{P}$ such that $\mathbf{A} = \mathbf{P}\mathbf{diag}(\mathbf{T}, \mathbf{N})\mathbf{P}^{-1}$, where $\mathbf{T}$ is non-singular and $\mathbf{N}$ is nilpotent.

Remark 2.2.25. Let $\mathbf{A}$ be an $n \times n$ matrix over $\mathbb{C}$. Then the algebraic multiplicity of the zero eigen-value of $\mathbf{A}$ is equal to its geometric multiplicity if and only if $\rho(\mathbf{A}) = \rho(\mathbf{A^2})$.

Definition 2.2.26. An $n \times n$ matrix $\mathbf{A}$ over $\mathbb{C}$ is said to be range-hermitian if $\mathcal{C}(\mathbf{A}) = \mathcal{C}(\mathbf{A}^\star)$ or equivalently if $\mathcal{N}(\mathbf{A}) = \mathcal{N}(\mathbf{A}^\star)$.

An $n \times n$ matrix $\mathbf{A}$ over $\mathbb{C}$ is said to be hermitian if $\mathbf{A} = \mathbf{A}^\star$.

Remark 2.2.27. Every hermitian matrix is range-hermitian. Every non-singular matrix is also range-hermitian.

Remark 2.2.28. If $\mathbf{A}$ is range-hermitian, then $\rho(\mathbf{A}) = \rho(\mathbf{A^2})$.

Theorem 2.2.29. *Let $\mathbf{A}$ be an $n \times n$ matrix over $\mathbb{C}$. Then $\mathbf{A}$ is range-hermitian if and only if there exists a unitary matrix $\mathbf{U}$ such that*

$$\mathbf{A} = \mathbf{U}\mathbf{diag}(\mathbf{T},\ \mathbf{0})\mathbf{U}^\star,$$

where $\mathbf{T}$ is non-singular.

Definition 2.2.30. A matrix $\mathbf{A}$ over $\mathbb{C}$ is said to be simple if all its eigen-values are distinct. $\mathbf{A}$ is said to be semi-simple, if the algebraic multiplicity for each of its distinct eigen-values equals its geometric multiplicity.

Theorem 2.2.31. *(Spectral Decomposition of a Semi-Simple Matrix) Let $\mathbf{A}$ be an $n \times n$ matrix over $\mathbb{C}$. Then $\mathbf{A}$ is semi-simple if and only if there exist matrices $\mathbf{E_1}, \ldots, \mathbf{E_s}$ of order $n \times n$ and distinct complex numbers $\lambda_1, \ldots, \lambda_s$ such that*

(i) $\mathbf{E_i^2} = \mathbf{E_i}$, $i = 1, \ldots s$

(ii) $\mathbf{E_i}\mathbf{E_j} = \mathbf{0}$ *whenever* $i \neq j$

(iii) $\mathbf{I} = \mathbf{E_1} + \ldots + \mathbf{E_s}$ *and*

(iv) $\mathbf{A} = \lambda_1\mathbf{E_1} + \ldots + \lambda_s\mathbf{E_s}$.

Further, the matrices $\mathbf{E_1}, \ldots, \mathbf{E_s}$ with these properties are unique.

Remark 2.2.32. Every simple matrix is semi-simple.

Remark 2.2.33. Let $\mathbf{E_i}$, $i = 1, \ldots, s$ as in Theorem 2.2.31. Then

$$n = \rho(\mathbf{I}) = \mathrm{tr}(\mathbf{I}) = \mathrm{tr}(\mathbf{E_1}) + \ldots + \mathrm{tr}(\mathbf{E_s}) = \rho(\mathbf{E_1}) + \ldots + \rho(\mathbf{E_s}).$$

Remark 2.2.34. Let $\mathbf{E_i}, i = 1, \ldots, s$ be as in Theorem 2.2.31. Let $(\mathbf{P_i}, \mathbf{Q_i})$ be a rank factorization of $\mathbf{E_i}$. Since $\mathbf{E_i^2} = \mathbf{E_i}$ for all i and $\mathbf{E_i}\mathbf{E_j} = \mathbf{0}$ whenever $i \neq j$, $\mathbf{Q_i}\mathbf{P_i} = \mathbf{I}$ for each i and $\mathbf{Q_i}\mathbf{P_j} = \mathbf{0}$ whenever $i \neq j$. Let $\mathbf{P} = (\mathbf{P_1}, \ldots, \mathbf{P_s})$ and $\mathbf{Q} = (\mathbf{Q_1^t}, \ldots, \mathbf{Q_s^t})^\mathbf{t}$. Then it follows that $\mathbf{P}$ and $\mathbf{Q}$ are square matrices and $\mathbf{QP} = \mathbf{I}$, so, $\mathbf{Q} = \mathbf{P^{-1}}$. Thus,

$$\mathbf{A} = \mathbf{P}\ \mathrm{diag}\ (\lambda_1, \ldots, \lambda_1,\ \lambda_2, \ldots, \lambda_2, \ldots, \lambda_s, \ldots, \lambda_s)\ \mathbf{P^{-1}}$$

where λ_i appear $\rho(\mathbf{E_i})$ times in the above diagonal matrix. Further, each column of $\mathbf{P}$ is a right eigen-vector of $\mathbf{A}$.

Remark 2.2.35. $\mathbf{A}$ is semi-simple if and only if $\mathbf{A}$ is similar to a diagonal matrix.

Remark 2.2.36. Let $\mathbf{A}$ be a semi-simple matrix and let $\mathbf{A} = \lambda_1\mathbf{E_1} + \ldots + \lambda_s\mathbf{E_s}$ be the spectral decomposition of $\mathbf{A}$. Then for each positive integer k, $\mathbf{A^k} = \lambda_1^\mathbf{k}\mathbf{E_1} + \ldots + \lambda_s^\mathbf{k}\mathbf{E_s}$ is the spectral decomposition of $\mathbf{A^k}$.

Definition 2.2.37. Let $\mathbf{A}$ be an $n \times n$ matrix over $\mathbb{C}$. Then $\mathbf{A}$ is called a normal matrix if $\mathbf{A}^\star\mathbf{A} = \mathbf{A}\mathbf{A}^\star$.

Theorem 2.2.38. *(Spectral Decomposition of a Normal Matrix) Let $\mathbf{A}$ be an $n \times n$ matrix over $\mathbb{C}$. Then $\mathbf{A}$ is normal if and only if there exists a unitary matrix $\mathbf{U}$ such that $\mathbf{A} = \mathbf{U}\boldsymbol{\Lambda}\mathbf{U}^\star$, where $\boldsymbol{\Lambda}$ is a diagonal matrix (possibly complex).*

Remark 2.2.39. Every normal matrix is range-hermitian.

Remark 2.2.40. The diagonal elements of $\boldsymbol{\Lambda}$ in Theorem 2.2.38 are eigen-values of $\mathbf{A}$ and columns of $\mathbf{U}$ are the ortho-normal eigen-vectors of $\mathbf{A}$.

Remark 2.2.41. Let $\mathbf{U} = (\mathbf{U_1}, \ldots, \mathbf{U_n})$ be a unitary matrix and let $\boldsymbol{\Lambda} = \mathbf{diag}(\lambda_1, \ldots, \lambda_\mathbf{n})$. Then the spectral decomposition $\mathbf{A} = \mathbf{U}\boldsymbol{\Lambda}\mathbf{U}^\star$ of $\mathbf{A}$ can also be written as

$$\mathbf{A} = \sum_{\mathbf{i=1}}^{\mathbf{n}} \lambda_\mathbf{i}\mathbf{U_i}\mathbf{U_i^\star}.$$

Remark 2.2.42. The above decomposition is in general not unique (unless all eigen-values are distinct). However, if $\mu_1, \ldots, \mu_s$ $(s \leq n)$ are distinct eigen-vectors of $\mathbf{A}$, then by appropriate pooling of the eigen-values we can write

$$\mathbf{U} = \sum_{\mathbf{i=1}}^{\mathbf{s}} \mu_\mathbf{i}\mathbf{V_i}\mathbf{V_i^\star}\ , \text{ where } \mathbf{V_i^\star}\mathbf{V_i} = \mathbf{I}\ ,\ \mathbf{V_i^\star}\mathbf{V_j} = \mathbf{0} \text{ and } \mathbf{I} = \sum \mathbf{V_i}\mathbf{V_i^\star}\ .$$

This decomposition is unique in the sense that the orthogonal projectors $\mathbf{V_i}\mathbf{V_i^\star}$ are unique.

Theorem 2.2.43. *(Spectral Decomposition of a Hermitian Matrix) Let $\mathbf{A}$ be an $n \times n$ matrix over $\mathbb{C}$. Then $\mathbf{A}$ is hermitian if and only if there exists a unitary matrix $\mathbf{U}$ such that $\mathbf{A} = \mathbf{U}\boldsymbol{\Lambda}\mathbf{U}^\star$, where Λ is a real diagonal matrix.*

Remark 2.2.44. Every hermitian matrix is normal with real eigen-values. Hence Remarks 2.2.39-2.2.42 are valid for hermitian matrices with obvious modifications.

Theorem 2.2.45. *(Singular Value Decomposition) Let $\mathbf{A}$ be an $n \times n$ matrix over $\mathbb{C}$ with rank $r > 0$. Then there exist unitary matrices $\mathbf{U}$ and $\mathbf{V}$ and a positive definite diagonal matrix $\boldsymbol{\Delta}$ of order $r \times r$ such that*

$$\mathbf{A} = \mathbf{U}\mathbf{diag}(\boldsymbol{\Delta}, \mathbf{0})\mathbf{V}^\star = \delta_1\mathbf{U_1}\mathbf{V_1^\star} + \ldots + \delta_\mathbf{r}\mathbf{U_r}\mathbf{V_r^\star},$$

where $\mathbf{U_i}$ and $\mathbf{V_i}$ are the ith columns of $\mathbf{U}$ and $\mathbf{V}$ respectively.

Remark 2.2.46. Let $\mathbf{A}, \mathbf{U}$ and $\mathbf{V}$ be as in Theorem 2.2.45. Then $\delta_1^2, \ldots, \delta_r^2$ are the non-zero eigen-values of the matrix $\mathbf{A}^\star\mathbf{A}$ (or equivalently of the matrix $\mathbf{A}\mathbf{A}^\star$) and $\mathbf{U_i}$ and $\mathbf{V_i}$ are eigen-vectors of the matrices $\mathbf{A}\mathbf{A}^\star$ and $\mathbf{A}^\star\mathbf{A}$ respectively corresponding to the eigen-value δ_i^2.

Remark 2.2.47. $\delta_1, \ldots, \delta_r$ in Remark 2.2.46 are called the singular values of $\mathbf{A}$ and $\mathbf{U_i}$ and $\mathbf{V_i}, i = 1, \ldots, r$, are called singular vectors of $\mathbf{A}$.

Remark 2.2.48. Compare Theorems 2.2.1 and 2.2.45. For matrices over complex field, we can use the unitary matrices to reduce $\mathbf{A}$ to a diagonal form. However, we cannot, in general reduce $\mathbf{A}$ to $\mathbf{diag}(\mathbf{I_r}, \mathbf{0})$ by using unitary matrices unless each singular value of $\mathbf{A}$ is 1 in which case $\mathbf{A}^\star\mathbf{A}$ and $\mathbf{A}\mathbf{A}^\star$ are orthogonal projectors.

Remark 2.2.49. In general a singular value decomposition as given in Theorem 2.2.45 is not unique. For a unique singular value decomposition, one has to pool the singular vectors corresponding to each distinct singular value. This is called the Penrose decomposition. For details see Chapter 12. See also [Rao and Bhimasankaram (2000)].

2.3 Generalized inverse of a matrix

Let $\mathbf{A}$ be a non-singular matrix. Then there exists a matrix $\mathbf{G}$ such that $\mathbf{AG} = \mathbf{GA} = \mathbf{I}$. Such a matrix $\mathbf{G}$ is unique. The matrix $\mathbf{G}$ is called the inverse of $\mathbf{A}$ and is denoted by $\mathbf{A}^{-1}$. If $\mathbf{A}$ is an $m \times n$ matrix ($m \neq n$) of rank m, there exists a matrix $\mathbf{G}$ such that $\mathbf{AG} = \mathbf{I}$. (However there is no matrix $\mathbf{H}$ such that $\mathbf{HA} = \mathbf{I}$.) Such a matrix $\mathbf{G}$, denoted by $\mathbf{A_R^{-1}}$, is called a right inverse of $\mathbf{A}$. Notice that $\mathbf{A_R^{-1}}$ is not unique unless $m = n$. Again, if $\mathbf{A}$ is an $m \times n$ matrix of rank n ($m \neq n$), there exists a matrix $\mathbf{G}$ such that $\mathbf{GA} = \mathbf{I}$ (note that there is no matrix $\mathbf{T}$ such that $\mathbf{AT} = \mathbf{I}$). Such a matrix $\mathbf{G}$ is called a left inverse of $\mathbf{A}$, denoted by $\mathbf{A_L^{-1}}$. Just as $\mathbf{A_R^{-1}}$ is not unique, so also $\mathbf{A_L^{-1}}$ is not unique unless $m = n$. In all the above cases, there exists a matrix $\mathbf{G}$ such that $\mathbf{AGA} = \mathbf{A}$. However, if $\mathbf{A}$ is an $m \times n$ matrix of rank $r < \min\{m, n\}$, then there is no matrix $\mathbf{R}$ such that $\mathbf{AR} = \mathbf{I}$ or $\mathbf{RA} = \mathbf{I}$. One can ask: does a matrix $\mathbf{G}$ satisfying $\mathbf{AGA} = \mathbf{A}$ exist? If this happens, $\mathbf{G}$ can be thought of as a generalization of the inverse/a right inverse/a left inverse.

From the utility point of view, if $\mathbf{A}$ is non-singular, then $\mathbf{Ax} = \mathbf{b}$ is consistent for all $\mathbf{b}$ and possesses unique solution given by $\mathbf{x} = \mathbf{Gb}$, where

$\mathbf{G}$ is the inverse of $\mathbf{A}$. Let $\mathbf{A}$ be an $m \times n$ matrix of rank m, $m \neq n$ and $\mathbf{G}$ be a right inverse of $\mathbf{A}$. Then $\mathbf{Ax} = \mathbf{b}$ is consistent for all $\mathbf{b}$ and $\mathbf{x} = \mathbf{Gb}$ is a solution. It should be noted that $\mathbf{Ax} = \mathbf{b}$ does not have a unique solution in this case. On the other hand if $\mathbf{A}$ is an $m \times n$ matrix of rank n and $\mathbf{G}$ is a left inverse of $\mathbf{A}$, then $\mathbf{Ax} = \mathbf{b}$ is not necessarily consistent for all b; but if it is consistent then it has a unique solution $\mathbf{x} = \mathbf{Gb}$ (interested readers may find examples). Now, let $\mathbf{A}$ be an $m \times n$ matrix of rank $r < \min\{m, n\}$. Does there exist a matrix $\mathbf{G}$ such that, whenever $\mathbf{Ax} = \mathbf{b}$ is consistent, $\mathbf{x} = \mathbf{Gb}$ is a solution to $\mathbf{Ax} = \mathbf{b}$? If so, such a matrix $\mathbf{G}$ can be thought of as a generalization of inverse/ right inverse/ left inverse.

We shall now establish the existence of a matrix $\mathbf{G}$ that will have most of these properties, there by answering both the questions posed in the preceding paragraphs.

Let $\mathbf{A}$ be an $m \times n$ matrix. If $\mathbf{A}$ is null, then $\mathbf{AGA} = \mathbf{A}$ for all matrices $\mathbf{G}$ of order $n \times m$. In this case, $\mathbf{Ax} = \mathbf{0}$ is the only consistent system and $\mathbf{x} = \mathbf{G0} = \mathbf{0}$ is a solution for each matrix $\mathbf{G}$ of order $n \times m$. If $\mathbf{A}$ is non-null, let $(\mathbf{P}, \mathbf{Q})$ be a rank factorization of $\mathbf{A}$. Then $\mathbf{P}$ has a left inverse $\mathbf{P}_{\mathbf{L}}^{-1}$ and $\mathbf{Q}$ has a right inverse $\mathbf{Q}_{\mathbf{R}}^{-1}$. Write $\mathbf{G} = \mathbf{Q}_{\mathbf{R}}^{-1}\mathbf{P}_{\mathbf{L}}^{-1}$. It is easy to check that $\mathbf{AGA} = \mathbf{A}$. Also, if $\mathbf{Ax} = \mathbf{b}$ is consistent, then $\mathbf{b} = \mathbf{Ac}$ for some vector $\mathbf{c}$. Now, $\mathbf{AGb} = \mathbf{AGAc} = \mathbf{Ac} = \mathbf{b}$, showing that $\mathbf{Gb}$ is a solution to $\mathbf{Ax} = \mathbf{b}$. Thus, for every matrix $\mathbf{A}$, there is a matrix $\mathbf{G}$ having both the properties. In fact a stronger result holds:

Theorem 2.3.1. *Let* $\mathbf{A}$ *is an* $m \times n$ *matrix and* $\mathbf{G}$ *be an* $n \times m$ *matrix. Then the following are equivalent:*

(i) $\mathbf{Gb}$ *is a solution to* $\mathbf{Ax} = \mathbf{b}$ *for all* $\mathbf{b} \in \mathcal{C}(\mathbf{A})$
(ii) $\mathbf{AGA} = \mathbf{A}$
(iii) $\mathbf{AG}$ *is idempotent and* $\rho(\mathbf{AG}) = \rho(\mathbf{A})$
(iv) $\mathbf{GA}$ *is idempotent and* $\rho(\mathbf{GA}) = \rho(\mathbf{A})$
(v) $\rho(\mathbf{I} - \mathbf{AG}) = \mathbf{m} - \rho(\mathbf{A})$ *and*
(vi) $\rho(\mathbf{I} - \mathbf{GA}) = \mathbf{n} - \rho(\mathbf{A})$.

Proof. (i) $\Leftrightarrow$ (ii)
$\mathbf{AGb} = \mathbf{b}$ for all $\mathbf{b} \in \mathcal{C}(\mathbf{A}) \Leftrightarrow \mathbf{AGAu} = \mathbf{Au}$ for all $\mathbf{u} \Leftrightarrow \mathbf{AGA} = \mathbf{A}$.
(ii) $\Rightarrow$ (iii) is trivial.
(iii) $\Rightarrow$ (ii)
Since $\mathcal{C}(\mathbf{AG}) \subseteq \mathcal{C}(\mathbf{A})$ and $\rho(\mathbf{AG}) = \rho(\mathbf{A})$, we have $\mathcal{C}(\mathbf{AG}) = \mathcal{C}(\mathbf{A})$. So, $\mathbf{A} = \mathbf{AGD}$ for some matrix $\mathbf{D}$. Further, since $\mathbf{AG}$ is idempotent, $\mathbf{AGAG} = \mathbf{AG}$. So, $\mathbf{AGAGD} = \mathbf{AGD}$ or $\mathbf{AGA} = \mathbf{A}$.

(ii) $\Leftrightarrow$ (iv) is established similarly.
(iv) $\Rightarrow$ (vi)
Since $\mathbf{GA}$ is idempotent, so is $\mathbf{I} - \mathbf{GA}$. Further, $\mathbf{I} = \mathbf{GA} + (\mathbf{I} - \mathbf{GA})$. Therefore,

$$n = \rho(\mathbf{I}) = \text{tr}(\mathbf{I}) = \text{tr}(\mathbf{GA}) + \text{tr}(\mathbf{I} - \mathbf{GA}) = \rho(\mathbf{GA}) + \rho(\mathbf{I} - \mathbf{GA}).$$

Since $\rho(\mathbf{GA}) = \rho(\mathbf{A})$, the result follows.
(vi) $\Rightarrow$ (iv)
Now, $\mathbf{I} = \mathbf{GA} + (\mathbf{I} - \mathbf{GA})$. So,

$$n \leq \rho(\mathbf{GA}) + \rho(\mathbf{I} - \mathbf{GA}) \leq \rho(\mathbf{A}) + \rho(\mathbf{I} - \mathbf{GA}) = n.$$

It follows that $\rho(\mathbf{GA}) = \rho(\mathbf{A})$. Again,

$$\rho(\mathbf{I}) = \rho(\mathbf{GA} + (\mathbf{I} - \mathbf{GA})) = \rho(\mathbf{GA}) + \rho(\mathbf{I} - \mathbf{GA}),$$

so, the column spaces of $\mathbf{GA}$ and $\mathbf{I} - \mathbf{GA}$ are virtually disjoint and so are the row spaces. Moreover, $\mathbf{GA}(\mathbf{I} - \mathbf{GA}) = (\mathbf{I} - \mathbf{GA})\mathbf{GA}$. Hence $\mathbf{GA}(\mathbf{I} - \mathbf{GA}) = \mathbf{0}$. So, $\mathbf{GA}$ is idempotent.

The proof for (iii) $\Leftrightarrow$ (v) can be completed along the lines of proof for (iv) $\Leftrightarrow$ (vi). □

Thus, we see that both the questions raised in the beginning of this section lead to the same class of solutions.

Definition 2.3.2. Let $\mathbf{A}$ be an $m \times n$ matrix. Then a matrix $\mathbf{G}$ is said to be a generalized inverse of $\mathbf{A}$ if it satisfies any one of the six equivalent conditions in Theorem 2.3.1. A generalized inverse (or in short a g-inverse) of $\mathbf{A}$ is denoted by $\mathbf{A}^-$ and the class of all generalized inverses of $\mathbf{A}$ is denoted by $\{\mathbf{A}^-\}$.

We shall now study the use of a g-inverse in solving linear equations.

Theorem 2.3.3. *Let $\mathbf{A}$ be an $m \times n$ matrix and let $\mathbf{G}$ be a g-inverse of $\mathbf{A}$. Then*

(i) $\mathbf{Ax} = \mathbf{b}$ *is consistent if and only if* $\mathbf{AGb} = \mathbf{b}$.
(ii) $\mathcal{N}(\mathbf{A})$, *the null space of* $\mathbf{A}$ *(or equivalently the class of all solutions to* $\mathbf{Ax} = \mathbf{0}$*) is given by* $\mathcal{C}(\mathbf{I} - \mathbf{GA})$ *(or equivalently,* $(\mathbf{I} - \mathbf{GA})\xi$, *where* ξ *is arbitrary).*
(iii) *Let* $\mathbf{Ax} = \mathbf{b}$ *be consistent. Then the class of all solutions to* $\mathbf{Ax} = \mathbf{b}$ *is given by* $\mathbf{Gb} + (\mathbf{I} - \mathbf{GA})\xi$, *where* ξ *is arbitrary.*

Proof. (i) 'If' part is obvious and the 'Only if' part follows from Theorem 2.3.1(i).
(ii) Since $\mathbf{A}(\mathbf{I}-\mathbf{GA}) = \mathbf{0}$, $\mathcal{C}(\mathbf{I}-\mathbf{GA}) \subseteq \mathcal{N}(\mathbf{A})$. Now, by Theorem 2.3.1(vi), $\rho(\mathbf{I}-\mathbf{GA}) = \mathbf{n} - \rho(\mathbf{A}) = \mathbf{d}(\mathcal{N}(\mathbf{A}))$, the dimension of the null space of $\mathbf{A}$, it follows that $\mathcal{C}(\mathbf{I}-\mathbf{GA}) = \mathcal{N}(\mathbf{A})$.
(iii) follows from (ii) and the fact that $\mathbf{Gb}$ is a solution to $\mathbf{Ax} = \mathbf{b}$. □

Thus, a g-inverse can be used to check the consistency of a given system of linear equations and also to obtain an expression for all possible solutions to a consistent system of linear equations.

Remark 2.3.4. Let $\mathbf{A}$ be an $m \times n$ matrix and $\mathbf{G}$ be a g-inverse of $\mathbf{A}$. Then for a matrix $\mathbf{B}$, $\mathcal{C}(\mathbf{B}) \subseteq \mathcal{C}(\mathbf{A})$ if and only if $\mathbf{B} = \mathbf{AGB}$.

Theorem 2.3.5. *Let* $\mathbf{Ax} = \mathbf{b}$ *be consistent and* $\mathbf{b} \neq \mathbf{0}$. *Then the class of all solutions of* $\mathbf{Ax} = \mathbf{b}$ *is* $\{\mathbf{Gb} : \mathbf{G} \in \{\mathbf{A}^-\}\}$.

Proof. Let $\mathbf{G}$ be a g-inverse of $\mathbf{A}$. Clearly, $\mathbf{AGb} = \mathbf{b}$, since $\mathbf{Ax} = \mathbf{b}$ is consistent. Hence, $\{\mathbf{Gb} : \mathbf{G} \in \{\mathbf{A}^-\}\}$ is a subset of the class of all solutions to $\mathbf{Ax} = \mathbf{b}$.

Now, let $\mathbf{u}$ be a solution to $\mathbf{Ax} = \mathbf{b}$ and $\mathbf{G}$ be a g-inverse of $\mathbf{A}$. Then by Theorem 2.3.3(iii), $\mathbf{u} = \mathbf{Gb} + (\mathbf{I}-\mathbf{GA})\xi$ for some ξ. Since $\mathbf{b} \neq \mathbf{0}$, there exists an integer j such that $b_j \neq 0$. Let $w_{ij} = \mathrm{\xi_i/b_j}$, $i = 1, \ldots, n$ and $w_{ik} = 0$, $k = 1, \ldots, n$ whenever $i \neq j$. Write $\mathbf{W} = (w_{pq})$. It is easy to see that $\mathbf{Wb} = \xi$. So,

$$\mathbf{u} = \mathbf{Gb} + (\mathbf{I}-\mathbf{GA})\xi = \mathbf{Gb} + (\mathbf{I}-\mathbf{GA})\mathbf{Wb} = (\mathbf{G} + (\mathbf{I}-\mathbf{GA})\mathbf{W})\mathbf{b}.$$

Now $\mathbf{A}(\mathbf{G}+(\mathbf{I}-\mathbf{GA})\mathbf{W})\mathbf{A} = \mathbf{AGA} = \mathbf{A}$. Hence, $\mathbf{G}+(\mathbf{I}-\mathbf{GA})\mathbf{W} \in \{\mathbf{A}^-\}$ and that $\mathbf{u} \in \{\mathbf{Gb} : \mathbf{G} \in \{\mathbf{A}^-\}\}$. □

We shall now obtain several expressions for the class of all g-inverses of a matrix. Notice that every matrix of order $n \times m$ is a g-inverse of the null matrix of order $m \times n$.

Theorem 2.3.6. *Let* $\mathbf{A}$ *be an* $m \times n$ *matrix of rank* $r(>0)$ *and let* $\mathbf{A} = \mathbf{P}\mathrm{\mathbf{diag}}(\mathbf{I},\ \mathbf{0})\mathbf{Q}$ *be a normal form of* $\mathbf{A}$ *(see Theorem 2.2.1). Then the class of all g-inverses of* $\mathbf{A}$ *is given by* $\mathbf{Q}^{-1}\begin{pmatrix}\mathbf{L_r} & \mathbf{L}\\ \mathbf{M} & \mathbf{N}\end{pmatrix}\mathbf{P}^{-1}$, *where* $\mathbf{L}, \mathbf{M}$ *and* $\mathbf{N}$ *are arbitrary.*

Proof. Straightforward. □

Remark 2.3.7. Let $\mathbf{A}$ be an $m \times n$ matrix. Let $\mathbf{P}$ and $\mathbf{Q}$ be non-singular matrices of order $m \times m$ and $n \times n$ respectively. Then $\mathbf{G}$ is a g-inverse of $\mathbf{A} \Leftrightarrow \mathbf{Q}^{-1}\mathbf{G}\mathbf{P}^{-1}$ is a g-inverse of $\mathbf{PAQ}$.

Theorem 2.3.8. *Let $\mathbf{A}$ be an $m \times n$ matrix and let $\mathbf{G}$ be a specific g-inverse of $\mathbf{A}$. Then the class of all g-inverses of $\mathbf{A}$ is given by*

$$\{\mathbf{A}^-\} = \{\mathbf{G} + \mathbf{U} - \mathbf{GAUAG},\ \mathbf{U} \textit{ arbitrary}\}$$

or equivalently

$$\{\mathbf{A}^-\} = \{\mathbf{G} + (\mathbf{I} - \mathbf{GA})\mathbf{V} + \mathbf{W}(\mathbf{I} - \mathbf{AG}),\ \mathbf{V},\ \mathbf{W} \textit{arbitrary}\}.$$

Proof. First notice that for all $\mathbf{U}$,

$$\begin{aligned}\mathbf{A}(\mathbf{G} + \mathbf{U} - \mathbf{GAUAG})\mathbf{A} &= \mathbf{AGA} + \mathbf{AUA} - \mathbf{AGAUAGA}\\ &= \mathbf{AGA}\\ &= \mathbf{A}.\end{aligned}$$

So, $\mathbf{G} + \mathbf{U} - \mathbf{GAUAG}$ is a g-inverse of $\mathbf{A}$ for all $\mathbf{U}$.

Let $\mathbf{G}_1$ be another g-inverse of $\mathbf{A}$. Choose $\mathbf{U} = \mathbf{G}_1 - \mathbf{G}$ and check that $\mathbf{G}_1 = \mathbf{G} + \mathbf{U} - \mathbf{GAUAG}$. Thus, the class of all g-inverses of $\mathbf{A}$ is given by

$$\{\mathbf{A}^-\} = \{\mathbf{G} + \mathbf{U} - \mathbf{GAUAG},\ \mathbf{U} \text{ arbitrary}\}.$$

Consider now $\mathbf{H} = \mathbf{G} + (\mathbf{I} - \mathbf{GA})\mathbf{V} + \mathbf{W}(\mathbf{I} - \mathbf{AG})$, where $\mathbf{V}$ and $\mathbf{W}$ are arbitrary matrices of appropriate order. It is easy to see that $\mathbf{H}$ is a g-inverse of $\mathbf{A}$ for all $\mathbf{V}$ and $\mathbf{W}$.

Choose and fix $\mathbf{V}$ and $\mathbf{W}$. Let $\mathbf{U} = (\mathbf{I} - \mathbf{GA})\mathbf{V} + \mathbf{W}(\mathbf{I} - \mathbf{AG})$. Then $\mathbf{AUA} = \mathbf{0}$. Therefore,

$$\mathbf{G} + (\mathbf{I} - \mathbf{GA})\mathbf{V} + \mathbf{W}(\mathbf{I} - \mathbf{AG}) = \mathbf{G} + \mathbf{U} - \mathbf{GAUAG},$$

so, $\mathbf{H} \in \{\mathbf{A}^-\}$ for all $\mathbf{V}$ and $\mathbf{W}$. Next, choose and fix $\mathbf{U}$. Then

$$\mathbf{G} + \mathbf{U} - \mathbf{GAUAG} = \mathbf{G} + (\mathbf{I} - \mathbf{GA})\mathbf{UAG} + \mathbf{U}(\mathbf{I} - \mathbf{AG}).$$

Let $\mathbf{V} = \mathbf{UAG}$ and $\mathbf{W} = \mathbf{U}$. Hence,

$$\begin{aligned}\{\mathbf{A}^-\} &= \{\mathbf{G} + \mathbf{U} - \mathbf{GAUAG},\ \mathbf{U} \text{ arbitrary}\}\\ &= \{\mathbf{G} + (\mathbf{I} - \mathbf{GA})\mathbf{V} + \mathbf{W}(\mathbf{I} - \mathbf{AG}),\ \mathbf{V},\ \mathbf{W} \text{ arbitrary}\}.\end{aligned}$$ □

We now give a basic lemma which will be needed in obtaining a useful result on the invariance of $\mathbf{BA}^-\mathbf{C}$ under the choice of g-inverses $\mathbf{A}^-$ of $\mathbf{A}$.

Lemma 2.3.9. *Let $\mathbf{a}$ and $\mathbf{c}$ be m-column and n-column vectors respectively such that $\mathbf{c} \neq \mathbf{0}$ and $\mathbf{a}^t\mathbf{Z}\mathbf{c} = \mathbf{0}$ for all $m \times n$ matrices $\mathbf{Z}$. Then $\mathbf{a} = \mathbf{0}$.*

Proof. If possible, let $\mathbf{a} \neq \mathbf{0}$. Then there exists an element say a_i of $\mathbf{a}$, which is non-null. Since $\mathbf{c} \neq \mathbf{0}$, there is an element c_j of $\mathbf{c}$ which is non-null. Take a matrix $\mathbf{Z}$ having the $(i, j)^{th}$ element $z_{ij} = 1$ and all other elements null. Then $\mathbf{a}^t\mathbf{Z}\mathbf{c} = a_i c_j \neq 0$, which is a contradiction. So, $\mathbf{a} = \mathbf{0}$. □

Remark 2.3.10. Similar result holds when $\mathbf{a}$ and $\mathbf{c}$ are matrices of suitable orders. More precisely, if $\mathbf{A}$ is a $m \times n$ matrix and $\mathbf{C}$ (non-null) is a $p \times q$ matrix such that $\mathbf{A}^t\mathbf{Z}\mathbf{C} = \mathbf{0}$ for all $n \times p$ matrices $\mathbf{Z}$, then $\mathbf{A} = \mathbf{0}$.

Theorem 2.3.11. *Let $\mathbf{A}$, $\mathbf{B}$ and $\mathbf{C}$ be matrices of appropriate orders such that the product $\mathbf{B}\mathbf{A}^-\mathbf{C}$ is defined. Then $\mathbf{B}\mathbf{A}^-\mathbf{C}$ is invariant under choices of $\mathbf{A}^-$ and is non-null if and only if $\mathbf{B}$ and $\mathbf{C}$ are non-null, $\mathcal{C}(\mathbf{C}) \subseteq \mathcal{C}(\mathbf{A})$ and $\mathcal{C}(\mathbf{B}^t) \subseteq \mathcal{C}(\mathbf{A}^t)$.*

Proof. 'If' part is trivial.
'Only if' part
Let $\mathbf{G}$ be a g-inverse of $\mathbf{A}$. Then $\mathbf{G}+(\mathbf{I}-\mathbf{G}\mathbf{A})\mathbf{V}+\mathbf{W}(\mathbf{I}-\mathbf{A}\mathbf{G})$ is a g-inverse of $\mathbf{A}$, for arbitrary matrices $\mathbf{V}$ and $\mathbf{W}$ of appropriate orders. Now, $\mathbf{B}\mathbf{G}\mathbf{C}$ is non-null $\Rightarrow$ $\mathbf{B}$ and $\mathbf{C}$ are non-null and

$$\mathbf{B}(\mathbf{G} + (\mathbf{I} - \mathbf{G}\mathbf{A})\mathbf{V} + \mathbf{W}(\mathbf{I} - \mathbf{A}\mathbf{G}))\mathbf{C} = \mathbf{B}\mathbf{G}\mathbf{C}$$

for all $\mathbf{V}$ and $\mathbf{W}$. So, $\mathbf{B}((\mathbf{I}-\mathbf{G}\mathbf{A})\mathbf{V}+\mathbf{W}(\mathbf{I}-\mathbf{A}\mathbf{G}))\mathbf{C} = \mathbf{0}$ for all $\mathbf{V}$ and $\mathbf{W}$. Taking $\mathbf{W} = \mathbf{0}$, we have $\mathbf{B}(\mathbf{I}-\mathbf{G}\mathbf{A})\mathbf{V}\mathbf{C} = \mathbf{0}$ for all $\mathbf{V}$. As $\mathbf{C}$ is non-null, by Lemma 2.3.10, $\mathbf{B}(\mathbf{I}-\mathbf{G}\mathbf{A}) = \mathbf{0}$. Thus, $\mathcal{C}(\mathbf{B}^t) \subseteq \mathcal{C}(\mathbf{A}^t)$. Similarly by taking $\mathbf{V} = \mathbf{0}$, we have $\mathcal{C}(\mathbf{C}) \subset \mathcal{C}(\mathbf{A})$. □

In the later sections we often need to get a g-inverse $\mathbf{G}$ of $\mathbf{A}$ such that column space or row space (of $\mathbf{G}$) is contained in a specified subspace. We shall now investigate when it is possible to get such a matrix $\mathbf{G}$. We shall also obtain an explicit expression for such a g-inverse.

Theorem 2.3.12. *Let $\mathbf{A}$ be an $m \times n$ matrix. Let r and s be positive integers and* $\mathbf{P}$ and $\mathbf{Q}$ *be matrices of orders $n \times s$ and $m \times r$ respectively. Then there exists a g-inverse $\mathbf{G}$ of $\mathbf{A}$ such that $\mathcal{C}(\mathbf{G}) \subseteq \mathcal{C}(\mathbf{P})$ and $\mathcal{C}(\mathbf{G}^t) \subseteq \mathcal{C}(\mathbf{Q}^t)$ if and only if $\rho(\mathbf{Q}\mathbf{A}\mathbf{P}) = \rho(\mathbf{A})$. If $\rho(\mathbf{Q}\mathbf{A}\mathbf{P}) = \rho(\mathbf{A})$, then $\mathbf{P}(\mathbf{Q}\mathbf{A}\mathbf{P})^-\mathbf{Q}$ is a g-inverse of $\mathbf{A}$. Further, if $\mathbf{P}\mathbf{C}\mathbf{Q}$ is a g-inverse of $\mathbf{A}$, then $\mathbf{C}$ is a g-inverse of $\mathbf{Q}\mathbf{A}\mathbf{P}$.*

Proof. If $\mathbf{G}$ is a matrix such that $\mathcal{C}(\mathbf{G}) \subseteq \mathcal{C}(\mathbf{P})$ and $\mathcal{C}(\mathbf{G}^t) \subseteq \mathcal{C}(\mathbf{Q}^t)$, then $\mathbf{G} = \mathbf{P}\mathbf{C}\mathbf{Q}$ for some matrix $\mathbf{C}$. Now, $\mathbf{A}\mathbf{G}\mathbf{A} = \mathbf{A} \Rightarrow \mathbf{A}\mathbf{P}\mathbf{C}\mathbf{Q}\mathbf{A} = \mathbf{A} \Rightarrow \rho(\mathbf{A}\mathbf{P}) = \rho(\mathbf{Q}\mathbf{A}) = \rho(\mathbf{A})$. Further, $\mathbf{Q}\mathbf{A}\mathbf{P}\mathbf{C}\mathbf{Q}\mathbf{A} = \mathbf{Q}\mathbf{A}$. Hence $\rho(\mathbf{Q}\mathbf{A}\mathbf{P}) = \rho(\mathbf{Q}\mathbf{A})$ and therefore, $\rho(\mathbf{Q}\mathbf{A}\mathbf{P}) = \rho(\mathbf{A})$.

Conversely, let $\rho(\mathbf{QAP}) = \rho(\mathbf{A})$. Then,

$$\rho(\mathbf{QAP}) = \rho(\mathbf{QA}) = \rho(\mathbf{AP}) = \rho(\mathbf{A}).$$

Consider $\mathbf{T} = \mathbf{P}(\mathbf{QAP})^-$. Now, $\mathbf{QAT} = \mathbf{QAP}(\mathbf{QAP})^-$ is idempotent. Also, $\rho(\mathbf{QAT}) = \rho(\mathbf{QAP}(\mathbf{QAP})^-) = \rho(\mathbf{QAP}) = \rho(\mathbf{QA})$. Therefore, $\mathbf{T}$ is a g-inverse of $\mathbf{QA}$, by Theorem 2.3.1(iii). Similarly, $\mathbf{TQ} = (\mathbf{QA})^-\mathbf{Q}$ is a g-inverse of $\mathbf{A}$. Thus, $\mathbf{P}(\mathbf{QAP})^-\mathbf{Q}$ is a g-inverse of $\mathbf{A}$.

Now, let $\mathbf{PCQ}$ be a g-inverse of $\mathbf{A}$. So, $\mathbf{APCQA} = \mathbf{A}$. Pre- and post-multiplying by $\mathbf{Q}$ and $\mathbf{P}$ respectively, we have $\mathbf{QAPCQAP} = \mathbf{QAP}$, showing $\mathbf{C}$ is a g-inverse of $\mathbf{QAP}$. □

We know that, if $\mathbf{G}$ is a g-inverse of $\mathbf{A}$, then $\mathbf{AGA} = \mathbf{A}$ and therefore, $\rho(\mathbf{G}) \geq \rho(\mathbf{A})$. Let $\rho(\mathbf{A}) = r < \min\{m, n\}$. Then by using Theorem 2.3.6, it can be shown that for each s such that $r \leq s \leq \min\{m, n\}$, there is a g-inverse $\mathbf{G}$ of $\mathbf{A}$ with rank s. Take for example $\mathbf{G} = \mathbf{Q}^{-1}\mathbf{diag}(\mathbf{I_s}, \mathbf{0})\mathbf{P}^{-1}$.

We now specialize to the class of g-inverses $\mathbf{G}$ of $\mathbf{A}$ such that $\rho(\mathbf{G}) = \rho(\mathbf{A})$.

Theorem 2.3.13. *Let* $\mathbf{A}$ *be an* $m \times n$ *matrix. Then* $\mathbf{G}$ *is a g-inverse of* $\mathbf{A}$ *such that* $\rho(\mathbf{G}) = \rho(\mathbf{A})$ *if only if* $\mathbf{AGA} = \mathbf{A}$ *and* $\mathbf{GAG} = \mathbf{G}$.

Proof. 'If' part
Since $\mathbf{AGA} = \mathbf{A}$, $\mathbf{G}$ is a g-inverse of $\mathbf{A}$ and $\rho(\mathbf{G}) \geq \rho(\mathbf{A})$. Moreover, $\mathbf{GAG} = \mathbf{G}$, so, $\rho(\mathbf{A}) \geq \rho(\mathbf{G})$. Hence $\rho(\mathbf{G}) = \rho(\mathbf{A})$.
'Only if' part
Since $\mathbf{G}$ is a g-inverse of $\mathbf{A}$ and $\rho(\mathbf{G}) = \rho(\mathbf{A})$, $\mathbf{AG}$ is idempotent and $\rho(\mathbf{AG}) = \rho(\mathbf{A}) = \rho(\mathbf{G})$. So, by (iv) of Theorem 2.3.1, $\mathbf{A}$ is a g-inverse of $\mathbf{G}$. Thus, $\mathbf{GAG} = \mathbf{G}$ and $\mathbf{G}$ being a g-inverse of $\mathbf{A}$, $\mathbf{AGA} = \mathbf{A}$. □

Definition 2.3.14. Let $\mathbf{A}$ be an $m \times n$ matrix. An $n \times m$ matrix $\mathbf{G}$ satisfying $\mathbf{AGA} = \mathbf{A}$ and $\mathbf{GAG} = \mathbf{G}$ is called a reflexive generalized inverse of $\mathbf{A}$, denoted by $\mathbf{A_r^-}$.

Remark 2.3.15. Notice the relationship $\mathbf{AGA} = \mathbf{A}$ and $\mathbf{GAG} = \mathbf{G}$ is one of symmetry between $\mathbf{A}$ and $\mathbf{G}$.

It is easy to see that $\{\mathbf{GAG} : \mathbf{G} \in \{\mathbf{A}^-\}\}$ is the class of all reflexive g-inverses of $\mathbf{A}$.

Remark 2.3.16. Analogous to Theorem 2.3.6, it is easy to show that if $\mathbf{A} = \mathbf{Pdiag}(\mathbf{I_r}, \mathbf{0})\mathbf{Q}$, then the class of all reflexive g-inverses of $\mathbf{A}$ is given by

$$\mathbf{A} = \mathbf{Q}^{-1}\begin{pmatrix} \mathbf{I_r} & \mathbf{L} \\ \mathbf{M} & \mathbf{ML} \end{pmatrix}\mathbf{P}^{-1},$$

where $\mathbf{L}$ and $\mathbf{M}$ are arbitrary.

For the null matrix of order $m \times n$, while every $n \times m$ matrix is a generalized inverse, there is only one reflexive g-inverse namely the null matrix.

We shall give below another characterization of reflexive g-inverses of a non-null matrix using rank factorization.

Theorem 2.3.17. *Let $\mathbf{A}$ be a non-null $m \times n$ matrix. Let $(\mathbf{P}, \mathbf{Q})$ be a rank factorization of $\mathbf{A}$. Then the class of all reflexive g-inverses of $\mathbf{A}$ is given by $\mathbf{Q_R^{-1}P_L^{-1}}$, where $\mathbf{P_L^{-1}}$ and $\mathbf{Q_R^{-1}}$ are arbitrary left and right inverses of $\mathbf{P}$ and $\mathbf{Q}$ respectively.*

Proof. Clearly, $\mathbf{G} = \mathbf{Q_R^{-1}P_L^{-1}}$ is a reflexive g-inverse of $\mathbf{A}$.

Conversely, let $\mathbf{G}$ be reflexive g-inverse of $\mathbf{A}$. Then $\mathbf{AGA} = \mathbf{A} \Rightarrow \mathbf{PQGPQ} = \mathbf{PQ} \Rightarrow \mathbf{QGP} = \mathbf{I}$, since $\mathbf{Q}$ has a right inverse and $\mathbf{P}$ has a left inverse. Clearly, a choice of $\mathbf{P_L^{-1}}$ is $\mathbf{QG}$ and a choice $\mathbf{Q_R^{-1}}$ is $\mathbf{GP}$.

Now, $\mathbf{GPQG} = \mathbf{GAG} = \mathbf{G}$. So, $\mathbf{G} = \mathbf{Q_R^{-1}P_L^{-1}}$. □

Let $\mathbf{A}$ be an $m \times n$ matrix of rank r $(< \min\{m, n\})$. Let s be an integer such that $r < s \le \min\{m, n\}$. We now obtain a characterization of all g-inverses of $\mathbf{A}$ with rank s.

Theorem 2.3.18. *Let $\mathbf{A}$ be a $m \times n$ matrix of rank r where $0 < r < \min\{m, n\}$. Let s be a positive integer such that $r < s \le \min\{m, n\}$. Then $\mathbf{G}$ is a g-inverse of $\mathbf{A}$ with $\rho(\mathbf{G}) = s$ if and only if then exist non-singular matrices $\mathbf{P}$ and $\mathbf{Q}$ of order $m \times m$ and $n \times n$ respectively such that $\mathbf{A} = \mathbf{P}\mathbf{diag}(\mathbf{I_r}, \mathbf{0})\mathbf{Q}$ and $\mathbf{G} = \mathbf{Q}^{-1}\mathbf{diag}(\mathbf{I_s}, \mathbf{0})\mathbf{P}^{-1}$.*

Proof. 'If' part is trivial.

'Only if' part

Let $\mathbf{A} = \mathbf{R_1}\mathbf{diag}(\mathbf{I_r}, \mathbf{0})\mathbf{R_2}$ be a normal form of $\mathbf{A}$ where $\mathbf{R_1}$ and $\mathbf{R_2}$ are non-singular. Since $\mathbf{G}$ is a g-inverse of $\mathbf{A}$,

$\mathbf{G} = \mathbf{R_2^{-1}}\begin{pmatrix} \mathbf{I_r} & \mathbf{L} \\ \mathbf{M} & \mathbf{N} \end{pmatrix}\mathbf{R_1^{-1}}$, for some matrices $\mathbf{L}, \mathbf{M}$ and $\mathbf{N}$.

Further, since $\mathbf{G}$ is of rank s, $\rho\begin{pmatrix} \mathbf{I_r} & \mathbf{L} \\ \mathbf{M} & \mathbf{N} \end{pmatrix} = s$. Now

$$\begin{pmatrix} \mathbf{I_r} & \mathbf{L} \\ \mathbf{M} & \mathbf{N} \end{pmatrix} = \begin{pmatrix} \mathbf{I_r} & \mathbf{0} \\ \mathbf{M} & \mathbf{I} \end{pmatrix}\begin{pmatrix} \mathbf{I_r} & \mathbf{0} \\ \mathbf{0} & \mathbf{N} - \mathbf{ML} \end{pmatrix}\begin{pmatrix} \mathbf{I_r} & \mathbf{L} \\ \mathbf{0} & \mathbf{I} \end{pmatrix}.$$

Hence $\rho(\mathbf{N}-\mathbf{ML}) = s-r$. Let $\mathbf{N}-\mathbf{ML} = \mathbf{R_3diag(I_{s-r},0)R_4}$ be a normal form of $\mathbf{N}-\mathbf{ML}$. Write

$$\mathbf{P} = \mathbf{R_1}\begin{pmatrix}\mathbf{I_r} & -\mathbf{L}\\ \mathbf{0} & \mathbf{I}\end{pmatrix}\begin{pmatrix}\mathbf{I_r} & \mathbf{0}\\ \mathbf{0} & \mathbf{R_4^{-1}}\end{pmatrix} \text{ and } \mathbf{Q} = \begin{pmatrix}\mathbf{I_r} & \mathbf{0}\\ \mathbf{0} & \mathbf{R_3^{-1}}\end{pmatrix}\begin{pmatrix}\mathbf{I_r} & \mathbf{0}\\ -\mathbf{M} & \mathbf{I}\end{pmatrix}\mathbf{R_2}.$$

It is easily verified that $\mathbf{A} = \mathbf{Pdiag(I_r,0)\,Q}$ and $\mathbf{G} = \mathbf{Q^{-1}diag(I_s,0)P^{-1}}$.

□

Corollary 2.3.19. *Consider the same setup as in Theorem 2.3.18. Then* $\mathbf{G}$ *is a g-inverse of* $\mathbf{A}$ *with rank* s *if and only if there exists a matrix* $\mathbf{B}$ *such that* (a) $\rho(\mathbf{A}+\mathbf{B}) = \rho(\mathbf{A})+\rho(\mathbf{B}) = s$ and (b) $\mathbf{G} = (\mathbf{A}+\mathbf{B})_\mathbf{r}^-$.

Proof. 'If' part

Let $(\mathbf{P_1},\mathbf{Q_1})$ and $(\mathbf{P_2},\mathbf{Q_2})$ be rank factorizations of $\mathbf{A}$ and $\mathbf{B}$ respectively. Since $\rho(\mathbf{A}+\mathbf{B}) = \rho(\mathbf{A})+\rho(\mathbf{B})$, $\left((\mathbf{P_1}:\mathbf{P_2}), \begin{pmatrix}\mathbf{Q_1}\\ \cdot\cdot\\ \mathbf{Q_2}\end{pmatrix}\right)$ is a rank factorization of $\mathbf{A}+\mathbf{B}$. Since $\mathbf{G}$ is a reflexive g-inverse of $\mathbf{A}+\mathbf{B}$, we have $\mathbf{G} = (\mathbf{R_1}:\mathbf{R_2})\begin{pmatrix}\mathbf{S_1}\\ \cdot\cdot\\ \mathbf{S_2}\end{pmatrix}$ where $(\mathbf{R_1}:\mathbf{R_2})$ is a right inverse of $\begin{pmatrix}\mathbf{Q_1}\\ \cdot\cdot\\ \mathbf{Q_2}\end{pmatrix}$ and $\begin{pmatrix}\mathbf{S_1}\\ \cdot\cdot\\ \mathbf{S_2}\end{pmatrix}$ is a left inverse of $(\mathbf{P_1}:\mathbf{P_2})$. Thus,

$$\mathbf{Q_1(R_1:R_2) = (I:0)} \text{ and } \begin{pmatrix}\mathbf{S_1}\\ \cdot\cdot\\ \mathbf{S_2}\end{pmatrix}\mathbf{P_1} = \begin{pmatrix}\mathbf{I}\\ \cdot\cdot\\ \mathbf{0}\end{pmatrix}.$$

Now,

$$\mathbf{AGA} = \mathbf{P_1Q_1(R_1:R_2)}\begin{pmatrix}\mathbf{S_1}\\ \cdot\cdot\\ \mathbf{S_2}\end{pmatrix}\mathbf{P_1Q_1} = \mathbf{P_1(I:0)}\begin{pmatrix}\mathbf{I}\\ \cdot\cdot\\ \mathbf{0}\end{pmatrix}\mathbf{Q_1} = \mathbf{P_1Q_1} = \mathbf{A}.$$

Thus, $\mathbf{G}$ is a g-inverse of $\mathbf{A}$. Further, $\rho(\mathbf{G}) = \rho(\mathbf{A}+\mathbf{B}) = s$.

'Only if' Part

By Theorem 2.3.18, $\mathbf{A} = \mathbf{Pdiag(I_r,0)Q}$ and $\mathbf{G} = \mathbf{Q^{-1}diag(I_s,0)P^{-1}}$ for some non-singular matrices $\mathbf{P}$ and $\mathbf{Q}$. Let $\mathbf{B} = \mathbf{Pdiag(0_r,I_{s-r},0)Q}$. Then $\mathbf{A}+\mathbf{B} = \mathbf{Pdiag(I_s,0)Q}$ and $\rho(\mathbf{A}+\mathbf{B}) = s = \rho(\mathbf{A})+\rho(\mathbf{B})$. Finally, $\mathbf{G} = \mathbf{Q^{-1}diag(I_s,0)P^{-1}}$ is $(\mathbf{A}+\mathbf{B})^-{}_r$. □

2.4 The group inverse

In the previous section, we studied one important sub-class of g-inverses of a matrix, namely, the class of reflexive g-inverses. In this section, we seek a g-inverse $\mathbf{G}$ of a square matrix $\mathbf{A}$ satisfying $\mathcal{C}(\mathbf{G}) \subseteq \mathcal{C}(\mathbf{A})$ or $\mathcal{C}(\mathbf{G^t}) \subseteq \mathcal{C}(\mathbf{A^t})$ or both. We shall see that not all matrices enjoy this property. It turns out that the same subclass of square matrices enjoys each of the properties mentioned above. We seek to identify this subclass of square matrices and for each matrix in this class, we characterize the sub-classes of all g-inverses with these properties. This leads us to the concept of the group inverse in a very natural manner. We study briefly the major properties of the group inverse. We also define the Drazin inverse for a general square matrix by extending the concept of the group inverse of a matrix of index 1. The group inverse is found useful in the analysis of Markov chains and the Drazin inverse is useful in solving differential equations and difference equations.

Definition 2.4.1. Let $\mathbf{A}$ be an $n \times n$ matrix. Then a g-inverse $\mathbf{G}$ of $\mathbf{A}$ such that $\mathcal{C}(\mathbf{G}) \subseteq \mathcal{C}(\mathbf{A})$ is called a χ-inverse of $\mathbf{A}$ and is denoted by $\mathbf{A}_\chi^-$.

A g-inverse such that $\mathcal{C}(\mathbf{G^t}) \subseteq \mathcal{C}(\mathbf{A^t})$ is called a ρ-inverse of $\mathbf{A}$ and is denoted by $\mathbf{A}_\rho^-$.

A g-inverse $\mathbf{G}$ of $\mathbf{A}$ such that $\mathcal{C}(\mathbf{G}) \subseteq \mathcal{C}(\mathbf{A})$ and $\mathcal{C}(\mathbf{G^t}) \subseteq \mathcal{C}(\mathbf{A^t})$ is called a $\rho\chi$-inverse of $\mathbf{A}$ and is denoted by $\mathbf{A}_{\rho\chi}^-$.

Theorem 2.4.2. *Let* $\mathbf{A}$ *be an* $n \times n$ *matrix. Then each of* $\mathbf{A}_\chi^-$, $\mathbf{A}_\rho^-$ *and* $\mathbf{A}_{\rho\chi}^-$ *exists if and only if* $\rho(\mathbf{A}) = \rho(\mathbf{A^2})$ *i.e.,* $\mathbf{A}$ *is of index not greater than 1.*

Proof. By Theorem 2.3.12, each of $\mathbf{A}_\chi^-$ and $\mathbf{A}_\rho^-$ exists $\Leftrightarrow \rho(\mathbf{A}) = \rho(\mathbf{A^2})$. Again by the same theorem, $\mathbf{A}_{\rho\chi}^-$ exists $\Leftrightarrow \rho(\mathbf{A}) = \rho(\mathbf{A^3})$. Since $\rho(\mathbf{A}) \geq \rho(\mathbf{A^2}) \geq \rho(\mathbf{A^3})$, $\mathbf{A}_{\rho\chi}^-$ exists $\Leftrightarrow \rho(\mathbf{A}) = \rho(\mathbf{A^2})$. □

Let $\mathbf{A}$ be an $n \times n$ matrix such that $\rho(\mathbf{A}) = \rho(\mathbf{A^2})$. Then by taking $\mathbf{Q} = \mathbf{I}$ and $\mathbf{P} = \mathbf{A}$ and applying Theorem 2.3.12, we see that a choice of $\mathbf{A}_\chi^-$ is $\mathbf{A}(\mathbf{A^2})^-$. Similarly, by taking $\mathbf{Q} = \mathbf{A}$, $\mathbf{P} = \mathbf{I}$, we see that a choice of $\mathbf{A}_\rho^-$ is $(\mathbf{A^2})^-\mathbf{A}$, where $(\mathbf{A^2})^-$ is any g-inverse of $\mathbf{A^2}$.

Theorem 2.4.3. *Let* $\mathbf{A}$ *be an* $n \times n$ *matrix such that* $\rho(\mathbf{A}) = \rho(\mathbf{A^2})$. *Then* $\mathbf{A}(\mathbf{A^3})^-\mathbf{A}$ *is an* $\mathbf{A}_{\rho\chi}^-$ *where* $(\mathbf{A^3})^-$ *is any g-inverse of* $\mathbf{A^3}$. *Further,* $\mathbf{A}_{\rho\chi}^-$ *is unique.*

Proof. By Theorem 2.3.12, it follows that $\mathbf{G} = \mathbf{A}(\mathbf{A^3})^-\mathbf{A}$ is a g-inverse of $\mathbf{A}$ such that $\mathcal{C}(\mathbf{G}) \subseteq \mathcal{C}(\mathbf{A})$ and $\mathcal{C}(\mathbf{G^t}) \subseteq \mathcal{C}(\mathbf{A^t})$. Further, if $\mathbf{ACA}$ is g-inverse of $\mathbf{A}$, then $\mathbf{C}$ must be of the form $(\mathbf{A^3})^-$ for some g- inverse $(\mathbf{A^3})^-$ of $\mathbf{A^3}$, again by Theorem 2.3.12. Since $\rho(\mathbf{A}) = \rho(\mathbf{A^2})$, $\mathcal{C}(\mathbf{A^3}) = \mathcal{C}(\mathbf{A})$, $\mathcal{C}((\mathbf{A^3})^\mathbf{t}) = \mathcal{C}(\mathbf{A^t})$. So, by Theorem 2.3.11, $\mathbf{G} = \mathbf{A}(\mathbf{A^3})^-\mathbf{A}$ is invariant under choices of g-inverses of $\mathbf{A^3}$. Thus $\mathbf{A}^-_{\rho\chi}$ exists, is unique and is equal to $\mathbf{A}(\mathbf{A^3})^-\mathbf{A}$. □

We shall now characterize the $\rho\chi$-inverse and the classes of all χ-inverses and all ρ-inverses using rank factorization.

Theorem 2.4.4. *Let* $\mathbf{A}$ *be an* $n \times n$ *matrix of rank* $r(>0)$. *Let* $(\mathbf{P}, \mathbf{Q})$ *be a rank factorization of* $\mathbf{A}$. *Then*

(i) $\rho(\mathbf{A}) = \rho(\mathbf{A^2})$ *if and only if* $\mathbf{QP}$ *is non-singular.*
(ii) *Let* $\rho(\mathbf{A}) = \rho(\mathbf{A^2})$. *The class of all* χ*-inverses is given by* $\{\mathbf{P}(\mathbf{QP})^{-1}\mathbf{P}_\mathbf{L}^{-1} : \mathbf{P}_\mathbf{L}^{-1}$ *is an arbitrary left inverse of* $\mathbf{P}\}$.
(iii) *Let* $\rho(\mathbf{A}) = \rho(\mathbf{A^2})$. *The class of all* ρ*-inverses is given by* $\{\mathbf{Q}_\mathbf{R}^{-1}(\mathbf{QP})^{-1}\mathbf{Q} : \mathbf{Q}_\mathbf{R}^{-1}$ is a right inverse of $\mathbf{Q}\}$.
(iv) *Let* $\rho(\mathbf{A}) = \rho(\mathbf{A^2})$. *Then the* $\rho\chi$*-inverse of* $\mathbf{A}$ *is* $\mathbf{P}((\mathbf{QP})^{-1})^\mathbf{2}\mathbf{Q}$.

Proof. (i) We have

$$\begin{aligned}\rho(\mathbf{A}) &= \rho(\mathbf{A^2})\\ &\Leftrightarrow \rho(\mathbf{PQ}) = \rho(\mathbf{PQPQ}) = \rho(\mathbf{QP})\\ &\Leftrightarrow \rho(\mathbf{QP}) = \rho(\mathbf{I}) = r.\end{aligned}$$

Thus, $\rho(\mathbf{A}) = \rho(\mathbf{A^2}) \Leftrightarrow \mathbf{QP}$ is non-singular.
(ii) Since $\rho(\mathbf{A}) = \rho(\mathbf{A^2})$, $\mathbf{A}^-_\chi$ exists. Clearly, $\rho(\mathbf{A}^-_\chi) \geq \rho(\mathbf{A})$. On the other hand, $\mathcal{C}(\mathbf{A}^-_\chi) \subseteq \mathcal{C}(\mathbf{A})$, so $\mathcal{C}(\mathbf{A}^-_\chi) = \mathcal{C}(\mathbf{A})$ and $\rho(\mathbf{A}^-_\chi) = \rho(\mathbf{A})$. Hence, $\mathbf{A}^-_\chi$ is a reflexive g-inverse of $\mathbf{A}$. By Theorem 2.3.15, every reflexive g-inverse is of the type $\mathbf{Q}_\mathbf{R}^{-1}\mathbf{P}_\mathbf{L}^{-1}$. Also, $\mathcal{C}(\mathbf{Q}_\mathbf{R}^{-1}\mathbf{P}_\mathbf{L}^{-1}) = \mathcal{C}(\mathbf{Q}_\mathbf{R}^{-1})$. Since $\mathcal{C}(\mathbf{A}^-\chi) = \mathcal{C}(\mathbf{A}) = \mathcal{C}(\mathbf{P})$, $\mathbf{Q}_\mathbf{R}^{-1} = \mathbf{PT}$ for some matrix $\mathbf{T}$. Further, $\mathbf{T}$ must satisfy $\mathbf{I} = \mathbf{QQ}_\mathbf{R}^{-1} = \mathbf{QPT}$. As $\mathbf{QP}$ is a square matrix, we have $\mathbf{T} = (\mathbf{QP})^{-1}$. Thus, every χ-inverse must be of the form $\mathbf{P}(\mathbf{QP})^{-1}\mathbf{P}_\mathbf{L}^{-1}$.

Conversely, $\mathbf{A}(\mathbf{P}(\mathbf{QP})^{-1}\mathbf{P}_\mathbf{L}^{-1})\mathbf{A} = \mathbf{PQP}(\mathbf{QP})^{-1}\mathbf{P}_\mathbf{L}^{-1}\mathbf{PQ} = \mathbf{PQ} = \mathbf{A}$ and $\mathcal{C}(\mathbf{P}(\mathbf{QP})^{-1}\mathbf{P}_\mathbf{L}^{-1}) = \mathcal{C}(\mathbf{P}) = \mathcal{C}(\mathbf{A})$, for each $\mathbf{P}_\mathbf{L}^{-1}$. So, $\mathbf{P}(\mathbf{QP})^{-1}\mathbf{P_L}^{-1}$ is a χ-inverse of $\mathbf{A}$.
Proof of (iii) is similar to proof of (ii).
(iv) follows form (ii) and (iii). □

Theorem 2.4.5. *Let* $\mathbf{A}$ *be an* $n \times n$ *matrix such that* $\rho(\mathbf{A}) = \rho(\mathbf{A^2})$. *Let* $\mathbf{A} = \mathbf{P}\mathbf{diag}(\mathbf{C}, \mathbf{0})\ \mathbf{P}^{-1}$, *where* $\mathbf{P}$ *and* $\mathbf{C}$ *are non-singular.*

(i) *The class of all* $\mathbf{A}_{\chi}^{-}$ *is given by*

$$\mathbf{P}\begin{pmatrix}\mathbf{C}^{-1} & \mathbf{L}\\ \mathbf{0} & \mathbf{0}\end{pmatrix}\mathbf{P}^{-1}\ ; \text{ where } \mathbf{L} \text{ is arbitrary.}$$

(ii) *The class of all* $\mathbf{A}_{\rho}^{-}$ *is given by*

$$\mathbf{P}\begin{pmatrix}\mathbf{C}^{-1} & \mathbf{0}\\ \mathbf{M} & \mathbf{0}\end{pmatrix}\mathbf{P}^{-1}\ ; \text{ where } \mathbf{M} \text{ is arbitrary.}$$

(iii) $\mathbf{G}$ *is* $\mathbf{A}_{\rho\chi}^{-}$ *if and only if* $\mathbf{G} = \mathbf{P}\mathbf{diag}(\mathbf{C}^{-1}, \mathbf{0})\ \mathbf{P}^{-1}$

Proof. Straightforward. □

Theorem 2.4.6. *Let* $\mathbf{A}$ *be an* $n \times n$ *matrix such that* $\rho(\mathbf{A}) = \rho(\mathbf{A}^2)$. *Then* $\mathbf{G}$ *is* $\mathbf{A}_{\rho\chi}^{-}$ *if and only if*

(i) $\mathbf{AGA} = \mathbf{A}$
(ii) $\mathbf{GAG} = \mathbf{G}$ *and*
(iii) $\mathbf{AG} = \mathbf{GA}$.

Proof. 'If' part
From (ii) and (iii) it follows that $\mathbf{AG^2} = \mathbf{G} = \mathbf{G^2A}$. So, $\mathcal{C}(\mathbf{G}) \subseteq \mathcal{C}(\mathbf{A})$ and $\mathcal{C}(\mathbf{G^t}) \subseteq \mathcal{C}(\mathbf{A^t})$. This together with (i) implies that $\mathbf{G}$ is $\mathbf{A}_{\rho\chi}^{-}$.

The 'only if' part can be easily verified using Theorem 2.4.4(iv). □

We formally define a commuting g-inverse of a square matrix in the following:

Definition 2.4.7. Let $\mathbf{A}$ be a square matrix. Then a square matrix $\mathbf{G}$ of the same order is called a commuting g-inverse of $\mathbf{A}$ if it satisfies the following:

(i) $\mathbf{AGA} = \mathbf{A}$, i.e. $\mathbf{G}$ is a g-inverse of $\mathbf{A}$ and
(ii) $\mathbf{AG} = \mathbf{GA}$.

Remark 2.4.8. A square matrix $\mathbf{A}$ has a commuting g-inverse if and only if $\rho(\mathbf{A}) = \rho(\mathbf{A}^2)$.

We now obtain the class of all commuting g-inverses of a square matrix.

Theorem 2.4.9. (a) Let $\mathbf{A}$ be a square matrix. Then $\mathbf{A}$ has a commuting g-inverse if and only if $\mathbf{A}$ is of index not greater than 1.
(b) Let $\mathbf{A}$ be of index not greater than 1 and let $\mathbf{A} = \mathbf{P}\mathbf{diag}(\mathbf{T}, \mathbf{0})\mathbf{P}^{-1}$, where $\mathbf{T}$ is non-singular. Then every commuting g-inverse of $\mathbf{A}$ is of the

form

$$\mathbf{G} = \mathbf{Pdiag}(\mathbf{T}^{-1},\ \mathbf{C})\mathbf{P}^{-1},$$

where $\mathbf{C}$ is arbitrary.

Proof. (a) 'If' part
Let $\mathbf{A}$ be of index not greater than 1. Then by Theorem 2.2.23, there exists a non-singular matrix $\mathbf{P}$ such that $\mathbf{A} = \mathbf{Pdiag}(\mathbf{T}, \mathbf{0})\mathbf{P}^{-1}$, where $\mathbf{T}$ is non-singular. Let $\mathbf{G} = \mathbf{Pdiag}(\mathbf{T}^{-1}, \mathbf{0})\mathbf{P}^{-1}$. Then $\mathbf{G}$ is a commuting g-inverse of $\mathbf{A}$.
'Only if' part
Let $\mathbf{A}$ have a commuting g-inverse $\mathbf{G}$. Then, $\mathbf{AGA} = \mathbf{A}^2\mathbf{G} = \mathbf{A}$. This implies $\rho(\mathbf{A}) \leq \rho(\mathbf{A}^2) \leq \rho(\mathbf{A})$, or equivalently, $\rho(\mathbf{A})^2 = \rho(\mathbf{A})$, so, $\mathbf{A}$ is index 1.
(b) As noted in the proof of (a), $\mathbf{G} = \mathbf{Pdiag}(\mathbf{T}^{-1}, \mathbf{0})\mathbf{P}^{-1}$ is a commuting g-inverse of $\mathbf{A}$. Let $\mathbf{H} = \mathbf{P}\begin{pmatrix}\mathbf{H_1} & \mathbf{H_2}\\ \mathbf{H_3} & \mathbf{H_4}\end{pmatrix}\mathbf{P}^{-1}$ be a commuting g-inverse of $\mathbf{A}$, partitioned conformably for multiplication with $\mathbf{A}$. Now, by direct computation, $\mathbf{H_1} = \mathbf{T}^{-1}, \mathbf{H_2}$ and $\mathbf{H_3}$ are null and $\mathbf{H_4}$ is arbitrary. □

Remark 2.4.10. Let $\mathbf{A}$ be a square matrix such that $\rho(\mathbf{A}) = \rho(\mathbf{A}^2)$. Then $\mathbf{A}^-_{\rho\chi}$ is the unique commuting reflexive g-inverse of $\mathbf{A}$.

Remark 2.4.11. In view of symmetry of (i)-(iii) in Theorem 2.4.6, $\mathbf{G} = \mathbf{A}^-_{\rho\chi}$ if and only if $\mathbf{A} = \mathbf{G}^-_{\rho\chi}$.

It is obvious that the set of all non-singular matrices of order $n \times n$ form a group under matrix multiplication with identity $\mathbf{I}$. The inverse for each matrix $\mathbf{A}$ in this group is $\mathbf{A}^{-1}$, the usual inverse of a non-singular matrix. We now show that the matrices satisfying $\rho(\mathbf{A}) = \rho(\mathbf{A}^2)$ also have similar property. Let $\mathbf{A}$ be an $n \times n$ matrix of rank r such that $\rho(\mathbf{A}) = \rho(\mathbf{A}^2)$. Consider the class of matrices defined as

$$\mathfrak{C}_{\mathbf{A}} = \{\mathbf{B} : \mathcal{C}(\mathbf{B}) = \mathcal{C}(\mathbf{A}) \text{ and } \mathcal{C}(\mathbf{B}^t) = \mathcal{C}(\mathbf{A}^t)\ ,\ \rho(\mathbf{B}) = \rho(\mathbf{B}^2)\}\ .$$

We shall show that $\mathfrak{C}_{\mathbf{A}}$ is a group under matrix multiplication.

Theorem 2.4.12. *Let $\mathbf{A}$ be an $n \times n$ matrix of rank r such that $\rho(\mathbf{A}) = \rho(\mathbf{A}^2)$. Let $\mathbf{A} = \mathbf{Pdiag}(\mathbf{T},\ \mathbf{0})\mathbf{P}^{-1}$, where $\mathbf{T}$ is an $r \times r$ non-singular matrix. Then*

(i) $\mathfrak{C}_{\mathbf{A}} = \{\mathbf{Pdiag}(\mathbf{R},\ \mathbf{0})\mathbf{P}^{-1}\}$, *where $\mathbf{R}$ is an arbitrary $r \times r$ non-singular matrix.*
(ii) $\mathfrak{C}_{\mathbf{A}}$ *forms a group under matrix multiplication.*

(iii) *For each* $\mathbf{B} \in \mathfrak{C}_{\mathbf{A}}$, $\mathbf{B}^{-}_{\rho\chi}$ *is the inverse of* $\mathbf{B}$ *in the group* $\mathfrak{C}_{\mathbf{A}}$.

Proof. Notice that for $\mathbf{B} \in \mathfrak{C}_{\mathbf{A}}$, since $\mathcal{C}(\mathbf{B}) = \mathcal{C}(\mathbf{A})$ and $\mathcal{C}(\mathbf{B^t}) = \mathcal{C}(\mathbf{A^t})$, we have

$$\mathcal{C}(\mathbf{PBP^{-1}}) = \mathcal{C}(\mathbf{PAP^{-1}}) \text{ and } \mathcal{C}(\mathbf{P^{-1}B^tP})^{\mathbf{t}} = \mathcal{C}(\mathbf{P^{-1}A^tP})^{\mathbf{t}}.$$

Now (i) follows easily.
(ii) Let each of $\mathbf{B} = \mathbf{P}\mathrm{\mathbf{diag}}(\mathbf{R}, \mathbf{0})\ \mathbf{P}^{-1}$ and $\mathbf{C} = \mathbf{P}\mathrm{\mathbf{diag}}(\mathbf{S}, \mathbf{0})\ \mathbf{P}^{-1}$ belong to $\mathfrak{C}_{\mathbf{A}}$, where $\mathbf{R}$ and $\mathbf{S}$ are non-singular matrices of order $r \times r$. Clearly, $\mathbf{BC} \in \mathfrak{C}_{\mathbf{A}}$. Thus, $\mathfrak{C}_{\mathbf{A}}$ is closed under multiplication. Associativity follows since the matrix multiplication is associative. It can be easily verified that $\mathbf{E} = \mathbf{P}\mathrm{\mathbf{diag}}(\mathbf{I_r}, \mathbf{0})\ \mathbf{P}^{-1}$ is the identity element and for each $\mathbf{B} = \mathbf{P}\mathrm{\mathbf{diag}}(\mathbf{R}, \mathbf{0})\mathbf{P}^{-1} \in \mathfrak{C}_{\mathbf{A}}$, its inverse is $\mathbf{P}\mathrm{\mathbf{daig}}(\mathbf{R}^{-1}, \mathbf{0})\mathbf{P}^{-1}$ which belongs to $\mathfrak{C}_{\mathbf{A}}$. Thus, $\mathfrak{C}_{\mathbf{A}}$ is a group.
(iii) is easy to check. □

In view of Theorem 2.4.12, we have

Theorem 2.4.13. *Let* $\mathbf{A}$ *be a matrix such that* $\rho(\mathbf{A}) = \rho(\mathbf{A^2})$. *Then the following statements are equivalent:*

(i) *A square matrix* $\mathbf{G}$ *is the unique g-inverse of* $\mathbf{A}$ *with* $\mathcal{C}(\mathbf{G}) \subseteq \mathcal{C}(\mathbf{A})$ *and* $\mathcal{C}(\mathbf{G^t}) \subseteq \mathcal{C}(\mathbf{A^t})$.
(ii) $\mathbf{G}$ *is the unique reflexive commuting g-inverse of* $\mathbf{A}$ *and*
(iii) $\mathbf{G}$ *is the inverse of* $\mathbf{A}$ *in the group*

$$\mathfrak{C}_{\mathbf{A}} = \{\mathbf{B} : \mathcal{C}(\mathbf{B}) = \mathcal{C}(\mathbf{A}) \textit{ and } \mathcal{C}(\mathbf{B^t}) = \mathcal{C}(\mathbf{A^t}),\ \rho(\mathbf{B}) = \rho(\mathbf{B^2})\}.$$

Remark 2.4.14. Henceforth, we shall refer to the matrix $\mathbf{G}$ satisfying any of the equivalent conditions in Theorem 2.4.13 as the **group inverse** of $\mathbf{A}$ and denote it by $\mathbf{A}^{\#}$. Thus, $\mathbf{A}^{\#}$ is the same as $\mathbf{A}^{-}_{\rho\chi}$.

If $\mathbf{A}$ is a non-singular matrix, then $\mathbf{A^k}$ is non-singular and $(\mathbf{A^k})^{-1}$ is $(\mathbf{A}^{-1})^{\mathbf{k}}$ for each (positive) integer k. We now establish a similar property for the group inverse.

Theorem 2.4.15. *Let* $\mathbf{A}$ *be a matrix such that* $\rho(\mathbf{A}) = \rho(\mathbf{A^2})$ *and let* $\mathbf{A}^{\#}$ *be the group inverse of* $\mathbf{A}$. *Then* $\rho(\mathbf{A^k}) = \rho((\mathbf{A^k})^2)$ *and* $(\mathbf{A^k})^{\#} = (\mathbf{A}^{\#})^{\mathbf{k}}$ *for each positive integer* k.

Proof. If $\rho(\mathbf{A}) = \rho(\mathbf{A^2})$, it is easy to establish $\rho(\mathbf{A}) = \rho(\mathbf{A^s})$ for each positive integer s. So, $\rho(\mathbf{A^k}) = \rho((\mathbf{A^k})^2) = \rho(\mathbf{A})$. The rest follows easily from Theorem 2.4.5(iii). □

For a non-singular matrix $\mathbf{A}$, if λ is an eigen-value of $\mathbf{A}$ with algebraic multiplicity k, then $1/\lambda$ is an eigen-value of $\mathbf{A}^{-1}$ with the same algebraic multiplicity k. The following is easy to establish.

Theorem 2.4.16. *Let* $\mathbf{A}$ *be a matrix of index* ≤ 1. *If* λ *is a non-null eigen-value of* $\mathbf{A}$ *with algebraic multiplicity* k, *then* $1/\lambda$ *is an eigen-value of* $\mathbf{A}^{\#}$ *with same algebraic multiplicity. Further, if zero is an eigen-value of* $\mathbf{A}$ *with algebraic multiplicity* t, *then zero is an eigen-value of* $\mathbf{A}^{\#}$ *with the same algebraic multiplicity* t.

Remark 2.4.17. A square matrix is of index not greater than 1 if and only if the algebraic and geometric multiplicities of its zero eigen-value are equal.

If $\mathbf{A}$ is a non-singular matrix, then it is well known that $\mathbf{A}^{-1}$ is a polynomial in $\mathbf{A}$. We now prove the following:

Theorem 2.4.18. *Let* $\mathbf{A}$ *be a square matrix. If* $\mathbf{A}$ *has a g-inverse* $\mathbf{G}$ *which is a polynomial in* $\mathbf{A}$, *then* $\mathbf{A}$ *has index not greater than* 1. *Moreover, if* $\rho(\mathbf{A}) = \rho(\mathbf{A^2})$, *then* $\mathbf{A}^{\#}$ *is a polynomial in* $\mathbf{A}$.

Proof. If $\mathbf{G}$ is a polynomial in $\mathbf{A}$, then $\mathbf{G}$ and $\mathbf{A}$ commute. So, $\mathbf{G}$ is a commuting g-inverse of $\mathbf{A}$. By Remark 2.4.8, it follows that $\rho(\mathbf{A}) = \rho(\mathbf{A^2})$.

Let $\rho(\mathbf{A}) = \rho(\mathbf{A^2})$. Write $\mathbf{A} = \mathbf{P}\mathbf{diag}(\mathbf{T},\mathbf{0})\mathbf{P}^{-1}$, where $\mathbf{P}$ and $\mathbf{T}$ are non-singular matrices. Since $\mathbf{T}$ is non-singular, $\mathbf{T}^{-1} = \sum_{j=0}^{k} c_j \mathbf{T}^{\mathbf{j}}$ for some positive integer k and some scalars $c_0, \ldots, c_k$. It is easy to check that $\mathbf{A}^{\#} = \sum_{j=0}^{k} c_j \mathbf{A}^{\mathbf{j}}$. □

The following Theorem is easy to establish.

Theorem 2.4.19. *Let* $\mathbf{A}$ *be an* $n \times n$ *matrix of index not greater than* 1. *Then the following hold:*

(i) $(\mathbf{A}^{\#})^{\#} = \mathbf{A}$.
(ii) $(\mathbf{A}^{\#})^{\mathbf{t}} = (\mathbf{A}^{\mathbf{t}})^{\#}$.
(iii) $F^n = \mathcal{C}(\mathbf{A}) \oplus \mathcal{N}(\mathbf{A})$.
(iv) *For each non-singular matrix* $\mathbf{P}$, $(\mathbf{PAP}^{-1})^{\#} = \mathbf{PA}^{\#}\mathbf{P}^{-1}$.

We now obtain the core-nilpotent decomposition (Theorem 2.2.21) for an $n \times n$ matrix $\mathbf{A}$ over a field $\mathbf{F}$.

Theorem 2.4.20. *Let $\mathbf{A}$ be an $n \times n$ matrix of index k. Then $\mathbf{A}$ can be written as $\mathbf{A} = \mathbf{A_1} + \mathbf{A_2}$ where*

(i) $\rho(\mathbf{A_1}) = \rho(\mathbf{A_1^2})$
(ii) $\mathbf{A_2}$ *is nilpotent* *and*
(iii) $\mathbf{A_1 A_2} = \mathbf{A_2 A_1} = \mathbf{0}$.

Further, such a decomposition is unique.

Proof. (i) If $k = 0$ or 1, take $\mathbf{A_1} = \mathbf{A}$ and $\mathbf{A_2} = \mathbf{0}$. Then $\mathbf{A_1}, \mathbf{A_2}$ satisfy (i), (ii) and (iii).

If $k > 1$, write $\mathbf{A_1} = \mathbf{A^k}(\mathbf{A^{2k-1}})^-\mathbf{A^k}$ and $\mathbf{A_2} = \mathbf{A} - \mathbf{A^k}(\mathbf{A^{2k-1}})^-\mathbf{A^k}$. Since $\rho(\mathbf{A^k}) = \rho(\mathbf{A^{k+1}}) = \rho(\mathbf{A^{2k-1}})$, by Theorem 2.3.11, $\mathbf{A_1}$ is invariant under choices of g-inverse of $\mathbf{A^{2k-1}}$. Now,

$$\mathbf{A_1^2} = \mathbf{A^k}(\mathbf{A^{2k-1}})^-\mathbf{A^k}\mathbf{A^k}(\mathbf{A^{2k-1}})^-\mathbf{A^k} = \mathbf{A^k}(\mathbf{A^{2k-1}})^-\mathbf{A^{k+1}}.$$

Also, $\mathbf{A_1^2}(\mathbf{A^{k+1}})^-\mathbf{A^k} = \mathbf{A^k}(\mathbf{A^{2k-1}})^-\mathbf{A^k} = \mathbf{A_1}$. Thus, $\rho(\mathbf{A_1}) = \rho(\mathbf{A_1^2})$.
(ii) It is easy to check that $\mathbf{A_2^r} = \mathbf{A^r} - \mathbf{A^{k+r-1}}(\mathbf{A^{2k-1}})^-\mathbf{A^k}$ for all $r \geq 1$. So, $\mathbf{A_2^k} = \mathbf{A^k} - \mathbf{A^{2k-1}}(\mathbf{A^{2k-1}})^-\mathbf{A^k} = \mathbf{A^k} - \mathbf{A^k} = \mathbf{0}$.
Thus $\mathbf{A_2}$ is nilpotent.
(iii) is easy to verify.

To prove uniqueness of the decomposition, we proceed as follows: Let $\mathbf{A} = \mathbf{B_1} + \mathbf{B_2}$ be another decomposition where $\mathbf{B_1}$ and $\mathbf{B_2}$ satisfy the conditions (i), (ii) and (iii). Then $\mathbf{A^k} = \mathbf{A_1^k} = \mathbf{B_1^k}$. So, $\mathcal{C}(\mathbf{A_1}) = \mathcal{C}(\mathbf{A_1^k}) = \mathcal{C}(\mathbf{B_1^k}) = \mathcal{C}(\mathbf{B_1})$. Similarly, $\mathcal{C}(\mathbf{A_1^t}) = \mathcal{C}(\mathbf{B_1^t})$.
Further, $\mathbf{A_1 A_2} = \mathbf{A_2 A_1} = \mathbf{B_1 B_2} = \mathbf{B_2 B_1} = \mathbf{0}$. Therefore, $\mathbf{0} = \mathbf{A_1 B_2} = \mathbf{B_2 A_1} = \mathbf{B_1 A_2} = \mathbf{A_2 B_1}$. Thus, $\mathbf{A_1}(\mathbf{A_1} - \mathbf{B_1}) = -\mathbf{A_1}(\mathbf{A_2} - \mathbf{B_2}) = \mathbf{0}$. Now, $\mathcal{C}(\mathbf{A_1} - \mathbf{B_1}) \subseteq \mathcal{C}(\mathbf{A_1}), (\mathbf{A_1} - \mathbf{B_1}) = \mathbf{A_1}(\mathbf{A_1^2})^-\mathbf{A_1}(\mathbf{A_1} - \mathbf{B_1}) = \mathbf{0}$. Hence, $\mathbf{A_1} = \mathbf{B_1}$ and $\mathbf{A_2} = \mathbf{B_2}$. □

In view of Theorem 2.2.23 and Remark 2.2.24, we can restate Theorem 2.4.20 as

Theorem 2.4.21. *Let $\mathbf{A}$ be an $n \times n$ matrix of index k and rank r. Then there exists a non-singular matrix $\mathbf{P}$ of order $n \times n$ such that $\mathbf{A} = \mathbf{P}\mathrm{\mathbf{diag}}(\mathbf{T}, \mathbf{N})\,\mathbf{P^{-1}}$, where $\mathbf{T}$ is non-singular and $\mathbf{N}$ is nilpotent such that $\mathbf{N^k} = \mathbf{0}$.*

Remark 2.4.22. The matrices $\mathbf{A_1}$ and $\mathbf{A_2}$ in Theorem 2.4.20 are called the **core** and the **nilpotent** parts respectively of the matrix $\mathbf{A}$. Further, $\rho(\mathbf{A_1})$ is called the **core rank** of $\mathbf{A}$. The core rank of $\mathbf{A}$ is less than or equal to the rank of $\mathbf{A}$. Also, k, the index of $\mathbf{A}$ is the least positive integer such that $\mathbf{A_2^k} = \mathbf{0}$ (called the index of nilpotency or simply **index** of $\mathbf{A_2}$). Notice that $\rho(\mathbf{A}) = \rho(\mathbf{A_1}) + \rho(\mathbf{A_2})$.

As noted earlier, a matrix does not have a commuting g-inverse unless it is of index not greater than 1. However, for a general square matrix not necessarily of index ≤ 1, we can define a type of commuting inverse (which may not be a generalized inverse) that is a generalization of the group inverse.

Definition 2.4.23. Let $\mathbf{A}$ be an $n \times n$ matrix. A matrix $\mathbf{G}$ is called the Drazin inverse of $\mathbf{A}$, denoted by $\mathbf{A^D}$, if $\mathbf{G}$ satisfies the following properties:

(i) $\mathbf{GAG} = \mathbf{G}$.
(ii) $\mathbf{A}^k = \mathbf{A}^{k+1}\mathbf{G}$ for some positive integer k and
(iii) $\mathbf{AG} = \mathbf{GA}$.

Remark 2.4.24. $\mathbf{A^D} = \mathbf{A^\#}$ if $k \leq 1$. If $\mathbf{A}$ is either null or nilpotent, then $\mathbf{A^D} = \mathbf{0}$.

Remark 2.4.25. Let k be the smallest positive integer such that $\mathbf{A^k} = \mathbf{A^{k+1}G}$. If $k \geq 2$, then $\mathbf{G}$ is not a g-inverse of $\mathbf{A}$.

We now prove the existence and uniqueness of the Drazin inverse for any square matrix.

Theorem 2.4.26. *Let* $\mathbf{A}$ *be an* $n \times n$ *matrix of index* k *and rank* r. *As in Theorem 2.4.21, let* $\mathbf{A} = \mathbf{Pdiag(T,\ N)P^{-1}}$, *where* $\mathbf{T}$ *and* $\mathbf{P}$ *are non-singular and* $\mathbf{N}$ *is nilpotent such that* $\mathbf{N^k} = \mathbf{0}$. *Then* $\mathbf{A^D} = \mathbf{Pdiag(T^{-1},\ 0)\ P^{-1}}$ *and* $\mathbf{A^D}$ *is unique.*

Proof. It is easy to check that $\mathbf{Pdiag(T^{-1},\ 0)P^{-1}}$ satisfies the conditions (i)-(iii) of Definition 2.4.23.

To show that $\mathbf{A^D}$ is unique, let $\mathbf{G_1}$ and $\mathbf{G_2}$ be any two choices for $\mathbf{A^D}$. Consider $\mathbf{G_1^{k+1}A^{2k+1}G_2^{k+1}}$. By repeated applications of the conditions (i)-(iii) of Definition 2.4.23, we have $\mathbf{G_1} = \mathbf{G_1^{k+1}A^{2k+1}G_2^{k+1}} = \mathbf{G_2}$. □

Remark 2.4.27. Consider the core-nilpotent decomposition as in Theorem 2.4.20. Then $\mathbf{A^D} = \mathbf{A_1^\#} = \mathbf{A^k(A^{2k+1})^-A^k}$.

Remark 2.4.28. Let $\mathbf{A}$ be a non-null matrix. Then $\mathbf{A}^{\mathrm{D}} = \mathbf{0} \Leftrightarrow \mathbf{A}$ is nilpotent.

The following theorem is easy to establish.

Theorem 2.4.29. *Let* $\mathbf{A}$ *be an* $n \times n$ *matrix of index* k. *Then*

(i) $((\mathbf{A}^{\mathrm{D}})^{\mathrm{D}})^{\mathrm{D}} = \mathbf{A}^{\mathrm{D}}$.
(ii) $(\mathbf{A}^{\mathrm{D}})^{\mathrm{t}} = (\mathbf{A}^{\mathrm{t}})^{\mathrm{D}}$.
(iii) $(\mathbf{PAP}^{-1})^{\mathrm{D}} = \mathbf{PA}^{\mathrm{D}}\mathbf{P}^{-1}$ *for each non-singular matrix* $\mathbf{P}$.
(iv) $\mathbf{F}^{\mathrm{n}} = \mathcal{C}(\mathbf{A}^{\mathrm{k}}) \oplus \mathcal{N}(\mathbf{A}^{\mathrm{k}})$.
(v) $\mathcal{C}(\mathbf{A}^{\mathrm{D}}) = \mathcal{C}(\mathbf{A}^{\mathrm{k}}), \mathcal{N}(\mathbf{A}^{\mathrm{D}}) = \mathcal{N}(\mathbf{A}^{\mathrm{k}})$.
(vi) $\mathbf{AA}^{\mathrm{D}} = \mathbf{A}^{\mathrm{D}}\mathbf{A}$ *is the projector* $\mathbf{P}_{\mathcal{C}(\mathbf{A}^{\mathrm{k}}),\ \mathcal{N}(\mathbf{A}^{\mathrm{k}})}$ *which projects vectors onto* $\mathcal{C}(\mathbf{A}^{\mathrm{k}})$ *along* $\mathcal{N}(\mathbf{A}^{\mathrm{k}})$.
(vii) $(\mathbf{I} - \mathbf{AA}^{\mathrm{D}}) = (\mathbf{I} - \mathbf{A}^{\mathrm{D}}\mathbf{A}) = \mathbf{P}_{\mathcal{N}(\mathbf{A}^{\mathrm{k}}),\ \mathcal{C}(\mathbf{A}^{\mathrm{k}})}$.

We now demonstrate that the Drazin inverse A^D too is a polynomial in A just as we showed that the group inverse of a matrix $\mathbf{B}$ when it exists, is a polynomial in $\mathbf{B}$.

Theorem 2.4.30. *Let* $\mathbf{A}$ *be an* $n \times n$ *matrix. Then* $\mathbf{A}^{\mathrm{D}}$ *is a polynomial in* $\mathbf{A}$.

Proof. Let k be the index of $\mathbf{A}$. Then as seen in Remark 2.4.27, $\mathbf{A}^{\mathrm{D}} = \mathbf{A}^{\mathrm{k}}(\mathbf{A}^{2\mathrm{k}+1})^{-}\mathbf{A}^{\mathrm{k}}$. Clearly, $\mathbf{A}^{\mathrm{k}}(\mathbf{A}^{2\mathrm{k}+1})^{-}\mathbf{A}^{\mathrm{k}}$ is invariant under choice of $(\mathbf{A}^{2\mathrm{k}+1})^{-}$ and $\rho(\mathbf{A}^{2\mathrm{k}+1}) = \rho((\mathbf{A}^{2\mathrm{k}+1})^2)$. So, $\mathbf{A}^{\mathrm{D}} = \mathbf{A}^{\mathrm{k}}(\mathbf{A}^{2\mathrm{k}+1})^{\#}\mathbf{A}^{\mathrm{k}}$. We know $(\mathbf{A}^{2\mathrm{k}+1})^{\#}$ is a polynomial in $\mathbf{A}^{2\mathrm{k}+1}$ and therefore, in $\mathbf{A}$. Hence, $\mathbf{A}^{\mathrm{D}}$ is a polynomial in $\mathbf{A}$. □

We now obtain a formula for the index and Drazin inverse using rank factorization.

Theorem 2.4.31. *Let* $\mathbf{A}$ *be a non-null* $n \times n$ *matrix. Write* $\mathbf{A}_0 = \mathbf{A}$. *Let*

$$\mathbf{A}_{\mathrm{i}+1} = \begin{cases} \mathbf{Q}_{\mathrm{i}}\mathbf{P}_{\mathrm{i}} & \text{if } \mathbf{A}_{\mathrm{i}} \neq \mathbf{0} \text{ and } (\mathbf{P}_{\mathrm{i}}, \mathbf{Q}_{\mathrm{i}}) \text{ is a rank factorization of } \mathbf{A}_{\mathrm{i}} \\ \mathbf{0} & \text{if } \mathbf{A}_{\mathrm{i}} = \mathbf{0} \end{cases}.$$

for $i = 1, 2, \ldots$. *Then the following hold:*

(i) $$\rho(\mathbf{A}^{\mathrm{i}+1}) = \begin{cases} \text{number of columns in } \mathbf{P}_{\mathrm{i}} & \text{if} \mathbf{A}_{\mathrm{i}} \neq \mathbf{0} \\ \mathbf{0} \text{ if } \mathbf{A}_{\mathrm{i}} = \mathbf{0} \end{cases}$$

(ii) *There exists an integer* $s \geq 0$ *such that* $\mathbf{A_s}$ *is either non-singular or null.*

Let k *be the smallest integer* ≥ 0 *such that* $\mathbf{A_k}$ *is either non-singular or null. Then*

(iii) $$\textit{index of } \mathbf{A} = \begin{cases} k \ \textit{ if } \mathbf{A_k} \textit{ is non-singular} \\ \\ k+1, \ \textit{if} \ \ \mathbf{A_k} = \mathbf{0} \end{cases}.$$

(iv) $\mathcal{C}(\mathbf{A^{i+1}}) = \mathcal{C}(\mathbf{P_0P_1 \dots P_i})$ *and* $\mathcal{N}(\mathbf{A^{i+1}}) = \mathcal{N}(\mathbf{Q_i \dots Q_1Q_0})$ *for* $i = 1, \dots k-1$ *if* $\mathbf{A_k}$ *is non-singular and for* $i = 1, \dots, k-2$ *if* $\mathbf{A_k}$ *is null.*

(v) $$\mathbf{A^D} = \begin{cases} \mathbf{0} \textit{ if } \mathbf{A_k} = \mathbf{0} \\ \\ \mathbf{P_0P_1 \dots P_{k-1}A_k^{-(k+1)}Q_{k-1} \dots Q_0} \text{ when } \mathbf{A_k} \textit{ is non-singular} \end{cases}.$$

Proof. (i) It is easy to see that

$$\begin{aligned} \mathbf{A^{i+1}} &= \mathbf{P_0P_1 \dots P_iQ_i \dots Q_1Q_0} \\ &= \mathbf{P_0P_1 \dots P_{i-1}A_iQ_{i-1} \dots Q_1Q_0}. \end{aligned}$$

If $\mathbf{A_i} \neq \mathbf{0}$, then $\mathbf{P_i}$ and $\mathbf{Q_i}$ are full rank matrices and $\rho(\mathbf{A^{i+1}}) = \rho(\mathbf{A_i}) = \rho(\mathbf{P_iQ_i})$ for each i. If $\mathbf{A_i}$ is null, so is $\mathbf{A^{i+1}}$, therefore $\rho(\mathbf{A^{i+1}}) = \mathbf{0}$ and if $\mathbf{A_i}$ is non-null, $\rho(\mathbf{A^{i+1}}) = \rho(\mathbf{A_i}) = r_i =$ number of columns in $\mathbf{P_i}$.
(ii) Suppose $\mathbf{A_j}$ is neither null nor non-singular for each j. Let the order of $\mathbf{A_j}$ be $r_j \times r_j$. Then $\mathbf{A_{j+1}}$ is of order $r_{j+1} \times r_{j+1}$, where $r_{j+1} < r_j$. Thus $\{r_j\}$ is a strictly decreasing sequence as long as $\{\mathbf{A_j}\}$ is a sequence of neither non-singular nor null matrices. But this is not possible since r_j can never be negative. This proves (ii). Suppose there exists $\mathbf{A_s}$ such that $\mathbf{A_s}$ is non-singular. Then it is of the same order as $\mathbf{A_t}$ for all $t \geq s$. Also, if $\mathbf{A_s} = \mathbf{0}$, then $\mathbf{A_t} = \mathbf{0}$ all $t \geq s$. Thus, there exists a smallest integer $k \geq 0$ such that $\mathbf{A_k}$ is either non-singular or null.
(iii) Let k be the smallest positive integer such that $\mathbf{A_k}$ is either non-singular or null. Then

$$\begin{aligned} \mathbf{A^{k+1}} &= \mathbf{P_0P_1 \dots P_{k-1}P_kQ_kQ_{k-1} \dots Q_1Q_0} \\ &= \mathbf{P_0P_1 \dots P_{k-1}A_kQ_{k-1} \dots Q_1Q_0}. \end{aligned}$$

Let $\mathbf{A_k}$ be null. Then, $\mathbf{A^{k+1}} = \mathbf{0}$, so, $\mathbf{A}$ is nilpotent. As k is the smallest positive integer for which $\mathbf{A_k}$ is null, $\mathbf{A}$ has index $k+1$.

Now, let $\mathbf{A_k}$ be non-singular. Then $\rho(\mathbf{A_k}) = r_k$, $\mathbf{A_{k+1}} = \mathbf{Q_kP_k}$ is of order $r_k \times r_k$. Therefore,

$\rho(\mathbf{A^{k+1}}) = \rho(\mathbf{A_k}) = r_k = \rho(\mathbf{A^k})$ and k is the smallest positive integer for which this is true. So, index of $\mathbf{A}$ is k.
(iv) Let $i = 1, \ldots, k-1$. From the proof of (ii) we have

$$\mathbf{A^{i+1}} = \mathbf{P_0P_1 \ldots P_{i-1}P_iQ_iQ_{i-1} \ldots Q_1Q_0},$$

where $\mathbf{P_j}$ has a left inverse and $\mathbf{Q_j}$ has a right inverse for $j = 0, 1, \ldots, i$. Hence $\mathcal{C}(\mathbf{A^{i+1}}) = \mathcal{C}(\mathbf{P_0P_1, \ldots P_{i-1}P_i})$ and $\mathcal{N}(\mathbf{A^{i+1}}) = \mathcal{N}(\mathbf{Q_i \ldots Q_1Q_0})$.

Let $\mathbf{A_k}$ be non-singular. Then we have

$$\mathbf{A^{k+1}} = \mathbf{P_0P_1 \ldots P_{k-1}P_kQ_kQ_{k-1} \ldots Q_1Q_0},$$

where $\mathbf{P_k}$ has a left inverse and $\mathbf{Q_k}$ has a right inverse. Hence,

$$\mathcal{C}(\mathbf{A^{k+1}}) = \mathcal{C}(\mathbf{P_0P_1 \ldots P_{k-1}P_k}) \text{and} \mathcal{N}(\mathbf{A^{k+1}}) = \mathcal{N}(\mathbf{Q_k \ldots Q_1Q_0}).$$

(v) If $\mathbf{A_k} = \mathbf{0}$, then $\mathbf{A}$ is nilpotent. By Remark 2.4.27, $\mathbf{A^D} = \mathbf{0}$. Let $\mathbf{A_k}$ be non-singular. Let $\mathbf{X} = \mathbf{P_0P_1 \ldots P_{k-1}A_k^{-(k+1)}Q_{k-1} \ldots Q_1Q_0}$. Then it is easy to check that $\mathbf{XAX} = \mathbf{X}$, $\mathbf{XA} = \mathbf{AX}$ and $\mathbf{XA^{k+1}} = \mathbf{A^k} = \mathbf{A^{k+1}X}$. Thus, $\mathbf{X}$ is the Drazin inverse of $\mathbf{A}$. □

Corollary 2.4.32. *Let* $\mathbf{A}$ *be an* $n \times n$ *matrix of index* ≤ 1 *and* $(\mathbf{P}, \mathbf{Q})$ *be a rank factorization of* $\mathbf{A}$. *Then*

$$\mathbf{A^{\#}} = \mathbf{P(QP)^{-2}Q}\ .$$

2.5 Moore-Penrose inverse

In this section we specialize to vectors and matrices over the field of complex numbers $\mathbb{C}$ and use the inner product $(\mathbf{x}, \mathbf{y}) = \mathbf{y^\star x}$ for $\mathbf{x}, \mathbf{y} \in \mathbb{C}^\mathbf{n}$. Let $\mathbf{Ax} = \mathbf{b}$ be a consistent system of linear equations for $\mathbf{A} \in \mathbb{C}^{\mathbf{m \times n}}$ and $\mathbf{b} \in \mathbb{C}^\mathbf{m}$. In Section 2.3, we noticed that, in general, a solution to $\mathbf{Ax} = \mathbf{b}$ is not unique. Since there are many solutions, we can look for solutions with some optimal property. One such is the minimum norm property as different solutions have possibly different norms. Does a solution with minimum norm exist? If so, does there exist a g-inverse $\mathbf{G}$ of $\mathbf{A}$ such that $\mathbf{Gb}$ is a solution to $\mathbf{Ax} = \mathbf{b}$ with minimum norm, whenever the system is consistent?

What if $\mathbf{Ax} = \mathbf{b}$ is not consistent? Does there exist a g-inverse $\mathbf{G}$ of $\mathbf{A}$ such that $\|\mathbf{AGb} - \mathbf{b}\| \leq \|\mathbf{Ax} - \mathbf{b}\|$ for each $\mathbf{x}$? This means that we are seeking a g-inverse $\mathbf{G}$ of $\mathbf{A}$ such that the vector $\mathbf{AGb}$ is the closest to $\mathbf{b}$.

Here we shall attempt to find answers to the above questions. It turns out that the answers to all of these are in affirmative. We shall see that the minimum norm solution is unique but a least squares solution, in general, is not unique. The next question that naturally arises is: Does there exist a g-inverse $\mathbf{G}$ of $\mathbf{A}$ such that $\mathbf{Gb}$ has the smallest norm in the class of all least squares solutions to $\mathbf{Ax} = \mathbf{b}$ for all column vectors $\mathbf{b}$? The answer again is in affirmative. In fact, we show that such a g-inverse is unique. It is this g-inverse that Moore and Penrose arrived at from different considerations. This g-inverse, named after them, is called the Moore-Penrose inverse. In this section, we shall characterize the subclasses of g-inverses mentioned in this and previous two paragraphs and study their properties.

In the previous section we studied g-inverse with specified column and row spaces. It turns out that the g-inverses studied in this section also have the column and row spaces contained in certain subspaces of interest which bear a striking similarity to those in the previous section. Thus, we notice that the g-inverses we study in this section have striking similarities to $\mathbf{A}_\chi^-$, $\mathbf{A}_\rho^-$ and $\mathbf{A}^\#$. While $\mathbf{A}_\chi^-$, $\mathbf{A}_\rho^-$ and $\mathbf{A}^\#$ exist for a matrix $\mathbf{A}$ of index ≤ 1 over a general field, the g-inverses that we are about to study in this section exist for every matrix over the field of complex numbers (or a subfield of it).

We first prove a lemma, which will be used frequently in this section.

Lemma 2.5.1. *Let* $\mathbf{A}$ *and* $\mathbf{B}$ *be* $m \times n$ *and* $m \times p$ *matrices respectively. Then* $\|\mathbf{Ax}\| \leq \|\mathbf{Ax} + \mathbf{By}\|$ *for all* $\mathbf{x}$ *and* $\mathbf{y}$ *if and only if* $\mathbf{A}^\star\mathbf{B} = \mathbf{0}$.

Proof. We first note that

$$\begin{aligned}\|\mathbf{Ax}\| \leq \|\mathbf{Ax} + \mathbf{By}\| &\Leftrightarrow (\mathbf{Ax}, \mathbf{Ax}) \leq (\mathbf{Ax} + \mathbf{By}, \mathbf{Ax} + \mathbf{By}) \\ &\Leftrightarrow \mathbf{y}^\star\mathbf{B}^\star\mathbf{Ax} + \mathbf{x}^\star\mathbf{A}^\star\mathbf{By} + \mathbf{y}^\star\mathbf{B}^\star\mathbf{By} \geq \mathbf{0} \ .\end{aligned}$$

'If' part follows trivially, since $\mathbf{B}^\star\mathbf{B}$ is *nnd* and $\mathbf{A}^\star\mathbf{B} = \mathbf{0}$ implies $(\mathbf{A}^\star\mathbf{B})^\star = \mathbf{B}^\star\mathbf{A} = \mathbf{0}$.

'Only if' part

Let if possible, $\mathbf{A}^\star\mathbf{B} \neq \mathbf{0}$. Then there exists a vector $\mathbf{y}$ necessarily non-null such that $\mathbf{A}^\star\mathbf{By} \neq \mathbf{0}$. So, $\mathbf{By} \neq \mathbf{0}$ and $\mathbf{y}^\star\mathbf{B}^\star\mathbf{By} > \mathbf{0}$. Let the jth component of $\mathbf{A}^\star\mathbf{By}$ be $a + ib \neq 0$. Choose x_j, the jth component of $\mathbf{x}$ equal to

$$-((a + ib)/(a^2 + b^2))(\mathbf{y}^\star\mathbf{B}^\star\mathbf{By})$$

and all other components of $\mathbf{x}$ as zero. Now,

$$\mathbf{x}^\star\mathbf{A}^\star\mathbf{By} = -((a - ib)/(a^2 + b^2))(\mathbf{y}^\star\mathbf{B}^\star\mathbf{By})(a + ib) = -\mathbf{y}^\star\mathbf{B}^\star\mathbf{By}.$$

Also, $\mathbf{y^\star B^\star Ax} = -(\mathbf{y^\star B^\star By})^\star = -\mathbf{y^\star B^\star By}$. So, for these choices of $\mathbf{x}$ and $\mathbf{y}$, we have

$$\mathbf{y^\star B^\star Ax + x^\star A^\star By + y^\star B^\star By} = -\mathbf{y^\star B^\star By} < 0,$$

which is a contradiction. So, $\mathbf{A^\star B = 0}$ □

Definition 2.5.2. Let $\mathbf{Ax = b}$ be consistent. Then a vector $\mathbf{x_0}$ is said to be a minimum norm solution to $\mathbf{Ax = b}$ if

(i) $\mathbf{Ax_0 = b}$ and
(ii) $\|\mathbf{x_0}\| \leq \|\mathbf{x}\|$ for every $\mathbf{x}$ such that $\mathbf{Ax = b}$.

Theorem 2.5.3. *Let* $\mathbf{A}$ *be an* $m \times n$ *matrix. Then there exists a g-inverse* $\mathbf{G}$ *of* $\mathbf{A}$ *such that* $\mathbf{Gb}$ *is a minimum norm solution to* $\mathbf{Ax = b}$ *whenever it is consistent.*

Proof. Consider $\mathbf{G} = \mathbf{A^\star(AA^\star)^-}$ where $\mathbf{(AA^\star)^-}$ is some g-inverse of $\mathbf{AA^\star}$. Since $\rho(\mathbf{AA^\star}) = \rho(\mathbf{A})$, by Theorem 2.3.12, $\mathbf{G}$ is a g-inverse of $\mathbf{A}$. So, $\mathbf{Gb}$ is a solution to $\mathbf{Ax = b}$ whenever it is consistent. Also, the class of all solutions to $\mathbf{Ax = b}$ is given by $\mathbf{Gb + (I - GA)}\xi$, ξ arbitrary. We now show that

$$\|\mathbf{Gb}\| \leq \|\mathbf{Gb + (I - GA)}\xi\| \text{ for all } \mathbf{b} \in \mathcal{C}(\mathbf{A}) \text{ and for all } \xi.$$

Now, $\mathbf{b} \in \mathcal{C}(\mathbf{A}) \Leftrightarrow \mathbf{b = Au}$ for some $\mathbf{u}$. Thus, we need to show

$$\|\mathbf{GAu}\| \leq \|\mathbf{GAu + (I - GA)}\xi\| \text{ for all } \mathbf{u} \text{ and } \xi.$$

By Lemma 2.5.1, the above happens if and only if $\mathbf{(GA)^\star(I - GA) = 0}$. Now,

$$\begin{aligned}\mathbf{(GA)^\star(I - GA)} &= \mathbf{(A^\star(AA^\star)^-A)^\star(I - A^\star(AA^\star)^-A)}\\ &= \mathbf{(A^\star((AA^\star)^-)^\star A)(I - A^\star(AA^\star)^-A)}\\ &= \mathbf{(A^\star((AA^\star)^-)^\star A) - (A^\star((AA^\star)^-)^\star)AA^\star(AA^\star)^-A}.\end{aligned}$$

Since $\mathcal{C}(\mathbf{A}) = \mathcal{C}(\mathbf{AA^\star})$, $\mathbf{A = AA^\star(AA^\star)^-A}$. Therefore,

$$\mathbf{(GA)^\star(I - GA) = (A^\star((AA^\star)^-)^\star A) - (A^\star((AA^\star)^-)^\star)A = 0}.$$

So, $\|\mathbf{GAu + (I - GA)}\xi\|^2 = \|\mathbf{GAu}\|^2 + \|\mathbf{(I - GA)}\xi\|^2$. Hence, $\mathbf{Gb}$ is a minimum norm solution to $\mathbf{Ax = b}$, whenever it is consistent. □

Definition 2.5.4. Let $\mathbf{A}$ be an $m \times n$ matrix. Then a matrix $\mathbf{G}$ of order $n \times m$ is said to be a minimum norm g-inverse of $\mathbf{A}$, if $\mathbf{Gb}$ provides a minimum norm solution to $\mathbf{Ax = b}$, whenever it is consistent. A minimum norm g-inverse of $\mathbf{A}$ is denoted by $\mathbf{A_m^-}$.

We shall now obtain some characterizations of $\mathbf{A}_{\mathbf{m}}^{-}$.

Theorem 2.5.5. *Let* $\mathbf{A}$ *be an* $m \times n$ *matrix. then* $\mathbf{G}$ *is a minimum norm g-inverse of* $\mathbf{A}$ *if and only if it satisfies any one of the following equivalent conditions:*

(i) $\mathbf{AGA} = \mathbf{A}$, $(\mathbf{GA})^\star = \mathbf{GA}$
(ii) $\mathbf{GA} = \mathbf{P}_{\mathbf{A}^\star}$, *the orthogonal projector onto* $\mathcal{C}(\mathbf{A}^\star)$ *and*
(iii) $\mathbf{GAA}^\star = \mathbf{A}^\star$.

Proof. In the proof of Theorem 2.5.3, we saw that $\mathbf{G}$ is a minimum norm g-inverse of $\mathbf{A}$ if and only if $\mathbf{AGA} = \mathbf{A}$ and $(\mathbf{GA})^\star(\mathbf{I}-\mathbf{GA}) = \mathbf{0}$. However, $(\mathbf{GA})^\star(\mathbf{I}-\mathbf{GA}) = \mathbf{0} \Rightarrow (\mathbf{GA})^\star = (\mathbf{GA})^\star\mathbf{GA} = \mathbf{GA}$, since $(\mathbf{GA})^\star\mathbf{GA}$ is hermitian.

Conversely, if $\mathbf{AGA} = \mathbf{A}$ and $(\mathbf{GA})^\star = \mathbf{GA}$, then

$$\begin{aligned}(\mathbf{GA})^\star(\mathbf{I}-\mathbf{GA}) &= (\mathbf{GA})^\star - (\mathbf{GA})^\star(\mathbf{GA}) = \mathbf{GA} - (\mathbf{GA})(\mathbf{GA}) \\ &= \mathbf{GA} - \mathbf{GAGA} = \mathbf{GA} - \mathbf{GA} = \mathbf{0}\,.\end{aligned}$$

So, $\mathbf{G}$ is a minimum norm g-inverse of $\mathbf{A}$ if and only if (i) holds

Therefore, it is enough to show (i) $\Rightarrow$ (ii) $\Rightarrow$ (iii) $\Rightarrow$ (i).

(i) $\Rightarrow$ (ii)

Let (i) hold. Clearly, $\mathbf{AGA} = \mathbf{A} \Rightarrow (\mathbf{GA})^2 = \mathbf{GA}$, so, $\mathbf{GA}$ is idempotent. Also, $(\mathbf{GA})^\star = \mathbf{GA}$ implies $\mathbf{GA}$ is hermitian. So, $\mathbf{GA}$ is the orthogonal projector onto $\mathcal{C}(\mathbf{GA})$. However, $\mathbf{GA} = (\mathbf{GA})^\star = \mathbf{A}^\star\mathbf{G}^\star$ and $\rho(\mathbf{A}^\star\mathbf{G}^\star) = \rho(\mathbf{GA}) = \rho(\mathbf{A}) = \rho(\mathbf{A}^\star)$. Therefore, $\mathcal{C}(\mathbf{GA}) = \mathcal{C}(\mathbf{A}^\star\mathbf{G}^\star) = \mathcal{C}(\mathbf{A}^\star)$, proving (ii) holds.

(ii) $\Rightarrow$ (iii) is trivial.

(iii) $\Rightarrow$ (i)

$\mathbf{GAA}^\star = \mathbf{A}^\star \Rightarrow \mathbf{GAA}^\star\mathbf{G}^\star = \mathbf{A}^\star\mathbf{G}^\star \Rightarrow \mathbf{GA}(\mathbf{GA})^\star = (\mathbf{GA})^\star$. Since $\mathbf{GA}(\mathbf{GA})^\star$ is hermitian, $(\mathbf{GA})^\star = \mathbf{GA}$. Also,

$$\mathbf{GAA}^\star = \mathbf{A}^\star \Rightarrow (\mathbf{GA})^\star\mathbf{A}^\star = \mathbf{A}^\star \Rightarrow \mathbf{AGA} = \mathbf{A}.$$

□

We now show that minimum norm solution is unique.

Theorem 2.5.6. *Let* $\mathbf{Ax} = \mathbf{b}$ *be consistent. Then minimum norm solution to* $\mathbf{Ax} = \mathbf{b}$ *is unique.*

Proof. Let $\mathbf{G}$ be one choice of $\mathbf{A}_{\mathbf{m}}^{-}$. Then $\mathbf{Gb}$ is a minimum norm solution to $\mathbf{Ax} = \mathbf{b}$. Every solution to $\mathbf{Ax} = \mathbf{b}$ is of the form $\mathbf{Gb} + (\mathbf{I}-\mathbf{GA})\xi$

for some vector ξ. Let, if possible, $\|\mathbf{Gb}\| = \|\mathbf{Gb} + (\mathbf{I} - \mathbf{GA})\xi\|$ for some ξ. Then

$$\begin{aligned}\|\mathbf{Gb}\|^2 &= \|\mathbf{Gb} + (\mathbf{I} - \mathbf{GA})\xi\|^2 \\ &= \|\mathbf{Gb}\|^2 + \|(\mathbf{I} - \mathbf{GA})\xi\|^2 + \mathbf{b}^\star\mathbf{G}^\star(\mathbf{I} - \mathbf{GA})\xi + \xi^\star(\mathbf{I} - \mathbf{GA})^\star\mathbf{Gb}.\end{aligned}$$

Since $\mathbf{b} \in \mathcal{C}(\mathbf{A})$, $\mathbf{b} = \mathbf{Au}$ for some $\mathbf{u}$. So,

$$\mathbf{b}^\star\mathbf{G}^\star(\mathbf{I} - \mathbf{GA})\xi = \mathbf{u}^\star\mathbf{A}^\star\mathbf{G}^\star(\mathbf{I} - \mathbf{GA})\xi = \mathbf{0}.$$

Also, $\xi^\star(\mathbf{I} - \mathbf{GA})^\star\mathbf{Gb} = (\mathbf{b}^\star\mathbf{G}^\star(\mathbf{I} - \mathbf{GA})\xi)^\star = \mathbf{0}$. Therefore, $\|\mathbf{Gb}\|^2 = \|\mathbf{Gb}\|^2 + \|(\mathbf{I} - \mathbf{GA})\xi\|^2$, which is possible only if $(\mathbf{I} - \mathbf{GA})\xi = \mathbf{0}$. Thus, minimum norm solution is unique. □

We now obtain the class of all minimum norm g-inverses of a matrix $\mathbf{A}$. Observe that for the null matrix of order $m \times n$, every matrix of order $n \times m$ is a minimum norm g-inverse.

Theorem 2.5.7. *Let* $\mathbf{A}$ *be an* $m \times n$ *matrix and* $\mathbf{G}$ *be a minimum norm g-inverse of* $\mathbf{A}$. *Then the class* $\{\mathbf{A}_\mathbf{m}^-\}$ *of all* $\mathbf{A}_\mathbf{m}^-$ *is given by*

$$\{\mathbf{G} + \mathbf{U}(\mathbf{I} - \mathbf{AG}),\ \mathbf{U}\ \textit{arbitrary}\}.$$

Proof. Let $\mathbf{U}$ be an arbitrary matrix. Then

$$(\mathbf{G} + \mathbf{U}(\mathbf{I} - \mathbf{AG}))\mathbf{AA}^\star = \mathbf{GAA}^\star = \mathbf{A}^\star.$$

So, $\mathbf{G} + \mathbf{U}(\mathbf{I} - \mathbf{AG})$ is a minimum norm g-inverse of $\mathbf{A}$ for all $\mathbf{U}$. Let $\mathbf{G_1}$ be any minimum norm g-inverse of $\mathbf{A}$. Then it is easy to check that $\mathbf{G_1} = \mathbf{G} + \mathbf{U}(\mathbf{I} - \mathbf{AG})$, where $\mathbf{U} = \mathbf{G_1} - \mathbf{G}$. Thus, the class of all minimum norm g-inverse is given by $\mathbf{G} + \mathbf{U}(\mathbf{I} - \mathbf{AG})$, where $\mathbf{U}$ is arbitrary. (Note that $\mathbf{GA} = \mathbf{P}_{\mathbf{A}^\star} = \mathbf{G_1A}$.)

□

Theorem 2.5.8. *Let* $\mathbf{A}$ *be an* $m \times n$ *matrix of rank* r (> 0). *Let* $\mathbf{U}\text{diag}(\mathbf{D}, \mathbf{0})\mathbf{V}^\star$ *be a singular value decomposition of* $\mathbf{A}$, *where* $\mathbf{U}$ *and* $\mathbf{V}$ *are unitary and* $\mathbf{D}$ *is a positive definite diagonal matrix of order* $r \times r$. *Then the class of all minimum norm g-inverses of* $\mathbf{A}$ *is given by*

$$\left\{\mathbf{V}\begin{pmatrix}\mathbf{D}^{-1} & \mathbf{L} \\ \mathbf{0} & \mathbf{N}\end{pmatrix}\mathbf{U}^\star : \mathbf{L},\ \mathbf{N}\ \textit{arbitrary}\right\}.$$

Proof. Notice that the class of all g-inverses of $\mathbf{A}$ is given by

$$\left\{\mathbf{V}\begin{pmatrix}\mathbf{D}^{-1} & \mathbf{L} \\ \mathbf{M} & \mathbf{N}\end{pmatrix}\mathbf{U}^\star;\ \mathbf{L},\ \mathbf{M} \text{ and } \mathbf{N}\ \textit{arbitrary}\right\}.$$

It is easy to check that $\mathbf{GAA}^\star = \mathbf{A}^\star$ if and only if $\mathbf{M} = \mathbf{0}$. □

Theorem 2.5.9. *Let* $\mathbf{A}$ *be an* $m \times n$ *matrix of rank* $r(>0)$. *The class* $\{\mathbf{A}_{\mathbf{mr}}^-\}$ *of all reflexive minimum norm g-inverses* $\mathbf{A}_{\mathbf{mr}}^-$ *of* $\mathbf{A}$ *is given by*

$$\{\mathbf{A}^\star(\mathbf{A}\mathbf{A}^\star)^- \ : \ (\mathbf{A}\mathbf{A}^\star)^- \text{ is an arbitrary g-inverse of } \mathbf{A}\mathbf{A}^\star\}.$$

Proof. In the proof of Theorem 2.5.3, we showed that $\mathbf{A}^\star(\mathbf{A}\mathbf{A}^\star)^-$ is a minimum norm g-inverse of $\mathbf{A}$. Notice that $\rho(\mathbf{A}^\star(\mathbf{A}\mathbf{A}^\star)^-) \leq \rho(\mathbf{A}^\star)$. However, $\rho(\mathbf{A}^\star(\mathbf{A}\mathbf{A}^\star)^-) \geq \rho(\mathbf{A}) = \rho(\mathbf{A}^\star)$ as $\mathbf{A}^\star(\mathbf{A}\mathbf{A}^\star)^-$ is a g-inverse of $\mathbf{A}$. So, $\rho(\mathbf{A}^\star(\mathbf{A}\mathbf{A}^\star)^-) = \rho(\mathbf{A})$. Hence, $\mathbf{A}^\star(\mathbf{A}\mathbf{A}^\star)^-$ is an $\mathbf{A}_{\mathbf{mr}}^-$.

Conversely, let $\mathbf{G} = \mathbf{A}_{\mathbf{mr}}^-$. As $\mathbf{G}$ is an $\mathbf{A}_{\mathbf{r}}^-$ and $\mathbf{GA}$ is hermitian, $\mathcal{C}(\mathbf{G}) = \mathcal{C}(\mathbf{GA}) = \mathcal{C}(\mathbf{A}^\star\mathbf{G}^\star) = \mathcal{C}(\mathbf{A}^\star)$. Thus $\mathbf{G} = \mathbf{A}^\star\mathbf{T}$ for some matrix $\mathbf{T}$. Now, $\mathbf{A}^\star = \mathbf{G}\mathbf{A}\mathbf{A}^\star = \mathbf{A}^\star\mathbf{T}\mathbf{A}\mathbf{A}^\star$. Therefore, $\mathbf{A}\mathbf{A}^\star = \mathbf{A}\mathbf{A}^\star\mathbf{T}\mathbf{A}\mathbf{A}^\star$, showing $\mathbf{T}$ is a g-inverse of $\mathbf{A}\mathbf{A}^\star$. □

Theorem 2.5.10. *Let* $(\mathbf{P}, \mathbf{Q})$ *be a rank factorization of an* $m \times n$ *matrix* $\mathbf{A}$ *of rank* $r(>0)$. *Then* $\mathbf{G}$ *is an* $\mathbf{A}_{\mathbf{mr}}^-$ *if and only if* $\mathbf{G} = \mathbf{Q}^\star(\mathbf{Q}\mathbf{Q}^\star)^{-1}\mathbf{P}_{\mathbf{L}}^{-1}$ *for some left inverse* $\mathbf{P}_{\mathbf{L}}^{-1}$ *of* $\mathbf{P}$.

Proof. We note that $\mathcal{C}(\mathbf{Q}^\star) = \mathcal{C}(\mathbf{A}^\star)$. The rest of the proof is similar to (i) of Theorem 2.4.4. □

Let us now turn our attention to a possibly inconsistent system of linear equations. We are given a system of linear equations $\mathbf{Ax} = \mathbf{b}$ which may or may not be consistent. In case it is consistent, we want to find a solution. Otherwise the next best thing to do is to find an approximate solution $\mathbf{x_0}$ such that $\mathbf{Ax_0}$ is the closest to $\mathbf{b}$ among all $\mathbf{Ax}$ i.e.

$$\|\mathbf{Ax_0} - \mathbf{b}\| \leq \|\mathbf{Ax} - \mathbf{b}\| \text{ for all } \mathbf{x}.$$

If such a vector exists, then for all $\mathbf{x}$,

$$(\mathbf{Ax_0} - \mathbf{b})^\star(\mathbf{Ax_0} - \mathbf{b}) \leq (\mathbf{Ax} - \mathbf{b})^\star(\mathbf{Ax} - \mathbf{b}).$$

Such an approximate solution is called a least (sum of) squares solution to $\mathbf{Ax} = \mathbf{b}$. We shall now explore the existence of a g-inverse $\mathbf{G}$ of $\mathbf{A}$ such that $\mathbf{x} = \mathbf{Gb}$ is a least squares solution to $\mathbf{Ax} = \mathbf{b}$. Notice that a least squares solution should be an exact solution if $\mathbf{Ax} = \mathbf{b}$ is consistent.

Theorem 2.5.11. *Let* $\mathbf{A}$ *be an* $m \times n$ *matrix. Then there exists a g-inverse* $\mathbf{G}$ *of* $\mathbf{A}$ *such that* $\mathbf{x} = \mathbf{Gb}$ *is a least squares solution to* $\mathbf{Ax} = \mathbf{b}$ *for all* $\mathbf{b} \in \mathbb{C}^{\mathbf{m}}$.

Proof. Consider $\mathbf{G} = (\mathbf{A}^\star\mathbf{A})^-\mathbf{A}^\star$. By Theorem 2.3.12, $\mathbf{G}$ is a g-inverse of $\mathbf{A}$. Now

$$\|\mathbf{Ax} - \mathbf{b}\| = \|\mathbf{Ax} - \mathbf{AGb} + \mathbf{AGb} - \mathbf{b}\|.$$

By Lemma 2.5.1,

$$\|\mathbf{AGb} - \mathbf{b}\| \leq \|\mathbf{AGb} - \mathbf{b} + \mathbf{A}(\mathbf{x} - \mathbf{Gb})\| \text{ for all } \mathbf{b} \text{ and for all } \mathbf{x}$$
$$\Leftrightarrow \mathbf{A}^{\star}(\mathbf{AG} - \mathbf{I}) = \mathbf{0}.$$

Moreover, $\mathbf{A}^{\star}\mathbf{AG} = \mathbf{A}^{\star}\mathbf{A}(\mathbf{A}^{\star}\mathbf{A})^{-}\mathbf{A}^{\star} = \mathbf{A}^{\star}$, since $\mathcal{C}(\mathbf{A}^{\star}\mathbf{A}) = \mathcal{C}(\mathbf{A}^{\star})$. Thus, $\mathbf{G} = (\mathbf{A}^{\star}\mathbf{A})^{-}\mathbf{A}^{\star}$ is a g-inverse of $\mathbf{A}$ such that $\mathbf{Gb}$ is a least squares solution to $\mathbf{Ax} = \mathbf{b}$ for all $\mathbf{b}$. □

Remark 2.5.12. If $\mathbf{Gb}$ has to be a least squares solution to $\mathbf{Ax} = \mathbf{b}$ for all $\mathbf{b}$, then $\mathbf{AGb} = \mathbf{b}$ for all $\mathbf{b} \in \mathcal{C}(\mathbf{A})$. So, $\mathbf{G}$ must be a g-inverse of $\mathbf{A}$.

Definition 2.5.13. Let $\mathbf{A}$ be an $m \times n$ matrix. Then a matrix $\mathbf{G}$ is said to be a least squares g-inverse of $\mathbf{A}$, denoted by $\mathbf{A}_{\ell}^{-}$, if $\mathbf{Gb}$ is a least squares solution to $\mathbf{Ax} = \mathbf{b}$ for all $\mathbf{b} \in \mathbb{C}^{\mathbf{m}}$.

We shall now obtain some characterizations of $\mathbf{A}_{\ell}^{-}$.

Theorem 2.5.14. *Let $\mathbf{A}$ be an $m \times n$ matrix. Then a matrix $\mathbf{G}$ is a least squares g-inverse of $\mathbf{A}$ if and only if it satisfies any one of the following equivalent conditions:*

(i) $\mathbf{AGA} = \mathbf{A}$, $(\mathbf{AG})^{\star} = \mathbf{AG}$
(ii) $\mathbf{AG} = \mathbf{P_A}$, *the orthogonal projector projecting vectors into* $\mathcal{C}(\mathbf{A})$.
(iii) $\mathbf{A}^{\star}\mathbf{AG} = \mathbf{A}^{\star}$.

Proof. In the proof of Theorem 2.5.11, we observed that $\mathbf{G}$ is a least squares g-inverse $\mathbf{G}$ of $\mathbf{A} \Leftrightarrow \mathbf{G}$ satisfies (iii) i.e. if $\mathbf{A}^{\star}\mathbf{AG} = \mathbf{A}^{\star}$.

Thus, it is enough to prove (iii) ⇒ (i) ⇒ (ii) ⇒ (iii).

Let $\mathbf{G}$ be a matrix satisfying (iii). Then, $\mathbf{A}^{\star}\mathbf{AGA} = \mathbf{A}^{\star}\mathbf{A}$. Now, $\mathcal{C}(\mathbf{A}^{\star}\mathbf{A}) = \mathcal{C}(\mathbf{A}^{\star})$, so, there exists a matrix $\mathbf{T}$ such that $\mathbf{A} = \mathbf{TA}^{\star}\mathbf{A}$. Hence, $\mathbf{A}^{\star}\mathbf{AGA} = \mathbf{A}^{\star}\mathbf{A} \Rightarrow \mathbf{TA}^{\star}\mathbf{AGA} = \mathbf{TA}^{\star}\mathbf{A} \Rightarrow \mathbf{AGA} = \mathbf{A}$. Also, $\mathbf{A}^{\star}\mathbf{AG} = \mathbf{A}^{\star} \Rightarrow \mathbf{G}^{\star}\mathbf{A}^{\star}\mathbf{AG} = \mathbf{G}^{\star}\mathbf{A}^{\star} = \mathbf{AG}$. So, $\mathbf{AG}$ is hermitian and therefore (i) holds.

Let (i) hold. Then $\mathbf{AG}$ is an orthogonal projector onto $\mathcal{C}(\mathbf{AG})$. But $\mathbf{AGA} = \mathbf{A} \Rightarrow \mathcal{C}(\mathbf{AG}) = \mathcal{C}(\mathbf{A})$. So, $\mathbf{AG} = \mathbf{P_A}$. Thus (ii) holds.
(ii)⇒ (iii) is trivial. □

Notice the striking similarity of conditions in Theorems 2.5.5 and 2.5.14. Consequently, we have the following important duality between minimum norm g-inverse and least squares g-inverses.

Theorem 2.5.15. *Let $\mathbf{A}$ be an $m \times n$ matrix. then $\mathbf{G}$ is an $\mathbf{A}_{\mathbf{m}}^{-}$ if and only if $\mathbf{G}^{\star}$ is an $(\mathbf{A}^{\star})_{\ell}^{-}$.*

Proof follows from Theorems 2.5.5 and 2.5.14, once we note that $\mathbf{G}$ is a g-inverse of $\mathbf{A} \Rightarrow \mathbf{G}^\star$ is a g-inverse of $\mathbf{A}^\star$.

Consider a system of linear equations $\mathbf{Ax} = \mathbf{b}$. Then $\mathbf{x_0}$ is a least squares solution to $\mathbf{Ax} = \mathbf{b}$ if and only if $\mathbf{Ax_0}$ is the orthogonal projection of $\mathbf{b}$ in $\mathcal{C}(\mathbf{A})$. Also, then $\mathbf{b} - \mathbf{Ax_0}$ is the orthogonal projection of $\mathbf{b}$ in $(\mathcal{C}(\mathbf{A}))^\perp$. Hence, even if $\mathbf{x_0}$ may not be unique, both the vectors $\mathbf{Ax_0}$ and $\mathbf{b} - \mathbf{Ax_0}$ are unique.

Theorem 2.5.16. *Let* $\mathbf{A}$ *be an* $m \times n$ *matrix and let* $\mathbf{G}$ *be a least squares g-inverse of* $\mathbf{A}$. *Then the class of all least squares solutions to* $\mathbf{Ax} = \mathbf{b}$ *is given by*

$$\{\mathbf{Gb} + (\mathbf{I} - \mathbf{GA})\xi,\ \xi \textit{ arbitrary}\}.$$

Proof. Clearly, $\mathbf{A}(\mathbf{Gb} + (\mathbf{I} - \mathbf{GA})\xi) = \mathbf{AGb}$ for all ξ. Further $\mathbf{Gb}$ is a least squares solution to $\mathbf{Ax} = \mathbf{b}$. Hence $\mathbf{Gb} + (\mathbf{I} - \mathbf{GA})\xi$ is a least squares solution to $\mathbf{Ax} = \mathbf{b}$ for all ξ.

Let $\mathbf{x_0}$ be a least squares solution to $\mathbf{Ax} = \mathbf{b}$. By the discussion just preceding this theorem, $\mathbf{Ax_0} = \mathbf{AGb}$. So, $\mathbf{A}(\mathbf{x_0} - \mathbf{Gb}) = \mathbf{0}$. Thus, $\mathbf{x_0} - \mathbf{Gb} = (\mathbf{I} - \mathbf{GA})\xi$ for some ξ. So, $\mathbf{x_0} = \mathbf{Gb} + (\mathbf{I} - \mathbf{GA})\xi$ for some ξ.

□

The following Theorems 2.5.17-2.5.19 and Theorem 2.5.21 can be established along the lines of Theorems 2.5.7-2.5.10.

Theorem 2.5.17. *Let* $\mathbf{A}$ *be an* $m \times n$ *matrix. Let* $\mathbf{G}$ *be a least squares g-inverse of* $\mathbf{A}$. *Then the class of all* $\mathbf{A}_\ell^-$ *is given by*

$$\{\mathbf{G} + (\mathbf{I} - \mathbf{GA})\mathbf{V} :\ \mathbf{V} \textit{ arbitrary}\}.$$

Theorem 2.5.18. *Let* $\mathbf{A}$ *be an* $m \times n$ *matrix of rank* $r(> 0)$. *Let* $\mathbf{U}\mathbf{diag}(\mathbf{D}, \mathbf{0})\mathbf{V}^\star$ *be a singular value decomposition of* $\mathbf{A}$, *where* $\mathbf{U}$ *and* $\mathbf{V}$ *are unitary and* $\mathbf{D}$ *is a positive definite diagonal matrix of order* $r \times r$. *Then the class* $\{\mathbf{A}_\ell^-\}$ *of all least squares g-inverse of* $\mathbf{A}$ *is given by*

$$\left\{\mathbf{V}\begin{pmatrix}\mathbf{D}^{-1} & \mathbf{0}\\ \mathbf{M} & \mathbf{N}\end{pmatrix}\mathbf{U}^\star :\ \mathbf{M} \textit{ and } \mathbf{N} \textit{ arbitrary}\right\}.$$

Theorem 2.5.19. *Let* $\mathbf{A}$ *be an* $m \times n$ *matrix of rank* $r(> 0)$. *The class* $\{\mathbf{A}_{\ell r}^-\}$ *of all reflexive least squares g-inverses* $\mathbf{A}_{\ell r}^-$ *of* $\mathbf{A}$ *is given by*

$$\{(\mathbf{A}^\star\mathbf{A})^-\mathbf{A}^\star\ :\ (\mathbf{A}^\star\mathbf{A})^- \textit{ is an arbitrary g-inverse of } \mathbf{A}^\star\mathbf{A}\}.$$

Remark 2.5.20. For a matrix $\mathbf{A}$, $\mathbf{G}$ is an $\mathbf{A}^-_{\ell \mathbf{r}}$ if and only if $\mathbf{G}$ is g-inverse of $\mathbf{A}$ such that $\mathcal{C}(\mathbf{G}^\star) = \mathcal{C}(\mathbf{A})$. Recall that $\mathbf{G}$ is an $\mathbf{A}^-_\rho$ if and only if $\mathcal{C}(\mathbf{G}^\star) = \mathcal{C}(\mathbf{A}^\star)$.

Theorem 2.5.21. *Let* $(\mathbf{P}, \mathbf{Q})$ *be a rank factorization of an* $m \times n$ *matrix* $\mathbf{A}$ *of rank* $r(> 0)$. *Then* $\mathbf{G}$ *is an* $\mathbf{A}^-_{\ell \mathbf{r}}$ *if and only if* $\mathbf{G} = \mathbf{Q}_{\mathbf{R}}^{-1}(\mathbf{P}^\star\mathbf{P})^{-1}\mathbf{P}^\star$ *for some right inverse* $\mathbf{Q}_{\mathbf{R}}^{-1}$ *of* $\mathbf{Q}$.

Notice the similarity between $\mathbf{A}^-_\chi$ and $\mathbf{A}^-_{\mathbf{mr}}$. Are there counterparts to $\mathbf{A}^-_{\mathbf{m}}$ and $\mathbf{A}^-_\ell$ for index-1 matrices? Indeed there are and we shall introduce them in Chapter 6.

We have seen in Theorem 2.5.16 that, in general, there are many least squares solutions to $\mathbf{Ax} = \mathbf{b}$. We shall now show that there is a g-inverse $\mathbf{G}$ of $\mathbf{A}$ such that $\mathbf{Gb}$ has the minimum norm among all least squares solutions to $\mathbf{Ax} = \mathbf{b}$ for all $\mathbf{b} \in \mathbb{C}^{\mathbf{m}}$.

Theorem 2.5.22. *Let* $\mathbf{A}$ *be an* $m \times n$ *matrix. Then, there exists a g-inverse* $\mathbf{G}$ *of* $\mathbf{A}$ *such that* $\mathbf{Gb}$ *has the minimum norm among all least squares solutions to* $\mathbf{Ax} = \mathbf{b}$ *for all* $\mathbf{b} \in \mathbb{C}^{\mathbf{m}}$.

Proof. Let $\mathbf{G}$ be a least squares g-inverse of $\mathbf{A}$. Then the class of all least squares solutions to $\mathbf{Ax} = \mathbf{b}$ is given by $\mathbf{Gb} + (\mathbf{I} - \mathbf{GA})\xi$, where ξ is arbitrary. Now, we are seeking a least squares g-inverse $\mathbf{G}$ such that $||\mathbf{Gb}|| \leq ||\mathbf{Gb} + (\mathbf{I} - \mathbf{GA})\xi||$ for all $\mathbf{b}$ and ξ. By Lemma 2.5.1,

$$||\mathbf{Gb}|| \leq ||\mathbf{Gb} + (\mathbf{I} - \mathbf{GA})\xi|| \Leftrightarrow \mathbf{G}^\star(\mathbf{I} - \mathbf{GA}) = \mathbf{0} \Leftrightarrow \mathbf{G}^\star\mathbf{GA} = \mathbf{G}^\star.$$

Consider $\mathbf{G} = \mathbf{A}^\star(\mathbf{A}^\star\mathbf{AA}^\star)^-\mathbf{A}^\star$. Then $\mathbf{A}^\star\mathbf{AG} = \mathbf{A}^\star\mathbf{AA}^\star(\mathbf{A}^\star\mathbf{AA}^\star)^-\mathbf{A}^\star = \mathbf{A}^\star$, since $\rho(\mathbf{A}^\star\mathbf{AA}^\star) = \rho(\mathbf{A}^\star)$. So $\mathbf{G}$ is a least square g-inverse of $\mathbf{A}$.

$$\begin{aligned} \text{Again } \mathbf{G}^\star\mathbf{GAA}^\star &= ((\mathbf{A}^\star\mathbf{A}^\star\mathbf{AA}^\star)^-\mathbf{A}^\star)^\star\mathbf{A}^\star(\mathbf{A}^\star\mathbf{AA}^\star)^-\mathbf{A}^\star\mathbf{AA}^\star \\ &= \mathbf{A}\{(\mathbf{A}^\star\mathbf{AA}^\star)^-\}^\star\mathbf{AA}^\star\ , \text{ since } \rho(\mathbf{A}^\star\mathbf{AA}^\star) = \rho(\mathbf{AA}^\star)\ . \end{aligned}$$

As $\rho(\mathbf{AA}^\star) = \rho(\mathbf{A})$, it follows that $\mathbf{G}^\star\mathbf{GA} = \mathbf{A}\{(\mathbf{A}^\star\mathbf{AA}^\star)^-\}^\star\mathbf{A} = \mathbf{G}^\star$.

Thus, $\mathbf{G} = \mathbf{A}^\star(\mathbf{A}^\star\mathbf{AA}^\star)^-\mathbf{A}^\star$ is a g-inverse with the desired properties. □

Remark 2.5.23. From the proof of Theorem 2.5.22, we see that $\mathbf{Gb}$ has the minimum norm among all least square solutions if and only if $\mathbf{A}^\star\mathbf{AG} = \mathbf{A}^\star$ and $\mathbf{G}^\star\mathbf{GA} = \mathbf{G}^\star$ or equivalently $\mathbf{G}$ is an $\mathbf{A}^-_\ell$ and $\mathbf{A}$ is a $\mathbf{G}^-_\ell$.

Theorem 2.5.24. *Let* $\mathbf{A}$ *be an* $m\times n$ *matrix. Then* $\mathbf{G}$ *is a matrix such that* $\mathbf{Gb}$ *has the minimum norm among all least squares solutions to* $\mathbf{Ax} = \mathbf{b}$ *for all* $\mathbf{b} \in \mathbb{C}^{\mathbf{m}}$ *if and only if any one of the following equivalent conditions holds:*

(i) $\mathbf{A}^\star\mathbf{AG} = \mathbf{A}^\star$ *and* $\mathbf{G}^\star\mathbf{GA} = \mathbf{G}^\star$
(ii) $\mathbf{AG} = \mathbf{P_A}$, $\mathbf{GA} = \mathbf{P_G}$ *and*
(iii) $\mathbf{AGA} = \mathbf{A}$, $\mathbf{GAG} = \mathbf{G}$, $(\mathbf{AG})^\star = \mathbf{AG}$, $(\mathbf{GA})^\star = \mathbf{GA}$.

Proof. (i) follows by Theorem 2.5.22. The equivalence of (i)-(iii) follows from Theorem 2.5.14. □

Remark 2.5.25. Let $\mathbf{Gb}$ have the minimum norm among all least squares solutions to $\mathbf{Ax} = \mathbf{b}$ for all $\mathbf{b} \in \mathbb{C}^{\mathbf{m}}$. Then (iii) of Theorem 2.5.24 shows that $\mathbf{G}$ is a minimum norm, least squares, reflexive g-inverse of $\mathbf{A}$.

Definition 2.5.26. Let $\mathbf{A}$ be an $m \times n$ matrix. Then an $n \times m$ matrix $\mathbf{G}$ satisfying any one of the equivalent conditions of Theorem 2.5.24 is called Moore-Penrose inverse of $\mathbf{A}$ and is denoted by $\mathbf{A}^\dagger$.

Theorem 2.5.27. *Let* $\mathbf{A}$ *be an* $m\times n$ *matrix. Then Moore-Penrose inverse* $\mathbf{A}^\dagger$ *of* $\mathbf{A}$ *is unique.*

Proof. Let $\mathbf{G_1}$ and $\mathbf{G_2}$ be two choices of Moore-Penrose inverse of $\mathbf{A}$. Then $\mathbf{G_1} = \mathbf{G_1AG_1} = \mathbf{G_1P_A} = \mathbf{G_1AG_2} = \mathbf{P_{A^\star}G_2} = \mathbf{G_2AG_2} = \mathbf{G_2}$. □

The following theorem is easy to prove.

Theorem 2.5.28. *Let* $\mathbf{A}$ *be an* $m \times n$ *matrix. Then the Moore-Penrose inverse* $\mathbf{A}^\dagger$ *has the following properties:*

(i) $\mathbf{A}^{\star\dagger} = \mathbf{A}^{\dagger\star}$.
(ii) $(\mathbf{A}^\dagger)^\dagger = \mathbf{A}$.
(iii) $(\mathbf{A}^\star\mathbf{A})^\dagger = \mathbf{A}^\dagger\mathbf{A}^{\dagger\star}$.
(iv) $\mathbf{A}^\dagger = (\mathbf{A}^\star\mathbf{A})^\dagger\mathbf{A}^\star$.
(v) *If* $\mathbf{R}$ *and* $\mathbf{S}$ *are unitary matrices such that the product* $\mathbf{RAS}$ *is defined, then* $(\mathbf{RAS})^\dagger = \mathbf{S}^\star\mathbf{A}^\dagger\mathbf{R}^\star$.

Remark 2.5.29. In Theorem 2.5.22, we gave an expression for $\mathbf{A}^\dagger$, namely $\mathbf{A}^\dagger = \mathbf{A}^\star(\mathbf{A}^\star\mathbf{A}\mathbf{A}^\star)^-\mathbf{A}^\star$. Thus, it follows that $\mathbf{A}^\dagger$ is the unique g-inverse $\mathbf{G}$ such that $\mathcal{C}(\mathbf{G}) = \mathcal{C}(\mathbf{A}^\star)$ and $\mathcal{C}(\mathbf{G}^\star) = \mathcal{C}(\mathbf{A})$.

Remark 2.5.30. Recall that $\mathbf{A}^\# = \mathbf{A}(\mathbf{A^3})^-\mathbf{A}$. Also, $\mathbf{A}^\#$ is the unique g-inverse of $\mathbf{A}$ such that $\mathcal{C}(\mathbf{G}) = \mathcal{C}(\mathbf{A})$ and $\mathcal{C}(\mathbf{G}^\star) = \mathcal{C}(\mathbf{A}^\star)$.

In view of this similarity, it is worth asking when $\mathbf{A}^\dagger$ and $\mathbf{A}^\#$ coincide.

Theorem 2.5.31. *Let* $\mathbf{A}$ *be an* $m \times m$ *matrix of index 1 over* $\mathbb{C}$. *Then* $\mathbf{A}^\dagger$ *and* $\mathbf{A}^\#$ *coincide if and only if* $\mathbf{A}$ *is a range-hermitian.*

Proof follows from Remark 2.5.29 and Remark 2.5.30.

The following characterizations of Moore-Penrose inverse are easy to establish.

Theorem 2.5.32. *Let* $\mathbf{A}$ *be an* $m \times n$ *matrix of rank* $(r > 0)$. *Let* $\mathbf{A} = \mathbf{Udaig(D,\ 0)V^\star}$ *be a singular value decomposition of* $\mathbf{A}$, *where* $\mathbf{U}$ *and* $\mathbf{V}$ *are unitary and* $\mathbf{D}$ *is a positive definite diagonal matrix of order* $r \times r$. *Then* $\mathbf{A}^\dagger = \mathbf{Vdiag(D^{-1},\ 0)U^\star}$. *(Notice that we have a singular value decomposition of* $\mathbf{A}^\dagger$.*)*

Theorem 2.5.33. *Let* $\mathbf{A}$ *be an* $m \times n$ *matrix of rank* $r(> 0)$. *Let* $(\mathbf{P}, \mathbf{Q})$ *be a rank factorization of* $\mathbf{A}$. *Then* $\mathbf{A}^\dagger = \mathbf{Q}^\star(\mathbf{QQ}^\star)^{-1}(\mathbf{P}^\star\mathbf{P})^{-1}\mathbf{P}^\star$.

We conclude this section by making a useful observation that, given an algorithm to compute a g-inverse of a matrix, we can compute various types of g-inverse of a given matrix using this algorithm.

Reflexive g-inverse	:	$\mathbf{GAG}$, if $\mathbf{G}$ is a g-inverse of $\mathbf{A}$
ρ-inverse	:	$\mathbf{A}(\mathbf{A}^2)^-$
χ-inverse	:	$(\mathbf{A}^2)^-\mathbf{A}$
Group inverse	:	$\mathbf{A}(\mathbf{A}^3)^-\mathbf{A}$
Least squares g-inverse	:	$(\mathbf{A}^\star\mathbf{A})^-\mathbf{A}^\star$
Minimum norm g-inverse	:	$\mathbf{A}^\star(\mathbf{AA}^\star)^-$
Moore-Penrose inverse	:	$\mathbf{A}^\star(\mathbf{A}^\star\mathbf{AA}^\star)^-\mathbf{A}^\star$

In each of the above cases, we apply the algorithm for computing a g-inverse to a suitable matrix (for example to $\mathbf{A}^\star\mathbf{AA}^\star$ in case of Moore-Penrose inverse) and make adjustment to it, be it a pre- and/or post-multiplication by a suitable matrix to get the required type of g-inverse of the given matrix.

2.6 Generalized inverses of modified matrices

In this section, we study the g-inverse of matrices that have been modified in some suitable ways. The need for a g-inverse of modified matrices arises in several situations. Consider a linear model $\mathbf{Y} = \mathbf{X}\boldsymbol{\beta} + \boldsymbol{\eta}$, where $\mathbf{X}$ is a known $m \times n$ matrix, $\boldsymbol{\beta}$ is an unknown but non-stochastic n-vector and $\boldsymbol{\eta}$ is a random m-vector such that its expected value $\mathbf{E}(\boldsymbol{\eta}) = \mathbf{0}$ and dispersion matrix $\mathbf{D}(\boldsymbol{\eta}) = \sigma^2\mathbf{I}$ where $\sigma^2(> 0)$ is an unknown constant. Then a least squares estimator of $\boldsymbol{\beta}$ is $\hat{\boldsymbol{\beta}} = \mathbf{X}_\ell^- \mathbf{Y} = (\mathbf{X}^t\mathbf{X})^-\mathbf{X}^t\mathbf{Y}$. If $\mathbf{p}^t\boldsymbol{\beta}$ is estimable, then $\mathbf{p}^t\hat{\boldsymbol{\beta}}$ is its BLUE with variance $V(\mathbf{p}^t\hat{\boldsymbol{\beta}}) = \mathbf{p}^t(\mathbf{X}^t\mathbf{X})^-\mathbf{p}\,\sigma^2$. Suppose one more uncorrelated observation is available, namely: $y_{m+1} = \mathbf{x}_{\mathbf{m+1}}^{\mathbf{t}}\boldsymbol{\beta} + \eta_{m+1}$, where $E(\eta_{m+1}) = 0, V(\boldsymbol{\eta}_{m+1}) = \sigma^2$ and $\mathrm{Cov}(\eta_{m+1}, \boldsymbol{\eta}) = 0$. Then a least squares estimator after incorporating the new observation may be obtained as

$$\begin{aligned}\tilde{\boldsymbol{\beta}} &= ((\mathbf{X}^t : \mathbf{x}_{\mathbf{m+1}}^{\mathbf{t}})^t)_\ell^-(\mathbf{Y}^t : y_{m+1})^t \\ &= (\mathbf{X}^t\mathbf{X} + \mathbf{x}_{\mathbf{m+1}}\mathbf{x}_{\mathbf{m+1}}^{\mathbf{t}})^-(\boldsymbol{X}^t\boldsymbol{Y} + y_{m+1}\mathbf{x}_{\mathbf{m+1}}).\end{aligned}$$

Also, the variance $V(\mathbf{p}^t\tilde{\boldsymbol{\beta}})$, of BLUE of $\mathbf{p}^t\boldsymbol{\beta}$ in the new model is $\mathbf{p}^t(\mathbf{X}^t\mathbf{X} + \mathbf{x}_{\mathbf{m+1}}\mathbf{x}_{\mathbf{m+1}}^{\mathbf{t}})^-\mathbf{p}\,\sigma^2$.

Thus after incorporating the new observation, we need

$$((\mathbf{X}^t : \mathbf{x}_{\mathbf{m+1}}^{\mathbf{t}})^t)_\ell^- \text{ and } (\mathbf{X}^t\mathbf{X} + \mathbf{x}_{\mathbf{m+1}}\mathbf{x}_{\mathbf{m+1}}^{\mathbf{t}})^-.$$

In order to assess the influence of an observation on the estimators of model parameters one needs to delete a row from $\mathbf{X}$ or subtract a rank 1 matrix from $\mathbf{X}^t\mathbf{X}$ and obtain a suitable g-inverse of the modified matrix.

When a new state is included in a Markov chain or a new port is introduced in an n-port network, modified matrices and their g-inverses of suitable type become important in the study of the effect of such modifications.

Let $\mathbf{A}$ be an $m \times n$ matrix and $\mathbf{G}$ be a g-inverse of $\mathbf{A}$. Here we will consider the following types of modifications

(i) appending/deleting a row or a column,
(ii) appending a row and a column, and
(iii) adding a rank 1 matrix of the same order.

In each case, we obtain a suitable modification of $\mathbf{G}$, so that it will be a g-inverse of the modified matrix.

The following lemma is useful in what follows:

Lemma 2.6.1. *Let* $\mathbf{A}$ *be an* $m \times n$ *matrix of rank* $r(> 0)$ *and let* $\mathbf{u_1}, \mathbf{u_2}, \ldots, \mathbf{u_k}$ *be* m*-vectors such that* $\rho(\mathbf{A} : \mathbf{u_1} : \ldots : \mathbf{u_k}) = \rho(\mathbf{A}) + k$, *where* $k \leq m - r$. *Then there exists a vector* $\mathbf{c}$ *such that* $\mathbf{c}^t\mathbf{A} = \mathbf{0}$, *and* $\mathbf{c}^t\mathbf{u_i} = 1$ *for* $i = 1, \ldots, k$.

Proof. Let $(\mathbf{P}, \mathbf{Q})$ be a rank factorization of $\mathbf{A}$. Then $(\mathbf{P} : \mathbf{u_1} : \ldots : \mathbf{u_k})$ is an $m \times (r+k)$ matrix of rank $r+k$. Let $(\mathbf{R^t} : \mathbf{v_1} : \mathbf{v_k})^\mathbf{t}$ be a left inverse of $(\mathbf{P} : \mathbf{u_1} : \ldots : \mathbf{u_k})$, where $\mathbf{R}$ is an $r \times m$ matrix. Then $\mathbf{v_i}^t\mathbf{P} = \mathbf{0}$ for $i = 1, \ldots, k$ and $\mathbf{v_i}^t\mathbf{u_j} = \delta_{\mathbf{ij}}, i, j = 1, \ldots, k$. Let $\mathbf{c} = \mathbf{v_1} + \mathbf{v_2} + \ldots + \mathbf{v_k}$. Clearly, $\mathbf{c^tP} = \mathbf{0}$ and $\mathbf{c^tu_i} = \mathbf{1}$ for $i = 1, \ldots, k$. □

Corollary 2.6.2. *If the matrices considered are over complex field, then there exists a vector* $\mathbf{c}$ *such that* $\mathbf{c^\star A} = \mathbf{0}$, $\mathbf{c^\star u_i} = 1$ *for all* $i = 1, \ldots, k$.

We shall start with the first type of modification.

Let $\mathbf{A}$ be an $m \times n$ matrix and a g-inverse $\mathbf{G}$ (of $\mathbf{A}$) of a suitable type be available. Let $\mathbf{a}$ be an m-column vector. Write $\mathbf{d} = \mathbf{Ga}$. We shall show in Theorems 2.6.3-2.6.10 that a corresponding type of g-inverse of $(\mathbf{A} : \mathbf{a})$ can be expressed in the form

$$\mathbf{X} = (\mathbf{G^t} - \mathbf{cd^t} : \mathbf{c})^\mathbf{t} \tag{2.6.1}$$

for some suitable $\mathbf{c}$.

Theorem 2.6.3. *Let* $\mathbf{A}$ *be an* $m \times n$ *matrix and* $\mathbf{G}$ *be a g-inverse of* $\mathbf{A}$. *Let* $\mathbf{a}$ *be an* m*-column vector and* $\mathbf{d} = \mathbf{Ga}$. *If* $\mathbf{a} \notin \mathcal{C}(\mathbf{A})$, $\mathbf{c}$ *is an* m*-column vector such that* $\mathbf{c^tA} = \mathbf{0}$, $\mathbf{c^ta} = 1$ *(such a* $\mathbf{c}$ *is guaranteed by Lemma 2.6.1), and* $\mathbf{X}$ *is as defined in (2.6.1), then*

(i) $\mathbf{X}$ *is* $(\mathbf{A} : \mathbf{a})^-$ *if* $\mathbf{G}$ *is* $\mathbf{A}^-$.
(ii) $\mathbf{X}$ *is* $(\mathbf{A} : \mathbf{a})_\mathbf{r}^-$ *if* $\mathbf{G}$ *is* $\mathbf{A}_\mathbf{r}^-$.
(iii) *Let all the vectors and matrices be over* $\mathbb{C}$. *Then* $\mathbf{X}$ *is* $(\mathbf{A} : \mathbf{a})_\mathbf{m}^-$ *if* $\mathbf{G}$ *is a* $\mathbf{A}_\mathbf{m}^-$.

Theorem 2.6.4. *Let* $\mathbf{A}, \mathbf{a}, \mathbf{G}$ *and* $\mathbf{d}$ *be as specified in Theorem 2.6.3, but with complex elements. Let* $\mathbf{a} \notin \mathcal{C}(\mathbf{A})$ *and* $\mathbf{c} = \dfrac{(\mathbf{I} - \mathbf{AG})\mathbf{a}}{\mathbf{a}^\star(\mathbf{I} - \mathbf{AG})\mathbf{a}}$. *Write* $\mathbf{X} = (\mathbf{G}^\star - \mathbf{cd}^\star : \mathbf{c})^\star$. *Then*

(i) $\mathbf{X}$ *is* $(\mathbf{A} : \mathbf{a})_\ell^-$ *if* $\mathbf{G}$ *is* $\mathbf{A}_\ell^-$.
(ii) $\mathbf{X}$ *is* $(\mathbf{A} : \mathbf{a})^\dagger$ *if* $\mathbf{G}$ *is* $\mathbf{A}^\dagger$.

Theorem 2.6.5. *Let* $\mathbf{A}, \mathbf{a}, \mathbf{G}, \mathbf{d}$ *be as in Theorem 2.6.3. Let* $\mathbf{a} \in \mathcal{C}(\mathbf{A})$. *Then*

(i) $\mathbf{X}$ *is* $(\mathbf{A} : \mathbf{a})^-$ *if* $\mathbf{G}$ *is an* $\mathbf{A}^-$ *and* $\mathbf{c}$ *is arbitrary.*
(ii) $\mathbf{X}$ *is* $(\mathbf{A} : \mathbf{a})_\mathbf{r}^-$ *if* $\mathbf{G}$ *is an* $\mathbf{A}_\mathbf{r}^-$ *and* $\mathbf{c} \in \mathcal{C}(\mathbf{G^t})$ *is arbitrary.*
(iii) *Assuming that the vectors and matrices are over* $\mathbb{C}, \mathbf{X}$ *is* $(\mathbf{A} : \mathbf{a})_\ell^-$ *if* $\mathbf{G}$ *is* $\mathbf{A}_\ell^-$ *and* $\mathbf{c}$ *is arbitrary.*

Theorem 2.6.6. *In the setup of Theorem 2.6.4, let* $\mathbf{c} = \dfrac{\mathbf{G^\star Ga}}{1+\mathbf{a^\star G^\star Ga}}$. *Then*

(i) $\mathbf{X}$ *is a* $(\mathbf{A}:\mathbf{a})^-_{\mathbf{m}}$ *if* $\mathbf{G}$ *is* $\mathbf{A}^-_{\mathbf{m}}$.
(ii) $\mathbf{X}$ *is* $(\mathbf{A}:\mathbf{a})^\dagger$ *if* $\mathbf{G}$ *is* $\mathbf{A}^\dagger$.

Suppose that a g-inverse $\mathbf{X} = (\mathbf{G^t} : \mathbf{b})^{\mathbf{t}}$ of a suitable type for $(\mathbf{A}:\mathbf{a})$ is available, where $\mathbf{G}$ is an $n\times m$ matrix and $\mathbf{b}$ is an m-column vector. We shall now find a modification of $\mathbf{X}$ so that it is a g-inverse of $\mathbf{A}$. Interestingly, $\mathbf{G}$ itself is a g-inverse of $\mathbf{A}$ when $\mathbf{a} \notin \mathcal{C}(\mathbf{A})$. We shall probe this further in the following

Lemma 2.6.7. *Let* $\mathbf{X} = (\mathbf{G^t} : \mathbf{b})^{\mathbf{t}}$ *be a g-inverse of* $(\mathbf{A}:\mathbf{a})$. *Then* $\mathbf{a} \notin \mathcal{C}(\mathbf{A})$ *if and only if* $\mathbf{b^t a} = 1$ *and* $\mathbf{b^t A} = \mathbf{0}$. *Further, if* $\mathbf{a} \notin \mathcal{C}(\mathbf{A})$, *then* $\mathbf{AGa} = \mathbf{0}$ *and* $\mathbf{X_1} = (\mathbf{G^t A^t G^t} : \mathbf{b})^{\mathbf{t}}$ *is a g-inverse of* $(\mathbf{A}:\mathbf{a})$. *Further, if the vectors and matrices are over* $\mathbb{C}$, *then the following holds:*

Let $\mathbf{X} = (\mathbf{G}^\star : \mathbf{b})^\star$ *be a g-inverse of* $(\mathbf{A}:\mathbf{a})$. *Then* $\mathbf{a} \notin \mathcal{C}(\mathbf{A})$ *if and only if* $\mathbf{b^\star a} = 1$ *and* $\mathbf{b^\star A} = \mathbf{0}$.

Theorem 2.6.8. *In the setup of Lemma 2.6.7, let* $\mathbf{a} \notin \mathcal{C}(\mathbf{A})$. *Then*

(i) $\mathbf{G}$ *is a* $\mathbf{A}^-$ *if* $(\mathbf{G^t}:\mathbf{b})^{\mathbf{t}}$ *is* $(\mathbf{A}:\mathbf{a})^-$.
(ii) $\mathbf{G}$ *is a* $\mathbf{A}^-_{\mathbf{r}}$ *if* $(\mathbf{G^t}:\mathbf{b})^{\mathbf{t}}$ *is* $(\mathbf{A}:\mathbf{a})^-_{\mathbf{r}}$.
Assume that the matrices and the vectors are over $\mathbb{C}$. *Then*
(iii) $\mathbf{G}$ *is a* $\mathbf{A}^-_{\mathbf{m}}$ *if* $(\mathbf{G}^\star:\mathbf{b})^\star$ *is* $(\mathbf{A}:\mathbf{a})^-_{\mathbf{m}}$.
(iv) $\mathbf{G}$ *is* $\mathbf{A}^-_{\ell}$ *if* $(\mathbf{G}^\star:\mathbf{b})^\star$ *is* $(\mathbf{A}:\mathbf{a})^-_{\ell}$.
(v) $\mathbf{G}$ *is* $\mathbf{A}^\dagger$ *if* $(\mathbf{G}^\star:\mathbf{b})^\star$ *is* $(\mathbf{A}:\mathbf{a})^\dagger$.

We shall now consider the case when $\mathbf{a} \in \mathcal{C}(\mathbf{A})$.

Theorem 2.6.9. *In the setup of Lemma 2.6.7, let* $\mathbf{X} = (\mathbf{G^t}:\mathbf{b})^{\mathbf{t}}$ *be a g-inverse of* $(\mathbf{A}:\mathbf{a})$ *and let* $\mathbf{b^t a} \neq 1$. *Write* $\mathbf{G_1} = \mathbf{G}\left(\mathbf{I} + \dfrac{\mathbf{ab^t}}{1-\mathbf{b^t a}}\right)$. *Then*

(i) $\mathbf{G_1}$ *is* $\mathbf{A}^-$.
(ii) $\mathbf{G_1}$ *is* $\mathbf{A_r}$ *if* $\mathbf{X}$ *is* $(\mathbf{A}:\mathbf{a})^-_{\mathbf{r}}$.
Let the matrices and the vectors be over $\mathbb{C}$. *Let* $\mathbf{X} = (\mathbf{G}^\star:\mathbf{b})^\star$ *be a g-inverse of* $(\mathbf{A}:\mathbf{a})$ *and let* $\mathbf{b^\star a} \neq 1$. *Write* $\mathbf{G_1} = \mathbf{G}\left(\mathbf{I} + \dfrac{\mathbf{ab}^\star}{1-\mathbf{b^\star a}}\right)$.
Then
(iii) $\mathbf{G_1}$ *is* $\mathbf{A}^-_{\ell}$ *if* $\mathbf{X}$ *is* $(\mathbf{A}:\mathbf{a})^-_{\ell}$.
(iv) $\mathbf{G_1}$ *is* $\mathbf{A}^-_{\mathbf{m}}$ *if* $\mathbf{X}$ *is* $(\mathbf{A}:\mathbf{a})^-_{\mathbf{m}}$.
(v) $\mathbf{G_1}$ *is* $\mathbf{A}^\dagger$ *if* $\mathbf{X}$ *is* $(\mathbf{A}:\mathbf{a})^\dagger$.

The case that is remaining is $\mathbf{b^t a} = \mathbf{1}$ and $\mathbf{b^t A} \neq \mathbf{0}$ over general field and $\mathbf{b^\star a} = \mathbf{1}$ and $\mathbf{b^\star A} = \mathbf{0}$ over $\mathbb{C}$. Since $\mathbf{b^t A} \neq \mathbf{0}$ (respectively $\mathbf{b^\star A} \neq \mathbf{0}$ over $\mathbb{C}$) there exists a positive integer j such that $\mathbf{b^t a_j} \neq \mathbf{0}$ (respectively $\mathbf{b^\star a_j} \neq \mathbf{0}$ over $\mathbb{C}$) where $\mathbf{a_j}$ is the jth column of $\mathbf{A}$. Define $\mathbf{c} = \mathbf{a} + \mathbf{a_j}$. Let $\mathbf{E}$ be the matrix obtained from $\mathbf{G}$ by replacing its jth row $\mathbf{g_j^t}$ ($\mathbf{g_j^\star}$ over $\mathbb{C}$) by $\mathbf{g_j^t} - \mathbf{b^t}$ (respectively $\mathbf{g_j^\star} - \mathbf{b^\star}$ over $\mathbb{C}$). Then clearly $(\mathbf{E^t} : \mathbf{b})^\mathbf{t}$ (respectively $(E^\star : \mathbf{b})^\star$ over $\mathbb{C}$) is a g-inverse of $(\mathbf{A} : \mathbf{c})$. Also $\mathbf{b^t c} = \mathbf{b^t a} + \mathbf{b^t a_j} \neq 1$ (respectively $\mathbf{b^\star c} = \mathbf{b^\star a} + \mathbf{b^\star a_j} \neq 1$ over $\mathbb{C}$). Thus, we have

Theorem 2.6.10. *In the setup of Lemma 2.6.7, let* $\mathbf{X} = (\mathbf{G^t} : \mathbf{b})^\mathbf{t}$ *be a g-inverse of* $(\mathbf{A} : \mathbf{a})$. *Moreover, let* $\mathbf{b^t a} = 1$ *and* $\mathbf{b^t A} \neq \mathbf{0}$. *Write* $\mathbf{G_2} = \mathbf{E}\left(\mathbf{I} - \dfrac{(\mathbf{a} + \mathbf{a_j})\mathbf{b^t}}{\mathbf{b^t a_j}}\right)$. *Then*

(i) $\mathbf{G_2}$ *is* $\mathbf{A}^-$.
(ii) $\mathbf{G_2}$ *is* $\mathbf{A}_\mathbf{r}^-$ *if* $\mathbf{X}$ *is* $(\mathbf{A} : \mathbf{a})_\mathbf{r}^-$.
Consider the matrices and the vectors over $\mathbb{C}$. *Let* $\mathbf{X} = (\mathbf{G^\star} : \mathbf{b})^\star$ *be a g-inverse of* $(\mathbf{A} : \mathbf{a})$ *and let* $\mathbf{b^\star a} = \mathbf{1}$ *and* $\mathbf{b^\star A} \neq \mathbf{0}$. *Write* $\mathbf{G_2} = \mathbf{E}\left(\mathbf{I} - \dfrac{(\mathbf{a} + \mathbf{a_j})\mathbf{b^\star}}{\mathbf{b^\star a_j}}\right)$. *Then*
(iii) $\mathbf{G_2}$ *is* $\mathbf{A}_\ell^-$ *if* $\mathbf{X}$ *is* $(\mathbf{A} : \mathbf{a})_\ell^-$.

Proofs of Theorems 2.6.3-2.6.10 are computational. For detailed proofs one may consult [Mitra and Bhimasankaram (1971)].

Remark 2.6.11. Consider matrices and vectors over $\mathbb{C}$. Let $(\mathbf{G^\star} : \mathbf{b})^\star$ be a minimum norm g-inverse of $(\mathbf{A} : \mathbf{a})$. Then $\mathbf{b^\star a} = 1$ if and only if $\mathbf{b^\star A} = \mathbf{0}$. This can be shown as follows:

Since $(\mathbf{G^\star} : \mathbf{b})^\star$ is $(\mathbf{A} : \mathbf{a})_\mathbf{m}^-$, we have $\mathbf{Ga} = \mathbf{A^\star b}$. Therefore,

$$\mathbf{b^\star a} = \mathbf{1} \Leftrightarrow \mathbf{AGa} = \mathbf{0} \Leftrightarrow \mathbf{AA^\star b} = \mathbf{0} \Leftrightarrow \mathbf{A^\star b} = \mathbf{0}.$$

Thus, $\mathbf{X} = (\mathbf{G^\star} : \mathbf{b})^\star$ can never be a minimum norm g-inverse or the Moore-Penrose inverse (left out in Theorem 2.6.10) in the set up considered in Theorem 2.6.10.

Remark 2.6.12. Notice that $\mathbf{G^t}$ is a g-inverse of $\mathbf{A^t}$ if $\mathbf{G}$ is a g-inverse of $\mathbf{A}$. Further, $\mathbf{G^\star}$ is a minimum norm g-inverse of $\mathbf{A^\star}$ if $\mathbf{G}$ is a least squares g-inverse of $\mathbf{A}$. With this observation one can deal with the case of g-inverse of $\mathbf{A}$ vs $(\mathbf{A^t} : \mathbf{b})^\mathbf{t}$ or $(\mathbf{A^\star} : \mathbf{b})^\star$ (in the case of vectors and matrices over $\mathbb{C}$).

We now consider g-inverse of $\mathbf{A}$ vs. $\begin{pmatrix} \mathbf{A} & \mathbf{u} \\ \mathbf{v^t} & a \end{pmatrix}$; where $\mathbf{u}$ and $\mathbf{v}$ are column vectors and a is scalar. Suppose a particular type of g-inverse $\mathbf{G}$ of $\mathbf{A}$ is a available to us and we want to compute a g-inverse of the same type of $\mathbf{B} = \begin{pmatrix} \mathbf{A} & \mathbf{u} \\ \mathbf{v}^t & a \end{pmatrix}$. Using the formulae so far developed, we can achieve this in two stages. First we compute a g-inverse $\mathbf{H}$ of $\mathbf{T} = (\mathbf{A} : \mathbf{u})$. The using $\mathbf{H}$, the formulae derived so far and the Remark 2.6.12, we compute a g-inverse of

$$\mathbf{B} = \begin{pmatrix} \mathbf{A} & \mathbf{u} \\ \mathbf{v}^t & a \end{pmatrix} = \begin{pmatrix} \mathbf{T} \\ \cdot\cdot \\ \mathbf{w} \end{pmatrix}, \text{ where } \mathbf{w} = (\mathbf{v}^t : \mathbf{a}) .$$

We give below another interesting way of obtaining g-inverse of $\mathbf{M} = \begin{pmatrix} \mathbf{A} & \mathbf{B} \\ \mathbf{C} & \mathbf{D} \end{pmatrix}$ in a single stage.

Let $\mathbf{A}$ be non-singular. Then the matrix $\mathbf{M} = \begin{pmatrix} \mathbf{A} & \mathbf{B} \\ \mathbf{C} & \mathbf{D} \end{pmatrix}$ is non-singular if and only if $\mathbf{F} = \mathbf{D} - \mathbf{C}\mathbf{A}^{-1}\mathbf{B}$ is non-singular. Also,

$$\begin{aligned} \mathbf{M}^{-1} &= \begin{pmatrix} \mathbf{A}^{-1} & \mathbf{0} \\ \mathbf{0} & \mathbf{0} \end{pmatrix} + \begin{pmatrix} \mathbf{A}^{-1}\mathbf{B} \\ \\ -\mathbf{I} \end{pmatrix} \mathbf{F}^{-1}(\mathbf{C}\mathbf{A}^{-1} : -\mathbf{I}) \\ &= \begin{pmatrix} \mathbf{A}^{-1} + \mathbf{A}^{-1}\mathbf{B}\mathbf{F}^{-1}\mathbf{C}\mathbf{A}^{-1} & -\mathbf{A}^{-}\mathbf{B}\mathbf{F}^{-1} \\ \\ -\mathbf{F}^{-1}\mathbf{C}\mathbf{A}^{-1} & \mathbf{F}^{-1} \end{pmatrix} . \end{aligned}$$

Remark 2.6.13. The matrix $\mathbf{F} = \mathbf{D} - \mathbf{C}\mathbf{A}^{-1}\mathbf{B}$ is known as the Schur complement of $\mathbf{A}$ in $\mathbf{M}$.

We shall now explore the availability of the above form for g- inverse of a partitioned matrix $\begin{pmatrix} \mathbf{A} & \mathbf{B} \\ \mathbf{C} & \mathbf{D} \end{pmatrix}$ when a g-inverse $\mathbf{G}$ of $\mathbf{A}$ is available.

Theorem 2.6.14. *Let a g-inverse* $\mathbf{G}$ *of* $\mathbf{A}$ *be available. Let*

$$\mathbf{M} = \begin{pmatrix} \mathbf{A} & \mathbf{B} \\ \mathbf{C} & \mathbf{D} \end{pmatrix} \text{ and } \mathbf{F} = \mathbf{D} - \mathbf{C}\mathbf{G}\mathbf{B}. \text{ Then}$$

$$\mathbf{R} = \begin{pmatrix} \mathbf{G} & 0 \\ \mathbf{0} & 0 \end{pmatrix} + \begin{pmatrix} \mathbf{G}\mathbf{B} \\ -\mathbf{I} \end{pmatrix} \mathbf{F}^{-}(\mathbf{C}\mathbf{G} : -\mathbf{I})$$

is a g-inverse of $\mathbf{M}$ *if and only if*

$$\left.\begin{aligned} \mathcal{C}(\mathbf{C}(\mathbf{I} - \mathbf{G}\mathbf{A})) &\subseteq \mathcal{C}(\mathbf{F}) \\ \mathcal{C}((\mathbf{I} - \mathbf{A}\mathbf{G})\mathbf{B})^t &\subseteq \mathcal{C}(\mathbf{F}^t) \ \ \textit{and} \\ (\mathbf{I} - \mathbf{A}\mathbf{G})\mathbf{B}\mathbf{F}^{-}\mathbf{C}(\mathbf{I} - \mathbf{G}\mathbf{A}) &= \mathbf{0} \end{aligned}\right\} \tag{2.6.2}$$

or equivalently

$$\rho(\mathbf{M}) = \rho(\mathbf{A}) + \rho(\mathbf{F}). \tag{2.6.3}$$

Proof. Proof of "$\mathbf{R}$ is M^- if and only if (2.6.2) holds" follows easily from the verification of the condition $\mathbf{MRM} = \mathbf{M}$.

To prove (2.6.2) and (2.6.3) are equivalent, we proceed as follows. Write

$$\mathbf{P} = \begin{pmatrix} \mathbf{I} & \mathbf{0} \\ -\mathbf{CG} & \mathbf{I} \end{pmatrix}, \quad \mathbf{Q} = \begin{pmatrix} \mathbf{I} & -\mathbf{GAGB} \\ \mathbf{0} & \mathbf{I} \end{pmatrix},$$

$$\mathbf{S} = \begin{pmatrix} \mathbf{I} & -(\mathbf{I}-\mathbf{AG})\mathbf{BF}^- \\ \mathbf{0} & \mathbf{I} \end{pmatrix} \text{ and}$$

$$\mathbf{T} = \begin{pmatrix} \mathbf{I} & \mathbf{0} \\ -\mathbf{F}-\mathbf{FF}^-\mathbf{C}(\mathbf{I}-\mathbf{GA}) & \mathbf{I} \end{pmatrix}.$$

Notice that $\mathbf{P}, \mathbf{Q}, \mathbf{S}$ and $\mathbf{T}$ are all non-singular matrices and straightforward computation shows that $\mathbf{SPMQT} = \mathbf{diag}(\mathbf{A}, \mathbf{F}) + \mathbf{Z}$, where

$$\mathbf{Z} = \begin{pmatrix} -(\mathbf{I}-\mathbf{AG})\mathbf{BF}^-\mathbf{C}(\mathbf{I}-\mathbf{GA}) & (\mathbf{I}-\mathbf{AG})\mathbf{B}(\mathbf{I}-\mathbf{F}^-\mathbf{F}) \\ (\mathbf{I}-\mathbf{FF}^-)\mathbf{C}(\mathbf{I}-\mathbf{GA}) & \mathbf{0} \end{pmatrix}.$$

It is easy to check that $\rho(\mathbf{M}) = \rho(\mathbf{SPMQT}) = \rho(\mathbf{diag}(\mathbf{A}, \mathbf{F}) + \mathbf{Z}) = \rho(\mathbf{diag}(\mathbf{A}, \mathbf{F})) + \rho(\mathbf{Z}) = \rho(\mathbf{A}) + \rho(\mathbf{F}) + \rho(\mathbf{Z})$. Hence, $\rho(\mathbf{M}) = \rho(\mathbf{A}) + \rho(\mathbf{F})$ if and only if $\mathbf{Z} = \mathbf{0}$, that is, if and only if (2.6.2) holds. □

Several remarks are in order.

Remark 2.6.15. The matrix $\mathbf{F} = \mathbf{D} - \mathbf{CGB}$ where $\mathbf{G}$ is a g-inverse of $\mathbf{A}$ is called a Schur complement of $\mathbf{A}$ in $\mathbf{M}$. Schur complements play an important role in the study of shorted operators. We shall study them in detail in Chapter 10.

Remark 2.6.16. If $\mathbf{G}$ is reflexive g-inverse of $\mathbf{A}$, and $\mathbf{F}^-$ is a reflexive g-inverse of $\mathbf{F}$, then $\mathbf{M}$ is always a g-inverse of $\mathbf{R}$.

Remark 2.6.17. If $\mathcal{C}(\mathbf{C}^t) \subseteq \mathcal{C}(\mathbf{A}^t)$ and $\mathcal{C}(\mathbf{B}) \subseteq \mathcal{C}(\mathbf{A})$, then (i) $\mathbf{F}$ is invariant under choices of g-inverse of $\mathbf{A}$ and (ii) $\mathbf{R}$ is a g-inverse of $\mathbf{M}$.

Remark 2.6.18. The following statement is equivalent to the statement of Remark 2.6.17:

If $\mathbf{M} = (\mathbf{X}_1^t, \mathbf{X}_2^t)^t(\mathbf{Y}_1, \mathbf{Y}_2)$, where $\rho(\mathbf{X}_1\mathbf{Y}_1) = \rho(\mathbf{X}_1) = \rho(\mathbf{Y}_1)$ (Here $\mathbf{A} = \mathbf{X}_1\mathbf{Y}_1, \mathbf{B} = \mathbf{X}_1\mathbf{Y}_2, \mathbf{C} = \mathbf{X}_2\mathbf{Y}_1$ and $\mathbf{D} = \mathbf{X}_2\mathbf{Y}_2$) then (i) $\mathbf{F}$ is invariant under the choices of g-inverses of $\mathbf{A}$ and (ii) $\mathbf{R}$ is a g-inverse of $\mathbf{M}$.

Theorem 2.6.19. *Let* $\mathbf{M}$ *be an nnd matrix over* $\mathbb{C}$. *Then* $\mathbf{R}$ *is a g-inverse of* $\mathbf{M}$.

Remark 2.6.20. Let $\mathbf{M}$ be as in Theorem 2.6.19.

(i) Let $\mathbf{G}$ be an $\mathbf{A}_\ell^-$ and $\mathbf{F}^-$ a least squares g-inverse of $\mathbf{F}$. Then $\mathbf{R}$ is an $\mathbf{M}_\ell^-$ if and only if $\mathcal{C}(\mathbf{CG}) \subseteq \mathcal{C}(\mathbf{F})$.
(ii) Let $\mathbf{G}$ be an $\mathbf{A}_\mathbf{m}^-$ and $\mathbf{F}^-$ a minimum norm g-inverse of $\mathbf{F}$. Then $\mathbf{R}$ is an $\mathbf{M}_\mathbf{m}^-$ if and only if $\mathcal{C}((\mathbf{GB})^t) \subseteq \mathcal{C}(\mathbf{F}^t)$.
(iii) Let $\mathbf{G}$ be $\mathbf{A}^\dagger$ and $\mathbf{F}^\dagger$ be the Moore-Penrose inverse of $\mathbf{F}$. Then $\mathbf{R}$ is $\mathbf{M}^\dagger$ if and only if $\mathcal{C}(\mathbf{CA}^\dagger) \subseteq \mathcal{C}(\mathbf{F})$ and $\mathcal{C}((\mathbf{A}^\dagger\mathbf{B})^\dagger) \subseteq \mathcal{C}(\mathbf{F}^t)$.

Obtaining a g-inverse of $\mathbf{M}$ in the most general case is a bit more complicated.

Theorem 2.6.21. *Let* $\mathbf{M} = \begin{pmatrix} \mathbf{A} & \mathbf{B} \\ \mathbf{C} & \mathbf{D} \end{pmatrix}$. *For a matrix* $\mathbf{W}$, *let us write* $\mathbf{E_w} = \mathbf{I} - \mathbf{WW}^-$ *and* $\mathbf{H_w} = \mathbf{I} - \mathbf{W}^-\mathbf{W}$, *where* $\mathbf{W}^-$ *is some g-inverse of* $\mathbf{W}$. *Let* $\mathbf{F} = \mathbf{D} - \mathbf{CA}^-\mathbf{B}$, $\mathbf{L} = \mathbf{E_A B}$, $\mathbf{K} = \mathbf{CH_A}$ *and* $\mathbf{N} = \mathbf{H_L}(\mathbf{E_K F H_L})^-\mathbf{E_K}$. *Then*

$$\mathbf{U} = \begin{pmatrix} \mathbf{A}^- - \mathbf{A}^-\mathbf{BL}^-\mathbf{E_A} - \mathbf{H_A K}^-\mathbf{CA}^- - \mathbf{H_A K}^-\mathbf{FL}^-\mathbf{E_A} & \mathbf{H_A K}^- \\ \mathbf{L}^-\mathbf{E_A} & \mathbf{0} \end{pmatrix}$$

$$+ \begin{pmatrix} \mathbf{H_A K}^-\mathbf{F} + \mathbf{A}^-\mathbf{B} \\ -\mathbf{I} \end{pmatrix} \mathbf{N}(\mathbf{FL}^-\mathbf{E_A} + \mathbf{CA}^- - \mathbf{I})$$

is a g-inverse of $\mathbf{M}$.

Proof. Check $\mathbf{MUM} = \mathbf{M}$. □

We shall now consider the third type of modification. Notice that a matrix is of rank 1 if and only if it is of the form $\mathbf{ab}^t$ for some non-null vectors $\mathbf{a}$ and $\mathbf{b}$.

Theorem 2.6.22. *Let* $\mathbf{A}$ *be an* $m \times n$ *matrix and* $\mathbf{A}^-$ *be a g-inverse of* $\mathbf{A}$. *Let* $\mathbf{a}$ *be an* m*-column vector and* $\mathbf{b}$ *an* n*-column vector. Let us write*

$\mathbf{E} = \mathbf{I} - \mathbf{A}\mathbf{A}^-, \mathbf{H} = \mathbf{I} - \mathbf{A}^-\mathbf{A}$ *and* $\beta = 1 + \mathbf{b}^t\mathbf{A}^-\mathbf{a}$. *If* $\mathbf{a} \notin \mathcal{C}(\mathbf{A})$, *let* $\mathbf{c}$ *be a column vector as per Lemma 2.6.1 so that* $\mathbf{c}^t\mathbf{A} = \mathbf{0}$ *and* $\mathbf{c}^t\mathbf{a} = 1$. *If* $\mathbf{b} \notin \mathcal{C}(\mathbf{A}^t)$, *let* $\mathbf{d}$ *be a column vector such that* $\mathbf{d}^t\mathbf{b} = 1$ *and* $\mathbf{A}\mathbf{d} = \mathbf{0}$. *Then the following hold:*

(i) $\mathbf{A}^- - \mathbf{A}^-\mathbf{a}\mathbf{c}^t - \mathbf{d}\mathbf{b}^t\mathbf{A}^- + \beta\mathbf{d}\mathbf{c}^t$ *is a g-inverse of* $\mathbf{A} + \mathbf{a}\mathbf{b}^t$, *when* $\mathbf{a} \notin \mathcal{C}(\mathbf{A})$ *and* $\mathbf{b} \notin \mathcal{C}(\mathbf{A}^t)$.

(ii) $\mathbf{A}^- - \beta^{-1}\mathbf{A}^-\mathbf{a}\mathbf{b}^t\mathbf{A}^-$ *is a g-inverse of* $\mathbf{A} + \mathbf{a}\mathbf{b}^t$, *when* $\beta \neq 0$, $\mathbf{a} \in \mathcal{C}(\mathbf{A})$ *or* $\mathbf{b} \in \mathcal{C}(\mathbf{A}^t)$ *or both.*

(iii) $\mathbf{A}^- - \mathbf{d}\mathbf{b}^t\mathbf{A}^-$ *is a g-inverse of* $\mathbf{A} + \mathbf{a}\mathbf{b}^t$, *when* $\beta = 0$, $\mathbf{a} \in \mathcal{C}(\mathbf{A})$ *and* $\mathbf{b} \notin \mathcal{C}(\mathbf{A}^t)$.

(iv) $\mathbf{A}^- - \mathbf{A}^-\mathbf{a}\mathbf{c}^t$ *is a g-inverse of* $\mathbf{A} + \mathbf{a}\mathbf{b}^t$, *when* $\beta = 0$, $\mathbf{a} \notin \mathcal{C}(\mathbf{A})$ *and* $\mathbf{b} \in \mathcal{C}(\mathbf{A}^t)$.

(v) $\mathbf{A}^-$ *is a g-inverse of* $\mathbf{A} + \mathbf{a}\mathbf{b}^t$, *when* $\beta = 0$, $\mathbf{a} \in \mathcal{C}(\mathbf{A})$ *and* $\mathbf{b} \in \mathcal{C}(\mathbf{A}^t)$.

Proof is straightforward.

We shall now give an expression for $(\mathbf{A} + \mathbf{a}\mathbf{b}^\star)^\dagger$ in terms of $\mathbf{A}^\dagger$ in various cases.

Theorem 2.6.23. *Let* $\mathbf{A}$ *be an* $m \times n$ *matrix over* $\mathbb{C}$ *and let* $\mathbf{A}^\dagger$ *be available. Let* $\mathbf{a}$ *be an* m*-column vector and* $\mathbf{b}$ *an* n*-column vector. Write* $\mathbf{k} = \mathbf{A}^\dagger\mathbf{a}$, $\mathbf{h} = \mathbf{A}^{\dagger\star}\mathbf{b}$, $\mathbf{E} = \mathbf{I} - \mathbf{A}\mathbf{A}^\dagger$, $\mathbf{F} = \mathbf{I} - \mathbf{A}^\dagger\mathbf{A}$ *and* $\beta = 1 + \mathbf{b}^\star\mathbf{A}^\dagger\mathbf{a}$. *Let* $\mathbf{u} = \mathbf{E}\mathbf{a}$, $\mathbf{v} = \mathbf{F}^\star\mathbf{b}$. *Then the following hold:*

(i) $\mathbf{A}^\dagger - \mathbf{k}\mathbf{u}^\dagger - \mathbf{v}^{\star\dagger}\mathbf{h}^\star + \beta\mathbf{v}^{\star\dagger}\mathbf{u}^\dagger$ *is* $(\mathbf{A} + \mathbf{a}\mathbf{b}^\star)^\dagger$, *when* $\mathbf{a} \notin \mathcal{C}(\mathbf{A})$ *and* $\mathbf{b} \notin \mathcal{C}(\mathbf{A}^\star)$.

(ii) $\mathbf{A}^\dagger - \mathbf{k}\mathbf{k}^\dagger\mathbf{A}^\dagger - \mathbf{v}^{\star\dagger}\mathbf{h}^\star$ *is* $(\mathbf{A} + \mathbf{a}\mathbf{b}^\star)^\dagger$, *when* $\beta = 0$, $\mathbf{a} \in \mathcal{C}(\mathbf{A})$ *and* $\mathbf{b} \notin \mathcal{C}(\mathbf{A}^\star)$.

(iii) $\mathbf{A}^\dagger + (1/\beta^\star)\mathbf{v}\mathbf{k}^\star\mathbf{A}^\dagger - (\beta^\star/\sigma_1)\mathbf{p_1}\mathbf{q_1^\star}$ *is* $(\mathbf{A} + \mathbf{a}\mathbf{b}^\star)^\dagger$, *where* $\sigma_1 = \|\mathbf{k}\|^2\|\mathbf{v}\|^2 + \beta^2$, $\mathbf{p_1} = ((\|\mathbf{k}\|^2/\beta^\star)\mathbf{v} + \mathbf{k})$, *and* $\mathbf{q_1} = ((\|\beta\mathbf{f}\mathbf{v}\|/\beta^\star)\,\mathbf{A}^{\dagger\star}\mathbf{k} + \mathbf{h})$, *when* $\beta \neq \mathbf{0}$, $\mathbf{a} \in \mathcal{C}(\mathbf{A})$.

(iv) $\mathbf{A}^\dagger + \mathbf{A}^\dagger\mathbf{h}^{\dagger\star}\mathbf{h}^\star - \mathbf{k}\mathbf{u}^\dagger$ *is* $(\mathbf{A} + \mathbf{a}\mathbf{b}^\star)^\dagger$, *when* $\beta = 0$, $\mathbf{a} \notin \mathcal{C}(\mathbf{A})$ *and* $\mathbf{b} \in \mathcal{C}(\mathbf{A}^\star)$.

(v) $\mathbf{A}^\dagger + (1/\beta^\star)\mathbf{A}^\dagger\mathbf{h}\mathbf{u}^\star - (\beta^\star/\sigma_2)\mathbf{p_2}\mathbf{q_2}^\star$ *is* $(\mathbf{A} + \mathbf{a}\mathbf{b}^\star)^\dagger$, *where* $\mathbf{p_2} = (1/\mathbf{b}^\star)\|\mathbf{u}\|^2\mathbf{A}^\dagger\mathbf{h} + \mathbf{k}$, $\mathbf{q_2} = (1/\beta^\star)\|\mathbf{h}\|^2\mathbf{u} + \mathbf{h}$ *and* $\sigma_2 = \|\mathbf{h}\|^2\|\mathbf{u}\|^2 + \beta^2$, *when* $\beta \neq 0$ *and* $\mathbf{b} \in \mathcal{C}(\mathbf{A}^\star)$.

(vi) $\mathbf{A}^\dagger - \mathbf{k}\mathbf{k}^\dagger\mathbf{A}^\dagger - \mathbf{A}^\dagger\mathbf{h}^{\dagger\star}\mathbf{h} + \mathbf{k}^\dagger\mathbf{A}^\dagger\mathbf{h}^{\dagger\star}\mathbf{k}\mathbf{h}^\star$ *is* $(\mathbf{A} + \mathbf{a}\mathbf{b}^\star)^\dagger$, *when* $\beta = 0$, $\mathbf{a} \in \mathcal{C}(\mathbf{A})$ *and* $\mathbf{b} \in \mathcal{C}(\mathbf{A}^\star)$.

Proof. See [Campbell and Meyer (1991)] for a proof. □

Group inverse of a modified matrix will be dealt with in detail in Chapter 13.

2.7 Simultaneous diagonalization

When we study matrix partial orders, parallel sums, and shorted operators in the later chapters of this monograph, we generally deal with a pair of matrices. Sometimes we study the common features of several matrices together as in Fisher-Cochran type theorems. Studying such relationships or common features is easy if the matrices involved are diagonal. In this section, we put together some results on simultaneous diagonalization of matrices by using (a) non-singular transformations, (b) contra-gradient transformations (c) unitary transformations and (d) similarity transformations. We also study the simultaneous diagonalization of a pair of (possibly) rectangular matrices using unitary transformations and simultaneous singular value decomposition of several (but finite in number) rectangular matrices. We then study the generalized singular value decomposition of a pair of matrices. Using the results mentioned above, we define generalized eigen-values and generalized singular values, which play a role in many applications. In this section, we work with matrices and vectors over the field of complex number $\mathbb{C}$.

Theorem 2.7.1. *Let $\mathbf{A_1}, \dots \mathbf{A_k}$ be hermitian matrices of the same order. Then there exists a unitary matrix $\mathbf{U}$ such that $\mathbf{U^\star A_j U}$ is diagonal for $j = 1, \dots, k$ if and only if $\mathbf{A_i}$ and $\mathbf{A_j}$ commute for $i, j = 1, \dots, k$ (or equivalently $\mathbf{A_i A_j}$ is hermitian for $i, j = 1, \dots, k$).*

Theorem 2.7.2. (a) *Let $\mathbf{A_1}$* and *$\mathbf{A_2}$ be hermitian matrices of the same order and let $\mathbf{A_1}$ be positive definite. Then there exists a non-singular matrix $\mathbf{P}$ such that $\mathbf{P^\star A_1 P} = \mathbf{I}$ and $\mathbf{P^\star A_2 P}$ is a real diagonal matrix.*
(b) *Let $\mathbf{A_1}, \mathbf{A_2}$ be hermitian matrices of the same order, $\mathbf{A_1}$ be non-negative definite and $\mathcal{C}(\mathbf{A_2}) \subseteq \mathcal{C}(\mathbf{A_1})$. Then there exists a non-singular matrix $\mathbf{P}$ such that $\mathbf{P^\star A_1 P} = \mathbf{diag}(\mathbf{I_r}, \mathbf{0})$ and $\mathbf{P^\star A_2 P} = \mathbf{diag}(\mathbf{D}, \mathbf{0})$; where $r = \rho(\mathbf{A_1})$ and $\mathbf{D}$ is a real diagonal matrix of order $r \times r$.*

Corollary 2.7.3. *Let $\mathbf{A_1}$ and $\mathbf{A_2}$ be nnd matrices of the same order. Then there exists a non-singular matrix $\mathbf{P}$ such that $\mathbf{P^\star A_1 P}$ and $\mathbf{P^\star A_2 P}$ are diagonal with non-negative diagonal elements.*

For proofs of Theorems 2.7.1, 2.7.2 and Corollary 2.7.3 see [Rao and Bhimasankaram (2000)].

Theorem 2.7.4. *Let* $\mathbf{A_1}$ and $\mathbf{A_2}$ *be hermitian matrices of the same order. Then there exists a non-singular matrix* $\mathbf{T}$ *such that the matrices* $\mathbf{T^{\star}A_1T}$ *and* $\mathbf{T^{-1}A_2(T^{-1})^{\star}}$ *are both diagonal if and only if* $\mathbf{A_1A_2}$ *is semi- simple with real eigen-values and* $\rho(\mathbf{A_1A_2}) = \rho(\mathbf{A_1A_2A_1})$.

Corollary 2.7.5. *Consider the same setup as in Theorem 2.7.4. Further, let* $\mathbf{A_1}$ *be nnd. Then there exists a non-singular matrix* $\mathbf{T}$ *such that matrices* $\mathbf{T^{\star}A_1T}$ *and* $\mathbf{T^{-1}A_2(T^{-1})^{\star}}$ *are both diagonal if and only if* $\rho(\mathbf{A_1A_2A_1}) = \rho(\mathbf{A_1A_2})$.

To prove Corollary 2.7.5, observe that $\mathbf{A_1}$ is *nnd* $\Rightarrow \mathbf{A_1} = \mathbf{QQ^{\star}}$ for some matrix $\mathbf{Q}$. Now, the non-null eigen-values of $\mathbf{A_1A_2}$ are the same as those of $\mathbf{Q^{\star}A_2Q}$ including the algebraic and geometric multiplicities. Further, since $\mathbf{Q^{\star}A_2Q}$ is hermitian, it is also semi-simple and has real eigen-values. If $\rho(\mathbf{A_1A_2}) = \rho(\mathbf{A_1A_2A_1})$, then $\rho(\mathbf{A_1A_2}) = \rho(\mathbf{A_1A_2A_1}) = \rho(\mathbf{A_1A_2A_1A_2})$ and so, the algebraic and geometric multiplicities of zero eigen-value of $\mathbf{A_1A_2}$ are equal. Hence, $\mathbf{A_1A_2}$ is semi-simple with real eigen-values. Now, the result follows by Theorem 2.7.4.

Corollary 2.7.6. *Let* $\mathbf{A_1}$ *and* $\mathbf{A_2}$ *be nnd matrices of the same order. Then there exists a non-singular matrix* $\mathbf{T}$ *such that the matrices* $\mathbf{T^{\star}A_1T}$ *and* $\mathbf{T^{-1}A_2(T^{-1})^{\star}}$ *are both diagonal.*

Theorem 2.7.7. *Let* $\mathbf{A_1}, \ldots, \mathbf{A_k}$ *be hermitian matrices of the same order and let* $\mathbf{A_1}$ *be non-singular. Then there exists a non-singular matrix* $\mathbf{P}$ *such that the matrix* $\mathbf{P^{\star}A_iP}$ *is diagonal for each* $i = 1, \ldots, k$ *if and only if* (i) $\mathbf{A_iA_1^{-1}}$ *is semi-simple with real eigen-values for* $i = 1, \ldots, k$ *and* (ii) $\mathbf{A_iA_1^{-1}A_j} = \mathbf{A_jA_1^{-1}A_i}$ *for* $i,\ j = 1, \ldots, k$.

Corollary 2.7.8. *Consider the same set up as in Theorem 2.7.7 and let* $\mathbf{A_1}$ *be positive definite. Then there exists a non-singular matrix* $\mathbf{P}$ *such that the matrix* $\mathbf{P^{\star}A_iP}$ *is diagonal for each* $i = 1, \ldots, k$ *if and only if* $\mathbf{A_iA_1^{-1}A_j} = \mathbf{A_jA_1^{-1}A_i}$ *for* $i,\ j = 1, \ldots, k$.

Corollary 2.7.9. *Let* $\mathbf{A_1}, \ldots, \mathbf{A_k}$ *be hermitian matrices of the same order such that* $\mathbf{B} = \mathbf{A_1} + \ldots + \mathbf{A_k}$ *is nnd. Also, let* $\mathcal{C}(\mathbf{A_i}) \subseteq \mathcal{C}(\mathbf{B})$ *for* $i = 1, \ldots, k$. *Then there exists a non-singular matrix* $\mathbf{P}$ *such that the matrix* $\mathbf{P^{\star}A_iP}$ *is diagonal for each* $i = 1, \ldots, k$ *if and only if* $\mathbf{A_iB^{-}A_j} = \mathbf{A_jB^{-}A_i}$ *for* $i,\ j = 1, \ldots k$, *where* $\mathbf{B^{-}}$ *is any g-inverse of* $\mathbf{B}$.

Corollary 2.7.10. *Let $\mathbf{A_1}, \ldots, \mathbf{A_k}$ be hermitian matrices of the same order such that $\mathcal{C}(\mathbf{A_i}) \subseteq \mathcal{C}(\mathbf{A_1})$ for $i = 1, \ldots, k$. Then there exists a non-singular matrix $\mathbf{P}$ such that $\mathbf{P^\star A_i P}$ is diagonal for $i = 1, \ldots, k$ if and only if* (i) *$\mathbf{A_i A_1^-}$ is semi-simple with real eigen-values and* (ii) *$\mathbf{A_i A_1^- A_j} = \mathbf{A_j A_1^- A_i}$ for each $i,\ j = 1, \ldots, k$, where $\mathbf{A_1^-}$ is any g-inverse of $\mathbf{A_1}$.*

For a proof of Theorem 2.7.7, see [Bhimasankaram (1971a)]. Corollaries 2.7.8-2.7.10 easily follow from Theorem 2.7.7.

Theorem 2.7.11. *(Similarity Transformation) Let $\mathbf{A_1}, \ldots, \mathbf{A_k}$ be semi-simple matrices of the same order. Then the following are equivalent:*

(i) *$\mathbf{A_1}, \ldots, \mathbf{A_k}$ commute pair-wise*
(ii) *$\mathbf{A_1}, \ldots, \mathbf{A_k}$ can be expressed as polynomials in a common semi-simple matrix $\mathbf{B}$ and*
(iii) *There exists a non-singular matrix $\mathbf{S}$ such that $\mathbf{S^{-1} A_i S}$ is diagonal for $i = 1, \ldots, k$.*

For a proof see [Bhimasankaram (1971a)] and [Ben-Israel and Greville (2001)].

We now study the simultaneous diagonalization of (possibly) rectangular matrices.

Theorem 2.7.12. *Let $\mathbf{A_1}, \mathbf{A_2}$ be $m \times n$ matrices. Then there exist unitary matrices $\mathbf{U}$ and $\mathbf{V}$ of order $m \times m$ and $n \times n$ respectively such that $\mathbf{U^\star A_i V}$ is diagonal with real diagonal elements, $i = 1, 2$ if and only if $\mathbf{A_1 A_2^\star}$ and $\mathbf{A_1^\star A_2}$ are hermitian.*

Proof. 'Only if' part is trivial.
'If' part
From Theorem 2.2.45, there exist unitary matrices $\mathbf{P}$ and $\mathbf{Q}$ such that $\mathbf{P^\star A_1 Q} = \mathbf{diag}(\mathbf{\Delta}\ , \mathbf{0})$, where $\mathbf{\Delta}$ is a positive definite diagonal matrix. Write $\mathbf{P^\star A_2 Q} = \begin{pmatrix} \mathbf{B} & \mathbf{C} \\ \mathbf{D} & \mathbf{E} \end{pmatrix}$, where the partitioning is conformable for addition of $\mathbf{P^\star A_1 Q}$ and $\mathbf{P^\star A_2 Q}$. Since $\mathbf{A_1 A_2^\star}$ is hermitian, so is $\mathbf{P^\star A_1 A_2^\star P}$.

Now,

$$\begin{aligned}\mathbf{P}^\star\mathbf{A_1}\mathbf{A_2^\star}\mathbf{P} &= \mathbf{P}^\star\mathbf{A_1}\mathbf{Q}\mathbf{Q}^\star\mathbf{A_2^\star}\mathbf{P}\\ &= \begin{pmatrix}\boldsymbol{\Delta} & \mathbf{0}\\ \mathbf{0} & \mathbf{0}\end{pmatrix}\begin{pmatrix}\mathbf{B}^\star & \mathbf{D}^\star\\ \mathbf{C}^\star & \mathbf{E}^\star\end{pmatrix}\\ &= \begin{pmatrix}\boldsymbol{\Delta}\mathbf{B}^\star & \boldsymbol{\Delta}\mathbf{D}^\star\\ \mathbf{0} & \mathbf{0}\end{pmatrix}.\end{aligned}$$

Thus, $\boldsymbol{\Delta}\mathbf{B}^\star$ is hermitian and $\mathbf{D} = \mathbf{0}$. Let $\boldsymbol{\Delta} = \mathbf{diag}(\delta_1, \ldots, \delta_\mathbf{r})$ and $\mathbf{B} = [b_{ij}]$. $\boldsymbol{\Delta}\mathbf{B}^\star = \mathbf{B}\boldsymbol{\Delta} \Rightarrow \delta_i b^\star_{ji} = b_{ij}\delta_j$ for all i and j. Similarly, '$\mathbf{A_1^\star A_2}$ is hermitian' implies that $\delta_i b_{ij} = \delta_j b^\star_{ji}$ for all i and j and and $\mathbf{C} = \mathbf{0}$. So, $b_{ij} = b^\star_{ji}$ for all i and j, and therefore, $\mathbf{B}$ is hermitian. Now, $\mathbf{P}^\star\mathbf{A_2}\mathbf{Q} = \mathbf{diag}(\mathbf{B},\ \mathbf{E})$, where $\mathbf{B}$ is hermitian and $\mathbf{B}$ and $\boldsymbol{\Delta}$ commute. So, by Theorem 2.7.1, there exists a unitary matrix $\mathbf{R}$ such that $\mathbf{R}^\star\boldsymbol{\Delta}\mathbf{R} = \boldsymbol{\Delta}$, $\mathbf{R}^\star\mathbf{B}\mathbf{R} = \boldsymbol{\Lambda}$ where $\boldsymbol{\Lambda}$ is a real diagonal matrix. Let $\mathbf{E} = \mathbf{X}\boldsymbol{\Gamma}\mathbf{Y}^\star$ be a singular value decomposition of $\mathbf{E}$ where $\mathbf{X}$ and $\mathbf{Y}$ are unitary and Γ is diagonal matrix with non-negative diagonal elements. Write $\mathbf{U} = \mathbf{P}\mathbf{diag}(\mathbf{R},\ \mathbf{X})$ and $\mathbf{V} = \mathbf{Q}\mathbf{diag}(\mathbf{R},\ \mathbf{Y})$. Clearly, $\mathbf{U}$ and $\mathbf{V}$ are unitary,

$$\mathbf{U}^\star\mathbf{A_1}\mathbf{V} = \mathbf{diag}(\mathbf{R}^\star, \mathbf{X}^\star)\ \mathbf{P}^\star\mathbf{A_1}\mathbf{Q}\ \mathbf{diag}(\mathbf{R}, \mathbf{Y}) = \mathbf{diag}(\boldsymbol{\Delta}, \mathbf{0})$$

$$\begin{aligned}\text{and } \mathbf{U}^\star\mathbf{A_2}\mathbf{V} &= \mathbf{diag}(\mathbf{R}^\star,\ \mathbf{X}^\star)\mathbf{P}^\star\mathbf{A_2}\mathbf{Q}\mathbf{diag}(\mathbf{R},\ \mathbf{Y})\\ &= \mathbf{diag}(\mathbf{R}^\star,\ \mathbf{X}^\star)\mathbf{diag}(\mathbf{B},\ \mathbf{E})\mathbf{diag}(\mathbf{R},\ \mathbf{Y})\\ &= \mathbf{diag}(\boldsymbol{\Lambda},\ \boldsymbol{\Gamma}).\end{aligned}$$

Thus, $\mathbf{U}^\star\mathbf{A_1}\mathbf{V}$ and $\mathbf{U}^\star\mathbf{A_2}\mathbf{V}$ are both diagonal and $\mathbf{U}, \mathbf{V}$ are unitary. □

Remark 2.7.13. In the proof of Theorem 2.7.12 we have $b^\star_{ji} = b_{ij}(\delta_i/\delta_j) = b_{ij}$. Thus, $\delta_i = \delta_j$ whenever $b_{ij} \neq 0$. Notice that all δ's are positive. So, it is easy to see that there exists a permutation matrix $\mathbf{T}$ such that $\mathbf{T}^\star\boldsymbol{\Delta}\mathbf{T} = \mathbf{diag}(\delta_1\mathbf{I}, \ldots, \delta_\mathbf{s}\mathbf{I})$ and $\mathbf{T}^\star\mathbf{B}\mathbf{T} = \mathbf{diag}(\mathbf{B_1}, \ldots, \mathbf{B_s})$ for some s where $\delta_i\mathbf{I}$ and $\mathbf{B_i}$ are of the same order for $i = 1, \ldots, s$ and the graph of the incidence matrix corresponding to $\mathbf{B_i}$ is connected for $i = 1, \ldots, s$.

We now prove a theorem which enables us to diagonalize several matrices simultaneously. Before we do so, we need the following lemmas:

Lemma 2.7.14. *Let* $\mathbf{a^t} = (a_1, \ldots, a_k)$ *and* $\mathbf{b^t} = (b_1, \ldots, b_k)$ *be complex vectors such that for* $i = 1, \ldots, k$ $\max\{|a_i|, |b_i|\} \neq 0$. *Then there exists a real scalar* θ *such that the vector* $\mathbf{c} = \mathbf{a} + \theta\mathbf{b}$ *has no null components i.e.* $c_i \neq 0$ *for all* i. *If vectors* $\mathbf{a}$ *and* $\mathbf{b}$ *are real, then* $\mathbf{c}$ *is also real.*

Proof. Let $\alpha = \max_i\{|a_i|\}$. If $\mathbf{b} = \mathbf{0}$, take $\theta = 0$. So $\mathbf{c} = \mathbf{a}$ has no null components. Now let $\mathbf{b} \neq \mathbf{0}$. Let $b_{i_1}, \ldots, b_{i_r}$ be the non-null components of $\mathbf{b}$. Write $\beta = \min_j |b_{i_j}|$. Let $\theta = \frac{\alpha+1}{\beta}$. Consider i such that $1 \leq i \leq k$. Let $b_i = 0$. Then $c_i = a_i + \theta b_i = a_i \neq 0$. Hence, $c_i \neq 0$. Now, let $b_i \neq 0$. Now, $c_i = a_i + \theta b_i = 0 \Rightarrow a_i = -\theta b_i = -\frac{\alpha+1}{\beta} b_i$. So, $|a_i| = \frac{\alpha+1}{\beta}|b_i| \geq \alpha + 1$, which is a contradiction, since $|a_i| \leq \alpha$. Hence, $c_i \neq 0$. □

Lemma 2.7.15. *Let $\mathbf{u_1}, \mathbf{u_2}, \ldots, \mathbf{u_k}$ be any finite number of complex vectors such that $\max_i\{|u_{ij}|\} \neq 0$ for all j, where u_{ij} is the jth component of $\mathbf{u_i}$. Then there exist real scalars $\theta_1, \theta_2, \ldots, \theta_k$ such that no component of the vector $\mathbf{v} = \theta_1\mathbf{u_1} + \theta_2\mathbf{u_2} + \ldots, \theta_k\mathbf{u_k}$ is null.*

Proof. Apply Lemma 2.7.14 recursively, noting that there is real linear combination of vectors $\mathbf{u_1}, \mathbf{u_2}$ no component of which is null. □

Theorem 2.7.16. *Let $\mathbf{A_1}, \ldots \mathbf{A_k}$ be a finite number of matrices of the same order $m \times n$. Then there exist unitary matrices $\mathbf{U}$ and $\mathbf{V}$ of order $m \times m$ and $n \times n$ respectively such that $\mathbf{U}^\star\mathbf{A_i}\mathbf{V}$ is diagonal with real diagonal elements if and only if $\mathbf{A_i^\star}\mathbf{A_j}$ and $\mathbf{A_i}\mathbf{A_j^\star}$ are hermitian for $i,\ j = 1, \ldots, k$.*

Proof. 'Only if' part is trivial.
'If' part
By Theorem 2.7.12, the result is true for $k = 2$. Let the result be true for $k = s$. So, there exist unitary matrices $\mathbf{P}$ and $\mathbf{Q}$ such that $\mathbf{D_i} = \mathbf{P}^\star\mathbf{A_i}\mathbf{Q}$ is real and diagonal for $i = 1, \ldots, s$. Without loss of generality we can assume $\mathbf{D_i} = \mathbf{diag}(\boldsymbol{\Delta}_\mathbf{i}, \mathbf{0}), \mathbf{i} = \mathbf{1}, \ldots, \mathbf{s}$ where $\boldsymbol{\Delta}_1 \ldots, \boldsymbol{\Delta}_s$ are of same order and $\max_i\{|\delta_j^{(i)}|\} \neq 0$ for all j (here $\delta_j^{(i)}$ is the jth diagonal element of $\boldsymbol{\Delta}_i$). By Lemma 2.7.15, there exists a real linear combination

$$\sum_{i=1}^{s} \mu_i \boldsymbol{\Delta}_i$$

which is a non-singular matrix. Write

$$\mathbf{P}^\star\mathbf{A_{s+1}}\mathbf{Q} = \begin{pmatrix} \mathbf{B} & \mathbf{C} \\ \mathbf{D} & \mathbf{E} \end{pmatrix}$$

where $\mathbf{B}$ is of the same order as $\boldsymbol{\Delta}_i$ (all $\boldsymbol{\Delta}_i$ are of same order). Since, $\mathbf{A_i^\star}\mathbf{A_{s+1}}$ and $\mathbf{A_i}\mathbf{A_{s+1}^\star}$ are hermitian for $i = 1, \ldots, s$, we have $(\sum_{i=1}^{s} \mu_i\mathbf{D}_i)\begin{pmatrix} \mathbf{B} & \mathbf{C} \\ \mathbf{D} & \mathbf{E} \end{pmatrix}^\star$ and $(\sum_{i=1}^{s} \mu_i\mathbf{D}_i)^\star\begin{pmatrix} \mathbf{B} & \mathbf{C} \\ \mathbf{D} & \mathbf{E} \end{pmatrix}$ are also hermitian. By

Theorem 2.7.12, it follows that $\mathbf{B}$ is hermitian, $\mathbf{C} = \mathbf{0}$ and $\mathbf{D} = \mathbf{0}$. Since $\mathbf{A_i^\star A_j}$ and $\mathbf{A_i A_j^\star}$ are hermitian for $i, j = 1, \ldots, s+1$, it follows that $\boldsymbol{\Delta}_1 \ldots, \boldsymbol{\Delta}_s$ and $\mathbf{B}$ commute pair-wise and are hermitian. So, there exists a unitary matrix $\mathbf{R}$ such that $\mathbf{R}^\star \boldsymbol{\Delta}_\mathbf{i} \mathbf{R}$ are diagonal with real diagonal elements $i = 1, \ldots, s$. Let $\mathbf{S}$ and $\mathbf{T}$ be unitary matrices such that $\mathbf{S}^\star \mathbf{E} \mathbf{T}$ is diagonal (singular value decomposition of $\mathbf{E}$). Write $\mathbf{U} = \mathbf{P}\,\mathrm{diag}(\mathbf{R}, \mathbf{S})$ and $\mathbf{V} = \mathbf{Q}\,\mathbf{diag}(\mathbf{R}, \mathbf{T})$. Clearly, $\mathbf{U}$ and $\mathbf{V}$ are unitary and $\mathbf{U}^\star \mathbf{A_i} \mathbf{V}$ is diagonal with real diagonal elements for $i = 1, \ldots, s+1$. The result now follows by induction on s. □

Definition 2.7.17. The matrices $\mathbf{A_1}, \ldots, \mathbf{A_k}$ of the same order are said to have simultaneous singular value decomposition if there exist unitary matrices $\mathbf{U}$ and $\mathbf{V}$ such that $\mathbf{U}^\star \mathbf{A_i} \mathbf{V}$ is a diagonal matrix with non-negative diagonal elements for $i = 1, \ldots, k$.

Theorem 2.7.18. *(Simultaneous Singular Value Decomposition)* (i) *Let $\mathbf{A_1}$ and $\mathbf{A_2}$ have simultaneous singular value decomposition. Then $\mathbf{A_1^\star A_2}$, $\mathbf{A_1 A_2^\star}$ are nnd.*
(ii) *Let $\mathbf{A_1}$ and $\mathbf{A_2}$ be matrices of same order such that $\mathbf{A_1^\star A_2}$ and $\mathbf{A_1 A_2^\star}$ are hermitian and at least one of them is nnd. Then $\mathbf{A_1}$ and $\mathbf{A_2}$ have simultaneous singular value decomposition.*

Proof. (i) is trivial.
(ii) Since $\mathbf{A_1 A_2^\star}$ and $\mathbf{A_1^\star A_2}$ are hermitian by Theorem 2.7.12, there exist unitary matrices $\mathbf{U}$ and $\mathbf{V}$ such that $\mathbf{A_i} = \mathbf{U} \boldsymbol{\Delta}_\mathbf{i} \mathbf{V}^\star$ where $\boldsymbol{\Delta}_i$ is a diagonal matrix with real diagonal elements δ_{ji}, $j = 1, \ldots,\ \min(m, n)$, $i = 1,\ 2$ where $\mathbf{A}$ and $\mathbf{B}$ are matrices of order $m \times n$. Let $\mathbf{A_1 A_2^\star}$ be *nnd*. Then $\boldsymbol{\Delta}_1 \boldsymbol{\Delta}_2$ is *nnd*. Clearly $\boldsymbol{\Delta}_2 \boldsymbol{\Delta}_1$ is also *nnd*. Thus, a diagonal element of $\boldsymbol{\Delta}_1$ is negative if and only if the corresponding diagonal element of $\boldsymbol{\Delta}_2$ is negative. If i^{th} diagonal element of $\boldsymbol{\Delta}_1$ is negative replace $\mathbf{u_i}$, the ith column of $\mathbf{U}$ by $-\mathbf{u_i}$ and δ_{1i} and δ_{2i} by $-\delta_{1i}$ and $-\delta_{2i}$. Repeat the same for all negative diagonal elements of $\boldsymbol{\Delta}_1$. Call the resultant matrices obtained from $\mathbf{U}$, $\boldsymbol{\Delta}_1$ and $\boldsymbol{\Delta}_2$ as $\mathbf{W}, \boldsymbol{\Lambda}_\mathbf{1}$ and $\boldsymbol{\Lambda}_\mathbf{2}$. Then $\mathbf{A_i} = \mathbf{W} \boldsymbol{\Lambda}_\mathbf{i} \mathbf{V}^\star$ when $\mathbf{W}$ as $\mathbf{V}$ are unitary, $\boldsymbol{\Lambda}_\mathbf{i}$ is diagonal *nnd*, $i = 1, 2$. Hence $\mathbf{A_i} = \mathbf{W} \boldsymbol{\Lambda}_\mathbf{i} \mathbf{V}^\star$ is a singular value decomposition for $i = 1, 2$. □

Theorem 2.7.19. *(Simultaneous Singular Value Decomposition for Several Matrices) Let $\mathbf{A_1}, \ldots, \mathbf{A_k}$ be matrices of the same order.*
(i) *If $\mathbf{A_1}, \ldots, \mathbf{A_k}$ have simultaneous singular value decomposition, then $\mathbf{A_i^\star A_j}$ and $\mathbf{A_i A_j^\star}$ are nnd for $i, j = 1, \ldots, k$.*
(ii) *Let $\mathbf{A_1}, \ldots, \mathbf{A_k}$ be matrices such that $\mathbf{A_i^\star A_j}$ and $\mathbf{A_i A_j^\star}$ are hermitian*

and at least one of them is nnd for $i, j = 1, \ldots, k$. *Then* $\mathbf{A_1}, \ldots, \mathbf{A_k}$ *have simultaneous singular value decomposition.*

Proof. We shall prove both (i) and (ii) by induction on k. The result is true for $k = 2$, by Theorem 2.7.18. Let it be true for $k = s$. Consider $\mathbf{A_1}, \ldots, \mathbf{A_{s+1}}$ such that $\mathbf{A_i^\star A_j}$ and $\mathbf{A_i A_j^\star}$ are hermitian and at least one of them is *nnd* for $i, j = 1, \ldots, s+1$. By induction hypothesis, $\mathbf{A_1}, \ldots, \mathbf{A_s}$ have simultaneous singular value decomposition. So, there exist unitary matrices $\mathbf{P}$ and $\mathbf{Q}$ such that $\mathbf{P^\star A_i Q} = \mathbf{D_i}$ where $\mathbf{D_i}$ is diagonal *nnd* for $i = 1, \ldots, s$. Without loss of generality, we can take $\mathbf{D_i} = \mathbf{diag}(\mathbf{\Delta_i},\ \mathbf{0})$, where each $\mathbf{\Delta_i}$ is *nnd* and $\sum_i^s \mathbf{\Delta_i}$ is positive definite. (This is achieved by permuting the rows and the same columns of each of $\mathbf{\Delta_i}$ - the same permutation transformation for each i - such that the null rows and columns are moved to be the last rows and last columns.)

Since, $\mathbf{A_i^\star A_{s+1}}$ and $\mathbf{A_i A_{s+1}^\star}$ are hermitian for each i, $(\sum_1^s \mathbf{A_i^\star})\mathbf{A_{s+1}}$ and $(\sum_1^s \mathbf{A_i})\mathbf{A_{s+1}^\star}$ are also hermitian. Moreover,

$$\mathbf{P}^\star\left(\sum \mathbf{A_i^\star}\right)\mathbf{Q} = \mathbf{diag}\left(\left(\sum \mathbf{\Delta_i}\right),\ \mathbf{0}\right).$$

Let $\mathbf{P^\star A_{s+1} Q} = \begin{pmatrix} \mathbf{B} & \mathbf{C} \\ \mathbf{F} & \mathbf{E} \end{pmatrix}$, where $\mathbf{B}$ has same order as $\sum \mathbf{\Delta_i}$, we notice that $\mathbf{B}$ is *nnd*, $\mathbf{C} = \mathbf{0}$ and $\mathbf{F} = \mathbf{0}$. Since $\mathbf{A_i^\star A_{s+1}^\star}$ are hermitian, $\mathbf{\Delta_i B}$ is hermitian, $\mathbf{\Delta_i}, \mathbf{B}$ commute for $i = 1, \ldots, s$. So, by Theorem 2.7.1, there exists a unitary matrix $\mathbf{R}$ such that $\mathbf{R^\star \Delta_i R}$ for $i = 1, \ldots, s$ and $\mathbf{R^\star B R}$ are each diagonal with real non-negative diagonal elements. Let $\mathbf{E} = \mathbf{S \Gamma T^\star}$ be singular value decomposition of $\mathbf{E}$. Now construct $\mathbf{U}$ and $\mathbf{V}$ as in Theorem 2.7.16 and check that $\mathbf{U^\star A_i V}$ is diagonal with real non-negative diagonal elements for $i = 1, \ldots, s+1$. □

Theorem 2.7.20. *(Generalized Singular Value Decomposition) Let* $\mathbf{A}$ *and* $\mathbf{B}$ *be matrices of orders* $m \times n$ *and* $p \times n$ *respectively. Let* $r = \rho(\mathbf{A^t} : \mathbf{B^t})^\mathbf{t}$ *and* $\max\{m, p\} \geq r$. *Then there exist unitary matrices* $\mathbf{U}$ *and* $\mathbf{V}$ *and a non-singular matrix* $\mathbf{Z}$ *such that* $\mathbf{A} = \mathbf{U \Delta_1 Z}$ *and* $\mathbf{B} = \mathbf{V \Delta_2 Z}$, *where* $\mathbf{\Delta_1}$ *and* $\mathbf{\Delta_2}$ *are diagonal matrices with non-negative diagonal elements.*

Before we prove the Theorem 2.7.20, we shall prove a lemma that is also of independent interest.

Lemma 2.7.21. *Let* $\mathbf{C}$ *and* $\mathbf{D}$ *be* $n \times m$ *matrices such that* $\mathbf{C^\star C} = \mathbf{D^\star D}$. *Then there exists a unitary matrix* $\mathbf{T}$ *such that* $\mathbf{D} = \mathbf{TC}$.

Proof. Let $\mathbf{C}^\star\mathbf{C} = \mathbf{D}^\star\mathbf{D} = \mathbf{P}\text{diag}(\mathbf{\Delta^2}, \mathbf{0})\mathbf{P}^\star$ be a spectral decomposition, where $\mathbf{\Delta}^2$ is a positive definite diagonal matrix and $\mathbf{P}$ is unitary. Then the singular value decompositions of $\mathbf{C}$ and $\mathbf{D}$ are given by $\mathbf{C} = \mathbf{Q}\text{diag}(\mathbf{\Delta}\ ,\mathbf{0})\mathbf{V}^\star$ and $\mathbf{D} = \mathbf{R}\text{diag}(\mathbf{\Delta}\ ,\mathbf{0})\ \mathbf{V}^\star$ where $\mathbf{Q}$ and $\mathbf{R}$ are unitary matrices. Let $\mathbf{T} = \mathbf{R}\mathbf{Q}^\star$. Then $\mathbf{T}$ is unitary and $\mathbf{D} = \mathbf{T}\mathbf{C}$. □

We now prove the Theorem 2.7.20.

Proof. Consider $\mathbf{A}^\star\mathbf{A}$ and $\mathbf{B}^\star\mathbf{B}$. By Corollary 2.7.3, there exists a non-singular matrix $\mathbf{Z}$ such that $\mathbf{A}^\star\mathbf{A} = \mathbf{Z}^\star\mathbf{\Gamma_1}\mathbf{Z}$ and $\mathbf{B}^\star\mathbf{B} = \mathbf{Z}^\star\mathbf{\Gamma_2}\mathbf{Z}$, where $\mathbf{\Gamma}_1$ and $\mathbf{\Gamma}_2$ are diagonal matrices with non-negative diagonal elements. Since $\rho(\mathbf{A}^\star\mathbf{A} + \mathbf{B}^\star\mathbf{B}) = \rho(\mathbf{A}^\star : \mathbf{B}^\star)^\star = r$, $\mathbf{\Gamma_1}+\mathbf{\Gamma_2}$ has exactly r diagonal elements which are positive. Let $\mathbf{\Gamma_1} = \mathbf{\Delta}_1^\star\mathbf{\Delta}_1$, $\mathbf{\Gamma_2} = \mathbf{\Delta}_2^\star\mathbf{\Delta_2}$ where $\mathbf{\Delta}_1, \mathbf{\Delta}_2$ are of orders $m \times n$ and $p \times n$ respectively, with all diagonal elements non-negative and all the off-diagonal elements zero. It is now clear that $\mathbf{A}^\star\mathbf{A} = \mathbf{Z}^\star\mathbf{\Delta}_1^\star\mathbf{\Delta}_1\mathbf{Z}$ and $\mathbf{B}^\star\mathbf{B} = \mathbf{Z}^\star\mathbf{\Delta}_2^\star\mathbf{\Delta}_2\mathbf{Z}$. So, by Lemma 2.7.21, there exist unitary matrices $\mathbf{U}$ and $\mathbf{V}$ such that $\mathbf{A} = \mathbf{U}\mathbf{\Delta}_1\mathbf{Z}$ and $\mathbf{B} = \mathbf{V}\mathbf{\Delta}_2\mathbf{Z}$. □

Remark 2.7.22. Suppose $r \leq \max\{m, p\}$ in Theorem 2.7.20. Let $m \geq p$. Then the matrix $\mathbf{\Delta}_1^\star\mathbf{\Delta}_1 + ({\mathbf{\Delta}_2}^\star : \mathbf{0})({\mathbf{\Delta}_2}^t : \mathbf{0})^\mathbf{t}$, has exactly r positive elements, where $({\mathbf{\Delta}_2}^t : \mathbf{0})^\mathbf{t}$ is of order $m \times n$. But this is not possible if $m < r$. The case where $m < p$ can be disposed of similarly. That is why we require the condition $\max\{m, p\} \geq r = \rho(\mathbf{A}^\mathbf{t} : \mathbf{B}^\mathbf{t})$.

We shall now present, in brief, the generalized eigen-values and generalized singular values.

Definition 2.7.23. We say a non-zero scalar λ is a generalized eigen-value of $\mathbf{A}$ with respect to $\mathbf{B}$ if there exists a non-null vector $\mathbf{x}$ such that $\mathbf{A}\mathbf{x} = \lambda\mathbf{B}\mathbf{x}$. Such a vector $\mathbf{x}$ is called a generalized eigen-vector of $\mathbf{A}$ with respect to $\mathbf{B}$ corresponding to the generalized eigen-value λ.

Let $\mathbf{A}$ and $\mathbf{B}$ be hermitian matrices of the same order with $\mathbf{B}$ positive definite. Then by Theorem 2.7.2, there exists a non-singular matrix $\mathbf{P}$ such that $\mathbf{P}^\star\mathbf{B}\mathbf{P} = \mathbf{I}$ and $\mathbf{P}^\star\mathbf{A}\mathbf{P} = \mathbf{\Lambda}$, a diagonal matrix. So, $\mathbf{A}\mathbf{P} = (\mathbf{P}^\star)^{-\mathbf{1}}\mathbf{\Lambda} = \mathbf{B}\mathbf{P}\mathbf{\Lambda}$. Let $\mathbf{P_i}$ be the ith column of $\mathbf{P}$. Then $\mathbf{A}\mathbf{P_i} = ((\mathbf{P}^\star)^{-\mathbf{1}}\mathbf{\Lambda})_\mathbf{i} = \lambda_i\mathbf{B}\mathbf{P_i}$, where λ_i's are diagonal elements of Λ. Thus, all the non-null diagonal elements of $\mathbf{\Lambda}$ are the generalized eigen-values of $\mathbf{A}$ with respect to $\mathbf{B}$ and the columns $\mathbf{P_i}$ of $\mathbf{P}$ are generalized eigen-vectors.

Notice that λ_i are the non-zero eigen values of $\mathbf{B}^{-\mathbf{1}}\mathbf{A}$ (and are same as non-null eigen-values of $\mathbf{A}\mathbf{B}^{-\mathbf{1}}$).

Definition 2.7.24. Consider the generalized singular value decomposition as in Theorem 2.7.20. Consider the common non-null diagonal elements of $\mathbf{\Delta}_1$ and $\mathbf{\Delta}_2$. Let them be $\delta_{1i_1}, \ldots, \delta_{1i_s}$ and $\delta_{2i_1}, \ldots, \delta_{2i_s}$. The scalars $\dfrac{\delta_{1i_j}}{\delta_{2i_j}}$, $j = 1, \ldots, s$ are called the generalized singular values of $\mathbf{A}$ with respect to $\mathbf{B}$.

Notice that $\mathbf{Z}^{-1}\mathbf{\Delta}_2^{-}\mathbf{V}^{\star}$ is a g-inverse of $\mathbf{B}$ where $\mathbf{\Delta}_2^{-}$ is any g-inverse of $\mathbf{\Delta}_2$. Now, construct the particular g-inverse $\mathbf{G}$ of $\mathbf{\Delta}_2$ as follows: $g_{ij} = 0$, whenever $i \neq j$ and $g_{ii} = \dfrac{1}{\delta_{2_{ii}}}$ if $\delta_{2_{ii}} \neq 0$ and 0 otherwise. Let $\mathbf{B}^{-} = \mathbf{Z}^{-1}\mathbf{G}\mathbf{V}^{\star}$. Then $\mathbf{U}\mathbf{\Delta}_1\mathbf{G}\mathbf{V}^{\star}$ is a singular value decomposition of $\mathbf{A}\mathbf{B}^{-}$, where $\mathbf{\Delta}_1\mathbf{G}$ is a diagonal *nnd* matrix. Further, the non-null diagonal elements of $\mathbf{\Delta}_1\mathbf{G}$ are precisely the generalized singular values of $\mathbf{A}$ with respect to $\mathbf{B}$. This shows the similarity between the concepts of generalized eigenvalues and generalized singular values. For more details on generalized eigen-values, see [Rao and Mitra (1971)]. For more details on generalized singular values, see [Golub and Van Loan (1996)].

2.8 Exercises

All matrices in the following exercises are over $\mathbb{C}$, the field of complex numbers.

(1) Let $\mathbf{A}$ be an idempotent matrix. Prove that $\rho(\mathbf{A}) = tr(\mathbf{A})$. However the converse is not true even if $\mathbf{A}$ is semi-simple.

(2) Let $\mathbf{E}$ be any idempotent matrix. Show that $\mathbf{E}^{\dagger}\mathbf{E}^{\star} = \mathbf{E}^{\star}\mathbf{E}^{\dagger} = \mathbf{E}^{\dagger}$.

(3) Let $\mathbf{A}$ be a square matrix of index 1 such that $\mathbf{A}^{\star} = \mathbf{A}^{\dagger}$. Then prove that

(i) $\mathbf{A}^{\#} = \mathbf{A}\mathbf{A}^{\star}\mathbf{A}^{\#} = \mathbf{A}^{\#}\mathbf{A}^{\star}\mathbf{A}$ and

(ii) $(\mathbf{A}^2\mathbf{A}^{\star})^{\dagger} = \mathbf{A}^{\#}\mathbf{A}\mathbf{A}^{\star}$ and $(\mathbf{A}^{\star}\mathbf{A}^2)^{\dagger} = \mathbf{A}^{\star}\mathbf{A}\mathbf{A}^{\#}\mathbf{A}^{\star}\mathbf{A} = \mathbf{A}^{\#}\mathbf{A}\mathbf{A}^{\star}$.

(iii) Is it necessary that $\mathbf{A}^{\#}$ is a partial isometry?

(4) Let $\mathbf{A}$ and $\mathbf{B}$ be matrices of the same order such that $\mathbf{A}$ ia range-hermitian and $\mathbf{B}$ is hermitian. Show that $\mathbf{A}\mathbf{B}^{\dagger}\mathbf{A} = \mathbf{A}$ implies $\mathbf{A}$ is hermitian.

(5) Let $\mathbf{P}$ and $\mathbf{Q}$ be two orthogonal projectors of order $n \times n$. Then prove the following statements:

(i) $\mathbf{PQ}$ is similar to a diagonal matrix.

(ii) $\mathbf{PQ}$ is an orthogonal projector if and only if $\mathbf{PQ}$ is range hermitian.

(iii) $n - \rho(\mathbf{I} - \mathbf{PQ}) = d(\mathcal{C}(\mathbf{P}) \cap \mathcal{C}(\mathbf{Q}))$.

(6) Let $\mathbf{A}$ be an $n \times n$ matrix. Then prove the following:

(i) $\mathbf{A}$ is range-hermitian $\Leftrightarrow \mathbf{A}^{\dagger} = \mathbf{A}^{\#}$.

(ii) $\mathbf{A}$ is range-hermitian $\Leftrightarrow \mathbf{A}\mathbf{A}^{\dagger} = \mathbf{A}^{\dagger}\mathbf{A}$.

(iii) $\mathbf{A}$ is of index $\leq 1 \Leftrightarrow \mathcal{C}(\mathbf{A})$ and $\mathcal{N}(\mathbf{A})$ are complementary.

(iv) Every range-hermitian matrix is of index 1.

(7) Let $\mathbf{A}$ be a partial isometry of index 1. Should $\mathbf{A}^{\#}$ be a partial isometry?

(8) Prove or disprove: $\mathbf{A}^{\#\dagger} = \mathbf{A}^{\dagger\#}$.

(9) Let $\mathbf{A}$ be an $n \times m$ matrix. Prove that $\mathbf{A}$ is unitarily similar to the matrix $\begin{pmatrix} \mathbf{\Sigma K} & \mathbf{\Sigma L} \\ \mathbf{0} & \mathbf{0} \end{pmatrix}$, where $\mathbf{K}\mathbf{K}^{\star} + \mathbf{L}\mathbf{L}^{\star} = \mathbf{I_r}$, $\mathbf{\Sigma} = diag(\sigma_1\mathbf{I_{r_1}}, \ldots, \sigma_t\mathbf{I_{r_t}})$; $r_1 + \ldots + r_t = r = \rho(\mathbf{A})$ and $\sigma_1 > \ldots \sigma_t > 0$. (Hint: Use Singular value decomposition.)

(10) Prove or disprove:

(i) A semi-simple matrix has index not greater than 1 and

(ii) A tripotent matrix is semi-simple.

(11) Let $\mathbf{A}$ be an $m \times n$ matrix. Then show that $\mathbf{A}$ is partial isometry $\Leftrightarrow \mathbf{A}\mathbf{A}^\star$ is an orthogonal projector $\Leftrightarrow \mathbf{A}^\star\mathbf{A}$ is an orthogonal projector.

(12) Let an $m \times n$ matrix $\mathbf{A}$ be a partial isometry. Then prove that $\mathbf{A}$ is normal $\Leftrightarrow \mathbf{A}$ is range hermitian.

(13) Prove or disprove: for a square matrix $\mathbf{A}$, both the matrices $\mathbf{A}$ and $\mathbf{A}^2$ are partial isometries implies $\mathbf{A}$ is normal.

(14) Show that an $m \times n$ matrix $\mathbf{A}$ is a partial isometry of rank $r \Leftrightarrow \mathbf{A}$ has exactly r singular values each equal to 1 and others are null.

(15) Show that a hermitian partial isometry is a tripotent matrix. Show the converse statement is not true. (Hint: Take $\mathbf{A} = \begin{pmatrix} 1 & 0 & 0 \\ 0 & -1 & 1 \\ 0 & 0 & 0 \end{pmatrix}$.)

(16) Show that an $n \times n$ matrix $\mathbf{A}$ is range-hermitian if and only if $\mathbf{A}^\sharp = \mathbf{A}^\dagger$ if and only if $\mathbf{A}^\dagger$ is a polynomial in $\mathbf{A}$.

All matrices in the following exercises are over F, an arbitrary field.

(17) Let $\mathbf{A}$ and $\mathbf{C}$ be any two matrices with $\mathbf{A}$ of index 1. Then prove the following:

(i) $\mathbf{A}^\#\mathbf{C} = \mathbf{0} \Leftrightarrow \mathbf{AC} = \mathbf{0}$ and

(ii) $\mathbf{CA}^\# = \mathbf{0} \Leftrightarrow \mathbf{CA} = \mathbf{0}$.

(18) Let $\mathbf{A}$ be a tripotent matrix. Then prove the following:

(i) $\mathbf{A}^2$ is idempotent.

(ii) $-\mathbf{A}$ is tripotent.

(iii) The only eigen-values of $\mathbf{A}$ are $-1, 0$ and 1.

(iv) $\rho(\mathbf{A}) = tr(\mathbf{A^2})$.

(v) $\mathbf{A}$ is its own g-inverse. Conversely, if $\mathbf{B}$ is its own inverse then $\mathbf{B}$ is tripotent.

(vi) For a tripotent matrix $\mathbf{A}$, $tr(\mathbf{A}^2 + \mathbf{A})$ = twice the number of positive eigen-values, $\text{tr}(\mathbf{A}^2 - \mathbf{A})$ = twice the number of negative eigenvalues and $\text{tr}(\mathbf{I} - \mathbf{A}^2)$ = the number of null eigen-values of $\mathbf{A}$.

(19) Prove that $\mathbf{A}$ is of index not greater than 1 and $\mathbf{A}^2$ is idempotent $\Leftrightarrow A$ is tripotent.

(20) Let $\mathbf{A}, \mathbf{B}$ be square matrices of the same order. Prove that $\mathbf{AB}(\mathbf{AB})^-\mathbf{A} = \mathbf{A}$ if and only if $\rho(\mathbf{AB}) = \rho(\mathbf{A})$ and $\mathbf{B}(\mathbf{AB})^-(\mathbf{AB}) = \mathbf{B}$ if and only if $\rho(\mathbf{AB}) = \rho(\mathbf{B})$.

(21) Let $\mathbf{A}$ be a square matrix and $(\mathbf{F}, \mathbf{G})$ be a rank factorization of $\mathbf{A}$. Then prove that $\mathbf{A}$ has a reflexive g-inverse $\mathbf{X}$ with $\mathcal{C}(\mathbf{X}) = \mathcal{C}(\mathbf{A})$ and $\mathcal{N}(\mathbf{X}) = \mathcal{N}(\mathbf{A})$ if and only if $\mathbf{GF}$ is invertible and if this is so, $\mathbf{X} = \mathbf{F}(\mathbf{GF})^{-2}\mathbf{G} = \mathbf{A}^\sharp$.

Chapter 3

The Minus Order

3.1 Introduction

After the first two introductory chapters, we are finally ready to explore the beautiful world of matrix partial orders. Historically these partial orders were defined through various generalized inverses. While it is true that the structure and mechanisms of these orders are more basic than the involvement of generalized inverse, we shall see that the use of generalized inverses makes the exposition elegant. Also, several properties can be expressed in a pleasant way in the process.

In Section 3.2, we begin with the study of one of the most basic matrix order relations, namely, the space pre-order. This is the stepping stone for almost all the partial orders that we shall study subsequently and therefore, we give a detailed account of its properties. This pre-order can be extended to a partial order with minor additions to its definition as we shall see in Section 3.3. Since this extension does not lend to any further properties than the space pre-order itself, we leave it at that and define a very important and the most fundamental partial order (that implies space pre-order and also the extension) called the minus order. We call it fundamental because the partial orders studied in the three subsequent chapters are built on this partial order by putting additional restrictions. We obtain several of its characterizations in Section 3.3. In Section 3.4, given a matrix $\mathbf{A}$, we obtain the class of all matrices $\mathbf{B}$ such that $\mathbf{A}$ is below $\mathbf{B}$ under the minus order (written as $\mathbf{A} <^{-} \mathbf{B}$). Assuming that $\mathbf{B}$ is given, we also obtain the class of all matrices $\mathbf{A}$ such that $\mathbf{A}$ is below $\mathbf{B}$ under the minus order. In Section 3.5, assuming that $\mathbf{A} <^{-} \mathbf{B}$, we characterize the class of all g-inverses of $\mathbf{A}$ such that $\mathbf{A}^{-}\mathbf{A} = \mathbf{A}^{-}\mathbf{B}$ and $\mathbf{A}\mathbf{A}^{-} = \mathbf{B}\mathbf{A}^{-}$. One of the characterising properties for $\mathbf{A}$ to be below $\mathbf{B}$ under minus order is $\mathbf{A} = \mathbf{P}\mathbf{B} = \mathbf{B}\mathbf{Q}$ for some projectors $\mathbf{P}$ and $\mathbf{Q}$. We also characterize the class of all projectors $\mathbf{P}$ and $\mathbf{Q}$ such that $\mathbf{A} = \mathbf{PB} = \mathbf{BQ}$ when $\mathbf{A} <^- \mathbf{B}$. Section 3.6 is devoted to the study of special properties of the minus order for idempotent matrices. In Section 3.7, we present some properties of the minus order for hermitian and *nnd* (complex) matrices.

We emphasize that all the matrices in this chapter are (a) possibly rectangular and (b) over an arbitrary field unless stated otherwise.

3.2 Space pre-order

A pre-order is a reflexive and transitive relation (see Appendix A). The space pre-order, which we define shortly, is the big daddy of (the most basic among) all the order relations that we study in this and subsequent chapters. We show that the space pre-order, as its name suggests, is a pre-order. We also obtain some characterizations of this pre-order which will be useful in what follows.

Definition 3.2.1. Let $\mathbf{A}$ and $\mathbf{B}$ be matrices (possibly rectangular) having the same order. Then $\mathbf{A}$ is said to be below $\mathbf{B}$ under the space pre-order, if $\mathcal{C}(\mathbf{A}) \subseteq \mathcal{C}(\mathbf{B})$ and $\mathcal{C}(\mathbf{A^t}) \subseteq \mathcal{C}(\mathbf{B^t})$. We denote the space pre-order by '$<^s$' and write $\mathbf{A} <^s \mathbf{B}$, whenever $\mathbf{A}$ is below $\mathbf{B}$ under '$<^s$'.

It is easy to check that $<^s$ is reflexive and transitive. Hence, it is a pre-order. However, it is not a partial order, for, if $\mathbf{A}$ and $\mathbf{B}$ are distinct non-singular matrices of the same order $n \times n$, then

$$\mathcal{C}(\mathbf{A}) = \mathcal{C}(\mathbf{B}) = \mathcal{C}(\mathbf{A^t}) = \mathcal{C}(\mathbf{B^t}) = \mathbf{F^n}.$$

More specifically, take

$$\mathbf{A} = \begin{pmatrix} 1 & 0 \\ 0 & 1 \end{pmatrix} \text{ and } \mathbf{B} = \begin{pmatrix} 1 & 1 \\ 0 & 1 \end{pmatrix}.$$

Thus $\mathbf{A} <^s \mathbf{B}$, $\mathbf{B} <^s \mathbf{A}$, but $\mathbf{A} \neq \mathbf{B}$, showing '$<^s$' is not anti-symmetric.

We give some more examples of the space pre-order.

Example 3.2.2. If $\mathbf{A} = \text{diag}(a_1, \ldots, a_n)$ and $\mathbf{B} = \text{diag}(b_1, \ldots, b_n)$, then $\mathbf{A} <^s \mathbf{B}$ if and only if $b_i \neq 0$ whenever $a_i \neq 0$.

Example 3.2.3. Let $\mathbf{A}$ and $\mathbf{B}$ be *nnd* matrices of the same order (over the field of complex numbers). Then $\mathbf{A} <^s \mathbf{A} + \mathbf{B}$.

Remark 3.2.4. Let $\mathbf{A}$ and $\mathbf{B}$ be matrices of the same order. In view of Remark 2.3.4, we have $\mathbf{A} <^s \mathbf{B} \Leftrightarrow \mathbf{A} = \mathbf{BB^-A} = \mathbf{AB^-B} = \mathbf{BB^-AB^-B}$

for all $\mathbf{B}^-$. So, $\mathbf{A} <^s \mathbf{B} \Leftrightarrow \mathbf{A} = \mathbf{BMB}$ for some matrix $\mathbf{M}$. Thus, $\rho(\mathbf{A}) \leq \rho(\mathbf{B})$ if $\mathbf{A} <^s \mathbf{B}$.

Remark 3.2.5. Let $\mathbf{A}$ and $\mathbf{B}$ be matrices of the same order. Let $\mathbf{P}$ and $\mathbf{Q}$ be non-singular matrices of appropriate orders so that the products $\mathbf{PAQ}$ and $\mathbf{PBQ}$ are defined. Then it is obvious that

$$\mathcal{C}(\mathbf{A}) \subseteq \mathcal{C}(\mathbf{B}) \Leftrightarrow \mathcal{C}(\mathbf{PAQ}) = \mathcal{C}(\mathbf{PA}) \subseteq \mathcal{C}(\mathbf{PB}) = \mathcal{C}(\mathbf{PBQ}).$$

Similarly, $\mathcal{C}(\mathbf{A}^t) \subseteq \mathcal{C}(\mathbf{B}^t) \Leftrightarrow \mathcal{C}((\mathbf{PBQ})^t) \subseteq \mathcal{C}((\mathbf{PBQ})^t)$. Thus, $\mathbf{A} <^s \mathbf{B} \Leftrightarrow \mathbf{PAQ} <^s \mathbf{PBQ}$, for all non-singular matrices $\mathbf{P}$ and $\mathbf{Q}$ of suitable orders. In other words, the space pre-order is invariant under equivalent transformations.

The following theorem gives some useful characterizations of the space pre-order.

Theorem 3.2.6. *Let $\mathbf{A}$ and $\mathbf{B}$ be matrices of the same order. The following are equivalent:*

(i) $\mathbf{A} <^s \mathbf{B}$,

(ii) *Let $(\mathbf{P}, \mathbf{Q})$ be a rank factorization of $\mathbf{B}$. Then $\mathbf{A} = \mathbf{PTQ}$ for some matrix $\mathbf{T}$ and*

(iii) *Let $(\mathbf{L}, \mathbf{M})$ be a rank factorization of $\mathbf{A}$. If $\rho(\mathbf{A}) = \rho(\mathbf{B})$, then $\mathbf{B} = \mathbf{LRM}$ for some non-singular matrix $\mathbf{R}$, and if $\rho(\mathbf{A}) < \rho(\mathbf{B})$, then there exist matrices $\mathbf{E}$ and $\mathbf{F}$ and a non-singular matrix $\mathbf{R}$ such that the matrices $(\mathbf{L} : \mathbf{E})$ and $\begin{pmatrix}\mathbf{M}\\ \cdot\cdot\\ \mathbf{F}\end{pmatrix}$ have full rank and $\mathbf{B} = (\mathbf{L} : \mathbf{E})\mathbf{R}\begin{pmatrix}\mathbf{M}\\ \cdot\cdot\\ \mathbf{F}\end{pmatrix}$.*

Proof. (i) $\Rightarrow$ (ii)
Let $(\mathbf{L}, \mathbf{M})$ be a rank factorization of $\mathbf{A}$. Since $\mathbf{A} <^s \mathbf{B}$, $\mathcal{C}(\mathbf{L}) = \mathcal{C}(\mathbf{A})$ and $\mathcal{C}(\mathbf{P}) = \mathcal{C}(\mathbf{B})$, we have $\mathcal{C}(\mathbf{L}) \subseteq \mathcal{C}(\mathbf{P})$. So, $\mathbf{L} = \mathbf{PT_1}$ for some matrix $\mathbf{T_1}$.

Similarly, $\mathcal{C}(\mathbf{M}^t) \subseteq \mathcal{C}(\mathbf{Q}^t)$ implies $\mathbf{M}^t = \mathbf{Q}^t\mathbf{T_2}$ for some matrix $\mathbf{T_2}$. Therefore, $\mathbf{A} = \mathbf{LM} = \mathbf{PT_1T_2^tQ} = \mathbf{PTQ}$, where $\mathbf{T} = \mathbf{T_1T_2^t}$.

(ii) $\Rightarrow$ (i)
Since $\mathbf{A} = \mathbf{PTQ}$, $\mathcal{C}(\mathbf{A}) \subseteq \mathcal{C}(\mathbf{P}) = \mathcal{C}(\mathbf{B})$ and $\mathcal{C}(\mathbf{A}^t) \subseteq \mathcal{C}(\mathbf{Q}^t) = \mathcal{C}(\mathbf{B}^t)$.

(i)$\Rightarrow$(iii)
Let $\rho(\mathbf{A}) = \rho(\mathbf{B})$. As in the proof of (i) $\Rightarrow$ (ii), it is easy to see that there exists a matrix $\mathbf{R}$ such that $\mathbf{B} = \mathbf{LRM}$. Let $\rho(\mathbf{A}) = a$. Then $\mathbf{R}$ is an $a \times a$ matrix. Further, since $\mathbf{L}$ has a left inverse and $\mathbf{M}$ has a right inverse, we have $a = \rho(\mathbf{A}) = \rho(\mathbf{B}) = \rho(\mathbf{LRM}) = \rho(\mathbf{R})$. Hence, $\mathbf{R}$ is non-singular.

Now, let $\rho(\mathbf{B}) > \rho(\mathbf{A})$. Since $\mathcal{C}(\mathbf{A}) \subseteq \mathcal{C}(\mathbf{B})$, there exists a full column rank matrix $\mathbf{E}$ such that columns of $(\mathbf{L} : \mathbf{E})$ form a basis of $\mathcal{C}(\mathbf{B})$. Similarly we can find a full row rank matrix $\mathbf{F}$ such that the rows of $\begin{pmatrix} \mathbf{M} \\ \cdot\cdot \\ \mathbf{F} \end{pmatrix}$ form a basis of the row space of $\mathbf{B}$. Let $(\mathbf{G}, \mathbf{H})$ be a rank factorization of $\mathbf{B}$. Then

$$\mathbf{G} = (\mathbf{L} : \mathbf{E})\mathbf{R_1},$$

where $\mathbf{R_1}$ is non-singular. Similarly,

$$\mathbf{H} = \mathbf{R_2} \begin{pmatrix} \mathbf{M} \\ \cdot\cdot \\ \mathbf{F} \end{pmatrix},$$

where $\mathbf{R_2}$ is non-singular. Hence

$$\mathbf{B} = \mathbf{GH} = (\mathbf{L} : \mathbf{E})\mathbf{R} \begin{pmatrix} \mathbf{M} \\ \cdot\cdot \\ F \end{pmatrix},$$

where $\mathbf{R} = \mathbf{R_1 R_2}$ is non-singular.

(iii) $\Rightarrow$ (i)

If $\rho(\mathbf{A}) = \rho(\mathbf{B})$, the proof is trivial. So, let $\rho(\mathbf{B}) > \rho(\mathbf{A})$. Let $\mathbf{E}$ and $\mathbf{F}$ be as in the proof of (i) $\Rightarrow$ (iii). Notice that $\mathcal{C}(\mathbf{A}) = \mathcal{C}(\mathbf{L}) \subseteq \mathcal{C}(\mathbf{L} : \mathbf{E}) = \mathcal{C}(\mathbf{B})$ and $\mathcal{C}(\mathbf{A^t}) = \mathcal{C}(\mathbf{M^t}) \subseteq \mathcal{C}((\mathbf{M^t} : \mathbf{F^t})) = \mathcal{C}(\mathbf{B^t})$. Therefore, $\mathbf{A} <^s \mathbf{B}$.

□

The characterizations of the space pre-order in Theorem 3.2.6 enable us to obtain the class of matrices, which are below a given matrix $\mathbf{B}$ under '$<^s$' (using (ii)) and the class of all matrices which are above a given matrix $\mathbf{A}$ under $<^s$ (using (iii)).

Remark 3.2.7. In Theorem 3.2.6, the statements (ii) and (iii) can be equivalently written as
(ii)$'$ Let $\mathbf{B} = \mathbf{P}\mathbf{diag}(\mathbf{I},\ \mathbf{0})\mathbf{Q}$; where $\mathbf{P}$ and $\mathbf{Q}$ are non-singular. Then $\mathbf{A} = \mathbf{P}\mathbf{diag}(\mathbf{T},\ \mathbf{0})\,\mathbf{Q}$ for some matrix $\mathbf{T}$
and
(iii)$'$ Let $\mathbf{A} = \mathbf{S}\mathbf{diag}(\mathbf{I_a}, \mathbf{0_{b-a}}, \mathbf{0})\mathbf{W}$ where $\mathbf{S}$ and $\mathbf{W}$ are non-singular. If $\rho(\mathbf{B}) = b \geq a$, then $\mathbf{B} = \mathbf{S}\mathbf{diag}(\mathbf{K},\ \mathbf{0})\mathbf{W}$, where $\mathbf{K}$ is a non-singular matrix of order $b \times b$ respectively.

Remark 3.2.8. Notice that $\mathcal{C}(\mathbf{A^t}) \subseteq \mathcal{C}(\mathbf{B^t})$ if and only if $\mathcal{N}(\mathbf{B}) \subseteq \mathcal{N}(\mathbf{A})$. Also $\mathcal{C}(\mathbf{A}) \subseteq \mathcal{C}(\mathbf{B})$ if and only if $\mathcal{N}(\mathbf{B^t}) \subseteq \mathcal{N}(\mathbf{A^t})$. Hence $\mathbf{A} <^s \mathbf{B}$, if and only if $\mathcal{N}(\mathbf{B}) \subseteq \mathcal{N}(\mathbf{A})$ and $\mathcal{N}(\mathbf{B^t}) \subseteq \mathcal{N}(\mathbf{A^t})$.

Theorem 3.2.9. *Let* $\mathbf{A}$ *and* $\mathbf{B}$ *be matrices having the same order. Then* $\mathbf{A} <^s \mathbf{B}$ *if and only if* $\mathbf{AB^-A}$ *is invariant under the choices of* $\mathbf{B}^-$.

Proof. If $\mathbf{A} = \mathbf{0}$, then result follows trivially.
If $\mathbf{A} \neq \mathbf{0}$, the result follows from Theorem 2.3.11. □

Let us now specialize to the case of square matrices of index 1. The following theorem can be proved along the lines of Theorem 3.2.6.

Theorem 3.2.10. *Let* $\mathbf{A}$ *and* $\mathbf{B}$ *be square matrices of order* $n \times n$ *with* $\rho(\mathbf{A}) = a$ *and* $\rho(\mathbf{B}) = b$. *Let each of* $\mathbf{A}$ *and* $\mathbf{B}$ *have index not greater than 1. Then the following are equivalent:*

(i) $\mathbf{A} <^s \mathbf{B}$,

(ii) *Let* $\mathbf{B} = \mathbf{P}\mathbf{diag}(\mathbf{D_b},\ \mathbf{0})\mathbf{P}^{-1}$, *where* $\mathbf{D_b}$ *is a* $b \times b$ *non-singular matrix and* $\mathbf{P}$ *is a non-singular matrix. Then* $\mathbf{A} = \mathbf{P}\mathbf{diag}(\mathbf{T}, \mathbf{0})\,\mathbf{P}^{-1}$, *where* $\mathbf{T}$ *is a* $b \times b$ *matrix of index not greater than 1 and*

(iii) *There exist non-singular matrices* $\mathbf{Q}$, $\mathbf{D_1}$ *and* $\mathbf{D_2}$ *of orders* $a \times a$ *and* $b \times b$ *such that*
$$\mathbf{A} = \mathbf{Q}\mathbf{diag}(\mathbf{D_1},\ \mathbf{0})\ \mathbf{Q}^{-1}, \text{ and } \mathbf{B} = \mathbf{Q}\mathbf{diag}(\mathbf{D_2},\ \mathbf{0})\,\mathbf{Q}^{-1}.$$

Remark 3.2.11. Let $\mathbf{A}$ and $\mathbf{B}$ be square matrices of index ≤ 1 such that $\mathbf{A} <^s \mathbf{B}$. Let $\mathbf{A} = \mathbf{Q}\mathbf{diag}(\mathbf{D_a},\ \mathbf{0})\ \mathbf{Q}^{-1}$, where $\mathbf{D_a}$ is non-singular. Then it is not necessarily true that $\mathbf{B} = \mathbf{Q}\mathbf{diag}(\mathbf{E}, \mathbf{0})\,\mathbf{Q}^{-1}$, where $\mathbf{E}$ is non-singular. However, if $\mathbf{A} <^s \mathbf{B}$, there exist nonsingular matrices $\mathbf{Q}$, $\mathbf{D}$ and $\mathbf{E}$ of appropriate orders such that $\mathbf{A} = \mathbf{Q}\ \mathbf{diag}(\mathbf{D}, \mathbf{0})\mathbf{Q}^{-1}$ and $\mathbf{B} = \mathbf{Q}\ \mathbf{diag}(\mathbf{E}, \mathbf{0})\mathbf{Q}^{-1}$.

We now consider matrices over the field of complex numbers $\mathbb{C}$ and give one more characterization of the space pre-order using singular value decomposition.

Theorem 3.2.12. *Let* $\mathbf{A}$ *and* $\mathbf{B}$ *be matrices in* $\mathbb{C}^{m\times n}$. *The following are equivalent:*

(i) $\mathbf{A} <^s \mathbf{B}$,

(ii) *Let* $\mathbf{B} = \mathbf{P}\mathbf{diag}(\mathbf{\Delta}, \mathbf{0})\mathbf{Q}^*$ *be a singular value decomposition of* $\mathbf{B}$, *where* $\mathbf{P}$ *and* $\mathbf{Q}$ *are unitary matrices of appropriate orders and* $\mathbf{\Delta}$ *is a positive definite diagonal matrix. Then* $\mathbf{A} = \mathbf{P}\mathbf{diag}(\mathbf{T}, \mathbf{0})\mathbf{Q}^*$ *for some matrix* $\mathbf{T}$ *of the same order as* $\mathbf{\Delta}$ *and*

(iii) *There exist unitary matrices* $\mathbf{U}$ *and* $\mathbf{V}$, *a positive definite diagonal matrix* $\mathbf{D}$ *and a nonsingular matrix* $\mathbf{S}$ *of order not less than that of* $\mathbf{D}$ *such that* $\mathbf{A} = \mathbf{U}\ \mathbf{diag}(\mathbf{D}, \mathbf{0})\mathbf{V}^*$ *is a singular value decomposition of* $\mathbf{A}$ *and* $\mathbf{B} = \mathbf{U}\ \mathbf{diag}(\mathbf{S}, \mathbf{0})\mathbf{V}^*$.

Proof can be given along the lines of Theorem 3.2.6.

3.3 Minus order - Some characterizations

In the previous section we studied an order relation which is a pre-order, namely, the space pre-order. We also noted that it is not anti-symmetric and hence is not a partial order. It is easy to see that the following minimal modification to the definition of the space pre-order yields a partial order. Define

$$\mathbf{A} <^{s+} \mathbf{B} \text{ if } \begin{cases} \mathbf{A} = \mathbf{B} & \text{if } \rho(\mathbf{A}) = \rho(\mathbf{B}) \\ \\ \mathbf{A} <^{s} \mathbf{B} & \text{if } \rho(\mathbf{A}) < \rho(\mathbf{B}) \end{cases}.$$

It is clear that '$<^{s+}$' is a partial order. This is the weakest partial order that implies the space pre-order. Unfortunately this partial order does not exhibit any special properties not possessed by the space pre-order. We now introduce a partial order which implies the space pre-order and also possesses interesting properties like the ones mentioned below.

Let us consider $\mathbf{A} = \mathbf{diag}(\mathbf{I_a},\ \mathbf{0},\ \mathbf{0})$ and $\mathbf{B} = \mathbf{diag}(\mathbf{I_a},\ \mathbf{I_c},\ \mathbf{0})$. It is reasonable to say $\mathbf{A}$ precedes $\mathbf{B}$ or $\mathbf{A}$ is below $\mathbf{B}$. The same status can be accorded to matrices $\mathbf{PAQ}$ and $\mathbf{PBQ}$, where $\mathbf{P}$ and $\mathbf{Q}$ are non-singular. Moreover, it is easy to see that $\mathbf{A} <^{s} \mathbf{B}$ and $\mathbf{B}$ is obtained from $\mathbf{A}$ by simply adding $\mathbf{C} = \mathbf{diag}(\mathbf{0},\ \mathbf{I_c},\ \mathbf{0})$. Notice that the column spaces of $\mathbf{A}$ and $\mathbf{C}$ are virtually disjoint and so are their row spaces. Thus, $\mathbf{B} = \mathbf{A} \oplus \mathbf{C}$. Such a matrix $\mathbf{A}$ can be thought as an independent section of $\mathbf{B}$ or we may say that $\mathbf{A}$ precedes $\mathbf{B}$ in some sense. In the previous section we saw that $\mathbf{A} <^{s} \mathbf{B}$ if and only if $\mathbf{AB^-A}$ is invariant under the choices of g-inverse $\mathbf{B}^-$ of $\mathbf{B}$. Further, if $\mathbf{AB^-A} = \mathbf{A}$, then $\{\mathbf{B}^-\}$, the set of the generalized inverses of $\mathbf{B}$ is a subset of $\{\mathbf{A}^-\}$. Thus, once again we can say $\mathbf{A}$ precedes $\mathbf{B}$ in some sense. In this section, we show that all the above requirements lead to an identical ordering between matrices of same order. In this chapter we make an in-depth study of this order. Consider the mathematically elegant relationship defined using generalized inverses as follows:

Definition 3.3.1. Let $\mathbf{A}$ and $\mathbf{B}$ be matrices of the same order. Then $\mathbf{A}$ is said to be below $\mathbf{B}$ under the minus order if there exist generalized inverses $\mathbf{G_1}$ and $\mathbf{G_2}$ of $\mathbf{A}$ such that $\mathbf{AG_1} = \mathbf{BG_1}$ and $\mathbf{G_2A} = \mathbf{G_2B}$.

When $\mathbf{A}$ is below $\mathbf{B}$ under the minus order, we write $\mathbf{A} <^{-} \mathbf{B}$.

We shall show that the relation in Definition 3.3.1 is identical with the various ways of $\mathbf{A}$ preceding $\mathbf{B}$ described in the second paragraph of this section. We shall also show that the minus order is indeed a partial order on $F^{m\times n}$. The minus order was first introduced by Hartwig on the set of regular elements of a semi-group. He developed the order relation as an extension of the standard or natural partial order on the set of idempotent elements of a semi-group and in particular, of elements of $F^{m\times n}$. This order also extends the natural partial order of Vagner on inverse semi-groups and the star order of Drazin (which is studied in detail in Chapter 5).

We first show that whenever $\mathbf{A} <^- \mathbf{B}$, there is no loss of generality in taking the g-inverses $\mathbf{G_1}$ and $\mathbf{G_2}$ of $\mathbf{A}$ to be the same in Definition 3.3.1.

Theorem 3.3.2. *Let* $\mathbf{A}$ and $\mathbf{B}$ *be matrices having the same order. Then* $\mathbf{A} <^- \mathbf{B}$ *if and only if there exists a g-inverse* $\mathbf{A}^-$ *of* $\mathbf{A}$ *such that* $\mathbf{A}^-\mathbf{A} = \mathbf{A}^-\mathbf{B}$ *and* $\mathbf{A}\mathbf{A}^- = \mathbf{B}\mathbf{A}^-$.

Proof. 'If' part is trivial.
'Only if' part
Let $\mathbf{G_1}$ and $\mathbf{G_2}$ to be the g-inverses of $\mathbf{A}$ such that $\mathbf{AG_1} = \mathbf{BG_1}$ and $\mathbf{G_2A} = \mathbf{G_2B}$. Let $\mathbf{G} = \mathbf{G_1AG_2}$. Then $\mathbf{G}$ is a g-inverse of $\mathbf{A}$ and satisfies $\mathbf{AG} = \mathbf{BG}$, $\mathbf{GA} = \mathbf{GB}$. □

Remark 3.3.3. (i) In view of Theorem 3.3.2, we may take $\mathbf{G_1}$ and $\mathbf{G_2}$ to be the same in Definition 3.3.1. Henceforth, we do so without making a special mention.

(ii) In the proof of Theorem 3.3.2, notice that $\mathbf{G} = \mathbf{G_1AG_2}$ is a reflexive g-inverse of $\mathbf{A}$. Hence, it is clear that if $\mathbf{A} <^- \mathbf{B}$, then there exists a reflexive g-inverse $\mathbf{A}_\mathbf{r}^-$ of $\mathbf{A}$ such that $\mathbf{A}\mathbf{A}_\mathbf{r}^- = \mathbf{B}\mathbf{A}_\mathbf{r}^-$ and $\mathbf{A}_\mathbf{r}^-\mathbf{A} = \mathbf{A}_\mathbf{r}^-\mathbf{B}$.

Clearly, $\mathbf{0} <^- \mathbf{B}$ for any matrix $\mathbf{B}$, where matrices $\mathbf{0}$ and $\mathbf{B}$ have the same order. Also, if $\mathbf{B} <^- \mathbf{0}$, then $\mathbf{B} = \mathbf{0}$. *Henceforth, we shall consider* $\mathbf{A}$ *and* $\mathbf{B}$ *to be non-null in our study of the minus order.*

Any $m \times n$ matrix is a representation of some linear transformation from F^n to F^m under specified bases for F^n and F^m. Thus, if $\mathbf{P}$ and $\mathbf{Q}$ are non-singular matrices of suitable orders so that the matrix $\mathbf{PAQ}$ is defined, then $\mathbf{A}$ and $\mathbf{PAQ}$ represent the same linear transformation. In view of Remark 3.2.5, the space pre-order is also a pre-order on the linear transformations. Our next Theorem shows the same is true for the minus order.

Theorem 3.3.4. *Let* $\mathbf{A}$ *and* $\mathbf{B}$ *be matrices of the same order and let* $\mathbf{P}$ *and* $\mathbf{Q}$ *be non-singular matrices of appropriate orders so that the matrix* $\mathbf{PAQ}$ *is defined. Then* $\mathbf{A} <^- \mathbf{B}$ *if and only if* $\mathbf{PAQ} <^- \mathbf{PBQ}$.

Proof. 'Only if' part

Let $\mathbf{A} <^- \mathbf{B}$. Then there exists a g-inverse $\mathbf{A}^-$ of $\mathbf{A}$ such that $\mathbf{A}^-\mathbf{A} = \mathbf{A}^-\mathbf{B}$ and $\mathbf{AA}^- = \mathbf{BA}^-$. In view of Remark 2.3.7, $\mathbf{C} = \mathbf{Q}^{-1}\mathbf{A}^-\mathbf{P}^{-1}$ is a g-inverse of $\mathbf{PAQ}$. Moreover, $\mathbf{CPAQ} = \mathbf{CPBQ}$ and $\mathbf{PAQC} = \mathbf{PBQC}$. Thus, we have $\mathbf{PAQ} <^- \mathbf{PBQ}$.

'If' part follows from the proof of 'only if' part once we observe that $\mathbf{A} = \mathbf{P}^{-1}(\mathbf{PAQ})\mathbf{Q}^{-1}$ and $\mathbf{B} = \mathbf{P}^{-1}(\mathbf{PBQ})\mathbf{Q}^{-1}$. □

We now establish the assertion made earlier about the equivalence of seemingly different relationships mentioned in the beginning of this section.

Theorem 3.3.5. *Let* $\mathbf{A}$ *and* $\mathbf{B}$ *be non-null matrices of the order* $m \times n$. *Let* $\rho(\mathbf{A}) = a$ *and* $\rho(\mathbf{B}) = b$. *Then the following are equivalent:*

(i) $\mathbf{A} <^- \mathbf{B}$
(ii) $\mathbf{A} <^s \mathbf{B}$ *and* $\mathbf{AA}^- = \mathbf{BA}^-$ *for some g-inverse* $\mathbf{A}^-$ *of* $\mathbf{A}$
(iii) $\mathbf{A} <^s \mathbf{B}$ *and* $\mathbf{AA}_r^- = \mathbf{BA}_r^-$ *for some reflexive g-inverse* $\mathbf{A}_r^-$ *of* $\mathbf{A}$
(iv) $\mathbf{AA}^- = \mathbf{BA}^-$ *for some g-inverse* $\mathbf{A}^-$ *of* $\mathbf{A}$ *and* $\mathcal{C}(\mathbf{A}^t) \subset \mathcal{C}(\mathbf{B}^t)$
(v) *If* $(\mathbf{P}, \mathbf{Q})$ *is a rank factorization of* $\mathbf{B}$, *then* $\mathbf{A} = \mathbf{PTQ}$ *for some idempotent matrix* $\mathbf{T}$
(vi) *There exist non-singular matrices* $\mathbf{R}$ *and* $\mathbf{S}$ *of orders* $m \times m$ *and* $n \times n$ *respectively such that*

$$\mathbf{A} = \mathbf{R}\mathbf{diag}(\mathbf{I_a},\ \mathbf{0},\ \mathbf{0})\mathbf{S} \text{ and } \mathbf{B} = \mathbf{R}\mathbf{diag}(\mathbf{I_a},\ \mathbf{I_{b-a}},\ \mathbf{0})\mathbf{S}$$

(vii) $\mathbf{B} = \mathbf{A} \oplus (\mathbf{B} - \mathbf{A})$ *and*
(viii) $\{\mathbf{B}^-\} \subseteq \{\mathbf{A}^-\}$.

Proof. (i)⇒(ii) and (ii)⇔(iii)⇔(iv) are trivial.

(iii)⇒(v)

Let $(\mathbf{P}, \mathbf{Q})$ be a rank factorization of $\mathbf{B}$. Since $\mathbf{A} <^s \mathbf{B}$, by Theorem 3.2.6(ii), we have $\mathbf{B} = \mathbf{PQ}$ and $\mathbf{A} = \mathbf{PTQ}$ for some matrix $\mathbf{T}$. Further, every reflexive g-inverse $\mathbf{A}_r^-$ of $\mathbf{A}$ is of the from $\mathbf{Q}_\mathbf{R}^{-1}\mathbf{T}_r^-\mathbf{P}_\mathbf{L}^{-1}$ for some right inverse $\mathbf{Q}_\mathbf{R}^{-1}$ of $\mathbf{Q}$, for some left inverse $\mathbf{P}_\mathbf{L}^{-1}$ of $\mathbf{P}$ and for some $\mathbf{T}_r^-$. Now $\mathbf{AA}_r^- = \mathbf{BA}_r^-$ for some reflexive g-inverse $\mathbf{A}_r^-$, so, $\mathbf{PTQQ}_\mathbf{R}^{-1}\mathbf{T}_r^-\mathbf{P}_\mathbf{L}^{-1} = \mathbf{PQQ}_\mathbf{R}^{-1}\mathbf{T}_r^-\mathbf{P}_\mathbf{L}^{-1}$ or $\mathbf{TT}_r^- = \mathbf{T}_r^-$. Clearly, $\mathbf{T} = \mathbf{T}_r^-\mathbf{T}$ is idempotent.

(v) ⇒ (vi)

Since $\mathbf{T}$ is idempotent with $\rho(\mathbf{T}) = a$, there exists a non-singular matrix $\mathbf{R}_1$ such that $\mathbf{T} = \mathbf{R}_1\mathbf{diag}(\mathbf{I_a},\ \mathbf{0})\mathbf{R}_1^{-1}$. So $\mathbf{B} = \mathbf{PQ} = \mathbf{PR}_1\mathbf{R}_1^{-1}\mathbf{Q}$, and $\mathbf{A} = \mathbf{PR}_1\mathbf{diag}(\mathbf{I_a},\ \mathbf{0})\mathbf{R}_1^{-1}\mathbf{Q}$. Clearly, $\rho(\mathbf{PR}_1) = \rho(\mathbf{R}_1^{-1}\mathbf{Q}) = b$. Also, $(\mathbf{PR}_1, \mathbf{R}_1^{-1}\mathbf{Q})$ is a rank factorization of $\mathbf{B}$. Hence, $\mathbf{PR}_1$ and $\mathbf{R}_1^{-1}\mathbf{Q}$ can be extended to non-singular matrices $\mathbf{R}$ and $\mathbf{S}$ respectively. It is easy to see

that $\mathbf{A} = \mathbf{R}\mathbf{diag}(\mathbf{I_a},\ \mathbf{0}, \mathbf{0})\mathbf{S}$ and $\mathbf{B} = \mathbf{R}\mathbf{diag}(\mathbf{I_a},\ \mathbf{I_{b-a}},\ \mathbf{0})\mathbf{S}$.
Proof for (vi) $\Rightarrow$ (vii) is trivial.
(vii) $\Rightarrow$ (viii)
Clearly, $\mathcal{C}(\mathbf{A}) \subseteq \mathcal{C}(\mathbf{B})$ and $\mathcal{C}(\mathbf{A^t}) \subseteq \mathcal{C}(\mathbf{B^t})$. So, for each $\mathbf{B}^-$,

$$\mathbf{A} = \mathbf{AB^-B} = \mathbf{AB^-A} + \mathbf{AB^-(B-A)} \text{ or } \mathbf{A} - \mathbf{AB^-A} = \mathbf{AB^-(B-A)}.$$

However,

$$\mathcal{C}(\mathbf{A} - \mathbf{AB^-A})^{\mathbf{t}} \subseteq \mathcal{C}(\mathbf{AB^-(B-A)})^{\mathbf{t}} \text{ and } \mathcal{C}(\mathbf{AB^-(B-A)})^{\mathbf{t}} \subseteq \mathcal{C}(\mathbf{B}-\mathbf{A})^{\mathbf{t}}.$$

$$\text{So, } \mathcal{C}(\mathbf{A} - \mathbf{AB^-A})^{\mathbf{t}} \cap \mathcal{C}(\mathbf{AB^-(B-A)})^{\mathbf{t}} \subseteq \mathcal{C}(\mathbf{A^t}) \cap \mathcal{C}(\mathbf{B}-\mathbf{A})^{\mathbf{t}} = \{\mathbf{0}\}.$$

Hence, $\mathbf{A} - \mathbf{AB^-A} = \mathbf{0}$ or equivalently $\mathbf{A} = \mathbf{AB^-A}$. Thus, $\mathbf{B}^- \in \{\mathbf{A}^-\}$.
(viii) $\Rightarrow$ (i)
Since $\mathbf{A} = \mathbf{AB^-A}$ for all $\mathbf{B}^-$, we have $\mathbf{A} <^s \mathbf{B}$. Let $\mathbf{B}^-$ and $\mathbf{A}^-$ be any g-inverses of $\mathbf{B}$ and $\mathbf{A}$ respectively. Write $\mathbf{G_1} = \mathbf{B^-AA^-}$ and $\mathbf{G_2} = \mathbf{A^-AB^-}$. Then it is easy to verify that $\mathbf{G_1}$ and $\mathbf{G_2}$ are g-inverse of $\mathbf{A}$ and $\mathbf{AG_1} = \mathbf{BG_1}$, $\mathbf{G_2A} = \mathbf{G_2B}$. □

Corollary 3.3.6. *If* $\mathbf{A}$ *and* $\mathbf{B}$ *are non-null matrices of the same order such that* $\mathbf{A} <^- \mathbf{B}$ *and* $\rho(\mathbf{A}) = \rho(\mathbf{B})$, *then* $\mathbf{A} = \mathbf{B}$.

Remark 3.3.7. Let $\mathbf{A}$ and $\mathbf{B}$ be non-null matrices of the same order such that $\mathbf{A}$ is a full rank matrix and $\mathbf{A} <^- \mathbf{B}$, then $\mathbf{A} = \mathbf{B}$. So, $\mathbf{A}$ is a maximal element of $F^{m \times n}$ under '$<^-$'.

Remark 3.3.8. The condition $\{\mathbf{B}^-\} \subseteq \{\mathbf{A}^-\}$ in (viii) of Theorem 3.3.5 can be replaced by $\{\mathbf{B}_\mathbf{r}^-\} \subseteq \{\mathbf{A}^-\}$.

Remark 3.3.9. Statement (vii) of Theorem 3.3.5 is equivalent to saying $\rho(\mathbf{B}) = \rho(\mathbf{A}) + \rho(\mathbf{B} - \mathbf{A})$.

Remark 3.3.10. Let $\mathbf{A}$ and $\mathbf{B}$ be non-null matrices of same order such that $\mathbf{A} <^- \mathbf{B}$. Then $\mathbf{A}$ and $\mathbf{B}$ have simultaneous normal form.

Remark 3.3.11. In Theorem 3.3.5, the statements (ii), (iii) and (iv) can equivalently be stated as

(ii)′ $\mathbf{A} <^s \mathbf{B}$ and $\mathbf{A^-A} = \mathbf{A^-B}$ for some g-inverse $\mathbf{A}^-$ of $\mathbf{A}$
(iii)′ $\mathbf{A} <^s \mathbf{B}$ and $\mathbf{A_r^-A} = \mathbf{A_r^-B}$ for some reflexive g-inverse $\mathbf{A}_\mathbf{r}^-$ of $\mathbf{A}$
(iv)′ $\mathbf{A^-A} = \mathbf{A^-B}$ for some g-inverse $\mathbf{A}^-$ of $\mathbf{A}$ and $\mathcal{C}(\mathbf{A}) \subset \mathcal{C}(\mathbf{B})$.

Remark 3.3.12. In Theorem 3.3.5, we proved that $\mathbf{A} <^s \mathbf{B}$ together with $\mathbf{AA}^- = \mathbf{BA}^-$ for some g-inverse $\mathbf{A}^-$ of $\mathbf{A}$ is equivalent to $\mathbf{A} <^- \mathbf{B}$. It is easy to see (in view of Remark 2.3.4) that $\mathbf{A} <^s \mathbf{B}$ together with $\mathbf{A} = \mathbf{AB}^-\mathbf{A}$ for some $\mathbf{B}^-$ is also equivalent to $\mathbf{A} <^- \mathbf{B}$.

Remark 3.3.13. Let $\mathbf{T_1}$ and $\mathbf{T_2}$ be linear transformations from F^n to F^m. Let $\rho(\mathbf{T_1}) = a$ and $\rho(\mathbf{T_2}) = b$, $b \geq a$. In view of (i) $\Leftrightarrow$ (vi) of Theorem 3.3.5, $\mathbf{T_1} <^- \mathbf{T_2}$ if and only if there exists bases $\{\mathbf{u_1}, \ldots, \mathbf{u_n}\}$ and $\{\mathbf{v_1}, \ldots, \mathbf{v_m}\}$ of F^n and F^m respectively such that

$$\mathbf{T_1 u_i} = \begin{cases} \mathbf{v_i} & 1 \leq i \leq a \\ \mathbf{0} & \text{otherwise} \end{cases}$$

and

$$\mathbf{T_2 u_i} = \begin{cases} \mathbf{v_i} & 1 \leq i \leq b \\ \mathbf{0} & \text{otherwise} \end{cases}.$$

We now show that the minus order is indeed a partial order.

Theorem 3.3.14. *The minus order is a partial order on* $F^{m \times n}$.

Proof. Reflexivity holds trivially. In view of Theorem 3.3.5 (viii), transitivity also holds. To prove anti-symmetry, observe that by Remark 3.3.9, $\mathbf{A} <^- \mathbf{B} \Rightarrow \rho(\mathbf{B}) = \rho(\mathbf{A}) + \rho(\mathbf{B} - \mathbf{A})$. So, $\rho(\mathbf{B}) \geq \rho(\mathbf{A})$. Similarly, we have $\mathbf{B} <^- \mathbf{A} \Rightarrow \rho(\mathbf{A}) \geq \rho(\mathbf{B})$. Combining the two inequalities, we have $\rho(\mathbf{B}) = \rho(\mathbf{A})$. Therefore, $\rho(\mathbf{B} - \mathbf{A}) = \mathbf{0}$ or equivalently $\mathbf{B} - \mathbf{A} = \mathbf{0}$. Hence, $\mathbf{B} = \mathbf{A}$. □

We now give some more characterizing properties of the minus order. Notice that if $\mathbf{A} <^- \mathbf{B}$, then there exists a g-inverse $\mathbf{A}^-$ of $\mathbf{A}$ such that $\mathbf{AA}^- = \mathbf{BA}^-$ and $\mathbf{A}^-\mathbf{A} = \mathbf{A}^-\mathbf{B}$. Hence $\mathbf{A} = \mathbf{AA}^-\mathbf{A} = \mathbf{AA}^-\mathbf{B} = \mathbf{BA}^-\mathbf{A}$. Thus, there are projectors $\mathbf{P}(= \mathbf{AA}^-)$ and $\mathbf{Q}(= \mathbf{A}^-\mathbf{A})$ such that $\mathbf{A} = \mathbf{PB} = \mathbf{BQ}$. It turns out that this is a characterizing property of the minus order. However, before we exhibit this, we prove the following:

Lemma 3.3.15. *Let* $\mathbf{A}, \mathbf{B}$ *be matrices such that the product* $\mathbf{AB}$ *is defined. Then* $\rho(\mathbf{AB}) = \rho(\mathbf{B}) - \mathbf{d}(\mathcal{C}(\mathbf{B}) \cap \mathcal{N}(\mathbf{A}))$.

Proof. Let $\rho(\mathbf{B}) = r$ and $d(\mathcal{C}(\mathbf{B}) \cap \mathcal{N}(\mathbf{A})) = s$. Let $\mathbf{C}$ and $\mathbf{D}$ be matrices of order $n \times s$ and $n \times (r-s)$ respectively such that $\mathcal{C}(\mathbf{C}) = \mathcal{C}(\mathbf{B}) \cap \mathcal{N}(\mathbf{A})$ and $\mathcal{C}(\mathbf{C} : \mathbf{D}) = \mathcal{C}(\mathbf{B})$. Clearly, $\mathbf{C}$ and $\mathbf{D}$ are matrices of full column rank. Since $\mathcal{C}(\mathbf{C}) = \mathcal{C}(\mathbf{B}) \cap \mathcal{N}(\mathbf{A})$, we have $\mathbf{AC} = \mathbf{0}$ and so, $\mathcal{C}(\mathbf{AD}) = \mathcal{C}(\mathbf{AB})$. Also,

$\mathcal{C}(\mathbf{C}) \cap \mathcal{C}(\mathbf{D}) = \{\mathbf{0}\}$. Further, if $\mathbf{ADu} = \mathbf{0}$, then $\mathbf{Du} \in \mathcal{C}(\mathbf{B}) \cap \mathcal{N}(\mathbf{A}) = \mathcal{C}(\mathbf{C})$. Conversely, if $\mathbf{Du} \in \mathcal{C}(\mathbf{B}) \cap \mathcal{N}(\mathbf{A}) = \mathcal{C}(\mathbf{C})$, then $\mathbf{Du} = \mathbf{Cv}$ for some $\mathbf{v}$. Hence $\mathbf{ADu} = \mathbf{ACv} = \mathbf{0}$. Now, $\mathbf{ADu} = \mathbf{0} \Leftrightarrow \mathbf{Du} \in \mathcal{C}(\mathbf{C})$. Hence, $\mathbf{Du} \in \mathcal{C}(\mathbf{C}) \cap \mathcal{C}(\mathbf{D}) = \{\mathbf{0}\}$, so, $\mathbf{Du} = \mathbf{0}$. Trivially, $\mathbf{Du} = \mathbf{0} \Rightarrow \mathbf{ADu} = \mathbf{0}$. Thus, $\rho(\mathbf{AD}) = \rho(\mathbf{D})$. As $\mathcal{C}(\mathbf{AD}) = \mathcal{C}(\mathbf{AB})$, $\rho(\mathbf{AB}) = \rho(\mathbf{D}) = r - s$. Hence, $\rho(\mathbf{AB}) = \rho(\mathbf{B}) - \mathbf{d}(\mathcal{C}(\mathbf{B}) \cap \mathcal{N}(\mathbf{A}))$.

(*Alternative Proof*): Consider the matrix $\mathbf{S} = \mathbf{B} - \mathbf{B}(\mathbf{AB})^-(\mathbf{AB})$. Clearly, for any g-inverse $\mathbf{B}^-$ of $\mathbf{B}$, $\mathbf{SB}^-\mathbf{S} = \mathbf{S}$, $\mathbf{SB}^-\mathbf{B} = \mathbf{S}$ and $\mathbf{BB}^-\mathbf{S} = \mathbf{S}$, so, $\mathbf{S} <^- \mathbf{B}$, by Remark 3.3.12. It is easy to see that $\mathcal{C}(\mathbf{S}) \subseteq \mathcal{C}(\mathbf{B}) \cap \mathcal{N}(\mathbf{A})$. Further, let $\mathbf{x} \in \mathcal{C}(\mathbf{B}) \cap \mathcal{N}(\mathbf{A})$. Then $\mathbf{x} = \mathbf{Bz}$ and $\mathbf{Ax} = \mathbf{0}$. So, $\mathbf{ABz} = \mathbf{0}$. Since $\mathbf{S} <^- \mathbf{B}$, we have $\mathbf{x} = \mathbf{Bz} = \mathbf{Sz} + (\mathbf{B}(\mathbf{AB})^-(\mathbf{AB}))\mathbf{z} = \mathbf{Sz}$ and so, $\mathbf{x} \in \mathcal{C}(\mathbf{S})$. Hence, $\mathcal{C}(\mathbf{S}) = \mathcal{C}(\mathbf{B}) \cap \mathcal{N}(\mathbf{A})$. Now,

$\rho(\mathbf{B}) = \rho(\mathbf{S}) + \rho(\mathbf{B}(\mathbf{AB})^-(\mathbf{AB})) = \mathbf{d}(\mathcal{C}(\mathbf{B}) \cap \mathcal{N}(\mathbf{A})) + \rho(\mathbf{AB})$, since $\mathbf{A}(\mathbf{B}(\mathbf{AB})^-(\mathbf{AB})) = \mathbf{AB}$. Thus, $\rho(\mathbf{AB}) = \rho(\mathbf{B}) - \mathbf{d}(\mathcal{C}(\mathbf{B}) \cap \mathcal{N}(\mathbf{A}))$.

□

Theorem 3.3.16. *Let* $\mathbf{A}$ and $\mathbf{B}$ *be matrices of the same order. Then the following are equivalent:*

(i) $\mathbf{A} <^- \mathbf{B}$,
(ii) *There exist projectors* $\mathbf{P}$ *and* $\mathbf{Q}$ *such that* $\mathbf{A} = \mathbf{PB} = \mathbf{BQ}$,
(iii) $\mathbf{B} - \mathbf{A} <^- \mathbf{B}$ *and*
(iv) $\rho(\mathbf{B} - \mathbf{A}) = \rho((\mathbf{I} - \mathbf{AA}^-)\mathbf{B}) = \rho(\mathbf{B}(\mathbf{I} - \mathbf{A}^-\mathbf{A}))$ *for any* $\mathbf{A}^-$.

Proof. (i) $\Rightarrow$ (ii)
Since $\mathbf{A} <^- \mathbf{B}$, there exists a g-inverse $\mathbf{A}^-$ of $\mathbf{A}$ such that $\mathbf{AA}^- = \mathbf{BA}^-$ and $\mathbf{A}^-\mathbf{A} = \mathbf{A}^-\mathbf{B}$. Choose $\mathbf{P} = \mathbf{AA}^-$ and $\mathbf{Q} = \mathbf{A}^-\mathbf{A}$. It is easy to check that $\mathbf{P}$ and $\mathbf{Q}$ are projectors.
(ii) $\Rightarrow$ (i)
For any g-inverse $\mathbf{B}^-$ of $\mathbf{B}$ we have, $\mathbf{AB}^-\mathbf{A} = \mathbf{PBB}^-\mathbf{BQ} = \mathbf{PBQ} = \mathbf{AQ} = \mathbf{BQ}^2 = \mathbf{BQ} = \mathbf{A}$. Thus, $\mathbf{AB}^-\mathbf{A} = \mathbf{A}$. This gives $\{\mathbf{B}^-\} \subseteq \{\mathbf{A}^-\}$ and therefore, $\mathbf{A} <^- \mathbf{B}$.

The equivalence of (i) and (iii) follows from the equivalence of (i) and (vii) of Theorem 3.3.5.
(iii) $\Leftrightarrow$ (iv) Notice that

$$\begin{aligned}\rho((\mathbf{I} - \mathbf{AA})^-)\mathbf{B} &= \rho((\mathbf{I} - \mathbf{AA}^-)(\mathbf{B} - \mathbf{A})) \\ &= \rho(\mathbf{B} - \mathbf{A}) - \mathbf{d}(\mathcal{C}(\mathbf{B} - \mathbf{A}) \cap \mathcal{N}(\mathbf{I} - \mathbf{AA}^-)) \\ &= \rho(\mathbf{B} - \mathbf{A}) - \mathbf{d}(\mathcal{C}(\mathbf{B} - \mathbf{A}) \cap \mathcal{C}(\mathbf{A})) \\ &= \rho(\mathbf{B} - \mathbf{A}) \Leftrightarrow \mathcal{C}(\mathbf{B} - \mathbf{A}) \cap \mathcal{C}(\mathbf{A}) = \{\mathbf{0}\}.\end{aligned}$$

Similarly, $\rho(\mathbf{B}(\mathbf{I}-\mathbf{A}^-\mathbf{A})) = \rho(\mathbf{B}-\mathbf{A}) \Leftrightarrow \mathcal{C}((\mathbf{B}-\mathbf{A})^t)\cap\mathcal{C}(\mathbf{A}^t) = \{\mathbf{0}\}$. □

Let $\mathbf{A}$ and $\mathbf{B}$ be matrices of the same order such that $\mathbf{A} <^- \mathbf{B}$. It is easy to see that there exist g-inverses $\mathbf{A}^-$ of $\mathbf{A}$ and $\mathbf{B}^-$ of $\mathbf{B}$ such that $\mathbf{A}^- <^- \mathbf{B}^-$ and also $\mathbf{B}^- <^- \mathbf{A}^-$. In fact, we can make a stronger statement as shown below.

Theorem 3.3.17. *Let* $\mathbf{A}$ *and* $\mathbf{B}$ *be matrices of the same order such that* $\mathbf{A} <^- \mathbf{B}$. *Let* $\mathbf{G}$ *be a g-inverse of* $\mathbf{B}$, $\rho(\mathbf{A}) = a$, $\rho(\mathbf{B}) = b$ *and* $\rho(\mathbf{G}) = g$. *Then there exists non-singular matrices* $\mathbf{P}$ *and* $\mathbf{Q}$ *such that*

$$\mathbf{A} = \mathbf{P}\mathbf{diag}(\mathbf{I_a},\ \mathbf{0},\ \mathbf{0},\ \mathbf{0})\mathbf{Q},\ \ \mathbf{B} = \mathbf{P}\mathbf{diag}(\mathbf{I_a},\ \mathbf{I_{b-a}},\ \mathbf{0},\mathbf{0})\mathbf{Q}$$

and $\mathbf{G} = \mathbf{Q}^{-1}\mathbf{diag}(\mathbf{I_a},\mathbf{I_{b-a}},\mathbf{I_{g-b}},\mathbf{0})\mathbf{P}^{-1}$.

Proof. By Theorem 3.3.5 ((i) ⇒ (vi)), there exist non-singular matrices $\mathbf{R}$, $\mathbf{S}$ such that $\mathbf{A} = \mathbf{R}\mathbf{diag}(\mathbf{I_a},\mathbf{0},\mathbf{0},\mathbf{0})\mathbf{S}$ and $\mathbf{B} = \mathbf{R}\mathbf{diag}(\mathbf{I_a},\mathbf{I_{b-a}},\mathbf{0},\mathbf{0})\mathbf{S}$. Since, $\mathbf{G}$ is a g-inverse of $\mathbf{B}$, we have

$$\mathbf{G} = \mathbf{S}^{-1}\begin{pmatrix}\mathbf{I_a} & \mathbf{0} & \mathbf{L_1}\\ \mathbf{0} & \mathbf{I_{b-a}} & \mathbf{L_2}\\ \mathbf{M_1} & \mathbf{M_2} & \mathbf{N}\end{pmatrix}\mathbf{R}^{-1}$$

for some matrices $\mathbf{L_1},\mathbf{L_2},\mathbf{M_1},\mathbf{M_2}$ and $\mathbf{N}$ of appropriate orders. Write

$$\mathbf{T} = \begin{pmatrix}\mathbf{I_a} & \mathbf{0} & \mathbf{0}\\ \mathbf{0} & \mathbf{I_{b-a}} & \mathbf{0}\\ \mathbf{M_1} & \mathbf{M_2} & \mathbf{I}\end{pmatrix} \text{ and } \mathbf{W} = \begin{pmatrix}\mathbf{I_a} & \mathbf{0} & \mathbf{L_1}\\ \mathbf{0} & \mathbf{I_{b-a}} & \mathbf{L_2}\\ \mathbf{0} & \mathbf{0} & \mathbf{I}\end{pmatrix}.$$

Clearly, $\mathbf{T}$ and $\mathbf{W}$ are non-singular matrices. Moreover, we have

$$\mathbf{G} = \mathbf{S}^{-1}\mathbf{T}\mathbf{diag}(\mathbf{I_a},\mathbf{I_{b-a}},\mathbf{N}-\mathbf{M_1}\mathbf{L_1}-\mathbf{M_2}\mathbf{L_2})\mathbf{W}\mathbf{R}^{-1}$$

and $\rho(\mathbf{N}-\mathbf{M_1}\mathbf{L_1}-\mathbf{M_2}\mathbf{L_2}) = g-b$. Let $\mathbf{X}^{-1}\mathbf{diag}(\mathbf{I_{g-b}},\mathbf{0})\mathbf{Y}^{-1}$ be a normal form of the matrix $\mathbf{N}-\mathbf{M_1}\mathbf{L_1}-\mathbf{M_2}\mathbf{L_2}$. Write, $\mathbf{P} = \mathbf{R}\mathbf{W}^{-1}\mathbf{diag}(\mathbf{I_b} : \mathbf{Y})$ and $\mathbf{Q} = \mathbf{diag}(\mathbf{I_b} : \mathbf{X})\mathbf{T}^{-1}\mathbf{S}$. Clearly, $\mathbf{A} = \mathbf{P}\mathbf{diag}(\mathbf{I_a},\ \mathbf{0},\ \mathbf{0},\ \mathbf{0})\mathbf{Q}$, $\mathbf{B} = \mathbf{P}\mathbf{diag}(\mathbf{I_a},\ \mathbf{I_{b-a}},\ \mathbf{0},\ \mathbf{0})\mathbf{Q}$ and $\mathbf{G} = \mathbf{Q}^{-1}\mathbf{diag}(\mathbf{I_a},\mathbf{I_{b-a}},\mathbf{I_{g-b}},\mathbf{0})\mathbf{P}^{-1}$. □

Theorem 3.3.18. *Let* $\mathbf{A}$ *and* $\mathbf{B}$ *be matrices of the same order such that* $\mathbf{A} <^- \mathbf{B}$. *Let* $\mathbf{G}$ *and* $\mathbf{H}$ *respectively be g-inverses of* $\mathbf{B}$ *and* $\mathbf{A}$ *such that* $\mathbf{G} <^- \mathbf{H}$. *Then* $\mathbf{H}$ *is a g-inverse of* $\mathbf{B}$.

Proof. In view of Theorem 3.3.17, there exist non-singular matrices $\mathbf{P}$ and $\mathbf{Q}$ such that $\mathbf{A} = \mathbf{P}\mathbf{diag}(\mathbf{I_a}, \mathbf{0}, \mathbf{0}, \mathbf{0})\mathbf{Q}$, $\mathbf{B} = \mathbf{P}\mathbf{diag}(\mathbf{I_a}, \mathbf{I_{b-a}}, \mathbf{0}, \mathbf{0})\mathbf{Q}$ and $\mathbf{G} = \mathbf{Q}^{-1}\mathbf{diag}(\mathbf{I_a}, \mathbf{I_{b-a}}, \mathbf{I_g}, \mathbf{0})\mathbf{P}^{-1}$, where $\rho(\mathbf{A}) = a, \rho(\mathbf{B}) = b$ and $\rho(\mathbf{G}) = g$. Let $\mathbf{V}$ be a g-inverse of $\mathbf{G}$ such that $\mathbf{GV} = \mathbf{HV}$ and $\mathbf{VG} = \mathbf{VH}$. Then $\mathbf{V}$ must be of the form

$$\mathbf{V} = \mathbf{P}\begin{pmatrix} \mathbf{I_a} & \mathbf{0} & \mathbf{0} & \mathbf{L_1} \\ \mathbf{0} & \mathbf{I_{b-a}} & \mathbf{0} & \mathbf{L_2} \\ \mathbf{0} & \mathbf{0} & \mathbf{I_{b-a}} & \mathbf{L_3} \\ \mathbf{M_1} & \mathbf{M_2} & \mathbf{M_3} & \mathbf{N} \end{pmatrix}\mathbf{Q}$$

for some matrices $\mathbf{L_1}, \mathbf{L_2}, \mathbf{L_3}, \mathbf{M_1}, \mathbf{M_2}, \mathbf{M_3}$ and $\mathbf{N}$ of appropriate orders. Proceeding as in Theorem 3.3.17, we can find non-singular matrices $\mathbf{R}, \mathbf{S}$ such that

$$\begin{aligned} &\mathbf{A} = \mathbf{R}\mathbf{diag}(\mathbf{I_a}, \mathbf{0}, \mathbf{0}, \mathbf{0}, \mathbf{0})\mathbf{S}, \mathbf{B} = \mathbf{R}\mathbf{diag}(\mathbf{I_a}, \mathbf{I_{b-a}}, \mathbf{0}, \mathbf{0}, \mathbf{0})\mathbf{S} \text{ and} \\ &\mathbf{G} = \mathbf{S}^{-1}\mathbf{diag}(\mathbf{I_a}, \mathbf{I_{b-a}}, \mathbf{I_{g-a}}, \mathbf{0}, \mathbf{0})\mathbf{R}^{-1} \text{ and} \\ &\mathbf{V} = \mathbf{S}^{-1}\mathbf{diag}(\mathbf{I_a}, \mathbf{I_{b-a}}, \mathbf{I_{g-b}}, \mathbf{I_{v-g}}, \mathbf{0})\mathbf{R}^{-1}, \end{aligned}$$

where $v = \rho(\mathbf{V})$. Since $\mathbf{H}$ is a g-inverse of $\mathbf{A}$, it must be of the form

$$\mathbf{H} = \mathbf{P}\begin{pmatrix} \mathbf{I_a} & \mathbf{D_{12}} & \mathbf{D_{13}} & \mathbf{D_{14}} & \mathbf{D_{15}} \\ \mathbf{D_{21}} & \mathbf{D_{22}} & \mathbf{D_{23}} & \mathbf{D_{24}} & \mathbf{D_{25}} \\ \mathbf{D_{31}} & \mathbf{D_{32}} & \mathbf{D_{33}} & \mathbf{D_{34}} & \mathbf{D_{35}} \\ \mathbf{D_{41}} & \mathbf{D_{42}} & \mathbf{D_{43}} & \mathbf{D_{44}} & \mathbf{D_{45}} \\ \mathbf{D_{51}} & \mathbf{D_{52}} & \mathbf{D_{53}} & \mathbf{D_{54}} & \mathbf{D_{55}} \end{pmatrix}\mathbf{Q}.$$

Since, $\mathbf{GV} = \mathbf{HV}$ and $\mathbf{VG} = \mathbf{VH}$, $\mathbf{D_{22}} = \mathbf{I_{b-a}}$, $\mathbf{D_{33}} = \mathbf{I_{g-b}}$, $\mathbf{D_{44}} = \mathbf{I_{v-g}}$ and $\mathbf{D_{ij}} = \mathbf{0}$ for all $i, j = 1, 2, 3, 4$ with $i \neq j$. Thus,

$$\mathbf{H} = \mathbf{P}\begin{pmatrix} \mathbf{I_a} & \mathbf{0} & \mathbf{0} & \mathbf{0} & \mathbf{D_{15}} \\ \mathbf{0} & \mathbf{I_{b-a}} & \mathbf{0} & \mathbf{0} & \mathbf{D_{25}} \\ \mathbf{D_{31}} & \mathbf{D_{32}} & \mathbf{I_{g-b}} & \mathbf{0} & \mathbf{D_{35}} \\ \mathbf{0} & \mathbf{0} & \mathbf{0} & \mathbf{I_{v-g}} & \mathbf{D_{45}} \\ \mathbf{0} & \mathbf{0} & \mathbf{0} & \mathbf{0} & \mathbf{D_{55}} \end{pmatrix}\mathbf{Q}.$$

It is now easy to check that $\mathbf{H}$ is a g-inverse of $\mathbf{B}$. □

We have seen in Theorem 3.3.5, that for any two matrices $\mathbf{A_1}$ and $\mathbf{A_2}$, $\mathbf{A_i} <^- (\mathbf{A_1} + \mathbf{A_2})$ if and only if $\mathbf{A_1}$ and $\mathbf{A_2}$ are disjoint matrices. We shall now extend this result for several matrices and their sum. We shall see in later chapters that this is a prelude to Fisher-Cochran type theorems on distribution of quadratic forms in normal variables.

Theorem 3.3.19. *Let* $\mathbf{A_1}, \mathbf{A_2}, \ldots, \mathbf{A_k}$ *be matrices of the same order. Write* $\mathbf{A} = \sum_{i=1}^{k} \mathbf{A_i}$. *Then the following are equivalent.*

(i) $\rho(\mathbf{A}) = \sum_{i=1}^{k} \rho(\mathbf{A_i})$
(ii) *There exist nonsingular matrices* $\mathbf{P}$ *and* $\mathbf{Q}$ *such that* $\mathbf{A_i} = \mathbf{P}\mathbf{diag}(\mathbf{0}\ldots,\mathbf{I},\mathbf{0}\ldots,\mathbf{0})\mathbf{Q}$, *where* $\mathbf{diag}(\mathbf{0},\ldots,\mathbf{I},\mathbf{0},\ldots,\mathbf{0})$ *has at least* $k+1$ *diagonal blocks and* $\mathbf{I}$ *occurs in the* i^{th} *block for* $i = 1,\ldots,k$
(iii) $\rho(\mathbf{A} - \mathbf{A_i}) = \rho(\mathbf{A}) - \rho(\mathbf{A_i})$, $i = 1,\ldots,k$ *and*
(iv) $\mathbf{A_i} <^- \mathbf{A}$, $i = 1,\ldots,k$.

Proof. (i) $\Rightarrow$ (ii)
Let $(\mathbf{P_i}, \mathbf{Q_i})$ be a rank factorization of $\mathbf{A_i}$, for each $i = 1,\ldots,k$. Since (i) holds, the columns of $(\mathbf{P_1} : \ldots : \mathbf{P_k})$ are linearly independent and so are the columns of $(\mathbf{Q^t} : \mathbf{Q_2^t} : \ldots : \mathbf{Q_k^t})$. Extend $(\mathbf{P_1} : \ldots : \mathbf{P_k})$ and $(\mathbf{Q_1^t} : \ldots : \mathbf{Q_k^t})$ to nonsingular matrices $\mathbf{P}$ and $\mathbf{Q^t}$ respectively. Now it is easy to see that

$$\mathbf{A_i} = \mathbf{P}\,\mathbf{diag}(\mathbf{0},\ldots,\mathbf{I},\mathbf{0}\ldots,\mathbf{0})\mathbf{Q}.$$

(ii) $\Rightarrow$ (iii) is trivial.
(iii) $\Rightarrow$ (iv) follows from Theorem 3.3.5.
(iv) $\Rightarrow$ (i)
Since (iv) holds, every $\mathbf{A^-} \in \{\mathbf{A^-}\}$ is a g-inverse of $\mathbf{A_i}$ for each i. Hence $\mathbf{A_iA^-}$ is idempotent and $\rho(\mathbf{A_iA^-}) = \rho(\mathbf{A_i})$ for all i. Now

$$\sum_{i=1}^{k}\rho(\mathbf{A_i}) = \sum_{i=1}^{k}\rho(\mathbf{A_iA^-}) = \sum_{i=1}^{k}\mathrm{tr}(\mathbf{A_iA^-}) = \mathrm{tr}\left(\left(\sum_{i=1}^{k}\mathbf{A_i}\right)\mathbf{A^-}\right)$$
$$= \mathrm{tr}(\mathbf{AA^-}) = \rho(\mathbf{A}).$$

□

Remark 3.3.20. It is interesting to note that $\mathbf{A_i}$ and $\mathbf{A} - \mathbf{A_i}$ are disjoint for all i implies that $\mathcal{C}(\mathbf{A_1}) + \mathcal{C}(\mathbf{A_2}) + \ldots + \mathcal{C}(\mathbf{A_k})$ and $\mathcal{C}(\mathbf{A_1^t}) + \ldots + \mathcal{C}(\mathbf{A_k^t})$ are actually direct sums.

Remark 3.3.21. Consider $\mathbf{A_1},\ldots,\mathbf{A_k}$ and $\mathbf{A}$ as in Theorem 3.3.19. If $\mathbf{A_i} <^- \mathbf{A}$, $i = 1,\ldots,k$, then $\mathbf{A_i} <^- \mathbf{A_{j_1}} + \ldots + \mathbf{A_{j_r}}$ for any sub-permutation $(j_1,\ldots,j_r)$ of $(1,\ldots,k)$ such that $i \in (j_1,\ldots,j_r)$.

3.4 Matrices above/below a given matrix under the minus order

Let one of the matrices $\mathbf{A}$ and $\mathbf{B}$ be given. In this section, we characterize the class of matrices for the other such that $\mathbf{A} <^- \mathbf{B}$ and obtain some of its formulations that are useful in different situations. However, before we actually do so, we show that every complement of $\mathcal{C}(\mathbf{A})$ is $\mathcal{C}(\mathbf{I} - \mathbf{AG})$ for some g-inverse $\mathbf{G}$ of $\mathbf{A}$.

Lemma 3.4.1. *Let* $\mathbf{A}$ *be a matrix of order* $m \times n$. *Then* $\mathcal{S}$ *is a complement of* $\mathcal{C}(\mathbf{A})$ *if and only if* $\mathcal{S} = \mathcal{C}(\mathbf{I} - \mathbf{AG})$ *for some g-inverse* $\mathbf{G}$ *of* $\mathbf{A}$.

Proof. 'If' part

We first notice that for any g-inverse $\mathbf{G}$ of $\mathbf{A}$, $\mathcal{C}(\mathbf{A}) \cap \mathcal{C}(\mathbf{I} - \mathbf{AG}) = \{\mathbf{0}\}$. Moreover, $\rho(\mathbf{I} - \mathbf{AG}) = \mathbf{m} - \rho(\mathbf{A}) = \mathbf{m} - \mathbf{d}(\mathcal{C}(\mathbf{A}))$. Thus, $\mathcal{C}(\mathbf{I} - \mathbf{AG})$ is a complement of $\mathcal{C}(\mathbf{A})$.

'Only if' part

Let $(\mathbf{P}_1, \mathbf{Q}_1)$ be a rank factorization of $\mathbf{A}$. Let the columns of the matrix $\mathbf{R}$ form a basis of $\mathcal{S}$ and $\mathbf{T}$ be a matrix such that $\begin{pmatrix} \mathbf{Q}_1 \\ \cdots \\ \mathbf{T} \end{pmatrix}$ is non-singular.

Then $\mathbf{A} = (\mathbf{P}_1 : \mathbf{R}) \begin{pmatrix} \mathbf{I} & \mathbf{0} \\ \mathbf{0} & \mathbf{0} \end{pmatrix} \begin{pmatrix} \mathbf{Q}_1 \\ \cdots \\ \mathbf{T} \end{pmatrix}$ and $\mathbf{G} = \begin{pmatrix} \mathbf{Q}_1 \\ \cdots \\ \mathbf{T} \end{pmatrix}^{-1} (\mathbf{P}_1 : \mathbf{R})^{-1}$ is a g-inverse of $\mathbf{A}$. Hence.

$$\mathbf{I} - \mathbf{AG} = (\mathbf{P}_1 : \mathbf{R}) \begin{pmatrix} \mathbf{0} & \mathbf{0} \\ \mathbf{0} & \mathbf{I} \end{pmatrix} (\mathbf{P}_1 : \mathbf{R})^{-1} \text{ and}$$

$$\mathcal{C}(\mathbf{I} - \mathbf{AG}) = \mathcal{C}\left((\mathbf{P}_1 : \mathbf{R}) \begin{pmatrix} \mathbf{0} & \mathbf{0} \\ \mathbf{0} & \mathbf{I} \end{pmatrix} \right) = \mathcal{C}(\mathbf{R}) = \mathcal{S}\ .$$

□

Theorem 3.4.2. *Let* $\mathbf{A}$ *be a matrix of order* $m \times n$ *such that* $\mathbf{0} < \rho(\mathbf{A}) < \min\{m, n\}$. *Let* $(\mathbf{P}_1, \mathbf{Q}_1)$ *be a rank factorization of* $\mathbf{A}$. *Then the class of all matrices* $\mathbf{B}$ *of order* $m \times n$ *such that* $\mathbf{A} <^- \mathbf{B}$ *is given by*

$$\{\mathbf{B} = \mathbf{A} \oplus \mathbf{C}, \mathbf{C} \textit{ arbitrary}\}$$

or equivalently by

$$\mathbf{B} = \mathbf{A} \text{ or } \mathbf{B} = \mathbf{P}_1\mathbf{Q}_1 + \mathbf{P}_2\mathbf{Q}_2,$$

where $\mathbf{P} = (\mathbf{P}_1 : \mathbf{P}_2)$ *and* $\mathbf{Q}^t = (\mathbf{Q}_1^t : \mathbf{Q}_2^t)$ *are arbitrary full rank extensions of* $\mathbf{P}_1$ *and* $\mathbf{Q}_1$ *respectively and involved partitions are conformable for matrix multiplication.*

Proof. This theorem is a restatement of Theorem 3.3.5(vii). □

Theorem 3.4.3. *Let* $\mathbf{A}$ *and* $\mathbf{B}$ *be matrices of the same order. Then*

$$\mathbf{A} <^{-} \mathbf{B} \text{ if and only if } \mathbf{B} = \mathbf{A} + (\mathbf{I} - \mathbf{A}\mathbf{A}^{-})\mathbf{W}(\mathbf{I} - \mathbf{A}^{-}\mathbf{A})$$

for some matrix $\mathbf{W}$ *and for some g-inverse* $\mathbf{A}^{-}$ *of* $\mathbf{A}$.

Proof. Proof follows from Theorem 3.4.2 and Lemma 3.4.1. □

Let $\mathbf{C}$ and $\mathbf{B}$ be matrices of the same order. We now consider a special form for the first matrix $\mathbf{C}$ and explore when $\mathbf{C} <^{-} \mathbf{B}$. This result will be useful later and is also of independent interest.

Theorem 3.4.4. *Let* $\mathbf{A}$ *be a non-singular matrix. Then*

$$\begin{pmatrix} \mathbf{A} & \mathbf{0} \\ \mathbf{0} & \mathbf{0} \end{pmatrix} <^{-} \mathbf{B} \textit{ if and only if } \mathbf{B} = \begin{pmatrix} \mathbf{A} & \mathbf{0} \\ \mathbf{0} & \mathbf{0} \end{pmatrix} + \begin{pmatrix} -\mathbf{AL} \\ \mathbf{0} \end{pmatrix} \mathbf{Z}(-\mathbf{MA} \ \mathbf{I})$$

for some matrices $\mathbf{L}, \mathbf{M}$ *and* $\mathbf{Z}$ *of appropriate orders, where the block matrix* $\mathrm{diag}(\mathbf{A}, \mathbf{0})$ *and the matrix* $\mathbf{B}$ *have the same order.*

Proof. Let $\mathbf{C} = \mathbf{diag}(\mathbf{A}\ \mathbf{0})$. In view of Theorem 3.4.3, $\mathbf{C} <^{-} \mathbf{B}$ if and only if $\mathbf{B} = \mathbf{C} + (\mathbf{I} - \mathbf{C}\mathbf{C}^{-})\mathbf{W}(\mathbf{I} - \mathbf{C}^{-}\mathbf{C})$ for some g-inverse $\mathbf{C}^{-}$ and some matrix $\mathbf{W}$. Since $\mathbf{A}$ is non-singular, every g-inverse of $\mathbf{C}$ is of the form $\mathbf{C}^{-} = \begin{pmatrix} \mathbf{A}^{-1} & \mathbf{L} \\ \mathbf{M} & \mathbf{N} \end{pmatrix}$, where $\mathbf{L}, \mathbf{M}, \mathbf{N}$ are arbitrary. Now,

$$\mathbf{I} - \mathbf{C}\mathbf{C}^{-} = \begin{pmatrix} \mathbf{0} & -\mathbf{AL} \\ \mathbf{0} & \mathbf{I} \end{pmatrix} \text{ and } \mathbf{I} - \mathbf{C}^{-}\mathbf{C} = \begin{pmatrix} \mathbf{0} & \mathbf{0} \\ -\mathbf{MA} & \mathbf{I} \end{pmatrix}.$$

Hence

$$\begin{aligned}(\mathbf{I} - \mathbf{C}\mathbf{C}^{-})\mathbf{W}(\mathbf{I} - \mathbf{C}^{-}\mathbf{C}) &= \begin{pmatrix} \mathbf{0} & -\mathbf{AL} \\ \mathbf{0} & \mathbf{I} \end{pmatrix} \begin{pmatrix} \mathbf{W}_{11} & \mathbf{W}_{12} \\ \mathbf{W}_{21} & \mathbf{W}_{22} \end{pmatrix} \begin{pmatrix} \mathbf{0} & \mathbf{0} \\ -\mathbf{MA} & \mathbf{I} \end{pmatrix}, \\ &= \begin{pmatrix} -\mathbf{AL} \\ \mathbf{I} \end{pmatrix} \mathbf{Z}(-\mathbf{MA}\ \mathbf{I})\end{aligned}$$

where $\mathbf{W}$ is partitioned as $\begin{pmatrix} \mathbf{W}_{11} & \mathbf{W}_{12} \\ \mathbf{W}_{21} & \mathbf{W}_{22} \end{pmatrix}$ conformable for the above matrix multiplication and $\mathbf{Z} = \mathbf{W}_{22}$. □

Remark 3.4.5. In Theorem 3.4.4, $\mathbf{B}$ is non-singular if and only if $\mathbf{Z}$ is non-singular.

We now ask the following question:
Given a matrix $\mathbf{B}$, what are all matrices $\mathbf{A}$ such that $\mathbf{A} <^{-} \mathbf{B}$?

The answer is actually contained in Theorem 3.3.5(v) but we restate for the sake of completeness. We note that if $\mathbf{B} = \mathbf{0}$, and $\mathbf{A} <^{-} \mathbf{B}$, then $\mathbf{A} = \mathbf{0}$. So, we shall take matrix $\mathbf{B}$ to be non-null.

Theorem 3.4.6. *Let* $\mathbf{B}$ *be a non-null matrix with a rank factorization* $(\mathbf{P}, \mathbf{Q})$. *Then the class of all matrices* $\mathbf{A}$ *such that* $\mathbf{A} <^{-} \mathbf{B}$ *is given by*

$$\{\mathbf{PTQ} : \mathbf{T} \textit{ idempotent}\}.$$

Theorem 3.4.7. *Let* $\mathbf{A}$ and $\mathbf{B}$ *be matrices of the same order and* $\mathbf{B}$ *be non-null. Then* $\mathbf{A} <^{-} \mathbf{B}$ *if and only if* $\mathbf{A} = \mathbf{B} - \mathbf{BR}(\mathbf{SBR})_r^{-}\mathbf{SB}$ *for some matrices* $\mathbf{S}$ *and* $\mathbf{R}$ *such that* $\mathbf{SBR}$ *is defined.*

Proof. 'If' part

Let $(\mathbf{P}, \mathbf{Q})$ be a rank factorization of $\mathbf{B}$. Then

$$\mathbf{A} = \mathbf{B} - \mathbf{BR}(\mathbf{SBR})_r^{-}\mathbf{SB} \Rightarrow \mathbf{A} = \mathbf{P}(\mathbf{I} - \mathbf{QR}(\mathbf{SPQR})_r^{-}\mathbf{SP})\mathbf{Q}\ .$$

It is easy to check that $\mathbf{QR}(\mathbf{SPQR})_r^{-}\mathbf{SP}$ and $\mathbf{I} - \mathbf{QR}(\mathbf{SPQR})_r^{-}\mathbf{SP}$ are idempotent. So, by Theorem 3.4.6, $\mathbf{A} <^{-} \mathbf{B}$.

'Only if' part

Let $\mathbf{A} <^{-} \mathbf{B}$ and $(\mathbf{P}, \mathbf{Q})$ be a rank factorization of $\mathbf{B}$. By Theorem 3.4.6 $\mathbf{A} = \mathbf{PTQ}$, where $\mathbf{T}$ is idempotent. Write

$$\begin{aligned}\mathbf{T} &= \mathbf{I} - (\mathbf{I} - \mathbf{T})\\ &= \mathbf{I} - \mathbf{QQ}_\mathbf{R}^{-1}(\mathbf{I} - \mathbf{T})((\mathbf{I} - \mathbf{T})\mathbf{P}_\mathbf{L}^{-1}\mathbf{PQQ}_\mathbf{R}^{-1}(\mathbf{I} - \mathbf{T}))_r^{-}(\mathbf{I} - \mathbf{T})\mathbf{P}_\mathbf{L}^{-1}\mathbf{P}.\end{aligned}$$

Let $\mathbf{R} = \mathbf{Q}_\mathbf{R}^{-1}(\mathbf{I} - \mathbf{T})$ and $\mathbf{S} = (\mathbf{I} - \mathbf{T})\mathbf{P}_\mathbf{L}^{-1}$. Then $\mathbf{A} = \mathbf{B} - \mathbf{BR}(\mathbf{SBR})_r^{-}\mathbf{SB}$. □

Let $\mathbf{B}$ be a given matrix. We now show that each matrix $\mathbf{A}$ of the form $\mathbf{B} - \mathbf{BR}(\mathbf{SBR})_r^{-}\mathbf{SB}$ for some matrices $\mathbf{S}$ and $\mathbf{R}$ is the unique matrix below $\mathbf{B}$ under the minus order with column and row spaces specified in an interesting way.

Theorem 3.4.8. *Let* $\mathbf{B}, \mathbf{R}$ *and* $\mathbf{S}$ *be matrices such that the product* $\mathbf{SBR}$ *is defined. Then there exists a matrix* $\mathbf{A}$ *satisfying '*$\mathcal{C}(\mathbf{A}) = \mathcal{N}(\mathbf{S}) \cap \mathcal{C}(\mathbf{B})$, $\mathcal{C}(\mathbf{A}^t) = \mathcal{N}(\mathbf{R}^t) \cap \mathcal{C}(\mathbf{B}^t)$ *and* $\mathbf{A} <^{-} \mathbf{B}$*' if and only if* $\rho(\mathbf{SB}) = \rho(\mathbf{SBR}) = \rho(\mathbf{BR})$. *If such a matrix* $\mathbf{A}$ *exists, then it is unique and is is of the form* $\mathbf{A} = \mathbf{B} - \mathbf{BR}(\mathbf{SBR})^{-}\mathbf{SB}$.

Proof. 'If' part

Let $\rho(\mathbf{SB}) = \rho(\mathbf{SBR}) = \rho(\mathbf{BR})$ hold and let $\mathbf{A} = \mathbf{B} - \mathbf{BR}(\mathbf{SBR})^{-}\mathbf{SB}$. Clearly, $\mathcal{C}(\mathbf{A}) \subseteq \mathcal{C}(\mathbf{B})$ and $\mathcal{C}(\mathbf{A}^t) \subseteq \mathcal{C}(\mathbf{B}^t)$. As $\mathcal{C}(\mathbf{SB}) = \mathcal{C}(\mathbf{SBR})$, we

have $\mathbf{SA} = \mathbf{SB} - \mathbf{SBR}(\mathbf{SBR})^-\mathbf{SB} = \mathbf{0}$. So, $\mathcal{C}(\mathbf{A}) \subseteq \mathcal{N}(\mathbf{S}) \cap \mathcal{C}(\mathbf{B})$. Let $\mathbf{x} \in \mathcal{N}(\mathbf{S}) \cap \mathcal{C}(\mathbf{B})$. Then $\mathbf{x} = \mathbf{Bu}$ and $\mathbf{Sx} = \mathbf{SBu} = \mathbf{0}$. Clearly, $\mathbf{Au} = \mathbf{Bu}$. Therefore, $\mathbf{x} \in \mathcal{C}(\mathbf{AS}) \subseteq \mathcal{C}(\mathbf{AS})$. Thus, $\mathcal{C}(\mathbf{A}) = \mathcal{N}(\mathbf{S}) \cap \mathcal{C}(\mathbf{B})$.
Similarly, we can prove that $\mathcal{C}(\mathbf{A^t}) = \mathcal{N}(\mathbf{R^t}) \cap \mathcal{C}(\mathbf{B^t})$.

Further, $\rho(\mathbf{SB}) = \rho(\mathbf{SBR}) = \rho(\mathbf{BR}) \Rightarrow \mathbf{BR}(\mathbf{SBR})^-\mathbf{SB}$ is invariant under choices of $(\mathbf{SBR})^-$. Hence by Theorem 3.4.7, $\mathbf{A} <^- \mathbf{B}$.
'Only if' part
Let $\mathbf{x} \in \mathcal{N}(\mathbf{R^tB^t})$. Then $\mathbf{B^tx} \in \mathcal{N}(\mathbf{R^t}) \cap \mathcal{C}(\mathbf{B^t}) = \mathcal{C}(\mathbf{A^t})$. We can write $\mathbf{B^tx} = \mathbf{A^tx} + (\mathbf{B} - \mathbf{A})^\mathbf{t}\mathbf{x}$ and also as $\mathbf{B^tx} = \mathbf{A^tu}$ for some vector $\mathbf{u}$, as $\mathbf{B^tx} \in \mathcal{C}(\mathbf{A^t})$. However, $\mathcal{C}(\mathbf{A^t}) \cap \mathcal{C}(\mathbf{B} - \mathbf{A})^\mathbf{t} = \{\mathbf{0}\}$, since $\mathbf{A} <^- \mathbf{B}$, we have $(\mathbf{B} - \mathbf{A})^\mathbf{t}\mathbf{x} = \mathbf{0}$. Therefore, $\mathcal{N}(\mathbf{R^tB^t}) \subseteq \mathcal{N}(\mathbf{B} - \mathbf{A})^\mathbf{t}$ or equivalently by Remark 3.2.8, $\mathcal{C}(\mathbf{B} - \mathbf{A}) \subseteq \mathcal{C}(\mathbf{BR})$.

Similarly, we can prove $\mathcal{C}(\mathbf{B} - \mathbf{A})^\mathbf{t} \subseteq \mathcal{C}(\mathbf{B^tS^t})$. So, $\mathbf{B} - \mathbf{A} = \mathbf{BRCSB}$ for some matrix $\mathbf{C}$. Since $\mathbf{SA} = \mathbf{0}$, we have $\mathbf{SB} = \mathbf{SBRCSB}$ or $\mathbf{C}$ is a g-inverse of $\mathbf{SBR}$, further, $\rho(\mathbf{SB}) = \rho(\mathbf{SBR})$ and $\rho(\mathbf{SBR}) = \rho(\mathbf{BR})$. Thus, $\mathbf{A} = \mathbf{B} - \mathbf{BRCSB} = \mathbf{B} - \mathbf{BR}(\mathbf{SBR})^-\mathbf{SB}$. Notice that $\mathbf{BR}(\mathbf{SBR})^-\mathbf{SB}$ is invariant under choices of $(\mathbf{SBR})^-$, it follows $\mathbf{A}$ is unique. □

Remark 3.4.9. Consider the setup of Theorem 3.4.8. In the 'only if' part we have shown that if $\mathcal{C}(\mathbf{A}) = \mathcal{N}(\mathbf{S}) \cap \mathcal{C}(\mathbf{B})$, $\mathcal{C}(\mathbf{A^t}) = \mathcal{N}(\mathbf{R^t}) \cap \mathcal{C}(\mathbf{B^t})$ and $\mathbf{A} <^- \mathbf{B}$, then $\mathcal{C}(\mathbf{B}-\mathbf{A}) \subseteq \mathcal{C}(\mathbf{BR})$ and $\mathcal{C}(\mathbf{B}-\mathbf{A})^\mathbf{t} \subseteq \mathcal{C}(\mathbf{B^tS^t})$. In fact, under this hypothesis one can show $\mathcal{C}(\mathbf{B}-\mathbf{A}) = \mathcal{C}(\mathbf{BR})$ and $\mathcal{C}(\mathbf{B}-\mathbf{A})^\mathbf{t} = \mathcal{C}(\mathbf{B^tS^t})$.

3.5 Subclass of g-inverses $\mathbf{A}^-$ of $\mathbf{A}$ such that $\mathbf{A}^-\mathbf{A} = \mathbf{A}^-\mathbf{B}$ and $\mathbf{AA}^- = \mathbf{BA}^-$ when $\mathbf{A} <^- \mathbf{B}$

Let $\mathbf{A}$ and $\mathbf{B}$ be matrices of the same order such that $\mathbf{A} <^- \mathbf{B}$. In this section we shall characterize the class of all g-inverses $\mathbf{A}^-$ of $\mathbf{A}$ such that $\mathbf{AA}^- = \mathbf{BA}^-$ and $\mathbf{A}^-\mathbf{A} = \mathbf{A}^-\mathbf{B}$. Let $\mathbf{A}^-$ be one such g-inverse, so that $\mathbf{AA}^- = \mathbf{BA}^-$ and $\mathbf{A}^-\mathbf{A} = \mathbf{A}^-\mathbf{B}$. We shall show that there exists a g-inverse $\mathbf{B}^-$ of $\mathbf{B}$ such that $\mathbf{AA}^- = \mathbf{AB}^-$ and $\mathbf{A}^-\mathbf{A} = \mathbf{B}^-\mathbf{A}$. We shall also characterize the class of all such $\mathbf{B}^-$ for which $\mathbf{AA}^- = \mathbf{AB}^-$ and $\mathbf{A}^-\mathbf{A} = \mathbf{B}^-\mathbf{A}$ holds. Let

$$\{\mathbf{A}^-\}_\mathbf{B} = \{\mathbf{G} : \mathbf{AGA} = \mathbf{A}, \mathbf{AG} = \mathbf{BG}, \mathbf{GA} = \mathbf{GB}\}.$$

We shall obtain an explicit representation of $\{\mathbf{A}^-\}_\mathbf{B}$. Let $\mathbf{A} <^- \mathbf{B}$. In view of Theorem 3.3.16, there exist projectors $\mathbf{P}$ and $\mathbf{Q}$ such that $\mathbf{A} = \mathbf{PB} = \mathbf{BQ}$. We shall also characterize the classes of all projectors $\mathbf{P}$ and $\mathbf{Q}$ such that $\mathbf{A} = \mathbf{PB} = \mathbf{BQ}$ when $\mathbf{A} <^- \mathbf{B}$. We begin with the following lemma, which is also of independent interest.

Lemma 3.5.1. *Let* $\mathbf{A} <^- \mathbf{B}$. *Then* $\mathbf{C}$ *is a reflexive g-inverse of* $\mathbf{B} - \mathbf{A}$ *such that* $\mathbf{AC} = \mathbf{0}$ *and* $\mathbf{CA} = \mathbf{0}$ *if and only if* $\mathbf{C} = \mathbf{B}^-(\mathbf{B}-\mathbf{A})\mathbf{B}^-$ *for some g-inverse* $\mathbf{B}^-$ *of* $\mathbf{B}$.

Proof. 'If' part
Let $\mathbf{B}^-$ be an arbitrary g-inverse of $\mathbf{B}$. Write $\mathbf{C} = \mathbf{B}^-(\mathbf{B} - \mathbf{A})\mathbf{B}^-$. Then $\mathbf{CA} = \mathbf{B}^-(\mathbf{B} - \mathbf{A})\mathbf{B}^-\mathbf{A} = \mathbf{B}^-(\mathbf{A} - \mathbf{A}) = \mathbf{0}$. Similarly, $\mathbf{AC} = \mathbf{0}$. Also, $(\mathbf{B} - \mathbf{A})\mathbf{C}(\mathbf{B} - \mathbf{A}) = (\mathbf{B} - \mathbf{A})\mathbf{B}^-(\mathbf{B} - \mathbf{A})\mathbf{B}^-(\mathbf{B} - \mathbf{A}) = \mathbf{B} - \mathbf{A}$, since $(\mathbf{B} - \mathbf{A}) <^- \mathbf{B}$. So, $\rho(\mathbf{C}) \geq \rho(\mathbf{B} - \mathbf{A})$. On the other hand $\rho(\mathbf{C}) = \rho(\mathbf{B}^-(\mathbf{B} - \mathbf{A})\mathbf{B}^-) \leq \rho(\mathbf{B} - \mathbf{A})$. Hence $\rho(\mathbf{C}) = \rho(\mathbf{B} - \mathbf{A})$. Therefore, $\mathbf{C}$ is a reflexive g-inverse of $\mathbf{B} - \mathbf{A}$.
'Only if' part

Since, $\mathbf{A} <^- \mathbf{B}$, there exist non-singular matrices $\mathbf{R}$ and $\mathbf{S}$ such that $\mathbf{A} = \mathbf{R}\mathbf{diag}(\mathbf{I_a},\ \mathbf{0},\ \mathbf{0})\mathbf{S}$ and $\mathbf{B} = \mathbf{R}\mathbf{diag}(\mathbf{I_a},\ \mathbf{I_{b-a}},\ \mathbf{0})\mathbf{S}$, where $a = \rho(\mathbf{A})$ and $b = \rho(\mathbf{B})$. It is easy to check that every reflexive g-inverse $\mathbf{C}$ of $\mathbf{B} - \mathbf{A}$ such that $\mathbf{AC} = \mathbf{0}$ and $\mathbf{CA} = \mathbf{0}$ is of the form

$$\mathbf{S}^{-1}\begin{pmatrix}\mathbf{0} & \mathbf{0} & \mathbf{0}\\ \mathbf{0} & \mathbf{I_{b-a}} & \mathbf{D_{23}}\\ \mathbf{0} & \mathbf{D_{32}} & \mathbf{D_{32}D_{23}}\end{pmatrix}\mathbf{R}^{-1},$$

where $\mathbf{D_{23}}$ and $\mathbf{D_{32}}$ are arbitrary. Choose and fix $\mathbf{D_{23}}$ and $\mathbf{D_{32}}$. It is easy to check that

$$\mathbf{T} = \mathbf{S}^{-1}\begin{pmatrix}\mathbf{I_a} & \mathbf{0} & \mathbf{0}\\ \mathbf{0} & \mathbf{I_{b-a}} & \mathbf{D_{23}}\\ \mathbf{0} & \mathbf{D_{32}} & \mathbf{T_{33}}\end{pmatrix}\mathbf{R}^{-1},$$

is a g-inverse of $\mathbf{B}$, where $\mathbf{T_{33}}$ is arbitrary and $\mathbf{C} = \mathbf{T}(\mathbf{B} - \mathbf{A})\mathbf{T}$ for a fixed choice of $\mathbf{T}$. □

Let $\mathbf{A} <^- \mathbf{B}$. In the following Lemma, we obtain a class of g-inverses of $\mathbf{A}$ which is a subset of $\{\mathbf{A}^-\}_{\mathbf{B}}$.

Lemma 3.5.2. *Let* $\mathbf{A}$ and $\mathbf{B}$ *be matrices having the same order such that* $\mathbf{A} <^- \mathbf{B}$ *and* $\mathbf{B}^-$ *be an arbitrary g-inverse of* $\mathbf{B}$. *Let* $\mathbf{C} = \mathbf{B}^-(\mathbf{B}-\mathbf{A})\mathbf{B}^-$ *be as in Lemma 3.5.1 and* $\mathbf{D_T} = \mathbf{B}^- - \mathbf{CTC}$, *where* $\mathbf{T}$ *is an arbitrary matrix of suitable order such that* $\mathbf{CTC}$ *is defined. Then the following hold:*

(i) $\mathbf{AD_T} = \mathbf{AB}^-,\ \mathbf{D_TA} = \mathbf{B}^-\mathbf{A}$, *for each* $\mathbf{T}$,
(ii) $\mathbf{AD_TA} = \mathbf{AD_TB} = \mathbf{BD_TA} = \mathbf{A}$, *for each* $\mathbf{T}$ *and*

(iii) $\mathbf{AD_T} = \mathbf{BD_T}$ *if and only if* $\mathbf{T}$ *is a g-inverse of* $\mathbf{C}$. *Further, if* $\mathbf{AD_T} = \mathbf{BD_T}$, *then* $\mathbf{D_TA} = \mathbf{D_TB}$.

Proof. Statement (i) follows from Lemma 3.5.1. Since $\mathbf{A} <^- \mathbf{B}$, (ii) is a simple consequence of (i).
(iii) $\mathbf{AD_T} = \mathbf{BD_T}$
$\Leftrightarrow \mathbf{A}(\mathbf{B}^- - \mathbf{CTC}) = \mathbf{B}(\mathbf{B}^- - \mathbf{CTC})$
$\Leftrightarrow \mathbf{AB}^- = \mathbf{B}(\mathbf{B}^- - (\mathbf{B}^-(\mathbf{B}-\mathbf{A})\mathbf{B}^-)\mathbf{TC})$, since $\mathbf{AC} = \mathbf{0}$ and $\mathbf{C} = \mathbf{B}^-(\mathbf{B}-\mathbf{A})\mathbf{B}^-$
$\Leftrightarrow \mathbf{AB}^- = \mathbf{BB}^-\mathbf{B}(\mathbf{B}^-(\mathbf{B}-\mathbf{A})\mathbf{B}^-)\mathbf{TC}$
$\qquad = \mathbf{BB}^- - (\mathbf{B}-\mathbf{A})\mathbf{B}^-\mathbf{TC}$
$\Leftrightarrow \quad (\mathbf{B}-\mathbf{A})\mathbf{B}^- = (\mathbf{B}-\mathbf{A})\mathbf{B}^-\mathbf{TC}$
$\Leftrightarrow \quad \mathbf{C} = \mathbf{B}^-(\mathbf{B}-\mathbf{A})\mathbf{B}^- = \mathbf{B}(\mathbf{B}-\mathbf{A})\mathbf{B}^-\mathbf{TC}$
$\Leftrightarrow \quad \mathbf{C} = \mathbf{CTC}$
$\Leftrightarrow \quad \mathbf{T}$ is a g-inverse of $\mathbf{C}$.

The remaining part of the statement is easy to check. □

Remark 3.5.3. Notice that Lemma 3.5.2(ii) and (iii) together imply that $\{\mathbf{B}^- - \mathbf{B}^-(\mathbf{B}-\mathbf{A})\mathbf{B}^- : \mathbf{B} \in \{\mathbf{B}^-\}\}$ is a subset of $\{\mathbf{A}^-\}_\mathbf{B}$. In the next theorem, we show that this subset is indeed the set $\{\mathbf{A}^-\}_\mathbf{B}$.

Theorem 3.5.4. *Let* $\mathbf{A}$ and $\mathbf{B}$ *be matrices of the same order such that* $\mathbf{A} <^- \mathbf{B}$. *Then*

(i) $\{\mathbf{A}^-\}_\mathbf{B} = \{\mathbf{B}^- - \mathbf{B}^-(\mathbf{B}-\mathbf{A})\mathbf{B}^- : \mathbf{B}^- \in \{\mathbf{B}^-\}\}$.
(ii) $\{\mathbf{A}_\mathbf{r}^-\}_\mathbf{B} = \{\mathbf{B}^-\mathbf{AB}^- : \mathbf{B}^- \in \{\mathbf{B}^-\}\}$.
(iii) $\{\mathbf{A}_\mathbf{r}^-\}_\mathbf{B} = \{\mathbf{B}_\mathbf{r}^-\mathbf{AB}_\mathbf{r}^- : \mathbf{B}_\mathbf{r}^- \in \{\mathbf{B}_\mathbf{r}^-\}\}$.

Proof. (i) We already proved that $\{\mathbf{B}^- - \mathbf{B}^-(\mathbf{B}-\mathbf{A})\mathbf{B}^- : \mathbf{B}^- \in \{\mathbf{B}^-\}\}$ is a subset of $\{\mathbf{A}^-\}_\mathbf{B}$ in Lemma 3.5.2. Let $\mathbf{A}^- \in \{\mathbf{A}^-\}_\mathbf{B}$. It is clear that $\mathbf{BA}^-\mathbf{B} = \mathbf{A}$. For any $\mathbf{B}^- \in \{\mathbf{B}^-\}$, we define $\mathbf{C} = \mathbf{B}^-(\mathbf{B}-\mathbf{A})\mathbf{B}^-$ and $\mathbf{E} = \mathbf{A}^- + \mathbf{C}$. It is easy to check that $\mathbf{E} \in \{\mathbf{B}^-\}$. By Lemma 3.5.1, $\mathbf{C}$ is a reflexive g-inverse of $\mathbf{B}-\mathbf{A}$ and $\mathbf{AC} = \mathbf{0}$, $\mathbf{CA} = \mathbf{0}$. Moreover, $\mathbf{A}^-\mathbf{B} = \mathbf{A}^-\mathbf{A}$ and $\mathbf{BA}^- = \mathbf{AA}^-$. Therefore, $\mathbf{E}(\mathbf{B}-\mathbf{A})\mathbf{E} = \mathbf{C}(\mathbf{B}-\mathbf{A})\mathbf{C} = \mathbf{C}$. So, $\mathbf{E} = \mathbf{A}^- + \mathbf{E}(\mathbf{B}-\mathbf{A})\mathbf{E}$ or $\mathbf{A}^- = \mathbf{E} - \mathbf{E}(\mathbf{B}-\mathbf{A})\mathbf{E}$ with $\mathbf{E} \in \{\mathbf{B}^-\}$.
(ii) Since $\mathbf{A} <^- \mathbf{B}$, we have $\mathbf{B}^-\mathbf{AB}^- \in \{\mathbf{A}_\mathbf{r}^-\}$. A routine check shows that $\mathbf{B}^-\mathbf{AB}^- \in \{\mathbf{A}_\mathbf{r}^-\}_\mathbf{B}$. To prove the other inclusion, we start with an $\mathbf{A}_\mathbf{r}^- \in \{\mathbf{A}_\mathbf{r}^-\}_\mathbf{B}$ and proceed along the same lines as in (i).
(iii) is proved in similar manner as (ii). □

Remark 3.5.5. Theorem 3.5.4(i) can be interpreted as follows:

Let $\mathbf{A} <^- \mathbf{B}$. Every g-inverse of $\mathbf{A}$ belonging to $\{\mathbf{A}^-\}_\mathbf{B}$ can be obtained by first choosing an arbitrary g-inverse $\mathbf{B}^-$ of $\mathbf{B}$ and subtracting from it a suitable reflexive g-inverse $\mathbf{H}$ of $\mathbf{B}-\mathbf{A}$, namely $\mathbf{H} = \mathbf{B}^-(\mathbf{B}-\mathbf{A})\mathbf{B}^-$, which also satisfies $\mathbf{AH} = \mathbf{0}$, $\mathbf{HA} = \mathbf{0}$. Thus,

$$\{\mathbf{A}^-\}_\mathbf{B} = \{\mathbf{B}^- - \mathbf{H}\ , \mathbf{B}^- \text{ arbitrary},\ \mathbf{AH} = \mathbf{0}, \mathbf{HA} - \mathbf{0}\}.$$

Let $\mathbf{A} <^- \mathbf{B}$. Let $\mathbf{A}^-$ belong to $\{\mathbf{A}^-\}_\mathbf{B}$. Does there exist a g-inverse $\mathbf{B}^-$ of $\mathbf{B}$ such that $\mathbf{B}^-\mathbf{A} = \mathbf{A}^-\mathbf{A}$ and $\mathbf{AB}^- = \mathbf{AA}^-$? The answer is in the affirmative and is contained in the following.

Theorem 3.5.6. *Let* $\mathbf{A} <^- \mathbf{B}$ *and let* $\mathbf{A}^-$ *be a g-inverse of* $\mathbf{A}$ *such that* $\mathbf{A}^-\mathbf{A} = \mathbf{A}^-\mathbf{B}$ *and* $\mathbf{AA}^- = \mathbf{BA}^-$. *Then there exists a g-inverse* $\mathbf{B}^-$ *of* $\mathbf{B}$ *such that* $\mathbf{B}^-\mathbf{A} = \mathbf{A}^-\mathbf{A}$ *and* $\mathbf{AB}^- = \mathbf{AA}^-$.

Proof. Let $a = \rho(\mathbf{A})$ and $b = \rho(\mathbf{B})$. Since $\mathbf{A} <^- \mathbf{B}$, there exist non-singular matrices $\mathbf{R}$ and $\mathbf{S}$ such that $\mathbf{A} = \mathbf{R}\mathbf{diag}(\mathbf{I_a}, \mathbf{0}, \mathbf{0})\mathbf{S}$ and $\mathbf{B} = \mathbf{R}\mathbf{diag}(\mathbf{I_a}, \mathbf{I_{b-a}}, \mathbf{0})\ \mathbf{S}$, by Theorem 3.3.5(vi). It is easy to check that the class of all g-inverses $\mathbf{A}^-$ of $\mathbf{A}$ such that $\mathbf{A}^-\mathbf{A} = \mathbf{A}^-\mathbf{B}$ and $\mathbf{AA}^- = \mathbf{BA}^-$ is given by

$$\mathbf{S}^{-1}\begin{pmatrix} \mathbf{I} & \mathbf{0} & \mathbf{C}_{13} \\ \mathbf{0} & \mathbf{0} & \mathbf{0} \\ \mathbf{C}_{31} & \mathbf{0} & \mathbf{C}_{33} \end{pmatrix}\mathbf{R}^{-1},$$

where $\mathbf{C}_{13}, \mathbf{C}_{31}, \mathbf{C}_{33}$ are arbitrary. Choose

$$\mathbf{A}^- = \mathbf{S}^{-1}\begin{pmatrix} \mathbf{I} & \mathbf{0} & \mathbf{C}_{13} \\ \mathbf{0} & \mathbf{0} & \mathbf{0} \\ \mathbf{C}_{31} & \mathbf{0} & \mathbf{C}_{33} \end{pmatrix}\mathbf{R}^{-1}$$

for a particular choice of $\mathbf{C}_{13}, \mathbf{C}_{31}, \mathbf{C}_{33}$. Then it is easy to check that one choice of required $\mathbf{B}^-$ is

$$\mathbf{B}^- = \mathbf{S}^{-1}\begin{pmatrix} \mathbf{I} & \mathbf{0} & \mathbf{C}_{13} \\ \mathbf{0} & \mathbf{I} & \mathbf{0} \\ \mathbf{C}_{31} & \mathbf{0} & \mathbf{C}_{33} \end{pmatrix}\mathbf{R}^{-1}.$$

In fact, the class of all such $\mathbf{B}^-$ is given by

$$\mathbf{S}^{-1}\begin{pmatrix} \mathbf{I} & \mathbf{0} & \mathbf{C}_{13} \\ \mathbf{0} & \mathbf{I} & \mathbf{0} \\ \mathbf{C}_{31} & \mathbf{0} & \mathbf{D} \end{pmatrix}\mathbf{R}^{-1},$$

where $\mathbf{D}$ is arbitrary. □

Corollary 3.5.7. *Let* $\mathbf{A}$ and $\mathbf{B}$ *be matrices of the same order such that* $\mathbf{A} <^- \mathbf{B}$ *and let* $\mathbf{A}^-$ *be a g-inverse of* $\mathbf{A}$ *such that* $\mathbf{A}^-\mathbf{A} = \mathbf{A}^-\mathbf{B}$ *and* $\mathbf{AA}^- = \mathbf{BA}^-$, *then for a g-inverse* $\mathbf{B}^-$ *of* $\mathbf{B}$,

(i) $\mathbf{AA}^- = \mathbf{AB}^- \Leftrightarrow \mathbf{AA}^-$ *and* $\mathbf{BB}^-$ *commute.*
(ii) $\mathbf{A}^-\mathbf{A} = \mathbf{B}^-\mathbf{A} \Leftrightarrow \mathbf{A}^-\mathbf{A}$ *and* $\mathbf{B}^-\mathbf{B}$ *commute.*

The following theorem deals with a question related to the one in Theorem 3.5.6 (from the point of view of a given g-inverse $\mathbf{B}^-$ of $\mathbf{B}$ namely: Given $\mathbf{A} <^- \mathbf{B}$ and any g-inverse $\mathbf{B}^-$ of $\mathbf{B}$, does there exist a g-inverse $\mathbf{A}^-$ of $\mathbf{A}$ such that $\mathbf{A}^-\mathbf{A} = \mathbf{A}^-\mathbf{B} = \mathbf{B}^-\mathbf{A}$ and $\mathbf{AA}^- = \mathbf{BA}^- = \mathbf{AB}^-$?).

Theorem 3.5.8. *Let* $\mathbf{A}$ and $\mathbf{B}$ *be matrices of the same order such that* $\mathbf{A} <^- \mathbf{B}$. *Then for any g-inverse* $\mathbf{B}^-$ *of* $\mathbf{B}$, *there exists a g-inverse* $\mathbf{A}^-$ *of* $\mathbf{A}$ *such that* (a) $\mathbf{AA}^- = \mathbf{BA}^-$, (b) $\mathbf{A}^-\mathbf{A} = \mathbf{A}^-\mathbf{B}$, (c) $\mathbf{AA}^- = \mathbf{AB}^-$, *and* (d) $\mathbf{A}^-\mathbf{A} = \mathbf{B}^-\mathbf{A}$.

Proof. Since $\mathbf{A} <^- \mathbf{B}$, there is a g-inverse $\mathbf{A}^-$ of $\mathbf{A}$ such that $\mathbf{A}^-\mathbf{A} = \mathbf{A}^-\mathbf{B}$ and $\mathbf{AA}^- = \mathbf{BA}^-$. Define $\mathbf{G} = \mathbf{B}^-\mathbf{BA}^-\mathbf{BB}^-$. It is easy to check that $\mathbf{G}$ is g-inverse of $\mathbf{A}$. Further, $\mathbf{AG} = \mathbf{AB}^-\mathbf{BA}^-\mathbf{BB}^- = \mathbf{AA}^-\mathbf{BB}^- = \mathbf{AA}^-\mathbf{AB}^- = \mathbf{AB}^-$. Also, $\mathbf{BG} = \mathbf{BB}^-\mathbf{BA}^-\mathbf{BB}^- = \mathbf{BA}^-\mathbf{BB}^- = \mathbf{AA}^-\mathbf{BB}^- = \mathbf{AA}^-\mathbf{AB}^- = \mathbf{AB}^-$. So, $\mathbf{G}$ satisfies (a).
The other equalities can be similarly established. □

While proving Theorem 3.5.8, we obtained another characterization of $\{\mathbf{A}^-\}_\mathbf{B}$. We state it below for completeness.

Theorem 3.5.9. *Let* $\mathbf{A}$ and $\mathbf{B}$ *be matrices of the same order such that* $\mathbf{A} <^- \mathbf{B}$ *and* $\mathbf{A} = \mathbf{R}\mathrm{diag}(\mathbf{I_a},\ \mathbf{0},\ \mathbf{0})\mathbf{S}$ *and* $\mathbf{B} = \mathbf{R}\mathrm{diag}(\mathbf{I_a},\ \mathbf{I_{b-a}},\ \mathbf{0})\ \mathbf{S}$. *Then*

$$\{\mathbf{A}^-\}_\mathbf{B} = \left\{\mathbf{S}^{-1}\begin{pmatrix}\mathbf{I_a} & \mathbf{0} & \mathbf{D_{13}}\\ \mathbf{0} & \mathbf{0} & \mathbf{0}\\ \mathbf{D_{31}} & \mathbf{0} & \mathbf{D_{33}}\end{pmatrix}\mathbf{R}^{-1}\right\},$$

where $\mathbf{D_{13}}, \mathbf{D_{31}}$ *and* $\mathbf{D_{33}}$ *are arbitrary.*

As observed in Theorem 3.3.16, for matrices $\mathbf{A}, \mathbf{B}$ of the same order whenever $\mathbf{A} <^- \mathbf{B}$, there exist projectors $\mathbf{P}$ and $\mathbf{Q}$ such that $\mathbf{A} = \mathbf{PB} = \mathbf{BQ}$. Let $\mathbf{A} <^- \mathbf{B}$. We shall now obtain explicitly the class of all projectors $\mathbf{P}$ and $\mathbf{Q}$ such that $\mathbf{A} = \mathbf{PB} = \mathbf{BQ}$. In the process we determine the possible ranks the projectors $\mathbf{P}$ and $\mathbf{Q}$ can have and obtain explicitly the subclass of the projectors $\mathbf{P}$ and $\mathbf{Q}$ of specified rank with the property $\mathbf{A} = \mathbf{PB} = \mathbf{BQ}$.

Lemma 3.5.10. *Let* $\mathbf{A}$ and $\mathbf{B}$ *be matrices of the same order and let* $\mathbf{P}$ *be a projector such that the product* $\mathbf{PB}$ *is defined. Then* $\mathbf{A} = \mathbf{PB}$ *if and only if* $\mathbf{PA} = \mathbf{A}$ *and* $\mathbf{P}(\mathbf{B} - \mathbf{A}) = \mathbf{0}$. *Also, for a projector* $\mathbf{Q}$ *such that the product* $\mathbf{BQ}$ *is defined,* $\mathbf{A} = \mathbf{BQ}$ *if and only if* $\mathbf{AQ} = \mathbf{A}$ *and* $(\mathbf{B} - \mathbf{A})\mathbf{Q} = \mathbf{0}$.

Proof. Proof is trivial. □

Theorem 3.5.11. *Let* $\mathbf{A}$ and $\mathbf{B}$ *be matrices of the same order* $m \times n$ *such that* $\mathbf{A} <^- \mathbf{B}$. *Let* $\mathbf{P}$ *be a projector such that the product* $\mathbf{PB}$ *is defined and has same order as* $\mathbf{B}$. *If* $\mathbf{A} = \mathbf{PB}$, *then*

$$\rho(\mathbf{A}) \leq \rho(\mathbf{P}) \leq \rho(\mathbf{A}) + \mathbf{m} - \rho(\mathbf{B}) \ .$$

Proof. In view of Lemma 3.5.10, $\mathbf{PA} = \mathbf{A}$ and $\mathbf{P}(\mathbf{B} - \mathbf{A}) = \mathbf{0}$. As $\mathbf{PA} = \mathbf{A}$, we have $\rho(\mathbf{A}) \leq \rho(\mathbf{P})$. Also,

$$\begin{aligned}\mathbf{P}(\mathbf{B} - \mathbf{A}) = \mathbf{0} &\Rightarrow \mathcal{C}(\mathbf{P^t}) \subseteq \mathcal{N}(\mathbf{B} - \mathbf{A})^{\mathbf{t}} \\ &\Rightarrow \rho(\mathbf{P}) = \rho(\mathbf{P^t}) = \mathbf{d}(\mathcal{C}(\mathbf{P^t})) \leq \mathbf{d}(\mathcal{N}(\mathbf{B} - \mathbf{A})^{\mathbf{t}}).\end{aligned}$$

Thus, $\rho(\mathbf{A}) \leq \rho(\mathbf{P}) \leq \rho(\mathbf{A}) \ + \ \mathbf{m} - \rho(\mathbf{B})$. □

Corollary 3.5.12. *Let* $\mathbf{A}, \mathbf{B}$ *be matrices of the same order* $m \times n$ *such that* $\mathbf{A} <^- \mathbf{B}$. *Let* $\mathbf{Q}$ *be a projector such that the product* $\mathbf{BQ}$ *is defined and is of the same order as* $\mathbf{B}$. *If* $\mathbf{A} = \mathbf{BQ}$, *then*

$$\rho(\mathbf{A}) \leq \rho(\mathbf{Q}) \leq \rho(\mathbf{A}) + \mathbf{n} - \rho(\mathbf{B}) \ .$$

Theorem 3.5.13. *Let* $\mathbf{A}$ and $\mathbf{B}$ *be matrices of the same order* $m \times n$ *with ranks 'a' and 'b' respectively. Let* $\mathbf{A} <^- \mathbf{B}$. *Let* $\mathbf{R}$ *and* $\mathbf{S}$ *be non-singular matrices such that* $\mathbf{A} = \mathbf{R}\mathbf{diag}(\mathbf{I_a},\ \mathbf{0},\ \mathbf{0})\mathbf{S}$ *and* $\mathbf{B} = \mathbf{R}\mathbf{diag}(\mathbf{I_a},\ \mathbf{I_{b-a}},\ \mathbf{0})\mathbf{S}$. *Then the class of all projectors* $\mathbf{P}$ *such that* $\mathbf{A} = \mathbf{PB}$ *is given by*

$$\mathbf{R}\begin{pmatrix}\mathbf{I_a} & \mathbf{0} & \mathbf{T_{13}} \\ \mathbf{0} & \mathbf{0} & \mathbf{T_{23}} \\ \mathbf{0} & \mathbf{0} & \mathbf{T_{33}}\end{pmatrix}\mathbf{R}^{-1} \ ,$$

where $\mathbf{T_{23}} = \mathbf{U}\mathbf{T_{33}}, \mathbf{T_{13}} = \mathbf{V}(\mathbf{I} - \mathbf{T_{33}})$ *and* $\mathbf{T_{33}}$ *is idempotent matrix of order* $(m - b) \times (m - b)$ *and matrices* $\mathbf{V}$ *and* $\mathbf{U}$ *are arbitrary.*

Proof. If $\mathbf{P}$ is an idempotent matrix of the given form, clearly, $\mathbf{A} = \mathbf{PB}$. Now, let $\mathbf{P}$ be projector such that $\mathbf{A} = \mathbf{PB}$. Write $\mathbf{P} = \mathbf{RTR}^{-1}$. Notice that $\mathbf{T}$ is idempotent since $\mathbf{P}$ is idempotent. Partition $\mathbf{T}$ as

$$\mathbf{T} = \begin{pmatrix}\mathbf{T_{11}} & \mathbf{T_{12}} & \mathbf{T_{13}} \\ \mathbf{T_{21}} & \mathbf{T_{22}} & \mathbf{T_{23}} \\ \mathbf{T_{31}} & \mathbf{T_{32}} & \mathbf{T_{33}}\end{pmatrix}$$

conformable for matrix multiplication with $\mathbf{diag(I_a, 0, 0)}$. Since, $\mathbf{A} = \mathbf{PB}$, $\mathbf{PA} = \mathbf{A}$ and $\mathcal{C}(\mathbf{A}) \subseteq \mathcal{C}(\mathbf{P})$. Now, $\mathbf{A} = \mathbf{PB} \Rightarrow \mathbf{T_{11}} = \mathbf{I_a}$, $\mathbf{T_{12}} = \mathbf{0}$, $\mathbf{T_{21}} = \mathbf{0}$, $\mathbf{T_{31}} = \mathbf{0}$, $\mathbf{T_{22}} = \mathbf{0}$ and $\mathbf{T_{32}} = \mathbf{0}$. As, $\mathbf{T}$ is idempotent, we have $\mathbf{T_{13}} + \mathbf{T_{13}T_{33}} = \mathbf{T_{13}}$, $\mathbf{T_{23}} = \mathbf{T_{23}T_{33}}$ and $\mathbf{T_{33}^2} = \mathbf{T_{33}}$.

So,

$$\mathbf{T} = \begin{pmatrix} \mathbf{I_a} & \mathbf{0} & \mathbf{T_{13}} \\ \mathbf{0} & \mathbf{0} & \mathbf{T_{23}} \\ \mathbf{0} & \mathbf{0} & \mathbf{T_{33}} \end{pmatrix}$$

where $\mathbf{T_{13}} = \mathbf{V(I - T_{33})}$, $\mathbf{T_{23}} = \mathbf{UT_{33}}$ for some matrices $\mathbf{U}$ and $\mathbf{V}$ and $\mathbf{T_{33}}$ is idempotent. Hence the result follows. □

Corollary 3.5.14. *Consider the setup of Theorem 3.5.13. Then the class of all projectors* $\mathbf{Q}$ *such that* $\mathbf{A} = \mathbf{BQ}$ *is given by*

$$\mathbf{R} \begin{pmatrix} \mathbf{I_a} & \mathbf{0} & \mathbf{0} \\ \mathbf{0} & \mathbf{0} & \mathbf{0} \\ \mathbf{T_{31}} & \mathbf{T_{32}} & \mathbf{T_{33}} \end{pmatrix} \mathbf{R^{-1}},$$

where $\mathbf{T_{13}} = \mathbf{(I - T_{33})Z}$, $\mathbf{T_{32}} = \mathbf{T_{33}W}$ *and* $\mathbf{T_{33}}$ *is an idempotent matrix of order* $(m - b) \times (m - b)$ *and* $\mathbf{Z}$ *and* $\mathbf{W}$ *are arbitrary.*

Theorem 3.5.15. *Let* $\mathbf{A}$ *and* $\mathbf{B}$ *be matrices of order* $m \times n$ *with ranks 'a' and 'b' respectively. Let* $\mathbf{A} <^{-} \mathbf{B}$ *and* $\mathbf{P}$ *be a projector of rank 'p' such that* $\mathbf{A} = \mathbf{PB}$. *Then there exist non-singular matrices* $\mathbf{X}$ *and* $\mathbf{Y}$ *such that* $\mathbf{A} = \mathbf{Xdiag(I_a,\ 0,\ 0)Y}$ *and* $\mathbf{B} = \mathbf{Xdiag(I_a,\ I_{b-a},\ 0,\ 0)Y}$ *and*

$$\mathbf{P} = \mathbf{X} \begin{pmatrix} \mathbf{I_a} & \mathbf{0} & \mathbf{0} & \mathbf{L} \\ \mathbf{0} & \mathbf{0} & \mathbf{0} & \mathbf{0} \\ \mathbf{0} & \mathbf{0} & \mathbf{I_{p-a}} & \mathbf{0} \\ \mathbf{0} & \mathbf{0} & \mathbf{0} & \mathbf{0} \end{pmatrix} \mathbf{X^{-1}} \text{ for some matrix } \mathbf{L}.$$

Proof. In view of Theorem 3.5.13, there exist non-singular matrices $\mathbf{R}$ and $\mathbf{S}$ such that $\mathbf{A} = \mathbf{Rdiag(I_a, 0, 0)S}$ and $\mathbf{B} = \mathbf{Rdiag(I_a, I_{b-a}, 0)S}$ and

$$\mathbf{P} = \mathbf{R} \begin{pmatrix} \mathbf{I_a} & \mathbf{0} & \mathbf{T_{13}} \\ \mathbf{0} & \mathbf{0} & \mathbf{T_{23}} \\ \mathbf{0} & \mathbf{0} & \mathbf{T_{33}} \end{pmatrix} \mathbf{R^{-1}},$$

where $\mathbf{T_{23}} = \mathbf{UT_{33}}$, $\mathbf{T_{13}} = \mathbf{V(I - T_{33})}$ for some matrices '$\mathbf{U}$ and $\mathbf{V}$' and $\mathbf{T_{33}}$ is idempotent matrix of order $(m - b) \times (m - b)$. Since $\rho(\mathbf{P}) =$

p, $\rho(\mathbf{T_{33}}) = p - a$. So, there exists a non-singular matrix $\mathbf{K}$ of order $(m-b)\times(m-b)$ such that $\mathbf{T_{33}} = \mathbf{K}\mathbf{diag}(\mathbf{I_{p-a}}, \mathbf{0})\mathbf{K}^{-1}$. Then

$$\mathbf{P} = \mathbf{R}\begin{pmatrix}\mathbf{I} & \mathbf{0} & \mathbf{0}\\ \mathbf{0} & \mathbf{I} & \mathbf{0}\\ \mathbf{0} & \mathbf{0} & \mathbf{K}\end{pmatrix}\begin{pmatrix}\mathbf{I_a} & \mathbf{0} & \mathbf{0} & \mathbf{L}\\ \mathbf{0} & \mathbf{0} & \mathbf{0} & \mathbf{0}\\ \mathbf{0} & \mathbf{0} & \mathbf{I_{p-a}} & \mathbf{0}\\ \mathbf{0} & \mathbf{0} & \mathbf{0} & \mathbf{0}\end{pmatrix}\begin{pmatrix}\mathbf{I} & \mathbf{0} & \mathbf{0}\\ \mathbf{0} & \mathbf{I} & \mathbf{0}\\ \mathbf{0} & \mathbf{0} & \mathbf{K}^{-1}\end{pmatrix}\mathbf{R}^{-1}.$$

Further,

$$\mathbf{P} = \mathbf{R}\begin{pmatrix}\mathbf{I_a} & \mathbf{0} & \mathbf{0}\\ \mathbf{0} & \mathbf{I_{b-a}} & \mathbf{0}\\ \mathbf{0} & \mathbf{0} & \mathbf{K}\end{pmatrix}\begin{pmatrix}\mathbf{I_a} & \mathbf{0} & \mathbf{0} & \mathbf{L}\\ \mathbf{0} & \mathbf{I_{b-a}} & \mathbf{W} & \mathbf{0}\\ \mathbf{0} & \mathbf{0} & \mathbf{I_{p-a}} & \mathbf{0}\\ \mathbf{0} & \mathbf{0} & \mathbf{0} & \mathbf{0}\end{pmatrix}\begin{pmatrix}\mathbf{I_a} & \mathbf{0} & \mathbf{0} & \mathbf{L}\\ \mathbf{0} & \mathbf{0} & \mathbf{0} & \mathbf{0}\\ \mathbf{0} & \mathbf{0} & \mathbf{I_{p-a}} & \mathbf{0}\\ \mathbf{0} & \mathbf{0} & \mathbf{0} & \mathbf{0}\end{pmatrix}$$

$$\times\begin{pmatrix}\mathbf{I_a} & \mathbf{0} & \mathbf{0} & \mathbf{L}\\ \mathbf{0} & \mathbf{I_{b-a}} & -\mathbf{W} & \mathbf{0}\\ \mathbf{0} & \mathbf{0} & \mathbf{I_{p-a}} & \mathbf{0}\\ \mathbf{0} & \mathbf{0} & \mathbf{0} & \mathbf{0}\end{pmatrix}\begin{pmatrix}\mathbf{I_a} & \mathbf{0} & \mathbf{0}\\ \mathbf{0} & \mathbf{I_{b-a}} & \mathbf{0}\\ \mathbf{0} & \mathbf{0} & \mathbf{K}^{-1}\end{pmatrix}\mathbf{R}^{-1}.$$

Let

$$\mathbf{X} = \mathbf{R}\begin{pmatrix}\mathbf{I_a} & \mathbf{0} & \mathbf{0}\\ \mathbf{0} & \mathbf{I_{b-a}} & \mathbf{0}\\ \mathbf{0} & \mathbf{0} & \mathbf{K}\end{pmatrix}\begin{pmatrix}\mathbf{I_a} & \mathbf{0} & \mathbf{0} & \mathbf{L}\\ \mathbf{0} & \mathbf{I_{b-a}} & \mathbf{W} & \mathbf{0}\\ \mathbf{0} & \mathbf{0} & \mathbf{I_{p-a}} & \mathbf{0}\\ \mathbf{0} & \mathbf{0} & \mathbf{0} & \mathbf{0}\end{pmatrix}$$

and $\mathbf{Y} = \mathbf{S}$. It can be checked that $\mathbf{A} = \mathbf{X}\mathbf{diag}(\mathbf{I_a}, \mathbf{0}, \mathbf{0}, \mathbf{0})\mathbf{Y}$, $\mathbf{B} = \mathbf{X}\mathbf{diag}(\mathbf{I_a}, \mathbf{I_{b-a}}, \mathbf{0}, \mathbf{0})\mathbf{Y}$ and

$$\mathbf{P} = \mathbf{X}\begin{pmatrix}\mathbf{I_a} & \mathbf{0} & \mathbf{0} & \mathbf{L}\\ \mathbf{0} & \mathbf{0} & \mathbf{0} & \mathbf{0}\\ \mathbf{0} & \mathbf{0} & \mathbf{I_{p-a}} & \mathbf{0}\\ \mathbf{0} & \mathbf{0} & \mathbf{0} & \mathbf{0}\end{pmatrix}\mathbf{X}^{-1}.$$

□

Corollary 3.5.16. *Consider the setup as in Theorem 3.5.15. Let* $\mathbf{Q}$ *be a projector of rank 'q' such that* $\mathbf{A} = \mathbf{BQ}$. *Then there exist non-singular matrices* $\mathbf{C}$ *and* $\mathbf{D}$ *such that* $\mathbf{A} = \mathbf{C}\mathbf{diag}(\mathbf{I_a}, \mathbf{0}, \mathbf{0}, \mathbf{0})\mathbf{D}$, $\mathbf{B} = \mathbf{C}\mathbf{diag}(\mathbf{I_{b-a}}, \mathbf{0}, \mathbf{0})\mathbf{D}$ *and*

$$\mathbf{Q} = \mathbf{C}\begin{pmatrix}\mathbf{I_a} & \mathbf{0} & \mathbf{0} & \mathbf{0}\\ \mathbf{0} & \mathbf{0} & \mathbf{0} & \mathbf{0}\\ \mathbf{0} & \mathbf{0} & \mathbf{I_{q-a}} & \mathbf{0}\\ \mathbf{M} & \mathbf{0} & \mathbf{0} & \mathbf{0}\end{pmatrix}\mathbf{C}^{-1}.$$

We now obtain the subclass of all projectors $\mathbf{P}$ of rank 'p' ($a \le p \le a+m-b$) such that $\mathbf{A} = \mathbf{PB}$, whenever $\mathbf{A} <^- \mathbf{B}$. From Theorem 3.5.13, it is clear that if $\mathbf{A} = \mathbf{Rdiag}(\mathbf{I_a}, \mathbf{0}, \mathbf{0})\ \mathbf{S}$ and $\mathbf{B} = \mathbf{Rdiag}(\mathbf{I_a}, \mathbf{I_{b-a}}, \mathbf{0})\ \mathbf{S}$, then every projector $\mathbf{P}$ such that $\mathbf{A} = \mathbf{PB}$ is of the form

$$\mathbf{P} = \mathbf{R}\begin{pmatrix} \mathbf{I_a} & \mathbf{0} & \mathbf{T_{13}} \\ \mathbf{0} & \mathbf{0} & \mathbf{T_{23}} \\ \mathbf{0} & \mathbf{0} & \mathbf{T_{33}} \end{pmatrix}\mathbf{R}^{-1},$$

where $\mathbf{T_{23}} = \mathbf{UT_{33}}, \mathbf{T_{13}} = \mathbf{V}(\mathbf{I} - \mathbf{T_{33}})$ for some matrices '$\mathbf{U}$ and $\mathbf{V}$' and $\mathbf{T_{33}}$ is idempotent matrix of order $(m-b) \times (m-b)$. Since $\mathbf{P}$ is idempotent and $\rho(\mathbf{P}) = p$, $\rho(\mathbf{P}) = \mathrm{tr}(\mathbf{P}) = \mathrm{tr}(\mathbf{I_a}) + \mathrm{tr}(\mathbf{T_{33}}) = a + \rho(\mathbf{T_{33}})$. Thus, the class of projectors $\mathbf{P}$ of given rank p is given by

$$\mathbf{P} = \mathbf{R}\begin{pmatrix} \mathbf{I_a} & \mathbf{0} & \mathbf{T_{13}} \\ \mathbf{0} & \mathbf{0} & \mathbf{T_{23}} \\ \mathbf{0} & \mathbf{0} & \mathbf{T_{33}} \end{pmatrix}\mathbf{R}^{-1},$$

where $\mathbf{T_{23}} = \mathbf{UT_{33}}, \mathbf{T_{13}} = \mathbf{V}(\mathbf{I} - \mathbf{T_{33}}), \mathbf{T_{33}}$ is idempotent matrix of order $(m-b) \times (m-b)$, $\rho(\mathbf{T_{33}}) = p - a$ and '$\mathbf{U}$ and $\mathbf{V}$' are arbitrary matrices.

The class of all projectors $\mathbf{P}$ of rank 'a' such that $\mathbf{A} = \mathbf{PB}$ deserves a special mention. It is clear that this class is given by

$$\mathbf{P} = \mathbf{R}\begin{pmatrix} \mathbf{I_a} & \mathbf{0} & \mathbf{T_{13}} \\ \mathbf{0} & \mathbf{0} & \mathbf{0} \\ \mathbf{0} & \mathbf{0} & \mathbf{0} \end{pmatrix}\mathbf{R}^{-1},$$

where $\mathbf{T_{13}}$ is arbitrary. Recall that in the proof of Theorem 3.5.6, the class of all g-inverse $\mathbf{A}^-$ such that $\mathbf{A}^-\mathbf{A} = \mathbf{A}^-\mathbf{B}$, $\mathbf{AA}^- = \mathbf{BA}^-$ is given by

$$\mathbf{P} = \mathbf{R}\begin{pmatrix} \mathbf{I_a} & \mathbf{0} & \mathbf{C_{13}} \\ \mathbf{0} & \mathbf{0} & \mathbf{0} \\ \mathbf{C_{31}} & \mathbf{0} & \mathbf{C_{33}} \end{pmatrix}\mathbf{R}^{-1}.$$

Hence for each $\mathbf{A}^-$ in that class,

$$\mathbf{AA}^- = \mathbf{R}\begin{pmatrix} \mathbf{I_a} & \mathbf{0} & \mathbf{C_{13}} \\ \mathbf{0} & \mathbf{0} & \mathbf{0} \\ \mathbf{0} & \mathbf{0} & \mathbf{0} \end{pmatrix}\mathbf{R}^{-1},$$

which is a projector of rank 'a'.

Conversely, each projector $\mathbf{P}$ of rank 'a' such that $\mathbf{A} = \mathbf{PB}$ is of the form

$$\mathbf{P} = \mathbf{R}\begin{pmatrix} \mathbf{I_a} & \mathbf{0} & \mathbf{T_{13}} \\ \mathbf{0} & \mathbf{0} & \mathbf{0} \\ \mathbf{0} & \mathbf{0} & \mathbf{0} \end{pmatrix}\mathbf{R}^{-1} = \mathbf{R}\begin{pmatrix} \mathbf{I_a} & \mathbf{0} & \mathbf{0} \\ \mathbf{0} & \mathbf{0} & \mathbf{0} \\ \mathbf{0} & \mathbf{0} & \mathbf{0} \end{pmatrix}\mathbf{SS}^{-1}\begin{pmatrix} \mathbf{I_a} & \mathbf{0} & \mathbf{T_{13}} \\ \mathbf{0} & \mathbf{0} & \mathbf{0} \\ \mathbf{0} & \mathbf{0} & \mathbf{0} \end{pmatrix}\mathbf{R}^{-1}$$

$= \mathbf{AA}^-$, where

$$\mathbf{A}^- = \mathbf{S}^{-1}\begin{pmatrix} \mathbf{I_a} & \mathbf{0} & \mathbf{T_{13}} \\ \mathbf{0} & \mathbf{0} & \mathbf{0} \\ \mathbf{0} & \mathbf{0} & \mathbf{0} \end{pmatrix}\mathbf{R}^{-1} \text{ satisfies } \mathbf{A}^-\mathbf{A} = \mathbf{A}^-\mathbf{B},\ \mathbf{AA}^- = \mathbf{BA}^-.$$

Thus, we have proved

Theorem 3.5.17. *Let* $\mathbf{A}$ and $\mathbf{B}$ *be matrices of the same order* $m \times n$ *with ranks 'a' and 'b' respectively. Let* $\mathbf{A} <^- \mathbf{B}$. *Then the class of all projectors* $\mathbf{P}$ *of rank 'a' such that* $\mathbf{A} = \mathbf{PB}$ *is the class of all* $\mathbf{AA}^-$ *satisfying* $\mathbf{A}^-\mathbf{A} = \mathbf{A}^-\mathbf{B}$, *and* $\mathbf{AA}^- = \mathbf{BA}^-$.

Theorem 3.5.18. *Let* $\mathbf{A}, \mathbf{B}$ *be matrices of the same order* $m \times n$ *with ranks 'a' and 'b' respectively. Let* $\mathbf{A} <^- \mathbf{B}$ *and* $\mathbf{P}$ *be a projector such that* $\mathbf{A} = \mathbf{PB}$. *Then* $\mathbf{P}$ *can be written as* $\mathbf{P} = \mathbf{P_1} + \mathbf{P_2}$ *where* $\mathbf{P_1}$ *is a projector of rank 'a' such that* $\mathbf{A} = \mathbf{P_1B}$ *and* $\mathbf{P_2}$ *is projector such that* $\mathbf{P_1P_2} = \mathbf{P_2P_1} = \mathbf{0}$, $\mathbf{P_2A} = \mathbf{0}$ *and* $\mathbf{P_2}(\mathbf{B} - \mathbf{A}) = \mathbf{0}$.

Proof. In the setup of Theorem 3.5.15, write

$$\mathbf{P_1} = \mathbf{X}\begin{pmatrix} \mathbf{I_a} & \mathbf{0} & \mathbf{0} & \mathbf{L} \\ \mathbf{0} & \mathbf{0} & \mathbf{0} & \mathbf{0} \\ \mathbf{0} & \mathbf{0} & \mathbf{0} & \mathbf{0} \\ \mathbf{0} & \mathbf{0} & \mathbf{0} & \mathbf{0} \end{pmatrix}\mathbf{X}^{-1}, \quad \mathbf{P_2} = \mathbf{X}\begin{pmatrix} \mathbf{0} & \mathbf{0} & \mathbf{0} & \mathbf{L} \\ \mathbf{0} & \mathbf{0} & \mathbf{0} & \mathbf{0} \\ \mathbf{0} & \mathbf{0} & \mathbf{I_{p-a}} & \mathbf{0} \\ \mathbf{0} & \mathbf{0} & \mathbf{0} & \mathbf{0} \end{pmatrix}\mathbf{X}^{-1}.$$

It is easy to check that $\mathbf{P_1}$ is a projector of rank 'a' such that $\mathbf{A} = \mathbf{P_1B}$ and $\mathbf{P_2}$ is idempotent such that $\mathbf{P_1P_2} = \mathbf{P_2P}_=\mathbf{0}$, $\mathbf{P_2A} = \mathbf{0}$ and $\mathbf{P_2}(\mathbf{B}-\mathbf{A}) = \mathbf{0}$. □

3.6 Minus order for idempotent matrices

Idempotent matrices (being projectors too) are a very important class of matrices in various applications such as in the theory of Linear Models. In this section, we study the minus order for this class. For an idempotent matrix $\mathbf{A}$, we find the class of all matrices $\mathbf{B}$ such that $\mathbf{A} <^- \mathbf{B}$. We then obtain the class of all idempotent matrices $\mathbf{B}$ such that $\mathbf{A} <^- \mathbf{B}$. Now, let $\mathbf{B}$ be an idempotent matrix. We show that if $\mathbf{A} <^- \mathbf{B}$, then $\mathbf{A}$ is necessarily idempotent. We also obtain the class of all $\mathbf{A}$ such that $\mathbf{A} <^- \mathbf{B}$.

Notice that for an idempotent matrix $\mathbf{E}$, $\mathbf{E}$ is a g-inverse of itself and $\rho(\mathbf{E}) = \mathrm{tr}(\mathbf{E})$. Let $(\mathbf{P}, \mathbf{Q})$ be a rank factorization of a non-null square matrix $\mathbf{A}$. Then $\mathbf{A}$ is idempotent if and only if $\mathbf{QP} = \mathbf{I}$. Moreover, for idempotent matrices $\mathbf{E}$ and $\mathbf{F}$ of the same order, it is trivial to check that $\mathbf{E} <^s \mathbf{F}$ if and only if $\mathbf{E} = \mathbf{FE} = \mathbf{EF}$ if and only if $\mathbf{E} <^- \mathbf{F}$. Thus, for any idempotent matrix $\mathbf{A}$, $\mathbf{A} <^- \mathbf{I}$, $\mathbf{A}$ and $\mathbf{I}$ being matrices of the same order.

Lemma 3.6.1. *Let* $\mathbf{E}$ *and* $\mathbf{F}$ *be idempotent matrices of the same order. Then* $\mathbf{E} <^- \mathbf{F}$ *if and only if* $\rho(\mathbf{F} - \mathbf{E}) = \rho(\mathbf{F} - \mathbf{EF}) = \rho(\mathbf{F} - \mathbf{FE})$.

Proof. Follows from Theorem 3.3.16(i) $\Leftrightarrow$ (iv). $\square$

We now prove a lemma that is also of independent interest.

Lemma 3.6.2. *Let* $\mathbf{E}$ *and* $\mathbf{F}$ *be idempotent matrices of the same order. Then* $\mathbf{F}-\mathbf{E}$ *is idempotent if and only if* $\mathbf{E}=\mathbf{FE}=\mathbf{EF}$.

Proof. 'If' part

Let $\mathbf{E}=\mathbf{FE}=\mathbf{EF}$. Then

$$\begin{aligned}(\mathbf{F}-\mathbf{E})^2 &= \mathbf{F}^2-\mathbf{FE}-\mathbf{EF}+\mathbf{E}^2\\ &= \mathbf{F}-\mathbf{FE}-\mathbf{EF}+\mathbf{E}=\mathbf{F}-\mathbf{E}.\end{aligned}$$

Hence, $\mathbf{F}-\mathbf{E}$ is idempotent.

'Only if' part

Write $\mathbf{F}=\mathbf{E}+(\mathbf{F}-\mathbf{E})$. Since $\mathbf{E},\mathbf{F}$ and $\mathbf{E}-\mathbf{F}$ are idempotent matrices,

$$\rho(\mathbf{F})=\mathrm{tr}(\mathbf{F})=\mathrm{tr}(\mathbf{E})+\mathrm{tr}(\mathbf{F}-\mathbf{E})=\rho(\mathbf{E})+\rho(\mathbf{F}-\mathbf{E}).$$

By Remark 3.3.9, we have $\mathbf{E}<^{-}\mathbf{F}$, and so, $\mathbf{E}=\mathbf{FE}=\mathbf{EF}$. $\square$

Theorem 3.6.3. *Let* $\mathbf{A}$ *be an idempotent matrix with a spectral decomposition* $\mathbf{A}=\mathbf{P}\mathbf{diag}(\mathbf{I},\mathbf{0})\mathbf{P}^{-1}$. *Then the class of all matrices* $\mathbf{B}$ *such that* $\mathbf{A}<^{-}\mathbf{B}$ *is given by* $\mathbf{A}+\mathbf{P}\begin{pmatrix}-\mathbf{L}\\ \mathbf{I}\end{pmatrix}\mathbf{W}(-\mathbf{M}\ \ \mathbf{I})\mathbf{P}^{-1}$ *where* $\mathbf{L},\mathbf{W}$ *and* $\mathbf{M}$ *are arbitrary.*

Proof follows from Theorem 3.4.4.

Theorem 3.6.4. *Let* $\mathbf{A}$ *be an idempotent matrix. Then the class of all idempotent matrices* $\mathbf{B}$ *such that* $\mathbf{A}<^{-}\mathbf{B}$ *is given by* $\mathbf{B}=\mathbf{A}+\mathbf{C}$ *where* $\mathbf{C}$ *is an idempotent matrix such that* $\mathbf{AC}=\mathbf{CA}=\mathbf{0}$.

Proof. Let $\mathbf{B}=\mathbf{A}+\mathbf{C}$, where $\mathbf{C}$ is an idempotent matrix such that $\mathbf{AC}=\mathbf{CA}=\mathbf{0}$. Then $\mathbf{B}^2=\mathbf{B}$ and trivially $\mathbf{A}<^{-}\mathbf{B}$.

Now, let $\mathbf{B}$ be an idempotent matrix such that $\mathbf{A}<^{-}\mathbf{B}$. Then by Theorem 3.3.5(vii), $\mathbf{B}=\mathbf{A}\oplus(\mathbf{B}-\mathbf{A})$. Since $\mathbf{B}=\mathbf{B}^2$, '$\mathcal{C}(\mathbf{A})$ and $\mathcal{C}(\mathbf{B}-\mathbf{A})$' are virtually disjoint and '$\mathcal{C}(\mathbf{A}^t)$ and $\mathcal{C}(\mathbf{B}-\mathbf{A})^t$' are virtually disjoint, it follows that $\mathbf{A}(\mathbf{B}-\mathbf{A})=\mathbf{0}$, $(\mathbf{B}-\mathbf{A})\mathbf{A}=\mathbf{0}$ and $\mathbf{B}-\mathbf{A}$ is idempotent. Let $\mathbf{C}=\mathbf{B}-\mathbf{A}$. Then $\mathbf{B}=\mathbf{A}+\mathbf{C}$ where $\mathbf{C}$ is an idempotent matrix such that $\mathbf{AC}=\mathbf{CA}=\mathbf{0}$. $\square$

Theorem 3.6.5. *Let* $\mathbf{A}$ *and* $\mathbf{B}$ *be square matrices of the same order with* $\mathbf{B}$ *idempotent. Then* $\mathbf{A}<^{-}\mathbf{B}$ *if and only if* $\mathbf{A}=\mathbf{A}^2=\mathbf{AB}=\mathbf{BA}$.

Proof. 'Only if' part
Since $\mathbf{A} <^- \mathbf{B}$ and $\mathbf{B}$ is idempotent, we have $\mathbf{A} = \mathbf{ABA} = \mathbf{AB} = \mathbf{BA}$. Therefore, $\mathbf{A^2} = (\mathbf{AB})(\mathbf{AB}) = (\mathbf{ABA})\mathbf{B} = \mathbf{AB} = \mathbf{A}$. Thus, $\mathbf{A} = \mathbf{A^2} = \mathbf{AB} = \mathbf{BA}$.
The proof of 'If' part is trivial. □

Theorem 3.6.6. *Let* $\mathbf{B}$ *be an idempotent matrix. Let* $(\mathbf{P}, \mathbf{Q})$ *be a rank factorization of* $\mathbf{B}$. *Then the class of all matrices* $\mathbf{A}$ *such that* $\mathbf{A} <^- \mathbf{B}$ *is given by* $\mathbf{A} = \mathbf{PTQ}$ *where* $\mathbf{T}$ *is an arbitrary idempotent matrix of appropriate order.*

Proof follows from Theorem 3.4.6.

Remark 3.6.7. The matrix $\mathbf{A}$ as in Theorem 3.6.6 is idempotent because $\mathbf{A^2} = \mathbf{PTQPTQ} = \mathbf{PT^2Q} = \mathbf{A}$.

Remark 3.6.8. If $\mathbf{A}$ and $\mathbf{B}$ are idempotent matrices such that $\mathbf{A} <^- \mathbf{B}$, then $\mathbf{A}^\#\mathbf{A} = \mathbf{A}^\#\mathbf{B}$ and $\mathbf{AA}^\# = \mathbf{BA}^\#$. This is so, as for an idempotent matrix $\mathbf{E}$, $\mathbf{E}^\# = \mathbf{E}$. (We shall make a detailed study of the relationship $\mathbf{A}^\#\mathbf{A} = \mathbf{A}^\#\mathbf{B}$ and $\mathbf{AA}^\# = \mathbf{BA}^\#$ for general square matrices in Chapter 4.)

3.7 Minus order for complex matrices

We shall now specialize to matrices over the field of complex numbers and study the special properties of the minus order. Let $\mathbf{A} <^- \mathbf{B}$. We obtain a canonical form for $\mathbf{A}$ and $\mathbf{B}$ in term of the singular value decomposition of the matrix $\mathbf{A}$.

Theorem 3.7.1. *Let* $\mathbf{A}$ *and* $\mathbf{B}$ *be nnd matrices of the same order. Then* $\mathbf{A} <^- \mathbf{B}$ *if and only if* $\mathbf{B} - \mathbf{A}$ *is an nnd matrix of rank equal to* $\rho(\mathbf{B}) - \rho(\mathbf{A})$.

Proof. 'If' part
Since $\rho(\mathbf{B} - \mathbf{A}) = \rho(\mathbf{B}) - \rho(\mathbf{A})$, we have $\mathbf{A} <^- \mathbf{B}$, by Remark 3.3.9.
'Only if' part
Clearly, $\rho(\mathbf{B} - \mathbf{A}) = \rho(\mathbf{B}) - \rho(\mathbf{A})$ as $\mathbf{A} <^- \mathbf{B}$. We also have $\mathbf{B} - \mathbf{A} <^- \mathbf{B}$ by Theorem 3.3.5. Since $\mathbf{B}$ is *nnd*, we take an *nnd* g-inverse $\mathbf{B}^-$ of $\mathbf{B}$. Then $(\mathbf{B} - \mathbf{A})\mathbf{B}^-(\mathbf{B} - \mathbf{A}) = \mathbf{B} - \mathbf{A}$ and hence, $\mathbf{B} - \mathbf{A}$ is *nnd*. □

Corollary 3.7.2. *Let* $\mathbf{A}$ and $\mathbf{B}$ *be matrices of the same order* $n \times n$. *Let* $\mathbf{B}$ *be an nnd matrix. If* $\mathbf{A} <^- \mathbf{B}$, *then* $\mathbf{A}$ *and* $\mathbf{B} - \mathbf{A}$ *are nnd.*

Proof. If $\mathbf{A} <^- \mathbf{B}$, then $\mathbf{AB}^\dagger\mathbf{A} = \mathbf{A}$. Since $\mathbf{B}$ is *nnd*, $\mathbf{B}^\dagger$ is also *nnd*. Hence $\mathbf{A}$ is *nnd*. Moreover, $\mathbf{A} <^- \mathbf{B} \Rightarrow \mathbf{B} - \mathbf{A} <^- \mathbf{B}$ and $\mathbf{B}$ be an *nnd* matrix, so, as in proof of 'Only if' part of Theorem 3.7.1, $\mathbf{B} - \mathbf{A}$ is *nnd*.□

By comparing Theorem 3.3.5(i) $\Leftrightarrow$ (vii) with Theorem 3.7.1, we notice that if $\mathbf{A}$ and $\mathbf{B}$ are *nnd* and $\mathbf{A} <^- \mathbf{B}$, $\mathbf{B} - \mathbf{A}$ is *nnd*. We now prove a result similar to Theorem 3.3.5(i) $\Leftrightarrow$ (vi) for complex matrices using unitary transformations.

Theorem 3.7.3. *Let* $\mathbf{A}$ *and* $\mathbf{B}$ *be matrices of the same order over the field of complex numbers with* $\rho(\mathbf{A}) = \mathbf{a}$, $\rho(\mathbf{B}) = \mathbf{b}$, *and* $b > a \geq 1$. *Then* $\mathbf{A} <^- \mathbf{B}$ *if and only if there exist unitary matrices* $\mathbf{U}$ *and* $\mathbf{V}$ *and non-singular matrices* $\mathbf{T_1}$ *and* $\mathbf{T_2}$ *such that*

$$\mathbf{A} = \mathbf{U}\mathbf{diag}(\mathbf{D_a}, \mathbf{0}, \mathbf{0})\mathbf{V}^\star = \mathbf{U}\mathbf{T_1}\mathbf{diag}(\mathbf{D_a}, \mathbf{0}, \mathbf{0})\mathbf{T_2}\mathbf{V}^\star$$

and $\mathbf{B} = \mathbf{U}\mathbf{T_1}\mathbf{diag}(\mathbf{D_a}, \mathbf{D_{b-a}}, \mathbf{0})\mathbf{T_2}\mathbf{V}^\star$, *where* $\mathbf{D_a}$ *and* $\mathbf{D_{b-a}}$ *are diagonal positive definite matrices.*

Proof. 'If' part follows form Theorem 3.3.4(vi)$\Rightarrow$ (i)
'Only if' part
Let $\mathbf{A} = \mathbf{U_1}\mathbf{D_a}\mathbf{V_1^\star}$ be a singular value decomposition of $\mathbf{A}$ where $\mathbf{U_1}$ and $\mathbf{V_1}$ are semi-unitary matrices and $\mathbf{D_a}$ is a diagonal positive definite matrix. Since, $\mathbf{A} <^- \mathbf{B}$ we have $\mathcal{C}(\mathbf{A}) \subseteq \mathcal{C}(\mathbf{B})$ and $\mathcal{C}(\mathbf{A}^\star) \subseteq \mathcal{C}(\mathbf{B}^\star)$. So, there exist semi-unitary matrices $\mathbf{U_2}$ and $\mathbf{V_2}$ such that the columns of $(\mathbf{U_1} : \mathbf{U_2})$ and $(\mathbf{V_1} : \mathbf{V_2})$ form ortho-normal basis of $\mathcal{C}(\mathbf{B})$ and $\mathcal{C}(\mathbf{B}^\star)$ respectively. Extend $(\mathbf{U_1} : \mathbf{U_2})$ and $(\mathbf{V_1} : \mathbf{V_2})$ to unitary matrices $\mathbf{U} = (\mathbf{U_1} : \mathbf{U_2} : \mathbf{U_3})$ and $\mathbf{V} = (\mathbf{V_1} : \mathbf{V_2} : \mathbf{V_3})$ respectively. Then we can write

$$\mathbf{A} = \mathbf{U}\mathbf{diag}(\mathbf{D_a}, \mathbf{0_{b-a}}, \mathbf{0})\,\mathbf{V}^\star \text{ and } \mathbf{B} = \mathbf{U}\mathbf{diag}(\mathbf{M}, \mathbf{0})\,\mathbf{V}^\star,$$

where $\mathbf{M}$ is a $b \times b$ non-singular matrix. (Compare with Theorem 3.3.5 (i) $\Rightarrow$ (vi).) Since, $\mathbf{A} <^- \mathbf{B}$, we have $\mathbf{diag}(\mathbf{D_a}, \mathbf{0}, \mathbf{0}) <^- \mathbf{diag}(\mathbf{M}, \mathbf{0})$ and $\mathbf{diag}(\mathbf{D_a}, \mathbf{0}) <^- \mathbf{M}$. Partition $\mathbf{M}$ as $\mathbf{M} = \begin{pmatrix} \mathbf{M_{11}} & \mathbf{M_{12}} \\ \mathbf{M_{21}} & \mathbf{M_{22}} \end{pmatrix}$, where $\mathbf{M_{11}}$ and $\mathbf{M_{22}}$ are matrices of orders $a \times a$ and $(b - a) \times (b - a)$ respectively. By Theorem 3.4.4 and Remark 3.4.5, $\mathbf{M_{22}}$ is non-singular and $\mathbf{M} = \begin{pmatrix} \mathbf{D_a} & \mathbf{0} \\ \mathbf{0} & \mathbf{0} \end{pmatrix} + \begin{pmatrix} -\mathbf{D_a}\mathbf{L_1} \\ \mathbf{I} \end{pmatrix} \mathbf{M_{22}}(-\mathbf{L_2}\mathbf{D_a},\ \mathbf{I})$ for some matrices $\mathbf{L_1}$ and $\mathbf{L_2}$. Notice that

$$\mathbf{M} = \mathbf{P}\mathbf{diag}(\mathbf{D_a}, \mathbf{M_{22}})\,\mathbf{Q} \text{ and } \mathbf{diag}(\mathbf{D_a}, \mathbf{0}) = \mathbf{P}\mathbf{diag}(\mathbf{D_a}, \mathbf{0})\mathbf{Q},$$

where $\mathbf{P} = \begin{pmatrix} \mathbf{I} & -\mathbf{D_a L_1} \\ \mathbf{0} & \mathbf{I} \end{pmatrix}$ and $\mathbf{Q} = \begin{pmatrix} \mathbf{I} & \mathbf{0} \\ -\mathbf{L_2 D_a} & \mathbf{I} \end{pmatrix}$ are non-singular matrices.

Now, let $\mathbf{M_{22}} = \mathbf{R D_{b-a} S}$ be a singular value decomposition of $\mathbf{M_{22}}$, where $\mathbf{R}$ and $\mathbf{S}$ are unitary and $\mathbf{D_{b-a}}$ is a diagonal positive definite matrix, as $\mathbf{M_{22}}$ is non-singular. Then it is easy to check that

$$\mathbf{A} = \mathbf{U}\mathbf{diag}(\mathbf{D_a} : \mathbf{0} : \mathbf{0})\ \mathbf{V}^{\star} = \mathbf{U T_1}\mathbf{diag}(\mathbf{D_a} : \mathbf{0} : \mathbf{0})\ \mathbf{T_2 V}^{\star} \text{ and}$$
$$\mathbf{B} = \mathbf{U T_1}\mathbf{diag}(\mathbf{D_a} : \mathbf{D_{b-a}} : \mathbf{0})\ \mathbf{T_2 V}^{\star}, \text{ where}$$

$$\mathbf{T_1} = \begin{pmatrix} \mathbf{P}\begin{pmatrix} \mathbf{I} & \mathbf{0} \\ \mathbf{0} & \mathbf{R} \end{pmatrix} & \mathbf{0} \\ \mathbf{0} & \mathbf{I} \end{pmatrix} \text{ and } \mathbf{T_2} = \begin{pmatrix} \begin{pmatrix} \mathbf{I} & \mathbf{0} \\ \mathbf{0} & \mathbf{R} \end{pmatrix}\mathbf{Q} & \mathbf{0} \\ \mathbf{0} & \mathbf{I} \end{pmatrix}.$$

Clearly $\mathbf{T_1}$ and $\mathbf{T_2}$ are non-singular. □

Corollary 3.7.4. *Let* $\mathbf{A}$ *and* $\mathbf{B} \in \mathbb{C}^{\mathbf{m}\times\mathbf{n}}$ *with* $\rho(\mathbf{A}) = a$, $\rho(\mathbf{B}) = b$, *and* $b > a \geq 1$. *Then* $\mathbf{A} <^{-} \mathbf{B}$ *if and only if there exist unitary matrices* $\mathbf{U}$ *and* $\mathbf{V}$, *diagonal positive definite matrix* $\mathbf{D_a}$ *and a* $b \times b$ *non-singular matrix* $\mathbf{M}$ *such that*

$$\mathbf{M} = \begin{pmatrix} \mathbf{D_a} & \mathbf{0} \\ \mathbf{0} & \mathbf{0} \end{pmatrix} + \begin{pmatrix} \mathbf{F_1} \\ \mathbf{I} \end{pmatrix} \mathbf{M_{22}}(\mathbf{F_2}, \mathbf{I})$$

such that $\mathbf{A} = \mathbf{U}\mathbf{diag}(\mathbf{D_a}, \mathbf{0_{b-a}}, \mathbf{0})\ \mathbf{V}^{\star}$ *and* $\mathbf{B} = \mathbf{U}(\mathbf{M}, \mathbf{0})\mathbf{V}^{\star}$, *where* $\mathbf{M_{22}}$ *is non-singular and* $\mathbf{F_1}$ *and* $\mathbf{F_2}$ *are some matrices of suitable orders.*

Corollary 3.7.5. *Let* $\mathbf{A}$ and $\mathbf{B} \in \mathbb{C}^{\mathbf{m}\times\mathbf{n}}$ *be nnd matrices with* $\rho(\mathbf{A}) = a$, $\rho(\mathbf{B}) = b$ *and* $b \geq a \geq 1$. *Then* $\mathbf{A} <^{-} \mathbf{B}$ *if and only if there exist a unitary matrix* $\mathbf{U}$, *an* $a \times a$ *positive definite diagonal matrix* $\mathbf{D_a}$ *and a* $b \times b$ *positive definite matrix* $\mathbf{M} = \begin{pmatrix} \mathbf{D_a} & \mathbf{0} \\ \mathbf{0} & \mathbf{0} \end{pmatrix} + \begin{pmatrix} \mathbf{F} \\ \mathbf{I} \end{pmatrix} \mathbf{M_{22}}(\mathbf{F}^{\star}, \mathbf{I})$ *such that* $\mathbf{A} = \mathbf{U}\mathbf{diag}(\mathbf{D_a}, \mathbf{0_{b-a}}, \mathbf{0})\ \mathbf{U}^{\star}$ *and* $\mathbf{B} = \mathbf{U}\mathbf{diag}(\mathbf{M}, \mathbf{0})\ \mathbf{U}^{\star}$; *where* $\mathbf{M_{22}}$ *is a positive definite matrix and* $\mathbf{F}$ *is some suitable matrix.*

Compare Theorem 3.7.3 with Theorem 3.3.5(i) ⇔ (vi). A vigilant reader might wonder: what is so special about Theorem 3.7.3 over Theorem 3.3.4(i) ⇒ (vi), since every non-singular matrix $\mathbf{L}$ can be written as $\mathbf{L} = \mathbf{UT}$ where $\mathbf{U}$ is unitary and $\mathbf{T}$ non-singular ($\mathbf{L} = \mathbf{U U^{\star} L}$ for any unitary matrix $\mathbf{U}$). In connection with Theorem 3.7.3, the point to be noted is that we can retain a singular value decomposition for $\mathbf{A}$, which is $\mathbf{A} = \mathbf{U}\mathbf{diag}(\mathbf{D_a}, \mathbf{0}, \mathbf{0})\mathbf{V}^{\star} = \mathbf{U T_1}\mathbf{diag}(\mathbf{D_a}, \mathbf{0}, \mathbf{0})\mathbf{T_2 V}^{\star}$. The importance of this result will be further clear when we deal with the star order.

3.8 Exercises

Unless otherwise stated matrices are over arbitrary field and may be square or rectangular.

(1) Let $\mathbf{A}$ and $\mathbf{B}$ be hermitian matrices of the same order over the complex field. Show that $\mathbf{A} <^s \mathbf{B}$ if and only if $\mathcal{C}(\mathbf{A}) \subset \mathcal{C}(\mathbf{B})$.

(2) If $\mathbf{A}$ and $\mathbf{B}$ are square matrices of the same order such that $\mathbf{A} <^s \mathbf{B}$ and $\mathbf{B}^2 = \mathbf{0}$, then show that $\mathbf{A}^2 = \mathbf{0}$. What happens if $\mathbf{B}^2 = \mathbf{B}$?

(3) Prove that $\mathbf{0} <^- \mathbf{B}$ for any matrix $\mathbf{B}$. Let $n > m$ and $\mathbf{I}$ be the identity matrix of order $m \times m$. Further let $\mathbf{B}$ be a matrix of order $n \times n$. Show that $(\mathbf{I}, \mathbf{0}) <^- \mathbf{B}$ if and only if $\mathbf{B} = (\mathbf{I}, \mathbf{0})$. State and prove the corresponding result for $m > n$ and for $m = n$.

(4) Let $\mathbf{A}$ and $\mathbf{B}$ be matrices of the same order such that $\mathbf{A} <^- \mathbf{B}$ and for a suitable matrix $\mathbf{C}, \mathbf{CA} = \mathbf{I}$. Prove that $\mathbf{A} = \mathbf{B}$. Is the conclusion still valid if there is a suitable matrix $\mathbf{D}$ such that $\mathbf{AD} = \mathbf{I}$? Justify your answer. Conclude that all invertible matrices are maximal under the minus-order.

(5) Let $\mathbf{A}$ and $\mathbf{B}$ be matrices of the same order such that $\mathbf{A} <^- \mathbf{B}$ and $(\mathbf{P}, \mathbf{Q})$ be a rank factorization of $\mathbf{A}$. Prove that (i) $\mathbf{P} = \mathbf{BGP}$ and (ii) $\mathbf{Q}^t = \mathbf{Q}^t\mathbf{GB}$ for some g-inverse $\mathbf{G}$ of $\mathbf{B}$. Also show that $\mathbf{A} = \mathbf{BHB}$ for some matrix $\mathbf{H}$.

(6) Prove that under the minus order every square matrix $\mathbf{A}$ lies below some invertible matrix. (Hint: Consider the normal form of $\mathbf{A}$.)

(7) Let $\mathbf{A}$ and $\mathbf{B}$ be square matrices of the same order such that $\mathbf{A} <^- \mathbf{B}$ and $\mathbf{AB} = \mathbf{BA}$. Show that if $\mathbf{B}$ is of index 1, then $\mathbf{A}$ is of index 1. What can be said about the index of $\mathbf{B}$ if index of $\mathbf{A}$ is 1?

(8) Let $\mathbf{A}$ and $\mathbf{B}$ be square matrices of the same order over the complex field such that $\mathbf{A} <^- \mathbf{B}$. Show that if $\mathbf{B}$ is range hermitian then $\mathbf{A}$ may not be range hermitian.

(9) Let $\mathbf{A}$ and $\mathbf{B}$ be square matrices of the same order over $\mathbb{C}$, the complex field such that $\mathbf{A} <^- \mathbf{B}$. Let $\mathbf{A}$ be normal and $\mathbf{B}$ be a hermitian matrix. Show that $\mathbf{A}$ is hermitian.

(10) Let $\mathbf{A} = \mathbf{AA}^-\mathbf{B} = \mathbf{BA}^-\mathbf{A}$ for some g-inverse $\mathbf{A}^-$ of $\mathbf{A}$. Show that there is a reflexive g-inverse $\mathbf{G}$ of $\mathbf{A}$ such that $\mathbf{A} = \mathbf{AGB} = \mathbf{BGA}$. Let $\mathbf{A} <^s \mathbf{B}$ and for some reflexive g-inverse $\mathbf{A}^-$ of A, $\mathbf{BA}^-\mathbf{A} = \mathbf{A}$. Show that $\mathbf{A} <^- \mathbf{B}$.

(11) Show that for matrices $\mathbf{A}$ and $\mathbf{B}$ over the field $\mathbb{C}$ of complex numbers (i) $\mathbf{A} <^- \mathbf{B} \Leftrightarrow \mathbf{A}^\star <^- \mathbf{B}^\star$.

(ii) $\mathbf{A} <^- \mathbf{B}$ and $\mathcal{C}(\mathbf{A}^\star) \subseteq \mathcal{C}(\mathbf{B}^\star)$. Then $\mathbf{A} = \mathbf{B}$.

(12) Give examples of matrices $\mathbf{A}$ and $\mathbf{B}$ over the field $\mathbb{C}$ such that $\mathbf{A} <^- \mathbf{B}$, but $\mathbf{A}^\dagger \not<^- \mathbf{B}^\dagger$. Further, show that if $\mathbf{A} <^- \mathbf{B}$, then $\mathbf{A}^\dagger <^- \mathbf{B}^\dagger \Leftrightarrow \mathbf{A}^\dagger\,\mathbf{B}\mathbf{A}^\dagger = \mathbf{A}^\dagger$.

(13) Show that the following hold:
(i) $\mathbf{A} <^- \mathbf{B} \Leftrightarrow \mathbf{A}^-\mathbf{B}\mathbf{A}^- = \mathbf{A}^-$ and $\mathbf{B}\mathbf{A}^-\mathbf{B} = \mathbf{A}$ for some $\mathbf{A}^- \in \{\mathbf{A}^-\}$.
(ii) $\mathbf{A} <^- \mathbf{B} \Leftrightarrow \{(A^- + \mathbf{B}^-)/2\} \subseteq \{\mathbf{A}^-\}$ for each $\mathbf{A}^- \in \{\mathbf{A}\}$ and for each $\mathbf{B}^- \in \{\mathbf{B}^-\}$.

(14) Show that $\mathbf{A} <^- \mathbf{B} \Rightarrow \{\mathbf{B}^-\} \subset \{\mathbf{A}^- + (\mathbf{B} - \mathbf{A})^-\}$. Is the converse true?

(15) Show that $\mathbf{A} <^- \mathbf{B} \Rightarrow \{\mathbf{B}_r^-\} \subset \{\mathbf{A}_r^- + (\mathbf{B} - \mathbf{A}_r^-)\}$. Is the converse true? Now suppose $\mathbf{A}$ and $\mathbf{B}$ are *nnd* matrices over the $\mathbb{C}$, the complex field. Show that $\{\mathbf{B}_r^-\} \subset \{\mathbf{A}_r^- + (\mathbf{B} - \mathbf{A})_r^-\} \Rightarrow \mathbf{A} <^- \mathbf{B}$.

(16) For an idempotent matrix $\mathbf{E}$, show that $\mathbf{E} <^- \mathbf{B} \Leftrightarrow \rho(\mathbf{B} - \mathbf{E}) = \rho(\mathbf{B} - \mathbf{B}\mathbf{E}) = \rho(\mathbf{B} - \mathbf{E}\mathbf{B})$.

(17) Let $\mathbf{A} <^- \mathbf{B}$ and one of $\mathcal{C}(\mathbf{B}) \subseteq \mathcal{C}(\mathbf{A})$ and $\mathcal{C}(\mathbf{B}^t) \subseteq \mathcal{C}(\mathbf{A}^t)$ and holds. Show that $\mathbf{A} = \mathbf{B}$. Show the same holds if $\mathbf{A} <^- \mathbf{B}$ and $\mathbf{A}\mathbf{X} = \mathbf{B}$ is consistent.

(18) Show that for square matrices $\mathbf{A}$ and $\mathbf{B}$, $\mathbf{A} <^- \mathbf{I} \Leftrightarrow \mathbf{A}^2 = \mathbf{A}$ and $\mathbf{I} <^- \mathbf{B} \Leftrightarrow \mathbf{B} = \mathbf{I}$.

(19) If for non-null matrix $\mathbf{A}$, $\mathbf{B}\mathbf{A}^-\mathbf{B} = \mathbf{A}$ for some $\mathbf{A}^- \in \{\mathbf{A}^-\}$, then show that $\mathbf{D} = \mathbf{A}\mathbf{B}^-\mathbf{A}$ is invariant under the choice of $\mathbf{B}^-$. Further show that $\mathbf{D}$ is the unique matrix such that $\mathbf{D} <^- \mathbf{B}$.

(20) For $m \times n$ matrices $\mathbf{A}$ and $\mathbf{B}$ over $\mathbb{C}$, show that $\mathbf{A} <^s \mathbf{B} \Leftrightarrow \mathbf{A}\mathbf{A}^\dagger <^- \mathbf{B}\mathbf{B}^\dagger$ and $\mathbf{A}^\dagger\mathbf{A} <^- \mathbf{B}^\dagger\mathbf{B}$.

(21) For $m \times n$ matrices $\mathbf{A}$ and $\mathbf{B}$ over $\mathbb{C}$, prove or disprove the following:
(i) $\mathbf{A} <^- \mathbf{B} \Rightarrow \mathbf{A}\mathbf{A}^\star <^- \mathbf{B}\mathbf{B}^\star$
(ii) $\mathbf{A} <^- \mathbf{B} \Rightarrow \mathbf{A}^\star\mathbf{A} <^- \mathbf{B}^\star\mathbf{B}$.

(22) Let $\mathbf{A}$ and $\mathbf{B}$ be $m \times n$ matrices. If $\mathbf{B}_1$ is an $r \times s$ matrix such that
$$\mathbf{B} = \begin{pmatrix} \mathbf{B}_1 & 0 \\ \mathbf{0} & \mathbf{0} \end{pmatrix},$$
where each $\mathbf{0}$ denotes a null matrix of appropriate order.
Then show that $\mathbf{0} <^- \mathbf{A} <^- \mathbf{B} \Leftrightarrow \mathbf{A} = \begin{pmatrix} \mathbf{A}_1 & 0 \\ \mathbf{0} & \mathbf{0} \end{pmatrix};$
where $\mathbf{A}_1$ an $r \times s$ matrix and $\mathbf{0} <^- \mathbf{A}_1 <^- \mathbf{B}_1$.

(23) Let $\mathbf{A}$ and $\mathbf{B}$ be normal matrices of order $n \times n$ over $\mathbb{C}$. Let $\rho(\mathbf{A}) = a$ and $\rho(\mathbf{B}) = b$, $1 \leq a < b \leq n$ with $m = b - a$. Then show that the following are equivalent:

(i) $\mathbf{A} <^- \mathbf{B}$

(ii) There exists a unitary matrix $\mathbf{U}$ such that $\mathbf{U}^\star\mathbf{AU} = \mathbf{diag(D, 0)}$ and

$$\mathbf{U}^\star\mathbf{BU} = \begin{pmatrix} \mathbf{D+RES} & \mathbf{RE} & \mathbf{0} \\ \mathbf{ES} & \mathbf{E} & \mathbf{0} \\ \mathbf{0} & \mathbf{0} & \mathbf{0} \end{pmatrix},$$

where $\mathbf{D}$ and $\mathbf{E}$ are respectively $a \times a$ and $m \times m$ non-singular diagonal matrices, $\mathbf{R}$ is of order $a \times m$ and $\mathbf{S}$ is order $m \times a$.

(iii) There exists a unitary matrix $\mathbf{U}$ such that $\mathbf{U}^\star\mathbf{AU} = \mathbf{diag(G, 0)}$ and

$$\mathbf{U}^\star\mathbf{BU} = \begin{pmatrix} \mathbf{G+RES} & \mathbf{RF} & \mathbf{0} \\ \mathbf{FS} & \mathbf{F} & \mathbf{0} \\ \mathbf{0} & \mathbf{0} & \mathbf{0} \end{pmatrix},$$

where $\mathbf{D}$ and $\mathbf{F}$ are $a \times a$ and $m \times m$ non-singular matrices, $\mathbf{R}$ is of order $a \times m$ and $\mathbf{S}$ is order $m \times a$.

(24) Let $\mathbf{Q}$ be a projector of order $m \times m$ and $\mathbf{B}$ be a $m \times n$ matrix. Show that the following are equivalent:

(i) $\mathbf{QB} <^- \mathbf{Q}$

(ii) $\{\mathbf{B}^-\} = \{\mathbf{B}^-\mathbf{Q}\} + \{\mathbf{B}^-(\mathbf{I}-\mathbf{Q})\}$, $\{\mathbf{B}^-\mathbf{Q}\} \subseteq \{(\mathbf{QB})^-\}$ and $\{\mathbf{B}^-(\mathbf{I}-\mathbf{Q})\} \subseteq \{(\mathbf{I}-\mathbf{QB})^-\}$.

(25) Let $\mathbf{A}, \mathbf{P}$ and $\mathbf{Q}$ be $m \times n$, $p \times m$ and $n \times q$ matrices respectively. Let matrices $\mathbf{T}$ and $\mathbf{R}$ be such that $\mathcal{N}(\mathbf{T}) = \mathcal{C}(\mathbf{AQ})$, $\mathcal{N}(\mathbf{PA}) = \mathcal{C}(\mathbf{R})$. Show that if $\mathbf{D} = \mathbf{AR}(\mathbf{TAR})_r^-\mathbf{TA}$, then $\mathcal{C}(\mathbf{D}) \subseteq \mathcal{N}(\mathbf{P})$, $\mathcal{C}(\mathbf{D}^t) \subseteq \mathcal{N}(\mathbf{Q}^t)$ and $\mathbf{D} <^- \mathbf{A}$.

(26) Let $\mathbf{A}, \mathbf{A}_1$ and $\mathbf{A}_2$ be $m \times n$ matrices. Then the following are equivalent:

(i) $\mathbf{A} = \mathbf{A}_1 \oplus \mathbf{A}_2$

(ii) $\mathbf{A}_1 <^- \mathbf{A}$, $\mathbf{A}_2 <^- \mathbf{A}$

(iii) $\mathbf{A}_1(\mathbf{A}_1 + \mathbf{A}_2)^-\mathbf{A}_2 = \mathbf{0}$

(iv) $\{\mathbf{A}^-\} \subseteq \{\mathbf{A}_1^-\} \cap \{\mathbf{A}_2^-\}$.

(27) Let $\mathbf{A}$ be an $m \times n$ matrix. Let $\mathbf{x} \in \mathcal{C}(\mathbf{A})$ and $\mathbf{y} \in \mathcal{C}(\mathbf{A^t})$. The vectors $\mathbf{x}$ and $\mathbf{y}$ are said to be separable if there exist disjoint matrices $\mathbf{A}_1$ and $\mathbf{A}_2$ such that

(a) $\mathbf{A} = \mathbf{A}_1 \oplus \mathbf{A}_2$ and

(b) $\mathbf{x} \in \mathcal{C}(\mathbf{A}_2)$ and $\mathbf{y} \in \mathcal{C}(\mathbf{A}_1^\mathbf{t})$.

Clearly, any pair of $\mathbf{x}, \mathbf{y}$ with $\mathbf{x} \in \mathcal{C}(\mathbf{A})$ and $\mathbf{y} \in \mathcal{C}(\mathbf{A}^t)$ is separable if at least one of $\mathbf{x}, \mathbf{y}$ is null. Prove that for a given matrix $\mathbf{A}$, $\mathbf{x} \in \mathcal{C}(\mathbf{A})$ and $\mathbf{y} \in \mathcal{C}(\mathbf{A}^t)$, the vectors $\mathbf{x}$ and $\mathbf{y}$ are separable if and only if $\mathbf{y}^t\mathbf{A}^-\mathbf{x} = \mathbf{0}$ for all g-inverses $\mathbf{A}^-$ of $\mathbf{A}$.

(28) Let $\mathbf{A}$ be an $m \times n$ matrix. Let $\mathbf{x} \in \mathcal{C}(\mathbf{A})$ and $\mathbf{y} \in \mathcal{C}(\mathbf{A}^t)$. Prove that $\mathbf{x}, \mathbf{y}$ are separable if and only if $\rho\begin{pmatrix} \mathbf{A} & \mathbf{x} \\ \mathbf{y}^t & \mathbf{0} \end{pmatrix} = \rho(\mathbf{A})$

(29) Let $\mathbf{A}$ and $\mathbf{B}$ be $m \times n$ matrices such that $\mathbf{A} <^- \mathbf{B}$. Suppose vectors $\mathbf{x}$, and $\mathbf{y}$ are separable with reference to $\mathbf{A}$. Show that they are also separable with reference to $\mathbf{B}$.

Chapter 4

The Sharp Order

4.1 Introduction

We know that if $\mathbf{A}$ is an idempotent matrix (that is $\mathbf{A^2} = \mathbf{A}$), then it is similar to a matrix of the form $\mathbf{diag(I, 0)}$. It also acts as an identity on the column space $\mathcal{C}(\mathbf{A})$ of the matrix $\mathbf{A}$. In the previous chapter, we saw that for two idempotent matrices $\mathbf{A}$ and $\mathbf{B}$ of the same order, $\mathbf{A} <^- \mathbf{B}$ if and only if $\mathbf{A} = \mathbf{AB} = \mathbf{BA} = \mathbf{ABA}$. Observe that an idempotent matrix is a matrix of index not exceeding 1 and is its unique commuting reflexive g-inverse. Let '$\mathcal{I}_1$' denotes the set of all matrices of index ≤ 1. The set '$\mathcal{I}_1$' contains a wide class of matrices. In fact, in addition to all idempotent matrices it includes the class of semi-simple matrices and the class of the range-hermitian matrices. Compare the following property of '$\mathcal{I}_1$' which is analogous to the property of the set of idempotent matrices mentioned above in this paragraph: each non-null matrix $\mathbf{A}$ in '$\mathcal{I}_1$' is similar to diag$(\mathbf{D}, \mathbf{0})$ for some non-singular matrix $\mathbf{D}$ and acts as a non-singular linear operator on $\mathcal{C}(\mathbf{A})$, that is, for each non-null $\mathbf{x} \in \mathcal{C}(\mathbf{A})$, $\mathbf{Ax} \neq \mathbf{0}$.

In this chapter, we define a partial order '$<^{\#}$' on $\mathcal{I}_1$, the set of square matrices of index ≤ 1. We shall show that under this order '$<^{\#}$' $\mathbf{A}$ is below $\mathbf{B}$ (that is $\mathbf{A} <^{\#} \mathbf{B}$) if and only if $\mathbf{A^2} = \mathbf{AB} = \mathbf{BA}$ and therefore, '$<^{\#}$' coincides with the minus order when restricted to idempotent matrices. A subclass of g-inverses called the commuting g-inverses plays an important role in developing this partial order. Recall from Chapter 2 that a matrix $\mathbf{G}$ is said to be a commuting g-inverse of a matrix $\mathbf{A}$ if (i) $\mathbf{AGA} = \mathbf{A}$ and (ii) $\mathbf{AG} = \mathbf{GA}$. Moreover, such a g-inverse for $\mathbf{A}$ exists if and only if $\mathbf{A}$ is of index ≤ 1. For a matrix $\mathbf{A}$ of index ≤ 1, the unique reflexive commuting g-inverse is called its group inverse and is denoted by $\mathbf{A}^{\#}$ Moreover, if $\mathbf{A} = \mathbf{Pdiag(D,\ 0)P^{-1}}$, where $\mathbf{P}$ and $\mathbf{D}$ are non-singular matrices, then every commuting g-inverse of $\mathbf{A}$ is of the form $\mathbf{P}\mathbf{diag}(\mathbf{D}^{-1},\ \mathbf{C})\mathbf{P}^{-1}$ where $\mathbf{C}$ is arbitrary and, in particular, $\mathbf{A}^{\#} = \mathbf{P}\mathbf{diag}(\mathbf{D}^{-1},\ \mathbf{0})\mathbf{P}^{-1}$.

In Section 4.2, we define a partial order, called the sharp order on matrices in $\mathcal{I}_{1,n}$, the set of $n \times n$ matrices of index 1 and obtain several characteristic properties of this order. In Section 4.3, we obtain the class of all matrices lying above (or below) a given matrix under the sharp order. We also specialize to some subclasses of $\mathcal{I}_{1,n}$ and obtain interesting properties of the sharp order. In Section 4.4, we extend the '$<^{\#}$' order to the class of general square matrices. Let $\mathbf{A} = \mathbf{A_1} + \mathbf{A_2}$ and $\mathbf{B} = \mathbf{B_1} + \mathbf{B_2}$ be the core-nilpotent decompositions of $\mathbf{A}$ and $\mathbf{B}$ respectively, where $\mathbf{A_1}$ and $\mathbf{B_1}$ are core parts and $\mathbf{A_2}$ and $\mathbf{B_2}$ are nilpotent parts. We define $\mathbf{A} <^{\mathbf{d}} \mathbf{B}$ if $\mathbf{A_1} <^{\#} \mathbf{B_1}$. (Thus, the nilpotent parts are ignored.) It turns out that $\mathbf{A} <^{\mathbf{d}} \mathbf{B}$ if and only if $\mathbf{A}\mathbf{A}^{\mathbf{D}} = \mathbf{B}\mathbf{A}^{\mathbf{D}} = \mathbf{A}^{\mathbf{D}}\mathbf{B}$, where $\mathbf{A}^{\mathbf{D}}$ is the Drazin inverse of $\mathbf{A}$. We show that '$<^{d}$' is a pre-order and study some of its interesting properties. We finally extend this pre-order to a partial order by involving the nilpotent parts also. We define: $\mathbf{A} <^{\#,-} \mathbf{B}$ if $\mathbf{A_1} <^{\#} \mathbf{B_1}$ and $\mathbf{A_2} <^{-} \mathbf{B_2}$. We also obtain a canonical form for the matrices $\mathbf{A}$ and $\mathbf{B}$ when $\mathbf{A} <^{\#,-} \mathbf{B}$.

4.2 Sharp order - Characteristic properties

The class of commuting g-inverses of a square matrix plays a vital role for the order relation we are about to study now. Recall from Chapter 2 that the group inverse of a square matrix is its unique commuting reflexive g-inverse and that a square matrix possesses a commuting g-inverse if and only if it is of index≤ 1. We begin this section by defining the Sharp order on $\mathcal{I}_{1,n}$, the set of all $n \times n$ matrices of index ≤ 1. We also obtain several characteristic properties of this order.

Definition 4.2.1. Let $\mathbf{A}, \mathbf{B} \in \mathcal{I}_{1,\mathbf{n}}$. $\mathbf{A}$ is said to be below $\mathbf{B}$ under the sharp order if there exist commuting g-inverses $\mathbf{G_1}$ and $\mathbf{G_2}$ of $\mathbf{A}$ such that $\mathbf{A}\mathbf{G_1} = \mathbf{B}\mathbf{G_1}$ and $\mathbf{G_2}\mathbf{A} = \mathbf{G_2}\mathbf{B}$.

When $\mathbf{A}$ is below $\mathbf{B}$ under the sharp order, we write $\mathbf{A} <^{\#} \mathbf{B}$.

Remark 4.2.2. It is obvious that $\mathbf{A} <^{\#} \mathbf{B} \Rightarrow \mathbf{A} <^{-} \mathbf{B}$.

Remark 4.2.3. In Definition 4.2.1, we actually do not require the matrix $\mathbf{B}$ to be of index ≤ 1. For example notice that the null matrix of order $n \times n$ belongs to $\mathcal{I}_{1,n}$ and $\mathbf{0} <^{\#} \mathbf{B}$, for each $n \times n$ matrix $\mathbf{B}$. We know that not every matrix is of index ≤ 1. we now give a non-trivial example.

Example 4.2.4. Consider

$$\mathbf{A} = \begin{pmatrix} 1 & 0 & 0 \\ 0 & 0 & 0 \\ 0 & 0 & 0 \end{pmatrix} \text{ and } \mathbf{B} = \begin{pmatrix} 1 & 0 & 0 \\ 0 & 0 & 1 \\ 0 & 0 & 0 \end{pmatrix}.$$

Clearly, $\mathbf{A^2} = \mathbf{AB} = \mathbf{BA}$, so, $\mathbf{A} <^{\#} \mathbf{B}$. Also, $\rho(\mathbf{B}) = \mathbf{2}$, $\rho(\mathbf{B^2}) = \mathbf{1}$. Thus, $\mathbf{B}$ is not a matrix of index ≤ 1.

However, in what follows, we take both $\mathbf{A}$ and $\mathbf{B}$ to be matrices of index ≤ 1. This is crucial for developing the properties of the sharp order. For example, in order to examine the transitivity of $<^{\#}$, we need to consider $\mathbf{A} <^{\#} \mathbf{B}$ and $\mathbf{B} <^{\#} \mathbf{C}$, where the matrix $\mathbf{B}$ must be of index ≤ 1.

In the previous chapter, we had shown that if $\mathbf{A} <^{-} \mathbf{B}$, then we can get the same g-inverse $\mathbf{G}$ of $\mathbf{A}$ such that $\mathbf{AG} = \mathbf{BG}$ and $\mathbf{GA} = \mathbf{GB}$. Is a similar thing possible for sharp order? The answer is in the affirmative and is contained in the following:

Theorem 4.2.5. *Let* $\mathbf{A}, \mathbf{B} \in \mathcal{I}_{\mathbf{1},\mathbf{n}}$. *Then* $\mathbf{A} <^{\#} \mathbf{B}$ *if and only if*

$$\mathbf{AA^{\#}} = \mathbf{BA^{\#}} \textit{ and } \mathbf{A^{\#}A} = \mathbf{A^{\#}B}.$$

Proof. 'If' part
Since $\mathbf{A^{\#}}$ is a commuting g-inverse of $\mathbf{A}$, the proof is trivial.
'Only if' part
Let $\mathbf{G_1}$ and $\mathbf{G_2}$ be commuting g-inverses of $\mathbf{A}$ such that $\mathbf{AG_1} = \mathbf{BG_1}$ and $\mathbf{G_2A} = \mathbf{G_2B}$. Now, $\mathbf{G_1AG_2}$ is a reflexive commuting g-inverse of $\mathbf{A}$. Therefore, $\mathbf{G_1AG_2} = \mathbf{A^{\#}}$ by uniqueness of the group inverse and $\mathbf{AA^{\#}} = \mathbf{BA^{\#}}$ and $\mathbf{A^{\#}A} = \mathbf{A^{\#}B}$. □

Remark 4.2.6. The sharp order was originally defined in [Mitra (1987)] using the group inverse.

Remark 4.2.7. Let $\mathbf{A}, \mathbf{B} \in \mathcal{I}_{\mathbf{1},\mathbf{n}}$ and $\mathbf{P}$ be an $n \times n$ non-singular matrix. Then $\mathbf{A} <^{\#} \mathbf{B}$ if and only if $\mathbf{PAP^{-1}} <^{\#} \mathbf{PBP^{-1}}$.

In the following theorem we obtain several characterizations of the sharp order.

Theorem 4.2.8. *Let* $\mathbf{A}, \mathbf{B} \in \mathcal{I}_{\mathbf{1},\mathbf{n}}$. *Then the following are equivalent:*

(i) $\mathbf{A} <^{\#} \mathbf{B}$

(ii) $\mathbf{A}^2 = \mathbf{AB} = \mathbf{BA}$ *or equivalently,* $\mathbf{A}(\mathbf{B}-\mathbf{A}) = (\mathbf{B}-\mathbf{A})\mathbf{A} = \mathbf{0}$.

(iii) *Let* $a = \rho(\mathbf{A})$ *and* $b = \rho(\mathbf{B})$. *Then there exists a non-singular matrix* $\mathbf{Q}$ *such that* $\mathbf{A} = \mathbf{Q}\mathrm{diag}(\mathbf{D_a}, \mathbf{0}, \mathbf{0})\,\mathbf{Q}^{-1}$ *and* $\mathbf{B} = \mathbf{Q}\mathrm{diag}(\mathbf{D_a}, \mathbf{D_{b-a}}, \mathbf{0})\,\mathbf{Q}^{-1}$, *where* $\mathbf{D_a}$ *and* $\mathbf{D_{b-a}}$ *are non-singular matrices of order* $a \times a$ *and* $(b-a) \times (b-a)$ *respectively*

(iv) $\{\mathbf{B}^-_{\mathbf{com}}\} \subseteq \{\mathbf{A}^-_{\mathbf{com}}\}$

(v) $\mathbf{B}^\# \in \{\mathbf{A}^-_{\mathbf{com}}\}$

(vi) *There exists a projector* $\mathbf{P}$ *onto* $\mathcal{C}(\mathbf{A})$ *such that* $\mathbf{A} = \mathbf{PB} = \mathbf{BP}$

(vii) $\mathbf{A}^\#\mathbf{A} = \mathbf{A}^\#\mathbf{B}$ *and* $(\mathbf{B}-\mathbf{A})\mathbf{A} = \mathbf{0}$

(viii) $\mathbf{AA}^\# = \mathbf{BA}^\#$ *and* $\mathbf{A}(\mathbf{B}-\mathbf{A}) = \mathbf{0}$ *and*

(ix) $\mathbf{B}-\mathbf{A} <^\# \mathbf{B}$.

Proof. (i) $\Rightarrow$ (ii)
Since $\mathbf{A} <^\# \mathbf{B}$, by Theorem 4.2.5, we have $\mathbf{AA}^\# = \mathbf{BA}^\#$ and $\mathbf{A}^\#\mathbf{A} = \mathbf{A}^\#\mathbf{B}$. Also, $\mathbf{A}^2\mathbf{A}^\# = \mathbf{A} = \mathbf{A}^\#\mathbf{A}^2$. So, $\mathbf{A}^2 = \mathbf{AA}^\#\mathbf{A}^2 = \mathbf{BA}^\#\mathbf{A}^2 = \mathbf{BA}$, since $\mathbf{AA}^\# = \mathbf{BA}^\#$. Similarly, $\mathbf{A}^2 = \mathbf{AB}$.
(ii) $\Rightarrow$ (i) follows from the equalities $(\mathbf{A}^\#)^2\mathbf{A} = \mathbf{A}^\# = \mathbf{A}(\mathbf{A}^\#)^2$.
(i) $\Rightarrow$ (iii)
Since $\mathbf{A}$ is of index ≤ 1, there exist non-singular matrices $\mathbf{R}$ and $\mathbf{D_a}$ such that $\mathbf{A} = \mathbf{R}\mathrm{diag}(\mathbf{D_a}, \mathbf{0}, \mathbf{0})\mathbf{R}^{-1}$ equivalently $\mathbf{R}^{-1}\mathbf{AR} = \mathrm{diag}(\mathbf{D_a}, \mathbf{0}, \mathbf{0})$. Let $\mathbf{R}^{-1}\mathbf{BR} = \mathbf{T}$. Partition $\mathbf{T}$ as $\mathbf{T} = \begin{pmatrix} \mathbf{T_{11}} & \mathbf{T_{12}} & \mathbf{T_{13}} \\ \mathbf{T_{21}} & \mathbf{T_{22}} & \mathbf{T_{23}} \\ \mathbf{T_{31}} & \mathbf{T_{32}} & \mathbf{T_{33}} \end{pmatrix}$ conformable for matrix multiplication with the partition of $\mathbf{R}^{-1}\mathbf{AR}$. Since $\mathbf{A}^2 = \mathbf{AB} = \mathbf{BA}$, we have $\mathbf{T_{11}} = \mathbf{D_a}$ and $\mathbf{T_{12}}, \mathbf{T_{13}}, \mathbf{T_{21}}$ and $\mathbf{T_{31}}$ are null matrices. Thus,

$$\mathbf{T} = \begin{pmatrix} \mathbf{D_a} & \mathbf{0} & \mathbf{0} \\ \mathbf{0} & \mathbf{T_{22}} & \mathbf{T_{23}} \\ \mathbf{0} & \mathbf{T_{32}} & \mathbf{T_{33}} \end{pmatrix}.$$

Since $\mathbf{B}$ is of index ≤ 1 and $\rho(\mathbf{B}) = \mathbf{b}$, the sub-matrix $\begin{pmatrix} \mathbf{T_{22}} & \mathbf{T_{23}} \\ \mathbf{T_{32}} & \mathbf{T_{33}} \end{pmatrix}$ is of index ≤ 1 with rank $b-a$. Hence there exists a non-singular matrix $\mathbf{S}$ such that $\begin{pmatrix} \mathbf{T_{22}} & \mathbf{T_{23}} \\ \mathbf{T_{32}} & \mathbf{T_{33}} \end{pmatrix} = \mathbf{S}\mathrm{diag}(\mathbf{D_{b-a}}, \mathbf{0})\,\mathbf{S}^{-1}$. Let $\mathbf{Q} = \mathbf{R}\mathrm{diag}(\mathbf{I_a}, \mathbf{S})$. Then $\mathbf{Q}$ is a non-singular matrix. Further,

$$\mathbf{A} = \mathbf{Q}\mathrm{diag}(\mathbf{D_a}, \mathbf{0}, \mathbf{0})\ \mathbf{Q}^{-1} \text{ and } \mathbf{B} = \mathbf{Q}\mathrm{diag}(\mathbf{D_a}, \mathbf{D_{b-a}}, \mathbf{0})\mathbf{Q}^{-1}.$$

(iii) $\Rightarrow$ (iv)
Let $\mathbf{G} \in \{\mathbf{B}^-_{\mathbf{com}}\}$. Then $\mathbf{G} = \mathbf{Q}\mathrm{diag}(\mathbf{D_a^{-1}}, \mathbf{D_{b-a}^{-1}}, \mathbf{C})\mathbf{Q}^{-1}$, where $\mathbf{C}$ is an

arbitrary matrix. It is easy to check that $\mathbf{G} \in \{\mathbf{A}^-_{\mathbf{com}}\}$.
(iv) $\Rightarrow$ (v) is trivial.
(v) $\Rightarrow$ (i)
Since $\mathbf{B}$ is of index ≤ 1, there exist non-singular matrices $\mathbf{Q}$ and $\mathbf{T}$ such that $\mathbf{B} = \mathbf{Q}\mathbf{diag}(\mathbf{T}, \mathbf{0})\ \mathbf{Q}^{-1}$. Notice that $\mathbf{B}^{\#} = \mathbf{Q}\mathbf{diag}(\mathbf{T}^{-1}, \mathbf{0})\mathbf{Q}^{-1}$. Partition $\mathbf{Q}^{-1}\mathbf{A}\mathbf{Q}$ conformably as $\mathbf{Q}^{-1}\mathbf{A}\mathbf{Q} = \begin{pmatrix} \mathbf{S_{11}} & \mathbf{S_{12}} \\ \mathbf{S_{21}} & \mathbf{S_{22}} \end{pmatrix}$. Since $\mathbf{B}^{\#}$ is a commuting g-inverse of $\mathbf{A}$, we have $\mathbf{A}\mathbf{B}^{\#} = \mathbf{B}^{\#}\mathbf{A}$. This yields $\mathbf{S_{12}} = \mathbf{0}$, $\mathbf{S_{21}} = \mathbf{0}$ and $\mathbf{S_{11}}\mathbf{T}^{-1} = \mathbf{T}^{-1}\mathbf{S_{11}}$. Further, $\mathbf{A}\mathbf{B}^{\#}\mathbf{A} = \mathbf{A}$ implies $\mathbf{S_{22}} = \mathbf{0}$ and $\mathbf{S_{11}}\mathbf{T}^{-1}\mathbf{S_{11}} = \mathbf{S_{11}}$. Now, let $\mathbf{G} = \mathbf{B}^{\#}\mathbf{A}\mathbf{B}^{\#}$. Then $\mathbf{G}$ is a reflexive commuting g-inverse of $\mathbf{A}$, so $\mathbf{G} = \mathbf{A}^{\#}$ by uniqueness of the group inverse and $\mathbf{GA} = \mathbf{GB}, \mathbf{AG} = \mathbf{BG}$.
(ii) $\Rightarrow$ (vi)
We have $\mathbf{A} = \mathbf{A}\mathbf{A}^{\#}\mathbf{A} = \mathbf{A}^{\#}\mathbf{A}^2 = \mathbf{A}^{\#}\mathbf{A}\mathbf{B}$. Let $\mathbf{P} = \mathbf{A}^{\#}\mathbf{A}$. Then $\mathbf{P}$ is a projector and $\mathbf{A} = \mathbf{PB}$. Now, $\mathbf{BP} = \mathbf{B}\mathbf{A}^{\#}\mathbf{A} = \mathbf{B}\mathbf{A}\mathbf{A}^{\#} = \mathbf{A}^2\mathbf{A}^{\#} = \mathbf{A}$.
(vi) $\Rightarrow$ (vii)
Notice that (vi) gives $\mathcal{C}(\mathbf{A}) \subseteq \mathcal{C}(\mathbf{P})$ and $\mathcal{C}(\mathbf{A}^t) \subseteq \mathcal{C}(\mathbf{P}^t)$ and therefore, $\mathbf{A} = \mathbf{AP} = \mathbf{PA}$. Since $\mathcal{C}(\mathbf{A}^{\#}) = \mathcal{C}(\mathbf{A})$, we have $\mathbf{A}^{\#} = \mathbf{A}^{\#}\mathbf{P} = \mathbf{P}\mathbf{A}^{\#}$. Now, $\mathbf{A}^{\#}\mathbf{A} = \mathbf{A}^{\#}\mathbf{PB} = \mathbf{A}^{\#}\mathbf{B}$. Again $\mathbf{A} = \mathbf{AP} = \mathbf{BP}^2 = \mathbf{BP}$. So, $(\mathbf{B} - \mathbf{A})\mathbf{P} = \mathbf{0}$. Hence, $(\mathbf{B} - \mathbf{A})\mathbf{A} = (\mathbf{B} - \mathbf{A})\mathbf{PB} = \mathbf{0}$.
(vii) $\Rightarrow$ (ii)
Since $(\mathbf{B} - \mathbf{A})\mathbf{A} = \mathbf{0}$, we have $\mathbf{BA} = \mathbf{A}^2$. Pre-multiplying $\mathbf{A}^{\#}\mathbf{A} = \mathbf{A}^{\#}\mathbf{B}$ by $\mathbf{A}^2$, we have $\mathbf{A}^2 = \mathbf{A}^2\mathbf{A}^{\#}\mathbf{A} = \mathbf{A}^2\mathbf{A}^{\#}\mathbf{B}$. Now, from $\mathbf{A}^2 = \mathbf{A}^2\mathbf{A}^{\#}\mathbf{B}$ we have $\mathbf{A}^2 = \mathbf{AB}$.
Proof for (ii) $\Rightarrow$ (vi) $\Rightarrow$ (viii) $\Rightarrow$ (ii) is similar and proof for (ix) $\Leftrightarrow$ (ii) follows from the equivalence of (i) and (ii). □

Remark 4.2.9. Let $\mathbf{A} \in \mathcal{I}_{\mathbf{1,n}}$ and $\mathbf{B}$ be a non-singular matrix. In view of Theorem 4.2.8, $\mathbf{A} <^{\#} \mathbf{B}$ if and only if $\mathbf{A}\mathbf{B}^{-1}\mathbf{A} = \mathbf{B}^{-1}\mathbf{A}\mathbf{B} = \mathbf{A}$.

Theorem 4.2.10. *The sharp order is a partial order on* $\mathcal{I}_{1,n}$.

Proof. The result follows trivially from the representations of $\mathbf{A}$ and $\mathbf{B}$ as given by Theorem 4.2.8 (iii) and (iv), when $\mathbf{A} <^{\#} \mathbf{B}$. □

As noted earlier in Remark 4.2.2, the sharp order implies the minus order. The following example shows that the converse is not true.

Example 4.2.11. Let

$$\mathbf{S} = \begin{pmatrix} 1 & 0 \\ 0 & 0 \end{pmatrix}, \ \mathbf{T} = \begin{pmatrix} 0 & 0 \\ 0 & 1 \end{pmatrix}, \ \mathbf{A} = \begin{pmatrix} \mathbf{I} & \mathbf{S} \\ \mathbf{T} & \mathbf{0} \end{pmatrix} \text{ and } \mathbf{B} = \begin{pmatrix} \mathbf{I} & \mathbf{S} \\ \mathbf{T} & -\mathbf{I} \end{pmatrix};$$

where $\mathbf{I}$ is a 2×2 unit matrix. Then $\mathbf{A} <^- \mathbf{B}$ but $\mathbf{A} \not<^\# \mathbf{B}$, since $\mathbf{A}^\# = \mathbf{A}$ and $\mathbf{A}(\mathbf{B}-\mathbf{A}) \neq \mathbf{0}$.

Notice that Example 4.2.11 shows that even on matrices of index ≤ 1 the minus order '$<^-$' does not imply the sharp order. Thus, for the minus order to imply the sharp order some additional conditions are required. The following theorem includes some of the equivalent additional conditions towards this end.

Theorem 4.2.12. *Let* $\mathbf{A}, \mathbf{B} \in \mathcal{I}_{\mathbf{1},\mathbf{n}}$ *such that* $\mathbf{A} <^- \mathbf{B}$. *Then the following are equivalent:*

(i) $\mathbf{A} <^\# \mathbf{B}$,
(ii) $\mathbf{A}$ *and* $\mathbf{B}$ *commute,*
(iii) $\mathbf{A}^\#$ *and* $\mathbf{B}$ *commute,*
(iv) $\mathbf{A}$ *and* $\mathbf{B}^\#$ *commute,*
(v) $\mathbf{A}^\#$ *and* $\mathbf{B}^\#$ *commute*
(vi) $\mathbf{B}^\#\mathbf{A}\mathbf{B}^\# = \mathbf{A}^\#$,
(vii) $\mathbf{B}\mathbf{A}^\#\mathbf{B} = \mathbf{A}$ *and*
(viii) $\mathcal{C}(\mathbf{B}\mathbf{A}^\#\mathbf{B}) \subseteq \mathcal{C}(\mathbf{A})$, $\mathcal{C}((\mathbf{B}\mathbf{A}^\#\mathbf{B})^\mathbf{t}) \subseteq \mathcal{C}(\mathbf{A}^\mathbf{t})$.

Proof. (i) $\Rightarrow$ (ii) follows from Theorem 4.2.8 (i) $\Rightarrow$ (ii).
(ii) $\Rightarrow$ (iii)
Since $\mathbf{A}$ and $\mathbf{B}$ commute and $\mathbf{A}^\#$ is a polynomial in $\mathbf{A}$, it follows that $\mathbf{A}^\#$ and $\mathbf{B}$ commute.
(iii) $\Rightarrow$ (iv)
Let $\mathbf{A}^\#$ and $\mathbf{B}$ commute. Since, $(\mathbf{A}^\#)^\# = \mathbf{A}$, $\mathbf{A}$ is a polynomial in $\mathbf{A}^\#$ and $\mathbf{B}^\#$ is a polynomial in $\mathbf{B}$, therefore $\mathbf{A}$ and $\mathbf{B}^\#$ commute.
(iv) $\Rightarrow$ (v)
Let $\mathbf{A}$ and $\mathbf{B}^\#$ commute. Since $\mathbf{A}^\#$ is a polynomial in $\mathbf{A}$, $\mathbf{A}^\#$ and $\mathbf{B}^\#$ commute.
(v) $\Rightarrow$ (vi)
Let $\mathbf{B} = \mathbf{P}\mathbf{diag}(\mathbf{T}, \mathbf{0})\,\mathbf{P}^{-1}$, where $\mathbf{T}$ is a non-singular matrix. As $\mathbf{A} <^- \mathbf{B}$, we have $\mathbf{A} = \mathbf{P}\mathbf{diag}(\mathbf{S}, \mathbf{0})\mathbf{P}^{-1}$ for some matrix $\mathbf{S}$ such that $\mathbf{S}\mathbf{T}^{-1}\mathbf{S} = \mathbf{S}$. Since $\mathbf{A}$ and $\mathbf{B}^\#$ commute, $\mathbf{S}$ and $\mathbf{T}^{-1}$ commute. Also, it is easy to check that $\mathbf{S}^\# = \mathbf{T}^{-1}\mathbf{S}\mathbf{T}^{-1}$ and hence $\mathbf{A}^\# = \mathbf{B}^\#\mathbf{A}\mathbf{B}^\#$.
(vi) $\Rightarrow$ (vii)
$\mathbf{B}\mathbf{A}^\#\mathbf{B} = \mathbf{B}\mathbf{B}^\#\mathbf{A}\mathbf{B}^\#\mathbf{B} = \mathbf{A}$ follows, as $\mathcal{C}(\mathbf{A}) \subseteq \mathcal{C}(\mathbf{B})$ and $\mathcal{C}(\mathbf{A}^\mathbf{t}) \subseteq \mathcal{C}(\mathbf{B}^\mathbf{t})$.
(vii) $\Rightarrow$ (viii) is obvious.

(viii) $\Rightarrow$ (i)
As $\mathbf{A} <^{-} \mathbf{B}$, $\mathbf{A} = \mathbf{AB}^{\#}\mathbf{A} = \mathbf{AB}^{\#}\mathbf{B} = \mathbf{BB}^{\#}\mathbf{A}$. Now,
$\mathcal{C}(\mathbf{BA}^{\#}\mathbf{B}) \subseteq \mathcal{C}(\mathbf{A}) \Rightarrow \mathbf{AA}^{\#}(\mathbf{BA}^{\#}\mathbf{B}) = \mathbf{BA}^{\#}\mathbf{B}$
$\Rightarrow \mathbf{AB}^{\#}(\mathbf{AA}^{\#}(\mathbf{BA}^{\#}\mathbf{B})) = \mathbf{AB}^{\#}(\mathbf{BA}^{\#}\mathbf{B})$
$\Rightarrow (\mathbf{AB}^{\#}\mathbf{A})(\mathbf{A}^{\#}(\mathbf{BA}^{\#}\mathbf{B})) = (\mathbf{AB}^{\#}\mathbf{B})\mathbf{A}^{\#}\mathbf{B}$
$\Rightarrow \mathbf{AA}^{\#}(\mathbf{BA}^{\#}\mathbf{B}) = \mathbf{AA}^{\#}\mathbf{B}$
$\Rightarrow \mathbf{BA}^{\#}\mathbf{B} = \mathbf{AA}^{\#}\mathbf{B}$.

Also, $\mathbf{A} <^{-} \mathbf{B} \Rightarrow \mathcal{C}(\mathbf{A}) \subseteq \mathcal{C}(\mathbf{B}) \Rightarrow \mathbf{A} = \mathbf{BU}$ for some $\mathbf{U}$. Therefore, $\mathbf{A} = \mathbf{AA}^{\#}\mathbf{A} = \mathbf{AA}^{\#}\mathbf{BU} = \mathbf{BA}^{\#}\mathbf{BU} = \mathbf{BA}^{\#}\mathbf{A}$. So, $\mathbf{AA}^{\#} = \mathbf{BA}^{\#}$.

Similarly, $\mathbf{A}^{\#}\mathbf{A} = \mathbf{A}^{\#}\mathbf{B}$. □

Corollary 4.2.13. *Let* $\mathbf{A}, \mathbf{B} \in \mathcal{I}_{1,\mathbf{n}}$. *Then the following are equivalent:*

(i) $\mathbf{A} <^{\#} \mathbf{B}$,
(ii) $\mathbf{A}^{\#}$ *and* $\mathbf{B}^{\#}$ *commute and*
(iii) $\mathbf{A}^{\#} <^{\#} \mathbf{B}^{\#}$.

It is important to note that for the equalities $\mathbf{AA}^{\#} = \mathbf{BA}^{\#}$ and $\mathbf{A}^{\#}\mathbf{A} = \mathbf{A}^{\#}\mathbf{B}$ to hold simultaneously, we require only $\mathbf{A}$ to be of index ≤ 1, whatever be the index of $\mathbf{B}$. In fact, it is easy to show that if $\mathbf{A} = \mathbf{P}\mathrm{diag}(\mathbf{D}, \mathbf{0})\mathbf{P}^{-1}$, $\mathbf{AA}^{\#} = \mathbf{BA}^{\#}$ and $\mathbf{A}^{\#}\mathbf{A} = \mathbf{A}^{\#}\mathbf{B}$, then $\mathbf{B} = \mathbf{P}\mathrm{diag}(\mathbf{D}, \mathbf{C})\mathbf{P}^{-1}$ for some arbitrary matrix $\mathbf{C}$. However, as mentioned earlier for the order relation '$<^{\#}$' to be a partial order, we do need $\mathbf{B}$ to be of index ≤ 1 and therefore we continue to take both $\mathbf{A}$ and $\mathbf{B}$ to be of index ≤ 1 in the rest of the chapter as well. We hasten to remark that some of the results proved above hold good even when $\mathbf{B}$ is not of index ≤ 1.

The reverse order law for the group inverse of a product of two (or a finite number) of matrices does not hold in general. However, in presence of the sharp order $\mathbf{A} <^{\#} \mathbf{B}$ on matrices $\mathbf{A}$ and $\mathbf{B}$, we have the following:

Theorem 4.2.14. *Let* $\mathbf{A}, \mathbf{B} \in \mathcal{I}_{1,\mathbf{n}}$ *such that* $\mathbf{A} <^{\#} \mathbf{B}$. *Then* $(\mathbf{AB})^{\#} = \mathbf{B}^{\#}\mathbf{A}^{\#} = \mathbf{A}^{\#}\mathbf{B}^{\#} = (\mathbf{A}^{\#})^{2}$.

Proof. The proof follows trivially in view of the representation of $\mathbf{A}$ and $\mathbf{B}$ in Theorem 4.2.8(iii). □

Theorem 4.2.15. *Let* $\mathbf{A}, \mathbf{B} \in \mathcal{I}_{1,\mathbf{n}}$ *such that* $\mathbf{A} <^{\#} \mathbf{B}$. *Then,* $\mathbf{A} <^{-} \mathbf{B}$ *and* $\mathbf{A}^{\mathbf{k}} <^{\#} \mathbf{B}^{\mathbf{k}}$ *hold for each positive integer k.*

Proof. The proof follows trivially in view of the representation of $\mathbf{A}$ and $\mathbf{B}$ in Theorem 4.2.8(iii). □

Remark 4.2.16. Let $\mathbf{A}, \mathbf{B} \in \mathcal{I}_{1,n}$. The conditions $\mathbf{A^k} <^- \mathbf{B^k}$ for each positive integer k and $\mathbf{A^k} <^\# \mathbf{B^k}$ for each positive integer $k \geq 2$ do not imply $\mathbf{A} <^\# \mathbf{B}$. For example, take $\mathbf{A}$ and $\mathbf{B}$ of Example 4.2.11.

4.3 Sharp order - Other properties

In this section, we obtain the class of all g-inverses $\mathbf{A}^-$ of $\mathbf{A}$ such that $\mathbf{A^-A} = \mathbf{A^\# A} = \mathbf{A^\# B} = \mathbf{A^- B}$ when $\mathbf{A} <^\# \mathbf{B}$. We also obtain the class of all matrices $\mathbf{B}$ lying above a given matrix $\mathbf{A}$ under the sharp order. We then obtain a solution to the dual problem of obtaining the class of all matrices $\mathbf{A}$ lying below a given matrix $\mathbf{B}$ under this order. We conclude the section with an extension of Fisher-Cochran Theorem for matrices of index ≤ 1. We begin with the following theorem which is analogous to Theorem 3.5.6 but is actually much stronger.

Theorem 4.3.1. *Let* $\mathbf{A}, \mathbf{B} \in \mathcal{I}_{1,n}$ *such that* $\mathbf{A} <^\# \mathbf{B}$. *Let* $\mathbf{A}^-_{\mathbf{com}}$ *be any commuting g-inverse of* $\mathbf{A}$ *such that* $\mathbf{A}^-_{\mathbf{com}}\mathbf{A} = \mathbf{A}^-_{\mathbf{com}}\mathbf{B}$ *and* $\mathbf{A}\mathbf{A}^-_{\mathbf{com}} = \mathbf{B}\mathbf{A}^-_{\mathbf{com}}$. *Then for every commuting g-inverse* $\mathbf{B}^-_{\mathbf{com}}$ *of* $\mathbf{B}$, *we have*

$$\mathbf{A}^-_{\mathbf{com}}\mathbf{A} = \mathbf{B}^-_{\mathbf{com}}\mathbf{A} \text{ and } \mathbf{A}\mathbf{A}^-_{\mathbf{com}} = \mathbf{A}\mathbf{B}^-_{\mathbf{com}}\ .$$

Proof. Since $\mathbf{A} <^\# \mathbf{B}$, by Theorem 4.2.8(iii), there exists a non-singular matrix $\mathbf{P}$ such that

$$\mathbf{A} = \mathbf{P}\mathrm{\mathbf{diag}}(\mathbf{D_a},\ \mathbf{0},\ \mathbf{0})\mathbf{P}^{-1} \text{ and } \mathbf{B} = \mathbf{P}\mathrm{\mathbf{diag}}(\mathbf{D_a},\ \mathbf{D_{b-a}},\ \mathbf{0})\mathbf{P}^{-1},$$

where $\mathbf{D_a}$ and $\mathbf{D_{b-a}}$ are non-singular matrices and $a = \rho(\mathbf{A})$ and $b = \rho(\mathbf{B})$. It is easy to check that the class of all commuting g-inverses $\mathbf{A}^-_{\mathbf{com}}$ of $\mathbf{A}$ such that $\mathbf{A}^-_{\mathbf{com}}\mathbf{A} = \mathbf{A}^-_{\mathbf{com}}\mathbf{B}$ and $\mathbf{A}\mathbf{A}^-_{\mathbf{com}} = \mathbf{B}\mathbf{A}^-_{\mathbf{com}}$ is given by $\mathbf{P}\mathrm{\mathbf{diag}}(\mathbf{D_a^{-1}},\ \mathbf{0_{b-a}},\ \mathbf{C})\mathbf{P}^{-1}$, where $\mathbf{C}$ is arbitrary. Also, we know that the class of all commuting g-inverses $\mathbf{B}^-_{\mathbf{com}}$ of $\mathbf{B}$ is given by $\mathbf{P}\mathrm{\mathbf{diag}}(\mathbf{D_a^{-1}},\ \mathbf{D_{b-a}^{-1}},\ \mathbf{E})\mathbf{P}^{-1}$, where $\mathbf{E}$ is arbitrary. The rest of the proof is computational. □

Corollary 4.3.2. *Let* $\mathbf{A}, \mathbf{B} \in \mathcal{I}_{1,n}$ *such that* $\mathbf{A} <^\# \mathbf{B}$ *and* $\mathbf{A}^-_{\mathbf{com}}$ *be any commuting g-inverse of* $\mathbf{A}$ *for which* $\mathbf{A}^-_{\mathbf{com}}\mathbf{A} = \mathbf{A}^-_{\mathbf{com}}\mathbf{B}$ *and* $\mathbf{A}\mathbf{A}^-_{\mathbf{com}} = \mathbf{B}\mathbf{A}^-_{\mathbf{com}}$. *Then for every commuting g-inverse* $\mathbf{B}^-_{\mathbf{com}}$ *of* $\mathbf{B}, \mathbf{A}^-_{\mathbf{com}}\mathbf{A}$ *and* $\mathbf{B}^-_{\mathbf{com}}\mathbf{B}$ *commute.*

Corollary 4.3.3. *Let* $\mathbf{A}, \mathbf{B} \in \mathcal{I}_{1,n}$ *such that* $\mathbf{A} <^\# \mathbf{B}$ *and* $\mathbf{A}^-_{\mathbf{com}}$ *be any commuting g-inverse of* $\mathbf{A}$ *for which* $\mathbf{A}^-_{\mathbf{com}}\mathbf{A} = \mathbf{A}^-_{\mathbf{com}}\mathbf{B}$ *and* $\mathbf{A}\mathbf{A}^-_{\mathbf{com}} = \mathbf{B}\mathbf{A}^-_{\mathbf{com}}$. *Then for any commuting g-inverse of* $\mathbf{B}^-_{\mathbf{com}}$ *of* $\mathbf{B}$ *we have*

$$\mathbf{A}^-_{\mathbf{com}}\mathbf{A} = \mathbf{A}^-_{\mathbf{com}}\mathbf{B} = \mathbf{B}^-_{\mathbf{com}}\mathbf{A} \text{ and } \mathbf{A}\mathbf{A}^-_{\mathbf{com}} = \mathbf{B}\mathbf{A}^-_{\mathbf{com}} = \mathbf{A}\mathbf{B}^-_{\mathbf{com}}.$$

While proving Theorem 4.3.1, we have already obtained a characterization of $\{\mathbf{A}^-_{\mathbf{com}}\}_\mathbf{B} = \{\mathbf{G} : \mathbf{AGA} = \mathbf{A}, \mathbf{AG} = \mathbf{GA}, \mathbf{AG} = \mathbf{BG}$ and $\mathbf{GA} = \mathbf{GB}\}$, when $\mathbf{A} <^{\#} \mathbf{B}$. We shall obtain another characterization below. First we prove

Lemma 4.3.4. *Let* $\mathbf{A}, \mathbf{B} \in \mathcal{I}_{1,n}$ *such that* $\mathbf{A} <^{\#} \mathbf{B}$. *Then*

$$(\mathbf{B} - \mathbf{A})^{\#} = \mathbf{B}^-_{\mathbf{com}}(\mathbf{B} - \mathbf{A})\mathbf{B}^-_{\mathbf{com}},$$

where $\mathbf{B}^-_{\mathbf{com}}$ *is any commuting g-inverse of* $\mathbf{B}$.

Proof. Since $\mathbf{A} <^{\#} \mathbf{B}$, we have $\mathbf{B} - \mathbf{A} <^{\#} \mathbf{B}$. By Theorem 4.2.8(ix), $\mathbf{B}^-_{\mathbf{com}}$ is a commuting g-inverse of $\mathbf{B} - \mathbf{A}$. So, $\mathbf{B}^-_{\mathbf{com}}(\mathbf{B} - \mathbf{A})\mathbf{B}^-_{\mathbf{com}}$, is the unique reflexive g-inverse of $\mathbf{B} - \mathbf{A}$ for each $\mathbf{B}^-_{\mathbf{com}}$. Therefore,

$$(\mathbf{B} - \mathbf{A})^{\#} = \mathbf{B}^-_{\mathbf{com}}(\mathbf{B} - \mathbf{A})\mathbf{B}^-_{\mathbf{com}}.$$

□

Theorem 4.3.5. *Let* $\mathbf{A}, \mathbf{B} \in \mathcal{I}_{1,n}$ *such that* $\mathbf{A} <^{\#} \mathbf{B}$. *Then*

$$\begin{aligned}\{\mathbf{A}^-_{\mathbf{com}}\}_\mathbf{B} &= \{\mathbf{B}^-_{\mathbf{com}} - \mathbf{B}^-_{\mathbf{com}}(\mathbf{B} - \mathbf{A})\mathbf{B}^-_{\mathbf{com}}\} \\ &= \{\mathbf{B}^-_{\mathbf{com}} - (\mathbf{B} - \mathbf{A})^{\#}\}.\end{aligned}$$

Proof. Follows easily from the proof of Theorem 4.3.1 and Lemma 4.3.4.

□

Corollary 4.3.6. *Let* $\mathbf{A}, \mathbf{B} \in \mathcal{I}_{1,n}$ *such that* $\mathbf{A} <^{\#} \mathbf{B}$. *Then* $\mathbf{A} <^{-} \mathbf{B}$ *and* $\mathbf{A}^{\#} = \mathbf{B}^{\#} - (\mathbf{B} - \mathbf{A})^{\#}$.

However, the converse of above the corollary is not true and can be shown by using the matrices $\mathbf{A}$ and $\mathbf{B}$ of Example 4.2.11.

Let $\mathbf{A} <^{\#} \mathbf{B}$. In view of Theorem 4.2.8, there exists a projector $\mathbf{P}$ such that $\mathbf{A} = \mathbf{PB} = \mathbf{BP}$. We now characterize the class of all projectors $\mathbf{P}$ such that $\mathbf{A} = \mathbf{PB} = \mathbf{BP}$, when $\mathbf{A} <^{\#} \mathbf{B}$.

Theorem 4.3.7. *Let* $\mathbf{A}, \mathbf{B} \in \mathcal{I}_{1,n}$ *such that* $\mathbf{A} <^{\#} \mathbf{B}$. *Then the class of all projectors* $\mathbf{P}$ *such that* $\mathbf{A} = \mathbf{PB} = \mathbf{BP}$ *is given by*

$$\mathbf{P} = \mathbf{P_A} + \mathbf{CTD},$$

where $\mathbf{P_A} = \mathbf{A}^{\#}\mathbf{A}$ *is the projector projecting vectors onto* $\mathcal{C}(\mathbf{A})$ *along* $\mathcal{N}(\mathbf{A})$, $(\mathbf{C}, \mathbf{D})$ *is a rank factorization of* $\mathbf{I} - \mathbf{B}^{\#}\mathbf{B}$ *and* $\mathbf{T}$ *is an arbitrary idempotent matrix of appropriate order.*

Proof. Notice that $\mathbf{A} <^{\#} \mathbf{B} \Rightarrow \mathbf{AA}^{\#} = \mathbf{BA}^{\#}$ and $\mathbf{A}^{\#}\mathbf{A} = \mathbf{A}^{\#}\mathbf{B}$. Also $\mathbf{I} - \mathbf{B}^{\#}\mathbf{B} = \mathbf{CD} \Rightarrow \mathbf{BC} = \mathbf{0}, \mathbf{DB} = \mathbf{0}$. So, $\mathbf{A} = \mathbf{P_A B} = \mathbf{BP_A}$. Thus, if $\mathbf{P}$ is a matrix of the form $\mathbf{P} = \mathbf{P_A} + \mathbf{CTD}$, then $\mathbf{B}(\mathbf{P_A} + \mathbf{CTD}) = \mathbf{BP_A} = \mathbf{A}$ and $(\mathbf{P_A} + \mathbf{CTD})\mathbf{B} = \mathbf{P_A B} = \mathbf{A}$ for all $\mathbf{T}$.

Let $\mathbf{P}$ be a projector such that $\mathbf{A} = \mathbf{PB} = \mathbf{BP}$. We show that $\mathbf{P}$ is of the given form. Let $\mathbf{P} = \mathbf{P_A} + \mathbf{Q}$. Since $\mathbf{A} = \mathbf{PB} = \mathbf{BP}$ and $\mathbf{A} = \mathbf{P_A B} = \mathbf{BP_A}$, so, $\mathbf{QB} = \mathbf{0}$, $\mathbf{BQ} = \mathbf{0}$. Also, $\mathbf{P_A P} = \mathbf{PP_A} = \mathbf{P_A}$, we have $\mathbf{P_A Q} = \mathbf{QP_A} = \mathbf{0}$. Thus, $\mathbf{Q}$ is idempotent. Let $(\mathbf{C}, \mathbf{D})$ be a rank factorization of $\mathbf{I} - \mathbf{B}^{\#}\mathbf{B}$. As $\mathbf{QB} = \mathbf{0}$, $\mathbf{BQ} = \mathbf{0}$, there exists a matrix $\mathbf{T}$ such that $\mathbf{Q} = \mathbf{CTD}$. Also, $\mathbf{CTD} = \mathbf{Q} = \mathbf{Q}^2 = \mathbf{CTDCTD} = \mathbf{CT}^2\mathbf{D}$, since $\mathbf{I} - \mathbf{B}^{\#}\mathbf{B}$ is an idempotent. Thus, $\mathbf{CTD} = \mathbf{CT}^2\mathbf{D}$. As $\mathbf{C}$ has a left inverse and $\mathbf{D}$ has a right inverse, we have $\mathbf{T} = \mathbf{T}^2$ i.e. $\mathbf{T}$ is idempotent.□

Given a matrix $\mathbf{A} \in \mathcal{I}_{1,n}$, we now characterize the class of all matrices $\mathbf{B} \in \mathcal{I}_{1,n}$ such that $\mathbf{A} <^{\#} \mathbf{B}$. First we prove a lemma.

Lemma 4.3.8. *Let* $\mathbf{A}$ and $\mathbf{B}$ *be matrices such that* $\mathbf{B} = \mathbf{A} \oplus (\mathbf{B} - \mathbf{A})$ *and* $\mathbf{B}^2 = \mathbf{A}^2 \oplus (\mathbf{B} - \mathbf{A})^2$. *If any two of* $\mathbf{A}, \mathbf{B}$ *and* $\mathbf{B} - \mathbf{A}$ *are of index* ≤ 1, *the the third is of index* ≤ 1.

Proof. Trivial. □

Theorem 4.3.9. *Let* $\mathbf{A} \in \mathcal{I}_{1,n}$. *Then for a matrix* $\mathbf{B} \in \mathcal{I}_{1,n}$, $\mathbf{A} <^{\#} \mathbf{B}$ *if and only if* $\mathbf{B} = \mathbf{A} + (\mathbf{I} - \mathbf{AA}^{\#})\mathbf{Z}(\mathbf{I} - \mathbf{A}^{\#}\mathbf{A})$ *for some* $\mathbf{Z}$ *such that* $(\mathbf{I} - \mathbf{AA}^{\#})\mathbf{Z}(\mathbf{I} - \mathbf{A}^{\#}\mathbf{A})$ *is of index* ≤ 1.

Proof. Notice that $\mathbf{B} = \mathbf{A} + (\mathbf{I} - \mathbf{AA}^{\#})\mathbf{Z}(\mathbf{I} - \mathbf{A}^{\#}\mathbf{A})$
$\Leftrightarrow \mathbf{A}(\mathbf{B} - \mathbf{A}) = (\mathbf{B} - \mathbf{A})\mathbf{A} = \mathbf{0}$
$\Leftrightarrow \mathbf{A} <^{\#} \mathbf{B}$.
Further, $\mathbf{B}^2 = \mathbf{A}^2 + \{((\mathbf{I} - \mathbf{AA}^{\#})\mathbf{Z}(\mathbf{I} - \mathbf{A}^{\#}\mathbf{A}))^2\}$. In view of the fact that the sums involved are direct, the result follows by Lemma 4.3.8. □

Corollary 4.3.10. *Let* $\mathbf{A}$ *be a non-singular matrix. Then* $\boldsymbol{diag}(\mathbf{A}, \mathbf{0}) <^{\#} \mathbf{B}$ *if and only if* $\mathbf{B} = \mathbf{diag}(\mathbf{A}, \mathbf{C})$ *for some matrix* $\mathbf{C}$ *of index* ≤ 1.

Corollary 4.3.11. *Let* $\mathbf{A} \in \mathcal{I}_{1,n}$. *The class of all matrices* $\mathbf{B} \in \mathcal{I}_{1,n}$ *such that* $\mathbf{A} <^{\#} \mathbf{B}$ *is given by* $\mathbf{B} = \mathbf{A} \oplus \mathbf{E}$ *where* $\mathbf{E}$ *is an arbitrary matrix of index* ≤ 1. *such that* $\mathbf{AE} = \mathbf{EA} = \mathbf{0}$.

Let $\mathbf{B} \in \mathcal{I}_{1,n}$ be a given matrix. We now obtain the class of all $\mathbf{A} \in \mathcal{I}_{1,n}$ such that $\mathbf{A} <^{\#} \mathbf{B}$.

Theorem 4.3.12. *Let* $\mathbf{B} \in \mathcal{I}_{1,n}$. *Then the class of all* $\mathbf{A} \in \mathcal{I}_{1,n}$ *such that* $\mathbf{A} <^{\#} \mathbf{B}$ *is given by* $\mathbf{A} = \mathbf{PCQ}$, *where* $(\mathbf{P}, \mathbf{Q})$ *is a rank factorization of* $\mathbf{B}$ *and* $\mathbf{C}$ *is an arbitrary idempotent matrix such that* $\mathbf{CQP} = \mathbf{QPC}$.

Proof. Let $(\mathbf{P}, \mathbf{Q})$ is a rank factorization of $\mathbf{B}$. Then $\mathbf{QP}$ is non-singular. Let $\mathbf{C}$ be an idempotent matrix such that $\mathbf{CQP} = \mathbf{QPC}$ and $\mathbf{A} = \mathbf{PCQ}$. Clearly, $\rho(\mathbf{A}) = \rho(\mathbf{C})$, as $\mathbf{P}$ and $\mathbf{Q}$ are full rank matrices. Further, $\mathbf{A^2} = \mathbf{PCQPCQ} = \mathbf{PC^2QPQ}$. Since $\mathbf{QP}$ is non-singular and $\mathbf{P}, \mathbf{Q}$ are full rank matrices, $\rho(\mathbf{A^2}) = \rho(\mathbf{C}) = \rho(\mathbf{A})$. Thus, $\mathbf{A} \in \mathcal{I}_{1,n}$. Also, $\mathbf{A^2} = \mathbf{PCQPCQ} = \mathbf{PC^2QPQ} = \mathbf{PCQPQ} = \mathbf{AB}$.

Similarly, we can show $\mathbf{A^2} = \mathbf{BA}$.

Next, let $\mathbf{A} \in \mathcal{I}_{1,n}$ such that $\mathbf{A} <^{\#} \mathbf{B}$. Then $\mathbf{A^2} = \mathbf{BA} = \mathbf{AB}$. Also, $\mathcal{C}(\mathbf{A}) \subseteq \mathcal{C}(\mathbf{B})$, and $\mathcal{C}(\mathbf{A^t}) \subseteq \mathcal{C}(\mathbf{B^t})$. So, $\mathbf{A} = \mathbf{PCQ}$, for some matrix $\mathbf{C}$. Since $\mathbf{A^2} = \mathbf{BA} = \mathbf{AB}$, we have $\mathbf{PCQPCQ} = \mathbf{PQPCQ} = \mathbf{PCQPQ}$. Since $\mathbf{P}$ and $\mathbf{Q}$ are full rank matrices, this further gives $\mathbf{CQPC} = \mathbf{QPC} = \mathbf{CQP}$. Now, $\mathbf{CQPC} = \mathbf{CQP}$ and $\mathbf{CQP} = \mathbf{QPC}$, so, $\mathbf{C^2QP} = \mathbf{CQP}$. As $\mathbf{QP}$ is non-singular, we have $\mathbf{C^2} = \mathbf{C}$. □

Let $\mathbf{A}, \mathbf{B}$ be square matrices of the same order over an algebraically closed field. Let $\mathbf{B}$ be of index ≤ 1 and $\mathbf{B} = \mathbf{P}\mathbf{diag}(\mathbf{J_1}, \ldots, \mathbf{J_r}, \mathbf{0})\mathbf{P^{-1}}$ be the Jordan decomposition of $\mathbf{B}$, where $\mathbf{J_1}, \ldots, \mathbf{J_r}$ are non-singular Jordan blocks and $\mathbf{P}$ is a non-singular matrix. Let $\mathbf{A} = \mathbf{P}\mathbf{diag}(\mathbf{D_1}, \ldots, \mathbf{D_r}, \mathbf{0})\,\mathbf{P^{-1}}$, where $\mathbf{D_{i_j}} = \mathbf{J_{i_j}}$, $i = 1, \ldots, s$ for sub-permutation $(i_1, \ldots, i_s)$ of $\{1, \ldots, r\}$ and $\mathbf{D_t} = \mathbf{0}$ for $t \in \{1, \ldots, r\} \cap \{i_1, \ldots, i_s\}^c$. It is easy to check that $\mathbf{A}$ is of index ≤ 1 and $\mathbf{A} <^{\#} \mathbf{B}$. We now prove the converse of this statement in a special case where there is exactly one Jordan block corresponding to each non-null eigen-value (or equivalently, the geometric multiplicity of each non-null eigen-value is 1).

Theorem 4.3.13. *Let* $\mathbf{A}, \mathbf{B}$ *be square matrices of same order over an algebraically closed field. Let* $\mathbf{B}$ *be of index* ≤ 1 *and* $\mathbf{B} = \mathbf{P}\mathbf{diag}(\mathbf{J_1}, \ldots, \mathbf{J_r}, \mathbf{0})\mathbf{P^{-1}}$ *be the Jordan decomposition of* $\mathbf{B}$, *where* $\mathbf{J_1}, \ldots, \mathbf{J_r}$ *are non-singular Jordan blocks corresponding to distinct eigen-values* $\lambda_1, \ldots, \lambda_r$ *and* $\mathbf{P}$ *is a non-singular matrix. Then* $\mathbf{A}$ *is of index* ≤ 1 *and* $\mathbf{A} <^{\#} \mathbf{B}$ *if and only if either* $\mathbf{A}$ *is a null matrix or* $\mathbf{A} = \mathbf{P}\mathbf{diag}(\mathbf{D_1}, \ldots, \mathbf{D_r}, \mathbf{0})\mathbf{P^{-1}}$, *where* $\mathbf{D_{i_j}} = \mathbf{J_{i_j}}$, $j = 1, \ldots, s$ *for some sub-permutation* $(i_1, \ldots, i_s)$ *of* $\{1, \ldots, r\}$ *and* $\mathbf{D_t} = \mathbf{0}$ *for* $t \in \{1, \ldots, r\} \cap \{i_1, \ldots, i_s\}^c$.

Proof. 'If' part is trivial.
'Only if' part

If $\mathbf{A}$ is null matrix, the result is trivial. So, let $\mathbf{A}$ be non-null. Notice that the following hold:

(i) For $\mathbf{C}, \mathbf{D} \in \mathcal{I}_{1,n}$ such that $\mathbf{C} <^{\#} \mathbf{D} \Leftrightarrow \mathbf{RCR^{-1}} <^{\#} \mathbf{RDR^{-1}}$.
(ii) If $\mathbf{J}$ is a non-singular Jordan block, then $\mathbf{C}$ is an index ≤ 1 matrix such that $\mathbf{C} <^{\#} \mathbf{J}$ if and only if either $\mathbf{C}$ is a null matrix or $\mathbf{C} = \mathbf{J}$.
(iii) If $\mathbf{J_1}, \mathbf{J_2}$ are non-singular Jordan blocks corresponding to distinct eigen-values, then $\mathbf{A} = \begin{pmatrix} \mathbf{A_{11}} & \mathbf{A_{12}} \\ \mathbf{A_{21}} & \mathbf{A_{22}} \end{pmatrix} <^{\#} \mathbf{diag(J_1, J_2)}$ if and only if $\mathbf{A_{12}} = \mathbf{0}$, $\mathbf{A_{21}} = \mathbf{0}$, '$\mathbf{A_{11}} = \mathbf{0}$ or $\mathbf{J_1}$' and '$\mathbf{A_{22}} = \mathbf{J_2}$ or $\mathbf{0}$'.

The proof follows from the above three steps and by induction on r. □

Conjecture: The conclusion of Theorem 4.3.13 remains valid even when some or all distinct non-null eigen-values are of geometric multiplicity exceeding 1.

We now prove

Theorem 4.3.14. *Let* $\mathbf{A}, \mathbf{B} \in \mathcal{I}_{1,n}$. *Then following are equivalent:*

(i) $\mathbf{A} = \mathbf{B}$,
(ii) $\mathbf{A}^{\#} \in \{\mathbf{B}^{-}_{\mathbf{com}}\}$ *and* $\mathbf{B}^{\#} \in \{\mathbf{A}^{-}\}$ *and*
(iii) $\mathbf{A}^{\#} \in \{\mathbf{B}^{-}_{\mathbf{com}}\}$ *and* $\rho(\mathbf{A}) = \rho(\mathbf{B})$.

Proof. (i) $\Rightarrow$ (ii) is trivial.
(ii) $\Rightarrow$ (iii)
As $\mathbf{A}^{\#} \in \{\mathbf{B}^{-}_{\mathbf{com}}\} \Rightarrow \rho(\mathbf{A}^{\#}) = \rho(\mathbf{B}^{-}_{\mathbf{com}})$ for some commuting g-inverse $\mathbf{B}^{-}_{\mathbf{com}}$ of $\mathbf{B}$. So, $\rho(\mathbf{A}) = \rho(\mathbf{A}^{\#}) \geq \rho(\mathbf{B})$. Also, $\mathbf{B}^{\#} \in \{\mathbf{A}^{-}\}$, therefore, $\rho(\mathbf{B}) = \rho(\mathbf{B}^{\#}) \geq \rho(\mathbf{A})$. Thus (iii) follows.
(iii) $\Rightarrow$ (i)
Since $\mathbf{A}^{\#} \in \{\mathbf{B}^{-}_{\mathbf{com}}\}$, $\mathbf{A}^{\#} = \mathbf{B}^{-}_{\mathbf{com}}$ for some B^{-}_{com}. Since $\rho(\mathbf{A}) = \rho(\mathbf{B})$, $\rho(\mathbf{A}^{\#}) = \rho(\mathbf{B})$ implying $\mathbf{A}^{\#}$ is the commuting reflexive g-inverse of $\mathbf{B}$. Thus, $\mathbf{A}^{\#} = \mathbf{B}^{\#}$. Since for any matrix $\mathbf{X}$ of index ≤ 1, $\mathbf{X} = (\mathbf{X}^{\#})^{\#}$, we have $\mathbf{A} = \mathbf{B}$. □

Let $\mathbf{A}, \mathbf{B} \in \mathcal{I}_{1,n}$ such that $\mathbf{A} <^{\#} \mathbf{B}$. It is easy to see that $\mathbf{A}^{\#} <^{\#} \mathbf{B}^{-}_{\mathbf{com}}$ for all $\mathbf{B}^{-}_{\mathrm{com}} \in \{\mathbf{B^{-}com}\}$. Since every $\mathbf{B}^{-}_{\mathbf{com}}$ is also an $\mathbf{A}^{-}_{\mathbf{com}}$, given a $\mathbf{B}^{-}_{\mathbf{com}}$ obviously there exists an $\mathbf{A}^{-}_{\mathbf{com}}$ such that $\mathbf{B}^{-}_{\mathrm{com}} <^{\#} \mathbf{A}^{-}_{\mathbf{com}}$. We now show that given commuting g-inverses $\mathbf{A}^{-}_{\mathbf{com}}$ and $\mathbf{B}^{-}_{\mathbf{com}}$ of $\mathbf{A}$ and $\mathbf{B}$ respectively such that $\mathbf{B}^{-}_{\mathbf{com}} <^{\#} \mathbf{A}^{-}_{\mathbf{com}}$ whenever $\mathbf{A} <^{\#} \mathbf{B}$, then $\mathbf{A}^{-}_{\mathbf{com}}$ is a commuting g-inverse of $\mathbf{B}$, a statement that is much stronger than the one in the previous sentence. (See also, Theorem 4.3.1.)

Theorem 4.3.15. *Let* $\mathbf{A}, \mathbf{B} \in \mathcal{I}_{1,n}$ *such that* $\mathbf{A} <^{\#} \mathbf{B}$. *Let* $\mathbf{A}^-_{com}$ *and* $\mathbf{B}^-_{com}$ *be any commuting g-inverses of* $\mathbf{A}$ *and* $\mathbf{B}$ *respectively satisfying* $\mathbf{B}^-_{com} <^{\#} \mathbf{A}^-_{com}$. *Then* $\mathbf{A}^-_{com}$ *is a commuting g-inverse of* $\mathbf{B}$.

Proof. Since $\mathbf{A} <^{\#} \mathbf{B}$, there exist non-singular matrices $\mathbf{P}, \mathbf{C}$ and $\mathbf{D}$ such that $\mathbf{A} = \mathbf{P}\mathbf{diag}(\mathbf{C},\ \mathbf{0}, \mathbf{0})\mathbf{P}^{-1}$ and $\mathbf{B} = \mathbf{P}\mathbf{diag}(\mathbf{C},\ \mathbf{D}, \mathbf{0})\mathbf{P}^{-1}$. Then $\mathbf{A}^-_{com} = \mathbf{P}\mathbf{diag}(\mathbf{C}^{-1},\ \mathbf{N})\mathbf{P}^{-1}$ and $\mathbf{B}^-_{com} = \mathbf{P}\mathbf{diag}(\mathbf{C}^{-1},\ \mathbf{D}^{-1},\ \mathbf{T})\mathbf{P}^{-1}$ for some matrices $\mathbf{N}$ and $\mathbf{T}$ of appropriate order. Since $\mathbf{B}^-_{com} <^{\#} \mathbf{A}^-_{com}$, $\mathbf{diag}(\mathbf{D}^{-1},\ \mathbf{T}) <^{\#} \mathbf{N}$, it follows that $\mathbf{N} = \mathbf{diag}(\mathbf{D}^{-1},\ \mathbf{S})$. Hence $\mathbf{A}^-_{com}$ is a commuting g-inverse of $\mathbf{B}$. □

As noted at the beginning of the chapter, every idempotent matrix is of index ≤ 1. We have the following simple but important result.

Theorem 4.3.16. *Let* $\mathbf{E}$ *and* $\mathbf{F}$ *be idempotent matrices of the same order. Then the following are equivalent:*

(i) $\mathbf{E} <^{s} \mathbf{F}$,
(ii) $\mathbf{E} <^{-} \mathbf{F}$ *and*
(iii) $\mathbf{E} <^{\#} \mathbf{F}$.

We now examine when each summand in a sum of matrices of index 1 lies below the sum. The following theorem is an analogue of Fisher-Cochran theorem for matrices of index ≤ 1.

Theorem 4.3.17. *Let* $\mathbf{A_1}, \ldots, \mathbf{A_k} \in \mathcal{I}_{1,n}$ *and* $\mathbf{A} = \sum_{i=1}^{k} \mathbf{A_i}$. *Consider the following statements:*

(i) $\mathbf{A_i}\mathbf{A_j} = \mathbf{0}$ *whenever* $i \neq j$,
(ii) *There exist non-singular matrices* $\mathbf{P}, \mathbf{D_1}, \ldots, \mathbf{D_k}$ *such that*

$$\mathbf{A_i} = \mathbf{P}\mathbf{diag}(\mathbf{0}, \ldots, \mathbf{0}, \mathbf{D_i}, \mathbf{0}, \ldots, \mathbf{0})\mathbf{P}^{-1} \text{ for each } i = 1, \ldots, k,$$

(iii) $\mathbf{A_i} <^{\#} \mathbf{A}$ *for each* $i = 1, \ldots, k$,
(iv) $\rho(\mathbf{A}) = \rho(\mathbf{A^2})$ *and*
(v) $\sum_{i=1}^{k} \rho(\mathbf{A_i}) = \rho(\mathbf{A})$.

Then any of (i) *and* (ii) *implies all the others from* (i) *to* (v).
(iii) *and* (v) *together imply all the others from* (i) *to* (iv).

Proof. (i) $\Rightarrow$ (ii)
Let $(\mathbf{E_i}, \mathbf{F_i})$ be a rank factorization of $\mathbf{A_i}$ for $i = 1, \ldots, k$. Since $\mathbf{A_i}$ for $i = 1, \ldots, k$ is of index ≤ 1, $\mathbf{F_iE_i}$ is non-singular for each $i = 1, \ldots, k$. Further, whenever $i \neq j$, $\mathbf{A_iA_j} = \mathbf{0}$ implies $\mathbf{E_iF_iE_jF_j} = \mathbf{0}$ and therefore, $\mathbf{F_iE_j} = \mathbf{0}$ as each $\mathbf{E_k}$ has a left inverse and each $\mathbf{F_i}$ has a right inverse. Write $\mathbf{E} = (\mathbf{E_1} : \ldots : \mathbf{E_k})$ and $\mathbf{F} = (\mathbf{F_1^t} : \ldots : \mathbf{F_k^t})^\mathbf{t}$. Let $\mathbf{E}$ be of order $n \times r$ and $\mathbf{F}$ of order $r \times n$. Now, $\mathbf{A} = \sum_\mathbf{i} \mathbf{E_iF_i} = \mathbf{EF}$. Notice that $\mathbf{FE}$ is a non-singular, since $\mathbf{F_iE_i}$ is non-singular for each i and $\mathbf{F_iE_j} = \mathbf{0}$ whenever $i \neq j$. So, $r = \rho(\mathbf{FE}) \leq \rho(\mathbf{E}) \leq r$. Hence $\rho(\mathbf{E}) = r$. Similarly, $\rho(\mathbf{F}) = r$. Thus, $(\mathbf{E}, \mathbf{F})$ is a rank factorization of $\mathbf{A}$. Let $\mathbf{E_{k+1}}$ be a matrix whose columns form a basis of the null space of $\mathbf{F}$. Then $\mathbf{E_{k+1}}$ is a matrix of order $n \times (n - r)$ of rank $n - r$. Now, if $\mathbf{E_{k+1}u} = \mathbf{Ev}$ for some vectors $\mathbf{u}$ and $\mathbf{v}$, then $\mathbf{0} = \mathbf{FE_{k+1}u} = \mathbf{FEv}$. As $\mathbf{FE}$ is non-singular, $\mathbf{v} = \mathbf{0}$. Thus $\mathcal{C}(\mathbf{E_{k+1}}) \cap \mathcal{C}(\mathbf{E}) = \{\mathbf{0}\}$. Therefore, $\mathbf{P} = (\mathbf{E} : \mathbf{E_{k+1}})$ is a non-singular matrix and $\mathbf{FE_{k+1}} = \mathbf{0}$. Similarly, we can extend $\mathbf{F}$ to a non-singular matrix $\mathbf{Q} = (\mathbf{F^t} : \mathbf{F_{k+1}^t})^\mathbf{t}$ such that $\mathbf{F_{k+1}E} = \mathbf{0}$. Now, $\mathbf{Q}, \mathbf{P}$ and $\mathbf{F_{k+1}E_{k+1}}$ are non-singular. Also $\mathbf{QP} = \mathbf{diag}(\mathbf{F_1E_1}, \ldots, \mathbf{F_kE_k}, \ \mathbf{F_{k+1}E_{k+1}})$ is non-singular. Notice that $\mathbf{P^{-1}} = (\mathbf{QP})^{-1}\mathbf{Q}$. Now, it is easy to check that for each $i = 1, \ldots, k$, $\mathbf{A_i} = \mathbf{Pdiag}(\mathbf{0}, \ldots, \mathbf{0}, \mathbf{F_iE_i}, \mathbf{0}, \ldots, \mathbf{0})\mathbf{P^{-1}}$. Let $\mathbf{D_i} = \mathbf{F_iE_i}$, for each $i = 1, \ldots, k$ and recall that $\mathbf{F_iE_i}$ is non-singular.

(ii) $\Rightarrow$ (i) and (i) $\Rightarrow$ (iii) are trivial.

(ii) $\Rightarrow$ (iv)
We have $\mathbf{A} = \mathbf{Pdiag}(\mathbf{D_1}, \ldots, \mathbf{D_k}, \mathbf{0})\mathbf{P^{-1}}$ where $\mathbf{D_1}, \ldots, \mathbf{D_k}$ are non-singular. Clearly, $\rho(\mathbf{A}) = \rho(\mathbf{A^2})$.

(ii) $\Rightarrow$ (v) is trivial.

(iii) and (v) $\Rightarrow$ (i)
Since $\rho(\mathbf{A}) = \sum_{\mathbf{i=1}}^{\mathbf{k}} \rho(\mathbf{A_i})$, the sums $\mathcal{C}(\mathbf{A_1}) + \ldots + \mathcal{C}(\mathbf{A_k})$ and $\mathcal{C}(\mathbf{A_1^t}) + \ldots + \mathcal{C}(\mathbf{A_k^t})$ are direct. Now, $\mathbf{AA_i} = \mathbf{A_i^2}$ implies $\sum_{i \neq j} \mathbf{A_jA_i} = \mathbf{0}$. As the sum $\mathcal{C}(\mathbf{A_1}) + \ldots + \mathcal{C}(\mathbf{A_k})$ is direct, so for each $m \neq i$, $\mathbf{A_mA_i} = \mathbf{0}$. □

Corollary 4.3.18. *Let* $\mathbf{A_1}, \ldots, \mathbf{A_k}$ *be square matrices of the same order and* $\mathbf{A} = \sum_{\mathbf{i=1}}^{\mathbf{k}} \mathbf{A_i}$. *Consider the following statements:*

(i) $\mathbf{A_iA_j} = \mathbf{0}$ *whenever* $i \neq j$ *and* $\rho(\mathbf{A_i}) = \rho(\mathbf{A_i^2})$ *for all* $i = 1, \ldots, k$.
(ii) $\mathbf{A_i} = \mathbf{A_i^2}$ *for all* $i = 1, \ldots, k$.
(iii) $\mathbf{A} = \mathbf{A^2}$.

(iv) $\rho(\mathbf{A}) = \sum_i \rho(\mathbf{A_i})$.
(v) *There exists a non-singular matrix* $\mathbf{P}$ *such that*

$$\mathbf{A_i} = \mathbf{P}\mathbf{diag}(\mathbf{0}, \ldots, \mathbf{0}, \mathbf{I}, \ \mathbf{0}, \ldots, \mathbf{0})\mathbf{P}^{-1},$$

where $\mathbf{I}$ *occurs in the* i^{th} *diagonal block for* $i = 1, \ldots k$.
Then (v) *implies all the others from* (i) *to* (iv) *and any two of* (i), (ii) *and* (iii) *imply the rest. Also (iii) and (iv) imply the rest.*

Proof. Exercise. □

Remark 4.3.19. The above corollary is a non-hermitian version of Fisher-Cochran theorem due to [Khatri (1968)].

4.4 Drazin order and an extension

In the last three sections, we studied the properties of the sharp order defined on the class '$\mathcal{I}_1$' of matrices of index 1. In this section we consider square matrices not necessarily of index ≤ 1. As noted in Theorem 2.2.21, every square matrix $\mathbf{A}$ has the Core-Nilpotent decomposition $\mathbf{A} = \mathbf{A_1} + \mathbf{A_2}$, where $\mathbf{A_1} \in \mathcal{I}_1$, $\mathbf{A_2}$ is nilpotent and $\mathbf{A_1}\mathbf{A_2} = \mathbf{A_2}\mathbf{A_1} = \mathbf{0}$. The matrix $\mathbf{A_1}$ is called the core part of $\mathbf{A}$ and $\mathbf{A_2}$ is called the nilpotent part of $\mathbf{A}$. To start with, we define an order relation using the core part of the matrices ignoring the nilpotent part altogether. We call this order relation as the Drazin order. In the process of studying the properties of this order, we show that the Drazin order is a pre-order and is in general different from the space pre-order. We obtain some characterizations of this order relation, one of which also justifies its name: the Drazin order. We also establish some interesting properties of Drazin order.

It is natural to ask is: Is there a way to extend the Drazin order so that it becomes a partial order? Indeed there is a way which involves the nilpotent part as well. We define this order and obtain a canonical form for the matrices under this order. We also show that this order is a partial order implying the minus order.

Definition 4.4.1. Let $\mathbf{A}$ and $\mathbf{B}$ be square matrices of the same order. Let $\mathbf{A} = \mathbf{A_1} + \mathbf{A_2}$ and $\mathbf{B} = \mathbf{B_1} + \mathbf{B_2}$ be the core-nilpotent decompositions of $\mathbf{A}$ and $\mathbf{B}$ respectively, where $\mathbf{A_1}$ is core part of $\mathbf{A}$, $\mathbf{B_1}$ is core part of $\mathbf{B}$, $\mathbf{A_2}$ is nilpotent part of $\mathbf{A}$ and $\mathbf{B}$ is nilpotent part of $\mathbf{B}$. The matrix $\mathbf{A}$ is said to be below the matrix $\mathbf{B}$ under the Drazin order if $\mathbf{A_1} <^{\#} \mathbf{B_1}$.

When this happens we write $\mathbf{A} <^{\mathbf{d}} \mathbf{B}$.

It is intuitive to define the order relation using Drazin inverse as follows:

$$\mathbf{A} <^{\mathbf{d}} \mathbf{B} \text{ if } \mathbf{A}^{\mathbf{D}}\mathbf{A} = \mathbf{A}^{\mathbf{D}}\mathbf{B}\ ,\ \mathbf{A}\mathbf{A}^{\mathbf{D}} = \mathbf{B}\mathbf{A}^{\mathbf{D}}\ .$$

In fact it turns out that this is equivalent to the Definition 4.4.1 as we shall see in Theorem 4.4.3.

Remark 4.4.2. Let $\mathbf{A}$ and $\mathbf{B}$ be square matrices with $\mathbf{A_1}$ and $\mathbf{B_1}$ as specified in Definition 4.4.1. the the following hold:

(a) $\mathbf{A}^{\mathbf{D}} = \mathbf{A_1}^{\#}$ and $(\mathbf{A}^{\mathbf{D}})^{\#} = \mathbf{A_1}$. Similarly, $(\mathbf{B}^{\mathbf{D}})^{\#} = \mathbf{B_1}$. Consequently,

$$\mathbf{A} <^{\mathbf{d}} \mathbf{B} \Leftrightarrow \mathbf{A}^{\mathbf{D}} <^{\#} \mathbf{B}^{\mathbf{D}}.$$

(b) If $\mathbf{A}$ is non-singular the only matrix $\mathbf{B}$ such that $\mathbf{A} <^{\mathbf{d}} \mathbf{B}$ is the matrix $\mathbf{A}$ itself.

(c) If $\mathbf{A}$ is nilpotent, then $\mathbf{A} <^{\mathbf{d}} \mathbf{B}$ for each matrix $\mathbf{B}$ of same order as that of $\mathbf{A}$.

(d) If $\mathbf{A}$ and $\mathbf{B}$ are of index ≤ 1, then $\mathbf{A} <^{\mathbf{d}} \mathbf{B} \Leftrightarrow \mathbf{A} <^{\#} \mathbf{B}$ and

(e) $\mathbf{A} <^{\mathbf{d}} \mathbf{B} \Leftrightarrow \mathbf{PAP}^{-1} <^{\mathbf{d}} \mathbf{PBP}^{-1}$ for each non-singular matrix $\mathbf{P}$.

Theorem 4.4.3. *Let* $\mathbf{A}$ *and* $\mathbf{B}$ *be square matrices of the same order. Let* $\mathbf{A} = \mathbf{A_1} + \mathbf{A_2}$ *and* $\mathbf{B} = \mathbf{B_1} + \mathbf{B_2}$ *be the core-nilpotent decompositions of* $\mathbf{A}$ *and* $\mathbf{B}$ *respectively, where* $\mathbf{A_1}$ *is core part of* $\mathbf{A}$, $\mathbf{B_1}$ *is core part of* $\mathbf{B}$, $\mathbf{A_2}$ *is nilpotent part of* $\mathbf{A}$ *and* $\mathbf{B_2}$ *is nilpotent part of* $\mathbf{B}$. *Then the following are equivalent:*

(i) $\mathbf{A} <^{\mathbf{d}} \mathbf{B}$

(ii) $\mathbf{A}\mathbf{A}^{\mathbf{D}} = \mathbf{B}\mathbf{A}^{\mathbf{D}} = \mathbf{A}^{\mathbf{D}}\mathbf{B} = \mathbf{A}^{\mathbf{D}}\mathbf{A}$

(iii) *There exists a non-singular matrix* $\mathbf{P}$ *such that* $\mathbf{A} = \mathbf{P}\mathbf{diag}(\mathbf{C_1}, \mathbf{N_1})\mathbf{P}^{-1}$ *and* $\mathbf{B} = \mathbf{P}\mathbf{diag}(\mathbf{C_1}, \mathbf{C_2}, \mathbf{N_2})\mathbf{P}^{-1}$, *where* $\mathbf{C_1}, \mathbf{C_2}$ *are non-singular and* $\mathbf{N_1}, \mathbf{N_2}$ *are nilpotent and*

(iv) $\mathbf{A}^{\mathbf{k}}\mathbf{B} = \mathbf{B}\mathbf{A}^{\mathbf{k}} = \mathbf{A}^{\mathbf{k+1}}$, *where* k *is the index of* $\mathbf{A}$.

Proof. (i) $\Rightarrow$ (ii)
Clearly, $\mathbf{A}\mathbf{A}^{\mathbf{D}} = \mathbf{A_1}\mathbf{A}_{\mathbf{1}}^{\#}$. Since, $\mathbf{A} <^{\mathbf{d}} \mathbf{B} \Rightarrow \mathbf{A_1} <^{\#} \mathbf{B_1}$, we have

$$\mathbf{A_1}^{\#}\mathbf{B_1} = \mathbf{A_1}^{\#}\mathbf{A_1} = \mathbf{A_1}\mathbf{A}_{\mathbf{1}}^{\#} = \mathbf{B_1}\mathbf{A_1}^{\#}.$$

Since, we also have $\mathcal{C}(\mathbf{A_1}) \subseteq \mathcal{C}(\mathbf{B_1})$ and $\mathcal{C}(\mathbf{A_2}) \subseteq \mathcal{C}(\mathbf{B_2})$, it follows that, $\mathbf{A_1}^{\#}\mathbf{B_2} = \mathbf{0} = \mathbf{B_2}\mathbf{A_1}^{\#} = \mathbf{0}$. Hence,

$$\mathbf{A}\mathbf{A}^{\mathbf{D}} = \mathbf{A_1}\mathbf{A_1}^{\#} = \mathbf{B_1}\mathbf{A_1}^{\#} = \mathbf{B}\mathbf{A_1}^{\#} = \mathbf{B}\mathbf{A}^{\mathbf{D}}.$$

Moreover,

$$\mathbf{A_1}^{\#}\mathbf{B_1} = \mathbf{A_1}^{\#}\mathbf{A_1} = \mathbf{A_1}\mathbf{A_1}^{\#} \Rightarrow \mathbf{A}^{\mathbf{D}}\mathbf{B} = \mathbf{A}^{\mathbf{D}}\mathbf{A} = \mathbf{A}\mathbf{A}^{\mathbf{D}}.$$

Hence, $\mathbf{A}\mathbf{A}^{\mathbf{D}} = \mathbf{B}\mathbf{A}^{\mathbf{D}} = \mathbf{A}^{\mathbf{D}}\mathbf{B} = \mathbf{A}^{\mathbf{D}}\mathbf{A}$.

(ii) ⇒ (iii)
By Remark 2.2.24, we can write $\mathbf{A} = \mathbf{Q}\mathbf{diag}(\mathbf{C_1}, \mathbf{M_1})\mathbf{Q}^{-1}$, where $\mathbf{C_1}$ is non-singular and $\mathbf{M_1}$ is nilpotent. Let $\mathbf{B} = \mathbf{Q}\begin{pmatrix} \mathbf{B_{11}} & \mathbf{B_{12}} \\ \mathbf{B_{21}} & \mathbf{B_{22}} \end{pmatrix}\mathbf{Q}^{-1}$, where partitioning of $\mathbf{B}$ is conformable with the partitioning of $\mathbf{A}$. Notice that $\mathbf{A^D} = \mathbf{Q}\mathbf{diag}(\mathbf{C_1^{-1}}, \mathbf{0})\mathbf{Q}^{-1}$. Since (ii) holds, we have

$$\begin{pmatrix} \mathbf{C_1} & \mathbf{0} \\ \mathbf{0} & \mathbf{M_1} \end{pmatrix}\begin{pmatrix} \mathbf{C_1^{-1}} & \mathbf{0} \\ \mathbf{0} & \mathbf{0} \end{pmatrix} = \begin{pmatrix} \mathbf{B_{11}} & \mathbf{B_{12}} \\ \mathbf{B_{21}} & \mathbf{B_{22}} \end{pmatrix}\begin{pmatrix} \mathbf{C_1^{-1}} & \mathbf{0} \\ \mathbf{0} & \mathbf{0} \end{pmatrix}$$
$$= \begin{pmatrix} \mathbf{C_1^{-1}} & \mathbf{0} \\ \mathbf{0} & \mathbf{0} \end{pmatrix}\begin{pmatrix} \mathbf{B_{11}} & \mathbf{B_{12}} \\ \mathbf{B_{21}} & \mathbf{B_{22}} \end{pmatrix}.$$

Or $\mathbf{B_{11}C_1^{-1}} = \mathbf{C_1^{-1}B_{11}} = \mathbf{I}$, $\mathbf{B_{21}C_1^{-1}} = \mathbf{0}$ and $\mathbf{C_1^{-1}B_{12}} = \mathbf{0}$. So, $\mathbf{B_{11}} = \mathbf{C_1}$, $\mathbf{B_{21}} = \mathbf{0}$ and $\mathbf{B_{12}} = \mathbf{0}$. Hence $\mathbf{B} = \mathbf{Q}\mathbf{diag}(\mathbf{C_1}, \mathbf{B_{22}})\ \mathbf{Q}^{-1}$. Let $\mathbf{B_{22}} = \mathbf{R}\mathbf{diag}(\mathbf{C_2}, \mathbf{N_2})\mathbf{R}^{-1}$, where $\mathbf{C_2}$ is non-singular and $\mathbf{N_2}$ is nilpotent. Write $\mathbf{P} = \mathbf{Q}\mathbf{diag}(\mathbf{I}, \mathbf{R})$. Then,

$$\mathbf{B} = \mathbf{P}\mathbf{diag}(\mathbf{C_1}, \mathbf{C_2}, \mathbf{N_2})\mathbf{P}^{-1} \text{ and } \mathbf{A} = \mathbf{P}\mathbf{diag}(\mathbf{C_1}, \mathbf{R^{-1}M_1R})\mathbf{P}^{-1}.$$

However, $\mathbf{M_1}$ is nilpotent implies $\mathbf{R^{-1}M_1R}$ is nilpotent. Let $\mathbf{N_1} = \mathbf{R^{-1}M_1R}$ and (iii) follows.
(iii) ⇒ (i) and (iii) ⇒ (iv) are trivial.
(iv) ⇒ (ii)
Since index of $\mathbf{A}$ is k, $\mathbf{A^k} = \mathbf{A^{k+1}A^D} = \mathbf{A^DA^{k+1}}$. So,

$$(\mathbf{A^D})^{\mathbf{k+1}}\mathbf{A^kB} = (\mathbf{A^D})^{\mathbf{k+1}}(\mathbf{A})^{\mathbf{k+1}} = \mathbf{A^DA}.$$

Also, $(\mathbf{A^D})^{\mathbf{k+1}}(\mathbf{A})^{\mathbf{k}} = \mathbf{A^D}$. So, we have $\mathbf{AA^D} = \mathbf{A^DA} = \mathbf{A^DB}$. Similarly, $\mathbf{BA^k} = \mathbf{A^{k+1}}$ yields $\mathbf{BA^D} = \mathbf{AA^D}$. □

Theorem 4.4.4. *The Drazin order is a pre-order on* $\mathbf{F^{n\times n}}$.

Proof. Reflexivity is trivial. To prove transitivity, let $\mathbf{A} <^{\mathbf{d}} \mathbf{B}$ and $\mathbf{B} <^{\mathbf{d}} \mathbf{C}$.
Let $\mathbf{A_1}$, $\mathbf{B_1}$ and $\mathbf{C_1}$ denote the core parts of $\mathbf{A}$, $\mathbf{B}$ and $\mathbf{C}$ respectively. Recall that $\mathbf{A} <^{\mathbf{d}} \mathbf{B} \Leftrightarrow \mathbf{A_1} <^{\#} \mathbf{B_1}$. So, $\mathbf{A} <^{\mathbf{d}} \mathbf{B}$ and $\mathbf{B} <^{\mathbf{d}} \mathbf{C} \Leftrightarrow \mathbf{A_1} <^{\#} \mathbf{B_1}$ and $\mathbf{B_1} <^{\#} \mathbf{C_1}$. However, '$<^{\#}$' is transitive, $\mathbf{A_1} <^{\#} \mathbf{C_1}$. and so, $\mathbf{A} <^{\mathbf{d}} \mathbf{C}$ in view of Theorem 4.4.3(iii). □

The following example shows that the Drazin order is not anti-symmetric.

Example 4.4.5. Let

$$\mathbf{A} = \begin{pmatrix} 1 & 0 & 0 \\ 0 & 0 & 1 \\ 0 & 0 & 0 \end{pmatrix}\ \mathbf{B} = \begin{pmatrix} 1 & 0 & 0 \\ 0 & 0 & 0 \\ 0 & 0 & 0 \end{pmatrix}.$$

Then $\mathbf{A} <^{\mathbf{d}} \mathbf{B}$ and $\mathbf{B} <^{\mathbf{d}} \mathbf{A}$. However, $\mathbf{A} \neq \mathbf{B}$.

Remark 4.4.6. Failure of anti-symmetry is due to the fact that the Drazin order ignores the nilpotent parts.

Remark 4.4.7. In general the Drazin order and the space pre-order are not comparable as the following examples show:

Example 4.4.8. Take

$$\mathbf{A} = \begin{pmatrix} 0 & 1 \\ 0 & 0 \end{pmatrix} \text{ and } \mathbf{B} = \begin{pmatrix} 0 & 0 \\ 0 & 0 \end{pmatrix}.$$

Then $\mathbf{A} <^{\mathbf{d}} \mathbf{B}$, but $\mathbf{A} \not<^{\mathbf{s}} \mathbf{B}$.

Example 4.4.9. Take

$$\mathbf{A} = \begin{pmatrix} 2 & 0 \\ 0 & 0 \end{pmatrix} \text{ and } \mathbf{B} = \begin{pmatrix} 1 & 0 \\ 0 & 0 \end{pmatrix}.$$

Then $\mathbf{A} <^{\mathbf{s}} \mathbf{B}$, but $\mathbf{A} \not<^{\mathbf{d}} \mathbf{B}$.

However, if $\mathbf{A}$ and $\mathbf{B}$ are index 1 matrices, then $\mathbf{A} <^{\mathbf{d}} \mathbf{B} \Rightarrow \mathbf{A} <^{\mathbf{s}} \mathbf{B}$.

We shall now obtain a characterization of the Drazin order $\mathbf{A} <^{\mathbf{d}} \mathbf{B}$ in terms of commutativity of $\mathbf{B}$ and $\mathbf{A}^{\mathbf{D}}$, the Drazin inverse of $\mathbf{A}$.

Theorem 4.4.10. *Let* $\mathbf{A}$ *and* $\mathbf{B}$ *be square matrices of the same order. Then* $\mathbf{A} <^{\mathbf{d}} \mathbf{B}$ *if and only if* $\mathbf{B}$ *is a commuting g-inverse of* $\mathbf{A}^{\mathbf{D}}$.

Proof. 'If' part
Let $\mathbf{A} = \mathbf{P}\mathbf{diag}(\mathbf{C}, \mathbf{N})\mathbf{P}^{-1}$, where $\mathbf{P}$ and $\mathbf{C}$ are non-singular and $\mathbf{N}$ is nilpotent. Then $\mathbf{A}^{\mathbf{D}} = \mathbf{P}\mathbf{diag}(\mathbf{C}, \mathbf{0})\mathbf{P}^{-1}$. Write $\mathbf{B} = \mathbf{P}\begin{pmatrix} \mathbf{B}_{11} & \mathbf{B}_{12} \\ \mathbf{B}_{21} & \mathbf{B}_{22} \end{pmatrix}\mathbf{P}^{-1}$.
Since, $\mathbf{A}^{\mathbf{D}}\mathbf{B}\mathbf{A}^{\mathbf{D}} = \mathbf{A}^{\mathbf{D}}$, we have

$$\begin{pmatrix} \mathbf{C}^{-1} & \mathbf{0} \\ \mathbf{0} & \mathbf{0} \end{pmatrix} = \begin{pmatrix} \mathbf{C}^{-1} & \mathbf{0} \\ \mathbf{0} & \mathbf{0} \end{pmatrix}\begin{pmatrix} \mathbf{B}_{11} & \mathbf{B}_{12} \\ \mathbf{B}_{21} & \mathbf{B}_{22} \end{pmatrix}\begin{pmatrix} \mathbf{C}^{-1} & \mathbf{0} \\ \mathbf{0} & \mathbf{0} \end{pmatrix} = \begin{pmatrix} \mathbf{C}^{-1}\mathbf{B}_{11}\mathbf{C}^{-1} & \mathbf{0} \\ \mathbf{0} & \mathbf{0} \end{pmatrix}.$$

So, $\mathbf{C} = \mathbf{B}_{11}$. Also, $\mathbf{A}^{\mathbf{D}}\mathbf{B} = \mathbf{B}\mathbf{A}^{\mathbf{D}}$, therefore,

$$\begin{pmatrix} \mathbf{C}^{-1}\mathbf{B}_{11} & \mathbf{C}^{-1}\mathbf{B}_{12} \\ \mathbf{0} & \mathbf{0} \end{pmatrix} = \begin{pmatrix} \mathbf{B}_{11}\mathbf{C}^{-1} & \mathbf{0} \\ \mathbf{C}^{-1}\mathbf{B}_{21} & \mathbf{0} \end{pmatrix}.$$

Moreover, $\mathbf{B}_{12} = \mathbf{0}$ and $\mathbf{B}_{21} = \mathbf{0}$. Thus $\mathbf{B} = \mathbf{P}\mathbf{diag}(\mathbf{C}, \mathbf{B}_{22})\mathbf{P}^{-1}$. It is easy to check that $\mathbf{A}\mathbf{A}^{\mathbf{D}} = \mathbf{B}\mathbf{A}^{\mathbf{D}} = \mathbf{A}^{\mathbf{D}}\mathbf{B} = \mathbf{P}\mathbf{diag}(\mathbf{I}, \mathbf{0})\mathbf{P}^{-1}$. It follows by Theorem 4.4.3, that $\mathbf{A} <^{\mathbf{d}} \mathbf{B}$.
'Only if' part follows from Theorem 4.4.3. □

Corollary 4.4.11. *Consider the same setup as in Theorem 4.4.10. Then*

(i) $\mathbf{A} <^{\mathbf{d}} \mathbf{B}$ *if* $\mathbf{B}(\mathbf{A}^{\mathbf{D}})^{\mathbf{r}+1} = (\mathbf{A}^{\mathbf{D}})^{\mathbf{r}+1}\mathbf{B} = (\mathbf{A}^{\mathbf{D}})^{\mathbf{r}}$ *for some positive integer* r.

(ii) *If* $\mathbf{A} <^{\mathbf{d}} \mathbf{B}$ *then* $\mathbf{B}(\mathbf{A}^{\mathbf{D}})^{\mathbf{r}+1} = (\mathbf{A}^{\mathbf{D}})^{\mathbf{r}+1}\mathbf{B} = (\mathbf{A}^{\mathbf{D}})^{\mathbf{r}}$ *for all positive integers* r.

The proof follows along the lines of Theorem 4.4.10.

Remark 4.4.12. In the set up of Theorem 4.4.10, if $\mathbf{A} <^{\mathbf{d}} \mathbf{B}$, then $\mathbf{A}^{\mathbf{D}}$ and $\mathbf{B}^{\mathbf{r}}$ commute for all positive integers r. Also, if $\mathbf{A} <^{\mathbf{d}} \mathbf{B}$, then $\mathbf{A}^{\mathbf{r}} <^{\mathbf{d}} \mathbf{B}^{\mathbf{r}}$ for each positive integer r. However, if $\mathbf{A}^{\mathbf{D}}$ and $\mathbf{B}^{\mathbf{r}}$ commute and $\mathbf{B}$ is a commuting g-inverse of $\mathbf{A}^{\mathbf{D}}$, it does not necessarily follow that $\mathbf{A} <^{\mathbf{d}} \mathbf{B}$. We give the following example. Incidentally, the same example shows, $\mathbf{A}^{\mathbf{r}} <^{\mathbf{d}} \mathbf{B}^{\mathbf{r}}$ for some positive integer $r > 1$ does not imply $\mathbf{A} <^{\mathbf{d}} \mathbf{B}$.

Example 4.4.13. Let

$$\mathbf{A} = \begin{pmatrix} 1 & 1 & 0 \\ 0 & 0 & 1 \\ 0 & 0 & 0 \end{pmatrix} \quad \mathbf{B} = \begin{pmatrix} 0 & 1 & 0 \\ 1 & 0 & 0 \\ 0 & 0 & 1 \end{pmatrix}.$$

Now, $\rho(\mathbf{A}) = \mathbf{2} > \rho(\mathbf{A}^2) = \rho(\mathbf{A}^3) = \mathbf{1}$. So index $\mathbf{A} = \mathbf{2}$. The Drazin inverse of $\mathbf{A}$ is $\mathbf{A}^{\mathbf{D}} = \begin{pmatrix} 1 & 1 & 1 \\ 0 & 0 & 0 \\ 0 & 0 & 0 \end{pmatrix}$. Clearly, $\mathbf{A}^{\mathbf{D}}\mathbf{B}\mathbf{A}^{\mathbf{D}} = \mathbf{A}^{\mathbf{D}}$, $\mathbf{A}^{\mathbf{D}}\mathbf{B}^2 = \mathbf{A}^{\mathbf{D}}\mathbf{B}^2 = \mathbf{A}^{\mathbf{D}}$, but $\mathbf{B}\mathbf{A}^{\mathbf{D}} \neq \mathbf{A}\mathbf{A}^{\mathbf{D}}$. Thus $\mathbf{A} \not<^{\mathbf{d}} \mathbf{B}$. Also, $\mathbf{A}^2 <^{\mathbf{d}} \mathbf{B}^2$. (In fact, $\mathbf{A}^2 <^{\#} \mathbf{B}^2$.)

Given a square matrix $\mathbf{A}$, we now obtain the class of all matrices $\mathbf{B}$ such that $\mathbf{A} <^{\mathbf{d}} \mathbf{B}$.

Theorem 4.4.14. *Let* $\mathbf{A}$ *be a square matrix. Let* $\mathbf{A} = \mathbf{P}\mathbf{diag}(\mathbf{C},\ \mathbf{N})\mathbf{P}^{-1}$, *where* $\mathbf{P}$ *and* $\mathbf{C}$ *are non-singular and* $\mathbf{N}$ *is nilpotent. Then the class of all matrices* $\mathbf{B}$ *such that* $\mathbf{A} <^{\mathbf{d}} \mathbf{B}$ *is given by*

$$\mathbf{B} = \mathbf{P}\mathbf{diag}(\mathbf{C},\ \mathbf{T})\mathbf{P}^{-1},$$

where $\mathbf{T}$ *arbitrary.*

Proof. If $\mathbf{A} <^{\mathbf{d}} \mathbf{B}$, then by Theorem 4.4.3(ii) $\Rightarrow$ (iii) $\mathbf{B}$ is of the form $\mathbf{B} = \mathbf{P}\mathbf{diag}(\mathbf{C}, \mathbf{T})\mathbf{P}^{-1}$ for some matrix $\mathbf{T}$.

Moreover, if $\mathbf{B} = \mathbf{P}\mathbf{diag}(\mathbf{C},\ \mathbf{T})\mathbf{P}^{-1}$, then we can check easily that $\mathbf{A}\mathbf{A}^{\mathbf{D}} = \mathbf{B}\mathbf{A}^{\mathbf{D}} = \mathbf{A}^{\mathbf{D}}\mathbf{B}$. □

Theorem 4.4.15. *Let* $\mathbf{A}$ *be a square matrix. Then the class of all matrices* $\mathbf{B}$ *such that* $\mathbf{A} <^{\mathrm{d}} \mathbf{B}$ *is given by* $\mathbf{B} = \mathbf{A} + \mathbf{C}$ *where* $\mathbf{C}$ *is arbitrary matrix such that* $\mathbf{A}^{\mathrm{D}}\mathbf{C} = \mathbf{C}\mathbf{A}^{\mathrm{D}} = \mathbf{0}$.

Proof. It is easy to see that $\mathbf{A}\mathbf{A}^{\mathrm{D}} = \mathbf{B}\mathbf{A}^{\mathrm{D}} = \mathbf{A}^{\mathrm{D}}\mathbf{B}$ if and only if $(\mathbf{B} - \mathbf{A})\mathbf{A}^{\mathrm{D}} = \mathbf{0}$, $\mathbf{A}^{\mathrm{D}}(\mathbf{B} - \mathbf{A}) = \mathbf{0}$. □

Remark 4.4.16. Let $\mathbf{A}$ be a square matrix of index k. Then the class of all matrices $\mathbf{B}$ such that $\mathbf{A} <^{\mathrm{d}} \mathbf{B}$ is given by

(i) $\mathbf{B} = \mathbf{A} + \mathbf{C}$ where $\mathbf{C}$ is arbitrary matrix such that $\mathbf{A}^{\mathrm{k}}\mathbf{C} = \mathbf{C}\mathbf{A}^{\mathrm{k}} = \mathbf{0}$. For, by Theorem 4.4.14, $\mathcal{C}(\mathbf{A}^{\mathrm{k}}) = \mathcal{C}(\mathbf{A}^{\mathrm{D}})$.

(ii) $\mathbf{B} = \mathbf{A} + \mathbf{C}$ where $\mathbf{C} = (\mathbf{I} - \mathbf{A}^{\mathrm{D}}\mathbf{A})\mathbf{Z}(\mathbf{I} - \mathbf{A}^{\mathrm{D}}\mathbf{A})$ and $\mathbf{Z}$ is arbitrary matrix.

We have already seen that the Drazin order is only a pre-order and not a partial order. We noticed that the main reason for this to happen is that the Drazin order does not take the nilpotent parts of the matrices into consideration. We modify the Drazin order so that the nilpotent part is appropriately involved and this leads to a partial order. We shall define this order and study some of its properties.

Definition 4.4.17. Let $\mathbf{A}$ and $\mathbf{B}$ be matrices of the same order. Let $\mathbf{A} = \mathbf{A}_1 + \mathbf{A}_2$ and $\mathbf{B} = \mathbf{B}_1 + \mathbf{B}_2$ be the core-nilpotent decompositions of $\mathbf{A}$ and $\mathbf{B}$ respectively, where $\mathbf{A}_1$ is core part of $\mathbf{A}$, $\mathbf{B}_1$ is core part of $\mathbf{B}$, $\mathbf{A}_2$ is nilpotent part of $\mathbf{A}$ and $\mathbf{B}$ is nilpotent part of $\mathbf{B}$. Then $\mathbf{A}$ is below $\mathbf{B}$ under $(\#, -)$ order if $\mathbf{A}_1 <^{\#} \mathbf{B}_1$ and $\mathbf{A}_2 <^{-} \mathbf{B}_2$.

Whenever this happens, we write $\mathbf{A} <^{\#,-} \mathbf{B}$.

We now obtain some characterizations of $(\#, -)$ order.

Theorem 4.4.18. *Let* $\mathbf{A}$ *and* $\mathbf{B}$ *be matrices of same order. Then the following are equivalent:*

(i) $\mathbf{A} <^{\#,-} \mathbf{B}$

(ii) $\mathbf{A} <^{\mathrm{d}} \mathbf{B}$, $\mathbf{A} - \mathbf{A}\mathbf{A}^{\mathrm{D}}\mathbf{A} <^{-} \mathbf{B} - \mathbf{B}\mathbf{B}^{\mathrm{D}}\mathbf{B}$ *and*

(iii) *There exists a non-singular matrix* $\mathbf{P}$ *such that*

$$\mathbf{A} = \mathbf{P}\mathrm{diag}(\mathbf{C}_1,\ \mathbf{0},\ \mathbf{N}_1)\mathbf{P}^{-1}, \text{ and } \mathbf{B} = \mathbf{P}\mathrm{diag}(\mathbf{C}_1,\ \mathbf{C}_2, \mathbf{N}_2)\mathbf{P}^{-1},$$

where $\mathbf{P}, \mathbf{C}_1$ *and* $\mathbf{C}_2$ *are non-singular and* $\mathbf{N}_1, \mathbf{N}_2$ *are nilpotent satisfying* $\mathbf{N}_1 <^{-} \mathbf{N}_2$.

Proof. (i) $\Leftrightarrow$ (ii) follows once we notice that $\mathbf{AA^DA}$ is the core part of $\mathbf{A}$ and $\mathbf{A} <^{\mathbf{d}} \mathbf{B} \Rightarrow \mathbf{A^D} <^{\mathbf{d}} \mathbf{B^D}$.

(ii) $\Rightarrow$ (iii)

Since $\mathbf{A} <^{\mathbf{d}} \mathbf{B}$, there exists a non-singular matrix $\mathbf{P}$ such that $\mathbf{A} = \mathbf{P}\mathbf{diag}(\mathbf{C_1}, \mathbf{M})\,\mathbf{P}^{-1}$ and $\mathbf{B} = \mathbf{P}\mathbf{diag}(\mathbf{C_1}, \mathbf{C_2}, \mathbf{N_2})\,\mathbf{P}^{-1}$, where '$\mathbf{C_1}$ and $\mathbf{C_2}$ are non-singular', and '$\mathbf{M}$ and $\mathbf{N_2}$ are nilpotent'.

Let $\mathbf{C_1}$ be of order $r \times r$, $\mathbf{C_2}$ on order $s \times s$ and $\mathbf{N_2}$ of order $(n-r-s) \times (n-r-s)$. Partition $\mathbf{P}$ and $\mathbf{P}^{-1}$ as $\mathbf{P} = (\mathbf{P_1} : \mathbf{P_2} : \mathbf{P_3})$ and $\mathbf{P}^{-1} = (\mathbf{P^{1t}} : \mathbf{P^{2t}} : \mathbf{P^{3t}})$, where $\mathbf{P_1}, \mathbf{P_2}$, $\mathbf{P_3}$ have $r, s, n-r-s$ columns respectively and $\mathbf{P^{1t}}, \mathbf{P^{2t}}, \mathbf{P^{3t}}$ have $r, s, n-r-s$ columns respectively.

Let $\mathbf{M} = \begin{pmatrix} \mathbf{M_{11}} & \mathbf{M_{12}} \\ \mathbf{M_{21}} & \mathbf{N_1} \end{pmatrix}$ and $\mathbf{N} = \begin{pmatrix} \mathbf{0} & \mathbf{0} \\ \mathbf{0} & \mathbf{N_2} \end{pmatrix}$ where $\mathbf{M}$ and $\mathbf{N}$ have conformable partitioning. Then $\mathbf{M} <^- \mathbf{N}$. So,

$$(\mathbf{P_2} : \mathbf{P_3})\mathbf{M} \begin{pmatrix} \mathbf{P^2} \\ \mathbf{P^3} \end{pmatrix} < (\mathbf{P_2} : \mathbf{P_3})\mathbf{N} \begin{pmatrix} \mathbf{P^2} \\ \mathbf{P^3} \end{pmatrix}.$$

It follows that $\mathbf{M_{11}}, \mathbf{M_{12}}$ and $\mathbf{M_{21}}$ are null and $\mathbf{P^3N_1P_3} <^- \mathbf{P^3N_2P_3}$. Since $\mathbf{P^3P_3} = \mathbf{I}$, we have $\mathbf{N_1} <^- \mathbf{N_2}$.

(iii) $\Rightarrow$ (i) is straightforward verification. □

Corollary 4.4.19. *The order relation* $(\#, -)$ *is a partial order of* $\mathbf{F^{n \times n}}$ *for each positive integer* n.

Proof. The Corollary follows from Theorem 4.4.18(iii). □

We now show that the order relation $(\#, -)$ implies '$<^-$'.

Theorem 4.4.20. *The order relation* $(\#, -)$ *on* $\mathbf{F^{n \times n}}$ *implies '*$<^-$*' i.e., if* $\mathbf{A} <^{\#,-} \mathbf{B}$, *then* $\mathbf{A} <^- \mathbf{B}$.

Proof. Let $\mathbf{A} <^{\#,-} \mathbf{B}$. Then by Theorem 4.4.18, there exists a non-singular matrix $\mathbf{P}$ such that $\mathbf{A} = \mathbf{P}\mathbf{diag}(\mathbf{C_1}, \mathbf{0}, \mathbf{N_1})\mathbf{P}^{-1}$, and $\mathbf{B} = \mathbf{P}\mathbf{diag}(\mathbf{C_1}, \mathbf{C_2}, \mathbf{N_2})\mathbf{P}^{-1}$ where $\mathbf{P}, \mathbf{C_1}$ and $\mathbf{C_2}$ are non-singular and $\mathbf{N_1}, \mathbf{N_2}$ are nilpotent satisfying $\mathbf{N_1} <^- \mathbf{N_2}$. Let $\mathbf{N_1^-}$ be a g-inverse of $\mathbf{N_1}$ such that $\mathbf{N_1^-N_1} = \mathbf{N_1^-N_2}$, $\mathbf{N_1N_1^-} = \mathbf{N_2N_1^-}$. Let $\mathbf{G} = \mathbf{P}\ \mathbf{diag}(\mathbf{C}^{-1}, \mathbf{0}, \mathbf{N_1^-})\mathbf{P}^{-1}$. Then, it is easy to check that $\mathbf{G}$ is a g-inverse of $\mathbf{A}$ for which $\mathbf{AG} = \mathbf{BG}$, $\mathbf{GA} = \mathbf{GB}$. Hence, $\mathbf{A} <^- \mathbf{B}$. □

4.5 Exercises

(1) Prove or disprove the following:
$\mathbf{A} <^{\#} \mathbf{B} \Rightarrow \mathbf{AA}^{\#} <^{\#} \mathbf{BB}^{\#}$.

(2) Prove that the conclusion of Theorem 4.3.15 is false if $\mathbf{A}^{\#} \in \{\mathbf{B}^{-}\}$, $\mathbf{B}^{\#} \in \{\mathbf{A}^{-}\}$, by giving a suitable example.

(3) Let $\mathbf{A}^{-}$ be any commuting g-inverse of a matrix $\mathbf{A}$ of index ≤ 1. Then show that $\mathbf{A}^{-}\mathbf{A} = \mathbf{A}^{\#}\mathbf{A}$.

(4) Prove the following statements:

(i) $\mathbf{A} <^{\#} \mathbf{B} \Leftrightarrow \mathbf{A}^{\#} <^{\#} \mathbf{B}^{\#}$
(ii) $\mathbf{A} <^{\#} \mathbf{B} \Rightarrow \mathbf{A} <^{-} \mathbf{B}$ and $\mathbf{A^2} <^{-} \mathbf{B^2}$.
Does the converse of (ii) hold?

(5) Let $\mathbf{A}$ be a matrix of index ≤ 1 and $(\mathbf{L}, \mathbf{R})$ be a rank factorization of $\mathbf{A}$. Suppose that $\mathbf{A} <^{\#} \mathbf{B}$. Then show that $\mathbf{L}(\mathbf{RL})^{-1}\mathbf{R} = \mathbf{L}(\mathbf{RL})^{-2}\mathbf{RB} = \mathbf{BL}(\mathbf{RL})^{-2}\mathbf{R}$.

(6) Let $\mathbf{A}$ and $\mathbf{B}$ be both of index ≤ 1 and satisfy (i) $\mathbf{AA}^{\#}\mathbf{BB}^{\#} = \mathbf{AB}^{\#}$ (ii) $\mathbf{A}^{\#}\mathbf{A} = \mathbf{A}^{\#}\mathbf{B}$ and (iii) $\mathcal{C}(\mathbf{A}) \subseteq \mathcal{C}(\mathbf{B})$. Then prove that $\mathbf{A} <^{\#} \mathbf{B}$.

(7) Let $\mathbf{A}^{\#} \in \{\mathbf{B}^{-}\}$, $\mathbf{A} <^{\mathbf{s}} \mathbf{B}$ and $\mathbf{B}^{\#}\mathbf{AB}^{\#} = \mathbf{B}^{\#}$. Show that $\mathbf{A} = \mathbf{B}$.

(8) Show that the relation $\preceq \mathbf{B}$ defined as $\mathbf{A} <^{\mathbf{s}} \mathbf{B}$ and and $\mathbf{A} = \mathbf{AB}^{\#}\mathbf{A}$ is a partial order on $\mathcal{I}_{1,n}$, the $n \times n$ matrices in $\mathcal{I}_1$. Show also, that this relation implies the minus order but is not the same as sharp order of $\mathcal{I}_{1,n}$.

(9) Prove or disprove: $\mathbf{A} <^{-} \mathbf{B} \Rightarrow \mathbf{A}^{\#} <^{-} \mathbf{B}^{\#}$.

(10) For $\mathbf{A}, \mathbf{B} \in \mathcal{I}_{\mathbf{1,n}}$ and for each $\lambda \neq 0$, define

$$\mathbf{G}(\lambda) = \begin{cases} \mathbf{B}^{\#} - \dfrac{\lambda}{\lambda - \mathbf{1}}\mathbf{A}^{\#} & \text{if } \lambda \neq 1 \\ \mathbf{B}^{\#} - \mathbf{A}^{\#} & \text{if } \lambda = 1 \end{cases}.$$

Let $\mathbf{A} <^{\#} \mathbf{B}$. Show that $\mathbf{A} <^{-} \mathbf{B}$ and for each $\lambda \neq 0$, $\mathbf{G}(\lambda) = (\mathbf{B} - \lambda\mathbf{A})^{\#}$, where $\mathbf{G}(\lambda)$ is as above.

(11) Let $\mathbf{A}$ and $\mathbf{B}$ be square matrices of same order. Let $\mathbf{A}$ be of index ≤ 1. Show that $\mathbf{A}^{\#}\mathbf{A} = \mathbf{A}^{\#}\mathbf{B}$, $\mathbf{AA}^{\#} = \mathbf{BA}^{\#} \Leftrightarrow \mathbf{A^2} = \mathbf{AB} = \mathbf{BA}$. Also, show that the relation $\mathbf{A} \ll \mathbf{B}$ defined by the equations $\mathbf{A}^{\#}\mathbf{A} = \mathbf{A}^{\#}\mathbf{B}$, $\mathbf{AA}^{\#} = \mathbf{BA}^{\#}$ is a partial order on F^n.

(12) Let $\mathbf{A} \in \mathcal{I}_\infty$. Then $\begin{pmatrix} \mathbf{A} & \mathbf{0} \\ \mathbf{0} & \mathbf{0} \end{pmatrix} <^{\#} \begin{pmatrix} \mathbf{B}_1 & \mathbf{B}_2 \\ \mathbf{B}_3 & \mathbf{B}_4 \end{pmatrix} = \mathbf{B} \Leftrightarrow \mathbf{B} = \begin{pmatrix} \mathbf{B}_1 & (\mathbf{I} - \mathbf{A}\mathbf{A}^{\#})\mathbf{S} \\ \mathbf{T}(\mathbf{I} - \mathbf{A}\mathbf{A}^{\#}) & \mathbf{B}_4 \end{pmatrix}$, where $\mathbf{A} <^{\#} \mathbf{B}_1$, $\mathbf{S}$, $\mathbf{T}$ and $\mathbf{B}_4$ are arbitrary and partitions are conformable for matrix multiplication.

Chapter 5

The Star Order

5.1 Introduction

The present chapter is devoted to the study of yet another partial order known as the star order. This order is defined for matrices $\mathbf{A}$ and $\mathbf{B}$ (possibly rectangular) over the field $\mathbb{C}$ of complex numbers. We say that a matrix $\mathbf{A}$ is below $\mathbf{B}$ under the star order if $\mathbf{AA}^\star = \mathbf{BA}^\star$ and $\mathbf{A}^\star\mathbf{A} = \mathbf{A}^\star\mathbf{B}$. If the matrices $\mathbf{A}$ and $\mathbf{B}$ are taken to be hermitian, these defining equalities read as $\mathbf{A^2} = \mathbf{AB} = \mathbf{BA}$. Recall from the previous chapter that a matrix $\mathbf{A}$ is below a matrix $\mathbf{B}$ under the sharp order if and only if $\mathbf{A^2} = \mathbf{AB} = \mathbf{BA}$. Thus, the star order and the sharp order coincide for the class of hermitian matrices. (Notice that a hermitioan matrix is a matrix of index ≤ 1.) In view of this, one can expect a number of results in this order to be parallel to those in the sharp order. This indeed is true as we shall see in due course.

A well explored partial order, the star order was first introduced by Drazin in a star semi-group with involution. While establishing several other properties of this order, he showed that the star order was a partial order. He further showed that in the defining equalities of this relation, one could actually replace '$\star$' by '$\dagger$', in other words the statement '$\mathbf{AA}^\star = \mathbf{BA}^\star$ and $\mathbf{A}^\star\mathbf{A} = \mathbf{A}^\star\mathbf{B}$' is equivalent to '$\mathbf{AA}^\dagger = \mathbf{BA}^\dagger$ and $\mathbf{A}^\dagger\mathbf{A} = \mathbf{A}^\dagger\mathbf{B}$'. However, M.Hestenes studied the properties of the matrices $\mathbf{A}$ and $\mathbf{B}$ satisfying the equalities $\mathbf{AA}^\star = \mathbf{BA}^\star$ and $\mathbf{A}^\star\mathbf{A} = \mathbf{A}^\star\mathbf{B}$ almost two decades earlier than Drazin. He called matrix $\mathbf{A}$, a section of the matrix $\mathbf{B}$ whenever $\mathbf{A}$ and $\mathbf{B}$ satisfied the equalities $\mathbf{AA}^\star = \mathbf{BA}^\star$, $\mathbf{A}^\star\mathbf{A} = \mathbf{A}^\star\mathbf{B}$. The term star order was coined by Drazin.

In Section 5.2, beginning with a formal definition of the star order, we obtain several of its characteristic properties. While studying the properties of the minus order and the sharp order, we found that the normal form and the reduced form (which is a special case of the core-nilpotent decomposition) $\mathbf{A} = \mathbf{P}\mathrm{diag}\left(\mathbf{D}\ ,\ \mathbf{0}\right)\mathbf{P}^{-1}$ for any matrix $\mathbf{A}$ of index ≤ 1, where $\mathbf{P}$ and $\mathbf{D}$ are non-singular matrices respectively proved useful. In the case of the star order, we shall see that the key reduced form is the singular value decomposition (SVD). In Section 5.3, apart from obtaining several other properties of the star order, we compare the star order with the other partial orders studied so far. We also identify the class of all matrices that lie above/below a given matrix under the star order. In Section 5.4, we specialize to some subclasses of matrices of index ≤ 1, which are important in applications. A detailed study of star order on these subclasses that include range-hermitian, normal, hermitian and *nnd* matrices, is undertaken here. Section 5.5 covers the star order on idempotent matrices. Idempotent matrices play a significant role in applications and for this reason a separate section on star order for the class of idempotent matrices is included. Finally in Section 5.6, Fisher-Cochran Type theorems are established for star order.

The matrices studied in this chapter are over $\mathbb{C}$ and the inner product of vectors $\mathbf{x}, \mathbf{y}$ in $\mathbb{C}^n$ will be the standard inner product, $(\mathbf{x}, \mathbf{y}) = \mathbf{y}^\star\mathbf{x}$, unless otherwise stated.

5.2 Star order - Characteristic properties

In this section we define the star order and obtain several of its characteristic properties and establish that the star order is a partial order on $\mathbb{C}^{m\times n}$. We show that if $\mathbf{A}$ is below $\mathbf{B}$ under the star order, then $\mathbf{A}$ and $\mathbf{B}$ have simultaneous singular value decomposition.

Definition 5.2.1. Let $\mathbf{A}$ and $\mathbf{B}$ be matrices of the same order $m \times n$. Then $\mathbf{A}$ is said to be below $\mathbf{B}$ under the star order if $\mathbf{A}\mathbf{A}^\star = \mathbf{B}\mathbf{A}^\star$ and $\mathbf{A}^\star\mathbf{A} = \mathbf{A}^\star\mathbf{B}$.
When $\mathbf{A}$ is below $\mathbf{B}$ under the star order, we write $\mathbf{A} <^\star \mathbf{B}$.

Remark 5.2.2. It is easy to show the following:
(i) $\mathbf{A} <^\star \mathbf{B} \Leftrightarrow \mathbf{A}^\star <^\star \mathbf{B}^\star$ and
(ii) If $\mathbf{U}$ and $\mathbf{V}$ are unitary matrices of orders $m \times m$ and $n \times n$ respectively, then $\mathbf{A} <^\star \mathbf{B} \Leftrightarrow \mathbf{UAV} <^\star \mathbf{UBV}$.

We start with the following theorem which shows that under the star order the matrices have a nice canonical form that is very useful in proving results.

Theorem 5.2.3. *Let* $\mathbf{A}$ *and* $\mathbf{B}$ *be matrices of the same order having ranks* a *and* b *respectively. Then the following are equivalent:*

(i) $\mathbf{A} <^\star \mathbf{B}$.

(ii) *There exist unitary matrices* $\mathbf{U}$ *and* $\mathbf{V}$, *positive definite diagonal matrices* $\mathbf{D_a}$ *and* $\mathbf{D_{b-a}}$ *of orders* $a \times a$ *and* $(b-a)\times(b-a)$ *respectively such that* $\mathbf{A} = \mathbf{U}\mathbf{diag}\left(\mathbf{D_a}, \mathbf{0}, \mathbf{0}\right)\mathbf{V}^\star$ *and* $\mathbf{B} = \mathbf{U}\mathbf{diag}\left(\mathbf{D_a}, \mathbf{D_{b-a}}, \mathbf{0}\right)\mathbf{V}^\star$ *and*

(iii) $\mathbf{B} - \mathbf{A} <^\star \mathbf{B}$.

Proof. (i) $\Rightarrow$ (ii)
Since $\mathbf{AA}^\star = \mathbf{BA}^\star$ and $\mathbf{A}^\star\mathbf{A} = \mathbf{A}^\star\mathbf{B}$, both $\mathbf{A}^\star\mathbf{B}$ and $\mathbf{BA}^\star$ are hermitian and *nnd*. So, by Theorem 2.7.18, there exist unitary matrices $\mathbf{U}$, $\mathbf{V}$ and *nnd* diagonal matrices $\mathbf{D_1}$ and $\mathbf{D_2}$ such that $\mathbf{A} = \mathbf{UD_1V}^\star$ and $\mathbf{B} = \mathbf{UD_2V}^\star$. Since $\mathbf{A} <^\star \mathbf{B}$, by Remark 5.2.2, we have $\mathbf{D_1} <^\star \mathbf{D_2}$. Without loss of generality, we can take $\mathbf{D_1} = \left(\mathbf{D_a}, \mathbf{0}, \mathbf{0}\right)$ and $\mathbf{D_2} = \left(\mathbf{D_a}, \mathbf{D_{b-a}}, \mathbf{0}\right)$, where $\mathbf{D_a}$ and $\mathbf{D_{b-a}}$ are positive definite diagonal matrices, and (ii) follows.
(ii) $\Rightarrow$ (i) is trivial.
(i) $\Leftrightarrow$ (iii) is easy to check. □

Remark 5.2.4. Let $\mathbf{U}$ and $\mathbf{V}$ be unitary matrices of orders $m \times m$ and $n \times n$ respectively such that $\mathbf{A} = \mathbf{U}\mathbf{diag}\left(\mathbf{D_a}, \mathbf{0}, \mathbf{0}\right)\mathbf{V}^\star$ and $\mathbf{B} = \mathbf{U}\mathbf{diag}\left(\mathbf{D_a}, \mathbf{D_{b-a}}, \mathbf{0}\right)\mathbf{V}^\star$, where $\mathbf{D_a}$ and $\mathbf{D_{b-a}}$ are positive definite diagonal matrices of order $a \times a$ and $(b-a)\times(b-a)$ respectively. Then the class of all

(a) least squares g-inverse of $\mathbf{A}$ and $\mathbf{B}$ are respectively given by

$$\{\mathbf{A}_\ell^-\} = \left\{\mathbf{V}\begin{pmatrix}\mathbf{D_a}^{-1} & \mathbf{0} & \mathbf{0}\\ \mathbf{M_1} & \mathbf{M_2} & \mathbf{M_3}\\ \mathbf{N_1} & \mathbf{N_2} & \mathbf{N_3}\end{pmatrix}\mathbf{U}^\star; \mathbf{M_i}\ \mathbf{N_i},\ \text{arbitrary } i = 1, 2, 3\right\},$$

and

$$\{\mathbf{B}_\ell^-\} = \left\{\mathbf{V}\begin{pmatrix}\mathbf{D_a}^{-1} & \mathbf{0} & \mathbf{0}\\ \mathbf{0} & \mathbf{D_{b-a}}^{-1} & \mathbf{0}\\ \mathbf{L_1} & \mathbf{L_2} & \mathbf{L_3}\end{pmatrix}\mathbf{U}^\star; \mathbf{L}_i,\ \text{arbitrary } i = 1,\ 2,\ 3\right\}.$$

(b) minimum norm g-inverses of $\mathbf{A}$ and $\mathbf{B}$ are respectively given by

$$\{\mathbf{A}_\mathbf{m}^-\} = \left\{\mathbf{V}\begin{pmatrix}\mathbf{D_a}^{-1} & \mathbf{S_2} & \mathbf{S_3}\\ \mathbf{0} & \mathbf{R_2} & \mathbf{R_3}\\ \mathbf{0} & \mathbf{T_2} & \mathbf{T_3}\end{pmatrix}\mathbf{U}^\star; \mathbf{S_i},\ \mathbf{R_i},\ \mathbf{T_i},\ \text{arbitrary } i = 2, 3\right\}$$

and

$$\{\mathbf{B}_{\mathbf{m}}^{-}\} = \left\{ \mathbf{V} \begin{pmatrix} \mathbf{D_a}^{-1} & \mathbf{0} & \mathbf{X_1} \\ \mathbf{0} & \mathbf{D_{b-a}}^{-1} & \mathbf{X_2} \\ \mathbf{0} & \mathbf{0} & \mathbf{X_3} \end{pmatrix} \mathbf{U}^{\star}; \mathbf{X_i}, \text{ arbitrary } i = 1,\ 2,\ 3 \right\}.$$

Remark 5.2.5. Let $\mathbf{A} <^{\star} \mathbf{B}$. Then
(i) $\mathbf{AB}^{\dagger}\mathbf{A} = \mathbf{BB}^{\dagger}\mathbf{A} = \mathbf{AB}^{\dagger}\mathbf{B} = \mathbf{A}$, or equivalently
(ii) $\mathbf{A} <^{-} \mathbf{B}$.

However, the converse is false as the following example shows.

Example 5.2.6. Let $\mathbf{A} = \begin{pmatrix} 1 & 1 \\ 0 & 0 \end{pmatrix}$ and $\mathbf{B} = \begin{pmatrix} 1 & 1 \\ 1 & 0 \end{pmatrix}$. Clearly, $\mathbf{A} <^{-} \mathbf{B}$.
As $\mathbf{AA}^{\star} = \begin{pmatrix} 2 & 0 \\ 0 & 0 \end{pmatrix}$ and $\mathbf{BA}^{\star} = \begin{pmatrix} 2 & 0 \\ 1 & 0 \end{pmatrix}$, $\mathbf{A} \not<^{\star} \mathbf{B}$.

It easy to see that each of the equalities in Remark 5.2.5(i) hold.

Theorem 5.2.7. *The star order is a partial order on* $\mathbb{C}^{m \times n}$.

Proof. Reflexivity is trivial and anti-symmetry follows by Remark 5.2.5. For transitivity, let $\mathbf{A} <^{\star} \mathbf{B}$ and $\mathbf{B} <^{\star} \mathbf{C}$. Then by Theorem 5.2.3, there exist unitary matrices $\mathbf{U}$ and $\mathbf{V}$ such that $\mathbf{A} = \mathbf{U}\,\mathbf{diag}\left(\mathbf{D_a}, \mathbf{0}, \mathbf{0}\right)\mathbf{V}^{\star}$ and $\mathbf{B} = \mathbf{U}\,\mathbf{diag}\left(\mathbf{D_a}, \mathbf{D_{b-a}}, \mathbf{0}\right)\mathbf{V}^{\star}$, where $\mathbf{D_a}$ and $\mathbf{D_{b-a}}$ are positive definite diagonal matrices of order $a \times a$ and $(b-a) \times (b-a)$ respectively. Let $\mathbf{C} = \mathbf{U} \begin{pmatrix} \mathbf{C_{11}} & \mathbf{C_{12}} & \mathbf{C_{13}} \\ \mathbf{C_{21}} & \mathbf{C_{22}} & \mathbf{C_{23}} \\ \mathbf{C_{31}} & \mathbf{C_{32}} & \mathbf{C_{33}} \end{pmatrix} \mathbf{V}^{\star}$, where partition involved is conformable for multiplication of $\mathbf{C}$ with $\mathbf{A}$ and $\mathbf{B}$. Since, $\mathbf{B} <^{\star} \mathbf{C}$, a simple computation shows that $\mathbf{C} = \mathbf{U} \begin{pmatrix} \mathbf{D_a} & \mathbf{0} & \mathbf{0} \\ \mathbf{0} & \mathbf{D_{b-a}} & \mathbf{0} \\ \mathbf{0} & \mathbf{0} & \mathbf{C_{33}} \end{pmatrix} \mathbf{V}^{\star}$. Clearly, $\mathbf{A} <^{\star} \mathbf{C}$. □

Theorem 5.2.8. *Let* $\mathbf{A}$ *and* $\mathbf{B}$ *be matrices of the same order. Then the following are equivalent:*

(i) $\mathbf{A}_{\ell \mathbf{m}}^{-}\mathbf{A} = \mathbf{A}_{\ell \mathbf{m}}^{-}\mathbf{B}$ *and* $\mathbf{AA}_{\ell \mathbf{m}}^{-} = \mathbf{BA}_{\ell \mathbf{m}}^{-}$ *for some minimum norm least squares g-inverse* $\mathbf{A}_{\ell \mathbf{m}}^{-}$ *of* $\mathbf{A}$
(ii) $\mathbf{A}^{\dagger}\mathbf{A} = \mathbf{A}^{\dagger}\mathbf{B}$ *and* $\mathbf{AA}^{\dagger} = \mathbf{BA}^{\dagger}$
(iii) $\mathbf{A} <^{\star} \mathbf{B}$
(iv) $\{\mathbf{B}_{\ell}^{-}\} \subseteq \{\mathbf{A}_{\ell}^{-}\}$ *and* $\{\mathbf{B}_{\mathbf{m}}^{-}\} \subseteq \{\mathbf{A}_{\mathbf{m}}^{-}\}$
(v) $\{\mathbf{B}_{\ell \mathbf{m}}^{-}\} \subseteq \{\mathbf{A}_{\ell \mathbf{m}}^{-}\}$
(vi) $\mathbf{A}^{\dagger}\mathbf{A} = \mathbf{B}^{\dagger}\mathbf{A}$ *and* $\mathbf{AA}^{\dagger} = \mathbf{AB}^{\dagger}$, *and*

(vii) *There exist orthogonal projectors* $\mathbf{P}$ *and* $\mathbf{Q}$ *such that* $\mathbf{A} = \mathbf{PB} = \mathbf{BQ}$.

Proof. (i) $\Rightarrow$ (ii) Since $\mathbf{A}^-_{\ell\mathbf{m}}\mathbf{A}\mathbf{A}^-_{\ell\mathbf{m}} = \mathbf{A}^\dagger$, the result follows.
(ii) $\Rightarrow$ (i)
Since $\mathbf{A}^\dagger$ is a minimum norm least squares g-inverse, the result follows.
(ii) $\Rightarrow$ (iii)
Notice that $\mathbf{A}^\star = \mathbf{A}^\dagger\mathbf{A}\mathbf{A}^\star = \mathbf{A}^\star\mathbf{A}\mathbf{A}^\dagger$. So, $\mathbf{A}\mathbf{A}^\dagger = \mathbf{B}\mathbf{A}^\dagger \Rightarrow \mathbf{A}\mathbf{A}^\dagger\mathbf{A}\mathbf{A}^\star = \mathbf{B}\mathbf{A}^\dagger\mathbf{A}\mathbf{A}^\star = \mathbf{B}(\mathbf{A}^\dagger\mathbf{A})^\star\mathbf{A}^\star \Rightarrow \mathbf{A}\mathbf{A}^\star = \mathbf{B}\mathbf{A}^\star$.
Similarly $\mathbf{A}^\star\mathbf{A} = \mathbf{A}^\star\mathbf{B}$.
(iii) $\Rightarrow$ (iv) follows from Theorem 5.2.3 in view of Remark 5.2.4.
(iv) $\Rightarrow$ (v)
Notice that for any $\mathbf{B}^-_{\ell\mathbf{m}}, \mathbf{B}^-_{\ell\mathbf{m}} \in \{\mathbf{B}^-_\ell\} \subseteq \{\mathbf{A}^-_\ell\}$ and $\mathbf{B}^-_{\ell\mathbf{m}} \in \{\mathbf{B}^-_\mathbf{m}\} \subseteq \{\mathbf{A}^-_\mathbf{m}\}$. Also, $\{\mathbf{A}^-_{\ell\mathbf{m}}\} = \{\mathbf{A}^-_\ell\} \cap \{\mathbf{A}^-_\mathbf{m}\}$. So, the result follows.
(v) $\Rightarrow$ (vi)
Since $\mathbf{B}^\dagger \in \{\mathbf{A}^-_{\ell\mathbf{m}}\}$, we have $\mathbf{A}\mathbf{B}^\dagger\mathbf{A} = \mathbf{A}$ and $\mathbf{A}\mathbf{B}^\dagger$ and $\mathbf{B}^\dagger\mathbf{A}$ are hermitian. So, $\mathbf{A}^\dagger\mathbf{A} = \mathbf{A}^\dagger\mathbf{A}\mathbf{B}^\dagger\mathbf{A} = (\mathbf{A}^\dagger\mathbf{A})^\star(\mathbf{B}^\dagger\mathbf{A})^\star = (\mathbf{B}^\dagger\mathbf{A}\mathbf{A}^\dagger\mathbf{A})^\star = (\mathbf{B}^\dagger\mathbf{A})^\star = \mathbf{B}^\dagger\mathbf{A}$.
Similarly, $\mathbf{A}\mathbf{A}^\dagger = \mathbf{A}\mathbf{B}^\dagger$.
(vi) $\Rightarrow$ (vii)
From Remark 5.2.5, we have $\mathbf{A} = \mathbf{A}\mathbf{B}^\dagger\mathbf{B} = \mathbf{A}\mathbf{A}^\dagger\mathbf{B}$, as $\mathbf{A}\mathbf{A}^\dagger = \mathbf{A}\mathbf{B}^\dagger$. Moreover, $\mathbf{A} = \mathbf{B}\mathbf{B}^\dagger\mathbf{A} = \mathbf{B}\mathbf{A}^\dagger\mathbf{A}$, since $\mathbf{A}^\dagger\mathbf{A} = \mathbf{B}^\dagger\mathbf{A}$. Let $\mathbf{P} = \mathbf{A}\mathbf{A}^\dagger$ and $\mathbf{Q} = \mathbf{A}^\dagger\mathbf{A}$. Clearly, $\mathbf{P}$ and $\mathbf{Q}$ are orthogonal projectors satisfying $\mathbf{A} = \mathbf{PB} = \mathbf{BQ}$.
(vii) $\Rightarrow$ (iii)
Now, $\mathbf{A} = \mathbf{PB} \Rightarrow \mathcal{C}(A) \subseteq \mathcal{C}(P) \Rightarrow \mathbf{PA} = \mathbf{A} \Rightarrow \mathbf{A}^\star\mathbf{P} = \mathbf{A}^\star$. Now $\mathbf{A} = \mathbf{PB} \Rightarrow \mathbf{A}^\star\mathbf{A} = \mathbf{A}^\star\mathbf{PB} = \mathbf{A}^\star\mathbf{B}$. Similarly, $\mathbf{A} = \mathbf{BQ} \Rightarrow \mathbf{A}\mathbf{A}^\star = \mathbf{B}\mathbf{A}^\star$.
(iii) $\Rightarrow$ (ii)
Since $\mathbf{A}^\dagger = \mathbf{A}^\dagger\mathbf{A}^{\dagger\star}\mathbf{A}^\star = \mathbf{A}^\star\mathbf{A}^{\dagger\star}\mathbf{A}^\dagger$, (ii) follows. □

Corollary 5.2.9. *Let* $\mathbf{A}$ *and* $\mathbf{B}$ *be matrices of the same order. Then the following are equivalent:*

(i) $\mathbf{A} <^\star \mathbf{B}$
(ii) $\mathbf{A}^\dagger <^\star \mathbf{B}^\dagger$
(iii) $\mathbf{A}\mathbf{A}^\dagger\mathbf{B} = \mathbf{A} = \mathbf{B}\mathbf{A}^\dagger\mathbf{A} = \mathbf{B}\mathbf{A}^\dagger\mathbf{B}$ *and*
(iv) $\mathbf{A}^\dagger\mathbf{A}\mathbf{B}^\dagger = \mathbf{A}^\dagger = \mathbf{B}^\dagger\mathbf{A}\mathbf{A}^\dagger = \mathbf{B}^\dagger\mathbf{A}\mathbf{B}^\dagger$.

Proof. By Theorem 5.2.3(ii) it is easy to see that (i) $\Leftrightarrow$ (ii), (i) $\Leftrightarrow$ (iii) and (i) $\Leftrightarrow$ (iv). □

We now explore the additional conditions required for the minus order to yield the star order.

Theorem 5.2.10. *Let* $\mathbf{A}$ *and* $\mathbf{B}$ *be matrices of the same order* $m \times n$ *such that* $\mathbf{A} <^{-} \mathbf{B}$. *Let* $\rho(\mathbf{A}) = a$, $\rho(\mathbf{B}) = b$ *and* k *be positive (integer or fraction).*

(i) *The following are equivalent:*

- (a) $\mathbf{A} <^{\star} \mathbf{B}$
- (b) $\mathbf{AB}^{\star}$ *and* $\mathbf{A}^{\star}\mathbf{B}$ *are hermitian*
- (c) $\mathbf{A}^{\dagger}\mathbf{B}$ *and* $\mathbf{BA}^{\dagger}$ *are hermitian* *and*
- (d) $\mathbf{AB}^{\dagger}$ *and* $\mathbf{B}^{\dagger}\mathbf{A}$ *are hermitian.*

(ii) *The following are equivalent:*

- (a_1) $\mathbf{A} <^{\star} \mathbf{B}$
- (b_1) $\mathbf{AA}^{\star} <^{\star} \mathbf{BB}^{\star}$
- (c_1) $(\mathbf{AA}^{\star})^{\mathbf{k}} <^{\star} (\mathbf{BB}^{\star})^{\mathbf{k}}$
- (d_1) $(\mathbf{AA}^{\star})^{\mathbf{k}}\mathbf{A} <^{\star} (\mathbf{BB}^{\star})^{\mathbf{k}}\mathbf{B}$
- (f_1) $\mathbf{BA}^{\dagger}\mathbf{B} = \mathbf{A}$ *and*
- (e_1) $\mathcal{C}(\mathbf{BA}^{\dagger}\mathbf{B}) \subseteq \mathcal{C}(\mathbf{A})$, $\mathcal{C}(\mathbf{BA}^{\dagger}\mathbf{B})^{\star} \subseteq \mathcal{C}(\mathbf{A})^{\star}$.

(iii) *The following are equivalent:*

- (a_2) $\mathbf{A} <^{\star} \mathbf{B}$
- (b_2) $\mathbf{A}^{\star}\mathbf{A} <^{\star} \mathbf{B}^{\star}\mathbf{B}$
- (c_2) $(\mathbf{A}^{\star}\mathbf{A})^{\mathbf{k}} <^{\star} (\mathbf{B}^{\star}\mathbf{B})^{\mathbf{k}}$
- (d_2) $(\mathbf{A}^{\star}\mathbf{A})^{\mathbf{k}}\mathbf{A}^{\star} <^{\star} (\mathbf{B}^{\star}\mathbf{B})^{\mathbf{k}}\mathbf{B}^{\star}$
- (f_2) $\mathbf{B}^{\dagger}\mathbf{AB}^{\dagger} = \mathbf{A}^{\dagger}$ *and*
- (e_2) $\mathcal{C}(\mathbf{B}^{\dagger}\mathbf{AB}^{\dagger}) \subseteq \mathcal{C}(\mathbf{A}^{\dagger})$.

Finally, all the conditions in anyone of (i), (ii) *and* (iii) *are equivalent to all conditions in any other.*

Proof. (i) (a) $\Rightarrow$ (b)
$\mathbf{A} <^{\star} \mathbf{B} \Rightarrow \mathbf{AA}^{\star} = \mathbf{BA}^{\star}$ and $\mathbf{A}^{\star}\mathbf{A} = \mathbf{A}^{\star}\mathbf{B}$. Also, $(\mathbf{BA}^{\star})^{\star} = \mathbf{AB}^{\star}$. It is now clear that $\mathbf{AB}^{\star}$ and $\mathbf{A}^{\star}\mathbf{B}$ are hermitian.
(b) $\Rightarrow$ (c), (d)
Since $\mathbf{AB}^{\star}$ and $\mathbf{A}^{\star}\mathbf{B}$ are hermitian, by Theorem 2.7.12, there exist unitary matrices $\mathbf{U}$, $\mathbf{V}$ and real diagonal matrices $\mathbf{D_1}$, $\mathbf{D_2}$ such that $\mathbf{A} = \mathbf{UD_1V}^{\star}$ and $\mathbf{B} = \mathbf{UD_2V}^{\star}$. Clearly, $\mathbf{A}^{\dagger}\mathbf{B}$ and $\mathbf{BA}^{\dagger}$ are hermitian and $\mathbf{AB}^{\dagger}$ and $\mathbf{B}^{\dagger}\mathbf{A}$ are hermitian.
(c) $\Rightarrow$ (b) follows along the lines of (b) $\Rightarrow$ (c).

Similarly, (d) $\Rightarrow$ (b).
(b) $\Rightarrow$ (a) Since $\mathbf{AB^\star}$ and $\mathbf{A^\star B}$ are hermitian, by Theorem 2.7.12, there exits unitary matrices $\mathbf{U}$, $\mathbf{V}$ and real diagonal matrices $\mathbf{D_1}$, $\mathbf{D_2}$ such that $\mathbf{A} = \mathbf{UD_1V^\star}$ and $\mathbf{B} = \mathbf{UD_2V^\star}$. As $\mathbf{A} <^- \mathbf{B}$, it follows that $\mathbf{D_1} <^- \mathbf{D_2}$. By suitable simultaneous permutations of rows and columns of $\mathbf{D_1}$ and $\mathbf{D_2}$ we have $\mathbf{PD_1Q} = \mathbf{diag(D_a\ , 0_{b-a}\ , 0)}$ and $\mathbf{PD_2Q} = \mathbf{diag(D_a\ , D_{b-a}\ , 0)}$, where $\mathbf{P}$ and $\mathbf{Q}$ are permutation matrices, $\mathbf{D_a}$ and $\mathbf{D_{b-a}}$ are non-singular diagonal matrices. Without loss of generality, we can take $\mathbf{D_a}$ and $\mathbf{D_{b-a}}$ as diagonal positive definite matrices. For, if some diagonal element d_i of $\mathbf{D_a}$ (or $\mathbf{D_{b-a}}$) is negative, then we can make it positive by multiplying it by -1 and multiplying the ith row of $\mathbf{V^\star}$ by -1. Thus,

$$\mathbf{A} = \mathbf{UP^\star diag(D_a\ , 0_{b-a}\ , 0)QV^\star} \text{ and } \mathbf{B} = \mathbf{UP^\star diag(D_a\ , D_{b-a}\ , 0)QV^\star}.$$

Clearly, $\mathbf{A} <^\star \mathbf{B}$.
(ii) $(a_1) \Rightarrow (b_1)$
Since $\mathbf{A} <^\star \mathbf{B}$, $\mathbf{AA^\star} = \mathbf{BA^\star}$ and $\mathbf{A^\star A} = \mathbf{A^\star B}$. Consider $\mathbf{AA^\star BB^\star} = \mathbf{AA^\star AB^\star} = \mathbf{AA^\star(AB^\star)} = \mathbf{AA^\star AA^\star}$, since $\mathbf{AA^\star} = \mathbf{BA^\star} = \mathbf{AB^\star}$ and $\mathbf{AA^\star}$ is hermitian.
Similarly, $\mathbf{AA^\star BB^\star} = \mathbf{AA^\star AA^\star}$.
$(b_1) \Rightarrow (c_1)$
If k is a positive integer, the result follows by $(a_1) \Rightarrow (b_1)$. If k is a fraction, then as $\mathbf{AA^\star}$ and $\mathbf{BB^\star}$ commute and are *nnd* matrices, therefore, there exists a unitary matrix $\mathbf{U}$ such that $\mathbf{A^\star A} = \mathbf{Udiag(D_1,\ 0,\ 0)U^\star}$ and $\mathbf{B^\star B} = \mathbf{Udiag(D_1,\ D_2,\ 0)U^\star}$. The result now follows by taking the kth powers.
$(c_1) \Rightarrow (b_1)$
Since $\mathbf{(A^\star A)^k} <^\star \mathbf{(B^\star B)^k}$, $\mathbf{(A^\star A)^k}$ and $\mathbf{(B^\star B)^k}$ commute and are *nnd* matrices. So, there exist unitary matrix $\mathbf{U}$ such that $\mathbf{(A^\star A)^k} = \mathbf{UD_1{}^kU^\star}$ and $\mathbf{(B^\star B)^k} = \mathbf{UD_2{}^kU^\star}$, where $\mathbf{D_1}$ and $\mathbf{D_2}$ are diagonal *nnd* matrices.Now, $\mathbf{(AA^\star)^k} <^\star \mathbf{(BB^\star)^k} \Rightarrow \mathbf{(D_1)^k} <^\star \mathbf{(D_2)^k}$. Let ${i_1}^{th}, {i_2}^{th} \ldots, {i_a}^{th}$ diagonal elements are non-null in $\mathbf{D_1}$. Then ${i_1}^{th}, {i_2}^{th} \ldots, {i_a}^{th}$ diagonal elements are same in both $\mathbf{D_1}$ and $\mathbf{D_2}$. So, $\mathbf{D_1} <^\star \mathbf{D_2}$ or $\mathbf{AA^\star} <^\star \mathbf{BB^\star}$.
$(b_1) \Rightarrow (a_1)$
Since $\mathbf{AA^\star} <^\star \mathbf{BB^\star}$, it follows that $\mathbf{AA^\star}$, $\mathbf{BB^\star}$ commute and are *nnd* matrices. Hence, there exists a unitary matrix $\mathbf{U}$, positive definite diagonal matrices $\mathbf{D_a}$ and $\mathbf{D_{b-a}}$ such that $\mathbf{AA^\star} = \mathbf{Udiag(D_a^2\ , 0_{b-a}\ , 0)U^\star}$ and $\mathbf{BB^\star} = \mathbf{Udiag(D_a^2\ , D_{b-a}^2\ , 0)U^\star}$. Partition $\mathbf{U} = (\mathbf{U_1} : \mathbf{U_2} : \mathbf{U_3})$, where $\mathbf{U_1}$ and $\mathbf{U_2}$ have a and $b - a$ columns respectively. Then $\mathbf{A} = \mathbf{U_1D_a^2L_1{}^\star}$ and $\mathbf{B} = (\mathbf{U_1} : \mathbf{U_2})\mathbf{diag(D_a^2\ , D_{b-a}^2)}(\mathbf{K_1} : \mathbf{K_2})^\star$ for some matrices $\mathbf{L_1}$, $\mathbf{K_1}$

and $\mathbf{K_2}$ such that $\mathbf{L_1L_1}^\star = \mathbf{I}$ and $(\mathbf{K_1} : \mathbf{K_2})^\star(\mathbf{K_1} : \mathbf{K_2}) = \mathbf{I}$. Since $\mathbf{A} <^- \mathbf{B}$, $\mathcal{C}(\mathbf{L_1}) \subseteq \mathcal{C}(\mathbf{A}^\star) \subseteq \mathcal{C}(\mathbf{B}^\star) = \mathcal{C}(\mathbf{K_1} : \mathbf{K_2})$. So, there exist matrices $\mathbf{F_1}$ and $\mathbf{F_2}$ such that $\mathbf{L_1} = (\mathbf{K_1} : \mathbf{K_2})(\mathbf{I} - \mathbf{F_1} : -\mathbf{F_2})^\star$. We can write $\mathbf{A} = (\mathbf{U_1} : \mathbf{U_2})\begin{pmatrix}\mathbf{D_a}(\mathbf{I}-\mathbf{F_1}) & -\mathbf{D_aF_2}\\ \mathbf{0} & \mathbf{0}\end{pmatrix}\begin{pmatrix}\mathbf{K_1}^\star\\ \mathbf{K_2}^\star\end{pmatrix}$. Hence, $\mathbf{B} - \mathbf{A} = (\mathbf{U_1} : \mathbf{U_2})\begin{pmatrix}\mathbf{D_aF_1} & \mathbf{D_aF_2}\\ \mathbf{0} & \mathbf{0}\end{pmatrix}\begin{pmatrix}\mathbf{K_1}^\star\\ \mathbf{K_2}^\star\end{pmatrix}$. As $\mathbf{A} <^- \mathbf{B}$, $\rho(\mathbf{B} - \mathbf{A}) = b - a$. So, $\mathbf{D_aF_1} = \mathbf{0}$ and therefore, $\mathbf{F_1} = \mathbf{0}$. Thus, $\mathbf{L_1} = \mathbf{K_1} - \mathbf{K_2F_2}^\star$. Now, $\mathbf{I} = \mathbf{L_1L_1}^\star = \mathbf{I} + \mathbf{F_2}^\star\mathbf{F_2}$. So, $\mathbf{F_2} = \mathbf{0}$. Thus, $\mathbf{L_1} = \mathbf{K_1}$ and $\mathbf{A} = \mathbf{U_1D_aK_1}$ and $\mathbf{B} = (\mathbf{U_1} : \mathbf{U_2})\mathbf{diag}(\mathbf{D_a^2}, \mathbf{D_{b-a}^2})(\mathbf{K_1} : \mathbf{K_2})^\star$. It is clear that $\mathbf{A} <^\star \mathbf{B}$.
$(c_1) \Rightarrow (d_1)$ is trivial.
$(d_1) \Rightarrow (c_1)$
From $(a_1) \Rightarrow (c_1)$, we have

$$\begin{aligned}(\mathbf{AA}^\star)^{\mathbf{k}}\mathbf{A} <^\star (\mathbf{BB}^\star)^{\mathbf{k}}\mathbf{B}&\\ &\Rightarrow (\mathbf{AA}^\star)^{\mathbf{k}}\mathbf{AA}^\star(\mathbf{AA}^\star)^{\mathbf{k}} <^\star (\mathbf{BB}^\star)^{\mathbf{k}}\mathbf{BB}^\star(\mathbf{BB}^\star)^{\mathbf{k}},\\ &\text{or } (\mathbf{AA}^\star)^{\mathbf{2k+1}} <^\star (\mathbf{BB}^\star)^{\mathbf{2k+1}}, \text{ so,}\\ &(\mathbf{AA}^\star)^{\mathbf{p}} <^\star (\mathbf{BB}^\star)^{\mathbf{p}}\end{aligned}$$

for some positive p.
$(a_1) \Rightarrow (f_1)$
By (i) $\mathbf{A}^\dagger\mathbf{A} = \mathbf{A}^\dagger\mathbf{B}$, $\mathbf{AA}^\dagger = \mathbf{BA}^\dagger$. So, $\mathbf{BA}^\dagger\mathbf{B} = \mathbf{AA}^\dagger\mathbf{A} = \mathbf{A}$.
$(f_1) \Rightarrow (e_1)$ is trivial
$(e_1) \Rightarrow (a_1)$
Now, $\mathcal{C}(\mathbf{BA}^\dagger\mathbf{B}) \subseteq \mathcal{C}(\mathbf{A}) \Rightarrow \mathbf{AA}^\dagger\mathbf{BA}^\dagger\mathbf{B} = \mathbf{BA}^\dagger\mathbf{B}$. Also, $\mathbf{A} <^- \mathbf{B}$, so, $\mathbf{A} = \mathbf{AB}^\dagger\mathbf{A} = \mathbf{AB}^\dagger\mathbf{B}$. Now, pre-multiplying and post-multiplying $\mathbf{AA}^\dagger\mathbf{BA}^\dagger\mathbf{B} = \mathbf{BA}^\dagger\mathbf{B}$ by $\mathbf{AB}^\dagger$ and $\mathbf{B}^\dagger\mathbf{AA}^\dagger$, we have $\mathbf{AB}^\dagger\mathbf{AA}^\dagger\mathbf{BA}^\dagger\mathbf{BB}^\dagger\mathbf{AA}^\dagger = \mathbf{AB}^\dagger\mathbf{BA}^\dagger\mathbf{BB}^\dagger\mathbf{AA}^\dagger$. As $\mathbf{A} = \mathbf{AB}^\dagger\mathbf{A} = \mathbf{AB}^\dagger\mathbf{B}$, we have $\mathbf{BA}^\dagger = \mathbf{AA}^\dagger$. Similarly, using $\mathcal{C}(\mathbf{BA}^\dagger\mathbf{B})^\star \subseteq \mathcal{C}(\mathbf{A})^\star$, we can establish $\mathbf{A}^\dagger\mathbf{A} = \mathbf{A}^\dagger\mathbf{B}$.

The proof for (iii) can be obtained by replacing $\mathbf{A}$ in (ii) by $\mathbf{A}^\star$ and noting that $\mathbf{A} <^\star \mathbf{B} \Leftrightarrow \mathbf{A}^\star <^\star \mathbf{B}^\star$. □

Theorem 5.2.11. *Let* $\mathbf{A}$ *and* $\mathbf{B}$ *be matrices of the same order with* $\rho(\mathbf{A}) = a$ *and* $\rho(\mathbf{B}) = b$. *Then the following are equivalent:*

(i) $\mathbf{A} <^\star \mathbf{B}$
(ii) $\mathbf{A} <^- \mathbf{B}$ *and* $(\mathbf{B} - \mathbf{A})^\dagger = \mathbf{B}^\dagger - \mathbf{A}^\dagger$
(iii) $\mathbf{A} <^- \mathbf{B}$ *and* $\mathbf{B}^\dagger - \mathbf{A}^\dagger \in \{(\mathbf{B} - \mathbf{A})_\ell^-\}$ *and*
(iv) $\mathbf{A} <^- \mathbf{B}$ *and* $\mathbf{B}^\dagger - \mathbf{A}^\dagger \in \{(\mathbf{B} - \mathbf{A})_{\mathbf{m}}^-\}$.

Proof. (i) $\Rightarrow$ (ii)
By Remark 5.2.5, $\mathbf{A} <^- \mathbf{B}$. By Theorem 5.2.3, there exist unitary matrices

$\mathbf{U}$ and $\mathbf{V}$ such that

$$\mathbf{A} = \mathbf{U}\mathbf{diag}\left(\mathbf{D_a}\ ,\ \mathbf{0}\ ,\ \mathbf{0}\right)\mathbf{V}^\star \text{ and } \mathbf{B} = \mathbf{U}\mathbf{diag}\left(\mathbf{D_a}\ ,\ \mathbf{D_{b-a}}\ ,\ \mathbf{0}\right)\mathbf{V}^\star,$$

where $\mathbf{D_a}$ and $\mathbf{D_{b-a}}$ are positive definite diagonal matrices of order $a \times a$ and $(b-a) \times (b-a)$ respectively. Clearly, $(\mathbf{B} - \mathbf{A})^\dagger = \mathbf{B}^\dagger - \mathbf{A}^\dagger$.

(ii) $\Rightarrow$ (iii) and (ii) $\Rightarrow$ (iv) are easy.

(iii) $\Rightarrow$ (i)

Since $\mathbf{A} <^- \mathbf{B}$, by Corollary 3.7.4, there exist unitary matrices $\mathbf{U}$, $\mathbf{V}$, a positive definite diagonal matrix $\mathbf{D_a}$ and a $b \times b$ non-singular matrix $\mathbf{M}$ given as

$$\mathbf{M} = \begin{pmatrix} \mathbf{D_a} & \mathbf{0} \\ \mathbf{0} & \mathbf{0} \end{pmatrix} + \begin{pmatrix} \mathbf{F_1} \\ \mathbf{I} \end{pmatrix} \mathbf{K}\left(\mathbf{F_2}\ ,\ \mathbf{I}\right)$$

such that $\mathbf{A} = \mathbf{U}\mathbf{diag}(\mathbf{D_a}, \mathbf{0_{b-a}}, \mathbf{0})\mathbf{V}^\star$ and $\mathbf{B} = \mathbf{U}(\mathbf{M}, \mathbf{0})\mathbf{V}^\star$, where $\mathbf{K}$ is non-singular and $\mathbf{F_1}$ and $\mathbf{F_2}$ are some matrices of suitable orders. Also, $\mathbf{B}^\dagger - \mathbf{A}^\dagger \in \{(\mathbf{B} - \mathbf{A})_\ell^-\}$, so, $(\mathbf{B} - \mathbf{A})^\star(\mathbf{B} - \mathbf{A})(\mathbf{B}^\dagger - \mathbf{A}^\dagger) = (\mathbf{B} - \mathbf{A})^\star$. However,

$$\mathbf{A}^\dagger = \mathbf{V}\mathbf{diag}(\mathbf{D_a}^{-1}, \mathbf{0_{b-a}}, \mathbf{0})\mathbf{U}^\star,\ \mathbf{B}^\dagger = \mathbf{V}(\mathbf{M}^{-1}, \mathbf{0})\mathbf{U}^\star,$$

where $\mathbf{M}^{-1} = \begin{pmatrix} \mathbf{D_a}^{-1} & -\mathbf{D_a}^{-1}\mathbf{F_1} \\ -\mathbf{F_2}\mathbf{D_a}^{-1} & \mathbf{K}^{-1} \end{pmatrix}$. Hence,

$$\begin{pmatrix} \mathbf{F_2^\star} \\ \mathbf{I} \end{pmatrix} \mathbf{K}^\star\left(\mathbf{F_1^\star}\ ,\ \mathbf{I}\right) \begin{pmatrix} \mathbf{F_1} \\ \mathbf{I} \end{pmatrix} \mathbf{K}\left(\mathbf{F_2}\ ,\ \mathbf{I}\right) = \begin{pmatrix} \mathbf{F_2^\star} \\ \mathbf{I} \end{pmatrix} \mathbf{K}^\star\left(\mathbf{F_1^\star}\ ,\ \mathbf{I}\right).$$

The matrix $\begin{pmatrix} \mathbf{F_2^\star} \\ \mathbf{I} \end{pmatrix}$ has a right inverse and $\mathbf{K}$ is invertible. So, we have

$$(\mathbf{F_1}\mathbf{F_1^\star} + \mathbf{I})\mathbf{K}\left(\mathbf{F_2}\ ,\ \mathbf{I}\right) \begin{pmatrix} \mathbf{0} & -\mathbf{D_a}^{-1}\mathbf{F_1} \\ -\mathbf{F_2}\mathbf{D_a}^{-1} & \mathbf{K}^{-1} \end{pmatrix} = \left(\mathbf{F_1^\star}\ ,\ \mathbf{I}\right). \quad (5.2.1)$$

Now, (5.2.1) $\Rightarrow$ $(\mathbf{F_1}\mathbf{F_1^\star} + \mathbf{I})\mathbf{K}(-\mathbf{F_2}\mathbf{D_a}^{-1}) = \mathbf{F_1^\star}$ and $\mathbf{F_1}\mathbf{F_1^\star} + (\mathbf{F_1}\mathbf{F_1^\star} + \mathbf{I})\mathbf{K}\mathbf{K}^{-1} = \mathbf{I}$. These give $\mathbf{F_1} = \mathbf{0}$ and $\mathbf{F_2} = \mathbf{0}$. It now follows that $\mathbf{A} <^\star \mathbf{B}$.

(iv) $\Rightarrow$ (i) follows along the same lines as the proof of (iii) $\Rightarrow$ (i). □

In Theorem 5.2.11, we have seen that if $\mathbf{A} <^- \mathbf{B}$ and $\mathbf{B}^\dagger - \mathbf{A}^\dagger$ is either a least squares g-inverse or a minimum norm g-inverse of $\mathbf{B} - \mathbf{A}$, then $\mathbf{A} <^\star \mathbf{B}$. One may be curious to know whether the condition '$\mathbf{A} <^- \mathbf{B}$ and $\mathbf{B}^\dagger - \mathbf{A}^\dagger$ is reflexive g-inverse of $\mathbf{B} - \mathbf{A}$' would also imply $\mathbf{A} <^\star \mathbf{B}$. The answer is in the negative as the following example shows:

Example 5.2.12. Let $\mathbf{A} = \begin{pmatrix} \mathbf{I} & \mathbf{0} \\ \mathbf{0} & \mathbf{0} \end{pmatrix}$ and $\mathbf{B} = \begin{pmatrix} \mathbf{I} & \mathbf{S} \\ \mathbf{T} & -\mathbf{I} \end{pmatrix}$, where $\mathbf{S}$ and $\mathbf{T}$ are matrices satisfying $\mathbf{ST} = \mathbf{0}, \mathbf{TS} = \mathbf{0}$. Then $\mathbf{A}^\dagger = \mathbf{A}$ and $\mathbf{B}^\dagger = \mathbf{B}^{-1} = \mathbf{B}$ and $\mathbf{B} - \mathbf{A}$ is its own reflexive g-inverse. Clearly, $\mathbf{A} \not<^\star \mathbf{B}$.

Incidentally, Example 5.2.12 also shows that $\mathbf{A} <^- \mathbf{B}$ and $\mathbf{A}^\dagger <^- \mathbf{B}^\dagger$ together do not imply $\mathbf{A} <^\star \mathbf{B}$.

5.3 Subclasses of g-inverses for which $\mathbf{A} <^\star \mathbf{B}$

In this section we characterize the class of all g-inverses $\mathbf{A}^-$ of a matrix $\mathbf{A}$ such that $\mathbf{A}^-\mathbf{A} = \mathbf{A}^\dagger\mathbf{A} = \mathbf{A}^\dagger\mathbf{B}$ and $\mathbf{A}\mathbf{A}^- = \mathbf{A}\mathbf{A}^\dagger = \mathbf{B}\mathbf{A}^\dagger$, when $\mathbf{A} <^\star \mathbf{B}$. We also obtain the class of all matrices that lie above/below a given matrix under the star order.

Theorem 5.3.1. *Let* $\mathbf{A}$ *and* $\mathbf{B}$ *be matrices of the same order such that* $\mathbf{A} <^\star \mathbf{B}$ *and let* $\mathbf{A}^-_{\ell\mathbf{m}}$ *be any minimum norm least squares g-inverse of* $\mathbf{A}$ *such that* $\mathbf{A}^-_{\ell\mathbf{m}}\mathbf{A} = \mathbf{A}^-_{\ell\mathbf{m}}\mathbf{B}$ *and* $\mathbf{A}\mathbf{A}^-_{\ell\mathbf{m}} = \mathbf{B}\mathbf{A}^-_{\ell\mathbf{m}}$. *Then for each minimum norm least squares g-inverse* $\mathbf{B}^-_{\ell\mathbf{m}}$ *of* $\mathbf{B}$, *we have* $\mathbf{A}^-_{\ell\mathbf{m}}\mathbf{A} = \mathbf{B}^-_{\ell\mathbf{m}}\mathbf{A} = \mathbf{A}^-_{\ell\mathbf{m}}\mathbf{B}$ *and* $\mathbf{A}\mathbf{A}^-_{\ell\mathbf{m}} = \mathbf{A}\mathbf{B}^-_{\ell\mathbf{m}} = \mathbf{B}\mathbf{A}^-_{\ell\mathbf{m}}$.

Proof. Proof follows along the lines of Theorem 4.3.1 by making use of Remark 5.2.4. □

Corollary 5.3.2. *Let* $\mathbf{A}$ *and* $\mathbf{B}$ *be matrices of the same order such that* $\mathbf{A} <^\star \mathbf{B}$ *and let* $\mathbf{A}^-_{\ell\mathbf{m}}$ *be any minimum norm least squares g-inverse of* $\mathbf{A}$ *such that* $\mathbf{A}^-_{\ell\mathbf{m}}\mathbf{A} = \mathbf{A}^-{}_{\ell\mathbf{m}}\mathbf{B}$ *and* $\mathbf{A}\mathbf{A}^-_{\ell\mathbf{m}} = \mathbf{B}\mathbf{A}^-_{\ell\mathbf{m}}$. *Then for each minimum norm least squares g-inverse* $\mathbf{B}^-_{\ell\mathbf{m}}$ *of* $\mathbf{B}$, $\mathbf{A}^-_{\ell\mathbf{m}}\mathbf{A}$ *and* $\mathbf{B}^-_{\ell\mathbf{m}}\mathbf{B}$ *commute. Further,* $\mathbf{A}\mathbf{A}^-_{\ell\mathbf{m}}$ *and* $\mathbf{B}\mathbf{B}^-_{\ell\mathbf{m}}$ *commute.*

Given matrices $\mathbf{A}$ and $\mathbf{B}$ of the same order, we shall now obtain a characterization of

$$\{\mathbf{A}^-_{\ell\mathbf{m}}\}_B = \{\mathbf{G} : \mathbf{A}^\star\mathbf{G}\mathbf{A} = \mathbf{A}^\star,\ \mathbf{A}\mathbf{G}\mathbf{A}^\star = \mathbf{A}^\star, \mathbf{A}\mathbf{G} = \mathbf{B}\mathbf{G}, \mathbf{G}\mathbf{A} = \mathbf{G}\mathbf{B}\},$$

the set of all minimum norm least squares g-inverses of $\mathbf{A}$ for which $\mathbf{A} <^\star \mathbf{B}$

Theorem 5.3.3. *Let* $\mathbf{A}$ *and* $\mathbf{B}$ *be matrices of the same order such that* $\mathbf{A} <^\star \mathbf{B}$. *Let* $\rho(\mathbf{A}) = a$ *and* $\rho(\mathbf{B}) = b$. *Then*

$$\{\mathbf{A}^-_{\ell\mathbf{m}}\}_\mathbf{B} = \{\mathbf{B}^-_{\ell\mathbf{m}} - \mathbf{B}^-_{\ell\mathbf{m}}(\mathbf{B}-\mathbf{A})\mathbf{B}^-_{\ell\mathbf{m}}\} = \{\mathbf{B}^-_{\ell\mathbf{m}} - (\mathbf{B}-\mathbf{A})^\dagger\}.$$

Proof. First observe that $\mathbf{B}^-_{\ell\mathbf{m}}(\mathbf{B}-\mathbf{A})\mathbf{B}^-_{\ell\mathbf{m}}$ is a minimum norm least squares g-inverse of $\mathbf{B}-\mathbf{A}$. Further, it is a reflexive g-inverse of $\mathbf{B}-\mathbf{A}$. Hence, for each $\mathbf{B}^-_{\ell\mathbf{m}}$, $\mathbf{B}^-_{\ell\mathbf{m}}(\mathbf{B}-\mathbf{A})\mathbf{B}^-_{\ell\mathbf{m}}$ is the Moore-Penrose inverse of $\mathbf{B}-\mathbf{A}$. Thus, $\{\mathbf{B}^-_{\ell\mathbf{m}} - \mathbf{B}^-_{\ell\mathbf{m}}(\mathbf{B}-\mathbf{A})\mathbf{B}^-_{\ell\mathbf{m}}\} = \{\mathbf{B}^-_{\ell\mathbf{m}} - (\mathbf{B}-\mathbf{A})^\dagger\}$. Using the representations of $\mathbf{A}$ and $\mathbf{B}$ as in Theorem 5.2.3, we can easily check that

$\mathbf{G} \in \{\mathbf{A}_{\ell m}^-\}$ if and only if $\mathbf{G} = \mathbf{V}\mathbf{diag}\left(\mathbf{D_a}^{-1}, \mathbf{0_{b-a}}, \mathbf{T}\right)\mathbf{U}^\star$ for some matrix $\mathbf{T}$. Also, each $\mathbf{B}_{\ell m}^-$ is of the form $\mathbf{V}\mathbf{diag}\left(\mathbf{D_a}^{-1}, \mathbf{D_{b-a}}^{-1}, \mathbf{T}\right)\mathbf{U}^\star$ for some matrix $\mathbf{S}$ and $(\mathbf{B}-\mathbf{A})^\dagger = \mathbf{V}\mathbf{diag}\left(\mathbf{0}, \mathbf{D_{b-a}}^{-1}, \mathbf{T}\right)\mathbf{U}^\star$. The rest of the proof is computational. □

We now obtain the class of all matrices that lie above a given matrix under the star order.

Theorem 5.3.4. *Let* $\mathbf{A}$ *and* $\mathbf{B}$ *be matrices having the same order. Then* $\mathbf{A} <^\star \mathbf{B}$ *if and only if* $\mathbf{B} = \mathbf{A} + (\mathbf{I} - \mathbf{A}\mathbf{A}_\ell^-)\mathbf{T}(\mathbf{I} - \mathbf{A}_\mathbf{m}^-\mathbf{A})$, *where* $\mathbf{A}_\ell^-$ *and* $\mathbf{A}_\mathbf{m}^-$ *are respectively arbitrary least squares g-inverse and minimum norm g-inverse of* $\mathbf{A}$ *and* $\mathbf{T}$ *is arbitrary.*

Proof follows by definition.

Our next theorem is yet another characterization of all matrices lying above a given matrix under the star order.

Theorem 5.3.5. *Let* $\mathbf{A}$ *be an* $m \times n$ *matrix and* $\mathbf{A} = \mathbf{U_1}\mathbf{D_1}\mathbf{V_1}^\star$ *be the singular value decomposition of* $\mathbf{A}$, *where* $\mathbf{U_1}$, $\mathbf{V_1}$ *are semi-unitary and* $\mathbf{D_1}$ *is a positive definite diagonal matrix. Then* $\mathbf{A} <^\star \mathbf{B}$ *if and only if* $\mathbf{B} = \mathbf{U}\mathbf{diag}\left(\mathbf{D_1}, \mathbf{D_2}\right)\mathbf{V}^\star$, *where* $\mathbf{U}$, $\mathbf{V}$ *are arbitrary extensions of* $\mathbf{U_1}$ *and* $\mathbf{V_1}$ *respectively to unitary matrices and* $\mathbf{D_2}$ *is arbitrary matrix of appropriate order.*

Proof. 'If' part is trivial.
'Only if' part
Extend $\mathbf{U_1}$ and $\mathbf{V_1}$ to unitary matrices $\mathbf{U} = (\mathbf{U_1} : \mathbf{U_2})$ and $\mathbf{V} = (\mathbf{V_1} : \mathbf{V_2})$. Then we can write $\mathbf{A} = \mathbf{U}\mathbf{diag}\left(\mathbf{D_1}, \mathbf{0}\right)\mathbf{V}^\star$. It is now easy to check that $\mathbf{B} = \mathbf{U}\mathbf{diag}\left(\mathbf{D_1}, \mathbf{D_2}\right)\mathbf{V}^\star$, for some matrix $\mathbf{D_2}$. □

In the next theorem we obtain the class of all matrices $\mathbf{A}$ that are below a given matrix $\mathbf{B}$ under the star order.

Theorem 5.3.6. *Let* $\mathbf{B}$ *an* $m \times n$ *matrix and* $\mathbf{B} = \mathbf{U_1}\mathbf{D}\mathbf{V_1}^\star$ *be the singular value decomposition of* $\mathbf{B}$, *where* $\mathbf{U_1}$, $\mathbf{V_1}$ *are semi-unitary and* $\mathbf{D} = [d_{ij}]$ *is a positive definite diagonal matrix. Then* $\mathbf{A} <^\star \mathbf{B}$ *if and only if* $\mathbf{A} = \mathbf{U_1}\mathbf{E}\mathbf{V_1}^\star$, *where* $\mathbf{E} = [e_{ij}]$ *is a diagonal matrix such that* $e_{ii} = d_{ii}$ *or* 0 *for each* i *and* $e_{ij} = 0$ *whenever* $i \neq j$.

Proof follows from Theorem 5.2.3.

Remark 5.3.7. The matrices $\mathbf{U_1}$ and $\mathbf{V_1}$ in Theorem 5.3.6 are in general not unique unless $\mathbf{D}$ has all diagonal elements distinct. Whenever $\mathbf{U_1}$ and $\mathbf{V_1}$ are not unique, we can vary them over all possible choices of $\mathbf{U_1}$ and $\mathbf{V_1}$ to obtain all $\mathbf{A}$ that are below $\mathbf{B}$ under $\mathbf{A} <^{\star} \mathbf{B}$.

5.4 Star order for special subclasses of matrices

In this section we specialize to some subclasses of matrices with index ≤ 1, namely, range-hermitian matrices, normal matrices, hermitian matrices and the *nnd* matrices and make a detailed study of the star order on these subclasses. We also study the relation between the minus order and the star order for several of these subclasses.

Theorem 5.4.1. *Let* $\mathbf{A}$ *and* $\mathbf{B}$ *be range-hermitian matrices of the same order having ranks* a *and* b $(b \geq a)$ *respectively. Then*

(i) $\mathbf{A} <^{-} \mathbf{B}$ *if and only if there exist a unitary matrix* $\mathbf{U}$, *non-singular matrices* $\mathbf{T}$ *and* $\mathbf{M}$, *where* $\mathbf{M}$ *is of the form*

$$\mathbf{M} = \begin{pmatrix} \mathbf{T} & \mathbf{0} \\ \mathbf{0} & \mathbf{0} \end{pmatrix} + \begin{pmatrix} \mathbf{F_1} \\ \mathbf{I} \end{pmatrix} \mathbf{K}(\mathbf{F_2}, \mathbf{I}),$$

where $\mathbf{K}$ *is a* $(b-a) \times (b-a)$ *non-singular matrix and* $\mathbf{F_1}$, $\mathbf{F_2}$ *are some matrices of appropriate order such that* $\mathbf{A} = \mathbf{U}\mathbf{diag}(\mathbf{T}, \mathbf{0})\mathbf{U}^{\star}$ *and* $\mathbf{B} = \mathbf{U}\mathbf{diag}(\mathbf{M}, \mathbf{0})\mathbf{U}^{\star}$.

(ii) $\mathbf{A} <^{\star} \mathbf{B}$ *if and only if* $\mathbf{A}$ *and* $\mathbf{B}$ *have the same form as in* (i) *with* $\mathbf{F_1} = \mathbf{0}$ *and* $\mathbf{F_2} = \mathbf{0}$.

Proof. (i) 'If' part is trivial.
'Only if' part
Since $\mathbf{A}$ is range-hermitian, by Theorem 2.2.9, there exists a unitary matrix $\mathbf{W}$ and a non-singular matrix $\mathbf{T}$ such that $\mathbf{A} = \mathbf{W}\mathbf{diag}(\mathbf{T}, \mathbf{0})\mathbf{W}^{\star}$. Also, $\mathbf{A}$ and $\mathbf{B}$ are range-hermitian matrices, so, $\mathcal{C}(\mathbf{A}^{\star}) = \mathcal{C}(\mathbf{A}) \subseteq \mathcal{C}(\mathbf{B}) = \mathcal{C}(\mathbf{B}^{\star})$. The remaining proof follows from Corollary 3.7.4.
(ii) The proof is trivial in view of (i) and the definition of star order. □

Let $\mathbf{A}$ and $\mathbf{B}$ be range-hermitian matrices such that $\mathbf{A} <^{-} \mathbf{B}$. What are the additional conditions that ensure $\mathbf{A} <^{\star} \mathbf{B}$? The answer is contained in the following theorem:

Theorem 5.4.2. *Let* $\mathbf{A}$ *and* $\mathbf{B}$ *be range-hermitian matrices of the same order such that* $\mathbf{A} <^{-} \mathbf{B}$. *Then the following are equivalent:*

(i) $\mathbf{A} <^{\star} \mathbf{B}$
(ii) $\mathbf{A}^{\star}\mathbf{B}$ *and* $\mathbf{B}\mathbf{A}^{\star}$ *are hermitian.*
(iii) $\mathbf{A}^{\dagger}\mathbf{B}$ *and* $\mathbf{B}\mathbf{A}^{\dagger}$ *are hermitian*
(iv) $\mathbf{A}\mathbf{B}^{\dagger}$ *and* $\mathbf{B}^{\dagger}\mathbf{A}$ *are hermitian*
(v) $\mathbf{A}^{\dagger}$ *and* $\mathbf{B}$ *commute*
(vi) $\mathbf{A}$ *and* $\mathbf{B}^{\dagger}$ *commute*
(vii) $\mathbf{A}$ *and* $\mathbf{B}$ *commute*
(viii) $\mathbf{B}\mathbf{A}^{\dagger}\mathbf{B}$ *is range-hermitian and* $\mathcal{C}(\mathbf{B}\mathbf{A}^{\dagger}\mathbf{B}) \in \mathcal{C}(\mathbf{A})$ *and*
(ix) $\mathbf{B}^{\dagger}\mathbf{A}\mathbf{B}^{\dagger}$ *is range-hermitian and* $\mathcal{C}(\mathbf{B}^{\dagger}\mathbf{A}\mathbf{B}^{\dagger}) \in \mathcal{C}(\mathbf{A})$.

Proof. Equivalence of (i)-(vii) follows from Theorem 5.4.1 and equivalence of (i), (viii) and (ix) follows from Theorem 5.2.10. □

Theorem 5.4.3. *Let* $\mathbf{A}$ *and* $\mathbf{B}$ *be matrices of the same order such that* $\mathbf{A} <^{\star} \mathbf{B}$. *The following are equivalent:*

(i) $\mathbf{A}$ *is range-hermitian*
(ii) $\mathbf{A}^{\dagger}$ *and* $\mathbf{B}$ *commute* *and*
(iii) $\mathbf{A}$ *and* $\mathbf{B}^{\dagger}$ *commute.*

Proof. (i) $\Leftrightarrow$ (ii) Since $\mathbf{A} <^{\star} \mathbf{B}$, $\mathbf{A}^{\dagger}\mathbf{A} = \mathbf{A}^{\dagger}\mathbf{B}$ and $\mathbf{A}\mathbf{A}^{\dagger} = \mathbf{B}\mathbf{A}^{\dagger}$. Now, $\mathbf{A}^{\dagger}$ and $\mathbf{B}$ commute $\Leftrightarrow$ $\mathbf{A}^{\dagger}\mathbf{B} = \mathbf{B}\mathbf{A}^{\dagger}$ $\Leftrightarrow$ $\mathbf{A}^{\dagger}\mathbf{A} = \mathbf{A}\mathbf{A}^{\dagger}$ $\Leftrightarrow$ $\mathbf{A}$ is range-hermitian.
(i) $\Leftrightarrow$ (iii) follows from Theorem 5.2.8. □

We now turn our attention to normal matrices.

Theorem 5.4.4. *Let* $\mathbf{A}$ *and* $\mathbf{B}$ *be matrices of the same order such that* $\mathbf{A} <^{\star} \mathbf{B}$. *Then the following are equivalent:*

(i) $\mathbf{A}$ *is normal*
(ii) $\mathbf{A}^{\star}$ *and* $\mathbf{B}$ *commute* *and*
(iii) $\mathbf{A}$ *and* $\mathbf{B}^{\star}$ *commute.*

The proof follows from definition of the star order and Remark 5.2.2(i).

Theorem 5.4.5. *Let* $\mathbf{A}$ *and* $\mathbf{B}$ *be normal matrices of the same order* $n \times n$ *having ranks* a *and* b $(b \geq a)$ *respectively. Then*

(i) $\mathbf{A} <^{-} \mathbf{B}$ *if and only if there exist a unitary matrix* $\mathbf{U}$, *a non-singular matrix* $\mathbf{D_a}$ *with possibly complex diagonal elements and a non-singular, normal matrix* $\mathbf{M}$ *of the form* $\mathbf{M} = \begin{pmatrix} \mathbf{D_a} & \mathbf{0} \\ \mathbf{0} & \mathbf{K} \end{pmatrix} + \begin{pmatrix} \mathbf{F_1} \\ \mathbf{I} \end{pmatrix} \mathbf{K}(\mathbf{F_2}, \mathbf{I})$, *where* $\mathbf{K}$ *is a* $(b-a) \times (b-a)$ *non-singular matrix and* $\mathbf{F_1}$, $\mathbf{F_2}$ *are some matrices of appropriate order such that* $\mathbf{A} = \mathbf{U}\mathbf{diag}(\mathbf{D_a},\ \mathbf{0_{b-a}},\ \mathbf{0})\mathbf{U}^\star$ *and* $\mathbf{B} = \mathbf{U}\mathbf{diag}(\mathbf{M}, \mathbf{0})\mathbf{U}^\star$.

(ii) $\mathbf{A} <^\star \mathbf{B}$ *if and only if* $\mathbf{A} = \mathbf{U}\mathbf{diag}(\mathbf{D_a},\ \mathbf{0_{b-a}},\ \mathbf{0})\mathbf{U}^\star$ *and* $\mathbf{B} = \mathbf{U}\mathbf{diag}(\mathbf{D_a},\ \mathbf{D_{b-a}},\ \mathbf{0})\mathbf{U}^\star$.

Proof. (a) Notice that a normal matrix is unitarily similar to a diagonal matrix with possibly complex diagonal elements and (b) $\mathbf{B}$ is normal if and only if $\mathbf{M}$ is normal. Proof now follows from Theorem 5.4.1. □

Theorem 5.4.6. *Let* $\mathbf{A}$ *and* $\mathbf{B}$ *be normal matrices of the same order such that* $\mathbf{A} <^- \mathbf{B}$. *Then the following are equivalent:*

(i) $\mathbf{A} <^\star \mathbf{B}$
(ii) $\mathbf{A}^\dagger\mathbf{B}$ *and* $\mathbf{B}\mathbf{A}^\dagger$ *are hermitian*
(iii) $\mathbf{B}^\dagger\mathbf{A}$ *and* $\mathbf{A}\mathbf{B}^\dagger$ *are hermitian*
(iv) $\mathbf{A}^\star\mathbf{B}$ *and* $\mathbf{B}\mathbf{A}^\star$ *are hermitian*
(v) $\mathbf{B}^\star\mathbf{A}$ *and* $\mathbf{A}\mathbf{B}^\star$ *are hermitian*
(vi) $\mathbf{A}$ *and* $\mathbf{B}$ *commute*
(vii) $\mathbf{A}^\star$ *and* $\mathbf{B}$ *commute*
(viii) $\mathbf{A}$ *and* $\mathbf{B}^\star$ *commute*
(ix) $\mathbf{A}^\star$ *and* $\mathbf{B}^\star$ *commute*
(x) $\mathbf{A}^\dagger$ *and* $\mathbf{B}$ *commute*
(xi) $\mathbf{A}$ *and* $\mathbf{B}^\dagger$ *commute*
(xii) $\mathbf{B}\mathbf{A}^\dagger\mathbf{B}$ *is normal and* $\mathcal{C}(\mathbf{B}\mathbf{A}^\dagger\mathbf{B}) \subseteq \mathcal{C}(\mathbf{A})$ *and*
(xiii) $\mathbf{B}^\dagger\mathbf{A}\mathbf{B}^\dagger$ *is normal and* $\mathcal{C}(\mathbf{B}^\dagger\mathbf{A}\mathbf{B}^\dagger) \subseteq \mathcal{C}(\mathbf{A})$.

In view of Theorem 5.4.5(i), proof follows from Theorem 5.2.10.

Remark 5.4.7. Let $\mathbf{A}$ and $\mathbf{B}$ be hermitian matrices of the same order. If we replace (xii) and (xiii) in Theorem 5.4.6 with (xii)$'$ $\mathcal{C}(\mathbf{B}\mathbf{A}^\dagger\mathbf{B}) \subseteq \mathcal{C}(\mathbf{A})$ and (xiii)$'$ $\mathcal{C}(\mathbf{B}^\dagger\mathbf{A}\mathbf{B}^\dagger) \subseteq \mathcal{C}(\mathbf{A})$ respectively, then the statements (i)-(xi), (xii)$'$ and (xiii)$'$ are equivalent.

Theorem 5.4.8. *Let* $\mathbf{A}$ *and* $\mathbf{B}$ *be matrices of the same order having ranks* a *and* b *respectively. Let* $\mathbf{A} <^- \mathbf{B}$. *Then* $\mathbf{A} <^\star \mathbf{B}$ *if and only if* $\mathbf{A}^\star\mathbf{B}$ *and* $\mathbf{A}\mathbf{B}^\star$ *are normal.*

Proof. 'If' part is trivial.
'Only if' part
By Corollary 3.7.4, there exist unitary matrices $\mathbf{U},\ \mathbf{V}$, a positive definite diagonal matrix $\mathbf{D_a}$, and a non-singular matrix $\mathbf{M}$ such that

$$\mathbf{A} = \mathbf{U}\mathbf{diag}(\mathbf{D_a},\ \mathbf{0_{b-a}},\ \mathbf{0})\mathbf{V}^\star \text{ and } \mathbf{B} = \mathbf{U}\mathbf{diag}(\mathbf{M}, \mathbf{0})\mathbf{V}^\star,$$

where $\mathbf{M} = \mathbf{diag}(\mathbf{D_a},\ \mathbf{0}) + \begin{pmatrix}\mathbf{F_1}\\ \mathbf{I}\end{pmatrix}\mathbf{K}(\mathbf{F_2},\mathbf{I})$, with $\mathbf{K}$ a $(b-a)\times(b-a)$ non-singular matrix and $\mathbf{F_1},\ \mathbf{F_2}$ some matrices of appropriate order.
Now, $\mathbf{A^\star B}$ is normal $\Rightarrow \mathbf{Vdiag}(\mathbf{D_a},\ \mathbf{0_{b-a}},\ \mathbf{0})\mathbf{diag}(\mathbf{M},\ \mathbf{0})\mathbf{V^\star}$ is normal

$$\Rightarrow \begin{pmatrix}\mathbf{D_a} & \mathbf{0} & \mathbf{0}\\ \mathbf{0} & \mathbf{0} & \mathbf{0}\\ \mathbf{0} & \mathbf{0} & \mathbf{0}\end{pmatrix}\begin{pmatrix}\mathbf{D_a}+\mathbf{F_1KF_2} & \mathbf{F_1K} & \mathbf{0}\\ \mathbf{KF_2} & \mathbf{K} & \mathbf{0}\\ \mathbf{0} & \mathbf{0} & \mathbf{0}\end{pmatrix} \text{ is normal.}$$

$$\Rightarrow \begin{pmatrix}\mathbf{D_a}^2+\mathbf{D_aF_1KF_2} & \mathbf{D_aF_1K} & \mathbf{0}\\ \mathbf{0} & \mathbf{0} & \mathbf{0}\\ \mathbf{0} & \mathbf{0} & \mathbf{0}\end{pmatrix}\begin{pmatrix}(\mathbf{D_a}^2+\mathbf{D_aF_1KF_2})^\star & \mathbf{0} & \mathbf{0}\\ (\mathbf{D_aF_1K})^\star & \mathbf{0} & \mathbf{0}\\ \mathbf{0} & \mathbf{0} & \mathbf{0}\end{pmatrix}$$

$$= \begin{pmatrix}(\mathbf{D_a}^2+\mathbf{D_aF_1KF_2})^\star & \mathbf{0} & \mathbf{0}\\ (\mathbf{D_aF_1K})^\star & \mathbf{0} & \mathbf{0}\\ \mathbf{0} & \mathbf{0} & \mathbf{0}\end{pmatrix}\begin{pmatrix}\mathbf{D_a}^2+\mathbf{D_aF_1KF_2} & \mathbf{D_aF_1K} & \mathbf{0}\\ \mathbf{0} & \mathbf{0} & \mathbf{0}\\ \mathbf{0} & \mathbf{0} & \mathbf{0}\end{pmatrix}$$

$$\Rightarrow \mathbf{F_1} = \mathbf{0}.$$

Similarly, $\mathbf{AB^\star}$ is normal $\Rightarrow \mathbf{F_2} = \mathbf{0}$. Therefore,

$$\mathbf{A} = \mathbf{Udiag}(\mathbf{D_a},\ \mathbf{0_{b-a}},\ \mathbf{0})\mathbf{V^\star}, \mathbf{B} = \mathbf{Udiag}(\mathbf{D_a},\ \mathbf{K},\ \mathbf{0})\mathbf{V^\star}.$$

Clearly, $\mathbf{A} <^\star \mathbf{B}$. □

Corollary 5.4.9. *Let* $\mathbf{A}$ *and* $\mathbf{B}$ *be matrices of the same order having ranks* a *and* b *respectively. Let* $\mathbf{A} <^- \mathbf{B}$. *Then the following are equivalent:*

(i) $\mathbf{A} <^\star \mathbf{B}$
(ii) $\mathbf{A^\dagger B}$ *and* $\mathbf{AB^\dagger}$ *are normal*
(iii) $\mathbf{B^\dagger A}$ *and* $\mathbf{BA^\dagger}$ *are normal*
(iv) $\mathbf{A^\star B}$ *and* $\mathbf{AB^\star}$ *are normal* *and*
(v) $\mathbf{B^\star A}$ *and* $\mathbf{BA^\star}$ *are normal.*

Theorem 5.4.10. *Let* $\mathbf{A}$ *and* $\mathbf{B}$ *be matrices of the same order having ranks* a *and* b $(b \geq a)$ *respectively. Let* $\mathbf{A} <^- \mathbf{B}$, $\mathbf{A}$ *be range-hermitian,* $\mathbf{B}$ *be hermitian and* $\mathbf{A^\star B}$ *be normal. Then* $\mathbf{A} <^\star \mathbf{B}$.

Proof. By Theorem 5.4.1,

$$\mathbf{A} = \mathbf{Udiag}(\mathbf{T},\mathbf{0})\mathbf{U^\star} \text{ and } \mathbf{B} = \mathbf{Udiag}(\mathbf{M},\mathbf{0})\mathbf{U^\star},$$

where $\mathbf{U}$ is unitary, $\mathbf{T}$ is non-singular matrix of order $a\times a$ and $\mathbf{M}$ is a $b\times b$ matrix of the form $\mathbf{M} = \begin{pmatrix}\mathbf{T} & \mathbf{0}\\ \mathbf{0} & \mathbf{0}\end{pmatrix} + \begin{pmatrix}\mathbf{F_1}\\ \mathbf{I}\end{pmatrix}\mathbf{K}(\mathbf{F_2},\mathbf{I})$, with $\mathbf{K}$ a $(b-a)\times(b-a)$ non-singular matrix and $\mathbf{F_1},\ \mathbf{F_2}$ are some matrices of appropriate order.

Since, $\mathbf{A}^\star\mathbf{B}$ is normal $\Rightarrow \begin{pmatrix}\mathbf{T} & \mathbf{0} & \mathbf{0}\\ \mathbf{0} & \mathbf{0} & \mathbf{0}\\ \mathbf{0} & \mathbf{0} & \mathbf{0}\end{pmatrix}\begin{pmatrix}\mathbf{T}+\mathbf{F_1KF_2} & \mathbf{F_1K} & \mathbf{0}\\ \mathbf{KF_2} & \mathbf{K} & \mathbf{0}\\ \mathbf{0} & \mathbf{0} & \mathbf{0}\end{pmatrix}$ is normal.
This gives $\mathbf{F_1} = \mathbf{0}$. As $\mathbf{B}$ is hermitian, $\mathbf{F_2} = \mathbf{0}$. So,

$$\mathbf{A} = \mathbf{U}\mathbf{diag}(\mathbf{T},\ \mathbf{0_{b-a}},\ \mathbf{0})\mathbf{V}^\star,\ \mathbf{B} = \mathbf{U}\mathbf{diag}(\mathbf{T},\ \mathbf{K},\ \mathbf{0})\mathbf{V}^\star.$$

Clearly, $\mathbf{A} <^\star \mathbf{B}$. □

Theorem 5.4.11. *Let* $\mathbf{A}$ *and* $\mathbf{B}$ *be nnd matrices of the same order. Consider the following statements:*

(i) $\mathbf{A} <^- \mathbf{B}$
(ii) $\mathbf{A^2} <^- \mathbf{B^2}$
(iii) $\mathbf{A} <^\star \mathbf{B}$
(iv) $\mathbf{A^2} <^\star \mathbf{B^2}$ *and*
(v) $\mathbf{AB} = \mathbf{BA} = \mathbf{A^2}$.

Then (i) *and* (ii) *imply all the others. Any of* (iii), (iv) *and* (v) *imply all others.*

Proof. (i) and (ii) $\Rightarrow$ (iii), (iv) and (v)
By Corollary 3.7.5, there exists a unitary matrix $\mathbf{U}$, an $a \times a$ positive definite diagonal matrix $\mathbf{D_a}$ and a $b \times b$ positive definite matrix $\mathbf{M} = \begin{pmatrix}\mathbf{D_a} & \mathbf{0}\\ \mathbf{0} & \mathbf{0}\end{pmatrix} + \begin{pmatrix}\mathbf{F}\\ \mathbf{I}\end{pmatrix}\mathbf{M_{22}}(\mathbf{F}^\star\ ,\mathbf{I})$ such that $\mathbf{A} = \mathbf{U}\mathbf{diag}(\mathbf{D_a}, \mathbf{0_{b-a}}, \mathbf{0})\ \mathbf{U}^*$ and $\mathbf{B} = \mathbf{U}\mathbf{diag}(\mathbf{M}, \mathbf{0})\ \mathbf{U}^\star$; where $\mathbf{M_{22}}$ is a positive definite matrix and $\mathbf{F}$ is some suitable matrix. As $\mathbf{A^2} <^- \mathbf{B^2}$, $\mathbf{A^2}(\mathbf{B}^\dagger)^\mathbf{2}\mathbf{A^2} = \mathbf{A^2}$. Therefore,

$$\begin{pmatrix}\mathbf{D_a}^2 & \mathbf{0} & \mathbf{0}\\ \mathbf{0} & \mathbf{0} & \mathbf{0}\\ \mathbf{0} & \mathbf{0} & \mathbf{0}\end{pmatrix}\left(\begin{pmatrix}\mathbf{D_a}+\mathbf{FKF}^\star & \mathbf{FK} & \mathbf{0}\\ \mathbf{F}^\star\mathbf{K} & \mathbf{K} & \mathbf{0}\\ \mathbf{0} & \mathbf{0} & \mathbf{0}\end{pmatrix}^\dagger\right)^2\begin{pmatrix}\mathbf{D_a}^2 & \mathbf{0} & \mathbf{0}\\ \mathbf{0} & \mathbf{0} & \mathbf{0}\\ \mathbf{0} & \mathbf{0} & \mathbf{0}\end{pmatrix} = \begin{pmatrix}\mathbf{D_a}^2 & \mathbf{0} & \mathbf{0}\\ \mathbf{0} & \mathbf{0} & \mathbf{0}\\ \mathbf{0} & \mathbf{0} & \mathbf{0}\end{pmatrix}$$

$\Rightarrow \mathbf{F} = \mathbf{0}$. So,(iii), (iv) and (v) all hold.
(v)$\Rightarrow$ (iii), (iii) $\Rightarrow$ (i) and (iv)$\Rightarrow$ (ii) are trivial.
(v)$\Rightarrow$ (ii) and (iv)
Let (v) hold. Then there exist *nnd* diagonal matrices $\mathbf{\Delta_1}, \mathbf{\Delta_2}$ and a unitary matrix $\mathbf{V}$ such that $\mathbf{A} = \mathbf{V\Delta_1V}^\star$ and $\mathbf{B} = \mathbf{V\Delta_2V}^\star$ with $\mathbf{\Delta_1\Delta_2} = \mathbf{\Delta_2\Delta_1} = \mathbf{\Delta_1^2}$. Hence, by a suitable permutation of rows and the corresponding columns of $\mathbf{\Delta_1}$, $\mathbf{\Delta_2}$, we have $\mathbf{P\Delta_1P}^\star = \mathbf{diag}(\mathbf{D_1}\ ,\mathbf{0}\ ,\mathbf{0})$ and $\mathbf{P\Delta_2P}^\star = \mathbf{diag}(\mathbf{D_1}\ ,\mathbf{D_2}\ ,\mathbf{0})$, where $\mathbf{D_1}\ ,\mathbf{D_2}$ are positive definite diagonal matrices and $\mathbf{P}$ is a permutation matrix. Let $\mathbf{U} = \mathbf{VP}$. Then $\mathbf{U}$ is unitary and $\mathbf{A} = \mathbf{U}\mathbf{diag}(\mathbf{D_1},\ \mathbf{0},\ \mathbf{0})\mathbf{U}^\star$ and $\mathbf{B} = \mathbf{U}\mathbf{diag}(\mathbf{D_1},\ \mathbf{D_2},\ \mathbf{0})\mathbf{U}^\star$.

Clearly, (ii) and (iv) hold.
(iv) $\Rightarrow$ (v)
Let (iv) hold. Clearly, $\mathbf{A^2}$ and $\mathbf{B^2}$ are *nnd* and therefore, $\mathbf{A^2B^2} = \mathbf{B^2A^2} = \mathbf{A^4}$. Thus, there exist *nnd* diagonal matrices $\mathbf{D_1}$, $\mathbf{D_2}$ and a unitary matrix $\mathbf{U}$ such that

$$\mathbf{A^2} = \mathbf{U}\mathbf{diag}(\mathbf{D_1}^2,\ \mathbf{0},\ \mathbf{0})\mathbf{U}^\star \text{ and } \mathbf{B^2} - \mathbf{U}\mathbf{diag}(\mathbf{D_1}^2,\ \mathbf{D_2}^2,\ \mathbf{0})\mathbf{U}^\star.$$

Since $\mathbf{A}$ and $\mathbf{B}$ are the unique *nnd* square roots of $\mathbf{A^2}$ and $\mathbf{B^2}$, we have $\mathbf{A} = \mathbf{U}\mathbf{diag}(\mathbf{D_1},\ \mathbf{0},\ \mathbf{0})\mathbf{U}^\star$ and $\mathbf{B^2} = \mathbf{U}\mathbf{diag}(\mathbf{D_1},\ \mathbf{D_2},\ \mathbf{0})\mathbf{U}^\star$. It is now clear that (v) holds. □

Corollary 5.4.12. *Let* $\mathbf{A}$ *and* $\mathbf{B}$ *be hermitian matrices of the same order. Consider the statements* (i)-(v) *in Theorem 5.4.11. Then* (i) *and* (ii) *imply all the others. Any of* (iii) *and* (v) *imply all others. Also,* (iv) *implies* (v) *if and only if* (a) *every negative eigen-value* λ *of* $\mathbf{A}$ *is an eigen-value of* $\mathbf{B}$ *and* (b) *the algebraic multiplicity of* λ *as an eigen value of* $\mathbf{A}$ *is* $\leq$ *the algebraic multiplicity of* λ *as an eigen value of* $\mathbf{B}$.

Theorem 5.4.13. *Let* $\mathbf{A}$ *and* $\mathbf{B}$ *be square matrices of the same order. Then the following hold:*

(i) *If* $\mathcal{C}(\mathbf{A}^\star) \subseteq \mathcal{C}(\mathbf{B})$, *then* $\mathbf{B}^\dagger\mathbf{A}^\dagger$ *is a reflexive least squares g-inverse of* $\mathbf{AB}$.
(ii) *If* $\mathbf{A} <^\star \mathbf{B}$ *and* $\mathbf{B}$ *is hermitian, then* $(\mathbf{AB})^\dagger = \mathbf{B}^\dagger\mathbf{A}^\dagger$.

Proof. (i) Since $\mathcal{C}(\mathbf{A}^\star) \subseteq \mathcal{C}(\mathbf{B})$, we have $\mathbf{A} = \mathbf{ABB}^\dagger$. Therefore, $\mathbf{AB}(\mathbf{B}^\dagger\mathbf{A}^\dagger)\mathbf{AB} = (\mathbf{ABB}^\dagger)\mathbf{A}^\dagger\mathbf{AB} = \mathbf{AB}$,

$$\begin{aligned}\mathbf{B}^\dagger\mathbf{A}^\dagger(\mathbf{AB})\mathbf{B}^\dagger\mathbf{A}^\dagger &= \mathbf{B}^\dagger\mathbf{A}^\dagger(\mathbf{ABB}^\dagger)\mathbf{A}^\dagger\\ &= \mathbf{B}^\dagger\mathbf{A}^\dagger\mathbf{AA}^\dagger = \mathbf{B}^\dagger\mathbf{A}^\dagger\end{aligned}$$

and $\mathbf{ABB}^\dagger\mathbf{A}^\dagger = \mathbf{AA}^\dagger$ is hermitian, as $\mathbf{AA}^\dagger$ is hermitian.
(ii) Since $\mathbf{A} <^\star \mathbf{B}$ and $\mathbf{B}$ is hermitian, in view of (i), we only need to show that $\mathbf{B}^\dagger\mathbf{A}^\dagger\mathbf{AB}$ is hermitian. However,
$\mathbf{A} <^\star \mathbf{B} \Rightarrow \mathbf{A}^\dagger\mathbf{A} = \mathbf{A}^\dagger\mathbf{B},\ \mathbf{AA}^\dagger = \mathbf{BA}^\dagger$ and $\mathbf{A}^\dagger\mathbf{A} = \mathbf{B}^\dagger\mathbf{A},\ \mathbf{AA}^\dagger = \mathbf{AB}^\dagger$, by Theorem 5.2.8. Therefore, $\mathbf{AA}^\dagger\mathbf{B} = \mathbf{A}$ and $\mathbf{A}^\dagger\mathbf{AB}^\star = \mathbf{A}^\star = \mathbf{B}^\star\mathbf{AA}^\dagger$. Hence, $\mathbf{B}^\dagger\mathbf{A}^\dagger\mathbf{AB} = (\mathbf{B}^\dagger\mathbf{A}^\dagger\mathbf{A})\mathbf{B} = (\mathbf{B}^\dagger\mathbf{A}^\star) = (\mathbf{AB}^\dagger)^\star = \mathbf{AB}^\dagger$. Since $\mathbf{AB}^\dagger$ is hermitian, $\mathbf{B}^\dagger\mathbf{A}^\dagger$ is the Moore-Penrose inverse of $\mathbf{AB}$. □

We observe that the hypothesis in Theorem 5.4.13(i) is not enough to guarantee $\mathbf{B}^\dagger\mathbf{A}^\dagger$ is the Moore-Penrose inverse of $\mathbf{AB}$. Also, we cannot replace the condition $\mathbf{A} <^\star \mathbf{B}$ in Theorem 5.4.13 (ii) by $\mathbf{A} <^- \mathbf{B}$, even when $\mathbf{A}$ and $\mathbf{B}$ are *nnd*. We give the following example:

Example 5.4.14. Let $\mathbf{A} = \begin{pmatrix} 1 & 1 \\ 1 & 1 \end{pmatrix}$ and $\mathbf{B} = \begin{pmatrix} 2 & 1 \\ 1 & 1 \end{pmatrix}$. The matrices $\mathbf{A}$ and $\mathbf{B}$ satisfy all the conditions in Theorem 5.4.13(i) and (ii).
Also, $\mathbf{A}^\dagger = \frac{1}{4}\begin{pmatrix} 1 & 1 \\ 1 & 1 \end{pmatrix}$ and $\mathbf{B}^\dagger = \begin{pmatrix} 1 & -1 \\ -1 & 2 \end{pmatrix}$. Now, $(\mathbf{AB})^\dagger = \frac{1}{26}\begin{pmatrix} 3 & 3 \\ 2 & 2 \end{pmatrix}$, and $\mathbf{B}^\dagger\mathbf{A}^\dagger = \frac{1}{4}\begin{pmatrix} 0 & 0 \\ 1 & 1 \end{pmatrix}$. Thus, $(\mathbf{AB})^\dagger \neq \mathbf{B}^\dagger\mathbf{A}^\dagger$.

Theorem 5.4.15. *Let* $\mathbf{A}$ *and* $\mathbf{B}$ *be range-hermitian matrices of the same order. Then* $\mathbf{A} <^\star \mathbf{B}$ *if and only if* $(\mathbf{AB})^\dagger = \mathbf{B}^\dagger\mathbf{A}^\dagger = \mathbf{A}^\dagger\mathbf{B}^\dagger = \mathbf{A}^{\dagger^2}$.

Proof. The 'only if' part follows from Theorem 4.2.8 as for any range-hermitian matrix $\mathbf{X}$, $\mathbf{X}^\# = \mathbf{X}^\dagger$ and $\mathbf{A} <^\star \mathbf{B}$ is equivalent to $\mathbf{A} <^\# \mathbf{B}$.
'If' part
Notice that $\mathbf{A}^{\dagger^2} = \mathbf{B}^\dagger\mathbf{A}^\dagger \Rightarrow \mathbf{A}^{\dagger^2}\mathbf{AA}^\star = \mathbf{B}^\dagger\mathbf{A}^\dagger\mathbf{AA}^\star = \mathbf{B}^\dagger\mathbf{A}^\star$ and $\mathbf{A}^{\dagger^2}\mathbf{AA}^\star = \mathbf{A}^\dagger\mathbf{A}^\star$, we have $\mathbf{A}^\dagger\mathbf{A}^\star = \mathbf{B}^\dagger\mathbf{A}^\star$. Since, $\mathbf{A}$ is range-hermitian, there is a matrix $\mathbf{D}$ such that $\mathbf{A} = \mathbf{A}^\star\mathbf{D}$. Therefore, $\mathbf{A}^\dagger\mathbf{A} = \mathbf{A}^\dagger\mathbf{A}^\star\mathbf{D} = \mathbf{B}^\dagger\mathbf{A}^\star\mathbf{D} = \mathbf{B}^\dagger\mathbf{A}$.

Similarly, using $\mathbf{A}^\dagger\mathbf{B}^\dagger = \mathbf{A}^{\dagger^2}$ we can establish $\mathbf{AA}^\dagger = \mathbf{BA}^\dagger$. Thus, $\mathbf{A} <^\star \mathbf{B}$. □

Recall that an $m \times n$ complex matrix $\mathbf{U}$ is a partial isometry if $\mathbf{UU}^\star\mathbf{U} = \mathbf{U}$ or equivalently, $\mathbf{U}^\dagger = \mathbf{U}^\star$. We can easily check that if $\mathbf{U}$ is a partial isometry, then for any unitary matrices $\mathbf{P}$, $\mathbf{Q}$ of appropriate order $\mathbf{PUQ}$ is also a partial isometry. We show that for partial isometries, the star order and the minus order coincide.

Theorem 5.4.16. *Let* $\mathbf{A}$ *and* $\mathbf{B}$ *be partial isometries of the same order* $m \times n$. *Then* $\mathbf{A} <^\star \mathbf{B}$ *if and only if* $\mathbf{A} <^- \mathbf{B}$.

Proof. 'Only if' part is trivial.
For 'If' part, notice that by Corollary 3.7.4, $\mathbf{A} <^- \mathbf{B} \Leftrightarrow$there exist unitary matrices $\mathbf{U}$, $\mathbf{V}$ positive definite diagonal matrix $\mathbf{D_a}$, and a non-singular matrix $\mathbf{M}$ such that

$$\mathbf{A} = \mathbf{Udiag}(\mathbf{D_a},\ \mathbf{0_{b-a}},\ \mathbf{0})\mathbf{V}^\star \text{ and } \mathbf{B} = \mathbf{Udiag}(\mathbf{M}, \mathbf{0})\mathbf{V}^\star,$$

where $\mathbf{M} = \mathbf{diag}(\mathbf{D_a},\ \mathbf{0}) + \begin{pmatrix} \mathbf{F_1} \\ \mathbf{I} \end{pmatrix}\mathbf{K}(\mathbf{F_2}, \mathbf{I})$, with $\mathbf{K}$ a $(b-a) \times (b-a)$ non-singular matrix and $\mathbf{F_1}$, $\mathbf{F_2}$ some matrices of appropriate order $\Leftrightarrow$ $\mathrm{diag}(\mathbf{D_a},\ \mathbf{0}) <^- \mathbf{M}$. Since $\mathbf{A}$ is a partial isometry, we have $\mathbf{D_a} = \mathbf{I}$ and $\mathbf{M}^\star = \mathbf{M}^\dagger = \mathbf{M}^{-1}$. It is easy to check that $\mathbf{F_1} = \mathbf{0}$ and $\mathbf{F_2} = \mathbf{0}$. It follows that $\mathbf{A} <^\star \mathbf{B}$. □

5.5 Star order and idempotent matrices

In this section we consider idempotent matrices and orthogonal projectors and study the star order for these matrices. Recall for idempotent matrices $\mathbf{E}$ and $\mathbf{F}$,

$$\mathbf{E} <^{-} \mathbf{F} \text{ if and only if } \mathbf{E} = \mathbf{EF} = \mathbf{FE}.$$

We have the following:

Theorem 5.5.1. *Let* $\mathbf{E}$ *and* $\mathbf{F}$ *be any idempotent matrices of the same order. Then* $\mathbf{E} <^{-} \mathbf{F}$ *if and only if* $\mathbf{E} <^{\star} \mathbf{E} + (\mathbf{I} - \mathbf{F})^{\star}$.

Proof. $\mathbf{E} <^{\star} \mathbf{E} + (\mathbf{I} - \mathbf{F})^{\star} \Leftrightarrow \mathbf{E}^{\star}\mathbf{E} = \mathbf{E}^{\star}(\mathbf{E} + (\mathbf{I} - \mathbf{F})^{\star})$,
$\mathbf{EE}^{\star} = (\mathbf{E} + (\mathbf{I} - \mathbf{F})^{\star})\mathbf{E}^{\star}$
$\Leftrightarrow \mathbf{E}^{\star}\mathbf{E} = \mathbf{E}^{\star}\mathbf{E} + \mathbf{E}^{\star} - \mathbf{E}^{\star}\mathbf{F}^{\star}$, $\mathbf{EE}^{\star} = \mathbf{EE}^{\star} + \mathbf{E}^{\star} - \mathbf{F}^{\star}\mathbf{E}^{\star}$
$\Leftrightarrow \mathbf{E} = \mathbf{EF} = \mathbf{FE} \Leftrightarrow \mathbf{E} <^{-} \mathbf{F}$. □

We now investigate when a given idempotent matrix $\mathbf{E}$ is below another under the star order.

Theorem 5.5.2. *Let* $\mathbf{E}$ *and* $\mathbf{F}$ *be idempotent matrices of the same order. Then the following are equivalent:*

(i) $\mathbf{E} <^{\star} \mathbf{F}$
(ii) $\mathbf{EE}^{\star} <^{\star} \mathbf{FF}^{\star}$ *and* $\mathbf{E}^{\star}\mathbf{E} <^{\star} \mathbf{F}^{\star}\mathbf{F}$
(iii) $\mathbf{EE}^{\star} <^{-} \mathbf{FF}^{\star}$ *and* $\mathbf{E}^{\star}\mathbf{E} <^{-} \mathbf{F}^{\star}\mathbf{F}$ *and*
(iv) $\mathbf{FF}^{\star} - \mathbf{EE}^{\star} = (\mathbf{F} - \mathbf{E})(\mathbf{F} - \mathbf{E})^{\star}$ *and* $\mathbf{F}^{\star}\mathbf{F} - \mathbf{E}^{\star}\mathbf{E} = (\mathbf{F} - \mathbf{E})^{\star}(\mathbf{F} - \mathbf{E})$.

Proof. (i) $\Rightarrow$ (ii)
Since $\mathbf{E} <^{\star} \mathbf{F}$, $\mathbf{E}^{\star}\mathbf{E} = \mathbf{E}^{\star}\mathbf{F}$ and $\mathbf{EE}^{\star} = \mathbf{FE}^{\star}$. Also, $\mathbf{EE}^{\star}\mathbf{EE}^{\star}$, $\mathbf{EE}^{\star}$ and $\mathbf{FF}^{\star}$ are all hermitian, therefore we have

$$\mathbf{EE}^{\star}\mathbf{EE}^{\star} = \mathbf{EE}^{\star}\mathbf{EF}^{\star} = \mathbf{EE}^{\star}\mathbf{FF}^{\star} = \mathbf{FF}^{\star}\mathbf{EE}^{\star}.$$

Thus, $\mathbf{EE}^{\star} <^{\star} \mathbf{FF}^{\star}$.
Similarly, $\mathbf{E}^{\star}\mathbf{E} <^{\star} \mathbf{F}^{\star}\mathbf{F}$.
(ii) $\Rightarrow$ (iii) is trivial.
(iii) $\Rightarrow$ (iv)
Since $\mathbf{EE}^{\star} <^{-} \mathbf{FF}^{\star}$, we have $\mathbf{FF}^{\star} - \mathbf{EE}^{\star} <^{-} \mathbf{FF}^{\star}$ and therefore, $(\mathbf{FF}^{\star} - \mathbf{EE}^{\star})(\mathbf{FF}^{\star})^{\dagger}(\mathbf{FF}^{\star} - \mathbf{EE}^{\star}) = \mathbf{FF}^{\star} - \mathbf{EE}^{\star}$. Also, $\mathbf{FF}^{\star}$ is *nnd*, and hence, $(\mathbf{FF}^{\star})^{\dagger}$ is *nnd*. It follows that $\mathbf{FF}^{\star} - \mathbf{EE}^{\star}$ is *nnd*. Moreover, $\mathcal{C}(\mathbf{E}) = \mathcal{C}(\mathbf{EE}^{\star}) \subseteq \mathcal{C}(\mathbf{FF}^{\star}) = \mathcal{C}(\mathbf{F})$ and $\mathcal{C}(\mathbf{E}^{\star}) = \mathcal{C}(\mathbf{E}^{\star}\mathbf{E}) \subseteq \mathcal{C}(\mathbf{F}^{\star}\mathbf{F}) = \mathcal{C}(\mathbf{F}^{\star})$. Hence, $\mathbf{E} = \mathbf{EFF} = \mathbf{FFE}$. So, $\mathbf{E} = \mathbf{EF} = \mathbf{FE}$. Now, $\mathbf{E}(\mathbf{FF}^{\star} - \mathbf{EE}^{\star})\mathbf{E}^{\dagger} = \mathbf{0}$,

and $\mathbf{FF}^\star - \mathbf{EE}^\star$ is *nnd*, we have $\mathbf{E}(\mathbf{FF}^\star - \mathbf{EE}^\star) = \mathbf{0}$, or $\mathbf{EF}^\star = \mathbf{EE}^\star = \mathbf{FE}^\star$. Hence, $\mathbf{FF}^\star - \mathbf{EE}^\star = (\mathbf{F} - \mathbf{E})(\mathbf{F} - \mathbf{E})^\star$.
Similarly, we can establish $\mathbf{F}^\star\mathbf{F} - \mathbf{E}^\star\mathbf{E} = (\mathbf{F} - \mathbf{E})^\star(\mathbf{F} - \mathbf{E})$.
(iv) $\Rightarrow$ (i)
Notice that $\mathbf{FF}^\star - \mathbf{EE}^\star = (\mathbf{F} - \mathbf{E})(\mathbf{F} - \mathbf{E})^\star \Rightarrow \mathbf{FF}^\star - \mathbf{EE}^\star$ is *nnd* and $\mathcal{C}(\mathbf{E}) = \mathcal{C}(\mathbf{EE}^\star) \subseteq \mathcal{C}(\mathbf{FF}^\star) = \mathcal{C}(\mathbf{F})$.

Similarly, $\mathbf{F}^\star\mathbf{F} - \mathbf{E}^\star\mathbf{E} = (\mathbf{F} - \mathbf{E})^\star(\mathbf{F} - \mathbf{E}) \Rightarrow \mathbf{F}^\star\mathbf{F} - \mathbf{E}^\star\mathbf{E}$ is *nnd* and $\mathcal{C}(\mathbf{E}^\star) \subseteq \mathcal{C}(\mathbf{F}^\star)$. Thus, $\mathbf{E} = \mathbf{EFF} = \mathbf{FFE}$ or $\mathbf{E} = \mathbf{EF} = \mathbf{FE}$. As in proof of (iii) $\Rightarrow$ (iv), we have $\mathbf{EF}^\star = \mathbf{EE}^\star = \mathbf{FE}^\star$. Similarly, $\mathbf{E}^\star\mathbf{E} = \mathbf{E}^\star\mathbf{F} = \mathbf{F}^\star\mathbf{E}$. So, $\mathbf{E} <^\star \mathbf{F}$. □

Let $\mathbf{E}$ and $\mathbf{F}$ be idempotent matrices. Then by Remark 3.6.8, we have $\mathbf{E} <^- \mathbf{F} \Leftrightarrow \mathbf{E} <^\# \mathbf{F}$. In fact we have the following:

Theorem 5.5.3. *Let* $\mathbf{E}$ *and* $\mathbf{F}$ *be idempotent matrices having the same order such that* $\mathbf{E} <^- \mathbf{F}$. *Let* $\rho(\mathbf{E}) = e$, $\rho(\mathbf{F}) = f$. *Then there exists a non-singular matrix* $\mathbf{P}$ *such that*

$$\mathbf{E} = \mathbf{P}\mathbf{diag}(\mathbf{I_e},\ \mathbf{0_{e-f}},\ \mathbf{0})\mathbf{P}^{-1} \text{ and } \mathbf{F} = \mathbf{P}\mathbf{diag}(\mathbf{I_e},\ \mathbf{I_{e-f}},\ \mathbf{0})\mathbf{P}^{-1}.$$

Proof. Since $\mathbf{E} <^- \mathbf{F}$, by Theorem 3.3.5(vi), there exists non-singular matrices $\mathbf{P}$, $\mathbf{D_e}$ and $\mathbf{D_{e-f}}$ such that $\mathbf{E} = \mathbf{P}\mathbf{diag}(\mathbf{D_e},\ \mathbf{0_{e-f}},\ \mathbf{0})\mathbf{P}^{-1}$ and $\mathbf{F} = \mathbf{P}\mathbf{diag}(\mathbf{D_e},\ \mathbf{D_{e-f}},\ \mathbf{0})\mathbf{P}^{-1}$. Since the only non-singular idempotent matrix is the identity matrix and $\mathbf{D_e}$ and $\mathbf{D_{e-f}}$ are non-singular, the result follows. □

Corollary 5.5.4. *Let* $\mathbf{E}$ *and* $\mathbf{F}$ *be idempotent matrices of the same order. Then* $\mathbf{E} <^- \mathbf{F}$ *if and only if* $\mathbf{F} - \mathbf{E}$ *is idempotent.*

Proof. 'If' part
Notice that $\rho(\mathbf{F}) = \mathbf{tr}(\mathbf{F}) = \mathbf{tr}(\mathbf{E}) + \mathbf{tr}(\mathbf{F} - \mathbf{E}) = \rho(\mathbf{E}) + \rho(\mathbf{F} - \mathbf{E})$. Therefore, by Remark 3.3.9, $\mathbf{E} <^- \mathbf{F}$.
'Only if' part follows by Theorem 5.5.3. □

Corollary 5.5.5. *Let* $\mathbf{E}$ *and* $\mathbf{F}$ *be any idempotent matrices of the same order. Then* $\mathbf{E} <^- \mathbf{F}$ *if and only if* $\mathbf{I} - \mathbf{F} <^- \mathbf{I} - \mathbf{E}$.

Remark 5.5.6. Corollary 5.5.5 is not true for star order as shown by the following example.

Example 5.5.7. Let $\mathbf{E} = \mathbf{diag}(\mathbf{0},\ \mathbf{0},\ \mathbf{1})$ and $\mathbf{F} = \begin{pmatrix} 1 & 1 & 0 \\ 0 & 0 & 0 \\ 0 & 0 & 1 \end{pmatrix}$. Then both $\mathbf{E}$, $\mathbf{F}$ are idempotent matrices satisfying Corollary 5.5.5. Clearly, $\mathbf{E} <^\star \mathbf{F}$, as $\mathbf{EE}^\star = \mathbf{EF}^\star = \mathbf{E}^\star\mathbf{E} = \mathbf{E}^\star\mathbf{F} = \mathbf{diag}(\mathbf{0},\ \mathbf{0},\ \mathbf{1})$. However, $\mathbf{I}-\mathbf{F}\not<^\star\mathbf{I}-\mathbf{E}$, Since $(\mathbf{I}-\mathbf{F})^\star(\mathbf{I}-\mathbf{F}) = \mathbf{diag}\,(0, 2, 0)$ and $(\mathbf{I}-\mathbf{F})^\star(\mathbf{I}-\mathbf{E}) = \begin{pmatrix} 0 & 0 & 0 \\ 1 & 1 & 0 \\ 0 & 0 & 0 \end{pmatrix}$.

Let $\mathbf{E}$ and $\mathbf{F}$ be idempotent matrices of the same order. We now investigate when both $\mathbf{E} <^\star \mathbf{F}$ and $\mathbf{I}-\mathbf{F} <^\star \mathbf{I}-\mathbf{E}$ hold. For this we need the following powerful result.

Theorem 5.5.8. *Let* $\mathbf{E}$ *and* $\mathbf{F} \in \mathbb{C}^{\mathbf{n}\times\mathbf{n}}$ *be idempotent matrices such that* $\mathbf{F}-\mathbf{E}$ *is a matrix of index* ≤ 1 *with all eigen-values real and non-negative. Then* $\mathbf{F}-\mathbf{E}$ *is idempotent.*

Proof. Notice that $\mathbf{I} = \mathbf{E} + (\mathbf{F}-\mathbf{E}) + (\mathbf{I}-\mathbf{F})$. The matrices $\mathbf{E}$ and $\mathbf{I}-\mathbf{F}$ are idempotent. Consider $\mathbf{F}-\mathbf{E}$ and let $\mathbf{F}-\mathbf{E} = \mathbf{Pdiag}(\mathbf{J_1},\ldots,\mathbf{J_p},\mathbf{J_{p+1}},\ldots,\mathbf{J_q})\mathbf{P}^{-1}$ be the Jordan decomposition, where the Jordan blocks $\mathbf{J_1},\ldots,\mathbf{J_p}$ correspond to eigen-value 1 and the Jordan blocks $\mathbf{J_{p+1}},\ldots,\mathbf{J_q}$ correspond to the other eigen-values. Let $\mathbf{Q_0} = \mathbf{Pdiag}(\mathbf{J_1},\ldots,\mathbf{J_p},\ \mathbf{0},\ldots,\mathbf{0})\mathbf{P}^{-1}$ and $\mathbf{Q_1} = \mathbf{Pdiag}(\mathbf{0},\ldots,\mathbf{0},\ \mathbf{J_{p+1}},\ldots,\mathbf{J_q})\mathbf{P}^{-1}$. Then $\mathbf{F}-\mathbf{E} = \mathbf{Q_0} + \mathbf{Q_1}$. Since $\mathbf{Q_1}$ does not have any eigen-value equal to 1, $\mathbf{I}-\mathbf{Q_1}$ is non-singular. So, $n = \rho(\mathbf{I}-\mathbf{Q_1}) = \rho(\mathbf{Q_0}+\mathbf{E}+\mathbf{I}-\mathbf{F}) \leq \rho(\mathbf{Q_0})+\rho(\mathbf{E})+\rho(\mathbf{I}-\mathbf{F}) = \rho(\mathbf{Q_0})+\mathbf{tr}(\mathbf{E})+\mathbf{tr}(\mathbf{I}-\mathbf{F}) \leq \mathbf{n}-\mathbf{tr}(\mathbf{Q_1})$. Thus, $\mathrm{tr}(\mathbf{Q_1}) = \mathbf{0}$. However, all eigen values of $\mathbf{Q_1}$ are non-negative, it follows that all eigen values of $\mathbf{Q_1}$ are null. As $\mathbf{F}-\mathbf{E}$ has index 1, it must have single Jordan block corresponding to null eigen-value, so, $\mathbf{Q_1} = \mathbf{0}$. Thus, $\rho(\mathbf{F}-\mathbf{E}) = \mathbf{tr}(\mathbf{F}-\mathbf{E}) = \mathbf{tr}(\mathbf{F}) - \mathbf{tr}(\mathbf{E}) = \mathbf{tr}(\rho\mathbf{F}) - \mathbf{tr}(\rho\mathbf{E})$. By Corollary 4.3.18, it follows that $\mathbf{F}-\mathbf{E}$ is idempotent. □

Theorem 5.5.9. *Let* $\mathbf{E}$ *and* $\mathbf{F}$ *be any idempotent matrices of the same order. Then the following are equivalent:*

(i) $\mathbf{E}-\mathbf{F}$ *is nnd*
(ii) $\mathbf{E}-\mathbf{F}$ *is hermitian and idempotent*
(iii) $\mathbf{E}-\mathbf{F} <^\star \mathbf{I}$
(iv) $\mathbf{E} <^- \mathbf{F}$ *and* $\mathbf{E}-\mathbf{F}$ *is hermitian and*
(v) $\mathbf{E} <^\star \mathbf{F}$ *and* $\mathbf{I}-\mathbf{F} <^\star \mathbf{I}-\mathbf{E}$.

Proof. (i) ⇔ (ii) ⇔ (iii) is easy.
(ii) ⇔ (iv) follows by Theorem 5.5.3.
(ii) ⇒ (v)
Since (ii) holds, $\mathbf{E} <^- \mathbf{F}$. Therefore, $\mathbf{E} = \mathbf{EF} = \mathbf{FE}$. Now, $\mathbf{0} =$

$\mathbf{FE} - \mathbf{E} = \mathbf{FE} - \mathbf{E^2} = (\mathbf{F} - \mathbf{E})\mathbf{E} = (\mathbf{F} - \mathbf{E})^\star\mathbf{E}$. Therefore, $\mathbf{F}^\star\mathbf{E} = \mathbf{E}^\star\mathbf{F}$ or $\mathbf{E}^\star\mathbf{F} = \mathbf{E}^\star\mathbf{F}$. Similarly, by using $\mathbf{E} = \mathbf{EF}$ we get $\mathbf{EE}^\star = \mathbf{FE}^\star$. Thus, $\mathbf{E} <^\star \mathbf{F}$. Since $(\mathbf{I} - \mathbf{E}) - (\mathbf{I} - \mathbf{F}) = \mathbf{F} - \mathbf{E}$ is hermitian idempotent, we have $\mathbf{I} - \mathbf{F} <^\star \mathbf{I} - \mathbf{E}$.
(v) $\Rightarrow$ (ii)
Since $\mathbf{E} <^\star \mathbf{F}$, we have $\mathbf{E} <^- \mathbf{F}$, so, $\mathbf{F} - \mathbf{E}$ is an idempotent. Consider $(\mathbf{F} - \mathbf{E})^\star(\mathbf{E} - \mathbf{F}) = \mathbf{F}^\star\mathbf{F} - \mathbf{E}^\star\mathbf{E}$, since $\mathbf{E} <^\star \mathbf{F}$. Also, since $\mathbf{I} - \mathbf{F} <^\star \mathbf{I} - \mathbf{E}$, $(\mathbf{I} - \mathbf{F})^\star\mathbf{I} - \mathbf{F} = \mathbf{I} - \mathbf{F}^\star\mathbf{I} - \mathbf{E}$ and therefore, $\mathbf{F} - \mathbf{E} = \mathbf{F}^\star\mathbf{F} - \mathbf{F}^\star\mathbf{E}$. Thus, $(\mathbf{F} - \mathbf{E})^\star = \mathbf{F}^\star\mathbf{F} - \mathbf{E}^\star\mathbf{F} = \mathbf{F}^\star\mathbf{F} - \mathbf{F}^\star\mathbf{E}$, as $\mathbf{E}^\star\mathbf{E} = \mathbf{E}^\star\mathbf{F} = \mathbf{F}^\star\mathbf{E}$. So, $(\mathbf{F} - \mathbf{E})^\star = (\mathbf{F} - \mathbf{E})$, showing $\mathbf{E} - \mathbf{F}$ is hermitian. □

Till now, we have been considering idempotent matrices $\mathbf{E}$ and $\mathbf{F}$ neither of which need be hermitian. We now take one of these to be hermitian and obtain some interesting results.

Theorem 5.5.10. *Let* $\mathbf{E}$ *and* $\mathbf{F}$ *be idempotent matrices of the same order and let* $\mathbf{E}$ *be hermitian. Then the following are equivalent:*

(i) $\mathbf{FF}^\star - \mathbf{E}$ *is nnd*
(ii) $\mathbf{E} = \mathbf{FE}$
(iii) $\mathbf{FF}^\star = \mathbf{E} + \mathbf{F}(\mathbf{I} - \mathbf{E})\mathbf{F}^\star$ *and*
(iv) $\mathbf{FF}^\star = \mathbf{E} + (\mathbf{F} - \mathbf{E})(\mathbf{E} - \mathbf{F})^\star$.

Proof. (i) $\Rightarrow$ (ii)
Since $\mathbf{FF}^\star - \mathbf{E} = \mathbf{FF}^\star - \mathbf{EE}^\star$ is *nnd*, we have $\mathcal{C}(\mathbf{E}) \subseteq \mathcal{C}(\mathbf{F})$. Hence, $\mathbf{FE} = \mathbf{FFE} = \mathbf{E}$.
(ii) $\Rightarrow$ (iii)
$\mathbf{E}+\mathbf{F}(\mathbf{I} - \mathbf{E})\mathbf{F}^\star = \mathbf{E}+\mathbf{FF}^\star-\mathbf{FEF}^\star = \mathbf{E}+\mathbf{FF}^\star-\mathbf{EF}^\star = \mathbf{E}+\mathbf{FF}^\star-\mathbf{E} = \mathbf{FF}^\star$.
(iii) $\Rightarrow$ (iv)
$(\mathbf{F} - \mathbf{E})(\mathbf{E} - \mathbf{F})^\star = \mathbf{F}(\mathbf{I} - \mathbf{E})(\mathbf{I} - \mathbf{E})^\star\mathbf{F}^\star = \mathbf{F}(\mathbf{I} - \mathbf{E})\mathbf{F}^\star = \mathbf{FF}^\star - \mathbf{E}$.
(iv) $\Rightarrow$ (i) is trivial. □

Theorem 5.5.11. *Let* $\mathbf{E}$ *and* $\mathbf{F}$ *be idempotent matrices of the same order and let* $\mathbf{E}$ *be hermitian. Then the following are equivalent:*

(i) $\mathbf{E} <^\star \mathbf{F}$
(ii) $\mathbf{E} <^- \mathbf{F}$
(iii) $\mathbf{E} <^\star \mathbf{FF}^\star$ *and*
(iv) $\mathbf{E} <^- \mathbf{FF}^\star$.

Proof. (i) $\Rightarrow$ (ii) is trivial.
(ii) $\Rightarrow$ (i)

Since $\mathbf{E} <^- \mathbf{F}$ we have $\mathbf{E} = \mathbf{EF} = \mathbf{FE}$. Therefore, $\mathbf{E} = \mathbf{E}^\star = \mathbf{EF}^\star = \mathbf{FE}^\star$. This further gives $\mathbf{E} = \mathbf{F}^\star\mathbf{E} = \mathbf{EF}^\star$. Thus, $\mathbf{E} = \mathbf{EE}^\star = \mathbf{F}^\star\mathbf{E}$ and $\mathbf{E} = \mathbf{E}^\star\mathbf{E} = \mathbf{EF}^\star$ and so, $\mathbf{E} <^\star \mathbf{F}$.

(i) $\Rightarrow$ (iii) and (i) $\Rightarrow$ (iv) follows by Theorem 5.5.2.

(iii) $\Rightarrow$ (iv) is trivial.

(iv) $\Rightarrow$ (ii)

Now, $\mathbf{E} <^- \mathbf{FF}^\star \Rightarrow \mathbf{FF}^\star - \mathbf{E} <^- \mathbf{FF}^\star$. Clearly, $\mathbf{FF}^\star - \mathbf{E}$ is *nnd*. Hence, by Theorem 5.5.10, we have $\mathbf{FF}^\star = \mathbf{E} + (\mathbf{F}-\mathbf{E})(\mathbf{F}-\mathbf{E})^\star$. This further gives $\mathbf{FF}^\star - \mathbf{EE}^\star = (\mathbf{F}-\mathbf{E})(\mathbf{F}-\mathbf{E})^\star$ and so, $\rho(\mathbf{FF}^\star - \mathbf{EE}^\star) = \rho((\mathbf{F}-\mathbf{E})(\mathbf{F}-\mathbf{E})^\star) = \rho(\mathbf{F}-\mathbf{E})$. Hence, $\rho(\mathbf{F}-\mathbf{E}) = \rho(\mathbf{FF}^\star - \mathbf{EE}^\star) = \rho(\mathbf{FF}^\star - \mathbf{E}) = \rho(\mathbf{FF}^\star) - \rho(\mathbf{E})$, as $\mathbf{E} <^- \mathbf{FF}^\star$. Finally, $\rho(\mathbf{F}-\mathbf{E}) = \rho(\mathbf{F}) - \rho(\mathbf{E})$ and so, $\mathbf{E} <^- \mathbf{F}$. □

Let $\mathbf{E}$ and $\mathbf{F}$ be idempotent matrices of the same order. Then we know that $\mathbf{E} <^- \mathbf{F}$ if and only if $\mathbf{E} = \mathbf{EF} = \mathbf{FE}$. Suppose one of the conditions, say, $\mathbf{E} = \mathbf{EF}$ holds. Further, let $\mathbf{E}$ be hermitian. Does that ensure $\mathbf{E} <^- \mathbf{F}$? The answer is in negative as the following example shows:

Example 5.5.12. Let $\mathbf{E} = \begin{pmatrix} 1 & 0 \\ 0 & 0 \end{pmatrix}$ and $\mathbf{F} = \begin{pmatrix} 1 & 1 \\ 0 & 0 \end{pmatrix}$. Then $\mathbf{E}$ and $\mathbf{F}$ are idempotents and $\mathbf{E}$ is hermitian. Also, $\mathbf{FE} = \mathbf{E}$. However, $\mathbf{EF} \neq \mathbf{E}$ and so, $\mathbf{E} \not<^- \mathbf{F}$.

Let $\mathbf{E}$ be a hermitian idempotent matrix. We shall now explore the additional conditions required so that the idempotent matrices $\mathbf{E}$ and $\mathbf{F}$ satisfy $\mathbf{E} <^\star \mathbf{F}$.

Theorem 5.5.13. *Let $\mathbf{E}$ and $\mathbf{F}$ be idempotent matrices of the same order and let $\mathbf{E}$ be hermitian. Then the following are equivalent:*

(i) $\mathbf{E} <^\star \mathbf{F}$
(ii) $\mathbf{E} = \mathbf{FE}$ *and* $\mathbf{F}^\dagger - \mathbf{E}$ *is a g-inverse of* $\mathbf{F} - \mathbf{E}$
(iii) $\mathbf{EF}^\dagger\mathbf{F} = \mathbf{FF}^\dagger\mathbf{E} = \mathbf{E}$ *or equivalently* $\mathbf{E} <^\star \mathbf{F}$ *and*
(iv) $\mathbf{EF} = (\mathbf{EF})^\star$.

Proof. (i)$\Rightarrow$ (ii)

As $\mathbf{E} <^\star \mathbf{F} \Rightarrow \mathbf{E} <^- \mathbf{F}$, we have $\mathbf{E} = \mathbf{FE}$. Now, $\mathbf{E}$ is hermitian idempotent, so, $\mathbf{E}^\dagger = \mathbf{E}$ and by Theorem 5.2.11, $\mathbf{F}^\dagger - \mathbf{E} = \mathbf{F}^\dagger - \mathbf{E}^\dagger = (\mathbf{E} - \mathbf{F})^\dagger$.

(ii) $\Rightarrow$ (iii) Since $\mathbf{E} = \mathbf{FE}$, we have $(\mathbf{F}-\mathbf{E}) = (\mathbf{F}-\mathbf{E})(\mathbf{F}^\dagger - \mathbf{E})(\mathbf{F}-\mathbf{E}) = (\mathbf{F}-\mathbf{E})\mathbf{F}^\dagger(\mathbf{F}-\mathbf{E})$. Hence, $\mathbf{EF}^\dagger\mathbf{F} + \mathbf{FF}^\dagger\mathbf{E} = \mathbf{EF}^\dagger\mathbf{E} + \mathbf{E}$. However, $\mathbf{FF}^\dagger\mathbf{E} = \mathbf{E}$, so $\mathbf{EF}^\dagger\mathbf{F} = \mathbf{FF}^\dagger\mathbf{E}$.

(iii)$\Rightarrow$ (iv)

Notice that $\mathbf{E} = \mathbf{FE} = \mathbf{FE}^\star = \mathbf{EF}^\star$. Now, $\mathbf{F}^\dagger\mathbf{E} = \mathbf{F}^\dagger\mathbf{FE} = \mathbf{EF}^\dagger\mathbf{F}$ is hermitian. Also, $\mathbf{EF} = (\mathbf{EF}^\star)\mathbf{F} = \mathbf{E}(\mathbf{F}\dagger\mathbf{FF})\star\mathbf{F} = \mathbf{F}\dagger\mathbf{FEF}\star\mathbf{F} = \mathbf{F}\dagger\mathbf{E}$. Therefore, $\mathbf{EF}$ is hermitian.

(iv) $\Rightarrow$ (i)

Since $\mathbf{EF}$ and $\mathbf{E}$ are hermitian, $\mathbf{EF} - \mathbf{E}$ is hermitian. So, $(\mathbf{EF} - \mathbf{E})^\star(\mathbf{EF} - \mathbf{E}) = (\mathbf{EF} - \mathbf{E})(\mathbf{EF} - \mathbf{E}) = \mathbf{EF} - \mathbf{E} - \mathbf{EF} + \mathbf{E} = \mathbf{0}$. Hence, $\mathbf{EF} = \mathbf{E} = \mathbf{FE}$. Thus, $\mathbf{E} <^- \mathbf{F}$ and by Theorem 5.5.11, $\mathbf{E} <^\star \mathbf{F}$. □

We can obtain similar theorems if instead of $\mathbf{E}$ we take $\mathbf{F}$ as hermitian. We conclude our discussion on partial orders on idempotent matrices with the following statement when both $\mathbf{E}$ and $\mathbf{F}$ are hermitian.

Theorem 5.5.14. *Let* $\mathbf{E}$ *and* $\mathbf{F}$ *be hermitian idempotent matrices of the same order. Then the following are equivalent:*

(i) $\mathbf{E} <^\star \mathbf{F}$

(ii) $\mathbf{E} <^- \mathbf{F}$

(iii) $\mathbf{E} = \mathbf{EFE}$

(iv) $(\mathbf{F} - \mathbf{E}) <^\star \mathbf{I}$

(v) $\mathbf{E} <^- \mathbf{EF}$

(vi) $\mathbf{I} - \mathbf{E} <^\star \mathbf{I} - \mathbf{F}$ *and*

(vii) $\mathbf{I} - \mathbf{E} <^- \mathbf{I} - \mathbf{F}$.

Proof is straightforward.

5.6 Fisher-Cochran type theorems

Recall that in Theorem 4.3.17, we had seen that if $\mathbf{A_1}, \ldots, \mathbf{A_k}$ are matrices each of index 1 satisfying $\mathbf{A_i}\mathbf{A_j} = \mathbf{0}$ whenever $i \neq j$ and $\mathbf{A} = \sum_{\mathbf{i}=1}^{\mathbf{k}} \mathbf{A_i}$, then each $\mathbf{A_i}$ lies below $\mathbf{A}$ under the sharp order. As a consequence, we had also deduced a non-hermitian version of Fisher-Cochran theorem for the idempotent matrices. In this section we study similar relationships for the range-hermitian matrices, normal matrices and hermitian matrices. We also remark that the last one is essentially an algebraic version of the Fisher Cochran theorem on the distribution of quadratic forms in independent standard normal variables.

Theorem 5.6.1. *Let* $\mathbf{A_1}, \ldots, \mathbf{A_k}$ *be range-hermitian matrices of the same order and let* $\mathbf{A} = \sum_{\mathbf{i}=1}^{\mathbf{k}} \mathbf{A_i}$. *Consider the following statements:*

(i) $\mathbf{A_i}\mathbf{A_j} = \mathbf{0}$ *for whenever* $i \neq j$

(ii) *there exist non-singular matrices* $\mathbf{D_1}, \dots, \mathbf{D_k}$, *and a unitary matrix* $\mathbf{U}$ *such that* $\mathbf{A_i} = \mathbf{U}\mathbf{diag}(\mathbf{0}, \dots, \mathbf{0}, \ \mathbf{D_i}, \ \mathbf{0}, \dots, \mathbf{0})\mathbf{U}^\star$, *for each* $i = 1, \dots, k$

(iii) $\mathbf{A_i} <^\star \mathbf{A}$

(iv) $\mathbf{A}$ *is range-hermitian* *and*

(v) $\rho(\mathbf{A}) = \sum_{\mathbf{i=1}}^{\mathbf{k}} \rho(\mathbf{A_i})$.

Then (i) *is equivalent to* (ii). *Any one of* (i) *and* (ii) *implies all others from* (i) *to* (v). (iii) *and* (iv) *together imply all others from* (i) *to* (v).

The proof follows along the lines of Theorem 4.3.17 using Theorem 2.2.9 and Theorem 5.4.1

The following theorems are now easy to prove.

Theorem 5.6.2. *Let* $\mathbf{A_1}, \dots, \mathbf{A_k}$ *be normal matrices of the same order and let* $\mathbf{A} = \sum_{\mathbf{i=1}}^{\mathbf{k}} \mathbf{A_i}$. *Consider the following statements:*

(i) $\mathbf{A_i}\mathbf{A_j} = \mathbf{0}$ *for whenever* $i \neq j$

(ii) *there exists non-singular diagonal matrices* $\mathbf{D_1}, \dots, \mathbf{D_k}$, *and a unitary matrix* $\mathbf{U}$ *such that* $\mathbf{A_i} = \mathbf{U}\mathbf{diag}(\mathbf{0}, \dots, \mathbf{0}, \ \mathbf{D_i}, \ \mathbf{0}, \dots, \mathbf{0})\mathbf{U}^\star$, *for each* $i = 1, \dots, k$

(iii) $\mathbf{A_i} <^\star \mathbf{A}$

(iv) $\mathbf{A}$ *is normal* *and*

(v) $\rho(\mathbf{A}) = \sum_{\mathbf{i=1}}^{\mathbf{k}} \rho(\mathbf{A_i})$.

Then (i) *is equivalent to* (ii). *Any one of* (i) *and* (ii) *implies all others from* (i) *to* (v). (iii) *and* (iv) *together imply all others from* (i) *to* (v).

Theorem 5.6.3. *Let* $\mathbf{A_1}, \dots, \mathbf{A_k}$ *be hermitian matrices of the same order and let* $\mathbf{A} = \sum_{\mathbf{i=1}}^{\mathbf{k}} \mathbf{A_i}$. *Consider the following statements:*

(i) $\mathbf{A^2} = \mathbf{A}$

(ii) $\mathbf{A_i}^2 = \mathbf{A_i}$ *for each* i

(iii) $\mathbf{A_i}\mathbf{A_j} = \mathbf{0}$ *whenever* $i \neq j$

(iv) $\rho(\mathbf{A}) = \sum_{\mathbf{i=1}}^{\mathbf{k}} \rho(\mathbf{A_i})$ *and*

(v) *there exists a unitary matrix* $\mathbf{U}$ *such that* $\mathbf{A_i} = \mathbf{U}\mathbf{diag}(\mathbf{0}, \dots, \mathbf{0}, \ \mathbf{I_i}, \ \mathbf{0}, \dots, \mathbf{0})\mathbf{U}^\star$, *for each* $i = 1, \dots, k$.

Then any two of (i), (ii) *and* (iv) *imply all others from* (i) *to* (v). *Any two of* (i), (ii) *and* (iii) *imply all others from* (i) *to* (v). (v) *implies all others from* (i) *to* (iv). *Further,* (i) *and* (iv) *imply all the others.*

5.7 Exercises

(1) Let $\mathbf{A}, \mathbf{B}$ be matrices of order $m \times n$. Let the symbol $<^?$ stand for $<^\star$ or $<^-$ or $<^s$. Then show that

(i) $\mathbf{A} <^? \mathbf{B} \Leftrightarrow \mathbf{B}^\star\mathbf{A} <^? \mathbf{B}^\star\mathbf{B}$ and $\mathbf{A}\mathbf{B}^\star <^? \mathbf{B}\mathbf{B}^\star$.

(ii) $\mathbf{A} <^? \mathbf{B} \Leftrightarrow \mathbf{B}^\dagger\mathbf{B}$ and $\mathbf{A}\mathbf{B}^\dagger <^? \mathbf{B}\mathbf{B}^\dagger$.

(2) Let $\mathbf{A}, \mathbf{B}$ be matrices of order $m \times n$. Show that $\mathbf{A}\mathbf{A}^\star <^\star \mathbf{B}\mathbf{B}^\star \Rightarrow \mathbf{A}\mathbf{A}^\dagger <^\star \mathbf{B}\mathbf{B}^\dagger$ and $\mathbf{A}^\star\mathbf{A} <^\star \mathbf{B}^\star\mathbf{B} \Rightarrow \mathbf{A}^\dagger\mathbf{A} <^\star \mathbf{B}^\dagger\mathbf{B}$.

(3) Let $\mathbf{A}, \mathbf{B}$ be matrices of order $m \times n$ and let scalars $a, b \in \mathbb{C}$. Suppose $\mathbf{A} \neq \mathbf{0}$ and $\mathbf{A} <^\star \mathbf{B}$. When does one have $a\mathbf{A} <^\star b\mathbf{B}$?

(4) For matrices $\mathbf{A}, \mathbf{B}$ of order $m \times n$, show that

(i) $\mathbf{A} <_\ell^- \mathbf{B} \Leftrightarrow \mathbf{A} = \mathbf{Q}\mathbf{B}$ where $\mathbf{Q}$ is projector onto a subspace of $\mathcal{C}(\mathbf{B})$ and

(ii) $\mathbf{A} <_m^- \mathbf{B} \Leftrightarrow \mathbf{A} = \mathbf{B}\mathbf{P}$ where $\mathbf{P}$ is projector onto a subspace of $\mathcal{C}(\mathbf{B})$.

(5) For matrices $\mathbf{A}, \mathbf{B}$ of order $m \times n$, show that the following are equivalent:

(i) $\mathbf{A} <^\star \mathbf{B}$;

(ii) $(\mathbf{B} - \mathbf{A}) <^\star \mathbf{B}$;

(iii) $(\mathbf{B} - \mathbf{A})^\dagger <^\star \mathbf{B}^\dagger$; and

(iv) $\mathbf{B}^\dagger - \mathbf{A}^\dagger <^\star \mathbf{B}^\dagger$.

(6) For matrices $\mathbf{A}, \mathbf{B}$ of order $m \times n$, show the following:

(i) $\mathbf{A} <^\star \mathbf{B} \Rightarrow \mathbf{B}^\star\mathbf{A} <^\star \mathbf{B}^\star\mathbf{B}$ and $\mathbf{A}\mathbf{B}^\star <^\star \mathbf{B}\mathbf{B}^\star$;

(ii) $\mathbf{A} <^\star \mathbf{B} \Rightarrow \mathbf{A}^\dagger\mathbf{A} <^\star \mathbf{B}^\dagger\mathbf{B}$ and $\mathbf{A}\mathbf{A}^\dagger <^\star \mathbf{B}\mathbf{B}^\dagger$;

(iii) If $m = n$, then $\mathbf{B}^\star\mathbf{B}^\dagger = \mathbf{B}^\dagger\mathbf{B}^\star$ and $\mathbf{A} <^\star \mathbf{B} \Rightarrow \mathbf{A}^\star\mathbf{A}^\dagger = \mathbf{A}^\dagger\mathbf{A}^\star$.

(7) Let $\mathbf{A}, \mathbf{B}$ be matrices of order $m \times n$ such that $\mathbf{A} <^\star \mathbf{B}$.

(i) Let $\mathbf{X} \in \mathbb{C}^{n \times m}$ be a matrix such that $\mathbf{B}\mathbf{A}^\dagger\mathbf{B} = \mathbf{A}\mathbf{X}\mathbf{A}$. Then prove that $\mathbf{A}\mathbf{X}\mathbf{A} = \mathbf{A}$.
If this is so, then show that $\mathbf{A} <^\star \mathbf{B} \Leftrightarrow \mathbf{A} <^- \mathbf{B}$, $\mathbf{B}\mathbf{A}^\dagger\mathbf{B} = \mathbf{A}$.

(ii) Let $\mathbf{X} \in \mathbb{C}^{n \times m}$ be a matrix such that $\mathbf{B}^\dagger\mathbf{A}\mathbf{X}\mathbf{A}\mathbf{B}^\dagger = \mathbf{A}^\dagger$, then show that $\mathbf{A}\mathbf{X}\mathbf{A} = \mathbf{A}$.
In such a case show that $\mathbf{A} <^\star \mathbf{B} \Leftrightarrow \mathbf{A} <^- \mathbf{B}$, $\mathbf{B}^\dagger\mathbf{A}\mathbf{X}\mathbf{A}\mathbf{B}^\dagger = \mathbf{A}^\dagger$.

(8) Let $\mathbf{A}, \mathbf{B}$ be matrices of order $n \times n$ such that $\mathbf{A} <^\star \mathbf{B}$. Then prove that $\mathbf{A}$ is Hermitian if and only if $\mathbf{A}, \mathbf{B}$ commute.

(9) Let $\mathbf{A} <^\star \mathbf{B}$. show that the following statements are equivalent:

(i) $\mathbf{A}$ is normal

(ii) $\mathbf{A}^\star, \mathbf{B}$ commute
(iii) $\mathbf{A}, \mathbf{B}^\star$ commute
(iv) $\mathbf{A}^\dagger, \mathbf{B}$ commute and
(v) $\mathbf{A}, \mathbf{B}^\dagger$ commute.

(10) Let $\mathbf{A}$ and $\mathbf{B}$ be square matrices of the same order such that $\mathbf{A}$ is range hermitian. Prove that (i) $\mathbf{A} <^\star \mathbf{B}$ implies $\mathbf{A}^2 <^\star \mathbf{B}^2$ and $\mathbf{AB} = \mathbf{BA}$ and (ii) $\mathbf{A} <^\star \mathbf{B}$ implies $\mathbf{A} <^\# \mathbf{B}$. Also show that the implication in (i) is not reversible.

(11) Let $\mathbf{A}$ and $\mathbf{B}$ be square matrices of the same order such that $\mathbf{A}$ is range hermitian and $\mathbf{B}$ is idempotent. Prove that $\mathbf{A} <^- \mathbf{B}$ if and only if $\mathbf{A} <^\star \mathbf{B}$.

(12) Let $\mathbf{A}$ and $\mathbf{B}$ be *nnd* matrices of the same order. Show that $\mathbf{A}^2 <^\star \mathbf{B}^2$ if and only if there exists an orthogonal projector $\mathbf{K}$ such that $\mathbf{A} = \mathbf{BK}$.

(13) Let $\mathbf{A}$ and $\mathbf{B}$ be hermitian matrices of the same order. Show that $\mathbf{A}^2 <^\star \mathbf{B}^2$ if and only if there exists a matrix $\mathbf{K}$ such that $\mathbf{A} = \mathbf{BK}$, $\mathcal{C}(\mathbf{K}) \subseteq \mathcal{C}(\mathbf{B})$ and $\mathbf{KK}^\star\mathbf{K} = \mathbf{K}$.

(14) Let $\mathbf{A}$ and $\mathbf{B}$ be partial isometries of the same order. Show that $\mathbf{A} <^\star \mathbf{B}$ if and only if $\mathbf{A} <^s \mathbf{B}$ and $\mathbf{AB}^\star = (\mathbf{AA}^\star)^{1/2}(\mathbf{BB}^\star)^{1/2}$.

(15) Let $\mathbf{A}$ and $\mathbf{B}$ be $m \times n$ matrices. Define $\mathbf{A} \ll \mathbf{B}$ if $\mathbf{AB}^\star\mathbf{A} = \mathbf{AA}^\star\mathbf{A}$. Show that '$\ll$' is a partial order and $\mathbf{A} \ll \mathbf{B}$ if and only if $\mathbf{A}^\dagger <^- \mathbf{B}^\dagger$. Show also that $\mathbf{A} \ll \mathbf{B}$ does not imply $\mathbf{A} <^- \mathbf{B}$.

(16) With notations as in Ex. 15, show that $\mathbf{A} <^\star \mathbf{B}$ if and only if $\mathbf{A} \ll \mathbf{B}$ $\mathbf{BA}^\dagger\mathbf{B} = \mathbf{B}$.

(17) Let $\mathbf{A}$ and $\mathbf{B}$ be orthogonal projectors. Should '$\mathbf{A} <^- \mathbf{B}$ if and only if $\mathbf{A} <^s \mathbf{B}$' hold? Justify.

(18) Let $\mathbf{A}, \mathbf{B}$ be complex matrices of index not greater than 1. Consider the following statements

(i) $\mathbf{A} <^\star \mathbf{B}$
(ii) $\mathbf{A}^2 <^\star \mathbf{B}^2$
(iii) $\mathbf{A} <^\# \mathbf{B}$.

Show that (i) and (ii) $\Rightarrow$ (iii) and (i) and (iii) $\Rightarrow$ (ii), but (ii) and (iii) $\not\Rightarrow$ (i).

(19) Let $\mathbf{A}$, $\mathbf{B}$ and $\mathbf{C}$ be matrices of the same order such that $\mathbf{A} <^\star \mathbf{B}$ and $\mathbf{B} <^\star \mathbf{C}$. Prove that $\mathbf{A}$, $\mathbf{B}$ and $\mathbf{C}$ have a simultaneous singular value decomposition.

Chapter 6

One-Sided Orders

6.1 Introduction

In the previous three chapters we studied the space pre-order and the minus, the sharp and the star partial orders. Recall that for matrices $\mathbf{A}$ and $\mathbf{B}$ of the same order

$\mathbf{A} <^- \mathbf{B} \Leftrightarrow \mathbf{A}\mathbf{A}^- = \mathbf{B}\mathbf{A}^-$ and $\mathbf{A}^-\mathbf{A} = \mathbf{A}^-\mathbf{B}$ for some $\mathbf{A}^- \in \{\mathbf{A}^-\}$,

$\mathbf{A} <^{\#} \mathbf{B} \Leftrightarrow \mathbf{A}\mathbf{B} = \mathbf{A}^2 = \mathbf{B}\mathbf{A}$, and

$\mathbf{A} <^{\star} \mathbf{B} \Leftrightarrow \mathbf{A}^{\star}\mathbf{A} = \mathbf{A}^{\star}\mathbf{B}$ and $\mathbf{A}\mathbf{A}^{\star} = \mathbf{B}\mathbf{A}^{\star}$.

Notice that there are two defining conditions for each of these partial orders. In this chapter, we shall examine the situations when only one of the two defining conditions is considered. Does one still get a partial order or at least a pre-order? If not, can one hope to get a partial order by addition of some milder condition(/s)? In case a partial order is obtained, should this new partial order be the same as the one whose defining conditions have been relaxed or does it result in a different partial order? In Chapter 3, we saw that if $\mathbf{A} <^{s} \mathbf{B}$ and $\mathbf{A}\mathbf{A}^- = \mathbf{B}\mathbf{A}^-$ (or $\mathbf{A}^-\mathbf{A} = \mathbf{A}^-\mathbf{B}$), then $\mathbf{A} <^- \mathbf{B}$. Suppose, we take the condition $\mathbf{A}\mathbf{A}^- = \mathbf{B}\mathbf{A}^-$ and add a condition as mild as $\mathcal{C}(\mathbf{A}^t) \subseteq \mathcal{C}(\mathbf{B}^t)$ to it, it is clear that we once again get $\mathbf{A} <^- \mathbf{B}$.

In Section 6.2, we study the condition $\mathbf{A}\mathbf{A}^- = \mathbf{B}\mathbf{A}^-$ (or $\mathbf{A}^-\mathbf{A} = \mathbf{A}^-\mathbf{B}$). We show that while this condition has some interesting characterizations, it does not even yield a pre-order. In Section 6.3, we relax one of the defining conditions of the sharp order and notice that in contrast to the minus order, addition of the condition $\mathcal{C}(\mathbf{A}) \subseteq \mathcal{C}(\mathbf{B})$ or $\mathcal{C}(\mathbf{A}^t) \subseteq \mathcal{C}(\mathbf{B}^t)$ results in a partial order distinct from the sharp order. In fact, for square matrices $\mathbf{A}$, $\mathbf{B}$ of the same order and of index ≤ 1, if we consider any one of the two equalities say, $\mathbf{A}\mathbf{B} = \mathbf{A}^2$ together with the condition $\mathcal{C}(\mathbf{A}) \subseteq \mathcal{C}(\mathbf{B})$

(or the condition $\mathbf{BA} = \mathbf{A^2}$ together with the condition $\mathcal{C}(\mathbf{A^t}) \subseteq \mathcal{C}(\mathbf{B^t})$), we obtain a partial order which we call the right sharp order (or the left sharp order respectively). We also obtain some conditions under which the one-sided sharp order and the sharp order coincide. In Section 6.4, we introduce two new classes of g-inverses of index 1 matrices. Their important properties are developed in the exercises at the end of this chapter. Our focus in this section is in exploring the role they play in the study of one-sided sharp orders. In fact, it turns out that the order relations defined through them coincide with the one-sided sharp orders. In Section 6.5, we study the order relations by taking one of the defining conditions of the star order say, $\mathbf{AA^\star} = \mathbf{BA^\star}$ along with $\mathcal{C}(\mathbf{A^\star}) \subseteq \mathcal{C}(\mathbf{B^\star})$(or the condition $\mathbf{A^\star A} = \mathbf{A^\star B}$ along with $\mathcal{C}(\mathbf{A}) \subseteq \mathcal{C}(\mathbf{B})$). We show that this leads us to a partial order that has properties very similar to the star order. The order relations thus obtained are known respectively as the right star and the left star orders. As seen in Chapter 5, for matrices $\mathbf{A}$ and $\mathbf{B}$ satisfying $\mathbf{A} <^\star \mathbf{B}$, the canonical form for $\mathbf{A}$ and $\mathbf{B}$ is the simultaneous singular value decomposition. We show that in the case of the right star (or the left star) order, the apt canonical form for the matrices $\mathbf{A}$ and $\mathbf{B}$ is a generalized singular value decomposition. We reiterate that all matrices in Sections 2, 3, and 4 are over a general field unless specified otherwise, but in Section 5, we deal with only complex matrices.

6.2 The condition $\mathbf{AA^-} = \mathbf{BA^-}$

We begin this section by studying the condition $\mathbf{AA^-} = \mathbf{BA^-}$ or ($\mathbf{A^-A} = \mathbf{A^-B}$). As we shall see, this condition by itself does not even define a pre-order, yet has some interesting properties. We have the following:

Theorem 6.2.1. *Let $\mathbf{A}$ and $\mathbf{B}$ be matrices of the same order. Then the following are equivalent:*

(i) $\mathbf{AA^-} = \mathbf{BA^-}$ *for some g-inverse $\mathbf{A^-}$ of $\mathbf{A}$*
(ii) $\mathbf{A} = \mathbf{BQ}$ *for some idempotent matrix $\mathbf{Q}$ and*
(iii) $\mathcal{C}(\mathbf{A})^{\mathbf{t}} \cap \mathcal{C}(\mathbf{B} - \mathbf{A})^{\mathbf{t}} = \{0\}$.

Proof. (i) $\Rightarrow$ (ii)
Since $\mathbf{AA^-} = \mathbf{BA^-}$, we have $\mathbf{A} = \mathbf{BA^-A} = \mathbf{BQ}$, where $\mathbf{Q} = \mathbf{A^-A}$ is idempotent.

(ii) $\Rightarrow$ (iii)
Let $\mathbf{A} = \mathbf{BQ}$. Then $\mathbf{AQ} = \mathbf{BQ}^2 = \mathbf{BQ} = \mathbf{A}$. Let $\mathbf{u} \in \mathcal{C}(\mathbf{A})^t \cap \mathcal{C}(\mathbf{B}-\mathbf{A})^t$. Then $\mathbf{u} = \mathbf{A}^t\mathbf{v} = (\mathbf{B}-\mathbf{A})^t\mathbf{w}$ for some column vectors $\mathbf{v}$ and $\mathbf{w}$. So, $\mathbf{u}^t = \mathbf{v}^t\mathbf{A} = \mathbf{w}^t(\mathbf{B}-\mathbf{A})$. This implies $\mathbf{u}^t\mathbf{Q} = \mathbf{v}^t\mathbf{AQ} = \mathbf{w}^t(\mathbf{BQ}-\mathbf{AQ})$. As $\mathbf{BQ} = \mathbf{A}$ and $\mathbf{AQ} = \mathbf{A}$, we have $\mathbf{u}^t\mathbf{Q} = \mathbf{v}^t\mathbf{AQ} = \mathbf{0}$ and so, $\mathbf{v}^t\mathbf{A} = \mathbf{0}$. It follows that $\mathbf{u} = \mathbf{0}$. Thus, $\mathcal{C}(\mathbf{A})^t \bigcap \mathcal{C}(\mathbf{B}-\mathbf{A})^t = \{0\}$.
(iii) $\rightarrow$ (i)
Let $(\mathbf{P}, \mathbf{T})$ and $(\mathbf{R}, \mathbf{S})$ be rank factorizations of $\mathbf{A}$ and $\mathbf{B}-\mathbf{A}$ respectively. So, $\mathbf{B} = \mathbf{A} + (\mathbf{B}-\mathbf{A}) = \mathbf{PT} + \mathbf{RS} = (\mathbf{P} : \mathbf{R})(\mathbf{T}^t : \mathbf{S}^t)^t$. Since $\mathcal{C}(\mathbf{A}^t) \cap \mathcal{C}((\mathbf{B}-\mathbf{A})^t) = \{\mathbf{0}\}$, the matrix $(\mathbf{T}^t : \mathbf{S}^t)^t$ has full row rank. Let $(\mathbf{L} : \mathbf{M})$ be a right inverse of it. Therefore, $\mathbf{TL} = \mathbf{I}$, $\mathbf{TM} = \mathbf{0}$; $\mathbf{SL} = \mathbf{0}$ and $\mathbf{SM} = \mathbf{I}$. As $\mathbf{P}$ has full column rank, $\mathbf{P}$ has a left inverse say $\mathbf{N}$. Then $\mathbf{G} = \mathbf{LN}$ is a g-inverse of $\mathbf{A}$ and $(\mathbf{B}-\mathbf{A})\mathbf{G} = \mathbf{RSLN} = \mathbf{0}$. Thus, $\mathbf{BG} = \mathbf{AG}$. □

Theorem 6.2.2. *Let* **A and B** *be matrices of the same order. Then the following are equivalent:*

(i) $\mathbf{A}^-\mathbf{A} = \mathbf{A}^-\mathbf{B}$ *for some g-inverse* $\mathbf{A}^-$ *of* $\mathbf{A}$
(ii) $\mathbf{A} = \mathbf{PB}$ *for some idempotent matrix* $\mathbf{P}$ *and*
(iii) $\mathcal{C}(\mathbf{A}) \cap \mathcal{C}(\mathbf{B}-\mathbf{A}) = 0$.

Proof is analogous to that of Theorem 6.2.1.

In the following theorem we obtain some equivalent conditions under which the condition $\mathbf{AA}^- = \mathbf{BA}^-$ yields the minus order.

Theorem 6.2.3. *Let* $\mathbf{A}$ *and* $\mathbf{B}$ *be matrices of the same order such that* $\mathbf{AA}^- = \mathbf{BA}^-$. *Then the following are equivalent:*

(i) $\mathbf{A}^-\mathbf{A} = \mathbf{A}^-\mathbf{B}$ *for some g-inverse* $\mathbf{A}^-$ *of* $\mathbf{A}$
(ii) $\mathbf{A} <^- \mathbf{B}$
(iii) $\mathcal{C}(\mathbf{A}^t) \subseteq \mathcal{C}(\mathbf{B}^t)$
(iv) $\mathcal{C}(\mathbf{A}) \cap \mathcal{C}(\mathbf{B}-\mathbf{A}) = 0$ *and*
(v) $\mathbf{A} = \mathbf{PB}$.

Proof. Use Theorems 6.2.1, 3.2.6 and 3.3.5. □

We now obtain a canonical form for matrices $\mathbf{A}$ and $\mathbf{B}$ when the condition $\mathbf{AA}^- = \mathbf{BA}^-$ holds.

Theorem 6.2.4. *Let* $\mathbf{A}$ *and* $\mathbf{B}$ *be matrices of order* $m \times n$. *Let* $\rho(\mathbf{B}) = s$, $\rho(\mathbf{A}) = r$ *and* $\rho(\mathbf{B}-\mathbf{A}) = t$. *Then* $\mathbf{AA}^- = \mathbf{BA}^-$ *if and only if there exist* *non-singular matrices* $\mathbf{P}$ *and* $\mathbf{Q}$ *such that*

$$\mathbf{A} = \mathbf{P}\begin{pmatrix} \mathbf{I_r} & \mathbf{0} & \mathbf{0} \\ \mathbf{0} & \mathbf{0} & \mathbf{0} \\ \mathbf{0} & \mathbf{0} & \mathbf{0} \end{pmatrix}\mathbf{Q} \text{ and } \mathbf{B} = \mathbf{P}\begin{pmatrix} \mathbf{I_r} & \mathbf{T_1} & \mathbf{0} \\ \mathbf{0} & \mathbf{T_2} & \mathbf{0} \\ \mathbf{0} & \mathbf{0} & \mathbf{0} \end{pmatrix}\mathbf{Q},$$

where $\begin{pmatrix} \mathbf{T_1} \\ \mathbf{T_2} \end{pmatrix}$ *an* $s \times t$ *matrix of rank* t.

Proof. 'If' part
Choose $\mathbf{A}^- = \mathbf{Q}^{-1}\mathbf{diag}\left(\mathbf{I_r}\ ,\mathbf{0}\ ,\mathbf{0}\right)\mathbf{P}^{-1}$. Then it is clear that

$$\mathbf{AA}^- = \mathbf{BA}^- = \mathbf{P}\left(\mathbf{I_r}\ ,\mathbf{0}\ ,\mathbf{0}\right)\mathbf{P}^{-1}.$$

'Only if' part
Let $(\mathbf{P_1}, \mathbf{Q_1})$ and $(\mathbf{P_2}, \mathbf{Q_2})$ be rank factorizations of $\mathbf{A}$ and $(\mathbf{B}-\mathbf{A})$ respectively. Since $\mathbf{AA}^- = \mathbf{BA}^-$, $\mathcal{C}(\mathbf{A^t}) \cap \mathcal{C}(\mathbf{B}-\mathbf{A})^\mathbf{t} = \mathbf{0}$. Hence, $\mathcal{C}(\mathbf{Q_1^t}) \cap \mathcal{C}(\mathbf{Q_2})^\mathbf{t} = 0$. By choosing a basis of $\mathcal{C}(\mathbf{P_1} : \mathbf{P_2})$ which includes the r columns of $\mathbf{P_1}$, we can find a matrix $\mathbf{R}$ of order $m \times (s-r)$ with $\rho(\mathbf{P_1} : \mathbf{R}) = s$ and matrices $\mathbf{T_1}$ and $\mathbf{T_2}$ such that the matrix $\begin{pmatrix} \mathbf{T_1} \\ \mathbf{T_2} \end{pmatrix}$ is of order $s \times t$ and rank t and $(\mathbf{P_1}, \mathbf{P_2}) = (\mathbf{P_1}, \mathbf{R})\begin{pmatrix} \mathbf{I} & \mathbf{T_1} \\ \mathbf{0} & \mathbf{T_2} \end{pmatrix}$. Since $(\mathbf{P_1}, \mathbf{R})$ is a matrix of order $m \times s$ with rank s and $\begin{pmatrix} \mathbf{Q_1} \\ \mathbf{Q_2} \end{pmatrix}$ is matrix of order $(r+t) \times n$ with rank $r+t$, we can find matrices $\mathbf{P_3}$ and $\mathbf{Q_3}$ such that $\mathbf{P} = (\mathbf{P_1}, \mathbf{R}, \mathbf{P_3})$ and $\mathbf{Q} = \begin{pmatrix} \mathbf{Q_1} \\ \mathbf{Q_2} \\ \mathbf{Q_3} \end{pmatrix}$ are non-singular. Now it is easy to see that $\mathbf{A} = \mathbf{P}\begin{pmatrix} \mathbf{I_r} & \mathbf{0} & \mathbf{0} \\ \mathbf{0} & \mathbf{0} & \mathbf{0} \\ \mathbf{0} & \mathbf{0} & \mathbf{0} \end{pmatrix}\mathbf{Q}$ and $\mathbf{B} = \mathbf{P}\begin{pmatrix} \mathbf{I_r} & \mathbf{T_1} & \mathbf{0} \\ \mathbf{0} & \mathbf{T_2} & \mathbf{0} \\ \mathbf{0} & \mathbf{0} & \mathbf{0} \end{pmatrix}\mathbf{Q}$. □

Given a g-inverse $\mathbf{A}^-$ of $\mathbf{A}$ such that $\mathbf{AA}^- = \mathbf{BA}^-$, we can now obtain the class of all g-inverses $\mathbf{A}^-$ of $\mathbf{A}$ for which $\mathbf{AA}^- = \mathbf{BA}^-$.

Theorem 6.2.5. *Let* $\mathbf{A}$ *and* $\mathbf{B}$ *be matrices of the same order such that* $\mathbf{AA}^- = \mathbf{BA}^-$. *Let* $\mathbf{A} = \mathbf{P}\begin{pmatrix} \mathbf{I_r} & \mathbf{0} & \mathbf{0} \\ \mathbf{0} & \mathbf{0} & \mathbf{0} \\ \mathbf{0} & \mathbf{0} & \mathbf{0} \end{pmatrix}\mathbf{Q}$ *and* $\mathbf{B} = \mathbf{P}\begin{pmatrix} \mathbf{I_r} & \mathbf{T_1} & \mathbf{0} \\ \mathbf{0} & \mathbf{T_2} & \mathbf{0} \\ \mathbf{0} & \mathbf{0} & \mathbf{0} \end{pmatrix}\mathbf{Q}$ *where* $\begin{pmatrix} \mathbf{T_1} \\ \mathbf{T_2} \end{pmatrix}$ *is of full column rank. Then for a g-inverse* $\mathbf{A}^-$ *of* $\mathbf{A}$, $\mathbf{AA}^- = \mathbf{BA}^-$

if and only if $\mathbf{A}^-$ *is of the form* $\mathbf{A}^- = \mathbf{Q}^{-1}\begin{pmatrix} \mathbf{I_r} & \mathbf{L_1} & \mathbf{L_2} \\ \mathbf{0} & \mathbf{0} & \mathbf{0} \\ \mathbf{N_1} & \mathbf{N_2} & \mathbf{N_3} \end{pmatrix}\mathbf{P}^{-1}$, *where* $\mathbf{L_1}, \mathbf{L_2}, \mathbf{N_1}, \mathbf{N_2}$, *and* $\mathbf{N_3}$ *are arbitrary.*

Remark 6.2.6. One is tempted to believe that the relation defined by the condition $\mathbf{AA}^- = \mathbf{BA}^-$ may be a pre-order. The belief, however, is put to rest by the following example:

Example 6.2.7. Let us take $\mathbf{A} = \begin{pmatrix} 1 & 0 \\ 0 & 0 \end{pmatrix}$, $\mathbf{B} = \begin{pmatrix} 2 & 1 \\ 0 & 0 \end{pmatrix}$ and $\mathbf{C} = \begin{pmatrix} 3 & 1 \\ 1 & 0 \end{pmatrix}$. We can easily check that $\mathbf{AA}^- = \mathbf{BA}^-$ for $\mathbf{A}^- = \begin{pmatrix} 1 & a \\ -1 & -a \end{pmatrix}$, where a is arbitrary and $\mathbf{BB}^- = \mathbf{CB}^-$ for $\mathbf{B}^- = \begin{pmatrix} 0 & 0 \\ 1 & 1 \end{pmatrix}$. However, $\mathbf{AA}^- = \mathbf{CA}^-$ for no g-inverse $\mathbf{A}^-$ of $\mathbf{A}$, so, the transitivity fails to hold.

Our next theorem shows that the condition $\mathbf{AA}^- = \mathbf{BA}^-$ is symmetric on the class of matrices that have same rank.

Theorem 6.2.8. *Let* $\mathbf{A}$ *and* $\mathbf{B}$ *be matrices of the same order. Consider the following:*

(i) $\mathbf{AA}^- = \mathbf{BA}^-$ *for some g-inverse* $\mathbf{A}^-$ *of* $\mathbf{A}$.
(ii) $\rho(\mathbf{A}) = \rho(\mathbf{B})$.
(iii) $\mathbf{BB}^- = \mathbf{AB}^-$ *for some g-inverse* $\mathbf{B}^-$ *of* $\mathbf{B}$.

Then any two of (i), (ii), (iii) *implies the third.*

Proof. (i) and (ii) $\Rightarrow$ (iii)
Let $\mathbf{A}^-$ be a g-inverse of $\mathbf{A}$ such that $\mathbf{AA}^- = \mathbf{BA}^-$. Then, $\mathbf{A} = \mathbf{AA}^-\mathbf{A} = \mathbf{BA}^-\mathbf{A}$. Further, $\rho(\mathbf{A}) = \rho(\mathbf{B})$, so, $\mathcal{C}(\mathbf{A}) = \mathcal{C}(\mathbf{B})$. So, $\mathbf{B} = \mathbf{AA}^-\mathbf{B} = \mathbf{BA}^-\mathbf{B}$. Hence $\mathbf{A}^-$ is a g-inverse of $\mathbf{B}$ and (iii) holds.
(i) and (iii)$\Rightarrow$(ii) is obvious.
Proof of (ii) and (iii) $\Rightarrow$ (i) is similar to the proof of (i) and (ii) $\Rightarrow$ (iii). □

Remark 6.2.9. In fact, we have proved a stronger statement than the one made in Theorem 6.2.8. If $\mathbf{AA}^- = \mathbf{BA}^-$ for some g-inverse $\mathbf{A}^-$ of $\mathbf{A}$ and $\rho(\mathbf{A}) = \rho(\mathbf{B})$, then $\mathbf{A}^-$ is a g-inverse of $\mathbf{B}$.

Remark 6.2.10. Let $\mathcal{C}(\mathbf{B}) = \mathcal{C}(\mathbf{A})$ and $\mathbf{AB}^-\mathbf{A} = \mathbf{A}$ for some g-inverse of $\mathbf{B}^-$ of $\mathbf{B}$. Then for any g-inverse $\mathbf{A}^-$ of $\mathbf{A}$, $\mathbf{G} = \mathbf{B}^-\mathbf{AA}^-$ is a g-inverse of $\mathbf{A}$. Also, $\mathbf{BG} = \mathbf{AA}^-$ and $\mathbf{BG} = \mathbf{AG}$, showing that the condition $\mathbf{AA}^- = \mathbf{BA}^-$ holds when $\mathbf{A}^- = \mathbf{G}$. Similarly, let $\mathcal{C}(\mathbf{B}^t) = \mathcal{C}(\mathbf{A}^t)$ and $\mathbf{AB}^-\mathbf{A} = \mathbf{A}$ for some g-inverse of $\mathbf{B}^-$ of $\mathbf{B}$. Then $\mathbf{GA} = \mathbf{GB}$ for some g-inverse $\mathbf{G}$ of $\mathbf{A}$.

6.3 One-sided sharp order

We begin this section by defining one-sided sharp orders, namely left sharp and right sharp orders. We show each of these relations defines a partial order which implies the minus order. We also obtain some conditions under which each of the left and the right sharp order is equivalent to the sharp order.

Definition 6.3.1. Let $\mathbf{A}$ and $\mathbf{B}$ be matrices of the same order and of index not greater than 1. We say $\mathbf{A}$ is below $\mathbf{B}$ under right sharp order, if $\mathbf{A}^2 = \mathbf{BA}$ and $\mathcal{C}(\mathbf{A}^t) \subseteq \mathcal{C}(\mathbf{B}^t)$. We write $\mathbf{A} < \# \mathbf{B}$, when this happens.

Similarly, if $\mathbf{A}^2 = \mathbf{AB}$ and $\mathcal{C}(\mathbf{A}) \subseteq \mathcal{C}(\mathbf{B})$, we say $\mathbf{A}$ is below $\mathbf{B}$ under the left sharp order. We write $\mathbf{A} \# < \mathbf{B}$, in case $\mathbf{A}$ is below $\mathbf{B}$ under the left sharp order.

Remark 6.3.2. Notice that in Definition 6.3.2, as in the case of the sharp order, we do not require matrix B to be of index ≤ 1. However for same reasons as mentioned in the case of the sharp order, we shall henceforth take both $\mathbf{A}$ and $\mathbf{B}$ to be matrices of index ≤ 1.

We shall now obtain a characterization of '$< \#$' using the matrix decompositions.

Theorem 6.3.3. *Let $\mathbf{A}$ and $\mathbf{B}$ be matrices of the same order and of index not greater than 1. Then $\mathbf{A} < \# \mathbf{B}$ if and only if there exists a non-singular matrix $\mathbf{P}$ such that*

$$\mathbf{A} = \mathbf{P}\begin{pmatrix} \mathbf{S} & \mathbf{0} & \mathbf{0} \\ \mathbf{0} & \mathbf{0} & \mathbf{0} \\ \mathbf{0} & \mathbf{0} & \mathbf{0} \end{pmatrix}\mathbf{P}^{-1} \text{ and } \mathbf{B} = \mathbf{P}\begin{pmatrix} \mathbf{S} & \mathbf{T}_{12} & \mathbf{0} \\ \mathbf{0} & \mathbf{T}_{22} & \mathbf{0} \\ \mathbf{0} & \mathbf{0} & \mathbf{0} \end{pmatrix}\mathbf{P}^{-1},$$

where $\mathbf{S}$ and $\mathbf{T}_{22}$ are non-singular.

Proof. 'If' part is trivial.
'Only if' part
Since $\mathbf{B}$ is of index ≤ 1, we can find non-singular matrices $\mathbf{R}$ and $\mathbf{T}$ such

that $\mathbf{B} = \mathbf{R}\mathbf{diag}(\mathbf{T}, \mathbf{0})\mathbf{R}^{-1}$. Write $\mathbf{A} = \mathbf{R}\begin{pmatrix} \mathbf{A}_{11} & \mathbf{A}_{12} \\ \mathbf{A}_{21} & \mathbf{A}_{22} \end{pmatrix}\mathbf{R}^{-1}$, partitioned in conformation for multiplication with $\mathbf{B}$. Since, $\mathcal{C}(\mathbf{A}^t) \subseteq \mathcal{C}(\mathbf{B}^t)$, we have $\mathbf{A}_{12} = \mathbf{0}$ and $\mathbf{A}_{22} = \mathbf{0}$. Now, $\mathbf{A}^2 = \mathbf{R}\begin{pmatrix} \mathbf{A}_{11}^2 & \mathbf{0} \\ \mathbf{A}_{21}\mathbf{A}_{11} & \mathbf{0} \end{pmatrix}\mathbf{R}^{-1}$ and $\mathbf{BA} = \mathbf{R}\begin{pmatrix} \mathbf{TA}_{11} & \mathbf{0} \\ \mathbf{0} & \mathbf{0} \end{pmatrix}\mathbf{R}^{-1}$. Since $\mathbf{A}^2 = \mathbf{BA}$, this gives $\mathbf{A}_{11}^2 = \mathbf{TA}_{11}$ and $\mathbf{A}_{21}\mathbf{A}_{11} = \mathbf{0}$. Further, $\rho(\mathbf{A}_{11}) = \rho(\mathbf{TA}_{11}) = \rho(\mathbf{BA}) = \rho(\mathbf{A}^2) = \rho(\mathbf{A})$, so, $\mathbf{A}_{21} = \mathbf{CA}_{11}$ and $\mathbf{CA}_{11}^2 = \mathbf{0}$. Also, $\rho(\mathbf{A}^2) = \rho(\mathbf{A}) \Rightarrow \mathcal{C}(\mathbf{A}^2) = \mathcal{C}(\mathbf{A})$. Thus, $\mathbf{A}_{11} = \mathbf{A}_{11}^2\mathbf{D}$ for some matrix $\mathbf{D}$. It follows that $\mathbf{A}_{21} = \mathbf{CA}_{11} = \mathbf{CA}_{11}^2\mathbf{D} = \mathbf{0}$. Now, index $\mathbf{A}_{11} \leq 1$, so, there exist non-singular matrices $\mathbf{S}$ and $\mathbf{Q}$ of suitable orders such that $\mathbf{A}_{11} = \mathbf{Q}\mathbf{diag}\begin{pmatrix} \mathbf{S} & \mathbf{0} \end{pmatrix}\mathbf{Q}^{-1}$. Write $\mathbf{T} = \mathbf{Q}\begin{pmatrix} \mathbf{T}_{11} & \mathbf{T}_{12} \\ \mathbf{T}_{21} & \mathbf{T}_{22} \end{pmatrix}\mathbf{Q}^{-1}$, where $\mathbf{T}_{11}$ and $\mathbf{S}$ are matrices of the same order. Now, $\mathbf{A}^2 = \mathbf{BA} \Rightarrow \mathbf{A}_{11} = \mathbf{TA}_{11}^2 \Rightarrow \begin{pmatrix} \mathbf{S}^2 & \mathbf{0} \\ \mathbf{0} & \mathbf{0} \end{pmatrix} = \begin{pmatrix} \mathbf{T}_{11}\mathbf{S} & \mathbf{0} \\ \mathbf{T}_{21}\mathbf{S} & \mathbf{0} \end{pmatrix}$. Therefore, $\mathbf{T}_{11} = \mathbf{S}$ and $\mathbf{T}_{21} = \mathbf{0}$. Moreover, $\mathbf{T}$ is non-singular $\Rightarrow \mathbf{T}_{22}$ is non-singular. Let $\mathbf{P} = \mathbf{R}\begin{pmatrix} \mathbf{Q} & \mathbf{0} \\ \mathbf{0} & \mathbf{I} \end{pmatrix}$ and notice that $\mathbf{A}$ and $\mathbf{B}$ have the desired forms. □

Theorem 6.3.4. *Let* $\mathbf{A}$ *and* $\mathbf{B}$ *be matrices of the same order and index not greater than 1. Then* $\mathbf{A} <\# \mathbf{B}$ *if and only if there exists a non-singular matrix* $\mathbf{P}$ *such that*

$$\mathbf{A} = \mathbf{P}\begin{pmatrix} \mathbf{S} & \mathbf{0} \\ \mathbf{0} & \mathbf{0} \end{pmatrix}\mathbf{P}^{-1} \text{ and } \mathbf{B} = \mathbf{P}\begin{pmatrix} \mathbf{S} & \mathbf{B}_{12} \\ \mathbf{0} & \mathbf{B}_{22} \end{pmatrix}\mathbf{P}^{-1},$$

where $\mathbf{S}$ *is non-singular,* $\mathbf{B}_{12}$ *is arbitrary and* $\mathbf{B}_{22}$ *is arbitrary matrix of index* ≤ 1.

Proof. Since $\mathbf{A}$ is of index ≤ 1, we can find non-singular matrices $\mathbf{P}$ and $\mathbf{S}$ such that $\mathbf{A} = \mathbf{P}\begin{pmatrix} \mathbf{S} & \mathbf{0} \\ \mathbf{0} & \mathbf{0} \end{pmatrix}\mathbf{P}^{-1}$. Write $\mathbf{B} = \mathbf{P}\begin{pmatrix} \mathbf{B}_{11} & \mathbf{B}_{12} \\ \mathbf{B}_{21} & \mathbf{B}_{22} \end{pmatrix}\mathbf{P}^{-1}$, partitioned in conformation with partitioning of $\mathbf{A}$. Since $\mathbf{A}^2 = \mathbf{BA}$, we have $\mathbf{A}^2 = \mathbf{P}\begin{pmatrix} \mathbf{S}^2 & \mathbf{0} \\ \mathbf{0} & \mathbf{0} \end{pmatrix}\mathbf{P}^{-1}$ and $\mathbf{BA} = \mathbf{P}\begin{pmatrix} \mathbf{B}_{11}\mathbf{S} & \mathbf{0} \\ \mathbf{B}_{21}\mathbf{S} & \mathbf{0} \end{pmatrix}\mathbf{P}^{-1}$. It follows that $\mathbf{B}_{11} = \mathbf{S}$ and $\mathbf{B}_{21} = \mathbf{0}$. Hence, $\mathbf{B} = \mathbf{P}\begin{pmatrix} \mathbf{S} & \mathbf{B}_{12} \\ \mathbf{0} & \mathbf{B}_{22} \end{pmatrix}\mathbf{P}^{-1}$, where $\mathbf{B}_{12}$ and $\mathbf{B}_{22}$ are arbitrary. Since $\mathbf{S}$ is non-singular, it is easy to check that $\rho(\mathbf{B}) = \rho(\mathbf{B}^2)$ if and only if $\rho(\mathbf{B}_{22}) = \rho(\mathbf{B}_{22}^2)$. □

Remark 6.3.5. Let $\mathbf{A}$ and $\mathbf{B}$ be matrices of the same order and index not greater than 1. If $\rho(\mathbf{A}) = \rho(\mathbf{B})$ and $\mathbf{A} < \# \mathbf{B}$, then $\mathbf{A} = \mathbf{B}$.

Theorem 6.3.6. *Let* $\mathbf{A}$ *and* $\mathbf{B}$ *be matrices of the same order and index not greater than 1. Then* $\mathbf{A} < \# \mathbf{B}$ *if and only if there exists an* $\mathbf{A}_\chi^-$ *such that* $\mathbf{A}\mathbf{A}_\chi^- = \mathbf{B}\mathbf{A}_\chi^-$ *and* $\mathbf{A}_\chi^-\mathbf{A} = \mathbf{A}_\chi^-\mathbf{B}$.

Proof. 'If' part
Let $\mathbf{A}\mathbf{A}_\chi^- = \mathbf{B}\mathbf{A}_\chi^-$ and $\mathbf{A}_\chi^-\mathbf{A} = \mathbf{A}_\chi^-\mathbf{B}$. Then $\mathbf{A}\mathbf{A}_\chi^- = \mathbf{B}\mathbf{A}_\chi^- \Rightarrow \mathbf{A}^2 = \mathbf{A}\mathbf{A}_\chi^-\mathbf{A}^2 = \mathbf{B}\mathbf{A}_\chi^-\mathbf{A}^2 = \mathbf{B}\mathbf{A}$. Also, $\mathbf{A}_\chi^-\mathbf{A} = \mathbf{A}_\chi^-\mathbf{B} \Rightarrow \mathbf{A} = \mathbf{A}\mathbf{A}_\chi^-\mathbf{A} = \mathbf{A}\mathbf{A}_\chi^-\mathbf{B} \Rightarrow \mathcal{C}(\mathbf{A}^t) \subseteq \mathcal{C}(\mathbf{B}^t)$.
'Only if' part
Let $\mathbf{A} < \# \mathbf{B}$. By Theorem 6.3.3, there exists a non-singular matrix $\mathbf{P}$ such that $\mathbf{A} = \mathbf{P}\begin{pmatrix}\mathbf{S} & \mathbf{0} & \mathbf{0}\\ \mathbf{0} & \mathbf{0} & \mathbf{0}\\ \mathbf{0} & \mathbf{0} & \mathbf{0}\end{pmatrix}\mathbf{P}^{-1}$ and $\mathbf{B} = \mathbf{P}\begin{pmatrix}\mathbf{S} & \mathbf{T_{12}} & \mathbf{0}\\ \mathbf{0} & \mathbf{T_{22}} & \mathbf{0}\\ \mathbf{0} & \mathbf{0} & \mathbf{0}\end{pmatrix}\mathbf{P}^{-1}$, where $\mathbf{S}$ and $\mathbf{T_{22}}$ are non-singular. Notice that any $\mathbf{A}_\chi^-$ is of the form $\mathbf{A}_\chi^- = \mathbf{P}\begin{pmatrix}\mathbf{S}^{-1} & \mathbf{L} & \mathbf{M}\\ \mathbf{0} & \mathbf{0} & \mathbf{0}\\ \mathbf{0} & \mathbf{0} & \mathbf{0}\end{pmatrix}\mathbf{P}^{-1}$ for some matrices $\mathbf{L}$ and $\mathbf{M}$. Also,

$\mathbf{A}\mathbf{A}_\chi^- = \mathbf{P}\begin{pmatrix}\mathbf{I} & \mathbf{SL} & \mathbf{SM}\\ \mathbf{0} & \mathbf{0} & \mathbf{0}\\ \mathbf{0} & \mathbf{0} & \mathbf{0}\end{pmatrix}\mathbf{P}^{-1}$ and $\mathbf{B}\mathbf{A}_\chi^- = \mathbf{P}\begin{pmatrix}\mathbf{I} & \mathbf{SL} & \mathbf{SM}\\ \mathbf{0} & \mathbf{0} & \mathbf{0}\\ \mathbf{0} & \mathbf{0} & \mathbf{0}\end{pmatrix}\mathbf{P}^{-1}$. Hence, for every $\mathbf{A}_\chi^-$, $\mathbf{A}\mathbf{A}_\chi^- = \mathbf{B}\mathbf{A}_\chi^-$. Moreover, $\mathbf{A}_\chi^-\mathbf{A} = \mathbf{P}\begin{pmatrix}\mathbf{I} & \mathbf{0} & \mathbf{0}\\ \mathbf{0} & \mathbf{0} & \mathbf{0}\\ \mathbf{0} & \mathbf{0} & \mathbf{0}\end{pmatrix}\mathbf{P}^{-1}$ and

$\mathbf{A}_\chi^-\mathbf{B} = \mathbf{P}\begin{pmatrix}\mathbf{I} & \mathbf{S}^{-1}\mathbf{T_{12}} + \mathbf{L}\mathbf{T_{22}} & \mathbf{0}\\ \mathbf{0} & \mathbf{0} & \mathbf{0}\\ \mathbf{0} & \mathbf{0} & \mathbf{0}\end{pmatrix}\mathbf{P}^{-1}$.

We take an $\mathbf{A}_\chi^-$ for which $\mathbf{L} = -\mathbf{S}^{-1}\mathbf{T_{12}}\mathbf{T_{22}^{-1}}$. Thus, there exists an $\mathbf{A}_\chi^-$ such that $\mathbf{A}_\chi^-\mathbf{A} = \mathbf{A}_\chi^-\mathbf{B}$. □

Remark 6.3.7. It is clear that if $\mathbf{T_{12}}$ and $\mathbf{L}$ are non-null, then we cannot have $\mathbf{A}\mathbf{A}^\# = \mathbf{B}\mathbf{A}^\#$ and $\mathbf{A}^\#\mathbf{A} = \mathbf{A}^\#\mathbf{B}$. In other words $\mathbf{A}$ cannot be below $\mathbf{B}$ under the sharp order.

Analogous to Theorem 6.3.6, it is easy to prove that the left sharp order is same the ρ-order.

Corollary 6.3.8. *Let* $\mathbf{A}$ *and* $\mathbf{B}$ *be matrices of order* $n \times n$ *and of index not greater than 1. Then* $\mathbf{A} \# < \mathbf{B}$ *if and only if there exists an* $\mathbf{A}_\rho^-$ *such that*

$\mathbf{AA}_\rho^- = \mathbf{BA}_\rho^-$ *and* $\mathbf{A}_\rho^-\mathbf{A} = \mathbf{A}_\rho^-\mathbf{B}$.

Corollary 6.3.9. *Let* $\mathbf{A}$ *and* $\mathbf{B}$ *be matrices of the same order and index not greater than 1. Then* $\mathbf{A} < \# \, \mathbf{B} \Rightarrow \mathbf{A} <^- \mathbf{B}$ *and* $\mathbf{A} \# < \mathbf{B} \Rightarrow \mathbf{A} <^- \mathbf{B}$.

Corollary 6.3.10. *The relation '$< \#$' (and also the relation '$\# <$') is a partial order on the set* $\mathcal{I}_1$ *of matrices of index* 1.

Proof. In view of Corollary 6.3.9, we only need to show that the relation '$< \#$' is transitive. Let $\mathbf{A}$, $\mathbf{B}$ and $\mathbf{C} \in \mathcal{I}_1$ such that $\mathbf{A} < \# \, \mathbf{B}$ and $\mathbf{B} < \# \, \mathbf{C}$. Then by definition, $\mathbf{A^2} = \mathbf{BA}$ and $\mathcal{C}(\mathbf{A^t}) \subseteq \mathcal{C}(\mathbf{B^t})$. and $\mathbf{B^2} = \mathbf{CB}$ and $\mathcal{C}(\mathbf{B^t}) \subseteq \mathcal{C}(\mathbf{C^t})$. Clearly, $\mathcal{C}(\mathbf{A^t}) \subseteq \mathcal{C}(\mathbf{C^t})$. To prove $\mathbf{A^2} = \mathbf{CA}$, we use Theorem 6.3.4. As $\mathbf{A} < \# \, \mathbf{B}$, there exists a non-singular matrix $\mathbf{P}$ such that $\mathbf{A} = \mathbf{P}\begin{pmatrix} \mathbf{S} & \mathbf{0} \\ \mathbf{0} & \mathbf{0} \end{pmatrix}\mathbf{P}^{-1}$ and $\mathbf{B} = \mathbf{P}\begin{pmatrix} \mathbf{S} & \mathbf{B}_{12} \\ \mathbf{0} & \mathbf{B}_{22} \end{pmatrix}\mathbf{P}^{-1}$, where $\mathbf{S}$ is non-singular and $\mathbf{B_{22}}$ is of index ≤ 1. Let $\mathbf{C} = \mathbf{P}\begin{pmatrix} \mathbf{C}_{11} & \mathbf{C}_{12} \\ \mathbf{C}_{21} & \mathbf{C}_{22} \end{pmatrix}\mathbf{P}^{-1}$. Since, $\mathbf{B^2} = \mathbf{CB}$, we have $\mathbf{C_{11}} = \mathbf{S}$ and $\mathbf{C_{21}} = \mathbf{0}$. Clearly, $\mathbf{CA} = \mathbf{P}\mathbf{diag}\left(\mathbf{S^2}, \ \mathbf{0}\right)\mathbf{P}^{-1} = \mathbf{A^2}$. Hence, $\mathbf{A} < \# \, \mathbf{C}$. □

Remark 6.3.11. Notice that in Theorem 6.3.6, the matrix $\mathbf{L}$ is unique. Thus, when $\mathbf{A} < \# \, \mathbf{B}$, there exists a unique $\mathbf{A}_\chi^-$ such that $\mathbf{AA}_\chi^- = \mathbf{BA}_\chi^-$ and $\mathbf{A}_\chi^-\mathbf{A} = \mathbf{A}_\chi^-\mathbf{B}$ and $\mathcal{C}({\mathbf{A}_\chi^-}^t) \subseteq \mathcal{C}(\mathbf{B^t})$.

Theorem 6.3.12. *Let* $\mathbf{A}$ *and* $\mathbf{B}$ *be matrices of the same order and of index not greater than 1. Then* $\mathbf{A} \# < \mathbf{B}$ *if and only* $\mathbf{A}^\# \# < \mathbf{B}^\#$.

Proof. Observe first that the group inverse of a block upper triangular matrix is also an upper triangular matrix. Now the proof follows by Theorem 6.3.3. □

Theorem 6.3.13. *Let* $\mathbf{A}$ *and* $\mathbf{B}$ *be matrices of the same order and of index not greater than 1. Then the following are equivalent:*

(i) $\mathbf{A} < \# \, \mathbf{B}$
(ii) $\mathbf{A} <^- \mathbf{B}$ *and* $\mathcal{C}(\mathbf{BA}) = \mathcal{C}(\mathbf{A})$
(iii) $\mathbf{A} <^- \mathbf{B}$ *and* $\mathcal{C}(\mathbf{BA}^\#) = \mathcal{C}(\mathbf{A})$
(iv) $\mathbf{A} <^- \mathbf{B}$ *and* $\mathcal{C}(\mathbf{B}^\#\mathbf{A}) = \mathcal{C}(\mathbf{A})$ *and*
(v) $\mathbf{A} <^- \mathbf{B}$ *and* $\mathcal{C}(\mathbf{B}^\#\mathbf{A}^\#) = \mathcal{C}(\mathbf{A})$.

Proof. (i) $\Rightarrow$ (ii)
Since (i) $\Rightarrow$ there exists an $\mathbf{A}_\chi^-$ such that $\mathbf{AA}_\chi^- = \mathbf{BA}_\chi^-$ and $\mathbf{A}_\chi^-\mathbf{A} = \mathbf{A}_\chi^-\mathbf{B}$,

it follows that $\mathbf{A} <^- \mathbf{B}$. Also, $\mathbf{A} < \# \mathbf{B} \Rightarrow \mathbf{A^2} = \mathbf{BA}$, so, $\mathcal{C}(\mathbf{BA}) = \mathcal{C}(\mathbf{A^2}) = \mathcal{C}(\mathbf{A})$. Thus, (ii) holds.

(ii) $\Rightarrow$ (i)

Since, $\mathbf{A} <^{\mathbf{s}} \mathbf{B}$ and each of $\mathbf{A}$ and $\mathbf{B}$ have index ≤ 1, by Theorem 3.2.10, there exist non-singular matrices $\mathbf{P}$, $\mathbf{S}$ and $\mathbf{T}$ of suitable order, such that $\mathbf{A} = \mathbf{P}\begin{pmatrix}\mathbf{S} & \mathbf{0} & \mathbf{0}\\ \mathbf{0} & \mathbf{0} & \mathbf{0}\\ \mathbf{0} & \mathbf{0} & \mathbf{0}\end{pmatrix}\mathbf{P}^{-1}$ and $\mathbf{B} = \mathbf{P}\begin{pmatrix}\mathbf{T_{11}} & \mathbf{T_{12}} & \mathbf{0}\\ \mathbf{T_{21}} & \mathbf{T_{22}} & \mathbf{0}\\ \mathbf{0} & \mathbf{0} & \mathbf{0}\end{pmatrix}\mathbf{P}^{-1}$ for some non-singular matrices $\mathbf{P}$, $\mathbf{S}$ and $\mathbf{T} = \begin{pmatrix}\mathbf{T_{11}} & \mathbf{T_{12}}\\ \mathbf{T_{21}} & \mathbf{T_{22}}\end{pmatrix}$. Since, $\mathcal{C}(\mathbf{BA}) = \mathcal{C}(\mathbf{A})$, it follows that $\mathbf{T_{21}} = \mathbf{0}$. Further, since $\mathbf{A} <^- \mathbf{B}$, $\mathbf{T_{11}} = \mathbf{S}$. So, (i) follows in view of Theorem 6.3.3.

(ii) $\Leftrightarrow$ (iii)

We have $\mathcal{C}(\mathbf{BA^\#}) = \mathcal{C}(\mathbf{BA(A^3)^-A}) \subseteq \mathcal{C}(\mathbf{BA}) \subseteq \mathcal{C}(\mathbf{BA^\#A^2}) \subseteq \mathcal{C}(\mathbf{BA^\#})$. So, $\mathcal{C}(\mathbf{BA^\#}) = \mathcal{C}(\mathbf{BA})$. It is now clear that (ii) $\Leftrightarrow$ (iii)

(i) $\Rightarrow$ (iv)

Since, $\mathbf{A} <^- \mathbf{B}$, we have $\mathbf{A} = \mathbf{AB^\#A} = \mathbf{AB^\#B} = \mathbf{BB^\#A}$. Also, $\mathbf{B^\#A} = \mathbf{B^\#AA^\#A} = \mathbf{B^\#A^2A^\#} = \mathbf{B^\#BAA^\#} = (\mathbf{B^\#BA})\mathbf{A^\#} = \mathbf{AA^\#}$. Therefore, $\mathcal{C}(\mathbf{B^\#A}) \subseteq \mathcal{C}(\mathbf{A})$. Also, $\mathcal{C}(\mathbf{A}) = \mathcal{C}(\mathbf{AA^\#A}) \subseteq \mathcal{C}(\mathbf{AA^\#}) \subseteq \mathcal{C}(\mathbf{B^\#A})$. So, $\mathcal{C}(\mathbf{B^\#A}) = \mathcal{C}(\mathbf{A})$.

(iv)$\Rightarrow$ (i) is similar to (ii)$\Rightarrow$ (i).

(iv)$\Leftrightarrow$ (v)

Notice that $\mathcal{C}(\mathbf{B^\#A}) = \mathcal{C}(\mathbf{B^\#A^\#})$. So, (iv)$\Leftrightarrow$ (v) holds. □

Corollary 6.3.14. *Let* $\mathbf{A}$ *and* $\mathbf{B}$ *be matrices of the same order and of index not greater than* 1. *Then the following are equivalent:*

(i) $\mathbf{A} <^\# \mathbf{B}$
(ii) $\mathbf{A} <^- \mathbf{B}$ *and* $\mathbf{A}$, $\mathbf{B}$ *commute*
(iii) $\mathbf{A} <^- \mathbf{B}$ *and* $\mathbf{A}$, $\mathbf{B^\#}$ *commute*
(iv) $\mathbf{A} <^- \mathbf{B}$ *and* $\mathbf{A^\#}$, $\mathbf{B}$ *commute* *and*
(v) $\mathbf{A} <^- \mathbf{B}$ *and* $\mathbf{A^\#}$, $\mathbf{B^\#}$ *commute.*

Proof. Proof follows by Theorem 6.3.13 and Theorem 4.2.12. □

Theorem 6.3.15. *Let* $\mathbf{A}$ *and* $\mathbf{B}$ *be matrices of the same order and of index not greater than 1 such that* $\mathbf{A} < \# \mathbf{B}$. *Then the following are equivalent:*

(i) $\mathbf{A} <^\# \mathbf{B}$
(ii) $\mathcal{C}(\mathbf{AB^t}) = \mathcal{C}(\mathbf{BA^t})$
(iii) $\mathcal{C}(\mathbf{AB^t}) = \mathcal{C}(\mathbf{A^t})$ *and*

(iv) $\mathbf{AB}^{\#} = \mathbf{AA}^{\#}$.

Proof. Since $\mathbf{A} < \# \, \mathbf{B}$, by Theorem 6.3.3, there exists a non-singular matrix $\mathbf{P}$ such that $\mathbf{A} = \mathbf{P}\begin{pmatrix} \mathbf{S} & \mathbf{0} & \mathbf{0} \\ \mathbf{0} & \mathbf{0} & \mathbf{0} \\ \mathbf{0} & \mathbf{0} & \mathbf{0} \end{pmatrix}$ and $\mathbf{B} = \mathbf{P}\begin{pmatrix} \mathbf{S} & \mathbf{T}_{12} & \mathbf{0} \\ \mathbf{0} & \mathbf{T}_{22} & \mathbf{0} \\ \mathbf{0} & \mathbf{0} & \mathbf{0} \end{pmatrix}\mathbf{P}^{-1}$, where $\mathbf{S}$ and $\mathbf{T}_{22}$ are non-singular.

(i) $\Rightarrow$ (ii)

Let (i) hold. So, $\mathbf{A}^2 = \mathbf{AB} = \mathbf{BA}$. Therefore,

$$\begin{pmatrix} \mathbf{S}^2 & \mathbf{0} & \mathbf{0} \\ \mathbf{0} & \mathbf{0} & \mathbf{0} \\ \mathbf{0} & \mathbf{0} & \mathbf{0} \end{pmatrix} = \begin{pmatrix} \mathbf{S}^2 & \mathbf{ST}_{12} & \mathbf{0} \\ \mathbf{0} & \mathbf{0} & \mathbf{0} \\ \mathbf{0} & \mathbf{0} & \mathbf{0} \end{pmatrix} = \begin{pmatrix} \mathbf{S}^2 & \mathbf{0} & \mathbf{0} \\ \mathbf{0} & \mathbf{0} & \mathbf{0} \\ \mathbf{0} & \mathbf{0} & \mathbf{0} \end{pmatrix}.$$

It follows that $\mathbf{T}_{12} = \mathbf{0}$. Clearly, $\mathcal{C}(\mathbf{AB}^t) = \mathcal{C}(\mathbf{BA}^t)$. Thus, (ii) holds.

(ii) $\Rightarrow$ (iii)

Next, let (ii) hold. So, $\mathbf{ST}_{12} = \mathbf{0}$. Since, $\mathbf{S}$ is non-singular $\mathbf{T}_{12} = \mathbf{0}$. Hence, $\mathcal{C}(\mathbf{AB}^t) = \mathcal{C}(\mathbf{A}^t)$ showing (iii) holds.

(iii) $\Rightarrow$ (i)

Now, let (iii) hold.

So, $\mathcal{C}(\mathbf{AB}^t) = \mathcal{C}(\mathbf{A}^t) = \mathcal{C}(\mathbf{A}^t)^2$. Once again $\mathbf{T}_{12} = \mathbf{0}$. So, $\mathbf{A}^2 = \mathbf{BA} = \mathbf{AB}$, which proves (i).

Finally for (i) $\Leftrightarrow$ (iv), use Theorem 6.3.3. □

Theorem 6.3.16. *Let* $\mathbf{A}$ *and* $\mathbf{B}$ *be matrices of the same order and of index not greater than 1. Let* $\mathbf{A}^2 = \mathbf{BA}$. *Then the following are equivalent:*

(i) $\mathcal{C}(\mathbf{A}^t) \subseteq \mathcal{C}(\mathbf{B}^t)$

(ii) $\mathcal{C}(\mathbf{A}) \cap \mathcal{C}(\mathbf{B} - \mathbf{A}) = \mathbf{0}$

(iii) $\mathbf{A}^-\mathbf{A} = \mathbf{A}^-\mathbf{B}$ *for some g-inverse* $\mathbf{A}^-$ *of* $\mathbf{A}$ *and*

(iv) $\mathbf{A} = \mathbf{PB}$ *for some projector* $\mathbf{P}$.

Proof. (i) $\Rightarrow$ (ii)

Since, $\mathbf{A}^2 = \mathbf{BA}$ and $\mathcal{C}(\mathbf{A}^t) \subseteq \mathcal{C}(\mathbf{B}^t)$, (ii) holds by Corollary 6.3.9 and Theorem 6.2.2.

(ii)$\Rightarrow$ (i)

Let $(\mathbf{P}, \mathbf{Q})$ and $(\mathbf{R}, \mathbf{S})$ be rank factorization of $\mathbf{A}$ and $\mathbf{B} - \mathbf{A}$ respectively. Then $\mathbf{B} = \mathbf{A} + (\mathbf{B} - \mathbf{A}) = \mathbf{PQ} + \mathbf{RS} = (\mathbf{P} \;\; \mathbf{R})\begin{pmatrix} \mathbf{Q} \\ \mathbf{S} \end{pmatrix}$. Notice that $(\mathbf{P} : \mathbf{R})$ is of full column rank, since $\mathcal{C}(\mathbf{A}) \cap \mathcal{C}(\mathbf{B} - \mathbf{A}) = \mathbf{0}$. So,

$$\mathcal{C}(\mathbf{A}^t) = \mathcal{C}(\mathbf{Q}^t) \subseteq \mathcal{C}(\mathbf{Q}^t : \mathbf{S}^t) = \mathcal{C}(\mathbf{B}^t).$$

(i)⇒ (iii) follows by Corollary 6.3.9.
(iii)⇒ (iv)
Since, $\mathbf{A}^-\mathbf{A} = \mathbf{A}^-\mathbf{B} \Rightarrow \mathbf{A} = \mathbf{A}\mathbf{A}^-\mathbf{A} = \mathbf{A}\mathbf{A}^-\mathbf{B}$, so, $\mathbf{A} = \mathbf{P}\mathbf{B}$, where $\mathbf{P} = \mathbf{A}\mathbf{A}^-$.
(iv)⇒ (i) is trivial. □

Theorem 6.3.17. *Let* $\mathbf{A}$ *and* $\mathbf{B}$ *be matrices of the same order and of index not greater than 1. The following are equivalent:*

(i) $\mathbf{A}\# < \mathbf{B}$ *and '*$\mathbf{A}\mathbf{A}^{\#}$ *and* $\mathbf{B}$*' commute and*
(ii) $\mathbf{A} <^{\#} \mathbf{B}$.

Proof. (i) ⇒ (ii)
Since $\mathbf{A} \# < \mathbf{B}$, we have $\mathbf{A}^2 = \mathbf{AB}$ and $\mathcal{C}(\mathbf{A}) \subseteq \mathcal{C}(\mathbf{B})$. Now, pre- multiplying and post multiplying $\mathbf{A}^2 = \mathbf{AB}$ by $\mathbf{A}^{\#}$ and $\mathbf{A}$ respectively, we have $\mathbf{A}^2 = \mathbf{A}^{\#}\mathbf{ABA} = \mathbf{AA}^{\#}\mathbf{BA}$. Since $\mathbf{AA}^{\#}$ and $\mathbf{B}$ commute, we obtain $\mathbf{A}^2 = \mathbf{BA}$. So, $\mathbf{A}^2 = \mathbf{AB} = \mathbf{BA}$ proving $\mathbf{A} <^{\#} \mathbf{B}$.
(ii) ⇒(i) is trivial. □

Remark 6.3.18. Notice that $\mathbf{A} < \# \mathbf{B} \Rightarrow (\mathbf{B} - \mathbf{A}) \# < \mathbf{B}$. Further, $\mathcal{C}(\mathbf{B} - \mathbf{A}) \subseteq \mathcal{C}(\mathbf{B})$, since $\mathbf{A} < \# \mathbf{B} \Rightarrow \mathbf{A} <^- \mathbf{B} \Rightarrow \mathbf{B} - \mathbf{A} <^- \mathbf{B}$.

Theorem 6.3.19. *Let* $\mathbf{A}$ *and* $\mathbf{B}$ *be matrices of the same order and of index not greater than 1. Then the following are equivalent:*

(i) $\mathbf{A} < \# \mathbf{B}$ *and* $\mathbf{B} - \mathbf{A} < \# \mathbf{B}$ *and*
(ii) $\mathbf{A} <^{\#} \mathbf{B}$.

Proof is easy.

Remark 6.3.20. Let $\mathbf{A}$ and $\mathbf{B}$ be matrices of the same order and of index not greater than 1. Then it is easy to see that
$\mathbf{A} < \# \mathbf{B} \Rightarrow \mathbf{A}^{\mathbf{k}} < \# \mathbf{B}^{\mathbf{k}}$ for each positive integer $k \geq 1$ and therefore, $\mathbf{A} < \# \mathbf{B} \Rightarrow \mathbf{A}^{\mathbf{k}} <^- \mathbf{B}^{\mathbf{k}}$ for each positive integer $k \geq 1$. A similar statement holds for '$\# <$'.

We can now show that one sided sharp order coincides with the sharp order for the class of range-hermitian matrices. In the rest of this section we consider matrices over $\mathbb{C}$, the field of complex numbers.

Theorem 6.3.21. *Let* $\mathbf{A}$ *and* $\mathbf{B}$ *be range-hermitian matrices of the same order. Then in notations of Corollary 3.7.4,* $\mathbf{A} < \# \mathbf{B}$ *if and only if*

$$\mathbf{A} = \mathbf{U}\mathrm{diag}\left(\mathbf{D_a},\ \mathbf{0_{b-a}},\ \mathbf{0}\right)\mathbf{U}^\star \text{ and } \mathbf{B} = \mathbf{U}\mathrm{diag}\left(\mathbf{M},\ \mathbf{0}\right)\mathbf{U}^\star,$$

where $\mathbf{U}$ *is a unitary matrix,* $\mathbf{D_a}$ *is positive definite matrix of order* $a \times a$ *and* $\mathbf{M}$ *is a non-singular matrix of order* $b \times b$ *of the form* $\mathbf{M} = \begin{pmatrix} \mathbf{D_a} & \mathbf{0} \\ \mathbf{0} & \mathbf{0} \end{pmatrix} + \begin{pmatrix} \mathbf{F_1} \\ \mathbf{I} \end{pmatrix} \mathbf{M_{22}} \begin{pmatrix} \mathbf{0} & \mathbf{I} \end{pmatrix}$, *for some matrix* $\mathbf{F_1}$ *of suitable order.*

Proof. 'If' part follows by direct verification.
'Only if' part
Since $\mathbf{A} < \# \mathbf{B} \Rightarrow \mathbf{A} <^- \mathbf{B}$, by Corollary 3.7.4,

$$\mathbf{A} = \mathbf{Udiag}\left(\mathbf{D_a},\ \mathbf{0_{b-a}},\ \mathbf{0}\right)\mathbf{U^\star} \text{ and } \mathbf{B} = \mathbf{Udiag}\left(\mathbf{M},\ \mathbf{0}\right)\mathbf{U^\star},$$

where $\mathbf{U}$ is a unitary matrix, $\mathbf{D_a}$ is positive definite matrix and $\mathbf{M}$ is a non-singular matrix of of order $a \times a$ and $b \times b$ respectively with $\mathbf{M}$ of the form $\mathbf{M} = \begin{pmatrix} \mathbf{D_a} & \mathbf{0} \\ \mathbf{0} & \mathbf{0} \end{pmatrix} + \begin{pmatrix} \mathbf{F_1} \\ \mathbf{I} \end{pmatrix} \mathbf{M_{22}} \begin{pmatrix} \mathbf{F_2} & \mathbf{I} \end{pmatrix}$, for some matrices $\mathbf{F_1}$ and $\mathbf{F_2}$ of suitable orders. Since, $\mathbf{A^2} = \mathbf{BA}$, it follows that $\mathbf{F_2} = \mathbf{0}$. □

Corollary 6.3.22. *Let* $\mathbf{A}$ *and* $\mathbf{B}$ *be matrices of the same order. Let* $\mathbf{A}$ *be range-hermitian and* $\mathbf{B}$ *be normal. Then in notations of Corollary 3.7.4,* $\mathbf{A} < \# \mathbf{B}$ *if and only if* $\mathbf{A} = \mathbf{Udiag}\left(\mathbf{D_a},\ \mathbf{0_{b-a}},\ \mathbf{0}\right)\mathbf{U^\star}$ *and* $\mathbf{B} = \mathbf{Udiag}\left(\mathbf{M},\ \mathbf{0}\right)\mathbf{U^\star}$, *where* $\mathbf{U}$ *is a unitary matrix,* $\mathbf{D_a}$ *is positive definite matrix and* $\mathbf{M}$ *is a non-singular matrix of of order* $a \times a$ *and* $b \times b$ *respectively with* $\mathbf{M}$ *of the form* $\mathbf{M} = \begin{pmatrix} \mathbf{D_a} & \mathbf{0} \\ \mathbf{0} & \mathbf{0} \end{pmatrix} + \begin{pmatrix} \mathbf{0} \\ \mathbf{I} \end{pmatrix} \mathbf{M_{22}} \begin{pmatrix} \mathbf{0} & \mathbf{I} \end{pmatrix}$, *where* $\mathbf{M_{22}}$ *is non-singular matrix of order* $(b-a) \times (b-a)$. *Moreover, in this case* $\mathbf{A} <^{\#} \mathbf{B}$.

Corollary 6.3.23. *Let* $\mathbf{A}$ *and* $\mathbf{B}$ *be range-hermitian matrices of the same order. Then* $\mathbf{A} < \# \mathbf{B}$ *if and only if* $\mathbf{A} <^{\#} \mathbf{B}$ *or equivalently if and only if* $\mathbf{A}$ *and* $\mathbf{B}$ *are normal.*

6.4 Roles of $\mathbf{A_c^-}$ and $\mathbf{A_a^-}$ in one-sided sharp order

In this section we first introduce briefly two new classes of g-inverses, namely: the class $\{\mathbf{A_c^-}\}$ and the class $\{\mathbf{A_a^-}\}$. Details of their properties are given in exercises at the end of this chapter. We then show that $\mathbf{A} < \# \mathbf{B}$ if and only if $\{\mathbf{B_c^-}\} \subseteq \{\mathbf{A_c^-}\}$ and $\mathbf{A} \# < \mathbf{B}$ if and only if $\{\mathbf{B_a^-}\} \subseteq \{\mathbf{A_a^-}\}$. We begin with the following:

Definition 6.4.1. Let $\mathbf{A}$ be a square matrix of index ≤ 1. A g-inverse $\mathbf{G}$ is called a c-inverse of $\mathbf{A}$ if $\mathcal{C}(\mathbf{GA}) = \mathcal{C}(\mathbf{A})$.

A c-inverse of $\mathbf{A}$ is denoted by $\mathbf{A}_{\mathbf{c}}^{-}$.

It is clear that every $\mathbf{A}_{\chi}^{-}$ is a c-inverse of $\mathbf{A}$. The following example shows that the converse is not true. However, every reflexive c-inverse is an $\mathbf{A}_{\chi}^{-}$.

Example 6.4.2. Let $\mathbf{A} = \begin{pmatrix} 1 & 0 \\ 0 & 0 \end{pmatrix}$. Then it is easy to check that $\mathbf{A}_{\chi}^{-} = \begin{pmatrix} 1 & \alpha \\ 0 & 0 \end{pmatrix}$ and $\mathbf{A}_{\mathbf{c}}^{-} = \begin{pmatrix} 1 & \alpha \\ 0 & \beta \end{pmatrix}$, where β is possibly a non-null scalar. It is now clear that a c-inverse may not be a χ-inverse.

Definition 6.4.3. Let $\mathbf{A}$ be a matrix of order $n \times n$ and of *index* ≤ 1. Let $\mathbf{Ax} = \mathbf{b}$ be a possibly inconsistent system of linear equations with $\mathbf{b} = \mathbf{c} + \mathbf{d}$, where $\mathbf{c} \in \mathcal{C}(\mathbf{A})$ and $\mathbf{d} \in \mathcal{N}(\mathbf{A})$. (Since $\mathbf{A}$ is of index ≤ 1, we can write $F^n = \mathcal{C}(\mathbf{A}) \oplus \mathcal{N}(\mathbf{A})$, so each $\mathbf{b} \in \mathbf{F^n}$ can be decomposed in the manner mentioned above.) We say $\mathbf{x_0}$ is a good approximate solution of $\mathbf{Ax} = \mathbf{b}$ if $\mathbf{Ax_0} = \mathbf{c}$.

Notice that in the setup of Definition 6.4.3, if $\mathbf{b} \in \mathcal{C}(\mathbf{A})$, then $\mathbf{x_0}$ is a good approximate solution of $\mathbf{Ax} = \mathbf{b}$ if and only if $\mathbf{x_0}$ is a solution of $\mathbf{Ax} = \mathbf{b}$ or equivalently $\mathbf{x_0}$ is a solution of $\mathbf{A^2x} = \mathbf{Ab}$.

Definition 6.4.4. Let $\mathbf{A}$ be a square matrix of index ≤ 1. A matrix $\mathbf{G}$ is said to be an $\mathbf{A}_{\mathbf{a}}^{-}$ if $\mathbf{Gb}$ is a good approximate solution of $\mathbf{Ax} = \mathbf{b}$ for each $\mathbf{b}$.

For a square matrix of index ≤ 1, a matrix $\mathbf{G}$ is an $\mathbf{A}_{\mathbf{c}}^{-}$ if and only if $\mathbf{G^t}$ is an $(\mathbf{A^t})_{\mathbf{a}}^{-}$.

Definition 6.4.5. Let $\mathbf{A}$ and $\mathbf{B}$ be matrices of the same order and of index not greater than 1. We say $\mathbf{A} <_{\mathbf{c}}^{-} \mathbf{B}$ if $\mathbf{AA}_{\mathbf{c}}^{-} = \mathbf{BA}_{\mathbf{c}}^{-}$ and $\mathbf{A}_{\mathbf{c}}^{-}\mathbf{A} = \mathbf{A}_{\mathbf{c}}^{-}\mathbf{B}$ for some $\mathbf{A}_{\mathbf{c}}^{-} \in \{\mathbf{A}_{\mathbf{c}}^{-}\}$.

Similarly we can define $\mathbf{A} <_{\mathbf{a}}^{-} \mathbf{B}$.

Theorem 6.4.6. *Let* $\mathbf{A}$ *and* $\mathbf{B}$ *be matrices of the same order and of index not greater than 1. Then* $\mathbf{A} <^{\#} \mathbf{B}$ *if and only if* $\mathbf{A} <_{\mathbf{c}}^{-} \mathbf{B}$.

Proof. Notice that for any $\mathbf{A}_{\mathbf{c}}^{-} \in \{\mathbf{A}_{\mathbf{c}}^{-}\}$, the matrix $\mathbf{G} = \mathbf{A}_{\mathbf{c}}^{-}\mathbf{AA}_{\mathbf{c}}^{-}$ is a $\mathbf{A}_{\chi}^{-}$. The theorem now follows from Theorem 6.3.6. □

Corollary 6.4.7. *Let* $\mathbf{A}$ *and* $\mathbf{B}$ *be matrices of the same order and of index not greater than 1. Then* $\mathbf{A} \# < \mathbf{B}$ *if and only if* $\mathbf{A} <_{\mathbf{a}}^{-} \mathbf{B}$.

Theorem 6.4.8. *Let* $\mathbf{A}$ *and* $\mathbf{B}$ *be matrices of the same order and of index not greater than 1. Then the following are equivalent:*

(i) $\mathbf{A} < \# \mathbf{B}$
(ii) $\{\mathbf{B}_{\mathbf{c}}^{-}\} \subseteq \{\mathbf{A}_{\mathbf{c}}^{-}\}$.

Proof. (i) $\Rightarrow$ (ii)
Since $\mathbf{A} < \# \mathbf{B}$, by Theorem 6.3.3 there exists a non-singular matrix $\mathbf{P}$ such that $\mathbf{A} = \mathbf{P}\begin{pmatrix} \mathbf{S} & \mathbf{0} & \mathbf{0} \\ \mathbf{0} & \mathbf{0} & \mathbf{0} \\ \mathbf{0} & \mathbf{0} & \mathbf{0} \end{pmatrix}$ and $\mathbf{B} = \mathbf{P}\begin{pmatrix} \mathbf{S} & \mathbf{T}_{12} & \mathbf{0} \\ \mathbf{0} & \mathbf{T}_{22} & \mathbf{0} \\ \mathbf{0} & \mathbf{0} & \mathbf{0} \end{pmatrix}\mathbf{P}^{-1}$, where $\mathbf{S}$ and $\mathbf{T}_{22}$ are non-singular. Also, $\mathbf{A} < \# \mathbf{B} \Rightarrow \mathbf{A} <^{-} \mathbf{B}$ and $\{\mathbf{B}_{\mathbf{c}}^{-}\} \subseteq \{\mathbf{B}^{-}\} \subseteq \{\mathbf{A}^{-}\}$. To show that $\{\mathbf{B}_{\mathbf{c}}^{-}\} \subseteq \{\mathbf{A}_{\mathbf{c}}^{-}\}$, we show each $\mathbf{B}_{\mathbf{c}}^{-}$ is an $\mathbf{A}_{\mathbf{c}}^{-}$. Any $\mathbf{B}_{\mathbf{c}}^{-}$ is of the form $\mathbf{P}\begin{pmatrix} \mathbf{S}^{-1} & -\mathbf{S}^{-1}\mathbf{T}_{12}\mathbf{T}_{22}{}^{-1} & \mathbf{L}_1 \\ \mathbf{0} & \mathbf{0} & \mathbf{L}_2 \\ \mathbf{0} & \mathbf{0} & \mathbf{L}_3 \end{pmatrix}\mathbf{P}^{-1}$. Clearly, $\mathbf{B}_{\mathbf{c}}^{-}\mathbf{A}^{2} = \mathbf{A}$. So, $\mathbf{B}_{\mathbf{c}}^{-} \in \{\mathbf{A}_{\mathbf{c}}^{-}\}$.
(ii) $\Rightarrow$ (i)
Since $\{\mathbf{B}_{\chi}^{-}\} \subseteq \{\mathbf{B}_{\mathbf{c}}^{-}\}$, it follows that $\{\mathbf{B}_{\chi}^{-}\} \subseteq \{\mathbf{A}_{\mathbf{c}}^{-}\}$. Since $\mathbf{B}$ is of *index* 1, there exist non-singular matrices $\mathbf{P}$ and $\mathbf{T}$ such that $\mathbf{B} = \mathbf{P}\begin{pmatrix} \mathbf{T} & \mathbf{0} \\ \mathbf{0} & \mathbf{0} \end{pmatrix}\mathbf{P}^{-1}$. By Theorem 2.4.5 any $\mathbf{B}_{\chi}^{-}$ is of the form $\mathbf{P}\begin{pmatrix} \mathbf{T}^{-1} & \mathbf{L} \\ \mathbf{0} & \mathbf{0} \end{pmatrix}\mathbf{P}^{-1}$, where $\mathbf{L}$ is arbitrary. Let $\mathbf{A} = \mathbf{P}\begin{pmatrix} \mathbf{A}_{11} & \mathbf{A}_{12} \\ \mathbf{A}_{21} & \mathbf{A}_{22} \end{pmatrix}\mathbf{P}^{-1}$ partitioned in conformation with the partitioning of $\mathbf{B}$. Since a $\mathbf{B}_{\chi}^{-}$ is an $\mathbf{A}_{\mathbf{c}}^{-}$, so, $\mathbf{B}_{\chi}^{-}\mathbf{A}^{2} = \mathbf{A}$. Therefore, $\mathbf{P}\begin{pmatrix} \mathbf{T}^{-1} & \mathbf{L} \\ \mathbf{0} & \mathbf{0} \end{pmatrix}\mathbf{P}^{-1}\mathbf{P}\begin{pmatrix} \mathbf{A}_{11} & \mathbf{A}_{12} \\ \mathbf{A}_{21} & \mathbf{A}_{22} \end{pmatrix}^{2}\mathbf{P}^{-1} = \mathbf{P}\begin{pmatrix} \mathbf{A}_{11} & \mathbf{A}_{12} \\ \mathbf{A}_{21} & \mathbf{A}_{22} \end{pmatrix}\mathbf{P}^{-1}$ for each $\mathbf{L}$ and in particular for $\mathbf{L} = \mathbf{0}$.
Therefore,

$$\mathbf{P}\begin{pmatrix} \mathbf{T}^{-1}\mathbf{A}_{11}^{2} + \mathbf{T}^{-1}\mathbf{A}_{12}\mathbf{A}_{21} & \mathbf{T}^{-1}\mathbf{A}_{12}\mathbf{A}_{21} + \mathbf{T}^{-1}\mathbf{A}_{12}\mathbf{A}_{22} \\ \mathbf{0} & \mathbf{0} \end{pmatrix}\mathbf{P}^{-1}$$

$$= \mathbf{P}\begin{pmatrix} \mathbf{A}_{11} & \mathbf{A}_{12} \\ \mathbf{A}_{21} & \mathbf{A}_{22} \end{pmatrix}\mathbf{P}^{-1}$$

$\Rightarrow$ $\mathbf{A}_{12} = \mathbf{0}, \mathbf{A}_{21} = \mathbf{0}, \mathbf{A}_{22} = \mathbf{0}$ and $\mathbf{T}^{-1}\mathbf{A}_{11}^2 = \mathbf{A}_{11}$.

So, $\mathbf{A} = \mathbf{P}\begin{pmatrix}\mathbf{A}_{11} & \mathbf{0}\\ \mathbf{0} & \mathbf{0}\end{pmatrix}\mathbf{P}^{-1}$. As $\mathbf{A}$ is of index 1, $\mathbf{A}_{11}$ is of *index* 1. Now, $\mathbf{A}_{11}$ is of index 1, so, there exists a non-singular matrix $\mathbf{R}$ and a non-singular matrix $\mathbf{C}$ such that $\mathbf{A}_{11} = \mathbf{R}\begin{pmatrix}\mathbf{C} & \mathbf{0}\\ \mathbf{0} & \mathbf{0}\end{pmatrix}\mathbf{R}^{-1}$. Let $\mathbf{Q} = \mathbf{P}\begin{pmatrix}\mathbf{R} & \mathbf{0}\\ \mathbf{0} & \mathbf{I}\end{pmatrix}$.

Then $\mathbf{A} = \mathbf{Q}\begin{pmatrix}\mathbf{C} & \mathbf{0} & \mathbf{0}\\ \mathbf{0} & \mathbf{0} & \mathbf{0}\\ \mathbf{0} & \mathbf{0} & \mathbf{0}\end{pmatrix}\mathbf{Q}^{-1}$ and $\mathbf{B} = \mathbf{Q}\begin{pmatrix}\mathbf{T}_{11} & \mathbf{T}_{12} & \mathbf{0}\\ \mathbf{T}_{21} & \mathbf{T}_{22} & \mathbf{0}\\ \mathbf{0} & \mathbf{0} & \mathbf{0}\end{pmatrix}\mathbf{Q}^{-1}$, where $\mathbf{RTR}^{-1} = \begin{pmatrix}\mathbf{T}_{11} & \mathbf{T}_{12}\\ \mathbf{T}_{21} & \mathbf{T}_{22}\end{pmatrix}$ and partitioning is in conformation with partitioning of $\mathbf{A}_{11}$. Now, $\mathbf{RTR}^{-1}$ is non-singular, let $(\mathbf{RTR}^{-1})^{-1} = \begin{pmatrix}\mathbf{T}^{11} & \mathbf{T}^{12}\\ \mathbf{T}^{21} & \mathbf{T}^{22}\end{pmatrix}$.

Then any $\mathbf{B}_{\chi}^{-}$ is of the form $\mathbf{P}\begin{pmatrix}\mathbf{T}^{11} & \mathbf{T}^{12} & \mathbf{L}_1\\ \mathbf{T}^{21} & \mathbf{T}^{22} & \mathbf{L}_2\\ \mathbf{0} & \mathbf{0} & \mathbf{0}\end{pmatrix}\mathbf{P}^{-1}$. Using the fact that a $\mathbf{B}_{\chi}^{-}$ is an $\mathbf{A}_{\mathbf{c}}^{-}$, we get $\mathbf{T}^{21} = \mathbf{0}$ and $\mathbf{T}^{11} = \mathbf{C}^{-1}$, so, $\mathbf{T}_{11} = \mathbf{C}, \mathbf{T}_{21} = \mathbf{0}$ and $\mathbf{T}_{22}$ is non-singular. Now, it is clear that $\mathbf{A}^2 = \mathbf{BA}$ and $\mathcal{C}(\mathbf{A}^t) \subseteq \mathcal{C}(\mathbf{B}^t)$. Thus, $\mathbf{A} < \# \mathbf{B}$. □

The following theorem can be proved analogously:

Theorem 6.4.9. *Let* $\mathbf{A}$ *and* $\mathbf{B}$ *be matrices of the same order and of index not greater than* 1. *Then the following are equivalent:*

(i) $\mathbf{A} \# < \mathbf{B}$
(ii) $\{\mathbf{B}_{\mathbf{a}}^{-}\} \subseteq \{\mathbf{A}_{\mathbf{a}}^{-}\}$.

Theorem 6.4.10. *Let* $\mathbf{A}$ *and* $\mathbf{B}$ *be matrices of the same order and index not greater than* 1. *Then the following are equivalent:*

(i) $\mathbf{A} <^{\#} \mathbf{B}$
(ii) $\{\mathbf{B}_{\mathbf{c}}^{-}\} \subseteq \{\mathbf{A}_{\mathbf{c}}^{-}\}$ *and* $\{\mathbf{B}_{\mathbf{a}}^{-}\} \subseteq \{\mathbf{A}_{\mathbf{a}}^{-}\}$
(iii) $\{\mathbf{B}_{\chi}^{-}\} \subseteq \{\mathbf{A}_{\mathbf{c}}^{-}\}$ *and* $\{\mathbf{B}_{\rho}^{-}\} \subseteq \{\mathbf{A}_{\mathbf{a}}^{-}\}$ *and*
(iv) $\mathbf{B}^{\#} \in \{\mathbf{A}_{\mathbf{ac}}^{-}\}$.

Proof. We only need to prove (i) $\Leftrightarrow$ (iv).
Since $\mathbf{B}^{\#} \in \{\mathbf{A}_{\mathbf{ac}}^{-}\} \Leftrightarrow \mathbf{B}^{\#} \in \{\mathbf{B}_{\mathbf{c}}^{-}\} \cap \{\mathbf{B}_{\mathbf{a}}^{-}\} \subseteq \{\mathbf{A}_{\mathbf{c}}^{-}\} \cap \{\mathbf{A}_{\mathbf{a}}^{-}\}$, the proof of (i) $\Rightarrow$ (iv) is clear.
(iv) $\Rightarrow$ (i)

it is easy to check that the class of all commuting g-inverses of $\mathbf{A}$ is precisely $\{\mathbf{A}_{\mathbf{ac}}^{-}\}$. Now, (iv) $\Rightarrow$ (i) follows from Theorem 4.2.8. $\square$

6.5 One-sided star order

The one-sided star orders maintain a similar relationship to one-sided sharp orders as did the star order to the sharp order. For hermitian matrices the right (left) sharp order and the right (left) star coincide. So, as expected, many theorems in this section will be similar to theorems in one-sided sharp order. However, a distinguishing feature is that there is no Fisher-Cochran type theorem for one-sided sharp orders, but for one-sided star order we show that it holds good. The important feature of this section is that: if $\mathbf{A}$ and $\mathbf{B}$ are matrices of the same order and $\mathbf{A}$ is below $\mathbf{B}$ under either the right star order or the left star order, then their reduced form (a canonical form) is generalized singular value decomposition of $\mathbf{A}$ with respect to $\mathbf{B}$. In this section all matrices are over the field $\mathbb{C}$ of the complex numbers.

Let us define for matrices $\mathbf{A}$ and $\mathbf{B}$ of the same order, $\mathbf{A} \prec \mathbf{B}$ if $\mathbf{AA}^\star = \mathbf{BA}^\star$ (or equivalently $\mathbf{AA}^\dagger = \mathbf{BA}^\dagger$). This relation is obviously reflexive. Moreover, if $\mathbf{A} \prec \mathbf{B}$ and $\mathbf{B} \prec \mathbf{A}$, then $(\mathbf{B}-\mathbf{A})\mathbf{A}^\star = \mathbf{0}$ and $(\mathbf{B}-\mathbf{A})\mathbf{B}^\star = \mathbf{0}$. Therefore, $(\mathbf{B}-\mathbf{A})(\mathbf{B}-\mathbf{A})^\star = \mathbf{0}$ and so, $\mathbf{A} = \mathbf{B}$. Thus, $\mathbf{A} \prec \mathbf{B}$ is anti-symmetric. However, this relation is not transitive as the following example shows:

Example 6.5.1. Consider matrices $\mathbf{A} = \begin{pmatrix}1 \;, -1\end{pmatrix}$, $\mathbf{B} = \begin{pmatrix}2 \;, 0\end{pmatrix}$ and $\mathbf{C} = \begin{pmatrix}2 \;, 1\end{pmatrix}$. Then $\mathbf{AA}^\star = \mathbf{2} = \mathbf{BA}^\star$ and $\mathbf{BB}^\star = \mathbf{4} = \mathbf{BC}^\star$, but $\mathbf{AA}^\star = \mathbf{2} \neq \mathbf{1} = \mathbf{AC}^\star$. Thus, the relation '$\prec$' is not transitive and so, is not even a pre-order.

Definition 6.5.2. Let $\mathbf{A}$ and $\mathbf{B}$ be matrices of same order. We define $\mathbf{A} < \star\, \mathbf{B}$ (read as $\mathbf{A}$ is below $\mathbf{B}$ under right star order) if $\mathbf{AA}^\star = \mathbf{BA}^\star$ and $\mathcal{C}(\mathbf{A}^\star) \subseteq \mathcal{C}(\mathbf{B}^\star)$ and $\mathbf{A} \star < \mathbf{B}$ (read as $\mathbf{A}$ is below $\mathbf{B}$ under left star order) if $\mathbf{A}^\star\mathbf{A} = \mathbf{A}^\star\mathbf{B}$ and $\mathcal{C}(\mathbf{A}) \subseteq \mathcal{C}(\mathbf{B})$.

Are the two relations defined above the same and the same as star order? Does each of them define a partial order? Or even a pre-order?

We give below an example which not only shows that none of them is same as the star order and that they themselves are different order relations.

Example 6.5.3. Let $\mathbf{A} = \begin{pmatrix} 1 & -1 \\ -1 & 1 \end{pmatrix}$ and $\mathbf{B} = \begin{pmatrix} 1 & -2 \\ -1 & 0 \end{pmatrix}$. Then clearly,

$\mathcal{C}(\mathbf{A}) \subseteq \mathcal{C}(\mathbf{B})$ and $\mathbf{A}^\star\mathbf{A} = \mathbf{A}^\star\mathbf{B} = \begin{pmatrix} 2 & -2 \\ -2 & 2 \end{pmatrix}$. However, $\mathbf{AA}^\star = \mathbf{A}^\star\mathbf{A}$ and $\mathbf{BA}^\star = \begin{pmatrix} 3 & -3 \\ -1 & 1 \end{pmatrix}$. Thus, $\mathbf{A} \star < \mathbf{B}$, $\mathbf{A} \not< \star\mathbf{B}$ and therefore, $\mathbf{A} \not<^\star \mathbf{B}$.

Remark 6.5.4. Notice that $\mathbf{A} < \star\, \mathbf{B} \Leftrightarrow \mathbf{AA}^\dagger = \mathbf{BA}^\dagger$ and $\mathcal{C}(\mathbf{A}^\star) \subseteq \mathcal{C}(\mathbf{B}^\star)$, since, $\mathbf{A}^\dagger = \mathbf{A}^\star(\mathbf{A}^\star\mathbf{AA}^\star)^-\mathbf{A}^\star$. Thus, $\mathbf{A} < \star\, \mathbf{B} \Rightarrow \mathbf{AA}^\dagger = \mathbf{BA}^\dagger$, and $\mathcal{C}(\mathbf{A}^\star) \subseteq \mathcal{C}(\mathbf{B}^\star)$ and this implies $\mathbf{A} <^- \mathbf{B}$, $\mathcal{C}(\mathbf{A}^\star) \subseteq \mathcal{C}(\mathbf{B}^\star)$, $\mathcal{C}(\mathbf{A}) \subseteq \mathcal{C}(\mathbf{B})$ and $\mathcal{C}(\mathbf{A}^\star) \perp \mathcal{C}(\mathbf{B}-\mathbf{A})^\star$. This is so because $\mathbf{AA}^\dagger = \mathbf{BA}^\dagger \Rightarrow \mathbf{A} = \mathbf{BA}^\dagger\mathbf{A}$ and therefore, $\mathbf{AB}^\dagger\mathbf{A} = \mathbf{AB}^\dagger\mathbf{BA}^\dagger\mathbf{A} = \mathbf{A}$; thus, $\mathbf{AB}^\dagger\mathbf{A} = \mathbf{A}$ and so, $\mathbf{A} <^- \mathbf{B}$. Further, $\mathbf{A}(\mathbf{B}^\star - \mathbf{A}^\star) = \mathbf{0}$.

Remark 6.5.5. $\mathbf{A} < \star\, \mathbf{B} \Rightarrow \mathbf{A} <^- \mathbf{B}$ and $\mathbf{A} \star < \mathbf{B} \Rightarrow \mathbf{A} <^- \mathbf{B}$.

We shall now first obtain a canonical form for matrices $\mathbf{A}$ and $\mathbf{B}$ when $\mathbf{A} < \star\, \mathbf{B}$.

Theorem 6.5.6. *Let* $\mathbf{A}$ *and* $\mathbf{B}$ *be matrices of the same order with ranks 'r' and 's' respectively. Then* $\mathbf{A} < \star\, \mathbf{B}$ *if and only if there exist a non-singular matrix* $\mathbf{P}$ *and a unitary matrix* $\mathbf{V}$ *such that* $\mathbf{A} = \mathbf{P}\mathbf{diag}\left(\mathbf{I_r}\ , \mathbf{0}\ , \mathbf{0}\right)\mathbf{V}^\star$ *and* $\mathbf{B} = \mathbf{P}\mathbf{diag}\left(\mathbf{I_r}, \mathbf{I_{r-s}}, \mathbf{0}\right)\mathbf{V}^\star$.

Proof. 'If' part is trivial.
'Only if' part
Let $\mathbf{A} < \star\, \mathbf{B}$. Then by Remark 6.5.4, $\mathbf{A} <^- \mathbf{B}$, $\mathcal{C}(\mathbf{A}^\star) \subseteq \mathcal{C}(\mathbf{B}^\star)$, $\mathcal{C}(\mathbf{A}) \subseteq \mathcal{C}(\mathbf{B})$ and $\mathcal{C}(\mathbf{A}^\star) \perp \mathcal{C}(\mathbf{B}-\mathbf{A})^\star$. Let $(\mathbf{P_1}, \mathbf{V_1^\star})$ and $(\mathbf{P_2}, \mathbf{V_2^\star})$ be rank factorizations of $\mathbf{A}$ and $\mathbf{B}-\mathbf{A}$ respectively, where $\mathbf{V_1^\star V_1} = \mathbf{I}$ and $\mathbf{V_2^\star V_2} = \mathbf{I}$. Then $\mathbf{B} = \mathbf{A} + (\mathbf{B}-\mathbf{A}) = \mathbf{P_1V_1^\star} + \mathbf{P_2V_2^\star} = (\mathbf{P_1} : \mathbf{P_2})(\mathbf{V_1} : \mathbf{V_2})^\star$. Since $\mathcal{C}(\mathbf{A}) \cap \mathcal{C}(\mathbf{B}-\mathbf{A}) = \{\mathbf{0}\}$, $(\mathbf{P_1} : \mathbf{P_2})$ is of full column rank 's'. Also, since $\mathcal{C}(\mathbf{A}^\star) \perp \mathcal{C}(\mathbf{B}-\mathbf{A})^\star$, so,$(\mathbf{V_1} : \mathbf{V_2})^\star(\mathbf{V_1} : \mathbf{V_2}) = \mathbf{I}$. Let $\mathbf{P_3}$ and $\mathbf{V_3^\star}$ be matrices such that $\mathbf{P} = (\mathbf{P_1} : \mathbf{P_2} : \mathbf{P_3})$ is non-singular and $\mathbf{V}^\star = (\mathbf{V_1} : \mathbf{V_2} : \mathbf{V_3})^\star$ is unitary. It is now easy to check that $\mathbf{A}$ and $\mathbf{B}$ have the desired forms.□

Remark 6.5.7. The canonical form of the matrices $\mathbf{A}$ and $\mathbf{B}$ in Theorem 6.5.6 is called a generalized singular value decomposition of $\mathbf{A}$ and $\mathbf{B}$.

Remark 6.5.8. Let $\mathbf{A}$ and $\mathbf{B}$ be matrices of the same order with ranks 'r' and 's' respectively. Then $\mathbf{A} \star < \mathbf{B}$ if and only if $\mathbf{A} = \mathbf{U}\mathbf{diag}\left(\mathbf{I_r}\ , \mathbf{0}\ , \mathbf{0}\right)\mathbf{Q}$ and $\mathbf{B} = \mathbf{U}\mathbf{diag}\left(\mathbf{I_r}\ , \mathbf{I_{r-s}}\ , \mathbf{0}\right)\mathbf{Q}$, for some non-singular matrix $\mathbf{Q}$ and unitary matrix $\mathbf{U}$.

We shall now obtain a characterization of $\mathbf{A}$ such that $\mathbf{A} < \star \mathbf{B}$ when a singular value decomposition of $\mathbf{B}$ is available.

Theorem 6.5.9. *Let* $\mathbf{A}$ *and* $\mathbf{B}$ *be matrices of the same order. Let* $\mathbf{B} = \mathbf{U}\mathbf{diag}\left(\mathbf{D}\ ,\ \mathbf{0}\right)\mathbf{V}^\star$, *where* $\mathbf{U}$, $\mathbf{V}$ *are unitary and* $\mathbf{D}$ *is a positive definite diagonal matrix. Then* $\mathbf{A} < \star \mathbf{B}$ *if and only if* $\mathbf{A} = \mathbf{U}\mathbf{diag}\left(\mathbf{T}\ ,\ \mathbf{0}\right)\mathbf{V}^\star$ *such that* $\mathbf{D}^{-1}$ *is a minimum norm g-inverse of* $\mathbf{T}$.

Proof. 'If' part is trivial.
'Only if' part
Let $\mathbf{B} = \mathbf{U}\mathbf{diag}\left(\mathbf{D}\ ,\ \mathbf{0}\right)\mathbf{V}^\star$. Write $\mathbf{A} = \mathbf{U}\begin{pmatrix}\mathbf{T} & \mathbf{E}\\ \mathbf{F} & \mathbf{G}\end{pmatrix}\mathbf{V}^\star$ where partitioning is in conformation with the partitioning of $\mathbf{B}$. Since $\mathcal{C}(\mathbf{A}^\star) \subseteq \mathcal{C}(\mathbf{B}^\star)$, we have $\mathbf{E} = \mathbf{0}$ and $\mathbf{G} = \mathbf{0}$. Also, $\mathbf{AA}^\star = \mathbf{BA}^\star$. So, $\mathbf{TT}^\star = \mathbf{DT}^\star$ and $\mathbf{FF}^\star = \mathbf{0}$. This further gives $\mathbf{D}^{-1}\mathbf{TT}^\star = \mathbf{T}^\star$ and $\mathbf{F} = \mathbf{0}$. Hence, $\mathbf{D}^{-1}$ is a minimum norm g-inverse of $\mathbf{T}$, showing that $\mathbf{A}$ has the desired form. □

Theorem 6.5.10. *Let* $\mathbf{A}$ *and* $\mathbf{B}$ *be matrices of the same order. Let* $\mathbf{A} = \mathbf{U}\mathbf{diag}\left(\mathbf{\Delta}\ ,\ \mathbf{0}\right)\mathbf{V}^\star$, *where* $\mathbf{U}$, $\mathbf{V}$ *are unitary and* $\mathbf{\Delta}$ *is a positive definite diagonal matrix. Then* $\mathbf{A} < \star \mathbf{B}$ *if and only if* $\mathbf{B} = \mathbf{U}\begin{pmatrix}\mathbf{\Delta} & \mathbf{T}_{12}\\ \mathbf{0} & \mathbf{T}_{22}\end{pmatrix}\mathbf{V}^\star$ *for some matrices* $\mathbf{T}_{12}$ *and* $\mathbf{T}_{22}$ *such that* $\mathcal{C}(\mathbf{T}_{12}) \subseteq \mathcal{C}(\mathbf{T}_{22})$.

Proof. Proof is similar to Theorem 6.5.9. □

Theorem 6.5.11. *The relation '$< \star$' is a partial order.*

Proof. The relation is trivially reflexive and as noted earlier it is also anti-symmetric. So, we only need to show that it is transitive. Since $\mathbf{A} < \star \mathbf{B} \Rightarrow \mathbf{AA}^\star = \mathbf{BA}^\star$ and $\mathcal{C}(\mathbf{A}^\star) \subseteq \mathcal{C}(\mathbf{B}^\star)$. Also, $\mathbf{B} < \star \mathbf{C} \Rightarrow \mathbf{BB}^\star = \mathbf{CB}^\star$ and $\mathcal{C}(\mathbf{B}^\star) \subseteq \mathcal{C}(\mathbf{C}^\star)$. Clearly, $\mathcal{C}(\mathbf{A}^\star) \subseteq \mathcal{C}(\mathbf{B}^\star)$ and $\mathcal{C}(\mathbf{B}^\star) \subseteq \mathcal{C}(\mathbf{C}^\star) \Rightarrow \mathcal{C}(\mathbf{A}^\star) \subseteq \mathcal{C}(\mathbf{C}^\star)$. Moreover, $\mathbf{AA}^\star = \mathbf{BA}^\star = \mathbf{AB}^\star = \mathbf{AB}^\star\mathbf{B}^{\dagger\star}\mathbf{B}^\star = \mathbf{AB}^\dagger\mathbf{BB}^\star = \mathbf{AB}^\dagger\mathbf{BC}^\star = \mathbf{AC}^\star = \mathbf{CA}^\star$. Therefore, $\mathbf{A} < \star \mathbf{C}$. □

Remark 6.5.12. The relation '$\star <$' is a partial order.

Remark 6.5.13. Notice that $\mathbf{A} < \star\mathbf{B} \Leftrightarrow \mathbf{A}^\star\star < \mathbf{B}^\star$

Theorem 6.5.14. *Let* $\mathbf{A}$ *and* $\mathbf{B}$ *be matrices of the same order. Then* $\mathbf{A} <^\star \mathbf{B}$ *if and only if* $\mathbf{A} < \star\mathbf{B}$ *and* $\mathbf{A} \star < \mathbf{B}$.

Theorem 6.5.15. *Let* $\mathbf{A}$ *and* $\mathbf{B}$ *be matrices of the same order such that* $\mathbf{AA}^\star = \mathbf{BA}^\star$. *Then the following are equivalent:*

(i) $\mathbf{A} < \star\, \mathbf{B}$
(ii) $\mathcal{C}(\mathbf{A}^\star) \subseteq \mathcal{C}(\mathbf{B}^\star)$
(iii) $\mathcal{C}(\mathbf{A}) \cap \mathcal{C}(\mathbf{B}-\mathbf{A}) = \{\mathbf{0}\}$
(iv) $\mathbf{A}^-\mathbf{A} = \mathbf{A}^-\mathbf{B}$ *for some g-inverse* $\mathbf{A}^-$ *of* $\mathbf{A}$ *and*
(v) $\mathbf{A} = \mathbf{PB}$ *for some projector* $\mathbf{P}$.

Proof follows along the lines of Theorem 6.3.16.
The following theorem follows trivially from Remark 6.5.13 and Theorem 6.5.15.

Theorem 6.5.16. *Let* $\mathbf{A}$ *and* $\mathbf{B}$ *be matrices of the same order such that* $\mathbf{A}^\star\mathbf{A} = \mathbf{A}^\star\mathbf{B}$. *Then the following are equivalent:*

(i) $\mathbf{A} \star\!< \mathbf{B}$
(ii) $\mathcal{C}(\mathbf{A}) \subseteq \mathcal{C}(\mathbf{B})$
(iii) $\mathcal{C}(\mathbf{A}^\star) \cap \mathcal{C}(\mathbf{B}-\mathbf{A})^\star = \{\mathbf{0}\}$
(iv) $\mathbf{AA}^- = \mathbf{BA}^-$ *for some g-inverse bf*A^- *of* $\mathbf{A}$ *and*
(v) $\mathbf{A} = \mathbf{BQ}$ *for some projector* $\mathbf{Q}$.

In Theorem 5.2.8, we saw that $\mathbf{A} <^\star \mathbf{B} \Leftrightarrow \{\mathbf{B}_\ell^-\} \subseteq \{\mathbf{A}_\ell^-\}$ and $\{\mathbf{B}_\mathbf{m}^-\} \subseteq \{\mathbf{A}_\mathbf{m}^-\} \Leftrightarrow \{\mathbf{B}_{\ell\mathbf{m}}^-\} \subseteq \{\mathbf{A}_{\ell\mathbf{m}}^-\} \Leftrightarrow \mathbf{B}^\dagger \in \{\mathbf{A}_{\ell\mathbf{m}}^-\}$. We shall now examine whether similar statements are true for $\mathbf{A} < \star\, \mathbf{B}$ and $\mathbf{A} \star\!< \mathbf{B}$.

Theorem 6.5.17. *Let* $\mathbf{A}$ *and* $\mathbf{B}$ *be matrices of the same order. Then* $\mathbf{A} < \star\, \mathbf{B}$ *if and only if* $\{\mathbf{B}_\mathbf{m}^-\} \subseteq \{\mathbf{A}_\mathbf{m}^-\}$.

Proof. 'If' part
Let $\mathbf{B} = \mathbf{U}\mathrm{diag}\begin{pmatrix}\mathbf{D} & \mathbf{0}\end{pmatrix}\mathbf{V}^\star$ be a singular value decomposition of $\mathbf{B}$, where $\mathbf{U}$ and $\mathbf{V}$ are unitary and $\mathbf{D}$ is a positive definite diagonal matrix. Then every $\mathbf{B}_\mathbf{m}^-$ is of the form $\mathbf{V}\begin{pmatrix}\mathbf{D}^{-1} & \mathbf{M}\\ \mathbf{0} & \mathbf{N}\end{pmatrix}\mathbf{U}^\star$, where $\mathbf{M}$ and $\mathbf{N}$ are arbitrary.

Write $\mathbf{A} = \mathbf{U}\begin{pmatrix}\mathbf{T}_{11} & \mathbf{T}_{12}\\ \mathbf{T}_{21} & \mathbf{T}_{22}\end{pmatrix}\mathbf{V}^\star$. Since, $\{\mathbf{B}_\mathbf{m}^-\} \subseteq \{\mathbf{A}_\mathbf{m}^-\}$, we have

$$\mathbf{V}\begin{pmatrix}\mathbf{D}^{-1} & \mathbf{M}\\ \mathbf{0} & \mathbf{N}\end{pmatrix}\mathbf{U}^\star\mathbf{U}\begin{pmatrix}\mathbf{T}_{11} & \mathbf{T}_{12}\\ \mathbf{T}_{21} & \mathbf{T}_{22}\end{pmatrix}\mathbf{V}^\star\mathbf{V}\begin{pmatrix}\mathbf{T}_{11}^\star & \mathbf{T}_{12}^\star\\ \mathbf{T}_{21}^\star & \mathbf{T}_{22}^\star\end{pmatrix}\mathbf{V}^\star = \mathbf{V}\begin{pmatrix}\mathbf{T}_{11}^\star & \mathbf{T}_{12}^\star\\ \mathbf{T}_{21}^\star & \mathbf{T}_{22}^\star\end{pmatrix}\mathbf{V}^\star.$$

In particular, by taking $\mathbf{M} = \mathbf{0}$ and $\mathbf{N} = \mathbf{0}$ and simplifying we have

$$\begin{pmatrix}\mathbf{D}^{-1}\mathbf{T}_{11}\mathbf{T}_{11}^\star + \mathbf{D}^{-1}\mathbf{T}_{12}\mathbf{T}_{12}^\star & \mathbf{D}^{-1}\mathbf{T}_{11}\mathbf{T}_{21}^\star + \mathbf{D}^{-1}\mathbf{T}_{12}\mathbf{T}_{22}^\star\\ \mathbf{0} & \mathbf{0}\end{pmatrix} = \begin{pmatrix}\mathbf{T}_{11}^\star & \mathbf{T}_{12}^\star\\ \mathbf{T}_{21}^\star & \mathbf{T}_{22}^\star\end{pmatrix}.$$

So, $\mathbf{T}_{12}^\star = \mathbf{0}, \mathbf{T}_{22}^\star = \mathbf{0}$ and $\mathbf{D}^{-1}\mathbf{T}_{11}\mathbf{T}_{11}^\star = \mathbf{T}_{11}^\star$, and $\mathbf{D}^{-1}\mathbf{T}_{11}\mathbf{T}_{21}^\star = \mathbf{T}_{21}^\star$

Next, by taking $\mathbf{M} = \mathbf{0}$ and simplifying we have $\mathbf{N}\mathbf{T}_{21}\mathbf{T}_{21}^{\star} = \mathbf{0}$ for all $\mathbf{N}$. So, $\mathbf{T}_{21} = \mathbf{0}$. Hence, $\mathbf{A} = \mathbf{U}\mathbf{diag}\begin{pmatrix}\mathbf{T}_{11} & \mathbf{0}\end{pmatrix}\mathbf{V}^{\star}$ such that $\mathbf{D}^{-1}$ is a minimum norm g-inverse of $\mathbf{T}_{11}$. Therefore by Theorem 6.5.9, we have $\mathbf{A} < \star\, \mathbf{B}$.

'Only if' part

Let $\mathbf{A} < \star\, \mathbf{B}$. Let $\mathbf{B} = \mathbf{U}\mathbf{diag}\begin{pmatrix}\mathbf{D} & \mathbf{0}\end{pmatrix}\mathbf{V}^{\star}$ be a singular value decomposition of $\mathbf{B}$, where $\mathbf{U}$ and $\mathbf{V}$ are unitary and $\mathbf{D}$ is a positive definite diagonal matrix. Then by Theorem 6.5.9, $\mathbf{A} = \mathbf{U}\mathbf{diag}\begin{pmatrix}\mathbf{T} & \mathbf{0}\end{pmatrix}\mathbf{V}^{\star}$ such that $\mathbf{D}^{-1}\mathbf{T}\mathbf{T}^{\star} = \mathbf{T}^{\star}$. The class all $\mathbf{B}_{\mathbf{m}}^{-}$ is given by $\mathbf{V}\begin{pmatrix}\mathbf{D}^{-1} & \mathbf{M}\\ \mathbf{0} & \mathbf{N}\end{pmatrix}\mathbf{U}^{\star}$, where $\mathbf{M}$ and $\mathbf{N}$ are arbitrary. Now, $\mathbf{B}_{\mathbf{m}}^{-}\mathbf{A}\mathbf{A}^{\star} = \mathbf{V}\begin{pmatrix}\mathbf{D}^{-1} & \mathbf{M}\\ \mathbf{0} & \mathbf{N}\end{pmatrix}\mathbf{U}^{\star}\mathbf{U}\begin{pmatrix}\mathbf{T} & \mathbf{0}\\ \mathbf{0} & \mathbf{0}\end{pmatrix}\mathbf{V}^{\star}\mathbf{V}\begin{pmatrix}\mathbf{T}^{\star} & \mathbf{0}\\ \mathbf{0} & \mathbf{0}\end{pmatrix}\mathbf{U}^{\star} = \mathbf{V}\begin{pmatrix}\mathbf{T}^{\star} & \mathbf{0}\\ \mathbf{0} & \mathbf{0}\end{pmatrix}\mathbf{U}^{\star} = \mathbf{A}^{\star}$.
Hence, $\{\mathbf{B}_{\mathbf{m}}^{-}\} \subseteq \{\mathbf{A}_{\mathbf{m}}^{-}\}$. □

Remark 6.5.18. In fact, we can prove that $\mathbf{A} < \star\, \mathbf{B} \Leftrightarrow \{\mathbf{B}_{\mathbf{mr}}^{-}\} \subseteq \{\mathbf{A}_{\mathbf{m}}^{-}\}$. The proof follows along the lines of Theorem 6.5.17. However, $\mathbf{B}^{\dagger} \in \{\mathbf{A}_{\mathbf{m}}^{-}\}$ does not imply $\mathbf{A} < \star\, \mathbf{B}$ as the following example shows.

Example 6.5.19. Let $\mathbf{A} = \begin{pmatrix}1 & 0\\ 1 & 0\end{pmatrix}$ and $\mathbf{B} = \begin{pmatrix}1 & 0\\ 0 & 0\end{pmatrix}$. Then $\mathbf{B}^{\dagger} = \mathbf{B}$ and $\mathbf{B}^{\dagger}\mathbf{A}\mathbf{A}^{\star} = \mathbf{A}^{\star}$. So, $\mathbf{B}^{\dagger} \in \{\mathbf{A}_{\mathbf{m}}^{-}\}$. However, $\mathbf{A}\mathbf{A}^{\star} = \begin{pmatrix}1 & 1\\ 1 & 1\end{pmatrix}$ and $\mathbf{B}\mathbf{A}^{\star} = \begin{pmatrix}1 & 1\\ 0 & 0\end{pmatrix}$. So, $\mathbf{A}\mathbf{A}^{\star} \neq \mathbf{B}\mathbf{A}^{\star}$. Thus, $\mathbf{A} \not< \star\, \mathbf{B}$.

Lemma 6.5.20. *Let* $\mathbf{C}$ *be a non-singular matrix. Then* $\mathbf{C}$ *is a minimum norm g-inverse of a matrix* $\mathbf{H}$ *if and only if* $\mathbf{C}^{-1}$ *is a minimum norm g-inverse of* $\mathbf{H}^{\dagger}$.

Proof. $\mathbf{C}$ is a minimum norm g-inverse of a matrix $\mathbf{H} \Leftrightarrow \mathbf{C}\mathbf{H}\mathbf{H}^{\star} = \mathbf{H}^{\star}$ $\Leftrightarrow \mathbf{C}\mathbf{H}\mathbf{H}^{\dagger} = \mathbf{H}^{\dagger} \Leftrightarrow \mathbf{H}\mathbf{H}^{\dagger} = \mathbf{C}^{-1}\mathbf{H}^{\dagger} \Leftrightarrow \mathbf{H}^{\dagger\star} = \mathbf{C}^{-1}\mathbf{H}^{\dagger}\mathbf{H}^{\dagger\star} \Leftrightarrow \mathbf{C}^{-1} \in \{(\mathbf{H}^{\dagger})_{\mathbf{m}}^{-}\}$. □

Theorem 6.5.21. *Let* $\mathbf{A}$ *and* $\mathbf{B}$ *be matrices of the same order. Then the following are equivalent:*

(i) $\mathbf{A}\mathbf{A}^{\star} = \mathbf{B}\mathbf{A}^{\star}$
(ii) $\mathbf{A}\mathbf{A}^{\dagger} = \mathbf{B}\mathbf{A}^{\dagger}$
(iii) $\mathbf{A} = \mathbf{B}\mathbf{L}$, *where* $\mathbf{L}$ *is an orthogonal projector on* $\mathcal{C}(\mathbf{A}^{\star})$
(iv) $\mathbf{A} = \mathbf{B}\mathbf{L}$, *where* $\mathbf{L}$ *is an orthogonal projector having range a subspace of* $\mathcal{C}(\mathbf{B}^{\star})$ *and*

(v) $\mathbf{A} = \mathbf{BL}$, *where* $\mathbf{L}$ *is an orthogonal projector.*

Proof. (i) $\Rightarrow$ (ii)
$\mathbf{AA}^\star = \mathbf{BA}^\star \Rightarrow \mathbf{AA}^\star\mathbf{A}^{\dagger\star}\mathbf{A}^\dagger = \mathbf{BA}^\star\mathbf{A}^{\dagger\star}\mathbf{A}^\dagger \Rightarrow \mathbf{AA}^\dagger = \mathbf{BA}^\dagger$.
(ii)$\Rightarrow$(iii)
$\mathbf{AA}^\dagger = \mathbf{BA}^\dagger \Rightarrow \mathbf{A} = \mathbf{AA}^\dagger\mathbf{A} = \mathbf{BA}^\dagger\mathbf{A}$. Take $\mathbf{L} = \mathbf{A}^\dagger\mathbf{A}$.
(iii) $\Rightarrow$ (iv) $\Rightarrow$(v) is trivial.
(v)$\Rightarrow$ (i)
Let $\mathbf{A} = \mathbf{BL}$, where $\mathbf{L}$ is an orthogonal projector. Then $\mathbf{A}^\star = \mathbf{LB}^\star$. So, $\mathcal{C}(\mathbf{A}^\star) \subseteq \mathcal{C}(\mathbf{L})$. This gives $\mathbf{A}^\star = \mathbf{LA}^\star$. Therefore, $\mathbf{AA}^\star = (\mathbf{BL})\mathbf{A}^\star = \mathbf{BA}^\star$. □

Remark 6.5.22. Any one of the five equivalent conditions in Theorem 6.5.21 together with $\mathcal{C}(\mathbf{A}^\star) \subseteq \mathcal{C}(\mathbf{B}^\star)$ implies $\mathbf{A} <\star \mathbf{B}$.

Remark 6.5.23. None of the five equivalent conditions in Theorem 6.5.21 implies $\mathcal{C}(\mathbf{A}^\star) \subseteq \mathcal{C}(\mathbf{B}^\star)$. For take $\mathbf{A} = \begin{pmatrix} 1 & 0 \\ 1 & 0 \end{pmatrix}$ and $\mathbf{B} = \begin{pmatrix} 1 & 1 \\ 1 & 1 \end{pmatrix}$.

Remark 6.5.24. Compare Theorem 6.5.15 and Theorem 6.5.21.

Theorem 6.5.25. *Let* $\mathbf{A}$ *and* $\mathbf{B}$ *be matrices of the same order. Then the following are equivalent:*

(i) $\mathbf{A} <\star \mathbf{B}$
(ii) $\mathbf{A} <^- \mathbf{B}$ *and* $\mathbf{BA}^\star$ *is hermitian*
(iii) $\mathbf{A} <^- \mathbf{B}$ *and* $\mathbf{BA}^\dagger$ *is hermitian* *and*
(iv) $\mathbf{A} <^- \mathbf{B}$ *and* $\mathbf{B}^\dagger\mathbf{A}$ *is hermitian.*

Proof. (i) $\Rightarrow$ (ii)
Proof is trivial in view of Remark 6.5.5 and definition of $\mathbf{A} <\star \mathbf{B}$.
(ii) $\Rightarrow$ (i) and (iv) $\Rightarrow$ (i)
Let $\mathbf{B} = \mathbf{U}\begin{pmatrix} \mathbf{D} & \mathbf{0} \\ \mathbf{0} & \mathbf{0} \end{pmatrix}\mathbf{V}^\star$ be a singular value decomposition of $\mathbf{B}$, where $\mathbf{U}$ and $\mathbf{V}$ are unitary and $\mathbf{D}$ is a positive definite diagonal matrix. In view of Theorem 6.5.9, the proof will be complete if we show that each of (ii) and (iv) implies that $\mathbf{A}$ is of the form $\mathbf{U}\begin{pmatrix} \mathbf{T} & \mathbf{0} \\ \mathbf{0} & \mathbf{0} \end{pmatrix}\mathbf{V}^\star$, where $\mathbf{D}^{-1}$ is a minimum norm g-inverse of $\mathbf{T}$. Let $\mathbf{A} = \mathbf{U}\begin{pmatrix} \mathbf{T} & \mathbf{E} \\ \mathbf{F} & \mathbf{G} \end{pmatrix}\mathbf{V}^\star$ partitioned in conformity with the partition of $\mathbf{B}$. Then $\mathbf{BA}^\star$ is hermitian $\Rightarrow$ $\mathbf{F} = \mathbf{0}$ and $\mathbf{A} <^- \mathbf{B} \Rightarrow \mathbf{G} = \mathbf{0}$ and $\mathbf{D}^{-1}$ is a g-inverse of $\mathbf{T}$ and $\mathbf{DT}$ is hermitian.

Since, $\mathcal{C}(\mathbf{A}^\star) \subseteq \mathcal{C}(\mathbf{B}^\star)$, we must have $\mathbf{E} = \mathbf{0}$. So, $\mathbf{D}^{-1}$ is a minimum norm g-inverse of $\mathbf{T}$.
Similarly, when $\mathbf{B}^\dagger\mathbf{A}$ is hermitian, $\mathbf{E} = \mathbf{0}$ and using $\mathcal{C}(\mathbf{A}) \subseteq \mathcal{C}(\mathbf{B})$, we have $\mathbf{F} = \mathbf{0}$ (i) $\Rightarrow$ (iii)
Proof is trivial in view of Remark 6.5.5 and definition of $\mathbf{A} < \star\, \mathbf{B}$.
(iii) $\Rightarrow$ (i)
Let $\mathbf{A} = \mathbf{U}\mathbf{diag}(\boldsymbol{\Delta}\ \mathbf{0})\mathbf{V}^\star$ be a singular value decomposition of $\mathbf{A}$. Then it is easy to show that $\mathbf{B} = \mathbf{U}\begin{pmatrix}\boldsymbol{\Delta} & \mathbf{T_{12}} \\ \mathbf{0} & \mathbf{T_{22}}\end{pmatrix}\mathbf{V}^\star$ for some matrices $\mathbf{T_{12}}$ and $\mathbf{T_{22}}$ such that $\mathcal{C}(\mathbf{T_{12}}^\star) \subseteq \mathcal{C}(\mathbf{T_{22}}^\star)$. So, (i) follows by Theorem 6.5.10. □

Theorem 6.5.26. *Let* $\mathbf{A}$ *and* $\mathbf{B}$ *be matrices of the same order. Then the following are equivalent:*

(i) $\mathbf{A} <^\star \mathbf{B}$
(ii) $\mathbf{A} < \star\, \mathbf{B}$ *and* $\mathbf{A}^\star\mathbf{B}$ *is hermitian*
(iii) $\mathbf{A} < \star\, \mathbf{B}$ *and* $\mathbf{A}^\dagger\mathbf{B}$ *is hermitian* *and*
(iv) $\mathbf{A} < \star\, \mathbf{B}$ *and* $\mathbf{A}\mathbf{B}^\dagger$ *is hermitian.*

Definition 6.5.27. Let $\mathbf{A}$ and $\mathbf{B}$ be matrices of same order. We say $\mathbf{A} <_\ell^- \mathbf{B}$ if $\mathbf{A}\mathbf{A}_\ell^- = \mathbf{B}\mathbf{A}_\ell^-$ and $\mathbf{A}_\ell^-\mathbf{A} = \mathbf{A}_\ell^-\mathbf{B}$ for some $\mathbf{A}_\ell^- \in \{\mathbf{A}_\ell^-\}$.
Also, $\mathbf{A} <_\mathbf{m}^- \mathbf{B}$ if $\mathbf{A}\mathbf{A}_\mathbf{m}^- = \mathbf{B}\mathbf{A}_\mathbf{m}^-$ and $\mathbf{A}_\mathbf{m}^-\mathbf{A} = \mathbf{A}_\mathbf{m}^-\mathbf{B}$ for some $\mathbf{A}_\mathbf{m}^- \in \{\mathbf{A}_\mathbf{m}^-\}$.

Theorem 6.5.28. *Let* $\mathbf{A}$ *and* $\mathbf{B}$ *be matrices of the same order. Then* $\mathbf{A} < \star\, \mathbf{B}$ *if and only if* $\mathbf{A} <_\mathbf{m}^- \mathbf{B}$.

Proof. 'If' part
Now, $\mathbf{A} <_\mathbf{m}^- \mathbf{B} \Rightarrow \mathbf{A}\mathbf{A}_\mathbf{m}^- = \mathbf{B}\mathbf{A}_\mathbf{m}^-$ for some $\mathbf{A}_\mathbf{m}^- \in \{\mathbf{A}_\mathbf{m}^-\}$. So, $\mathbf{A}\mathbf{A}^\star = \mathbf{A}\mathbf{A}_\mathbf{m}^-\mathbf{A}\mathbf{A}^\star = \mathbf{B}\mathbf{A}_\mathbf{m}^-\mathbf{A}\mathbf{A}^\star = \mathbf{B}\mathbf{A}^\star$ Also, $\mathbf{A} <_\mathbf{m}^- \mathbf{B} \Rightarrow \mathbf{A} <^- \mathbf{B} \Rightarrow \mathcal{C}(\mathbf{A}^\star) \subseteq \mathcal{C}(\mathbf{B}^\star)$, hence, $\mathbf{A} < \star\, \mathbf{B}$.
'Only if' part
We first note that $\mathbf{A}^\star(\mathbf{A}\mathbf{A}^\star)^-$ is a minimum norm g-inverse of $\mathbf{A}$. Since, $\mathbf{A} < \star\, \mathbf{B} \Rightarrow \mathbf{A}\mathbf{A}^\star = \mathbf{B}\mathbf{A}^\star \Rightarrow \mathbf{A}\mathbf{A}^\star(\mathbf{A}\mathbf{A}^\star)^- = \mathbf{B}\mathbf{A}^\star(\mathbf{A}\mathbf{A}^\star)^-$ for any g-inverse $(\mathbf{A}\mathbf{A}^\star)^-$ of $\mathbf{A}\mathbf{A}^\star$. Therefore, $\mathbf{A}\mathbf{A}_\mathbf{m}^- = \mathbf{B}\mathbf{A}_\mathbf{m}^-$ for $\mathbf{A}_\mathbf{m}^- = \mathbf{A}^\star(\mathbf{A}\mathbf{A}^\star)^-$. Also, $\mathbf{A} < \star\, \mathbf{B} \Rightarrow \mathbf{A} <^- \mathbf{B} \Rightarrow \mathbf{A} = \mathbf{A}\mathbf{B}^\dagger\mathbf{A}$. Let $\mathbf{A}^\sim = \mathbf{A}^\dagger\mathbf{A}\mathbf{B}^\dagger$. Then $\mathbf{A}^\sim\mathbf{A}\mathbf{A}^\star = \mathbf{A}^\dagger\mathbf{A}\mathbf{B}^\dagger\mathbf{A}\mathbf{A}^\star = \mathbf{A}^\dagger\mathbf{A}\mathbf{A}^\star = \mathbf{A}^\sim$. Hence, $\mathbf{A}^\sim$ is a minimum norm g-inverse of $\mathbf{A}$. Now, $\mathbf{A}^\sim\mathbf{B} = \mathbf{A}^\dagger\mathbf{A}\mathbf{B}^\dagger\mathbf{B} = \mathbf{A}^\dagger\mathbf{A} = \mathbf{A}^\sim\mathbf{A}$. Thus, $\mathbf{A} <_\mathbf{m}^- \mathbf{B}$.
□

Theorem 6.5.29. *Let* $\mathbf{A}$ *and* $\mathbf{B}$ *be range-hermitian matrices of same order. Then in notations of Theorem 5.4.1,* $\mathbf{A} < \star\, \mathbf{B}$ *if and only if*

$\mathbf{A} = \mathbf{U}\mathrm{\mathbf{diag}}\left(\mathbf{D_a},\ \mathbf{0_{b-a}},\ \mathbf{0}\right)\mathbf{U}^\star$ *and* $\mathbf{B} = \mathbf{U}\mathrm{\mathbf{diag}}\left(\mathbf{M},\ \mathbf{0}\right)\mathbf{U}^\star$, *where* $\mathbf{U}$ *is unitary,* $\mathbf{D_a}$ *and* $\mathbf{M}$ *are non-singular matrices of order* $a \times a$ *and* $b \times b$ *respectively with* $\mathbf{M}$ *of the form* $\mathbf{M} = \begin{pmatrix} \mathbf{D_a}, & \mathbf{0} \\ \mathbf{0}, & \mathbf{0} \end{pmatrix} + \begin{pmatrix} \mathbf{F_1} \\ \mathbf{I} \end{pmatrix} \mathbf{M_{22}} \left(\mathbf{0}\ \ \mathbf{I}\right)$, *for some matrix* $\mathbf{F_1}$ *and a non-singular matrix* $\mathbf{M_b}$ *of order* $(b-a) \times (b-a)$.

Proof. Proof follows from Theorem 5.4.1(i) and Definition of $\mathbf{A} < \star\, \mathbf{B}$. □

We now show that a one-sided star order is the star order if and only if the matrices involved are both normal. Thus, the class of the normal matrices is precisely the class of matrices for which the one-sided star order coincides with the star order.

Remark 6.5.30. Let $\mathbf{A}$ and $\mathbf{B}$ be normal matrices of same order. Then $\mathbf{A} < \star\, \mathbf{B}$ if and only if $\mathbf{A} = \mathbf{U}\mathrm{\mathbf{diag}}\left(\mathbf{D_a}\ \ \mathbf{0_{b-a}}\ \ \mathbf{0}\right)\mathbf{U}^\star$ and $\mathbf{B} = \mathbf{U}\mathrm{\mathbf{diag}}\left(\mathbf{M_b}\ \ \mathbf{0}\right)\mathbf{U}^\star$, where $\mathbf{U}$ is unitary, $\mathbf{D_a}$ and $\mathbf{M_b}$ are non-singular matrices of order $a \times a$ $b \times b$ respectively with $\mathbf{M_b}$ of the form $\mathbf{M_b} = \begin{pmatrix} \mathbf{D_a} & \mathbf{0} \\ \mathbf{0} & \mathbf{0} \end{pmatrix} + \begin{pmatrix} \mathbf{0} \\ \mathbf{I} \end{pmatrix} \mathbf{M_{22}} \left(\mathbf{0},\ \mathbf{I}\right)$, for some non-singular matrix $\mathbf{M_b}$ of order $(b-a) \times (b-a)$. Thus, it follows that $\mathbf{A} <^\star\ \mathbf{B}$.

Theorem 6.5.31. *Let* $\mathbf{A}$ *and* $\mathbf{B}$ *be range-hermitian matrices of the same order such that* $\mathbf{A} < \star\, \mathbf{B}$. *Then* $\mathbf{A}^\star\mathbf{A} = \mathbf{A}^\star\mathbf{B}$ *implies* $\mathbf{A}$ *and* $\mathbf{B}$ *are normal.*

Proof. Proof follows immediately from Theorem 6.5.29 and the fact that $\mathbf{A}^\star\mathbf{A} = \mathbf{A}^\star\mathbf{B} \Rightarrow \mathbf{F_1} = \mathbf{0}$ and therefore, both $\mathbf{A}$ and $\mathbf{B}$ are normal. □

We have the following:

Fisher-Cochran type theorem for one-sided star order

Theorem 6.5.32. *Let* $\mathbf{A_1}, \mathbf{A_2}, \ldots, \mathbf{A_k}$ *be any* $m \times n$ *matrices such that* $\mathbf{A} = \mathbf{A_1} + \mathbf{A_2} + \ldots + \mathbf{A_k}$. *Consider the following statements:*

(i) $\mathbf{A_i}\mathbf{A_j}^\star = \mathbf{0}$, *whenever* $i \neq j$
(ii) $\mathbf{A}\mathbf{A_i}^\star = \mathbf{A_i}\mathbf{A_i}^\star$
(iii) $\mathbf{A}\mathbf{A_i}^\dagger = \mathbf{A_i}\mathbf{A_i}^\dagger$
(iv) $\mathbf{A}\mathbf{A_i}^\star\mathbf{A_i} = \mathbf{A_i}$
(v) $\mathcal{C}(\mathbf{A_i}) \subseteq \mathcal{C}(\mathbf{A})$
(vi) $\rho(\mathbf{A}) = \sum \rho(\mathbf{A_i})$
(vii) $\mathcal{C}(\mathbf{A_i}) \cap \mathcal{C}(\mathbf{A_j}^\star) = \{\mathbf{0}\}$ *for* $i \neq j$

(viii) $\mathbf{A}\mathbf{A_i}^\star = \mathbf{A_i}\mathbf{A}^\star$

Then (i) $\Rightarrow$ (ii) $\Rightarrow$ (iii) $\Rightarrow$ (iv) $\Rightarrow$ (v). (i) $\Rightarrow$ (vii), (iv) $\Rightarrow$ (vi) *and* (i) $\Rightarrow$ (viii).

Proof. Proof is straightforward. (Recall $\mathbf{A}^\dagger = \mathbf{A}^\star(\mathbf{A}^\star\mathbf{A}\mathbf{A}^\star)^-\mathbf{A}^\star$.) □

6.6 Exercises

(1) Let $\mathbf{A}, \mathbf{B}$ be range-hermitian matrices. Show that if either $\mathbf{A}^2 = \mathbf{BA}$ or $\mathbf{A}^2 = \mathbf{AB}$ and $\mathbf{B}$ is an orthogonal projector, then $\mathbf{A}$ is also an orthogonal projector.

(2) Show that for square matrices $\mathbf{A}$ and $\mathbf{B}$ of the same order and of index ≤ 1, $\mathbf{A} < \#\mathbf{B} \not\Leftrightarrow \{\mathbf{B}_{\chi}^-\} \subseteq \{\mathbf{A}_{\chi}^-\}$ and $\mathbf{A}\# < \mathbf{B} \not\Leftrightarrow \{\mathbf{B}_{\rho}^-\} \subseteq \{\mathbf{A}_{\rho}^-\}$.

(3) Let $\mathbf{A}$ and $\mathbf{B}$ be square matrices of the same order and of index ≤ 1 such that $\mathbf{AA}^- = \mathbf{BA}^-$ (or $\mathbf{A}^-\mathbf{A} = \mathbf{A}^-\mathbf{B}$) for some g-inverse $\mathbf{A}^-$ of $\mathbf{A}$ and $\mathbf{AB} = \mathbf{BA}$. Show that $\mathbf{A} <^{\#} \mathbf{B}$.

(4) Let $\mathbf{A}$ and $\mathbf{B}$ be square matrices of the same order and of index ≤ 1 such that $\mathbf{AA}^- = \mathbf{BA}^-$ (or $\mathbf{A}^-\mathbf{A} = \mathbf{A}^-\mathbf{B}$) for some g-inverse $\mathbf{A}^-$ of $\mathbf{A}$ and $\mathbf{A}^2 = \mathbf{AB}$ (respectively $\mathbf{A}^2 = \mathbf{BA}$). Then show that $\mathbf{A} <^{\#} \mathbf{B}$.

(5) Let $\mathbf{A}$ and $\mathbf{B}$ be matrices of the same order $m \times n$ over an arbitrary field. Prove that $\mathbf{AA}^- = \mathbf{BA}^-$ (or $\mathbf{A}^-\mathbf{A} = \mathbf{A}^-\mathbf{B}$) is invariant under choice of g-inverse $\mathbf{A}^-$ if and only if $\mathcal{C}(\mathbf{B}^t) \subseteq \mathcal{C}(\mathbf{A}^t)$ (or $\mathcal{C}(\mathbf{B}) \subseteq \mathcal{C}(\mathbf{B})$). Show that if $\mathbf{A} <^- \mathbf{B}$, the condition that each of the defining conditions is invariant under choices of $\mathbf{A}^-$ implies $\mathbf{A} = \mathbf{B}$.

(6) Let $\mathbf{A}$ and $\mathbf{B}$ be square matrices of the same order and of index ≤ 1 such that (i) $\mathbf{A}^2 = \mathbf{BA}$ (or $\mathbf{A}^2 = \mathbf{AB}$) and (ii) $\mathbf{BA}^{\#}\mathbf{B} = \mathbf{A}$. Then prove that $\mathbf{A} <^{\#} \mathbf{B}$. What happens if the condition (ii) is replaced by (ii)$'$ $\mathbf{B}^{\#}\mathbf{AB}^{\#} = \mathbf{A}^{\#}$? Should the same result hold if the group inverse (i) and replaced by a commuting g-inverse?

(7) Let $\mathbf{A}$ and $\mathbf{B}$ be matrices of the same order $m \times n$ over the field of complex numbers such that (i) $\mathbf{A}^-\mathbf{A} = \mathbf{A}^-\mathbf{B}$ and (ii) $\mathbf{BA}^{\dagger}\mathbf{B} = \mathbf{A}$. Prove that $\mathbf{A} <^{\star} \mathbf{B}$. Assume now (i) $\mathbf{AA}^- = \mathbf{BA}^-$ and (ii) $\mathbf{B}^{\dagger}\mathbf{AB}^{\dagger} = \mathbf{A}^{\dagger}$. Show that $\mathbf{A} <^{\star} \mathbf{B}$.

(8) Let $\mathbf{A}$ and $\mathbf{B}$ be square matrices with index $\mathbf{A} = 1$. We say $\mathbf{A} <^{-\rho} \mathbf{B}$ if for some $\mathbf{A}_{\rho}^- \in \{\mathbf{A}_{\rho}^-\}$, $\mathbf{A}_{\rho}^-\mathbf{A} = \mathbf{A}_{\rho}^-\mathbf{B}$ and $\mathbf{AA}_{\rho}^- = \mathbf{BA}_{\rho}^-$. Similarly one defines $\mathbf{A} <^{-\chi} \mathbf{B}$. Prove that
$\mathbf{A} <^{\#} \mathbf{B} \Leftrightarrow \mathbf{A} <^{-\rho} \mathbf{B}$ and $\mathbf{A} <^{-\chi} \mathbf{B}$.

(9) Let $\mathbf{A}$ be an $n \times n$ complex matrix of index 1. Show that an $\mathbf{A}_{\rho}^-$ is an $\mathbf{A}_{\ell r}^-$ if and only if $\mathbf{A}$ is range-hermitian. Thus, for square matrices $\mathbf{A}$ and $\mathbf{B}$, show that $\mathbf{A} <^{-\rho} \mathbf{B} \Leftrightarrow \mathbf{A}^{\star} < \mathbf{B}$ holds $\mathbf{A}$ if and only if is range-hermitian. Similarly, $\mathbf{A} <^{-\chi} \mathbf{B} \Leftrightarrow \mathbf{A} <^{\star} \mathbf{B} \Leftrightarrow \mathbf{A}$ is range-hermitian.

(10) Let $\mathbf{A}$ and $\mathbf{B}$ be square matrices of the same order, $\mathbf{A}$ be a range hermitian matrix and $\mathbf{B}$ be idempotent. Then show that the following are equivalent:

(i) $\mathbf{A} <^{-} \mathbf{B}$
(ii) $\mathbf{A}^{\star} < \mathbf{B}$
(iii) $\mathbf{A} < \star\mathbf{B}$
(iv) $\mathbf{A} <^{\star} \mathbf{B}$

(11) Let $\mathbf{A}$ and $\mathbf{B}$ be square matrices of the same order, $\mathbf{A}$ be a range-hermitian matrix and $\mathbf{B}$ a normal matrix. Show that (i) $\mathbf{A} < \star\mathbf{B}$ if and only if $\mathbf{A} <^{\star} \mathbf{B}$. (ii) $\mathbf{A}^{\star} < \mathbf{B}$ if and only if $\mathbf{A} <^{\star} \mathbf{B}$.

(12) Let $\mathbf{A}$ and $\mathbf{B}$ be matrices of the same order $m \times n$. Show the following: (i) $\mathbf{A} < \star\mathbf{B}$ if and only if $\mathbf{A} <^{-} \mathbf{B}$ and (ii) $\mathbf{A}^{\star}\mathbf{A} <^{L} \mathbf{B}^{\star}\mathbf{B}$ if and only if $\mathbf{A} <^{-} \mathbf{B}$ and $\mathbf{A}^{\star}\mathbf{A} <^{-} \mathbf{B}^{\star}\mathbf{B}$.

(13) Let $\mathbf{A}$ and $\mathbf{B}$ be matrices of the same order $m \times n$. Show the following: (i) $\mathbf{A} < \star\mathbf{B}$ if and only if $\mathbf{A} <^{-} \mathbf{B}$ and (ii) $\mathbf{A}\mathbf{A}^{\star} <^{L} \mathbf{B}^{\star}\mathbf{B}$ if and only if $\mathbf{A} <^{-} \mathbf{B}$ and $\mathbf{A}\mathbf{A}^{\star} <^{-} \mathbf{B}^{\star}\mathbf{B}$.

The following set of exercises gives some properties of the g-inverses introduced in Section 4 of the chapter.

(14) Let $\mathbf{A}$ be a square matrix of index ≤ 1. Write $\mathbf{A} = \mathbf{P}\begin{pmatrix}\mathbf{S} & \mathbf{0}\\ \mathbf{0} & \mathbf{0}\end{pmatrix}\mathbf{P}^{-1}$ where $\mathbf{S}$ and $\mathbf{P}$ are non-singular. Then show that $\{\mathbf{A}_c^{-}\}$, the class of all c-inverses is given by $\mathbf{P}\begin{pmatrix}\mathbf{S}^{-1} & \mathbf{L}\\ \mathbf{0} & \mathbf{N}\end{pmatrix}\mathbf{P}^{-1}$, where $\mathbf{L}$ and $\mathbf{N}$ are arbitrary. Hence conclude that every $\mathbf{A}_{\chi}^{-}$ is a c-inverse of $\mathbf{A}$. Notice that every reflexive c-inverse is an $\mathbf{A}_{\chi}^{-}$. What is the class of all $\mathbf{A}_a^{-}$? Conclude further that $\mathbf{G}$ is a c-inverse if and only if $\mathbf{G}^t$ is $(\mathbf{A}^t)_a^{-}$.

(15) Let $\mathbf{A}$ be a square matrix of index 1. Prove that $\mathbf{AGA} = \mathbf{A}$, $\mathcal{C}(\mathbf{GA}) \subseteq \mathcal{C}(\mathbf{A})$ if and only if $\mathbf{GA} = \mathbf{P}$, where $\mathbf{P}$ is a projector which projects vectors into $\mathcal{C}(\mathbf{A})$ along $\mathcal{N}(\mathbf{A})$ if and only if $\mathbf{GA}^2 = A$.

(16) Let $\mathbf{A}$ be a square matrix of index 1. Show that for every consistent system $\mathbf{Ax} = \mathbf{b}$, $\mathbf{Gb} \in \mathcal{C}(\mathbf{A})$ if and only if $\mathbf{G} \in \{\mathbf{A}_c^{-}\}$. Show further that $\mathbf{Ax} = \mathbf{b}$ has a unique solution $\mathbf{x}_0$ belonging to $\mathcal{C}(\mathbf{A})$.

(17) Let $\mathbf{A}$ be a square matrix of index ≤ 1. Let $\mathbf{G} \in \{\mathbf{A}_a^{-}\}$. Prove that the class of all good approximate solutions of $\mathbf{Ax} = \mathbf{b}$ is given by $\mathbf{Gb} + (\mathbf{I} - \mathbf{GA})\xi$, where ξ is arbitrary.

(18) Let $\mathbf{A}$ be a square matrix of index ≤ 1. Show that the following are equivalent:

(i) $\mathbf{G}$ is a $\mathbf{A}_a^{-}$
(ii) $\mathbf{A}^2\mathbf{G} = \mathbf{A}$
(iii) $\mathbf{AGA} = \mathbf{A}$ $\mathcal{C}(\mathbf{G}^t\mathbf{A}^t) = \mathcal{C}(\mathbf{A}^t)$ and

(iv) $\mathbf{G}^t\mathbf{A}^t$ is a projector that projects vectors into $\mathcal{C}(\mathbf{A}^t)$ along $\mathcal{N}(\mathbf{A}^t)$.

(19) Find the class of all $\mathbf{A}_a^-$ for the matrix $\mathbf{A}$ in question no. 15.

Chapter 7

Unified Theory of Matrix Partial Orders through Generalized Inverses

7.1 Introduction

In Chapters 3-5, we studied three major matrix partial orders namely: the minus, star and sharp orders. In each of these cases, we saw that they possess a similar characterizing property in terms of a suitable subclass of g-inverses: a matrix $\mathbf{A}$ is below a matrix $\mathbf{B}$ if $\mathbf{AG} = \mathbf{BG}$ and $\mathbf{GA} = \mathbf{GB}$ for some matrix $\mathbf{G}$ belonging to a suitable subclass of g-inverses of $\mathbf{A}$. For the minus, star and sharp orders, the subclasses are $\{\mathbf{A}^-\}$, $\{\mathbf{A}^-_{\ell\mathbf{m}}\}$ and $\{\mathbf{A}^-_{\mathbf{ac}}\}$ respectively. A vigilant reader must have noticed that the three partial orders mentioned above have a number of characterizing properties which are analogous. We recall a couple of them here.

If $\mathbf{A} <^- \mathbf{B}$, then
(i) $\{\mathbf{B}^-\} \subseteq \{\mathbf{A}^-\}$ and
(ii) there exist projectors $\mathbf{P}$ and $\mathbf{Q}$ such that $\mathbf{A} = \mathbf{PB} = \mathbf{BQ}$.

If $\mathbf{A} <^{\#} \mathbf{B}$, then
(i) $\{\mathbf{B}^-_{\mathbf{ac}}\} \subseteq \{\mathbf{A}^-_{\mathbf{ac}}\}$ and
(ii) there exists a projector $\mathbf{P}$ onto $\mathcal{C}(\mathbf{A})$ such that $\mathbf{A} = \mathbf{BP} = \mathbf{PB}$.

Also, if $\mathbf{A} <^{\star} \mathbf{B}$, then
(i) $\{\mathbf{B}^-_{\ell\mathbf{m}}\} \subseteq \{\mathbf{A}^-_{\ell\mathbf{m}}\}$ and
(ii) there exist orthogonal projectors $\mathbf{P}$ and $\mathbf{Q}$ such that

$$\mathbf{A} = \mathbf{PB} = \mathbf{BQ}.$$

In Chapter 6, we studied one-sided partial orders which also exhibited some similar properties.

In the present chapter, we first develop a unified theory of matrix partial orders that are defined via g-inverses. We then attempt to obtain characterizing properties similar to the ones mentioned earlier under the unified theory. We also develop a unified theory based on subclasses of outer inverses. We finally consider some extensions of the above mentioned unifications.

In Section 7.2, we define g-maps and a $\mathcal{G}$-based order relation (which includes as special cases the minus, the star and the sharp orders) on matrices based on subclasses of generalized inverses. We study the conditions under which a $\mathcal{G}$-based order relation is a partial order. We also show that not every partial order defined using subclasses of g-inverses is a $\mathcal{G}$-based order. The order relation '$<^{\#,-}$' discussed in Section 4 of Chapter 4 provides a nontrivial example of the same. Section 7.3 is devoted to the study of orders based on the subclasses of outer inverses. We also study when an order relation based on a subclass of g-inverses coincides with an order relation based on a subclass of outer inverses. In Section 7.4, we develop a unified theory of one-sided orders. In Section 7.5, we study some characterizing properties of $\mathcal{G}$-based partial orders similar to those mentioned in the first paragraph above. Theorem 7.5.5 provides us a true unification of almost all the partial orders via g-inverses that we studied so far. Section 7.6 deals with extensions of $\mathcal{G}$-based orders to accommodate some orders that are not $\mathcal{G}$-based.

7.2 $\mathcal{G}$-based order relations: Definitions and preliminaries

Let '$<$' denote any of the partial orders studied in earlier chapters. Recall that they all shared a common property: $\mathbf{A} < \mathbf{B} \Leftrightarrow$ some specific subset of g-inverses of $\mathbf{B}$ is a subset of a suitable subset of g-inverses of $\mathbf{A}$. Keeping this in view, we introduce in this section the notion of a $\mathcal{G}$-map and a $\mathcal{G}$-based order relation. We study the properties of this order relation and examine when it defines a partial order at least on its support to be defined shortly. We also give appropriate interpretations of the $\mathcal{G}$-based order relation for most of the partial orders studied earlier.

Definition 7.2.1. Let $\mathfrak{P}(\mathbf{F^{n\times m}})$ denote the power set (class of all subsets) of $\mathbf{F^{n\times m}}$. A g-map is a map

$$\mathcal{G} : \mathbf{F^{m\times n}} \longrightarrow \mathfrak{P}(\mathbf{F^{n\times m}})$$

such that for each $\mathbf{A} \in \mathbf{F^{m\times n}}$, $\mathcal{G}(\mathbf{A})$ is a certain subset (possibly non-empty) of $\{\mathbf{A}^-\}$ and the set

$$\Omega_{\mathcal{G}} = \{\mathbf{A} \in \mathbf{F}^{\mathbf{m}\times\mathbf{n}} : \mathcal{G}(\mathbf{A}) \neq \emptyset\}$$

is called the support of the g-map $\mathcal{G}$.

Remark 7.2.2. Notice that the set $\mathcal{G}(\mathbf{A})$ is not always non-empty. For instance, let $\mathcal{G}(\mathbf{A}) = \{\mathbf{A}^{\#}\}$ for each $\mathbf{A} \in \mathbf{F}^{\mathbf{n}\times\mathbf{n}}$. If $\mathbf{A}$ is not of index 1, then $\mathcal{G}(\mathbf{A}) = \emptyset$.

Definition 7.2.3. Let $\mathcal{G} : \mathbf{F}^{\mathbf{m}\times\mathbf{n}} \longrightarrow \mathfrak{P}(\mathbf{F}^{\mathbf{n}\times\mathbf{m}})$ be a g-map. For $\mathbf{A}$, $\mathbf{B} \in \mathbf{F}^{\mathbf{m}\times\mathbf{n}}$, we say $\mathbf{A} <^{\mathcal{G}} \mathbf{B}$ if there exists a $G \in \mathcal{G}(\mathbf{A})$ such that $\mathbf{GA} = \mathbf{GB},\ \mathbf{AG} = \mathbf{BG}$.

The order relation '$<^{\mathcal{G}}$' is called $\mathcal{G}$-based order relation. We give below some examples

Example 7.2.4. Let $\mathcal{G}(\mathbf{A}) = \{\mathbf{A}^-\}$. Then the order relation '$<^{\mathcal{G}}$' is the minus order.

Remark 7.2.5. The order relation '$<^{\mathcal{G}}$' implies '$<^-$'.

Example 7.2.6. Let $\mathcal{G}(\mathbf{A}) = \{\mathbf{A}^{\dagger}\}$ or $\{\mathbf{A}^-_{\ell,\mathbf{m}}\}$. Then the order relation '$<^{\mathcal{G}}$' is the star order.

Similarly, if $\mathcal{G}(\mathbf{A}) = \{\mathbf{A}^-_{\mathbf{m}}\}$, (or $\{\mathbf{A}^-_{\ell}\}$,) then this order relation is right star (or left star) order respectively. For $\mathbf{A}, \mathbf{B} \in \mathcal{I}_1$, if we take $\mathcal{G}(\mathbf{A}) = \{\mathbf{A}^-_{\mathrm{com}}\}$ or $\{\mathbf{A}^-_{\chi}\}$ or $\{\mathbf{A}^-_{\rho}\}$, then we get the sharp order or right sharp or left sharp orders respectively.

Thus, almost all partial orders appear to be $\mathcal{G}$-based order relations. Are there some known partial orders that are not $\mathcal{G}$-based order relations? The Löwner order as defined in Section 1.1, is clearly not a $\mathcal{G}$-based order relation. We give a non-trivial example of order relation that is not $\mathcal{G}$-based.

Example 7.2.7. Consider the order relation '$<^{\#,-}$' of Definition 4.4.17. We show that this order relation is not a $\mathcal{G}$-based an order relation. Let

$$\mathbf{A} = \begin{pmatrix} 0 & 0 & 0 & 0 \\ 1 & 1 & 1 & 0 \\ -1 & -1 & -1 & 0 \\ 1 & 1 & 1 & 0 \end{pmatrix}, \mathbf{B} = \begin{pmatrix} 0 & 0 & 0 & 0 \\ 1 & 0 & 0 & 0 \\ 0 & 1 & 0 & 0 \\ 0 & 0 & 1 & 0 \end{pmatrix} \in \mathbb{R}^{\mathbf{4}\times\mathbf{4}}.$$

Clearly, $\mathbf{A^2} = \mathbf{0}$ and $\mathbf{B^4} = \mathbf{0}$, hence, $\mathbf{A}$ and $\mathbf{B}$ are nilpotent. Also, $\rho(\mathbf{A}) = \mathbf{1},\ \rho(\mathbf{B}) = 3,$ and $\rho(\mathbf{B} - \mathbf{A}) = 2$. So, $\mathbf{A} <^- \mathbf{B}$. Since $\mathbf{A}$ and $\mathbf{B}$ are nilpotent, it follows that $\mathbf{A} <^{\#,-} \mathbf{B}$.

If the relation $<^{\#,-}$ were a $\mathcal{G}$-based order relation, then there exists a $\mathbf{G} \in \mathcal{G}(\mathbf{A})$, such that $\mathbf{GA} = \mathbf{GB}$, $\mathbf{AG} = \mathbf{BG}$. Let $\mathbf{T} = \mathbf{2B} - \mathbf{A}$, then $\mathbf{G}(\mathbf{T} - \mathbf{A}) = \mathbf{0}$, $(\mathbf{T} - \mathbf{A})\mathbf{G} = \mathbf{0}$. So, $\mathbf{A}<^{\#,-}\mathbf{T}$. However, $\mathbf{T} = \begin{pmatrix} 0 & 0 & 0 & 0 \\ 1 & -1 & -1 & 0 \\ 1 & 3 & 1 & 0 \\ -1 & -1 & 1 & 0 \end{pmatrix}$, $\rho(\mathbf{T}) = \mathbf{3}$, $\rho(\mathbf{T^2}) = \mathbf{2} = \rho(\mathbf{T^3})$. Thus, $\mathbf{T}$ is of index 2, so, $\mathbf{C_T}$, the core part is of rank 2 and $\mathbf{N_T}$, the nilpotent part is of rank 1. Since $\mathbf{A}<^{\#,-}\mathbf{T}$, we have $\mathbf{0} = \mathbf{C_A} <^{\#} \mathbf{C_T}$ and $\mathbf{A} = \mathbf{N_A} <^{-} \mathbf{N_T}$. It follows that $\mathbf{A} = \mathbf{N_A} = \mathbf{N_T}$, which is impossible. Since $\mathbf{T} - \mathbf{A} = \mathbf{C_T}$, so, $\mathbf{AC_T} = \mathbf{AT}$ and $\mathbf{C_TA} = \mathbf{TA}$ should both be null. However, an easy computation shows that neither is a null matrix.

In view of Remarks 7.2.5, it is clear that the $\mathcal{G}$-based order relation is always anti-symmetric and reflexive on its support. Since our main objective here is to study partial orders on matrices, it is very natural to ask if a $\mathcal{G}$-based order relation is transitive at least on its support $\Omega_{\mathcal{G}}$, if not on its domain. We remark that if the support, $\Omega_{\mathcal{G}}$ of a g-map consists of exactly two elements, then the g-map always induces a partial order on $\Omega_{\mathcal{G}}$. However, if $\Omega_{\mathcal{G}}$ contains three or more elements, the answer is no as the following example shows:

Example 7.2.8. Let $\mathcal{G} : \mathbb{C}^{3\times3} \longrightarrow \mathfrak{P}(\mathbb{C}^{3\times3})$ be the map

$$\mathcal{G}(\mathbf{A}) = \begin{cases} \mathbf{A}^{\dagger}, & \text{if } \rho(\mathbf{A}) = \mathbf{1} \\ \mathbf{A}^{-}, & \text{otherwise.} \end{cases}$$

$$\text{Thus, } \mathbf{A} <^{\mathcal{G}} \mathbf{B} \Leftrightarrow \begin{cases} \mathbf{A} <^{\star} \mathbf{B}, & \text{if } \rho(\mathbf{A}) = \mathbf{1} \\ \mathbf{A} <^{-} \mathbf{B}, & \text{otherwise.} \end{cases}$$

Now take $\mathbf{A} = \begin{pmatrix} 1 & 0 & 0 \\ 0 & 0 & 0 \\ 0 & 0 & 0 \end{pmatrix}$, $\mathbf{B} = \begin{pmatrix} 1 & 0 & 0 \\ 0 & 1 & 0 \\ 0 & 0 & 0 \end{pmatrix}$ and $\mathbf{C} = \begin{pmatrix} 1 & 0 & 1 \\ 0 & 1 & 0 \\ 0 & 0 & 1 \end{pmatrix}$.

Notice that $\rho(\mathbf{A}) = \mathbf{1}, \rho(\mathbf{B}) = \mathbf{2}$ and $\rho(\mathbf{C}) = \mathbf{3}$. Further, $\mathbf{A} <^{\star} \mathbf{B}$ and $\mathbf{B} <^{-} \mathbf{C}$. So, $\mathbf{A} <^{\mathcal{G}} \mathbf{B}$ and $\mathbf{B} <^{\mathcal{G}} \mathbf{C}$. However, $\mathbf{A} \not<^{\star} \mathbf{C}$, so, $\mathbf{A} \not<^{\mathcal{G}} \mathbf{C}$. Hence the transitivity fails to hold for the relation $<^{\mathcal{G}}$. As a result $<^{\mathcal{G}}$ is not a partial order.

Thus, a $\mathcal{G}$-based order relation on $\mathbf{F^{m\times n}}$ may fail to be either reflexive or transitive. We will now study the conditions under which a $\mathcal{G}$-based order relation becomes a partial order. Our next theorem gives a sufficient condition under which it becomes a partial order. But before this we need to give some definitions.

Definition 7.2.9. For a g-map $\mathcal{G} : \mathbf{F}^{\mathbf{m}\times\mathbf{n}} \longrightarrow \mathfrak{P}(\mathbf{F}^{\mathbf{n}\times\mathbf{m}})$, $\mathbf{A} \in \mathbf{F}^{\mathbf{m}\times\mathbf{n}}$ is said to be maximal with respect to the order relation '$<^{\mathcal{G}}$' ($\mathcal{G}$-maximal) if for any matrix $\mathbf{D}$, $\mathbf{A} <^{\mathcal{G}} \mathbf{D}$, then $\mathbf{A} = \mathbf{D}$.

Similarly, $\mathbf{A} \in \mathbf{F}^{\mathbf{m}\times\mathbf{n}}$ is said to be minimal with respect to the order relation '$<^{\mathcal{G}}$' ($\mathcal{G}$-minimal) if for any matrix $\mathbf{C}$, $\mathbf{C} <^{\mathcal{G}} \mathbf{A}$, then $\mathbf{C} = \mathbf{A}$. Thus, any matrix with full rank (row or column) is $\mathcal{G}$-maximal and the null matrix $\mathbf{0}$ is $\mathcal{G}$-minimal.

We have seen in Example 7.2.8, the order relation '$<^{\mathcal{G}}$' may fail to satisfy transitivity. So, we require some additional condition/s to make it a partial order.

We now define a condition to be satisfied by the order relation '$<^{\mathcal{G}}$' and call it the (T)-condition, (where T is first letter of the word transitive). This condition, as we shall shortly see is only a sufficient condition for an order relation to be a partial order. We shall, in due course find a necessary and sufficient condition that will make the order relation '$<^{\mathcal{G}}$' a partial order.

Definition 7.2.10. Let $\mathcal{G} : \mathbf{F}^{\mathbf{m}\times\mathbf{n}} \longrightarrow \mathfrak{P}(\mathbf{F}^{\mathbf{n}\times\mathbf{m}})$ be a g-map. Then for a pair of matrices $\mathbf{A}, \mathbf{B} \in \mathbf{F}^{\mathbf{m}\times\mathbf{n}}$, let $\mathcal{G}(\mathbf{A}, \mathbf{B}) = \{\mathbf{B}^-\mathbf{A}\mathbf{B}^- : \mathbf{B}^- \in \mathcal{G}(\mathbf{A})\}$. A pair $(\mathbf{A}, \mathbf{B})$ is said to satisfy the (T)-condition if $\mathcal{G}(\mathbf{A}, \mathbf{B}) \subseteq \mathcal{G}(\mathbf{A})$. Furthermore, if for each pair of matrices $\mathbf{A}, \mathbf{B}$, the pair $(\mathbf{A}, \mathbf{B})$ satisfies the (T)-condition, we say the g-map $\mathcal{G}$ satisfies the (T)-condition.

We first prove the following:

Theorem 7.2.11. *Let* $\mathcal{G} : \mathbf{F}^{\mathbf{m}\times\mathbf{n}} \longrightarrow \mathfrak{P}(\mathbf{F}^{\mathbf{n}\times\mathbf{m}})$ *be a g-map. Suppose for the matrices* $\mathbf{A}, \mathbf{B} \in \mathbf{\Omega}_{\mathcal{G}}$, *the pair* $(\mathbf{A}, \mathbf{B})$ *satisfies the (T)-condition. Then* $\mathbf{A} <^{\mathcal{G}} \mathbf{B}$ *if and only if* $\mathbf{A} <^- \mathbf{B}$.

Proof. We only need to prove the 'if' part. So, let $\mathbf{A} <^- \mathbf{B}$. Since $\mathbf{B} \in \mathbf{\Omega}_{\mathcal{G}}$, $\mathcal{G}(\mathbf{B}) \neq \emptyset$. Let $\mathbf{H} \in \mathcal{G}(\mathbf{B})$. Consider $\mathbf{G} = \mathbf{HAH}$. As the pair $(\mathbf{A}, \mathbf{B})$ satisfies the (T)-condition, $\mathbf{G} \in \mathcal{G}(\mathbf{A})$ and by Theorem 3.3.2, we have $(\mathbf{B} - \mathbf{A})\mathbf{G} = \mathbf{BHAH} - \mathbf{AHAH} = \mathbf{0}$. So, $\mathbf{BH} = \mathbf{AH}$. Similarly, $\mathbf{HB} = \mathbf{HA}$. Hence, $\mathbf{A} <^{\mathcal{G}} \mathbf{B}$. □

Remark 7.2.12. Let $\mathcal{G} : \mathbb{C}^{m\times n} \longrightarrow \mathfrak{P}(\mathbb{C}^{n\times m})$ be the g-map $\mathcal{G}(\mathbf{A}) = \{\mathbf{A}^\dagger\}$ for each $\mathbf{A}$. Then a pair $(\mathbf{A}, \mathbf{B})$ satisfies the (T)-condition is equivalent to saying that the pair $(\mathbf{A}, \mathbf{B})$ satisfies the condition $\mathbf{B}^\dagger\mathbf{A}\mathbf{B}^\dagger = \mathbf{A}^\dagger$. So, in view of Theorem 5.2.10, $\mathbf{A} <^{\mathcal{G}} \mathbf{B}$ if and only if $\mathbf{A} <^{-} \mathbf{B}$, where $\mathbf{A} <^{\mathcal{G}} \mathbf{B}$ means $\mathbf{A} <^{\star} \mathbf{B}$.

Similarly, if $\mathcal{G} : \mathbf{F^{n\times n}} \longrightarrow \mathfrak{P}(\mathbf{F^{n\times n}})$ be the g-map given by $\mathcal{G}(\mathbf{A}) = \{\mathbf{A}^{\#}\}$ for each $\mathbf{A}$, then in view of Theorem 4.2.12, $\mathbf{A} <^{\#} \mathbf{B}$ if and only if $\mathbf{A} <^{-} \mathbf{B}$ for matrices $\mathbf{A}$ and $\mathbf{B}$ of index ≤ 1.

We are now ready to show that under the (T)-condition, the relation '$<^{\mathcal{G}}$' defines a partial order on $\Omega_{\mathcal{G}}$, the support of the g-map $\mathcal{G}$.

Theorem 7.2.13. *Let $\mathcal{G} : \mathbf{F^{m\times n}} \longrightarrow \mathfrak{P}(\mathbf{F^{n\times m}})$ be a g-map with support $\Omega_{\mathcal{G}}$. Let $\mathbf{A} <^{\mathcal{G}} \mathbf{B}$, $\mathbf{B}$ not $\mathcal{G}$-maximal imply the pair $(\mathbf{A}, \mathbf{B})$ satisfies the (T)-condition. Then the order relation $<^{\mathcal{G}}$ is transitive and therefore defines a partial order on $\Omega_{\mathcal{G}}$.*

Proof. Let $\mathbf{A} <^{\mathcal{G}} \mathbf{B}$ and $\mathbf{B} <^{\mathcal{G}} \mathbf{C}$. If $\mathbf{B}$ is $\mathcal{G}$-maximal, then $\mathbf{B} <^{\mathcal{G}} \mathbf{C}$ implies $\mathbf{B} = \mathbf{C}$ and trivially $<^{\mathcal{G}}$ is transitive. So, let $\mathbf{B}$ be not $\mathcal{G}$-maximal. Now, $\mathbf{B} <^{\mathcal{G}} \mathbf{C} \Rightarrow$ there exists a $\mathbf{H} \in \mathcal{G}(\mathbf{B})$ such that $\mathbf{BH} = \mathbf{CH}$ and $\mathbf{HB} = \mathbf{HC}$. As the pair $(\mathbf{A}, \mathbf{B})$ satisfies the (T)-condition, $\mathbf{HAH} \in \mathcal{G}(\mathbf{A})$. Let $\mathbf{G} = \mathbf{HAH}$. Then $\mathbf{AG} = \mathbf{AHAH} = \mathbf{AH} = \mathbf{BHAH} = \mathbf{CHAH} = \mathbf{CG}$. Similarly, $\mathbf{GA} = \mathbf{GC}$. Thus, $\mathbf{A} <^{\mathcal{G}} \mathbf{C}$. As already noted that $<^{\mathcal{G}}$ is reflexive and anti-symmetric on the support $\Omega_{\mathcal{G}}$, it follows that the order relation is a partial order. □

Remark 7.2.14. The condition '$\mathbf{A} <^{\mathcal{G}} \mathbf{B} \Rightarrow$ the pair $(\mathbf{A}, \mathbf{B})$ satisfies the (T)-condition' is true for the minus, the star and the sharp orders and as seen already, they are all partial orders.

As pointed out in the opening para of the section, the (T)-condition just defined above is only a sufficient condition and not necessary for the order relation $<^{\mathcal{G}}$ to become a partial order. We now give an example to show that an order relation $<^{\mathcal{G}}$ can still be a partial order even when $\mathbf{A} <^{\mathcal{G}} \mathbf{B}$ does not imply that the pair $(\mathbf{A}, \mathbf{B})$ satisfies the (T)-condition.

Example 7.2.15. Consider the matrices $\mathbf{A} = \begin{pmatrix} 1 & 0 & 0 \\ 0 & 0 & 0 \\ 0 & 0 & 0 \end{pmatrix}, \mathbf{B} = \begin{pmatrix} 1 & 0 & 0 \\ 0 & 1 & 0 \\ 0 & 0 & 0 \end{pmatrix}$ and $\mathbf{G} = \begin{pmatrix} 1 & 0 & 0 \\ 0 & 1 & 0 \\ 1 & 0 & 1 \end{pmatrix} \in \mathbb{R}^{\mathbf{3\times 3}}$. Then $\mathbf{G}$ is a g-inverse of $\mathbf{B}$ and the matrices $\mathbf{H_a} = \begin{pmatrix} 1 & 0 & a \\ 0 & 0 & 0 \\ 0 & 0 & a \end{pmatrix}$, $0 \neq a \in \mathbb{R}$ are g-inverses of $\mathbf{A}$.

Let $\mathcal{G} : \mathbb{R}^{3\times 3} \longrightarrow \mathfrak{P}(\mathbb{R}^{3\times 3})$ be the g-map defined as

$$\mathcal{G}(\mathbf{X}) = \begin{cases} \{\mathbf{H_a}\}, & \text{for } \mathbf{X} = \mathbf{A} \\ \{\mathbf{G}\}, & \text{for } \mathbf{X} = \mathbf{B} \\ \emptyset & \text{otherwise.} \end{cases}$$

Then it is easy to verify that the relation '$<^{\mathcal{G}}$' is a partial order on its support $\Omega_{\mathcal{G}} = \{\mathbf{A},\ \mathbf{B}\}$. However, the pair $(\mathbf{A}, \mathbf{B})$ does not satisfy the (T)-condition, since $\mathbf{GAG} = \begin{pmatrix} 1 & 0 & 0 \\ 0 & 0 & 0 \\ 1 & 0 & 0 \end{pmatrix}$ and it is clearly not of the form $\mathbf{H_a}$ for any non-null real number $\mathbf{a}$ and therefore does not belong to $\mathcal{G}(\mathbf{A})$.

Thus, the (T)-condition is not a necessary condition to obtain a partial order defined by a g-map. We also notice that for the star and the sharp orders this condition is not only sufficient but is necessary as well. Therefore, there must be some underlying property that both these relations satisfy that is not inherent in the definition of a $\mathcal{G}$-based order relation. We try to capture this property. We first give two definitions for this purpose.

Definition 7.2.16. Let $\mathcal{G} : \mathbf{F}^{\mathbf{m}\times\mathbf{n}} \longrightarrow \mathfrak{P}(\mathbf{F}^{\mathbf{n}\times\mathbf{m}})$ be a g-map. For a matrix $\mathbf{A} \in \mathbf{F}^{\mathbf{m}\times\mathbf{n}}$, the class

$$\widetilde{\mathcal{G}(\mathbf{A})} = \{\mathbf{G} \in \{\mathbf{A}^-\} : \mathbf{GA} = \mathbf{HA},\ \mathbf{AG} = \mathbf{AH} \text{ for some } \mathrm{H} \in \mathcal{G}(\mathbf{A})\}$$

is called the completion of $\mathcal{G}(\mathbf{A})$. We say $\mathcal{G}(\mathbf{A})$ is complete if $\mathcal{G}(\mathbf{A}) = \widetilde{\mathcal{G}(\mathbf{A})}$. Further, if for each $\mathbf{A} \in \mathbf{F}^{\mathbf{m}\times\mathbf{n}}$, $\mathcal{G}(\mathbf{A})$ is complete, we say the g-map is complete.

For example, the completion of $\{\mathbf{A}_\mathbf{r}^-\}$, the class of reflexive g-inverses of $\mathbf{A}$ is $\{\mathbf{A}^-\}$. Similarly, the completion of $\{\mathbf{A}^\dagger\}$ is the class $\{\mathbf{A}_{\ell\mathbf{m}}^-\}$. The classes $\{\mathbf{A}^-\}$ and $\{\mathbf{A}_{\ell\mathbf{m}}^-\}$ are complete for each $\mathbf{A} \in \mathbf{F}^{\mathbf{m}\times\mathbf{n}}$. However, not each subclass of $\{\mathbf{A}^-\}$ is complete, for as noted $\{\mathbf{A}_\mathbf{r}^-\}$, the class of reflexive g-inverses is not complete. In fact, for an $\mathbf{A} \in \mathbf{F}^{\mathbf{m}\times\mathbf{n}}$, if $\{\mathbf{A}_\mathbf{r}^-\} \subset \mathcal{G}(\mathbf{A}) \subset \{\mathbf{A}^-\}$, $\mathcal{G}(\mathbf{A})$ is not complete. Similarly, for $\mathbf{A}, \mathbf{B} \in \mathbf{F}^{\mathbf{m}\times\mathbf{n}}$ such that $\mathbf{A} <^- \mathbf{B}$, the class $\{\mathbf{G} : \mathbf{AG} = \mathbf{BG},\ \mathbf{GA} = \mathbf{GB}\}$ is also incomplete.

Definition 7.2.17. For a g-map $\mathcal{G} : \mathbf{F}^{\mathbf{m}\times\mathbf{n}} \longrightarrow \mathfrak{P}(\mathbf{F}^{\mathbf{n}\times\mathbf{m}})$ and $\mathbf{A} \in \mathbf{F}^{\mathbf{m}\times\mathbf{n}}$, the set $\mathcal{G}(\mathbf{A})$ is said to be semi-complete (complete with respect to reflexive g-inverses), if $\widetilde{\mathcal{G}(\mathbf{A})} \cap \{\mathbf{A}_\mathbf{r}^-\} = \mathcal{G}(\mathbf{A}) \cap \{\mathbf{A}_\mathbf{r}^-\}$ and if for each $\mathbf{A} \in \mathbf{F}^{\mathbf{m}\times\mathbf{n}}$, $\mathcal{G}(\mathbf{A})$ is semi-complete, we say the g-map $\mathcal{G}$ is semi-complete.

We shall denote the set $\mathcal{G}(\mathbf{A}) \cap \{\mathbf{A}_\mathbf{r}^-\}$ by $\mathcal{G}_r(\mathbf{A})$ and $\widetilde{\mathcal{G}(\mathbf{A})} \cap \{\mathbf{A}_\mathbf{r}^-\}$ by $\tilde{\mathcal{G}}_r(\mathbf{A})$. Thus, $\mathcal{G}(\mathbf{A})$ is semi-complete if and only if $\mathcal{G}_r(\mathbf{A}) = \tilde{\mathcal{G}}_\mathbf{r}(\mathbf{A})$. Also, each complete set is semi-complete and the same is true for g-maps.

Remark 7.2.18. For a g-map $\mathcal{G} : \mathbf{F}^{\mathbf{m}\times\mathbf{n}} \longrightarrow \mathfrak{P}(\mathbf{F}^{\mathbf{n}\times\mathbf{m}})$ and $\mathbf{A} \in \mathbf{F}^{\mathbf{m}\times\mathbf{n}}$, the set $\mathcal{G}(\mathbf{A})$ such that $\{\mathbf{A}_\mathbf{r}^-\} \subseteq \mathcal{G}(\mathbf{A}) \subset \{\mathbf{A}^-\}$ is semi-complete.

Theorem 7.2.19. *Let $\mathcal{G} : \mathbf{F}^{\mathbf{m}\times\mathbf{n}} \longrightarrow \mathfrak{P}(\mathbf{F}^{\mathbf{n}\times\mathbf{m}})$ be g-map and let $\mathbf{A} \in \mathbf{F}^{\mathbf{m}\times\mathbf{n}}$. Then $\mathcal{G}(\mathbf{A})$ is semi-complete if and only if $\mathcal{G}(\mathbf{A},\mathbf{A}) \subseteq \mathcal{G}(\mathbf{A})$.*

Proof. 'Only if' part
Let $\mathcal{G}(\mathbf{A})$ be semi-complete and $\mathbf{GAG} \in \mathcal{G}(\mathbf{A},\mathbf{A})$ for $\mathbf{G} \in \mathcal{G}(\mathbf{A})$. Clearly, $\mathbf{GAG} \in \widetilde{\mathcal{G}(\mathbf{A})}$ and is a reflexive g-inverse of $\mathbf{A}$. Therefore,

$$\mathbf{GAG} \in \widetilde{\mathcal{G}(\mathbf{A})} \cap \{\mathbf{A}_\mathbf{r}^-\} = \mathcal{G}(\mathbf{A}) \cap \{\mathbf{A}_\mathbf{r}^-\}$$

and hence $\mathcal{G}(A,A) \subseteq \mathcal{G}(\mathbf{A})$.
'If' part
Let $\mathcal{G}(A,A) \subseteq \mathcal{G}(\mathbf{A})$. As $\mathcal{G}(\mathbf{A}) \cap \{\mathbf{A}_\mathbf{r}^-\} \subseteq \widetilde{\mathcal{G}(\mathbf{A})} \cap \{\mathbf{A}_\mathbf{r}^-\}$, we only need to show that $\widetilde{\mathcal{G}(\mathbf{A})} \cap \{\mathbf{A}_\mathbf{r}^-\} \subseteq \mathcal{G}(\mathbf{A}) \cap \{\mathbf{A}_\mathbf{r}^-\}$. So, let $\mathbf{H} \in \widetilde{\mathcal{G}(\mathbf{A})} \cap \{\mathbf{A}_\mathbf{r}^-\}$. Then there exists a $\mathbf{G} \in \mathcal{G}(\mathbf{A})$ such that $\mathbf{GAG} = \mathbf{HAH} = \mathbf{H}$, as $\mathbf{H}$ is a reflexive g-inverse of $\mathbf{A}$. But then $\mathbf{GAG} \in \mathcal{G}(\mathbf{A},\mathbf{A}) \subseteq \mathcal{G}(\mathbf{A})$, so $\mathbf{H} \in \{\mathbf{A}_\mathbf{r}^-\} \subseteq \mathcal{G}(\mathbf{A})$. Hence, $\mathcal{G}(\mathbf{A})$ is semi-complete. □

Corollary 7.2.20. *A g-map satisfying the (T)-condition is semi-complete.*

Theorem 7.2.21. *Let $\mathcal{G} : \mathbf{F}^{\mathbf{m}\times\mathbf{n}} \longrightarrow \mathfrak{P}(\mathbf{F}^{\mathbf{n}\times\mathbf{m}})$ be a semi-complete g-map. Then for $\mathbf{A},\mathbf{B} \in \mathbf{\Omega}_\mathcal{G}$, the pair $(\mathbf{A},\mathbf{B})$ satisfies the (T)-condition if and only if $\mathcal{G}(\mathbf{B}) \subseteq \widetilde{\mathcal{G}(\mathbf{A})}$.*

Proof. 'If' part
Let $\mathcal{G}(\mathbf{B}) \subseteq \widetilde{\mathcal{G}(\mathbf{A})}$ and $\mathbf{H} \in \mathcal{G}(\mathbf{B})$ such that $\mathbf{HAH} \in \mathcal{G}(\mathbf{A},\mathbf{B})$. There exists a $\mathbf{G} \in \mathcal{G}(\mathbf{A})$ such that $\mathbf{AH} = \mathbf{AG}$, $\mathbf{HA} = \mathbf{GA}$. As $\mathcal{G}$ is semi-complete we have $\mathbf{HAH} = \mathbf{GAG} \in \mathcal{G}(\mathbf{A},\mathbf{A}) \subseteq \mathcal{G}(\mathbf{A})$. So, $\mathbf{HAH} \in \mathcal{G}(\mathbf{A})$. Thus, the pair $(\mathbf{A},\mathbf{B})$ satisfies the (T)-condition.
'Only if' part
Let $\mathbf{H} \in \mathcal{G}(\mathbf{B})$. Then $\mathbf{HAH} \in \mathcal{G}(\mathbf{A},\mathbf{B})$ and the pair $(\mathbf{A},\mathbf{B})$ satisfies the (T)-condition, so, there exists a $\mathbf{G} \in \mathcal{G}(\mathbf{A})$ such that $\mathbf{HAH} = \mathbf{G}$. It is now clear that $\mathbf{H} \in \widetilde{\mathcal{G}(\mathbf{A})}$. □

Our next theorem is a step in the direction of unification of these host of partial orders that we have studied in the earlier chapters.

Theorem 7.2.22. *Let* $\mathcal{G} : \mathbf{F}^{\mathbf{m}\times\mathbf{n}} \longrightarrow \mathfrak{P}(\mathbf{F}^{\mathbf{n}\times\mathbf{m}})$ *be a complete g-map and* $\mathbf{A}, \mathbf{B} \in \mathbf{\Omega}_{\mathcal{G}}$. *Then the following are equivalent:*

(i) $\mathbf{A} <^{-} \mathbf{B}$ *and the pair* $(\mathbf{A}, \mathbf{B})$ *satisfies the (T)-condition* *and*
(ii) $\mathcal{G}(\mathbf{B}) \subseteq \mathcal{G}(\mathbf{A})$ *and* $\mathbf{A} <^{\mathbf{s}} \mathbf{B}$.

Proof. In view of Theorem 7.2.21, (i) $\Rightarrow$ (ii) follows from Theorem 3.3.5.
(ii) $\Rightarrow$ (i)
We only need to prove that $\mathbf{AB^{-}A} = \mathbf{A}$ for some $\mathbf{B}^{-} \in \{\mathbf{B}^{-}\}$. Let $\mathbf{B}^{-} \in \mathcal{G}(\mathbf{B})$. As $\mathcal{G}(\mathbf{B}) \subseteq \mathcal{G}(\mathbf{A})$, $\mathbf{B}^{-} \in \mathcal{G}(\mathbf{A})$, so, $\mathbf{B}^{-}$ is g-inverse of $\mathbf{A}$. Thus, $\mathbf{AB^{-}A} = \mathbf{A}$ and by Remark 3.3.12, $\mathbf{A} <^{-} \mathbf{B}$. □

Theorem 7.2.23. *Let* $\mathcal{G} : \mathbf{F}^{\mathbf{m}\times\mathbf{n}} \longrightarrow \mathfrak{P}(\mathbf{F}^{\mathbf{n}\times\mathbf{m}})$ *be a semi-complete g-map and* $\mathbf{A}, \mathbf{B} \in \mathbf{\Omega}_{\mathcal{G}}$. *Then the following are equivalent:*

(i) $\mathbf{A} <^{-} \mathbf{B}$ *and the pair* $(\mathbf{A}, \mathbf{B})$ *satisfies the (T)-condition*
(ii) $\widetilde{\mathcal{G}(\mathbf{B})} \subseteq \widetilde{\mathcal{G}(\mathbf{A})}$ *and* $\mathbf{A} <^{\mathbf{s}} \mathbf{B}$ *and*
(iii) $\mathcal{G}(\mathbf{B}) \subseteq \widetilde{\mathcal{G}(\mathbf{A})}$ *and* $\mathbf{A} <^{\mathbf{s}} \mathbf{B}$.

Proof. (i) $\Rightarrow$ (ii)
We first note that the set $\widetilde{\mathcal{G}(\mathbf{A}, \mathbf{B})} = \{\mathbf{HAH} : \mathbf{H} \in \widetilde{\mathcal{G}(\mathbf{B})}\} \subseteq \widetilde{\mathcal{G}(\mathbf{A})}$. To see this, let $\mathbf{HAH} \in \widetilde{\mathcal{G}(\mathbf{A}, \mathbf{B})}$ with $\mathbf{H} \in \widetilde{\mathcal{G}(\mathbf{B})}$. As $\mathbf{H} \in \widetilde{\mathcal{G}(\mathbf{B})}$, there exists a $\mathbf{G} \in \mathcal{G}(\mathbf{B})$ such that $\mathbf{HB} = \mathbf{GB}$, $\mathbf{BH} = \mathbf{BG}$. So, $\mathbf{HBH} = \mathbf{GBG}$. Since $\mathbf{A} <^{-} \mathbf{B}$, we have

$$\mathbf{A} = \mathbf{AGA} = \mathbf{AGB} = \mathbf{BGA} = \mathbf{AHB} = \mathbf{BHA}.$$

Hence, $\mathbf{HAH} = \mathbf{H(AHB)H} = \mathbf{H((BHA)HB)H} = \mathbf{(HBH)A(HBH)} = \mathbf{(GBG)A(GBG)} = \mathbf{GAG}$. Thus, $\mathbf{HAH} \in \mathcal{G}(\mathbf{A}, \mathbf{B}) \subseteq \mathcal{G}(\mathbf{A}) \subseteq \widetilde{\mathcal{G}(\mathbf{A})}$. Since, $\tilde{\mathcal{G}}$ is a complete map, by Theorem 7.2.22, (ii) follows.
(ii) $\Rightarrow$ (iii) is trivial.
(iii) $\Rightarrow$ (i)
By Theorem 7.2.21, the pair $(\mathbf{A}, \mathbf{B})$ satisfies the (T)-condition. Further, $\mathbf{A} <^{-} \mathbf{B}$ follows along the lines of (ii) $\Rightarrow$ (i) of Theorem 7.2.22. □

Remark 7.2.24. Notice that the completion of $\{\mathbf{A}^{\#}\}$ is $\{\mathbf{A}^{-}_{\mathrm{com}}\}$, the class of commuting g-inverses of $\mathbf{A}$ and the completion of $\{\mathbf{A}^{\dagger}\}$ is $\{\mathbf{A}^{-}_{\ell \mathbf{m}}\}$, the class of least squares minimum norm g-inverses of $\mathbf{A}$. Compare now Theorem 4.2.8(i) $\Rightarrow$ (iv) $\Rightarrow$ (v) and Theorem 5.2.8 (i) $\Rightarrow$ (v) with Theorems 7.2.22 and 7.2.23.

Corollary 7.2.25. *Let $\mathcal{G} : \mathbf{F}^{\mathbf{m}\times\mathbf{n}} \longrightarrow \mathfrak{P}(\mathbf{F}^{\mathbf{n}\times\mathbf{m}})$ be a g-map. Then the order relation defined as '$\mathbf{A} <_{\mathcal{G}} \mathbf{B}$ if $\mathbf{A} <^{-} \mathbf{B}$ and the pair $(\mathbf{A}, \mathbf{B})$ satisfies the (T)-condition' defines a partial order if and only if $\mathcal{G}$ is semi-complete.*

Proof. 'If' part follows by reflexivity and the (T)-condition.
'Only if' part
We only need to show that transitivity holds, which follows from Theorem 7.2.23 and the fact that the completion of the completion is the completion itself. □

Remark 7.2.26. Theorems 7.2.13 and Corollary 7.2.25 together are analogues of Theorem 4.2.12(i) ⇔ (vi) and Theorem 5.2.10(iii) $(a_2) \Leftrightarrow (f_2)$.

Remark 7.2.27. The hypothesis $\mathbf{A} <^{\mathbf{s}} \mathbf{B}$ in Theorems 7.2.22 and 7.2.23 cannot be dropped as the following example shows:

Example 7.2.28. Let $\mathbf{A}, \mathbf{B} \in \mathbf{F}^{\mathbf{m}\times\mathbf{n}}$, $m > n$ be matrices having same rank and a common left inverse $\mathbf{C}$. (For example take $\begin{pmatrix} 1 & 1 \\ 0 & 0 \\ 1 & 2 \end{pmatrix}$, and $\begin{pmatrix} 1 & 1 \\ 1 & 1 \\ 1 & 2 \end{pmatrix}$) Define $\mathcal{G} : \mathbf{F}^{\mathbf{m}\times\mathbf{n}} \longrightarrow \mathfrak{P}(\mathbf{F}^{\mathbf{n}\times\mathbf{m}})$ as follows:

$$\mathcal{G}(\mathbf{X}) = \begin{cases} \mathbf{C}, & \text{for } \mathbf{X} = \mathbf{A} \\ \mathbf{C}, & \text{for } \mathbf{X} = \mathbf{B} \\ \mathbf{X}^{-}, & \text{for all other } \mathbf{X} \in \mathbf{F}^{\mathbf{m}\times\mathbf{n}}. \end{cases}$$

Then $\mathcal{G}$ is a complete map. If Theorem 7.2.23 were to be true, we would have $\mathbf{A} <^{-} \mathbf{B}$. Since $\mathbf{A}$ and $\mathbf{B}$ have same rank, this would mean $\mathbf{A} = \mathbf{B}$ and that is not true.

Corollary 7.2.29. *Let $\mathcal{G} : \mathbf{F}^{\mathbf{m}\times\mathbf{n}} \longrightarrow \mathfrak{P}(\mathbf{F}^{\mathbf{n}\times\mathbf{m}})$ be a g-map defined as $\mathcal{G}(\mathbf{A}) = \{\mathbf{A}^{+}\}$, where $\mathbf{A}^{+} \in \{\mathbf{A}_{\mathbf{r}}^{-}\}$ for each $\mathbf{A} \in \mathbf{F}^{\mathbf{m}\times\mathbf{n}}$. Then the following are equivalent:*

(i) *The order relation $\mathbf{A} <^{\mathcal{G}} \mathbf{B}$ is a partial order on $\Omega_{\mathcal{G}}$ and*
(ii) *$\mathbf{A} <^{\mathcal{G}} \mathbf{B} \Rightarrow$ the pair $(\mathbf{A}, \mathbf{B})$ satisfies the (T)-condition.*

Proof. Notice that $\mathcal{G}$ is semi-complete. Now, (ii) ⇒ (i) follows by Theorem 7.2.13.
(i) ⇒ (ii)
By Theorem 7.2.23, we have to show that $\mathcal{G}(\mathbf{B}) \subseteq \widetilde{\mathcal{G}(\mathbf{A})}$ or equivalently $\mathbf{B}^{+}\mathbf{A} = \mathbf{A}^{+}\mathbf{A}$ and $\mathbf{A}\mathbf{B}^{+} = \mathbf{A}\mathbf{A}^{+}$. This is clear, since $\mathbf{A} <^{\mathcal{G}} \mathbf{B} \Rightarrow \mathbf{A} <^{-} \mathbf{B}$.

Hence, $\mathbf{B}^+$ is a g-inverse of $\mathbf{A}$. As $\mathcal{G}$ is semi-complete, $\mathbf{B}^+\mathbf{A}\mathbf{B}^+ \in \mathcal{G}(\mathbf{A})$. Thus, $\mathbf{B}^+\mathbf{A}\mathbf{B}^+ = \mathbf{A}^+$. □

Recall that $\mathbf{A} <^- \mathbf{B} \Leftrightarrow \mathbf{B} - \mathbf{A} <^- \mathbf{B}$ and similar results hold for the star and the sharp order. However, $\mathbf{A} <^{\mathcal{G}} \mathbf{B} \not\Rightarrow \mathbf{B} - \mathbf{A} <^{\mathcal{G}} \mathbf{B}$. Similarly, $\mathbf{A} <_{\mathcal{G}} \mathbf{B} \not\Rightarrow \mathbf{B} - \mathbf{A} <_{\mathcal{G}} \mathbf{B}$. We give the following example:

Example 7.2.30. Let $\mathcal{G} : \mathbb{C}^{4\times 3} \longrightarrow \mathfrak{P}(\mathbb{R}^{3\times 4})$ be the g-map defined as

$$\mathcal{G}(\mathbf{A}) = \begin{cases} \mathbf{A}^-, & \text{if } \rho(\mathbf{A}) < \delta \\ \mathbf{A}^\dagger, & \text{otherwise} \end{cases}$$

where $\delta = 2$. Thus, if $\rho(\mathbf{A}) < \delta$, $\mathbf{A} <^{\mathcal{G}} \mathbf{B} \Leftrightarrow \mathbf{A} <^- \mathbf{B}$ and if $\rho(\mathbf{A}) \geq \delta$, then $\mathbf{A} <^{\mathcal{G}} \mathbf{B} \Leftrightarrow \mathbf{A} <^\star \mathbf{B}$. Clearly, the order relation '$<^{\mathcal{G}}$' is reflexive and anti-symmetric. We only need to check the transitivity. So, let $\mathbf{A} <^{\mathcal{G}} \mathbf{B}$ and $\mathbf{B} <^{\mathcal{G}} \mathbf{C}$. We consider the following cases:

(i) Let $\rho(\mathbf{A}) < \delta$ and $\rho(\mathbf{B}) < \delta$, or $\rho(\mathbf{A}) < \delta$ and $\rho(\mathbf{B}) \geq \delta$. Since the minus order is transitive, in either case '$<^{\mathcal{G}}$' is transitive.
(ii) Let $\rho(\mathbf{A}) \geq \delta$. Consequently, $\rho(\mathbf{B}) \geq \delta$ and $\rho(\mathbf{C}) \geq \delta$. So,$\mathbf{A} <^\star \mathbf{B}$ and $\mathbf{B} <^\star \mathbf{C}$. As the star order is transitive, '$<^{\mathcal{G}}$' is transitive.

Trivially, the map $\mathcal{G}$ satisfies (T)-condition. Let $\mathbf{K} = \begin{pmatrix} 1 & 0 & 1 \\ 0 & 1 & 0 \\ 0 & 0 & 0 \\ 0 & 0 & 0 \end{pmatrix}$. Then

$\rho(\mathbf{K}) = 2$ and $\mathbf{K}^\dagger = \begin{pmatrix} 1/2 & 0 & 0 & 0 \\ 0 & 1 & 0 & 0 \\ 1/2 & 0 & 0 & 0 \end{pmatrix}$. Choose $\mathbf{u_1} = \begin{pmatrix} 1 \\ 0 \\ 0 \\ 0 \end{pmatrix}$, $\mathbf{u_2} = \begin{pmatrix} 0 \\ 0 \\ 0 \\ 1 \end{pmatrix}$,

$\mathbf{v_1} = \begin{pmatrix} 1/2 \\ 0 \\ 1/2 \end{pmatrix}$ and $\mathbf{v_2} = \begin{pmatrix} 0 \\ 0 \\ -1 \end{pmatrix}$. Also, $\mathbf{u_1} \in \mathcal{C}(\mathbf{K})$, $\mathbf{u_2} \in \mathcal{C}(\mathbf{I} - \mathbf{K}\mathbf{K}^\dagger)$, $\mathbf{v_1} \in \mathcal{C}(\mathbf{K}^\dagger)$ and $\mathbf{v_2} \in \mathcal{C}(\mathbf{I} - \mathbf{K}^\dagger\mathbf{K})$. Let $\mathbf{u} = \mathbf{u_1} + \mathbf{u_2}$, $\mathbf{v} = \mathbf{v_1} + \mathbf{v_2}$, and

$$\mathbf{A} = \mathbf{u}\mathbf{v}^\star = \begin{pmatrix} 1/2 & 0 & -1/2 \\ 0 & 0 & 0 \\ 0 & 0 & 0 \\ 1/2 & 0 & 1/2 \end{pmatrix}, \quad \mathbf{B} = \mathbf{K} + \mathbf{A} = \begin{pmatrix} 3/2 & 0 & -1/2 \\ 0 & 1 & 0 \\ 0 & 0 & 0 \\ 3/2 & 0 & -1/2 \end{pmatrix}.$$

Note that $\mathbf{B} = \mathbf{A} \oplus (\mathbf{B} - \mathbf{A})$. Further, $\mathbf{A}^\star(\mathbf{B} - \mathbf{A}) \neq \mathbf{0}$ and also, $(\mathbf{B} - \mathbf{A})\mathbf{A}^\star \neq \mathbf{0}$. Thus, $\mathbf{B} - \mathbf{A} \not<^\star \mathbf{B}$ implying $\mathbf{B} - \mathbf{A} \not<^{\mathcal{G}} \mathbf{B}$.

We have seen that the (T)-condition plays a vital role in proving transitivity of an order relation. What about in reverse direction, i.e. if an order relation is transitive, whether the g-map defining it satisfies the (T)-condition? In our next theorem we show that the answer is in affirmative when the matrices are square and the g-map defining it is semi-complete. We also note that there is no loss of generality, if the g-map is taken to be complete.

Theorem 7.2.31. *Let $\mathcal{G} : \mathbf{F^{n\times n}} \longrightarrow \mathfrak{P}(\mathbf{F^{n\times n}})$ be a complete g-map. If the order relation '$<^{\mathcal{G}}$' is transitive, then $\mathbf{A} <^{\mathcal{G}} \mathbf{B}$, $\mathbf{B}$ not maximal implies that the pair $(\mathbf{A}, \mathbf{B})$ satisfies the (T)-condition.*

Proof. Let $\mathbf{A}, \mathbf{B} \in \mathbf{F^{n\times n}}$ such that $\mathbf{A} <^{\mathcal{G}} \mathbf{B}$, $\mathbf{B}$ not maximal. Let $\mathbf{GAG} \in \mathcal{G}(\mathbf{A}, \mathbf{B})$ with $\mathbf{G} \in \mathcal{G}(\mathbf{B})$. Write $\mathbf{H} = \mathbf{GBG}$. Then $\mathbf{GAG} = \mathbf{HAH}$ and $\mathbf{H} \in \widetilde{\mathcal{G}(\mathbf{B})} = \mathcal{G}(\mathbf{B})$. Also, $\mathbf{H}$ being reflexive g-inverse of $\mathbf{B}$, $\mathbf{H} \in \mathcal{G}_{\mathbf{r}}(\mathbf{B})$. Let $\rho(\mathbf{B}) = r$. Then $\mathbf{HB}$ and $\mathbf{BH}$ are idempotent of rank r and therefore, $\mathbf{I} - \mathbf{HB}$ and $\mathbf{I} - \mathbf{BH}$ are idempotent of rank $n - r$. Let $j_1, j_2, \ldots, j_{n-r}$ be the $n - r$ linearly independent columns of $\mathbf{I} - \mathbf{BH}$ and $\mathbf{U_1}$ be the matrix formed by the corresponding columns of $\mathbf{I_n}$. Similarly, let $i_1, i_2, \ldots, i_{n-r}$ be the $n - r$ linearly independent rows of $\mathbf{I} - \mathbf{BH}$ and $\mathbf{U_2}$ be the matrix formed by the corresponding rows of $\mathbf{I_n}$. Then $\mathbf{U_1}$ and $\mathbf{U_2}$ are full rank matrices. Let $\mathbf{U} = \mathbf{U_1U_2}$. The matrix $(\mathbf{I} - \mathbf{BH})\mathbf{U}(\mathbf{I} - \mathbf{HB})$ has rank $n - r$. Let $\mathbf{V_1}$ be a right inverse of $\mathbf{U_2}(\mathbf{I} - \mathbf{HB})$, $\mathbf{V_2}$ be a left inverse of $(\mathbf{I} - \mathbf{BH})\mathbf{U_1}$ and $\mathbf{V} = \mathbf{V_1V_2}$. Write $\mathbf{C} = \mathbf{B} + (\mathbf{I} - \mathbf{BH})\mathbf{U}(\mathbf{I} - \mathbf{HB})$ and note that $\mathbf{C}$ is non-singular with $\mathbf{C}^{-1} = \mathbf{H} + (\mathbf{I} - \mathbf{HB})\mathbf{V}(\mathbf{I} - \mathbf{BH})$. Moreover, $\mathbf{HC} = \mathbf{HB}$, $\mathbf{CH} = \mathbf{BH}$ and $\mathbf{H} \in \mathcal{G}_{\mathbf{r}}(\mathbf{B})$. Therefore, $\mathbf{B} <^{\mathcal{G}} \mathbf{C}$. Since '$<^{\mathcal{G}}$' is transitive, $\mathbf{A} <^{\mathcal{G}} \mathbf{C}$. So, there exists a $\mathbf{G_0} \in \mathcal{G}(\mathbf{A})$ such that $\mathbf{G_0A} = \mathbf{G_0C}$, $\mathbf{AG_0} = \mathbf{CG_0}$. As $\mathcal{G}$ is semi-complete, $\mathbf{G_0AG_0} \in \mathcal{G}(\mathbf{A})$ and is a reflexive g-inverse of $\mathbf{A}$. So, we can take $\mathbf{G_0} \in \mathcal{G}_{\mathbf{r}}(\mathbf{A})$. Clearly, $\mathbf{A} = \mathbf{CG_0C}$ or $\mathbf{C}^{-1}\mathbf{AC}^{-1} = \mathbf{G_0}$. As $\mathbf{C}^{-1}\mathbf{AC}^{-1} = \mathbf{HAH}$, therefore, $\mathbf{GAG} = \mathbf{HAH} \in \mathcal{G}_{\mathbf{r}}(\mathbf{A}) \subseteq \mathcal{G}(\mathbf{A})$. Thus, the pair $(\mathbf{A}, \mathbf{B})$ satisfies the (T)-condition. □

Remark 7.2.32. (i) Notice that $<^{\mathcal{G}} \Rightarrow <^{\tilde{\mathcal{G}}}$.
(ii) For any g-map $\mathcal{G}$ on $\mathbf{F^{m\times n}}$, $\tilde{\mathcal{G}}(\mathbf{A}, \mathbf{B}) = \mathcal{G}(\mathbf{A}, \mathbf{B})$ and if $\mathcal{G}$ is semi-complete, then $\tilde{\mathcal{G}}(\mathbf{A}, \mathbf{B}) \subseteq \tilde{\mathcal{G}}(\mathbf{A}) \Rightarrow \mathcal{G}(\mathbf{A}, \mathbf{B}) \subseteq \mathcal{G}(\mathbf{A})$.

Corollary 7.2.33. *Let $\mathcal{G} : \mathbf{F^{n\times n}} \longrightarrow \mathfrak{P}(\mathbf{F^{n\times n}})$ be a complete or semi-complete g-map. Then the following are equivalent:*

(i) *$<^{\mathcal{G}}$ is a partial order*

(ii) $\mathbf{A} <^{\mathcal{G}} \mathbf{B}$, $\mathbf{B}$ *not maximal implies* $\mathcal{G}_r(\mathbf{A}, \mathbf{B}) \subseteq \mathcal{G}_{\mathbf{r}}(\mathbf{A})$ *and*

(iii) $\mathbf{A} <^{\mathcal{G}} \mathbf{B}$, $\mathbf{B}$ *not maximal implies the pair* $(\mathbf{A}, \mathbf{B})$ *satisfies the (T)-condition.*

Proof. Proof for (i) $\Rightarrow$ (iii) follows by Theorem 7.2.31 and for (i) $\Rightarrow$ (ii) follows along the lines of Theorem 7.2.31. For (iii) $\Rightarrow$ (i), and (ii) $\Rightarrow$ (i) use Theorem 7.2.19. □

Before we prove the last result of this section, we need another definition.

Definition 7.2.34. Let '$<^{\mathcal{G}}$' and '$<^{\mathcal{G}_0}$' be two order relations on $\mathbf{F}^{\mathbf{m} \times \mathbf{n}}$. Then we say the order relation '$<^{\mathcal{G}}$' is finer than the order relation '$<^{\mathcal{G}_0}$' if $<^{\mathcal{G}} \Rightarrow <^{\mathcal{G}_0}$. Said otherwise, $\mathcal{G}(\mathbf{A}) \subseteq \mathcal{G}_{\mathbf{0}}(\mathbf{A})$ for each $\mathbf{A} \in \mathbf{F}^{\mathbf{m} \times \mathbf{n}}$.

For example, each of the star order and the sharp order are finer than the minus order. Any $\mathcal{G}$-based order relation is finer than the minus order. Our next theorem is a very powerful result, for it shows how two completely unrelated partial orders influence each others presence.

Theorem 7.2.35. *Let* $<^{\mathcal{G}}$ *and* $<^{\mathcal{G}^o}$ *be two distinct* $\mathcal{G}$*-based partial orders on* $\mathbf{F}^{\mathbf{m} \times \mathbf{n}}$. *If* $\mathbf{A} <^{\mathcal{G}} \mathbf{C}$ *and* $\mathbf{A} <^{\mathcal{G}^o} \mathbf{B} <^{\mathcal{G}^o} \mathbf{C}$, *then* $\mathbf{A} <^{\mathcal{G}} \mathbf{B}$.

Proof. The statement is clearly true if $<^{\mathcal{G}^o}$ is finer than $<^{\mathcal{G}}$. Let us suppose the order relations are not related so. Since $\mathbf{A} <^{\mathcal{G}^o} \mathbf{B} <^{\mathcal{G}^o} \mathbf{C}$, we have $\mathbf{A} <^- \mathbf{B} <^- \mathbf{C}$. However, $\mathbf{A} <^- \mathbf{B} \Rightarrow \mathbf{B} - \mathbf{A} <^- \mathbf{B}$, $\mathbf{A} <^- \mathbf{C} \Rightarrow \mathbf{C} - \mathbf{A} <^- \mathbf{C}$ and $\mathbf{B} <^- \mathbf{C} \Rightarrow \mathbf{B} - \mathbf{C} <^- \mathbf{C}$. We can write $\mathbf{C} - \mathbf{A} = \mathbf{C} - \mathbf{B} + \mathbf{B} - \mathbf{A}$. Moreover,

$$\mathcal{C}(\mathbf{C} - \mathbf{B}) \cap \mathcal{C}(\mathbf{B} - \mathbf{A}) \subseteq \mathcal{C}(\mathbf{C} - \mathbf{B}) \cap \mathcal{C}(\mathbf{B}) = \mathbf{0},$$
$$\mathcal{C}(\mathbf{C} - \mathbf{B})^{\mathbf{t}} \cap \mathcal{C}(\mathbf{B} - \mathbf{A})^{\mathbf{t}} \subseteq \mathcal{C}(\mathbf{C} - \mathbf{B})^{\mathbf{t}} \cap \mathcal{C}(\mathbf{B})^{\mathbf{t}} = \mathbf{0}.$$

So, $\mathbf{C} - \mathbf{A} = (\mathbf{C} - \mathbf{B}) \oplus (\mathbf{B} - \mathbf{A})$. Since $\mathbf{A} <^{\mathcal{G}} \mathbf{C}$, there exists a $\mathbf{G} \in \mathcal{G}(\mathbf{A})$ such that $\mathbf{G}(\mathbf{C} - \mathbf{A}) = \mathbf{0}, (\mathbf{C} - \mathbf{A})\mathbf{G} = \mathbf{0}$. Therefore, $\mathbf{G}(\mathbf{C} - \mathbf{B}) = \mathbf{0}$, $\mathbf{G}(\mathbf{B} - \mathbf{A}) = \mathbf{0}$, $(\mathbf{C} - \mathbf{B})\mathbf{G} = \mathbf{0}$ and $(\mathbf{B} - \mathbf{A})\mathbf{G} = \mathbf{0}$. Thus, $\mathbf{A} <^{\mathcal{G}} \mathbf{B}$. □

7.3 $\mathcal{O}$-based order relations and their properties

Until now we have been concentrating on mainly the partial orders induced by g-inverses of matrices, since our main aim in this chapter is to unify various partial orders that are generalizations of the minus order. In this section we change our focus from the order relations defined through g-inverses to the order relations defined by the outer inverses. One reason that can be given for this is that in Chapter 4, we saw that the Drazin inverse (which is an outer inverse) was used to define a pre-order and later modified to give a partial order. This gives a clear indication that outer inverses can be used to obtain new partial orders. The other reason, as seen in Example 7.2.7, is that not all partial orders are $\mathcal{G}$-based order relations. So, where do such kind of order relations fit in the present scenario? We examine the possibility of getting an answer in partial orders that can be induced by selecting a suitable subset of outer inverses.

We shall denote an outer inverse of any matrix $\mathbf{A} \in \mathbf{F}^{\mathbf{m}\times\mathbf{n}}$ by $\mathbf{A}_-$ and the set all outer inverses of $\mathbf{A}$ by $\{\mathbf{A}_-\}$. So,

$$\{\mathbf{A}_-\} = \{\mathbf{G} : \mathbf{GAG} = \mathbf{G}\}.$$

We begin by giving some properties of the outer inverses.

Theorem 7.3.1. *Let* $\mathbf{B} \in \mathbf{F}^{\mathbf{m}\times\mathbf{n}}$ *and* $\mathbf{A} = \mathbf{BB}_-\mathbf{B}$ *for some* $\mathbf{B}_- \in \{\mathbf{B}_-\}$. *Then the following hold:*

(i) $\mathbf{B}_-$ *is reflexive g-inverse of* $\mathbf{A}$.
(ii) $\mathbf{A} <^- \mathbf{B}$.
(iii) $\mathbf{BA}_-\mathbf{B}$ and $\mathbf{B}_-\mathbf{BA}_-\mathbf{BB}_- \in \{\mathbf{B}_-\}$ *for all* $\mathbf{A}_- \in \{\mathbf{A}_-\}$.
(iv) $\mathbf{B}_- = \mathbf{B}_\mathbf{r}^-\mathbf{BB}_\mathbf{r}^-$ *for some* $\mathbf{B}_\mathbf{r}^- \in \{\mathbf{B}_\mathbf{r}^-\}$.
(v) $\mathbf{A} = \mathbf{BB}_\mathbf{r}^-\mathbf{AB}_\mathbf{r}^-\mathbf{B}$ *for some* $\mathbf{B}_\mathbf{r}^- \in \{\mathbf{B}_\mathbf{r}^-\}$.

Proof. (i) Since

$$\begin{aligned}\mathbf{AB}_-\mathbf{A} &= (\mathbf{BB}_-\mathbf{B})\mathbf{B}_-(\mathbf{BB}_-\mathbf{B}) = \mathbf{B}((\mathbf{B}_-\mathbf{BB}_-)(\mathbf{BB}_-)\mathbf{B})\\ &= (\mathbf{BB}_-\mathbf{B}) = \mathbf{A}\end{aligned}$$

and $\mathbf{B}_-\mathbf{AB}_- = \mathbf{B}_-(\mathbf{BB}_-\mathbf{B})\mathbf{B}_- = (\mathbf{B}_-\mathbf{BB}_-)\mathbf{BB}_- = \mathbf{B}_-$. So, the result follows.

(ii) Notice that $\mathbf{A} = \mathbf{BB}_-\mathbf{B} \Rightarrow \mathbf{B}_-\mathbf{A} = \mathbf{B}_-\mathbf{B}$ and $\mathbf{AB}_- = \mathbf{BB}_-$. By (i) $\mathbf{B}_-$ is reflexive g-inverse of $\mathbf{A}$, hence, $\mathbf{A} <^- \mathbf{B}$.

(iii) is easy.

(iv) Proof follows from(i), (ii) and Remark 3.3.3(ii).

(v) follows from (i) and (iii). □

We now give definition of an o-map and an $\mathcal{O}$-based order relation in analogy with a g-map.

Definition 7.3.2. An o-map is a map $\mathcal{O} : \mathbf{F}^{\mathbf{m}\times\mathbf{n}} \longrightarrow \mathfrak{P}(\mathbf{F}^{\mathbf{n}\times\mathbf{m}})$ such that for each $\mathbf{A} \in \mathbf{F}^{\mathbf{n}\times\mathbf{m}}$, $\mathcal{O}(\mathbf{A})$ is certain specified (possibly non-empty) subset of $\{\mathbf{A}_-\}$, the set all outer inverses of $\mathbf{A}$.

Definition 7.3.3. Let $\mathcal{O} : \mathbf{F}^{\mathbf{m}\times\mathbf{n}} \longrightarrow \mathfrak{P}(\mathbf{F}^{\mathbf{n}\times\mathbf{m}})$ be an o-map. We define an order relation '$<^{\mathcal{O}}$' on $\mathbf{F}^{\mathbf{n}\times\mathbf{m}}$ as follows:

For $\mathbf{A}, \mathbf{B} \in \mathbf{F}^{\mathbf{m}\times\mathbf{n}}$, $\mathbf{A} <^{\mathcal{O}} \mathbf{B}$ if $\mathbf{A} = \mathbf{BB}_-\mathbf{B}$ for some $\mathbf{B}_- \in \mathcal{O}(\mathbf{B})$

and call it an $\mathcal{O}$-based order relation.

Remark 7.3.4. (i) Notice that $\mathbf{A} = \mathbf{B}\mathbf{B}_-\mathbf{B}$ for some $\mathbf{B}_- \in \{\mathbf{B}_-\}$ implies $\mathbf{B}_-\mathbf{A} = \mathbf{B}_-\mathbf{B}$ and $\mathbf{A}\mathbf{B}_- = \mathbf{B}\mathbf{B}_-$. However, the converse is not true. For example, let $\mathbf{A}, \mathbf{B} \in \mathbf{F}^{\mathbf{n}\times\mathbf{m}}$ be distinct matrices such that $\mathbf{B} <^{\star} \mathbf{A}$. Then $\mathbf{B}^{\dagger}\mathbf{B} = \mathbf{B}^{\dagger}\mathbf{A}$, $\mathbf{B}\mathbf{B}^{\dagger} = \mathbf{A}\mathbf{B}^{\dagger}$, but $\mathbf{A} \neq \mathbf{B}\mathbf{B}^{\dagger}\mathbf{B}$.
(ii) If $\mathcal{O}(\mathbf{A}) = \{\mathbf{A}_-\}$, then by Theorem 7.3.1(ii), the order relation '$<^{\mathcal{O}}$' is just the minus order.
(iii) If $\mathcal{O}(\mathbf{A}) = \{\mathbf{0}\}$, then $\mathbf{A} = \mathbf{0}$.

In view of Theorem 7.3.1 and Remark 7.3.4(ii), we have

Theorem 7.3.5. *Let* $\mathcal{O} : \mathbf{F}^{\mathbf{m}\times\mathbf{n}} \longrightarrow \mathfrak{P}(\mathbf{F}^{\mathbf{n}\times\mathbf{m}})$ *be an o-map and* $\mathbf{A}, \mathbf{B} \in \mathbf{F}^{\mathbf{n}\times\mathbf{m}}$. *Then the following hold:*

(i) $\mathbf{A} <^{\mathcal{O}} \mathbf{B} \Rightarrow \mathbf{A} <^{-} \mathbf{B}$.
(ii) $\mathcal{O}(\mathbf{A}) = \{\mathbf{A}_-\} \Rightarrow \mathbf{A} <^{\mathcal{O}} \mathbf{B} \Leftrightarrow \mathbf{A} <^{-} \mathbf{B}$.
(iii) $\mathbf{A} <^{\mathcal{O}} \mathbf{B} \Rightarrow \mathbf{B}_-\mathbf{B}\mathbf{A}_-\mathbf{B}\mathbf{B}_- <^{\mathcal{O}} \mathbf{B}_-$.

Proof. (ii) We only need to prove the $\mathbf{A} <^{-} \mathbf{B} \Rightarrow \mathbf{A} <^{\mathcal{O}} \mathbf{B}$. Notice that for each $\mathbf{B}^{-} \in \{\mathbf{B}^{-}\}, \mathbf{B}^{-}\mathbf{A}\mathbf{B}^{-} \in \mathcal{O}(\mathbf{B})$ and $\mathbf{B}\mathbf{B}^{-}\mathbf{A}\mathbf{B}^{-}\mathbf{B} = \mathbf{A}$.
(iii) We first note that $\mathbf{B}\mathbf{A}_-\mathbf{B}$ is an outer inverse of $\mathbf{B}_-$ and $\mathbf{B}_-\mathbf{B}\mathbf{A}_-\mathbf{B}\mathbf{B}_- = \mathbf{B}_-(\mathbf{B}\mathbf{A}_-\mathbf{B})\mathbf{B}_-$, so (iii) holds. □

Even before we dwell upon the question: whether or not an $\mathcal{O}$-based order relation is a partial order, we show that any partial order finer than the minus order is an $\mathcal{O}$-based order relation.

Theorem 7.3.6. *Let '$<$' be a partial order on* $\mathbf{F}^{\mathbf{m}\times\mathbf{n}}$ *that is finer than '$<^{-}$' the minus order. Then '$<$' is an* $\mathcal{O}$*-based order relation.*

Proof. Let $\mathbf{B} \in \mathbf{F}^{\mathbf{m}\times\mathbf{n}}$. Write $\mathbf{D}_{\mathbf{B}} = \{\mathbf{C} : \mathbf{C} < \mathbf{B}\}$. Then $\mathbf{D}_{\mathbf{B}}$ is non-empty. Choose and fix a reflexive g-inverse $\mathbf{G}$ of $\mathbf{B}$ and notice that $\mathbf{G}$ is g-inverse of $\mathbf{C}$, since, $\mathbf{C} < \mathbf{B} \Rightarrow \mathbf{C} <^{-} \mathbf{B}$. Clearly, $\mathbf{GCG}$ is an outer inverse of $\mathbf{B}$ for each $\mathbf{C} \in \mathbf{D}_{\mathbf{B}}$. Let $\mathcal{O}(\mathbf{B}) = \{\mathbf{GCG} : \mathbf{C} \in \mathbf{D}_{\mathbf{B}}\}$. We show that $\mathbf{A} < \mathbf{B} \Leftrightarrow \mathbf{A} <^{\mathcal{O}} \mathbf{B}$. So, let $\mathbf{A} < \mathbf{B}$. Since $<$ is finer than $<^{-}$, $\mathbf{A} <^{-} \mathbf{B}$. Also, $\mathbf{GAG}$ is a reflexive g-inverse of $\mathbf{A}$ and $\mathbf{A} = \mathbf{BGAGB}$. So, $\mathbf{A} <^{\mathcal{O}} \mathbf{B}$. Next, let $\mathbf{A} <^{\mathcal{O}} \mathbf{B}$. Then for some $\mathbf{C} \in \mathbf{D}_{\mathbf{B}}$, $\mathbf{GCG}$ is an outer inverse of $\mathbf{B}$, so, $\mathbf{A} = \mathbf{BGCGB} = \mathbf{C}$. Hence, $\mathbf{A} < \mathbf{B}$. □

Remark 7.3.7. Thus, the partial order in Example 7.2.7 is an $\mathcal{O}$-based order relation. It will be interesting to work out the classes $\mathcal{O}(\mathbf{A})$, $\mathcal{O}(\mathbf{B})$ and $\mathcal{O}(\mathbf{T})$ for $\mathbf{A}$, $\mathbf{B}$ and $\mathbf{T}$ for this example.

We now study when an $\mathcal{O}$-base order relation is a partial order. Since $\mathbf{A} <^{\mathcal{O}} \mathbf{B} \Rightarrow \mathbf{A} <^{-} \mathbf{B}$, this order relation is always anti-symmetric. It is also clear that this relation will be reflexive if for each $\mathbf{A}$, $\mathcal{O}(\mathbf{A})$ contains at least one reflexive g-inverse of $\mathbf{A}$. For transitivity, we have the following:

Theorem 7.3.8. *Let $\mathcal{O} : \mathbf{F}^{\mathbf{m}\times\mathbf{n}} \longrightarrow \mathfrak{P}(\mathbf{F}^{\mathbf{n}\times\mathbf{m}})$ be an o-map. Then the order relation $<^{\mathcal{O}}$ is transitive, if for each $\mathbf{C}_{=} \in \mathcal{O}(\mathbf{C})$ and $\mathbf{B} = \mathbf{C}\mathbf{C}_{=}\mathbf{C}$, the matrix $\mathbf{C}_{=}\mathbf{C}\mathbf{B}_{=}\mathbf{C}\mathbf{C}_{=} \in \mathcal{O}(\mathbf{C})$ for all $\mathbf{B}_{=} \in \mathcal{O}(\mathbf{B})$.*

Proof. Let $\mathbf{A} <^{\mathcal{O}} \mathbf{B}$, and $\mathbf{B} <^{\mathcal{O}} \mathbf{C}$. Then there exist $\mathbf{B}_{=} \in \mathcal{O}(\mathbf{B})$ and $\mathbf{C}_{=} \in \mathcal{O}(\mathbf{C})$ such that $\mathbf{A} = \mathbf{B}\mathbf{B}_{=}\mathbf{B}$, and $\mathbf{B} = \mathbf{C}\mathbf{C}_{=}\mathbf{C}$. Consider $\mathbf{C}\mathbf{C}_{=}\mathbf{C}\mathbf{B}_{=}\mathbf{C}\mathbf{C}_{=}\mathbf{C} = \mathbf{B}\mathbf{B}_{=}\mathbf{B} = \mathbf{A}$. As $\mathbf{C}_{=}\mathbf{C}\mathbf{B}_{=}\mathbf{C}\mathbf{C}_{=} \in \mathcal{O}(\mathbf{C})$, we have $\mathbf{A} <^{\mathcal{O}} \mathbf{C}$. □

Remark 7.3.9. (i) The matrix $\mathbf{C}_{=}\mathbf{C}\mathbf{B}_{=}\mathbf{C}\mathbf{C}_{=}$ is an outer inverse of $\mathbf{C}$.
(ii) Let $\mathbf{A}, \mathbf{B} \in \mathbf{F}^{\mathbf{m}\times\mathbf{n}}$ be such that $\mathbf{A} <^{-} \mathbf{B}$ and $\mathbf{G} = \mathbf{H}\mathbf{A}\mathbf{H}$ for some reflexive g-inverse $\mathbf{H}$ of $\mathbf{B}$. Then $\mathbf{G}$ is an outer inverse of $\mathbf{B}$ and $\mathbf{A} = \mathbf{B}\mathbf{G}\mathbf{B}$. This is a dual statement of Theorem 7.3.1(i) and (v).

The following theorem is comparable to Theorem 3.5.4(i):

Theorem 7.3.10. *Let $\mathbf{A}, \mathbf{B} \in \mathbf{F}^{\mathbf{m}\times\mathbf{n}}$ be such that $\mathbf{A} <^{-} \mathbf{B}$. Then for any $\mathbf{B}_{=} \in \mathcal{O}(\mathbf{B})$ for which $\mathbf{A} = \mathbf{B}\mathbf{B}_{=}\mathbf{B}$, $\mathbf{G} = \mathbf{B}_{=} + \mathbf{B}^{-}(\mathbf{B} - \mathbf{A})\mathbf{B}^{-}$ is a reflexive g-inverse of $\mathbf{B}$ for each $\mathbf{B}^{-} \in \{\mathbf{B}^{-}\}$.*

Proof is by direct verification.

Remark 7.3.11. Recall that $\mathbf{A} <^{-} \mathbf{B} \Leftrightarrow \{\mathbf{B}^{-}\} \subseteq \{\mathbf{A}^{-}\}$. Imitating this statement for outer inverses, one wonders if the order relation defined as: $\mathbf{A} < \mathbf{B} \Rightarrow \{\mathbf{A}_{=}\} \subseteq \{\mathbf{B}_{=}\}$ gives a partial order. Trivially, this relation is reflexive and transitive. The anti-symmetry can be checked as follows: Let $\mathbf{A} < \mathbf{B}$ and $\mathbf{B} < \mathbf{A}$. Then $\{\mathbf{A}_{=}\} = \{\mathbf{B}_{=}\}$. So, $\mathbf{A} = \mathbf{A}\mathbf{B}_{=}\mathbf{A}$ and $\mathbf{B} = \mathbf{B}\mathbf{A}_{=}\mathbf{B}$ for each $\mathbf{B}_{=}$ and for each $\mathbf{A}_{=}$. So, $\rho(\mathbf{A}) \leq \rho(\mathbf{B}_{=}) \leq \rho(\mathbf{B}) \leq \rho(\mathbf{A}_{=}) \leq \rho(\mathbf{A}) \Rightarrow \rho(\mathbf{A}) = \rho(\mathbf{B})$. It follows that $\{\mathbf{A}_{\mathbf{r}}^{-}\} = \{\mathbf{B}_{\mathbf{r}}^{-}\}$. Thus, $\mathbf{A} = \mathbf{B}$.

We conclude this section by giving relations between various $\mathcal{G}$-based order relations and $\mathcal{O}$-based order relations. The various generalized inverses satisfy suitable subsets the following set of equations:

(1) $\mathbf{A}\mathbf{X}\mathbf{A} = \mathbf{A}$
(2) $\mathbf{X}\mathbf{A}\mathbf{X} = \mathbf{A}$
(3) $(\mathbf{A}\mathbf{X})^{\mathbf{t}} = \mathbf{A}\mathbf{X}$

(4) $(\mathbf{XA})^{\mathbf{t}} = \mathbf{XA}$
(5) $\mathbf{AX} = \mathbf{XA}$
(6) $\mathcal{C}(\mathbf{X}) \subseteq \mathcal{C}(\mathbf{AX})$
(7) $\mathcal{C}(\mathbf{X}^{\mathbf{t}}) \subseteq \mathcal{C}(\mathbf{AX})^{\mathbf{t}}$

where '$\mathbf{t}$' denotes operation of taking transpose, if matrices are over arbitrary field and taking conjugate transpose if matrices are over complex field.

We denote by $\mathcal{S}$ any subset of $\{1, 2, 3, 4, 5, 6, 7\}$ that contains 2. Let $\mathcal{S}_1$ denote the subset obtained from $\mathcal{S}$ by replacing 2 by 1. We define $\overline{\mathcal{S}}$ as follows:

$$\overline{\mathcal{S}} = \begin{cases} \mathcal{S}, & \text{if } 1 \in \mathcal{S} \\ \mathcal{S}_1, & \text{otherwise.} \end{cases}$$

We write $\mathbf{X}(\mathcal{S})$ to be the set of all outer inverses of $\mathbf{X}$ that satisfy the equations corresponding to elements of $\mathcal{S}$, whenever meaningful. For example, if $\mathcal{S} = \{1, 2, 3\}$, the set $\mathbf{X}(\mathcal{S})$ is the set of all outer inverses of $\mathbf{X}$ that satisfy equations (1), (2) and (3). For $\mathcal{S} = \{6, 7\}$, the set $\mathbf{X}(\mathcal{S})$ is not defined. Similarly, the set $X(\overline{\mathcal{S}})$ will denote the set of generalized inverses of $\mathbf{X}$ that satisfy the equations corresponding to elements of $\overline{\mathcal{S}}$. With these notations in place we now prove

Theorem 7.3.12. *Let* $\mathbf{B} \in \mathbf{F}^{\mathbf{m}\times\mathbf{n}}$. *Let* $\mathbf{A} = \mathbf{B}\mathbf{B}_-\mathbf{B}$ *for some* $\mathbf{B}_- \in \mathbf{B}(\mathcal{S})$. *Then the condition:*

$$(T)_o \quad \mathbf{B}_-\mathbf{B}\mathbf{A}_-\mathbf{B}\mathbf{B}_- \in \mathbf{B}(\mathcal{S}) \text{ for each } \mathbf{A}_- \in \mathbf{A}(\mathcal{S})$$

is satisfied. Further, if $\mathcal{O}(\mathbf{B}) = \mathbf{B}(\mathcal{S})$ *and* $\mathcal{G}(\mathbf{A}) = \mathbf{A}(\overline{\mathcal{S}})$, *then the two order relations '*$<^{\mathcal{G}}$*' and '*$<^{\mathcal{O}}$*' are identical.*

Proof. Notice that by Theorem 7.3.1, if $\mathbf{B}_- \in \mathbf{B}(\mathcal{S})$ and $\mathbf{A} = \mathbf{B}\mathbf{B}_-\mathbf{B}$, then $\mathbf{B}_-$ is a reflexive g-inverse of $\mathbf{A}$ and by Remark 7.3.4(i), we have $\mathbf{B}_-\mathbf{A} = \mathbf{B}_-\mathbf{B}$ and $\mathbf{A}\mathbf{B}_- = \mathbf{B}\mathbf{B}_-$. Thus, $\mathbf{B}_-\mathbf{B}$ (or $\mathbf{B}\mathbf{B}_-$) is hermitian implies $\mathbf{B}_-\mathbf{A}$ (or $\mathbf{B}\mathbf{B}_-$ respectively) is hermitian and vice versa. If $\mathbf{B}_-$ and $\mathbf{B}$ commute, so do $\mathbf{B}_-$ and $\mathbf{A}$. Similarly, if $\mathcal{C}(\mathbf{B}_-) \subseteq \mathcal{C}(\mathbf{B}\mathbf{B}_-)$, then $\mathcal{C}(\mathbf{B}_-) \subseteq \mathcal{C}(\mathbf{A}\mathbf{B}_-)$ and so on. Thus, the last part of the statement holds. For the first half, we verify some cases in detail and leave the remaining for the reader to supply the details.
(i) Let $\mathcal{S} = \{2\}$. Then $\mathbf{B}(\mathcal{S}) = \{\mathbf{B}_-\}$. Let $\mathbf{A} = \mathbf{B}\mathbf{B}_-\mathbf{B}$ for some $\mathbf{B}_- \in \mathbf{B}(\mathcal{S})$ and $\mathbf{A}_- \in \mathbf{A}(\mathcal{S})$ be arbitrary. Consider,

$$\mathbf{B}_-\mathbf{B}\mathbf{A}_-\mathbf{B}\mathbf{B}_-\mathbf{B}\mathbf{B}_-\mathbf{B}\mathbf{A}_-\mathbf{B}\mathbf{B}_- = \mathbf{B}_-\mathbf{B}\mathbf{A}_-\mathbf{B}\mathbf{B}_-\mathbf{B}\mathbf{A}_-\mathbf{B}\mathbf{B}_-$$

$$= \mathbf{B_-BA_-AA_-BB_-} = \mathbf{B_-BABB_-} \in \mathbf{B}(\mathcal{S}).$$

Therefore, the condition T_o is satisfied. As $\mathbf{A}(\mathcal{S}) = \{\mathbf{A_-}\}$, the later statement is clear and the order relation '$<^{\mathcal{O}}$' is the minus order.
(ii) We next verify the statements in case $\mathcal{S} = \{2, 3\}$. Notice that the condition T_o is satisfied as in earlier case. We only need to check $\mathbf{BB_-BA_-BB_-}$ is hermitian. We know that $\mathbf{AA_-}$ and $\mathbf{BB_-}$ are hermitian. However, $\mathbf{BB_-BA_-BB_-} = \mathbf{AA_-BB_-} = (\mathbf{AA_-})^\star(\mathbf{BB_-})^\star = (\mathbf{A_-})^\star\mathbf{A}^\star(\mathbf{B_-})^\star\mathbf{B}^\star = (\mathbf{A_-})^\star\mathbf{B}^\star(\mathbf{B_-})^\star\mathbf{B}^\star(\mathbf{B_-})^\star\mathbf{B}^\star = (\mathbf{A_-})^\star\mathbf{B}^\star(\mathbf{B_-})^\star\mathbf{B}^\star = (\mathbf{A_-})^\star\mathbf{A}^\star = \mathbf{AA_-}$ and $\mathbf{AA_-}$ is hermitian. Hence, $\mathbf{BB_-BA_-BB_-}$ is hermitian and therefore belongs to $\mathbf{B}(\mathcal{S})$. Notice that the order relation '$<^{\mathcal{O}}$' is the left star order.
(iii) In case $\mathcal{S} = \{2, 5\}$, one obtains the order relation '$<^{\mathcal{O}}$' as the sharp order. □

Remark 7.3.13. Let $\mathbf{A}, \mathbf{B} \in \mathbf{\Omega}_{\mathcal{G}}$ and $\mathbf{A} <^{\mathcal{G}} \mathbf{B} \Rightarrow$ the pair $(\mathbf{A}, \mathbf{B})$ satisfies the (T)-condition. Let $\mathbf{G}$ be a fixed reflexive g-inverse of $\mathbf{B}$, and

$$\mathbf{D_B} = \{\mathbf{C} : \mathbf{C} <^{\mathcal{G}} \mathbf{B}\} \text{ and } \mathcal{O}(\mathbf{B}) = \{\mathbf{GCG} : \mathbf{C} \in \mathbf{D_B}\},$$

be the sets as in Theorem 7.3.6. Then $\mathbf{A} <^{\mathcal{O}} \mathbf{B} \Leftrightarrow \mathbf{A} <^{\mathcal{G}} \mathbf{B}$. Thus, it also follows that the pair $(\mathbf{A}, \mathbf{B})$ satisfies the $(T)_o$-condition of Theorem 7.3.12 as well.

7.4 One-sided $\mathcal{G}$-based order relations

In this section we take up the case of one-sided $\mathcal{G}$-based order relations defined by g-maps in Section 2 and examine the conditions under which these will define a partial order.

Definition 7.4.1. Let $\mathcal{G} : \mathbf{F^{m\times n}} \longrightarrow \mathfrak{P}(\mathbf{F^{m\times n}})$ be a g-map. For any $\mathbf{A}, \mathbf{B} \in \mathbf{F^{m\times n}}$, we say $\mathbf{A}\ \mathcal{G} < \mathbf{B}$, if there exists a $\mathbf{G} \in \mathcal{G}(\mathbf{A})$ such that $\mathbf{GA} = \mathbf{GB}$, and $\mathcal{C}(\mathbf{A}) \subseteq \mathcal{C}(\mathbf{B})$. We call this order relation as left $\mathcal{G}$-order.

Similarly, we say $\mathbf{A} < \mathcal{G}\ \mathbf{B}$ if there exists a $\mathbf{G} \in \mathcal{G}(\mathbf{A})$ such that $\mathbf{AG} = \mathbf{BG}$, and $\mathcal{C}(\mathbf{A^t}) \subseteq \mathcal{C}(\mathbf{B^t})$. We call this order relation right $\mathcal{G}$-order.

Remark 7.4.2. If $\mathbf{A}\ \mathcal{G} < \mathbf{B}$ or $\mathbf{A} < \mathcal{G}\ \mathbf{B}$ holds, then it ia clear that $\mathbf{A} <^- \mathbf{B}$.

Remark 7.4.3. If $\mathcal{G}(\mathbf{A}) = \{\mathbf{A}^-\}$, then both left $\mathcal{G}$-order and right $\mathcal{G}$-order are the same as the minus order.

In view of the Remark 7.4.2, both left $\mathcal{G}$-order and right $\mathcal{G}$-order are reflexive and anti-symmetric on the support $\Omega_{\mathcal{G}}$ of the g-map $\mathcal{G}$. In absence of any further conditions on the g-map neither of the left $\mathcal{G}$-order and the right $\mathcal{G}$-order defines a partial order. One obvious choice of the additional condition is the (T)-condition and this makes both left $\mathcal{G}$-order and right $\mathcal{G}$-order into partial orders as the following theorem shows:

Theorem 7.4.4. *Let* $\mathcal{G} : \mathbf{F}^{\mathbf{m}\times\mathbf{n}} \longrightarrow \mathfrak{P}(\mathbf{F}^{\mathbf{m}\times\mathbf{n}})$ *be a g-map. Then*

(i) *If* $\mathbf{A}\ \mathcal{G} < \mathbf{B} \Rightarrow$ *the pair* $(\mathbf{A}, \mathbf{B})$ *satisfies the (T)-condition, then the left* $\mathcal{G}$*-order is a partial order on the support* $\Omega_{\mathcal{G}}$.
(ii) *If* $\mathbf{A} < \mathcal{G}\ \mathbf{B} \Rightarrow$ *the pair* $(\mathbf{A}, \mathbf{B})$ *satisfies the (T)-condition, then the right* $\mathcal{G}$*-order is a partial order on the support* $\Omega_{\mathcal{G}}$.

Proof is easy.

Is there any other additional condition that ensures that both left $\mathcal{G}$-order and right $\mathcal{G}$-order induce partial orders on the support $\Omega_{\mathcal{G}}$? The answer is contained in the following theorem:

Theorem 7.4.5. *Let* $\mathcal{G} : \mathbf{F}^{\mathbf{m}\times\mathbf{n}} \longrightarrow \mathfrak{P}(\mathbf{F}^{\mathbf{n}\times\mathbf{m}})$ *be a g-map. Then*

(i) *If '*$\mathbf{A}\ \mathcal{G} < \mathbf{B} \Rightarrow$ *the pair* $(\mathbf{A}, \mathbf{B})$ *satisfies* $\mathcal{G}(\mathbf{A})\mathbf{A}\mathcal{G}(\mathbf{B}) \subseteq \mathcal{G}(\mathbf{A})$*', then the left* $\mathcal{G}$*-order is a partial order on the support* $\Omega_{\mathcal{G}}$.
(ii) *If '*$\mathbf{A} < \mathcal{G}\ \mathbf{B} \Rightarrow$ *the pair* $(\mathbf{A}, \mathbf{B})$ *satisfies* $\mathcal{G}(\mathbf{B})\mathbf{A}\mathcal{G}(\mathbf{A}) \subseteq \mathcal{G}(\mathbf{A})$*', then the right* $\mathcal{G}$*-order is a partial order on the support* $\Omega_{\mathcal{G}}$.

Proof. We prove that the left $\mathcal{G}$-order is transitive and the proof for right $\mathcal{G}$-order is similar.

So, let $\mathbf{A}\ \mathcal{G} < \mathbf{B}$ and $\mathbf{B}\ \mathcal{G} < \mathbf{C}$. Now, $\mathbf{A}\ \mathcal{G} < \mathbf{B} \Rightarrow$ there exists a $\mathbf{G} \in \mathcal{G}(\mathbf{A})$ such that $\mathbf{GA} = \mathbf{GB}$ and $\mathcal{C}(\mathbf{A}) \subseteq \mathcal{C}(\mathbf{B})$. Also $\mathbf{B}\ \mathcal{G} < \mathbf{C} \Rightarrow$ there exists an $\mathbf{H} \in \mathcal{G}(\mathbf{B})$ such that $\mathbf{HB} = \mathbf{HC}$ and $\mathcal{C}(\mathbf{B}) \subseteq \mathcal{C}(\mathbf{C})$. Let $\mathbf{H_1} = \mathbf{GAH}$. Then $\mathbf{H_1} \in \mathcal{G}(\mathbf{A})$ and $\mathbf{H_1A} = \mathbf{GAHA} = \mathbf{GA}$, as $\mathbf{A}\ \mathcal{G} < \mathbf{B} \Rightarrow \mathbf{A} <^{-} \mathbf{B}$. Similarly, $\mathbf{H_1C} = \mathbf{GAHC} = \mathbf{GAHB} = \mathbf{GA}$. Also, $\mathcal{C}(\mathbf{A}) \subseteq \mathcal{C}(\mathbf{C})$. Hence, $\mathbf{A}\ \mathcal{G} < \mathbf{C}$. □

Remark 7.4.6. (i) The conditions in Theorem 7.4.5 (i) and (ii) are in general, different from the (T)-condition.
(ii) The left $\mathcal{G}$-order and the right $\mathcal{G}$-order can be partial orders even when defining g-map does not satisfy the respective conditions.
(iii) The left $\mathcal{G}$-order and the right $\mathcal{G}$-order can be partial orders even when

the order relation $<^{\mathcal{G}}$ is not a partial order. We give an example to justify the statements (i), (ii) and (iii).

Example 7.4.7. Let $\mathcal{G} : \mathbb{C}^{m\times n} \longrightarrow \mathfrak{P}(\mathcal{C}^{n\times m})$ be the g-map defined as:

$$\text{For each } \mathbf{A} \in \mathbb{C}^{\mathbf{m}\times\mathbf{n}},\ \mathcal{G}(\mathbf{A}) = (\{\mathbf{A}_\ell^-\} \cup \{\mathbf{A}_\mathbf{m}^-\}) \setminus \{\mathbf{A}_{\ell\mathbf{m}}^-\}.$$

Let $\mathbf{A}, \mathbf{B} \in \mathbb{C}^{\mathbf{m}\times\mathbf{n}}$. Then
(i) It is obvious that $\mathcal{G}$ is a complete g-map. For, if $\mathbf{H} \in \widetilde{\mathcal{G}(\mathbf{A})}$, then there exists a $\mathbf{G} \in \mathcal{G}(\mathbf{A})$ such that $\mathbf{HA} = \mathbf{GA},\ \mathbf{AH} = \mathbf{AG}$. Thus, if $\mathbf{G} \in \{\mathbf{A}_\ell^-\}$, then so is $\mathbf{H} \in \{\mathbf{A}_\ell^-\}$ and if $\mathbf{G} \in \{\mathbf{A}_\mathbf{m}^-\}$, then so is $\mathbf{H} \in \{\mathbf{A}_\mathbf{m}^-\}$.
(ii) We show that $\mathbf{A}\ \mathcal{G} < \mathbf{B}$ and $\mathbf{A} < \mathcal{G}\ \mathbf{B}$ are each equivalent to $\mathbf{A} <^- \mathbf{B}$. Since, $\mathbf{A}\ \mathcal{G} < \mathbf{B} \Rightarrow \mathbf{A} <^- \mathbf{B}$ and $\mathbf{A} < \mathcal{G}\ \mathbf{B} \Rightarrow \mathbf{A} <^- \mathbf{B}$, is always true, we must show that $\mathbf{A} <^- \mathbf{B} \Rightarrow \mathbf{A}\ \mathcal{G} < \mathbf{B}$ and $\mathbf{A} <^- \mathbf{B} \Rightarrow \mathbf{A} < \mathcal{G}\ \mathbf{B}$. So, let $\mathbf{A} <^- \mathbf{B}$. Then there exists a $\mathbf{G} \in \{\mathbf{A}^-\}$ such that $\mathbf{AG} = \mathbf{BG},\ \mathbf{GA} = \mathbf{GB}$. If $\mathbf{G} \in \{\mathbf{A}_\ell^-\} \setminus \{\mathbf{A}_{\ell\mathbf{m}}^-\}$, or $\mathbf{G} \in \{\mathbf{A}_\mathbf{m}^-\} \setminus \{\mathbf{A}_{\ell\mathbf{m}}^-\}$, then in view of the fact that $\mathbf{A} <^- \mathbf{B} \Rightarrow \mathbf{A} <^\mathbf{s} \mathbf{B}$, we have $\mathbf{A} <^- \mathbf{B} \Rightarrow \mathbf{A}\ \mathcal{G} < \mathbf{B}$ and also $\mathbf{A} <^- \mathbf{B} \Rightarrow \mathbf{A} < \mathcal{G}\ \mathbf{B}$. Let $\mathbf{G}$ does not belong to $\{\mathbf{A}_\ell^-\}$ or does not belong to $\{\mathbf{A}_\mathbf{m}^-\}$. Then for an arbitrary $\mathbf{A}_\ell^-$, $\mathbf{GAA}_\ell^- \in \mathcal{G}(\mathbf{A})$ and for an arbitrary $\mathbf{A}_\mathbf{m}^-$, $\mathbf{A}_\mathbf{m}^-\mathbf{AG} \in \mathcal{G}(\mathbf{A})$, as $\mathcal{G}$ is a complete g-map. Moreover, neither of $\mathbf{GAA}_\ell^-$ and $\mathbf{A}_\mathbf{m}^-\mathbf{AG}$ is in $\{\mathbf{A}_{\ell\mathbf{m}}^-\}$, because $\mathbf{A}(\mathbf{GAA}_\ell^-)$ and $(\mathbf{A}_\mathbf{m}^-\mathbf{AG})\mathbf{A}$ are not hermitian. Also, $\mathbf{A}(\mathbf{GAA}_\ell^-) = \mathbf{B}(\mathbf{GAA}_\ell^-) = \mathbf{AA}_\ell^-$ and $\mathbf{A}_\mathbf{m}^-\mathbf{AGA} = \mathbf{A}_\mathbf{m}^-\mathbf{AGB} = \mathbf{A}_\mathbf{m}^-\mathbf{A}$. So, $\mathbf{A}\ \mathcal{G} < \mathbf{B}$ and $\mathbf{A} < \mathcal{G}\ \mathbf{B}$ hold. Note that it is quite possible that

$$\{\mathbf{A}^- : \mathbf{A}^-(\mathbf{B} - \mathbf{A}) = \mathbf{0},\ (\mathbf{B} - \mathbf{A})\mathbf{A}^- = \mathbf{0}\} = \{\mathbf{A}_{\ell\mathbf{m}}^-\}.$$

If this is true, then $\{\mathbf{B}^-\mathbf{AB}^-, \mathbf{B}^- \text{ arbitrary}\} = \{\mathbf{A}^\dagger\}$. So, for each $\mathbf{B}^-$, $\mathbf{B}^-\mathbf{AB}^-\mathbf{A} = \mathbf{A}^\dagger\mathbf{A}$ and $\mathbf{AB}^-\mathbf{AB}^- = \mathbf{AA}^\dagger$. Thus, $\mathbf{B}^-\mathbf{A} = \mathbf{A}^\dagger\mathbf{A}$ and $\mathbf{AB}^- = \mathbf{AA}^\dagger$. This means, $\mathbf{B}^-\mathbf{A}$ and $\mathbf{AB}^-$ are invariant under choice of $\mathbf{B}^-$. Therefore, $\mathbb{C}^n = \mathcal{C}(\mathbf{I_n}) \subseteq \mathcal{C}(\mathbf{B^t})$ and $\mathbb{C}^m = \mathcal{C}(\mathbf{I_m}) \subseteq \mathcal{C}(\mathbf{B})$, which is not possible if $m \neq n$.
(iii) We now show that the conditions in Theorem 7.4.5 (i) and (ii) are not satisfied. Let $\mathbf{A}, \mathbf{B} \in \mathbb{C}^{\mathbf{m}\times\mathbf{n}}$ such that $\mathbf{A} <^{-\ell} \mathbf{B}$. Then $\{\mathbf{B}_\ell^-\} \subseteq \{\mathbf{A}_\ell^-\}$. As, $\mathbf{A} <^{-\ell} \mathbf{B} \Rightarrow \mathbf{A} <^- \mathbf{B}$, we have $\mathbf{A}\ \mathcal{G} < \mathbf{B}$. Also, $\mathbf{A}^\dagger = \mathbf{A}_\mathbf{m}^-\mathbf{AB}_\ell^- \in \mathcal{G}(\mathbf{A})\mathbf{A}\mathcal{G}(\mathbf{B})$, and $\mathbf{A}^\dagger$ does not belong to $\mathcal{G}(\mathbf{A})$. So, $\mathcal{G}(\mathbf{A})\mathbf{A}\mathcal{G}(\mathbf{B}) \nsubseteq \mathcal{G}(\mathbf{A})$. Similarly, $\mathcal{G}(\mathbf{B})\mathbf{A}\mathcal{G}(\mathbf{A}) \nsubseteq \mathcal{G}(\mathbf{A})$.
(iv) We now show that the (T)-condition does not hold.
For this, let $\mathbf{A}, \mathbf{B} \in \mathbb{C}^{\mathbf{m}\times\mathbf{n}}$ such that $\mathbf{A} <^{-\ell} \mathbf{B}$. Let $\mathbf{B}_\ell^-\mathbf{AB}_\ell^- \in \mathcal{G}(\mathbf{A}, \mathbf{B})$. If $\mathbf{B}_\ell^-\mathbf{AB}_\ell^- \in \mathcal{G}(\mathbf{A})$, then $\mathbf{B}_\ell^-\mathbf{AB}_\ell^- \in \{\mathbf{A}_\ell^-\}$ or $\mathbf{B}_\ell^-\mathbf{AB}_\ell^- \in \{\mathbf{A}_\mathbf{m}^-\}$. Let $\mathbf{B}_\ell^-\mathbf{AB}_\ell^- \in \{\mathbf{A}_\ell^-\}$. So, $\mathbf{AB}_\ell^-\mathbf{AB}_\ell^- = \mathbf{AB}_\ell^-$ is hermitian. Choose a g-inverse $\mathbf{B}^-$ not in $(\{\mathbf{B}_\ell^-\} \cup \{\mathbf{B}_\mathbf{m}^-\})$. Now, $\mathbf{AB}^-{}_\ell = \mathbf{BB}^-\mathbf{AB}_\ell^-$ is hermitian. So, $\mathbf{B}$

and $\mathbf{B}^-\mathbf{A}\mathbf{B}_\ell^-$ commute. Therefore, $\mathbf{A}\mathbf{B}_\ell^- = \mathbf{B}^-\mathbf{A}\mathbf{B}_\ell^-\mathbf{B} = \mathbf{B}^-\mathbf{A}$ should be hermitian, which is not true.
Let $\mathbf{B}_\ell^-\mathbf{A}\mathbf{B}_\ell^- \in \{\mathbf{A}_\mathbf{m}^-\}$. Then $\mathbf{B}_\ell^-\mathbf{A}\mathbf{B}_\ell^-\mathbf{A} = \mathbf{B}_\ell^-\mathbf{A}$ should be hermitian, which is again not true. Hence, (T)-condition does not hold and the $<^{\mathcal{G}}$ cannot be a partial order.

We now consider another condition (that is stronger than semi-completeness) under which the left $\mathcal{G}$ order and right $\mathcal{G}$ order are equivalent to the order relation $<^{\mathcal{G}}$.

Definition 7.4.8. Let $\mathcal{G} : \mathbf{F}^{\mathbf{m}\times\mathbf{n}} \longrightarrow \mathfrak{P}(\mathbf{F}^{\mathbf{n}\times\mathbf{m}})$ be a g-map and $\mathbf{A} \in \mathbf{F}^{\mathbf{m}\times\mathbf{n}}$. If $\mathcal{G}(\mathbf{A})\mathbf{A}\mathcal{G}(\mathbf{A}) \subseteq \mathcal{G}(\mathbf{A})$, then we say $\mathcal{G}(\mathbf{A})$ is strongly semi-complete. If for each $\mathbf{A}$, $\mathcal{G}(\mathbf{A})$ is strongly semi-complete, we say the g-map $\mathcal{G}$ is strongly semi-complete. A g-map that is complete as well as strongly semi-complete is called strongly complete.

Remark 7.4.9. A strongly semi-complete map is semi-complete, but not conversely.

Lemma 7.4.10. *Let $\mathcal{G} : \mathbf{F}^{\mathbf{m}\times\mathbf{n}} \longrightarrow \mathfrak{P}(\mathbf{F}^{\mathbf{n}\times\mathbf{m}})$ be a g-map and $\mathbf{A} \in \mathbf{F}^{\mathbf{m}\times\mathbf{n}}$. If $\mathcal{G}(\mathbf{A})$ is strongly semi-complete, then for $\mathbf{B} \in \mathbf{F}^{\mathbf{m}\times\mathbf{n}}$, '$\mathcal{G}(\mathbf{A})\mathbf{A}\mathcal{G}(\mathbf{B}) \subseteq \mathcal{G}(\mathbf{A})$ and $\mathcal{G}(\mathbf{B})\mathbf{A}\mathcal{G}(\mathbf{A}) \subseteq \mathcal{G}(\mathbf{A})$' $\Rightarrow \mathcal{G}(\mathbf{A},\mathbf{B}) \subseteq \mathcal{G}(\mathbf{A})$.*

Proof. Let $\mathbf{HAH} \in \mathcal{G}(\mathbf{A},\mathbf{B})$. Then $\mathbf{HAH} = (\mathbf{HAG})\mathbf{A}(\mathbf{GAH})$ for each $\mathbf{G} \in \mathcal{G}(\mathbf{A})$. Since, $\mathbf{HAG} \in \mathcal{G}(\mathbf{B})\mathbf{A}\mathcal{G}(\mathbf{A})$ and $\mathbf{GAH} \in \mathcal{G}(\mathbf{A})\mathbf{A}\mathcal{G}(\mathbf{B})$ and $\mathcal{G}(\mathbf{A})$ is strongly complete, it follows that $\mathbf{HAH} \in \mathcal{G}(\mathbf{A})$. □

Theorem 7.4.11. *Let $\mathcal{G} : \mathbf{F}^{\mathbf{m}\times\mathbf{n}} \longrightarrow \mathfrak{P}(\mathbf{F}^{\mathbf{n}\times\mathbf{m}})$ be a strongly complete g-map satisfying both $\mathcal{G}(\mathbf{A})\mathbf{A}\mathcal{G}(\mathbf{B}) \subseteq \mathcal{G}(\mathbf{A})$ and $\mathcal{G}(\mathbf{B})\mathbf{A}\mathcal{G}(\mathbf{A}) \subseteq \mathcal{G}(\mathbf{A})$, for $\mathbf{A},\mathbf{B} \in \mathbf{F}^{\mathbf{m}\times\mathbf{n}}$. Then $\mathbf{A}\ \mathcal{G} < \mathbf{B}$ and $\mathbf{A} < \mathcal{G}\ \mathbf{B} \Leftrightarrow \mathbf{A} <^{\mathcal{G}} \mathbf{B}$. Further, the order relation $<^{\mathcal{G}}$ is a partial order.*

Proof. We only need to prove $\mathbf{A}\ \mathcal{G} < \mathbf{B}$ and $\mathbf{A} < \mathcal{G}\ \mathbf{B} \Rightarrow \mathbf{A} <^{\mathcal{G}} \mathbf{B}$. Choose and fix $\mathbf{H} \in \mathcal{G}(\mathbf{B})$. Then by Lemma 7.4.10, $\mathbf{H_1} = \mathbf{HAH} \in \mathcal{G}(\mathbf{A})$. Also, $\mathbf{H_1A} = \mathbf{HA}, \mathbf{H_1B} = \mathbf{HAHB} = \mathbf{HA}$. So, $\mathbf{H_1A} = \mathbf{H_1B}$. Similarly, $\mathbf{AH_1} = \mathbf{BH_1}$. □

7.5 Properties of $\mathcal{G}$-based order relations

We now study some further properties of $\mathcal{G}$-based order relations, yet another step to unify the theory of matrix partial orders.

Theorem 7.5.1. *Let* $\mathcal{G} : F^{n\times n} \longrightarrow \mathfrak{P}(F^{n\times n})$ *be a g-map. Let* $\mathbf{A}, \mathbf{B} \in \mathbf{F}^{\mathbf{n}\times\mathbf{n}}$. *Then* $\mathbf{A} <^{\mathcal{G}} \mathbf{B}$ *if and only if there exist projectors* $\mathbf{P}$ and $\mathbf{Q}$ *such that* $\mathbf{A} = \mathbf{BP} = \mathbf{QB}$ *and* $\mathcal{C}(\mathbf{P}) = \mathcal{C}(\mathbf{G})$ and $\mathcal{C}(\mathbf{Q^t}) = \mathcal{C}(\mathbf{G^t})$ *for some* $\mathbf{G} \in \mathcal{G}(\mathbf{A})$.

Proof. 'If' part
Since $\mathcal{C}(\mathbf{P}) = \mathcal{C}(\mathbf{G})$ and $\mathcal{C}(\mathbf{Q^t}) = \mathcal{C}(\mathbf{G^t})$ for some $\mathbf{G} \in \mathcal{G}(\mathbf{A})$, we have $\mathbf{P} = \mathbf{GT}$ for some matrix $\mathbf{T}$ and $\mathbf{Q} = \mathbf{SG}$ for some matrix $\mathbf{S}$. Consider $\mathbf{GA} = \mathbf{GQB} = \mathbf{GSGB}$. Since $\mathbf{Q}$ is idempotent, $\rho(\mathbf{SG}) = \rho(\mathbf{SGSG}) \leq \rho(\mathbf{GSG}) \leq \rho(\mathbf{SG})$. So, $\rho(\mathbf{Q}) = \rho(\mathbf{SG}) = \rho(\mathbf{GSG})$. Therefore, $\mathbf{S}$ is a g-inverse of $\mathbf{G}$ and hence, $\mathbf{GA} = \mathbf{GB}$.
Similarly, $\mathbf{AG} = \mathbf{BG}$.
'Only if' part
Let $\mathbf{A} <^{\mathcal{G}} \mathbf{B}$. Then there exists a g-inverse $\mathbf{G} \in \mathcal{G}(\mathbf{A})$ such that $\mathbf{GA} = \mathbf{GB}$, $\mathbf{AG} = \mathbf{BG}$. So, $\mathbf{A} = \mathbf{AGB} = \mathbf{BGA}$. Let $\rho(\mathbf{G}) = \mathbf{s}$. Then by Corollary 2.3.19, there exists a matrix $\mathbf{D}$ such that

$$\rho(\mathbf{A}+\mathbf{D}) = \rho(\mathbf{A}) + \rho(\mathbf{D}) = \mathbf{s} \text{ and } \mathbf{G} = (\mathbf{A}+\mathbf{D})_{\mathbf{r}}^{-}.$$

Let $\mathbf{C} = \mathbf{A}+\mathbf{D}$. Then $\mathbf{CGB} = (\mathbf{A}+\mathbf{D})\mathbf{GB} = (\mathbf{A}+\mathbf{D})\mathbf{GA} = \mathbf{A}$. Since $\mathbf{AGA} = \mathbf{A}$, so, $\mathbf{DGA} = \mathbf{0}$. Thus, $\mathbf{CGB} = \mathbf{A}$. Now, take $\mathbf{Q} = \mathbf{CG}$. Clearly, $\mathbf{Q}$ is idempotent and $\rho(\mathbf{CG}) = \rho(\mathbf{G})$, so, $\mathcal{C}(\mathbf{Q^t}) = \mathcal{C}(\mathbf{G^t})$. Similarly, we can obtain a projector $\mathbf{P}$ such that $\mathcal{C}(\mathbf{P}) = \mathcal{C}(\mathbf{G})$. □

Theorem 7.5.2. *Let m, n, p and q be positive integers and* $\mathbf{A} \in \mathbf{F}^{\mathbf{m}\times\mathbf{n}}$, $\mathbf{P} \in \mathbf{F}^{\mathbf{n}\times\mathbf{p}}$ and $\mathbf{Q} \in \mathbf{F}^{\mathbf{q}\times\mathbf{m}}$. *Then the following hold:*

(i) $\mathbf{G}$ *is a g-inverse of* $\mathbf{A}$ *if and only if there exist matrices* $\mathbf{P}$ *and* $\mathbf{Q}$ *such that* (a) $\rho(\mathbf{QAP}) = \rho(\mathbf{A})$ *and* (b) $\mathbf{G} = \mathbf{P}(\mathbf{QAP})^{-}\mathbf{Q}$.

(ii) $\mathbf{H}$ *is an outer inverse of* $\mathbf{A}$ *if and only if there exist matrices* $\mathbf{P}$ *and* $\mathbf{Q}$ *such that* $\mathbf{H} = \mathbf{P}(\mathbf{QAP})_{\mathbf{r}}^{-}\mathbf{Q}$.

(iii) $\mathbf{H}$ *is an outer inverse of* $\mathbf{A}$ *if and only if there exist matrices* $\mathbf{P}$ *and* $\mathbf{Q}$ *such that*

$$\text{(a) } \rho(\mathbf{QAP}) = \rho(\mathbf{P}) = \rho(\mathbf{Q}) \text{ and (b) } \mathbf{H} = \mathbf{P}(\mathbf{QAP})^{-}\mathbf{Q}.$$

Proof. (i) 'If' part is trivial and for 'Only if' part take $\mathbf{P} = \mathbf{I_n}$ and $\mathbf{Q} = \mathbf{I_m}$.
(ii) We only prove the 'Only if' part. Let $\mathbf{A}_{\mathbf{r}}^{-}$ be any reflexive g-inverse of $\mathbf{A}$. Consider $\mathbf{H} = \mathbf{A}_{\mathbf{r}}^{-}$. It is clear that $\mathbf{H}$ is an outer inverse of $\mathbf{A}$. So, for $\mathbf{P} = \mathbf{I_n}$ and $\mathbf{Q} = \mathbf{I_m}$, $\mathbf{H}$ is of the desired form.
(iii) 'If' part
Let there exist matrices $\mathbf{P}$ and $\mathbf{Q}$ such that (a) $\rho(\mathbf{QAP}) = \rho(\mathbf{P}) = \rho(\mathbf{Q})$

and (b) $\mathbf{H} = \mathbf{P}(\mathbf{QAP})^-\mathbf{Q}$. The condition (a) implies $\mathcal{C}(\mathbf{QAP}) = \mathcal{C}(\mathbf{Q})$ and $\mathcal{C}(\mathbf{QAP})^{\mathbf{t}} = \mathcal{C}(\mathbf{P})^{\mathbf{t}}$. Further, $\mathcal{C}(\mathbf{QAP}) = \mathcal{C}(\mathbf{Q}) \Rightarrow$ there exists a matrix $\mathbf{U}$ such that $\mathbf{Q} = \mathbf{QAPU}$. So, $\mathbf{HAH} = \mathbf{P}(\mathbf{QAP})^-\mathbf{QAP}(\mathbf{QAP})^-\mathbf{Q} = \mathbf{P}(\mathbf{QAP})^-\mathbf{QAP}(\mathbf{QAP})^-\mathbf{QAPU} = \mathbf{P}(\mathbf{QAP})^-\mathbf{QAPU} = \mathbf{P}(\mathbf{QAP})^-\mathbf{Q} = \mathbf{H}$. Thus, $\mathbf{H}$ is an outer inverse of $\mathbf{A}$.

'Only if' part

Let $\mathbf{H}$ be an outer inverse of $\mathbf{A}$. Take $\mathbf{P} = \mathbf{H}$, $\mathbf{Q} = \mathbf{H}$. Then $\mathbf{H} = \mathbf{H}(\mathbf{HAH})^-\mathbf{H}$ is in the desired form. □

Given a matrix Theorem 7.5.3 determines the class of all g-inverses as the class of all outer inverses with specified row and column spaces. Theorem 7.5.5 unifies several known partial orders as demonstrated in Remark 7.5.6.

Theorem 7.5.3. *Let m, n, p and q be positive integers and $\mathbf{A} \in \mathbf{F}^{\mathbf{m}\times\mathbf{n}}$, $\mathbf{P} \in \mathbf{F}^{\mathbf{n}\times\mathbf{p}}$ and $\mathbf{Q} \in \mathbf{F}^{\mathbf{q}\times\mathbf{m}}$. Let $\mathbf{X} = \mathbf{P}(\mathbf{QAP})^-\mathbf{Q}$. Then the following hold:*

(i) *$\mathbf{X}$ is a g-inverse of $\mathbf{A}$ if and only if $\rho(\mathbf{QAP}) = \rho(\mathbf{A})$.*
(ii) *$\mathbf{X}$ is an outer inverse of $\mathbf{A}$ with $\mathcal{C}(\mathbf{X}) = \mathcal{C}(\mathbf{P})$ and $\mathcal{C}(\mathbf{X})^{\mathbf{t}} = \mathcal{C}(\mathbf{Q})^{\mathbf{t}}$ if and only if $\rho(\mathbf{QAP}) = \rho(\mathbf{P}) = \rho(\mathbf{Q})$.*

Proof. (i) 'If' part

Since $\rho(\mathbf{QAP}) = \rho(\mathbf{A})$, it follows that $\rho(\mathbf{QAP}) = \rho(\mathbf{AP}) = \rho(\mathbf{A})$ and therefore, $\mathcal{C}(\mathbf{AP}) = \mathcal{C}(\mathbf{A})$. So, there exists a matrix $\mathbf{U}$ such that $\mathbf{A} = \mathbf{APU}$. Consider $\mathbf{AXA} = \mathbf{AXAXA} = \mathbf{AP}(\mathbf{QAP})^-\mathbf{QAP}(\mathbf{QAP})^-\mathbf{QAPU} = \mathbf{AP}(\mathbf{QAP})^-\mathbf{QAPU} = \mathbf{APU}$, since $\mathcal{C}(\mathbf{P^t}) \subseteq \mathcal{C}(\mathbf{QAP^t})$. So, $\mathbf{AXA} = \mathbf{A}$. Thus $\mathbf{X}$ is a g-inverse of $\mathbf{A}$.

'Only if' part

If $\mathbf{X} = \mathbf{P}(\mathbf{QAP})^-\mathbf{Q}$ is a g-inverse of $\mathbf{A}$, then $\mathbf{A} = \mathbf{AXA} = \mathbf{AXAXA} = \mathbf{AP}(\mathbf{QAP})^-\mathbf{QAP}(\mathbf{QAP})^-\mathbf{QA} \Rightarrow \rho(\mathbf{A}) \leq \rho(\mathbf{QAP}) \leq \rho(\mathbf{A})$. So, $\rho(\mathbf{QAP}) = \rho(\mathbf{A})$.

(ii) 'If' part follows along the same lines as in Theorem 7.5.2 (iii).

'Only if' part

Let $\mathbf{X}$ is an outer inverse of $\mathbf{A}$ with $\mathcal{C}(\mathbf{X}) = \mathcal{C}(\mathbf{P})$ and $\mathcal{C}(\mathbf{X})^{\mathbf{t}} = \mathcal{C}(\mathbf{Q})^{\mathbf{t}}$. Then $\mathbf{X} = \mathbf{XAX} = \mathbf{XAXAX} \Rightarrow \rho(\mathbf{X}) \leq \rho(\mathbf{QAP}) \leq \rho(\mathbf{AP}) \leq \rho(\mathbf{P})$ and $\rho(\mathbf{X}) \leq \rho(\mathbf{QAP}) \leq \rho(\mathbf{QA}) \leq \rho(\mathbf{Q})$. Since, $\mathcal{C}(\mathbf{X}) = \mathcal{C}(\mathbf{P})$ and $\mathcal{C}(\mathbf{X})^{\mathbf{t}} = \mathcal{C}(\mathbf{Q})^{\mathbf{t}}$, it follows that $\rho(\mathbf{QAP}) = \rho(\mathbf{P})$ and $\rho(\mathbf{QAP}) = \rho(\mathbf{Q})$. □

Remark 7.5.4. Thus, as $\mathbf{P}$ and $\mathbf{Q}$ vary in Theorem 7.5.2(i), we get all g-inverses of $\mathbf{A}$ and in (ii) all outer inverses. Similarly, Theorem 7.5.3, gives us a characterization of all g-inverses and outer inverses with a specified column space and a specified row space.

Theorem 7.5.5. *Let* m, n, p and q *be positive integers and* $\mathbf{A} \in \mathbf{F}^{\mathbf{m}\times\mathbf{n}}$, $\mathbf{P} \in \mathbf{F}^{\mathbf{n}\times\mathbf{p}}$ and $\mathbf{Q} \in \mathbf{F}^{\mathbf{q}\times\mathbf{m}}$. *Let* $\mathcal{G} : \mathbf{F}^{\mathbf{m}\times\mathbf{n}} \longrightarrow \mathfrak{P}(\mathbf{F}^{\mathbf{n}\times\mathbf{m}})$ *be a g-map defined as follows:*
For each $\mathbf{A} \in \mathbf{F}^{\mathbf{m}\times\mathbf{n}}$, $\mathcal{G}(\mathbf{A}) = \{\mathbf{P}(\mathbf{QAP})^{-}\mathbf{Q} : \rho(\mathbf{QAP}) = \rho(\mathbf{A})\}$.
Then the following hold:

(i) $\mathcal{G}$ *satisfies the (T)-condition.*
(ii) $\mathcal{G}$ *is semi-complete.*
(iii) *The order relation '*$<^{\mathcal{G}}$*' defines a partial order on* $\mathbf{F}^{\mathbf{m}\times\mathbf{n}}$.

Proof. As $\mathbf{A} <^{\mathcal{G}} \mathbf{B} \Rightarrow \mathbf{A} <^{-} \mathbf{B}$, each g-inverse $\mathbf{P}(\mathbf{QBP})^{-}\mathbf{Q}$ of $\mathbf{B}$ is a g-inverse of $\mathbf{A}$. So, $\mathbf{AP}(\mathbf{QBP})^{-}\mathbf{QA} = \mathbf{A}$. Pre-multiplying by $\mathbf{Q}$ and post-multiplying $\mathbf{P}$ respectively implies $\mathbf{QAP}(\mathbf{QBP})^{-}\mathbf{QAP} = \mathbf{QAP}$. Thus, $(\mathbf{QBP})^{-}$ is a g-inverse of $\mathbf{QAP}$.
(i) We show that the pair $(\mathbf{A}, \mathbf{B})$ for $\mathbf{A}, \mathbf{B} \in \mathbf{F}^{\mathbf{m}\times\mathbf{n}}$ satisfies the (T)-condition. Let $\mathbf{P}(\mathbf{QBP})^{-}\mathbf{QAP}(\mathbf{QBP})^{-}\mathbf{Q} \in \mathcal{G}(\mathbf{A}, \mathbf{B})$. The matrix $(\mathbf{QBP})^{-}\mathbf{QAP}(\mathbf{QBP})^{-}$ is a reflexive g-inverse of $\mathbf{QAP}$. So,

$$\rho((\mathbf{QBP})^{-}\mathbf{QAP}(\mathbf{QBP})^{-}) = \rho(\mathbf{QAP}) = \rho(\mathbf{A})$$

and hence

$$\mathbf{P}(\mathbf{QBP})^{-}\mathbf{QAP}(\mathbf{QBP})^{-}\mathbf{Q} \in \mathcal{G}(\mathbf{A}).$$

(ii) Let $\mathbf{A} \in \mathbf{F}^{\mathbf{m}\times\mathbf{n}}$. We show that $\mathcal{G}(\mathbf{A})$ is semi-complete.
So, let $\mathbf{P}(\mathbf{QAP})^{-}\mathbf{Q} \in \mathcal{G}(\mathbf{A})$ and consider $\mathbf{P}(\mathbf{QAP})^{-}\mathbf{QAP}(\mathbf{QAP})^{-}\mathbf{Q}$. As $(\mathbf{QAP})^{-}\mathbf{QAP}(\mathbf{QAP})^{-}$ is a reflexive g-inverse of $\mathbf{QAP}$, we have

$$\rho((\mathbf{QAP})^{-}\mathbf{QAP}(\mathbf{QAP})^{-}) = \rho(\mathbf{QAP}) = \rho(\mathbf{A}).$$

Hence, $\mathbf{P}(\mathbf{QAP})^{-}\mathbf{QAP}(\mathbf{QAP})^{-}\mathbf{Q} \in \mathcal{G}(\mathbf{A})$, showing $\mathcal{G}(\mathbf{A})$ is semi-complete. Thus, $\mathcal{G}$ is semi-complete.
(iii) Reflexivity is trivial.
For anti-symmetry, let $\mathbf{A} <^{\mathcal{G}} \mathbf{B}$ and $\mathbf{B} <^{\mathcal{G}} \mathbf{A}$. Then there exist a $\mathbf{G} \in \mathcal{G}(\mathbf{A})$ and an $\mathbf{H} \in \mathcal{G}(\mathbf{B})$ such that $\mathbf{AG} = \mathbf{BG}$, $\mathbf{GA} = \mathbf{GB}$ and $\mathbf{HB} = \mathbf{HA}$, $\mathbf{BH} = \mathbf{AH}$. Since $\mathbf{G} = \mathbf{P}(\mathbf{QAP})^{-}\mathbf{Q}$ and $\mathbf{H} = \mathbf{P}(\mathbf{QBP})^{-}\mathbf{Q}$ for some choice of $(\mathbf{QAP})^{-}$ and $(\mathbf{QBP})^{-}$. Now,

$$\mathbf{GA} = \mathbf{GB} \Rightarrow \mathbf{P}(\mathbf{QAP})^{-}\mathbf{QA} = \mathbf{P}(\mathbf{QAP})^{-}\mathbf{QB}$$

$$\Rightarrow \mathbf{QAP}(\mathbf{QAP})^{-}\mathbf{QAP} = \mathbf{QAP}(\mathbf{QAP})^{-}\mathbf{QBP}$$

$$\Rightarrow \mathbf{QAP} = \mathbf{QBP}.$$

Therefore, $\rho(\mathbf{QAP}) = \rho(\mathbf{QBP})$ implies $\rho(\mathbf{A}) = \rho(\mathbf{B})$. As $\mathbf{A} <^{\mathcal{G}} \mathbf{B} \Rightarrow \mathbf{A} <^{-} \mathbf{B}$, it follows $\mathbf{A} = \mathbf{B}$.

We now show that the relation is transitive as well. So, let $\mathbf{A} <^{\mathcal{G}} \mathbf{B}$ and $\mathbf{B} <^{\mathcal{G}} \mathbf{C}$. Since $\mathbf{B} <^{\mathcal{G}} \mathbf{C}$, there exists an $\mathbf{H} \in \mathcal{G}(\mathbf{B})$ such that $\mathbf{HB} = \mathbf{HC}$, $\mathbf{BH} = \mathbf{CH}$. As $\mathbf{A} <^{\mathcal{G}} \mathbf{B} \Rightarrow \mathbf{A} <^{-} \mathbf{B}$, it follows that $\mathbf{A} = \mathbf{AHB} = \mathbf{HBA} = \mathbf{AHA}$. The matrix $\mathbf{HAH}$ is a reflexive g-inverse of $\mathbf{A}$, so, $\rho(\mathbf{HAH}) = \rho(\mathbf{A})$. Hence, $\mathbf{HAH} \in \mathcal{G}(\mathbf{A})$. Further, $\mathbf{C}(\mathbf{HAH}) = \mathbf{BHAH} = \mathbf{AH} = \mathbf{A}(\mathbf{HAH})$, $(\mathbf{HAH})\mathbf{C} = \mathbf{HAHB} = \mathbf{HA} = (\mathbf{HAH})\mathbf{A}$. Thus, $\mathbf{A} <^{\mathcal{G}} \mathbf{C}$.

□

Remark 7.5.6. The following (a)-(g) show that Theorem 7.5.5 is very a strong result.

(a) If we take $\mathbf{P} = \mathbf{I_m}$ and $\mathbf{Q} = \mathbf{I_n}$, we have $\mathbf{P}(\mathbf{QAP})^{-}\mathbf{Q} = \mathbf{A}^{-}$ and the order relation $<^{\mathcal{G}}$ reduces to the minus order.

(b) If we take $\mathbf{P} = \mathbf{A}^{\star}$ and $\mathbf{Q} = \mathbf{I_n}$, we have $\mathbf{P}(\mathbf{QAP})^{-}\mathbf{Q} = \mathbf{A}^{\star}(\mathbf{AA}^{\star})^{-}$, which is minimum norm g-inverse of $\mathbf{A}$. The resulting order relation $<^{\mathcal{G}}$ is the right star order.

(c) If we take $\mathbf{P} = \mathbf{I_m}$ and $\mathbf{Q} = \mathbf{A}^{\star}$, we have $\mathbf{P}(\mathbf{QAP})^{-}\mathbf{Q} = (\mathbf{A}^{\star}\mathbf{A})^{-}\mathbf{A}^{\star}$. Thus, $\mathcal{G}(\mathbf{A})$ is the class of least squares g-inverses and the resulting order relation is the left star order.

(d) Take $\mathbf{P} = \mathbf{A}^{\star}$ and $\mathbf{Q} = \mathbf{A}^{\star}$, or $\mathbf{P} = \mathbf{A}^{\star}\mathbf{A}$ and $\mathbf{Q} = \mathbf{A}^{\star}$, then $\mathcal{G}(\mathbf{A}) = \{\mathbf{A}^{\dagger}\}$. the order relation $<^{\mathcal{G}}$ is the star order.

Now let $\mathbf{A}$ be an $n \times n$ matrix of index 1 and $\mathcal{G}$ to be a g-map on $F^{n \times n}$.

(e) If we take $\mathbf{P} = \mathbf{A}$ and $\mathbf{Q} = \mathbf{A}$, we have $\mathbf{P}(\mathbf{QAP})^{-}\mathbf{Q} = \mathbf{A}(\mathbf{A^3})^{-}\mathbf{A}$ and the order relation $<^{\mathcal{G}}$ is nothing but the sharp order.

(f) If we take $\mathbf{P} = \mathbf{A}$ and $\mathbf{Q} = \mathbf{I}$, we have $\mathbf{P}(\mathbf{QAP})^{-}\mathbf{Q} = \mathbf{A}(\mathbf{A^2})^{-}$, which is a $\mathbf{A}_{\chi}^{-}$ -inverse of $\mathbf{A}$, so, the order relation is right sharp order and finally

(g) if we take $\mathbf{P} = \mathbf{I}$ and $\mathbf{Q} = \mathbf{A}$, we have $\mathbf{P}(\mathbf{QAP})^{-}\mathbf{Q} = (\mathbf{A^2})^{-}\mathbf{A}$, which is a $\mathbf{A}_{\rho}^{-}$ -inverse of $\mathbf{A}$ and the order relation $<^{\mathcal{G}}$ is nothing but the left sharp order.

In this way various partial orders are obtained from order relation in Theorem 7.5.5 by a suitable choice of the matrices $\mathbf{P}$ and $\mathbf{Q}$. This is a single theorem that unifies the partial orders through g-inverses.

7.6 $\mathcal{G}$-based extensions

In Section 2, we have seen that a $\mathcal{G}$-based order relation '$<^{\mathcal{G}}$ need not always correspond to a partial order. Even when it defines a partial order it may not support full $\mathbf{F}^{\mathbf{m}\times\mathbf{n}}$. For example as seen in Theorem 4.2.10, the sharp order is a partial order only on $\mathcal{I}_1$, the index 1 matrices and not on $\mathbf{F}^{\mathbf{n}\times\mathbf{n}}$. Thus, a natural question for one to ask is: Is it possible to modify the g-map $\mathcal{G}$ on the complement of the its support $\Omega_{\mathcal{G}}$ so that the modified g-map say $\hat{\mathcal{G}}$ continues to induce a partial order? We shall see that there are many trivial ways of extending a $\mathcal{G}$-based partial order relation to the domain of g-map $\mathcal{G}$. We study some non-trivial extensions of the sharp order and refer the reader to [Jain, Mitra and Werner (1996)] for other details. We begin with the definition of an extension.

Definition 7.6.1. Let $\mathcal{G} : \mathbf{F}^{\mathbf{m}\times\mathbf{n}} \longrightarrow \mathfrak{P}(\mathbf{F}^{\mathbf{m}\times\mathbf{n}})$ be a g-map with support $\Omega_{\mathcal{G}} \neq \mathbf{F}^{\mathbf{m}\times\mathbf{n}}$. Then a g-map $\hat{\mathcal{G}}$ is called an extension of $\mathcal{G}$ if $\hat{\mathcal{G}}(\mathbf{A}) = \mathcal{G}(\mathbf{A})$ whenever $\mathbf{A} \in \mathbf{\Omega}_{\mathcal{G}}$. Further, the order relation '$<^{\hat{\mathcal{G}}}$' defined by the g-map $\hat{\mathcal{G}}$ is called $\mathcal{G}$-based extension of the order relation '$<^{\mathcal{G}}$'.

Theorem 7.6.2. *Let $\mathcal{G} : \mathbf{F}^{\mathbf{m}\times\mathbf{n}} \longrightarrow \mathfrak{P}(\mathbf{F}^{\mathbf{m}\times\mathbf{n}})$ be a g-map and let the order relation $<^{\mathcal{G}}$ define a partial order on the support $\Omega_{\mathcal{G}} \neq \mathbf{F}^{\mathbf{m}\times\mathbf{n}}$. Then there exists g-map $\hat{\mathcal{G}}$ with support $\mathbf{F}^{\mathbf{m}\times\mathbf{n}}$ such that* (i) $<^{\hat{\mathcal{G}}}$ *is a partial order and* (ii) *is a $\mathcal{G}$-based extension of the order relation $<^{\mathcal{G}}$.*

Proof. Define $\hat{\mathcal{G}} : \mathbf{F}^{\mathbf{m}\times\mathbf{n}} \longrightarrow \mathfrak{P}(\mathbf{F}^{\mathbf{m}\times\mathbf{n}})$ as follows:

$$\hat{\mathcal{G}}(\mathbf{A}) = \begin{cases} \mathcal{G}(\mathbf{A}), & \text{if } \mathbf{A} \in \mathbf{\Omega}_{\mathcal{G}} \\ \{\mathbf{A}^-_{\mathbf{max}}\}, & \text{otherwise} \end{cases}$$

where $\mathbf{A}^-_{\mathbf{max}}$ is a fixed full rank g-inverse of $\mathbf{A}$ (the existence of such a g-inverse is guaranteed by Theorem 2.3.18). We show that if $\mathbf{A} \notin \mathbf{\Omega}_{\mathcal{G}}$, then $\mathbf{A}$ is $\hat{\mathcal{G}}$-maximal. So, let $\mathbf{A} \notin \mathbf{\Omega}_{\mathcal{G}}$ and $\mathbf{B}$ be such that $\mathbf{A} <^{\hat{\mathcal{G}}} \mathbf{B}$. Then $(\mathbf{B}-\mathbf{A})\mathbf{A}^-_{\mathbf{max}} = \mathbf{0}$ and $\mathbf{A}^-_{\mathbf{max}}(\mathbf{B}-\mathbf{A}) = \mathbf{0}$. Let $\{\mathbf{A}^-_{\mathbf{max}}\}$ be full column rank matrix. Then $\mathbf{A}^-_{\mathbf{max}}(\mathbf{B}-\mathbf{A}) = \mathbf{0} \Rightarrow \mathbf{A} = \mathbf{B}$.
It is easy to check that $<^{\hat{\mathcal{G}}}$ is a partial order on $\mathbf{F}^{\mathbf{m}\times\mathbf{n}}$. □

Remark 7.6.3. Theorem 7.6.2 is one of many trivial ways to obtain an extension of a g-map with support a proper subset of its domain. Another such extension arises if we take $\hat{\mathcal{G}}(\mathbf{A}) = \{\mathbf{A}^-\}$ when $\mathbf{A}$ does not belong to $\Omega_{\mathcal{G}}$.

We conclude this chapter with the study of the non-trivial partial order extensions of the sharp order. We begin with a lemma that provides us with a necessary condition for a partial order to be an extension of the sharp order.

Lemma 7.6.4. *Let $<^t$ be any partial order extension to $F^{n\times n}$ of the sharp order. Then for* $\mathbf{A}, \mathbf{B} \in \mathbf{F}^{\mathbf{n}\times\mathbf{n}}$, $\mathbf{A} <^{\mathbf{t}} \mathbf{B} \Rightarrow \mathbf{C_A} <^{\#} \mathbf{C_B}$, *where* $\mathbf{C_X}, \mathbf{N_X}$ *denote respectively the core part and the nilpotent part of the square matrix* $\mathbf{X}$.

Proof. Write $\mathbf{A} = \mathbf{C_A} + \mathbf{N_A}$ with $\mathbf{C_A}\mathbf{N_A} = \mathbf{0}$, $\mathbf{N_A}\mathbf{C_A} = \mathbf{0}$ and $\mathbf{B} = \mathbf{C_B} + \mathbf{N_B}$ with $\mathbf{C_B}\mathbf{N_B} = \mathbf{0}$, $\mathbf{N_B}\mathbf{C_B} = \mathbf{0}$. Clearly, $\mathbf{C_A} <^{\#} \mathbf{A}$ and therefore, $\mathbf{C_A} <^{\mathbf{t}} \mathbf{A}$. So, $\mathbf{C_A} <^{\mathbf{t}} \mathbf{B}$, by transitivity. Moreover, $\mathbf{C_A} <^{\#} \mathbf{B}$, so, we can write $\mathbf{B} = \mathbf{C_A} \oplus (\mathbf{B} - \mathbf{C_A})$. Let $\mathbf{X} = \mathbf{B} - \mathbf{C_A}$. Then $\mathbf{X}\mathbf{C_A} = \mathbf{0} = \mathbf{C_A}\mathbf{X}$. As $\mathbf{C_A}$ is of index 1, there exist non-singular matrices $\mathbf{P}$ and $\mathbf{T}$ such that $\mathbf{C_A} = \mathbf{P}\mathbf{diag}\,(\mathbf{T} \;\; \mathbf{0})\,\mathbf{P}^{-1}$. Let $\mathbf{X} = \mathbf{P}\begin{pmatrix}\mathbf{X_1} & \mathbf{X_2}\\ \mathbf{X_3} & \mathbf{X_4}\end{pmatrix}\mathbf{P}^{-1}$, partitioning of $\mathbf{X}$ being in conformation with partitioning of $\mathbf{C_A}$. Now, $\mathbf{X}\mathbf{C_A} = \mathbf{0} = \mathbf{C_A}\mathbf{X}$ implies $\mathbf{X_i} = \mathbf{0}$, for $i = 1, 2, 3$. Hence, $\mathbf{B} = \mathbf{P}\begin{pmatrix}\mathbf{T} & \mathbf{0}\\ \mathbf{0} & \mathbf{X_4}\end{pmatrix}\mathbf{P}^{-1}$. Let $\mathbf{X_4} = \mathbf{Q}\begin{pmatrix}\mathbf{U} & \mathbf{0}\\ \mathbf{0} & \mathbf{N_1}\end{pmatrix}\mathbf{Q}^{-1}$ for some non-singular matrices $\mathbf{Q}$ and $\mathbf{U}$. Let us take $\mathbf{R} = \mathbf{P}\begin{pmatrix}\mathbf{I} & \mathbf{0}\\ \mathbf{0} & \mathbf{Q}\end{pmatrix}$. Then $\mathbf{R}$ is non-singular and

$$\mathbf{B} = \mathbf{R}\begin{pmatrix}\mathbf{T} & \mathbf{0}\\ \mathbf{0} & \begin{pmatrix}\mathbf{U} & \mathbf{0}\\ \mathbf{0} & \mathbf{N_1}\end{pmatrix}\end{pmatrix}\mathbf{R}^{-1}.$$

Thus, $\mathbf{C_B} = \mathbf{C_A} + \mathbf{C_X}$, $\mathbf{C_A}\mathbf{C_X} = \mathbf{0}$, $\mathbf{C_X}\mathbf{C_A} = \mathbf{0}$. So, $\mathbf{C_A} <^{\#} \mathbf{C_B}$. □

Definition 7.6.5. Let $\mathbf{A}, \mathbf{B} \in \mathbf{F}^{\mathbf{n}\times\mathbf{n}}$. We say $\mathbf{A}$ is below $\mathbf{B}$ under $<^t$, if $\mathbf{A} <^{-} \mathbf{B}$ and $\mathbf{C_A} <^{\#} \mathbf{C_B}$. We write $\mathbf{A} <^{\mathbf{t}} \mathbf{B}$ if $\mathbf{A} <^{-} \mathbf{B}$ and $\mathbf{C_A} <^{\#} \mathbf{C_B}$.

Lemma 7.6.6. *The order relation $<^{\mathbf{t}}$ is a partial order on $\mathbf{F}^{\mathbf{n}\times\mathbf{n}}$ that extends the sharp order.*

Proof. The order relation $<^{\mathbf{t}}$ is a partial order follows from the uniqueness of the core-nilpotent decomposition and the fact that both the minus order and the sharp order are partial orders.

To show that $<^{\mathbf{t}}$ extends $<^{\#}$, let $\mathbf{A}, \mathbf{B} \in \mathbf{F}^{\mathbf{n}\times\mathbf{n}}$ be each of index ≤ 1 such that $\mathbf{A} <^{\mathbf{t}} \mathbf{B}$. It is enough to show that $\mathbf{A}$ and $\mathbf{B}$ commute. Since $\mathbf{A} <^{-} \mathbf{B}$, there exist non-singular matrices $\mathbf{P}$, $\mathbf{T}$ and a matrix $\mathbf{S}$ of order same as of $\mathbf{T}$ such that $\mathbf{A} = \mathbf{P}^{-1}\mathbf{diag}\begin{pmatrix}\mathbf{S} & \mathbf{0}\end{pmatrix}\mathbf{P}$ and $\mathbf{B} = \mathbf{P}^{-1}\mathbf{diag}\begin{pmatrix}\mathbf{T} & \mathbf{0}\end{pmatrix}\mathbf{P}$. As $\mathbf{A}$ is of index ≤ 1, $\mathbf{S}$ is of index ≤ 1. Now, $\mathbf{C_A} <^{\#} \mathbf{C_B}$, so, $\mathbf{C_A}$ and $\mathbf{C_B}$ commute and $\mathbf{C_A}^2 = \mathbf{C_A}\mathbf{C_B}$. So, $\mathbf{T}$ and $\mathbf{S}$ commute. Hence, $\mathbf{A}$ and $\mathbf{B}$ commute. □

We now show that the order relation $<^t$ is a $\mathcal{G}$-based extension of $<^{\#}$.

Let $\mathcal{G}^0 : F^{n\times n} \longrightarrow \mathfrak{P}(F^{n\times n})$ be a g-map defined as

$$\mathcal{G}^0(\mathbf{A}) = \{\mathbf{A}^- : \mathcal{C}(\mathbf{C_A}) \subseteq \mathcal{C}(\mathbf{A}^-),\ \mathcal{C}(\mathbf{C_A}^{\mathbf{t}}) \subseteq \mathcal{C}(\mathbf{A}^{-\mathbf{t}})\}.$$

Notice that $\mathcal{G}^0(\mathbf{A})$ is non-empty, as we can find a non-singular g-inverse of $\mathbf{A}$ and such a g-inverse lies in $\mathcal{G}^0(\mathbf{A})$. Let $\mathbf{P_A} = \mathbf{C_A}\mathbf{C_A^{\#}} = \mathbf{C_A^{\#}}\mathbf{C_A}$. We have the following:

Lemma 7.6.7. *In the setup of the preceding para, the following hold:*

(i) *For each choice of* $\mathbf{N_A^-}$*, g-inverse of* $\mathbf{N_A}$*, the nilpotent part of* $\mathbf{A}$*, the matrix* $\mathbf{C_A^{\#}} + (\mathbf{I} - \mathbf{P_A})\mathbf{N_A^-}(\mathbf{I} - \mathbf{P_A}) \in \mathcal{G}^0(\mathbf{A})$.
(ii) *For each choice of* $\mathbf{A}^-$, $\mathbf{C_A^{\#}} + (\mathbf{I} - \mathbf{P_A})\mathbf{A}^-(\mathbf{I} - \mathbf{P_A}) \in \mathcal{G}^0(\mathbf{A})$.

Proof. If $\mathbf{A} = \mathbf{P}\mathbf{diag}\begin{pmatrix}\mathbf{C}\ , \mathbf{N}\end{pmatrix}\mathbf{P}^{-1}$ be core nilpotent decomposition of $\mathbf{A}$, then $\mathbf{P_A} = \mathbf{P}\mathbf{diag}\begin{pmatrix}\mathbf{I}\ , \mathbf{0}\end{pmatrix}\mathbf{P}^{-1}$. Also, $\mathbf{A}^- = \mathbf{P}\begin{pmatrix}\mathbf{C}^{-1} & \mathbf{L} \\ \mathbf{M} & \mathbf{N}^-\end{pmatrix}\mathbf{P}^{-1}$, where $\mathbf{LN} = \mathbf{0}$, $\mathbf{NM} = \mathbf{0}$ and $\mathbf{N}^-$ is a g-inverse of $\mathbf{N}$. So, $\mathbf{C_A^{\#}} + (\mathbf{I} - \mathbf{P_A})\mathbf{N_A^-}(\mathbf{I} - \mathbf{P_A})$ and $\mathbf{C_A^{\#}} + (\mathbf{I} - \mathbf{P_A})\mathbf{A}^-(\mathbf{I} - \mathbf{P_A})$ are each equal to $\mathbf{P}\mathbf{diag}\begin{pmatrix}\mathbf{C}^{-1}\ , \mathbf{N}^-\end{pmatrix}\mathbf{P}^{-1}$.
(i) Clearly, $\mathbf{C_A^{\#}} + (\mathbf{I} - \mathbf{P_A})\mathbf{N_A^-}(\mathbf{I} - \mathbf{P_A})$ is g-inverse of $\mathbf{A}$ and that $\mathcal{C}(\mathbf{C_A}) \cap \mathcal{C}(\mathbf{I} - \mathbf{P_A})\mathbf{A}^-(\mathbf{I} - \mathbf{P_A}) = \{\mathbf{0}\}$, so, $\mathcal{C}(\mathbf{C_A}) \subseteq \mathcal{C}(\mathbf{G})$.
Similarly, $\mathcal{C}(\mathbf{C_A^t}) \subseteq \mathcal{C}(\mathbf{G^t})$. Thus, $\mathbf{G} \in \mathcal{G}^0(\mathbf{A})$.
(ii) Similar to (i). □

Remark 7.6.8. The two classes

$$\mathcal{G}^1(\mathbf{A}) = \{\mathbf{C_A^{\#}} + (\mathbf{I} - \mathbf{P_A})\mathbf{N_A^-}(\mathbf{I} - \mathbf{P_A}) : \mathbf{N_A^-} \text{ is arbitrary}\}$$

and

$$\mathcal{G}^2(\mathbf{A}) = \{\mathbf{C_A^{\#}} + (\mathbf{I} - \mathbf{P_A})\mathbf{A}^-(\mathbf{I} - \mathbf{P_A}) : \mathbf{A}^- \text{ is arbitrary}\}$$

are identical.

Theorem 7.6.9. *In the setup of Lemma 7.6.7, for the order relation* $<^{\mathcal{G}^0}$ *defined by* $\mathcal{G}^0$ *on* $F^{n\times n}$, $\mathbf{A} <^{\mathcal{G}^0} \mathbf{B}$ *if and only if* $\mathbf{A} <^{\mathbf{t}} \mathbf{B}$.

Proof. 'Only if' part
Let $\mathbf{A} <^{\mathcal{G}^0} \mathbf{B}$. So, there exists a $\mathbf{A}^- \in \mathcal{G}^0(\mathbf{A})$ such that $\mathbf{B} = \mathbf{A} + \mathbf{X}$, $\mathbf{XA}^- = \mathbf{0}$, and $\mathbf{A}^-\mathbf{X} = \mathbf{0}$. Since $\mathbf{A}^- \in \mathcal{G}^0(\mathbf{A})$, $\mathcal{C}(\mathbf{C_A}) \subseteq \mathcal{C}(\mathbf{A}^-)$, and $\mathcal{C}(\mathbf{C_A^t}) \subseteq \mathcal{C}(\mathbf{A}^{-t})\}$. So, $\mathbf{XC_A} = \mathbf{0}$, $\mathbf{C_AX} = \mathbf{0}$. Rewrite $\mathbf{B} = \mathbf{C_A} + (\mathbf{N_A} + \mathbf{X})$. We have $\mathbf{C_A}(\mathbf{N_A} + \mathbf{X}) = \mathbf{0}, (\mathbf{N_A} + \mathbf{X})\mathbf{C_A} = \mathbf{0}$. By Lemma 7.6.6, we have $\mathbf{C_A} + \mathbf{C_Y} = \mathbf{C_B}$, where $\mathbf{Y} = \mathbf{N_A} + \mathbf{X}$ and $\mathbf{C_YC_A} = \mathbf{0} = \mathbf{C_AC_Y}$. So, $\mathbf{C_A} <^{\#} \mathbf{C_B}$. However, $\mathbf{A} <^{\mathcal{G}^0} \mathbf{B} \Rightarrow \mathbf{A} <^- \mathbf{B}$. So, $\mathbf{A} <^{\mathbf{t}} \mathbf{B}$.
'If' part
Let $\mathbf{A} <^{\mathbf{t}} \mathbf{B}$. So, $\mathbf{A} <^- \mathbf{B}$. Let $\mathbf{A}^-$ be such that $\mathbf{A}^-(\mathbf{B} - \mathbf{A}) = \mathbf{0}$ and $(\mathbf{B} - \mathbf{A})\mathbf{A}^- = \mathbf{0}$. Now, $\mathbf{G} = \mathbf{C_A^{\#}} + (\mathbf{I} - \mathbf{P_A})\mathbf{A}^-(\mathbf{I} - \mathbf{P_A}) \in \mathcal{G}^0(\mathbf{A})$, by Lemma 7.6.7(ii) and $\mathbf{C_A} = \mathbf{C_A}(\mathbf{C_B^{\#}})\mathbf{C_B}$ and $\mathbf{C_A} = \mathbf{C_B}(\mathbf{C_B^{\#}})\mathbf{C_A}$, as $\mathbf{C_A} <^{\#} \mathbf{C_B}$. So, $\mathbf{C_AN_B} = \mathbf{0} = \mathbf{N_BC_A}$. Consider $\mathbf{C_A}(\mathbf{B} - \mathbf{A})$. We can rewrite $\mathbf{C_A}(\mathbf{B} - \mathbf{A}) = \mathbf{C_A}(\mathbf{C_B} - \mathbf{C_A}) + \mathbf{C_A}(\mathbf{N_B} - \mathbf{N_A}) = \mathbf{0}$, since $\mathbf{C_AN_A} = \mathbf{0}$, $\mathbf{C_AN_B} = \mathbf{0}$ and $\mathbf{C_A} <^{\#} \mathbf{C_B}$. Similarly, $(\mathbf{B} - \mathbf{A})\mathbf{C_A} = \mathbf{0}$. Therefore, $\mathbf{C_A^{\#}}(\mathbf{B} - \mathbf{A}) = \mathbf{0}$, $(\mathbf{B} - \mathbf{A})\mathbf{C_A^{\#}} = \mathbf{0}$. Also, $(\mathbf{I} - \mathbf{P_A})(\mathbf{B} - \mathbf{A}) = (\mathbf{B} - \mathbf{A})$. Thus, $\mathbf{G}(\mathbf{B} - \mathbf{A}) = \mathbf{0}$.
Similarly, $(\mathbf{B} - \mathbf{A})\mathbf{G} = \mathbf{0}$. Hence, $\mathbf{A} <^{\mathcal{G}^0} \mathbf{B}$. □

Remark 7.6.10. (i) Clearly, for each $\mathbf{A}$, the class $\mathcal{G}^0(\mathbf{A})$ contains the class $\mathcal{G}^1(\mathbf{A})$ and therefore, $\mathcal{G}^2(\mathbf{A})$.
(ii) A $\mathcal{G}$-based extension of the sharp order need not be unique.
(iii) A $\mathcal{G}$-based extension of the sharp order will have to be finer than the order relation $<^{\mathcal{G}^0}$, by Lemma 7.6.4.

7.7 Exercises

(1) Prove that if $\mathcal{G}(\mathbf{A}) = \{\mathbf{A}_r^-\}$, then the relation $<^{\mathcal{G}}$ is the minus-order.

(2) Let $\mathcal{G}$ be a complete g-map on $F^{m\times n}$ such that whenever $\mathbf{A} \in \Omega_{\mathcal{G}}$, then $\mathcal{G}(\mathbf{A}) \cap \{\mathbf{A}_r^-\} \neq \emptyset$. Then any $\mathbf{A} \in \Omega_{\mathcal{G}}$ is maximal $\Leftrightarrow$ $\rho(\mathbf{A}) = \min\{m, n\}$. Prove that the same conclusion holds when the map is semi-complete.

(3) Let $\mathcal{G}$ be a g-map of $F^{m\times n}$ with support $\Omega_{\mathcal{G}}$ and $\Omega \subset \Omega_{\mathcal{G}}$. Let for each $\mathbf{A} \in \Omega$, the set $\mathcal{G}(\mathbf{A})$ is replaced by $\emptyset$. If the relation $<^{\mathcal{G}}$ is a partial order, then show that the modified map also induces a partial order. What is the support of the new g-map?

(4) Let $\mathcal{G}$ be a semi-complete g-map on $\mathbb{C}^{m\times n}$ that satisfies the condition $(\mathbf{T})$. Suppose $\mathbf{A}, \mathbf{B}, \mathbf{C} \in \mathbb{C}^{m\times n}$ such that $\mathbf{A} \neq \mathbf{B}$, $\mathbf{B} \neq \mathbf{C}$ and $\mathbf{A} <^{\mathcal{G}} \mathbf{B}$, $\mathbf{B} <^{\mathcal{G}} \mathbf{C}$. Then show that $\mathbf{A} <^{\mathcal{G}} \mathbf{C}$. Hence, conclude that '$<^{\mathcal{G}}$' is a partial order on $\Omega_{\mathcal{G}}$.

(5) If $\mathbf{A} <^- \mathbf{B}$ and the pair $(\mathbf{A}, \mathbf{B})$ satisfies condition $(\mathbf{T})$, then show that it also satisfies the following conditions:

$$\mathcal{G}_r(\mathbf{A}|\mathbf{B}) \subseteq \mathcal{G}_r(\mathbf{A}) \text{ and } \subseteq \tilde{\mathcal{G}}(\mathbf{A}).$$

(6) Let $\mathbf{A}$ and $\mathbf{C} \in \mathbb{C}^{m\times n}$ be non-null and $\mathcal{G}$ be a g-map on $\mathbb{C}^{m\times n}$. Let $\mathcal{G}(\mathbf{A}) = Ann(\mathbf{C}) \cap \{\mathbf{A}^-\} \neq \emptyset$. Show that $\mathcal{G}(\mathbf{A})$ is complete if and only if $Ann(\mathbf{A}) \subset Ann(\mathbf{C})$, where $Ann(\mathbf{X})$ denotes that annihilator of the matrix $\mathbf{X}$.

(7) Let $\mathcal{G} : F^{m\times n} \to \mathcal{P}(F^{m\times n})$ be defined as

$$\mathcal{G}(\mathbf{A}) = \begin{cases} \mathbf{A}^{\#}, & \text{if Index } \mathbf{A} = 1 \\ \mathbf{A}^{-}, & \text{if } \mathbf{A} \text{ is nilpotent} \\ \emptyset, & \text{otherwise.} \end{cases}$$

Then show that the order relation $<^{\mathbf{G}}$ is a partial order that extends $<^{\#}$.

(8) Let $\mathcal{G} : F^{m\times n} \to \mathcal{P}(F^{n\times m})$ be a one-one map. Further, suppose $\{\mathbf{A}^+ : \mathbf{A} \in F^{m\times n}\}$, where $\mathbf{A}^+$ is a specified reflexive g-inverse of $\mathbf{A}$ is equal to $F^{n\times m}$. Define $\mathcal{G}^{-1}$, the inverse map $\mathcal{G}^{-1}(\mathbf{D}) = \mathbf{A}$, if $\mathcal{G}(\mathbf{A}) = \mathbf{D}$. In the setup of Corollary 7.2.29, assume $\mathcal{G}$ is one-one and $<^{\mathbf{G}}$ is a partial order. Then, show that $\mathbf{A} <^{\mathcal{G}} \mathbf{B}$ implies $\mathbf{A}^+ <^{\mathcal{G}^{-1}} \mathbf{B}^+$ and if in addition $\mathbf{C} < \mathcal{G}^{-1}\mathbf{E}$ implies $\mathbf{D}^+ <^{s} \mathbf{E}^+$, the order relation $<^{\mathcal{G}^{-1}}$ is a partial order.

(9) In the setup of Lemma 7.6.7 show that
$\mathcal{G}^1(\mathbf{A}) = \{\mathbf{C}_A^{\#} + (\mathbf{I} - \mathbf{P}_A)\mathbf{N}_A^-(\mathbf{I} - \mathbf{P}_A) : \mathbf{N}_A^-$ is arbitrary$\}$ and
$\mathcal{G}^2(\mathbf{A}) = \{\mathbf{C}_A^{\#} + (\mathbf{I} - \mathbf{P}_A)\mathbf{A}_A^-(\mathbf{I} - \mathbf{P}_A) : \mathbf{A}^-$ is arbitrary$\}$ are the same

sets i.e., $\mathcal{G}^1(\mathbf{A}) = \mathcal{G}^2(\mathbf{A})$.

(10) Let $\mathcal{G} : F^{m\times n} \to \mathcal{P}(F^{n\times m})$ and $\mathcal{G}^0 : F^{m\times n} \to \mathcal{P}(F^{n\times m})$ be two g-maps. Write $\mathbf{A} <_{\mathcal{G},\mathcal{G}^0} \mathbf{B}$, if $\mathbf{A} <^{\mathcal{G}^0} \mathbf{B}$ and $\mathcal{G}(\mathbf{A},\mathbf{B}) \subseteq \mathcal{G}(\mathbf{A})$. When does '$<_{\mathcal{G},\mathcal{G}^0}$' define a partial order? Is it always different from the order relations '$<_{\mathcal{G}}$' and '$<_{\mathcal{G}^0}$'?

(11) Take $\mathcal{G}(\mathbf{A}) = \{\mathbf{A}^\dagger\}$ and $\mathcal{G}^0(\mathbf{A}) = \{\mathbf{A}^-\}$ in Ex. 10. What is the order defined by '$<_{\mathcal{G},\mathcal{G}^0}$'. Identify the orders $<_{\mathcal{G},\mathcal{G}^0}$ when (i) $\mathcal{G}(\mathbf{A}) = \{\mathbf{A}^\#\}$ and $\mathcal{G}^0(\mathbf{A}) = \{\mathbf{A}^-\}$ (ii) $\mathcal{G}(\mathbf{A}) = \{\mathbf{A}_r^-\}$ and $\mathcal{G}^0(\mathbf{A}) = \{\mathbf{A}^-\}$.

(12) Let $\mathcal{G} : \mathbb{C}^{n\times n} \to \mathcal{P}(\mathbb{C}^{n\times n})$ be defined as $\mathcal{G}(\mathbf{A}) = \{\mathbf{A}^\dagger\mathbf{A}\mathbf{A}^\#\}$. Show that the map $\mathcal{G}$ gives a $\mathcal{G}$-based partial order.

(13) Open Problem: Find a necessary and sufficient condition that a g-map $\mathcal{G} : F^{m\times n} \to \mathcal{P}(F^{n\times m})$ defines a partial order on its support?

The following exercises use systems of matrix equations to define partial orders.

(14) Let $\mathbf{A},\mathbf{B} \in F^{m\times n}$. Show that the following are equivalent:

(i) $\mathbf{A} <^- \mathbf{B}$
(ii) $\mathbf{AXB} = \mathbf{A}$ and $\mathbf{BXB} = \mathbf{A}$ are jointly consistent;
(iii) $\mathbf{BXA} = \mathbf{A}$ and $\mathbf{BXB} = \mathbf{A}$ are jointly consistent;
(iv) The matrix equation $\begin{pmatrix}\mathbf{A} & \mathbf{A}\\ \mathbf{A} & \mathbf{A}\end{pmatrix} = \begin{pmatrix}\mathbf{A}\\ \mathbf{B}\end{pmatrix}\mathbf{X}(\mathbf{A}\ \mathbf{B})$ is consistent.

(15) Let $\mathbf{A},\mathbf{B} \in \mathbb{C}^{m\times n}$. Show that the following are equivalent:

(i) $\mathbf{A} \star\!< \mathbf{B}$
(ii) $\mathbf{A}^\star\mathbf{AX} = \mathbf{A}^\star$, $\mathbf{AXB} = \mathbf{A}$ and $\mathbf{BXB} = \mathbf{A}$ are jointly consistent.

(16) Let $\mathbf{A},\mathbf{B} \in F^{m\times n}$. Show that the following are equivalent:

(i) $\mathbf{A} <\star\ \mathbf{B}$ and
(ii) $\mathbf{XAA}^\star = \mathbf{A}^\star$, $\mathbf{AXB} = \mathbf{A}$ and $\mathbf{BXB} = \mathbf{A}$ are jointly consistent.

(17) Let $\mathbf{A},\mathbf{B} \in F^{m\times n}$. Then the following are equivalent:

(i) $\mathbf{A} <^0 \mathbf{B}$
(ii) $\mathbf{XBX} = \mathbf{X}$ and $\mathbf{BXB} = \mathbf{A}$ are jointly consistent;
(iii) $\mathbf{BXB} = \mathbf{A}$, $\mathbf{BXBXB} = \mathbf{BXB}$ are jointly consistent and
(iv) $\mathbf{XBXBX} = \mathbf{XBX}$ and $\mathbf{BXB} = \mathbf{A}$ are jointly consistent.

(18) Let $\mathbf{A},\mathbf{B} \in F^{m\times n}$. Consider the matrix equations $\mathbf{AXA} = \mathbf{A}$ and $\mathbf{BXB} = \mathbf{A}$. Define an order relation as follows: $\mathbf{A} \ll \mathbf{B}$ if the equa-

tions $\mathbf{AXA} = \mathbf{A}$ and $\mathbf{BXB} = \mathbf{A}$ are jointly consistent. Check if this order relation is a partial order.

Chapter 8

The Löwner Order

8.1 Introduction

The Löwner order is one of the oldest and most widely used matrix partial orders. However, the properties and the results concerning the Löwner order do not seem to have been compiled together in one source and are rather scattered. In this chapter, we attempt to give a comprehensive exposition of the Löwner order. In Section 8.2, we define the Löwner order on the class of hermitian matrices. The class of *nnd* matrices is an important subclass of the class of hermitian matrices. We obtain several important properties of the Löwner order on *nnd* matrices and discuss some of the possible extensions to hermitian matrices and limitations when this is not possible. In Section 8.3, we study the relationship of the Löwner order on a pair of hermitian matrices with that on their powers. We also study the relationship of the Löwner order with some of the other partial orders that we studied earlier. In Section 8.4, we study the ordering properties of g-inverses of matrices with respect to the Löwner order. Section 8.5 contains two generalizations of the Löwner order, one for hermitian matrices and other for arbitrary rectangular matrices. We make a comparison of these extensions on the class of hermitian matrices.

8.2 Definition and basic properties

In this section, we formally define the Löwner order and show that it is a partial order. We also derive some of its useful and interesting properties. Recall that $\mathfrak{H}_n$ denotes the class of all $n \times n$ hermitian matrices.

We give below two well-known results in Linear Algebra which will be needed subsequently.

Theorem 8.2.1. *(Fisher-Courant) Let* $\mathbf{A} \in \mathfrak{H}_n$. *Let* $\lambda_1 \geq \lambda_2 \geq \ldots \lambda_n$ *be the eigen-values of* $\mathbf{A}$ *and* $\mathbf{u_1}, \mathbf{u_2}, \ldots, \mathbf{u_n}$ *be the corresponding ortho-normal eigen vectors. (Notice that all eigen-values of* $\mathbf{A}$ *are real.) Then*

(i) $\max_{\|x\|=1} \mathbf{x}^\star \mathbf{A}\mathbf{x} = \max\limits_{\mathbf{y} \neq \mathbf{0}} \dfrac{\mathbf{y}^\star \mathbf{A}\mathbf{y}}{\mathbf{y}^\star \mathbf{y}} = \lambda_1$,

$\min_{\|x\|=1} \mathbf{x}^\star \mathbf{A}\mathbf{x} = \min\limits_{\mathbf{y} \neq \mathbf{0}} \dfrac{\mathbf{y}^\star \mathbf{A}\mathbf{y}}{\mathbf{y}^\star \mathbf{y}} = \lambda_\mathbf{n}$.

(ii) $\max\{\dfrac{\mathbf{y}^\star \mathbf{A}\mathbf{y}}{\mathbf{y}^\star \mathbf{y}} : \mathbf{u_i^\star y} = 0,\ i = 1, \ldots, k-1\} = \lambda_k$.

(iii) *Let* k *be a fixed integer such that* $2 \leq k \leq n$ *and let* $\mathbf{B}$ *denote an* $n \times (k-1)$ *matrix. Then*

$$\inf_{\mathbf{B}} \sup_{\mathbf{B}^\star \mathbf{y}=\mathbf{0}} \left(\frac{\mathbf{y}^\star \mathbf{A}\mathbf{y}}{\mathbf{y}^\star \mathbf{y}}\right) = \lambda_k.$$

Proof. (i) and (ii) are easy.

(iii) Let $\mathbf{u_1}, \mathbf{u_2}, \ldots, \mathbf{u_n}$ be the ortho-normal eigen vectors of $\mathbf{A}$ corresponding to the eigen-values $\lambda_1, \lambda_2, \ldots, \lambda_n$ respectively. Write $\mathbf{U} = (\mathbf{u_1} : \cdots : \mathbf{u_n})$. Clearly, $\mathbf{U}$ is a unitary matrix such that $\mathbf{U}^\star \mathbf{A}\mathbf{U} = \mathbf{diag}\,(\lambda_1 \ \cdots \ \lambda_n)$. Let $\mathbf{B}$ be an $n \times (k-1)$ matrix. Write $\mathbf{y} = \mathbf{U}\mathbf{z}$. Then

$$\sup_{\mathbf{B}^\star \mathbf{y}=\mathbf{0}} \left(\frac{\mathbf{y}^\star \mathbf{A}\mathbf{y}}{\mathbf{y}^\star \mathbf{y}}\right) = \sup\nolimits_{\mathbf{B}^\star \mathbf{U}\mathbf{z}=\mathbf{0}} \left(\frac{\mathbf{z}^\star \mathbf{U}^\star \mathbf{A}\mathbf{U}\mathbf{z}}{\mathbf{z}^\star \mathbf{z}}\right) = \sup_{\mathbf{B}^\star \mathbf{U}\mathbf{z}=\mathbf{0}} \left(\frac{\sum_{i=1}^n \lambda_i {\mathbf{z}_i}^2}{\sum_{i=1}^n {\mathbf{z}_i}^2}\right).$$

Since $\mathbf{B}^\star \mathbf{U}$ is of order $(k-1) \times n$, there exists a non-null vector $\mathbf{w}$ such that $\mathbf{B}^\star \mathbf{U}\mathbf{w} = \mathbf{0}$ and $w_{k+1} = \ldots = w_n = 0$. So,

$$\sup_{\mathbf{B}^\star \mathbf{U}\mathbf{z}=\mathbf{0}} \left(\frac{\sum_{i=1}^n \lambda_i {\mathbf{z}_i}^2}{\sum_{i=1}^n {\mathbf{z}_i}^2}\right) \geq \left(\frac{\sum_{i=1}^k \lambda_i {\mathbf{w_i}}^2}{\sum_{i=1}^k {\mathbf{w_i}}^2}\right). \tag{8.2.1}$$

Since, (8.2.1) holds for all matrices $\mathbf{B}$ of order $(k-1) \times n$, we have

$$\inf_{\mathbf{B}} \sup_{\mathbf{B}^\star \mathbf{y}=\mathbf{0}} \left(\frac{\mathbf{y}^\star \mathbf{A}\mathbf{y}}{\mathbf{y}^\star \mathbf{y}}\right) \geq \lambda_k. \tag{8.2.2}$$

Taking $\mathbf{B} = (\mathbf{u_1} : \cdots : \mathbf{u_{k-1}})$, it follows from (ii) that equality holds in (8.2.2).

□

Theorem 8.2.2. *Let* $\mathbf{A} = \begin{pmatrix} \mathbf{A_{11}} & \mathbf{A_{12}} \\ \mathbf{A_{12}^\star} & \mathbf{A_{22}} \end{pmatrix}$ *be an nnd matrix of order* $n \times n$, *where* $\mathbf{A_{11}}, \mathbf{A_{12}}$ *and* $\mathbf{A_{22}}$ *are of orders* $r \times r$, $r \times (n-r)$ *and* $(n-r) \times (n-r)$ *respectively. Then* $\mathbf{A}$ *is nnd if and only if*

(i) $\mathbf{A_{11}}$ *is nnd*
(ii) $\mathcal{C}(\mathbf{A_{12}}) \subseteq \mathcal{C}(\mathbf{A_{11}})$ *and*
(iii) $\mathbf{A_{22}} - \mathbf{A_{12}^{\star}A_{11}^{-}A_{12}}$ *is nnd, where* $\mathbf{A_{11}^{-}}$ *is an arbitrary g-inverse of* $\mathbf{A_{11}}$.

Proof. Notice first that we can write

$$\mathbf{A} = \begin{pmatrix} \mathbf{I} & \mathbf{0} \\ \mathbf{A_{12}^{\star}A_{11}^{-}} & \mathbf{I} \end{pmatrix} \begin{pmatrix} \mathbf{A_{11}} & \mathbf{0} \\ \mathbf{0} & \mathbf{A_{22}} - \mathbf{A_{12}^{\star}A_{11}^{-}A_{12}} \end{pmatrix} \begin{pmatrix} \mathbf{I} & \mathbf{A_{11}^{-}A_{12}} \\ \mathbf{0} & \mathbf{I} \end{pmatrix}.$$

The matrices $\begin{pmatrix} \mathbf{I} & \mathbf{A_{11}^{-}A_{12}} \\ \mathbf{0} & \mathbf{I} \end{pmatrix}$ and $\begin{pmatrix} \mathbf{I} & \mathbf{0} \\ \mathbf{A_{12}^{\star}A_{11}^{-}} & \mathbf{I} \end{pmatrix}$ are non-singular and are hermitian conjugates of each other. Moreover, under both 'if' part as well as 'only if' part, we have $\mathcal{C}(\mathbf{A_{12}}) \subseteq \mathcal{C}(\mathbf{A_{11}})$. So, $\mathbf{A_{12}^{\star}A_{11}^{-}A_{12}}$ is invariant under choices of g-inverses $\mathbf{A_{11}^{-}}$ of $\mathbf{A_{11}}$. Therefore, $\mathbf{A}$ is *nnd* if and only if $\mathbf{A_{11}}$ and $\mathbf{A_{22}} - \mathbf{A_{12}^{\star}A_{11}^{-}A_{12}}$ are *nnd*. □

Corollary 8.2.3. *Consider the same setup as in Theorem 8.2.2. Then* $\mathbf{A}$ *is nnd if and only if*

(i) $\mathbf{A_{22}}$ *is nnd,*
(ii) $\mathcal{C}(\mathbf{A_{12}^{\star}}) \subseteq \mathcal{C}(\mathbf{A_{22}})$ *and*
(iii) $\mathbf{A_{11}} - \mathbf{A_{12}^{\star}A_{22}^{-}A_{12}}$ *is nnd, where* $\mathbf{A_{22}^{-}}$ *is a g-inverse of* $\mathbf{A_{22}}$.

Theorem 8.2.4. *Let* $\mathbf{A} \in \mathfrak{H}_\mathbf{n}$. *Then there exist nnd matrices* $\mathbf{A_1}$ *and* $\mathbf{A_2}$ *such that*

(i) $\mathbf{A} = \mathbf{A_1} - \mathbf{A_2}$ *and*
(ii) $\mathbf{A_1A_2} = \mathbf{0}$.

Further, such a decomposition of $\mathbf{A}$ *is unique (i.e. if* $\mathbf{A} = \mathbf{B_1} - \mathbf{B_2}$, *where* $\mathbf{B_1}$ *and* $\mathbf{B_2}$ *are nnd and* $\mathbf{B_1B_2} = \mathbf{0}$, *then* $\mathbf{A_1} = \mathbf{B_1}$ *and* $\mathbf{A_2} = \mathbf{B_2}$*).*

Proof. Let $\lambda_1 > \lambda_2 > \ldots \lambda_s = 0 > \lambda_{s+1} > \ldots \lambda_t$ be distinct eigen-values of $\mathbf{A}$. Since every hermitian matrix has real eigen values it follows from Remark 2.2.42, that $\mathbf{A} = \sum_{\mathbf{i=1}}^{\mathbf{t}} \lambda_\mathbf{i}\mathbf{T_i}$, where $\mathbf{T_i}$ is hermitian and idempotent for $i = 1, \ldots, t$ and $\mathbf{T_iT_j} = \mathbf{0}$ whenever $i \neq j$. Take $\mathbf{A_1} = \sum_{\mathbf{i=1}}^{\mathbf{s-1}} \lambda_\mathbf{i}\mathbf{T_i}$ and $\mathbf{A_2} = -(\sum_{\mathbf{i=s+1}}^{\mathbf{t}} \lambda_\mathbf{i}\mathbf{T_i})$. Clearly, $\mathbf{A_1}$ and $\mathbf{A_2}$ are *nnd*, $\mathbf{A_1A_2} = \mathbf{0}$ and $\mathbf{A} = \mathbf{A_1} + \mathbf{A_2}$.

We now show that this decomposition is unique. Let $\mathbf{A} = \mathbf{B_1} - \mathbf{B_2}$, where $\mathbf{B_1}$, $\mathbf{B_2}$ are *nnd*, and $\mathbf{B_1B_2} = \mathbf{0}$. It is easy to see that (see Theorem 2.7.1) $\mathbf{B_1} = \mathbf{Udiag}\left(\mathbf{\Delta_1}\ ,\ \mathbf{0}\ ,\ \mathbf{0}\right)\mathbf{U^\star}$ and $\mathbf{B_2} = \mathbf{Udiag}\left(\mathbf{0}\ ,\ \mathbf{\Delta_2}\ ,\ \mathbf{0}\right)\mathbf{U^\star}$, where $\mathbf{\Delta_1}$, $\mathbf{\Delta_2}$ are positive definite diagonal matrices with their diagonal elements arranged in decreasing and increasing orders respectively and $\mathbf{U}$ is a unitary matrix. Now, $\mathbf{A} = \mathbf{B_1} - \mathbf{B_2} = \mathbf{U}\mathbf{diag}\left(\mathbf{\Delta_1}\ , -\mathbf{\Delta_2}\ , \mathbf{0}\right)\mathbf{U}^\star$, is a spectral decomposition of $\mathbf{A}$ for which the distinct diagonal elements must be precisely $\lambda_1 > \lambda_2 > \ldots > \lambda_t$. By appropriate pooling of the terms we have $\mathbf{A} = \sum_{\mathbf{i}=1}^{\mathbf{t}} \lambda_\mathbf{i}\mathbf{S_i}$. Thus, $\mathbf{B_1} = \sum_{\mathbf{i}=1}^{\mathbf{s}-1} \lambda_\mathbf{i}\mathbf{S_i}$ and $\mathbf{B_2} = -(\sum_{\mathbf{i}=\mathbf{s}+1}^{\mathbf{t}} \lambda_\mathbf{i}\mathbf{S_i})$. Once again by Remark 2.2.42, $\mathbf{S_i} = \mathbf{T_i}$. So, $\mathbf{A_1} = \mathbf{B_1}$ and $\mathbf{A_2} = \mathbf{B_2}$. □

We now define the Löwner order on the class of hermitian matrices $\mathfrak{H}_n$ of order $n \times n$.

Definition 8.2.5. Let $\mathbf{A}, \mathbf{B} \in \mathfrak{H}_n$. Then $\mathbf{A}$ is below $\mathbf{B}$ under the Löwner order if $\mathbf{B} - \mathbf{A}$ is *nnd*. When this happens, we write $\mathbf{A} <^\mathbf{L} \mathbf{B}$.

The following two theorems are easy to prove:

Theorem 8.2.6. *The Löwner order is a partial order on* $\mathfrak{H}_n$.

Theorem 8.2.7. *Let* $\mathbf{A}$, $\mathbf{B}$, $\mathbf{C} \in \mathfrak{H}_n$ *and let* $\mathbf{A} <^\mathbf{L} \mathbf{B}$. *Then the following hold:*

(i) $\mathbf{A} + \mathbf{C} <^\mathbf{L} \mathbf{B} + \mathbf{C}$.
(ii) $\lambda\mathbf{A} <^\mathbf{L} \lambda\mathbf{B}$ *for all* $\lambda \geq 0$.
(iii) $\mathbf{TAT}^\star <^\mathbf{L} \mathbf{TBT}^\star$ *for every matrix* $\mathbf{T}$ *such that* $\mathbf{TAT}^\star$ *is defined.*

Remark 8.2.8. Let $\mathbf{A}, \mathbf{B} \in \mathfrak{H}_n$ and $\mathbf{T}$ be a matrix of rank n. Then $\mathbf{TAT}^\star <^\mathbf{L} \mathbf{TBT}^\star$ implies $\mathbf{A} <^\mathbf{L} \mathbf{B}$.

We now prove some useful and non-trivial properties of the Löwner order.

Theorem 8.2.9. *Let* $\mathbf{A}, \mathbf{B} \in \mathfrak{H}_n$ *and* $\mathbf{A} <^\mathbf{L} \mathbf{B}$. *For a matrix* $\mathbf{C} \in \mathfrak{H}_n$, *let* $\lambda_1(\mathbf{C}), \lambda_2(\mathbf{C}), \ldots, \lambda_\mathbf{n}(\mathbf{C})$ *denote the eigen-values of* $\mathbf{C}$ *such that* $\lambda_1(\mathbf{C}) \geq \lambda_2(\mathbf{C}) \geq \ldots \lambda_\mathbf{n}(\mathbf{C})$. *Then* $\lambda_k(\mathbf{A}) \leq \lambda_\mathbf{k}(\mathbf{B})$ *for* $k = 1, \ldots, n$.

Proof. Let $k \geq 2$ and consider matrices $\mathbf{T}$ of order $n \times (k-1)$. By Theorem 8.2.1 (iii),

$$
\begin{aligned}
\inf_\mathbf{T}\sup_{\mathbf{T^\star x}=\mathbf{0}}\left\{\frac{\mathbf{x^\star Bx}}{\mathbf{x^\star x}}\right\} &= \inf_\mathbf{T}\sup_{\mathbf{T^\star x}=\mathbf{0}}\left\{\frac{\mathbf{x}^\star(\mathbf{A}+\mathbf{B}-\mathbf{A})\mathbf{x}}{\mathbf{x^\star x}}\right\} \\
&= \inf_\mathbf{T}\sup_{\mathbf{T^\star x}=\mathbf{0}}\left\{\frac{\mathbf{x^\star Ax}}{\mathbf{x^\star x}} + \frac{\mathbf{x}^\star(\mathbf{B}-\mathbf{A})\mathbf{x}}{\mathbf{x^\star x}}\right\} \\
&\geq \inf_\mathbf{T}\sup_{\mathbf{T^\star x}=\mathbf{0}}\left\{\frac{\mathbf{x^\star Ax}}{\mathbf{x^\star x}} + \lambda_n(\mathbf{B}-\mathbf{A})\right\} \\
&= \lambda_k(\mathbf{A}) + \lambda_\mathbf{n}(\mathbf{B}-\mathbf{A}) \geq \lambda_\mathbf{k}(\mathbf{A}),
\end{aligned}
$$

since $\mathbf{B} - \mathbf{A}$ is *nnd*. Therefore, $\lambda_k(\mathbf{B}) \geq \lambda_{\mathbf{k}}(\mathbf{A})$ for each $k \geq 2$. Also,

$$\mathbf{A} <^{\mathbf{L}} \mathbf{B} \Rightarrow \frac{\mathbf{x}^\star\mathbf{A}\mathbf{x}}{\mathbf{x}^\star\mathbf{x}} \leq \frac{\mathbf{x}^\star\mathbf{B}\mathbf{x}}{\mathbf{x}^\star\mathbf{x}}$$

for each $\mathbf{x} \neq \mathbf{0}$. Hence,

$$\lambda_1(\mathbf{A}) = \max_{\mathbf{x}\neq\mathbf{0}} \frac{\mathbf{x}^\star\mathbf{A}\mathbf{x}}{\mathbf{x}^\star\mathbf{x}} < \max_{\mathbf{x}\neq\mathbf{0}} \frac{\mathbf{x}^\star\mathbf{B}\mathbf{x}}{\mathbf{x}^\star\mathbf{x}} = \lambda_1(\mathbf{B}).$$

□

Corollary 8.2.10. *Let* $\mathbf{A}, \mathbf{B} \in \mathfrak{H}_n$ *and* $\mathbf{A} <^{\mathbf{L}} \mathbf{B}$. *Then the following hold:*

(i) $tr(\mathbf{A}) \leq tr(\mathbf{B})$.
(ii) *If* $\mathbf{A}$ *is nnd, then* $\det(\mathbf{A}) \leq \det(\mathbf{B})$.
(iii) *If* $\mathbf{A}$ *is not nnd, then* $\det(\mathbf{A}) \leq \det(\mathbf{B})$ *need not be true.*
(iv) $\mathbf{A}$ *has at least as many negative eigen-values (including multiplicities) as* $\mathbf{B}$.

As an example for (iii), take $\mathbf{A} = \mathbf{diag}\left(-1\ ,\ -1\right)$ and $\mathbf{B} = \mathbf{0}$.

We now obtain necessary and sufficient conditions for $\mathbf{A} <^{\mathbf{L}} \mathbf{B}$, when $\mathbf{A}, \mathbf{B}$ are *nnd*.

Theorem 8.2.11. *Let* $\mathbf{B}$ *be an nnd matrix of order* $n \times n$ *and* $\mathbf{C}$ *a matrix of order* $m \times n$. *Then the following are equivalent:*

(i) $\mathbf{C}\mathbf{C}^\star <^{\mathbf{L}} \mathbf{B}$
(ii) $\mathcal{C}(\mathbf{C}) \subseteq \mathcal{C}(\mathbf{B})$ *and* $\lambda_{max}(\mathbf{C}^\star\mathbf{B}^-\mathbf{C}) \leq \mathbf{1}$, *where* $\mathbf{B}^-$ *is any g-inverse of* $\mathbf{B}$ *and*
(iii) $\mathcal{C}(\mathbf{C}) \subseteq \mathcal{C}(\mathbf{B})$ *and* $\mathbf{C}^\star\mathbf{B}^-\mathbf{C} <^{\mathbf{L}} \mathbf{I}$.

Proof. We first note that $\mathbf{B}$ *nnd* and $\mathcal{C}(\mathbf{C}) \subseteq \mathcal{C}(\mathbf{B})$ together imply $\mathbf{C}^\star\mathbf{B}^-\mathbf{C}$ is invariant under the choices of $\mathbf{B}^-$. Since $\mathbf{B}$ is *nnd*, $\mathbf{B}$ has an *nnd* g-inverse. Hence, $\mathbf{C}^\star\mathbf{B}^-\mathbf{C}$ is *nnd* for all $\mathbf{B}^-$ when $\mathcal{C}(\mathbf{C}) \subseteq \mathcal{C}(\mathbf{B})$. Thus, (iii) $\Rightarrow$ (ii) is clear by Theorem 8.2.9. (ii) $\Rightarrow$ (iii) follows by Theorem 8.2.1.

Notice that $\mathbf{C}\mathbf{C}^\star <^{\mathbf{L}} \mathbf{B}$ and $\mathbf{B}$ is *nnd* $\Leftrightarrow$ the matrix $\mathbf{M} = \begin{pmatrix} \mathbf{I} & \mathbf{C}^\star \\ \mathbf{C} & \mathbf{B} \end{pmatrix}$ is *nnd*, by Theorem 8.2.2. By Corollary 8.2.3, $\mathbf{M}$ is *nnd* $\Leftrightarrow \mathbf{I} - \mathbf{C}^\star\mathbf{B}^-\mathbf{C}$ is *nnd* and $\mathcal{C}(\mathbf{C}) \subseteq \mathcal{C}(\mathbf{B}) \Leftrightarrow \mathbf{C}^\star\mathbf{B}^-\mathbf{C} <^{\mathbf{L}} \mathbf{I}$ and $\mathcal{C}(\mathbf{C}) \subseteq \mathcal{C}(\mathbf{B})$. Hence, (i) $\Leftrightarrow$ (iii). □

Corollary 8.2.12. *Let* $\mathbf{A}$ *and* $\mathbf{B}$ *be nnd matrices. Then the following are equivalent:*

(i) $\mathbf{A} <^{\mathbf{L}} \mathbf{B}$

(ii) $\mathcal{C}(\mathbf{A}) \subseteq \mathcal{C}(\mathbf{B})$ *and* $\lambda_{max}(\mathbf{B}^-\mathbf{A}) \leq 1$, *where* $\mathbf{B}^-$ *is any g-inverse of* $\mathbf{B}$ *and*

(iii) $\mathcal{C}(\mathbf{A}) \subseteq \mathcal{C}(\mathbf{B})$ *and* $\mathbf{AB}^-\mathbf{A} <^{\mathbf{L}} \mathbf{A}$, *where* $\mathbf{B}^-$ *is any g-inverse of* $\mathbf{B}$.

Proof. Write $\mathbf{A} = \mathbf{CC}^\star$. Then the non-null eigen-values of $\mathbf{C}^\star\mathbf{B}^-\mathbf{C}$ and $\mathbf{B}^-\mathbf{A}$ are the same. Further, $\mathbf{C}^\star\mathbf{B}^-\mathbf{C} <^{\mathbf{L}} \mathbf{I}$
$\Rightarrow \mathbf{CC}^\star\mathbf{B}^-\mathbf{CC}^\star <^{\mathbf{L}} \mathbf{CC}^\star$, by Theorem 8.2.7 (iii)
$\Rightarrow \mathbf{AB}^-\mathbf{A} <^{\mathbf{L}} \mathbf{A}$.
Also, $\mathbf{AB}^-\mathbf{A} <^{\mathbf{L}} \mathbf{A} \Rightarrow \mathbf{CC}^\star\mathbf{B}^-\mathbf{CC}^\star <^{\mathbf{L}} \mathbf{CC}^\star$
$\Rightarrow \mathbf{C}^\star\mathbf{B}^-\mathbf{C} = \mathbf{C}^\dagger\mathbf{CC}^\star\mathbf{B}^-\mathbf{CC}^\star\mathbf{C}^{\dagger\star} <^{\mathbf{L}} \mathbf{C}^\dagger\mathbf{CC}^\star\mathbf{C}^{\dagger\star} = \mathbf{C}^\dagger\mathbf{C} <^{\mathbf{L}} \mathbf{I}$.
Corollary now follows by Theorem 8.2.11. □

Remark 8.2.13. Notice that if $\mathbf{A}, \mathbf{B}$ are *nnd* matrices, then $\mathbf{A} <^- \mathbf{B} \Leftrightarrow$ '$\mathcal{C}(\mathbf{A}) \subseteq \mathcal{C}(\mathbf{B})$, and $\mathbf{AB}^-\mathbf{A} = \mathbf{A}$'. Now compare this with (iii) of Corollary 8.2.12.

We now show that Corollary 8.2.12 can be generalized to the wider class of hermitian matrices that are not necessarily *nnd* matrices. We start with a few preliminaries:

Definition 8.2.14. Let $\mathbf{A} \in \mathfrak{H}_n$. The inertia of $\mathbf{A}$ is the ordered triplet $(\pi(\mathbf{A}), \nu(\mathbf{A}), \eta(\mathbf{A}))$, where $\pi(\mathbf{A})$, $\nu(\mathbf{A})$, and $\eta(\mathbf{A})$ denote respectively the number of positive, negative and zero eigen-values of $\mathbf{A}$. ($\eta(\mathbf{A})$ is also the nullity of $\mathbf{A}$.) The inertia of $\mathbf{A}$ is denoted by $In(\mathbf{A})$.

The following is a well known theorem known as Silvester's Law of inertia.

Theorem 8.2.15. *Let* $\mathbf{A} \in \mathfrak{H}_n$. *Then* $In(\mathbf{A})$ *is invariant under similarity transformations i.e. for any non-singular matrix* $\mathbf{P}$, $In(\mathbf{P}^\star\mathbf{AP}) = In(\mathbf{A})$.

For a proof see [Rao and Bhimasankaram (2000)]

Corollary 8.2.16. *If* $\mathbf{M}_1 = \begin{pmatrix} \mathbf{A} & \mathbf{AB} \\ \mathbf{BA} & \mathbf{B} \end{pmatrix}$, *then*

$$In(\mathbf{M}_1) = In(\mathbf{A}) + In(\mathbf{B} - \mathbf{BAB}).$$

Proof. Let $\mathbf{A}^-$ be any g-inverse of $\mathbf{A}$ and $\mathbf{T} = \begin{pmatrix} \mathbf{I} & \mathbf{0} \\ \mathbf{BAA}^- & \mathbf{I} \end{pmatrix}$. Then we can write $\mathbf{M}_1 = \mathbf{T}\mathbf{diag}\left(\mathbf{A}\ ,\ \mathbf{B} - \mathbf{BAB}\right)\mathbf{T}^\star$. Since $\mathbf{T}$ is non-singular, the result follows. □

Corollary 8.2.17. *Let* $\mathbf{M_1}$ *be the matrix as in Corollary 8.2.16 and* $\mathbf{M_2} = \begin{pmatrix} \mathbf{B} & \mathbf{BA} \\ \mathbf{AB} & \mathbf{A} \end{pmatrix}$. *Then* $In(\mathbf{M_1}) = In(\mathbf{M_2})$.

Proof. Let $\mathbf{U} = \begin{pmatrix} \mathbf{0} & \mathbf{I} \\ \mathbf{I} & \mathbf{0} \end{pmatrix}$ partitioned conformably for multiplication with $\mathbf{M_1}$. Then $\mathbf{M_2} = \mathbf{UM_1U^\star}$ and by Silvester's Law of inertia the result follows. □

We now prove

Theorem 8.2.18. *Let* $\mathbf{A} \in \mathfrak{H}_\mathbf{n}$ *and* $\mathbf{L}$ *be an* $n \times n$ *matrix. Then the following are equivalent:*

(i) $In(\mathbf{LAL^\star}) = In(\mathbf{A})$
(ii) $\rho(\mathbf{LAL^\star}) = \rho(\mathbf{A})$ *and*
(iii) $\mathcal{C}(\mathbf{A}) \cap \mathcal{N}(\mathbf{L}) = \{\mathbf{0}\}$.

Proof. (i) ⇒ (ii) is trivial.
(ii) ⇔ (iii)
We first note that $\mathbf{A}$ hermitian $\Rightarrow \rho(\mathbf{AL^\star}) = \rho(\mathbf{LA})$. Now, by Lemma 3.3.15,

$$\begin{aligned} \rho(\mathbf{LAL^\star}) &= \rho(\mathbf{AL^\star}) - \mathbf{d}(\mathcal{C}(\mathbf{AL^\star})) \cap \mathcal{N}(\mathbf{L}) \\ &= \rho(\mathbf{LA}) - \mathbf{d}(\mathcal{C}(\mathbf{AL^\star})) \cap \mathcal{N}(\mathbf{L}) \\ \text{and } \ \rho(\mathbf{LA}) &= \rho(\mathbf{A}) - \mathbf{d}(\mathcal{C}(\mathbf{A}) \cap \mathcal{N}(\mathbf{L})). \end{aligned}$$

Hence, $\rho(\mathbf{LAL^\star}) = \rho(\mathbf{A}) - \mathbf{d}(\mathcal{C}(\mathbf{A}) \cap \mathcal{N}(\mathbf{L})) - \mathbf{d}(\mathcal{C}(\mathbf{AL^\star}) \cap \mathcal{N}(\mathbf{L}))$. So, $\rho(\mathbf{LAL^\star}) = \rho(\mathbf{A}) \Leftrightarrow \mathbf{d}(\mathcal{C}(\mathbf{A}) \cap \mathcal{N}(\mathbf{L})) = \mathbf{0}$ and $d(\mathcal{C}(\mathbf{AL^\star}) \cap \mathcal{N}(\mathbf{L})) = \mathbf{0}$. Since, $\mathcal{C}(\mathbf{AL^\star}) \subseteq \mathcal{C}(\mathbf{A})$, we have $\rho(\mathbf{LAL^\star}) = \rho(\mathbf{A}) \Leftrightarrow \mathbf{d}(\mathcal{C}(\mathbf{A}) \cap \mathcal{N}(\mathbf{L})) = \mathbf{0}$.
(ii) ⇒ (i)
Let $\mathbf{A} = \mathbf{P}\left(\mathbf{\Delta_1}, -\mathbf{\Delta_2}, \mathbf{0}\right)\mathbf{P^\star}$ be a spectral decomposition of $\mathbf{A}$, where $\mathbf{P}$ is unitary and $\mathbf{\Delta_1}$, $\mathbf{\Delta_2}$ are positive definite diagonal matrices. We can write $\mathbf{A} = \mathbf{P_1}\left(\mathbf{\Delta_1}, -\mathbf{\Delta_2}\right)\mathbf{P_1^\star}$, where $\mathbf{P_1^\star P_1} = \mathbf{I}$ by partitioning $\mathbf{P}$ as $\mathbf{P} = (\mathbf{P_1}, \mathbf{P_2})$, where $\mathbf{P_1}$ is of appropriate order. Now, $\mathbf{LAL^\star} = \mathbf{LP_1diag}\left(\Delta_1, -\Delta_2\right)\mathbf{P_1^\star L^\star}$. Since $\rho(\mathbf{LAL^\star}) = \rho(\mathbf{A})$, we have $\mathcal{C}(\mathbf{A}) \cap \mathcal{N}(\mathbf{L}) = \{\mathbf{0}\}$. Also, $\mathcal{C}(\mathbf{A}) = \mathcal{C}(\mathbf{P_1})$. So, by Lemma 3.3.15, $\rho(\mathbf{LP_1}) = \rho(\mathbf{P_1})$ = number of columns in $\mathbf{P_1}$. Now, $\mathbf{LP_1}$ has full column rank and can be extended to a non-singular matrix $\mathbf{Q} = \left(\mathbf{LP_1} : \mathbf{T}\right)$. Thus, $\mathbf{LP_1diag}\left(\mathbf{\Delta_1}, -\mathbf{\Delta_2}\right)\mathbf{P_1^\star L^\star} = \mathbf{Qdiag}\left(\mathbf{\Delta_1}, -\mathbf{\Delta_2}, \mathbf{0}\right)\mathbf{Q^\star}$. It follows by Theorem 8.2.15 that $In(\mathbf{LAL^\star}) = In(\mathbf{A})$. □

Theorem 8.2.19. *Let* $\mathbf{A}, \mathbf{B} \in \mathfrak{H}_\mathbf{n}$. *Then* $In(\mathbf{A} - \mathbf{A}\mathbf{B}^\dagger\mathbf{A}) = In(\mathbf{B} - \mathbf{A}) - (In(\mathbf{B}) - In(\mathbf{A}))$ *if and only if* $\mathcal{C}(\mathbf{B} - \mathbf{A}) \cap \mathcal{N}(\mathbf{B}) = \{\mathbf{0}\}$.

Proof. We have $In(\mathbf{B}^\dagger) + In(\mathbf{A} - \mathbf{A}\mathbf{B}^\dagger\mathbf{A}) = In(\mathbf{A}) + In(\mathbf{B}^\dagger - \mathbf{B}^\dagger\mathbf{A}\mathbf{B}^\dagger)$, by Corollaries 8.2.16 and 8.2.17. Since, $In(\mathbf{B}^\dagger) = In(\mathbf{B})$ and $\mathbf{B}^\dagger - \mathbf{B}^\dagger\mathbf{A}\mathbf{B}^\dagger = \mathbf{B}^\dagger(\mathbf{B} - \mathbf{A})\mathbf{B}^\dagger$, it follows that

$$In(\mathbf{A} - \mathbf{A}\mathbf{B}^\dagger\mathbf{A}) = In(\mathbf{B}^\dagger(\mathbf{B} - \mathbf{A})\mathbf{B}^\dagger) - (In(\mathbf{B}) - In(\mathbf{A})).$$

Now by Theorem 8.2.18, $In(\mathbf{B}^\dagger(\mathbf{B} - \mathbf{A})\mathbf{B}^\dagger) = In(\mathbf{B} - \mathbf{A})$ if and only if $\mathcal{C}(\mathbf{B} - \mathbf{A}) \cap \mathcal{N}(\mathbf{B}^\dagger) = \mathbf{0}$. Since, $\mathcal{N}(\mathbf{B}^\dagger) = \mathcal{N}(\mathbf{B})$, the result follows. □

Corollary 8.2.20. *Let* $\mathbf{A}, \mathbf{B} \in \mathfrak{H}_\mathbf{n}$ *such that* $\nu(\mathbf{A}) = \nu(\mathbf{B})$ *and* $\mathbf{A} <^\mathbf{s} \mathbf{B}$. *Then* $\nu(\mathbf{A} - \mathbf{A}\mathbf{B}^-\mathbf{A}) = \nu(\mathbf{B} - \mathbf{A})$ *for every g-inverse* $\mathbf{B}^-$ *of* $\mathbf{B}$.

Proof. Since $\mathbf{A} <^\mathbf{s} \mathbf{B}$, and $\nu(\mathbf{A}) = \nu(\mathbf{B})$, we have $\mathcal{N}(\mathbf{A}) = \mathcal{N}(\mathbf{B})$. So, $\mathcal{C}(\mathbf{B} - \mathbf{A}) \cap \mathcal{N}(\mathbf{B}) = \{\mathbf{0}\}$. Also, as $\nu(\mathbf{A}) = \nu(\mathbf{B})$, so, $\nu(\mathbf{A} - \mathbf{A}\mathbf{B}^\dagger\mathbf{A}) = \nu(\mathbf{B} - \mathbf{A})$, by Theorem 8.2.19. Further, $\mathbf{A} <^\mathbf{s} \mathbf{B}$, gives $\mathbf{A}\mathbf{B}^\dagger\mathbf{A} = \mathbf{A}\mathbf{B}^-\mathbf{A}$, for each g-inverse $\mathbf{B}^-$ of $\mathbf{B}$, so, the result follows. □

We are now ready to give the promised generalization of Corollary 8.2.12.

Theorem 8.2.21. *Let* $\mathbf{A}$, $\mathbf{B} \in \mathfrak{H}_n$ *such that* $\nu(\mathbf{A}) = \nu(\mathbf{B})$. *Then* $\mathbf{A} <^\mathbf{L} \mathbf{B}$ *if and only if* '$\mathbf{A} <^\mathbf{s} \mathbf{B}$ *(or equivalently* $\mathcal{C}(\mathbf{A}) \subseteq \mathcal{C}(\mathbf{B})$*) and* $\mathbf{A}\mathbf{B}^-\mathbf{A} <^\mathbf{L} \mathbf{A}$' *for an arbitrary choice of g-inverse* $\mathbf{B}^-$ *of* $\mathbf{B}$.

Proof. 'If' part follows from Corollary 8.2.20.
'Only if' part
In view of Theorem 8.2.4, we can write $\mathbf{A} = \mathbf{A}_1 - \mathbf{A}_2$ and $\mathbf{B} = \mathbf{B}_1 - \mathbf{B}_2$, where $\mathbf{A}_1$, $\mathbf{A}_2, \mathbf{B}_1$ and $\mathbf{B}_2$ are *nnd* matrices such that $\mathbf{A}_1\mathbf{A}_2 = \mathbf{0}$ and $\mathbf{B}_1\mathbf{B}_2 = \mathbf{0}$. Now,

$$\nu(\mathbf{A}) = \nu(\mathbf{B}) \Leftrightarrow \rho(\mathbf{A}_2) = \rho(\mathbf{B}_2) \text{ and } \mathbf{A} <^\mathbf{L} \mathbf{B} \Leftrightarrow \mathbf{A}_1 + \mathbf{B}_2 <^\mathbf{L} \mathbf{A}_2 + \mathbf{B}_1.$$

Since $\mathbf{A}_1, \mathbf{A}_2, \mathbf{B}_1$ and $\mathbf{B}_2$ are *nnd*, it follows that $\mathcal{C}(\mathbf{A}_1 + \mathbf{B}_2) \subseteq \mathcal{C}(\mathbf{A}_2 + \mathbf{B}_1)$ or equivalently

$$\mathcal{C}(\mathbf{A}_1 : \mathbf{B}_2) \subseteq \mathcal{C}(\mathbf{A}_2 : \mathbf{B}_1) = \mathcal{C}(\mathbf{A}_2 : \mathbf{B}_1 : \mathbf{B}_2). \tag{8.2.3}$$

Now,

$$\begin{aligned}\rho(\mathbf{A}_2 : \mathbf{B}_1 : \mathbf{B}_2) &= d(\mathcal{C}(\mathbf{A}_2 : \mathbf{B}_1 : \mathbf{B}_2)) \\ &= d(\mathcal{C}(\mathbf{A}_2)) + \mathbf{d}(\mathcal{C}(\mathbf{B}_1 : \mathbf{B}_2)) - \mathbf{d}(\mathcal{C}(\mathbf{A}_2) \cap \mathcal{C}(\mathbf{B}_1 : \mathbf{B}_2)) \\ &= \rho(\mathbf{A}_2) + \rho(\mathbf{B}_1) + \rho(\mathbf{B}_2) - \mathbf{d}(\mathcal{C}(\mathbf{A}_2) \cap \mathcal{C}(\mathbf{B}_1 : \mathbf{B}_2)).\end{aligned} \tag{8.2.4}$$

and

$$\rho(\mathbf{A_2} : \mathbf{B_1}) = \rho(\mathbf{A_2}) + \rho(\mathbf{B_1}) - \mathbf{d}(\mathcal{C}(\mathbf{A_2}) \cap \mathcal{C}(\mathbf{B_1})). \tag{8.2.5}$$

Since $\rho(\mathbf{A_2}) = \rho(\mathbf{B_2})$, Equations (8.2.3), (8.2.4) and (8.2.5) together give

$$\rho(\mathbf{A_2}) - \mathbf{d}(\mathcal{C}(\mathbf{A_2}) \cap \mathcal{C}(\mathbf{B_1} : \mathbf{B_2})) = -\mathbf{d}(\mathcal{C}(\mathbf{A_2}) \cap \mathcal{C}(\mathbf{B_1})).$$

Moreover, $\mathcal{C}(\mathbf{A_2}) \cap \mathcal{C}(\mathbf{B_1} : \mathbf{B_2}) \subseteq \mathcal{C}(\mathbf{A_2})$, the left hand of the above equality is a positive number, whereas the right hand is a negative number, so, either side is null. This further gives $\mathcal{C}(\mathbf{A_2}) \cap \mathcal{C}(\mathbf{B_1}) = \{\mathbf{0}\}$ and $\mathcal{C}(\mathbf{A_2}) = \mathcal{C}(\mathbf{A_2}) \cap \mathcal{C}(\mathbf{B_1} : \mathbf{B_2})$. It follows that $\mathcal{C}(\mathbf{A_2}) \subseteq \mathcal{C}(\mathbf{B_1} : \mathbf{B_2}) = \mathcal{C}(\mathbf{B})$ and so, $\mathcal{C}(\mathbf{A_1} : \mathbf{B_2}) \subseteq \mathcal{C}(\mathbf{A_2} : \mathbf{B_1}) \subseteq \mathcal{C}(\mathbf{B})$. Hence, $\mathcal{C}(\mathbf{A_1}) \subseteq \mathcal{C}(\mathbf{B})$ and therefore $\mathcal{C}(\mathbf{A}) \subseteq \mathcal{C}(\mathbf{B})$. By Corollary 8.2.20, we have $\mathbf{AB^-A} <^{\mathbf{L}} \mathbf{A}$. □

We note that for $\mathbf{A}, \mathbf{B} \in \mathfrak{H}_n$ such that $\nu(\mathbf{A}) = \nu(\mathbf{B})$, the condition '$\mathbf{A} <^{\mathbf{s}} \mathbf{B}$ and $\mathbf{AB^-A} <^{\mathbf{L}} \mathbf{A}$' need not imply $\lambda_{max}(\mathbf{B^-A}) \leq \mathbf{1}$ as the following example shows.

Example 8.2.22. Let $\mathbf{A} = \mathbf{diag}\left(-1\ ,\ -2\right)$ and $\mathbf{B} = \mathbf{diag}\left(-0.5\ ,\ -1\right)$. Clearly, $\mathbf{A} <^{\mathbf{L}} \mathbf{B}$ and also $\mathbf{A} <^{\mathbf{s}} \mathbf{B}$. However, $\mathbf{B^-A} = \mathbf{diag}\left(2\ ,\ 2\right)$ and $\lambda_{max}(\mathbf{B^-A}) \geq \mathbf{1}$.

Thus, the hypothesis $\mathbf{A}$ and $\mathbf{B}$ are *nnd* can not be dropped from Corollary 8.2.12. In fact, if $\mathbf{A}, \mathbf{B} \in \mathfrak{H}_{\mathbf{n}}$ such that $\nu(\mathbf{A}) = \nu(\mathbf{B}) \neq \mathbf{0}$ and $\mathbf{A} <^{\mathbf{L}} \mathbf{B}$, then it can be shown that $\lambda_{max}(\mathbf{B^-A}) \geq \mathbf{1}$. (See Exercise 3.)

We remark that Theorem 8.2.11 is the key to some very interesting results that we obtain here in what follows and also in the next section. We start with

Theorem 8.2.23. *Let* $\mathbf{A}$ *and* $\mathbf{B}$ *be matrices of the same order with* $\mathbf{B}$ *an nnd matrix. Let* $\mathbf{B^-}$ *is some g-inverse of* $\mathbf{B}$. *Then the following are equivalent:*

(i) $\mathbf{BA} = \mathbf{BA^2}$, $\mathcal{C}(\mathbf{A^\star B}) \subseteq \mathcal{C}(\mathbf{B})$ *and* $\lambda_{max}(\mathbf{AB^-A^\star B}) \leq \mathbf{1}$
(ii) $\mathbf{BA} = \mathbf{BA^2}$, $\mathcal{C}(\mathbf{A^\star B}) \subseteq \mathcal{C}(\mathbf{B})$ *and* $tr(\mathbf{AB^-A^\star B}) \leq \rho(\mathbf{BA})$
(iii) $\mathbf{BA} = \mathbf{BA^2}$, $\mathbf{A^\star BA} <^{\mathbf{L}} \mathbf{B}$ *and*
(iv) $\mathbf{BA} = \mathbf{A^\star BA}$.

Proof. Write $\mathbf{B} = \mathbf{CC^\star}$ and let $\mathbf{S} = \mathbf{C^\dagger A^\star C}$.
(i) ⇒ (ii)
First note that $\lambda_{max}(\mathbf{AB^-A^\star B}) = \lambda_{max}(\mathbf{C^\star AB^-A^\star C})$. Since $\mathcal{C}(\mathbf{A^\star C}) = \mathcal{C}(\mathbf{A^\star B}) \subseteq \mathcal{C}(\mathbf{B})$, $\mathbf{C^\star AB^-A^\star C}$ is invariant under the choice of g-inverses of

$\mathbf{B}$. Hence, we can use $\mathbf{B}^\dagger$ for a g-inverse of $\mathbf{B}$ in $\mathbf{C}^\star\mathbf{A}\mathbf{B}^-\mathbf{A}^\star\mathbf{C}$. Since $\mathbf{S}^\star\mathbf{S} = \mathbf{C}^\star\mathbf{A}\mathbf{C}^{\star\dagger}\mathbf{C}^\dagger\mathbf{A}^\star\mathbf{C} = \mathbf{C}^\star\mathbf{A}\mathbf{B}^\dagger\mathbf{A}^\star\mathbf{C}$, the non-null eigen-values of $\mathbf{A}\mathbf{B}^-\mathbf{A}^\star\mathbf{B}$ and $\mathbf{S}^\star\mathbf{S}$ are the same. Also, $\lambda_{max}(\mathbf{A}\mathbf{B}^-\mathbf{A}^\star\mathbf{B}) \leq \mathbf{1} \Rightarrow \mathbf{tr}(\mathbf{S}^\star\mathbf{S}) \leq \rho(\mathbf{S})$, since the number of non-null eigen-values of $\mathbf{S}^\star\mathbf{S}$ (including the algebraic multiplicity) is $\rho(\mathbf{S})$. Now, $\mathcal{C}(\mathbf{A}^\star\mathbf{C}) = \mathcal{C}(\mathbf{A}^\star\mathbf{B}) \subseteq \mathcal{C}(\mathbf{B}) = \mathcal{C}(\mathbf{C})$, we have $\mathbf{S^2} = \mathbf{C}^\dagger\mathbf{A}^\star\mathbf{C}\mathbf{C}^\dagger\mathbf{A}^\star\mathbf{C} = \mathbf{C}^\dagger\mathbf{A}^{2\star}\mathbf{C}$. Since $\mathbf{BA} = \mathbf{BA^2}$, it follows that $\mathbf{C}^\dagger\mathbf{A}^{2\star}\mathbf{C} = \mathbf{C}^\dagger\mathbf{A}^\star\mathbf{C} = \mathbf{S}$. Thus, $\mathbf{S}$ is an idempotent. Also, $\mathbf{S}$ idempotent and $\mathcal{C}(\mathbf{A}^\star\mathbf{C}) \subseteq \mathcal{C}(\mathbf{C})$ together imply that $\mathbf{A}^\star\mathbf{C}\mathbf{C}^\dagger$ is an idempotent. Therefore,

$$\rho(\mathbf{S}) = tr(\mathbf{S}) = tr(\mathbf{A}^\star\mathbf{C}\mathbf{C}^\dagger) = \rho(\mathbf{A}^\star\mathbf{C}\mathbf{C}^\dagger) = \rho(\mathbf{A}^\star\mathbf{C}\mathbf{C}^\star) = \rho(\mathbf{BA}).$$

(ii) $\Rightarrow$ (iv)

$\mathbf{BA} = \mathbf{BA^2}$ and $\mathcal{C}(\mathbf{A}^\star\mathbf{B}) \subseteq \mathcal{C}(\mathbf{B}) \Rightarrow tr(\mathbf{S}^\star\mathbf{S}) \leq \rho(\mathbf{S}) = tr(\mathbf{S^2})$ as shown in (i) $\Rightarrow$ (ii). However, $0 \leq tr(\mathbf{S} - \mathbf{S}^\star)^\star(\mathbf{S} - \mathbf{S}^\star) = \mathbf{2}(tr(\mathbf{S}^\star\mathbf{S}) - tr(\mathbf{S^2}))$, so, $\mathbf{S} = \mathbf{S}^\star$. Also, $\mathbf{CSC}^\star = \mathbf{C}\mathbf{C}^\dagger\mathbf{A}^\star\mathbf{C}\mathbf{C}^\star = \mathbf{A}^\star\mathbf{B}^\star = \mathbf{A}^\star\mathbf{B}$. Since $\mathbf{CSC}^\star = \mathbf{C}\mathbf{S}^\star\mathbf{C}^\star = (\mathbf{CSC}^\star)^\star$, we have $\mathbf{A}^\star\mathbf{B} = \mathbf{BA}$. Hence, $\mathbf{BA} = \mathbf{BA^2} = \mathbf{A}^\star\mathbf{BA}$.

(iv) $\Rightarrow$ (i)

$\mathbf{BA} = \mathbf{A}^\star\mathbf{BA} = \mathbf{A}^\star\mathbf{B}$. So, $\mathbf{BA} = \mathbf{BA^2}$ and $\mathcal{C}(\mathbf{A}^\star\mathbf{B}) \subseteq \mathcal{C}(\mathbf{B})$. Moreover, $\mathbf{C}\mathbf{C}^\star\mathbf{A}\mathbf{C}^{\dagger\star}\mathbf{C}^\dagger\mathbf{A}^\star\mathbf{C} = \mathbf{BA}\mathbf{B}^\dagger\mathbf{A}^\star\mathbf{C} = \mathbf{A}^{\star 2}\mathbf{C} = \mathbf{A}^\star\mathbf{C} = \mathbf{C}\mathbf{C}^\dagger\mathbf{A}^\star\mathbf{C}$. Hence, $\mathbf{S} = \mathbf{S}^\star\mathbf{S}$. So, $\mathbf{S}$ is hermitian and idempotent. From the proof above of (i) $\Rightarrow$ (ii), we have non-null eigen-values of $\mathbf{A}\mathbf{B}^-\mathbf{A}^\star\mathbf{B}$ and $\mathbf{S}^\star\mathbf{S}$ are same. Hence, $\lambda_{max}(\mathbf{A}\mathbf{B}^-\mathbf{A}^\star\mathbf{B}) = \mathbf{1}$.

(i) $\Leftrightarrow$ (iii) follows from Theorem 8.2.11. □

Remark 8.2.24. In the setup of Theorem 8.2.23, the proofs of the equivalences (i) $\Leftrightarrow$ (ii) $\Leftrightarrow$ (iv) $\Leftrightarrow$ (i) of Theorem 8.2.23, show that $\mathbf{BA} = \mathbf{BA^2}$, $\mathcal{C}(\mathbf{A}^\star\mathbf{B}) \subseteq \mathcal{C}(\mathbf{B})$ and $\lambda_{max}(\mathbf{A}\mathbf{B}^-\mathbf{A}^\star\mathbf{B}) \leq \mathbf{1} \Rightarrow \lambda_{max}(\mathbf{A}\mathbf{B}^-\mathbf{A}^\star\mathbf{B}) = \mathbf{1}$.

Similarly, $\mathbf{BA} = \mathbf{BA^2}$, $\mathcal{C}(\mathbf{A}^\star\mathbf{B}) \subseteq \mathcal{C}(\mathbf{B})$ and $tr(\mathbf{A}\mathbf{B}^-\mathbf{A}^\star\mathbf{B}) \leq \rho(\mathbf{BA})$ together imply $tr(\mathbf{A}\mathbf{B}^-\mathbf{A}^\star\mathbf{B}) = \rho(\mathbf{BA})$.

Theorem 8.2.25. *Let* $\mathbf{A}, \mathbf{B}$ *be nnd matrices such that* $\mathbf{A} <^{\mathbf{L}} \mathbf{B}$. *Then,* $\mathbf{0} <^{\mathbf{L}} \mathbf{A}\mathbf{B}^\dagger\mathbf{A} <^{\mathbf{L}} \mathbf{A} <^{\mathbf{L}} \mathbf{A} + \mathbf{B}((\mathbf{I} - \mathbf{A}\mathbf{A}^\dagger)\mathbf{B}(\mathbf{I} - \mathbf{A}^\dagger\mathbf{A}))^\dagger\mathbf{B} <^{\mathbf{L}} \mathbf{B}$.

Proof. Since $\mathbf{B}$ is *nnd*, $\mathbf{B}^\dagger$ is *nnd*. So, $\mathbf{0} <^{\mathbf{L}} \mathbf{A}\mathbf{B}^\dagger\mathbf{A}$. By Corollary 8.2.12(iii), $\mathbf{A}\mathbf{B}^\dagger\mathbf{A} <^{\mathbf{L}} \mathbf{A}$. Notice that $\mathbf{I} - \mathbf{A}\mathbf{A}^\dagger = \mathbf{I} - \mathbf{A}^\dagger\mathbf{A}$, since $\mathbf{A}$ is hermitian. Therefore, $\mathbf{C} = (\mathbf{I} - \mathbf{A}\mathbf{A}^\dagger)\mathbf{B}(\mathbf{I} - \mathbf{A}^\dagger\mathbf{A})$ is also *nnd* and hence $\mathbf{C}^\dagger$ is *nnd*. So, $\mathbf{B}\mathbf{C}^\dagger\mathbf{B}$ is *nnd*. Thus, $\mathbf{A} <^{\mathbf{L}} \mathbf{A} + \mathbf{B}((\mathbf{I} - \mathbf{A}\mathbf{A}^\dagger)\mathbf{B}(\mathbf{I} - \mathbf{A}^\dagger\mathbf{A}))^\dagger\mathbf{B}$. Now,

$$\begin{aligned}\mathcal{C}(((\mathbf{I} - \mathbf{A}\mathbf{A}^\dagger)\mathbf{B}(\mathbf{I} - \mathbf{A}^\dagger\mathbf{A}))^\dagger) &= \mathcal{C}((\mathbf{I} - \mathbf{A}\mathbf{A}^\dagger)\mathbf{B}(\mathbf{I} - \mathbf{A}^\dagger\mathbf{A})) \\ &\subseteq \mathcal{C}(\mathbf{I} - \mathbf{A}\mathbf{A}^\dagger) = \mathcal{C}(\mathbf{I} - \mathbf{A}^\dagger\mathbf{A}).\end{aligned}$$

So, $\mathbf{B}[(\mathbf{I}-\mathbf{A}\mathbf{A}^\dagger)\mathbf{B}(\mathbf{I}-\mathbf{A}^\dagger\mathbf{A})]^\dagger\mathbf{B} = \mathbf{B}(\mathbf{I}-\mathbf{A}^\dagger\mathbf{A})[(\mathbf{I}-\mathbf{A}\mathbf{A}^\dagger)\mathbf{B}(\mathbf{I}-\mathbf{A}^\dagger\mathbf{A})]^\dagger$ $(\mathbf{I}-\mathbf{A}\mathbf{A}^\dagger)\mathbf{B} = (\mathbf{B}-\mathbf{A})(\mathbf{I}-\mathbf{A}^\dagger\mathbf{A})[(\mathbf{I}-\mathbf{A}\mathbf{A}^\dagger)\mathbf{B}(\mathbf{I}-\mathbf{A}^\dagger\mathbf{A})]^\dagger(\mathbf{I}-\mathbf{A}\mathbf{A}^\dagger)$ $(\mathbf{B}-\mathbf{A})$. Consider the matrix

$$\mathbf{M} = \begin{pmatrix} \mathbf{B}-\mathbf{A} & (\mathbf{B}-\mathbf{A})(\mathbf{I}-\mathbf{A}^\dagger\mathbf{A}) \\ (\mathbf{I}-\mathbf{A}\mathbf{A}^\dagger)(\mathbf{B}-\mathbf{A}) & (\mathbf{I}-\mathbf{A}\mathbf{A}^\dagger)(\mathbf{B}-\mathbf{A})(\mathbf{I}-\mathbf{A}^\dagger\mathbf{A}) \end{pmatrix}.$$

By Theorem 8.2.2 it follows that $\mathbf{M}$ is *nnd*. So, by Corollary 8.2.3, we have

$$(\mathbf{B}-\mathbf{A}) - \mathbf{B}[(\mathbf{I}-\mathbf{A}\mathbf{A}^\dagger)\mathbf{B}(\mathbf{I}-\mathbf{A}^\dagger\mathbf{A})]^-\mathbf{B} =$$
$$(\mathbf{B}-\mathbf{A}) - (\mathbf{B}-\mathbf{A})(\mathbf{I}-\mathbf{A}^\dagger\mathbf{A})[(\mathbf{I}-\mathbf{A}\mathbf{A}^\dagger)\mathbf{B}(\mathbf{I}-\mathbf{A}^\dagger\mathbf{A})]^\dagger(\mathbf{I}-\mathbf{A}\mathbf{A}^\dagger)(\mathbf{B}-\mathbf{A})$$

is *nnd*. □

Let $\mathbf{A}$ and $\mathbf{B}$ be matrices of the same order (possibly rectangular). Then the matrix $\mathbf{A}\mathbf{B}^\star + \mathbf{B}\mathbf{A}^\star$ is hermitian. When is this matrix *nnd*? We answer in the following:

Theorem 8.2.26. *Let* $\mathbf{A}$ *and* $\mathbf{B}$ *be matrices of order* $m \times n$. *Then the following are equivalent:*

(i) $\mathbf{0} <^{\mathbf{L}} \mathbf{A}\mathbf{B}^\star + \mathbf{B}\mathbf{A}^\star$
(ii) $\lambda_{max}((\mathbf{A}-\mathbf{B})^\star(\mathbf{A}\mathbf{A}^\star+\mathbf{B}\mathbf{B}^\star)^-(\mathbf{A}-\mathbf{B})) \leq \mathbf{1}$
(iii) $\lambda_{max}[(\mathbf{A}:\mathbf{B})^\star\{(\mathbf{A}+\mathbf{B})(\mathbf{A}+\mathbf{B})^-\}(\mathbf{A}:\mathbf{B})] \leq \mathbf{1}$ *and* $\mathcal{C}(\mathbf{A}:\mathbf{B}) = \mathcal{C}(\mathbf{A}+\mathbf{B})$,

where all g-inverses involved in (ii) and (iii) are arbitrary.

Proof. Notice that (a) $(\mathbf{A}-\mathbf{B})(\mathbf{A}-\mathbf{B})^\star = \mathbf{A}\mathbf{A}^\star+\mathbf{B}\mathbf{B}^\star-(\mathbf{A}\mathbf{B}^\star+\mathbf{B}\mathbf{A}^\star)$, (b) $(\mathbf{A}+\mathbf{B})(\mathbf{A}+\mathbf{B})^\star = \mathbf{A}\mathbf{A}^\star+\mathbf{B}\mathbf{B}^\star+(\mathbf{A}\mathbf{B}^\star+\mathbf{B}\mathbf{A}^\star)$ and (c) $(\mathbf{A}:\mathbf{B})(\mathbf{A}:\mathbf{B})^\star = \mathbf{A}\mathbf{A}^\star+\mathbf{B}\mathbf{B}^\star$. Hence

$$\begin{aligned} &\mathbf{0} <^{\mathbf{L}} \mathbf{A}\mathbf{B}^\star + \mathbf{B}\mathbf{A}^\star \\ &\Leftrightarrow (\mathbf{A}-\mathbf{B})(\mathbf{A}-\mathbf{B})^\star <^{\mathbf{L}} (\mathbf{A}:\mathbf{B})(\mathbf{A}:\mathbf{B})^\star \\ &\Leftrightarrow (\mathbf{A}:\mathbf{B})(\mathbf{A}:\mathbf{B})^\star <^{\mathbf{L}} (\mathbf{A}+\mathbf{B})(\mathbf{A}+\mathbf{B})^\star. \end{aligned}$$

Since $\mathcal{C}(\mathbf{A}-\mathbf{B}) \subseteq \mathcal{C}(\mathbf{A}:\mathbf{B})$ holds trivially, we have by Theorem 8.2.11, $(\mathbf{A}-\mathbf{B})(\mathbf{A}-\mathbf{B})^\star <^{\mathbf{L}} (\mathbf{A}:\mathbf{B})(\mathbf{A}:\mathbf{B})^\star \Leftrightarrow \lambda_{max}((\mathbf{A}-\mathbf{B})^\star(\mathbf{A}\mathbf{A}^\star+\mathbf{B}\mathbf{B}^\star)^-(\mathbf{A}-\mathbf{B})) \leq \mathbf{1}$. Now, it follows that (i) and (ii) are equivalent. Again, by Theorem 8.2.11 we have

$$\begin{aligned} &(\mathbf{A}:\mathbf{B})(\mathbf{A}:\mathbf{B})^\star <^{\mathbf{L}} (\mathbf{A}+\mathbf{B})(\mathbf{A}+\mathbf{B})^\star \\ &\Leftrightarrow \mathcal{C}(\mathbf{A}:\mathbf{B}) \subseteq \mathcal{C}(\mathbf{A}+\mathbf{B}) \text{ and} \end{aligned}$$

$$\lambda_{max}((\mathbf{A}:\mathbf{B})^{\star}\{(\mathbf{A}+\mathbf{B})(\mathbf{A}+\mathbf{B})^{-}\}(\mathbf{A}:\mathbf{B})) \leq \mathbf{1}.$$

However, $\mathcal{C}(\mathbf{A}+\mathbf{B}) \subseteq \mathcal{C}(\mathbf{A}:\mathbf{B})$, so, (i) $\Leftrightarrow$ (iii). □

Theorem 8.2.27. *Let* $\mathbf{A}, \mathbf{B} \in \mathfrak{H}_{\mathbf{n}}$ *and* $\mathbf{B}$ *be an nnd matrix such that* $\mathcal{C}(\mathbf{A}) \subseteq \mathcal{C}(\mathbf{B})$. *Further, let* $\mathbf{AB}+\mathbf{BA}$ *be nnd. Then* $\mathbf{A}$ *is nnd.*

Proof. Let $\mathbf{A} = \mathbf{U}\begin{pmatrix}\mathbf{\Delta} & \mathbf{0}\\ \mathbf{0} & \mathbf{0}\end{pmatrix}\mathbf{U}^{\star}$ be a spectral decomposition of $\mathbf{A}$ and $\mathbf{\Delta}$ is a non-singular real diagonal matrix. Write $\mathbf{B} = \mathbf{U}\begin{pmatrix}\mathbf{C_{11}} & \mathbf{C_{12}}\\ \mathbf{C_{12}}^{\star} & \mathbf{C_{22}}\end{pmatrix}\mathbf{U}^{\star}$, where $\mathbf{C_{11}}$ and $\mathbf{\Delta}$ have same order. Since, $\mathbf{AB}+\mathbf{BA}$ is *nnd*, we have

$$\begin{pmatrix}\mathbf{\Delta} & \mathbf{0}\\ \mathbf{0} & \mathbf{0}\end{pmatrix}\begin{pmatrix}\mathbf{C_{11}} & \mathbf{C_{12}}\\ \mathbf{C_{12}}^{\star} & \mathbf{C_{22}}\end{pmatrix} + \begin{pmatrix}\mathbf{C_{11}} & \mathbf{C_{12}}\\ \mathbf{C_{12}}^{\star} & \mathbf{C_{22}}\end{pmatrix}\begin{pmatrix}\mathbf{\Delta} & \mathbf{0}\\ \mathbf{0} & \mathbf{0}\end{pmatrix} = \begin{pmatrix}\mathbf{\Delta C_{11}}+\mathbf{C_{11}\Delta} & \mathbf{\Delta C_{12}}\\ \mathbf{C_{11}}^{\star}\mathbf{\Delta} & \mathbf{0}\end{pmatrix}$$

is *nnd*. Hence, $\mathbf{\Delta C_{11}}+\mathbf{C_{11}\Delta}$ is *nnd* and $\mathbf{C_{12}} = \mathbf{0}$. Since, $\mathcal{C}(\mathbf{A}) \subseteq \mathcal{C}(\mathbf{B})$, we have $\mathbf{C_{11}}$ is positive definite. The ith diagonal element of $\mathbf{\Delta C_{11}}+\mathbf{C_{11}\Delta}$ is $\mathbf{\Delta}_{ii}(\mathbf{C_{11}})_{ii} + (\mathbf{C_{11}})_{ii}\mathbf{\Delta}_{ii} = 2(\mathbf{\Delta}_{ii}(\mathbf{C_{11}})_{ii})$. Since, $\mathbf{\Delta C_{11}}$ and $\mathbf{C_{11}\Delta}$ are *nnd* and $\mathbf{C_{11}}$ is positive definite, we have $\mathbf{\Delta}_{ii} \geq 0$ for all i. Since $\mathbf{\Delta}$ is non-singular, we have $\mathbf{\Delta}_{ii} > 0$. Hence, $\mathbf{A}$ is *nnd*. □

In the case of the minus order, if $\mathbf{A} <^{-} \mathbf{B}$ and $\rho(\mathbf{A}) = \rho(\mathbf{B})$, then $\mathbf{A} = \mathbf{B}$. Same conclusion holds even when $\mathbf{A} <^{\star} \mathbf{B}$ and $\rho(\mathbf{A}) = \rho(\mathbf{B})$ or $\mathbf{A} <^{\#} \mathbf{B}$ and $\rho(\mathbf{A}) = \rho(\mathbf{B})$. However, this is not the case with the Löwner order. For example, we can take $\mathbf{A} = \mathbf{I}, \mathbf{B} = \mathbf{2I}$. Then $\mathbf{A} <^{\mathbf{L}} \mathbf{B}$ and $\rho(\mathbf{A}) = \rho(\mathbf{B})$, but $\mathbf{A} \neq \mathbf{B}$.

8.3 Löwner order on powers and its relation with other partial orders

Let $\mathbf{A}$ and $\mathbf{B}$ be *nnd* matrices of the same order. In this section we study the Löwner order for powers of $\mathbf{A}$ and $\mathbf{B}$ when $\mathbf{A} <^{\mathbf{L}} \mathbf{B}$. While we prove $\mathbf{A^2} <^{\mathbf{L}} \mathbf{B^2}$ implies $\mathbf{A} <^{\mathbf{L}} \mathbf{B}$, the implication in general is not reversible. We show that the same can be obtained by taking an additional condition that $\mathbf{A}$ and $\mathbf{B}$ commute when $\mathbf{A} <^{\mathbf{L}} \mathbf{B}$. We also study the Löwner order in conjunction with some of the earlier partial orders namely, the minus and the star order. Finally, we consider $\mathbf{A}$ and $\mathbf{B}$ to be hermitian matrices that not are necessarily *nnd* and obtain some of the results on the Löwner order as an extension of the corresponding results on the Löwner order for *nnd* matrices. We begin with a lemma that we need for our results in this and later sections of this chapter.

Lemma 8.3.1. *Let* $\mathbf{A}$ *be an nnd matrix. Let* λ_1 *and* σ_1 *respectively be the largest eigen-value and the largest singular value of* $\mathbf{A}$. *Then* $|\lambda_1| < \sigma_1$.

Proof. Since σ_1 is the largest singular value, ${\sigma_1}^2$ is the largest eigen-value of $\mathbf{A}^\star\mathbf{A}$. Hence by Theorem 8.2.1 (a), we have

$$ {\sigma_1}^2 = \max_{\mathbf{y}\neq\mathbf{0}} \frac{\mathbf{y}^\star\mathbf{A}^\star\mathbf{A}\mathbf{y}}{\mathbf{y}^\star\mathbf{y}}. $$

Let $\mathbf{u}$ be the eigen vector of $\mathbf{A}$ corresponding to the eigen-value λ_1. So,

$$ |\lambda_1|^2 = \lambda_1\overline{\lambda_1} = \frac{\mathbf{u}^\star\mathbf{A}^\star\mathbf{A}\mathbf{u}}{\mathbf{u}^\star\mathbf{u}} \leq \max_{\mathbf{y}\neq\mathbf{0}} \frac{\mathbf{y}^\star\mathbf{A}^\star\mathbf{A}\mathbf{y}}{\mathbf{y}^\star\mathbf{y}} = {\sigma_1}^2. $$

□

We now prove the following:

Theorem 8.3.2. *Let A and* $\mathbf{B}$ *be nnd matrices of the same order. Consider the following statements:*

(i) $\mathbf{A^2} <^{\mathbf{L}} \mathbf{B^2}$
(ii) $\mathbf{A} <^{\mathbf{L}} \mathbf{B}$ *and*
(iii) $\mathbf{AB} = \mathbf{BA}$.

Then (i) $\Rightarrow$ (ii) and '(ii) and (iii)' $\Rightarrow$ (i).

Proof. (i) $\Rightarrow$ (ii)
Notice first that any eigen-value of $(\mathbf{B^2})^\dagger\mathbf{A^2}$ is a singular value of $\mathbf{B}^\dagger\mathbf{A}$, as $\mathbf{A}$ and $\mathbf{B}$ are *nnd*. So, by Corollary 8.2.12, $\mathbf{A^2} <^{\mathbf{L}} \mathbf{B^2} \Leftrightarrow \mathcal{C}(\mathbf{A}) \subseteq \mathcal{C}(\mathbf{B})$ and $\sigma_1(\mathbf{B}^\dagger\mathbf{A}) \leq \mathbf{1}$, where σ_1 is the largest singular value of $\mathbf{B}^\dagger\mathbf{A}$. Hence, by Lemma 8.3.1, $\lambda_1(\mathbf{B}^\dagger\mathbf{A}) \leq \mathbf{1}$. Since $\mathbf{A}$ and $\mathbf{B}$ are *nnd*, the eigen-values of $\mathbf{B}^\dagger\mathbf{A}$ are real and non-negative. Once again invoking Corollary 8.2.12, we have $\mathbf{A} <^{\mathbf{L}} \mathbf{B}$.
(ii) and (iii) $\Rightarrow$ (i)
Since $\mathbf{A}$ and $\mathbf{B}$ commute, by Theorem 2.7.1, we have $\mathbf{A} = \mathbf{P\Delta_1P^\star}$ and $\mathbf{B} = \mathbf{P\Delta_2P^\star}$ for some unitary matrix $\mathbf{P}$ and diagonal *nnd* matrices $\mathbf{\Delta_1}$ and $\mathbf{\Delta_2}$. $\mathbf{A} <^{\mathbf{L}} \mathbf{B} \Rightarrow \mathbf{\Delta_1} <^{\mathbf{L}} \mathbf{\Delta_2} \Rightarrow \mathbf{\Delta_1^2} <^{\mathbf{L}} \mathbf{\Delta_2}^{\mathbf{2}}$, since $\mathbf{\Delta_1}$ and $\mathbf{\Delta_2}$ are diagonal and *nnd*, so, $\mathbf{A^2} = \mathbf{P\Delta_1^2P^\star} <^{\mathbf{L}} \mathbf{P\Delta_2^2P^\star} = \mathbf{B^2}$. □

Remark 8.3.3. Let $\mathbf{A}$ and $\mathbf{B}$ be *nnd* matrices of the same order such that $\mathbf{A} <^{\mathbf{L}} \mathbf{B}$. We need not have $\mathbf{A^2} <^{\mathbf{L}} \mathbf{B^2}$.
For, let $\mathbf{A} = \begin{pmatrix} 1 & 1 \\ 1 & 1 \end{pmatrix}, \mathbf{B} = \begin{pmatrix} 1.4 & 0 \\ 0 & 4 \end{pmatrix}$.

Remark 8.3.4. Let $\mathbf{A}$ and $\mathbf{B}$ be *nnd* matrices of the same order such that $\mathbf{A} <^{\mathbf{L}} \mathbf{B}$ and $\mathbf{A^2} <^{\mathbf{L}} \mathbf{B^2}$. However $\mathbf{A}$ and $\mathbf{B}$ need not commute.
Let $\mathbf{A} = \begin{pmatrix} 1 & 1 \\ 1 & 1 \end{pmatrix}, \mathbf{B} = \begin{pmatrix} 2 & 0 \\ 0 & 4 \end{pmatrix}$. Clearly, $\mathbf{A} <^{\mathbf{L}} \mathbf{B}$ and $\mathbf{A^2} <^{\mathbf{L}} \mathbf{B^2}$. But $\mathbf{AB} = \begin{pmatrix} 2 & 4 \\ 2 & 4 \end{pmatrix}$ and $\mathbf{BA} = \begin{pmatrix} 2 & 2 \\ 4 & 4 \end{pmatrix}$.

Remark 8.3.5. From the proof of Theorem 8.3.2 (i) $\Rightarrow$ (ii), it is clear that if $\mathbf{A}$ and $\mathbf{B}$ are *nnd* matrices of the same order such that $\mathbf{A} <^{\mathbf{L}} \mathbf{B}$ and the largest eigen-value $\lambda_1(\mathbf{B}^\dagger\mathbf{A})$ of $\mathbf{B}^\dagger\mathbf{A}$ is the same as the largest singular value $\sigma_1(\mathbf{B}^\dagger\mathbf{A})$ of $\mathbf{B}^\dagger\mathbf{A}$, then $\mathbf{A^2} <^{\mathbf{L}} \mathbf{B^2}$.

We now study the inter-relationship of the Löwner order with other partial orders studied in earlier chapters.

Theorem 8.3.6. *Let* $\mathbf{A}, \mathbf{B} \in \mathfrak{H}_n$. *Let* $\mathbf{A} <^\star \mathbf{B}$. *Then* $\mathbf{A} <^{\mathbf{L}} \mathbf{B}$ *if and only if the number of negative eigen-values (including algebraic multiplicities) of* $\mathbf{A}$ *and* $\mathbf{B}$ *are equal.*

Proof. Since $\mathbf{A} <^\star \mathbf{B}$ and $\mathbf{A}$ and $\mathbf{B}$ are hermitian, it follows that $\mathbf{A}$ and $\mathbf{B}$ commute. It is easy to see that there exists a unitary matrix $\mathbf{P}$ and non-singular diagonal matrices $\mathbf{D_1}$ and $\mathbf{D_2}$ such that

$$\mathbf{A} = \mathbf{P}\textbf{diag}\left(\mathbf{D_1}, \mathbf{0}, \mathbf{0}\right)\mathbf{P}^\star \text{ and } \mathbf{B} = \mathbf{P}\textbf{diag}\left(\mathbf{D_1}, \mathbf{D_2}, \mathbf{0}\right)\mathbf{P}^\star.$$

So, $\mathbf{A} <^{\mathbf{L}} \mathbf{B}$ if and only if $\mathbf{D_2}$ is positive definite. Thus, the negative eigen-values are the same in $\mathbf{A}$ and $\mathbf{B}$. □

Corollary 8.3.7. *If* $\mathbf{A}$ *and* $\mathbf{B}$ *be nnd matrices of the same order, then* $\mathbf{A} <^\star \mathbf{B}$ *implies* $\mathbf{A} <^{\mathbf{L}} \mathbf{B}$.

Notice that $\mathbf{A} <^\star \mathbf{B}$ implies $\mathbf{A} <^- \mathbf{B}$. Let $\mathbf{A}$ and $\mathbf{B}$ be *nnd*. Should $\mathbf{A} <^- \mathbf{B}$ imply $\mathbf{A} <^{\mathbf{L}} \mathbf{B}$? The answer is contained in the following:

Theorem 8.3.8. *Let* $\mathbf{A}$ *and* $\mathbf{B}$ *be nnd matrices of the same order. Then* $\mathbf{A} <^- \mathbf{B}$ *implies* $\mathbf{A} <^{\mathbf{L}} \mathbf{B}$.

Proof. Since $\mathbf{A} <^- \mathbf{B} \Rightarrow \mathbf{B} - \mathbf{A} <^- \mathbf{B}$, we have $(\mathbf{B} - \mathbf{A})\mathbf{B}^-(\mathbf{B} - \mathbf{A}) = \mathbf{B} - \mathbf{A}$ for each g-inverse $\mathbf{B}^-$ of $\mathbf{B}$. Moreover, $\mathbf{B}$ being *nnd*, we can choose an *nnd* g-inverse of $\mathbf{B}$. It follows that $\mathbf{B} - \mathbf{A}$ is *nnd*, so, $\mathbf{A} <^{\mathbf{L}} \mathbf{B}$. □

Remark 8.3.9. Observe that in Theorem 8.3.8, we can actually assume $\mathbf{A}$ to be just a hermitian matrix.

Theorem 8.3.10. *Let* $\mathbf{A}, \mathbf{B} \in \mathfrak{H}_n$. *Then '*$\mathbf{A} <^- \mathbf{B}$ *and* $\mathbf{A}^2 <^L \mathbf{B}^2$*' if and only if* $\mathbf{A} <^\star \mathbf{B}$.

Proof. 'If' part
Since $\mathbf{A} <^\star \mathbf{B}$, we have $\mathbf{A} <^- \mathbf{B}$. Moreover, as mentioned earlier, there exists a unitary matrix $\mathbf{U}$ such that $\mathbf{A} = \mathbf{U}\mathbf{diag}\left(\mathbf{D_a}, \mathbf{0}, \mathbf{0}\right)\mathbf{U}^\star$ and $\mathbf{B} = \mathbf{U}\mathbf{diag}\left(\mathbf{D_a}, \mathbf{D_{b-a}}, \mathbf{0}\right)\mathbf{U}^\star$, where $\mathbf{D_a}$, $\mathbf{D_{b-a}}$ are real non-singular diagonal matrices. Clearly, $\mathbf{A}^2 <^L \mathbf{B}^2$.
'Only if' part
Since $\mathbf{B}$ is hermitian, $\mathbf{B}^\dagger\mathbf{B} = \mathbf{B}\mathbf{B}^\dagger$. As, $\mathbf{A}^2 <^L \mathbf{B}^2$, $\mathbf{B}^\dagger\mathbf{A}^2\mathbf{B}^\dagger <^L \mathbf{B}^\dagger\mathbf{B}^2\mathbf{B}^\dagger$ or equivalently $\mathbf{B}^\dagger\mathbf{A}(\mathbf{B}^\dagger\mathbf{A})^\star <^L \mathbf{B}^\dagger\mathbf{B}$, since $\mathbf{B}^\dagger\mathbf{B} = \mathbf{B}\mathbf{B}^\dagger$. Since $\mathbf{A} <^- \mathbf{B}$, and $\mathbf{A}$ and $\mathbf{B}$ are hermitian, it follows from Corollary 3.7.5 that there is a unitary matrix $\mathbf{U}$ such that $\mathbf{A} = \mathbf{U}\mathbf{diag}\left(\mathbf{D_a}, \mathbf{0}, \mathbf{0}\right)\mathbf{U}^\star$ and $\mathbf{B} = \mathbf{U}\ \mathbf{diag}\left(\mathbf{M}, \mathbf{0}\right)\mathbf{U}^\star$, where $\mathbf{M} = \begin{pmatrix} \mathbf{D_a} + \mathbf{F}\mathbf{M_{22}}\mathbf{F}^\star & \mathbf{F}\mathbf{M_{22}} \\ \mathbf{M_{22}}\mathbf{F}^\star & \mathbf{M_{22}} \end{pmatrix}$ for some matrix $\mathbf{F}$, and for some non-singular hermitian matrices $\mathbf{D_a}$ and $\mathbf{M_{22}}$. Therefore, $\mathbf{B}^\dagger = \mathbf{U}\begin{pmatrix} \mathbf{M}^{-1} & \mathbf{0} \\ \mathbf{0} & \mathbf{0} \end{pmatrix}\mathbf{U}^\star$, $\mathbf{B}^\dagger\mathbf{B} = \mathbf{U}\mathbf{diag}\left(\mathbf{I}, \mathbf{I}, \mathbf{0}\right)\mathbf{U}^\star$ and $\mathbf{B}^\dagger\mathbf{A} = \mathbf{U}\begin{pmatrix} \mathbf{I} & \mathbf{0} & \mathbf{0} \\ -\mathbf{F}^\star & \mathbf{0} & \mathbf{0} \\ \mathbf{0} & \mathbf{0} & \mathbf{0} \end{pmatrix}\mathbf{U}^\star$. Also, $\mathbf{B}^\dagger\mathbf{A}(\mathbf{B}^\dagger\mathbf{A})^\star = \mathbf{U}\begin{pmatrix} \mathbf{I} & -\mathbf{F} & \mathbf{0} \\ -\mathbf{F}^\star & \mathbf{F}^\star\mathbf{F} & \mathbf{0} \\ \mathbf{0} & \mathbf{0} & \mathbf{0} \end{pmatrix}\mathbf{U}^\star$.
Since $\mathbf{B}^\dagger\mathbf{A}(\mathbf{B}^\dagger\mathbf{A})^\star <^L \mathbf{B}^\dagger\mathbf{B}$, we have $\mathbf{B}^\dagger\mathbf{B} - \mathbf{B}^\dagger\mathbf{A}(\mathbf{B}^\dagger\mathbf{A})^\star$ is *nnd*. However, $\mathbf{B}^\dagger\mathbf{B} - \mathbf{B}^\dagger\mathbf{A}(\mathbf{B}^\dagger\mathbf{A})^\star = \mathbf{U}\begin{pmatrix} \mathbf{0} & \mathbf{F} & \mathbf{0} \\ \mathbf{F}^\star & \mathbf{I} - \mathbf{F}^\star\mathbf{F} & \mathbf{0} \\ \mathbf{0} & \mathbf{0} & \mathbf{0} \end{pmatrix}\mathbf{U}^\star$, so, $\mathbf{F} = \mathbf{0}$. Thus, $\mathbf{A} = \mathbf{U}\mathbf{diag}\left(\mathbf{D_a}, \mathbf{0}, \mathbf{0}\right)\mathbf{U}^\star$ and $\mathbf{B} = \mathbf{U}\mathbf{diag}\left(\mathbf{D_a}, \mathbf{M_{22}}, \mathbf{0}\right)\mathbf{U}^\star$. It is now easy to see that $\mathbf{A} <^\star \mathbf{B}$. □

For a generalization of Theorem 8.3.10, we need the following

Lemma 8.3.11. *Let* $\mathbf{A}$ *and* $\mathbf{B}$ *be nnd matrices of the same order such that* $\mathbf{A} <^L \mathbf{B}$. *Then* $\mathbf{A}^r <^L \mathbf{B}^r$ *for every* r *such that* $0 < r \leq 1$.

For a proof of Lemma 8.3.11, we refer the reader to [Löwner (1934)] or [Bhagwat and Subramanian (1978)]. We now prove the following:

Theorem 8.3.12. *Let* $\mathbf{A}$ *and* $\mathbf{B}$ *be nnd matrices of the same order. Then for any integer* $m \geq 2$, $\mathbf{A} <^- \mathbf{B}$ *and* $\mathbf{A}^m <^L \mathbf{B}^m$ *if and only if* $\mathbf{A} <^\star \mathbf{B}$.

Proof. 'If' part follows easily along the lines of the proof of 'if' part of Theorem 8.3.10. In view of Lemma 8.3.11, if for some integer $m \geq 2$, we have $\mathbf{A}^m <^L \mathbf{B}^m$, then $\mathbf{A}^2 <^L \mathbf{B}^2$. The rest follows from Theorem 8.3.10. □

What if $\mathbf{A}$ and $\mathbf{B}$ are just hermitian and not *nnd* matrices? Should Theorem 8.3.12 still hold? We have the following:

Theorem 8.3.13. *Let* $\mathbf{A}, \mathbf{B} \in \mathfrak{H}_{\mathbf{n}}$. *Then* $\mathbf{A} <^{\star} \mathbf{B}$ *if and only if* $\mathbf{A} <^{-} \mathbf{B}$ *and* $\mathbf{A}^{\mathbf{2m}} <^L \mathbf{B}^{\mathbf{2m}}$ *where m is a positive integer.*

Proof. 'If' part follows along the lines of the 'if' part of Theorem 8.2.11. 'Only if' part follows from Theorem 8.2.11, since $\mathbf{A}^{\mathbf{2}}$ and $\mathbf{B}^{\mathbf{2}}$ are *nnd*. □

8.4 Löwner order on generalized inverses

Let $\mathbf{A}$ be an *nnd* matrix. Let $\mathbf{G}$ be an *nnd* g-inverse of $\mathbf{A}$. We obtain a characterization of *nnd* g-inverses $\mathbf{H}$ of $\mathbf{A}$ with a specified rank each of which dominates $\mathbf{G}$ or is dominated by $\mathbf{G}$ under the Löwner order. We also study a similar problem for *nnd* outer inverses. We then consider *nnd* matrices $\mathbf{A}$ and $\mathbf{B}$ such that $\mathbf{A} <^{\mathbf{L}} \mathbf{B}$. Let $\mathbf{H}$ be an *nnd* g-inverse of $\mathbf{B}$. We characterize the class of g-inverses $\mathbf{G}$ of $\mathbf{A}$ such that $\mathbf{H} <^{\mathbf{L}} \mathbf{G}$ (antitonicity property of g-inverse) and also consider some related problems. Finally we conclude the section by studying the monotonicity property of Moore-Penrose inverse of *nnd* matrices.

We start with a few characterizations of the *nnd* inverses (g-inverses/outer inverses) of an *nnd* matrix.

Theorem 8.4.1. *Let* $\mathbf{A}$ *be an nnd matrix of order* $n \times n$ *with rank* r. *Let* $\mathbf{A} = \mathbf{P}\mathrm{\mathbf{diag}}\,(\boldsymbol{\Delta}\ ,\ \mathbf{0})\,\mathbf{P}^{\star}$ *be spectral decomposition of* $\mathbf{A}$, *where* $\boldsymbol{\Delta}$ *is a positive definite diagonal matrix of order* $r \times r$ *and* $\mathbf{P}$ *is a unitary matrix. Let* s *be an integer such that* $r \leq s \leq n$. *Then an* $n \times n$ *matrix* $\mathbf{G}$ *is an nnd g-inverse of* $\mathbf{A}$ *with rank* s *if and only if* $\mathbf{G} = \mathbf{P}\begin{pmatrix} \boldsymbol{\Delta}^{-1} & \mathbf{L} \\ \mathbf{L}^{\star} & \mathbf{L}^{\star}\boldsymbol{\Delta}\mathbf{L} + \mathbf{S} \end{pmatrix}\mathbf{P}^{\star}$ *for some matrix* $\mathbf{L}$ *and some nnd matrix* $\mathbf{S}$ *of order* $(n-r) \times (n-r)$ *with rank* $s - r$. *Further, for a given g-inverse* $\mathbf{G}$ *of* $\mathbf{A}$, *the matrices* $\mathbf{S}$ *and* $\mathbf{L}$ *are uniquely determined.*

Proof is easy.

Corollary 8.4.2. *Let* $\mathbf{A}$ *be an nnd matrix of order* $n \times n$ *with rank* r. *Let* $\mathbf{A} = \mathbf{Q}\,\mathbf{diag}\,(\mathbf{I_r}\ ,\ \mathbf{0})\,\mathbf{Q}^{\star}$, *where* $\mathbf{Q}$ *is non-singular. Let* s *be an integer such that* $r \leq s \leq n$. *Then an* $n \times n$ *matrix* $\mathbf{G}$ *is an nnd g-inverse of* $\mathbf{A}$ *with* *rank s if and only if* $\mathbf{G} = (\mathbf{Q}^\star)^{-1}\begin{pmatrix}\mathbf{I_r} & \mathbf{L} \\ \mathbf{L}^\star & \mathbf{L}^\star\mathbf{L}+\mathbf{S}\end{pmatrix}\mathbf{Q}^{-1}$ *for some matrix* $\mathbf{L}$ *and for some nnd matrix* $\mathbf{S}$ *of order* $(n-r)\times(n-r)$ *with rank* $s-r$. *Further, for a given g-inverse* $\mathbf{G}$ *of* $\mathbf{A}$, *the matrices* $\mathbf{S}$ *and* $\mathbf{L}$ *are uniquely determined.*

Corollary 8.4.3. *Let* $\mathbf{A}$ *be an nnd matrix of order* $n \times n$ *with rank* r. *Let* s *be an integer such that* $r \le s \le n$. *Then an* $n\times n$ *matrix* $\mathbf{G}$ *is an nnd g-inverse of* $\mathbf{A}$ *with rank* s *if and only if there exists a non-singular matrix* $\mathbf{T}$ *such that* $\mathbf{A} = \mathbf{T}\mathbf{diag}\left(\mathbf{I_r}\ ,\ \mathbf{0}\right)\mathbf{T}^\star$, $\mathbf{G} = (\mathbf{T}^\star)^{-1}\mathbf{diag}\left(\mathbf{I_s}\ ,\ \mathbf{0}\right)\mathbf{T}^{-1}$.

Corollary 8.4.4. *Let* $\mathbf{A}$ *be an nnd matrix of order* $n \times n$ *with rank* r. *Let* s *be an integer such that* $r \le s \le n$. *Then an* $n\times n$ *matrix* $\mathbf{G}$ *is an nnd g-inverse of* $\mathbf{A}$ *with rank* s *if and only if* $\mathbf{G} = \mathbf{A_r^-} + \mathbf{R}$ *for some nnd reflexive g-inverse* $\mathbf{A_r^-}$ *of* $\mathbf{A}$ *and for some matrix* $\mathbf{R}$ *such that* $\mathbf{AR} = \mathbf{0}$ *and* $\rho(\mathbf{R}) = \mathbf{s}-\mathbf{r}$. *Further given* $\mathbf{G}$, *such a decomposition is unique. (Given* $\mathbf{G}$, *the unique reflexive g-inverse mentioned above is* $\mathbf{GAG}$.*)*

Corollary 8.4.5. *Let* $\mathbf{A}$ *be an nnd matrix of order* $n \times n$ *with rank* r. *Let* s *be an integer such that* $r \le s \le n$. *Then an* $n\times n$ *matrix* $\mathbf{G}$ *is an nnd g-inverse of* $\mathbf{A}$ *with rank* s *if and only if* $\mathbf{G}$ *is an nnd reflexive g-inverse of* $\mathbf{A}+\mathbf{B}$ *for some nnd matrix* $\mathbf{B}$ *such that* $\rho(\mathbf{A}+\mathbf{B}) = \rho(\mathbf{A}) + \rho(\mathbf{B})$.

We now divert our attention to the outer inverses of an *nnd* matrix.

Theorem 8.4.6. *Let* $\mathbf{A}$ *be an nnd matrix of order* $n \times n$ *with rank* r. *Let* $(\mathbf{P}, \mathbf{P}^\star)$ *be a rank factorization of* $\mathbf{A}$. *Then* $\mathbf{G}$ *is an nnd outer inverse of* $\mathbf{A}$ *with rank* s, *where* $0 \le s \le r$ *if and only if* $\mathbf{G} = (\mathbf{P_L^{-1}})^\star\mathbf{T}\mathbf{P_L^{-1}}$ *for some left inverse* $\mathbf{P_L^{-1}}$ *of* $\mathbf{P}$ *and for some idempotent matrix* $\mathbf{T}$ *with rank* s.

Proof. 'If' part is trivial.
'Only if' part
Since $\mathbf{P}$ is a full column rank matrix, there exists a matrix $\mathbf{Q}$ such that the matrix $\mathbf{R} = (\mathbf{P} : \mathbf{Q})$ is non-singular. So, $\mathbf{A} = \mathbf{R}\mathbf{diag}\left(\mathbf{I_r}\ ,\ \mathbf{0}\right)\mathbf{R}^\star$. Let $\mathbf{G}$ be an outer inverse of $\mathbf{A}$ with rank s. Since, $\mathbf{GAG} = \mathbf{G}$, it follows that $\mathbf{G} = \mathbf{R}^{\star-1}\begin{pmatrix}\mathbf{B_{11}} & \mathbf{B_{12}} \\ \mathbf{B_{12}^\star} & \mathbf{B_{12}B_{12}^\star}\end{pmatrix}\mathbf{R}^{-1}$, where $\mathbf{B_{11}} = \mathbf{B_{11}B_{11}^\star}$, $\rho(\mathbf{B_{11}}) = s$ and $\mathbf{B_{11}B_{12}} = \mathbf{B_{12}}$. We can rewrite $\mathbf{G}$ as $\mathbf{G} = \mathbf{R}^{\star-1}\mathbf{S}^\star\mathbf{diag}\left(\mathbf{B_{11}}\ ,\ \mathbf{0}\right)\mathbf{S}\mathbf{R}^{-1}$, where $\mathbf{S} = \begin{pmatrix}\mathbf{I} & \mathbf{B_{12}} \\ \mathbf{0} & \mathbf{I}\end{pmatrix}$. It is easy to see that $(\mathbf{I} : -\mathbf{B_{12}})\mathbf{R}^{-1}$ is a left inverse of $\mathbf{P}$. By choosing $\mathbf{T} = \mathbf{B_{11}}$ and $\mathbf{P_L} = (\mathbf{I} : -\mathbf{B_{12}})\mathbf{R}^{-1}$, the result follows.

□

Corollary 8.4.7. *Let* $\mathbf{A}$ *be an nnd matrix of order* $n \times n$ *with rank* r. *Then* $\mathbf{G}$ *is an nnd outer inverse of* $\mathbf{A}$ *with rank* s, *where* $0 \le s \le r$ *if and only if there exists a full column rank matrix* $\mathbf{S}$ *of order* $n \times r$ *such that* $\mathbf{A} = \mathbf{S}\mathbf{S}^\star$ *and* $\mathbf{G} = (\mathbf{S}_\mathbf{L}^{-1})^\star \mathbf{diag}\,(\mathbf{I_s}\ , \mathbf{0})\, \mathbf{S}_\mathbf{L}^{-1}$.

We are now ready for exploring the Löwner ordering properties of *nnd* g-inverses and outer inverses of an *nnd* matrix. We begin with g-inverses.

Theorem 8.4.8. *Let* $\mathbf{A}$ *be an* $n \times n$ *nnd matrix with rank* r. *Let* $\mathbf{G}$ *be an nnd g-inverse of* $\mathbf{A}$ *with rank* s, *where* $r \le s \le n$. *Let* $\mathbf{H}$ *be an* $n \times n$ *matrix of rank* t, *where* $s \le t \le n$. *Then the following are equivalent:*

(i) $\mathbf{H}$ *is an nnd g-inverse of* $\mathbf{A}$ *and* $\mathbf{G} <^\mathbf{L} \mathbf{H}$
(ii) $\mathbf{H} = \mathbf{G} + (\mathbf{I} - \mathbf{GA})\mathbf{Z}(\mathbf{I} - \mathbf{AG})$ *for some nnd matrix* $\mathbf{Z}$ *and*
(iii) $\mathbf{H} = \mathbf{GAG} + \mathbf{S}$ *for some nnd matrix* $\mathbf{S}$ *such that* $\mathbf{G} - \mathbf{GAG} <^\mathbf{L} \mathbf{S}$ *and* $\mathbf{AS} = \mathbf{0}$.

Proof. (i) $\Rightarrow$ (ii)
Let $\mathbf{H} - \mathbf{G} = \mathbf{S}$. Since $\mathbf{G} <^\mathbf{L} \mathbf{H}$, $\mathbf{S}$ is *nnd* and $\mathbf{ASA} = \mathbf{0}$. But then it follows that $\mathbf{AS} = \mathbf{0}$. This further gives $\mathbf{S} = (\mathbf{I} - \mathbf{GA})\mathbf{Z}(\mathbf{I} - \mathbf{AG})$ for some *nnd* matrix $\mathbf{Z}$.
(ii) $\Rightarrow$ (iii)
Let $\mathbf{H} = \mathbf{G} + (\mathbf{I} - \mathbf{GA})\mathbf{Z}(\mathbf{I} - \mathbf{AG})$ for some *nnd* matrix $\mathbf{Z}$. Then $\mathbf{H}$ is a g-inverse of $\mathbf{A}$ with $\mathbf{AH} = \mathbf{AG}$ and $\mathbf{HAH} = \mathbf{GAG}$. By Corollary 8.4.4, there is an *nnd* matrix $\mathbf{R}$ such that $\mathbf{G} = \mathbf{GAG} + \mathbf{R}$ and $\mathbf{AR} = \mathbf{0}$. Let $\mathbf{S} = \mathbf{R} + (\mathbf{I} - \mathbf{GA})\mathbf{Z}(\mathbf{I} - \mathbf{AG})$. Then $\mathbf{H} = \mathbf{GAG} + \mathbf{S}$. As '$\mathbf{Z}$ and $\mathbf{R}$' are *nnd* and '$\mathbf{A}$ are $\mathbf{G}$' are hermitian, $\mathbf{S}$ is *nnd* and $\mathbf{AS} = \mathbf{0}$. Moreover, $\mathbf{G} - \mathbf{GAG} = \mathbf{R}$, $\mathbf{H} - \mathbf{GAG} = \mathbf{S}$ and $\mathbf{S} - \mathbf{R}$ is *nnd*, so, (iii) holds.
(iii) $\Rightarrow$ (i)
Since, $\mathbf{A}$, $\mathbf{G}$, $\mathbf{S}$ are *nnd*, $\mathbf{H}$ is *nnd*. Also, $\mathbf{AHA} = \mathbf{AGAGA} + \mathbf{ASA} = \mathbf{A}$, giving $\mathbf{H}$ is a g-inverse of $\mathbf{A}$ and $\mathbf{G} - \mathbf{GAG} <^\mathbf{L} \mathbf{S} \Rightarrow \mathbf{S} + \mathbf{GAG} - \mathbf{G} = \mathbf{H} - \mathbf{G}$ is *nnd*. Thus (i) holds. □

Theorem 8.4.9. *Let* $\mathbf{A}$ *be an* $n \times n$ *nnd matrix with rank* r. *Let* $\mathbf{G}$ *be an nnd g-inverse of* $\mathbf{A}$ *with rank* s, *where* $r \le s \le n$. *Let* $\mathbf{H}$ *be an* $n \times n$ *matrix of rank* t, *where* $s \le t \le n$. *Then the following are equivalent:*

(i) $\mathbf{H}$ *is an nnd g-inverse of* $\mathbf{A}$ *and* $\mathbf{H} <^\mathbf{L} \mathbf{G}$
(ii) $\mathbf{H} = \mathbf{G} - (\mathbf{I} - \mathbf{GA})\mathbf{Z}(\mathbf{I} - \mathbf{AG})$ *for some nnd matrix* $\mathbf{Z}$ *and*
(iii) $\mathbf{H} = \mathbf{GAG} + \mathbf{U}$ *for some nnd matrix* $\mathbf{U}$ *such that* $\mathbf{U} <^\mathbf{L} \mathbf{G} - \mathbf{GAG}$.

Proof follows along similar lines to Theorem 8.4.8.
Note that in Theorem 8.4.9, for the matrix $\mathbf{U}$, $\mathbf{U} <^{\mathbf{L}} \mathbf{G} - \mathbf{GAG} \Rightarrow \mathbf{AU} = \mathbf{0}$.

Theorem 8.4.10. *Let* $\mathbf{A}$ *be an* $n \times n$ *nnd matrix with rank* r. *Let* $0 \leq s \leq t \leq r$. *Let* $\mathbf{G}$, $\mathbf{H}$ *be nnd outer inverses of* $\mathbf{A}$ *with ranks* s *and* t *respectively. Then the following are equivalcnt:*

(i) $\mathbf{G} <^{\mathbf{L}} \mathbf{H}$
(ii) $\mathcal{C}(\mathbf{G}) \subseteq \mathcal{C}(\mathbf{H})$
(iii) $\mathbf{H} - \mathbf{G}$ *is an outer inverse of* $\mathbf{A}$ *and*
(iv) $\mathbf{G} <^{-} \mathbf{H}$.

Proof. (i) $\Rightarrow$ (ii) is trivial.
(ii) $\Rightarrow$ (iii)
Since $\mathbf{A}$ is a g-inverse of $\mathbf{H}$, we have $\mathbf{HAG} = \mathbf{G} = \mathbf{GAH}$. Hence, $\mathbf{H} - \mathbf{GAH} - \mathbf{G} = \mathbf{H} - \mathbf{G}$.
(iii) $\Rightarrow$ (iv)
Since $\mathbf{H} - \mathbf{G}$ is an outer inverse of $\mathbf{A}$, $\rho(\mathbf{H} - \mathbf{G}) = tr((\mathbf{H} - \mathbf{G})\mathbf{A}) = tr(\mathbf{HA}) - tr(\mathbf{GA}) = \rho(\mathbf{H}) - \rho(\mathbf{G})$. Hence, $\mathbf{G} <^{-} \mathbf{H}$.
(iv) $\Rightarrow$ (i) follows from Theorem 8.3.8. □

How does one obtain the class of all *nnd* outer inverses of a specified rank dominating (or dominated by) a given outer inverse of a matrix? We describe below a constructive procedure for this:

Let $\mathbf{A}$ be an $n \times n$ *nnd* matrix with rank r. Let $\mathbf{G}$ be an *nnd* outer inverse of $\mathbf{A}$ with rank $s (s \leq r)$. Then by Corollary 8.4.7, there exists a full column rank matrix $\mathbf{S}$ such that $\mathbf{A} = \mathbf{SS}^{\star}$ and $\mathbf{G} = (\mathbf{S}_{\mathbf{L}}^{-1})^{\star}\mathbf{diag}\left(\mathbf{I}\ ,\ \mathbf{0}\right) \mathbf{S}_{\mathbf{L}}^{-1^{-1}}$.

(i) The class of all *nnd* outer inverses $\mathbf{H}$ of $\mathbf{A}$ with rank t $(s \leq t \leq r)$ such that $\mathbf{G} <^{\mathbf{L}} \mathbf{H}$ is given by
$$\mathbf{H} = \{\mathbf{S_L}^{-1} + \mathbf{Z}(\mathbf{I} - \mathbf{SS_L}^{-1})\}^{\star}\mathbf{T}\{\mathbf{S_L}^{-1} + \mathbf{Z}(\mathbf{I} - \mathbf{SS_L}^{-1})\},$$
where
   (a) $\mathbf{T} = \mathbf{diag}\left(\mathbf{I_s}\ ,\ \mathbf{T_{22}}\right)$ with $\mathbf{T_{22}}$ arbitrary hermitian idempotent such that $\rho(\mathbf{T_{22}}) = t - s$ and
   (b) $\mathbf{Z} = \begin{pmatrix} \mathbf{US}_{\mathbf{L}}^{-1} \\ .. \\ \mathbf{Z_2} \end{pmatrix}$ with $\mathbf{U}$ an $s \times r$ matrix and $\mathbf{Z_2}$ arbitrary.

(ii) The class of all *nnd* outer inverses $\mathbf{H}$ of $\mathbf{A}$ with rank t $(0 \leq t \leq s)$ such that $\mathbf{H} <^{\mathbf{L}} \mathbf{G}$ is given by $\mathbf{H} = \mathbf{P}_{\mathbf{L}}^{-1^{\star}}\mathbf{UP_L}^{-1}$, where $\mathbf{U} = \left(\mathbf{U_1}\ ,\ \mathbf{0}\right)$, with $\mathbf{U_1}$ an arbitrary idempotent matrix of order $s \times s$ and rank t.

Let $\mathbf{A}$ and $\mathbf{B}$ be positive definite matrices of the same order. Then it is well known that $\mathbf{A} <^{\mathbf{L}} \mathbf{B}$ if and only if $\mathbf{B}^{-1} <^{\mathbf{L}} \mathbf{A}^{-1}$, which essentially follows from Theorem 2.7.2. We now consider two *nnd* matrices $\mathbf{A}$ and $\mathbf{B}$ of the same order such that $\mathbf{A} <^{\mathbf{L}} \mathbf{B}$. Let an *nnd* g-inverse $\mathbf{H}$ of $\mathbf{B}$ with specified rank be given. We shall characterize the class of all *nnd* g-inverses $\mathbf{G}$ of $\mathbf{A}$ such that $\mathbf{H} <^{\mathbf{L}} \mathbf{G}$.

Theorem 8.4.11. *Let* $\mathbf{A}$ *and* $\mathbf{B}$ *be nnd matrices of the same order with ranks* r *and* s *respectively such that* $\mathbf{A} <^{\mathbf{L}} \mathbf{B}$. *Let* u *and* t *be positive integers such that* $r \leq s \leq t \leq u$.

(i) *Let* $\mathbf{H}$ *be an nnd g-inverse of* $\mathbf{B}$. *Let* $\left(\mathbf{P_1} : \mathbf{P_2} : \mathbf{P_3}\right)$ *be a non-singular matrix such that* $\mathbf{B} = \mathbf{P_1}\mathbf{P_1^\star} + \mathbf{P_2}\mathbf{P_2^\star}, \mathbf{A} = \mathbf{P_1}\mathbf{\Lambda}\mathbf{P_1^\star}$ *and* $\mathbf{H} = \mathbf{Q}\ \mathbf{diag}\left(\mathbf{I_t}\ , \mathbf{0}\right)\mathbf{Q}^\star$, *where* $\mathbf{Q} = (\mathbf{P}^\star)^{-1}$ *and* $\mathbf{\Lambda}$ *is a* $r \times r$ *positive definite matrix such that* $\mathbf{I} - \mathbf{\Lambda}$ *is nnd. Then an nnd g-inverse* $\mathbf{G}$ *of* $\mathbf{A}$ *with rank* u *satisfies* $\mathbf{H} <^{\mathbf{L}} \mathbf{G}$ *if and only if*

$$\mathbf{G} = \mathbf{Q}\begin{pmatrix} \mathbf{\Lambda}^{-1} & \mathbf{S_{12}} & \mathbf{S_{13}} \\ \mathbf{S_{12}}^\star & \mathbf{S_{22}} & \mathbf{S_{23}} \\ \mathbf{S_{13}}^\star & \mathbf{S_{23}}^\star & \mathbf{S_{33}} \end{pmatrix}\mathbf{Q}^\star,$$

where $\mathbf{S_{12}}, \mathbf{S_{13}}, \mathbf{S_{22}}, \mathbf{S_{23}},$ *and* $\mathbf{S_{33}}$ *are arbitrary matrices satisfying*

(a) $\mathbf{S_{12}} = (\mathbf{\Lambda}^{-1} - \mathbf{I})\mathbf{Z}$, *where* $\mathbf{Z}$ *is arbitrary*

(b) $\mathbf{S_{22}} = \mathbf{I} + \mathbf{S_{12}^\star}(\mathbf{\Lambda}^{-1} - \mathbf{I})^{-}\mathbf{S_{12}} + \mathbf{V}$, *where* $\mathbf{V}$ *is an arbitrary nnd matrix*

(c) $\begin{pmatrix} \mathbf{S_{13}} \\ \mathbf{S_{23}} \end{pmatrix} = \begin{pmatrix} \mathbf{\Lambda}^{-1} - \mathbf{I} & \mathbf{S_{12}} \\ \mathbf{S_{12}^\star} & \mathbf{S_{22}} - \mathbf{I} \end{pmatrix}\begin{pmatrix} \mathbf{U_1} \\ \mathbf{U_2} \end{pmatrix}$, *where* $\mathbf{U_1}$ *and* $\mathbf{U_2}$ *are arbitrary and*

(d) $\mathbf{S_{33}} = \left(\mathbf{S_{13}^\star}\ , \mathbf{S_{23}^\star}\right)\begin{pmatrix} \mathbf{\Lambda}^{-1} & \mathbf{S_{12}} \\ \mathbf{S_{12}^\star} & \mathbf{S_{22}} \end{pmatrix}^{-1}\begin{pmatrix} \mathbf{S_{13}} \\ \mathbf{S_{23}} \end{pmatrix} + \mathbf{W}$, *where* $\mathbf{W}$ *is an arbitrary nnd matrix such that* $\rho(\mathbf{W}) = u - t$.

(ii) *Let* $r = s$. *Let* $\mathbf{G}$ *be a given nnd g-inverse of* $\mathbf{A}$ *with rank* u. *Let* $\mathbf{P}$ *be a non-singular matrix such that* $\mathbf{A} = \mathbf{P}\ \mathbf{diag}\left(\mathbf{I_r}\ , \mathbf{0}\right)\mathbf{P}^\star$, $\mathbf{G} = \mathbf{Q}\ \mathbf{diag}\left(\mathbf{I_u}\ , \mathbf{0}\right)\mathbf{Q}^\star$ *and* $\mathbf{B} = \mathbf{P}\ \mathbf{diag}\left(\mathbf{\Delta}\ , \mathbf{0}\right)\mathbf{P}^\star$, *where* Δ *is a positive definite matrix such that* $\mathbf{I} <^{\mathbf{L}} \mathbf{\Delta}$ *and* $\mathbf{Q} = (\mathbf{P}^\star)^{-1}$. *Then*

$$\mathbf{H} <^{\mathbf{L}} \mathbf{G} \text{ if and only if } \mathbf{H} = \mathbf{Q}\begin{pmatrix} \mathbf{\Lambda}^{-1} & \mathbf{L} & \mathbf{0} \\ \mathbf{L}^\star & \mathbf{M} & \mathbf{0} \\ \mathbf{0} & \mathbf{0} & \mathbf{0} \end{pmatrix}\mathbf{Q}^\star,$$

where $\mathbf{L}, \mathbf{M}$ *are arbitrary matrices satisfying*

(a) $\mathcal{C}(\mathbf{L}) \subseteq \mathcal{C}(\mathbf{I} - \mathbf{\Delta}^{-1})$
(b) $\mathbf{L}^\star\mathbf{\Delta L} <^{\mathbf{L}} \mathbf{M} <^{\mathbf{L}} \mathbf{I} - \mathbf{L}^\star(\mathbf{\Lambda}^{-1} - \mathbf{I})^{-}\mathbf{L}$ *and*
(c) $\rho(\mathbf{M} - \mathbf{L}^\star\mathbf{\Delta L}) = \mathbf{t} - \mathbf{r}$.

Proof follows by repeated applications of Theorem 8.2.2.

Remark 8.4.12. For an interesting characterization of matrices satisfying (a)-(c) of Theorem 8.4.11, see [Bhimasankaram and Mathew Thomas (1993)].

In Theorem 8.4.11 (ii), we considered a special case when $\rho(\mathbf{A}) = \rho(\mathbf{B})$. However, if $\mathbf{A} <^{\mathbf{L}} \mathbf{B}$ and $\rho(\mathbf{A}) < \rho(\mathbf{B})$, then given an *nnd* g-inverse $\mathbf{G}$ of $\mathbf{A}$, it may not be possible to find an *nnd* g-inverse $\mathbf{H}$ of $\mathbf{B}$ such that $\mathbf{H} <^{\mathbf{L}} \mathbf{G}$ as shown by the following example:

Example 8.4.13. Let $\mathbf{A} = \begin{pmatrix} 1 & 0 \\ 0 & 0 \end{pmatrix}$ and $\mathbf{B} = \begin{pmatrix} 2 & 1 \\ 1 & 2 \end{pmatrix}$. Then it is clear that $\mathbf{A} <^{\mathbf{L}} \mathbf{B}$. Now, $\mathbf{B}^{-1} = \frac{1}{3}\begin{pmatrix} 2 & -1 \\ -1 & 2 \end{pmatrix}$. Every *nnd* g-inverse of $\mathbf{A}$ is of the form $\begin{pmatrix} 1 & c \\ c & d \end{pmatrix}$, where $d - c^2 \geq 0$. However, $\mathbf{B}^{-1} - \begin{pmatrix} 1 & c \\ c & d \end{pmatrix}$ is not *nnd* for any c, d which satisfy the condition $d - c^2 \geq 0$.

Theorem 8.4.14. *Let* $\mathbf{A}$ *and* $\mathbf{B}$ *be nnd matrices of the same order with ranks* r *and* s *respectively such that* $\mathbf{A} <^{\mathbf{L}} \mathbf{B}$. *Let* $\mathbf{A} <^{-} \mathbf{B}$. *Then for a given nnd g-inverse* $\mathbf{G}$ *of* $\mathbf{A}$ *with rank* u $(u \geq s)$, *there exists an nnd g-inverse* $\mathbf{H}$ *of* $\mathbf{B}$ *such that* $\mathbf{H} <^{\mathbf{L}} \mathbf{G}$ *if and only if* $\mathbf{B} - \mathbf{A} <^{\mathbf{L}} (\mathbf{B} - \mathbf{A})\mathbf{G}(\mathbf{B} - \mathbf{A})$ *and* $\mathbf{AG}(\mathbf{B} - \mathbf{A}) = \mathbf{0}$.

Proof. Since $\mathbf{A}$ and $\mathbf{B}$ are *nnd* and $\mathbf{A} <^{-} \mathbf{B}$, there exists a non-singular matrix $\mathbf{P}$ such that

$$\mathbf{A} = \mathbf{P}\text{diag}\left(\mathbf{I_r}\ ,\ \mathbf{0}\ ,\ \mathbf{0}\right)\mathbf{P}^\star \text{ and } \mathbf{B} = \mathbf{P}\ \text{diag}\left(\mathbf{I_r}\ ,\ \mathbf{I_{s-a}}\ ,\ \mathbf{0}\right)\mathbf{P}^\star.$$

So, $\mathbf{G} = \mathbf{Q}\begin{pmatrix} \mathbf{I} & \mathbf{S_{12}} & \mathbf{S_{13}} \\ \mathbf{S_{12}^\star} & \mathbf{S_{22}} & \mathbf{S_{23}} \\ \mathbf{S_{13}^\star} & \mathbf{S_{23}^\star} & \mathbf{S_{33}} \end{pmatrix}\mathbf{Q}^\star$, where $\mathbf{Q} = (\mathbf{P}^{-1})^\star$, and $\mathbf{S_{12}}$, $\mathbf{S_{13}}$, $\mathbf{S_{22}}$, $\mathbf{S_{23}}$, and $\mathbf{S_{33}}$ are such that $\mathbf{G}$ is an *nnd* matrix of rank u. Any *nnd* g-inverse $\mathbf{H}$ of $\mathbf{B}$ must be of the form $\mathbf{H} = \mathbf{Q}\begin{pmatrix} \mathbf{I} & \mathbf{0} & \mathbf{T_{13}} \\ \mathbf{0} & \mathbf{I} & \mathbf{T_{23}} \\ \mathbf{T_{13}^\star} & \mathbf{T_{23}^\star} & \mathbf{T_{33}} \end{pmatrix}\mathbf{Q}^\star$ for some $\mathbf{T_{13}}$, $\mathbf{T_{23}}$, and $\mathbf{T_{33}}$ such that $\mathbf{T_{33}} - \mathbf{T_{13}^\star T_{13}} - \mathbf{T_{23}^\star T_{23}}$ is *nnd*.

Clearly, there exist $\mathbf{T_{13}}$, $\mathbf{T_{23}}$, and $\mathbf{T_{33}}$ such that $\mathbf{H} <^{\mathbf{L}} \mathbf{G}$ if and only if $\mathbf{S_{12}} = \mathbf{0}$ and $\mathbf{I} <^{\mathbf{L}} \mathbf{S_{22}}$ or equivalently if and only if $\mathbf{AG}(\mathbf{B}-\mathbf{A}) = \mathbf{0}$ and $\mathbf{B}-\mathbf{A} <^{\mathbf{L}} (\mathbf{B}-\mathbf{A})\mathbf{G}(\mathbf{B}-\mathbf{A})$. □

Remark 8.4.15. The necessary and sufficient condition in Theorem 8.4.14 for $\mathbf{H} <^{\mathbf{L}} \mathbf{G}$ can be equivalently stated as $\mathbf{BGA} = \mathbf{A}$, $\mathbf{B} <^{\mathbf{L}} \mathbf{BGB}$.

Remark 8.4.16. Let $\mathbf{B} = \mathbf{A} + \mathbf{xx}^\star$, where $\mathbf{A}$ is *nnd* and $\mathbf{x} \notin \mathcal{C}(\mathbf{A})$. (So, $\rho(\mathbf{B}) = r+1$.) For an *nnd* g-inverse $\mathbf{G}$ of $\mathbf{A}$ with rank $u \geq r+1$, there exists an *nnd* g-inverse $\mathbf{H}$ of $\mathbf{B}$ such that $\mathbf{H} <^{\mathbf{L}} \mathbf{G}$ if and only if $\mathbf{x}^\star\mathbf{Gx} \geq \mathbf{1}$ and $\mathbf{AGx} = \mathbf{0}$.

Let $\mathbf{A}$ and $\mathbf{B}$ be *nnd* matrices of the same order such that $\mathbf{A} <^{\mathbf{L}} \mathbf{B}$. If there exist reflexive *nnd* g-inverses $\mathbf{G}$ of $\mathbf{A}$ and $\mathbf{H}$ of $\mathbf{B}$ such that $\mathbf{H} <^{\mathbf{L}} \mathbf{G}$, then $\rho(\mathbf{A}) \leq \rho(\mathbf{B}) = \rho(\mathbf{H}) \leq \rho(\mathbf{G}) = \rho(\mathbf{A})$. Thus, $\rho(\mathbf{A}) = \rho(\mathbf{B})$. We show in the following theorem that if $\mathbf{A} <^{\mathbf{L}} \mathbf{B}$ and $\rho(\mathbf{A}) = \rho(\mathbf{B})$, then given any reflexive *nnd* g-inverse $\mathbf{H}$ of $\mathbf{B}$, there exists a unique reflexive g-inverse $\mathbf{G}$ of $\mathbf{A}$ such that $\mathbf{H} <^{\mathbf{L}} \mathbf{G}$.

Theorem 8.4.17. *Let $\mathbf{A}$ and $\mathbf{B}$ be nnd matrices of the same order and of the same rank r such that $\mathbf{A} <^{\mathbf{L}} \mathbf{B}$. Then for any reflexive nnd g-inverse $\mathbf{H}$ of $\mathbf{B}$, there exists a unique nnd reflexive g-inverse $\mathbf{G}$ of $\mathbf{A}$ such that $\mathbf{H} <^{\mathbf{L}} \mathbf{G}$.*

Proof. Let $\mathbf{P}$ be matrix of full column rank r such that $\mathbf{A} = \mathbf{PP}^\star$ and $\mathbf{B} = \mathbf{P}\boldsymbol{\Delta}^{-1}\mathbf{P}^\star$, where $\boldsymbol{\Delta}$ is a positive definite diagonal matrix such that $\mathbf{I} <^{\mathbf{L}} \boldsymbol{\Delta}$. Let $\mathbf{H} = \mathbf{Q}\boldsymbol{\Delta}^{-1}\mathbf{Q}^\star$ be a given *nnd* reflexive g-inverse of $\mathbf{B}$, where $\mathbf{Q}^\star$ is a g-inverse of $\mathbf{P}$. Every left inverse of $\mathbf{P}$ is given by $\mathbf{Q}^\star + \mathbf{Z}^\star(\mathbf{I}-\mathbf{PQ}^\star)$ for some $\mathbf{Z}$. So, every *nnd* reflexive g-inverse of $\mathbf{A}$ is given by $\mathbf{TT}^\star$, where $\mathbf{T} = \mathbf{Q} + (\mathbf{I}-\mathbf{QP}^\star)\mathbf{Z}$. Now, $\mathbf{H} <^{\mathbf{L}} \mathbf{G} \Rightarrow \mathbf{Q} = \mathbf{TW}$ for some $\mathbf{W} \Rightarrow \mathbf{W} = \mathbf{I}$ and $\mathbf{Q} + (\mathbf{I}-\mathbf{QP}^\star)\mathbf{Z} = \mathbf{0}$. Let $\mathbf{G} = \mathbf{QQ}^\star$. Clearly, $\mathbf{G}$ is the unique g-inverse of $\mathbf{A}$ such that $\mathbf{H} <^{\mathbf{L}} \mathbf{G}$. □

Corollary 8.4.18. *Let $\mathbf{A}$ and $\mathbf{B}$ be nnd matrices of the same order and of the same rank r such that $\mathbf{A} <^{\mathbf{L}} \mathbf{B}$. Let $\mathbf{G}$ and $\mathbf{H}$ be two given nnd reflexive g-inverses of $\mathbf{A}$ and $\mathbf{B}$ respectively. Then $\mathbf{H} <^{\mathbf{L}} \mathbf{G}$ if and only if $\mathbf{AG} = \mathbf{BH}$.*

Proof. Let $\mathbf{P}$ be a non-singular matrix such that

$$\mathbf{A} = \mathbf{P}\ \mathrm{\mathbf{diag}}\left(\mathbf{I_r}, \mathbf{0}\right)\mathbf{P}^\star,\ \mathbf{B} = \mathbf{P}\ \mathrm{\mathbf{diag}}\left(\boldsymbol{\Delta}, \mathbf{0}\right)\mathbf{P}^\star \text{ and } \mathbf{G} = \mathbf{Q}\begin{pmatrix}\mathbf{I_r} & \mathbf{0}\\ \mathbf{0} & \mathbf{0}\end{pmatrix}\mathbf{Q}^\star,$$

where $\mathbf{Q} = (\mathbf{P}^{-1})^\star$, $\mathbf{\Delta}$ is positive definite matrix satisfying $\mathbf{I} <^{\mathbf{L}} \mathbf{\Delta}$. Clearly, $\mathbf{H} = \mathbf{Q}\begin{pmatrix}\mathbf{\Delta}^{-1} & \mathbf{0} \\ \mathbf{0} & \mathbf{0}\end{pmatrix}\mathbf{Q}^\star$ is an *nnd* g-inverse of $\mathbf{B}$ such that $\mathbf{H} <^{\mathbf{L}} \mathbf{G}$. But by Theorem 8.4.17, such a $\mathbf{G}$ is unique. Also, $\mathbf{AT} = \mathbf{BH}$ for some reflexive *nnd* g-inverse $\mathbf{T}$ of $\mathbf{A}$ if and only if $\mathbf{T} = \mathbf{G}$. □

Corollary 8.4.19. *Let* $\mathbf{A}$ *and* $\mathbf{B}$ *be nnd matrices of the same order such that* $\mathbf{A} <^{\mathbf{L}} \mathbf{B}$. *Then* $\mathbf{B}^\dagger <^{\mathbf{L}} \mathbf{A}^\dagger$ *if and only if* $\rho(\mathbf{A}) = \rho(\mathbf{B})$.

Proof. 'Only if' part is trivial.
'If' part
$\mathbf{A} <^{\mathbf{L}} \mathbf{B}$ and $\rho(\mathbf{A}) = \rho(\mathbf{B})$. So, $\mathbf{AA}^\dagger = \mathbf{P_A} = \mathbf{P_B} = \mathbf{BB}^\dagger$. So, by Corollary 8.4.18, we have $\mathbf{B}^\dagger <^{\mathbf{L}} \mathbf{A}^\dagger$. □

Let $\mathbf{A}$ and $\mathbf{B}$ be *nnd* matrices of the same order such that $\mathbf{A} <^{\mathbf{L}} \mathbf{B}$. Given an *nnd* outer inverse $\mathbf{G}$ of $\mathbf{A}$, what is the class of all *nnd* outer inverses $\mathbf{H}$ of $\mathbf{B}$ such that $\mathbf{H} <^{\mathbf{L}} \mathbf{G}$? For solutions to these and related problems we refer the readers to [Bhimasankaram and Mathew Thomas (1993)].

So far we have been studying the antitonicity property of g-inverses of *nnd* matrices (i.e. when does $\mathbf{A} <^{\mathbf{L}} \mathbf{B}$ imply $\mathbf{B}^- <^{\mathbf{L}} \mathbf{A}^-$ for *nnd* g-inverses $\mathbf{A}^-$ of $\mathbf{A}$ and $\mathbf{B}^-$ of $\mathbf{B}$?) We conclude this section with the following theorem showing the monotonicity of Moore-Penrose inverse:

Theorem 8.4.20. *Let* $\mathbf{A}$ *and* $\mathbf{B}$ *be nnd matrices of the same order such that* $\mathbf{A} <^{\mathbf{L}} \mathbf{B}$. *Then* $\mathbf{A}^\dagger <^{\mathbf{L}} \mathbf{B}^\dagger$ *if and only if* $\mathbf{A}^\dagger <^\star \mathbf{B}^\dagger$.

Proof. 'If'part is trivial.
'Only if' part
Since $\mathbf{A}^\dagger <^{\mathbf{L}} \mathbf{B}^\dagger$, we have $\mathbf{B}^\dagger = \mathbf{A}^\dagger + \mathbf{CC}^\star$ for some matrix $\mathbf{C}$. Therefore, $\mathbf{AB}^\dagger\mathbf{A} = \mathbf{A} + \mathbf{ACC}^\star\mathbf{A}^\star$. Thus, $\mathbf{A} <^{\mathbf{L}} \mathbf{AB}^\dagger\mathbf{A}$. But then by Corollary 8.2.12, $\mathbf{AB}^\dagger\mathbf{A} <^{\mathbf{L}} \mathbf{A}$, so, $\mathbf{AB}^\dagger\mathbf{A} = \mathbf{A}$. Thus, $\mathbf{AC} = \mathbf{0}$ and so, $\mathbf{AB}^\dagger = \mathbf{AA}^\dagger$. As $\mathbf{A}$ is hermitian and $\mathbf{B}^\dagger\mathbf{A} = \mathbf{A}^\dagger\mathbf{A} + \mathbf{CC}^\star\mathbf{A}$, we have $\mathbf{B}^\dagger\mathbf{A} = \mathbf{A}^\dagger\mathbf{A}$. Hence, $\mathbf{A}^\dagger <^\star \mathbf{B}^\dagger$. □

Corollary 8.4.21. *Let* $\mathbf{A}$ *and* $\mathbf{B}$ *be nnd matrices of the same order such that* $\mathbf{A} <^{\mathbf{L}} \mathbf{B}$. *Then* $\mathbf{A}^\dagger <^{\mathbf{L}} \mathbf{B}^\dagger$ *if and only if* $\mathbf{A} <^\star \mathbf{B}$.

Remark 8.4.22. Theorem 8.4.20 holds even if we replace $\mathbf{A}^\dagger$ and $\mathbf{B}^\dagger$ by *nnd* reflexive g-inverses $\mathbf{G}$ and $\mathbf{H}$ of $\mathbf{A}$ and $\mathbf{B}$ respectively. The proof also follows along the same lines.

8.5 Generalizations of the Löwner order

In this section we give two generalizations of the Löwner order, one for the class of hermitian matrices and the other for arbitrary matrices, square or rectangular. We study their properties briefly and compare the two on the class of hermitian matrices.

Let $\mathbf{A}$ and $\mathbf{B}$ be hermitian matrices of the same order. As we have seen in Theorem 8.2.21, the statements '$\mathbf{A} <^{\mathbf{L}} \mathbf{B}$' and '$\mathcal{C}(\mathbf{A}) \subseteq \mathcal{C}(\mathbf{B})$ and $\mathbf{AB}^-\mathbf{A} <^{\mathbf{L}} \mathbf{A}$' are equivalent whenever $\nu(\mathbf{A}) = \nu(\mathbf{B})$. We now show that the condition '$\mathcal{C}(\mathbf{A}) \subseteq \mathcal{C}(\mathbf{B})$ and $\mathbf{AB}^-\mathbf{A} <^{\mathbf{L}} \mathbf{A}$' can be used to define a new order relation on the class of hermitian matrices, $\mathfrak{H}_n$. It turns out that it is a partial order on $\mathfrak{H}_n$ that is equivalent to the Löwner order when matrices under consideration have the same number of non-negative eigen-values. Before studying the properties of this order relation we formally define this in

Definition 8.5.1. Let $\mathbf{A}, \mathbf{B} \in \mathfrak{H}_n$. We say $\mathbf{A}$ lies below $\mathbf{B}$ under '$<^o$' if $\mathcal{C}(\mathbf{A}) \subseteq \mathcal{C}(\mathbf{B})$ and $\mathbf{AB}^-\mathbf{A} <^{\mathbf{L}} \mathbf{A}$ for some g-inverse $\mathbf{B}^-$ of $\mathbf{B}$.

When this is so, we write $\mathbf{A} <^{\mathbf{o}} \mathbf{B}$.

Remark 8.5.2. If $\mathbf{A}$ and $\mathbf{B}$ be hermitian matrices of the same order such that $\nu(\mathbf{A}) = \nu(\mathbf{B})$, then $\mathbf{A} <^{\mathbf{o}} \mathbf{B} \Leftrightarrow \mathbf{A} <^{\mathbf{L}} \mathbf{B}$.

We now show that the order relation '$<^o$' defines a partial order on $\mathfrak{H}_n$.

Theorem 8.5.3. *The order relation '$<^o$' is a partial order on $\mathfrak{H}_n$.*

Proof. Reflexivity is trivial.
We next show that the relation is transitive. Let $\mathbf{A} <^{\mathbf{o}} \mathbf{B}$ and $\mathbf{B} <^{\mathbf{o}} \mathbf{C}$. Then $\mathcal{C}(\mathbf{A}) \subseteq \mathcal{C}(\mathbf{B})$ and $\mathcal{C}(\mathbf{B}) \subseteq \mathcal{C}(\mathbf{C})$. Therefore, $\mathcal{C}(\mathbf{A}) \subseteq \mathcal{C}(\mathbf{C})$. Since, $\mathcal{C}(\mathbf{A}) \subseteq \mathcal{C}(\mathbf{B})$, $\mathbf{AB}^-\mathbf{A} = \mathbf{AB}^\dagger\mathbf{A}$ for all g-inverses $\mathbf{B}^-$ of $\mathbf{B}$. Similarly, $\mathcal{C}(\mathbf{B}) \subseteq \mathcal{C}(\mathbf{C}) \Rightarrow \mathbf{BC}^-\mathbf{B} = \mathbf{BC}^\dagger\mathbf{B}$ for all g-inverses $\mathbf{C}^-$ of $\mathbf{C}$. Thus both the matrices $\mathbf{A}(\mathbf{A}^\dagger - \mathbf{B}^\dagger)\mathbf{A}$ and $\mathbf{B}(\mathbf{B}^\dagger - \mathbf{C}^\dagger)\mathbf{B}$ are *nnd*. Pre- and post-multiplying $\mathbf{B}(\mathbf{B}^\dagger - \mathbf{C}^\dagger)\mathbf{B}$ with $\mathbf{AB}^\dagger$ and $(\mathbf{AB}^\dagger)^\star$ respectively, we have $\mathbf{AB}^\dagger\mathbf{B}(\mathbf{B}^\dagger - \mathbf{C}^\dagger)\mathbf{B}(\mathbf{AB}^\dagger)^\star = \mathbf{A}(\mathbf{B}^\dagger - \mathbf{C}^\dagger)\mathbf{A}$ is *nnd*. Thus, $\mathbf{AC}^\dagger\mathbf{A} <^{\mathbf{L}} \mathbf{AB}^\dagger\mathbf{A} <^{\mathbf{L}} \mathbf{A}$ and the transitivity holds.
For anti-symmetry, let $\mathbf{A} <^{\mathbf{o}} \mathbf{B}$ and $\mathbf{B} <^{\mathbf{o}} \mathbf{A}$. Then $\mathcal{C}(\mathbf{A}) \subseteq \mathcal{C}(\mathbf{B}) \subseteq \mathcal{C}(\mathbf{A})$, so, $\mathcal{C}(\mathbf{A}) = \mathcal{C}(\mathbf{B})$. It follows that $\mathbf{P_A} = \mathbf{AA}^\dagger = \mathbf{A}^\dagger\mathbf{A} = \mathbf{A}^\dagger\mathbf{AB}^\dagger\mathbf{B} = \mathbf{P_A P_B} = \mathbf{P_B} = \mathbf{B}^\dagger\mathbf{B} = \mathbf{BB}^\dagger$. Also, since $\mathbf{A}(\mathbf{A}^\dagger - \mathbf{B}^\dagger)\mathbf{A}$ and $\mathbf{B}(\mathbf{B}^\dagger - \mathbf{A}^\dagger)\mathbf{B}$ are *nnd*, so, $\mathbf{P_A}(\mathbf{A}^\dagger - \mathbf{B}^\dagger)\mathbf{P_A} = \mathbf{A}^\dagger - \mathbf{B}^\dagger$ and $\mathbf{P_B}(\mathbf{B}^\dagger - \mathbf{A}^\dagger)\mathbf{P_B} = \mathbf{B}^\dagger - \mathbf{A}^\dagger$ are *nnd*. Hence $\mathbf{A}^\dagger = \mathbf{B}^\dagger$ or $\mathbf{A} = \mathbf{B}$. □

Theorem 8.5.4. *Let* $\mathbf{A}, \mathbf{B} \in \mathfrak{H}_\mathbf{n}$. *Then* $\mathbf{A} <^{\mathbf{o}} \mathbf{B}$ *if and only if* $\mathcal{C}(\mathbf{A}) \subseteq \mathcal{C}(\mathbf{B})$ *and* $\nu(\mathbf{B}-\mathbf{A}) = \nu(\mathbf{B}) - \nu(\mathbf{A})$.

Proof. Clearly, the condition $\mathcal{C}(\mathbf{A}) \subseteq \mathcal{C}(\mathbf{B}) \Rightarrow \mathcal{C}(\mathbf{B}-\mathbf{A}) \cap \mathcal{N}(\mathbf{B}) = \mathbf{0}$. So, by Theorem 8.2.19 we have $\nu(\mathbf{A} - \mathbf{A}\mathbf{B}^{-}\mathbf{A}) = \nu(\mathbf{B}-\mathbf{A}) - \nu(\mathbf{B}) - \nu(\mathbf{A})$. Hence the result follows. □

Remark 8.5.5. If $\mathbf{A}$ and $\mathbf{B}$ are hermitian matrices of the same order, then $\mathbf{A} <^{-} \mathbf{B} \Rightarrow \mathbf{A} <^{\mathbf{o}} \mathbf{B}$.

However, the converse is not true as the following example shows:

Example 8.5.6. Let $\mathbf{A} = \begin{pmatrix} 1 & 1 \\ 1 & 1 \end{pmatrix}$ and $\mathbf{B} = \begin{pmatrix} 2 & 2 \\ 2 & -2 \end{pmatrix}$. It is easy to see that $\mathbf{A} <^{\mathbf{o}} \mathbf{B}$. However, $\rho(\mathbf{B}-\mathbf{A}) = \mathbf{2}$. So, $\rho(\mathbf{B}-\mathbf{A}) + \rho(\mathbf{A}) \neq \rho(\mathbf{B})$. Thus, $\mathbf{A} \not<^{-} \mathbf{B}$.

We now show that neither of the two partial orders $\mathbf{A} <^{\mathbf{o}} \mathbf{B}$ and $\mathbf{A} <^{\mathbf{L}} \mathbf{B}$ implies the other. For $\mathbf{A} <^{\mathbf{o}} \mathbf{B} \not\Rightarrow \mathbf{A} <^{\mathbf{L}} \mathbf{B}$ we can take matrices as in Example 8.5.6.
For $\mathbf{A} <^{\mathbf{L}} \mathbf{B} \not\Rightarrow \mathbf{A} <^{\mathbf{o}} \mathbf{B}$, we give the following example:

Example 8.5.7. Let $\mathbf{A} = \mathbf{diag}\,(0\ , \ \ -2)$ and $\mathbf{B} = \mathbf{I}$.

The following theorem establishes a nice connection between the order relations '$<^{L}$' and '$<^{o}$.'

Theorem 8.5.8. *Let* $\mathbf{A}, \mathbf{B} \in \mathfrak{H}_\mathbf{n}$ *such that* $\mathcal{C}(\mathbf{A}) = \mathcal{C}(\mathbf{B})$. *Then* $\mathbf{A} <^{\mathbf{o}} \mathbf{B}$ *if and only if* $\mathbf{B}^{\dagger} <^{\mathbf{L}} \mathbf{A}^{\dagger}$.

Proof. 'If' part is trivial.
'Only if'part
Let $\mathbf{A} <^{\mathbf{o}} \mathbf{B}$. Then $\mathbf{A} - \mathbf{A}\mathbf{B}^{\dagger}\mathbf{A}$ is *nnd*. Hence, $\mathbf{A}^{\dagger}(\mathbf{A} - \mathbf{A}\mathbf{B}^{\dagger}\mathbf{A})\mathbf{A}^{\dagger}$ is *nnd*. Notice that $\mathbf{A}^{\dagger}\mathbf{A}\mathbf{B}^{\dagger} = \mathbf{B}^{\dagger} = \mathbf{B}^{\dagger}\mathbf{A}\mathbf{A}^{\dagger}$, since $\mathcal{C}(\mathbf{A}) = \mathcal{C}(\mathbf{B})$. So,

$$\mathbf{A}^{\dagger}(\mathbf{A} - \mathbf{A}\mathbf{B}^{\dagger}\mathbf{A})\mathbf{A}^{\dagger} = \mathbf{P}_{\mathbf{A}}(\mathbf{A}^{\dagger} - \mathbf{B}^{\dagger})\mathbf{P}_{\mathbf{A}} \Rightarrow \mathbf{A}^{\dagger} - \mathbf{B}^{\dagger} \text{ is } nnd.$$ □

We now consider another generalization of the Löwner order for arbitrary matrices square or rectangular. Given a matrix $\mathbf{A}$ of order $m \times n$, this ordering makes use of the unique *nnd* square root of $\mathbf{A}\mathbf{A}^{\star}$, denoted by $(\mathbf{A}\mathbf{A}^{\star})^{\frac{1}{2}}$ for this purpose.

Definition 8.5.9. Let $\mathbf{A}$ and $\mathbf{B}$ be $m \times n$ matrices. The matrix $\mathbf{A}$ is said to be below the matrix $\mathbf{B}$ under the GL-ordering (Generalized Löwner order) if $(\mathbf{AA}^\star)^{\frac{1}{2}} <^{\mathbf{L}} (\mathbf{BB}^\star)^{\frac{1}{2}}$, $\mathcal{C}(\mathbf{A}^\star) \subseteq \mathcal{C}(\mathbf{B}^\star)$ and $\mathbf{AB}^\star = (\mathbf{AA}^\star)^{\frac{1}{2}}(\mathbf{BB}^\star)^{\frac{1}{2}}$. We write $\mathbf{A} <^{\mathbf{GL}} \mathbf{B}$ if $\mathbf{A}$ is below the matrix $\mathbf{B}$ under the GL-ordering.

One easily sees that if $\mathbf{A}$ and $\mathbf{B}$ are *nnd* matrices, $\mathbf{A} <^{\mathbf{GL}} \mathbf{B} \Rightarrow \mathbf{A} <^{\mathbf{L}} \mathbf{B}$. In fact, we shall soon see that the GL-ordering is actually the Löwner ordering on *nnd* matrices. But before we prove this, we prove several other results leading to this. We first show

Theorem 8.5.10. *GL-ordering is a partial order on* $\mathbb{C}^{m\times n}$.

Proof. Reflexivity is trivial.
To see that the relation is antisymmetric, let $\mathbf{A} <^{\mathbf{GL}} \mathbf{B}$ and $\mathbf{B} <^{\mathbf{GL}} \mathbf{A}$. Notice that $(\mathbf{A}-\mathbf{B})(\mathbf{A}-\mathbf{B})^\star = \mathbf{0}$, since $(\mathbf{AA}^\star)^{\frac{1}{2}} = (\mathbf{BB}^\star)^{\frac{1}{2}}$. So, $\mathbf{A} = \mathbf{B}$.
For transitivity, let $\mathbf{A} <^{\mathbf{GL}} \mathbf{B}$ and $\mathbf{B} <^{\mathbf{GL}} \mathbf{C}$. Clearly, $(\mathbf{AA}^\star)^{\frac{1}{2}} <^{\mathbf{L}} (\mathbf{CC}^\star)^{\frac{1}{2}}$ and $\mathcal{C}(\mathbf{A}^\star) \subseteq \mathcal{C}(\mathbf{C}^\star)$. Further, since $\mathcal{C}(\mathbf{A}^\star) \subseteq \mathcal{C}(\mathbf{B}^\star)$, we have

$$\begin{aligned}\mathbf{AC}^\star &= \mathbf{AB}^\dagger\mathbf{BC}^\star = \mathbf{AB}^\star\mathbf{B}^{\dagger\star}\mathbf{C}^\star\\ &= (\mathbf{AA}^\star)^{\frac{1}{2}}(\mathbf{BB}^\star)^{\frac{1}{2}}\mathbf{B}^{\dagger\star}\mathbf{C}^\star = (\mathbf{AA}^\star)^{\frac{1}{2}}(\mathbf{BB}^\star)^{\frac{1}{2}}(\mathbf{BB}^\star)^\dagger\mathbf{BC}^\star\\ &= (\mathbf{AA}^\star)^{\frac{1}{2}}(\mathbf{BB}^\star)^{\frac{1}{2}}(\mathbf{BB}^\star)^\dagger(\mathbf{BB}^\star)^{\frac{1}{2}}(\mathbf{CC}^\star)^{\frac{1}{2}} = (\mathbf{AA}^\star)^{\frac{1}{2}}(\mathbf{CC}^\star)^{\frac{1}{2}},\end{aligned}$$

since $\mathcal{C}(\mathbf{A}^\star) \subseteq \mathcal{C}(\mathbf{B}^\star)$, we can write $(\mathbf{AA}^\star)^{\frac{1}{2}} = \mathbf{k}(\mathbf{BB}^\star)^{\frac{1}{2}}$. Thus, $\mathbf{A} <^{\mathbf{GL}} \mathbf{C}$, showing that transitivity holds. □

Suppose $\mathbf{A} <^{\mathbf{GL}} \mathbf{B}$. Do $\mathbf{A}^\star$ and $\mathbf{B}^\star$ have similar relationship? The answer is contained in the following:

Theorem 8.5.11. *Let* $\mathbf{A}$ *and* $\mathbf{B}$ *be matrices of the same order. Then* $\mathbf{A} <^{\mathbf{GL}} \mathbf{B}$ *if and only if* $\mathbf{A}^\star <^{\mathbf{GL}} \mathbf{B}^\star$.

Proof. 'Only if' part
Let $\mathbf{A} = \mathbf{U}_1\mathbf{\Lambda}_1\mathbf{V}_1^\star$ and $\mathbf{B} = \mathbf{U}_2\mathbf{\Lambda}_2\mathbf{V}_2^\star$ be singular value decompositions of $\mathbf{A}$ and $\mathbf{B}$ respectively, where $\mathbf{\Lambda}_1$ and $\mathbf{\Lambda}_2$ are positive definite diagonal matrices and $\mathbf{U}_1, \mathbf{U}_2, \mathbf{V}_1, \mathbf{V}_2$ are matrices such that $\mathbf{U}_1^\star\mathbf{U}_1 = \mathbf{I}, \mathbf{U}_2^\star\mathbf{U}_2 = \mathbf{I}, \mathbf{V}_1^\star\mathbf{V}_1 = \mathbf{I}$ and $\mathbf{V}_2^\star\mathbf{V}_2 = \mathbf{I}$. Then $\mathbf{A} <^{\mathbf{GL}} \mathbf{B}$ if and only if

(i) $\mathbf{U}_1\mathbf{\Lambda}_1\mathbf{V}_1^\star <^{\mathbf{L}} \mathbf{U}_2\mathbf{\Lambda}_2\mathbf{V}_2{}^\star$
(ii) $\mathbf{V}_1 = \mathbf{V}_2\mathbf{V}_2^\star\mathbf{V}_1$ and
(iii) $\mathbf{U}_1^\star\mathbf{U}_2 = \mathbf{V}_1^\star\mathbf{V}_2$.

(i) and (iii) $\Rightarrow \mathbf{V}_2^\star\mathbf{V}_1\mathbf{\Lambda}_1\mathbf{V}_1^\star\mathbf{V}_2 <^{\mathbf{L}} \mathbf{\Lambda}_2 \Rightarrow \mathbf{V}_1\mathbf{\Lambda}_1\mathbf{V}_1^\star <^{\mathbf{L}} \mathbf{V}_2\mathbf{\Lambda}_1\mathbf{V}_2^\star$, since $\mathbf{V}_2^\star\mathbf{V}_2\mathbf{V}_1 = \mathbf{V}_1$. It now follows that $\mathbf{A}^\star <^{\mathbf{GL}} \mathbf{B}^\star$.

'If' part follows from only if part by replacing $\mathbf{A}$ with $\mathbf{A}^\star$ and $\mathbf{B}$ with $\mathbf{A}^\star$. □

Corollary 8.5.12. *Let* $\mathbf{A}$ *and* $\mathbf{B}$ *be matrices of the same order.*
If $\mathbf{A} <^{\mathbf{GL}} \mathbf{B}$, *then* $\mathbf{A} <^{\mathbf{s}} \mathbf{B}$.

Corollary 8.5.13. *Let* $\mathbf{A}$ *and* $\mathbf{B}$ *be matrices of the same order such that* $\mathbf{A} <^{\mathbf{GL}} \mathbf{B}$. *Then* $\mathbf{B}^{\dagger} <^{\mathbf{L}} \mathbf{A}^{\dagger}$ *if and only if* $\rho(\mathbf{A}) = \rho(\mathbf{B})$.

Proof follows from Definition 8.5.9, Theorem 8.5.11 and Corollary 8.4.19.

We now give a characterization of the GL-ordering similar to the one for Löwner ordering in Corollary 8.2.12 (i) $\Leftrightarrow$ (ii).

Theorem 8.5.14. *Let* $\mathbf{A}$ *and* $\mathbf{B}$ *be matrices of the same order. Then* $\mathbf{A} <^{\mathbf{GL}} \mathbf{B}$ *if and only if* $\lambda_{max}(\mathbf{B}^{\dagger}\mathbf{A}) \leq 1, \mathbf{A} <^{\mathbf{s}} \mathbf{B}$ *and* $\mathbf{AB}^{\star} = (\mathbf{AA}^{\star})^{\frac{1}{2}}(\mathbf{BB}^{\star})^{\frac{1}{2}}$.

Proof. In view of Corollary 8.2.12, $(\mathbf{AA}^{\star})^{\frac{1}{2}} <^{\mathbf{L}} (\mathbf{BB}^{\star})^{\frac{1}{2}}$ if and only if $\lambda_{\mathbf{max}}((\mathbf{BB}^{\star})^{\frac{1}{2}})^{\dagger}(\mathbf{AA}^{\star})^{\frac{1}{2}} \leq 1$ and $\mathcal{C}(\mathbf{A}) \subseteq \mathcal{C}(\mathbf{B})$. However, we can write $((\mathbf{BB}^{\star})^{\frac{1}{2}})^{\dagger} = (\mathbf{BB}^{\star})^{\dagger}((\mathbf{BB}^{\star})^{\frac{1}{2}})$, so,

$$\begin{aligned}((\mathbf{BB}^{\star})^{\frac{1}{2}})^{\dagger}(\mathbf{AA}^{\star})^{\frac{1}{2}} &= ((\mathbf{BB}^{\star})^{\dagger}(\mathbf{BB}^{\star})^{\frac{1}{2}})(\mathbf{AA}^{\star})^{\frac{1}{2}}\\ &= (\mathbf{BB}^{\star})^{\dagger}((\mathbf{BB}^{\star})^{\frac{1}{2}}(\mathbf{AA}^{\star})^{\frac{1}{2}}) = ((\mathbf{BB}^{\star})^{\dagger})\mathbf{BA}^{\star}\\ &= \mathbf{B}^{\dagger^{\star}}\mathbf{A}^{\star}.\end{aligned}$$

Hence,

$$\begin{aligned}\lambda_{max}((\mathbf{BB}^{\star})^{\frac{1}{2}})^{\dagger}(\mathbf{AA}^{\star})^{\frac{1}{2}} &= \lambda_{max}(\mathbf{B}^{\dagger^{\star}}\mathbf{A}^{\star})\\ &= \lambda_{max}(\mathbf{AB}^{\dagger})\\ &= \lambda_{max}(\mathbf{B}^{\dagger}\mathbf{A}).\end{aligned}$$

Rest of the proof follows by Definition 8.5.9. □

Theorem 8.5.15. *Let* $\mathbf{A}$ *and* $\mathbf{B}$ *be nnd matrices. Then* $\mathbf{A} <^{\mathbf{GL}} \mathbf{B}$ *if and only if* $\mathbf{A} <^{\mathbf{L}} \mathbf{B}$.

Proof is trivial.

Remark 8.5.16. Let $\mathbf{A}$ and $\mathbf{B}$ be matrices of the same order such that $\mathbf{A} <^{\star} \mathbf{B}$. Then $\mathbf{A} <^{\mathbf{GL}} \mathbf{B}$.

Consider the following version of the polar decomposition of a rectangular matrix as given in [Ben-Israel and Greville (2001)].

Theorem 8.5.17. *(Polar Decomposition) Let* $\mathbf{A}$ *be a matrix of order* $m \times n$. *Then there exists an nnd matrix* $\mathbf{H}$ *and a partial isometry* $\mathbf{W}$

(i.e. $\mathbf{W}^\dagger = \mathbf{W}^\star$*) such that* $\mathbf{A} = \mathbf{HW}$*. Further the matrices* $\mathbf{H}$ *and* $\mathbf{W}$ *are uniquely determined by* $\mathcal{C}(\mathbf{H}) = \mathcal{C}(\mathbf{W})$ *in which case* $\mathbf{H^2} = \mathbf{AA}^\star$ *and* $\mathbf{W} = \mathbf{H}^\dagger\mathbf{A}$.

For a proof see page 220 of [Ben-Israel and Greville (2001)].

We now obtain a characterization of GL-ordering in terms of polar decomposition.

Theorem 8.5.18. *Let* $\mathbf{A}$ *and* $\mathbf{B}$ *be matrices of order* $m \times n$*. Let* $\mathbf{A} = \mathbf{H_1W_1}$ *and* $\mathbf{B} = \mathbf{H_2W_2}$ *be the polar decomposition of* $\mathbf{A}$ *and* $\mathbf{B}$ *respectively with* $\mathcal{C}(\mathbf{H_i}) = \mathcal{C}(\mathbf{W_i})$*, for* $i = 1, 2$*. Then* $\mathbf{A} <^{\mathbf{GL}} \mathbf{B}$ *if and only if* $\mathbf{H_1} <^{\mathbf{L}} \mathbf{H_2}$ *and* $\mathbf{W_1} <^\star \mathbf{W_2}$.

Proof. Notice that $\mathbf{P_A} = \mathbf{P_{H_1}} = \mathbf{P_{W_1}}$ and $\mathbf{P_B} = \mathbf{P_{H_2}} = \mathbf{P_{W_2}}$. Thus, $\mathbf{P_A} = \mathbf{H_1H_1^\dagger} = \mathbf{H_1^\dagger H_1} = \mathbf{W_1W_1^\star}$ and $\mathbf{P_B} = \mathbf{H_2H_2^\dagger} = \mathbf{H_2^\dagger H_2} = \mathbf{W_2W_2^\star}$. Further, $\mathbf{A} <^{\mathbf{GL}} \mathbf{B}$ if and only if $\mathbf{H_1} <^{\mathbf{L}} \mathbf{H_2}, \mathcal{C}(\mathbf{W_1})^\star \subseteq \mathcal{C}(\mathbf{W_2})^\star$ and $\mathbf{H_1W_1W_2{}^\star H_2} = \mathbf{H_1H_2}$. Now,

$$\begin{aligned}\mathbf{H_1W_1W_2^\star H_2} &= \mathbf{H_1H_2}\\ &\Leftrightarrow \mathbf{H_1^\dagger H_1W_1W_2^\star H_2H_2^\dagger} = \mathbf{H_1^\dagger H_1H_2H_2^\dagger}\\ &\Leftrightarrow \mathbf{W_1W_1^\star W_1W_2^\star W_2W_2^\star} = \mathbf{W_1W_1^\star W_2W_2^\star}\\ &\Leftrightarrow \mathbf{W_1W_2^\star} = \mathbf{W_1W_1^\star W_2W_2^\star}.\end{aligned}$$

Thus, $\mathbf{A} <^{\mathbf{GL}} \mathbf{B}$
$\Leftrightarrow \mathbf{H_1} <^{\mathbf{L}} \mathbf{H_2}$, $\mathcal{C}(\mathbf{W_1^\star}) \subseteq \mathcal{C}(\mathbf{W_2^\star})$ and $\Leftrightarrow \mathbf{W_1W_2^\star} = \mathbf{W_1W_1^\star W_2W_2^\star}$
$\Leftrightarrow \mathbf{H_1} <^{\mathbf{L}} \mathbf{H_2}$, $\mathcal{C}(\mathbf{W_1})^\star \subseteq \mathcal{C}(\mathbf{W_2})^\star$ and $\mathbf{W_1W_2^\star} = \mathbf{W_1W_1^\star}$
$\Leftrightarrow \mathbf{H_1} <^{\mathbf{L}} \mathbf{H_2}, \mathbf{W_2^\star W_2W_1^\star W_1} = \mathbf{W_1{}^\star W_1}$ and $\mathbf{W_1W_2^\star} = \mathbf{W_1W_1^\star}$
$\Leftrightarrow \mathbf{H_1} <^{\mathbf{L}} \mathbf{H_2}, \mathbf{W_2^\star W_1} = \mathbf{W_1^\star W_1}$ and $\mathbf{W_1W_2^\star} = \mathbf{W_1W_1^\star}$
$\Leftrightarrow \mathbf{H_1} <^{\mathbf{L}} \mathbf{H_2}$ and $\mathbf{W_1} <^\star \mathbf{W_2}$. □

Let us now compare the two order relations '$<^o$' and '$<^{GL}$' on the class of all hermitian matrices of the same order. Using Example 8.5.6, one can show that $\mathbf{A} <^{\mathbf{o}} \mathbf{B}$ does not imply $\mathbf{A} <^{\mathbf{GL}} \mathbf{B}$. Next take $\mathbf{C} = \mathbf{diag}\,(0\ ,\ -1)$ and $\mathbf{D} = \mathbf{diag}\,(-1\ ,\ -2)$. it is easy to see that $\mathbf{C} <^{\mathbf{GL}} \mathbf{D}$. However, $\mathbf{C} \not<^{\mathbf{o}} \mathbf{D}$. Thus the two orderings '$<^o$' and '$<^{GL}$' coincide for the class of *nnd* matrices, but are in general, not comparable in the class of hermitian matrices. However, both orderings imply the space pre-order.

8.6 Exercises

(1) Let $\mathbf{A}$ and $\mathbf{B}$ be *nnd* matrices of the same order such that $\mathbf{B} <^{\mathbf{L}} \mathbf{A}$ and $\tilde{\mathbf{A}}$ and $\tilde{\mathbf{B}}$ be hermitian reflexive g-inverses of $\mathbf{A}$ and $\mathbf{B}$ respectively. Show that $\tilde{\mathbf{A}} <^{\mathbf{L}} \tilde{\mathbf{B}}$ if and only if $\mathcal{C}(\tilde{\mathbf{A}}) = \mathcal{C}(\tilde{\mathbf{B}})$.

(2) Let $\mathbf{A}$ and $\mathbf{B}$ be *nnd* matrices of the same order and $\tilde{\mathbf{A}}$ and $\tilde{\mathbf{B}}$ be hermitian reflexive g-inverses of $\mathbf{A}$ and $\mathbf{B}$ respectively. Then the following are equivalent:

   (i) $\mathbf{B} <^{\mathbf{L}} \mathbf{A}$
   (ii) $\tilde{\mathbf{A}} <^{\mathbf{L}} \tilde{\mathbf{B}}$
   (iii) $\mathcal{C}(\mathbf{A}) = \mathcal{C}(\mathbf{B})$, $\mathcal{C}(\tilde{\mathbf{A}}) = \mathcal{C}(\tilde{\mathbf{B}})$.

(3) Let $\mathbf{A}$ and $\mathbf{B} \in \mathfrak{H}_n$, $\nu(\mathbf{A}) = \nu(\mathbf{B}) \neq 0$ and $\mathbf{A} <^{\mathbf{L}} \mathbf{B}$. Show that $\lambda_{max}(\mathbf{B}^-\mathbf{A}) \geq \mathbf{1}$, where $\mathbf{B}^-$ is any g-inverse of $\mathbf{B}$.

(4) For a matrix $\mathbf{A}$, let $\sigma(\mathbf{A})$ denote the set of all non-zero singular values of $\mathbf{A}$. Define $\mathbf{A} <^{\sigma} \mathbf{B}$ if $\sigma(\mathbf{A}) \subseteq \sigma(\mathbf{B})$. Show the following hold:

   (i) '$<^{\sigma}$' is a pre-order on $\mathbb{C}^{m\times n}$.
   (ii) '$\mathbf{B}^{\dagger} = \mathbf{B}^{\star}$ and $\mathbf{A} <^{\sigma} \mathbf{B}$' implies $\mathbf{A}^{\dagger} = \mathbf{A}^{\star}$.
   (iii) '$\mathbf{B}\mathbf{B}^{\star} <^{\mathbf{L}} \mathbf{I}$ and $\mathbf{A} <^{\sigma} \mathbf{B}$' implies $\mathbf{A}\mathbf{A}^{\star} <^{\mathbf{L}} \mathbf{I}$.

(5) Let $\mathbf{A}$ and $\mathbf{B} \in \mathfrak{H}_n$ such that $\nu(\mathbf{A}) = \nu(\mathbf{B})$. Show that (i) $\mathbf{A} <^{\mathbf{L}} \mathbf{B}$ and $\mathbf{A}^{\dagger} <^{\mathbf{L}} \mathbf{B}^{\dagger} \Leftrightarrow \mathbf{A} <^{-} \mathbf{B}$ and $\mathbf{A}^{\dagger} <^{-} \mathbf{B}^{\dagger}$ and (ii) $\mathbf{A} <^{\star} \mathbf{B} \Rightarrow \mathbf{A} <^{\mathbf{L}} \mathbf{B}$ and $\mathbf{A}^{\dagger} <^{\mathbf{L}} \mathbf{B}^{\dagger}$.
Show also that $\mathbf{A} <^{\mathbf{L}} \mathbf{B}$ and $\mathbf{A}^{\dagger} <^{\mathbf{L}} \mathbf{B}^{\dagger} \Rightarrow \mathbf{A} <^{\star} \mathbf{B}$ for *nnd* matrices $\mathbf{A}$ and $\mathbf{B}$.

Definition 8.6.1. Let $\mathbf{A} = [a_{ij}]$ and $\mathbf{B} = [b_{kt}]$ be matrices of orders $m \times n$ and $r \times s$ respectively. Then the Kronecker product $\mathbf{A} \otimes \mathbf{B}$ of $\mathbf{A}$ and $\mathbf{B}$ is defined to be the $mr \times ns$ matrix

$$\begin{pmatrix} a_{11}\mathbf{B} & a_{12}\mathbf{B} & \dots & a_{1n}\mathbf{B} \\ a_{21}\mathbf{B} & a_{22}\mathbf{B} & \dots & a_{2n}\mathbf{B} \\ \vdots & & \dots & \vdots \\ a_{m1}\mathbf{B} & a_{m2}\mathbf{B} & \dots & a_{mn}\mathbf{B} \end{pmatrix}.$$

If $\mathbf{A}$ and $\mathbf{B}$ are of same order, then the Schur product $\mathbf{A} \odot \mathbf{B}$ of $\mathbf{A}$ and $\mathbf{B}$ is the matrix $\mathbf{A} \odot \mathbf{B} = [a_{ij}b_{ij}]$.
In particular, if $\mathbf{u} = (u_1, u_2, \dots, u_n)^t$ and $\mathbf{v} = (v_1, v_2, \dots, v_n)^t$, then $\mathbf{u} \otimes \mathbf{v} = (u_1v_1, u_1v_2, \dots u_1v_n, \dots, u_nv_1, \dots, u_nv_n)^t$ and $\mathbf{u} \odot \mathbf{v} = (u_1v_1, u_2v_2, \dots, u_nv_n)^t$.

(6) Let $\mathbf{A}$ and $\mathbf{B}$ be *nnd* matrices of the same order $n \times n$ and $\mathbf{C}$ and $\mathbf{D}$ be *nnd* matrices of the same order $m \times m$. Show the following:

(i) $\mathbf{A} \otimes \mathbf{C}$ is *nnd*.

(ii) $\mathbf{A} \otimes \mathbf{C} <^{\mathbf{L}} \mathbf{B} \otimes \mathbf{D} \Leftrightarrow$(a) $\mathcal{C}(\mathbf{A}) \subseteq \mathcal{C}(\mathbf{B}), \mathcal{C}(\mathbf{C}) \subseteq \mathcal{C}(\mathbf{D})$ and (b)$\lambda_1(\mathbf{B}^-\mathbf{A})\lambda_1(\mathbf{D}^-\mathbf{C}) \leq 1$, where $\mathbf{B}^-$ is any g-inverse of $\mathbf{B}$ and $\mathbf{D}^-$ is any g-inverse of $\mathbf{D}$ respectively.

(7) Let $\mathbf{A}$, $\mathbf{B}$, $\mathbf{C}$ and $\mathbf{D}$ be *nnd* matrices of the same order $n \times n$. Show the following:

(i) $\mathbf{A} \otimes \mathbf{B}$ is *nnd*.

(ii) If $\mathbf{A} <^{\mathbf{L}} \mathbf{B}$ and $\mathbf{C} <^{\mathbf{L}} \mathbf{D}$, then $\mathbf{A} \otimes \mathbf{C} <^{\mathbf{L}} \mathbf{B} \otimes \mathbf{D}$.

Chapter 9

Parallel Sums

9.1 Introduction

The present chapter is a prelude to the study of one of many applications of g-inverses and matrix partial orders. Here we study parallel sums which originally arose in the study of network synthesis. The concept of parallel sum is analogous to the concept of connecting resistors either in series or in parallel, a basic concept in elementary network theory. If two resistors having resistances $\mathbf{R_1}$ and $\mathbf{R_2}$ are connected in series, their joint resistance is $\mathbf{R_1} + \mathbf{R_2}$ and if they are connected in parallel then their joint resistance $\mathbf{R}$ is

$$\mathbf{R} = \mathbf{R_1^{-1}} + \mathbf{R_2^{-1}} = \frac{\mathbf{R_1 R_2}}{\mathbf{R_1} + \mathbf{R_2}} = \mathbf{R_1}(\mathbf{R_1} + \mathbf{R_2})^{-1}\mathbf{R_2}$$

and is called parallel sum of $\mathbf{R_1}$ and $\mathbf{R_2}$. Notice that while giving the total resistance of these resistors in parallel, we have tacitly assumed the resistances to be positive numbers. However, if $\mathbf{R_1}$ and/or $\mathbf{R_2}$ are zero, we say the joint resistance is zero.

As we shall see in Chapter 10, the impedance matrix of a reciprocal resistive n-port network is an *nnd* matrix and that *nnd* matrices can be considered a generalization of non-negative real numbers. Further, if two such n-port networks $\mathbf{N_1}$ and $\mathbf{N_2}$ with impedance matrices $\mathbf{Z_1}$ and $\mathbf{Z_2}$ respectively are connected in parallel then the impedance matrix of their parallel connection is $\mathbf{Z_1}(\mathbf{Z_1}+\mathbf{Z_2})^{\dagger}\mathbf{Z_2}$. Following the discussion in the opening para, one definition of the parallel sum of two *nnd* matrices $\mathbf{A}$ and $\mathbf{B}$ can be $\mathbf{A}(\mathbf{A}+\mathbf{B})^{-1}\mathbf{B}$. This will certainly have a meaning if $\mathbf{A}+\mathbf{B}$ is non-singular. If both $\mathbf{A}$ and $\mathbf{B}$ are non-singular, this parallel sum can be written as $(\mathbf{A}^{-1}+\mathbf{B}^{-1})^{-1}$. If the sum $\mathbf{A}+\mathbf{B}$ is singular, then suitable modifications are required to give a meaningful definition of the parallel sum. Since $(\mathbf{A}+\mathbf{B})^\dagger$, the Moore-Penrose inverse of $\mathbf{A}+\mathbf{B}$ always exists and is unique, a possible definition of the parallel sum of $\mathbf{A}$ and $\mathbf{B}$ is the matrix $\mathbf{A}(\mathbf{A}+\mathbf{B})^\dagger\mathbf{B}$. Following what we just said for non-singular matrices, one may then be tempted to express this parallel sum as $(\mathbf{A}^\dagger+\mathbf{B}^\dagger)^\dagger$. This, however, is far from true as can seen by taking $\mathbf{A}=\mathbf{I_2}$, the 2×2 identity matrix and $\mathbf{B}=\begin{pmatrix}0 & 0\\ 0 & 1\end{pmatrix}$. In this chapter we begin by studying the parallel sum and its properties in a more general context, where the latter definition is the guiding source. Parallel sums of *nnd* matrices are discussed simultaneously.

In Section 9.2, we define the notion of parallel sum for arbitrary rectangular matrices using generalized inverses and study in detail the various properties of these sums. The corresponding theorems for *nnd* matrices, whenever possible have been included too. We also obtain a solution for matrix equation $\mathcal{P}(\mathbf{A},\mathbf{X})=\mathbf{C}$, where $\mathcal{P}(\mathbf{A},\mathbf{X})$ denotes that parallel sum of matrices $\mathbf{A}$ and $\mathbf{X}$. In Section 9.3, we study the inter-relationship of parallel sums and various partial orders studied in earlier chapters. We also prove that the set of *nnd* matrices form a partially ordered abelian group with respect to parallel addition as its binary operation. Section 9.4 contains results about continuity of parallel sums and their index. Drazin inverse of parallel sums is also discussed towards the end of this section. As has been our tradition, all matrices are over arbitrary field unless otherwise stated.

9.2 Definition and properties

In this section, we first define the parallel sum of two arbitrary matrices (possibly rectangular) over a general field. Traditionally we should begin by defining the parallel sum of two *nnd* matrices as the concept of parallel sums first came into being for these matrices only. However, we choose to start by defining the parallel sum of two arbitrary rectangular matrices over a general field and as we go along studying the various properties, we also develop similar results for parallel sums of two *nnd* matrices.

Definition 9.2.1. Let $\mathbf{A}$ and $\mathbf{B}$ be any two $m\times n$ matrices. We say $\mathbf{A}$ and $\mathbf{B}$ are parallel summable if $\mathbf{A}(\mathbf{A}+\mathbf{B})^-\mathbf{B}$ is invariant under the choice of g-inverses $(\mathbf{A}+\mathbf{B})^-$ of $\mathbf{A}+\mathbf{B}$.

When this is so, the common value of $\mathbf{A}(\mathbf{A}+\mathbf{B})^-\mathbf{B}$ is called the parallel sum of $\mathbf{A}$ and $\mathbf{B}$. We denote the parallel sum by $\mathcal{P}(\mathbf{A},\mathbf{B})$ and write $\mathcal{P}(\mathbf{A},\mathbf{B})=\mathbf{A}(\mathbf{A}+\mathbf{B})^-\mathbf{B}$.

Before going any further, we make some useful observations in the following remarks.

Remark 9.2.2. If either $\mathbf{A}$ or $\mathbf{B}$ is null, then $\mathbf{A}$ and $\mathbf{B}$ are trivially parallel summable and their parallel sum is the null matrix. Thus, to have anything interesting about parallel sums one must take only non-null matrices.

Remark 9.2.3. If the non-null matrices $\mathbf{A}$ and $\mathbf{B}$ are such that $\mathbf{A}+\mathbf{B}$ is null, their parallel sum is not defined, since $\mathbf{A}(\mathbf{A}+\mathbf{B})^{-}\mathbf{B}$ remains no longer invariant under the choice of g-inverses of $\mathbf{A}+\mathbf{B}$. Thus, over any field, the matrices $\mathbf{A}(\neq \mathbf{0})$ and $-\mathbf{A}$ are not parallel summable.

We now give a characterization and its various formulations for parallel summability of two matrices.

Theorem 9.2.4. *Let* $\mathbf{A}$ *and* $\mathbf{B}$ *be two matrices of the same order. Then the following are equivalent:*

(i) $\mathbf{A}$ *and* $\mathbf{B}$ *are parallel summable*

(ii) $\mathcal{C}(\mathbf{A}) \subseteq \mathcal{C}(\mathbf{A}+\mathbf{B})$ *and* $\mathcal{C}(\mathbf{A^t}) \subseteq \mathcal{C}(\mathbf{A}+\mathbf{B})^{\mathbf{t}}$

(iii) $\mathcal{C}(\mathbf{B}) \subseteq \mathcal{C}(\mathbf{A}+\mathbf{B})$ *and* $\mathcal{C}(\mathbf{B^t}) \subseteq \mathcal{C}(\mathbf{A}+\mathbf{B})^{\mathbf{t}}$

(iv) $\mathcal{C}(\mathbf{A})+\mathcal{C}(\mathbf{B}) = \mathcal{C}(\mathbf{A}+\mathbf{B})$ *and* $\mathcal{C}(\mathbf{A^t})+\mathcal{C}(\mathbf{B^t}) = \mathcal{C}(\mathbf{A}+\mathbf{B})^{\mathbf{t}}$

(v) *Let* $\rho(\mathbf{A}+\mathbf{B}) = r$. *There exist non-singular matrices* $\mathbf{P}$ *and* $\mathbf{Q}$ *of orders* $m\times m$ *and* $n\times n$ *respectively and an* $r\times r$ *matrix* $\mathbf{T}$ *such that* $\mathbf{A} = \mathbf{P}\ \mathbf{diag}\left(\mathbf{T}\ ,\ \mathbf{0}\right)\mathbf{Q}$ *and* $\mathbf{B} = \mathbf{P}\ \mathbf{diag}\left(\mathbf{I}-\mathbf{T}\ ,\ \mathbf{0}\right)\mathbf{Q}$

(vi) *Let* $\rho(\mathbf{A}+\mathbf{B}) = r$. *There exist non-singular matrices* $\mathbf{R},\mathbf{S}$ *of orders* $m\times m$ *and* $n\times n$ *respectively and an* $r\times r$ *matrix* $\mathbf{W}$ *such that* $\mathbf{A} = \mathbf{R}\ \mathbf{diag}\left(\mathbf{I}-\mathbf{W}\ ,\ \mathbf{0}\right)\mathbf{S}$ *and* $\mathbf{B} = \mathbf{R}\ \mathbf{diag}\left(\mathbf{W}\ ,\ \mathbf{0}\right)\mathbf{S}$

(vii) *Let* $\rho(\mathbf{A}) = a$ *and* $\rho(\mathbf{A}+\mathbf{B}) = r$. *There exist non-singular matrices* $\mathbf{P}',\mathbf{Q}'$ *of orders* $m\times m$ *and* $n\times n$ *respectively and an* $r\times r$ *non-singular matrix* $\mathbf{K} = \begin{pmatrix}\mathbf{K_{11}} & \mathbf{K_{12}}\\ \mathbf{K_{21}} & \mathbf{K_{22}}\end{pmatrix}$ *such that* $\mathbf{A} = \mathbf{P}'\mathbf{diag}\left(\mathbf{I_a}\ ,\ \mathbf{0_{r-a}}\ ,\ \mathbf{0}\right)\mathbf{Q}'$ *and* $\mathbf{B} = \mathbf{P}'\begin{pmatrix}\mathbf{K_{11}}-\mathbf{I_a} & \mathbf{K_{12}} & \mathbf{0}\\ \mathbf{K_{21}} & \mathbf{K_{22}} & \mathbf{0}\\ \mathbf{0} & \mathbf{0} & \mathbf{0}\end{pmatrix}\mathbf{Q}'$, *where* $\mathbf{K_{11}}$ *is* $a\times a$ *matrix and*

(viii) *Let* $\rho(\mathbf{B}) = b$ *and* $\rho(\mathbf{A}+\mathbf{B}) = r$. *There exist non-singular matrices* $\mathbf{R}',\mathbf{S}'$ *of orders* $m\times m$ *and* $n\times n$ *respectively and an* $r\times r$ *non-singular matrix* $\mathbf{W} = \begin{pmatrix}\mathbf{W_{11}} & \mathbf{W_{12}}\\ \mathbf{W_{21}} & \mathbf{W_{22}}\end{pmatrix}$ *such that*

$$\mathbf{A} = \mathbf{R}' \begin{pmatrix} \mathbf{W_{11}} - \mathbf{I_b} & \mathbf{W_{12}} & \mathbf{0} \\ \mathbf{W_{21}} & \mathbf{W_{22}} & \mathbf{0} \\ \mathbf{0} & \mathbf{0} & \mathbf{0} \end{pmatrix} \mathbf{S}' \text{ and } \mathbf{B} = \mathbf{R}'\mathrm{diag}\left(\mathbf{I_b}\ ,\ \mathbf{0_{r-b}}\ ,\ \mathbf{0}\right)\mathbf{S}',$$

where $\mathbf{W_{11}}$ *is a* $b \times b$ *matrix.*

Proof. (i)⇒(ii)
By Theorem 2.3.11, $\mathcal{C}(\mathbf{B}) \subseteq \mathcal{C}(\mathbf{A}+\mathbf{B})$ and $\mathcal{C}(\mathbf{A^t}) \subseteq \mathcal{C}(\mathbf{A}+\mathbf{B})^{\mathbf{t}}$. However, $\mathcal{C}(\mathbf{A}) = \mathcal{C}(\mathbf{A}+\mathbf{B}-\mathbf{B}) \subseteq \mathcal{C}(\mathbf{A}+\mathbf{B}) + \mathcal{C}(\mathbf{B}) \subseteq \mathcal{C}(\mathbf{A}+\mathbf{B}) + \mathcal{C}(\mathbf{A}+\mathbf{B}) = \mathcal{C}(\mathbf{A}+\mathbf{B})$, as $\mathcal{C}(\mathbf{A}+\mathbf{B})$ is a subspace. So, (ii) holds.
(ii) ⇒ (i)
Once again by Theorem 2.3.11, $\mathbf{A}(\mathbf{A}+\mathbf{B})^{-}\mathbf{A}$ is invariant under choice of g-inverses $(\mathbf{A}+\mathbf{B})^{-}$. But $\mathbf{A}(\mathbf{A}+\mathbf{B})^{-}\mathbf{A} = \mathbf{A}(\mathbf{A}+\mathbf{B})^{-}(\mathbf{A}+\mathbf{B}-\mathbf{B}) = \mathbf{A}(\mathbf{A}+\mathbf{B})^{-}(\mathbf{A}+\mathbf{B}) - \mathbf{A}(\mathbf{A}+\mathbf{B})^{-}\mathbf{B} = \mathbf{A} - \mathbf{A}(\mathbf{A}+\mathbf{B})^{-}\mathbf{B}$, therefore, $\mathbf{A}(\mathbf{A}+\mathbf{B})^{-}\mathbf{B}$ is invariant under choices of $(\mathbf{A}+\mathbf{B})^{-}$ proving $\mathbf{A}$ and $\mathbf{B}$ are parallel summable.
Proof of (i) ⇔ (iii) is similar. (i) ⇔ (iv) follows from the proof of (i) ⇔ (ii) and (i) ⇔ (iii). Proofs of (ii) ⇔ (v) ⇔ (vii) and (iii) ⇔ (vi) ⇔ (viii) follow by Remark 3.2.7. □

Corollary 9.2.5. *Let* $\mathbf{A}$ *and* $\mathbf{B}$ *be two matrices of the same order. Then*

(i) $\mathbf{A}$ *and* $\mathbf{B}$ *are parallel summable if and only if* $\mathbf{B}$ *and* $\mathbf{A}$ *are parallel summable. Further,* $\mathcal{P}(\mathbf{A},\mathbf{B}) = \mathcal{P}(\mathbf{B},\mathbf{A})$.
(ii) $\mathbf{A}$ *and* $\mathbf{B}$ *are parallel summable if and only if* $\mathbf{A^t}$ *and* $\mathbf{B^t}$ *are parallel summable. Further,* $\mathcal{P}(\mathbf{A^t},\mathbf{B^t}) = (\mathcal{P}(\mathbf{B},\mathbf{A}))^{\mathbf{t}}$.
(iii) *If* $\mathbf{A},\mathbf{B} \in \mathbb{C}^{\mathbf{n}\times\mathbf{n}}$ *are hermitian and* $\mathbf{A}$ *and* $\mathbf{B}$ *are parallel summable, then* $\mathcal{P}(\mathbf{A},\mathbf{B})$ *is hermitian.*

Proof. (i) First half of the statement follows from Theorem 9.2.4. For the later half, note that

$$\begin{aligned} \mathbf{A}(\mathbf{A}+\mathbf{B})^{-}\mathbf{B} &= \mathbf{A}(\mathbf{A}+\mathbf{B})^{-}(\mathbf{A}+\mathbf{B}-\mathbf{A}) \\ &= \mathbf{A}(\mathbf{A}+\mathbf{B})^{-}(\mathbf{A}+\mathbf{B}) - \mathbf{A}(\mathbf{A}+\mathbf{B})^{-}\mathbf{A} \\ &= \mathbf{A} - \mathbf{A}(\mathbf{A}+\mathbf{B})^{-}\mathbf{A} \\ &= (\mathbf{A}+\mathbf{B})(\mathbf{A}+\mathbf{B})^{-}\mathbf{A} - \mathbf{A}(\mathbf{A}+\mathbf{B})^{-}\mathbf{A} \\ &= \mathbf{B}(\mathbf{A}+\mathbf{B})^{-}\mathbf{A}. \end{aligned}$$

Hence, $\mathcal{P}(\mathbf{A},\mathbf{B}) = \mathcal{P}(\mathbf{B},\mathbf{A})$.
(ii) First part of the statement follows from Theorem 9.2.4. For the second part, notice that a choice of g-inverse of $\mathbf{A^t} + \mathbf{B^t}$ is $((\mathbf{A}+\mathbf{B})^{-})^{\mathbf{t}}$.
(iii) follows in view of (ii). □

Remark 9.2.6. Even when at least one of $\mathbf{A}$ and $\mathbf{B}$ is a full rank matrix, the matrices $\mathbf{A}$ and $\mathbf{B}$ may not be parallel summable. A simple example is to take any non-singular matrix $\mathbf{A}$ and $\mathbf{B} = -\mathbf{A}$.

We next show that if the matrices $\mathbf{A}$ and $\mathbf{B}$ are *nnd*, then they are always parallel summable and that their parallel sum is below each of $\mathbf{A}$ and $\mathbf{B}$ under the Löwner order.

Theorem 9.2.7. *Let* $\mathbf{A}$, $\mathbf{B} \in \mathbb{C}^{\mathbf{n}\times\mathbf{n}}$ *be nnd matrices. Then the following hold:*

(i) $\mathcal{C}(\mathbf{A}) + \mathcal{C}(\mathbf{B}) = \mathcal{C}(\mathbf{A}+\mathbf{B})$
(ii) $\mathbf{A}$ *and* $\mathbf{B}$ *are parallel summable* *and*
(iii) $\mathcal{P}(\mathbf{A},\mathbf{B}) <^{\mathbf{L}} \mathbf{A}$ *and* $\mathcal{P}(\mathbf{A},\mathbf{B}) <^{\mathbf{L}} \mathbf{B}$.

Proof. Let $(\mathbf{P},\mathbf{P}^\star)$ be a rank factorization of $\mathbf{A}$ and $(\mathbf{R},\mathbf{R}^\star)$ be a rank factorization of $\mathbf{B}$. Then $\mathcal{C}(\mathbf{A}) = \mathcal{C}(\mathbf{P})$ and $\mathcal{C}(\mathbf{B}) = \mathcal{C}(\mathbf{R})$ and $\mathbf{A}+\mathbf{B} = \left(\mathbf{P}:\mathbf{R}\right)\left(\mathbf{P}:\mathbf{R}\right)^\star$. Therefore,

$$\mathcal{C}(\mathbf{A}+\mathbf{B}) = \mathcal{C}(\left(\mathbf{P}:\mathbf{R}\right)) = \mathcal{C}(\mathbf{P}) + \mathcal{C}(\mathbf{R}) = \mathcal{C}(\mathbf{A}) + \mathcal{C}(\mathbf{B}).$$

(ii) follows from (i) and Theorem 9.2.4.
(iii) Notice that $\mathbf{A} - \mathcal{P}(\mathbf{A},\mathbf{B}) = \mathbf{A}(\mathbf{A}+\mathbf{B})^-\mathbf{A}$. Since $\mathbf{A}+\mathbf{B}$ is *nnd*, we can choose an *nnd* g-inverse of $\mathbf{A}+\mathbf{B}$ and $\mathbf{A}$ is hermitian, so, $\mathbf{A} - \mathcal{P}(\mathbf{A},\mathbf{B})$ is *nnd*.
Similarly, $\mathbf{B} - \mathcal{P}(\mathbf{A},\mathbf{B})$ is *nnd*. □

Remark 9.2.8. For an alternate proof of (iii) of Theorem 9.2.7, use Theorem 9.2.4 (v) or (vi).

It is pertinent to ask, if a theorem like Theorem 9.2.7 is valid for the class of hermitian matrices. The answer is contained in the following theorem:

Theorem 9.2.9. *Let* $\mathbf{A},\mathbf{B} \in \mathfrak{H}_n$. *Then* $\mathbf{A}$ *and* $\mathbf{B}$ *are parallel summable and* $\mathcal{P}(\mathbf{A},\mathbf{B})$ *is nnd if and only if* $\mathcal{C}(\mathbf{A}) \subseteq \mathcal{C}(\mathbf{A}+\mathbf{B})$ *and* $\mathbf{A}(\mathbf{A}+\mathbf{B})^-\mathbf{A} <^{\mathbf{L}} \mathbf{A}$.

Proof. By Theorem 9.2.2 and the hypothesis that $\mathbf{A}$ and B are hermitian $\Rightarrow$ $\mathbf{A}$ and B are parallel summable if and only if $\mathcal{C}(\mathbf{A}) \subseteq \mathcal{C}(\mathbf{A}+\mathbf{B})$. We can rewrite $\mathcal{P}(\mathbf{A},\mathbf{B}) = \mathbf{A} - \mathbf{A}(\mathbf{A}+\mathbf{B})^-\mathbf{A}$, so, the result follows. □

We next show that if $\mathbf{A}$ and $\mathbf{B}$ are parallel summable matrices, then matrices equivalent to $\mathbf{A}$ and $\mathbf{B}$ respectively are also parallel summable and have parallel sum equivalent to the parallel sum of $\mathbf{A}$ and $\mathbf{B}$.

Theorem 9.2.10. *Let* $\mathbf{A}, \mathbf{B}$ *be* $m \times n$ *matrices. Let* $\mathbf{C}$ *and* $\mathbf{D}$ *be matrices of order* $p \times m$ *and* $n \times q$ *respectively. Further let* $\rho(\mathbf{C}) = m$ *and* $\rho(\mathbf{D}) = n$. *Then the following hold:*

(i) $\mathbf{A}$ *and* $\mathbf{B}$ *are parallel summable if and only if* $\mathbf{CAD}$ *and* $\mathbf{CBD}$ *are parallel summable.*
(ii) *When* $\mathcal{P}(\mathbf{A}, \mathbf{B})$ *exists,* $\mathcal{P}(\mathbf{CAD}, \mathbf{CBD}) = \mathbf{C}\mathcal{P}(\mathbf{A}, \mathbf{B})\mathbf{D}$.

Proof. (i) Observe that $\mathcal{C}(\mathbf{CAD}) = \mathcal{C}(\mathbf{CA})$ and $\mathcal{C}(\mathbf{CAD})^{\mathbf{t}} = \mathcal{C}(\mathbf{AD})^{\mathbf{t}}$, since $\mathbf{C}$ has a left inverse and $\mathbf{D}$ has a right inverse. Further, for any matrices $\mathbf{R}$, $\mathbf{S}$ and $\mathbf{T}$, if $\mathcal{C}(\mathbf{R}) \subseteq \mathcal{C}(\mathbf{R}+\mathbf{S})$, then $\mathcal{C}(\mathbf{TR}) \subseteq \mathcal{C}(\mathbf{T}(\mathbf{R}+\mathbf{S}))$. Now, it follows immediately that $\mathcal{C}(\mathbf{A}) \subseteq \mathcal{C}(\mathbf{A}+\mathbf{B}) \Leftrightarrow \mathcal{C}(\mathbf{CA}) \subseteq \mathcal{C}(\mathbf{C}(\mathbf{A}+\mathbf{B}))$, since $\mathbf{C}$ has a left inverse. Similarly, $\mathcal{C}(\mathbf{A}^{\mathbf{t}}) \subseteq \mathcal{C}(\mathbf{A}+\mathbf{B})^{\mathbf{t}} \Leftrightarrow \mathcal{C}(\mathbf{AD})^{\mathbf{t}} \subseteq \mathcal{C}((\mathbf{A}+\mathbf{B})\mathbf{D})^{\mathbf{t}}$. Thus, (i) holds.
(ii) Let $\mathcal{P}(\mathbf{A}, \mathbf{B})$ exist. Notice that $\mathbf{D}_{\mathbf{R}}^{-1}(\mathbf{A}+\mathbf{B})^{-}\mathbf{C}_{\mathbf{L}}^{-1}$ is a g-inverse of $\mathbf{C}(\mathbf{A}+\mathbf{B})\mathbf{D}$. Hence, $\mathcal{P}(\mathbf{CAD}, \mathbf{CBD}) = \mathbf{C}\mathcal{P}(\mathbf{A}, \mathbf{B})\mathbf{D}$. □

Corollary 9.2.11. *Let* $\mathbf{A}, \mathbf{B}$ *be* $m \times n$ *matrices. Let* $\mathbf{P}$ *and* $\mathbf{Q}$ *respectively be* $m \times m$ *and* $n \times n$ *non-singular matrices. Then* $\mathbf{A}$, $\mathbf{B}$ *are parallel summable if and only if* $\mathbf{PAQ}$, $\mathbf{PBQ}$ *are parallel summable. Furthermore,*

$$\mathcal{P}(\mathbf{PAQ}, \mathbf{PBQ}) = \mathbf{P}\mathcal{P}(\mathbf{A}, \mathbf{B})\mathbf{Q}.$$

Remark 9.2.12. If $\mathbf{A}$ and $\mathbf{B}$ are parallel summable, it need not follow that $\mathbf{A}^{\dagger}$ and $\mathbf{B}^{\dagger}$ are parallel summable. For example take $\mathbf{A} = \begin{pmatrix} 3 & -5 \\ -1 & 2 \end{pmatrix}$ and $\mathbf{B} = \begin{pmatrix} -\frac{1}{5} & 0 \\ -\frac{2}{5} & 0 \end{pmatrix}$. Then $\mathbf{A}^{\dagger} = \mathbf{A}^{-1} = \begin{pmatrix} 2 & 5 \\ 1 & 3 \end{pmatrix}$, $\mathbf{B}^{\dagger} = \begin{pmatrix} -1 & -2 \\ 0 & 0 \end{pmatrix}$. Trivially $\mathbf{A}$ and $\mathbf{B}$ are parallel summable. But $\mathbf{A}^{\dagger} + \mathbf{B}^{\dagger} = \begin{pmatrix} 1 & 3 \\ 1 & 3 \end{pmatrix}$ is a singular matrix where as $\mathbf{A}^{\dagger}$ is non-singular. So, $\mathbf{A}^{\dagger}$ and $\mathbf{B}^{\dagger}$ cannot be parallel summable.

Thus, one may ask the following:

When are the generalized inverses of parallel summable matrices themselves parallel summable?

We do not know the answer yet. However see Ex. 13 at the end of the chapter.

We now obtain the class of all generalized inverses of the parallel sum of two parallel summable matrices. From the utility point of view this is an important result. However, we need the following lemma to prove it.

Lemma 9.2.13. *Let* $\mathbf{A}$ *and* B *be parallel summable. Then*

$$\mathcal{N}(\mathcal{P}(\mathbf{A}, \mathbf{B})) = \mathcal{N}(\mathbf{A}) + \mathcal{N}(\mathbf{B}) \text{ and } \mathcal{N}(\mathcal{P}(\mathbf{A}, \mathbf{B}))^{\mathbf{t}} = \mathcal{N}(\mathbf{A}^{\mathbf{t}}) + \mathcal{N}(\mathbf{B}^{\mathbf{t}}).$$

Proof. Clearly, $\mathcal{N}(\mathbf{A})+\mathcal{N}(\mathbf{B}) \subseteq \mathcal{N}(\mathcal{P}(\mathbf{A},\mathbf{B}))$. Further, $\mathcal{N}(\mathbf{A}) \cap \mathcal{N}(\mathbf{B}) \subseteq \mathcal{N}(\mathbf{A})+\mathcal{N}(\mathbf{B})$ holds always. Let $\mathbf{x} \in \mathcal{N}(\mathcal{P}(\mathbf{A},\mathbf{B}))$. So,

$$\begin{aligned}\mathcal{P}(\mathbf{A},\mathbf{B})\mathbf{x} = \mathbf{0} &\Rightarrow \mathcal{P}(\mathbf{B},\mathbf{A})\mathbf{x} = \mathbf{0}\\ &\Rightarrow (\mathbf{A}+\mathbf{B})^-\mathbf{B}\mathbf{x} \in \mathcal{N}(\mathbf{A}) \text{ and } (\mathbf{A}+\mathbf{B})^-\mathbf{A}\mathbf{x} \in \mathcal{N}(\mathbf{B}).\end{aligned}$$

Therefore, $(\mathbf{A}+\mathbf{B})^-(\mathbf{A}+\mathbf{B})\mathbf{x} \in \mathcal{N}(\mathbf{A})+\mathcal{N}(\mathbf{B})$. Since $\mathcal{N}(\mathbf{A})+\mathcal{N}(\mathbf{B}) \subseteq \mathcal{N}(\mathbf{A}+\mathbf{B})$, we have $(\mathbf{A}+\mathbf{B})(\mathbf{A}+\mathbf{B})^-(\mathbf{A}+\mathbf{B})\mathbf{x} = \mathbf{0}$. So, $(\mathbf{A}+\mathbf{B})\mathbf{x} = \mathbf{0}$. As $\mathcal{C}(\mathbf{A}^t) \subseteq \mathcal{C}(\mathbf{A}+\mathbf{B})^t$ and $\mathcal{C}(\mathbf{B}^t) \subseteq \mathcal{C}(\mathbf{A}+\mathbf{B})^t$ we have, $\mathbf{A}\mathbf{x} = \mathbf{0}$ and $\mathbf{B}\mathbf{x} = \mathbf{0}$. Hence, $\mathbf{x} \in \mathcal{N}(\mathbf{A}) \cap \mathcal{N}(\mathbf{B}) \Rightarrow \mathbf{x} \in \mathcal{N}(\mathbf{A})+\mathcal{N}(\mathbf{B})$. Thus, $\mathcal{N}(\mathcal{P}(\mathbf{A},\mathbf{B})) = \mathcal{N}(\mathbf{A})+\mathcal{N}(\mathbf{B})$. Similarly, we can prove $\mathcal{N}(\mathcal{P}(\mathbf{A},\mathbf{B}))^t = \mathcal{N}(\mathbf{A}^t)+\mathcal{N}(\mathbf{B}^t)$. □

Theorem 9.2.14. *Let* $\mathbf{A}$ *and* $\mathbf{B}$ *be parallel summable matrices of order* $m \times n$. *Then the class of all generalized inverses of* $\mathcal{P}(\mathbf{A},\mathbf{B})$ *is given by*

$$\{\mathcal{P}(\mathbf{A},\mathbf{B})^-\} = \{\mathbf{A}^-\}+\{\mathbf{B}^-\}.$$

Conversely, if $\mathbf{C}$ *is a non-null matrix of order* $m \times n$ *such that* $\{\mathbf{C}^-\} = \{\mathbf{A}^-\}+\{\mathbf{B}^-\}$, *then* $\mathbf{A}$ *and* $\mathbf{B}$ *are parallel summable and* $\mathbf{C} = \mathcal{P}(\mathbf{A},\mathbf{B})$.

Proof. Let $\mathbf{A}$ and $\mathbf{B}$ be parallel summable and $\mathbf{A}^-$ and $\mathbf{B}^-$ be any g-inverses of $\mathbf{A}$ and $\mathbf{B}$ respectively. Consider

$$\begin{aligned}\mathcal{P}(\mathbf{A},\mathbf{B})(\mathbf{A}^-+\mathbf{B}^-)\mathcal{P}(\mathbf{A},\mathbf{B}) &= \mathcal{P}(\mathbf{A},\mathbf{B})\mathbf{A}^-\mathcal{P}(\mathbf{A},\mathbf{B})+\mathcal{P}(\mathbf{A},\mathbf{B})\mathbf{B}^-\mathcal{P}(\mathbf{A},\mathbf{B})\\ &= \mathbf{B}(\mathbf{A}+\mathbf{B})^-\mathbf{A}\mathbf{A}^-\mathbf{A}(\mathbf{A}+\mathbf{B})^-\mathbf{B}+\mathbf{A}(\mathbf{A}+\mathbf{B})^-\mathbf{B}\mathbf{B}^-\mathbf{B}(\mathbf{A}+\mathbf{B})^-\mathbf{A}\\ &= \mathbf{B}(\mathbf{A}+\mathbf{B})^-\mathbf{A}(\mathbf{A}+\mathbf{B})^-\mathbf{B}+\mathbf{A}(\mathbf{A}+\mathbf{B})^-\mathbf{B}(\mathbf{A}+\mathbf{B})^-\mathbf{A}\\ &= \mathcal{P}(\mathbf{A},\mathbf{B})(\mathbf{A}+\mathbf{B})^-\mathbf{B}+\mathcal{P}(\mathbf{A},\mathbf{B})(\mathbf{A}+\mathbf{B})^-\mathbf{A}\\ &= \mathcal{P}(\mathbf{A},\mathbf{B})(\mathbf{A}+\mathbf{B})^-(\mathbf{A}+\mathbf{B})\\ &= \mathcal{P}(\mathbf{A},\mathbf{B}),\end{aligned}$$

since $\mathcal{C}(\mathcal{P}(\mathbf{A},\mathbf{B})) \subseteq \mathcal{C}(\mathbf{A}) \cap \mathcal{C}(\mathbf{B}) \subseteq \mathcal{C}(\mathbf{A}+\mathbf{B})$. Thus, $\mathbf{A}^-+\mathbf{B}^-$ is a g-inverse of $\mathcal{C}(\mathbf{A}+\mathbf{B})$, showing that $\{\mathbf{A}^-\}+\{\mathbf{B}^-\} \subseteq \{\mathcal{P}(\mathbf{A},\mathbf{B})^-\}$.

To prove $\{\mathcal{P}(\mathbf{A},\mathbf{B})^-\} \subseteq \{\mathbf{A}^-\}+\{\mathbf{B}^-\}$, note that as seen in the above proof, a choice for a g-inverse of $\mathcal{P}(\mathbf{A},\mathbf{B})$ is $\mathbf{A}^-+\mathbf{B}^-$ for some g-inverse $\mathbf{A}^-$ of $\mathbf{A}$ and $\mathbf{B}^-$ of $\mathbf{B}$. So, every g-inverse of $\mathcal{P}(\mathbf{A},\mathbf{B})$ is of the form $\mathbf{G} = \mathbf{A}^-+\mathbf{B}^-+(\mathbf{I}-(\mathcal{P}(\mathbf{A},\mathbf{B}))^-\mathcal{P}(\mathbf{A},\mathbf{B}))\mathbf{W}+\mathbf{Z}(\mathbf{I}-\mathcal{P}(\mathbf{A},\mathbf{B})(\mathcal{P}(\mathbf{A},\mathbf{B}))^-)$ and $(\mathbf{I}-(\mathcal{P}(\mathbf{A},\mathbf{B}))^-\mathcal{P}(\mathbf{A},\mathbf{B})) \in \mathcal{N}(\mathcal{P}(\mathbf{A},\mathbf{B}))$. However, by Lemma 9.2.13, $\mathcal{N}(\mathcal{P}(\mathbf{A},\mathbf{B})) = \mathcal{N}(\mathbf{A})+\mathcal{N}(\mathbf{B})$. So, $(\mathbf{I}-(\mathcal{P}(\mathbf{A},\mathbf{B}))^-\mathcal{P}(\mathbf{A},\mathbf{B})) = (\mathbf{I}-\mathbf{A}^-\mathbf{A})\mathbf{W}_1+(\mathbf{I}-\mathbf{B}^-\mathbf{B})\mathbf{W}_2$ for some matrices $\mathbf{W}_1$ and $\mathbf{W}_2$. Similarly, $\mathbf{Z}(\mathbf{I}-\mathcal{P}(\mathbf{A},\mathbf{B})(\mathcal{P}(\mathbf{A},\mathbf{B}))^-) = \mathbf{Z}_1(\mathbf{I}-\mathbf{A}\mathbf{A}^-)+\mathbf{Z}_2(\mathbf{I}-\mathbf{B}\mathbf{B}^-)$ for some matrices $\mathbf{Z}_1$ and $\mathbf{Z}_2$. Therefore, $\mathbf{G} = \mathbf{A}^-+\mathbf{B}^-+(\mathbf{I}-\mathbf{A}^-\mathbf{A})\mathbf{W}_1+(\mathbf{I}-\mathbf{B}^-\mathbf{B})\mathbf{W}_2+\mathbf{Z}_1(\mathbf{I}-\mathbf{A}\mathbf{A}^-)+\mathbf{Z}_2(\mathbf{I}-\mathbf{B}\mathbf{B}^-) = \mathbf{A}^-+(\mathbf{I}-\mathbf{A}^-\mathbf{A})\mathbf{W}_1+\mathbf{Z}_1(\mathbf{I}-\mathbf{A}\mathbf{A}^-)+\mathbf{B}^-+$

$(\mathbf{I}-\mathbf{B}^-\mathbf{B})\mathbf{W_2}+\mathbf{Z_2}(\mathbf{I}-\mathbf{BB}^-)$. Clearly, $\mathbf{A}^-+(\mathbf{I}-\mathbf{A}^-\mathbf{A})\mathbf{W_1}+\mathbf{Z_1}(\mathbf{I}-\mathbf{AA}^-)$ is a g-inverse of $\mathbf{A}$ and $\mathbf{B}^-+(\mathbf{I}-\mathbf{B}^-\mathbf{B})\mathbf{W_2}+\mathbf{Z_2}(\mathbf{I}-\mathbf{BB}^-)$ is a g-inverse of $\mathbf{B}$. Thus, $\mathbf{G}$ is a sum of a g-inverse of $\mathbf{A}$ and a g-inverse of $\mathbf{B}$ proving $\{\mathcal{P}(\mathbf{A},\mathbf{B})^-\}\subseteq\{\mathbf{A}^-\}+\{\mathbf{B}^-\}$. Hence, $\{\mathcal{P}(\mathbf{A},\mathbf{B})^-\}=\{\mathbf{A}^-\}+\{\mathbf{B}^-\}$.

For converse, we first show that $\mathbf{A}$ and $\mathbf{B}$ are parallel summable. Let $(\mathbf{C_1},\mathbf{D_1})$ be a rank factorization of $\mathbf{C}$. Since, $\{\mathbf{C}^-\}=\{\mathbf{A}^-\}+\{\mathbf{B}^-\}$, for each $\mathbf{G}\in\{\mathbf{A}^-\}$ and for each $\mathbf{H}\in\{\mathbf{B}^-\}$ as seen above $\mathbf{G}+(\mathbf{I}-\mathbf{GA})\mathbf{W_1}+\mathbf{Z_1}(\mathbf{I}-\mathbf{AG})+\mathbf{H}+(\mathbf{I}-\mathbf{HB})\mathbf{W_2}+\mathbf{Z_2}(\mathbf{I}-\mathbf{BH})$ is a g-inverse of $\mathbf{C}$. So, $\mathbf{C_1D_1}(\mathbf{G}+(\mathbf{I}-\mathbf{GA})\mathbf{W_1}+\mathbf{Z_1}(\mathbf{I}-\mathbf{AG})+\mathbf{H}+(\mathbf{I}-\mathbf{HB})\mathbf{W_2}+\mathbf{Z_2}(\mathbf{I}-\mathbf{BH}))\mathbf{C_1D_1}=\mathbf{C_1D_1}$, equivalently, $\mathbf{C_1D_1}((\mathbf{I}-\mathbf{GA})\mathbf{W_1}+\mathbf{Z_1}(\mathbf{I}-\mathbf{AG})+(\mathbf{I}-\mathbf{HB})\mathbf{W_2}+\mathbf{Z_2}(\mathbf{I}-\mathbf{BH}))\mathbf{C_1D_1}=\mathbf{0}$. As $\mathbf{C_1}$ and $\mathbf{D_1}$ are full rank matrices this further gives $\mathbf{D_1}((\mathbf{I}-\mathbf{GA})\mathbf{W_1}+\mathbf{Z_1}(\mathbf{I}-\mathbf{AG})+(\mathbf{I}-\mathbf{HB})\mathbf{W_2}+\mathbf{Z_2}(\mathbf{I}-\mathbf{BH}))\mathbf{C_1}=\mathbf{0}$. Since the last equation holds for all $\mathbf{W_1},\mathbf{W_2},\mathbf{Z_1},\mathbf{Z_2}$, we have $\mathbf{D_1}(\mathbf{I}-\mathbf{GA})=\mathbf{0}$, $(\mathbf{I}-\mathbf{AG})\mathbf{C_1}=\mathbf{0},\mathbf{D_1}(\mathbf{I}-\mathbf{HB})=\mathbf{0},(\mathbf{I}-\mathbf{BH})\mathbf{C_1}=\mathbf{0}$. It follows that $\mathcal{C}(\mathbf{C_1})\subseteq\mathcal{C}(\mathbf{A})\cap\mathcal{C}(\mathbf{B})$ and $\mathcal{C}(\mathbf{D_1})\subseteq\mathcal{C}(\mathbf{A^t})\cap\mathcal{C}(\mathbf{B^t})$. Since $\mathbf{C_1}$ has full column rank $=\rho(\mathbf{C})$ and $\mathbf{D_1}$ has full row rank $=\rho(\mathbf{C})$, it follows that $d(\mathcal{C}(\mathbf{A})\cap\mathcal{C}(\mathbf{B}))=\mathbf{d}(\mathcal{C}(\mathbf{A^t})\cap\mathcal{C}(\mathbf{B^t}))=\rho(\mathbf{C})$. Let $\mathbf{C_2}$ and $\mathbf{C_3}$ be matrices such that the columns of the matrix $(\mathbf{C_1}:\mathbf{C_2})$ form a basis of $\mathcal{C}(\mathbf{A})$ and that of $(\mathbf{C_1}:\mathbf{C_3})$ form a basis of $\mathcal{C}(\mathbf{B})$. Similarly, let $\mathbf{D_2}$ and $\mathbf{D_3}$ be matrices such that the rows of $(\mathbf{D_1}:\mathbf{D_2})^t$ form a basis of $\mathcal{C}(\mathbf{A^t})$ and rows of $(\mathbf{D_2}:\mathbf{D_3})^t$ form a basis of $\mathcal{C}(\mathbf{B^t})$. Then $\mathbf{C}=(\mathbf{C_1}:\mathbf{C_2}:\mathbf{C_3})\begin{pmatrix}\mathbf{I}&0&0\\0&0&0\\0&0&0\end{pmatrix}\begin{pmatrix}\mathbf{D_1}\\\mathbf{D_2}\\\mathbf{D_3}\end{pmatrix}$. Extend $(\mathbf{C_1}:\mathbf{C_2}:\mathbf{C_3})$ to a non-singular matrix $\mathbf{P}=(\mathbf{C_1}:\mathbf{C_2}:\mathbf{C_3}:\mathbf{C_4})$ and $\begin{pmatrix}\mathbf{D_1}\\\mathbf{D_2}\\\mathbf{D_3}\end{pmatrix}$ to a non-singular matrix $\mathbf{Q}=\begin{pmatrix}\mathbf{D_1}\\\mathbf{D_2}\\\mathbf{D_3}\\\mathbf{D_4}\end{pmatrix}$, so, $\mathbf{C}=\mathbf{P}\begin{pmatrix}\mathbf{I}&0&0&0\\0&0&0&0\\0&0&0&0\\0&0&0&0\end{pmatrix}\mathbf{Q}$.

Also, $\mathbf{A}=\mathbf{P}\begin{pmatrix}\mathbf{R_{11}}&\mathbf{R_{12}}&0&0\\\mathbf{R_{21}}&\mathbf{R_{22}}&0&0\\0&0&0&0\\0&0&0&0\end{pmatrix}\mathbf{Q}$ and $\mathbf{B}=\mathbf{P}\begin{pmatrix}\mathbf{S_{11}}&0&\mathbf{S_{13}}&0\\0&0&0&0\\\mathbf{S_{31}}&0&\mathbf{S_{33}}&0\\0&0&0&0\end{pmatrix}\mathbf{Q}$,

where $\begin{pmatrix}\mathbf{R_{11}}&\mathbf{R_{12}}\\\mathbf{R_{12}}&\mathbf{R_{22}}\end{pmatrix}$ and $\begin{pmatrix}\mathbf{S_{11}}&\mathbf{S_{13}}\\\mathbf{S_{31}}&\mathbf{S_{33}}\end{pmatrix}$ are non-singular matrices and let

$\begin{pmatrix} \mathbf{R}^{11} & \mathbf{R}^{12} \\ \mathbf{R}^{12} & \mathbf{R}^{22} \end{pmatrix}$ and $\begin{pmatrix} \mathbf{S}^{11} & \mathbf{S}^{13} \\ \mathbf{S}^{31} & \mathbf{S}^{33} \end{pmatrix}$ be their respective g-inverses.

Any g-inverse of $\mathbf{C}$ is of the form $\mathbf{C}^- = \mathbf{Q}^{-1} \begin{pmatrix} \mathbf{I} & X & X & X \\ X & X & X & X \\ X & X & X & X \\ X & X & X & X \end{pmatrix} \mathbf{P}^{-1}$.

Similarly, g-inverses of $\mathbf{A}$ and $\mathbf{B}$ are respectively of the form

$$\mathbf{A}^- = \mathbf{Q}^{-1} \begin{pmatrix} \mathbf{R}^{11} & X & X & X \\ X & X & X & X \\ X & X & X & X \\ X & X & X & X \end{pmatrix} \mathbf{P}^{-1} \text{ and } \mathbf{B}^- = \mathbf{Q}^{-1} \begin{pmatrix} \mathbf{S}^{11} & X & X & X \\ X & X & X & X \\ X & X & X & X \\ X & X & X & X \end{pmatrix} \mathbf{P}^{-1}.$$

Notice that here X's represent some suitable matrices not of interest. Now,

$$\mathbf{A}+\mathbf{B} = \mathbf{P} \begin{pmatrix} \mathbf{R}_{11}+\mathbf{S}_{11} & \mathbf{R}_{12} & \mathbf{S}_{13} & \mathbf{0} \\ \mathbf{R}_{21} & \mathbf{R}_{22} & \mathbf{0} & \mathbf{0} \\ \mathbf{S}_{31} & \mathbf{0} & \mathbf{S}_{33} & \mathbf{0} \\ \mathbf{0} & \mathbf{0} & \mathbf{0} & \mathbf{0} \end{pmatrix} \mathbf{Q} = \mathbf{P} \begin{pmatrix} \Delta & \mathbf{0} \\ \mathbf{0} & \mathbf{0} \end{pmatrix} \mathbf{Q}.$$

We show that $\mathbf{\Delta} = \begin{pmatrix} \mathbf{R}_{11}+\mathbf{S}_{11} & \mathbf{R}_{12} & \mathbf{S}_{13} \\ \mathbf{R}_{21} & \mathbf{R}_{22} & \mathbf{0} \\ \mathbf{S}_{31} & \mathbf{0} & \mathbf{S}_{33} \end{pmatrix}$ is a non-singular matrix.

Let

$$\mathbf{F}_1 = \begin{pmatrix} \mathbf{R}^{11} & \mathbf{R}^{12} & \mathbf{0} \\ \mathbf{R}^{12} & \mathbf{R}^{22} & \mathbf{0} \\ \mathbf{0} & \mathbf{0} & \mathbf{I} \end{pmatrix}, \quad \mathbf{F}_2 = \begin{pmatrix} \mathbf{S}^{11} & \mathbf{0} & \mathbf{S}^{13} \\ \mathbf{0} & \mathbf{I} & \mathbf{0} \\ \mathbf{S}^{31} & \mathbf{0} & \mathbf{S}^{33} \end{pmatrix},$$

where $\mathbf{I}$ is an identity matrix of suitable order. Then $\mathbf{F}_1$ and $\mathbf{F}_2$ are non-singular

and $\mathbf{F}_1\mathbf{\Delta}\mathbf{F}_2 = \begin{pmatrix} \mathbf{I} & \mathbf{0} & \mathbf{S}^{13} \\ \mathbf{R}^{21} & \mathbf{I} & \mathbf{0} \\ \mathbf{0} & \mathbf{0} & \mathbf{I} \end{pmatrix}$. Clearly, $\mathbf{F}_1\mathbf{\Delta}\mathbf{F}_2$ is non-singular. Therefore, $\mathbf{\Delta}$ is a non-singular matrix. Also, a direct computation shows that $(\mathbf{A}+\mathbf{B})(\mathbf{A}+\mathbf{B})^-\mathbf{A} = \mathbf{A}$ and $\mathbf{A}(\mathbf{A}+\mathbf{B})^-(\mathbf{A}+\mathbf{B}) = \mathbf{A}$. So, $\mathcal{C}(\mathbf{A}) \subseteq \mathcal{C}(\mathbf{A}+\mathbf{B})$ and $\mathcal{C}(\mathbf{A}^t) \subseteq \mathcal{C}(\mathbf{A}+\mathbf{B})^t$ i.e., $\mathbf{A}$ and $\mathbf{B}$ are parallel summable. By first part $\mathbf{C}^- = \mathcal{P}(\mathbf{A},\mathbf{B})^-$, so, by Theorem 2.4.2 [Rao, Mitra and Bhimasankaram (1972)], $\mathbf{C} = \mathcal{P}(\mathbf{A},\mathbf{B})$ □

Remark 9.2.15. The converse of Theorem 9.2.14 is not true if $\mathbf{C} = \mathbf{0}$. We give the following example: Let $\mathbf{A} = \begin{pmatrix} 1 & 1 \\ 0 & 0 \end{pmatrix}, \mathbf{B} = \begin{pmatrix} 0 & 1 \\ 0 & 0 \end{pmatrix}$. Then

$\{\mathbf{A}^-\} = \left\{\begin{pmatrix} a & b \\ 1-a & d \end{pmatrix}\right\}$ and $\{\mathbf{B}^-\} = \left\{\begin{pmatrix} x & y \\ 1 & w \end{pmatrix}\right\}$, where a, b, d, x, y, w are arbitrary. Clearly, **A** and **B** are not parallel summable. However, for $\mathbf{C} = \mathbf{0}$, $\{\mathbf{C}^-\} = \mathbb{C}^{\mathbf{2}\times\mathbf{2}} = \{\mathbf{A}^-\} + \{\mathbf{B}^-\}$.

Theorem 9.2.16. *Let* **A** *and B be parallel summable. Then*

$$\mathcal{C}(\mathcal{P}(\mathbf{A},\mathbf{B})) = \mathcal{C}(\mathbf{A}) \cap \mathcal{C}(\mathbf{B}) \text{ and } \mathcal{C}(\mathcal{P}(\mathbf{A},\mathbf{B}))^{\mathbf{t}} = \mathcal{C}(\mathbf{A}^{\mathbf{t}}) \cap \mathcal{C}(\mathbf{B}^{\mathbf{t}}).$$

Proof. In view of Corollary 9.2.5, clearly we have

$$\mathcal{C}(\mathcal{P}(\mathbf{A},\mathbf{B})) \subseteq \mathcal{C}(\mathbf{A}) \cap \mathcal{C}(\mathbf{B}),\ \mathcal{C}(\mathcal{P}(\mathbf{A},\mathbf{B}))^{\mathbf{t}} \subseteq \mathcal{C}(\mathbf{A}^{\mathbf{t}}) \cap \mathcal{C}(\mathbf{B}^{\mathbf{t}}).$$

So, let $\mathbf{x} \in \mathcal{C}(\mathbf{A}) \cap \mathcal{C}(\mathbf{B})$. Let $\mathbf{A}^-, \mathbf{B}^-$ and $(\mathbf{A}+\mathbf{B})^-$ be any g-inverses of $\mathbf{A}, \mathbf{B}$ and $\mathbf{A}+\mathbf{B}$ respectively. Then $\mathbf{A}^- + \mathbf{B}^-$ is a g-inverse of $\mathcal{P}(\mathbf{A},\mathbf{B})$ and

$$\begin{aligned}\mathcal{P}(\mathbf{A},\mathbf{B})(\mathbf{A}^- + \mathbf{B}^-)\mathbf{x} &= \mathcal{P}(\mathbf{B},\mathbf{A})\mathbf{A}^-\mathbf{x} + \mathcal{P}(\mathbf{A},\mathbf{B})\mathbf{B}^-\mathbf{x} \\ &= \mathbf{B}(\mathbf{A}+\mathbf{B})^-\mathbf{A}\mathbf{A}^-\mathbf{x} + \mathbf{A}(\mathbf{A}+\mathbf{B})^-\mathbf{B}\mathbf{B}^-\mathbf{x}.\end{aligned}$$

Since $\mathbf{x} \in \mathcal{C}(\mathbf{A}) \cap \mathcal{C}(\mathbf{B})$, we have $\mathbf{A}\mathbf{A}^-\mathbf{x} = \mathbf{x}$, $\mathbf{B}\mathbf{B}^-\mathbf{x} = \mathbf{x}$. Moreover, $\mathbf{x} \in \mathcal{C}(\mathbf{A}+\mathbf{B})$, so,

$$\begin{aligned}\mathcal{P}(\mathbf{A},\mathbf{B})(\mathbf{A}^- + \mathbf{B}^-)\mathbf{x} &= \mathbf{B}(\mathbf{A}+\mathbf{B})^-\mathbf{x} + \mathbf{A}(\mathbf{A}+\mathbf{B})^-\mathbf{x} \\ &= (\mathbf{A}+\mathbf{B})(\mathbf{A}+\mathbf{B})^-\mathbf{x} = \mathbf{x}.\end{aligned}$$

Hence, $\mathbf{x} \in \mathcal{C}(\mathcal{P}(\mathbf{A},\mathbf{B}))$, and so, $\mathcal{C}(\mathbf{A}) \cap \mathcal{C}(\mathbf{B}) \subseteq \mathcal{C}(\mathcal{P}(\mathbf{A},\mathbf{B}))$.
The proof of the other part is similar. □

Remark 9.2.17. If **A** and **B** are disjoint matrices, then they are parallel summable with parallel sum, the null matrix.

Theorem 9.2.18. *Let* $\mathbf{A}, \mathbf{B} \in \mathbb{C}^{\mathbf{n}\times\mathbf{n}}$ *be nnd matrices. Then their parallel sum* $\mathcal{P}(\mathbf{A},\mathbf{B})$ *is nnd.*

Proof. By Theorem 9.2.7, **A** and **B** are parallel summable. Let $\mathbf{z} \in \mathbb{C}^{\mathbf{n}}$ and $\mathbf{x} = \mathcal{P}(\mathbf{A},\mathbf{B})\mathbf{z}$. Then $\mathcal{P}(\mathbf{A},\mathbf{B})(\mathbf{A}^\dagger + \mathbf{B}^\dagger)\mathbf{x} = \mathbf{x}$, since $\mathbf{A}^\dagger + \mathbf{B}^\dagger$ is a g-inverse of $\mathcal{P}(\mathbf{A},\mathbf{B})$. So,

$$\begin{aligned}(\mathcal{P}(\mathbf{A},\mathbf{B})\mathbf{z},\mathbf{z}) &= (\mathbf{x},\mathbf{z}) = (\mathcal{P}(\mathbf{A},\mathbf{B})(\mathbf{A}^\dagger + \mathbf{B}^\dagger)\mathbf{x},\mathbf{z}) \\ &= ((\mathbf{A}^\dagger + \mathbf{B}^\dagger)\mathbf{x}), \mathcal{P}(\mathbf{A},\mathbf{B})^\star\mathbf{z} = ((\mathbf{A}^\dagger + \mathbf{B}^\dagger)\mathbf{x}, \mathcal{P}(\mathbf{A},\mathbf{B})\mathbf{z}),\end{aligned}$$

since, $\mathcal{P}(\mathbf{A},\mathbf{B})$ is hermitian. Thus, $(\mathcal{P}(\mathbf{A},\mathbf{B})\mathbf{z},\mathbf{z}) = ((\mathbf{A}^\dagger + \mathbf{B}^\dagger)\mathbf{x},\mathbf{x}) \geq \mathbf{0}$, as $\mathbf{A}^\dagger + \mathbf{B}^\dagger$ is *nnd*. Thus, $\mathcal{P}(\mathbf{A},\mathbf{B})$ is *nnd*. □

Corollary 9.2.19. *Let* $\mathbf{A}, \mathbf{B} \in \mathbb{C}^{\mathbf{n}\times\mathbf{n}}$ *be range-hermitian matrices. Then their parallel sum* $\mathcal{P}(\mathbf{A}, \mathbf{B})$ *is range-hermitian.*

Proof follows from (ii) of Corollary 9.2.5 and Theorem 9.2.16.

Theorem 9.2.20. *Let* $\mathbf{A}, \mathbf{B} \in \mathbb{C}^{\mathbf{n}\times\mathbf{n}}$. *Then the following hold:*

(i) *Let* $\mathbf{P}$, *and* $\mathbf{Q}$ *denote the orthogonal projectors onto* $\mathcal{C}(\mathbf{A}) \cap \mathcal{C}(\mathbf{B})$ *and* $\mathcal{C}(\mathbf{A}^\star) \cap \mathcal{C}(\mathbf{B}^\star)$ *respectively. Then* $\mathcal{P}(\mathbf{A}, \mathbf{B})^\dagger = \mathbf{Q}(\mathbf{A}^- + \mathbf{B}^-)\mathbf{P}$, *where* $\mathbf{A}^-$ *and* $\mathbf{B}^-$ *are any g-inverses of* $\mathbf{A}$ *and* $\mathbf{B}$ *respectively. In particular, the expression* $\mathbf{Q}(\mathbf{A}^- + \mathbf{B}^-)\mathbf{P}$ *is independent of choice of* $\mathbf{A}^-$ *and* $\mathbf{B}^-$.

(ii) *Let* $\mathbf{P_1}$ *and* $\mathbf{P_2}$ *be orthogonal projectors onto* $\mathcal{C}(\mathbf{A})$ *and* $\mathcal{C}(\mathbf{B})$ *respectively. Then the orthogonal projector onto* $\mathcal{C}(\mathbf{A}) \cap \mathcal{C}(\mathbf{B})$ *is* $2(\mathcal{P}(\mathbf{P_1}, \mathbf{P_2}))$.

Proof. (i) Since $\mathcal{C}(\mathcal{P}(\mathbf{A}, \mathbf{B})) = \mathcal{C}(\mathbf{A}) \cap \mathcal{C}(\mathbf{B})$, $\mathbf{P} = (\mathcal{P}(\mathbf{A}, \mathbf{B}))(\mathcal{P}(\mathbf{A}, \mathbf{B}))^\dagger$ and $\mathbf{Q} = (\mathcal{P}(\mathbf{A}, \mathbf{B}))^\dagger(\mathcal{P}(\mathbf{A}, \mathbf{B}))$. So,

$$\mathbf{Q}(\mathbf{A}^- + \mathbf{B}^-)\mathbf{P} = (\mathcal{P}(\mathbf{A}, \mathbf{B}))^\dagger(\mathcal{P}(\mathbf{A}, \mathbf{B}))(\mathbf{A}^- + \mathbf{B}^-)(\mathcal{P}(\mathbf{A}, \mathbf{B}))(\mathcal{P}(\mathbf{A}, \mathbf{B}))^\dagger.$$

As $(\mathbf{A}^- + \mathbf{B}^-)$ is a g-inverse of $\mathcal{P}(\mathbf{A}, \mathbf{B})$, by Theorem 9.2.14, we have $\mathbf{Q}(\mathbf{A}^- + \mathbf{B}^-)\mathbf{P} = (\mathcal{P}(\mathbf{A}, \mathbf{B}))^\dagger$.
(ii) Since $\mathcal{C}(\mathbf{P_1}) = \mathcal{C}(\mathbf{A})$ and $\mathcal{C}(\mathbf{P_2}) = \mathcal{C}(\mathbf{B})$ and $\mathbf{P_1}, \mathbf{P_2}$ being hermitian and idempotent are parallel summable by Theorem 9.2.7. Also, $\mathcal{P}(\mathbf{P_1}, \mathbf{P_2})$ is hermitian. A simple computation shows that $2(\mathcal{P}(\mathbf{P_1}, \mathbf{P_2}))$ is idempotent. So, the result follows. □

Remark 9.2.21. Let $\mathbf{A}, \mathbf{B}$ be parallel summable matrices. For any g-inverse $(\mathcal{P}(\mathbf{A}, \mathbf{B}))^-$ of $\mathcal{P}(\mathbf{A}, \mathbf{B})$ and projectors $\mathbf{P} = (\mathcal{P}(\mathbf{A}, \mathbf{B}))(\mathcal{P}(\mathbf{A}, \mathbf{B}))^-$ and $\mathbf{Q} = (\mathcal{P}(\mathbf{A}, \mathbf{B}))^-(\mathcal{P}(\mathbf{A}, \mathbf{B}))$, the matrix $\mathbf{Q}(\mathbf{A}^- + \mathbf{B}^-)\mathbf{P}$ is a reflexive g-inverse of $\mathcal{P}(\mathbf{A}, \mathbf{B})$. It will be interesting to know what are all the reflexive g-inverses of $\mathcal{P}(\mathbf{A}, \mathbf{B})$.

We now give all possible reflexive generalized inverses of the parallel sum $\mathcal{P}(\mathbf{A}, \mathbf{B})$ of matrices $\mathbf{A}, \mathbf{B}$ whenever it is defined.

Theorem 9.2.22. *Let* $\mathbf{A}$ *and* $\mathbf{B}$ *be two parallel summable matrices of order* $m \times n$. *Let* $\rho(\mathbf{A}) = a$ *and* $\rho(\mathbf{A} + \mathbf{B}) = r$. *Further, let* $\mathbf{P}, \mathbf{Q}$ *be non-singular matrices of orders* $m \times m$ *and* $n \times n$ *respectively and* $\mathbf{K} = \begin{pmatrix} \mathbf{K}_{11} & \mathbf{K}_{12} \\ \mathbf{K}_{21} & \mathbf{K}_{22} \end{pmatrix}$ *be an* $r \times r$ *non-singular matrix with inverse* $\mathbf{K}^{-1} = \begin{pmatrix} \mathbf{K}^{11} & \mathbf{K}^{12} \\ \mathbf{K}^{21} & \mathbf{K}^{22} \end{pmatrix}$, *where* $\mathbf{K}_{11}$ *is* $a \times a$ *matrix such that* $\mathbf{A} = \mathbf{P}\ \mathbf{diag}\left(\mathbf{I_a}\ , \mathbf{0_{r-a}}\ , \mathbf{0}\right)\mathbf{Q}$ *and* $\mathbf{B} =$

$\mathbf{P}\begin{pmatrix}\mathbf{K_{11}}-\mathbf{I_a} & \mathbf{K_{12}} & \mathbf{0}\\ \mathbf{K_{21}} & \mathbf{K_{22}} & \mathbf{0}\\ \mathbf{0} & \mathbf{0} & \mathbf{0}\end{pmatrix}\mathbf{Q}$. *Then the class of all reflexive g- inverses of* $\mathcal{P}(\mathbf{A},\mathbf{B})$ *consist of matrices* $\mathbf{X}$ *of the form* $\mathbf{X} = \mathbf{Q}^{-1}\begin{pmatrix}\mathbf{T} & \mathbf{L}\\ \mathbf{M} & \mathbf{ML}\end{pmatrix}\mathbf{P}^{-1}$, *where* $\mathbf{T} = (\mathbf{I_a}-\mathbf{K^{11}})_\mathbf{r}$ *is any reflexive g-inverse of* $\mathbf{I_a}-\mathbf{K^{11}}$ *and* $\mathbf{L},\mathbf{M}$ *are arbitrary.*

Proof is by direct verification.

In general, we may not have any of $\mathcal{P}(\mathbf{A}+\mathbf{C}|\mathbf{B}) = \mathcal{P}(\mathbf{A},\mathbf{B}) + \mathcal{P}(\mathbf{C}|\mathbf{B})$ or $\mathcal{P}(\mathbf{A}|\mathbf{B}+\mathbf{C}) = \mathcal{P}(\mathbf{A},\mathbf{B}) + \mathcal{P}(\mathbf{A},\mathbf{C})$, even when all parallel sums are defined. However, we have the following:

Theorem 9.2.23. *Let* $\mathbf{A},\mathbf{B}$, *and* $\mathbf{C}$ *be* $m\times n$ *matrices. If* $\mathbf{A},\mathbf{B}$ *are parallel summable,* $\mathbf{A}+\mathbf{B}$, $\mathbf{C}$ *are parallel summable and* $\mathcal{P}(\mathbf{A}+\mathbf{B}|\mathbf{C}) = \mathbf{0}$, *then the following hold:*

(i) $\mathbf{A}+\mathbf{C}$, $\mathbf{B}$ *are parallel summable*
(ii) $\mathbf{A}$, $\mathbf{B}+\mathbf{C}$ *are parallel summable and*
(iii) $\mathcal{P}(\mathbf{A}+\mathbf{C},\mathbf{B}) = \mathcal{P}(\mathbf{A},\mathbf{B}) = \mathcal{P}(\mathbf{A},\mathbf{B}+\mathbf{C})$.

Proof. Since $\mathbf{A},\mathbf{B}$ are parallel summable and $\mathbf{A}+\mathbf{B},\mathbf{C}$ are parallel summable, we have

$$\mathcal{C}(\mathbf{A}) \subseteq \mathcal{C}(\mathbf{A}+\mathbf{B}) \subseteq \mathcal{C}(\mathbf{A}+\mathbf{B}+\mathbf{C}),\ \mathcal{C}(\mathbf{A^t}) \subseteq \mathcal{C}(\mathbf{A}+\mathbf{B})^\mathbf{t} \subseteq \mathcal{C}(\mathbf{A}+\mathbf{B}+\mathbf{C})^\mathbf{t} \tag{9.2.1}$$

or equivalently

$$\mathcal{C}(\mathbf{B}) \subseteq \mathcal{C}(\mathbf{A}+\mathbf{B}) \subseteq \mathcal{C}(\mathbf{A}+\mathbf{B}+\mathbf{C}),\ \mathcal{C}(\mathbf{B^t}) \subseteq \mathcal{C}(\mathbf{A}+\mathbf{B})^\mathbf{t} \subseteq \mathcal{C}(\mathbf{A}+\mathbf{B}+\mathbf{C})^\mathbf{t}. \tag{9.2.2}$$

By (9.2.1) and (9.2.1), both (i) and (ii) follow.
(iii) $\mathcal{P}(\mathbf{A}+\mathbf{B},\mathbf{C}) = \mathbf{0} \Rightarrow (\mathbf{A}+\mathbf{B})(\mathbf{A}+\mathbf{B}+\mathbf{C})^-\mathbf{C} = \mathbf{0} \Rightarrow \mathbf{B}(\mathbf{A}+\mathbf{B}+\mathbf{C})^-\mathbf{C} = \mathbf{0}$, since $\mathcal{C}(\mathbf{B})^\mathbf{t} \subseteq \mathcal{C}(\mathbf{A}+\mathbf{B})^\mathbf{t}$. Similarly, $\mathbf{C}(\mathbf{A}+\mathbf{B}+\mathbf{C})^-\mathbf{B} = \mathbf{0}$. So, $\mathcal{P}(\mathbf{A}+\mathbf{C},\mathbf{B}) = (\mathbf{A}+\mathbf{C})(\mathbf{A}+\mathbf{B}+\mathbf{C})^-\mathbf{B} = \mathbf{A}(\mathbf{A}+\mathbf{B}+\mathbf{C})^-\mathbf{B}$. Again, $\mathcal{P}(\mathbf{A}+\mathbf{B},\mathbf{C}) = \mathbf{0} \Rightarrow \rho(\mathbf{A}+\mathbf{B}+\mathbf{C}) = \rho(\mathbf{A}+\mathbf{B}) + \rho(\mathbf{C})$. This gives $(\mathbf{A}+\mathbf{B}+\mathbf{C})^-$ is a g-inverse of $\mathbf{A}+\mathbf{B}$. Hence,

$$\mathcal{P}(\mathbf{A}+\mathbf{C},\mathbf{B}) = \mathbf{A}(\mathbf{A}+\mathbf{B})^-\mathbf{B} = \mathcal{P}(\mathbf{A},\mathbf{B}).$$

Similarly, $\mathcal{P}(\mathbf{A},\mathbf{B}) = \mathcal{P}(\mathbf{B}+\mathbf{C},\mathbf{A}) = \mathcal{P}(\mathbf{A},\mathbf{B}+\mathbf{C})$. □

Remark 9.2.24. Let $\mathbf{A}$ and $\mathbf{B}$ be any $m \times n$ matrices such that $\mathbf{A} <^{s} \mathbf{B}$. Does there exist a matrix $\mathbf{C}$ such that $\mathbf{B}$ and $\mathbf{C}$ are parallel summable and $\mathcal{P}(\mathbf{B}, \mathbf{C}) = \mathbf{A}$. The answer is in the negative and we give below an example:

Example 9.2.25. Let $\mathbf{A} = \begin{pmatrix} 1 & 1 & 0 \\ 0 & 0 & 0 \\ 0 & 0 & 0 \end{pmatrix}$ and $\mathbf{B} = \begin{pmatrix} 1 & 0 & 0 \\ 0 & 1 & 0 \\ 0 & 0 & 0 \end{pmatrix}$. Then there exists no matrix $\mathbf{C}$ such that $\mathbf{B}$ and $\mathbf{C}$ are parallel summable and $\mathcal{P}(\mathbf{B}, \mathbf{C}) = \mathbf{A}$.

An affirmative answer for the above question in special cases is contained in the following theorems:

Theorem 9.2.26. *Let* $\mathbf{A}$ *and* $\mathbf{B}$ *be any* $m \times n$ *matrices such that* $\mathbf{A}$, $\mathbf{B}$ *are parallel summable. Then the following hold:*

(i) *If either* $\mathcal{C}(\mathbf{B}) \subseteq \mathcal{C}(\mathbf{A})$ or $\mathcal{C}(\mathbf{B}^{t}) \subseteq \mathcal{C}(\mathbf{A}^{t})$, *then both the inclusions hold and* $\rho(\mathbf{A} - \mathcal{P}(\mathbf{A}, \mathbf{B})) = \rho(\mathbf{A})$.

(ii) *Let* $\rho(\mathbf{A} - \mathcal{P}(\mathbf{A}, \mathbf{B})) = \rho(\mathbf{A})$. *Then* $\mathbf{A}$, $-\mathcal{P}(\mathbf{A}, \mathbf{B})$ *are parallel summable and* $\mathbf{B} = -\mathcal{P}(\mathbf{A}, -\mathcal{P}(\mathbf{A}, \mathbf{B})) + \mathbf{W}$, *for some* $\mathbf{W}$ *such that* $\mathbf{A}$ *and* $\mathbf{W}$ *are disjoint matrices.*

Proof. (i) Since $\mathbf{A}$, $\mathbf{B}$ are parallel summable, $\mathcal{C}(\mathbf{A}) + \mathcal{C}(\mathbf{B}) = \mathcal{C}(\mathbf{A} + \mathbf{B})$ and $\mathcal{C}(\mathbf{A}^{t}) + \mathcal{C}(\mathbf{B}^{t}) = \mathcal{C}(\mathbf{A} + \mathbf{B})^{t}$.
Assume first that $\mathcal{C}(\mathbf{B}) \subseteq \mathcal{C}(\mathbf{A})$. Then

$$\mathcal{C}(\mathbf{A} + \mathbf{B}) = \mathcal{C}(\mathbf{A}) + \mathcal{C}(\mathbf{B}) \subseteq \mathcal{C}(\mathbf{A}) \subseteq \mathcal{C}(\mathbf{A} + \mathbf{B}) \Rightarrow \rho(\mathbf{A} + \mathbf{B}) = \rho(\mathbf{A}).$$

This further gives $\mathcal{C}(\mathbf{A}^{t}) = \mathcal{C}(\mathbf{A} + \mathbf{B})^{t}$. Hence, $\mathcal{C}(\mathbf{B}^{t}) \subseteq \mathcal{C}(\mathbf{A}^{t})$.
Similarly, if $\mathcal{C}(\mathbf{B}^{t}) \subseteq \mathcal{C}(\mathbf{A}^{t})$, then $\mathcal{C}(\mathbf{B}) \subseteq \mathcal{C}(\mathbf{A})$.
Notice that $\mathbf{A} - \mathcal{P}(\mathbf{A}, \mathbf{B}) = \mathbf{A}(\mathbf{A} + \mathbf{B})^{-}\mathbf{A}$. So, $\rho(\mathbf{A} - \mathcal{P}(\mathbf{A}, \mathbf{B})) \leq \rho(\mathbf{A})$. Let $\mathcal{C}(\mathbf{B}) \subseteq \mathcal{C}(\mathbf{A})$. Then $\mathbf{B} = \mathbf{AU}$ for some matrix $\mathbf{U}$. Denote $\mathcal{P}(\mathbf{A}, \mathbf{B})$ by $\mathbf{D}$. As $\mathcal{C}(\mathbf{A}) \subseteq \mathcal{C}(\mathbf{A} + \mathbf{B})$, we have $\mathbf{A} = \mathbf{A}(\mathbf{A} + \mathbf{B})^{-}(\mathbf{A} + \mathbf{B}) = (\mathbf{A} - \mathbf{D})(\mathbf{I} + \mathbf{U})$. So, $\rho(\mathbf{A}) \leq \rho(\mathbf{A} - \mathbf{D}) = \rho(\mathbf{A} - \mathcal{P}(\mathbf{A}, \mathbf{B}))$.

(ii) Now, $\rho(\mathbf{A} - \mathcal{P}(\mathbf{A}, \mathbf{B})) = \rho(\mathbf{A})$ and $\mathcal{P}(\mathbf{A}, \mathbf{B}) <^{s} \mathbf{A}$. So, $\mathbf{A}$ and $-\mathcal{P}(\mathbf{A}, \mathbf{B}) = -\mathbf{D}$ are parallel summable. Let $\mathbf{C}$ denote the parallel sum of $\mathbf{A}$ and $-\mathbf{D}$. Clearly, $\mathbf{A} + \mathbf{C} = \mathbf{A}(\mathbf{A} - \mathbf{D})^{-}\mathbf{A}$ and by part (i) $\rho(\mathbf{A} + \mathbf{C}) = \rho(\mathbf{A})$. Thus, $\mathbf{A}$ and $\mathbf{C}$ are also parallel summable. A g-inverse of $\mathbf{A} + \mathbf{C}$ is $\mathbf{A}^{-}(\mathbf{A} - \mathbf{D})\mathbf{A}^{-}$, since

$$(\mathbf{A} + \mathbf{C})\mathbf{A}^{-}(\mathbf{A} - \mathbf{D})\mathbf{A}^{-}(\mathbf{A} + \mathbf{C}) = \mathbf{A}(\mathbf{A} - \mathbf{D})^{-}\mathbf{A}(\mathbf{A}^{-}(\mathbf{A} - \mathbf{D})\mathbf{A}^{-})\mathbf{A}(\mathbf{A} - \mathbf{D})^{-}\mathbf{A}$$

$$= \mathbf{A}(\mathbf{A}-\mathbf{D})^{-}\mathbf{A}(\mathbf{A}-\mathbf{D})(\mathbf{A}-\mathbf{D})^{-}\mathbf{A} = \mathbf{A}(\mathbf{A}-\mathbf{D})^{-}\mathbf{A} = \mathbf{A}+\mathbf{C}.$$

So, the parallel sum of $\mathbf{A}$ and $\mathbf{C}$ is

$$\mathcal{P}(\mathbf{A},\mathbf{C}) = \mathbf{A}\mathbf{A}^{-}(\mathbf{A}-\mathbf{D})\mathbf{A}^{-}\mathbf{C} = \mathbf{A}-\mathbf{D}\mathbf{A}^{-}\mathbf{A}(\mathbf{A}-\mathbf{D})^{-}\mathbf{D} = \mathbf{D} = \mathcal{P}(\mathbf{A},\mathbf{B}).$$

This further implies $\mathbf{A}(\mathbf{A}+\mathbf{C})^{-}\mathbf{A} = \mathbf{A}(\mathbf{A}+\mathbf{B})^{-}\mathbf{A}$ and that every g-inverse of $\mathbf{A}+\mathbf{B}$ is a g-inverse of $\mathbf{A}+\mathbf{C}$. For, if $\mathbf{G}$ is a g-inverse of $\mathbf{A}+\mathbf{B}$, then

$$\begin{aligned}(\mathbf{A}+\mathbf{C})\mathbf{G}(\mathbf{A}+\mathbf{C}) &= \mathbf{A}(\mathbf{A}-\mathbf{D})^{-}\mathbf{A}\mathbf{G}\mathbf{A}(\mathbf{A}-\mathbf{D})^{-}\mathbf{A}\\ &= \mathbf{A}(\mathbf{A}-\mathbf{D})^{-}\mathbf{A}(\mathbf{A}+\mathbf{C})^{-}\mathbf{A}(\mathbf{A}-\mathbf{D})^{-}\mathbf{A}\\ &= (\mathbf{A}+\mathbf{C})(\mathbf{A}+\mathbf{C})^{-}(\mathbf{A}+\mathbf{C}) = \mathbf{A}+\mathbf{C}.\end{aligned}$$

So, $\mathbf{A}+\mathbf{C} <^{-} \mathbf{A}+\mathbf{B}$, and hence $\mathbf{A}+\mathbf{B} = \mathbf{A}+\mathbf{C}+\mathbf{W}$, for some matrix $\mathbf{W}$ disjoint with $\mathbf{A}+\mathbf{B}$ and therefore with $\mathbf{A}$. □

Remark 9.2.27. Given matrices $\mathbf{A}$ and $\mathbf{C}$ of the same order such that $\mathbf{C} <^{s} \mathbf{A}$ and $\rho(\mathbf{A}-\mathbf{C}) = \rho(\mathbf{A})$, there exists a unique matrix $\mathbf{B}$, such that $\mathbf{A}$ and $\mathbf{B}$ are parallel summable and their parallel sum is $\mathbf{C}$. If matrices $\mathbf{A}$ and $\mathbf{C}$ are *nnd*, then so is the matrix $\mathbf{B}$.

Theorem 9.2.28. *Let* $\mathbf{A}$, *and* $\mathbf{B}$ *be* $m \times n$ *matrices such that* $\mathbf{A}$, $\mathbf{B}$ *are parallel summable. Let* $\mathbf{C} = \mathcal{P}(\mathbf{A},\mathbf{B})$. *Then* $\rho(\mathbf{A}-\mathbf{C}) \geq 2\rho(\mathbf{A}) - \rho(\mathbf{A}+\mathbf{B})$. *Conversely, let* $\mathbf{A}$, *and* $\mathbf{B}$ *be any* $m \times n$ *matrices such that* (i) $\mathbf{C} <^{s} \mathbf{A}$ *and* (ii) $\rho(\mathbf{A}-\mathbf{C}) \geq 2\rho(\mathbf{A}) - \min\{m,n\}$. *Then there exists a matrix* $\mathbf{X}$ *of order* $m \times n$ *such that* $\mathbf{A}$ *and* $\mathbf{X}$ *are parallel summable and their parallel sum is* $\mathbf{C}$.

Proof. Let $\rho(\mathbf{A}+\mathbf{B}) = r$ and $(\mathbf{L},\mathbf{R})$ be a rank factorization of $\mathbf{A}+\mathbf{B}$. Since $\mathbf{A}$, $\mathbf{B}$ are parallel summable, there exists an $r \times r$ matrix $\mathbf{D}$ such that $\mathbf{A} = \mathbf{L}\mathbf{D}\mathbf{R}$. Notice that $\rho(\mathbf{A}) = \rho(\mathbf{D})$. Clearly, $\mathbf{B} = \mathbf{L}(\mathbf{I}-\mathbf{D})\mathbf{R}$ and $\mathbf{C} = \mathbf{L}\mathbf{D}(\mathbf{I}-\mathbf{D})\mathbf{R}$. Also, $\mathbf{A}-\mathbf{C} = \mathbf{L}\mathbf{D}^{2}\mathbf{R}$ and $\rho(\mathbf{A}-\mathbf{C}) = \rho(\mathbf{D}^{2})$. By applying Fröbenius inequality to $\mathbf{D}$, $\mathbf{I}$ and $\mathbf{D}$, we have $\rho(\mathbf{D}^{2}) + \rho(\mathbf{I}) \geq \rho(\mathbf{D}) + \rho(\mathbf{D}) = 2\rho(\mathbf{A})$ and $\rho(\mathbf{A}-\mathbf{C}) = \rho(\mathbf{D}^{2}) \geq 2\rho(\mathbf{A}) - r = 2\rho(\mathbf{A}) - \rho(\mathbf{A}+\mathbf{B})$.

(*Converse*). Let $\rho(\mathbf{A}) = r$ and $(\mathbf{P},\mathbf{Q})$ be a rank factorization of $\mathbf{A}$. Since $\mathbf{C} <^{s} \mathbf{A}$, we have $\mathbf{A}-\mathbf{C} <^{s} \mathbf{A}$, so, we can write $\mathbf{A}-\mathbf{C} = \mathbf{P}\mathbf{\Lambda}\mathbf{Q}$. Note that $\rho(\mathbf{A}-\mathbf{C}) = \rho(\mathbf{\Lambda}) = q$ *(say)*. Let p be an integer such that $p \leq \min\{m,n\}$ and $\rho(\mathbf{A}-\mathbf{C}) \geq 2\rho(\mathbf{A}) - p$. Let $\mathbf{F}$ be a matrix of order $p \times r$ and rank r and $\mathbf{E}$ a matrix of $r \times p$ and rank r such that $\mathbf{E}\mathbf{F} = \mathbf{\Lambda}$. Let $\mathbf{L}$ and $\mathbf{R}$ be matrices of full rank such that $\mathbf{L}\mathbf{F} = \mathbf{P}$ and $\mathbf{E}\mathbf{R} = \mathbf{Q}$. Then $\mathbf{A} = \mathbf{P}\mathbf{Q} = \mathbf{L}\mathbf{E}\mathbf{F}\mathbf{R}$ and

$\mathbf{A} - \mathbf{C} = \mathbf{L}(\mathbf{EF})^2\mathbf{R}$. Let $\mathbf{X} = \mathbf{LR} - \mathbf{A}$. Then clearly, $\mathbf{A}$ and $\mathbf{X}$ are parallel summable and $\mathcal{P}(\mathbf{A}, \mathbf{X}) = \mathbf{C}$. □

9.3 Parallel sums and partial orders

In this section we study the parallel sums in relation to the some of the partial orders studied in earlier chapters. We first begin with space pre-order.

Theorem 9.3.1. *Let* $\mathbf{A}$, $\mathbf{B}$ *be* $m \times n$ *matrices. Then the following are equivalent:*

(i) $\mathbf{A}, \mathbf{B}$ *are parallel summable*
(ii) $\mathbf{A} <^{\mathbf{s}} \mathbf{A} + \mathbf{B}$ *and*
(iii) $\mathbf{B} <^{\mathbf{s}} \mathbf{A} + \mathbf{B}$.

Proof follows by Theorem 9.2.7.

Theorem 9.3.2. *Let* $\mathbf{A}, \mathbf{B}$, *be parallel summable matrices of order* $m \times n$. *Then the following hold:*

(i) $\mathcal{P}(\mathbf{A}, \mathbf{B}) <^{\mathbf{s}} \mathbf{A}$
(ii) $\mathcal{P}(\mathbf{A}, \mathbf{B}) <^{\mathbf{s}} \mathbf{B}$ *and*
(iii) *If* $\mathbf{C}$ *is any* $m \times n$ *matrix such that* $\mathbf{C} <^{\mathbf{s}} \mathbf{A}$ *and* $\mathbf{C} <^{\mathbf{s}} \mathbf{B}$ *then* $\mathbf{C} <^{\mathbf{s}} \mathcal{P}(\mathbf{A}, \mathbf{B})$. *Thus, for parallel summable matrices, their parallel sum is a maximal element below each of* $\mathbf{A}, \mathbf{B}$ *under space pre-order.*

Proof. (i) and (ii) are trivial in view of Theorem 9.2.16.
(iii) Since $\mathbf{C} <^{\mathbf{s}} \mathbf{A}, \mathbf{C} <^{\mathbf{s}} \mathbf{B}$, we have $\mathcal{C}(\mathbf{C}) \subseteq \mathcal{C}(\mathbf{A})$, $\mathcal{C}(\mathbf{C}) \subseteq \mathcal{C}(\mathbf{B})$ and $\mathcal{C}(\mathbf{C}^{\mathbf{t}}) \subseteq \mathcal{C}(\mathbf{A}^{\mathbf{t}})$, $\mathcal{C}(\mathbf{C}^{\mathbf{t}}) \subseteq \mathcal{C}(\mathbf{B}^{\mathbf{t}})$. So, $\mathcal{C}(\mathbf{C}) \subseteq \mathcal{C}(\mathbf{A}) \cap \mathcal{C}(\mathbf{B}) = \mathcal{C}(\mathcal{P}(\mathbf{A}, \mathbf{B}))$ and $\mathcal{C}(\mathbf{C}^{\mathbf{t}}) \subseteq \mathcal{C}(\mathbf{A}^{\mathbf{t}}) \cap \mathcal{C}(\mathbf{B}^{\mathbf{t}}) \subseteq \mathcal{C}(\mathbf{B}^{\mathbf{t}}) = \mathcal{C}(\mathcal{P}(\mathbf{A}, \mathbf{B}))^{\mathbf{t}}$. □

Theorem 9.3.3. *Let* $\mathbf{A}, \mathbf{B}$ *and* $\mathbf{C}$ *be matrices of the same order such that*

(i) $\mathbf{A} <^{\mathbf{s}} \mathbf{B}$
(ii) $\mathbf{A}, \mathbf{C}$ *are parallel summable* *and*
(iii) $\mathbf{B}, \mathbf{C}$ *are parallel summable.*
Then $\mathcal{P}(\mathbf{A}, \mathbf{C}) <^{\mathbf{s}} \mathcal{P}(\mathbf{B}, \mathbf{C})$.

Proof is trivial.

Theorem 9.3.4. *Let* $\mathbf{A}$, $\mathbf{B}$ *be* $m \times n$ *matrices. Then the following are equivalent:*

(i) $\mathbf{A} <^- \mathbf{B}$
(ii) $\mathbf{A}, \mathbf{B} - \mathbf{A}$ *are parallel summable and* $\mathcal{P}(\mathbf{A}, \mathbf{B} - \mathbf{A}) = \mathbf{0}$ *and*
(iii) $\mathbf{B} - \mathbf{A} <^- \mathbf{B}$.

Proof. (i) $\Leftrightarrow$ (iii) by Remark 3.3.12.
(i) $\Leftrightarrow$ (ii)
Since $\mathbf{A} <^- \mathbf{B} \Rightarrow \mathbf{B} = \mathbf{A} \oplus \mathbf{B} - \mathbf{A} \Leftrightarrow \mathbf{A}, \mathbf{B} - \mathbf{A}$ are parallel summable and $\mathcal{P}(\mathbf{A}, \mathbf{B} - \mathbf{A}) = \mathbf{0}$. □

Remark 9.3.5. Notice that Theorem 9.3.4 remains valid for any partial order finer than minus order.

We have seen in Theorem 9.2.7 (iii) that for *nnd* matrices $\mathbf{A}$ and $\mathbf{B}$, while $\mathcal{P}(\mathbf{A}, \mathbf{B}) <^{\mathbf{L}} \mathbf{A}$ (and $\mathcal{P}(\mathbf{A}, \mathbf{B}) <^{\mathbf{L}} \mathbf{B}$), the same is not true for minus order as the following example shows:

Example 9.3.6. Let $\mathbf{A} = \begin{pmatrix} 1 & 0 \\ 0 & 0 \end{pmatrix}$ and $\mathbf{B} = \begin{pmatrix} 1 & 0 \\ 0 & 1 \end{pmatrix}$. Then $\mathbf{A}$ and $\mathbf{B}$ are trivially parallel summable and $\mathbf{A} <^- \mathbf{B}$, Now, $\mathcal{P}(\mathbf{A}, \mathbf{B}) = \begin{pmatrix} \frac{1}{2} & 0 \\ 0 & 0 \end{pmatrix}$ and $\mathcal{P}(\mathbf{A}, \mathbf{B}) \not<^- \mathbf{A}$.

In fact it is the same with any order finer than the minus order.

Theorem 9.3.7. *Let* $\mathbf{A}$, $\mathbf{B}$ *be parallel summable matrices of the same order and* $\mathcal{C}(\mathbf{B}) \subseteq \mathcal{C}(\mathbf{A})$. *Then* $\mathcal{P}(\mathbf{A}, \mathbf{B}) <^- \mathbf{A} \Leftrightarrow \mathcal{P}(\mathbf{A}, \mathbf{B}) = \mathbf{0}$.

Proof is trivial in view of Theorem 9.2.26.

Theorem 9.3.8. *Let* $\mathbf{A}, \mathbf{B}$ *and* $\mathbf{C}$ *be matrices of the same order such that*

(i) $\mathbf{A} <^- \mathbf{B}$
(ii) $\mathbf{A}, \mathbf{C}$ *are parallel summable and*
(iii) $\mathbf{B}$, $\mathbf{C}$ *are parallel summable.*

Then $\mathcal{P}(\mathbf{A}, \mathbf{C}) <^- \mathcal{P}(\mathbf{B}, \mathbf{C})$. *Thus parallel sums respect the minus order.*

Proof. By Theorem 9.2.16, we have

$$\mathcal{C}(\mathcal{P}(\mathbf{A}, \mathbf{C})) = \mathcal{C}(\mathbf{A}) \cap \mathcal{C}(\mathbf{C}),\ \mathcal{C}(\mathcal{P}(\mathbf{A}, \mathbf{C}))^{\mathbf{t}} = \mathcal{C}(\mathbf{A}^{\mathbf{t}}) \cap \mathcal{C}(\mathbf{C}^{\mathbf{t}})$$

and $\mathcal{C}(\mathcal{P}(\mathbf{B}, \mathbf{C})) = \mathcal{C}(\mathbf{B}) \cap \mathcal{C}(\mathbf{C}),\ \mathcal{C}(\mathcal{P}(\mathbf{B}, \mathbf{C}))^{\mathbf{t}} = \mathcal{C}(\mathbf{B}^{\mathbf{t}}) \cap \mathcal{C}(\mathbf{C}^{\mathbf{t}})$.

Since $\mathbf{A} <^- \mathbf{B} \Rightarrow \mathbf{A} <^s \mathbf{B}$, so, by Theorem 9.3.3, $\mathcal{P}(\mathbf{A}, \mathbf{C}) <^s \mathcal{P}(\mathbf{B}, \mathbf{C})$. It remains to show that $\mathcal{P}(\mathbf{A}, \mathbf{C})[\mathcal{P}(\mathbf{B}, \mathbf{C})]^- \mathcal{P}(\mathbf{A}, \mathbf{C}) = \mathcal{P}(\mathbf{A}, \mathbf{C})$ for some g-inverse $[\mathcal{P}(\mathbf{B}, \mathbf{C})]^-$ of $\mathcal{P}(\mathbf{B}, \mathbf{C})$. By Theorem 9.2.14 any g-inverse of $\mathcal{P}(\mathbf{B}, \mathbf{C})$ is of the form $\mathbf{B}^- + \mathbf{C}^-$ for some choice of $\mathbf{B}^-$ and $\mathbf{C}^-$. Also, since (i) holds we have $\{\mathbf{B}^-\} \subseteq \{\mathbf{A}^-\}$. So, $\mathbf{B}^- + \mathbf{C}^-$ is also a g-inverse of $\mathcal{P}(\mathbf{A}, \mathbf{C})$. Therefore, the result follows. □

Corollary 9.3.9. *Let* $\mathbf{A}, \mathbf{B}$ *be any* $m \times n$ *matrices such that* $\mathbf{A} <^\star \mathbf{A} + \mathbf{B}$. *Then* $\mathbf{A}, \mathbf{B}$ *are parallel summable and* $\mathcal{P}(\mathbf{A}, \mathbf{B}) = \mathbf{0}$.

The proof follows by Theorems 9.3.2 and 9.3.4. However, an independent proof using Theorem 5.3.3 will be more interesting.

Remark 9.3.10. A theorem similar to Theorem 9.3.8 does not hold for star order as the following example shows.

Example 9.3.11. Let $\mathbf{A} = \begin{pmatrix} 1 & 0 \\ 0 & 0 \end{pmatrix}, \mathbf{B} = \begin{pmatrix} 1 & 0 \\ 0 & 1 \end{pmatrix}$ and $\mathbf{C} = \begin{pmatrix} 1 & 1 \\ 1 & 0 \end{pmatrix}$. Clearly, $\mathbf{A} <^\star \mathbf{B}$, $\mathbf{A}, \mathbf{C}$ and $\mathbf{B}, \mathbf{C}$ are parallel summable. But then $\mathcal{P}(\mathbf{A}, \mathbf{C}) = \mathbf{A}, \mathcal{P}(\mathbf{B}, \mathbf{C}) = \begin{pmatrix} 0 & 1 \\ 1 & -1 \end{pmatrix}$, and $\mathcal{P}(\mathbf{A}, \mathbf{C}) \not<^\star \mathcal{P}(\mathbf{B}, \mathbf{C})$.

We now give two interesting results for matrices in the class of hermitian matrices. For the first result which shows parallel sums respect the Löwner order, we give two proofs. The first proof has an advantage of obtaining the conclusion quickly. The second proof follows after of a series of theorems which are themselves useful in obtaining several other useful properties of parallel sums in this and in the next section.

Theorem 9.3.12. *Let* $\mathbf{A}, \mathbf{B}$ *and* $\mathbf{C}$ *be nnd matrices of the same order such that* $\mathbf{A} <^{\mathbf{L}} \mathbf{B}$. *Then* $\mathcal{P}(\mathbf{A}, \mathbf{C}) <^{\mathbf{L}} \mathcal{P}(\mathbf{B}, \mathbf{C})$.

Proof. Notice that $\mathbf{A}, \mathbf{C}$ and $\mathbf{B}, \mathbf{C}$ are both parallel summable pairs by Theorem 9.2.7. Since $\mathbf{A} <^{\mathbf{L}} \mathbf{B} \Rightarrow \mathbf{A} <^s \mathbf{B}$, by Theorem 9.3.3, we have $\mathcal{P}(\mathbf{A}, \mathbf{C}) <^s \mathcal{P}(\mathbf{B}, \mathbf{C})$. Also, $\mathcal{P}(\mathbf{A}, \mathbf{C})$ and $\mathcal{P}(\mathbf{B}, \mathbf{C})$ are *nnd* and $\mathcal{C}(\mathcal{P}(\mathbf{A}, \mathbf{C})) \subseteq \mathcal{C}(\mathcal{P}(\mathbf{A}, \mathbf{C}))$, so, by Corollary 8.2.12, it is enough to prove $\mathcal{P}(\mathbf{A}, \mathbf{C})[\mathcal{P}(\mathbf{B}, \mathbf{C})]^- \mathcal{P}(\mathbf{A}, \mathbf{C}) <^{\mathbf{L}} \mathcal{P}(\mathbf{A}, \mathbf{C})$ for any g-inverse $[\mathcal{P}(\mathbf{B}, \mathbf{C})]^-$ of $\mathcal{P}(\mathbf{B}, \mathbf{C})$. We can choose $\mathbf{B}^\dagger + \mathbf{C}^\dagger$ for a g-inverse of $\mathcal{P}(\mathbf{B}, \mathbf{C})$ as $\mathbf{B}$ and $\mathbf{C}$ are *nnd*. Consider

$\mathcal{P}(\mathbf{A}, \mathbf{C}) - \mathcal{P}(\mathbf{A}, \mathbf{C})[\mathcal{P}(\mathbf{B}, \mathbf{C})]^- \mathcal{P}(\mathbf{A}, \mathbf{C}) = \mathcal{P}(\mathbf{A}, \mathbf{C}) - \mathcal{P}(\mathbf{A}, \mathbf{C})(\mathbf{B}^\dagger + \mathbf{C}^\dagger)\mathcal{P}(\mathbf{A}, \mathbf{C})$
$= \mathbf{C}(\mathbf{A} + \mathbf{C})^\dagger \mathbf{A} - \mathbf{C}(\mathbf{A} + \mathbf{C})^\dagger \mathbf{A}(\mathbf{B}^\dagger + \mathbf{C}^\dagger)\mathbf{C}(\mathbf{A} + \mathbf{C})^\dagger \mathbf{A} = \mathbf{C}(\mathbf{A} + \mathbf{C})^\dagger \mathbf{A}$

$- \mathbf{C}(\mathbf{A}+\mathbf{C})^{\dagger}\mathbf{A}\mathbf{B}^{\dagger}\mathbf{C}(\mathbf{A}+\mathbf{C})^{\dagger}\mathbf{A} - \mathbf{A}(\mathbf{A}+\mathbf{C})^{\dagger}\mathbf{C}\mathbf{C}^{\dagger}\mathbf{C}(\mathbf{A}+\mathbf{C})^{\dagger}\mathbf{A}.$

However,

$$\begin{aligned}\mathbf{A}(\mathbf{A}+\mathbf{C})^{\dagger}\mathbf{C}\mathbf{C}^{\dagger}\mathbf{C}(\mathbf{A}+\mathbf{C})^{\dagger}\mathbf{A} &= (\mathbf{A}+\mathbf{C}-\mathbf{C})(\mathbf{A}+\mathbf{C})^{\dagger}\mathbf{C}(\mathbf{A}+\mathbf{C})^{\dagger}\mathbf{A}\\ &= (\mathbf{A}+\mathbf{C})(\mathbf{A}+\mathbf{C})^{\dagger}\mathbf{C}(\mathbf{A}+\mathbf{C})^{\dagger}\mathbf{A} - \mathbf{C}(\mathbf{A}+\mathbf{C})^{\dagger}\mathbf{C}(\mathbf{A}+\mathbf{C})^{\dagger}\mathbf{A}\\ &= \mathbf{C}(\mathbf{A}+\mathbf{C})^{\dagger}\mathbf{A} - \mathbf{C}(\mathbf{A}+\mathbf{C})^{\dagger}\mathbf{C}(\mathbf{A}+\mathbf{C})^{\dagger}\mathbf{A}\\ &= \mathbf{C}(\mathbf{A}+\mathbf{C})^{\dagger}\mathbf{A} - \mathbf{C}(\mathbf{A}+\mathbf{C})^{\dagger}\mathbf{A}(\mathbf{A}+\mathbf{C})^{\dagger}\mathbf{C}.\end{aligned}$$

Therefore,

$$\mathcal{P}(\mathbf{A},\mathbf{C}) - \mathcal{P}(\mathbf{A},\mathbf{C})[\mathcal{P}(\mathbf{B},\mathbf{C})]^{-}\mathcal{P}(\mathbf{A},\mathbf{C}) = \mathbf{C}(\mathbf{A}+\mathbf{C})^{\dagger}(\mathbf{A}-\mathbf{A}\mathbf{B}^{\dagger}\mathbf{A})(\mathbf{A}+\mathbf{C})^{\dagger}\mathbf{A}.$$

Now, since $\mathbf{A} <^{\mathbf{L}} \mathbf{B}$, we have $\mathbf{A}\mathbf{B}^{\dagger}\mathbf{A} <^{\mathbf{L}} \mathbf{A}$. Therefore,

$$\mathbf{C}(\mathbf{A}+\mathbf{C})^{\dagger}(\mathbf{A}-\mathbf{A}\mathbf{B}^{\dagger}\mathbf{A})(\mathbf{A}+\mathbf{C})^{\dagger}\mathbf{A} \geq \mathbf{0}.$$

Hence the result follows. □

We now give the second proof of Theorem 9.3.12. We first prove a lemma.

Lemma 9.3.13. *Let* $\mathbf{A}$, $\mathbf{B} \in \mathbb{C}^{\mathbf{n}\times\mathbf{n}}$ *be nnd matrices. Then for any n-vectors* $\mathbf{x},\mathbf{y},\mathbf{z}$ *such that* $\mathbf{x}+\mathbf{y}=\mathbf{z}$, $(\mathcal{P}(\mathbf{A},\mathbf{B})\mathbf{z},\mathbf{z}) \leq (\mathbf{A}\mathbf{x},\mathbf{x}) + (\mathbf{B}\mathbf{y},\mathbf{y})$ *and equality holds if* $\mathbf{z} \in \mathcal{C}(\mathbf{A})+\mathcal{C}(\mathbf{B})$ *and* $\mathbf{z} = \mathbf{x_0}+\mathbf{y_0}$, *where* $\mathbf{x_0} = (\mathbf{A}+\mathbf{B})^{\dagger}\mathbf{B}\mathbf{z}$ *and* $\mathbf{y_0} = (\mathbf{A}+\mathbf{B})^{\dagger}\mathbf{A}\mathbf{z}$.

Proof. First notice that $\mathbf{A}\mathbf{x_0} = \mathcal{P}(\mathbf{A},\mathbf{B})\mathbf{z}$, $\mathbf{B}\mathbf{y_0} = \mathcal{P}(\mathbf{A},\mathbf{B})\mathbf{z}$ and $\mathbf{x_0} + \mathbf{y_0} = (\mathbf{A}+\mathbf{B})^{\dagger}(\mathbf{A}+\mathbf{B})\mathbf{z} = (\mathbf{A}+\mathbf{B})(\mathbf{A}+\mathbf{B})^{\dagger}\mathbf{z}$. As $\mathbf{z} \in \mathcal{C}(\mathbf{A})+\mathcal{C}(\mathbf{B})$ and $\mathbf{A},\mathbf{B}$ are parallel summable, $\mathcal{C}(\mathbf{A})+\mathcal{C}(\mathbf{B}) = \mathcal{C}(\mathbf{A}+\mathbf{B})$. So, $\mathbf{x_0}+\mathbf{y_0} = \mathbf{z}$. Also,

$$\begin{aligned}(\mathcal{P}(\mathbf{A},\mathbf{B})\mathbf{z},\mathbf{z}) &= \mathbf{z}^{\star}\mathcal{P}(\mathbf{A},\mathbf{B})\mathbf{z} = \mathbf{z}^{\star}\mathcal{P}(\mathbf{A},\mathbf{B})\mathbf{x_0} + \mathbf{z}^{\star}\mathcal{P}(\mathbf{A},\mathbf{B})\mathbf{y_0}\\ &= (\mathbf{A}\mathbf{x_0},\mathbf{x_0}) + (\mathbf{B}\mathbf{y_0},\mathbf{y_0}).\end{aligned}$$

Thus, the equality holds.

Now, for the first part, let $\mathbf{x^1} = \mathbf{P}_{(\mathbf{A}+\mathbf{B})}\mathbf{x}$, $\mathbf{y^1} = \mathbf{P}_{(\mathbf{A}+\mathbf{B})}\mathbf{y}$ and $\mathbf{z^1} = \mathbf{P}_(\mathbf{A}+\mathbf{B})\mathbf{z}$, where $\mathbf{P}_{(\mathbf{A}+\mathbf{B})}$ is the orthogonal projection onto $\mathcal{C}(\mathbf{A}+\mathbf{B})$. Write $\mathbf{x_0^1} = (\mathbf{A}+\mathbf{B})^{\dagger}\mathbf{B}\mathbf{z^1}, \mathbf{y_0^1} = (\mathbf{A}+\mathbf{B})^{\dagger}\mathbf{A}\mathbf{z^1}$. Note that $\mathbf{x_0^1} + \mathbf{y_0^1} = \mathbf{z^1} = \mathbf{x^1}+\mathbf{y^1}$. We can take $\mathbf{x^1} = \mathbf{x_0^1}+\mathbf{t}, \mathbf{y^1} = \mathbf{y_0^1}-\mathbf{t}$. Then $(\mathbf{A}\mathbf{x^1},\mathbf{x^1}) = (\mathbf{A}\mathbf{x_0^1},\mathbf{x_0^1}) + \mathbf{2Re}(\mathbf{A}\mathbf{x_0^1},\mathbf{t}) + (\mathbf{A}\mathbf{t},\mathbf{t})$ and $(\mathbf{B}\mathbf{y^1},\mathbf{y^1}) = (\mathbf{A}\mathbf{y_0^1},\mathbf{y_0^1}) - \mathbf{2Re}(\mathbf{B}\mathbf{y_0^1},\mathbf{t}) + (\mathbf{B}\mathbf{t},\mathbf{t})$. Since, $\mathbf{A}\mathbf{x_0^1} = \mathbf{B}\mathbf{y_0^1} = \mathcal{P}(\mathbf{A},\mathbf{B})\mathbf{z^1}$, therefore,

$$\begin{aligned}(\mathbf{A}\mathbf{x^1},\mathbf{x^1}) + (\mathbf{B}\mathbf{y^1},\mathbf{y^1}) &= (\mathbf{A}\mathbf{x_0^1},\mathbf{x_0^1}) + (\mathbf{A}\mathbf{t},\mathbf{t}) + (\mathbf{A}\mathbf{y_0^1},\mathbf{y_0^1}) + (\mathbf{B}\mathbf{t},\mathbf{t})\\ &\geq (\mathcal{P}(\mathbf{A},\mathbf{B})\mathbf{z^1},\mathbf{z^1}).\end{aligned}$$

However, $(\mathbf{A}\mathbf{x^1},\mathbf{x^1}) = (\mathbf{A}\mathbf{P}_{(\mathbf{A}+\mathbf{B})}\mathbf{x},\mathbf{P}_{(\mathbf{A}+\mathbf{B})}\mathbf{x}) = (\mathbf{A}\mathbf{x},\mathbf{x}), (\mathbf{B}\mathbf{y^1},\mathbf{y^1}) = (\mathbf{B}\mathbf{y},\mathbf{y})$ and $(\mathcal{P}(\mathbf{A},\mathbf{B})\mathbf{z^1},\mathbf{z^1}) = (\mathcal{P}(\mathbf{A},\mathbf{B})\mathbf{z},\mathbf{z})$, so the result follows. □

Corollary 9.3.14. *If* $\mathbf{A}$ *and* $\mathbf{B}$ *are nnd matrices of the same order, then* $(\mathcal{P}(\mathbf{A},\mathbf{B})\mathbf{z},\mathbf{z}) \le \mathcal{P}((\mathbf{Az},\mathbf{z}),(\mathbf{Bz},\mathbf{z}))$.

Proof. If $(\mathbf{Az},\mathbf{z}) + (\mathbf{Bz},\mathbf{z}) = \mathbf{0}$, $\mathcal{P}((\mathbf{Az},\mathbf{z}),(\mathbf{Bz},\mathbf{z}))$ is not defined. So, we may assume that $(\mathbf{Az},\mathbf{z}) + (\mathbf{Bz},\mathbf{z}) \ne \mathbf{0}$. Let

$$\mathbf{x} - \frac{(\mathbf{Bz},\mathbf{z})}{(\mathbf{A}+\mathbf{Bz},\mathbf{z})}\mathbf{z} \text{ and } \mathbf{y} = \frac{(\mathbf{Az},\mathbf{z})}{(\mathbf{A}+\mathbf{Bz},\mathbf{z})}\mathbf{z}.$$

Then $\mathbf{z} = \mathbf{x}+\mathbf{y}$. By Lemma 9.3.13, $(\mathcal{P}(\mathbf{A},\mathbf{B})\mathbf{z},\mathbf{z}) \le (\mathbf{Ax},\mathbf{x}) + (\mathbf{By},\mathbf{y})$. However,

$$(\mathbf{Ax},\mathbf{x}) + (\mathbf{By},\mathbf{y}) = \frac{(\mathbf{Az},\mathbf{z})(\mathbf{Bz},\mathbf{z})}{((\mathbf{A}+\mathbf{B})\mathbf{z},\mathbf{z})} = \mathcal{P}((\mathbf{Az},\mathbf{z}),(\mathbf{Bz},\mathbf{z})),$$

the result follows. □

In fact, we have a more general result.

Theorem 9.3.15. *Let* $\mathbf{A},\mathbf{B}$ *be nnd matrices of order* $n \times n$ *and let* $\mathbf{Z} \in \mathbb{C}^{\mathbf{n}\times\mathbf{n}}$ *be an arbitrary matrix. Then* $\mathbf{Z}^\star\mathcal{P}(\mathbf{A},\mathbf{B})\mathbf{Z} <^{\mathbf{L}} \mathcal{P}(\mathbf{Z}^\star\mathbf{AZ},\mathbf{Z}^\star\mathbf{BZ})$.

Proof. Let $\mathbf{Q}$ be an orthogonal projection onto $\mathcal{C}(\mathbf{A}+\mathbf{B})$. Then, $(\mathbf{QZ})^\star(\mathbf{A}+\mathbf{B})\mathbf{QZ} = \mathbf{Z}^\star\mathbf{Q}(\mathbf{A}+\mathbf{B})\mathbf{QZ} = \mathbf{Z}^\star(\mathbf{A}+\mathbf{B})\mathbf{Z}$, so, we may assume that $\mathbf{QZ} = \mathbf{Z}$, i.e. $\mathbf{Z} \in \mathcal{C}(\mathbf{A}+\mathbf{B})$. Let $\mathbf{X_0} = (\mathbf{A}+\mathbf{B})^\dagger\mathbf{BZ}$ and $\mathbf{Y_0} = (\mathbf{A}+\mathbf{B})^\dagger\mathbf{AZ}$. Then $\mathbf{X_0}+\mathbf{Y_0} = \mathbf{Z}$ and $\mathbf{X_0^\star AX_0} + \mathbf{Y_0^\star BY_0} = \mathbf{Z}^\star\mathcal{P}(\mathbf{A},\mathbf{B})\mathbf{Z}$. For any $\mathbf{X},\mathbf{Y}$ such that $\mathbf{Z} = \mathbf{X}+\mathbf{Y}$, let $\mathbf{X} = \mathbf{X_0} - \mathbf{T}$, $\mathbf{Y} = \mathbf{Y_0} + \mathbf{T}$. Now,
$\mathbf{X^\star AX} + \mathbf{Y^\star BY} = (\mathbf{X_0}-\mathbf{T})^\star\mathbf{A}(\mathbf{X_0}-\mathbf{T}) + (\mathbf{Y_0}+\mathbf{T})^\star\mathbf{B}(\mathbf{X_0}+\mathbf{T})$
$= \mathbf{X_0^\star AX_0} + \mathbf{Y_0^\star BY_0} + \mathbf{T^\star AT} + \mathbf{T^\star BT} - \mathbf{T^\star AX_0} - \mathbf{X_0^\star AT} + \mathbf{T^\star BY_0} + \mathbf{Y_0^\star BT}$
$= \mathbf{X_0^\star AX_0} + \mathbf{Y_0^\star BY_0} + \mathbf{T^\star AT} + \mathbf{T^\star BT}$, since, $\mathbf{AX_0} = \mathbf{BY_0}$.
Therefore, $\mathbf{Z}^\star\mathcal{P}(\mathbf{A},\mathbf{B})\mathbf{Z} <^{\mathbf{L}} \mathbf{X^\star AX} + \mathbf{Y^\star BY}$.
Now, let $\mathbf{X} = \mathbf{Z}(\mathbf{Z}^\star(\mathbf{A}+\mathbf{B})\mathbf{Z})^\dagger\mathbf{Z}^\star\mathbf{BZ}$ and $\mathbf{Y} = \mathbf{Z}(\mathbf{Z}^\star(\mathbf{A}+\mathbf{B})\mathbf{Z})^\dagger\mathbf{Z}^\star\mathbf{BZ}$. Clearly, $\mathbf{X}+\mathbf{Y} = \mathbf{Z}$, as $\mathbf{QZ} = \mathbf{Z}$. It is easy to check now that

$$\mathbf{X^\star AX} + \mathbf{Y^\star BY} = \mathcal{P}(\mathbf{Z^\star AZ}|\mathbf{Z^\star BZ}).$$

□

Lemma 9.3.16. *Let* $\mathbf{A},\mathbf{B},\mathbf{C}$ *and* $\mathbf{D}$ *be nnd matrices of the same order. Then* $\mathcal{P}(\mathbf{A},\mathbf{C}) + \mathcal{P}(\mathbf{B},\mathbf{D}) <^{\mathbf{L}} \mathcal{P}(\mathbf{A}+\mathbf{B},\mathbf{C}+\mathbf{D})$.

Proof. Notice that the column space of any of the involved parallel sums is contained in $\mathcal{C}(\mathbf{A}+\mathbf{B}+\mathbf{C}+\mathbf{D})$. So, we take $\mathbf{z} \in \mathcal{C}(\mathbf{A}+\mathbf{B}+\mathbf{C}+\mathbf{D})$. By Theorem 9.2.18, for suitable $\mathbf{x_0},\mathbf{y_0}$ we have $(\mathcal{P}(\mathbf{A}+\mathbf{B}|\mathbf{C}+\mathbf{D})\mathbf{z},\mathbf{z}) = (\mathbf{A}+\mathbf{Bx_0},\mathbf{x_0}) + (\mathbf{C}+\mathbf{Dy_0},\mathbf{y_0}) = (\mathbf{Ax_0},\mathbf{x_0}) + (\mathbf{Bx_0},\mathbf{x_0}) + (\mathbf{Cy_0},\mathbf{y_0}) + (\mathbf{Dy_0},\mathbf{y_0}) \ge (\mathcal{P}(\mathbf{A},\mathbf{C})\mathbf{z},\mathbf{z}) + (\mathcal{P}(\mathbf{B},\mathbf{D})\mathbf{z},\mathbf{z})$. □

And finally now, we give the second proof of Theorem 9.3.12. Corollary 9.3.17 below is a restatement of Theorem 9.3.12.

Corollary 9.3.17. *Let* $\mathbf{A}, \mathbf{B}$ *and* $\mathbf{C}$ *be nnd matrices of the same order such that* $\mathbf{A} <^{\mathbf{L}} \mathbf{B}$. *Then* $\mathcal{P}(\mathbf{A}, \mathbf{C}) <^{\mathbf{L}} \mathcal{P}(\mathbf{B}, \mathbf{C})$.

Proof. Let $\mathbf{D} = \mathbf{B} - \mathbf{A}$. Then $\mathbf{D}$ is *nnd* and by Lemma 9.3.16,
$\mathcal{P}(\mathbf{B}, \mathbf{C}) = \mathcal{P}(\mathbf{A} + \mathbf{D}, \mathbf{C} + \mathbf{0}) \geq \mathcal{P}(\mathbf{A}, \mathbf{C}) + \mathcal{P}(\mathbf{D}, \mathbf{0}) = \mathcal{P}(\mathbf{A}, \mathbf{C})$, □

Corollary 9.3.18. *Let* $\mathbf{A}$ *and* $\mathbf{B}$ *be nnd matrices of the same order such that* $\mathbf{A} <^{\mathbf{L}} \mathbf{B}$. *Then* $\mathbf{A} <^{\mathbf{L}} 2\mathcal{P}(\mathbf{A}, \mathbf{B}) <^{\mathbf{L}} \mathbf{B}$.

Remark 9.3.19. The result in Lemma 9.3.16 is called series-parallel inequality. This corresponds to the physical fact that the impedance on the right hand side(RHS) of the inequality is bigger than the one on the left hand side(LHS) as the network corresponding to the RHS has more paths than the one corresponding to the LHS.

Remark 9.3.20. Notice that from Theorems 9.2.7(i), 9.2.16(i) and 9.3.12 and Ex. 10, it follows that the class of all *nnd* matrices forms a partially ordered abelian semi-group with respect to parallel addition as its binary operation.

9.4 Continuity and index of parallel sums

In this section we first give the error bounds on the matrix $\mathcal{P}(\mathbf{A} + \mathbf{X}, \mathbf{B} + \mathbf{Y}) - \mathcal{P}(\mathbf{A}, \mathbf{B})$, which is the difference of the the parallel sums of matrices $\mathbf{A}$, $\mathbf{B}$ and their perturbations $\mathbf{A} + \mathbf{X}$, $\mathbf{B} + \mathbf{Y}$ and use the same to discuss the continuity of parallel sums. We begin with

Theorem 9.4.1. *If* $\mathbf{A}, \mathbf{B}$ *be nnd matrices of the same order, then* $\|\mathcal{P}(\mathbf{A}, \mathbf{B})\| \leq \mathcal{P}(\|\mathbf{A}\|, \|\mathbf{B}\|)$.

Proof. We first note that for an *nnd* matrix $\mathbf{A}$, we have for each $\mathbf{x}$, $(\mathbf{Ax}, \mathbf{Ax}) \leq \|\mathbf{A}\|(\mathbf{Ax}, \mathbf{x})$. The inequality is trivial if the matrix $\mathbf{A} = \mathbf{0}$. So, let $\mathbf{A} \neq \mathbf{0}$. Let ϵ be any real positive number. Consider $\|\mathbf{A}\| - \epsilon$. There exists a non-null vector $\mathbf{x_0}$ such that

$$\frac{(\mathbf{Ax_0}, \mathbf{Ax_0})}{(A\mathbf{x_0}, \mathbf{x_0})} \geq \|\mathbf{A}\| - \epsilon.$$

So, by above observation, as $\mathcal{P}(\mathbf{A},\mathbf{B})$ is *nnd*, for each positive real number ϵ there exists an $\mathbf{x}$ such that

$$\frac{(\mathcal{P}(\mathbf{A},\mathbf{B})\mathbf{x},\mathcal{P}(\mathbf{A},\mathbf{B})\mathbf{x})}{(\mathcal{P}(\mathbf{A},\mathbf{B})\mathbf{x},\mathbf{x})} \geq \|\mathcal{P}(\mathbf{A},\mathbf{B})\| - \epsilon.$$

Let $\mathbf{y} = \mathcal{P}(\mathbf{A},\mathbf{B})\mathbf{x}$. Then $\mathcal{P}(\mathbf{A},\mathbf{B})(\mathbf{A}^\dagger + \mathbf{B}^\dagger)\mathbf{y} = \mathbf{y}$ and $(\mathcal{P}(\mathbf{A},\mathbf{B})\mathbf{x},\mathbf{x}) = (\mathbf{y},\mathbf{x}) = (\mathcal{P}(\mathbf{A},\mathbf{B})((\mathbf{A}^\dagger + \mathbf{B}^\dagger)\mathbf{y},\mathbf{x}) = (\mathbf{A}^\dagger + \mathbf{B}^\dagger)\mathbf{y},\mathcal{P}(\mathbf{A},\mathbf{B})\mathbf{x}) = ((\mathbf{A}^\dagger + \mathbf{B}^\dagger)\mathbf{y},\mathbf{y}) = (\mathbf{A}^\dagger\mathbf{y},\mathbf{y}) + (\mathbf{B}^\dagger\mathbf{y},\mathbf{y})$. Since $\mathbf{y} \in \mathcal{C}(\mathcal{P}(\mathbf{A},\mathbf{B}))$, let $\mathbf{y} = \mathbf{A}\mathbf{u} = \mathbf{B}\mathbf{v}$. Then $(\mathbf{A}^\dagger\mathbf{y},\mathbf{y}) = (\mathbf{u},\mathbf{A}\mathbf{u})$ and $(\mathbf{B}^\dagger\mathbf{y},\mathbf{y}) = (\mathbf{v},\mathbf{B}\mathbf{v})$. So,

$$\begin{aligned}(\frac{1}{\|\mathbf{A}\|} + \frac{1}{\|\mathbf{B}\|}) &\leq \frac{(\mathbf{u},\mathbf{A}\mathbf{u})}{(\mathbf{A}\mathbf{u},\mathbf{A}\mathbf{u})} + \frac{(\mathbf{v},\mathbf{B}\mathbf{v})}{(\mathbf{B}\mathbf{v},\mathbf{B}\mathbf{v})} = \frac{(\mathcal{P}(\mathbf{A},\mathbf{B})\mathbf{x},\mathbf{x})}{(\mathcal{P}(\mathbf{A},\mathbf{B})\mathbf{x},\mathcal{P}(\mathbf{A},\mathbf{B})\mathbf{x})}\\ &\leq \frac{1}{\|\mathcal{P}(\mathbf{A},\mathbf{B})\| - \epsilon}\end{aligned}$$

or equivalently $\mathcal{P}(\|\mathbf{A}\|,\|\mathbf{B}\|) \geq \|\mathcal{P}(\mathbf{A},\mathbf{B})\| - \epsilon$. as ϵ is arbitrary, result follows. □

Remark 9.4.2. We note that in Theorem 9.4.1, equality occurs if there exists a vector $\mathbf{y}$ such that $\mathbf{A}\mathbf{y} = \|\mathbf{A}\|\mathbf{y}$ and $\mathbf{B}\mathbf{y} = \|\mathbf{B}\|\mathbf{y}$. It also implies the continuity of the operator $\mathcal{P}(\mathbf{A},\mathbf{B})$ about '$\mathbf{0}$'.

Theorem 9.4.3. *Let* $\mathbf{A},\mathbf{B}$ *and* $\mathbf{X}$ *be nnd matrices the same order and* $\mathbf{G}$ *denote the matrix* $\mathcal{P}(\mathbf{A},\mathbf{B}+\mathbf{X}) - \mathcal{P}(\mathbf{A},\mathbf{B})$. *Then* $\mathbf{G}$ *is nnd and if* $\mathbf{C} = \mathbf{A}+\mathbf{B}$, $\mathbf{G} = \mathbf{A}\mathbf{C}^\dagger(\mathcal{P}(\mathbf{C},\mathbf{X}))\mathbf{C}^\dagger\mathbf{A}$ *and* $\|\mathbf{G}\| \leq \|\mathbf{C}^\dagger\mathbf{A}\|^2\|\mathbf{X}\|$.

Proof. Notice that

$$\begin{aligned}\mathbf{G} &= \mathbf{A}(\mathbf{A}+\mathbf{B}+\mathbf{X})^\dagger\mathbf{B}+\mathbf{X} - \mathbf{A}(\mathbf{A}+\mathbf{B})^\dagger\mathbf{B}\\ &= \mathbf{A}(\mathbf{C}+\mathbf{X})^\dagger\mathbf{B}+\mathbf{X} - \mathbf{A}\mathbf{C}^\dagger\mathbf{B}\\ &= \mathbf{A}((\mathbf{C}+\mathbf{X})^\dagger - \mathbf{C}^\dagger)\mathbf{B} + \mathbf{A}(\mathbf{C}+\mathbf{X})^\dagger\mathbf{X}.\end{aligned}$$

If $\mathbf{P_C}$ is the orthogonal projector onto $\mathcal{C}(\mathbf{C})$, then $\mathbf{P_C}\mathbf{B} = \mathbf{B}$. So,

$$\begin{aligned}\mathbf{A}((\mathbf{C}+\mathbf{X})^\dagger - \mathbf{C}^\dagger)\mathbf{B} &= \mathbf{A}((\mathbf{C}+\mathbf{X})^\dagger - \mathbf{C}^\dagger)\mathbf{C}\mathbf{C}^\dagger\mathbf{B}\\ &= \mathbf{A}((\mathbf{C}+\mathbf{X})^\dagger)\mathbf{C}+\mathbf{X} - \mathbf{X}\mathbf{C}^\dagger\mathbf{B} - \mathbf{A}\mathbf{C}^\dagger\mathbf{C}\mathbf{C}^\dagger\mathbf{B}\\ &= \mathbf{A}(\mathbf{C}^\dagger\mathbf{B} - \mathbf{C}^\dagger\mathbf{B}) - \mathbf{A}(\mathbf{C}+\mathbf{X})^\dagger\mathbf{X}\mathbf{C}^\dagger\mathbf{B}.\end{aligned}$$

Hence, $\mathbf{G} = -\mathbf{A}(\mathbf{C}+\mathbf{X})^\dagger\mathbf{X}\mathbf{C}^\dagger\mathbf{B} + \mathbf{A}(\mathbf{C}+\mathbf{X})^\dagger\mathbf{X}$. Since, $\mathcal{C}(\mathbf{A}) \subseteq \mathcal{C}(\mathbf{C})$, $\mathbf{P_C}\mathbf{G} = \mathbf{G}$, and $\mathbf{G}$ being hermitian, $\mathbf{G}\mathbf{P_C} = \mathbf{G}$. So,

$$\begin{aligned}\mathbf{G} = \mathbf{G}\mathbf{P_C} &= \mathbf{A}(\mathbf{C}+\mathbf{X})^\dagger\mathbf{X}\mathbf{P_C}(\mathbf{I} - \mathbf{C}^\dagger\mathbf{B})\\ &= \mathbf{A}(\mathbf{C}+\mathbf{X})^\dagger\mathbf{X}\mathbf{P_C}\mathbf{C}^\dagger\mathbf{A}\\ &= \mathbf{A}\mathbf{C}^\dagger\mathbf{C}(\mathbf{C}+\mathbf{X})^\dagger\mathbf{X}\mathbf{P_C}\mathbf{C}^\dagger\mathbf{A}\\ &= \mathbf{A}\mathbf{C}^\dagger\mathcal{P}(\mathbf{C}|\mathbf{X})\mathbf{C}^\dagger\mathbf{A}.\end{aligned}$$

Since $\mathcal{P}(\mathbf{C},\mathbf{X})$ is hermitian, $\mathbf{G}$ is hermitian. Moreover,

$$\|\mathbf{G}\| \leq \|\mathbf{A}\mathbf{C}^\dagger\|\|\mathcal{P}(\mathbf{C},\mathbf{X})\|\|\mathbf{C}^\dagger\mathbf{A}\| \leq \|\mathbf{A}\mathbf{C}^\dagger\|^2\|\mathbf{X}\|,$$

for by Theorem 9.4.1,

$$\|\mathcal{P}(\mathbf{C},\mathbf{X})\| \leq \frac{\|\mathbf{C}\|\|\mathbf{X}\|}{\|\mathbf{C}\|+\|\mathbf{X}\|}.$$ □

Lemma 9.4.4. *If* $\mathbf{A},\mathbf{B}$ *and* $\mathbf{X}$ *be nnd matrices of the same order, then* $2\mathcal{P}(\mathbf{A}+\mathbf{X},\mathbf{B}+\mathbf{X})+\mathcal{P}(\mathbf{A}+\mathbf{B},2\mathbf{X}) = 2\mathcal{P}(\mathbf{A}+\mathbf{B}+\mathbf{X},\mathbf{X})+\mathcal{P}(\mathbf{A},\mathbf{B}+2\mathbf{X})$ $+\mathcal{P}(\mathbf{A}+2\mathbf{X},\mathbf{B})$.

Proof. Let $\mathbf{A}+\mathbf{B}+2\mathbf{X}=\mathbf{D}$. Then

$$\begin{aligned}&2\mathcal{P}(\mathbf{A}+\mathbf{X},\mathbf{B}+\mathbf{X})+\mathcal{P}(\mathbf{A}+\mathbf{B},2\mathbf{X})\\ &=2\mathcal{P}(\mathbf{A}+\mathbf{X},\mathbf{B}+\mathbf{X})+\mathcal{P}(\mathbf{A}+\mathbf{B},2\mathbf{X})\\ &=2(\mathbf{A}+\mathbf{X})\mathbf{D}^{\dagger}\mathbf{B}+\mathbf{X}+2(\mathbf{A}+\mathbf{B})\mathbf{D}^{\dagger}\mathbf{X}\\ &=2\{\mathbf{A}\mathbf{D}^{\dagger}\mathbf{B}+2\mathbf{A}\mathbf{D}^{\dagger}\mathbf{X}+\mathbf{X}\mathbf{D}^{\dagger}\mathbf{B}+\mathbf{X}\mathbf{D}^{\dagger}\mathbf{X}\}.\end{aligned}$$

Also,

$$\begin{aligned}&2\mathcal{P}(\mathbf{A}+\mathbf{B}+\mathbf{X},\mathbf{X})+\mathcal{P}(\mathbf{A},\mathbf{B}+2\mathbf{X})\\ &=2\mathcal{P}(\mathbf{A}+\mathbf{B}+\mathbf{X},\mathbf{X})+\mathcal{P}(\mathbf{A},\mathbf{B}+2\mathbf{X})+\mathcal{P}(\mathbf{A}+2\mathbf{X},\mathbf{B})\\ &=2(\mathbf{A}+\mathbf{B}+\mathbf{X})\mathbf{D}^{\dagger}\mathbf{X}+(\mathbf{A})\mathbf{D}^{\dagger}\mathbf{B}-2\mathbf{X}+(\mathbf{A}+2\mathbf{X})\mathbf{D}^{\dagger}\mathbf{B}\\ &=2\{\mathbf{A}\mathbf{D}^{\dagger}\mathbf{B}+2\mathbf{A}\mathbf{D}^{\dagger}\mathbf{X}++\mathbf{D}^{\dagger}\mathbf{B}+\mathbf{X}\mathbf{D}^{\dagger}\mathbf{X}\}.\end{aligned}$$

So, the two sides are equal. □

Lemma 9.4.5. *If* $\mathbf{A},\mathbf{B}$ *and* $\mathbf{X}$ *be nnd matrices of the same order and* $\mathbf{H}=\mathcal{P}(\mathbf{A}+\mathbf{X},\mathbf{B}+\mathbf{X})-\mathcal{P}(\mathbf{A},\mathbf{B})-\mathcal{P}(\mathbf{X},\mathbf{X})$, *then* $\mathbf{H}$ *is hermitian and the matrix* $2\mathbf{H}=\mathbf{A}\mathbf{C}^{\dagger}(\mathcal{P}(\mathbf{C},2\mathbf{X}))\mathbf{C}^{\dagger}\mathbf{A}+\mathbf{B}\mathbf{C}^{\dagger}(\mathcal{P}(\mathbf{C},2\mathbf{X}))\mathbf{C}^{\dagger}\mathbf{B}-\frac{1}{2}\mathcal{P}(\mathbf{C},2\mathbf{X})$, *where* $\mathbf{C}=\mathbf{A}+\mathbf{B}$. *Furthermore,* $\|\mathbf{H}\|\leq 2(\|\mathbf{C}^{\dagger}\mathbf{A}\|^2+\|\mathbf{C}^{\dagger}\mathbf{B}\|^2)\|\mathbf{X}\|$.

Proof. First notice that by Lemma 9.4.4,

$$\begin{aligned}2\mathbf{H}&=[\mathcal{P}(\mathbf{A},\mathbf{B}+2\mathbf{X})-\mathcal{P}(\mathbf{A},\mathbf{B})]+[\mathcal{P}(\mathbf{B},\mathbf{A}+2\mathbf{X})-\mathcal{P}(\mathbf{B},\mathbf{A})]\\ &\quad+2[\mathcal{P}(\mathbf{X},\mathbf{A}+\mathbf{B}+\mathbf{X})-\mathcal{P}(\mathbf{X},\mathbf{X})]-\mathcal{P}(\mathbf{A}+\mathbf{B},2\mathbf{X}).\end{aligned}$$

Consider $2[\mathcal{P}(\mathbf{X},\mathbf{A}+\mathbf{B}+\mathbf{X})-\mathcal{P}(\mathbf{X},\mathbf{X})]-\mathcal{P}(\mathbf{A}+\mathbf{B},2\mathbf{X})$

$$\begin{aligned}&=2[\mathcal{P}(\mathbf{X},\mathbf{C}+\mathbf{X})-\mathcal{P}(\mathbf{X},\mathbf{X})]-\mathcal{P}(\mathbf{C},2\mathbf{X})\\ &=2[(\mathbf{C}+\mathbf{X})(\mathbf{C}+2\mathbf{X})^{\dagger}\mathbf{X}]-2\mathbf{X}(2\mathbf{X})^{\dagger}\mathbf{X}-2\mathbf{C}(\mathbf{C}+2\mathbf{X})^{\dagger}\mathbf{X}\\ &=2\mathbf{X}(\mathbf{C}+2\mathbf{X})^{\dagger}\mathbf{X}-\mathbf{X}\\ &=-\mathbf{C}(\mathbf{C}+2\mathbf{X})^{\dagger}\mathbf{X}\\ &=-\tfrac{1}{2}\mathcal{P}(\mathbf{C},2\mathbf{X}),\end{aligned}$$

since, $\mathbf{X}(\mathbf{C}+2\mathbf{X})^{\dagger}(\mathbf{C}+2\mathbf{X})=\mathbf{X}$. Now, using Theorem 9.4.3, for $[\mathcal{P}(\mathbf{A},\mathbf{B}+2\mathbf{X})-\mathcal{P}(\mathbf{A},\mathbf{B})]$ and $[\mathcal{P}(\mathbf{B},\mathbf{A}+2\mathbf{X})-\mathcal{P}(\mathbf{B},\mathbf{A})]$, we have $2\mathbf{H}=\mathbf{A}\mathbf{C}^{\dagger}(\mathcal{P}(\mathbf{C},2\mathbf{X}))\mathbf{C}^{\dagger}\mathbf{A}+\mathbf{B}\mathbf{C}^{\dagger}(\mathcal{P}(\mathbf{C},2\mathbf{X}))\mathbf{C}^{\dagger}\mathbf{B}-\frac{1}{2}\mathcal{P}(\mathbf{C}|2\mathbf{X})$. It follows that $\mathbf{A}\mathbf{C}^{\dagger}(\mathcal{P}(\mathbf{C}|2\mathbf{X}))\mathbf{C}^{\dagger}\mathbf{A}+\mathbf{B}\mathbf{C}^{\dagger}(\mathcal{P}(\mathbf{C},2\mathbf{X}))\mathbf{C}^{\dagger}\mathbf{B}-2\mathbf{H}=\frac{1}{2}\mathcal{P}(\mathbf{C},2\mathbf{X})$, which is an *nnd* matrix. So,

$$2\mathbf{H}<^{\mathbf{L}}\mathbf{A}\mathbf{C}^{\dagger}(\mathcal{P}(\mathbf{C},2\mathbf{X}))\mathbf{C}^{\dagger}\mathbf{A}+\mathbf{B}\mathbf{C}^{\dagger}\mathcal{P}(\mathbf{C},2\mathbf{X})\mathbf{C}^{\dagger}\mathbf{B}.$$

The last statement now follows by Theorem 9.4.3. □

Theorem 9.4.6. *If* $\mathbf{A}, \mathbf{B}, \mathbf{X}$ *and* $\mathbf{Y}$ *be nnd matrices of the same order, then* $\|\mathcal{P}(\mathbf{A}+\mathbf{X}, \mathbf{B}+\mathbf{Y}) - \mathcal{P}(\mathbf{A}, \mathbf{B})\|$
$\leq \mathbf{2}(\|(\mathbf{A}+\mathbf{B})^{\dagger}\mathbf{A}\|^{\mathbf{2}} + \|(\mathbf{A}+\mathbf{B})^{\dagger}\mathbf{B}\|^{\mathbf{2}})\|\mathbf{X}+\mathbf{Y}\|^{\mathbf{2}} + \frac{\mathbf{1}}{\mathbf{2}}\|\mathbf{X}+\mathbf{Y}\|^{\mathbf{2}}$.

Proof. Since $\mathbf{A}, \mathbf{B}, \mathbf{X}$ and $\mathbf{Y}$ are all *nnd*, we have

$$\mathbf{A} <^{\mathbf{L}} \mathbf{A}+\mathbf{X} <^{\mathbf{L}} \mathbf{A}+\mathbf{X}+\mathbf{Y}, \mathbf{B}+\mathbf{Y} <^{\mathbf{L}} \mathbf{B}+\mathbf{X}+\mathbf{Y},$$

so, $\mathcal{P}(\mathbf{A}+\mathbf{X}, \mathbf{B}+\mathbf{Y}) <^{\mathbf{L}} \mathcal{P}(\mathbf{A}+\mathbf{X}+\mathbf{Y}, \mathbf{B}+\mathbf{Y})$.
As $\mathcal{P}(\mathbf{A}+\mathbf{X}+\mathbf{Y}, \mathbf{B}+\mathbf{Y}) = \mathcal{P}(\mathbf{B}+\mathbf{Y}, \mathbf{A}+\mathbf{X}+\mathbf{Y})$, $\mathcal{P}(\mathbf{B}+\mathbf{Y}, \mathbf{A}+\mathbf{X}+\mathbf{Y}) <^{\mathbf{L}} \mathcal{P}(\mathbf{B}+\mathbf{X}+\mathbf{Y}, \mathbf{A}+\mathbf{X}+\mathbf{Y})$ and $\mathcal{P}(\mathbf{B}+\mathbf{X}+\mathbf{Y}, \mathbf{A}+\mathbf{X}+\mathbf{Y}) = \mathcal{P}(\mathbf{A}+\mathbf{X}+\mathbf{Y}, \mathbf{B}+\mathbf{X}+\mathbf{Y})$.
Therefore,
$\mathcal{P}(\mathbf{A}+\mathbf{X}, \mathbf{B}+\mathbf{Y}) - \mathcal{P}(\mathbf{A}, \mathbf{B}) \leq \mathcal{P}(\mathbf{B}+\mathbf{X}+\mathbf{Y}, \mathbf{A}+\mathbf{X}+\mathbf{Y}) - \mathcal{P}(\mathbf{B}, \mathbf{A})$.
By Lemma 9.4.5,

$$\begin{aligned}
&\|\mathcal{P}(\mathbf{B}+\mathbf{X}+\mathbf{Y}, \mathbf{A}+\mathbf{X}+\mathbf{Y}) - \mathcal{P}(\mathbf{B}, \mathbf{A})\| \\
&\leq \|\mathcal{P}(\mathbf{B}+\mathbf{X}+\mathbf{Y}, \mathbf{A}+\mathbf{X}+\mathbf{Y}) - \mathcal{P}(\mathbf{B}, \mathbf{A}) \\
&- \mathcal{P}(\mathbf{X}+\mathbf{Y}, \mathbf{X}+\mathbf{Y})\| + \|\mathcal{P}(\mathbf{X}+\mathbf{Y}, \mathbf{X}+\mathbf{Y})\| \\
&\leq \mathbf{2}(\|(\mathbf{A}+\mathbf{B})^{\dagger}\mathbf{A}\|^{\mathbf{2}} + \|(\mathbf{A}+\mathbf{B})^{\dagger}\mathbf{B}\|^{\mathbf{2}})\|(\mathbf{X}+\mathbf{Y})\| + \|(\tfrac{\mathbf{1}}{\mathbf{2}}(\mathbf{X}+\mathbf{Y}))\| \\
&= \mathbf{2}(\|(\mathbf{A}+\mathbf{B})^{\dagger}\mathbf{A}\|^{\mathbf{2}} + \|(\mathbf{A}+\mathbf{B})^{\dagger}\mathbf{B}\|^{\mathbf{2}} + \tfrac{\mathbf{1}}{\mathbf{4}})\|\mathbf{X}+\mathbf{Y}\|^{\mathbf{2}}.
\end{aligned}$$

□

Corollary 9.4.7. *In the set up of Theorem 9.4.6, let* $\mathbf{X} \to \mathbf{0}, \mathbf{Y} \to \mathbf{0}$. *Then* $\mathcal{P}(\mathbf{A}+\mathbf{X}, \mathbf{B}+\mathbf{Y}) \to \mathcal{P}(\mathbf{A}, \mathbf{B})$.

Given matrices $\mathbf{A}$, $\mathbf{B}$, we shall now obtain the rank and index of their parallel sum. We begin first with the following remark:

Remark 9.4.8. We know that $\mathcal{C}(\mathcal{P}(\mathbf{A}, \mathbf{B})) = \mathcal{C}(\mathbf{A}) \cap \mathcal{C}(\mathbf{B})$ and $\rho(\mathbf{A}+\mathbf{B}) = \rho(\mathbf{A}) + \rho(\mathbf{B}) - \mathbf{d}(\mathcal{C}(\mathbf{A}) \cap \mathcal{C}(\mathbf{B})) = \rho(\mathbf{A}) + \rho(\mathbf{B}) - \rho(\mathcal{P}(\mathbf{A}, \mathbf{B}))$. Thus, $\rho(\mathbf{A}+\mathbf{B}) = \rho(\mathbf{A}) \Rightarrow \rho(\mathbf{B}) = \rho(\mathcal{P}(\mathbf{A}, \mathbf{B}))$. Similarly, $\rho(\mathbf{A}+\mathbf{B}) = \rho(\mathbf{B}) \Rightarrow \rho(\mathbf{A}) = \rho(\mathcal{P}(\mathbf{A}, \mathbf{B}))$. Moreover, if any three of $\mathbf{A}, \mathbf{B}, \mathbf{A}+\mathbf{B}$ and $\mathcal{P}(\mathbf{A}, \mathbf{B})$ have same rank then all four have the same rank.

Theorem 9.4.9. *Let* $\mathbf{A}, \mathbf{B}$ *be parallel summable matrices of order* $m \times n$ *and* $\rho(\mathbf{A}+\mathbf{B}) = r$. *Then* $\rho(\mathcal{P}(\mathbf{A}, \mathbf{B})) = \mathbf{r}$ *if and only if* $\rho(\mathbf{A}) = \rho(\mathbf{B}) = r$.

Proof. Let $(\mathbf{L}, \mathbf{R})$ be a rank factorization of $\mathbf{A}+\mathbf{B}$. Then there exists an $r \times r$ matrix $\mathbf{D}$ unique up to similarity such that $\mathbf{A} = \mathbf{LDR}$, $\mathbf{A} = \mathbf{L}(\mathbf{I}-\mathbf{D})\mathbf{R}$ and $\mathcal{P}(\mathbf{A}, \mathbf{B}) = \mathbf{LD}(\mathbf{I}-\mathbf{D})\mathbf{R}$. Note that $\rho(\mathbf{A}) = \rho(\mathbf{D})$, $\rho(\mathbf{B}) = \rho(\mathbf{I}-\mathbf{D})$ and $\rho(\mathcal{P}(\mathbf{A}, \mathbf{B})) = \rho(\mathbf{D}(\mathbf{I}-\mathbf{D}))$. Now,

$$\rho(\mathcal{P}(\mathbf{A}, \mathbf{B})) = \mathbf{r} \Leftrightarrow \rho(\mathbf{D}(\mathbf{I}-\mathbf{D})) = \mathbf{r}$$

$\Leftrightarrow \mathbf{D}(\mathbf{I}-\mathbf{D})$ is a non-singular matrix of rank r
$\Leftrightarrow \rho(\mathbf{D}) = r = \rho(\mathbf{A}),\ \rho(\mathbf{I}-\mathbf{D}) = r = \rho(\mathbf{B}).$ □

Theorem 9.4.10. *Let* $\mathbf{A}, \mathbf{B}$ *be parallel summable matrices of order* $n \times n$ *and any three of* $\rho(\mathbf{A}+\mathbf{B}),\ \rho(\mathbf{A}),\ \rho(\mathbf{B})$ *and* $\rho(\mathcal{P}(\mathbf{A}, \mathbf{B}))$ *are equal. Then* $\mathbf{A}+\mathbf{B}$ *has index* 1 *if and only if* $\mathcal{P}(\mathbf{A}, \mathbf{B})$ *has index* 1.

Proof. 'Only if' part
Let $\rho(\mathbf{A}+\mathbf{B}) = r$ and $(\mathbf{L}, \mathbf{R})$ be a rank factorization of $\mathbf{A}+\mathbf{B}$. Then as seen in Theorem 9.4.9, $\mathbf{A} = \mathbf{LDR}$, $\mathbf{A} = \mathbf{L}(\mathbf{I}-\mathbf{D})\mathbf{R}$ and $\mathcal{P}(\mathbf{A}, \mathbf{B}) = \mathbf{LD}(\mathbf{I}-\mathbf{D})\mathbf{R}$. Also, $(\mathcal{P}(\mathbf{A}, \mathbf{B}))^2 = \mathbf{LD}(\mathbf{I}-\mathbf{D})\mathbf{RLD}(\mathbf{I}-\mathbf{D})\mathbf{R}$. So, $\rho((\mathcal{P}(\mathbf{A}, \mathbf{B}))^2) = \rho(\mathbf{LD}(\mathbf{I}-\mathbf{D})\mathbf{RLD}(\mathbf{I}-\mathbf{D})\mathbf{R}) = \rho(\mathbf{D}(\mathbf{I}-\mathbf{D})\mathbf{RLD}(\mathbf{I}-\mathbf{D}))$. As $\mathbf{A}+\mathbf{B}$ has index 1, the matrix $\mathbf{RL}$ has order $r \times r$ and is invertible. By Theorem 9.4.6, $\rho(\mathcal{P}(\mathbf{A}, \mathbf{B})) = \rho(\mathbf{D}(\mathbf{I}-\mathbf{D})) = \mathbf{r}$. Therefore, $\mathbf{D}(\mathbf{I}-\mathbf{D})$ is invertible and so, $\mathbf{D}(\mathbf{I}-\mathbf{D})\mathbf{RLD}(\mathbf{I}-\mathbf{D})$ is invertible matrix of rank r showing $\rho(\mathcal{P}(\mathbf{A}, \mathbf{B}))^{\mathbf{2}} = \rho(\mathcal{P}(\mathbf{A}, \mathbf{B}))$, i.e. $\mathcal{P}(\mathbf{A}, \mathbf{B})$ has index 1.
'If' part
Let $\mathcal{P}(\mathbf{A}, \mathbf{B})$ has index 1. Then from proof of 'Only if' part $\rho(\mathbf{D}(\mathbf{I}-\mathbf{D})\mathbf{RLD}(\mathbf{I}-\mathbf{D})) = \rho(\mathbf{D}(\mathbf{I}-\mathbf{D}))$. Since $\rho(\mathbf{A}) = \rho(\mathbf{B}) = r \Rightarrow \rho(\mathbf{D}) = \rho(\mathbf{I}-\mathbf{D}) = r$, both $\mathbf{D}$ and $\mathbf{I}-\mathbf{D}$ are invertible. So, $\mathbf{D}(\mathbf{I}-\mathbf{D})$ is invertible. Thus, $\rho(\mathbf{D}(\mathbf{I}-\mathbf{D})\mathbf{RLD}(\mathbf{I}-\mathbf{D})) = r$ and therefore, is invertible. It follows that $\mathbf{RL}$ is invertible. Thus, $\mathbf{A}+\mathbf{B}$ has index 1. □

Corollary 9.4.11. *In the set up of Theorem 9.4.10, The matrix* $\mathbf{A}+\mathbf{B}$ *has the group inverse if and only if the parallel sum* $\mathcal{P}(\mathbf{A}, \mathbf{B})$ *has the group inverse.*

Lemma 9.4.12. *Let* $\mathbf{A}$ *be a matrix of index* k *and* $(\mathbf{L}, \mathbf{R})$ *be a rank factorization of* $\mathbf{A}$. *Then the Drazin inverse of* $\mathbf{A}$ *is given by* $\mathbf{A}^{\mathbf{D}} = \mathbf{L}((\mathbf{RL})^{\mathbf{D}})^{\mathbf{2}}\mathbf{R}$.

Proof. As $\mathbf{L}, \mathbf{R}$ are full rank matrices, Index of $\mathbf{A}$ is $k \Leftrightarrow \rho(\mathbf{A}^{k+1}) = \rho(\mathbf{A}^k) \Leftrightarrow \rho((\mathbf{LR})^{k+1}) = \rho((\mathbf{LR})^k) \Leftrightarrow \rho((\mathbf{RL})^k) = \rho((\mathbf{RL})^{k-1}) \Leftrightarrow$ index of $\mathbf{RL}$ is $k-1$. Let $\mathbf{X} = \mathbf{L}((\mathbf{RL})^{\mathbf{D}})^{\mathbf{2}}\mathbf{R}$. It can be easily verified that $\mathbf{X}$ satisfies Definition 2.4.23. □

Theorem 9.4.13. *Let* $\mathbf{A}, \mathbf{B}$ *be parallel summable matrices of order* $n \times n$ *with index of* $\mathbf{A}+\mathbf{B}$ *equal to* 1 *and* $\rho(\mathbf{A}+\mathbf{B}) = r$. *Let* $(\mathbf{L}, \mathbf{R})$ *be a rank factorization of* $\mathbf{A}+\mathbf{B}$ *and* $\mathbf{D}$ *be an* $r \times r$ *matrix such that* $\mathbf{A} = \mathbf{LDR}, \mathbf{B} + \mathbf{L}(\mathbf{I}-\mathbf{D})\mathbf{R}$. *If* $\mathbf{D}$ *and* $\mathbf{RL}$ *commute, then*

(i) *Index of* $\mathcal{P}(\mathbf{A}, \mathbf{B})$ *is* k *if and only if index of* $\mathbf{D}(\mathbf{I}-\mathbf{D})$ *equals* $k-1$.

(ii) *If in addition any two of* $\rho(\mathbf{A}), \rho(\mathbf{B})$ *and* $\rho(\mathcal{P}(\mathbf{A},\mathbf{B}))$ *are each equal to* r, *then the Drazin inverse of* $\mathcal{P}(\mathbf{A},\mathbf{B})$ *is given by* $(\mathcal{P}(\mathbf{A},\mathbf{B}))^{\mathbf{D}} = \mathbf{LD}((\mathbf{I}-\mathbf{D})\mathbf{RLD})^{2}(\mathbf{I}-\mathbf{D})\mathbf{R}$.

Proof. (i) Since $\mathbf{D}$ and $\mathbf{RL}$ commute, we have $(\mathcal{P}(\mathbf{A},\mathbf{B}))^{\mathbf{k}+1} = \mathbf{L}(\mathbf{D}(\mathbf{I}-\mathbf{D}))^{\mathbf{k}}\mathbf{RLR}$. Also, $\mathbf{L},\mathbf{R}$ being full rank matrices, we have $\rho(\mathcal{P}(\mathbf{A},\mathbf{B}))^{\mathbf{k}-1} = \rho((\mathbf{D}(\mathbf{I}-\mathbf{D}))^{\mathbf{k}}\mathbf{RL})$. Now, $\mathbf{\Lambda}$ is of index 1 $\Rightarrow$ $\mathbf{RL}$ is invertible, so, $\rho(\mathcal{P}(\mathbf{A},\mathbf{B}))^{\mathbf{k}+1} = \rho((\mathbf{D}(\mathbf{I}-\mathbf{D}))^{\mathbf{k}})$. Hence, (i) follows.
(ii) We first note that in this case $(\mathbf{LD},(\mathbf{I}-\mathbf{D})\mathbf{R})$ is a rank factorization of $\mathcal{P}(\mathbf{A},\mathbf{B})$. Let index of $\mathcal{P}(\mathbf{A},\mathbf{B})$ be k. Then by (i) above index of $\mathbf{D}(\mathbf{I}-\mathbf{D})$ is $k-1$. By Lemma 9.4.12, the Drazin inverse of $\mathcal{P}(\mathbf{A},\mathbf{B})$ is $(\mathcal{P}(\mathbf{A},\mathbf{B}))^{\mathbf{D}} = \mathbf{LD}((\mathbf{I}-\mathbf{D})\mathbf{RLD})^{2}(\mathbf{I}-\mathbf{D})\mathbf{R}$. □

9.5 Exercises

(1) Let $\mathbf{A}$, $\mathbf{B}$ and $\mathbf{C} \in \mathrm{F}^{\mathrm{m}\times\mathrm{n}}$. Show that $\mathcal{P}(\mathcal{P}(\mathbf{A},\mathbf{B}),\mathbf{C}) = \mathcal{P}(\mathbf{A},\mathcal{P}(\mathbf{B},\mathbf{C}))$, whenever all parallel sums involved are defined.

(2) Let $\mathbf{A},\mathbf{B}$ and $\mathbf{C} \in \mathrm{F}^{\mathrm{m}\times\mathrm{n}}$ such that (i) $\mathbf{CA} = \mathbf{AC}$, $\mathbf{CB} = \mathbf{BC}$ (ii) $\mathbf{A}+\mathbf{B}$ and $\mathbf{C}$ each have index 1 and (iii) $\mathbf{A},\mathbf{B}$ are parallel summable. Prove that $\mathbf{AC},\mathbf{BC}$ are parallel summable with $\mathcal{P}(\mathbf{AC},\mathbf{BC}) = \mathcal{P}(\mathbf{A},\mathbf{B})\mathbf{C} = \mathbf{C}\mathcal{P}(\mathbf{A},\mathbf{B})$.

(3) Let $\mathbf{A},\mathbf{B} \in \mathrm{F}^{\mathrm{m}\times\mathrm{n}}$. Prove that $\rho(\mathbf{A}+\mathbf{B}) = \rho(\mathbf{A}) + \rho(\mathbf{B}) \Rightarrow \mathbf{A},\mathbf{B}$ are parallel summable. Is the converse true?

(4) Let $\mathbf{A},\mathbf{B}$ and $\mathbf{C} \in \mathbb{C}^{m\times n}$ such that (i) $\mathbf{A},\mathbf{B}$ are parallel summable with either $\mathcal{C}(\mathbf{B}) \subseteq \mathcal{C}(\mathbf{A})$ or $\mathcal{C}(\mathbf{B}^t) \subseteq \mathcal{C}(\mathbf{A}^t)$ (ii) $\mathbf{A}$ and $\mathbf{C}$ are range-Hermitian (or of index 1) and (iii) $\mathbf{CA} = \mathbf{AC}$, $\mathbf{CB} = \mathbf{BC}$. Prove that $\mathbf{AC},\mathbf{BC}$ are parallel summable with $\mathcal{P}(\mathbf{AC},\mathbf{BC}) = \mathcal{P}(\mathbf{A},\mathbf{B})\mathbf{C} = \mathbf{C}\mathcal{P}(\mathbf{A},\mathbf{B})$.

(5) Let $\mathbf{A},\mathbf{B} \in \mathrm{F}^{\mathrm{m}\times\mathrm{n}}$ be parallel summable and $\mathbf{x} \in F^n$ such that $\mathbf{Ax} = a\mathbf{x}$ and $\mathbf{Bx} = b\mathbf{x}$, where a, b are both non-null scalars. Show that $\mathcal{P}(\mathbf{A},\mathbf{B})\mathbf{x} = \mathcal{P}(\mathbf{A},\mathbf{B})\mathbf{x}$, provided $a+b$ is non-null.

(6) Let $\mathbf{A},\mathbf{B} \in \mathrm{F}^{\mathrm{m}\times\mathrm{n}}$ be parallel summable. Prove that $\mathcal{C}(\mathbf{B}) \subseteq \mathcal{C}(\mathbf{A})$ if and only if $\mathcal{C}(\mathbf{B}^t) \subseteq \mathcal{C}(\mathbf{A}^t)$. Conclude that if $\mathbf{A},\mathbf{B}$ are parallel summable and either $\mathcal{C}(\mathbf{B}) \subseteq \mathcal{C}(\mathbf{A})$ or $\mathcal{C}(\mathbf{B}^t) \subseteq \mathcal{C}(\mathbf{A}^t)$, then $\mathbf{A} <^s \mathbf{B}$.

(7) Let $\mathbf{A},\mathbf{B} \in \mathbb{C}^{n\times n}$ be parallel summable. If any two of $\mathbf{A},\mathbf{B}$ and $\mathbf{A}+\mathbf{B}$ are range-hermitian then so the third.

(8) Let $\mathbf{A},\mathbf{B} \in \mathbb{C}^{n\times n}$ be *nnd* matrices. We can define the matrix of $\mathcal{P}(\mathbf{A},\mathbf{B})$ as follows:
Since $\mathbf{A}+\mathbf{B}$ is hermitian we can find a unitary matrix $\mathbf{U}$ such that $\mathbf{U}^\star(\mathbf{A}+\mathbf{B})\mathbf{U}$ is diagonal, say $\mathbf{C} = [c_{ij}]$. Let $\mathbf{U}^\star\mathbf{AU} = [a_{ij}]$, $\mathbf{U}^\star\mathbf{BU} = [b_{ij}]$ and $\mathbf{U}^\star\mathcal{P}(\mathbf{A},\mathbf{B})\mathbf{U} = [d_{ij}]$. Notice that if $a_{kk} = \mathbf{0}$ for some k, then $a_{i,k} = \mathbf{0}$ for each k. With the convention $\frac{0}{0} = \mathbf{0}$, and ignoring the unitary matrix i.e. $\mathbf{A}+\mathbf{B} = \mathbf{A}$, $\mathbf{A} = [a_{ij}]$ etc. we can write
$$d_{ij} = \sum_k a_{i,k} \sum_m c_{k,m} + b_{m,j} = \sum_k \frac{a_{i,k}b_{k,j}}{a_{i,k}+b_{k,j}}$$
Prove that (i) $tr(\mathcal{P}(\mathbf{A},\mathbf{B})) \le tr(\mathbf{A})tr(\mathbf{B})$ and (ii) $\det(\mathcal{P}(\mathbf{A},\mathbf{B})) \le \det(\mathbf{A})\det(\mathbf{B})$.

(9) Let $\mathbf{A},\mathbf{B}$ and $\mathbf{C} \in \mathbb{C}^{n\times n}$ be *nnd* matrices. Prove that $9\mathcal{P}(\mathcal{P}(\mathbf{A},\mathbf{B}),\mathbf{C}) <^L \mathbf{A}+\mathbf{B}+\mathbf{C}$ (Hint: Use Lemma 9.3.12).

(10) Let $\mathbf{A},\mathbf{B},\mathbf{X}$ and $\mathbf{Y} \in \mathbb{C}^{\mathrm{n}\times\mathrm{n}}$ be *nnd* matrices. Let a_1 and a_2 be non-negative real numbers such that $a_1 + a_2 = 1$. Then prove that
$$\mathcal{P}(a_1\mathbf{A}_1 + a_2\mathbf{A}_2, a_1\mathbf{B}_1 + a_2\mathbf{B}_2) = a_1\mathcal{P}(\mathbf{A}_1,\mathbf{B}_1) + a_2\mathcal{P}(\mathbf{A}_2,\mathbf{B}_2).$$

(11) Let $\mathbf{A}, \mathbf{B} \in \mathbb{C}^{m \times n}$. Prove that $\mathbf{A}, \mathbf{B}$ are parallel summable if and only if $\mathbf{AA}^{\dagger} <^{\star} (\mathbf{A}+\mathbf{B})(\mathbf{A}+\mathbf{B})^{\dagger}$, $\mathbf{A}^{\dagger}\mathbf{A} <^{\star} (\mathbf{A}+\mathbf{B})^{\dagger}(\mathbf{A}+\mathbf{B})$ if and only if $\mathbf{BB}^{\dagger} <^{\star} (\mathbf{A}+\mathbf{B})(\mathbf{A}+\mathbf{B})^{\dagger}$, $\mathbf{B}^{\dagger}\mathbf{B} <^{\star} (\mathbf{A}+\mathbf{B})^{\dagger}(\mathbf{A}+\mathbf{B})$.

(12) Let $\mathbf{A}, \mathbf{B}$ be parallel summable matrices such that $\mathbf{A} <^{-} \mathbf{B}$. Prove that $\mathbf{A} <^{-} 2\mathbf{P}(\mathbf{A}, \mathbf{B})$.

(13) Let $\mathbf{A}, \mathbf{B}$ be parallel summable matrices of index not greater than 1 such that (i) $\mathbf{A}+\mathbf{B}$ is of index 1 and (ii) $(\mathbf{A}+\mathbf{B})^{\#} = \mathbf{A}^{\#} + \mathbf{B}^{\#}$. Show that $\mathbf{A}^{\#}$ and $\mathbf{B}^{\#}$ are parallel summable.

Chapter 10

Schur Complements and Shorted Operators

10.1 Introduction

Schur complements are an interesting class of operators having a wide range of applications. For example, these operators play an important role in determining the shorted operator of the impedance matrix of an n-port electrical network some of whose ports have been shorted. These operators also play an important role in multivariate statistics.

In this chapter, we shall define this class of operators and study its properties. We start with providing some motivation for the shorted operators via electrical network theory and statistics in Section 10.2. We formally define the concept of Schur complement as the shorted operator of an *nnd* matrix in Section 10.3 and identify this as the supremum of a certain class of *nnd* operators under the Löwner order. In Section 10.4, we obtain the shorted operator/Schur complement of an *nnd* matrix as the limit of a sequence of parallel sums (which provides a base for obtaining the shorted matrix of a given matrix with respect to another matrix as the limit of a sequence of parallel sums in the next chapter). In Section 10.5, we extend the concept of Schur complements/shorted operator to general matrices (possibly rectangular) over a general field. We define the complementability of a matrix with respect to a pair of projectors or a pair of subspaces and give a necessary and sufficient condition for complementability of a matrix. We also show that the Schur complement is the supremum of a certain class of matrices under the minus order. Moreover, we show that Schur compression of a matrix with respect to a pair of projectors is indeed the Schur complement of the matrix with respect to a suitable pair of projectors. This chapter in fact, serves as a prelude to a more detailed study of the shorted operator in the next chapter.

In Sections 10.1–10.4 we consider matrices over the field of complex numbers. In Section 10.5 we consider matrices over a general field.

10.2 Shorted operator - A motivation

Electrical networks are a great source of motivating the shorted operator. We use n-port electrical networks of specialized nature for this purpose. We consider current through any terminal of a port as an input and the resulting potential (voltage) as output. We assume that the network has the properties of homogeneity and superposition. By homogeneity, we mean that if we multiply the inputs by some factor, the outputs get multiplied by the same factor. (Thus, the scale of measurement does not affect the relationship.) A network has the property of superposition if the output of the sum of several inputs is the sum of the outputs of each individual input. Further, we consider the networks that are resistive and reciprocal in nature. A network is called resistive if it is composed of only resistors and is reciprocal if the relationship between inputs and outputs is unchanged when input and output terminals are interchanged. Throughout this section, we consider the networks with these properties only i.e., they are resistive and reciprocal in nature and have the properties of homogeneity and superposition and no further mention will be made of this when dealing with a network.

Consider an n-port network. Let v_j be the potential across the port j and let i_j be the current through the terminals of the jth port. The vectors $\mathbf{i} = \begin{pmatrix} i_1 \\ \vdots \\ i_n \end{pmatrix}$ and $\mathbf{v} = \begin{pmatrix} v_1 \\ \vdots \\ v_n \end{pmatrix}$ are respectively called the current and voltage vectors to the network. Thus, $\mathbf{i}$ is the input vector and $\mathbf{v}$ is the output vector. Since the network is homogeneous and has the property of superposition, v_r is a linear combination of $i_1, \ldots, i_n$ for $r = 1, \ldots n$. Thus, we have $v_r = \sum_{j=1}^{n} z_{rj} i_j$, for $r = 1, \ldots n$ or equivalently $\mathbf{v} = \mathbf{Z}\mathbf{i}$, where $\mathbf{Z} = \big(z_{ij}\big)$ is a matrix of order $n \times n$. The matrix $\mathbf{Z}$ is called the **impedance matrix**. The elements z_{ij} can be interpreted as follows:
If we apply a current of strength i_j across a terminal in port j leaving the other ports open and measure the voltage across the port r, then $v_r = z_{rj} i_j$ or $z_{rj} = \dfrac{v_r}{i_j}$. Since the network is reciprocal and resistive, the impedance matrix is an *nnd* matrix.

Let us take an n-port network N with impedance matrix $\mathbf{Z}$. Suppose we short out the last $n-r$ ports to obtain a new n-port network N_* with impedance matrix $\mathbf{Z}_*$. We assume that this new network has all the properties of the original network. Then the voltage across the last $n-r$ ports of the network N_* is zero irrespective of inputs and therefore, the last $n-r$ rows of impedance matrix $\mathbf{Z}_*$ are null. Also, since the network is assumed to be reciprocal, the last $n-r$ elements of each of first r rows are also null. Thus, we can write the impedance matrix $\mathbf{Z}_* = \begin{pmatrix} \mathbf{Z}_{*11} & \mathbf{0} \\ \mathbf{0} & \mathbf{0} \end{pmatrix}$, where $\mathbf{Z}_{*11}$ is an *nnd* matrix of order $r \times r$. We obtain the relation between $\mathbf{Z}_{*11}$ and $\mathbf{Z}$ in the following theorem:

Theorem 10.2.1. *Let N be an n-port network with impedance matrix $\mathbf{Z}$. Let $\mathbf{Z} = \begin{pmatrix} \mathbf{Z}_{11} & \mathbf{Z}_{12} \\ \mathbf{Z}_{21} & \mathbf{Z}_{22} \end{pmatrix}$, where $\mathbf{Z}_{11}$ is of order $r \times r$ and $\mathbf{Z}_{22}$ is of order $(n-r) \times (n-r)$. Let $\mathbf{Z}_{*11} = (\mathbf{Z}_{11} - \mathbf{Z}_{12}\mathbf{Z}_{22}^{-}\mathbf{Z}_{21})$. Then $\mathbf{Z}_* = \begin{pmatrix} \mathbf{Z}_{*11} & \mathbf{0} \\ \mathbf{0} & \mathbf{0} \end{pmatrix}$ is the impedance matrix of the network N_* obtained from N by shorting the last $n-r$ ports.*

Proof. Write $\mathbf{i} = \begin{pmatrix} \mathbf{i}_1 \\ \mathbf{i}_2 \end{pmatrix}$ and $\mathbf{v} = \begin{pmatrix} \mathbf{v}_1 \\ \mathbf{v}_2 \end{pmatrix}$, where $\mathbf{i}_1$ and $\mathbf{v}_1$ are r-vectors. Since $\mathbf{v} = \mathbf{Z}\mathbf{i}$, we have

$$\mathbf{v}_1 = \mathbf{Z}_{11}\mathbf{i}_1 + \mathbf{Z}_{12}\mathbf{i}_2 \text{ and } \mathbf{v}_2 = \mathbf{Z}_{21}\mathbf{i}_1 + \mathbf{Z}_{22}\mathbf{i}_2.$$

Since the last $n-r$ ports are shorted, we have $\mathbf{v}_2 = \mathbf{0}$. Hence, we must determine the matrix $\mathbf{Z}_{*11}$ such that $\mathbf{v}_1 = \mathbf{Z}_{*11}\mathbf{i}_1$, given that $\mathbf{v}_2 = \mathbf{0}$. Now, $\mathbf{v}_2 = \mathbf{0} \Rightarrow \mathbf{Z}_{22}\mathbf{i}_2 = -\mathbf{Z}_{21}\mathbf{i}_1$. Since $\mathbf{Z}$ is *nnd*, we have $\mathcal{C}(\mathbf{Z}_{21}) \subseteq \mathcal{C}(\mathbf{Z}_{22})$. Therefore, $\mathbf{i}_2 = -\mathbf{Z}_{22}^{-}\mathbf{Z}_{21}\mathbf{i}_1 + (\mathbf{I} - \mathbf{Z}_{22}^{-}\mathbf{Z}_{22})\eta$ for some η and arbitrary g-inverse $\mathbf{Z}_{22}^{-}$ of $\mathbf{Z}_{22}$. Thus,

$$\mathbf{v}_1 = \mathbf{Z}_{11}\mathbf{i}_1 + \mathbf{Z}_{12}\mathbf{i}_2 = \mathbf{Z}_{11}\mathbf{i}_1 + \mathbf{Z}_{12}(-\mathbf{Z}_{22}^{-}\mathbf{Z}_{21}\mathbf{i}_i + (\mathbf{I} - \mathbf{Z}_{22}^{-}\mathbf{Z}_{22})\eta).$$

Since $\mathbf{Z}_{12}(\mathbf{I} - \mathbf{Z}_{22}^{-}\mathbf{Z}_{22}) = \mathbf{0}$, we have

$$\mathbf{v}_1 = \mathbf{Z}_{11}\mathbf{i}_1 - (\mathbf{Z}_{12}\mathbf{Z}_{22}^{-}\mathbf{Z}_{21})\mathbf{i}_i = (\mathbf{Z}_{11} - \mathbf{Z}_{12}\mathbf{Z}_{22}^{-}\mathbf{Z}_{21})\mathbf{i}_i.$$

Thus, $\mathbf{Z}_{*11} = \mathbf{Z}_{11} - \mathbf{Z}_{12}\mathbf{Z}_{22}^{-}\mathbf{Z}_{21}$. □

Remark 10.2.2. The matrix $\mathbf{Z}_*$ is referred to as the **shorted operator** of $\mathbf{Z}$ corresponding to the shorting of last $n-r$ ports of the network N.

Remark 10.2.3. It is easy to check that $\mathbf{0} <^{\mathbf{L}} \mathbf{Z}_* <^{\mathbf{L}} \mathbf{Z}$. Thus a short circuit can only reduce the resistance of a network and can never increase the same.

Remark 10.2.4. Notice that the matrix $\mathbf{Z_{*11}}$ is a **Schur complement** of $\mathbf{Z_{22}}$ in $\mathbf{Z}$.

Let $\mathbf{e_1}, \ldots, \mathbf{e_n}$ be the standard basis of $\mathbb{C}^n$, where for each $j = 1, \ldots, n$ the vector $\mathbf{e_j^\star}$ is defined as $\mathbf{e_j^\star} = (0, \ldots 0, 1, 0 \ldots, 0)$ with 1 at the jth position. Then $\mathbf{v} = \mathbf{v_1 e_1} + \ldots + \mathbf{v_n e_n}$ and $\mathbf{i} = \mathbf{i_1 e_1} + \ldots + \mathbf{i_n e_n}$. When we short the last $n - r$ ports of the network N, the coefficients of $\mathbf{e_{r+1}}, \ldots, \mathbf{e_n}$ become zero. Thus, shorting the last $n - r$ ports can be viewed as shorting with respect to the subspace $\mathcal{S}$, spanned by $\mathbf{e_{r+1}}, \ldots, \mathbf{e_n}$ and $\mathbf{Z_{*11}}$ as the shorted operator of $\mathbf{Z}$ with respect this subspace $\mathcal{S}$.

Yet another motivation for shorted operator comes from multivariate statistics.

Let $\mathbf{Y}$ follow an n-variate normal distribution with dispersion matrix $\mathbf{\Sigma}$. Write $\mathbf{Y} = \begin{pmatrix} \mathbf{Y_1} \\ \mathbf{Y_2} \end{pmatrix}$ and $\mathbf{\Sigma} = \begin{pmatrix} \mathbf{\Sigma_{11}} & \mathbf{\Sigma_{12}} \\ \mathbf{\Sigma_{21}} & \mathbf{\Sigma_{22}} \end{pmatrix}$, where $\mathbf{Y_1}$ has r components and $\mathbf{\Sigma_{11}}$ and $\mathbf{\Sigma_{22}}$ are of orders $r \times r$ and $(n-r) \times (n-r)$ respectively. Then it is well known that the conditional dispersion matrix of $\mathbf{Y_1}$ when $\mathbf{Y_2}$ is fixed is given as: $\mathbf{\Sigma_{11}} - \mathbf{\Sigma_{12}\Sigma_{22}^{-}\Sigma_{21}}$, which is the Schur complement of $\mathbf{\Sigma_{22}}$ in $\mathbf{\Sigma}$. Also, the dispersion matrix of $(\mathbf{Y_1} \mid \mathbf{Y_2} = \mathbf{c})$ together with $\mathbf{Y_2} = \mathbf{c}$ is the shorted operator of $\mathbf{\Sigma}$, namely $\begin{pmatrix} \mathbf{\Sigma_{11}} - \mathbf{\Sigma_{12}\Sigma_{22}^{-}\Sigma_{21}} & \mathbf{0} \\ \mathbf{0} & \mathbf{0} \end{pmatrix}$.

10.3 Generalized Schur complement and shorted operator

Let $\mathbf{A} = \begin{pmatrix} \mathbf{A_{11}} & \mathbf{A_{12}} \\ \mathbf{A_{21}} & \mathbf{A_{22}} \end{pmatrix}$ be the impedance matrix of a reciprocal, resistive electrical network with n-ports, where $\mathbf{A_{11}}$ is an $r \times r$ matrix and $\mathbf{A_{22}}$ is an $(n-r) \times (n-r)$ matrix. In the previous section, we saw that if the last $n - r$ ports are shorted, the resulting impedance matrix is $\begin{pmatrix} \mathbf{A_{11}} - \mathbf{A_{12}A_{22}^{-}A_{21}} & \mathbf{0} \\ \mathbf{0} & \mathbf{0} \end{pmatrix}$. This can be viewed as shorting the network with respect to the subspace $\mathcal{S} = \mathcal{C}(\mathbf{e_1} : \cdots : \mathbf{e_r})$. This is so because we are shorting the last $n-r$ ports. In this chapter, we consider shorting of a given matrix $\mathbf{A}$ with respect to a general subspace $\mathcal{S}$ and obtain the expression for the impedance matrix, which we call as the shorted operator $\mathbf{A}_{\mathcal{S}}$. We emphasize that the subspace $\mathcal{S}$ relates to unshorted ports.

Let $\mathbf{A}$ be an *nnd* matrix of order $n \times n$. Let $\mathcal{S}$ be a subspace of $\mathbb{C}^n$ with an ortho-normal basis $\{\mathbf{s_1}, \mathbf{s_2}, \ldots, \mathbf{s_r}\}$. Extend $\mathbf{s_1}, \mathbf{s_2}, \ldots, \mathbf{s_r}$ to an ortho-normal basis $\mathbf{s_1}, \mathbf{s_2}, \ldots, \mathbf{s_n}$ of $\mathbb{C}^n$. Then $\mathbf{s_{r+1}}, \mathbf{s_{r+2}}, \ldots, \mathbf{s_n}$ forms a basis of

$\mathcal{S}^{\perp}$, the orthogonal complement of $\mathcal{S}$. Let $\mathbf{S} = (\mathbf{S_1} : \mathbf{S_2})$, where $\mathbf{S_1} = (\mathbf{s_1} : \cdots : \mathbf{s_r})$ and $\mathbf{S_2} = (\mathbf{s_{r+1}} : \cdots : \mathbf{s_n})$. Then $\mathbf{S}$ is a unitary matrix and the representation of $\mathbf{A}$ with respect to the ortho-normal basis $\mathbf{s_1}, \mathbf{s_2}, \ldots, \mathbf{s_n}$ is $\mathbf{S^\star A S} = \begin{pmatrix} \mathbf{S_1^\star A S_1} & \mathbf{S_1^\star A S_2} \\ \mathbf{S_2^\star A S_1} & \mathbf{S_2^\star A S_2} \end{pmatrix}$. In this case we call the matrix

$$\begin{pmatrix} \mathbf{S_1^\star A S_1} - \mathbf{S_1^\star A S_2}(\mathbf{S_2^\star A S_2})^{-}\mathbf{S_2^\star A S_1} & \mathbf{0} \\ \mathbf{0} & \mathbf{0} \end{pmatrix}$$

as the Generalized Schur Complement of the matrix $\mathbf{S_2^\star A S_2}$ in the matrix $\mathbf{S^\star A S}$. (Notice that $\mathcal{C}(\mathbf{S_2^\star A S_1}) \subseteq \mathcal{C}(\mathbf{S_2^\star A}) = \mathcal{C}(\mathbf{S_2^\star A S_2})$. Hence, the quantity $\mathbf{S_1^\star A S_2}(\mathbf{S_2^\star A S_2})^{-}\mathbf{S_2^\star A S_1}$ is invariant under the choices of g-inverses of $\mathbf{S_2^\star A S_2}$.) The generalized Schur complement with respect to the standard basis has the representation

$$\mathbf{S}\begin{pmatrix} \mathbf{S_1^\star A S_1} - \mathbf{S_1^\star A S_2}(\mathbf{S_2^\star A S_2})^{-}\mathbf{S_2^\star A S_1} & \mathbf{0} \\ \mathbf{0} & \mathbf{0} \end{pmatrix}\mathbf{S^\star} \tag{10.3.1}$$

or equivalently,

$$\mathbf{S_1 S_1^\star A S_1 S_1^\star} - \mathbf{S_1 S_1^\star A S_2}(\mathbf{S_2^\star A S_2})^{-}\mathbf{S_2^\star A S_1 S_1^\star}. \tag{10.3.2}$$

Definition 10.3.1. Let $\mathbf{A}$ be an $n \times n$ *nnd* matrix and let $\mathcal{S}$ be a subspace of $\mathbb{C}^n$. Let $\mathbf{S} = (\mathbf{S_1} : \mathbf{S_2})$ be a unitary matrix where $\mathcal{C}(\mathbf{S_1}) = \mathcal{S}$. Then the shorted operator $\mathbf{A}_{\mathcal{S}}$ of $\mathbf{A}$ with respect to $\mathcal{S}$ is defined by (10.3.1) or equivalently by (10.3.2).

We now show that the shorted operator $\mathbf{A}_{\mathcal{S}}$ depends only on $\mathcal{S}$ and not on a particular basis of $\mathcal{S}$ and the orthogonal complement $\mathcal{S}^{\perp}$ of $\mathcal{S}$ chosen the for representation of $\mathbf{A}$. We will use the machinery and notations developed above in this entire section without making a repeated mention of the same.

Theorem 10.3.2. *Let* $\mathbf{A}$ *be an* $n \times n$ *nnd matrix and let* $\mathcal{S}$ *be a subspace of* $\mathbb{C}^n$*. Then*

$$\mathcal{C}(\mathbf{A}_{\mathcal{S}}) \subseteq \mathcal{S} \text{ and } \mathbf{A}_{\mathcal{S}} = \mathbf{A} - \mathbf{A P}_{\mathcal{S}^\perp}(\mathbf{P}_{\mathcal{S}^\perp}\mathbf{A P}_{\mathcal{S}^\perp})^{-}\mathbf{P}_{\mathcal{S}^\perp}\mathbf{A}$$

where $\mathbf{P}_{\mathcal{S}}$ and $\mathbf{P}_{\mathcal{S}^\perp}$ *are orthogonal projectors onto* $\mathcal{S}$ *and* $\mathcal{S}^{\perp}$ *respectively.Thus,* $\mathbf{A}_{\mathcal{S}}$ *depends on the subspace* $\mathcal{S}$ *and not on a particular basis of* $\mathcal{S}$ *and its orthogonal complement* $\mathcal{S}^{\perp}$ *chosen for representing* $\mathbf{A}_{\mathcal{S}}$*.*

Proof. Consider the form (10.3.2) for $\mathbf{A}_{\mathcal{S}}$. As noted earlier the quantity $\mathbf{S_1^\star A S_2}(\mathbf{S_2^\star A S_2})^{-}\mathbf{S_2^\star A S_1}$ ia always invariant under the choices of g-inverse of $\mathbf{S_2^\star A S_2}$. Since $\rho(\mathbf{S_2 S_2^\star A S_2 S_2^\star}) = \rho(\mathbf{S_2^\star A S_2})$, we have

$\mathbf{S}_2^\star(\mathbf{S}_2\mathbf{S}_2^\star\mathbf{A}\mathbf{S}_2\mathbf{S}_2^\star)^-\mathbf{S}_2$ is a g-inverse of $\mathbf{S}_2^\star\mathbf{A}\mathbf{S}_2$ for each choice of g-inverse $(\mathbf{S}_2\mathbf{S}_2^\star\mathbf{A}\mathbf{S}_2\mathbf{S}_2^\star)^-$ of $\mathbf{S}_2\mathbf{S}_2^\star\mathbf{A}\mathbf{S}_2\mathbf{S}_2^\star$. So, we can write (10.3.2) as

$$\begin{aligned}\mathbf{A}_{\mathcal{S}} &= \mathbf{S}_1\mathbf{S}_1^\star\mathbf{A}\mathbf{S}_1\mathbf{S}_1^\star - \mathbf{S}_1\mathbf{S}_1^\star\mathbf{A}\mathbf{S}_2\mathbf{S}_2^\star(\mathbf{S}_2\mathbf{S}_2^*\mathbf{A}\mathbf{S}_2\mathbf{S}_2^*)^-\mathbf{S}_2\mathbf{S}_2^*\mathbf{A}\mathbf{S}_2\mathbf{S}_1^\star\\ &= \mathbf{P}_{\mathcal{S}}\mathbf{A}\mathbf{P}_{\mathcal{S}} - \mathbf{P}_{\mathcal{S}}\mathbf{A}\mathbf{P}_{\mathcal{S}^\perp}(\mathbf{P}_{\mathcal{S}^\perp}\mathbf{A}\mathbf{P}_{\mathcal{S}^\perp})^-\mathbf{P}_{\mathcal{S}^\perp}\mathbf{A}\mathbf{P}_{\mathcal{S}}\\ &= \mathbf{P}_{\mathcal{S}}(\mathbf{A} - \mathbf{A}\mathbf{P}_{\mathcal{S}^\perp}(\mathbf{P}_{\mathcal{S}^\perp}\mathbf{A}\mathbf{P}_{\mathcal{S}^\perp})^-\mathbf{P}_{\mathcal{S}^\perp}\mathbf{A})\mathbf{P}_{\mathcal{S}}.\end{aligned}$$

So, $\mathcal{C}(\mathbf{A}_{\mathcal{S}}) \subseteq \mathcal{S}$. Notice that $\mathbf{P}_{\mathcal{S}^\perp}(\mathbf{P}_{\mathcal{S}^\perp}\mathbf{A}\mathbf{P}_{\mathcal{S}^\perp})^-$ is a g-inverse of $\mathbf{P}_{\mathcal{S}^\perp}\mathbf{A}$ and therefore, $\mathbf{P}_{\mathcal{S}^\perp}(\mathbf{A} - \mathbf{A}\mathbf{P}_{\mathcal{S}^\perp}(\mathbf{P}_{\mathcal{S}^\perp}\mathbf{A}\mathbf{P}_{\mathcal{S}^\perp})^-\mathbf{P}_{\mathcal{S}^\perp}\mathbf{A}) = \mathbf{0}$. This further gives that $\mathcal{C}(\mathbf{A} - \mathbf{A}\mathbf{P}_{\mathcal{S}^\perp}(\mathbf{P}_{\mathcal{S}^\perp}\mathbf{A}\mathbf{P}_{\mathcal{S}^\perp})^-\mathbf{P}_{\mathcal{S}^\perp}\mathbf{A}) \subseteq \mathcal{S}$. Hence,

$$\mathbf{A}_{\mathcal{S}} = \mathbf{A} - \mathbf{A}\mathbf{P}_{\mathcal{S}^\perp}(\mathbf{P}_{\mathcal{S}^\perp}\mathbf{A}\mathbf{P}_{\mathcal{S}^\perp})^-\mathbf{P}_{\mathcal{S}^\perp}\mathbf{A}.$$ □

Corollary 10.3.3. *Let* $\mathbf{Q}$ *be a matrix such that* $\mathcal{C}(\mathbf{Q}) = \mathcal{S}^\perp$. *Then*

$$\mathbf{A}_{\mathcal{S}} = \mathbf{A} - \mathbf{A}\mathbf{Q}(\mathbf{Q}^\star\mathbf{A}\mathbf{Q})^-\mathbf{Q}^\star\mathbf{A},$$

where $(\mathbf{Q}^\star\mathbf{A}\mathbf{Q})^-$ *is any g-inverse of* $\mathbf{Q}^\star\mathbf{A}\mathbf{Q}$.

Proof. Clearly, $\mathbf{P}_{\mathcal{S}^\perp} = \mathbf{Q}(\mathbf{Q}^\star\mathbf{Q})^-\mathbf{Q}^\star$. Since $\rho(\mathbf{Q}^\star\mathbf{A}\mathbf{Q}) = \rho(\mathbf{P}_{\mathcal{S}^\perp}\mathbf{A}\mathbf{P}_{\mathcal{S}^\perp})$, the matrix $(\mathbf{Q}^\star\mathbf{Q})^-\mathbf{Q}^\star(\mathbf{Q}(\mathbf{Q}^\star\mathbf{Q})^-\mathbf{Q}^\star\mathbf{A}\mathbf{Q}(\mathbf{Q}^\star\mathbf{Q})^-\mathbf{Q}^\star)^-\mathbf{Q}(\mathbf{Q}^\star\mathbf{Q})^-$ is a g-inverse of $\mathbf{Q}^\star\mathbf{A}\mathbf{Q}$. So,

$$\mathbf{A}_{\mathcal{S}} = \mathbf{A} - \mathbf{A}\mathbf{P}_{\mathcal{S}^\perp}(\mathbf{P}_{\mathcal{S}}^\perp\mathbf{A}\mathbf{P}_{\mathcal{S}}^\perp)^-\mathbf{P}_{\mathcal{S}^\perp}\mathbf{A} = \mathbf{A} - \mathbf{A}\mathbf{Q}(\mathbf{Q}^\star\mathbf{A}\mathbf{Q})^-\mathbf{Q}^\star\mathbf{A}$$

for some g-inverse $(\mathbf{Q}^\star\mathbf{A}\mathbf{Q})^-$ of $\mathbf{Q}^\star\mathbf{A}\mathbf{Q}$. Notice that $\mathbf{A}\mathbf{Q}(\mathbf{Q}^\star\mathbf{A}\mathbf{Q})^-\mathbf{Q}^\star\mathbf{A}$ is invariant under choices of g-inverse of $\mathbf{Q}^\star\mathbf{A}\mathbf{Q}$, since $\mathcal{C}(\mathbf{Q}^\star\mathbf{A}) = \mathcal{C}(\mathbf{Q}^\star\mathbf{A}\mathbf{Q})$. Now, the result follows. □

Remark 10.3.4. Write $\tilde{\mathbf{A}} = \begin{pmatrix}\mathbf{A} & \mathbf{A}\mathbf{Q}\\ \mathbf{Q}^\star\mathbf{A} & \mathbf{Q}^\star\mathbf{A}\mathbf{Q}\end{pmatrix}$. Then, $\tilde{\mathbf{A}}$ is a $2n \times 2n$ matrix. It is easy to see that $\mathbf{A}_{\mathcal{S}}$ is the usual Schur complement of $\mathbf{Q}^\star\mathbf{A}\mathbf{Q}$ in $\tilde{\mathbf{A}}$.

In the setup of Theorem 10.3.2, write $\tilde{\mathbf{A}} = \begin{pmatrix}\mathbf{P}_{\mathcal{S}}\mathbf{A}\mathbf{P}_{\mathcal{S}} & \mathbf{P}_{\mathcal{S}}\mathbf{A}\mathbf{P}_{\mathcal{S}^\perp}\\ \mathbf{P}_{\mathcal{S}^\perp}\mathbf{A}\mathbf{P}_{\mathcal{S}} & \mathbf{P}_{\mathcal{S}^\perp}\mathbf{A}\mathbf{P}_{\mathcal{S}^\perp}\end{pmatrix}$. Let $\mathbf{B} = \mathbf{P}_{\mathcal{S}}\mathbf{A}\mathbf{P}_{\mathcal{S}}$, $\mathbf{C} = \mathbf{P}_{\mathcal{S}}\mathbf{A}\mathbf{P}_{\mathcal{S}^\perp}$, $\mathbf{E} = \mathbf{P}_{\mathcal{S}^\perp}\mathbf{A}\mathbf{P}_{\mathcal{S}}$ and $\mathbf{H} = \mathbf{P}_{\mathcal{S}^\perp}\mathbf{A}\mathbf{P}_{\mathcal{S}^\perp}$. Thus, $\tilde{\mathbf{A}} = \begin{pmatrix}\mathbf{B} & \mathbf{C}\\ \mathbf{E} & \mathbf{H}\end{pmatrix}$. Further, $\mathbf{P}_{\mathcal{S}} + \mathbf{P}_{\mathcal{S}^\perp} = \mathbf{I}$. Hence,

$$\begin{aligned}\mathbf{A} &= (\mathbf{P}_{\mathcal{S}} + \mathbf{P}_{\mathcal{S}^\perp})\mathbf{A}(\mathbf{P}_{\mathcal{S}} + \mathbf{P}_{\mathcal{S}^\perp})\\ &= \mathbf{P}_{\mathcal{S}}\mathbf{A}\mathbf{P}_{\mathcal{S}} + \mathbf{P}_{\mathcal{S}}\mathbf{A}\mathbf{P}_{\mathcal{S}^\perp} + \mathbf{P}_{\mathcal{S}^\perp}\mathbf{A}\mathbf{P}_{\mathcal{S}} + \mathbf{P}_{\mathcal{S}^\perp}\mathbf{A}\mathbf{P}_{\mathcal{S}^\perp}\\ &= \mathbf{B} + \mathbf{C} + \mathbf{E} + \mathbf{H}.\end{aligned}$$

The expression $\mathbf{A} = \mathbf{B} + \mathbf{C} + \mathbf{E} + \mathbf{H}$ is called the **Orthogonal Projective decomposition** of $\mathbf{A}$. The matrix $\mathbf{A}_{\mathcal{S}} = \mathbf{B} - \mathbf{C}\mathbf{H}^{-}\mathbf{E}$ is called the **Schur Complement** of $\mathbf{A}$ with respect to the subspace $\mathcal{S}$ and the matrix $\mathbf{A} - \mathbf{A}_{\mathcal{S}}$ is called the **Schur Compression** of $\mathbf{A}$ with respect to the subspace $\mathcal{S}$.

We now show that the shorted operator $\mathbf{A}_{\mathcal{S}}$ is the supremum of a particular set of *nnd* matrices.

Theorem 10.3.5. *Let* $\mathbf{A}$ *be an nnd matrix of order* $n \times n$ *and let* $\mathcal{S}$ *be a subspace of* $\mathbb{C}^n$. *Then*

$$\mathbf{A}_{\mathcal{S}} = \max\{\mathbf{D} : \mathbf{0} <^{\mathbf{L}} \mathbf{D} <^{\mathbf{L}} \mathbf{A}, \mathcal{C}(\mathbf{D}) \subseteq \mathcal{S}\}.$$

Proof. By Corollary 8.2.3, it follows that $\mathbf{0} <^{\mathbf{L}} \mathbf{A}_{\mathcal{S}}$ and

$$\mathbf{A} - \mathbf{A}_{\mathcal{S}} = \mathbf{S}\begin{pmatrix} \mathbf{S}_1^{\star}\mathbf{A}\mathbf{S}_2(\mathbf{S}_2^{\star}\mathbf{A}\mathbf{S}_2)^{-}\mathbf{S}_2^{\star}\mathbf{A}\mathbf{S}_1 & \mathbf{S}_1^{\star}\mathbf{A}\mathbf{S}_2 \\ \mathbf{S}_2^{\star}\mathbf{A}\mathbf{S}_1 & \mathbf{S}_2^{\star}\mathbf{A}\mathbf{S}_2 \end{pmatrix}\mathbf{S}^{\star} \tag{10.3.3}$$

is *nnd*. Thus, $\mathbf{0} <^{\mathbf{L}} \mathbf{A}_{\mathcal{S}} <^{\mathbf{L}} \mathbf{A}$. From Theorem 10.3.2, it is clear that $\mathcal{C}(\mathbf{A}_{\mathcal{S}}) \subseteq \mathcal{C}(\mathbf{P}_{\mathcal{S}}) = \mathcal{S}$, where $\mathbf{P}_{\mathcal{S}}$ is orthogonal projector on $\mathcal{S}$. Hence, $\mathbf{A}_{\mathcal{S}} \in \{\mathbf{D} : \mathbf{0} <^{\mathbf{L}} \mathbf{D} <^{\mathbf{L}} \mathbf{A}, \mathcal{C}(\mathbf{D}) \subseteq \mathcal{S}\}$.

Now, let $\mathbf{D}$ be an *nnd* matrix such that $\mathcal{C}(\mathbf{D}) \subseteq \mathcal{S}$ and $\mathbf{D} <^{\mathbf{L}} \mathbf{A}$. Then $\mathbf{D} = \mathbf{S}\mathbf{S}^{\star}\mathbf{D}\mathbf{S}\mathbf{S}^{\star} = \mathbf{S}\begin{pmatrix} \mathbf{S}_1^{\star}\mathbf{D}\mathbf{S}_1 & \mathbf{0} \\ \mathbf{0} & \mathbf{0} \end{pmatrix}\mathbf{S}^{\star} = \mathbf{S}_1\mathbf{S}_1^{\star}\mathbf{D}\mathbf{S}_1\mathbf{S}_1^{\star}$. As $\mathbf{D} <^{\mathbf{L}} \mathbf{A}$, the matrix $\mathbf{A} - \mathbf{D} = \mathbf{S}\begin{pmatrix} \mathbf{S}_1^{\star}\mathbf{A}\mathbf{S}_1 - \mathbf{S}_1^{\star}\mathbf{D}\mathbf{S}_1 & \mathbf{S}_1^{\star}\mathbf{A}\mathbf{S}_2 \\ \mathbf{S}_2^{\star}\mathbf{A}\mathbf{S}_1 & \mathbf{S}_2^{\star}\mathbf{A}\mathbf{S}_2 \end{pmatrix}\mathbf{S}^{\star}$ is *nnd*. It now follows that $\mathbf{S}_1\{\mathbf{S}_1^{\star}\mathbf{A}\mathbf{S}_1 - \mathbf{S}_1^{\star}\mathbf{D}\mathbf{S}_1 - \mathbf{S}_1^{\star}\mathbf{A}\mathbf{S}_2(\mathbf{S}_2^{\star}\mathbf{A}\mathbf{S}_2)^{-}\mathbf{S}_2^{\star}\mathbf{A}\mathbf{S}_1\}\mathbf{S}_1^{\star} = \mathbf{A}_{\mathcal{S}} - \mathbf{D}$ is *nnd*. So, $\mathbf{D} <^{\mathbf{L}} \mathbf{A}_{\mathcal{S}}$. □

Corollary 10.3.6. *Let* $\mathbf{A}, \mathbf{B}$ *be nnd matrices of the same order such that* $\mathbf{B} <^{\mathbf{L}} \mathbf{A}$. *Then* $\mathbf{B}_{\mathcal{S}} <^{\mathbf{L}} \mathbf{A}_{\mathcal{S}}$ *for each subspace* $\mathcal{S}$ *of* $\mathbb{C}^n$.

Proof. By Theorem 10.3.2, $\mathcal{C}(\mathbf{B}_{\mathcal{S}}) \subseteq \mathcal{S}$. Also, by Theorem 10.3.5, we have $\mathbf{B}_{\mathcal{S}} <^{\mathbf{L}} \mathbf{B}$, so, $\mathbf{B}_{\mathcal{S}} <^{\mathbf{L}} \mathbf{B} <^{\mathbf{L}} \mathbf{A}$. Another application of Theorem 10.3.5 yields the result. □

Corollary 10.3.7. *Let* $\mathbf{A}$ *be an nnd matrix of order* $n \times n$ *and let* $\mathcal{S}$, *and* $\mathcal{T}$ *be subspaces of* $\mathbb{C}^n$. *Then* $(\mathbf{A}_{\mathcal{S}})_{\mathcal{T}} = (\mathbf{A}_{\mathcal{T}})_{\mathcal{S}} = \mathbf{A}_{\mathcal{S}\cap\mathcal{T}}$.

Proof. Clearly, $\mathbf{0} <^{\mathbf{L}} (\mathbf{A}_{\mathcal{S}})_{\mathcal{T}} <^{\mathbf{L}} \mathbf{A}_{\mathcal{S}} <^{\mathbf{L}} \mathbf{A}$. So, $\mathcal{C}((\mathbf{A}_{\mathcal{S}})_{\mathcal{T}}) \subseteq \mathcal{S}$. However, $\mathcal{C}((\mathbf{A}_{\mathcal{S}})_{\mathcal{T}}) \subseteq \mathcal{T}$, so, $\mathcal{C}((\mathbf{A}_{\mathcal{S}})_{\mathcal{T}}) \subseteq \mathcal{S} \cap \mathcal{T}$. Further, since $\mathcal{C}((\mathbf{A}_{\mathcal{S}})_{\mathcal{T}}) \subseteq \mathcal{S} \cap \mathcal{T}$ and $\mathbf{0} <^{\mathbf{L}} (\mathbf{A}_{\mathcal{S}})_{\mathcal{T}} <^{\mathbf{L}} \mathbf{A}$, it follows that $(\mathbf{A}_{\mathcal{S}})_{\mathcal{T}} <^{\mathbf{L}} \mathbf{A}_{\mathcal{S}\cap\mathcal{T}}$. On the other hand, $\mathbf{0} <^{\mathbf{L}} \mathbf{A}_{\mathcal{S}\cap\mathcal{T}} <^{\mathbf{L}} \mathbf{A}$ and $\mathcal{C}(\mathbf{A}_{\mathcal{S}\cap\mathcal{T}}) \subseteq \mathcal{S}$. Hence, $\mathbf{0} <^{\mathbf{L}} \mathbf{A}_{\mathcal{S}\cap\mathcal{T}} <^{\mathbf{L}} \mathbf{A}_{\mathcal{S}}$. Also, $\mathcal{C}(\mathbf{A}_{\mathcal{S}\cap\mathcal{T}}) \subseteq \mathcal{T}$. Therefore, by Theorem 10.3.5, $\mathbf{A}_{\mathcal{S}\cap\mathcal{T}} <^{\mathbf{L}} (\mathbf{A}_{\mathcal{S}})_{\mathcal{T}}$ and hence, $(\mathbf{A}_{\mathcal{S}})_{\mathcal{T}} = \mathbf{A}_{\mathcal{S}\cap\mathcal{T}}$. Interchanging the roles of $\mathcal{S}$, and $\mathcal{T}$ we get the $(\mathbf{A}_{\mathcal{T}})_{\mathcal{S}} = \mathbf{A}_{\mathcal{S}\cap\mathcal{T}}$. □

Corollary 10.3.8. *Let* $\mathbf{A}$ *be an nnd matrix of order* $n \times n$ *and let* $\mathcal{S}$ *be a subspace of* $\mathbb{C}^n$. *Then* $\mathcal{C}(\mathbf{A}_{\mathcal{S}}) = \mathcal{C}(\mathbf{A}) \cap \mathcal{S}$.

Proof. Let $\mathbf{P}_{\mathcal{S}}$ be the orthogonal projector onto $\mathcal{S}$. Then $\mathbf{A}$ and $\mathbf{P}_{\mathcal{S}}$ are parallel summable, so, by Theorem 9.2.16, $\mathcal{C}(\mathcal{P}(\mathbf{A}, \mathbf{P}_{\mathcal{S}})) = \mathcal{C}(\mathbf{A}) \cap \mathcal{S}$ and $\mathcal{P}(\mathbf{A}, \mathbf{P}_{\mathcal{S}}) <^{\mathbf{L}} \mathbf{A}$ since $\mathbf{A} - \mathcal{P}(\mathbf{A}, \mathbf{P}_{\mathcal{S}}) = \mathbf{P}_{\mathcal{S}}(\mathbf{A} + \mathbf{P}_{\mathcal{S}})^{-}\mathbf{P}_{\mathcal{S}}$ is *nnd*. So, by Theorem 10.3.5, $\mathcal{P}(\mathbf{A}, \mathbf{P}_{\mathcal{S}}) <^{\mathbf{L}} \mathbf{A}_{\mathcal{S}}$. Since $\mathbf{A}_{\mathcal{S}} <^{\mathbf{L}} \mathbf{A}$ we have $\mathcal{C}(\mathbf{A}_{\mathcal{S}}) \subseteq \mathcal{C}(\mathbf{A})$. Also, $\mathcal{C}(\mathbf{A}_{\mathcal{S}}) \subseteq \mathcal{S}$. Thus, $\mathcal{C}(\mathbf{A}_{\mathcal{S}}) \subseteq \mathcal{C}(\mathbf{A}) \cap \mathcal{S}$. Now, $\mathcal{C}(\mathbf{A}) \cap \mathcal{S} = \mathcal{C}(\mathcal{P}(\mathbf{A}, \mathbf{P}_{\mathcal{S}})) \subseteq \mathcal{C}(\mathbf{A}_{\mathcal{S}}) \subseteq \mathcal{C}(\mathbf{A}) \cap \mathcal{S}$. Hence, $\mathcal{C}(\mathbf{A}_{\mathcal{S}}) = \mathcal{C}(\mathbf{A}) \cap \mathcal{S}$. □

Corollary 10.3.9. *If* $\mathbf{A}$ *is an orthogonal projector, then* $\mathbf{A}_{\mathcal{S}}$ *is the orthogonal projector onto* $\mathcal{C}(\mathbf{A}) \cap \mathcal{S}$.

Proof. Let $\mathbf{Q}$ be the orthogonal projector onto $\mathcal{C}(\mathbf{A}) \cap \mathcal{S}$. Since $\mathbf{A}$ is the orthogonal projector onto $\mathcal{C}(\mathbf{A})$, we have $\mathbf{Q} <^{\mathbf{L}} \mathbf{A}$. Also, $\mathcal{C}(\mathbf{Q}) \subseteq \mathcal{S}$. Further, $\mathbf{QAQ} = \mathbf{Q}$. Let $\mathbf{0} <^{\mathbf{L}} \mathbf{D} <^{\mathbf{L}} \mathbf{A}$ and $\mathcal{C}(\mathbf{D}) \subseteq \mathcal{S}$. Then $\mathcal{C}(\mathbf{D}) \subseteq \mathcal{C}(\mathbf{A}) \cap \mathcal{S}$. Hence, $\mathbf{QDQ} = \mathbf{D}$. Since $\mathbf{0} <^{\mathbf{L}} \mathbf{D} <^{\mathbf{L}} \mathbf{A}$, we have $\mathbf{0} <^{\mathbf{L}} \mathbf{QDQ} <^{\mathbf{L}} \mathbf{QAQ} \Rightarrow \mathbf{O} <^{\mathbf{L}} \mathbf{D} <^{\mathbf{L}} \mathbf{Q}$. Thus $\mathbf{Q} = \mathbf{A}_{\mathcal{S}}$. □

Theorem 10.3.10. *(Schur Decomposition) Let* $\mathbf{A}$ *be an nnd matrix of order* $n \times n$ *and let* $\mathcal{S}$ *be a subspace of* $\mathbb{C}^n$. *Then there exists a unique nnd matrix* $\mathbf{F}$ *of order* $n \times n$ *such that*

(i) $\mathbf{F} <^{-} \mathbf{A}$
(ii) $\mathcal{C}(\mathbf{F}) \subseteq \mathcal{S}$ *and*
(iii) $\mathcal{C}(\mathbf{A} - \mathbf{F}) \cap \mathcal{S} = \mathbf{0}$.
Moreover, the matrix satisfying the above properties is precisely the shorted operator $\mathbf{A}_{\mathcal{S}}$.

Proof. We first show that $\mathbf{A}_{\mathcal{S}}$ satisfies (i)-(iii) and is *nnd*. We already know that $\mathbf{A}_{\mathcal{S}}$ is *nnd* in view of Theorem 10.3.5 and that $\mathcal{C}(\mathbf{A}_{\mathcal{S}}) \subseteq \mathcal{S}$. Moreover, $\mathbf{A}_{\mathcal{S}} <^{\mathbf{L}} \mathbf{A}$, so, $\mathbf{A} - \mathbf{A}_{\mathcal{S}}$ is *nnd*. We show that $\mathcal{C}(\mathbf{A} - \mathbf{A}_{\mathcal{S}}) \cap \mathcal{S} = \{\mathbf{0}\}$. Let us write

$$\begin{pmatrix} \mathbf{S}_1^{\star}\mathbf{A}\mathbf{S}_2(\mathbf{S}_2^{\star}\mathbf{A}\mathbf{S}_2)^{-}\mathbf{S}_2^{\star}\mathbf{A}\mathbf{S}_1 & \mathbf{S}_1^{\star}\mathbf{A}\mathbf{S}_2 \\ \mathbf{S}_2^{\star}\mathbf{A}\mathbf{S}_1 & \mathbf{S}_2^{\star}\mathbf{A}\mathbf{S}_2 \end{pmatrix} = \mathbf{X}$$

and

$$\begin{pmatrix} \mathbf{S}_1^{\star}\mathbf{A}\mathbf{S}_1 - \mathbf{S}_1^{\star}\mathbf{A}\mathbf{S}_2(\mathbf{S}_2^{\star}\mathbf{A}\mathbf{S}_2)^{-}\mathbf{S}_2^{\star}\mathbf{A}\mathbf{S}_1 & \mathbf{0} \\ \mathbf{0} & \mathbf{0} \end{pmatrix} = \mathbf{Y}.$$

Notice that

$$\rho\begin{pmatrix}\mathbf{S}_1^\star\mathbf{AS}_1 & \mathbf{S}_1^\star\mathbf{AS}_2\\ \mathbf{S}_2^\star\mathbf{AS}_1 & \mathbf{S}_2^\star\mathbf{AS}_2\end{pmatrix} = \rho(\mathbf{S}_2^\star\mathbf{AS}_2) + \rho(\mathbf{S}_1^\star\mathbf{AS}_1 - \mathbf{S}_1^\star\mathbf{AS}_2(\mathbf{S}_2^\star\mathbf{AS}_2)^-\mathbf{S}_2^\star\mathbf{AS}_1).$$

$$= \rho(\mathbf{X}) + \rho(\mathbf{Y}).$$

Hence, $\mathcal{C}(\mathbf{X}) \cap \mathcal{C}(\mathbf{Y}) = \{\mathbf{0}\}$. Now, $\mathcal{S} = \mathcal{C}(\mathbf{S}_1)$, so any element in $\mathcal{S}$ is of the form $\mathbf{S}_1(\mathbf{u}) = \mathbf{S}\begin{pmatrix}\mathbf{u}\\ \mathbf{0}\end{pmatrix}$. Consider a vector $\mathbf{x} \in \mathcal{S} \cap \mathcal{C}(\mathbf{A} - \mathbf{A}_\mathcal{S})$. So, $\mathbf{x} = \mathbf{S}_1(\mathbf{u}) = \mathbf{S}\begin{pmatrix}\mathbf{u}\\ \mathbf{0}\end{pmatrix} = \mathbf{S}\begin{pmatrix}\mathbf{S}_1^\star\mathbf{AS}_2(\mathbf{S}_2^\star\mathbf{AS}_2)^-\mathbf{S}_2^\star\mathbf{AS}_1 & \mathbf{S}_1^\star\mathbf{AS}_2\\ \mathbf{S}_2^\star\mathbf{AS}_1 & \mathbf{S}_2^\star\mathbf{AS}_2\end{pmatrix}\mathbf{S}^\star\begin{pmatrix}\mathbf{v}\\ \mathbf{w}\end{pmatrix}$ for some $\mathbf{u}$, $\mathbf{v}$ and $\mathbf{w}$. Therefore,

$$\mathbf{u} = \mathbf{S}_1^\star\mathbf{AS}_2(\mathbf{S}_2^\star\mathbf{AS}_2)^-\mathbf{S}_2^\star\mathbf{AS}_1\mathbf{S}^\star\mathbf{v} + \mathbf{S}_1^\star\mathbf{AS}_2\mathbf{S}^\star\mathbf{w} \tag{10.3.4}$$

and

$$\mathbf{0} = \mathbf{S}_2^\star\mathbf{AS}_1\mathbf{S}^\star\mathbf{v} + \mathbf{S}_2^\star\mathbf{AS}_2\mathbf{S}^\star\mathbf{w}. \tag{10.3.5}$$

Pre-multiplying (10.3.5) by $\mathbf{S}_1^\star\mathbf{AS}_2(\mathbf{S}_2^\star\mathbf{AS}_2)^-$, we have

$$\mathbf{S}_1^\star\mathbf{AS}_2(\mathbf{S}_2^\star\mathbf{AS}_2)^-\mathbf{S}_2^\star\mathbf{AS}_1\mathbf{S}^\star\mathbf{v} + \mathbf{S}_1^\star\mathbf{AS}_2\mathbf{S}^\star\mathbf{w} = \mathbf{0}$$

Using (10.3.4) we have $\mathbf{u} = \mathbf{0}$. Thus, $\mathcal{C}(\mathbf{A} - \mathbf{A}_\mathcal{S}) \cap \mathcal{S} = \mathbf{0}$. Thus, $\mathbf{F} = \mathbf{A}_\mathcal{S}$ satisfies all the properties (i)-(iii).

We now show $\mathbf{F}$ is unique. Let $\mathbf{F}$ be any *nnd* matrix satisfying (i)-(iii). Clearly, $\mathbf{A} - \mathbf{F}$ is *nnd*. So, $\mathbf{F} <^{\mathbf{L}} \mathbf{A}$. Further, $\mathcal{C}(\mathbf{F}) \subseteq \mathcal{S}$. Therefore, $\mathbf{F} <^{\mathbf{L}} \mathbf{A}_\mathcal{S}$. We can write $\mathbf{A} = \mathbf{F} + (\mathbf{A}_\mathcal{S} - \mathbf{F}) + (\mathbf{A} - \mathbf{A}_\mathcal{S})$. So, $\mathcal{C}(\mathbf{A} - \mathbf{F}) = \mathcal{C}(\mathbf{A}_\mathcal{S} - \mathbf{F}) + \mathcal{C}(\mathbf{A} - \mathbf{A}_\mathcal{S})$. However, $\mathcal{C}(\mathbf{A}_\mathcal{S} - \mathbf{F}) \subseteq \mathcal{S}$. Hence, in view of (iii), $\mathbf{A}_\mathcal{S} - \mathbf{F}$ must be the null matrix. In other words $\mathbf{F} = \mathbf{A}_\mathcal{S}$.

□

Theorem 10.3.11. *Let* $\mathbf{A}$ *be an nnd matrix of order* $n \times n$ *and let* $\mathcal{S}$ *be a subspace of* $\mathbb{C}^n$. *Then* $\mathbf{A}_\mathcal{S} = \max\{\mathbf{D} : \mathbf{D} \in \mathfrak{C}\}$, *where*

$$\mathfrak{C} = \{\mathbf{D} : \mathbf{D} \text{ is nnd, } \mathbf{D} <^- \mathbf{A},\ \mathcal{C}(\mathbf{D}) \subseteq \mathcal{S}\}.$$

Proof. By Theorem 10.3.5, $\mathbf{A}_\mathcal{S} <^- \mathbf{A}$ and $\mathbf{A}_\mathcal{S}$ is *nnd*. Therefore, $\mathbf{A}_\mathcal{S} \in \mathfrak{C}$. Let $\mathbf{D} \in \mathfrak{C}$. Write $\mathbf{A} - \mathbf{D} = (\mathbf{A} - \mathbf{A}_\mathcal{S}) + (\mathbf{A}_\mathcal{S} - \mathbf{D})$. This sum is direct as $\mathcal{C}(\mathbf{A}_\mathcal{S} - \mathbf{D}) \subseteq \mathcal{S}$ and $(\mathbf{A} - \mathbf{A}_\mathcal{S}) \cap \mathcal{S} = \{\mathbf{0}\}$. Thus, $\mathbf{A} - \mathbf{A}_\mathcal{S} <^- \mathbf{A} - \mathbf{D}$ and $\mathbf{A}_\mathcal{S} - \mathbf{D} <^- \mathbf{A} - \mathbf{D}$. Since, $\mathbf{D} <^- \mathbf{A}$, by Theorem 8.3.8, $\mathbf{A} - \mathbf{D}$ is *nnd*. Also $\mathbf{A} - \mathbf{A}_\mathcal{S}$ is *nnd*, by Theorem 10.3.5. So, from $\mathbf{A} - \mathbf{A}_\mathcal{S} <^- \mathbf{A} - \mathbf{D}$, and Theorem 8.3.8, $\mathbf{A} - \mathbf{A}_\mathcal{S} <^{\mathbf{L}} \mathbf{A} - \mathbf{D}$. Hence, $\mathbf{A}_\mathcal{S} - \mathbf{D}$ is *nnd*. Now, $\mathbf{A}_\mathcal{S} - \mathbf{D} <^- \mathbf{A} - \mathbf{D} <^- \mathbf{A}$. Since $\mathbf{A}_\mathcal{S} <^- \mathbf{A}$, we have $\rho(\mathbf{A}) = \rho(\mathbf{A} - \mathbf{A}_\mathcal{S}) + \rho(\mathbf{A}_\mathcal{S})$. Similarly $\rho(\mathbf{A} - \mathbf{D}) = \rho(\mathbf{A}_\mathcal{S} - \mathbf{D}) + \rho(\mathbf{A} - \mathbf{A}_\mathcal{S})$ and $\rho(\mathbf{A}) = \rho(\mathbf{A} - \mathbf{D}) + \rho(\mathbf{D})$. From these it follows that $\rho(\mathbf{A}_\mathcal{S}) - \rho(\mathbf{D}) = \rho(\mathbf{A}_\mathcal{S} - \mathbf{D})$. So, by Remark 3.3.9 we have $\mathbf{D} <^- \mathbf{A}_\mathcal{S}$.

□

Theorem 10.3.12. *Let* $\mathbf{A}, \mathbf{B}$ *be nnd matrices of the same order and let* $\mathcal{S}$ *be a subspace of* $\mathbb{C}^n$. *Then* $\mathbf{A}_{\mathcal{S}} + \mathbf{B}_{\mathcal{S}} <^{\mathbf{L}} (\mathbf{A}+\mathbf{B})_{\mathcal{S}}$. *Further the equality holds if and only if* $\mathcal{C}(\mathbf{A} - \mathbf{A}_{\mathcal{S}} + \mathbf{B} - \mathbf{B}_{\mathcal{S}}) \cap \mathcal{S} = \{\mathbf{0}\}$.

Proof. Clearly, $\mathbf{A}_{\mathcal{S}} + \mathbf{B}_{\mathcal{S}} <^{\mathbf{L}} \mathbf{A}+\mathbf{B}$, since $\mathbf{A}_{\mathcal{S}} <^{\mathbf{L}} \mathbf{A}$ and $\mathbf{B}_{\mathcal{S}} <^{\mathbf{L}} \mathbf{B}$. Also, $\mathcal{C}(\mathbf{A}_{\mathcal{S}}) \subseteq \mathcal{S}$ and $\mathcal{C}(\mathbf{B}_{\mathcal{S}}) \subseteq \mathcal{S}$, so, $\mathcal{C}(\mathbf{A}_{\mathcal{S}}+\mathbf{B}_{\mathcal{S}}) \subseteq \mathcal{S}$. Therefore, we have $\mathbf{A}_{\mathcal{S}}+\mathbf{B}_{\mathcal{S}} <^{\mathbf{L}} (\mathbf{A}+\mathbf{B})_{\mathcal{S}}$. Write $\mathbf{A}+\mathbf{B} = (\mathbf{A}_{\mathcal{S}}+\mathbf{B}_{\mathcal{S}}) + (\mathbf{A}-\mathbf{A}_{\mathcal{S}}+\mathbf{B}-\mathbf{B}_{\mathcal{S}})$. In view of Theorem 10.3.10,

$$\mathbf{A}_{\mathcal{S}}+\mathbf{B}_{\mathcal{S}} = (\mathbf{A}+\mathbf{B})_{\mathcal{S}} \Leftrightarrow \mathcal{C}(\mathbf{A}-\mathbf{A}_{\mathcal{S}}+\mathbf{B}-\mathbf{B}_{\mathcal{S}}) \cap \mathcal{S} = \{\mathbf{0}\}. \qquad \square$$

Corollary 10.3.13. *Let* $\mathbf{A}, \mathbf{B}$ *be positive definite matrices of the same order* $n \times n$ *and let* $\mathcal{S}$ *be a subspace of* $\mathbb{C}^n$. *Then* $\mathbf{A}_{\mathcal{S}}+\mathbf{B}_{\mathcal{S}} = (\mathbf{A}+\mathbf{B})_{\mathcal{S}}$ *if and only if* $\mathcal{C}(\mathbf{A}-\mathbf{A}_{\mathcal{S}}) = \mathcal{C}(\mathbf{B}-\mathbf{B}_{\mathcal{S}})$.

Proof. Since $\mathbf{A}$ is positive definite and $\mathbf{A}_{\mathcal{S}} <^{-} \mathbf{A}$, we have

$$\rho(\mathbf{A}_{\mathcal{S}}) + \rho(\mathbf{A}-\mathbf{A}_{\mathcal{S}}) = n.$$

If possible, let $\mathcal{C}(\mathbf{A}-\mathbf{A}_{\mathcal{S}}) \neq \mathcal{C}(\mathbf{A}-\mathbf{B}_{\mathcal{S}})$. As $\mathbf{A}-\mathbf{A}_{\mathcal{S}}$ and $\mathbf{B}-\mathbf{B}_{\mathcal{S}}$ are *nnd*, we have $\mathcal{C}(\mathbf{A}-\mathbf{A}_{\mathcal{S}}) + \mathcal{C}(\mathbf{B}-\mathbf{B}_{\mathcal{S}}) = \mathcal{C}(\mathbf{A}-\mathbf{A}_{\mathcal{S}}+\mathbf{B}-\mathbf{B}_{\mathcal{S}})$ and $\mathcal{C}(\mathbf{A}-\mathbf{A}_{\mathcal{S}})$ is its proper subspace. So, $\rho(\mathbf{A}-\mathbf{A}_{\mathcal{S}}+\mathbf{B}-\mathbf{B}_{\mathcal{S}}) > \rho(\mathbf{A}-\mathbf{A}_{\mathcal{S}}) = n - \rho(\mathbf{A}_{\mathcal{S}})$. Further, $\mathcal{C}(\mathbf{A}-\mathbf{A}_{\mathcal{S}}+\mathbf{B}-\mathbf{B}_{\mathcal{S}}) \cap \mathcal{S} \neq \{\mathbf{0}\}$. Hence, $\mathbf{A}_{\mathcal{S}}+\mathbf{B}_{\mathcal{S}} \neq (\mathbf{A}+\mathbf{B})_{\mathcal{S}}$.

Converse follows from Theorem 10.3.12. $\square$

Theorem 10.3.14. *Let* $\mathbf{A}$ *be nnd matrix of the order* $n \times n$ *and let* $\mathcal{S}$ *be a subspace of* $\mathbb{C}^n$. *Then there exist matrices* $\mathbf{M_r}$ *and* $\mathbf{M}_\ell$ *such that*

(i) $\mathbf{P}_{\mathcal{S}^\perp}\mathbf{M_r} = \mathbf{M_r}$
(ii) $\mathbf{M}_\ell\mathbf{P}_{\mathcal{S}^\perp} = \mathbf{M}_\ell$
(iii) $\mathbf{P}_{\mathcal{S}^\perp}\mathbf{A}\mathbf{M_r} = \mathbf{P}_{\mathcal{S}^\perp}\mathbf{A}$ *and*
(iv) $\mathbf{M}_\ell\mathbf{A}\mathbf{P}_{\mathcal{S}^\perp} = \mathbf{A}\mathbf{P}_{\mathcal{S}^\perp}$

As a consequence, $\mathbf{A}\mathbf{M_r} = \mathbf{M}_\ell\mathbf{A}\mathbf{M_r} = \mathbf{M}_\ell\mathbf{A}$. *Also, for given matrix* $\mathbf{A}$ *and subspace* $\mathcal{S}$, *the matrices* $\mathbf{A}\mathbf{M_r}$ *and* $\mathbf{M}_\ell\mathbf{A}$ *are unique and* $\mathbf{A}_{\mathcal{S}} = \mathbf{A} - \mathbf{A}\mathbf{M_r}$.

Proof. Let $\mathbf{S_1}$ and $\mathbf{S_2}$ be the matrices as in Definition 10.3.1. Then we can write $\mathbf{A} = \mathbf{S}\begin{pmatrix} \mathbf{S_1^\star A S_1} & \mathbf{S_1^\star A S_2} \\ \mathbf{S_2^\star A S_1} & \mathbf{S_2^\star A S_2} \end{pmatrix}\mathbf{S}^\star$. It is easy to see that the matrices

$$\mathbf{M_r} = \mathbf{S}\begin{pmatrix} \mathbf{0} & \mathbf{0} \\ (\mathbf{S_2^\star A S_2})^-\mathbf{S_2^\star A S_1} & (\mathbf{S_2^\star A S_2})^-\mathbf{S_2^\star A S_2} \end{pmatrix}\mathbf{S}^\star$$

and $\mathbf{M}_\ell = \mathbf{S}\begin{pmatrix} \mathbf{0} & \mathbf{S_1^\star A S_2}(\mathbf{S_2^\star A S_2})^- \\ \mathbf{0} & \mathbf{S_2^\star A S_2}(\mathbf{S_2^\star A S_2})^- \end{pmatrix}\mathbf{S}^\star$ satisfy (i)-(iv).

Now, $\mathbf{AM_r} = \mathbf{A}(\mathbf{P}_{\mathcal{S}^\perp}\mathbf{M_r}) = (\mathbf{AP}_{\mathcal{S}^\perp})\mathbf{M_r} = \mathbf{M}_\ell\mathbf{AP}_{\mathcal{S}^\perp}\mathbf{M_r} = \mathbf{M}_\ell\mathbf{AM_r}$. Similarly, $\mathbf{M}_\ell\mathbf{A} = \mathbf{M}_\ell\mathbf{AP}_{\mathcal{S}^\perp}\mathbf{M_r} = \mathbf{M}_\ell\mathbf{AM_r}$.
Also, it is easy to check that $\mathbf{AM_r} = \mathbf{A} - \mathbf{A}_\mathcal{S} = \mathbf{M}_\ell\mathbf{A}$. Thus, the matrices $\mathbf{AM_r}$ and $\mathbf{M}_\ell\mathbf{A}$ do not depend on choice of matrices $\mathbf{M_r}$ and $\mathbf{M}_\ell$ respectively and so are unique. □

10.4 Shorted operator via parallel sums

In this section we obtain the shorted operator of an *nnd* matrix with respect to a given subspace as the limit of a certain sequence of parallel sums. For this we first give the following:

Lemma 10.4.1. *Let* $\mathbf{C}$ *be an nnd matrix of order* $n \times n$ *and* $\mathbf{B}$ *any matrix of order* $m \times n$ *such that* $\mathcal{C}(\mathbf{B}^\star) \subseteq \mathcal{C}(\mathbf{C})$. *Then*
$\lim_{\epsilon\to 0}\mathbf{B}(\mathbf{C}+\epsilon\mathbf{I})^{-1}\mathbf{B}^\star = \mathbf{BC}^\dagger\mathbf{B}^\star = \mathbf{BC}^-\mathbf{B}^\star$,
where $\mathbf{C}^-$ *is an arbitrary g-inverse of* $\mathbf{C}$.

Proof. If $\mathbf{C}$ is non-singular, the result follows trivially. So, let $\mathbf{C}$ be singular and $\mathbf{C} = (\mathbf{P_1} : \mathbf{P_2})\mathbf{diag}\,(\Delta\ \ \mathbf{0})\begin{pmatrix}\mathbf{P}_1^\star\\ \mathbf{P}_2^\star\end{pmatrix}$ be a spectral decomposition of $\mathbf{C}$, where $\mathbf{P} = (\mathbf{P_1} : \mathbf{P_2})$ is unitary, Δ is positive definite diagonal matrix and partitioning is such that $\mathbf{P_1\Delta}$ is defined. Clearly, $\mathcal{C}(\mathbf{C}) = \mathcal{C}(\mathbf{P_1})$. Since, $\mathcal{C}(\mathbf{B}^\star) \subseteq \mathcal{C}(\mathbf{C})$ we can write $\mathbf{B}^\star = \mathbf{P_1D} = \mathbf{P}\begin{pmatrix}\mathbf{D}\\ \mathbf{0}\end{pmatrix}$ for some matrix $\mathbf{D}$. Now,

$$\mathbf{B}(\mathbf{C}+\epsilon\mathbf{I})^{-1}\mathbf{B}^\star = (\mathbf{D}^\star : \mathbf{0})\begin{pmatrix}\Delta+\epsilon\mathbf{I} & \mathbf{0}\\ \mathbf{0} & \epsilon\mathbf{I}\end{pmatrix}^{-1}\begin{pmatrix}\mathbf{D}\\ \mathbf{0}\end{pmatrix}$$
$$= \mathbf{D}^\star(\mathbf{\Delta}+\epsilon\mathbf{I})^{-1}\mathbf{D}.$$

Taking limit as $\epsilon \to 0$, the result follows. □

We are now ready to obtain the shorted operator of an *nnd* matrix as the limit of a sequence of parallel sums. We prove

Theorem 10.4.2. *Let* $\mathbf{A}$ *be nnd matrix of the order* $m \times m$ *and let* $\mathcal{S}$ *be a subspace of* $\mathbb{C}^m$. *Let* $\mathbf{P}_\mathcal{S}$ *be the orthogonal projector onto* $\mathcal{S}$. *Then* $\mathbf{A}_\mathcal{S} = \lim_{n\to\infty}\mathcal{P}(\mathbf{A}, n\mathbf{P}_\mathcal{S})$.

Proof. As seen in the previous section, $\mathbf{A} = \mathbf{S}\begin{pmatrix}\mathbf{S}_1^\star\mathbf{AS}_1 & \mathbf{S}_1^\star\mathbf{AS}_2\\ \mathbf{S}_2^\star\mathbf{AS}_1 & \mathbf{S}_2^\star\mathbf{AS}_2\end{pmatrix}\mathbf{S}^\star$, where $\mathbf{S} = (\mathbf{S_1} : \mathbf{S_2})$ is unitary and $\mathcal{C}(\mathbf{S_1}) = \mathcal{S}$. Also, $\mathbf{P}_\mathcal{S} = \mathbf{S}_1\mathbf{S}_1^\star =$

$\mathbf{S}\text{diag}\,(\mathbf{I}\ \ \mathbf{0})\,\mathbf{S}^\star$. Write $\mathbf{A}_\epsilon = \mathbf{S}\begin{pmatrix}\mathbf{S}_1^\star\mathbf{A}\mathbf{S}_1 & \mathbf{S}_1^\star\mathbf{A}\mathbf{S}_2\\ \mathbf{S}_2^\star\mathbf{A}\mathbf{S}_1 & \mathbf{S}_2^\star\mathbf{A}\mathbf{S}_2+\epsilon\mathbf{I}\end{pmatrix}\mathbf{S}^\star$. Then $\mathbf{A}_\epsilon + n\mathbf{P}_\mathcal{S} = \mathbf{S}\begin{pmatrix}\mathbf{S}_1^\star\mathbf{A}\mathbf{S}_1+n\mathbf{I} & \mathbf{S}_1^\star\mathbf{A}\mathbf{S}_2\\ \mathbf{S}_2^\star\mathbf{A}\mathbf{S}_1 & \mathbf{S}_2^\star\mathbf{A}\mathbf{S}_2+\epsilon\mathbf{I}\end{pmatrix}\mathbf{S}^\star$, which is a positive definite matrix whenever $n>0$ and $\epsilon>0$. Clearly, $(\mathbf{A}_\epsilon+n\mathbf{P}_\mathcal{S})^{-1} = \mathbf{S}\mathbf{X}\mathbf{S}^\star$, where $\mathbf{X} = \begin{pmatrix}(n\mathbf{I}+\mathbf{S}_1^\star\mathbf{A}\mathbf{S}_1-\mathbf{S}_1^\star\mathbf{A}\mathbf{S}_2(\mathbf{S}_2^\star\mathbf{A}\mathbf{S}_2+\epsilon\mathbf{I})^{-1}\mathbf{S}_2^\star\mathbf{A}\mathbf{S}_1)^{-1} & \bullet\\ \bullet & \bullet\end{pmatrix}$ and $\bullet$ stand for some suitable matrices not of interest to us. Now,

$$\begin{aligned}\mathcal{P}(\mathbf{A}_\epsilon, n\mathbf{P}_\mathcal{S}) &= n\mathbf{P}_\mathcal{S}(\mathbf{A}_\epsilon+n\mathbf{P}_\mathcal{S})^{-1}\mathbf{A}_\epsilon\\ &= n\mathbf{P}_\mathcal{S} - n\mathbf{P}_\mathcal{S}((\mathbf{A}_\epsilon+n\mathbf{P}_\mathcal{S})^{-1})n\mathbf{P}_\mathcal{S}\\ &= \mathbf{S}\begin{pmatrix}n\mathbf{I} & \mathbf{0}\\ \mathbf{0} & \mathbf{0}\end{pmatrix}\mathbf{X}\begin{pmatrix}n\mathbf{I} & \mathbf{0}\\ \mathbf{0} & \mathbf{0}\end{pmatrix}\mathbf{S}^\star\\ &= \mathbf{S}\begin{pmatrix}\mathbf{Z} & \mathbf{0}\\ \mathbf{0} & \mathbf{0}\end{pmatrix}\mathbf{S}^\star,\end{aligned}$$

where $\mathbf{Z} = n\mathbf{I} - n^2(n\mathbf{I}+\mathbf{S}_1^\star\mathbf{A}\mathbf{S}_1-\mathbf{S}_1^\star\mathbf{A}\mathbf{S}_2(\mathbf{S}_2^\star\mathbf{A}\mathbf{S}_2+\epsilon\mathbf{I})^{-1}\mathbf{S}_2^\star\mathbf{A}\mathbf{S}_1)^{-1}$. Let $\mathbf{Q}_\epsilon = \mathbf{S}_1^\star\mathbf{A}\mathbf{S}_1-\mathbf{S}_1^\star\mathbf{A}\mathbf{S}_2(\mathbf{S}_2^\star\mathbf{A}\mathbf{S}_2+\epsilon\mathbf{I})^{-1}\mathbf{S}_2^\star\mathbf{A}\mathbf{S}_1$ and $N(\epsilon) = \|\mathbf{Q}_\epsilon\| = \|(\mathbf{A}_\epsilon)_\mathcal{S}\|$ and n be a positive integer such that $n > N(\epsilon)$. Then $(\mathbf{I}+\frac{1}{n}\mathbf{Q}_\epsilon)$ is non-singular and $(\mathbf{I}+\frac{1}{n}\mathbf{Q}_\epsilon)^{-1} = \mathbf{I}-\frac{1}{\mathbf{n}}\mathbf{Q}_\epsilon+\frac{1}{\mathbf{n}^2}\mathbf{Q}_\epsilon{}^{\mathbf{2}}\cdots$.
Hence,

$$\begin{aligned}\mathcal{P}(\mathbf{A}_\epsilon, n\mathbf{P}_\mathcal{S}) &= \mathbf{S}\begin{pmatrix}\mathbf{Q}_\epsilon-\frac{1}{n}\mathbf{Q}_\epsilon{}^2\cdots & \mathbf{0}\\ \mathbf{0} & \mathbf{0}\end{pmatrix}\mathbf{S}^\star\\ &= (\mathbf{A}_\epsilon)_\mathcal{S} - \frac{1}{n}(\mathbf{A}_\epsilon)_\mathcal{S}^2+\cdots\\ &= (\mathbf{A}_\epsilon)_\mathcal{S} - \frac{1}{n}(\mathbf{A}_\epsilon)_\mathcal{S}^2(\mathbf{I}+\frac{1}{n}(\mathbf{A}_\epsilon)_\mathcal{S})^{-\mathbf{1}}.\end{aligned}$$

By Lemma 10.4.1, $\mathbf{A}_\mathcal{S} = \lim_{\epsilon\to\mathbf{0}}(\mathbf{A}_\epsilon)_\mathcal{S}$. So,

$$\lim_{n\to\infty}\lim_{\epsilon\to 0}\mathcal{P}(\mathbf{A}_\epsilon, n\mathbf{P}_\mathcal{S}) = \lim_{n\to\infty}\lim_{\epsilon\to 0}((\mathbf{A}_\epsilon)_\mathcal{S} - \frac{1}{n}(\mathbf{A}_\epsilon)_\mathcal{S}^2(\mathbf{I}+\frac{1}{n}(\mathbf{A}_\epsilon)_\mathcal{S})^{-1})$$

$$= \mathbf{A}_\mathcal{S} - \lim_{n\to\infty}\frac{1}{n}([\mathbf{A^2}_\mathcal{S}](I+\frac{1}{n}\mathbf{A}_\mathcal{S})^{-1}) = \mathbf{A}_\mathcal{S}.$$

□

Corollary 10.4.3. *Let* $\mathbf{A}, \mathbf{B}$ *be nnd matrices of same the order* $m\times m$ *and let* $\mathcal{S}$ *be a subspace of* $\mathbb{C}^m$. *Then* $\mathcal{P}(\mathbf{A},\mathbf{B})_\mathcal{S} = \mathcal{P}(\mathbf{A}_\mathcal{S},\mathbf{B}) = \mathcal{P}(\mathbf{A}_\mathcal{S},\mathbf{B}_\mathcal{S})$.

Proof. Observe that$\mathcal{P}(2n\mathbf{P}_{\mathcal{S}}, 2n\mathbf{P}_{\mathcal{S}}) = n\mathbf{P}_{\mathcal{S}}$. Now,

$$\mathcal{P}(\mathcal{P}(\mathbf{A},\mathbf{B}), n\mathbf{P}_{\mathcal{S}}) = \mathcal{P}(\mathcal{P}(\mathbf{A}, n\mathbf{P}_{\mathcal{S}}), \mathbf{B}) = \mathcal{P}(\mathcal{P}(\mathbf{A}, 2n\mathbf{P}_{\mathcal{S}}), \mathcal{P}(\mathbf{B}, 2n\mathbf{P}_{\mathcal{S}})),$$

in view of commutativity and associativity of parallel sums. The result now follows by taking limits. □

10.5 Generalized Schur complement and shorted operator of a matrix over general field

In Section 10.3, we defined the shorted operator of an *nnd* matrix $\mathbf{A}$ with respect to a given subspace $\mathcal{S}$, which we denoted by $\mathbf{A}_{\mathcal{S}}$. We also saw that it was a Schur complement of $\mathbf{A}$ in a suitably constructed matrix. (See Remark 10.3.4). For this purpose we used the orthogonal projectors onto $\mathcal{S}$ and the orthogonal complement $\mathcal{S}^{\perp}$ or equivalently an ortho-normal basis of $\mathcal{S}$ and an ortho-normal basis of the orthogonal complement $\mathcal{S}^{\perp}$. We showed that given any subspace $\mathcal{S}$, the matrix $\mathbf{A}$ has a shorted operator $\mathbf{A}_{\mathcal{S}}$ indexed by subspace $\mathcal{S}$.

For compatibility with notation for shorted operator in general case and for convenience, henceforth we use the notation $\mathbf{S}(\mathbf{A}|\mathcal{S})$ to denote the shorted operator of the nnd matrix $\mathbf{A}$ indexed by $\mathcal{S}$ which we have so far denoted as $\mathbf{A}_{\mathcal{S}}$. Thus, $\mathbf{A}_{\mathcal{S}}$ and $\mathbf{S}(\mathbf{A}|\mathcal{S})$ mean the same.

In this section, we consider matrices (possibly rectangular) over an arbitrary field F and define the notions of complementability and shorted operator for a given matrix $\mathbf{A}$ of order $m \times n$. In this case we deal with two subspaces V_m and V_n of F^m and F^n respectively. In the absence of the distinct advantage of orthogonal complements and ortho-normal basis as in case of complex matrices, we choose complements W_m of V_m and W_n of V_n in F^m and F^n respectively. Using these we construct projectors $\mathbf{P}_{\ell}$ and $\mathbf{P}_{\mathbf{r}}$ of orders $m \times m$ and $n \times n$ onto V_m along W_m and V_n along W_n respectively. Employing conditions similar to those in Theorem 10.3.14 in terms of $\mathbf{P}_{\ell}$ and $\mathbf{P}_{\mathbf{r}}$, we define complementability of a matrix and a shorted operator of a matrix indexed by V_m, W_m, V_n and W_n (or equivalently indexed by $\mathbf{P}_{\ell}$ and $\mathbf{P}_{\mathbf{r}}$). We then investigate the existence of the shorted operator defined thus. We show that this shorted operator whenever it exists depends only on the subspaces V_m and V_n and does not depend on the choice of complements W_m and W_n. We finally show that shorted operator of a given matrix $\mathbf{A}$ indexed by V_m and V_n (dropping the subspaces W_m and W_n in view of the last statement), whenever it exists is a solution to a maximization problem similar to the one considered in the Theorem 10.3.11.

Let V_m, V_n, W_m, W_n, $\mathbf{P}_\ell$ and $\mathbf{P}_\mathrm{r}$ be as in the preceding para. One nice way to construct $\mathbf{P}_\ell$ is as follows:

Let $\mathbf{P_1}$ and $\mathbf{P_2}$ be matrices whose columns form a bases of V_m and W_m respectively. Let $\begin{pmatrix}\mathbf{P}^1\\ \mathbf{P}^2\end{pmatrix}$ be the inverse of $(\mathbf{P_1} : \mathbf{P_2})$. Then $\mathbf{P}_\ell = \mathbf{P_1 P^1}$. Similarly, let $\mathbf{P_3}$ and $\mathbf{P_4}$ be matrices whose rows form a bases of V_n and W_n respectively. Let $(\mathbf{P^3} : \mathbf{P^4})$ be inverse of $\begin{pmatrix}\mathbf{P_3}\\ \mathbf{P_4}\end{pmatrix}$, then $\mathbf{P}_\mathrm{r}^\mathrm{t} = \mathbf{P^3 P_3}$. Write $\mathbf{Q}_\ell = \mathbf{I} - \mathbf{P}_\ell$ and $\mathbf{Q}_\mathrm{r} = \mathbf{I} - \mathbf{P}_\mathrm{r}$. Let $\mathbf{A}$ be an $m \times n$ matrix. Then $\mathbf{A} = (\mathbf{P}_\ell + \mathbf{Q}_\ell)\mathbf{A}(\mathbf{P}_\mathrm{r}^\mathrm{t} + \mathbf{Q}_\mathrm{r}^\mathrm{t}) = \mathbf{P}_\ell\mathbf{A}\mathbf{P}_\mathrm{r}^\mathrm{t} + \mathbf{P}_\ell\mathbf{A}\mathbf{Q}_\mathrm{r}^\mathrm{t} + \mathbf{Q}_\ell\mathbf{A}\mathbf{P}_\mathrm{r}^\mathrm{t} + \mathbf{Q}_\ell\mathbf{A}\mathbf{Q}_\mathrm{r}^\mathrm{t}$. Let $\mathbf{B} = \mathbf{P}_\ell\mathbf{A}\mathbf{P}_\mathrm{r}^\mathrm{t}$, $\mathbf{C} = \mathbf{P}_\ell\mathbf{A}\mathbf{Q}_\mathrm{r}^\mathrm{t}, \mathbf{E} = \mathbf{Q}_\ell\mathbf{A}\mathbf{P}_\mathrm{r}^\mathrm{t}$ and $\mathbf{H} = \mathbf{Q}_\ell\mathbf{A}\mathbf{Q}_\mathrm{r}^\mathrm{t}$. Thus, $\mathbf{A} = \mathbf{B} + \mathbf{C} + \mathbf{E} + \mathbf{H}$, which is analogous to the orthogonal projective decomposition of Remark 10.3.4. We call this decomposition as ***projective decomposition*** of $\mathbf{A}$ with respect to the pair $(\mathbf{P}_\ell, \mathbf{P}_\mathrm{r})$ of projections.

Definition 10.5.1. Let $\mathbf{A}$ be an $m \times n$ matrix. Let $\mathbf{P}_\ell$ and $\mathbf{P}_\mathrm{r}$ be projectors of order $m \times m$ and $n \times n$ respectively. Then $\mathbf{A}$ is said to be complementable with respect to the pair $(\mathbf{P}_\ell, \mathbf{P}_\mathrm{r})$ or equivalently $\mathbf{A}$ is said to be $(\mathbf{P}_\ell, \mathbf{P}_\mathrm{r})$-complementable if there exist matrices $\mathbf{M}_\ell$ and $\mathbf{M}_\mathrm{r}$ of orders $m \times m$ and $n \times n$ respectively such that

(i) $\mathbf{P}_\mathrm{r}^\mathrm{t}\mathbf{M}_\mathrm{r} = \mathbf{M}_\mathrm{r}$
(ii) $\mathbf{M}_\ell\mathbf{P}_\ell = \mathbf{M}_\ell$
(iii) $\mathbf{P}_\ell\mathbf{A}\mathbf{M}_\mathrm{r} = \mathbf{P}_\ell\mathbf{A}$ and
(iv) $\mathbf{M}_\ell\mathbf{A}\mathbf{P}_\mathrm{r}^\mathrm{t} = \mathbf{A}\mathbf{P}_\mathrm{r}^\mathrm{t}$.

Remark 10.5.2. Notice the similarity of conditions (i)-(iv) of Definition 10.5.1 and the conditions (i)- (iv) of Theorem 10.3.14. In Theorem 10.3.14, we have $\mathbf{P}_\mathrm{r} = \mathbf{P}_\ell = \mathbf{P}_{\mathcal{S}^\perp}$. Thus, an *nnd* matrix over $\mathbb{C}$ is complementable with respect to the pair $(\mathbf{P}_{\mathcal{S}^\perp}, \mathbf{P}_{\mathcal{S}^\perp})$, where $\mathcal{S}$ is an arbitrary subspace of $\mathbb{C}^n$.

Analogous to the *nnd* case, we now define Schur complement and Schur compression in general case.

Definition 10.5.3. Let $\mathbf{A}$ be complementable with respect to the pair $(\mathbf{P}_\ell, \mathbf{P}_\mathrm{r})$ and $\mathbf{M}_\ell$, $\mathbf{M}_\mathrm{r}$ be matrices satisfying (i)-(iv) of Definition 10.5.1. Then $\mathbf{A}_{(\mathbf{P}_\ell,\mathbf{P}_\mathrm{r})} = \mathbf{M}_\ell\mathbf{A}\mathbf{M}_\mathrm{r} = \mathbf{M}_\ell\mathbf{A} = \mathbf{A}\mathbf{M}_\mathrm{r}$ is called the **Schur compression** of $\mathbf{A}$ with respect to the pair $(\mathbf{P}_\ell, \mathbf{P}_\mathrm{r})$ and $\mathbf{A}/_{(\mathbf{P}_\ell,\mathbf{P}_\mathrm{r})} = \mathbf{A} - \mathbf{A}_{(\mathbf{P}_\ell,\mathbf{P}_\mathrm{r})}$ is called the **Schur complement** of $\mathbf{A}$ with respect to the pair $(\mathbf{P}_\ell, \mathbf{P}_\mathbf{r})$. We call $\mathbf{A}/_{(\mathbf{P}_\ell,\mathbf{P}_\mathbf{r})}$ as the shorted matrix indexed by $\mathcal{C}(\mathbf{I}-\mathbf{P}_\ell)$ and $\mathcal{C}(\mathbf{I}-\mathbf{P}_\mathbf{r}^\mathbf{t})$.

Several questions arise. Why is the name 'Schur complement' justified in this case? Does it confirm to the usual definition of Schur complement? How relevant is the choice of the complements W_m of V_m and W_n of V_n? Does the shorted operator defined above have optimality properties similar to those of the shorted operator in the *nnd* case? Before we address all these questions, we establish the following necessary and sufficient conditions for the complementability of a matrix with respect to a pair of projections.

Theorem 10.5.4. *Let* $\mathbf{A}$ *be an* $m \times n$ *matrix over* F. *Let* $\mathbf{P}_\ell$ *and* $\mathbf{P}_\mathbf{r}$ *be projectors of order* $m \times m$ *and* $n \times n$ *respectively. Let* $\mathbf{A} = \mathbf{B}+\mathbf{C}+\mathbf{E}+\mathbf{H}$ *be projective decomposition of* $\mathbf{A}$ *with respect to the pair* $(\mathbf{P}_\ell, \mathbf{P}_\mathbf{r})$ *as given in opening para of the section (before Definition 10.5.1). Also, let* $\mathbf{P}_\ell = \mathbf{P}_1\mathbf{P}^1$ *and* $\mathbf{P}_\mathbf{r}^\mathbf{t} = \mathbf{P}^3\mathbf{P}_3$, *where* $\mathbf{P}^1$ *and* $\mathbf{P}^3$ *are as constructed in the para before Definition 10.5.1. Then the following are equivalent:*

(i) $\mathbf{A}$ *is* $(\mathbf{P}_\ell, \mathbf{P}_\mathbf{r})$*-complementable*
(ii) $\mathcal{C}(\mathbf{E}^\mathbf{t}) \subseteq \mathcal{C}(\mathbf{B}^\mathbf{t})$, $\mathcal{C}(\mathbf{C}) \subseteq \mathcal{C}(\mathbf{B})$
(iii) $\mathcal{N}(\mathbf{B}) \subseteq \mathcal{N}(\mathbf{B}+\mathbf{E})$, $\mathcal{N}(\mathbf{B}^\mathbf{t}) \subseteq \mathcal{N}(\mathbf{B}+\mathbf{C})^\mathbf{t}$
(iv) $\mathcal{C}(\mathbf{T}_{12}) \subseteq \mathcal{C}(\mathbf{T}_{11})$, $\mathcal{C}(\mathbf{T}_{21}^\mathbf{t}) \subseteq \mathcal{C}(\mathbf{T}_{11}^\mathbf{t})$,
where $\mathbf{A} = (\mathbf{P}_1 : \mathbf{P}_2)\begin{pmatrix}\mathbf{T}_{11} & \mathbf{T}_{12}\\ \mathbf{T}_{21} & \mathbf{T}_{22}\end{pmatrix}\begin{pmatrix}\mathbf{P}_3\\ \mathbf{P}_4\end{pmatrix}$ *and*
(v) $\rho(\mathbf{P}_\ell\mathbf{A}\mathbf{P}_\mathbf{r}^\mathbf{t}) = \rho(\mathbf{P}_\ell\mathbf{A}) = \rho(\mathbf{A}\mathbf{P}_\mathbf{r}^\mathbf{t})$.

Proof. (ii)$\Leftrightarrow$ (iii)
Notice that $\mathcal{C}(\mathbf{E}^\mathbf{t}) \subseteq \mathcal{C}(\mathbf{B}^\mathbf{t}) \Leftrightarrow \mathcal{N}(\mathbf{B}) \subseteq \mathcal{N}(\mathbf{E}) \Leftrightarrow \mathcal{N}(\mathbf{B}) = \mathcal{N}(\mathbf{B}+\mathbf{E})$, since $\mathcal{C}(\mathbf{B}) \cap \mathcal{C}(\mathbf{E}) = \{\mathbf{0}\}$.
Similarly, $\mathcal{C}(\mathbf{C}) \subseteq \mathcal{C}(\mathbf{B}) \Leftrightarrow \mathcal{N}(\mathbf{B}^\mathbf{t}) \subseteq \mathcal{N}(\mathbf{B}+\mathbf{C})^\mathbf{t}$.
(ii)$\Leftrightarrow$ (iv)
Notice that

$$\mathbf{A} = (\mathbf{P}_1 : \mathbf{P}_2)\begin{pmatrix}\mathbf{P}^1\\ \mathbf{P}^2\end{pmatrix}\mathbf{A}\,(\mathbf{P}^3 : \mathbf{P}^4)\begin{pmatrix}\mathbf{P}_3\\ \mathbf{P}_4\end{pmatrix}.$$

Thus,

$$\mathbf{T}_{11} = \mathbf{P}^1\mathbf{A}\mathbf{P}^3,\ \mathbf{T}_{12} = \mathbf{P}^1\mathbf{A}\mathbf{P}^4,\ \mathbf{T}_{21} = \mathbf{P}^2\mathbf{A}\mathbf{P}^3 \text{ and } \mathbf{T}_{22} = \mathbf{P}^2\mathbf{A}\mathbf{P}^4.$$

Further,

$$\mathbf{B} = \mathbf{P}_1\mathbf{T}_{11}\mathbf{P}_3,\ \mathbf{C} = \mathbf{P}_1\mathbf{T}_{12}\mathbf{P}_4,\ \mathbf{E} = \mathbf{P}_2\mathbf{T}_{21}\mathbf{P}_3,$$

and $\mathbf{H} = \mathbf{P_2T_{22}P_4}$. Since, $\mathbf{P_1}$ has a left inverse and $\mathbf{P_3}$ has a right inverse, $\mathcal{C}(\mathbf{C}) \subseteq \mathcal{C}(\mathbf{B})$ if and only if $\mathcal{C}(\mathbf{T_{12}}) \subseteq \mathcal{C}(\mathbf{T_{11}})$.
Similarly, $\mathcal{C}(\mathbf{E^t}) \subseteq \mathcal{C}(\mathbf{B^T})$ if and only if $\mathcal{C}(\mathbf{T_{21}^t}) \subseteq \mathcal{C}(\mathbf{T_{11}^t})$.
(i) $\Rightarrow$ (ii)
Since $\mathbf{A}$ is $(\mathbf{P_\ell}, \mathbf{P_r})$-complementable, there exist matrices $\mathbf{M_\ell}$, $\mathbf{M_r}$ satisfying (i)-(iv) of Definition 10.5.1. So,

$$\mathbf{M_\ell B} = \mathbf{M_\ell P_\ell AP_r^t} = \mathbf{M_\ell AP_r^t} = \mathbf{AP_r^t} = \mathbf{B} + \mathbf{E}.$$

Hence, $\mathcal{C}(\mathbf{E^t}) \subseteq \mathcal{C}(\mathbf{B^t})$.
Similarly, $\mathbf{BM_r} = \mathbf{B} + \mathbf{C}$. Thus, $\mathcal{C}(\mathbf{C}) \subseteq \mathcal{C}(\mathbf{B})$.
(ii) $\Rightarrow$ (i)
Let (ii) hold. Choose $\mathbf{M_\ell} = (\mathbf{B}+\mathbf{E})\mathbf{B^-P_\ell}$ and $\mathbf{M_r} = \mathbf{P_r^tB^-}(\mathbf{B}+\mathbf{C})$, where $\mathbf{B^-}$ is any g-inverse of $\mathbf{B}$. Clearly, $\mathbf{P_r^tM_r} = \mathbf{M_r}$ and $\mathbf{M_\ell P_\ell} = \mathbf{M_\ell}$. Also, $\mathbf{M_\ell AP_r^t} = (\mathbf{B}+\mathbf{E})\mathbf{B^-P_\ell AP_r^t} = (\mathbf{B}+\mathbf{E})\mathbf{B^-B} = \mathbf{B}+\mathbf{E} = \mathbf{AP_r^t}$. Similarly, we can show that $\mathbf{P_\ell AM_r} = \mathbf{P_\ell A}$.
(ii) $\Rightarrow$ (v)
From the proof of (ii) $\Rightarrow$ (i), ii follows that there exist matrices $\mathbf{D_1}$ and $\mathbf{D_2}$ such that $\mathbf{D_1P_\ell AP_r^t} = \mathbf{AP_r^t}$ and $\mathbf{P_\ell AP_r^tD_2} = \mathbf{P_\ell A}$. So,

$$\rho(\mathbf{P_\ell AP_r^t}) = \rho(\mathbf{P_\ell A}) = \rho(\mathbf{AP_r^t}).$$

(v) $\Rightarrow$ (i)
Since (v) holds, there exist matrices $\mathbf{D_1}$ and $\mathbf{D_2}$ such that $\mathbf{D_1P_\ell AP_r^t} = \mathbf{AP_r^t}$ and $\mathbf{P_\ell AP_r^tD_2} = \mathbf{P_\ell A}$. Let $\mathbf{M_\ell} = \mathbf{D_1P_\ell}$ and $\mathbf{M_r} = \mathbf{P_r^tD_2}$. Now, it is easy to see that conditions (i)-(iv) of Definition 10.5.1 hold, so (i) holds for this choice of matrices $\mathbf{M_\ell}$ and $\mathbf{M_r}$. □

We next obtain expressions for the Schur compression, Schur complement and and shorted operator. Let any one of equivalent conditions (ii)-(v) of Theorem 10.5.4 hold. Then $\mathbf{A}$ is complementable with respect to the pair $(\mathbf{P_\ell}, \mathbf{P_r})$. Let $\mathbf{M_\ell}$ and $\mathbf{M_r}$ be matrices as chosen in the proof of Theorem 10.5.4 (ii) $\Rightarrow$ (i). The Schur compression $\mathbf{A}_{(\mathbf{P_\ell},\mathbf{P_r})} = \mathbf{M_\ell AM_r} = \mathbf{M_\ell A} = (\mathbf{B}+\mathbf{E})\mathbf{B^-P_\ell A} = (\mathbf{B}+\mathbf{E})\mathbf{B^-}(\mathbf{B}+\mathbf{C}) = (\mathbf{B}+\mathbf{C}) + \mathbf{E} + \mathbf{EB^-C}$. Notice that $\mathbf{EB^-C}$ is invariant under choices of $\mathbf{B^-}$ in view of (ii) of Theorem 10.5.4. Further,

$$\begin{aligned}\mathbf{A}_{/(\mathbf{P_\ell},\mathbf{P_r})} &= \mathbf{A} - \mathbf{A}_{(\mathbf{P_\ell},\mathbf{P_r})} = \mathbf{B}+\mathbf{C}+\mathbf{E}+\mathbf{H} - (\mathbf{B}+\mathbf{C}+\mathbf{E}+\mathbf{EB^-C})\\ &= \mathbf{H} - \mathbf{EB^-C} = (\mathbf{I}-\mathbf{P_\ell})(\mathbf{A} - \mathbf{AP_r^t}(\mathbf{P_\ell AP_r^t})\mathbf{P_\ell A})(\mathbf{I}-\mathbf{P_r^t})\\ &= \mathbf{A} - \mathbf{AP_r^t}(\mathbf{P_\ell AP_r^t})\mathbf{P_\ell A}, \end{aligned} \tag{10.5.1}$$

as $\rho(\mathbf{P_\ell AP_r^t}) = \rho(\mathbf{P_\ell A}) = \rho(\mathbf{AP_r^t})$.
Now, $\mathbf{AP_r^t} = (\mathbf{P_1T_{11}} + \mathbf{P_2T_{21}})\mathbf{P_3}$, $\mathbf{P_\ell A} = \mathbf{P_1}(\mathbf{T_{11}P_3} + \mathbf{T_{12}P_4})$

and $\mathbf{P}_\ell\mathbf{AP}_\mathrm{r}^\mathrm{t} = \mathbf{P}_1\mathbf{T}_{11}\mathbf{P}_3$. Clearly, $\rho(\mathbf{P}_\ell\mathbf{AP}_\mathrm{r}^\mathrm{t}) = \rho(\mathbf{T}_{11})$. Hence, $\mathbf{P}_3(\mathbf{P}_1\mathbf{T}_{11}\mathbf{P}_3)^-\mathbf{P}_1$ is a g-inverse of $\mathbf{T}_{11}$. It can be easily seen that

$$\mathbf{A}_{/(\mathbf{P}_\ell,\mathbf{P}_\mathrm{r})} = \mathbf{A} - \mathbf{AP}_\mathrm{r}^\mathrm{t}(\mathbf{P}_\ell\mathbf{AP}_\mathrm{r}^\mathrm{t})\mathbf{P}_\ell\mathbf{A} = \mathbf{P}_2(\mathbf{T}_{22} - \mathbf{T}_{21}\mathbf{T}_{11}^-\mathbf{T}_{12})\mathbf{P}_4.$$

Thus, the shorted operator of $\mathbf{A}$ indexed by W_m and W_n written as $\mathbf{S}(\mathbf{A}|W_m, W_n, V_m, V_n)$ is given by:

$$\mathbf{S}(\mathbf{A}|W_m, W_n, V_m, V_n) = \mathbf{P}_2(\mathbf{T}_{22} - \mathbf{T}_{21}\mathbf{T}_{11}^-\mathbf{T}_{12})\mathbf{P}_4. \tag{10.5.2}$$

Accordingly

$$\mathbf{S}(\mathbf{A}|V_m, V_n, W_m, W_n) = \mathbf{P}_1(\mathbf{T}_{11} - \mathbf{T}_{12}\mathbf{T}_{22}^-\mathbf{T}_{21})\mathbf{P}_3. \tag{10.5.3}$$

So, the Schur compression of $\mathbf{A}$ with respect to the pair $(\mathbf{P}_\ell, \mathbf{P}_\mathrm{r})$ is

$$\mathbf{A}_{(\mathbf{P}_\ell,\mathbf{P}_\mathrm{r})} = \mathbf{A} - \mathbf{A}_{/(\mathbf{P}_\ell,\mathbf{P}_\mathrm{r})} = \mathbf{AP}_\mathrm{r}^\mathrm{t}(\mathbf{P}_\ell\mathbf{AP}_\mathrm{r}^\mathrm{t})\mathbf{P}_\ell\mathbf{A}. \tag{10.5.4}$$

We have taken a specific choice of $\mathbf{M}_\ell$ and $\mathbf{M}_\mathrm{r}$, namely, $\mathbf{M}_\ell = (\mathbf{B}+\mathbf{E})\mathbf{B}^-\mathbf{P}_\ell$ and $\mathbf{M}_\mathrm{r} = \mathbf{P}_\mathrm{r}^\mathrm{t}\mathbf{B}^-(\mathbf{B}+\mathbf{C})$, while arriving at the expressions (10.5.1)-(10.5.4). To show these expressions are independent of any particular choice of $\mathbf{M}_\ell$ and $\mathbf{M}_\mathrm{r}$, we first obtain the class of all $\mathbf{M}_\ell$ and $\mathbf{M}_\mathrm{r}$, satisfying the conditions (i)-(iv) of Definition 10.5.1 under the assumption that $\mathbf{A}$ is $(\mathbf{P}_\ell, \mathbf{P}_\mathrm{r})$-complementable.

Assume that $\mathbf{A}$ is $(\mathbf{P}_\ell, \mathbf{P}_\mathrm{r})$-complementable. Then

$$\rho(\mathbf{P}_\ell\mathbf{AP}_\mathrm{r}^\mathrm{t}) = \rho(\mathbf{P}_\ell\mathbf{A}) = \rho(\mathbf{AP}_\mathrm{r}^\mathrm{t}).$$

Since, $\mathbf{M}_\ell\mathbf{P}_\ell = \mathbf{M}_\ell$, $\mathcal{C}(\mathbf{M}_\ell^\mathrm{t}) \subseteq \mathcal{C}(\mathbf{P}_\ell^\mathrm{t})$, we have, $\mathbf{M}_\ell = \mathbf{D}_1\mathbf{P}_\ell$ for some $\mathbf{D}_1$. Also, $\mathbf{M}_\ell\mathbf{AP}_\mathrm{r}^\mathrm{t} = \mathbf{AP}_\mathrm{r}^\mathrm{t}$, so, we have, $\mathbf{D}_1\mathbf{P}_\ell\mathbf{AP}_\mathrm{r}^\mathrm{t} = \mathbf{AP}_\mathrm{r}^\mathrm{t}$. This equation in $\mathbf{D}_1$ is evidently consistent, so,

$$\mathbf{D}_1 = \mathbf{AP}_\mathrm{r}^\mathrm{t}(\mathbf{P}_\ell\mathbf{AP}_\mathrm{r}^\mathrm{t})^- + \mathbf{Z}(\mathbf{I} - (\mathbf{P}_\ell\mathbf{AP}_\mathrm{r}^\mathrm{t})(\mathbf{P}_\ell\mathbf{AP}_\mathrm{r}^\mathrm{t})^-),$$

where $\mathbf{Z}$ is arbitrary.

Hence, the class of all $\mathbf{M}_\ell$ is given by:

$$\mathbf{M}_\ell = \mathbf{AP}_\mathrm{r}^\mathrm{t}(\mathbf{P}_\ell\mathbf{AP}_\mathrm{r}^\mathrm{t})^-\mathbf{P}_\ell + \mathbf{Z}_1(\mathbf{I} - (\mathbf{P}_\ell\mathbf{AP}_\mathrm{r}^\mathrm{t})(\mathbf{P}_\ell\mathbf{AP}_\mathrm{r}^\mathrm{t})^-)\mathbf{P}_\ell, \tag{10.5.5}$$

where $\mathbf{Z}_1$ is arbitrary.
Similarly, the class of all $\mathbf{M}_\mathrm{r}$ is given as:

$$\mathbf{M}_\mathrm{r} = \mathbf{P}_\mathrm{r}^\mathrm{t}(\mathbf{P}_\ell\mathbf{AP}_\mathrm{r}^\mathrm{t})^-\mathbf{P}_\ell\mathbf{A} + (\mathbf{I} - (\mathbf{P}_\ell\mathbf{AP}_\mathrm{r}^\mathrm{t})^-(\mathbf{P}_\ell\mathbf{AP}_\mathrm{r}^\mathrm{t}))\mathbf{W}, \tag{10.5.6}$$

where $\mathbf{W}$ is arbitrary.
Consider

$$\begin{aligned}\mathbf{M}_\ell\mathbf{AM}_\mathrm{r} &= \mathbf{AP}_\mathrm{r}^\mathrm{t}(\mathbf{P}_\ell\mathbf{AP}_\mathrm{r}^\mathrm{t})^-\mathbf{P}_\ell\mathbf{AM}_\mathrm{r} + \mathbf{Z}_1(\mathbf{I} - (\mathbf{P}_\ell\mathbf{AP}_\mathrm{r}^\mathrm{t})(\mathbf{P}_\ell\mathbf{AP}_\mathrm{r}^\mathrm{t})^-)\mathbf{P}_\ell\mathbf{AM}_\mathrm{r}\\ &= \mathbf{AP}_\mathrm{r}^\mathrm{t}(\mathbf{P}_\ell\mathbf{AP}_\mathrm{r}^\mathrm{t})^-\mathbf{P}_\ell\mathbf{AM}_\mathrm{r},\end{aligned}$$

since $\rho(\mathbf{P}_\ell\mathbf{A}\mathbf{P}_\mathbf{r}^\mathbf{t}) = \rho(\mathbf{P}_\ell\mathbf{A})$. So, $\mathbf{M}_\ell\mathbf{A}\mathbf{M}_\mathbf{r} = \mathbf{A}\mathbf{P}_\mathbf{r}^\mathbf{t}(\mathbf{P}_\ell\mathbf{A}\mathbf{P}_\mathbf{r}^\mathbf{t})^-\mathbf{P}_\ell\mathbf{A}$, which is independent of $\mathbf{M}_\ell$ and $\mathbf{M}_\mathbf{r}$. Thus, the expressions (10.5.1)-(10.5.4) are independent of choices of $\mathbf{M}_\ell$ and $\mathbf{M}_\mathbf{r}$.

Recall that $\mathbf{P}_\ell = \mathbf{P}_1\mathbf{P}^1$ and $\mathbf{P}_\mathbf{r}^\mathbf{t} = \mathbf{P}^3\mathbf{P}_3$. Hence the Schur compression is $\mathbf{A}_{(\mathbf{P}_\ell,\mathbf{P}_\mathbf{r})} = \mathbf{A}\mathbf{P}_\mathbf{r}^\mathbf{t}(\mathbf{P}_\ell\mathbf{A}\mathbf{P}_\mathbf{r}^\mathbf{t})^-\mathbf{P}_\ell\mathbf{A} = \mathbf{A}\mathbf{P}^3\mathbf{P}_3(\mathbf{P}_1\mathbf{P}^1\mathbf{A}\mathbf{P}^3\mathbf{P}_3)^-\mathbf{P}_1\mathbf{P}^1\mathbf{A}$. Since, $\rho(\mathbf{P}_\ell\mathbf{A}\mathbf{P}_\mathbf{r}^\mathbf{t}) = \rho(\mathbf{P}_1\mathbf{P}^1\mathbf{A}\mathbf{P}^3\mathbf{P}_3) = \rho(\mathbf{P}^1\mathbf{A}\mathbf{P}^3)$, we have $\mathbf{A}_{(\mathbf{P}_\ell,\mathbf{P}_\mathbf{r})} = \mathbf{A}\mathbf{P}^3(\mathbf{P}^1\mathbf{A}\mathbf{P}^3)^-\mathbf{P}^1\mathbf{A}$.

Clearly, $\mathbf{A}\mathbf{P}^3(\mathbf{P}^1\mathbf{A}\mathbf{P}^3)^-\mathbf{P}^1\mathbf{A}$ is invariant under choices of g-inverses of $\mathbf{P}^1\mathbf{A}\mathbf{P}^3$. Also note that $\mathcal{C}(\mathbf{P}^1) = \mathcal{N}(\mathbf{P}_2)$ and $\mathcal{C}(\mathbf{P}^{3\mathbf{t}}) = \mathcal{N}(\mathbf{P}_4^\mathbf{t})$. Choice of matrices $\mathbf{P}_2$ and $\mathbf{P}_4$ is arbitrary subject to the conditions: columns of $\mathbf{P}_2$ form a basis of W_m and rows of $\mathbf{P}_4$ form a basis of W_n. Hence, the Schur compression and Schur complement can be obtained as follows:

Let columns of $\mathbf{P}_2$ and rows of $\mathbf{P}_4$ form a basis of W_m and W_n respectively. Let $\mathbf{P}_1$ and $\mathbf{P}_3$ be matrices such that $\mathcal{C}(\mathbf{P}^1)$ and $\mathcal{C}(\mathbf{P}^{3\mathbf{t}})$ form the null spaces of $\mathbf{P}_2$ and $\mathbf{P}_4^\mathbf{t}$ respectively. Then the Schur compression of $\mathbf{A}$ with respect to the pair $(\mathbf{P}_\ell, \mathbf{P}_\mathbf{r})$ is given as:

$$\mathbf{A}_{(\mathbf{P}_\ell,\mathbf{P}_\mathbf{r})} = \mathbf{A}\mathbf{P}^3(\mathbf{P}^1\mathbf{A}\mathbf{P}^3)^-\mathbf{P}^1\mathbf{A}, \tag{10.5.7}$$

and Schur complement of $\mathbf{A}$ with respect to the pair $(\mathbf{P}_\ell, \mathbf{P}_\mathbf{r})$ is given as:

$$\mathbf{A}_{/(\mathbf{P}_\ell,\mathbf{P}_\mathbf{r})} = \mathbf{A} - \mathbf{A}\mathbf{P}^3(\mathbf{P}^1\mathbf{A}\mathbf{P}^3)^-\mathbf{P}^1\mathbf{A}. \tag{10.5.8}$$

A careful reader will notice that we have arrived at (10.5.7) and (10.5.8) when the columns of $\mathbf{P}^1$ and rows of $\mathbf{P}^3$ form bases of the null spaces of $\mathbf{P}_2$ and $\mathbf{P}_4^\mathbf{t}$ respectively. However, both these hold even when the columns of $\mathbf{P}^1$ and rows of $\mathbf{P}^3$ are not necessarily linearly independent.

Let us write $\tilde{\mathbf{A}} = \begin{pmatrix} \mathbf{A} & \mathbf{A}\mathbf{P}^3 \\ \mathbf{P}^1\mathbf{A} & \mathbf{P}^1\mathbf{A}\mathbf{P}^3 \end{pmatrix}$. Then $\mathbf{A}_{/(\mathbf{P}_\ell,\mathbf{P}_\mathbf{r})}$ is the Schur complement of $\mathbf{A}$ in $\tilde{\mathbf{A}}$, answering the first question before Theorem 10.5.4.

Now, let the shorted operator $\mathbf{S}(\mathbf{A}|V_m, V_n, W_m, W_n)$ exist and U_m, U_n be complements of V_m and V_n respectively (possibly different from W_m and W_n). We show that, if $\mathbf{S}(\mathbf{A}|V_m, V_n, W_m, W_n)$ exists, then $\mathbf{S}(\mathbf{A}|V_m, V_n, U_m, U_n)$ also exists and the two are equal. Notice that every basis of U_m is the columns of a matrix of the form $\mathbf{P}_1\mathbf{L}_1 + \mathbf{P}_2\mathbf{L}_2$ for some matrices $\mathbf{L}_1$ and $\mathbf{L}_2$ such that $\mathbf{L}_2$ is non-singular. Similarly, every basis of U_n is the columns of a matrix of the form $\mathbf{P}_3^\mathbf{t}\mathbf{L}_3 + \mathbf{P}_4^\mathbf{t}\mathbf{L}_4$ for some matrices $\mathbf{L}_3$ and $\mathbf{L}_4$ such that $\mathbf{L}_4$ is non-singular. Since $\mathbf{S}(\mathbf{A}|V_m, V_n, W_m, W_n)$ exists, we have

$$\mathbf{A} = (\mathbf{P}_1 : \mathbf{P}_2)\begin{pmatrix} \mathbf{T}_{11} & \mathbf{T}_{12} \\ \mathbf{T}_{21} & \mathbf{T}_{22} \end{pmatrix}\begin{pmatrix} \mathbf{P}_3 \\ \mathbf{P}_4 \end{pmatrix},$$

where $\mathcal{C}(\mathbf{T_{21}}) \subseteq \mathcal{C}(\mathbf{T_{22}})$, $\mathcal{C}(\mathbf{T_{12}^t}) \subseteq \mathcal{C}(\mathbf{T_{22}^t})$. We can write

$$\begin{aligned}\mathbf{A} &= (\mathbf{P_1} : \mathbf{P_2})\begin{pmatrix}\mathbf{T_{11}} & \mathbf{T_{12}}\\ \mathbf{T_{21}} & \mathbf{T_{22}}\end{pmatrix}\begin{pmatrix}\mathbf{P_3}\\ \mathbf{P_4}\end{pmatrix}\\ &= (\mathbf{P_1} : \mathbf{P_2})\mathbf{X}\begin{pmatrix}\mathbf{P_3}\\ \mathbf{P_4}\end{pmatrix},\\ &= (\mathbf{P_1} : \mathbf{P_1L_1} + \mathbf{P_2L_2})\mathbf{Y}\begin{pmatrix}\mathbf{P_3}\\ \mathbf{L_3P_3} + \mathbf{L_4P_4}\end{pmatrix}\end{aligned}$$

where

$$\mathbf{X} = \begin{pmatrix}\mathbf{I} & \mathbf{L_1}\\ \mathbf{0} & \mathbf{L_2}\end{pmatrix}\begin{pmatrix}\mathbf{I} & -\mathbf{L_1L_2^{-1}}\\ \mathbf{0} & \mathbf{L_2^{-1}}\end{pmatrix}\begin{pmatrix}\mathbf{T_{11}} & \mathbf{T_{12}}\\ \mathbf{T_{21}} & \mathbf{T_{22}}\end{pmatrix}\begin{pmatrix}\mathbf{I} & \mathbf{0}\\ -\mathbf{L_4^{-1}L_3} & \mathbf{L_4^{-1}}\end{pmatrix}\begin{pmatrix}\mathbf{I} & \mathbf{0}\\ \mathbf{L_3} & \mathbf{L_4}\end{pmatrix},$$

$$\mathbf{Y} = \begin{pmatrix}\mathbf{K} & \mathbf{T_{12}L_4^{-1}} - \mathbf{L_1L_2^{-1}T_{22}L_4^{-1}}\\ \mathbf{L_2^{-1}T_{21}} - \mathbf{L_2^{-1}T_{22}L_4^{-1}L_3} & \mathbf{L_2^{-1}T_{22}L_4^{-1}}\end{pmatrix}$$

and $\mathbf{K} = \mathbf{T_{11}} - \mathbf{L_1L_2^{-1}T_{21}} - \mathbf{T_{12}L_4^{-1}L_3} + \mathbf{L_1L_2^{-1}T_{22}L_4^{-1}L_3}$.
Since $\mathcal{C}(\mathbf{T_{21}}) \subseteq \mathcal{C}(\mathbf{T_{22}})$, we have

$$\mathcal{C}(\mathbf{L_2^{-1}T_{21}} - \mathbf{L_2^{-1}T_{22}L_4^{-1}L_3}) \subseteq \mathcal{C}(\mathbf{L_2^{-1}T_{22}L_4^{-1}}).$$

Also, since $\mathcal{C}(\mathbf{T_{12}^t}) \subseteq \mathcal{C}(\mathbf{T_{22}^t})$, we have

$$\mathcal{C}(\mathbf{T_{12}L_4^{-1}} - \mathbf{L_1L_2^{-1}T_{22}L_4^{-1}})^{\mathbf{t}} \subseteq \mathcal{C}(\mathbf{L_2^{-1}T_{22}L_4^{-1}})^{\mathbf{t}}.$$

Hence, $\mathbf{S}(\mathbf{A}|V_m, V_n, U_m, U_n)$ exists by Theorem 10.5.4(iv).
It is computational to show that
$\mathbf{T_{11}} - \mathbf{T_{12}T_{22}^{-}T_{21}} = \mathbf{T_{11}} - \mathbf{L_1L_2^{-1}T_{21}} - \mathbf{T_{12}L_4^{-1}L_3} + \mathbf{L_1L_2^{-1}T_{22}L_4^{-1}L_3} - (\mathbf{T_{12}L_4^{-1}} - \mathbf{L_1L_2^{-1}T_{22}L_4^{-1}})(\mathbf{L_2^{-1}T_{22}L_4^{-1}})^{-}(\mathbf{L_2^{-1}T_{22}L_4^{-1}})$.
Thus, $\mathbf{S}(\mathbf{A}|V_m, V_n, U_m, U_n) = \mathbf{S}(\mathbf{A}|V_m, V_n, W_m, W_n)$. As the choice of U_m, U_n is arbitrary, it follows that the shorted operator does not depend on the choice of the complements W_m and W_n of V_m and V_n respectively. Henceforth, we shall write $\mathbf{S}(\mathbf{A}|V_m, V_n, W_m, W_n)$ simply as $\mathbf{S}(\mathbf{A}|V_m, V_n)$. Also, $\mathbf{S}(\mathbf{A}|V_m, V_n)$ when it exists is given as:

$$\mathbf{P_1}(\mathbf{T_{11}} - \mathbf{T_{12}T_{22}^{-}T_{21}})\mathbf{P_3}. \tag{10.5.9}$$

We now show that the shorted operator $\mathbf{S}(\mathbf{A}|V_m, V_n)$ when exists is also the solution to the maximization problem analogous to the one considered in Theorem 10.3.11.

Theorem 10.5.5. *Let* $\mathbf{A}$ *be an* $m \times n$ *matrix and let* V_m *and* V_n *be subspaces of* F^m *and* F^n *respectively such that* $\mathbf{S}(\mathbf{A}|V_m, V_n)$ *exists. Then* $\mathbf{S}(\mathbf{A}|V_m, V_n) = \max\{\mathbf{D} : \mathbf{D} <^{-} \mathbf{A},\ \mathcal{C}(\mathbf{D}) \subseteq V_m,\ \mathcal{C}(\mathbf{D^t}) \subseteq V_n\}$.

Proof. Since $\mathbf{S}(\mathbf{A}|V_m, V_n) = \mathbf{P}_1(\mathbf{T}_{11} - \mathbf{T}_{12}\mathbf{T}_{22}^{-}\mathbf{T}_{21})\mathbf{P}_3$, it follows that $\mathcal{C}(\mathbf{S}(\mathbf{A}|V_m, V_n)) \subseteq V_m$ and $\mathcal{C}(\mathbf{S}(\mathbf{A}|V_m, V_n))^t \subseteq V_n$. Also $\mathbf{A} = (\mathbf{P}_1 : \mathbf{P}_2)\begin{pmatrix}\mathbf{T}_{11} & \mathbf{T}_{12}\\ \mathbf{T}_{21} & \mathbf{T}_{22}\end{pmatrix}\begin{pmatrix}\mathbf{P}_3\\ \mathbf{P}_4\end{pmatrix}$ and we can write

$$\begin{aligned}\mathbf{S}(\mathbf{A}|V_m, V_n) &= \mathbf{P}_1(\mathbf{T}_{11} - \mathbf{T}_{12}\mathbf{T}_{22}^{-}\mathbf{T}_{21})\mathbf{P}_3\\ &= (\mathbf{P}_1 : \mathbf{P}_2)\begin{pmatrix}\mathbf{T}_{11} - \mathbf{T}_{12}\mathbf{T}_{22}^{-}\mathbf{T}_{21} & \mathbf{0}\\ \mathbf{0} & \mathbf{0}\end{pmatrix}\begin{pmatrix}\mathbf{P}_3\\ \mathbf{P}_4\end{pmatrix}.\end{aligned}$$

Now, $\begin{pmatrix}\mathbf{T}_{11} & \mathbf{T}_{12}\\ \mathbf{T}_{21} & \mathbf{T}_{22}\end{pmatrix} - \begin{pmatrix}\mathbf{T}_{11} - \mathbf{T}_{12}\mathbf{T}_{22}^{-}\mathbf{T}_{21} & \mathbf{0}\\ \mathbf{0} & \mathbf{0}\end{pmatrix} = \begin{pmatrix}\mathbf{T}_{12}\mathbf{T}_{22}^{-}\mathbf{T}_{21} & \mathbf{T}_{12}\\ \mathbf{T}_{21} & \mathbf{T}_{22}\end{pmatrix}$ and

$$\begin{aligned}\rho\begin{pmatrix}\mathbf{T}_{11} & \mathbf{T}_{12}\\ \mathbf{T}_{21} & \mathbf{T}_{22}\end{pmatrix} &= \rho(\mathbf{T}_{11} - \mathbf{T}_{12}\mathbf{T}_{22}^{-}\mathbf{T}_{21}) + \rho(\mathbf{T}_{22})\\ &= \rho\begin{pmatrix}\mathbf{T}_{11} - \mathbf{T}_{12}\mathbf{T}_{22}^{-}\mathbf{T}_{21} & \mathbf{0}\\ \mathbf{0} & \mathbf{0}\end{pmatrix} + \rho\begin{pmatrix}\mathbf{T}_{12}\mathbf{T}_{22}^{-}\mathbf{T}_{21} & \mathbf{T}_{12}\\ \mathbf{T}_{21} & \mathbf{T}_{22}\end{pmatrix}.\end{aligned}$$

Hence,

$$\begin{pmatrix}\mathbf{T}_{11} - \mathbf{T}_{12}\mathbf{T}_{22}^{-}\mathbf{T}_{21} & \mathbf{0}\\ \mathbf{0} & \mathbf{0}\end{pmatrix} <^{-} \begin{pmatrix}\mathbf{T}_{11} & \mathbf{T}_{12}\\ \mathbf{T}_{21} & \mathbf{T}_{22}\end{pmatrix}$$

and thus, $\mathbf{S}(\mathbf{A}|V_m, V_n) <^{-} \mathbf{A}$.

Let $\mathbf{D}$ be an arbitrary matrix such that $\mathbf{D} <^{-} \mathbf{A}$, $\mathcal{C}(\mathbf{D}) \subseteq V_m$ and $\mathcal{C}(\mathbf{D^t}) \subseteq V_n$. Then we can write $\mathbf{D} = (\mathbf{P}_1 : \mathbf{P}_2)\begin{pmatrix}\mathbf{D}_1 & \mathbf{0}\\ \mathbf{0} & \mathbf{0}\end{pmatrix}\begin{pmatrix}\mathbf{P}_3\\ \mathbf{P}_4\end{pmatrix}$ for some matrix $\mathbf{D}_1$. Since $\mathbf{D} <^{-} \mathbf{A}$, we have $\begin{pmatrix}\mathbf{D}_1 & \mathbf{0}\\ \mathbf{0} & \mathbf{0}\end{pmatrix} <^{-} \begin{pmatrix}\mathbf{T}_{11} & \mathbf{T}_{12}\\ \mathbf{T}_{21} & \mathbf{T}_{22}\end{pmatrix}$ and therefore, $\mathbf{D}_1 <^{-} \mathbf{T}_{11} - \mathbf{T}_{12}\mathbf{T}_{22}^{-}\mathbf{T}_{21}$. So,

$$\mathbf{D} = \mathbf{P}_1\mathbf{D}_1\mathbf{P}_3 <^{-} \mathbf{P}_1(\mathbf{T}_{11} - \mathbf{T}_{12}\mathbf{T}_{22}^{-}\mathbf{T}_{21})\mathbf{P}_3 = \mathbf{S}(\mathbf{A}|V_m, V_n).$$

Hence, the result follows. □

Corollary 10.5.6. *Let* $\mathbf{A}$ *and* $\mathbf{B}$ *be matrices such that* $\mathbf{S}(\mathbf{A}|V_m, V_n)$ *and* $\mathbf{S}(\mathbf{B}|V_m, V_n)$ *exist for some subspaces* V_m *and* V_n *of* F^m *and* F^n *respectively. If* $\mathbf{A} <^{-} \mathbf{B}$ *then* $\mathbf{S}(\mathbf{A}|V_m, V_n) <^{-} \mathbf{S}(\mathbf{B}|V_m, V_n)$.

10.6 Exercises

(1) Let $\mathbf{A} = \begin{pmatrix} \mathbf{A_{11}} & \mathbf{A_{12}} \\ \mathbf{A_{21}} & \mathbf{A_{22}} \end{pmatrix}$ be an *nnd* matrix. Define $\mathbf{A}^{\mathcal{S}} = \begin{pmatrix} \mathbf{A_{11}} & \mathbf{0} \\ \mathbf{0} & \mathbf{0} \end{pmatrix}$.
Prove the following:

(i) $\mathbf{A}_{\mathcal{S}}^{\dagger} = \mathbf{A}^{\dagger\mathcal{S}}$.
(ii) $\mathcal{P}(\mathbf{A},\mathbf{B})^{\mathcal{S}} <^{L} \mathcal{P}(\mathbf{A}^{\mathcal{S}},\mathbf{B}^{\mathcal{S}})$.
(iii) $\mathbf{A}^{\mathcal{S}} + \mathbf{B}^{\mathcal{S}} = (\mathbf{A}+\mathbf{B})^{\mathcal{S}}$

The matrices $\mathbf{A}_{\mathcal{S}}$ and $\mathbf{A}^{\mathcal{S}}$ can be thought as dual to each other and corresponds to the the fact that theorems about electrical networks have dual when one interchanges the current and the voltage variables and replaces the resistance with conductance.

(2) Let $\mathbf{A}$ be an *nnd* matrix of order $n \times n$ and $\mathcal{S}$ and $\mathcal{T}$ be subspaces of $\mathbb{C}^n$. If $\mathcal{S} \subseteq \mathcal{T}$, then show that $\mathbf{A}_{\mathcal{S}} <^{L} \mathbf{A}_{\mathcal{T}}$.

(3) Show that for an *nnd* matrix $\mathbf{A}$ the following hold:
(i) $\mathbf{A}_{\mathcal{S}} = \inf\{\mathbf{QAQ}^{\star} : \mathbf{Q^2} = \mathbf{Q}, \mathcal{C}(\mathbf{Q}) = \mathcal{S}\}$ and
(ii) $\mathbf{A}_{\mathcal{S}}^{\mathbf{2}} <^{L} (\mathbf{A}_{\mathcal{S}})^{\mathbf{2}}$.

(4) Let $\mathbf{A_n}$, $n \in \mathbb{N}$ be a sequence of positive definite matrices of the same order such that $\mathbf{A_n} \downarrow \mathbf{A}$ and $\mathcal{S}$ be a subspace of $\mathbb{C}^n$. Prove that $(\mathbf{A_n})_{\mathcal{S}} \downarrow \mathbf{A}_{\mathcal{S}}$.

(5) Using notation and conclusions of the exercises 2 and 3, prove that for each real number $t \geq 1$, $(\mathbf{A}_{\mathcal{S}}^{\mathbf{t}})^{\frac{1}{t}} <^{L} \mathbf{A}_{\mathcal{S}}$. Also, prove that for $1 \leq s \leq t$, $(\mathbf{A}_{\mathcal{S}}^{\mathbf{t}})^{\frac{1}{t}} <^{L} (\mathbf{A}_{\mathcal{S}}^{\mathbf{s}})^{\frac{1}{s}}$.

(6) A matrix $\mathbf{A} \in \mathbb{C}^{n\times n}$ is said to be almost definite if, for any $\mathbf{x} \in \mathbb{C}^n$,

$$\mathbf{x}^{\star}\mathbf{A}\mathbf{x} = 0 \Rightarrow \mathbf{A}\mathbf{x} = \mathbf{0}.$$

Consider a matrix $\mathbf{A} \in \mathbb{C}^{n\times n}$ and $\mathcal{S}$, a subspace of $\mathbb{C}^n$. Prove that the following are equivalent:

(i) $\mathbf{A}$ is $\mathcal{S}$-complementable for each subspace $\mathcal{S}$ of $\mathbb{C}^n$
(ii) $\mathbf{A}$ is $\mathcal{S}$-complementable for each 1-dimensional subspace $\mathcal{S}$ of $\mathbb{C}^n$
(iii) for any $\mathbf{x} \in \mathbb{C}^n$, $\mathbf{x}^{\star}\mathbf{A}\mathbf{x} = 0 \Rightarrow \mathbf{A}\mathbf{x} = \mathbf{0}$ and
(iv) for any $\mathbf{x} \in \mathbb{C}^n$, $\mathbf{x}^{\star}\mathbf{A}\mathbf{x} = 0 \Rightarrow \mathbf{A}\mathbf{x} = \mathbf{0}$ and $\mathbf{x}^{\star}\mathbf{A}\mathbf{x} = \mathbf{0}$.

(7) Let $\mathbf{A} = \begin{pmatrix} \mathbf{A_{11}} & \mathbf{A_{12}} \\ \mathbf{A_{21}} & \mathbf{A_{22}} \end{pmatrix}$ be an $m \times n$ matrix over a field F, with $\mathbf{A_{11}}$ as a square matrix of order $r \times r$. Let $\mathbf{P}_{\ell}$ and $\mathbf{P_r}$ be projector of order $m \times m$ and $n \times n$ respectively. Then prove that $\mathbf{A}$ is $\mathbf{P}_{\ell}$, $\mathbf{P_r}$-complementable if and only if $\mathcal{C}(\mathbf{A_{12}}) \subseteq \mathcal{C}(\mathbf{A_{11}})$ and $\mathcal{C}(\mathbf{A_{21}^t}) \subseteq \mathcal{C}(\mathbf{A_{11}^t})$. Also, find the matrices $\mathbf{M}_{\ell}$ and $\mathbf{M_r}$.

(8) Let $\mathbf{A}$, $\mathbf{B}$, $\mathbf{C}$ and $\mathbf{D}$ be *nnd* matrices such that $\mathbf{A} <^L \mathbf{C}$ and $\mathbf{B} <^L \mathbf{D}$. Check whether or not any of the following hold:
(i) $\mathbf{S}(\mathbf{A}|\mathbf{B}) <^L \mathbf{S}(\mathbf{C}|\mathbf{B})$ and
(ii) $\mathbf{S}(\mathbf{A}|\mathbf{B}) <^L \mathbf{S}(\mathbf{C}|\mathbf{D})$.

Chapter 11

Shorted Operators - Other Approaches

11.1 Introduction

Motivated by electrical network theory and statistics, we introduced in the previous chapter the concept of shorted operator and studied some of its properties. We saw that the shorted matrix we defined is also a suitable Schur complement. Moreover, the shorted operator $\mathbf{S}(\mathbf{A}|V_m, V_n)$ of a given matrix $\mathbf{A}$ indexed by the given subspaces V_m and V_n, if exists, is the closest to $\mathbf{A}$ in the class of matrices

$$\mathfrak{M} = \{\mathbf{D} : \mathbf{D} <^- \mathbf{A},\ \mathcal{C}(\mathbf{D}) \subseteq V_m,\ \mathcal{C}(\mathbf{D}^t) \subseteq V_n\}$$

in the sense that $\mathbf{S}(\mathbf{A}|V_m, V_n) \in \mathfrak{M}$, and $\mathbf{D} <^- \mathbf{S}(\mathbf{A}|V_m, V_n)$ for each $\mathbf{D} \in \mathfrak{M}$. We also saw that the shorted operator of an *nnd* matrix indexed by a given subspace is the limit of a sequence of parallel sums of suitable matrices. In the present chapter, we give some more equivalent definitions for the shorted operator and alongside develop methods to compute this operator.

In Section 11.2, we extend the concept of shorted operator as a limit of a sequence of parallel sums to rectangular matrices over the field of complex numbers. We investigate the conditions under which the shorted operator thus defined exists. We obtain a formula for computing the shorted operator when it exists. We also establish several of its interesting properties which include the closeness property mentioned above. Another closeness property of the shorted operator to the given matrix is in terms of rank. More precisely, we consider the class $\mathfrak{M}$ as in the preceding paragraph. The shorted operator $\mathbf{S}(\mathbf{A}|V_m, V_n)$, when it exists is the unique matrix such that $\mathbf{S}(\mathbf{A}|V_m, V_n) \in \mathfrak{M}$ and for each matrix $\mathbf{D} \in \mathfrak{M}$, $\rho(\mathbf{A} - \mathbf{S}(\mathbf{A}|\mathbf{V_m}, \mathbf{V_n})) \leq \rho(\mathbf{A} - \mathbf{D})$.

In Section 11.3, we consider matrices over a general field. We define the shorted operator using the closeness property in terms of rank and investigate the conditions under which this operator exists. We show the equivalence of the various definitions given in this and the preceding chapter.

Section 11.4 contains one more method to compute the shorted operator. We also compare different computational methods in terms of the labour involved for computing them. No comparisons are made in terms of error analysis.

11.2 Shorted operator as the limit of parallel sums - General matrices

In this section, we consider the matrices over $\mathbb{C}$, the field of complex numbers. In Section 10.4, we obtained the shorted operator $\mathbf{S}(\mathbf{A}|\mathcal{S})$ of *nnd* matrix $\mathbf{A}$ of order $m \times m$ with respect to a subspace $\mathcal{S}$ of $\mathbb{C}^m$ as the limit of a sequence of parallel sums, $\lim_{n\to\infty} \mathcal{P}(\mathbf{A}, n\mathbf{P}_{\mathcal{S}})$, where $\mathbf{P}_{\mathcal{S}}$ is the orthogonal projector onto $\mathcal{S}$. With obvious modifications in Lemma 10.4.1 and Theorem 10.4.2, it can be shown that $\mathbf{S}(\mathbf{A}|\mathcal{S}) = \lim_{n\to\infty} \mathcal{P}(\mathbf{A}, n\mathbf{B})$, where $\mathbf{B}$ is an arbitrary *nnd* matrix such that $\mathcal{C}(\mathbf{B}) = \mathcal{S}$. It is easy to see that $\mathbf{S}(\mathbf{A}|\mathcal{S}) = \frac{1}{\lambda} \lim_{\lambda\to 0} \mathcal{P}(\lambda\mathbf{A}, \mathbf{B})$. Thus, $\mathbf{S}(\mathbf{A}|\mathcal{S})$ may be called the shorted operator of $\mathbf{A}$ shorted by $\mathbf{B}$ and is denoted by $\mathbf{S}(\mathbf{A}|\mathbf{B})$. As seen in Chapter 10, in the case of *nnd* matrices, as seen already, the shorted operator $\mathbf{S}(\mathbf{A}|\mathbf{B})$ always exists. In this section, we extend this concept of shorted operator to arbitrary matrices (possibly rectangular) by defining it via the limit of suitable sequence of parallel sums. We explore the conditions under which $\mathbf{S}(\mathbf{A}|\mathbf{B})$ exists for arbitrary matrices $\mathbf{A}$ and $\mathbf{B}$ of the same order. We also obtain several properties of this operator whenever it exists. We begin by proving a lemma that will be used in the sequel.

Lemma 11.2.1. *Let* $\mathbf{A}$, $\mathbf{B}$ *be matrices of the same order* $m \times n$ *such that* $\mathbf{A}$ *and* $\alpha\mathbf{B}$ *are parallel summable for some complex number* $\alpha(\neq 0)$. *Then there exists a complex number* θ_0 *such that for every complex number* θ *such that* $|\theta| \geq |\theta_0|$, $\mathbf{A}$ *and* $\theta\mathbf{B}$ *are parallel summable.*

Proof. Since $\mathbf{A}$ and $\alpha\mathbf{B}$ are parallel summable, $\mathbf{A} <^s \mathbf{A} + \alpha\mathbf{B}$. Let $\rho(\mathbf{A} + \alpha\mathbf{B}) = r$. By Remark 3.2.7, there exist non-singular matrices $\mathbf{P}$, $\mathbf{Q}$ and an $r \times r$ matrix $\mathbf{T}$ such that

$$\mathbf{A} = \mathbf{P}\mathrm{diag}\,(\mathbf{T},\ \mathbf{0})\,\mathbf{Q} \text{ and } \mathbf{A} + \alpha\mathbf{B} = \mathbf{P}\mathrm{diag}\,(\mathbf{I}_r,\ \mathbf{0})\,\mathbf{Q}.$$

Thus, $\alpha\mathbf{B} = \mathbf{P}\mathbf{diag}\left(\mathbf{I_r} - \mathbf{T},\ \mathbf{0}\right)\mathbf{Q}$ or $\mathbf{B} = \mathbf{P}\mathbf{diag}\left(\frac{1}{\alpha}(\mathbf{I_r} - \mathbf{T}),\ \mathbf{0}\right)\mathbf{Q}$. Consider $\mathbf{A} + \theta\mathbf{B} = \mathbf{P}\mathbf{diag}\left(\mathbf{T} + \frac{\theta}{\alpha}(\mathbf{I_r} - \mathbf{T}),\ \mathbf{0}\right)\mathbf{Q}$. If each eigen-value of $\mathbf{T}$ is 1, then it is easy to see that $(\mathbf{T} + \frac{\theta}{\alpha}(\mathbf{I_r} - \mathbf{T}))$ is non-singular for all θ. Suppose at least one eigen-value of $\mathbf{T}$ is different from 1. Then we can rewrite $\mathbf{T} + \frac{\theta}{\alpha}(\mathbf{I_r} - \mathbf{T}) = \frac{\theta}{\alpha}\mathbf{I_r} + (1 - \frac{\theta}{\alpha})\mathbf{T}$. Let $\theta_0 = \mid \alpha \mid \max_i\{\mid \frac{\lambda_i}{1-\lambda_i} \mid + 1\}$, where $\lambda_i' s$ are the eigen-values of $\mathbf{T}$ not equal to 1. It is easy to see that $\mathbf{T} + \frac{\theta}{\alpha}(\mathbf{I_r} - \mathbf{T})$ is non-singular whenever $\mid \theta \mid \ \geq\ \mid \theta_0 \mid$. So, $\mathbf{T} <^{\mathbf{s}} \mathbf{T} + \frac{\theta}{\alpha}(\mathbf{I_r} - \mathbf{T})$ and therefore, $\mathbf{A} <^{\mathbf{s}} \mathbf{A} + \theta\mathbf{B}$ for each θ such that $\mid \theta \mid \ \geq\ \mid \theta_0 \mid$. □

Corollary 11.2.2. *Let* $\mathbf{A}$, $\mathbf{B}$ *be matrices of the same order such that* $\mathbf{A}$ *and* $\alpha\mathbf{B}$ *are parallel summable for some complex number* $\alpha(\neq 0)$. *Then there exists a positive integer* m *such that* $\mathbf{A}$ *and* $m\mathbf{B}$ *are parallel summable.*

Remark 11.2.3. If $\mathbf{A}$, $\mathbf{B}$ are matrices of the same order such that $\mathbf{A}$ and $\alpha\mathbf{B}$ are parallel summable for some complex number $\alpha(\neq 0)$, then there exists a complex number γ_0 such that for every complex number γ such that $\mid \gamma \mid \ \leq\ \mid \gamma_0 \mid$, $\gamma\mathbf{A}$ and $\mathbf{B}$ are parallel summable.

We now define the shorted operator.

Definition 11.2.4. Let $\mathbf{A}$, $\mathbf{B}$ be matrices of the same order $m \times n$ such that $\mathbf{A}$ and $\alpha\mathbf{B}$ are parallel summable for some complex number α. If the limit

$$\lim_{\lambda \to 0} \mathbf{A}(\lambda\mathbf{A} + \mathbf{B})^{-}\mathbf{B}$$

exists and is finite, then the limit is called the shorted operator of $\mathbf{A}$ shorted by $\mathbf{B}$ and is denoted by $\mathbf{S}(\mathbf{A}|\mathbf{B})$.

Clearly, $\lim_{\lambda \to 0} \mathbf{A}(\lambda\mathbf{A} + \mathbf{B})^{-}\mathbf{B} = \lim_{\lambda \to \mathbf{0}} \frac{\mathbf{1}}{\lambda}\mathcal{P}(\lambda\mathbf{A}, \mathbf{B})$ and if $\mathbf{A}$ or $\mathbf{B}$ is a null matrix, then so is $\mathbf{S}(\mathbf{A}|\mathbf{B})$. Henceforth we consider only non-null matrices $\mathbf{A}$ and $\mathbf{B}$.

Remark 11.2.5. In view of Corollary 11.2.2 and Remark 11.2.3, we notice that $\mathbf{S}(\mathbf{A}|\mathbf{B})$, whenever it exists, is given by

$$\mathbf{S}(\mathbf{A}|\mathbf{B}) = \lim_{\mathbf{n} \to \infty} \mathbf{A}(\frac{\mathbf{1}}{\mathbf{n}}\mathbf{A} + \mathbf{B})^{-}\mathbf{B}.$$

We now give some elementary properties of the shorted operator when it exists. These properties will be used in exploring the conditions under which the shorted operator exists. For the sake of convenience we shall write $\mathbf{S} = \mathbf{S}(\mathbf{A}|\mathbf{B})$.

Theorem 11.2.6. *Let* $\mathbf{A}$ and $\mathbf{B}$ *be matrices of the same order for which* $\mathbf{S}$ *exists. Then the following hold:*

(i) $\mathcal{C}(\mathbf{S}) \subseteq \mathcal{C}(\mathbf{A}) \cap \mathcal{C}(\mathbf{B})$ *and* $\mathcal{C}(\mathbf{S}^\star) \subseteq \mathcal{C}(\mathbf{A}^\star) \cap \mathcal{C}(\mathbf{B}^\star)$.
(ii) $\mathcal{C}(\mathbf{A}-\mathbf{S}) \cap \mathcal{C}(\mathbf{B}) = \{\mathbf{0}\}$ *and* $\mathcal{C}(\mathbf{A}-\mathbf{S})^\star \cap \mathcal{C}(\mathbf{B}^\star) = \{\mathbf{0}\}$.
(iii) $\mathbf{S} <^- \mathbf{A}$.
(iv) $\mathcal{C}(\mathbf{S}) = \mathcal{C}(\mathbf{A}) \cap \mathcal{C}(\mathbf{B})$ *and* $\mathcal{C}(\mathbf{S}^\star) = \mathcal{C}(\mathbf{A}^\star) \cap \mathcal{C}(\mathbf{B}^\star)$. *Thus,* $\mathbf{S} <^{\mathbf{s}} \mathbf{B}$.

Proof. (i) Let $\mathbf{u} \in \mathcal{C}(\mathbf{S})$. Then $\mathbf{u} = \mathbf{S}\mathbf{y} = \lim_{\mathbf{s}\to\infty}(\frac{\mathbf{A}}{\mathbf{s}} + \mathbf{B})^-\mathbf{B}\mathbf{y}$ for some vector $\mathbf{y}$. For each s, $\mathbf{A}(\frac{\mathbf{A}}{\mathbf{s}} + \mathbf{B})^-\mathbf{B}\mathbf{y} \in \mathcal{C}(\mathbf{A})$. Since $\mathcal{C}(\mathbf{A})$ is a closed subspace of $\mathbb{C}^m$ and for each s, $\mathbf{A}(\frac{\mathbf{A}}{\mathbf{s}} + \mathbf{B})^-\mathbf{B} = \mathbf{B}(\frac{\mathbf{A}}{\mathbf{s}} + \mathbf{B})^-\mathbf{A}$, so, $\mathbf{u} \in \mathcal{C}(\mathbf{A})$ and $\mathbf{u} \in \mathcal{C}(\mathbf{B})$. Hence, $\mathcal{S} \subseteq \mathcal{C}(\mathbf{A}) \cap \mathcal{C}(\mathbf{B})$.
Similarly we can prove that $\mathcal{C}(\mathbf{S}^\star) \subseteq \mathcal{C}(\mathbf{A}^\star) \cap \mathcal{C}(\mathbf{B}^\star)$.
(ii) Let $\mathbf{u} \in \mathcal{C}(\mathbf{A}-\mathbf{S}) \cap \mathcal{C}(\mathbf{B})$. Then $\mathbf{u} = (\mathbf{A}-\mathbf{S})\mathbf{v} = \mathbf{B}\mathbf{w}$ for for some vectors $\mathbf{v}$ and $\mathbf{w}$. So,

$$\begin{aligned}\mathbf{A}\mathbf{v} - \mathbf{B}\mathbf{w} = \mathbf{S}\mathbf{v} &= \lim_{s\to\infty}\mathbf{B}(\frac{\mathbf{A}}{\mathbf{s}} + \mathbf{B})^-\mathbf{A}\mathbf{v} \\ &= \lim_{s\to\infty}\mathbf{B}(\frac{\mathbf{A}}{\mathbf{s}} + \mathbf{B})^-(\mathbf{B}\mathbf{w} + \mathbf{S}\mathbf{v}).\end{aligned}$$

Since $\mathcal{C}(\mathbf{S}) \subseteq \mathcal{C}(\mathbf{A})$ and $\lim_{s\to\infty}\mathbf{B}(\frac{\mathbf{A}}{\mathbf{s}} + \mathbf{B})^-\mathbf{A}$ exists, it follows that $\lim_{s\to\infty}\mathbf{B}(\frac{\mathbf{A}}{\mathbf{s}} + \mathbf{B})^-\mathbf{S}$ exists. So, $\lim_{s\to\infty}\mathbf{B}(\frac{\mathbf{A}}{\mathbf{s}} + \mathbf{B})^-\mathbf{B}\mathbf{w}$ exists and is equal to $\mathbf{S}\mathbf{v} - \lim_{\mathbf{s}\to\infty}\mathbf{B}(\frac{\mathbf{A}}{\mathbf{s}} + \mathbf{B})^-\mathbf{S}\mathbf{v}$. Now,

$$\begin{aligned}\mathbf{S} - \mathbf{B}(\frac{\mathbf{A}}{\mathbf{s}} + \mathbf{B})^-\mathbf{S} &= \mathbf{S} - (\frac{\mathbf{A}}{\mathbf{s}} + \mathbf{B} - \frac{\mathbf{A}}{\mathbf{s}})(\frac{\mathbf{A}}{\mathbf{s}} + \mathbf{B})^-\mathbf{S} \\ &= \mathbf{S} - \mathbf{S} + \frac{\mathbf{A}}{\mathbf{s}}(\frac{\mathbf{A}}{\mathbf{s}} + \mathbf{B})^-\mathbf{S},\end{aligned}$$

since $\mathcal{C}(\mathbf{S}) \subseteq \mathcal{C}(\mathbf{B}) \subseteq \mathcal{C}(\frac{\mathbf{A}}{\mathbf{s}} + \mathbf{B})$. Hence,

$$\lim_{s\to\infty}\mathbf{B}(\frac{\mathbf{A}}{\mathbf{s}} + \mathbf{B})^-\mathbf{B}\mathbf{w} = \lim_{\mathbf{s}\to\infty}\frac{\mathbf{A}}{\mathbf{s}}(\frac{\mathbf{A}}{\mathbf{s}} + \mathbf{B})^-\mathbf{S}\mathbf{v} = \mathbf{0}.$$

However, $\mathbf{B}(\frac{\mathbf{A}}{\mathbf{s}}+\mathbf{B})^{-}\mathbf{Bw} = \mathbf{Bw} - \frac{\mathbf{A}}{\mathbf{s}}(\frac{\mathbf{A}}{\mathbf{s}}+\mathbf{B})^{-}\mathbf{Bw}$, so, $\lim_{s\to\infty}\mathbf{B}(\frac{\mathbf{A}}{\mathbf{s}}+\mathbf{B})^{-}\mathbf{Bw} = \mathbf{Bw}$. Therefore, $\mathbf{Bw} = \mathbf{0}$ i.e., $\mathbf{v} = \mathbf{0}$, proving $\mathcal{C}(\mathbf{A}-\mathbf{S}) \cap \mathcal{C}(\mathbf{B}) = \{\mathbf{0}\}$.
Similarly, $\mathcal{C}(\mathbf{A}-\mathbf{S})^\star \cap \mathcal{C}(\mathbf{B})^\star = \{\mathbf{0}\}$.
(iii) Since $\mathcal{C}(\mathbf{S}) \subseteq \mathcal{C}(\mathbf{B})$ by (i) and $\mathcal{C}(\mathbf{A}-\mathbf{S}) \cap \mathcal{C}(\mathbf{B}) = \{\mathbf{0}\}$ by (ii), we have $\mathcal{C}(\mathbf{A}-\mathbf{S}) \cap \mathcal{C}(\mathbf{S}) = \{\mathbf{0}\}$. Similarly, $\mathcal{C}(\mathbf{A}-\mathbf{S})^\star \cap \mathcal{C}(\mathbf{S}^\star) = \{\mathbf{0}\}$. Hence, $\mathbf{S} <^- \mathbf{A}$.
(iv) By (i), we have $\mathcal{C}(\mathbf{S}) \subseteq \mathcal{C}(\mathbf{A}) \cap \mathcal{C}(\mathbf{B})$ and $\mathcal{C}(\mathbf{S}^\star) \subseteq \mathcal{C}(\mathbf{A}^\star) \cap \mathcal{C}(\mathbf{B}^\star)$. We now show that the reverse inclusions hold. Let $\mathbf{x} \in \mathcal{C}(\mathbf{A}) \cap \mathcal{C}(\mathbf{B})$. By (iii), $\mathcal{C}(\mathbf{A}) = \mathcal{C}(\mathbf{A}-\mathbf{S}) \oplus \mathcal{C}(\mathbf{S})$, so, we can write $\mathbf{x} = \mathbf{x_1} + \mathbf{x_2}$, where $\mathbf{x_1} \in \mathcal{C}(\mathbf{S})$ and $\mathbf{x_2} \in \mathcal{C}(\mathbf{A}-\mathbf{S})$. Now, $\mathcal{C}(\mathbf{S}) \subseteq \mathcal{C}(\mathbf{B})$, so, $\mathbf{x} - \mathbf{x_1} = \mathbf{x_2} \in \mathcal{C}(\mathbf{B})$. By (ii), $\mathbf{x_2} = \mathbf{0}$. Thus, $\mathbf{x} = \mathbf{x_1} \in \mathcal{C}(\mathbf{S})$.
Similarly, $\mathcal{C}(\mathbf{A}^\star) \cap \mathcal{C}(\mathbf{B}^\star) \subseteq \mathcal{C}(\mathbf{S}^\star)$. Hence, the result follows. □

Corollary 11.2.7. *Let* $\mathbf{A}$, $\mathbf{B}$ *be matrices of the same order. Then the following hold:*

(i) *If* $\rho(\mathbf{A}+\mathbf{B}) = \rho(\mathbf{A}) + \rho(\mathbf{B})$, *then* $\mathbf{S} = \mathbf{0}$.
(ii) *If* $\mathbf{B}$ *is non-singular, then* $\mathbf{S} = \mathbf{A}$.
(iii) *If* $\mathbf{A}$ *and* $\mathbf{B}$ *are range-hermitian, then so is* $\mathbf{S}$.

Remark 11.2.8. In Theorem 11.2.6 the statement (i) is redundant in view of statement (iv). However, we need (i) to prove statements (ii) and (iii) which in turn are required to prove (iv). Notice that (i) implies that $\mathbf{S} <^{\mathbf{s}} \mathbf{A}$ and $\mathbf{S} <^{\mathbf{s}} \mathbf{B}$. Also, if $\mathbf{S_0}$ is a matrix of the same order as $\mathbf{A}$ such that $\mathbf{S_0} <^{\mathbf{s}} \mathbf{A}$ and $\mathbf{S_0} <^{\mathbf{s}} \mathbf{B}$, then $\mathbf{S_0} <^{\mathbf{s}} \mathbf{S}$.

Remark 11.2.9. Let $\mathbf{A}$, $\mathbf{B}$ be matrices of the same order such that $\mathbf{S} = \mathbf{S}(\mathbf{A}|\mathbf{B})$ exists. Then $\mathbf{S}(\mathbf{A}|\mathbf{cB})$ exists for each non-null complex number c and is also equal to $\mathbf{S}$. So, if $\mathbf{S} = \mathbf{S}(\mathbf{A}|\mathbf{B})$ exists, we can assume that $\mathbf{A}$ and $\mathbf{B}$ are parallel summable.

Remark 11.2.10. Let $\mathbf{A}$, $\mathbf{B}$ be matrices of the same order for which $\mathbf{S}(\mathbf{A}|\mathbf{B})$ exists and $\mathbf{L}$ and $\mathbf{R}$ be matrices such that $\mathbf{LAR}$ is defined. Let $\mathbf{L}$ have a left inverse and $\mathbf{R}$ have a right inverse. Then $\mathbf{S}(\mathbf{LAR}|\mathbf{LBR})$ exists and is equal to $\mathbf{LS}(\mathbf{A}|\mathbf{B})\mathbf{R}$.

We are now ready to obtain necessary and sufficient conditions for the existence of the shorted matrix $\mathbf{S}(\mathbf{A}|\mathbf{B})$.

Theorem 11.2.11. *Let* $\mathbf{A}$, $\mathbf{B}$ *be parallel summable matrices of the same order. Let* $(\mathbf{L}, \mathbf{R})$ *be a rank factorization of* $\mathbf{A}+\mathbf{B}$ *and* $\mathbf{D}$ *be a matrix such that* $\mathbf{A} = \mathbf{LDR}$. *Then the following are equivalent:*

(i) $\mathbf{S} := \mathbf{S}(\mathbf{A}|\mathbf{B})$ *exists*
(ii) *Index of* $\mathbf{I}-\mathbf{D}$ *is* ≤ 1
(iii) $\rho(\mathbf{B}(\mathbf{A}+\mathbf{B})^{-}\mathbf{B}) = \rho(\mathbf{B})$.

Proof. (ii) $\Leftrightarrow$ (iii)
Let $\mathbf{L}_\mathbf{L}^{-1}$ be a left inverse of $\mathbf{L}$ and $\mathbf{R}_\mathbf{R}^{-1}$ a right inverse of $\mathbf{R}$. Then $\mathbf{R}_\mathbf{R}^{-1}\mathbf{L}_\mathbf{L}^{-1}$ is a g-inverse of $\mathbf{A}+\mathbf{B}$. Since

$$\mathbf{B}(\mathbf{A}+\mathbf{B})^{-}\mathbf{B} = \mathbf{L}(\mathbf{I}-\mathbf{D})\mathbf{R}\mathbf{R}_\mathbf{R}^{-1}\mathbf{L}_\mathbf{L}^{-1}\mathbf{L}(\mathbf{I}-\mathbf{D})\mathbf{R} = \mathbf{L}(\mathbf{I}-\mathbf{D})^2\mathbf{R},$$

it follows that $\rho(\mathbf{B}(\mathbf{A}+\mathbf{B})^{-}\mathbf{B}) = \rho(\mathbf{I}-\mathbf{D})^2$. Also, $\mathbf{B} = \mathbf{L}(\mathbf{I}-\mathbf{D})\mathbf{R}$, so, we have $\rho(\mathbf{B}) = \rho(\mathbf{I}-\mathbf{D})$. Thus,
$\rho(\mathbf{B}(\mathbf{A}+\mathbf{B})^{-}\mathbf{B}) = \rho(\mathbf{B}) \Leftrightarrow \rho(\mathbf{I}-\mathbf{D})^2 = \rho(\mathbf{I}-\mathbf{D}) \Leftrightarrow \mathbf{I}-\mathbf{D}$ is of index not exceeding 1.
(i) $\Rightarrow$ (iii)
Let $\mathbf{S} := \mathbf{S}(\mathbf{A}|\mathbf{B})$ exist. Write $\mathbf{A}+\mathbf{B} = \mathbf{A}-\mathbf{S}+\mathbf{B}+\mathbf{S}$. Since $\mathbf{S} <^{s} \mathbf{B}$, we have $\mathbf{S}+\mathbf{B} <^{s} \mathbf{B}$. Since $\mathcal{C}(\mathbf{A}-\mathbf{S}) \cap \mathcal{C}(\mathbf{S}+\mathbf{B}) \subseteq \mathcal{C}(\mathbf{A}-\mathbf{S}) \cap \mathcal{C}(\mathbf{B})$ and $\mathcal{C}(\mathbf{A}-\mathbf{S})^\star \cap \mathcal{C}(\mathbf{S}+\mathbf{B})^\star \subseteq \mathcal{C}(\mathbf{A}-\mathbf{S})^\star \cap \mathcal{C}(\mathbf{B})^\star$, we have by Theorem 11.2.6(ii), $\mathcal{C}(\mathbf{A}-\mathbf{S}) \cap \mathcal{C}(\mathbf{S}+\mathbf{B}) = \{\mathbf{0}\}$ and $\mathcal{C}(\mathbf{A}-\mathbf{S})^\star \cap \mathcal{C}(\mathbf{S}+\mathbf{B})^\star = \{\mathbf{0}\}$. Therefore, $\mathbf{A}+\mathbf{B} = (\mathbf{A}-\mathbf{S}) \oplus (\mathbf{B}+\mathbf{S})$, or equivalently $\mathbf{S}+\mathbf{B} <^{-} \mathbf{A}+\mathbf{B}$. We now show that $\mathcal{C}(\mathbf{B}) = \mathcal{C}(\mathbf{S}+\mathbf{B})$ and $\mathcal{C}(\mathbf{B}^\star) = \mathcal{C}(\mathbf{S}+\mathbf{B})^\star$. Since $\mathbf{S}+\mathbf{B} <^{s} \mathbf{B}$, we already have $\mathcal{C}(\mathbf{S}+\mathbf{B}) \subseteq \mathcal{C}(\mathbf{B})$ and $\mathcal{C}(\mathbf{S}+\mathbf{B})^\star \subseteq \mathcal{C}(\mathbf{B}^\star)$. Let $\mathbf{x} \in \mathcal{C}(\mathbf{B})$. Then $\mathbf{x} = \mathbf{Bu}$ for some $\mathbf{u}$. Now, $\mathbf{A}$, $\mathbf{B}$ be parallel summable, so, $\mathcal{C}(\mathbf{B}) \subseteq \mathcal{C}(\mathbf{A}+\mathbf{B})$. Therefore, there exists a vector $\mathbf{v}$ such that $\mathbf{x} = \mathbf{Bu} = (\mathbf{A}+\mathbf{B})\mathbf{v} = (\mathbf{A}-\mathbf{S})\mathbf{v} + (\mathbf{S}+\mathbf{B})\mathbf{v}$. Therefore, $\mathbf{Bu} - (\mathbf{S}+\mathbf{B})\mathbf{v} = (\mathbf{A}-\mathbf{S})\mathbf{v}$. Since $\mathbf{Bu} - (\mathbf{S}+\mathbf{B})\mathbf{v} \in \mathcal{C}(\mathbf{B})$, $(\mathbf{A}-\mathbf{S})\mathbf{v} \in \mathcal{C}(\mathbf{A}-\mathbf{S})$ and $\mathcal{C}(\mathbf{A}-\mathbf{S}) \cap \mathcal{C}(\mathbf{B}) = \{\mathbf{0}\}$, so, $\mathbf{Bu} = (\mathbf{S}+\mathbf{B})\mathbf{v}$. But then $\mathbf{x} = \mathbf{Bu}$, we have $\mathbf{x} \in \mathcal{C}(\mathbf{S}+\mathbf{B})$, giving $\mathcal{C}(\mathbf{B}) \subseteq \mathcal{C}(\mathbf{S}+\mathbf{B})$ and therefore $\mathcal{C}(\mathbf{B}) = \mathcal{C}(\mathbf{S}+\mathbf{B})$. Similarly, $\mathcal{C}(\mathbf{B}^\star) = \mathcal{C}(\mathbf{S}+\mathbf{B})^\star$.

Now, $\mathcal{C}(\mathbf{B}) = \mathcal{C}(\mathbf{S}+\mathbf{B})$, and $\mathcal{C}(\mathbf{B}^\star) = \mathcal{C}(\mathbf{S}+\mathbf{B})^\star \Rightarrow$ there exist matrices $\mathbf{T}_1$ and $\mathbf{T}_2$ such that $\mathbf{B} = \mathbf{T}_1(\mathbf{S}+\mathbf{B})$ and $\mathbf{BT}_2 = \mathbf{S}+\mathbf{B}$. Also, since $(\mathbf{S}+\mathbf{B}) <^{-} (\mathbf{A}+\mathbf{B}) \Rightarrow \mathbf{S}+\mathbf{B} = (\mathbf{S}+\mathbf{B})(\mathbf{A}+\mathbf{B})^{-}(\mathbf{S}+\mathbf{B})$. Hence, $\mathbf{T}_1(\mathbf{S}+\mathbf{B})(\mathbf{A}+\mathbf{B})^{-}(\mathbf{S}+\mathbf{B}) = \mathbf{T}_1(\mathbf{S}+\mathbf{B})(\mathbf{A}+\mathbf{B})^{-}\mathbf{BT}_2 = \mathbf{T}_1(\mathbf{S}+\mathbf{B})$, or equivalently, $\mathbf{B}(\mathbf{A}+\mathbf{B})^{-}\mathbf{BT}_2 = \mathbf{B}$. It now follows that

$$\rho(\mathbf{B}) = \rho(\mathbf{B}(\mathbf{A}+\mathbf{B})^{-}\mathbf{BT}_2) \leq \rho(\mathbf{B}(\mathbf{A}+\mathbf{B})^{-}\mathbf{B}) \leq \rho(\mathbf{B})$$

and hence, $\rho(\mathbf{B}(\mathbf{A}+\mathbf{B})^{-}\mathbf{B}) = \rho(\mathbf{B})$.
(ii) $\Rightarrow$ (i)

Let us first consider the case when $\mathbf{D}$ does not possess 1 as an eigen-value. In this case $\mathbf{I}-\mathbf{D}$ is non-singular, so, $\mathbf{S}(\mathbf{D}|\mathbf{I}-\mathbf{D})$ exists and is equal to $\mathbf{D}$, by Corollary 11.2.7. Hence, $\mathbf{S}(\mathbf{A}|\mathbf{B}) = \mathbf{S}(\mathbf{LDR}|\mathbf{L}(\mathbf{I}-\mathbf{D})\mathbf{R})$ exists and is equal to $\mathbf{LDR} = \mathbf{A}$.

Now, let 1 be an eigen-value of $\mathbf{D}$. Since $\mathbf{I}-\mathbf{D}$ is of index 1, the algebraic multiplicity of 1 as an eigen-value of $\mathbf{D}$ is same as its geometric multiplicity. So, there exists a non-singular matrix $\mathbf{Q}$ and a matrix $\mathbf{D_1}$ such that $\mathbf{I}-\mathbf{D_1}$ is non-singular and $\mathbf{I}-\mathbf{D} = \mathbf{Q}\mathbf{diag}(\mathbf{I}-\mathbf{D_1},\ \mathbf{0})\mathbf{Q}^{-1}$. So, $\mathbf{D} = \mathbf{Q}\mathbf{diag}(\mathbf{D_1},\ \mathbf{I})\mathbf{Q}^{-1}$. Consider

$$\mathbf{diag}\left(\mathbf{D_1},\mathbf{I}\right)\left(\mathbf{diag}\left(\frac{1}{n}\mathbf{D_1},\ \frac{1}{n}\mathbf{I}\right)+\mathbf{diag}\left(\mathbf{I}-\mathbf{D_1},\mathbf{0}\right)\right)^{-}\mathbf{diag}\left(\mathbf{I}-\mathbf{D_1},\mathbf{0}\right)$$

$$=\mathbf{diag}\left(\mathbf{D_1},\mathbf{I}\right)\left(\mathbf{diag}\left(\frac{1}{n}\mathbf{D_1}+(\mathbf{I}-\mathbf{D_1}),\ \frac{1}{n}\mathbf{I}\right)\right)^{-}\mathbf{diag}\left(\mathbf{I}-\mathbf{D_1},\mathbf{0}\right).$$

It is easy to check that for all $n > n_0 = \max_i\{\frac{d_i}{d_i-1}\}$, where d_i are the eigen-values of $\mathbf{D_1}$, the matrix $\frac{1}{n}\mathbf{D_1}+\mathbf{I}-\mathbf{D_1}$ is non-singular.
Thus, for all $n > n_0$,

$$\mathbf{diag}(\mathbf{D_1},\ \mathbf{I})\left(\mathbf{diag}(\tfrac{1}{n}\mathbf{D_1}+\mathbf{I}-\mathbf{D_1},\ \tfrac{1}{n}\mathbf{I})\right)^{-1}\mathbf{diag}(\mathbf{I}-\mathbf{D_1},\ \mathbf{0})$$
$$=\mathbf{diag}(\mathbf{D_1}(\tfrac{1}{\mathbf{n}}\mathbf{D_1}+\mathbf{I}-\mathbf{D_1})^{-1}(\mathbf{I}-\mathbf{D_1}),\mathbf{0}).$$

It follows that,

$$\lim_{n\to\infty}\mathbf{diag}(\mathbf{D_1},\ \mathbf{I})\mathbf{diag}((\tfrac{1}{n}\mathbf{D_1}+\mathbf{I}-\mathbf{D_1},\ \tfrac{1}{n}\mathbf{I},\ \mathbf{nI})^{-1},\mathbf{nI})\mathbf{diag}(\mathbf{I}-\mathbf{D_1},\ \mathbf{0})$$
$$=\mathbf{diag}(\lim_{n\to\infty}(\mathbf{D_1}(\frac{1}{\mathbf{n}}\mathbf{D_1}+\mathbf{I}-\mathbf{D_1})^{-1}(\mathbf{I}-\mathbf{D_1})),\mathbf{0}) = \mathbf{diag}(\mathbf{D_1},\mathbf{0}).$$

So, $\mathbf{S}(\mathbf{D}|\mathbf{I}-\mathbf{D})$ exists and $\mathbf{S}(\mathbf{D}|\mathbf{I}-\mathbf{D}) = \mathbf{Q}\mathbf{diag}(\mathbf{D_1},\mathbf{0})\mathbf{Q}^{-1}$.
As a consequence, we have $\mathbf{S}(\mathbf{A}|\mathbf{B})$ exists and by Remark 11.2.10, it is equal to $\mathbf{LQ}\mathbf{diag}(\mathbf{D_1},\mathbf{0})\mathbf{Q}^{-1}\mathbf{R}$. □

Corollary 11.2.12. *Let* $\mathbf{A}$ *and* $\mathbf{B}$ *be parallel summable matrices of the same order. Then in the notations of Theorem 11.2.11, the following are equivalent:*

(i) $\mathbf{S} = \mathbf{S}(\mathbf{B}|\mathbf{A})$ *exists*
(ii) *Index of* $\mathbf{D}$ *is* ≤ 1,
(iii) $\rho(\mathbf{A}(\mathbf{A}+\mathbf{B})^{-}\mathbf{A}) = \rho(\mathbf{A})$.

Corollary 11.2.13. *If* $\mathbf{A}$ *and* $\mathbf{B}$ *are parallel summable matrices of the same order and either* $\mathcal{C}(\mathbf{A}) \subseteq \mathcal{C}(\mathbf{B})$ *or* $\mathcal{C}(\mathbf{B}) \subseteq \mathcal{C}(\mathbf{A})$, *then* $\mathbf{S}(\mathbf{A}|\mathbf{B})$ *exists.*

Proof. $\mathbf{S}(\mathbf{A}|\mathbf{B})$ exists by Theorem 9.2.26(i) and Theorem 11.2.11. □

Theorem 11.2.14. *Let* $\mathbf{A}$ *and* $\mathbf{B}$ *be matrices of the same order such that* $\mathbf{S} = \mathbf{S}(\mathbf{A}|\mathbf{B})$ *exists. Then the following hold:*

(i) $\mathbf{S} = \lim_{\lambda \to 0} \frac{1}{\lambda}\mathcal{P}(\lambda\mathbf{A}, \mathbf{B})$ *and*
(ii) $\mathbf{S}(\mathbf{A}^\star|\mathbf{B}^\star)$ *exists and equals* $(\mathbf{S}(\mathbf{A}|\mathbf{B}))^\star$.

Proof is trivial.

Corollary 11.2.15. *Let* $\mathbf{A}$ and $\mathbf{B}$ *be hermitian matrices of the same order such that* $\mathbf{S}(\mathbf{A}|\mathbf{B})$ *exists. Then* $\mathbf{S}(\mathbf{A}|\mathbf{B})$ *is hermitian.*

Remark 11.2.16. Let $\mathbf{A}$, $\mathbf{B}$ be square matrices of the same order such that $\mathbf{S}(\mathbf{A}|\mathbf{B})$ exists and is hermitian. It does not necessarily follow that $\mathbf{A}$ and $\mathbf{B}$ are hermitian. For, let $\mathbf{A}$ be hermitian and $\mathbf{B}$ non-hermitian invertible matrix such that $\mathbf{S}(\mathbf{A}|\mathbf{B})$ exists. Then by Corollary 11.2.7 (ii), $\mathbf{S}(\mathbf{A}|\mathbf{B}) = \mathbf{A}$ is hermitian, but $\mathbf{B}$ is not. We now give an example where neither of the two matrices is hermitian, yet $\mathbf{S}(\mathbf{A}|\mathbf{B})$ is hermitian.

Example 11.2.17. Let $\mathbf{A} = \begin{pmatrix} 0 & 0 \\ 1 & 1 \end{pmatrix}$ and $\mathbf{B} = \begin{pmatrix} 1 & 2 \\ 0 & 0 \end{pmatrix}$. Then $\mathbf{A}, \mathbf{B}$ are non-hermitian. Since $\mathcal{C}(\mathbf{A}) \cap \mathcal{C}(\mathbf{B}) = \{\mathbf{0}\}$ and $\mathcal{C}(\mathbf{A}^\star) \cap \mathcal{C}(\mathbf{B}^\star) = \{\mathbf{0}\}$ we have $\dfrac{\mathbf{A}}{\mathbf{n}}, \mathbf{B}$ are parallel summable for each positive integer n. Therefore, $\mathbf{S}$ exists. Also, for each positive integer n, $\dfrac{\mathbf{A}}{\mathbf{n}} + \mathbf{B} = \begin{pmatrix} 1 & 2 \\ \frac{1}{n} & \frac{1}{n} \end{pmatrix}$ is invertible with inverse $\begin{pmatrix} -1 & 2n \\ 1 & -n \end{pmatrix}$. So,

$$\mathbf{A}(\frac{\mathbf{A}}{\mathbf{n}} + \mathbf{B})^- \mathbf{B} = \begin{pmatrix} 0 & 0 \\ 1 & 1 \end{pmatrix} \begin{pmatrix} -1 & 2n \\ 1 & -n \end{pmatrix} \begin{pmatrix} 1 & 2 \\ 0 & 0 \end{pmatrix} = \begin{pmatrix} 0 & 0 \\ 0 & 0 \end{pmatrix}.$$

Thus, $\mathbf{S}$ is hermitian but neither of $\mathbf{A}, \mathbf{B}$ is.

Let $\mathbf{A}$ be a given matrix. In Theorem 10.3.5 and Theorem 10.5.5, we had obtained a matrix $\mathbf{S}$ which has its row space and column space contained in specified subspaces such that $\mathbf{S}$ is the closest to $\mathbf{A}$ in a certain sense. We show that the shorted operator, we have defined above in this section, has a similar property. But before we exhibit this, we prove the following useful theorem:

Theorem 11.2.18. *Let* $\mathbf{P}$, $\mathbf{Q}$ *and* $\mathbf{R}$ *be matrices of the same order such that* $\mathbf{P} <^{\mathbf{s}} \mathbf{Q}$, $\mathbf{P} <^- \mathbf{R}$ *and* $\mathbf{Q} <^- \mathbf{R}$. *Then* $\mathbf{P} <^- \mathbf{Q}$.

Proof. Since $\mathbf{P} <^- \mathbf{R}$, we have $\mathbf{PP}^- = \mathbf{RP}^-$ and $\mathbf{P}^-\mathbf{P} = \mathbf{P}^-\mathbf{R}$ for some g-inverse $\mathbf{P}^-$ of $\mathbf{P}$, and $\mathbf{Q} <^- \mathbf{R} \Rightarrow \{\mathbf{R}^-\} \subseteq \{\mathbf{Q}^-\}$. Let $\mathbf{R}^-$ be a g-inverse of $\mathbf{R}$. Then $\mathbf{PP}^- = \mathbf{RP}^- \Rightarrow \mathbf{QR}^-\mathbf{PP}^- = \mathbf{QR}^-\mathbf{RP}^-$. Since $\mathbf{P} <^{\mathbf{s}} \mathbf{Q}$, $\mathcal{C}(\mathbf{P}) \subseteq \mathcal{C}(\mathbf{Q})$ and $\mathcal{C}(\mathbf{P}^\star) \subseteq \mathcal{C}(\mathbf{Q}^\star)$. Therefore, $\mathbf{QR}^-\mathbf{P} = \mathbf{P}$ and $\mathbf{QR}^-\mathbf{R} = \mathbf{Q}$. As consequence, $\mathbf{PP}^- = \mathbf{QP}^-$. Similarly, $\mathbf{P}^-\mathbf{P} = \mathbf{P}^-\mathbf{Q}$ and hence, $\mathbf{P} <^- \mathbf{Q}$. □

Theorem 11.2.19. *Let* $\mathbf{A}$, $\mathbf{B}$ *be square matrices of the same order such that* $\mathbf{S} = \mathbf{S}(\mathbf{A}|\mathbf{B})$ *exists. Then*

$$\mathbf{S} = \mathbf{S}(\mathbf{A}|\mathbf{B}) = \max\{\mathbf{D} : \mathbf{D} <^- \mathbf{A},\ \mathcal{C}(\mathbf{D}) \subseteq \mathcal{C}(\mathbf{B}),\ \mathcal{C}(\mathbf{D}^\star) \subseteq \mathcal{C}(\mathbf{B}^\star).\}$$

Proof. Let $\mathfrak{C} = \{\mathbf{D} : \mathbf{D} <^- \mathbf{A},\ \mathcal{C}(\mathbf{D}) \subseteq \mathcal{C}(\mathbf{B}),\ \mathcal{C}(\mathbf{D}^\star) \subseteq \mathcal{C}(\mathbf{B}^\star)\}$. By Theorem 11.2.6 (iii) and (iv), $\mathbf{S} \in \mathfrak{C}$. Let $\mathbf{D}$ be an arbitrary matrix in $\mathfrak{C}$. Clearly, $\mathcal{C}(\mathbf{D}) \subseteq \mathcal{C}(\mathbf{A}) \cap \mathcal{C}(\mathbf{B}) = \mathcal{C}(\mathbf{S})$ and $\mathcal{C}(\mathbf{D}) \subseteq \mathcal{C}(\mathbf{A}) \cap \mathcal{C}(\mathbf{B}) = \mathcal{C}(\mathbf{S}^\star)$. So, $\mathbf{D} <^{\mathbf{s}} \mathbf{S}$. Now, $\mathbf{D} <^{\mathbf{s}} \mathbf{S}$, $\mathbf{D} <^- \mathbf{A}$ and $\mathbf{S} <^- \mathbf{A}$, so, the result follows by Theorem 11.2.18. □

We now obtain another property related to the closeness of the shorted operator to the matrix $\mathbf{A}$ according to somewhat different looking criterion. We say that it is somewhat different looking criterion, since we shall later prove that these two different criteria lead to the same optimizing matrix whenever it exists.

Theorem 11.2.20. *Let* $\mathbf{A}$, $\mathbf{B}$ *be square matrices of the same order such that* $\mathbf{S} = \mathbf{S}(\mathbf{A}|\mathbf{B})$ *exists. Then*

$$\mathbf{S} = \arg\min_{\mathbf{D}}\{\rho(\mathbf{A} - \mathbf{D}) : \mathcal{C}(\mathbf{D}) \subseteq \mathcal{C}(\mathbf{B}),\ \mathcal{C}(\mathbf{D}^\star) \subseteq \mathcal{C}(\mathbf{B}^\star)\}.$$

Proof. Let $\mathfrak{F} = \{\mathbf{D} : \mathcal{C}(\mathbf{D}) \subseteq \mathcal{C}(\mathbf{B}),\ \mathcal{C}(\mathbf{D}^\star) \subseteq \mathcal{C}(\mathbf{B}^\star)\}$. Notice that $\mathbf{S} \in \mathfrak{F}$. Let $\mathbf{D}$ be an arbitrary element of $\mathfrak{F}$. Write $\mathbf{A} - \mathbf{D} = (\mathbf{A} - \mathbf{S}) + (\mathbf{S} - \mathbf{D})$. Since $\mathcal{C}(\mathbf{S}) \subseteq \mathcal{C}(\mathbf{B})$, $\mathcal{C}(\mathbf{S} - \mathbf{D}) \subseteq \mathcal{C}(\mathbf{B})$, so, $\mathcal{C}(\mathbf{A} - \mathbf{S}) \cap \mathcal{C}(\mathbf{S} - \mathbf{D}) = \{\mathbf{0}\}$. Similarly, $\mathcal{C}(\mathbf{A} - \mathbf{S})^\star \cap \mathcal{C}(\mathbf{S} - \mathbf{D})^\star = \{\mathbf{0}\}$. So, $\mathbf{A} - \mathbf{D} = (\mathbf{A} - \mathbf{S}) \oplus (\mathbf{S} - \mathbf{D})$. In view of Remark 3.3.9, we have $\rho(\mathbf{A} - \mathbf{S}) \leq \rho(\mathbf{A} - \mathbf{D})$. Further, the equality sign holds if and only if $\rho(\mathbf{S} - \mathbf{D}) = \mathbf{0}$ or $\mathbf{D} = \mathbf{S}$. □

Let $\mathbf{C} <^- \mathbf{A}$. Does $\mathbf{S}(\mathbf{A}|\mathbf{C})$ exist? If so, what is it? The answer to all these questions is contained in the following:

Theorem 11.2.21. *Let* $\mathbf{A}$ *and* $\mathbf{C}$ *be matrices such that* $\mathbf{C} <^- \mathbf{A}$. *Then* $\mathbf{S}(\mathbf{A}|\mathbf{C})$ *and* $\mathbf{S}(\mathbf{C}|\mathbf{A})$ *exist and are each equal to* $\mathbf{C}$.

Proof. Since $\mathbf{C} <^- \mathbf{A}$, there exist non-singular matrices $\mathbf{P}$, $\mathbf{Q}$ such that

$$\mathbf{C} = \mathbf{P}\mathbf{diag}\left(\mathbf{I_c},\ \mathbf{0},\ \mathbf{0}\right)\mathbf{Q} \text{ and } \mathbf{A} = \mathbf{P}\mathbf{diag}\left(\mathbf{I_c},\ \mathbf{I_{a-c}},\ \mathbf{0}\right)\mathbf{Q},$$

where $c = \rho(\mathbf{C})$ and $a = \rho(\mathbf{A})$. Notice that $\mathbf{A}$ and $\mathbf{C}$ are parallel summable. Write $\mathbf{P} = (\mathbf{P_1}\ :\mathbf{P_2}\ :\mathbf{P_3})$ and $\mathbf{Q^t} = (\mathbf{Q^t}_1\ :\mathbf{Q^t}_2\ :\mathbf{Q^t}_3)$, where '$\mathbf{P_1}$ and $\mathbf{Q^t}_1$' each have c number of columns and '$\mathbf{P_2}$ and $\mathbf{Q^t}_2$' each have $a-c$ number of columns. Now,

$$\mathbf{A}+\mathbf{C} = (\sqrt{2}\mathbf{P_1}\ :\mathbf{P_2}\ :\mathbf{P_3})\mathbf{diag}\left(\mathbf{I_c},\ \mathbf{I_{a-c}},\ \mathbf{0}\right)(\sqrt{2}\mathbf{Q^t}_1\ :\mathbf{Q^t}_2\ :\mathbf{Q^t}_3)^{\mathbf{t}}$$

$$\text{and } \mathbf{C} = (\sqrt{2}\mathbf{P_1}\ :\mathbf{P_2}\ :\mathbf{P_3})\mathbf{diag}\left(\frac{\mathbf{I_c}}{2},\ \mathbf{0},\ \mathbf{0}\right)(\sqrt{2}\mathbf{Q^t}_1\ :\mathbf{Q^t}_2\ :\mathbf{Q^t}_3)^{\mathbf{t}}.$$

Further, both $(\sqrt{2}\mathbf{P_1}\ :\mathbf{P_2}\ :\mathbf{P_3})$ and $(\sqrt{2}\mathbf{Q^t}_1\ :\mathbf{Q^t}_2\ :\mathbf{Q^t}_3)$ are non-singular. Under the notations of Theorem 11.2.11, $\mathbf{I}-\mathbf{D} = \mathbf{diag}\left(\frac{\mathbf{I}}{2}\ ,\mathbf{0}\right)$ and $\mathbf{D} = \mathbf{diag}\left(\frac{\mathbf{I}}{2}\ ,\mathbf{I}\right)$. Clearly each of $\mathbf{D}$ and $\mathbf{I}-\mathbf{D}$ are of index 1. Thus $\mathbf{S}(\mathbf{A}|\mathbf{C})$ and $\mathbf{S}(\mathbf{C}|\mathbf{A})$ exist. Following the proof of Theorem 11.2.11, we can show that $\mathbf{S}(\mathbf{A}|\mathbf{C}) = \mathbf{C}$. Since $\mathbf{C} <^- \mathbf{A}$, using the decompositions of $\mathbf{C}$ and $\mathbf{A}$ as above it follows that $\mathbf{S}(\mathbf{C}|\mathbf{A}) = \mathbf{C}$.

□

Corollary 11.2.22. *Let* $\mathbf{A}$ *and* $\mathbf{B}$ *be matrices of the same order such that* $\mathbf{S}(\mathbf{A}|\mathbf{B})$ *exists. Then* $\mathbf{S}(\mathbf{A}|\mathbf{S}(\mathbf{A}|\mathbf{B}))$ *and* $\mathbf{S}(\mathbf{S}(\mathbf{A}|\mathbf{B})|\mathbf{A})$ *exist and are each equal to* $\mathbf{S}(\mathbf{A}|\mathbf{B})$.

We now obtain the class of all g-inverses of $\mathbf{S}(\mathbf{A}|\mathbf{B})$.

Theorem 11.2.23. *Let* $\mathbf{A}$ *and* $\mathbf{B}$ *be matrices of the same order such that* $\mathbf{S} = \mathbf{S}(\mathbf{A}|\mathbf{B})$ *exists. Then*

$$\{\mathbf{S}(\mathbf{A}|\mathbf{B})^-\} = \{\mathbf{A}^- + \mathbf{X} : \mathbf{A}^- \in \{\mathbf{A}^-\} \text{ and } \mathbf{BXB} = \mathbf{0}\}.$$

Proof. We first note that $\mathbf{S}(\mathbf{A}|\mathbf{B}) <^- \mathbf{A} \Rightarrow \{\mathbf{A}^-\} \subseteq \{\mathbf{S}(\mathbf{A}|\mathbf{B})^-\}$. Let $\mathbf{BXB} = \mathbf{0}$ for some matrix $\mathbf{X}$ of suitable order. Since $\mathcal{C}(\mathbf{S}(\mathbf{A}|\mathbf{B})) \subseteq \mathcal{C}(\mathbf{B})$ and $\mathcal{C}(\mathbf{S}(\mathbf{A}|\mathbf{B})^\star) \subseteq \mathcal{C}(\mathbf{B}^\star)$, it follows that $\mathbf{S}(\mathbf{A}|\mathbf{B})\mathbf{X}\mathbf{S}(\mathbf{A}|\mathbf{B}) = \mathbf{0}$. Hence, $\mathbf{S}(\mathbf{A}|\mathbf{B})(\mathbf{A}^- + \mathbf{X})\mathbf{S}(\mathbf{A}|\mathbf{B}) = \mathbf{S}(\mathbf{A}|\mathbf{B})$ for each $\mathbf{A}^- \in \{\mathbf{A}^-\}$. Thus, $\mathbf{A}^- + \mathbf{X}$ is a g-inverse of $\mathbf{S}(\mathbf{A}|\mathbf{B})$ for each $\mathbf{A}^- \in \{\mathbf{A}^-\}$.

Conversely, let $\mathbf{A}^- + \mathbf{X}$ be a g-inverse of $\mathbf{S}(\mathbf{A}|\mathbf{B})$ for some matrix $\mathbf{X}$ of appropriate order. Then, $\mathbf{S}(\mathbf{A}|\mathbf{B})\mathbf{X}\mathbf{S}(\mathbf{A}|\mathbf{B}) = \mathbf{0}$. As $\mathcal{C}(\mathbf{S}) = \mathcal{C}(\mathbf{A}) \cap \mathcal{C}(\mathbf{B})$, and $\mathcal{C}(\mathbf{S}^\star) = \mathcal{C}(\mathbf{A}^\star) \cap \mathcal{C}(\mathbf{B}^\star)$, it follows that $\mathbf{X} = \mathbf{X_a} + \mathbf{X_b}$, where $\mathbf{A}\mathbf{X_a}\mathbf{A} = \mathbf{0}$ and $\mathbf{B}\mathbf{X_a}\mathbf{B} = \mathbf{0}$. Notice that $\mathbf{A}^- + \mathbf{X_a} \in \{\mathbf{A}^-\}$, so, $\mathbf{A}^- + \mathbf{X} = \mathbf{G} + \mathbf{X_b}$ with $\mathbf{G} \in \{\mathbf{A}^-\}$ and $\mathbf{B}\mathbf{X_b}\mathbf{B} = \mathbf{0}$. Hence, the matrix $\mathbf{A}^- + \mathbf{X}$ belongs to $\{\mathbf{A}^- + \mathbf{X} : \mathbf{A}^- \in \{\mathbf{A}^-\} \text{ and } \mathbf{BXB} = \mathbf{0}\}$ and so,

$$\{\mathbf{S}(\mathbf{A}|\mathbf{B})^-\} = \{\mathbf{A}^- + \mathbf{X} : \mathbf{A}^- \in \{\mathbf{A}^-\} \text{ and } \mathbf{BXB} = \mathbf{0}\}.$$

□

11.3 Rank minimization problem and shorted operator

In Section 10.5 we defined the shorted operator of a matrix over a general field, indexed by some specified subspaces in terms of Schur complements extending the approach followed in Section 10.3 for *nnd* matrices over $\mathbb{C}$. In Section 11.2, we defined the shorted operator of a matrix over $\mathbb{C}$ as the limit of a sequence of parallel sums of suitable matrices. In both the cases the shorted operator whenever it exists has the optimality property of closeness in the following sense: Let $\mathbf{A}$ be an $m \times n$ matrix over a field F. Let V_m and V_n be subspaces of F^m and F^n respectively. Then the shorted operator $\mathbf{S}(\mathbf{A}|V_m, V_n)$, whenever it exists has the property:

$$\mathbf{S}(\mathbf{A}|V_m, V_n) = \max\{\mathbf{D} : \mathbf{D} <^- \mathbf{A},\ \mathcal{C}(\mathbf{D}) \subseteq V_m,\ \mathcal{C}(\mathbf{D^t}) \subseteq V_n\}.$$

In Section 11.2, under the same setup as above, we also obtained another optimality property of closeness for $\mathbf{S}(\mathbf{A}|V_m, V_n)$ namely:

$$\mathbf{S}(\mathbf{A}|V_m,\ V_n) = \arg\min_D\{\rho(\mathbf{A} - \mathbf{D}) : \mathcal{C}(\mathbf{D}) \subseteq V_m,\ \mathcal{C}(\mathbf{D^t}) \subseteq V_n\},$$

whenever it exists. We call this later property as the rank minimization property. When this property is satisfied by a unique matrix, we show that it is equivalent to the definition of the shorted operator via parallel sums and is also equivalent to the definition via optimality property of maximization as mentioned above.

All matrices in this section are over an arbitrary field F unless stated otherwise.

Definition 11.3.1. Let $\mathbf{A}$ be an $m \times n$ matrix. Let V_m and V_n be subspaces of F^m and F^n respectively. Then an $m \times n$ matrix $\mathbf{S}$ is said to have property $(*)$ with respect to the triple $(\mathbf{A}, V_m, V_n)$ if

$$\rho(\mathbf{A} - \mathbf{S}) = \min_D\{\rho(\mathbf{A} - \mathbf{D}) : \mathcal{C}(\mathbf{D}) \subseteq V_m, \mathcal{C}(\mathbf{D^t}) \subseteq V_n\}.$$

In general, $\mathbf{S}$ as defined in Definition 11.3.1 is not unique as the following example shows:

Example 11.3.2. Let $\mathbf{A} = \begin{pmatrix} 1 & 0 & 0 \\ 1 & 1 & 1 \\ 0 & 1 & 0 \end{pmatrix}$. Let $V_m = V_n = \mathcal{C}\begin{pmatrix} 1 & 0 \\ 0 & 1 \\ 0 & 0 \end{pmatrix}$. Now it is easy to see that for arbitrary scalars a and b, the matrix $\mathbf{S_{a,b}} = \begin{pmatrix} 1 & 0 & 0 \\ a & b & 0 \\ 0 & 0 & 0 \end{pmatrix}$ has the property $(*)$ with respect to the triple $(\mathbf{A}, V_m, V_n)$.

When is the matrix $\mathbf{S}$ having property $(*)$ with respect to the the triple $(\mathbf{A}, V_m, V_n)$ as in Definition 11.3.1 unique? Interestingly it turns out that $\mathbf{S}$ is unique if and only if the shorted operator $\mathbf{S}(\mathbf{A}|V_m, V_n)$ as in Section 10.5 exists and is equal to $\mathbf{S}(\mathbf{A}|V_m, V_n)$. We show this in the following theorem:

Theorem 11.3.3. *Let* $\mathbf{A}$ *be an* $m \times n$ *matrix. Let* V_m *and* V_n *be subspaces of* F^m *and* F^n *respectively. Then an* $m \times n$ *matrix* $\mathbf{S}$ *having property* $(*)$ *with respect to the the triple* $(\mathbf{A}, V_m, V_n)$ *exists and is unique if and only if* $\mathbf{S}(\mathbf{A}|V_m, V_n)$ *as defined in Section 10.5 exists. Further,* $\mathbf{S} = \mathbf{S}(\mathbf{A}|V_m, V_n)$.

Proof. 'If' part
Let $\mathbf{S}(\mathbf{A}|V_m, V_n)$ exist. We write $\mathbf{S_1} = \mathbf{S}(\mathbf{A}|V_m, V_n)$. Clearly, $\mathcal{C}(\mathbf{S_1}) \subseteq V_m$ and $\mathcal{C}(\mathbf{S_1^t}) \subseteq V_n$. Now, $\mathbf{S_1} = \mathbf{P_1}(\mathbf{T_{11}} - \mathbf{T_{12}}\mathbf{T_{22}^-}\mathbf{T_{21}})\mathbf{P_3}$, by (10.5.9). It is easy to check that

$$\begin{aligned}\mathbf{A} - \mathbf{S_1} &= (\mathbf{P_1}\mathbf{T_{12}}\mathbf{T_{22}^-} + \mathbf{P_2})(\mathbf{T_{21}}\mathbf{P_3} + \mathbf{T_{22}}\mathbf{P_4}) \\ &= (\mathbf{P_1}\mathbf{T_{12}} + \mathbf{P_2}\mathbf{T_{22}})(\mathbf{P_4} + \mathbf{T_{22}^-}\mathbf{T_{21}}\mathbf{P_3}).\end{aligned}$$

We show that $\mathcal{C}(\mathbf{A} - \mathbf{S_1}) \cap V_m = \{\mathbf{0}\}$. Recall that $V_m = \mathcal{C}(\mathbf{P_1})$. Let $(\mathbf{A} - \mathbf{S_1})\mathbf{x} = \mathbf{P_1}\mathbf{y}$. But $(\mathbf{A} - \mathbf{S_1})\mathbf{x} = (\mathbf{P_1}\mathbf{T_{12}} + \mathbf{P_2}\mathbf{T_{22}})\mathbf{u}$ for some $\mathbf{u}$. So, $\mathbf{P_1}\mathbf{T_{12}}\mathbf{u} + \mathbf{P_2}\mathbf{T_{22}}\mathbf{u} = \mathbf{P_1}\mathbf{y}$ or equivalently, $\mathbf{P_2}\mathbf{T_{22}}\mathbf{u} = \mathbf{P_1}\mathbf{y} - \mathbf{P_1}\mathbf{T_{12}}\mathbf{u} = \mathbf{0}$, since $\mathcal{C}(\mathbf{P_1}) \cap \mathcal{C}(\mathbf{P_2}) = \{\mathbf{0}\}$. Also, $\mathbf{P_2}\mathbf{T_{22}}\mathbf{u} = \mathbf{0} \Rightarrow \mathbf{T_{22}}\mathbf{u} = \mathbf{0}$, as $\mathbf{P_2}$ has a left inverse. Therefore, $\mathbf{T_{12}}\mathbf{u} = \mathbf{0}$, since $\mathbf{S}(\mathbf{A}|V_m, V_n)$ exists. Thus, $(\mathbf{A} - \mathbf{S_1})\mathbf{x} = \mathbf{0}$. Likewise we can prove $\mathcal{C}(\mathbf{A} - \mathbf{S_1})^{\mathbf{t}} \cap V_n = \{\mathbf{0}\}$.
Now, let $\mathbf{D}$ be a matrix such that $\mathcal{C}(\mathbf{D}) \subseteq V_m$ and $\mathcal{C}(\mathbf{D^t}) \subseteq V_n$. Write $\mathbf{A} - \mathbf{D} = \mathbf{A} - \mathbf{S_1} + \mathbf{S_1} - \mathbf{D}$. Since $\mathcal{C}(\mathbf{S_1}) \subseteq V_m$ and $\mathcal{C}(\mathbf{D}) \subseteq V_m$, we have $\mathcal{C}(\mathbf{S_1} - \mathbf{D}) \subseteq V_m$. So, $\mathcal{C}(\mathbf{A} - \mathbf{S_1}) \cap \mathcal{C}(\mathbf{S_1} - \mathbf{D}) = \{\mathbf{0}\}$. Similarly, $\mathcal{C}(\mathbf{A} - \mathbf{S_1})^{\mathbf{t}} \cap \mathcal{C}(\mathbf{S_1} - \mathbf{D})^{\mathbf{t}} = \{\mathbf{0}\}$. Hence, $\rho(\mathbf{A} - \mathbf{D}) = \rho(\mathbf{A} - \mathbf{S_1}) + \rho(\mathbf{S_1} - \mathbf{D})$. It follows that $\rho(\mathbf{A} - \mathbf{S_1}) \leq \rho(\mathbf{A} - \mathbf{D})$ and the equality holds if and only if $\mathbf{S_1} = \mathbf{D}$. Thus, $\mathbf{S_1}$ is the unique matrix satisfying the property $(*)$.
'Only if' part
Suppose there exists a unique matrix $\mathbf{S}$ satisfying the property $(*)$ with respect to the triple $(\mathbf{A}, V_m, V_n)$. Then $\mathcal{C}(\mathbf{A} - \mathbf{S}) \cap V_m = \{\mathbf{0}\}$ and $\mathcal{C}(\mathbf{A} - \mathbf{S})^t \cap V_n = \{\mathbf{0}\}$. Let $(\mathbf{L_{11}}, \mathbf{R_{11}})$ and $(\mathbf{L_{21}}, \mathbf{R_{21}})$ be rank factorizations of $\mathbf{S}$ and $\mathbf{A} - \mathbf{S}$ respectively. We clearly have, $\mathcal{C}(\mathbf{L_{11}}) \cap \mathcal{C}(\mathbf{L_{21}}) = \{\mathbf{0}\}$ and $\mathcal{C}(\mathbf{R_{11}^t}) \cap \mathcal{C}(\mathbf{R_{21}^t}) = \{\mathbf{0}\}$. Let the columns of $\mathbf{L_1} = \begin{pmatrix}\mathbf{L_{11}} : \mathbf{L_{12}}\end{pmatrix}$ form a basis for V_m and the rows of $\mathbf{R_1} = \begin{pmatrix}\mathbf{R_{11}} \\ \mathbf{R_{21}}\end{pmatrix}$ form a basis of V_n. Let $\mathbf{L_2}$ and $\mathbf{R_2}$ be matrices such that $\begin{pmatrix}\mathbf{L_1} : \mathbf{L_2}\end{pmatrix}$ and $\begin{pmatrix}\mathbf{R_1} \\ \mathbf{R_2}\end{pmatrix}$ are non-singular. Then $\mathbf{S} = \mathbf{L_1}\mathbf{T_{11}}\mathbf{R_1}$ and

$\mathbf{A}-\mathbf{S} = \mathbf{L_2T_{22}R_2}$. Thus, $\mathbf{A} = (\mathbf{L_1} : \mathbf{L_2})\begin{pmatrix}\mathbf{T_{11}} & \mathbf{0}\\ \mathbf{0} & \mathbf{T_{22}}\end{pmatrix}\begin{pmatrix}\mathbf{R_1}\\ \mathbf{R_2}\end{pmatrix}$. Clearly, $\mathbf{S}(\mathbf{A}|V_m, V_n)$ exists by Theorem 10.5.4 and is equal to $\mathbf{S}$. □

Thus, there is a unique matrix $\mathbf{S}$ having the property $(*)$ with respect to the triple $(\mathbf{A}, V_m, V_n)$ if and only if $\mathbf{S}(\mathbf{A}|V_m, V_n)$ exists and in such a case $\mathbf{S} = \mathbf{S}(\mathbf{A}|V_m, V_n)$. Henceforth, we consider the situation when $\mathbf{S}$ in Definition 11.3.1 exists and is unique. We identify $\mathbf{S}$ with $\mathbf{S}(\mathbf{A}|V_m, V_n)$.

Given a matrix $\mathbf{A}$ and subspaces V_m and V_n of F^m and F^n respectively, let $\mathbf{B}$ be a matrix such that $\mathcal{C}(\mathbf{B}) = \mathbf{V_m}$ and $\mathcal{C}(\mathbf{B^t}) = \mathbf{V_n}$. Our next object is to show that $\mathbf{S}(\mathbf{A}|\mathbf{B})$ as defined in Section 11.2 and $\mathbf{S}(\mathbf{A}|V_m, V_n)$ as defined above are the same when matrices are over the complex field $\mathbb{C}$.

Theorem 11.3.4. *Let* $\mathbf{A}$ *and* $\mathbf{B}$ *be matrices of the same order over a field* F *of characteristic* 0. *Let* $(\mathbf{P_1}, \mathbf{P_3})$ *be a rank factorization of* $\mathbf{B}$ *and let* $\mathbf{R} = \begin{pmatrix}\mathbf{A} & \mathbf{P_1}\\ \mathbf{P_3} & \mathbf{0}\end{pmatrix}$. *Let* $\mathbf{P_2}$ *and* $\mathbf{P_4}$ *be matrices such that* $(\mathbf{P_1} : \mathbf{P_2})$ *and* $\begin{pmatrix}\mathbf{P_3}\\ \mathbf{P_4}\end{pmatrix}$ *are non-singular matrices and* $\mathbf{A} = (\mathbf{P_1} : \mathbf{P_2})\begin{pmatrix}\mathbf{T_{11}} & \mathbf{T_{12}}\\ \mathbf{T_{21}} & \mathbf{T_{22}}\end{pmatrix}\begin{pmatrix}\mathbf{P_3}\\ \mathbf{P_4}\end{pmatrix}$. *Then the following are equivalent:*

(i) *There exists a non-zero scalar* c *such that* $\mathbf{A}$ *and* $c\mathbf{B}$ *are parallel summable and* $\rho(\mathbf{B}(\mathbf{A}+\mathbf{B})^-\mathbf{B}) = \rho(\mathbf{B})$

(ii) $\rho(\mathbf{R}) = \rho(\mathbf{A} : \mathbf{P_1}) + \rho(\mathbf{P_3}) = \rho\begin{pmatrix}\mathbf{A}\\ \mathbf{P_3}\end{pmatrix} + \rho(\mathbf{P_1})$

(iii) $\mathcal{C}\begin{pmatrix}\mathbf{A}\\ \mathbf{0}\end{pmatrix} \subseteq \mathcal{C}(\mathbf{R})$, $\mathcal{C}\begin{pmatrix}\mathbf{A^t}\\ \mathbf{0}\end{pmatrix} \subseteq \mathcal{C}(\mathbf{R^t})$ *and*

(iv) $\mathcal{C}(\mathbf{T_{21}}) \subseteq \mathcal{C}(\mathbf{T_{22}})$ *and* $\mathcal{C}(\mathbf{T_{12}^t}) \subseteq \mathcal{C}(\mathbf{T_{22}^t})$.

Proof. (i) ⇒ (ii)

Since $\mathbf{A}$ and $c\mathbf{B}$ are parallel summable, we have $\mathbf{M} = \begin{pmatrix}\mathbf{A}+\mathbf{cB} & \mathbf{P_1}\\ \mathbf{P_3} & \mathbf{0}\end{pmatrix} = \begin{pmatrix}\mathbf{I} & \mathbf{0}\\ \mathbf{P_3}(\mathbf{A}+\mathbf{cB})^- & \mathbf{0}\end{pmatrix}\begin{pmatrix}\mathbf{A}+\mathbf{cB} & \mathbf{0}\\ \mathbf{0} & \mathbf{P_3}(\mathbf{A}+\mathbf{cB})^-\mathbf{P_1}\end{pmatrix}\begin{pmatrix}\mathbf{I} & (\mathbf{A}+\mathbf{cB})^-\mathbf{P_1}\\ \mathbf{0} & \mathbf{I}\end{pmatrix}$.

So,

$$\begin{aligned}\rho(\mathbf{M}) &= \rho(\mathbf{A}+\mathbf{cB}) + \rho(\mathbf{P_3}(\mathbf{A}+\mathbf{cB})^-\mathbf{P_1})\\ &= \rho(\mathbf{A}+\mathbf{cB}) + \rho(\mathbf{B}(\mathbf{A}+\mathbf{cB})^-\mathbf{B}) = \rho(\mathbf{A}+\mathbf{cB}) + \rho(\mathbf{B}).\end{aligned}$$

However,

$$\mathbf{M}\begin{pmatrix}\mathbf{I} & \mathbf{0}\\ -c\mathbf{P_3} & \mathbf{I}\end{pmatrix} = \begin{pmatrix}\mathbf{A} & \mathbf{P_1}\\ \mathbf{P_3} & \mathbf{0}\end{pmatrix} = \mathbf{R}.$$

Hence, $\rho(\mathbf{R}) = \rho(\mathbf{M}) = \rho(\mathbf{A}+\mathbf{cB}) + \rho(\mathbf{B})$. Since $\mathbf{A}$ and $c\mathbf{B}$ are parallel summable, $\rho(\mathbf{A}+\mathbf{cB}) = \rho(\mathbf{A}:\mathbf{B}) = \begin{pmatrix}\mathbf{A}\\ \mathbf{B}\end{pmatrix} = \rho(\mathbf{A}:\mathbf{P_1}) = \begin{pmatrix}\mathbf{A}\\ \mathbf{P_3}\end{pmatrix}$. Also, $\rho(\mathbf{B}) = \rho(\mathbf{P_1}) = \rho(\mathbf{P_3})$. Thus,

$$\rho(\mathbf{R}) = \rho(\mathbf{A}:\mathbf{P_1}) + \rho(\mathbf{P_3}) = \rho\begin{pmatrix}\mathbf{A}\\ \mathbf{P_3}\end{pmatrix} + \rho(\mathbf{P_1}).$$

(ii) $\Rightarrow$ (i)

Since (ii) holds, we have $\rho\begin{pmatrix}\mathbf{A} & \mathbf{B}\\ \mathbf{B} & \mathbf{0}\end{pmatrix} = \rho\begin{pmatrix}\mathbf{A}\\ \mathbf{B}\end{pmatrix} + \rho(\mathbf{B}) = \rho(\mathbf{A}:\mathbf{B}) + \rho(\mathbf{B})$. Thus, $\begin{pmatrix}\mathbf{A} & \mathbf{B}\\ \mathbf{0} & \mathbf{0}\end{pmatrix} <^- \begin{pmatrix}\mathbf{A} & \mathbf{B}\\ \mathbf{B} & \mathbf{0}\end{pmatrix}$ and $\begin{pmatrix}\mathbf{A} & \mathbf{0}\\ \mathbf{B} & \mathbf{0}\end{pmatrix} <^- \begin{pmatrix}\mathbf{A} & \mathbf{B}\\ \mathbf{B} & \mathbf{0}\end{pmatrix}$. Let $\begin{pmatrix}\mathbf{F}^1 & \mathbf{F}^2\\ \mathbf{F}^3 & \mathbf{F}^4\end{pmatrix}$ be a g-inverse of $\begin{pmatrix}\mathbf{A} & \mathbf{B}\\ \mathbf{B} & \mathbf{0}\end{pmatrix}$. Then

$$\mathbf{B} = \mathbf{BF^2B} = \mathbf{BF^3B},\ \mathbf{AF^2B} = \mathbf{BF^4B} = \mathbf{BF^3A}.$$

Choose a scalar c such that $\mathbf{F^4B}+\mathbf{cI}$ is non-singular. We claim that $\mathbf{A}-\mathbf{BF^4B}$ and $\mathbf{B}$ are disjoint. Let $\mathbf{x} \in \mathcal{C}(\mathbf{A}-\mathbf{BF^4B}) \cap \mathcal{C}(\mathbf{B})$. Then there exist vectors $\mathbf{u}$ and $\mathbf{v}$ such that $\mathbf{x} = (\mathbf{A}-\mathbf{BF^4B})\mathbf{u} = \mathbf{Bv}$. Pre-multiplying by $\mathbf{BF^3}$, we have $\mathbf{BF^3x} = \mathbf{BF^3}(\mathbf{A}-\mathbf{BF^4B})\mathbf{u} = \mathbf{BF^3Bv}$. Since $\mathbf{B} = \mathbf{BF^3B}$, $\mathbf{BF^4B} = \mathbf{BF^3A}$, we have $\mathbf{Bv} = \mathbf{0}$. Therefore, $\mathbf{x} = \mathbf{0}$, so, $\mathcal{C}(\mathbf{A}-\mathbf{BF^4B}) \cap \mathcal{C}(\mathbf{B}) = \{\mathbf{0}\}$. Similarly, $\mathcal{C}(\mathbf{A}-\mathbf{BF^4B})^{\mathbf{t}} \cap \mathcal{C}(\mathbf{B^t}) = \{\mathbf{0}\}$. So, $\mathbf{A}-\mathbf{BF^4B}$ and $\mathbf{B}$ are disjoint and therefore, $\mathbf{A}-\mathbf{BF^4B}$ and $\mathbf{BF^4B}$ are disjoint. Thus,

$$\begin{aligned}\mathcal{C}(\mathbf{A}) &= \mathcal{C}(\mathbf{A}-\mathbf{BF^4B}) + \mathcal{C}(\mathbf{BF^4B})\\ &\subseteq \mathcal{C}(\mathbf{A}-\mathbf{BF^4B}) + \mathcal{C}(\mathbf{cB}+\mathbf{BF^4B})\\ &= \mathcal{C}(\mathbf{A}+\mathbf{cB}).\end{aligned}$$

Similarly, $\mathcal{C}(\mathbf{A^t}) \subseteq \mathcal{C}(\mathbf{A}+\mathbf{cB})^{\mathbf{t}}$. Hence, $\mathbf{A}$ and $c\mathbf{B}$ are parallel summable for some non-null scalar c. The proof of $\rho(\mathbf{B}(\mathbf{A}+\mathbf{B})^-\mathbf{B}) = \rho(\mathbf{B})$ follows from the proof of (i) $\Rightarrow$ (ii).

(ii) $\Rightarrow$ (iii)

Since $\rho(\mathbf{R}) = \rho(\mathbf{A}:\mathbf{P_1}) + \rho(\mathbf{P_3}) = \rho\begin{pmatrix}\mathbf{A}\\ \mathbf{P_3}\end{pmatrix} + \rho(\mathbf{P_1})$, we have

$$\begin{pmatrix}\mathbf{A} & \mathbf{P_1}\\ \mathbf{0} & \mathbf{0}\end{pmatrix} <^- \mathbf{R} \text{ and } \begin{pmatrix}\mathbf{A} & \mathbf{0}\\ \mathbf{P_3} & \mathbf{0}\end{pmatrix} <^- \mathbf{R}.$$

Hence, $\mathcal{C}\begin{pmatrix}\mathbf{A}\\ \mathbf{0}\end{pmatrix} \subseteq \mathcal{C}(\mathbf{R})$ and $\mathcal{C}\begin{pmatrix}\mathbf{A^t}\\ \mathbf{0}\end{pmatrix} \subseteq \mathcal{C}(\mathbf{R^t})$.

(iii) $\Rightarrow$ (iv) Let $\mathbf{P}^1$ be a left inverse of $\mathbf{P_1}$ and $\mathbf{P}^3$ be a right inverse of $\mathbf{P_3}$. Then it is easy to see that $\begin{pmatrix} \mathbf{0} & \mathbf{P^3} \\ \mathbf{P^1} & -\mathbf{P^1AP^3} \end{pmatrix}$ is a g-inverse of $\mathbf{R}$. Since $\mathcal{C}\begin{pmatrix} \mathbf{A} \\ \mathbf{0} \end{pmatrix} \subseteq \mathcal{C}(\mathbf{R})$, $\mathbf{A} = \mathbf{P_1P^1A}$. Thus, $\mathcal{C}(\mathbf{A}) \subseteq \mathcal{C}(\mathbf{P_1})$. Now, $\mathbf{A} = \mathbf{P_1}(\mathbf{T_{11}P_3} + \mathbf{T_{12}P_4}) + \mathbf{P_2}(\mathbf{T_{21}P_3} + \mathbf{T_{22}P_4})$. Pre-multiplying by $\mathbf{P_1P^1}$ and using $\mathbf{A} = \mathbf{P_1P^1A}$, we have $\mathbf{P_2}(\mathbf{T_{21}P_3} + \mathbf{T_{22}P_4}) = \mathbf{0}$ or equivalently $\mathbf{T_{21}P_3} + \mathbf{T_{22}P_4} = \mathbf{0}$, since $\mathbf{P_2}$ has a left inverse. So, $\mathbf{T_{21}} = -\mathbf{T_{22}P_4P^3}$. This shows that $\mathcal{C}(\mathbf{T_{21}}) \subseteq \mathcal{C}(\mathbf{T_{22}})$. Similarly, we can show that $\mathcal{C}(\mathbf{T_{12}^t}) \subseteq \mathcal{C}(\mathbf{T_{22}^t})$.

(iv) $\Rightarrow$ (ii)

As noted earlier in Theorem 11.3.3, we have

$$\mathbf{A} = \mathbf{P_1}(\mathbf{T_{11}} - \mathbf{T_{12}T_{22}^{-}T_{21}})\mathbf{P_3} + (\mathbf{P_1T_{12}} + \mathbf{P_2T_{22}})(\mathbf{P_4} + \mathbf{T_{22}^{-}T_{21}P_3}).$$

We now show that $\mathcal{C}\begin{pmatrix} \mathbf{A} \\ \mathbf{P_3} \end{pmatrix} \cap \mathcal{C}\begin{pmatrix} \mathbf{P_1} \\ \mathbf{0} \end{pmatrix} = \{\mathbf{0}\}$ and $\mathcal{C}\begin{pmatrix} \mathbf{A} \\ \mathbf{P_3} \end{pmatrix}^t \cap \mathcal{C}\begin{pmatrix} \mathbf{P_1} \\ \mathbf{0} \end{pmatrix}^t = \{\mathbf{0}\}$.

Let $\begin{pmatrix} \mathbf{A} \\ \mathbf{P_3} \end{pmatrix}\mathbf{Y} = \begin{pmatrix} \mathbf{P_1} \\ \mathbf{0} \end{pmatrix}\mathbf{Z}$. Then $\mathbf{AY} = \mathbf{P_1Z}$ and $\mathbf{P_3Y} = \mathbf{0}$. Now,

$$\mathbf{P_3Y} = \mathbf{0} \Rightarrow \mathbf{AY} = (\mathbf{P_1T_{12}} + \mathbf{P_2T_{22}})(\mathbf{P_4Y}) = \mathbf{P_1Z}.$$

Since $\mathbf{P_2}$ has a left inverse and $\mathbf{Z} = \mathbf{T_{12}P_4Y}$, we obtain $\mathbf{T_{22}}(\mathbf{P_4Y}) = \mathbf{0}$. Since $\mathcal{C}(\mathbf{T_{12}^t}) \subseteq \mathcal{C}(\mathbf{T_{22}^t})$, so, $\mathbf{Z} = \mathbf{0}$. Thus, $\mathcal{C}\begin{pmatrix} \mathbf{A} \\ \mathbf{P_3} \end{pmatrix} \cap \mathcal{C}\begin{pmatrix} \mathbf{P_1} \\ \mathbf{0} \end{pmatrix} = \{\mathbf{0}\}$.

Similarly, $\mathcal{C}\begin{pmatrix} \mathbf{A} \\ \mathbf{P_3} \end{pmatrix}^t \cap \mathcal{C}\begin{pmatrix} \mathbf{P_1} \\ \mathbf{0} \end{pmatrix}^t = \{\mathbf{0}\}$. Hence, $\rho(\mathbf{R}) = \rho\begin{pmatrix} \mathbf{A} \\ \mathbf{P_3} \end{pmatrix} + \rho(\mathbf{P_1})$.
Similarly, we can prove that $\rho(\mathbf{R}) = \rho(\mathbf{A} : \mathbf{P_1}) + \rho(\mathbf{P_3})$. □

A few remarks are in order.

Remark 11.3.5. Let V_m and V_n be subspaces of F^m and F^n respectively. Let $\mathbf{X}$ and $\mathbf{Y}$ be matrices such that $V_m = \mathcal{C}(\mathbf{X})$ and $V_n = \mathcal{C}(\mathbf{Y^t})$. Then it can be easily established that (ii)-(iv) of Theorem 11.3.4 are equivalent if we replace $\mathbf{P_1}$ and $\mathbf{P_3}$ by $\mathbf{X}$ and $\mathbf{Y}$ respectively. Thus, the definitions of $\mathbf{S}(\mathbf{A}|V_m, V_n)$ as in Section 10.5 and the one in this section are equivalent.

Remark 11.3.6. Let $\mathbf{A}$ and $\mathbf{B}$ be matrices of the same order over a field of characteristic 0. Let $V_m = \mathcal{C}(\mathbf{B})$ and $V_n = \mathcal{C}(\mathbf{B^t})$. We proved that $\mathbf{S}(\mathbf{A}|V_m,\ V_n)$ exists if and only if $\mathbf{A}$ and $c\mathbf{B}$ are parallel summable for some scalar c and $\rho(\mathbf{B}(\mathbf{A}+\mathbf{B})^{-}\mathbf{B}) = \rho(\mathbf{B})$. As seen in Section 11.2, when these condition hold, $\mathbf{S}(\mathbf{A}|\mathbf{B}) = \mathbf{LQdiag}(\mathbf{D},\ \mathbf{0})\mathbf{Q^{-1}R}$, where $(\mathbf{L}, \mathbf{R})$ is a rank factorization of $\mathbf{A}+\mathbf{B}$, $\mathbf{A} = \mathbf{LDR}$ and $\mathbf{I}-\mathbf{D} = \mathbf{Qdiag}(\mathbf{I}-\mathbf{D_1},\ \mathbf{0})\mathbf{Q^{-1}}$ for some non-singular matrices $\mathbf{Q}$ and $\mathbf{I}-\mathbf{D_1}$.

Remark 11.3.7. The definitions of a shorted operator as given in Sections 10.5, 11.2 and 11.3 all coincide when matrices are taken over the field of the complex numbers.

Remark 11.3.8. The definition of the shorted operator $\mathbf{S}(\mathbf{A}|\mathbf{B})$ as the limit of a sequence of parallel sums appears to be more restrictive than that of $\mathbf{S}(\mathbf{A}|V_m, V_n)$ for the following reason: In definition of $\mathbf{S}(\mathbf{A}|\mathbf{B})$, the spaces $\mathcal{C}(\mathbf{B})$ and $\mathcal{C}(\mathbf{B}^t)$ are necessarily of the same dimension where as V_m and V_n can be of different dimensions.

Remark 11.3.9. The statements (i)-(iv) in Theorem 11.3.4 are also equivalent over a general field.

11.4 Computation of shorted operator

We gave quite a few definitions of shorted operator in this chapter as well as the previous one and proved their equivalence in the last section. (See Theorem 11.3.4, Remarks 11.3.5 and 11.3.6.) We investigated the conditions under which the shorted operator exists and provided some formulae for computing the shorted operator whenever it exists. In this section, we first develop one more method for computing the shorted operator and then briefly compare the various different computational methods. We first prove the following useful lemma which is also of independent interest:

Lemma 11.4.1. *Let* $\mathbf{A}$ *be an* $m \times n$ *matrix over a field* F *and* V_m *and* V_n *be subspaces of* F^m *and* F^n *respectively. Let* $\mathbf{X}$ *and* $\mathbf{Y}$ *be matrices such that* $V_m = \mathcal{C}(\mathbf{X})$ *and* $V_n = \mathcal{C}(\mathbf{Y}^t)$. *Denote by* $\mathbf{R}$, *the matrix* $\begin{pmatrix} \mathbf{A} & \mathbf{X} \\ \mathbf{Y} & \mathbf{0} \end{pmatrix}$. *Let* $\mathbf{G} = \begin{pmatrix} \mathbf{G}_{11} & \mathbf{G}_{12} \\ \mathbf{G}_{21} & -\mathbf{G}_{22} \end{pmatrix}$ *be a g-inverse of* $\mathbf{R}$. *If*

$$\mathcal{C}\begin{pmatrix} \mathbf{A} \\ \mathbf{0} \end{pmatrix} \subseteq \mathcal{C}(\mathbf{R}) \text{ and } \mathcal{C}\begin{pmatrix} \mathbf{A}^t \\ \mathbf{0} \end{pmatrix} \subseteq \mathcal{C}(\mathbf{R}^t),$$

then the following hold:

(i) $\mathbf{X}\mathbf{G}_{21}\mathbf{X} = \mathbf{X}, \ \mathbf{Y}\mathbf{G}_{12}\mathbf{Y} = \mathbf{Y}$.

(ii) $\mathbf{Y}\mathbf{G}_{11}\mathbf{X} = \mathbf{0}, \ \mathbf{A}\mathbf{G}_{11}\mathbf{X} = \mathbf{0}$ *and* $\mathbf{Y}\mathbf{G}_{11}\mathbf{A} = \mathbf{0}$.

(iii) $\mathbf{A}\mathbf{G}_{12}\mathbf{Y} = \mathbf{X}\mathbf{G}_{21}\mathbf{A} = \mathbf{X}\mathbf{G}_{22}\mathbf{Y}$.

(iv) $\mathbf{A}\mathbf{G}_{11}\mathbf{A}\mathbf{G}_{11}\mathbf{A} = \mathbf{A}\mathbf{G}_{11}\mathbf{A}$, $\operatorname{tr}(\mathbf{A}\mathbf{G}_{11}) = \rho(\mathbf{A} : \mathbf{X}) - \rho(\mathbf{X}) = \rho\begin{pmatrix} \mathbf{A} \\ \mathbf{Y} \end{pmatrix} - \rho(\mathbf{Y})$.

(v) $\begin{pmatrix} \mathbf{G}_{11} \\ \mathbf{G}_{21} \end{pmatrix}$ *and* $(\mathbf{G}_{11} : \mathbf{G}_{21})$ *are g-inverses of* $(\mathbf{A} : \mathbf{X})$ *and* $\begin{pmatrix} \mathbf{A} \\ \mathbf{Y} \end{pmatrix}$ *respectively.*

Proof. We first notice that $\mathcal{C}\begin{pmatrix} \mathbf{A} \\ \mathbf{0} \end{pmatrix} \subseteq \mathcal{C}(\mathbf{R}) \Rightarrow \mathbf{RG}\begin{pmatrix} \mathbf{A} \\ \mathbf{0} \end{pmatrix} = \begin{pmatrix} \mathbf{A} \\ \mathbf{0} \end{pmatrix}$ and $\mathcal{C}\begin{pmatrix} \mathbf{A}^t \\ \mathbf{0} \end{pmatrix} \subseteq \mathcal{C}(\mathbf{R}^t) \Rightarrow (\mathbf{A} : \mathbf{0})\mathbf{GR} = (\mathbf{A} : \mathbf{0})$. Further,

$$\mathbf{RG}\begin{pmatrix} \mathbf{A} \\ \mathbf{0} \end{pmatrix} = \begin{pmatrix} \mathbf{A} \\ \mathbf{0} \end{pmatrix} \Rightarrow \begin{cases} \mathbf{AG}_{11}\mathbf{A} + \mathbf{XG}_{21}\mathbf{A} = \mathbf{A} & (11.4.1) \\ \mathbf{YG}_{11}\mathbf{A} = \mathbf{0}. & (11.4.2) \end{cases}$$

$$(\mathbf{A} : \mathbf{0})\mathbf{GR} = (\mathbf{A} : \mathbf{0}) \Rightarrow \begin{cases} \mathbf{AG}_{11}\mathbf{A} + \mathbf{AG}_{12}\mathbf{Y} = \mathbf{A} & (11.4.3) \\ \mathbf{AG}_{11}\mathbf{X} = \mathbf{0}. & (11.4.4) \end{cases}$$

Also, $\mathbf{RGR} = \mathbf{R}$

$$\Rightarrow \begin{cases} \mathbf{AG}_{11}\mathbf{A} + \mathbf{XG}_{21}\mathbf{A} + \mathbf{AG}_{12}\mathbf{Y} - \mathbf{XG}_{22}\mathbf{Y} = \mathbf{A} & (11.4.5) \\ \mathbf{AG}_{11}\mathbf{X} + \mathbf{XG}_{21}\mathbf{X} = \mathbf{X} & (11.4.6) \\ \mathbf{YG}_{11}\mathbf{A} + \mathbf{YG}_{21}\mathbf{Y} = \mathbf{Y} & (11.4.7) \\ \mathbf{YG}_{11}\mathbf{X} = \mathbf{0}. & (11.4.8) \end{cases}$$

Now, (i) follows from (11.4.2), (11.4.4), (11.4.6) and (11.4.7), (ii) follows from (11.4.2), (11.4.4) and (11.4.8) and (iii) follows from (11.4.1), (11.4.3) and (11.4.5).
For (iv), (11.4.1) and (11.4.4) imply $\mathbf{AG}_{11}\mathbf{AG}_{11}\mathbf{A} = \mathbf{AG}_{11}\mathbf{A}$. Further,

$$\begin{aligned} \rho(\mathbf{R}) &= \mathrm{tr}(\mathbf{RG}) = \mathrm{tr}(\mathbf{AG}_{11}) + \mathrm{tr}(\mathbf{XG}_{21}) + \mathrm{tr}(\mathbf{YG}_{12}) \\ &= \mathrm{tr}(\mathbf{AG}_{11}) + \rho(\mathbf{X}) + \rho(\mathbf{Y}). \end{aligned}$$

Hence,

$$\mathrm{tr}(\mathbf{AG}_{11}) = \rho(\mathbf{R}) - \rho(\mathbf{X}) - \rho(\mathbf{Y}) = \rho(\mathbf{A} : \mathbf{X}) - \rho(\mathbf{X}) = \rho\begin{pmatrix} \mathbf{A} \\ \mathbf{Y} \end{pmatrix} - \rho(\mathbf{Y}),$$

by using (ii) $\Leftrightarrow$ (iii) of Theorem 11.3.4 and Remark 11.3.5.
(v) follows from (11.4.1), (11.4.3), (11.4.6) and (11.4.7). □

We are now ready to obtain the shorted operator $\mathbf{S}(\mathbf{A}|V_m, V_n)$.

Theorem 11.4.2. *Consider the same setup as in Lemma 11.4.1. Then the matrix* $\mathbf{AG}_{12}\mathbf{Y} = \mathbf{XG}_{21}\mathbf{A} = \mathbf{XG}_{22}\mathbf{Y}$ *is the shorted operator* $\mathbf{S}(\mathbf{A}|V_m, V_n)$.

Proof. Let us write $\mathbf{S} = \mathbf{AG}_{12}\mathbf{Y} = \mathbf{XG}_{21}\mathbf{A} = \mathbf{XG}_{22}\mathbf{Y}$. Clearly, we have $\mathcal{C}(\mathbf{S}) \subseteq \mathcal{C}(\mathbf{X}) = V_m$ and $\mathcal{C}(\mathbf{S}^t) \subseteq \mathcal{C}(\mathbf{Y}^t) = V_n$. We next show that

$$\mathcal{C}(\mathbf{A} - \mathbf{S}) \cap V_m = \{\mathbf{0}\} \text{ and } \mathcal{C}(\mathbf{A} - \mathbf{S})^t \cap V_n = \{\mathbf{0}\}.$$

We have from (11.4.1), $\mathbf{A} - \mathbf{S} = \mathbf{A}\mathbf{G}_{11}\mathbf{A}$. Let $\mathbf{A}\mathbf{G}_{11}\mathbf{A}\mathbf{u} = \mathbf{X}\mathbf{v}$ for some vectors $\mathbf{u}$ and $\mathbf{v}$. Then by Lemma 11.4.1 (iv) and (ii), $\mathbf{A}\mathbf{G}_{11}\mathbf{A}\mathbf{G}_{11}\mathbf{A}\mathbf{u} = \mathbf{A}\mathbf{G}_{11}\mathbf{X}\mathbf{v} = \mathbf{0}$. So, $\mathbf{A}\mathbf{G}_{11}\mathbf{A}\mathbf{G}_{11}\mathbf{A}\mathbf{u} = \mathbf{A}\mathbf{G}_{11}\mathbf{A}\mathbf{u} = \mathbf{X}\mathbf{v} = \mathbf{0}$. Hence, $\mathcal{C}(\mathbf{A} - \mathbf{S}) \cap V_m = \mathcal{C}(\mathbf{A} - \mathbf{S}) \cap \mathcal{C}(\mathbf{X}) = \{\mathbf{0}\}$.
Similarly, we can show that $\mathcal{C}(\mathbf{A} - \mathbf{S})^{\mathbf{t}} \subseteq \mathcal{C}(\mathbf{Y}^{\mathbf{t}}) = \{\mathbf{0}\}$.

Let $\mathbf{E}$ be a matrix such that $\mathcal{C}(\mathbf{E}) \subseteq V_m$ and $\mathcal{C}(\mathbf{E}^{\mathbf{t}}) \subseteq V_n$. We can write $\mathbf{A} - \mathbf{E} = \mathbf{A} - \mathbf{S} + \mathbf{S} - \mathbf{E}$. Since $\mathcal{C}(\mathbf{S} - \mathbf{E}) \subseteq V_m$ and $\mathcal{C}(\mathbf{S} - \mathbf{E})^{\mathbf{t}} \subseteq V_n$, it follows that $\mathcal{C}(\mathbf{A} - \mathbf{S}) \cap \mathcal{C}(\mathbf{S} - \mathbf{E}) = \{\mathbf{0}\}$ and $\mathcal{C}(\mathbf{A} - \mathbf{S})^{\mathbf{t}} \cap \mathcal{C}(\mathbf{S} - \mathbf{E})^{\mathbf{t}} = \{\mathbf{0}\}$. Hence, $\rho(\mathbf{A} - \mathbf{E}) \geq \rho(\mathbf{A} - \mathbf{S})$ and the equality holds if and only if $\mathbf{E} = \mathbf{S}$. Thus, by Definition 11.3.1, $\mathbf{S} = \mathbf{S}(\mathbf{A}|V_m, V_n)$. □

We now discuss different methods of computing $\mathbf{S}(\mathbf{A}|V_m, V_n)$.

Let $\mathbf{A}$ be an $m \times n$ matrix over a field F and V_m and V_n be subspaces of F^m and F^n respectively. Let $\mathbf{P_1}$ be matrix whose columns form a basis of V_m and $\mathbf{P_3}$ be a matrix rows of which form a basis of V_n and W_m and W_n denote complements of V_m and V_n in F^m and F^n respectively. Let the columns of a matrix $\mathbf{P_2}$ for a basis of W_m and the rows of a matrix $\mathbf{P_4}$ form a basis for W_n. Let $\mathbf{X}$ and $\mathbf{Y}$ be matrices such that $V_m = \mathcal{C}(\mathbf{X})$ and $V_n = \mathcal{C}(\mathbf{Y}^{\mathbf{t}})$. Denote by $\mathbf{R}$, the matrix $\begin{pmatrix} \mathbf{A} & \mathbf{X} \\ \mathbf{Y} & \mathbf{0} \end{pmatrix}$ and let $\mathbf{G} = \begin{pmatrix} \mathbf{G}_{11} & \mathbf{G}_{12} \\ \mathbf{G}_{21} & -\mathbf{G}_{22} \end{pmatrix}$ be a g-inverse of $\mathbf{R}$. Write $(\mathbf{P_1} : \mathbf{P_2})^{-1} = \begin{pmatrix} \mathbf{P}^1 \\ \mathbf{P}^2 \end{pmatrix}$ and $\begin{pmatrix} \mathbf{P_3} \\ \mathbf{P_4} \end{pmatrix}^{-1} = (\mathbf{P}^3 : \mathbf{P}^4)$.

Method 11.4.3. Check whether $\mathbf{R}\mathbf{G}\begin{pmatrix} \mathbf{A} \\ \mathbf{0} \end{pmatrix} = \begin{pmatrix} \mathbf{A} \\ \mathbf{0} \end{pmatrix}$ and $(\mathbf{A} : \mathbf{0})\mathbf{G}\mathbf{R} = (\mathbf{A} : \mathbf{0})$. If yes, the shorted operator $\mathbf{S}(\mathbf{A}|V_m, V_n)$ exists and is given by $\mathbf{S}(\mathbf{A}|V_m, V_n) = \mathbf{X}\mathbf{G}_{11}\mathbf{Y} = \mathbf{A}\mathbf{G}_{11}\mathbf{Y} = \mathbf{X}\mathbf{G}_{11}\mathbf{A}$. If no, the shorted operator does not exist.

Method 11.4.4. Write $\begin{pmatrix} \mathbf{P}^1 \\ \mathbf{P}^2 \end{pmatrix}\mathbf{A}(\mathbf{P}^3 : \mathbf{P}^4) = \begin{pmatrix} \mathbf{T}_{11} & \mathbf{T}_{12} \\ \mathbf{T}_{21} & \mathbf{T}_{22} \end{pmatrix}$, where $\mathbf{T}_{11} = \mathbf{P}^1\mathbf{A}\mathbf{P}^3$, $\mathbf{T}_{12} = \mathbf{P}^1\mathbf{A}\mathbf{P}^4$, $\mathbf{T}_{21} = \mathbf{P}^2\mathbf{A}\mathbf{P}^3$, and $\mathbf{T}_{22} = \mathbf{P}^2\mathbf{A}\mathbf{P}^4$. Check, if $\mathcal{C}(\mathbf{T}_{21}) \subseteq \mathcal{C}(\mathbf{T}_{22})$ and $\mathcal{C}(\mathbf{T}_{12}^{\mathbf{t}}) \subseteq \mathcal{C}(\mathbf{T}_{22}^{\mathbf{t}})$. If so, $\mathbf{S}(\mathbf{A}|V_m, V_n)$ exists and is given by $\mathbf{S}(\mathbf{A}|V_m, V_n) = \mathbf{P}_1(\mathbf{T}_{11} - \mathbf{T}_{12}\mathbf{T}_{22}^{-}\mathbf{T}_{21})\mathbf{P}_3$, where $\mathbf{T}_{22}^{-}$ is an arbitrary g-inverse of $\mathbf{T}_{22}$. If no, the shorted operator $\mathbf{S}(\mathbf{A}|V_m, V_n)$ does not exist.

Equivalently,

Method 11.4.4*. Check whether $\rho(\mathbf{P}^2\mathbf{A}\mathbf{P}^4) = \rho(\mathbf{P}^2\mathbf{A}) = \rho(\mathbf{A}\mathbf{P}^4)$. If yes, $\mathbf{S}(\mathbf{A}|V_m, V_n)$ exists and is given by $\mathbf{A} - \mathbf{A}\mathbf{P}^4(\mathbf{P}^2\mathbf{A}\mathbf{P}^4)^{-}\mathbf{P}^2\mathbf{A}$, where $(\mathbf{P}^2\mathbf{A}\mathbf{P}^4)^{-}$ is any g-inverse of $\mathbf{P}^2\mathbf{A}\mathbf{P}^4$.

Method 11.4.5. This is a special case of Method 11.4.3. Let F be a field of characteristiç 0 and $\mathbf{A}$, $\mathbf{B}$ be $m \times n$ matrices over F. Let $V_m = \mathcal{C}(\mathbf{B})$ and $V_n = \mathcal{C}(\mathbf{B^t})$. Notice that $d(V_m) = d(V_n)$. In this case, check if

$$\mathcal{C}\begin{pmatrix} \mathbf{A} \\ \mathbf{0} \end{pmatrix} \subseteq \mathcal{C}\begin{pmatrix} \mathbf{A} & \mathbf{B} \\ \mathbf{B} & \mathbf{0} \end{pmatrix} \text{ and } \mathcal{C}\begin{pmatrix} \mathbf{A} \\ \mathbf{0} \end{pmatrix}^t \subseteq \mathcal{C}\begin{pmatrix} \mathbf{A} & \mathbf{B} \\ \mathbf{B} & \mathbf{0} \end{pmatrix}^t.$$

If so, let $(\mathbf{L},\ \mathbf{R})$ be a rank factorization of $\mathbf{A} + \mathbf{B}$. Let $\mathbf{D}$ be a square matrix such that $\mathbf{A} = \mathbf{LDR}$ and $\mathbf{I} - \mathbf{D} = \mathbf{Q}\mathbf{diag}\left(\mathbf{I} - \mathbf{D_1} \ \ \mathbf{0}\right)\mathbf{Q}^{-1}$, where $\mathbf{Q}$ and $\mathbf{I} - \mathbf{D_1}$ are non-singular. Then $\mathbf{S}(\mathbf{A}|\mathbf{B})$ exists and is given by $\mathbf{S}(\mathbf{A}|\mathbf{B}) = \mathbf{LQ}\mathbf{diag}\left(\mathbf{D_1} \ \ \mathbf{0}\right)\mathbf{Q}^{-1}\mathbf{R}$. If no, then $\mathbf{S}(\mathbf{A}|\mathbf{B})$ does not exist.

Let us now compare the methods.

For **Method 11.4.3**, we need to compute a g-inverse of $\begin{pmatrix} \mathbf{A} & \mathbf{X} \\ \mathbf{Y} & \mathbf{0} \end{pmatrix}$. Once we have this, it is very easy to check if $\mathbf{S}(\mathbf{A}|V_m, V_n)$ exists and then obtain it. However, if the orders of the matrices $\mathbf{X}$ and $\mathbf{Y}$ are $m \times s$ and $r \times n$ respectively, then we need to compute a g-inverse of a matrix of order $(m + r) \times (s + n)$ which is a matrix of much larger order.

For **Method 11.4.4**, we need to know a basis of W_m and and a basis W_n, where W_m and W_n denote the complements of V_m and V_n respectively. We also need to know the matrix $(\mathbf{P_1} : \mathbf{P_2})^{-1}\mathbf{A}\begin{pmatrix} \mathbf{P^3} \\ \mathbf{P^4} \end{pmatrix}^{-1}$. One way to obtain this matrix is as follows:

Form the matrix $(\mathbf{P_1} \ : \ \mathbf{P_2} \ : \ \mathbf{A})$ and reduce it to the form $(\mathbf{I} \ : \ (\mathbf{P_1} \ : \ \mathbf{P_2})^{-1}\mathbf{A})$ by elementary row operations. This reduces $(\mathbf{P_1} : \mathbf{P_2})$ to $\mathbf{I}$. Now consider $\left((\mathbf{P_1} : \mathbf{P_2})^{-1}\mathbf{A} : \begin{pmatrix} \mathbf{P_3} \\ \mathbf{P_4} \end{pmatrix}\right)$ and reduce it to $\left((\mathbf{P_1} : \mathbf{P_2})^{-1}\mathbf{A}\begin{pmatrix} \mathbf{P_3} \\ \mathbf{P_4} \end{pmatrix}^{-1} : \mathbf{I}\right)$ by elementary column operations. This reduces $\begin{pmatrix} \mathbf{P_3} \\ \mathbf{P_4} \end{pmatrix}$ to $\mathbf{I}$. Thus, we have the matrix $\begin{pmatrix} \mathbf{T_{11}} & \mathbf{T_{12}} \\ \mathbf{T_{21}} & \mathbf{T_{22}} \end{pmatrix}$. If this matrix can be further reduced to $\begin{pmatrix} \mathbf{H} & \mathbf{0} \\ \mathbf{0} & \mathbf{T_{22}} \end{pmatrix}$ by elementary row and column operations, we have $\mathcal{C}(\mathbf{T_{21}}) \subseteq \mathcal{C}(\mathbf{T_{22}})$, $\mathcal{C}(\mathbf{T^t_{12}}) \subseteq \mathcal{C}(\mathbf{T^t_{22}})$ and $\mathbf{H} = \mathbf{T_{11}} - \mathbf{T_{12}}\mathbf{T^-_{22}}\mathbf{T_{21}}$. Further, $\mathbf{S}(\mathbf{A}|V_m, V_n) = \mathbf{P_1}\mathbf{H}\mathbf{P_3}$. If the last reduction is not possible, then $\mathbf{S}(\mathbf{A}|V_m, V_n)$ does not exist. If the matrices $\mathbf{P_2}$ and $\mathbf{P_4}$ can be easily obtained, then this method has an edge over the method offered by Lemma 11.4.1. It may be noted that for any full column rank matrix $\mathbf{P_1}$, there exists a sub-permutation $(i_1, i_2, \ldots, i_{m-t})$ of $(1, 2, \ldots, m)$ such that

$\mathbf{P_2} = (\mathbf{e_{i_1}} : \mathbf{e_{i_2}} : ..\mathbf{e_{i_{m-t}}})$. However this may not be very easy to obtain algorithmically.

Thus, the Methods 11.4.3 and 11.4.4 are of comparable complexity.

For **Method 11.4.5**, one has to check $\mathcal{C}\begin{pmatrix}\mathbf{A}\\ \mathbf{0}\end{pmatrix} \subseteq \mathcal{C}\begin{pmatrix}\mathbf{A} & \mathbf{B}\\ \mathbf{B} & \mathbf{0}\end{pmatrix}$ and $\mathcal{C}\begin{pmatrix}\mathbf{A^t}\\ \mathbf{0}\end{pmatrix} \subseteq \mathcal{C}\begin{pmatrix}\mathbf{A} & \mathbf{B}\\ \mathbf{B} & \mathbf{0}\end{pmatrix}^t$, which is a task by itself. Even after this, there is significant amount of computation left to be done that leads to the computation of $\mathbf{S}(\mathbf{A}|\mathbf{B})$. This method appears to be computationally more expensive as compared to other methods.

11.5 Exercises

Unless stated otherwise all matrices are over a general field F.

(1) Let $\mathbf{A}$, $\mathbf{B}$ and $\mathbf{C}$ be matrices of the same order such that $\mathbf{C} \neq \mathbf{0}$ and $\{\mathbf{C}^-\} = \{\mathbf{A}^- + \mathbf{X} : \mathbf{A}^- \in \{\mathbf{A}^-\}$ and $\mathbf{BXB} = \mathbf{0}\}$. Show that $\mathbf{S} = \mathbf{S}(\mathbf{A}|\mathbf{B})$ exists and $\mathbf{S} = \mathbf{C}$.

(2) Let $\mathbf{A}$ and $\mathbf{B}$ be matrices of the same order such that $\mathbf{S}(\mathbf{A}|\mathbf{B})$ and $\mathbf{S}(\mathbf{B}|\mathbf{A})$ both exist. Then show that $\mathbf{S}(\mathbf{A}|\mathbf{B})$ and $\mathbf{S}(\mathbf{B}|\mathbf{A})$ are parallel summable with $\mathcal{P}(\mathbf{S}(\mathbf{A}|\mathbf{B}), \mathbf{S}(\mathbf{B}|\mathbf{A})) = \mathcal{P}(\mathbf{A}, \mathbf{B})$.

(3) Let $\mathbf{A}$ and $\mathbf{B}$ be matrices of the same order such that $\mathbf{S}(\mathbf{A}|\mathbf{B})$ and $\mathbf{S}(\mathbf{B}|\mathbf{A})$ both exist. Then show that $\{(\mathbf{S}(\mathbf{A}|\mathbf{B}))^- + (\mathbf{S}(\mathbf{B}|\mathbf{A}))^-\} = \{\mathbf{A}^-\} + \{\mathbf{B}^-\}$.

(4) Let $\mathbf{A}$, $\mathbf{B}$ and $\mathbf{C}$ be matrices of the same order such that (i) $\mathbf{A} <^- \mathbf{B}$ and (ii) the shorted matrices $\mathbf{S}(\mathbf{A}|\mathbf{C})$ and $\mathbf{S}(\mathbf{B}|\mathbf{C})$ exist. Show that $\mathbf{S}(\mathbf{A}|\mathbf{C}) <^- \mathbf{S}(\mathbf{B}|\mathbf{C})$. Verify the validity of the conclusion if '$<^-$' is replaced by '$<^\star$'

(5) Let $\mathbf{A}$ be an $m \times n$ matrix and $\mathcal{S}$ and $\mathcal{T}$ be subspaces of F^m and F^n. Let $\mathcal{S}' \subseteq \mathcal{S}$ and $\mathcal{T}' \subseteq \mathcal{T}$. Show that $\mathbf{S}(\mathbf{A}|\mathcal{S}', \mathcal{T}') <^- \mathbf{S}(\mathbf{A}|\mathcal{S}, \mathcal{T})$ whenever the shorted matrices exist.

(6) Let $\mathbf{A}$ be an $m \times n$ matrix, $\mathcal{S}$ and $\mathcal{S}'$ be subspaces of F^m and $\mathcal{T}$ and $\mathcal{T}'$ be subspaces of F^n such that $\mathcal{S} \cap \mathcal{S}' \neq \{\mathbf{0}\}$ and $\mathcal{T} \cap \mathcal{T}' \neq \{\mathbf{0}\}$. Assume $\mathbf{S}(\mathbf{A}|\mathcal{S}, \mathcal{T})$ exists. Show that $\mathbf{S}(\mathbf{A}|\mathcal{S} \cap \mathcal{S}', \mathcal{T} \cap \mathcal{T}')$ exists if and only if $\mathbf{S}(\mathbf{S}(\mathbf{A}|\mathcal{S}, \mathcal{T})|\mathcal{S}', \mathcal{T}')$ exists, in which case they are equal.

(7) Let the matrix $\mathbf{A}$ be idempotent of order $n \times n$. Prove that $\mathbf{S}(\mathbf{A}|V_n, V_n)$, if it exists, is also idempotent where V_n is a subspace of F^n.

Chapter 12

Lattice Properties of Partial Orders

12.1 Introduction

We studied the shorted operators in Chapters 10 and 11 in some detail. Their characterizing property as a maximal element of a certain collection of matrices also sets up the tone to consider an associated natural question namely, if a class of matrices is equipped with any of the partial orders on it, whether or not the set becomes a lattice under the partial order. If not, whether or not, it is a semi-lattice. In this chapter we discuss this problem for the three major partial orders namely the minus, the star and the sharp orders.

Let $\mathfrak{U}$ be a set of matrices equipped with a partial order '$\prec$'. Let $\mathbf{A},\ \mathbf{B} \in \mathfrak{U}$. We consider the following two classes of matrices

$$\mathfrak{C}_1 = \{\mathbf{C} : \mathbf{C} \prec \mathbf{A},\ \mathbf{C} \prec \mathbf{B}\}$$

and

$$\mathfrak{C}_2 = \{\mathbf{C} : \mathbf{A} \prec \mathbf{C},\ \mathbf{B} \prec \mathbf{C}\}.$$

The class $\mathfrak{C}_1$ is non- empty as the null matrix is always below every matrix under '$\prec$'. However, the class $\mathfrak{C}_2$ may well be empty in general. One instance when it happens is when the matrices $\mathbf{A}$ and $\mathbf{B}$ are maximal with respect to the given partial order '$\prec$'. Let us remind ourselves of the following:

Definition 12.1.1. The unique maximal element of $\mathfrak{C}_1$, if it exists is called the infimum of the matrices $\mathbf{A},\ \mathbf{B}$ under '$\prec$' and is denoted by $\mathbf{A} \wedge \mathbf{B}$.

Similarly, if $\mathfrak{C}_2$ is non-empty, the unique minimal element of $\mathfrak{C}_2$, if it exists, is called the supremum of the matrices $\mathbf{A},\ \mathbf{B}$ under '$\prec$' and is denoted by $\mathbf{A} \vee \mathbf{B}$.

Section 2 gives a detailed account of finding the infimum and supremum of a given pair of matrices under the minus order. In general neither the supremum nor the infimum of a given pair of elements may exist. We study the conditions under which either one or both of infimum and supremum of a given pair of matrices under the minus order exist. Section 3 deals with the same problem for the star order. Here we show that the infimum of any pair of elements always exists. In Section 4, we study the conditions under which a pair of matrices has an infimum under the sharp order. The related problem for associated one sided orders for the star order can be found in exercises at the end of the chapter.

We begin with the minus order.

12.2 Supremum and infimum of a pair of matrices under the minus order

Let $\mathbf{A}$ and $\mathbf{B}$ be matrices of the same order $m \times n$ over an arbitrary field F. For the minus order, we shall write $\underline{\mathfrak{C}}$ for the class $\mathfrak{C}_1$ and $\overline{\mathfrak{C}}$ for the class $\mathfrak{C}_2$. Thus

$$\underline{\mathfrak{C}} = \{\mathbf{C} : \mathbf{C} <^- \mathbf{A},\ \mathbf{C} <^- \mathbf{B}\}$$
$$\text{and } \overline{\mathfrak{C}} = \{\mathbf{C} : \mathbf{A} <^- \mathbf{C},\ \mathbf{B} <^- \mathbf{C}\}.$$

Also as noted earlier $\underline{\mathfrak{C}} \neq \emptyset$ and $\overline{\mathfrak{C}}$ may be empty. We begin by giving an example of a pair of matrices for which neither $\mathbf{A} \wedge \mathbf{B}$ nor $\mathbf{A} \vee \mathbf{B}$ exists.

Example 12.2.1. Let $\mathbf{A} = \begin{pmatrix} 1 & 0 & 1 \\ 0 & 1 & 0 \\ 0 & 0 & 0 \end{pmatrix}$ and $\mathbf{B} = \begin{pmatrix} 1 & 0 & 0 \\ 0 & 1 & 0 \\ 0 & 0 & 0 \end{pmatrix}$.
Notice that any non-null matrix $\mathbf{C}$ such that $\mathbf{C} <^- \mathbf{A},\ \mathbf{C} <^- \mathbf{B}$ must have rank 1. Now each of $\mathbf{C_1} = \begin{pmatrix} 1 & 0 & 0 \\ 0 & 0 & 0 \\ 0 & 0 & 0 \end{pmatrix}$ and $\mathbf{C_2} = \begin{pmatrix} 0 & 0 & 0 \\ 0 & 1 & 0 \\ 0 & 0 & 0 \end{pmatrix} \in \underline{\mathfrak{C}}$. However, both $\mathbf{C_1}$ and $\mathbf{C_2}$ are non-comparable under the minus order. Hence, $\mathbf{A} \wedge \mathbf{B}$ does not exist.

Moreover, the class $\overline{\mathfrak{C}} \neq \emptyset$, since both the matrices $\mathbf{D_1} = \begin{pmatrix} 1 & 0 & 1 \\ 0 & 1 & 1 \\ 0 & 0 & 1 \end{pmatrix}$ and $\mathbf{D_2} = \begin{pmatrix} 1 & 0 & 0 \\ 0 & 1 & 0 \\ 0 & 0 & 1 \end{pmatrix} \in \overline{\mathfrak{C}}$. Clearly, each has rank 3 and are not comparable under the minus order. Thus, $\mathbf{A} \vee \mathbf{B}$ also does not exist.

Thus, in general, $F^{m\times n}$ may fail to be a lower semi-lattice as well as an upper semi-lattice under the minus order. What conditions will ensure that this is possible? The obvious thing to do is to find out when can $\overline{\mathfrak{C}}$ have the supremum and when can $\underline{\mathfrak{C}}$ have the infimum. We first study the conditions under which the supremum of any two matrices in $F^{m\times n}$ exists. The following theorem gives a necessary and sufficient condition such that $\overline{\mathfrak{C}} \neq \emptyset$ when the two matrices possess certain properties.

Theorem 12.2.2. *Let* $\mathbf{A}$ *and* $\mathbf{B}$ *be matrices of the same order. Write* $\mathcal{S}_a = \mathcal{C}(\mathbf{A})$, $\mathcal{T}_a = \mathcal{C}(\mathbf{A}^t)$, $\mathcal{S}_\mathbf{b} = \mathcal{C}(\mathbf{B})$ *and* $\mathcal{T}_b = \mathcal{C}(\mathbf{B}^t)$. *Further, assume that the shorted operators* $\mathbf{S}(\mathbf{A}|\mathcal{S}_\mathbf{b}, \mathcal{T}_\mathbf{b})$ *and* $\mathbf{S}(\mathbf{B}|\mathcal{S}_\mathbf{a}, \mathcal{T}_\mathbf{a})$ *both exist. Then* $\overline{\mathfrak{C}} \neq \emptyset$ *if and only if* $\mathbf{S}(\mathbf{A}|\mathcal{S}_\mathbf{b}, \mathcal{T}_\mathbf{b}) = \mathbf{S}(\mathbf{B}|\mathcal{S}_\mathbf{a}, \mathcal{T}_\mathbf{a})$.

Proof. We note that if $\mathbf{S}(\mathbf{A}|\mathcal{S}_\mathbf{b}, \mathcal{T}_\mathbf{b})$ exists and $\begin{pmatrix}\mathbf{F}^1 & \mathbf{F}^2\\ \mathbf{F}^3 & \mathbf{F}^4\end{pmatrix}$ is a g-inverse of $\begin{pmatrix}\mathbf{A} & \mathbf{B}\\ \mathbf{B} & \mathbf{0}\end{pmatrix}$, then $\mathbf{B} = \mathbf{B}\mathbf{F}^2\mathbf{B} = \mathbf{B}\mathbf{F}^3\mathbf{B}$, $\mathbf{A}\mathbf{F}^2\mathbf{B} = \mathbf{B}\mathbf{F}^4\mathbf{B} = \mathbf{B}\mathbf{F}^3\mathbf{A}$, and the shorted operator $\mathbf{S}(\mathbf{A}|\mathcal{S}_\mathbf{b}, \mathcal{T}_\mathbf{b}) = \mathbf{B}\mathbf{F}^3\mathbf{A}$. (See Theorem 11.4.2.)

'If' part

Let $\mathbf{S}(\mathbf{A}|\mathcal{S}_\mathbf{b}, \mathcal{T}_\mathbf{b}) = \mathbf{S}(\mathbf{B}|\mathcal{S}_\mathbf{a}, \mathcal{T}_\mathbf{a})$. We show that $\mathbf{A}+\mathbf{B} - \mathbf{S}(\mathbf{A}|\mathcal{S}_\mathbf{b}, \mathcal{T}_\mathbf{b}) \in \overline{\mathfrak{C}}$. Note that $\mathbf{A} <^- \mathbf{A}+\mathbf{B} - \mathbf{S}(\mathbf{A}|\mathcal{S}_\mathbf{b}, \mathcal{T}_\mathbf{b}) \Leftrightarrow \mathbf{B} - \mathbf{S}(\mathbf{A}|\mathcal{S}_\mathbf{b}, \mathcal{T}_\mathbf{b})$ and $\mathbf{A}$ are disjoint matrices. Therefore, we must show that

$$\mathcal{C}(\mathbf{B} - \mathbf{S}(\mathbf{A}|\mathcal{S}_\mathbf{b}, \mathcal{T}_\mathbf{b})) \cap \mathcal{C}(\mathbf{A}) = \{\mathbf{0}\} \text{ and } \mathcal{C}(\mathbf{B} - \mathbf{S}(\mathbf{A}|\mathcal{S}_\mathbf{b}, \mathcal{T}_\mathbf{b}))^t \cap \mathcal{C}(\mathbf{A})^t = \{\mathbf{0}\}.$$

Let $\mathbf{z} = (\mathbf{B} - \mathbf{S}(\mathbf{A}|\mathcal{S}_\mathbf{b}, \mathcal{T}_\mathbf{b}))\mathbf{x} = \mathbf{A}\mathbf{y}$. As noted above, $\mathbf{S}(\mathbf{A}|\mathcal{S}_\mathbf{b}, \mathcal{T}_\mathbf{b}) = \mathbf{B}\mathbf{F}^3\mathbf{A}$, so, $\mathbf{z} \in \mathcal{C}(\mathbf{A}) \cap \mathcal{C}(\mathbf{B}) = \mathcal{C}(\mathbf{S}(\mathbf{A}|\mathcal{S}_\mathbf{b}, \mathcal{T}_\mathbf{b})) = \mathcal{C}(\mathbf{S}(\mathbf{B}|\mathcal{S}_\mathbf{a}, \mathcal{T}_\mathbf{a}))$. Therefore, $\mathbf{z} \in \mathcal{C}(\mathbf{B} - \mathbf{S}(\mathbf{B}|\mathcal{S}_\mathbf{a}, \mathcal{T}_\mathbf{a})) \cap \mathcal{C}(\mathbf{S}(\mathbf{B}|\mathcal{S}_\mathbf{a}, \mathcal{T}_\mathbf{a})) = \{\mathbf{0}\}$. Since $\mathbf{S}(\mathbf{B}|\mathcal{S}_\mathbf{a}, \mathcal{T}_\mathbf{a}) <^- \mathbf{B}$, we have $\mathcal{C}(\mathbf{B} - \mathbf{S}(\mathbf{A}|\mathcal{S}_\mathbf{b}, \mathcal{T}_\mathbf{b})) \cap \mathcal{C}(\mathbf{A}) = \{\mathbf{0}\}$. Hence, $\mathbf{z} = \mathbf{0}$. This proves $\mathcal{C}(\mathbf{B} - \mathbf{S}(\mathbf{B}|\mathcal{S}_\mathbf{a}, \mathcal{T}_\mathbf{a})) \cap \mathcal{C}(\mathbf{A}) = \{\mathbf{0}\}$. Similarly, we can show that $\mathcal{C}(\mathbf{B} - \mathbf{S}(\mathbf{B}|\mathcal{S}_\mathbf{a}, \mathcal{T}_\mathbf{a}))^t \cap \mathcal{C}(\mathbf{A})^t = \{\mathbf{0}\}$.

We can prove that $\mathbf{B} <^- \mathbf{A}+\mathbf{B} - \mathbf{S}(\mathbf{A}|\mathcal{S}_\mathbf{b}, \mathcal{T}_\mathbf{b})$ along the same lines. Thus, $\mathbf{A}+\mathbf{B} - \mathbf{S}(\mathbf{A}|\mathcal{S}_\mathbf{b}, \mathcal{T}_\mathbf{b}) \in \overline{\mathfrak{C}}$ and so, $\overline{\mathfrak{C}} \neq \emptyset$.

'Only if' part

Let $\overline{\mathfrak{C}} \neq \emptyset$ and $\mathbf{C} \in \overline{\mathfrak{C}}$. Then $\mathbf{A} <^- \mathbf{C}$ and $\mathbf{B} <^- \mathbf{C}$. By Ex.6 of Chapter 11, we have

$$\begin{aligned}\mathbf{S}(\mathbf{S}(\mathbf{C}|\mathcal{S}_a, \mathcal{T}_a)|\mathcal{S}_b, \mathcal{T}_b) &= \mathbf{S}(\mathbf{S}(\mathbf{C}|\mathcal{S}_b, \mathcal{T}_b)|\mathcal{S}_a, \mathcal{T}_a)\\ &= \mathbf{S}(\mathbf{C}|\mathcal{S}_a \cap \mathcal{S}_b, \mathcal{T}_a \cap \mathcal{T}_b). \qquad (12.2.1)\end{aligned}$$

Now, by Theorem 10.3.5

$$\mathbf{S}(\mathbf{C}|\mathcal{S}_\mathbf{a}, \mathcal{T}_\mathbf{a}) = \max\{\mathbf{Z} : \mathbf{Z} <^- \mathbf{C},\ \mathcal{C}(\mathbf{X}) \subseteq \mathcal{C}(\mathbf{A}).\ \mathcal{C}(\mathbf{X}^t) \subseteq \mathcal{C}(\mathbf{A}^t)\}.$$

Clearly, $\mathbf{A} \in \{\mathbf{Z} : \mathbf{Z} <^- \mathbf{C},\ \mathcal{C}(\mathbf{X}) \subseteq \mathcal{C}(\mathbf{A}).\ \mathcal{C}(\mathbf{X^t}) \subseteq \mathcal{C}(\mathbf{A^t})\}$, and therefore, $\mathbf{A} <^- \mathbf{S}(\mathbf{C}|\mathcal{S}_\mathbf{a}, \mathcal{T}_\mathbf{a})$. Since $\mathbf{S}(\mathbf{A}|\mathcal{S}_\mathbf{b}, \mathcal{T}_\mathbf{b}) <^- \mathbf{A}$, $\mathcal{C}(\mathbf{S}(\mathbf{A}|\mathcal{S}_\mathbf{b}, \mathcal{T}_\mathbf{b})) = \mathcal{C}(\mathbf{A}) \cap \mathcal{C}(\mathbf{B}) \subseteq \mathcal{C}(\mathbf{B})$ and $\mathcal{C}(\mathbf{S}(\mathbf{A}|\mathcal{S}_\mathbf{b}, \mathcal{T}_\mathbf{b}))^\mathbf{t} = \mathcal{C}(\mathbf{A^t}) \cap \mathcal{C}(\mathbf{B^t}) \subseteq \mathcal{C}(\mathbf{B^t})$, so, $\mathbf{S}(\mathbf{A}|\mathcal{S}_\mathbf{b}, \mathcal{T}_\mathbf{b}) <^- \mathbf{S}(\mathbf{S}(\mathbf{C}|\mathcal{S}_\mathbf{a}, \mathcal{T}_\mathbf{a})|\mathcal{S}_\mathbf{b}, \mathcal{T}_\mathbf{b})$. However, both the matrices $\mathbf{S}(\mathbf{S}(\mathbf{C}|\mathcal{S}_\mathbf{a}, \mathcal{T}_\mathbf{a})|\mathcal{S}_\mathbf{b}, \mathcal{T}_\mathbf{b})$ and $\mathbf{S}(\mathbf{A}|\mathcal{S}_\mathbf{b}, \mathcal{T}_\mathbf{b})$ have same column space $\mathcal{C}(\mathbf{A}) \cap \mathcal{C}(\mathbf{B})$ and same row space $\mathcal{C}(\mathbf{A^t}) \cap \mathcal{C}(\mathbf{B^t})$, therefore have same rank and so,

$$\mathbf{S}(\mathbf{A}|\mathcal{S}_\mathbf{b}, \mathcal{T}_\mathbf{b}) = \mathbf{S}(\mathbf{S}(\mathbf{C}|\mathcal{S}_\mathbf{a}, \mathcal{T}_\mathbf{a})|\mathcal{S}_\mathbf{b}, \mathcal{T}_\mathbf{b}). \tag{12.2.2}$$

Similarly,

$$\mathbf{S}(\mathbf{B}|\mathcal{S}_\mathbf{a}, \mathcal{T}_\mathbf{a}) = \mathbf{S}(\mathbf{S}(\mathbf{C}|\mathcal{S}_\mathbf{b}, \mathcal{T}_\mathbf{b})|\mathcal{S}_\mathbf{a}, \mathcal{T}_\mathbf{a}). \tag{12.2.3}$$

From (12.2.1)-(12.2.3), we have $\mathbf{S}(\mathbf{A}|\mathcal{S}_\mathbf{b}, \mathcal{T}_\mathbf{b}) = \mathbf{S}(\mathbf{B}|\mathcal{S}_\mathbf{a}, \mathcal{T}_\mathbf{a})$. □

Remark 12.2.3. It is possible that the class $\overline{\mathfrak{C}}$ be non-empty even when $\mathbf{S}(\mathbf{A}|\mathcal{S}_\mathbf{b}, \mathcal{T}_\mathbf{b})$ and /or $\mathbf{S}(\mathbf{B}|\mathcal{S}_\mathbf{a}, \mathcal{T}_\mathbf{a})$ are not defined as the following example shows:

Example 12.2.4. Let $\mathbf{A}$, $\mathbf{B}$ be the matrices as in Example 12.2.1. Consider the matrix $\mathbf{C} = \begin{pmatrix} 1 & 0 & a \\ 0 & 1 & b \\ 0 & 0 & c \end{pmatrix}$, where a, b, c are arbitrary and $c \neq 0$. Then $\mathbf{A} <^- \mathbf{C}$ and $\mathbf{B} <^- \mathbf{C}$. Also, $\mathcal{C}(\mathbf{A}) \cap \mathcal{C}(\mathbf{B}) = \mathcal{C}(\mathbf{A}) = \mathcal{C}(\mathbf{B})$ and $\mathcal{C}(\mathbf{A^t}) \cap \mathcal{C}(\mathbf{B^t}) = \mathcal{C}\begin{pmatrix} 0 \\ 1 \\ 0 \end{pmatrix}$. Thus, $d(\mathcal{C}(\mathbf{A}) \cap \mathcal{C}(\mathbf{B})) = 2$ and $d(\mathcal{C}(\mathbf{A^t}) \cap \mathcal{C}(\mathbf{B^t})) = 1$. Hence, both the shorted matrices $\mathbf{S}(\mathbf{A}|\mathcal{S}_\mathbf{b}, \mathcal{T}_\mathbf{b})$ and $\mathbf{S}(\mathbf{B}|\mathcal{S}_\mathbf{a}, \mathcal{T}_\mathbf{a})$ do not exist, yet $\mathbf{C} \in \overline{\mathfrak{C}}$, showing $\overline{\mathfrak{C}} \neq \emptyset$.

Theorem 12.2.5. *Let* $\mathbf{A}$ *and* $\mathbf{B}$ *be matrices of the same order. Write* $\mathcal{S}_a = \mathcal{C}(\mathbf{A})$, $\mathcal{T}_a = \mathcal{C}(\mathbf{A^t})$, $\mathcal{S}_\mathbf{b} = \mathcal{C}(\mathbf{B})$ *and* $\mathcal{T}_b = \mathcal{C}(\mathbf{B^t})$. *Further, assume that the shorted operators* $\mathbf{S}(\mathbf{A}|\mathcal{S}_\mathbf{b}, \mathcal{T}_\mathbf{b})$ *and* $\mathbf{S}(\mathbf{B}|\mathcal{S}_\mathbf{a}, \mathcal{T}_\mathbf{a})$ *both exist and are equal. Then* $\mathbf{A} \vee \mathbf{B}$ *exists if and only if at least one of* $\mathbf{A} - \mathbf{S}(\mathbf{A}|\mathcal{S}_\mathbf{b}, \mathcal{T}_\mathbf{b})$ *and* $\mathbf{B} - \mathbf{S}(\mathbf{B}|\mathcal{S}_\mathbf{a}, \mathcal{T}_\mathbf{a})$ *is null or equivalently at least one of* $\mathbf{A} <^- \mathbf{B}$ *or* $\mathbf{B} <^- \mathbf{A}$ *holds.*

Proof. The matrix $\mathbf{A} + \mathbf{B} - \mathbf{S}(\mathbf{A}|\mathcal{S}_\mathbf{b}, \mathcal{T}_\mathbf{b}) \in \overline{\mathfrak{C}}$, as shown in Theorem 12.2.2. We show that $\mathbf{A} \vee \mathbf{B} = \mathbf{A} + \mathbf{B} - \mathbf{S}(\mathbf{A}|\mathcal{S}_\mathbf{b}, \mathcal{T}_\mathbf{b}) \in \overline{\mathfrak{C}}$. Let $\mathbf{C} \in \overline{\mathfrak{C}}$. Then $\mathbf{A} <^- \mathbf{C}$ and $\mathbf{B} <^- \mathbf{C}$. So, $\mathcal{C}(\mathbf{A} + \mathbf{B}) = \mathcal{C}(\mathbf{A}) + \mathcal{C}(\mathbf{B}) \subseteq \mathcal{C}(\mathbf{C})$. Hence,

$$\begin{aligned} \rho(\mathbf{C}) &= d(\mathcal{C}(\mathbf{A}) + \mathcal{C}(\mathbf{B})) \\ &= \rho(\mathbf{A}) + \rho(\mathbf{B}) - \mathbf{d}(\mathcal{C}(\mathbf{A}) \cap \mathcal{C}(\mathbf{B})) \\ &\geq \rho(\mathbf{A} + \mathbf{B} - \mathbf{S}(\mathbf{A}|\mathcal{S}_\mathbf{b}, \mathcal{T}_\mathbf{b})), \end{aligned}$$

so, $\mathbf{C} \not<^- \mathbf{A} + \mathbf{B} - \mathbf{S}(\mathbf{A}|\mathcal{S}_\mathbf{b}, \mathcal{T}_\mathbf{b})$ under '$<^-$'.

Let none of the matrices $\mathbf{A} - \mathbf{S}(\mathbf{A}|\mathcal{S}_\mathbf{b}, \mathcal{T}_\mathbf{b})$ and $\mathbf{B} - \mathbf{S}(\mathbf{B}|\mathcal{S}_\mathbf{a}, \mathcal{T}_\mathbf{a})$ be null. Consider the matrix

$$\mathbf{X_K} = \mathbf{A} + \mathbf{B} - \mathbf{S}(\mathbf{A}|\mathcal{S}_\mathbf{b}, \mathcal{T}_\mathbf{b}) + (\mathbf{A} - \mathbf{S}(\mathbf{A}|\mathcal{S}_\mathbf{b}, \mathcal{T}_\mathbf{b}))\mathbf{K}(\mathbf{B} - \mathbf{S}(\mathbf{B}|\mathcal{S}_\mathbf{a}, \mathcal{T}_\mathbf{a})),$$

where $\mathbf{K}$ is an arbitrary matrix of appropriate order. We show that for each such matrix $\mathbf{K}$, the matrix $\mathbf{X_K}$ is a minimal element of $\overline{\mathfrak{C}}$. We first prove that $\mathbf{X_K} \in \overline{\mathfrak{C}}$ or equivalently, $\mathbf{A} <^- \mathbf{X_K}$ and $\mathbf{B} <^- \mathbf{X_K}$ which is further equivalent to establishing that the matrices '$\mathbf{A}$ and $\mathbf{X_K} - \mathbf{A}$' are disjoint and '$\mathbf{B}$ and $\mathbf{X_K} - \mathbf{B}$' are disjoint. Consider the matrix $\mathbf{X_K} - \mathbf{A}$.

$$\begin{aligned}\mathbf{X_K} - \mathbf{A} &= \mathbf{B} - \mathbf{S}(\mathbf{A}|\mathcal{S}_\mathbf{b}, \mathcal{T}_\mathbf{b}) + (\mathbf{A} - \mathbf{S}(\mathbf{A}|\mathcal{S}_\mathbf{b}, \mathcal{T}_\mathbf{b}))\mathbf{K}(\mathbf{B} - \mathbf{S}(\mathbf{B}|\mathcal{S}_\mathbf{a}, \mathcal{T}_\mathbf{a}))\\ &= \mathbf{B} - \mathbf{S}(\mathbf{B}|\mathcal{S}_\mathbf{a}, \mathcal{T}_\mathbf{a}) + (\mathbf{A} - \mathbf{S}(\mathbf{A}|\mathcal{S}_\mathbf{b}, \mathcal{T}_\mathbf{b}))\mathbf{K}(\mathbf{B} - \mathbf{S}(\mathbf{B}|\mathcal{S}_\mathbf{a}, \mathcal{T}_\mathbf{a})).\end{aligned}$$

Let $\mathbf{z} = \mathbf{Ax} = (\mathbf{X_K} - \mathbf{A})\mathbf{y}$. Clearly, $\mathbf{z} \in \mathcal{C}(\mathbf{A}) \cap \mathcal{C}(\mathbf{B}) = \mathcal{C}(\mathbf{S}(\mathbf{B}|\mathcal{S}_\mathbf{a}, \mathcal{T}_\mathbf{a}))$. So, $\mathbf{z} \in \mathcal{C}(\mathbf{B} - \mathbf{S}(\mathbf{B}|\mathcal{S}_\mathbf{a}, \mathcal{T}_\mathbf{a})) \cap \mathcal{C}(\mathbf{S}(\mathbf{B}|\mathcal{S}_\mathbf{a}, \mathcal{T}_\mathbf{a})) = \{\mathbf{0}\}$, as $\mathbf{S}(\mathbf{B}|\mathcal{S}_\mathbf{a}, \mathcal{T}_\mathbf{a}) <^- \mathbf{B}$. Therefore, $\mathbf{z} = \mathbf{0}$ implying $\mathcal{C}(\mathbf{A}) \cap \mathcal{C}(\mathbf{X_K} - \mathbf{A}) = \{\mathbf{0}\}$.

Similarly, $\mathcal{C}(\mathbf{A}^\mathbf{t}) \cap \mathcal{C}(\mathbf{X_K} - \mathbf{A})^\mathbf{t} = \{\mathbf{0}\}$. Thus, $\mathbf{A} <^- \mathbf{X_K}$. Similar argument shows that $\mathbf{B} <^- \mathbf{X_K}$. Moreover,

$$\begin{aligned}\rho(\mathbf{X_K}) &= \rho(\mathbf{A} + (\mathbf{B} - \mathbf{S}(\mathbf{A}|\mathcal{S}_\mathbf{b}, \mathcal{T}_\mathbf{b})))(\mathbf{I} + \mathbf{K}(\mathbf{A} - \mathbf{S}(\mathbf{A}|\mathcal{S}_\mathbf{b}, \mathcal{T}_\mathbf{b})))\\ &\leq \rho(\mathbf{A}) + \rho((\mathbf{B} - \mathbf{S}(\mathbf{A}|\mathcal{S}_\mathbf{b}, \mathcal{T}_\mathbf{b}))(\mathbf{I} + \mathbf{K}(\mathbf{A} - \mathbf{S}(\mathbf{A}|\mathcal{S}_\mathbf{b}, \mathcal{T}_\mathbf{b}))))\\ &\leq \rho(\mathbf{A}) + \rho((\mathbf{B} - \mathbf{S}(\mathbf{A}|\mathcal{S}_\mathbf{b}, \mathcal{T}_\mathbf{b}))).\end{aligned}$$

Thus, $\mathbf{X_K}$ is a minimal element for each choice of $\mathbf{K}$, showing $\mathbf{A} \vee \mathbf{B}$ does not exist. Thus, either $\mathbf{A} - \mathbf{S}(\mathbf{A}|\mathcal{S}_\mathbf{b}, \mathcal{T}_\mathbf{b})$ or $\mathbf{B} - \mathbf{S}(\mathbf{B}|\mathcal{S}_\mathbf{a}, \mathcal{T}_\mathbf{a})$ is a null matrix or equivalently, either $\mathbf{A} <^- \mathbf{B}$ or $\mathbf{B} <^- \mathbf{A}$ holds.

'If' part

Assume at least one of $\mathbf{A} - \mathbf{S}(\mathbf{A}|\mathcal{S}_\mathbf{b}, \mathcal{T}_\mathbf{b})$ or $\mathbf{B} - \mathbf{S}(\mathbf{B}|\mathcal{S}_\mathbf{a}, \mathcal{T}_\mathbf{a})$ be null, say $\mathbf{A} - \mathbf{S}(\mathbf{A}|\mathcal{S}_\mathbf{b}, \mathcal{T}_\mathbf{b}) = \{\mathbf{0}\}$. Then $\mathbf{A} = \mathbf{S}(\mathbf{A}|\mathcal{S}_\mathbf{b}, \mathcal{T}_\mathbf{b}) = \mathbf{S}(\mathbf{B}|\mathcal{S}_\mathbf{a}, \mathcal{T}_\mathbf{a}) <^- \mathbf{B}$. So, $\mathbf{A} \vee \mathbf{B} = \mathbf{A}$.

Similarly, if $\mathbf{B} - \mathbf{S}(\mathbf{B}|\mathcal{S}_\mathbf{a}, \mathcal{T}_\mathbf{a}) = \{\mathbf{0}\}$, then $\mathbf{A} \vee \mathbf{B} = \mathbf{B}$. □

Remark 12.2.6. In the notations of Theorem 12.2.5, when $\mathbf{S}(\mathbf{A}|\mathcal{S}_\mathbf{b}, \mathcal{T}_\mathbf{b})$ and $\mathbf{S}(\mathbf{B}|\mathcal{S}_\mathbf{a}, \mathcal{T}_\mathbf{a})$ both exist, for $\mathbf{A} \vee \mathbf{B}$ to exist, either $\mathbf{A} <^- \mathbf{B}$ or $\mathbf{B} <^- \mathbf{A}$.

Our next theorem shows that the only instance when $\mathbf{A} \vee \mathbf{B}$ exists is provided by conditions of Theorem 12.2.5.

Theorem 12.2.7. *Let* $\mathbf{A}$ *and* $\mathbf{B}$ *be matrices of the same order. Write* $\mathcal{S}_a = \mathcal{C}(\mathbf{A})$, $\mathcal{T}_a = \mathcal{C}(\mathbf{A}^\mathbf{t})$, $\mathcal{S}_\mathrm{b} = \mathcal{C}(\mathbf{B})$ *and* $\mathcal{T}_b = \mathcal{C}(\mathbf{B}^\mathbf{t})$. *Further, assume that either of the shorted operators* $\mathbf{S}(\mathbf{A}|\mathcal{S}_\mathbf{b}, \mathcal{T}_\mathbf{b})$ *or* $\mathbf{S}(\mathbf{B}|\mathcal{S}_\mathbf{a}, \mathcal{T}_\mathbf{a})$ *does not exist. Then* $\mathbf{A} \vee \mathbf{B}$ *does not exist.*

Proof. Let us suppose that $\mathbf{S}(\mathbf{A}|\mathcal{S}_\mathbf{b}, \mathcal{T}_\mathbf{b})$ does not exist, yet $\mathbf{A} \vee \mathbf{B}$ exists. Let $\mathbf{C} = \mathbf{A} \vee \mathbf{B}$ and $(\mathbf{P}, \mathbf{Q})$ be a rank factorization of $\mathbf{C}$. As $\mathbf{A} <^- \mathbf{C}$ and $\mathbf{B} <^- \mathbf{C}$, by Theorem 3.3.5(v), we can write $\mathbf{B} = \mathbf{P}\mathrm{diag}(\mathbf{I_r},\ \mathbf{0})\mathbf{Q}$ and $\mathbf{A} = \mathbf{P}\begin{pmatrix}\mathbf{H_{11}} & \mathbf{H_{12}}\\ \mathbf{H_{21}} & \mathbf{H_{22}}\end{pmatrix}\mathbf{Q}$, where $r = \rho(\mathbf{B})$ and $\mathbf{H} = \begin{pmatrix}\mathbf{H_{11}} & \mathbf{H_{12}}\\ \mathbf{H_{21}} & \mathbf{H_{22}}\end{pmatrix}$ is an idempotent matrix.

Let $\lambda \neq 0, \neq 1$ be a scalar such that $\det(\mathbf{I} + (\lambda - 1)(\mathbf{I} - \mathbf{H_{22}})) \neq \mathbf{0}$. If $\mathbf{I} = \mathbf{H_{22}}$, let

$$\mathbf{T} = \begin{pmatrix}\mathbf{I} & -(\lambda-1)\mathbf{H_{12}}\\ \mathbf{0} & \mathbf{I} + (\lambda-1)(\mathbf{I} - \mathbf{H_{22}})\end{pmatrix} = \mathbf{T_1}$$

or

$$\mathbf{T} = \begin{pmatrix}\mathbf{I} & \mathbf{0}\\ -(\lambda-1)\mathbf{H_{21}} & \mathbf{I} + (\lambda-1)(\mathbf{I} - \mathbf{H_{22}})\end{pmatrix} = \mathbf{T_2}$$

according as $\mathbf{H_{12}}$ is non-null or $\mathbf{H_{21}}$ is non-null. If $\mathbf{I} \neq \mathbf{H_{22}}$, then either choice is permissible. Clearly, $\det(\mathbf{T}) \neq \mathbf{0}$. Hence, $\rho(\mathbf{C}) = \rho(\mathbf{PTQ})$. We now show that $\mathbf{A} <^- \mathbf{PTQ}$ and $\mathbf{B} <^- \mathbf{PTQ}$. Consider $\mathbf{T_1} - \mathbf{H} = \begin{pmatrix}\mathbf{I} - \mathbf{H_{11}} & -\lambda\mathbf{H_{12}}\\ -\mathbf{H_{21}} & \lambda(\mathbf{I} - \mathbf{H_{22}})\end{pmatrix}$. We have $\mathcal{C}(\mathbf{T_1} - \mathbf{H}) = \mathcal{C}(\mathbf{I} - \mathbf{H})$ and $\mathbf{I} - \mathbf{H}$ is an idempotent. So, $\rho(\mathbf{T_1} - \mathbf{H}) = \rho(\mathbf{I} - \mathbf{H})$ and $\mathbf{T_1} = \mathbf{H} \oplus (\mathbf{T_1} - \mathbf{H})$. Similarly, $\mathbf{T_2} = \mathbf{H} \oplus (\mathbf{T_2} - \mathbf{H})$. Therefore, in either case $\mathbf{A} <^- \mathbf{PTQ}$. However, $\mathbf{B} <^- \mathbf{PTQ}$ holds trivially. This contradicts that $\mathbf{A} \vee \mathbf{B}$ exists.

In case $\mathbf{I} = \mathbf{H_{22}}$, $\mathbf{H_{12}} = \mathbf{0}$ and $\mathbf{H_{21}} = \mathbf{0}$, we have $\mathbf{A} = \mathbf{P}\begin{pmatrix}\mathbf{H_{11}} & \mathbf{0}\\ \mathbf{0} & \mathbf{I}\end{pmatrix}\mathbf{Q}$. Consider the matrix $\begin{pmatrix}\mathbf{A} & \mathbf{B}\\ \mathbf{B} & \mathbf{0}\end{pmatrix}$. By Theorem 11.4.2, $\mathbf{S}(\mathbf{A}|\mathcal{S}_\mathbf{b}, \mathcal{T}_\mathbf{b})$ exists and $\mathbf{S}(\mathbf{A}|\mathcal{S}_\mathbf{b}, \mathcal{T}_\mathbf{b}) = \mathbf{A}$, a clear contradiction to the supposition that $\mathbf{S}(\mathbf{A}|\mathcal{S}_\mathbf{b}, \mathcal{T}_\mathbf{b})$ does not exist.

The case when $\mathbf{S}(\mathbf{B}|\mathcal{S}_\mathbf{a}, \mathcal{T}_\mathbf{a})$ does not exist can be completed in the same manner. □

Remark 12.2.8. In the setup of Theorem 12.2.7, $\mathbf{A} \vee \mathbf{B}$ does not exist. We show that the same is not the case with infimum. Even when one of the shorted operators $\mathbf{S}(\mathbf{A}|\mathcal{S}_\mathbf{b}, \mathcal{T}_\mathbf{b})$ or $\mathbf{S}(\mathbf{B}|\mathcal{S}_\mathbf{a}, \mathcal{T}_\mathbf{a})$ does not exist, the infimum $\mathbf{A} \wedge \mathbf{B}$ may still exist.

We now study the conditions under which the infimum $\mathbf{A} \wedge \mathbf{B}$ of the matrices $\mathbf{A}$ and $\mathbf{B}$ of the same order exists. We prove the following:

Theorem 12.2.9. *Let* $\mathbf{A}$ *and* $\mathbf{B}$ *be matrices of the same order. Let* $\mathcal{S}_a = \mathcal{C}(\mathbf{A})$, $\mathcal{T}_a = \mathcal{C}(\mathbf{A}^\mathbf{t})$, $\mathcal{S}_\mathbf{b} = \mathcal{C}(\mathbf{B})$ *and* $\mathcal{T}_b = \mathcal{C}(\mathbf{B}^\mathbf{t})$. *Further, assume* *that the shorted operators* $\mathbf{S}(\mathbf{A}|\mathcal{S}_\mathbf{b}, \mathcal{T}_\mathbf{b})$ *and* $\mathbf{S}(\mathbf{B}|\mathcal{S}_\mathbf{a}, \mathcal{T}_\mathbf{a})$ *both exist and the field* F *has characteristic different from 2. Then the following hold:*

(i) *The pairs '*$\mathbf{A}$ *and* $\mathbf{B}$*,' and '*$\mathbf{S}(\mathbf{A}|\mathcal{S}_\mathbf{b}, \mathcal{T}_\mathbf{b})$ *and* $\mathbf{S}(\mathbf{B}|\mathcal{S}_\mathbf{a}, \mathcal{T}_\mathbf{a})$*' are parallel summable and* $\mathcal{P}(\mathbf{S}(\mathbf{A}, \mathcal{S}_\mathbf{b}, \mathcal{T}_\mathbf{b})|\mathbf{S}(\mathbf{B}|\mathcal{S}_\mathbf{a}, \mathcal{T}_\mathbf{a})) = \mathcal{P}(\mathbf{A}, \mathbf{B})$.

(ii) *If* $\mathbf{S}(\mathbf{A}|\mathcal{S}_\mathbf{b}, \mathcal{T}_\mathbf{b}) = \mathbf{S}(\mathbf{B}|\mathcal{S}_\mathbf{a}, \mathcal{T}_\mathbf{a})$*, then* $\mathbf{A} \wedge \mathbf{B}$ *exists and is given by* $2\mathcal{P}(\mathbf{A}|\mathbf{B})$.

Proof. (i) Proof follows by Ex. 11.2, Ex. 11.3 and Theorem 9.2.14.
(ii) If $\mathbf{S}(\mathbf{A}|\mathcal{S}_\mathbf{b}, \mathcal{T}_\mathbf{b}) = \mathbf{S}(\mathbf{B}|\mathcal{S}_\mathbf{a}, \mathcal{T}_\mathbf{a})$, then

$$\mathcal{P}(\mathbf{S}(\mathbf{A}|\mathcal{S}_\mathbf{b}, \mathcal{T}_\mathbf{b})|\mathbf{S}(\mathbf{B}|\mathcal{S}_\mathbf{a}, \mathcal{T}_\mathbf{a})) = \frac{\mathbf{S}(\mathbf{A}|\mathcal{S}_\mathbf{b}, \mathcal{T}_\mathbf{b})}{\mathbf{2}}.$$

So by (i), we have $\frac{\mathbf{S}(\mathbf{A}|\mathcal{S}_\mathbf{b}, \mathcal{T}_\mathbf{b})}{2} = \mathcal{P}(\mathbf{A}|\mathbf{B})$. Clearly, $\mathbf{S}(\mathbf{A}|\mathcal{S}_\mathbf{b}, \mathcal{T}_\mathbf{b}) \in \mathfrak{C}$, as $\mathbf{S}(\mathbf{A}|\mathcal{S}_\mathbf{b}, \mathcal{T}_\mathbf{b}) <^- \mathbf{A}$, $\mathbf{S}(\mathbf{B}|\mathcal{S}_\mathbf{a}, \mathcal{T}_\mathbf{a}) <^- \mathbf{B}$ and $\mathbf{S}(\mathbf{A}|\mathcal{S}_\mathbf{b}, \mathcal{T}_\mathbf{b}) = \mathbf{S}(\mathbf{B}|\mathcal{S}_\mathbf{a}, \mathcal{T}_\mathbf{a})$. If $\mathbf{C} \in \mathfrak{C}$, then $\mathbf{C} <^- \mathbf{A}$ and $\mathbf{C} <^- \mathbf{B}$. So, $\mathbf{C} <^- \mathbf{A}$, $\mathcal{C}(\mathbf{C}) \subseteq \mathcal{C}(\mathbf{B})$, and $\mathcal{C}(\mathbf{C}^\mathbf{t}) \subseteq \mathcal{C}(\mathbf{B}^\mathbf{t})$. So, $\mathbf{C} <^- \mathbf{S}(\mathbf{A}|\mathcal{S}_\mathbf{b}, \mathcal{T}_\mathbf{b})$. Hence, $\mathbf{A} \wedge \mathbf{B} = \mathbf{S}(\mathbf{A}|\mathcal{S}_\mathbf{b}, \mathcal{T}_\mathbf{b}) = \mathbf{2}\mathcal{P}(\mathbf{A}|\mathbf{B})$. □

Theorem 12.2.10. *Let* $\mathbf{A}$ *and* $\mathbf{B}$ *be matrices of the same order. Let* $\mathcal{S}_a = \mathcal{C}(\mathbf{A})$, $\mathcal{T}_a = \mathcal{C}(\mathbf{A}^\mathbf{t})$, $\mathcal{S}_\mathbf{b} = \mathcal{C}(\mathbf{B})$ *and* $\mathcal{T}_b = \mathcal{C}(\mathbf{B}^\mathbf{t})$*. Further, assume that the shorted operators* $\mathbf{S}(\mathbf{A}|\mathcal{S}_\mathbf{b}, \mathcal{T}_\mathbf{b})$ *and* $\mathbf{S}(\mathbf{B}|\mathcal{S}_\mathbf{a}, \mathcal{T}_\mathbf{a})$ *both exist and are not equal. Then either* $\mathbf{A} \wedge \mathbf{B}$ *does not exist or if it exists, it must be the null matrix.*

Proof. We first show that when $\mathbf{S}(\mathbf{A}|\mathcal{S}_\mathbf{b}, \mathcal{T}_\mathbf{b})$ and $\mathbf{S}(\mathbf{B}|\mathcal{S}_\mathbf{a}, \mathcal{T}_\mathbf{a})$ exist, then $\mathbf{A} \wedge \mathbf{B} = \mathbf{S}(\mathbf{A}|\mathcal{S}_\mathbf{b}, \mathcal{T}_\mathbf{b}) \wedge \mathbf{S}(\mathbf{B}|\mathcal{S}_\mathbf{a}, \mathcal{T}_\mathbf{a})$. Recall that $\mathbf{S}(\mathbf{A}|\mathcal{S}_\mathbf{b}, \mathcal{T}_\mathbf{b})$ is the unique maximal element of $\{\mathbf{C} : \mathbf{C} <^- \mathbf{A},\ \mathcal{C}(\mathbf{C}) \subseteq \mathcal{S}_\mathbf{b},\ \mathcal{C}(\mathbf{C}^\mathbf{t}) \subseteq \mathcal{T}_\mathbf{b}\}$ and $\mathbf{S}(\mathbf{B}|\mathcal{S}_\mathbf{a}, \mathcal{T}_\mathbf{a})$ is that of $\{\mathbf{C} : \mathbf{C} <^- \mathbf{B},\ \mathcal{C}(\mathbf{C}) \subseteq \mathcal{S}_\mathbf{a},\ \mathcal{C}(\mathbf{C}^\mathbf{t}) \subseteq \mathcal{T}_\mathbf{a}\}$. Also, $\mathbf{A} \wedge \mathbf{B}$ is the unique maximal element of $\mathfrak{C}$. Let $\mathbf{D} = \mathbf{A} \wedge \mathbf{B}$. Clearly, $\mathbf{D} <^- \mathbf{S}(\mathbf{A}|\mathcal{S}_\mathbf{b}, \mathcal{T}_\mathbf{b})$ and $\mathbf{D} <^- \mathbf{S}(\mathbf{B}|\mathcal{S}_\mathbf{a}, \mathcal{T}_\mathbf{a})$, since $\mathbf{D} <^- \mathbf{A}$ and $\mathbf{D} <^- \mathbf{B}$. Therefore, $\mathbf{D} <^- \mathbf{S}(\mathbf{A}|\mathcal{S}_\mathbf{b}, \mathcal{T}_\mathbf{b}) \wedge \mathbf{S}(\mathbf{B}|\mathcal{S}_\mathbf{a}, \mathcal{T}_\mathbf{a})$.

However, $\mathbf{S}(\mathbf{A}|\mathcal{S}_\mathbf{b}, \mathcal{T}_\mathbf{b}) \wedge \mathbf{S}(\mathbf{B}|\mathcal{S}_\mathbf{a}, \mathcal{T}_\mathbf{a}) <^- \mathbf{D}$, so,

$$\mathbf{S}(\mathbf{A}|\mathcal{S}_\mathbf{b}, \mathcal{T}_\mathbf{b}) \wedge \mathbf{S}(\mathbf{B}|\mathcal{S}_\mathbf{a}, \mathcal{T}_\mathbf{a}) = \mathbf{D} = \mathbf{A} \wedge \mathbf{B}.$$

Let $(\mathbf{P}, \mathbf{Q})$ be a rank factorization of $\mathbf{S}(\mathbf{A}|\mathcal{S}_\mathbf{b}, \mathcal{T}_\mathbf{b})$. Since, $\mathcal{C}(\mathbf{S}(\mathbf{A}|\mathcal{S}_\mathbf{b}, \mathcal{T}_\mathbf{b})) = \mathcal{C}(\mathbf{S}(\mathbf{B}|\mathcal{S}_\mathbf{a}, \mathcal{T}_\mathbf{a})) = \mathcal{C}(\mathbf{A}) \cap \mathcal{C}(\mathbf{B})$ and $\mathcal{C}(\mathbf{S}(\mathbf{A}|\mathcal{S}_\mathbf{b}, \mathcal{T}_\mathbf{b}))^\mathbf{t} = \mathcal{C}(\mathbf{S}(\mathbf{B}|\mathcal{S}_\mathbf{a}, \mathcal{T}_\mathbf{a}))^\mathbf{t} = \mathcal{C}(\mathbf{A}^\mathbf{t}) \cap \mathcal{C}(\mathbf{B}^\mathbf{t})$, by Theorem 3.2.6(iii), there exists a non-singular matrix $\mathbf{T}$ such that $\mathbf{S}(\mathbf{B}|\mathcal{S}_\mathbf{a}, \mathcal{T}_\mathbf{a}) = \mathbf{PTQ}$ for some non-singular matrix $\mathbf{T}$. It follows that without any loss of generality we may take $\mathbf{A} = \mathbf{I}$ and $\mathbf{B}$ to be any non-singular matrix. Let $\mathbf{C}$ be a non-null matrix such that $\mathbf{C} <^- \mathbf{I}$ and

$\mathbf{C} <^- \mathbf{B}$. We consider two cases namely: (i) when $\mathbf{B}$ and $\mathbf{C}$ do not commute and (ii) when they commute.

Case 1: Let $\mathbf{B}$ and $\mathbf{C}$ do not commute. Then $\mathbf{C_0} = \mathbf{BCB^{-1}}$ and $\mathbf{C}$ are distinct matrices with same rank and satisfy $\mathbf{C_0} <^- \mathbf{I}$ and $\mathbf{C_0} <^- \mathbf{B}$. As $\mathbf{C_0}$ and $\mathbf{C}$ are non-comparable, $\underline{\mathfrak{C}}$ can not have a unique maximal element.

Case 2: Let $\mathbf{B}$ and $\mathbf{C}$ commute and $\mathbf{X}$ be a matrix of appropriate order such that $(\mathbf{I}-\mathbf{C})\mathbf{XC} \neq \mathbf{0}$ and $\det(\mathbf{I}+(\mathbf{I}-\mathbf{C})\mathbf{X}) \neq \mathbf{0}$. Now, let $\mathbf{C_0} = \mathbf{C}+(\mathbf{I}-\mathbf{C})\mathbf{XC} = (\mathbf{I}+(\mathbf{I}-\mathbf{C})\mathbf{X})\mathbf{C}$. Clearly, $\rho(\mathbf{C_0}) = \rho(\mathbf{C})$. Also, $\mathbf{C_0}$ is idempotent as $\mathbf{C}$ is idempotent. So, $\mathbf{C_0} <^- \mathbf{I}$. It is, now, easy to see that $\mathbf{C_0} = \mathbf{C_0B^{-1}C_0}$, and therefore, $\mathbf{C_0} <^- \mathbf{B}$. Also, $\mathbf{C_0}$ and $\mathbf{C}$ are non-comparable. Thus, once again $\underline{\mathfrak{C}}$ can not have a unique maximal element. □

We next show that the only situation in which $\mathbf{A} \wedge \mathbf{B}$ is non-null occurs when $\mathbf{S}(\mathbf{A}|\mathcal{S}_\mathbf{b}, \mathcal{T}_\mathbf{b})$ and $\mathbf{S}(\mathbf{B}|\mathcal{S}_\mathbf{a}, \mathcal{T}_\mathbf{a})$ are multiples of each other with multiplication factor $\neq 1$. Once again, we take $\mathbf{A} = \mathbf{I}$ and $\mathbf{B}$ to be any non-singular matrix. The following theorem identifies the situation when $\mathbf{A} \wedge \mathbf{B}$ is a non-null matrix.

Theorem 12.2.11. *Let* $\mathbf{B}$ *be a square matrix such that either* (a) $\mathbf{B}$ *is a non-null idempotent or* (b) $\mathbf{B^2} \neq \mathbf{kB}$, *for any scalar k. Then there exists a non-null idempotent matrix* $\mathbf{H}$ *such that* $\mathbf{H} <^- \mathbf{B}$.

Proof. (a) If $\mathbf{B}$ is a non-null idempotent, take $\mathbf{H} = \mathbf{B}$ and we are done.
(b) Let $\mathbf{B^2} \neq \mathbf{kB}$, for any scalar k i.e., $\mathbf{B}$ is not an idempotent. We shall actually manufacture a non-null idempotent matrix $\mathbf{H}$ of rank 1 such that $\mathbf{H} <^- \mathbf{B}$.
Let $\mathbf{H} = \mathbf{xy^t}$ be rank 1 matrix such that $\mathbf{H} <^- \mathbf{B}$. Clearly, $\mathbf{xy^tB^-xy^t} = \mathbf{xy^t}$ for each g-inverse $\mathbf{B^-}$ of $\mathbf{B}$. It follows that $\mathbf{y^tB^-x} = 1$ for each g-inverse $\mathbf{B^-}$ of $\mathbf{B}$. This is possible when $\mathbf{x} \in \mathcal{C}(\mathbf{B})$ and $\mathbf{y} \in \mathcal{C}(\mathbf{B^t})$ and $\mathbf{y^tB^-x} = 1$. Let $\mathbf{x} = \mathbf{Bu}$ and $\mathbf{y^t} = \mathbf{v^tB}$. Then $1 = \mathbf{y^tB^-x} = \mathbf{v^tBu}$. Also, $\mathbf{H^2} = \mathbf{H} \Rightarrow \mathbf{y^tx} = \mathbf{1} \Rightarrow \mathbf{v^tB^2u} = 1$. If $\mathbf{B^2} \neq \mathbf{kB}$, the equations $\mathbf{v^tBu} = 1$ and $\mathbf{v^tB^2u} = 1$ clearly posses a solution for $\mathbf{u}, \mathbf{v}$. Let $\mathbf{H} = \mathbf{Buv^tB}$. □

Thus, if $\mathbf{S}(\mathbf{A}|\mathcal{S}_\mathbf{b}, \mathcal{T}_\mathbf{b})$ and $\mathbf{S}(\mathbf{B}|\mathcal{S}_\mathbf{a}, \mathcal{T}_\mathbf{a})$ exist and unless they are multiples of each other with multiplication factor $\neq 1$, we can always find a non-null matrix that is dominated by $\mathbf{A}$ and $\mathbf{B}$ under the minus order. This guarantees the non-null $\mathbf{A} \wedge \mathbf{B}$.

We now turn our attention to the case when one of $\mathbf{S}(\mathbf{A}|\mathcal{S}_{\mathbf{b}}, \mathcal{T}_{\mathbf{b}})$ and $\mathbf{S}(\mathbf{B}|\mathcal{S}_{\mathbf{a}}, \mathcal{T}_{\mathbf{a}})$ does not exist. We will show that $\mathbf{A} \wedge \mathbf{B}$ exists but under some conditions. Before we can proceed to do so, we need some preparation.

Lemma 12.2.12. *Let $(\mathbf{L}, \mathbf{R})$ be a rank factorization of a square matrix $\mathbf{A}$. Then $\mathbf{A}$ is idempotent if and only if $\mathbf{RL} = \mathbf{I}$.*

Proof is trivial.

Theorem 12.2.13. *Let $\mathbf{E_1}$ be a matrix of order $k \times m$ and rank r. Let $(\mathbf{L_1}, \mathbf{R})$ be a rank factorization of $\mathbf{E_1}$ and $\mathbf{R} = (\mathbf{R_1} : \mathbf{R_2})$ be a partition of $\mathbf{R}$, where $\mathbf{R_1}$ is of order $r \times k$. then there exists a matrix $\mathbf{E_2}$ of order $(m-k) \times m$ such that $\mathbf{E} = \begin{pmatrix} \mathbf{E_1} \\ \mathbf{E_2} \end{pmatrix}$ is idempotent of rank r if and only if $\mathcal{C}(\mathbf{I} - \mathbf{R_1 L_1}) \subseteq \mathcal{C}(\mathbf{R_2})$.*

Proof. 'If' part
Let $\mathcal{C}(\mathbf{I} - \mathbf{R_1 L_1}) \subseteq \mathcal{C}(\mathbf{R_2})$, so, $\mathbf{I} - \mathbf{R_1 L_1} = \mathbf{R_1 X}$ for some matrix X.
Let $\mathbf{E_2} = \mathbf{XR}$ and $\mathbf{E} = \begin{pmatrix} \mathbf{E_1} \\ \mathbf{E_2} \end{pmatrix}$. Then

$$\mathbf{E}^2 = \begin{pmatrix} \mathbf{L_1} \\ \mathbf{X} \end{pmatrix} \mathbf{R} \begin{pmatrix} \mathbf{L_1} \\ \mathbf{X} \end{pmatrix} \mathbf{R} = \begin{pmatrix} \mathbf{L_1} \\ \mathbf{X} \end{pmatrix} (\mathbf{R_1} : \mathbf{R_2}) \begin{pmatrix} \mathbf{L_1} \\ \mathbf{X} \end{pmatrix} (\mathbf{R_1} : \mathbf{R_2}) = \mathbf{E},$$

since, $(\mathbf{R_1} : \mathbf{R_2}) \begin{pmatrix} \mathbf{L_1} \\ \mathbf{X} \end{pmatrix} = \mathbf{I}$.
'Only if' part
Let $\mathbf{E} = \begin{pmatrix} \mathbf{E_1} \\ \mathbf{E_2} \end{pmatrix}$ be an idempotent matrix of rank r. We first notice that $\mathbf{E_1} = (\mathbf{L_1 R_1} : \mathbf{L_1 R_2})$. Partition the matrix $\mathbf{E_2}$ as $\mathbf{E_2} = (\mathbf{E_2'} : \mathbf{E_2''})$. Then the matrix $\mathbf{E} = \begin{pmatrix} \mathbf{L_1 R_1} & \mathbf{L_1 R_2} \\ \mathbf{E_2'} & \mathbf{E_2''} \end{pmatrix}$ is idempotent. So,

$$\mathbf{L_1 R_1 E_2'} = \mathbf{L_1}(\mathbf{I} - \mathbf{R_1 L_1})\mathbf{R_1}, \ \mathbf{L_1 R_2 E_2''} = \mathbf{L_1}(\mathbf{I} - \mathbf{R_1 L_1})\mathbf{R_2},$$

$$\mathbf{E_2' L_1 R_1} + \mathbf{E_2'' E_2'} = \mathbf{E_2'} \text{ and } \mathbf{E_2' L_1 R_2} + \mathbf{E_2'' E_2''} = \mathbf{E_2''}.$$

Thus, $\mathbf{L_1 R_2 E_2} = \mathbf{L_1}(\mathbf{I} - \mathbf{R_1 L_1})\mathbf{R}$. As $\mathbf{L_1}$ has a left inverse and $\mathbf{R}$ has a right inverse, we see that $\mathcal{C}(\mathbf{I} - \mathbf{R_1 L_1}) \subseteq \mathcal{C}(\mathbf{R_2})$. □

Theorem 12.2.14. *In the setup of Theorem 12.2.13, there exists a matrix $\mathbf{E_2}$ of order $(m-k) \times m$ such that $\mathbf{E} = \begin{pmatrix} \mathbf{E_1} \\ \mathbf{E_2} \end{pmatrix}$ is idempotent of rank $r + s$, $s > 0$ into which $\mathbf{E_1}$ is embedded if and only if $\mathcal{C}(\mathbf{I} - \mathbf{R_1 L_1}) \subseteq \mathcal{C}(\mathbf{R_2})$ and $s \leq (m-k) - \rho(\mathbf{R_2})$.*

Proof. 'Only if' part

Let there exists a matrix $\mathbf{E_2}$ of order $(m-k) \times m$ such that $\mathbf{E} = \begin{pmatrix} \mathbf{E_1} \\ \mathbf{E_2} \end{pmatrix}$ is idempotent of rank $r+s,\ s > 0$ into which $\mathbf{E_1}$ is embedded. We will construct a rank factorization $(\mathbf{S}, \mathbf{T})$ of $\mathbf{E}$ using the rank factorization $(\mathbf{L_1}, \mathbf{R})$ of $\mathbf{E_1}$ as in Theorem 12.2.13. Since $\mathbf{E}$ has rank $r+s,\ \mathcal{C}(\mathbf{R^t}) \neq \mathcal{C}(\mathbf{E^t})$. So, we must add additional rows to $\mathbf{R}$ so that $\mathcal{C}(\mathbf{R^t}) = \mathcal{C}(\mathbf{E^t})$. We add terminal s rows forming a matrix say, $\mathbf{Y}$ to $\mathbf{R}$ such that the matrix $\begin{pmatrix} \mathbf{R} \\ \mathbf{Y} \end{pmatrix}$ is a matrix of rank $r+s$ and is the right factor in our proposed factorization. Since the first k rows of $\mathbf{E}$ are also the first k rows of $\mathbf{E_1}$, the rows of $\mathbf{R}$ form a basis of the row span of these k rows of $\mathbf{E}$. So, in the left factor of $\mathbf{E}$ we can take the first k rows of $\mathbf{L_1}$ followed by s null rows. We can write $\mathbf{E} = \begin{pmatrix} \mathbf{L_1} & \mathbf{L_2} \\ \begin{pmatrix} \mathbf{0} \\ \circ \end{pmatrix} & \begin{pmatrix} \mathbf{0} \\ \circ \end{pmatrix} \end{pmatrix} \begin{pmatrix} \mathbf{R_1} & \mathbf{R_2} \\ \mathbf{Y_1} & \mathbf{Y_2} \end{pmatrix}$, where $\mathbf{L_2}$ is some $k \times s$ matrix and $\begin{pmatrix} \mathbf{0} \\ \circ \end{pmatrix}$ is some $(m-k) \times s$ matrix $\circ$ being some appropriate matrix. Since $\mathbf{E}$ is an idempotent by Lemma 12.2.12, $\begin{pmatrix} \mathbf{R_1} & \mathbf{R_2} \\ \mathbf{Y_1} & \mathbf{Y_2} \end{pmatrix} \begin{pmatrix} \mathbf{L_1} & \mathbf{L_2} \\ \begin{pmatrix} \mathbf{0} \\ \circ \end{pmatrix} & \begin{pmatrix} \mathbf{0} \\ \circ \end{pmatrix} \end{pmatrix} = \mathbf{I_{r+s}}$. So,

$$\mathbf{R_1L_1} + \mathbf{R_2}\begin{pmatrix} \mathbf{0} \\ \circ \end{pmatrix} = \mathbf{I_r},\ \mathbf{Y_1L_2} + \mathbf{Y_2}\begin{pmatrix} \mathbf{0} \\ \circ \end{pmatrix} = \mathbf{I_s},\ \mathbf{R}\begin{pmatrix} \mathbf{L_2} \\ \begin{pmatrix} \mathbf{0} \\ \circ \end{pmatrix} \end{pmatrix} = \mathbf{0}$$

and

$$\mathbf{Y}\begin{pmatrix} \mathbf{L_1} \\ \begin{pmatrix} \mathbf{0} \\ \circ \end{pmatrix} \end{pmatrix} = \mathbf{0}.$$

Now, $\mathbf{R_1L_1} + \mathbf{R_2}\begin{pmatrix} \mathbf{0} \\ \circ \end{pmatrix} = \mathbf{I_r} \Rightarrow \mathcal{C}(\mathbf{I} - \mathbf{R_1L_1}) \subseteq \mathcal{C}(\mathbf{R_2})$ and $\mathbf{R}\begin{pmatrix} \mathbf{L_2} \\ \begin{pmatrix} \mathbf{0} \\ \circ \end{pmatrix} \end{pmatrix} = \mathbf{0}$ $\Rightarrow$ the last s columns of the left factor are in null space of $\mathbf{R}$. Therefore, $s \leq d(\mathcal{N}(\mathbf{R_2})) = (m-k) - \rho(\mathbf{R_2})$ and so, we can take the last s columns of the left factor as $\begin{pmatrix} \mathbf{0} \\ \mathbf{Z} \end{pmatrix}$, where $\mathbf{0}$ is the null matrix of order $k \times s$ and $\mathbf{Z}$ is the matrix of order $(m-k) \times s$, whose columns are linearly independent vectors from the null space of $\mathbf{R_2}$. Thus, $\mathbf{S} = \begin{pmatrix} \mathbf{L_1} & \mathbf{0} \\ \mathbf{X} & \mathbf{Z} \end{pmatrix}$ for some matrix $\mathbf{X}$.

'If' part

Let $\mathcal{C}(\mathbf{I} - \mathbf{R_1L_1}) \subseteq \mathcal{C}(\mathbf{R_2})$ and $s \leq (m-k) - \rho(\mathbf{R_2})$ hold. By Theorem 12.2.13, $\mathbf{E_1}$ can be embedded in an idempotent of rank r and $\mathbf{R_2K} = \mathbf{I} - \mathbf{R_1L_1}$ for some matrix $\mathbf{K}$ of order $(m-k) \times r$. Let $\mathbf{Y}$ be some matrix of order $s \times m$ to be determined, so that $\mathbf{T} = \begin{pmatrix} \mathbf{R} \\ \mathbf{Y} \end{pmatrix}$ is of order $(r+s) \times m$ and of rank $r+s$. Choose s columns from a basis of $\mathcal{N}(\mathbf{R_2})$ and let $\mathbf{Z}$ be a matrix of order $(m-k) \times s$ formed by them. Let $\mathbf{S} = \begin{pmatrix} \mathbf{L_1} & \mathbf{0} \\ \mathbf{K} & \mathbf{Z} \end{pmatrix}$. Then $\mathbf{E} = \mathbf{ST}$ is of order $m \times m$. Clearly, $\rho(\mathbf{S}) = r+s$. By Lemma 12.2.12, $\mathbf{E}$ is idempotent $\Leftrightarrow \mathbf{TS} = \mathbf{I_{r+s}} \Leftrightarrow \begin{pmatrix} \mathbf{R} \\ \mathbf{Y} \end{pmatrix} \begin{pmatrix} \mathbf{L_1} & \mathbf{0} \\ \mathbf{K} & \mathbf{Z} \end{pmatrix} = \mathbf{I_{r+s}}$. Partition $\mathbf{Y}$ as $(\mathbf{Y_1} : \mathbf{Y_2})$, where $\mathbf{Y_1}$ is of order $s \times k$ and $\mathbf{Y_2}$ is of order $s \times (m-k)$. Now,

$$\begin{aligned} &\begin{pmatrix} \mathbf{R} \\ \mathbf{Y} \end{pmatrix} \begin{pmatrix} \mathbf{L_1} & \mathbf{0} \\ \mathbf{K} & \mathbf{Z} \end{pmatrix} = \mathbf{I_{r+s}} \\ \Leftrightarrow &\begin{pmatrix} \mathbf{R_1} & \mathbf{R_2} \\ \mathbf{Y_1} & \mathbf{Y_2} \end{pmatrix} \begin{pmatrix} \mathbf{L_1} & \mathbf{0} \\ \mathbf{K} & \mathbf{Z} \end{pmatrix} = \mathbf{I_{r+s}} \\ \Leftrightarrow &\,\mathbf{R_2K} = \mathbf{I_r} - \mathbf{R_1L_1},\ \mathbf{R_2Z} = \mathbf{0}, \\ &\mathbf{Y_1L_1} + \mathbf{Y_2K} = \mathbf{0} \text{ and } \mathbf{Y_2Z} = \mathbf{I_s}. \end{aligned}$$

The equation $\mathbf{Y_1L_1} + \mathbf{Y_2K} = \mathbf{0}$ is equivalent to $\mathbf{Y}\begin{pmatrix} \mathbf{L_1} \\ \mathbf{K} \end{pmatrix} = \mathbf{0}$. Both the equations $\mathbf{Y}\begin{pmatrix} \mathbf{L_1} \\ \mathbf{K} \end{pmatrix} = \mathbf{0}$ and $\mathbf{Y}\begin{pmatrix} \mathbf{0} \\ \mathbf{Z} \end{pmatrix} = \mathbf{I}$ are consistent individually and jointly, being linearly independent, hence form a solution for $\mathbf{Y}$. Thus, with this solution as $\mathbf{Y}$, the matrix $\mathbf{T}$ is the right factor and $\mathbf{E} = \mathbf{ST}$ is idempotent of rank $r+s$ in which $\mathbf{E_1}$ is embedded. □

Finally we are ready to study the conditions under which $\mathbf{A} \wedge \mathbf{B}$ exists.

Theorem 12.2.15. *Let* $\mathbf{A}$ *and* $\mathbf{B}$ *be matrices of the same order. Let* $\mathcal{S}_a = \mathcal{C}(\mathbf{A})$, $\mathcal{T}_a = \mathcal{C}(\mathbf{A^t})$, $\mathcal{S}_b = \mathcal{C}(\mathbf{B})$ *and* $\mathcal{T}_b = \mathcal{C}(\mathbf{B^t})$. *Further, assume that the shorted operator* $\mathbf{S}(\mathbf{B}|\mathcal{S}_\mathbf{a}, \mathcal{T}_\mathbf{a})$ *does not exist. Then the following hold:*

(i) *There exists a matrix* $\mathbf{C}$ *such that* $\mathbf{C} <^- \mathbf{B}$
(ii) *There exists an idempotent* $\mathbf{C}$ *such that* $\mathbf{C} <^- \mathbf{B}$ *and* $\mathbf{C} <^- \mathbf{A}$ *and*
(iii) *If there exists a unique idempotent* $\mathbf{C}$ *such that* $\mathbf{C} <^- \mathbf{B}$ *and* $\mathbf{C} <^- \mathbf{A}$, *then* $\mathbf{C} = \mathbf{A} \wedge \mathbf{B}$, *the infimum of* $\mathbf{A}$, $\mathbf{B}$.

Proof. There is no loss of generality if we take $\mathbf{A} = \begin{pmatrix} \mathbf{I_m} & \mathbf{0} \\ \mathbf{0} & \mathbf{0} \end{pmatrix}$ and $\mathbf{B} = \begin{pmatrix} \mathbf{B_{11}} & \mathbf{B_{12}} \\ \mathbf{B_{21}} & \mathbf{B_{22}} \end{pmatrix}$. Since, $\mathbf{S}(\mathbf{B}|\mathcal{S}_\mathbf{a}, \mathcal{T}_\mathbf{a})$ exists $\Leftrightarrow \mathcal{C}(\mathbf{B_{21}}) \subseteq \mathcal{C}(\mathbf{B_{22}})$ and

$\mathcal{C}(\mathbf{B}^t_{12}) \subseteq \mathcal{C}(\mathbf{B}^t_{22})$, so, either of the two conditions must fail to hold. Let us assume $\mathcal{C}(\mathbf{B}_{21}) \subseteq \mathcal{C}(\mathbf{B}_{22})$ but $\mathcal{C}(\mathbf{B}^t_{12}) \nsubseteq \mathcal{C}(\mathbf{B}^t_{22})$. Now,

$$\mathcal{C}(\mathbf{B}_{21}) \subseteq \mathcal{C}(\mathbf{B}_{22}) \Leftrightarrow \mathcal{C}(\mathbf{B}_{21},\ \mathbf{B}_{22})^t \cap \mathcal{C}(\mathbf{I},\ \mathbf{0})^t = \{\mathbf{0}\}.$$

For, if $(\mathbf{B}_{21},\ \mathbf{B}_{22})^t\mathbf{x} = (\mathbf{I},\ \mathbf{0})^t\mathbf{z}$, then $\mathbf{B}^t_{21}\mathbf{x} = \mathbf{z}$ and $\mathbf{B}^t_{22}\mathbf{x} = \mathbf{0}$. However, $\mathbf{B}^t_{22}\mathbf{x} = \mathbf{0} \Rightarrow \mathbf{B}^t_{21}\mathbf{B}^{-t}_{22}\mathbf{B}^t_{22}\mathbf{x} = (\mathbf{B}_{22}\mathbf{B}^-_{22}\mathbf{B}_{21})^t\mathbf{x} = \mathbf{0} \Rightarrow \mathbf{B}^t_{21}\mathbf{x} = \mathbf{0} \Rightarrow \mathbf{z} = \mathbf{0}$. Thus, $(\mathbf{B}_{21},\ \mathbf{B}_{22})^t\mathbf{x} = \mathbf{0}$ and $(\mathbf{I},\ \mathbf{0})^t\mathbf{z} = \mathbf{0}$.
Conversely, let $\mathcal{C}(\mathbf{B}_{21},\ \mathbf{B}_{22})^t \cap \mathcal{C}(\mathbf{I},\ \mathbf{0})^t = \{\mathbf{0}\}$. So, $(\mathbf{B}_{21},\ \mathbf{B}_{22})^t\mathbf{x} = (\mathbf{I},\ \mathbf{0})^t\mathbf{z} \Rightarrow (\mathbf{B}_{21},\ \mathbf{B}_{22})^t\mathbf{x} = \mathbf{0}$ and $(\mathbf{I},\ \mathbf{0})^t\mathbf{z} = \mathbf{0}$. For an arbitrary vector $\mathbf{u}$, let $\mathbf{x} = \begin{pmatrix}\mathbf{I}-\mathbf{B}_{22}\mathbf{B}^-_{22}\\ \mathbf{I}-\mathbf{B}_{22}\mathbf{B}^-_{22}\end{pmatrix}^t\mathbf{u}$ and $\mathbf{z} = \begin{pmatrix}\mathbf{B}_{21}(\mathbf{I}-\mathbf{B}_{22}\mathbf{B}^-_{22})\\ \mathbf{0}\end{pmatrix}^t\mathbf{u}$.
Then $(\mathbf{B}_{21},\ \mathbf{B}_{22})^t\mathbf{x} = (\mathbf{I},\ \mathbf{0})^t\mathbf{z}$, so, $(\mathbf{I}-\mathbf{B}_{22}\mathbf{B}^-_{22})^t\mathbf{B}^t_{21}\mathbf{u} = \mathbf{0}$. This gives $(\mathbf{I}-\mathbf{B}_{22}\mathbf{B}^-_{22})^t\mathbf{B}^t_{21} = \mathbf{0}$ or equivalently, $\mathcal{C}(\mathbf{B}_{21}) \subseteq \mathcal{C}(\mathbf{B}_{22})$.

Let $\mathbf{x}_{i2}$, $i = 1, \ldots, \ell$ be a basis of of $\mathcal{N}(\mathbf{B}^t_{21})$. Each $\mathbf{x}_{i2}$ is an $n-m$-vector. Then for each $i = 1, \ldots, \ell$, the vectors $\mathbf{x}_i = \begin{pmatrix}\mathbf{0}\\ \mathbf{x}_{i2}\end{pmatrix} \in \mathcal{N}(\mathbf{B}^t_{12},\ \mathbf{B}^t_{22})$, where $\mathbf{0}$ is a null m-vector. Extend these ℓ vectors to a basis of $\mathcal{N}(\mathbf{B}^t_{12},\ \mathbf{B}^t_{22})$, say by taking additional k vectors $\mathbf{x}_i = \begin{pmatrix}\mathbf{x}_{i1}\\ \mathbf{x}_{i2}\end{pmatrix}$, $i = \ell+1, \ldots, \ell+k$. Since $\mathbf{x}_i$, $i = \ell+1, \ldots, \ell+k$ are linearly independent, $\mathbf{x}_{i1}$, $i = \ell+1, \ldots, \ell+k$ are also linearly independent. Choose $\mathbf{x}_{i1}$, $i = \ell+k+1, \ldots, \ell+m$ such that $\mathbf{x}_i$, $i = \ell+1, \ldots, \ell+m$ form a basis of F^m. Set $\mathbf{x}^t_{i1}\mathbf{B}_{11} + \mathbf{x}^t_{i2}\mathbf{B}_{21} = \mathbf{z}^t_{i-\ell}$ for $i = \ell+1, \ldots, \ell+k$. Note that each $\mathbf{z}^t_i$ is an m-vector and are k in number. We define a matrix $\mathbf{C}_{11}$ as: $\mathbf{x}^t_{i1}\mathbf{C}_{11} = \mathbf{z}^t_{i-\ell}$ for $i = \ell+1, \ldots, \ell+k$. Such a matrix $\mathbf{C}_{11}$ exists. Let $\mathbf{C} = \begin{pmatrix}\mathbf{C}_{11} & \mathbf{0}\\ \mathbf{0} & \mathbf{0}\end{pmatrix}$. We show $\mathcal{C}(\mathbf{B}-\mathbf{C})^t \cap \mathcal{C}(\mathbf{I},\ \mathbf{0})^t = \{\mathbf{0}\}$.

Let $\mathbf{x} = (\mathbf{B}-\mathbf{C})^t\mathbf{y} = (\mathbf{I},\ \mathbf{0})^t\mathbf{z}$. Partition $\mathbf{y} = \begin{pmatrix}\mathbf{y}_1\\ \mathbf{y}_2\end{pmatrix}$ and $\mathbf{z}$ as $\mathbf{z} = \begin{pmatrix}\mathbf{z}_1\\ \mathbf{z}_2\end{pmatrix}$.
Then $(\mathbf{B}_{11}-\mathbf{C}_{11})^t\mathbf{y}_1 + \mathbf{B}^t_{21}\mathbf{y}_2 = \mathbf{z}_1$ and $\mathbf{B}^t_{21}\mathbf{y}_2 + \mathbf{B}^t_{22}\mathbf{y}_2 = \mathbf{0}$. So, $\mathbf{y} \in \mathcal{N}(\mathbf{B}^t_{21},\ \mathbf{B}^t_{22})$. Let $\mathbf{y} = \Sigma^{\ell+k}_{i=1}\alpha_i\mathbf{x}_i$. Then $(\mathbf{B}_{11}-\mathbf{C}_{11})^t\mathbf{y}_1 + \mathbf{B}^t_{21}\mathbf{y}_2 = \mathbf{z}_1$

$$\Rightarrow \begin{pmatrix}(\mathbf{B}_{11}-\mathbf{C}_{11})^t\\ \mathbf{B}^t_{21}\end{pmatrix}\mathbf{y} = \begin{pmatrix}\mathbf{I}\\ \mathbf{0}\end{pmatrix}\mathbf{z} \Rightarrow \Sigma^{\ell+k}_{i=1}\alpha_i\begin{pmatrix}(\mathbf{B}_{11}-\mathbf{C}_{11})^t\\ \mathbf{B}^t_{21}\end{pmatrix}\mathbf{x}_i = \begin{pmatrix}\mathbf{I}\\ \mathbf{0}\end{pmatrix}\mathbf{z}$$

$$\Rightarrow \Sigma^{\ell}_{i=1}\alpha_i\begin{pmatrix}(\mathbf{B}_{11}-\mathbf{C}_{11})^t\\ \mathbf{B}^t_{21}\end{pmatrix}\mathbf{x}_i + \Sigma^{\ell+k}_{i=\ell+1}\alpha_i\begin{pmatrix}(\mathbf{B}_{11}-\mathbf{C}_{11})^t\\ \mathbf{B}^t_{21}\end{pmatrix}\mathbf{x}_i = \begin{pmatrix}\mathbf{I}\\ \mathbf{0}\end{pmatrix}\mathbf{z}$$

$$\Rightarrow \Sigma^{\ell}_{i=1}\alpha_i\begin{pmatrix}(\mathbf{B}_{11}-\mathbf{C}_{11})^t\\ \mathbf{B}^t_{21}\end{pmatrix}\begin{pmatrix}\mathbf{0}\\ \mathbf{x}_{i2}\end{pmatrix} + \Sigma^{\ell+k}_{i=\ell+1}\alpha_i\begin{pmatrix}(\mathbf{B}_{11}-\mathbf{C}_{11})^t\\ \mathbf{B}^t_{21}\end{pmatrix}\begin{pmatrix}\mathbf{x}_{i1}\\ \mathbf{x}_{i2}\end{pmatrix}$$

$$= \begin{pmatrix}\mathbf{I}\\ \mathbf{0}\end{pmatrix}\mathbf{z}.$$

It can now be seen that this gives $\mathbf{x} = \mathbf{0}$.

Similarly, $\mathcal{C}(\mathbf{B}-\mathbf{C})^{\mathbf{t}} \cap \mathcal{C}(\mathbf{C_{11}},\ \mathbf{0})^{\mathbf{t}} = \{\mathbf{0}\}$.

Let $\mathbf{Z}$ be a matrix whose rows are the k $\mathbf{z_i^t}$ in natural order as defined and $r = \rho(\mathbf{Z})$. Then $\rho(\mathbf{B}) = n = (\ell+k)+r$, for $\mathbf{x_i^t B} = \mathbf{0}$ for $i = 1,\ldots,\ell$. Also, $\mathbf{x_i^t B} = (\mathbf{x_{i1}^t},\ \mathbf{x_{i2}^t}) \begin{pmatrix} \mathbf{B_{11}} & \mathbf{B_{12}} \\ \mathbf{B_{21}} & \mathbf{B_{22}} \end{pmatrix} = (\mathbf{z_{i-\ell}^t},\ \mathbf{0})$ for $i = \ell+1,\ldots,\ell+m$. Moreover, $\rho(\mathbf{Z}) = r$, $d(\mathcal{N}(\mathbf{Z})) = k-r$, so, $\mathbf{x_i^t B} = \mathbf{0}$ for exactly those i for which $\mathbf{z_{i-\ell}^t}$ is a linear combination of r of linearly independent $\mathbf{z_j^t}$. Such $\mathbf{z_{i-\ell}^t}$ are $k-r$ in number and form a basis of $\mathcal{N}(\mathbf{Z})$. So, there are $\ell+k-r$ linearly independent vectors $\mathbf{x_i^t}$ for which $\mathbf{x_i^t B} = \mathbf{0}$. Thus, $d(\mathcal{N}(\mathbf{B^t})) = \ell+k-r$, which gives $\rho(\mathbf{B}) = \rho(\mathbf{B^t}) = n-(\ell+k)+r$. Notice that

$\mathbf{x_i^t}(\mathbf{B}-\mathbf{C}) = (\mathbf{x_{i1}^t},\ \mathbf{x_{i2}^t}) \begin{pmatrix} \mathbf{B_{11}}-\mathbf{C_{11}} & \mathbf{B_{12}} \\ \mathbf{B_{21}} & \mathbf{B_{22}} \end{pmatrix} = \mathbf{0}$ for $i = \ell+1,\ldots,\ell+m$.

So, $d(\mathcal{N}(\mathbf{B^t}-\mathbf{C^t})) = \ell+k$, and therefore, $\rho(\mathbf{B}-\mathbf{C}) = n-(\ell+k)$. Clearly, $\rho(\mathbf{C_{11}}) \geq r$. We can choose $\mathbf{C_{11}}$ of rank r. (Since exactly r of $\mathbf{z_{i-\ell}^t}$ are linearly independent, we take corresponding $\mathbf{x_{i1}^t}$ and choose $\mathbf{C_{11}}$ which will be of rank r.) Then $\mathbf{B} = \mathbf{C} \oplus (\mathbf{B}-\mathbf{C})$, so, $\mathbf{C} <^- \mathbf{B}$.

(ii) We first show that the choice of $\mathbf{C_{11}}$ is independent of choice of a basis of $\mathcal{N}(\mathbf{B_{12}^t},\ \mathbf{B_{22}^t})$. Let $\mathbf{y_i}$, $i = 1,\ldots,\ell+k$ be another basis of $\mathcal{N}(\mathbf{B_{12}^t},\ \mathbf{B_{22}^t})$, and $\mathbf{y_i} = \begin{pmatrix} \mathbf{0} \\ \mathbf{y_{i2}} \end{pmatrix}$, $i = 1,\ldots,\ell$. Then each $\mathbf{y_i}$ is a linear combination of $\mathbf{x_i}$, $i = 1,\ldots,\ell+k$. So, $\mathbf{y_{\ell+1}} = \mathbf{\Sigma_{i=1}^{\ell+k}}\beta_\mathbf{i}\mathbf{x_i}$ for some scalars β_i. Consider

$$\mathbf{w_1^t} = \mathbf{y_{(\ell+1)1}^t B_{11}} + \mathbf{y_{(\ell+1)2}^t B_{21}} = \mathbf{\Sigma_{i=1}^{\ell+k}}\beta_\mathbf{i}(\mathbf{x_{i1}^t B_{11}} + \mathbf{x_{i2}^t B_{21}})$$
$$= \Sigma_{i=\ell+1}^{\ell+k}\beta_i(\mathbf{x_{i1}^t B_{11}} + \mathbf{x_{i2}^t B_{21}}) = \mathbf{\Sigma_{i=\ell+1}^{\ell+k}}\beta_\mathbf{i}\mathbf{z_{i-\ell}^t},$$

as $\mathcal{C}(\mathbf{B_{21}}) \subseteq \mathcal{C}(\mathbf{B_{22}})$. Thus,

$$\mathbf{w_1^t} = \mathbf{\Sigma_{i=\ell+1}^{\ell+k}}\beta_\mathbf{i}\mathbf{z_{i-\ell}^t} = \mathbf{\Sigma_{i=\ell+1}^{\ell+k}}\beta_\mathbf{i}\mathbf{x_{i1}^t C_{11}} = \mathbf{y_{\ell+1}^t C_{11}}.$$

Therefore, $\mathbf{C_{11}}$ satisfies the same conditions in terms of $\mathbf{y_{i1}^t}$ as it did in terms of $\mathbf{x_{i1}^t}$.

We next show that we can choose $\mathbf{C_{11}}$ to be an idempotent of rank r. This will show $\mathbf{C} <^- \mathbf{A}$. As $\mathbf{C} <^- \mathbf{B}$ and $\mathbf{C}$ is unique, whence (iii) follows. So, let the vectors $\mathbf{x_{i1}^t}$ be arranged in their natural order and form the rows of of a matrix $\mathbf{M}$. Clearly, $\mathbf{M}$ is non-singular as $\mathbf{x_{i1}^t}$ for $i = \ell+1,\ldots,\ell+m$ form a basis of F^m. Let $\mathbf{T}$ be a matrix such that $\mathbf{MC_{11}} = \begin{pmatrix} \mathbf{Z} \\ \mathbf{T} \end{pmatrix}$. We have to find a matrix $\mathbf{T}$ such that $\begin{pmatrix} \mathbf{Z} \\ \mathbf{T} \end{pmatrix}$ is a square matrix of rank r and $\mathbf{M^{-1}} \begin{pmatrix} \mathbf{Z} \\ \mathbf{T} \end{pmatrix}$ is idempotent say $\mathbf{E}$. Notice that $\begin{pmatrix} \mathbf{Z} \\ \mathbf{T} \end{pmatrix} \mathbf{E} = \begin{pmatrix} \mathbf{Z} \\ \mathbf{T} \end{pmatrix} = \mathbf{E} \begin{pmatrix} \mathbf{Z} \\ \mathbf{T} \end{pmatrix}$. Having found a matrix $\mathbf{T}$, we let $(\mathbf{L_1}, \mathbf{R})$ be a rank factorization of $\mathbf{ZM^{-1}} = \mathbf{Z_1}$ and

$\mathbf{R} = (\mathbf{R_1} : \mathbf{R_2})$ be a partition of $\mathbf{R}$, where $\mathbf{R_1}$ is of order $r \times (m-k)$ and has full column rank. By Theorem 12.2.14, there exists a unique $\mathbf{T_1}$ such that $\mathbf{E} = \begin{pmatrix} \mathbf{Z_1} \\ \mathbf{T_1} \end{pmatrix}$ is idempotent of rank r. Choose $\mathbf{C_{11}}$ as $\mathbf{M^{-1}EM}$. Then $\mathbf{C_{11}}$ is the unique idempotent of rank r. Now (iii) follows. □

We now give an example of a pair of matrices $\mathbf{A}$ and $\mathbf{B}$ to show when $\mathbf{S}(\mathbf{A}|\mathcal{S}_\mathbf{b}, \mathcal{T}_\mathbf{b})$ and $\mathbf{S}(\mathbf{B}|\mathcal{S}_\mathbf{a}, \mathcal{T}_\mathbf{a})$ do not exist, the infimum $\mathbf{A} \wedge \mathbf{B}$ also does not exist.

Example 12.2.16. Let $\mathbf{A} = \begin{pmatrix} 1 & 0 & 1 \\ 0 & 1 & 0 \\ 0 & 0 & 0 \end{pmatrix}$ and $\mathbf{B} = \begin{pmatrix} 1 & 0 & 0 \\ 0 & 1 & 0 \\ 0 & 0 & 0 \end{pmatrix}$ be the matrices as in Example 12.2.1. Then $\mathbf{S}(\mathbf{A}|\mathcal{S}_\mathbf{b}, \mathcal{T}_\mathbf{b})$ and $\mathbf{S}(\mathbf{B}|\mathcal{S}_\mathbf{a}, \mathcal{T}_\mathbf{a})$ both do not exist. Also, $\mathbf{C} = \begin{pmatrix} 1 & 0 & 0 \\ 0 & 0 & 0 \\ 0 & 0 & 0 \end{pmatrix}$ and $\mathbf{D} = \begin{pmatrix} 0 & 0 & 0 \\ 0 & 1 & 0 \\ 0 & 0 & 0 \end{pmatrix}$ are such that $\mathbf{C} <^- \mathbf{A}$, $\mathbf{C} <^- \mathbf{B}$, $\mathbf{D} <^- \mathbf{A}$, and $\mathbf{D} <^- \mathbf{B}$. But $\mathbf{C}$ and $\mathbf{D}$ are not comparable under the minus order. So, the infimum $\mathbf{A} \wedge \mathbf{B}$, of $\mathbf{A}$ and $\mathbf{B}$ also does not exist.

The following theorem gives an upper bound for $\mathbf{A} \wedge \mathbf{B}$, the infimum of $\mathbf{A}$ and $\mathbf{B}$.

Theorem 12.2.17. *Let* $\mathbf{A}$ *and* $\mathbf{B}$ *be matrices of the same order. If for some* $\gamma \neq 0, \neq 1$ $\gamma^{-1}\mathbf{A}$ *and* $(1-\gamma)^{-1}\mathbf{B}$ *are parallel summable, then*

$$\mathbf{A} \wedge \mathbf{B} <^- \mathcal{P}(\gamma^{-1}\mathbf{A}, (1-\gamma)^{-1}\mathbf{B}).$$

Proof. Let $\mathbf{C} = \mathbf{A} \wedge \mathbf{B}$. Then $\mathbf{C} <^- \mathbf{A}$ and $\mathbf{C} <^- \mathbf{B}$. So, $\{\mathbf{A}^-\} \subseteq \{\mathbf{C}^-\}$ and $\{\mathbf{B}^-\} \subseteq \{\mathbf{C}^-\}$. By Theorem 9.2.14, $\{\mathcal{P}(\gamma^{-1}\mathbf{A}, (1-\gamma)^{-1}\mathbf{B})^-\} = \{\mathbf{A}^-\} + \{\mathbf{B}^-\} \subseteq \{\mathbf{C}^-\}$. Hence, $\mathbf{C} = \mathbf{A} \wedge \mathbf{B} <^- \mathcal{P}(\gamma^{-1}\mathbf{A}, (1-\gamma)^{-1}\mathbf{B})$, by using Theorem 3.3.5. □

12.3 Supremum and infimum under the star order

We have seen in the last section that under the minus order $F^{m\times n}$ fails to be even a semi-lattice in general. However, the same is not the case with $\mathbb{C}^{m\times n}$ under the star order. We show here that under star order, $\mathbb{C}^{m\times n}$ is a lower semi-lattice but is not an upper semi-lattice for, if $\mathbf{A}$, $\mathbf{B}$ is a pair of matrices, with $\mathbf{A}$ or $\mathbf{B}$, a maximal element of $\mathbb{C}^{m\times n}$ under the star order, then there is no matrix $\mathbf{C}$ that can dominate $\mathbf{A}$ and $\mathbf{B}$ and hence $\mathbf{A} \vee \mathbf{B}$ cannot exist.

We first note that for any two orthogonal projectors $\mathbf{E}$ and $\mathbf{F}$, $\mathbf{E} \vee \mathbf{F}$ and $\mathbf{E} \wedge \mathbf{F}$ always exist. (See Ex. 12.4.) To show the existence of the infimum of any two matrices, we first proceed to show the existence of the infimum when both matrices are partial isometries. Then by using a special type of representation of matrices known as the Penrose decomposition, we establish the existence of infimum of arbitrary matrices. All matrices in this section are complex matrices unless stated otherwise. We start with the following lemmas:

Lemma 12.3.1. *Let* $\mathbf{A}$ *and* $\mathbf{B} = \begin{pmatrix} \mathbf{L} & \mathbf{0} \\ \mathbf{0} & \mathbf{0} \end{pmatrix}$ *be matrices of the same order. Then* $\mathbf{A} <^\star \mathbf{B}$ *if and only if* $\mathbf{A} = \begin{pmatrix} \mathbf{K} & \mathbf{0} \\ \mathbf{0} & \mathbf{0} \end{pmatrix}$ *with* $\mathbf{K} <^\star \mathbf{L}$, *and all partitions involved are conformable for matrix operations.*

Proof. 'If' part is trivial.
'Only if' part
Let $\mathbf{A} = \begin{pmatrix} \mathbf{K} & \mathbf{M} \\ \mathbf{N} & \mathbf{T} \end{pmatrix}$ be the partition of $\mathbf{A}$ conformable with the given partition of $\mathbf{B}$. Then $\mathbf{A} <^\star \mathbf{B} \Rightarrow \mathbf{A} <^{\mathbf{s}} \mathbf{B} \Rightarrow \mathbf{M} = \mathbf{0}$, $\mathbf{N} = \mathbf{0}$ and $\mathbf{T} = \mathbf{0}$. Now, the result follows immediately. □

Lemma 12.3.2. *Let* $\mathbf{B_i}$ *be matrices of order* $m_i \times n$ *for each* $i = 1, \ldots, s$ *and* $\mathfrak{C} = \{\mathbf{E} \in \mathbb{C}^{\mathbf{n}\times\mathbf{n}} : \mathbf{E}$ *is an orthogonal projector and* $\mathbf{B_i E} = \mathbf{0}$ *for each* $\mathbf{i}\}$. *Then* $\mathfrak{C}$ *has the infimum given by an orthogonal projector* $\mathbf{F}$ *on the space* $\cap_{i=1}^{s} \mathcal{N}(\mathbf{B_i})$.

Proof. Since $\mathbf{B_i E} = \mathbf{0} \Rightarrow \mathcal{C}(\mathbf{E}) \subseteq \mathcal{N}(\mathbf{B_i})$ for each i, so, $\mathbf{E} \in \mathfrak{C} \Leftrightarrow \mathcal{C}(\mathbf{E}) \subseteq \cap_{\mathbf{i}=1}^{\mathbf{s}} \mathcal{N}(\mathbf{B_i})$. Let $\mathbf{F}$ be a projection on the space $\cap_{i=1}^{s} \mathcal{N}(\mathbf{B_i})$. Clearly, $\mathbf{F} \in \mathfrak{C}$ and for each $\mathbf{E} \in \mathfrak{C}, \mathcal{C}(\mathbf{E}) \subseteq \mathcal{C}(\mathbf{F})$. Now, $\mathbf{E} <^\star \mathbf{F} \Leftrightarrow \mathcal{C}(\mathbf{E}) \subseteq \mathcal{C}(\mathbf{F})$, so, $\mathbf{F}$ is an upper bound of $\mathfrak{C}$. It is trivial to see that $\mathbf{F}$ is the unique maximal element of $\mathfrak{C}$. □

Recall a partial isometry is a matrix $\mathbf{A} \in \mathbb{C}^{\mathbf{m}\times\mathbf{n}}$ if $\mathbf{A}\mathbf{A}^\star\mathbf{A} = \mathbf{A}$, equivalently $\mathbf{A}^\star = \mathbf{A}^\dagger$. If $\mathbf{A}$ is a partial isometry, then its singular value decomposition is $\mathbf{X}^\star\mathbf{A}\mathbf{Y} = \begin{pmatrix} \mathbf{I_r} & \mathbf{0} \\ \mathbf{0} & \mathbf{0} \end{pmatrix}$, where $r = \rho(\mathbf{A})$, matrices $\mathbf{X}$ and $\mathbf{Y}$ are unitary.

Lemma 12.3.3. *Let* $\mathbf{A}$ *and* $\mathbf{B}$ *be matrices of the same order such that*

$\mathbf{A}$ *is an isometry. Let* $\mathbf{X}^\star\mathbf{A}\mathbf{Y} = \begin{pmatrix} \mathbf{I_r} & \mathbf{0} \\ \mathbf{0} & \mathbf{0} \end{pmatrix}$, *where* $r = \rho(\mathbf{A})$, $\mathbf{X}$ *and* $\mathbf{Y}$ *are unitary. Further, let* $\mathbf{X}^\star\mathbf{B}\mathbf{Y} = \begin{pmatrix} \mathbf{L} & \mathbf{M} \\ \mathbf{N} & \mathbf{T} \end{pmatrix}$, *where partitions are conformable for matrix operations. Then* $\mathbf{A} \wedge \mathbf{B}$ *exists and is given by* $\mathbf{A} \wedge \mathbf{B} = \mathbf{X}\begin{pmatrix} \mathbf{Q} & \mathbf{0} \\ \mathbf{0} & \mathbf{0} \end{pmatrix}\mathbf{Y}^\star$, *where* $\mathbf{Q}$ *is the orthogonal projector onto* $\mathcal{N}(\mathbf{I}-\mathbf{L}) \cap \mathcal{N}(\mathbf{I}-\mathbf{L})^\star \cap \mathcal{N}(\mathbf{M})^\star \cap \mathcal{N}(\mathbf{N})$.

Proof. Now, $\mathbf{A} <^\star \mathbf{B} \Leftrightarrow \mathbf{X}^\star\mathbf{A}\mathbf{Y} <^\star \mathbf{X}^\star\mathbf{B}\mathbf{Y}$, since $\mathbf{X}$ and $\mathbf{Y}$ are unitary. So, we may take $\mathbf{A} = \begin{pmatrix} \mathbf{I_r} & \mathbf{0} \\ \mathbf{0} & \mathbf{0} \end{pmatrix}$ and $\mathbf{B} = \begin{pmatrix} \mathbf{L} & \mathbf{M} \\ \mathbf{N} & \mathbf{T} \end{pmatrix}$. By Lemma 12.3.1,
$\mathbf{C} <^\star \begin{pmatrix} \mathbf{I_r} & \mathbf{0} \\ \mathbf{0} & \mathbf{0} \end{pmatrix} \Leftrightarrow \mathbf{C} = \begin{pmatrix} \mathbf{E} & \mathbf{0} \\ \mathbf{0} & \mathbf{0} \end{pmatrix}$ for some projector $\mathbf{E}$ and $\mathbf{C} <^\star \begin{pmatrix} \mathbf{L} & \mathbf{M} \\ \mathbf{N} & \mathbf{T} \end{pmatrix}$
$\Leftrightarrow \mathbf{LE} = \mathbf{E} = \mathbf{EL}$, $\mathbf{EM} = \mathbf{0}$ and $\mathbf{NE} = \mathbf{0}$
$\Leftrightarrow \mathcal{C}(\mathbf{E}) \subseteq \mathcal{N}(\mathbf{I}-\mathbf{L}) \cap \mathcal{N}(\mathbf{I}-\mathbf{L})^\star \cap \mathcal{N}(\mathbf{M})^\star \cap \mathcal{N}(\mathbf{N})$.
Let $\mathbf{Q}$ be the orthogonal projector on

$$\mathcal{N}(\mathbf{I}-\mathbf{L}) \cap \mathcal{N}(\mathbf{I}-\mathbf{L})^\star \cap \mathcal{N}(\mathbf{M})^\star \cap \mathcal{N}(\mathbf{N}).$$

Clearly, $\mathcal{C}(\mathbf{E}) \subseteq \mathcal{C}(\mathbf{Q})$. Thus, $\mathbf{A} \wedge \mathbf{B}$ exists and is given by

$$\mathbf{A} \wedge \mathbf{B} = \mathbf{X}\begin{pmatrix} \mathbf{Q} & \mathbf{0} \\ \mathbf{0} & \mathbf{0} \end{pmatrix}\mathbf{Y}^\star.$$

□

We now study the existence of the infimum for two arbitrary matrices. For this, we need a representation of matrices known as the Penrose decomposition. We describe it below.

Let $\mathbf{A}$ be an $m \times n$ matrix. Consider the singular value decomposition $\mathbf{A} = \mathbf{X}\begin{pmatrix} \mathbf{D} & \mathbf{0} \\ \mathbf{0} & \mathbf{0} \end{pmatrix}\mathbf{Y}^\star$, where $\mathbf{X} \in \mathbb{C}^{\mathbf{m}\times\mathbf{m}}$, $\mathbf{Y} \in \mathbb{C}^{\mathbf{n}\times\mathbf{n}}$ are unitary, $\mathbf{D} = \mathbf{diag}(\sigma_1\mathbf{I_{r_1}}, \ldots, \sigma_\mathbf{k}\mathbf{I_{r_k}}) \in \mathbb{C}^{\mathbf{r}\times\mathbf{r}}$ and σ_i are distinct eigen-values of $\mathbf{A}$ such that $r = r_1 + \ldots r_k = \rho(\mathbf{A})$. Partition $\mathbf{X}$, $\mathbf{Y}$ conformably with the block diagonal decomposition of $\mathbf{D}$, say $\mathbf{X} = (\mathbf{X_1}, \ldots, \mathbf{X_{k+1}})$ and $\mathbf{Y} = (\mathbf{Y_1}, \ldots, \mathbf{Y_{k+1}})$. Define $\mathbf{U}_{\sigma_\mathbf{i}} = \mathbf{X}\mathbf{diag}(\mathbf{0}, \ldots, \mathbf{0}, \mathbf{I_{r_i}}, \mathbf{0}, \ldots, \mathbf{0})$ for each $i = 1, \ldots, k$, and $\mathbf{U}_\alpha = \mathbf{0}$ for each positive real α, if α is not a singular value of $\mathbf{A}$. Then we can write

$$\mathbf{A} = \mathbf{\Sigma}_{\mathbf{i}=1}^{\mathbf{k}}\sigma_\mathbf{i}\mathbf{X_i}\mathbf{Y_i}^\star = \mathbf{\Sigma}_{\sigma>0}\alpha\mathbf{U}_\alpha \tag{12.3.1}$$

where $\mathbf{U}_\alpha$ are non-zero only for finitely many values of α and are pairwise $\star$-orthogonal partial isometries i.e., $\mathbf{U}_\alpha\mathbf{U}^\star{}_\beta\mathbf{U}_\alpha = \mathbf{U}_\alpha$, $\mathbf{U}_\alpha\mathbf{U}^\star{}_\beta = \mathbf{0}$ and

$\mathbf{U^\star}_\alpha \mathbf{U}_\beta = \mathbf{0}$ for $\alpha \neq \beta$.
The representation in (12.3.1) is called the Penrose decomposition of $\mathbf{A}$.

Theorem 12.3.4. *Let* $\mathbf{A}$ *and* $\mathbf{B}$ *be matrices of the same order having Penrose decompositions* $\mathbf{A} = \Sigma_{\alpha>0}\alpha\mathbf{U}_\alpha$ *and* $\mathbf{B} = \Sigma_{\beta>0}\beta\mathbf{V}_\beta$ *respectively. Consider the following statements:*

(i) $\mathbf{A} <^\star \mathbf{B}$,
(ii) $\mathbf{U}_\alpha <^\star \mathbf{V}_\alpha$ *for each positive real* α *and*
(iii) *For all* α, β, $\alpha \neq \beta$, $\mathbf{U^\star}_\alpha \mathbf{V}_\beta = \mathbf{0}$ *and* $\mathbf{U}_\alpha \mathbf{V^\star}_\beta = \mathbf{0}$, *i.e.,* $\mathbf{U}_\alpha$ *and* $\mathbf{V}_\beta$ *are* $\star-$*orthogonal.*

Then (i) $\Leftrightarrow$ (ii). *Also, any of* (i) *or* (ii) $\Rightarrow$ (iii).

Proof. We first prove (i) $\Rightarrow$ (iii).
By (i), $\mathbf{A^\star A} = \mathbf{A^\star B}$ and $\mathbf{AA^\star} = \mathbf{BA^\star}$.
$\mathbf{A^\star A} = \mathbf{A^\star B} \Rightarrow \mathbf{U}_\alpha\mathbf{A^\star A} = \mathbf{U}_\alpha\mathbf{A^\star B} \Rightarrow \alpha^2\mathbf{U}_\alpha = \alpha\mathbf{U}_\alpha\mathbf{U}^\star_\alpha\mathbf{B}$.
Also, $\mathbf{BV}^\star_\beta = \beta\mathbf{V}_\beta\mathbf{V}^\star_\beta$. Therefore, $\alpha\mathbf{U}_\alpha\mathbf{V}^\star_\beta = \beta\mathbf{U}_\alpha\mathbf{U}^\star_\alpha\mathbf{V}_\beta\mathbf{V}^\star_\beta$.
Similarly, $\alpha\mathbf{U^\star V}_\beta = \beta\mathbf{U}^\star_\alpha\mathbf{U}_\alpha\mathbf{V}^\star_\beta\mathbf{V}_\beta$. So,

$$\begin{aligned}\alpha^2\mathbf{U}^\star_\alpha\mathbf{V}_\beta &= \beta\mathbf{U}^\star_\alpha(\alpha\mathbf{U}_\alpha\mathbf{V}^\star_\beta)\mathbf{V}_\beta\\ &= \beta^2\mathbf{U}^\star_\alpha(\mathbf{U}_\alpha\mathbf{U}^\star_\alpha\mathbf{V}_\beta\mathbf{V}^\star_\beta)\mathbf{V}_\beta\\ &= \beta^2\mathbf{U}^\star_\alpha\mathbf{V}_\beta.\end{aligned}$$

Since α, β are positive real numbers and $\alpha \neq \beta$, we have $\mathbf{U}^\star_\alpha\mathbf{V}_\beta = \mathbf{0}$. Similarly, $\mathbf{U}_\alpha\mathbf{V}^\star_\beta = \mathbf{0}$.
(i) $\Rightarrow$ (ii).
From (i) $\Rightarrow$ (iii), we have $\alpha\mathbf{U}_\alpha = \mathbf{U}_\alpha\mathbf{U}^\star_\alpha\mathbf{B}$. Pre-multiplying by $\mathbf{U}^\star_\alpha$ and using (iii), $\mathbf{U}^\star_\alpha\alpha\mathbf{U}_\alpha = \mathbf{U}^\star_\alpha\mathbf{U}_\alpha\mathbf{U}^\star_\alpha\mathbf{B} = \mathbf{U}^\star_\alpha\mathbf{B} = \alpha\mathbf{U}^\star_\alpha\mathbf{V}_\alpha$. As $\alpha \neq 0$, we have $\mathbf{U}^\star_\alpha\mathbf{U}_\alpha = \mathbf{U}^\star_\alpha\mathbf{V}_\alpha$. Similarly, $\mathbf{U}_\alpha\mathbf{U}^\star_\alpha = \mathbf{V}_\alpha\mathbf{U}^\star_\alpha$. Therefore, $\mathbf{U}_\alpha <^\star \mathbf{V}_\alpha$.
(ii) $\Rightarrow$ (i) is easy. □

Corollary 12.3.5. *Let* $\mathbf{A} = \Sigma_{\alpha>0}\alpha\mathbf{U}_\alpha$ *and* $\mathbf{B}$ *be matrices of the same order. Then* $\mathbf{A} <^\star \mathbf{B}$ *if and only if* $\alpha(\mathbf{U}_\alpha <^\star \mathbf{B})$ *for each* α.

Theorem 12.3.6. *Let* $\mathbf{A}$ *and* $\mathbf{B}$ *be matrices of the same order having Penrose decompositions* $\mathbf{A} = \Sigma_{\alpha>0}\alpha\mathbf{U}_\alpha$ *and* $\mathbf{B} = \Sigma_{\beta>0}\beta\mathbf{V}_\beta$ *respectively. Then* $\mathbf{A} \wedge \mathbf{B}$ *exists and is given by* $\mathbf{A} \wedge \mathbf{B} = \Sigma_{\alpha>0}\alpha\mathbf{U}_\alpha \wedge \mathbf{V}_\alpha$.

Proof. Let $\mathbf{W}_\alpha = \mathbf{U}_\alpha \wedge \mathbf{V}_\alpha$. Each $\mathbf{W}_\alpha$ exists by Lemma 12.3.2 and is a partial isometry. Also, $\mathbf{W}_\alpha <^\star \mathbf{U}_\alpha$ and $\mathbf{U}_\alpha$ is a partial isometry for each α. Let $\mathbf{D} = \Sigma_{\alpha>0}\alpha\mathbf{W}_\alpha$. We show that $\mathbf{D}$ is the infimum of $\mathbf{A}$ and $\mathbf{B}$. Since for each α, $\mathbf{W}_\alpha <^\star \mathbf{U}_\alpha$, we have $\mathbf{W}_\alpha^\star \mathbf{W}_\alpha = \mathbf{W}_\alpha^\star \mathbf{U}_\alpha$, $\mathbf{W}_\alpha \mathbf{W}_\alpha^\star = \mathbf{U}_\alpha \mathbf{W}_\alpha^\star$. So, $\mathbf{W}_\alpha^\star \mathbf{W}_\beta = \mathbf{W}_\alpha^\star \mathbf{W}_\alpha \mathbf{W}_\alpha^\star \mathbf{W}_\beta \mathbf{W}_\beta^\star \mathbf{W}_\beta = \mathbf{W}_\alpha^\star \mathbf{W}_\alpha \mathbf{U}_\alpha^\star \mathbf{U}_\beta \mathbf{W}_\beta^\star \mathbf{W}_\beta = \mathbf{0}$ for $\alpha \neq \beta$, since $\mathbf{U}_\alpha^\star \mathbf{U}_\beta = \mathbf{0}$. Similarly, $\mathbf{W}_\alpha \mathbf{W}_\beta^\star = \mathbf{0}$. Let $\mathbf{C} = \Sigma \alpha \mathbf{T}_\alpha$ be an $m \times n$ matrix such that $\mathbf{C} <^\star \mathbf{A}$ and $\mathbf{C} <^\star \mathbf{B}$. Then by Theorem 12.3.4, $\mathbf{T}_\alpha <^\star \mathbf{U}_\alpha$, $\mathbf{T}_\alpha <^\star \mathbf{V}_\alpha$ for each α. So, $\mathbf{T}_\alpha <^\star \mathbf{W}_\alpha$ for each α. Once again by Theorem 12.3.4, $\mathbf{C} <^\star \mathbf{D}$, and hence $\mathbf{D} = \mathbf{A} \wedge \mathbf{B}$. □

As an immediate consequence of Theorem 12.3.6, we have

Theorem 12.3.7. *Let* $\mathbf{A}$ *and* $\mathbf{B}$ *be matrices of the same order. Then*

(i) $(\mathbf{A} \wedge \mathbf{B})^\star = \mathbf{A}^\star \wedge \mathbf{B}^\star$,
(ii) $(\mathbf{A} \wedge \mathbf{B})^\dagger = \mathbf{A}^\dagger \wedge \mathbf{B}^\dagger$ *and*
(iii) $\alpha(\mathbf{A} \wedge \mathbf{B}) = \alpha\mathbf{A} \wedge \alpha\mathbf{B}$.

Thus, it follows that the infimum of two hermitian matrices is hermitian.

We have expressed the infimum of given matrices in terms of some partial isometries. The following theorems express $\mathbf{A} <^\star \mathbf{B}$ in terms of $\mathbf{A}$, $\mathbf{B}$, $\mathbf{A}^\star$ and $\mathbf{B}^\star$.

Theorem 12.3.8. *Let* $\mathbf{A}$ *and* $\mathbf{B}$ *be matrices of the same order. Let*

$$\mathbf{E} = \mathbf{I} - (\mathbf{I} - \mathbf{A}^\dagger\mathbf{B})(\mathbf{I} - \mathbf{A}^\dagger\mathbf{B})^\dagger \text{ and } \mathbf{F} = \mathbf{I} - (\mathbf{I} - \mathbf{B}\mathbf{A}^\dagger)^\dagger(\mathbf{I} - \mathbf{B}\mathbf{A}^\dagger).$$

Then the following are equivalent:

(i) $\mathbf{AE} = \mathbf{FA}$
(ii) $\mathbf{A}(\mathbf{I} - \mathbf{E}) = (\mathbf{I} - \mathbf{F})\mathbf{A}$ *and*
(iii) $\mathcal{C}(\mathbf{A}(\mathbf{I} - \mathbf{A}^\dagger\mathbf{B})) \subseteq \mathcal{C}(\mathbf{I} - \mathbf{B}\mathbf{A}^\dagger)^\star$ and
$\mathcal{C}((\mathbf{I} - \mathbf{B}\mathbf{A}^\dagger)\mathbf{A})^\star \subseteq \mathcal{C}(\mathbf{I} - \mathbf{A}^\dagger\mathbf{B})$.

When this is so, $\mathbf{A} \wedge \mathbf{B} = \mathbf{AE} = \mathbf{FA}$.

Proof. Notice that both $\mathbf{E}$ and $\mathbf{F}$ are projectors.
(i) ⇔ (ii) is easy.
(ii) ⇒ (iii)
Since $\mathbf{A}(\mathbf{I} - \mathbf{E}) = (\mathbf{I} - \mathbf{F})\mathbf{A}$, we have

$$\mathbf{A}(\mathbf{I} - \mathbf{A}^\dagger\mathbf{B})(\mathbf{I} - \mathbf{A}^\dagger\mathbf{B})^\dagger = (\mathbf{I} - \mathbf{B}\mathbf{A}^\dagger)^\dagger(\mathbf{I} - \mathbf{B}\mathbf{A}^\dagger)\mathbf{A}. \tag{12.3.2}$$

Post-multiplying (12.3.2) by $\mathbf{I} - \mathbf{A}^\dagger\mathbf{B}$, we have

$$\mathbf{A}(\mathbf{I} - \mathbf{A}^\dagger\mathbf{B}) = (\mathbf{I} - \mathbf{B}\mathbf{A}^\dagger)^\dagger(\mathbf{I} - \mathbf{B}\mathbf{A}^\dagger)\mathbf{A}(\mathbf{I} - \mathbf{A}^\dagger\mathbf{B}).$$

Thus,

$$\mathcal{C}(\mathbf{A}(\mathbf{I}-\mathbf{A}^{\dagger}\mathbf{B})) = \mathcal{C}((\mathbf{I}-\mathbf{B}\mathbf{A}^{\dagger})^{\dagger}(\mathbf{I}-\mathbf{B}\mathbf{A}^{\dagger})\mathbf{A}(\mathbf{I}-\mathbf{A}^{\dagger}\mathbf{B})) \subseteq \mathcal{C}(\mathbf{I}-\mathbf{B}\mathbf{A}^{\dagger})^{\star}.$$

Similarly, pre-multiplying (12.3.2) by $\mathbf{I}-\mathbf{B}\mathbf{A}^{\dagger}$ and taking transposes we have $\mathcal{C}((\mathbf{I}-\mathbf{B}\mathbf{A}^{\dagger})\mathbf{A})^{\star} \subseteq \mathcal{C}(\mathbf{I}-\mathbf{A}^{\dagger}\mathbf{B})$.
(iii) $\Rightarrow$ (ii)
Now,

$$\begin{aligned}\mathcal{C}(\mathbf{A}(\mathbf{I}-\mathbf{A}^{\dagger}\mathbf{B})) &\subseteq \mathcal{C}(\mathbf{I}-\mathbf{B}\mathbf{A}^{\dagger})^{\star}\\ &\Rightarrow \mathbf{A}(\mathbf{I}-\mathbf{A}^{\dagger}\mathbf{B}) = (\mathbf{I}-\mathbf{B}\mathbf{A}^{\dagger})^{\star}(\mathbf{I}-\mathbf{B}\mathbf{A}^{\dagger})^{\star\dagger}\mathbf{A}(\mathbf{I}-\mathbf{A}^{\dagger}\mathbf{B})\\ &= ((\mathbf{I}-\mathbf{B}\mathbf{A}^{\dagger})^{\dagger}(\mathbf{I}-\mathbf{B}\mathbf{A}^{\dagger}))^{\star}\mathbf{A}(\mathbf{I}-\mathbf{A}^{\dagger}\mathbf{B})\\ &= (\mathbf{I}-\mathbf{B}\mathbf{A}^{\dagger})^{\dagger}(\mathbf{I}-\mathbf{B}\mathbf{A}^{\dagger})\mathbf{A}(\mathbf{I}-\mathbf{A}^{\dagger}\mathbf{B}).\end{aligned}$$

Post-multiplying by $(\mathbf{I}-\mathbf{A}^{\dagger}\mathbf{B})^{\dagger}$, we have

$$\mathbf{A}(\mathbf{I}-\mathbf{A}^{\dagger}\mathbf{B})(\mathbf{I}-\mathbf{A}^{\dagger}\mathbf{B})^{\dagger} = (\mathbf{I}-\mathbf{B}\mathbf{A}^{\dagger})^{\dagger}(\mathbf{I}-\mathbf{B}\mathbf{A}^{\dagger})\mathbf{A}(\mathbf{I}-\mathbf{A}^{\dagger}\mathbf{B})(\mathbf{I}-\mathbf{A}^{\dagger}\mathbf{B})^{\dagger}.$$

Also, $(\mathbf{I}-\mathbf{B}\mathbf{A}^{\dagger})\mathbf{A} = (\mathbf{I}-\mathbf{B}\mathbf{A}^{\dagger})\mathbf{A}(\mathbf{I}-\mathbf{A}^{\dagger}\mathbf{B})(\mathbf{I}-\mathbf{A}^{\dagger}\mathbf{B})^{\dagger}$, since

$$\mathcal{C}((\mathbf{I}-\mathbf{B}\mathbf{A}^{\dagger})\mathbf{A})^{\star} \subseteq \mathcal{C}(\mathbf{I}-\mathbf{A}^{\dagger}\mathbf{B}).$$

So,

$$\mathbf{A}(\mathbf{I}-\mathbf{A}^{\dagger}\mathbf{B})(\mathbf{I}-\mathbf{A}^{\dagger}\mathbf{B})^{\dagger} = (\mathbf{I}-\mathbf{B}\mathbf{A}^{\dagger})^{\dagger}(\mathbf{I}-\mathbf{B}\mathbf{A}^{\dagger})\mathbf{A}$$

or equivalently, $\mathbf{A}(\mathbf{I}-\mathbf{E}) = (\mathbf{I}-\mathbf{F})\mathbf{A}$. Thus, (ii) holds.
It is easy to see that $\mathbf{AE} <^{\star} \mathbf{A}$ and $\mathbf{AE} <^{\star} \mathbf{B}$. To show that $\mathbf{A} \wedge \mathbf{B} = \mathbf{AE}$, let $\mathbf{C} <^{\star} \mathbf{A}$ and $\mathbf{C} <^{\star} \mathbf{B}$. Since $\mathbf{C} <^{\star} \mathbf{A}$, we have $\mathbf{C}^{\dagger} <^{\star} \mathbf{A}^{\dagger} \Rightarrow \mathbf{C}\mathbf{C}^{\dagger} = \mathbf{C}\mathbf{A}^{\dagger}$ and $\mathbf{C}^{\dagger}\mathbf{C} = \mathbf{A}^{\dagger}\mathbf{C}$. Similarly, $\mathbf{C}\mathbf{C}^{\dagger} = \mathbf{C}\mathbf{B}^{\dagger}$ and $\mathbf{C}^{\dagger}\mathbf{C} = \mathbf{B}^{\dagger}\mathbf{C}$. So, $\mathbf{C}\mathbf{A}^{\dagger}\mathbf{B} = \mathbf{C}\mathbf{C}^{\dagger}\mathbf{B} = \mathbf{C}\mathbf{C}^{\dagger}\mathbf{C} = \mathbf{C}$. Similarly, $\mathbf{B}\mathbf{A}^{\dagger}\mathbf{C} = \mathbf{C}$. Therefore, $\mathbf{CE} = \mathbf{C} = \mathbf{FC}$, so, $\mathbf{C}\mathbf{C}^{\dagger}\mathbf{AE} = \mathbf{CE} = \mathbf{C} = \mathbf{FC} = \mathbf{FA}\mathbf{C}^{\dagger}\mathbf{C} = \mathbf{EA}\mathbf{C}^{\dagger}\mathbf{C}$. Thus, $\mathbf{C} <^{\star} \mathbf{AE}$, hence the result. □

Theorem 12.3.9. *Let* $\mathbf{A}$ *and* $\mathbf{B}$ *be matrices of the same order. Let* $\mathbf{G} = \mathbf{I}-(\mathbf{A}-\mathbf{B})^{\dagger}(\mathbf{A}-\mathbf{B})$ *and* $\mathbf{H} = \mathbf{I}-(\mathbf{A}-\mathbf{B})(\mathbf{A}-\mathbf{B})^{\dagger}$. *Then the following are equivalent:*

(i) $\mathbf{AG} = \mathbf{HA}$
(ii) $\mathbf{A}(\mathbf{I}-\mathbf{G}) = (\mathbf{I}-\mathbf{H})\mathbf{A}$
(iii) $\mathcal{C}(\mathbf{A}(\mathbf{A}^{\star}-\mathbf{B}^{\star})) \subseteq \mathcal{C}(\mathbf{A}-\mathbf{B})$ $\mathcal{C}((\mathbf{A}^{\star}-\mathbf{B}^{\star})\mathbf{A})^{\star} \subseteq \mathcal{C}(\mathbf{A}-\mathbf{B})^{\star}$ *and*
(iv) $\mathbf{A}(\mathbf{A}-\mathbf{B})^{\dagger}\mathbf{B} = \mathbf{B}(\mathbf{A}-\mathbf{B})^{\dagger}\mathbf{A}$.

When this is so, $\mathbf{A} \wedge \mathbf{B} = \mathbf{AG} = \mathbf{HA}$.

Proof. (i) $\Leftrightarrow$ (ii) is easy.
(ii) $\Leftrightarrow$ (iv) is obvious.
(ii) $\Rightarrow$ (iii)
Since (ii) holds, $\mathbf{A}(\mathbf{A}-\mathbf{B})^{\dagger}(\mathbf{A}-\mathbf{B}) = (\mathbf{A}-\mathbf{B})(\mathbf{A}-\mathbf{B})^{\dagger}\mathbf{A}$.
Post-multiplying by $(\mathbf{A}-\mathbf{B})^{\star}$, we have

$$\mathbf{A}(\mathbf{A}-\mathbf{B})^{\dagger}(\mathbf{A}-\mathbf{B})(\mathbf{A}-\mathbf{B})^{\star} = (\mathbf{A}-\mathbf{B})(\mathbf{A}-\mathbf{B})^{\dagger}\mathbf{A}(\mathbf{A}-\mathbf{B})^{\star}$$

and therefore,

$$\mathbf{A}(\mathbf{A}-\mathbf{B})^{\star} = (\mathbf{A}-\mathbf{B})(\mathbf{A}-\mathbf{B})^{\dagger}\mathbf{A}(\mathbf{A}-\mathbf{B})^{\star}$$

$$\Rightarrow \mathcal{C}(\mathbf{A}(\mathbf{A}^{\star}-\mathbf{B}^{\star})) \subseteq \mathcal{C}(\mathbf{A}-\mathbf{B}).$$

Now,

$$\begin{aligned}\mathcal{C}((\mathbf{A}^{\star}-\mathbf{B}^{\star})\mathbf{A})^{\star} &= \mathcal{C}(\mathbf{A}^{\star}(\mathbf{A}-\mathbf{B}))\\ &= \mathcal{C}(\mathbf{A}^{\star}(\mathbf{A}-\mathbf{B})(\mathbf{A}-\mathbf{B})^{\dagger}(\mathbf{A}-\mathbf{B}))\\ &= \mathcal{C}(\mathbf{A}^{\star}((\mathbf{A}-\mathbf{B})(\mathbf{A}-\mathbf{B})^{\dagger})^{\star}(\mathbf{A}-\mathbf{B}))\\ &= \mathcal{C}((((\mathbf{A}-\mathbf{B})(\mathbf{A}-\mathbf{B})^{\dagger})\mathbf{A})^{\star}(\mathbf{A}-\mathbf{B}) \subseteq \mathcal{C}(\mathbf{A}-\mathbf{B})^{\star}.\end{aligned}$$

Remaining proof can be completed along the lines of the proof in Theorem 12.3.8. □

We have studied the shorted matrix defined through minus partial order in Chapter 11 at length. There is really nothing particularly special about the minus order. As a matter of fact one can use any of the partial orders studied till now and define a shorted matrix. In the remainder of this section we briefly study the shorted matrix defined through the star order.

Definition 12.3.10. Let $\mathbf{A}$ be an $m \times n$ matrix. Let $\mathcal{S}$ and $\mathcal{T}$ be subspaces of $\mathbb{C}^n$ and $\mathbb{C}^m$ respectively. Let

$$\mathfrak{D} = \{\mathbf{C} \in \mathbb{C}^{\mathbf{m}\times\mathbf{n}} : \mathbf{C} <^{\star} \mathbf{A},\ \mathcal{C}(\mathbf{C}) \subseteq \mathcal{S} \text{ and } \mathcal{C}(\mathbf{C}^{\star}) \subseteq \mathcal{T}\}.$$

The unique maximal element of $\mathfrak{D}$ is called the shorted matrix of $\mathbf{A}$ relative to the subspaces $\mathcal{S}$ and $\mathcal{T}$. We denote this shorted matrix by $\mathbf{S}_{\star}(\mathbf{A}|\mathcal{S},\mathcal{T})$.

Recall that $\mathbf{S}(\mathbf{A}|\mathcal{S},\mathcal{T})$, the shorted matrix of $\mathbf{A}$ relative the subspaces $\mathcal{S}$ and $\mathcal{T}$ under minus order does not always exist. However, we show in our next theorem that $\mathbf{S}_{\star}(\mathbf{A}|\mathcal{S},\mathcal{T})$ always exists.

Theorem 12.3.11. *Let $\mathbf{A}$ be an $m \times n$ matrix. Let $\mathcal{S}$ and $\mathcal{T}$ be subspaces of $\mathbb{C}^n$ and $\mathbb{C}^m$ respectively. Then the shorted matrix $\mathbf{S}_{\star}(\mathbf{A}|\mathcal{S},\mathcal{T})$ exists.*

Proof. Let $\mathfrak{D} = \{\mathbf{C} \in \mathbb{C}^{\mathbf{m}\times\mathbf{n}} : \mathbf{C} <^{\star} \mathbf{A},\ \mathcal{C}(\mathbf{C}) \subseteq \mathcal{S} \text{ and } \mathcal{C}(\mathbf{C}^{\star}) \subseteq \mathcal{T}\}$. and $\mathbf{C} \in \mathfrak{D}$. Since $\mathbf{C} <^{\star} \mathbf{A}$, there exist unitary matrices $\mathbf{X}$ and $\mathbf{Y}$

such that $\mathbf{C} = \mathbf{X}\mathbf{diag}(\mathbf{D_a},\ \mathbf{0},\ \mathbf{0})\mathbf{Y}^\star$ and $\mathbf{A} = \mathbf{X}\mathbf{diag}(\mathbf{D_a},\ \mathbf{D_b},\ \mathbf{0})\mathbf{Y}^\star$, where $a = \rho(\mathbf{C})$, $a + b = \rho(\mathbf{A})$, $\mathbf{D_a}$ and $\mathbf{D_b}$ are positive definite diagonal matrices. Also, $\mathbf{C} = \mathbf{A}\mathbf{C}^\dagger\mathbf{C} = \mathbf{C}\mathbf{C}^\dagger\mathbf{A}$. Let $\mathbf{P} = \mathbf{C}^\dagger\mathbf{C}$ and $\mathbf{Q} = \mathbf{C}\mathbf{C}^\dagger$. Then $\mathbf{P}$ and $\mathbf{Q}$ are orthogonal projectors onto $\mathcal{C}(\mathbf{C}^\star)$ and $\mathcal{C}(\mathbf{C})$ respectively. Write $\mathbf{D_a} = \mathbf{diag}(\sigma_1\mathbf{I_{r_1}}, \ldots, \sigma_\mathbf{m}\mathbf{I_{r_m}})$ and $\mathbf{diag}(\mathbf{D_a}, \mathbf{D_b}) = \mathbf{diag}(\sigma_1\mathbf{I_{r_1}}, \ldots, \sigma_\mathbf{k}\mathbf{I_{r_k}})$, where $a = r_1 + \ldots + r_m$ and $a + b = r_1 + \ldots + r_k$. Partition $\mathbf{X}$, $\mathbf{Y}$ conformably with the block diagonal decomposition of $diag(\mathbf{D_a},\ \mathbf{D_b})$, say $\mathbf{X} = (\mathbf{X_1}, \ldots, \mathbf{X_{k+1}})$ and $\mathbf{Y} = (\mathbf{Y_1}, \ldots, \mathbf{Y_{k+1}})$. Then $\mathbf{A} = \Sigma_{\mathbf{i}=1}^{\mathbf{k}}\sigma_\mathbf{i}\mathbf{X_i}\mathbf{Y_i^\star}$, for each i, $\mathbf{X_i^\star X_i} = \mathbf{I} = \mathbf{Y_i^\star Y_i}$ and for each $i \neq j$, $\mathbf{X_i^\star X_j} = \mathbf{0} = \mathbf{Y_i^\star Y_j}$. Also, if $\mathbf{X_i'} = (\mathbf{0}, \ldots, \mathbf{X_i}, \ldots, \mathbf{0})$, then each $\mathbf{X_i'} <^\star \mathbf{X}$. Since $\mathbf{P} = \mathbf{C}^\dagger\mathbf{C}$, we can write $\mathbf{P} = \Sigma_{\mathbf{i}=1}^{\mathbf{k}}\mathbf{Y_i P_i Y_i^\star}$, where $\mathbf{P_i}$ are orthogonal projectors. This gives $\mathbf{C} = \mathbf{AP} = \Sigma_{\mathbf{i}=1}^{\mathbf{k}}\sigma_\mathbf{i}\mathbf{X_i P_i Y_i^\star}$. For each i, we define projectors $\mathbf{E_i}$ and $\mathbf{F_i}$ as follows:

Clearly, $\mathbf{CY_i} = \sigma_\mathbf{i}\mathbf{X_i P_i}$ and $\mathcal{C}(\mathbf{X_i P_i}) = \mathcal{C}(\sigma_\mathbf{i}\mathbf{X_i P_i}) = \mathcal{C}(\mathbf{CY_i}) \subseteq \mathcal{C}(\mathbf{C}) \subseteq \mathcal{S}$. Thus, there exists a projector $\mathbf{P_i}$ such that $\mathcal{C}(\mathbf{X_i P_i}) \subseteq \mathcal{S}$. So, let $\mathbf{E_i}$ be the maximal projector such that $\mathcal{C}(\mathbf{X_i E_i}) \subseteq \mathcal{S}$. Similarly, let $\mathbf{F_i}$ be the maximal projector such that $\mathcal{C}(\mathbf{Y_i F_i}) \subseteq \mathcal{T}$. The maximal $\mathbf{C}$ is the one with the $\mathbf{P_i}$'s selected as large as possible such that $\mathbf{P_i} <^\star \mathbf{E_i}$ and $\mathbf{P_i} <^\star \mathbf{F_i}$. Since $\mathbf{P_i}$, $\mathbf{E_i}$ and $\mathbf{F_i}$ are projectors, $\mathcal{C}(\mathbf{P_i}) \subseteq \mathcal{C}(\mathbf{E_i})$ and $\mathcal{C}(\mathbf{P_i}) \subseteq \mathcal{C}(\mathbf{F_i})$. Therefore, the maximal $\mathbf{P_i}$ that can be selected is $\mathbf{E_i} \wedge \mathbf{F_i} = \mathbf{2}\mathcal{P}(\mathbf{E_i}|\mathbf{F_i})$. We show that $\mathbf{Z} = \mathbf{2}\Sigma_{\mathbf{i}=1}^{\mathbf{k}}\sigma_\mathbf{i}\mathbf{X_i}\mathcal{P}(\mathbf{E_i}|\mathbf{F_i})\mathbf{Y_i^\star}$ is the maximal element of $\mathfrak{D}$. Clearly, $\mathcal{C}(\mathbf{Z}) \subseteq \mathcal{S}$ and $\mathcal{C}(\mathbf{Z}^\star) \subseteq \mathcal{T}$. Also, $\mathbf{ZZ}^\star = \mathbf{2}\Sigma_{\mathbf{i}=1}^{\mathbf{k}}\sigma^\mathbf{2}{}_\mathbf{i}\mathbf{X_i}\mathcal{P}(\mathbf{E_i}|\mathbf{F_i})\mathbf{Y_i^\star}$, since, $2\mathcal{P}(\mathbf{E_i}|\mathbf{F_i})$ is a projector onto $\mathcal{C}(\mathbf{E_i}) \cap \mathcal{C}(\mathbf{F_i})$. Moreover, $\mathbf{AZ}^\star = \mathbf{2}\Sigma_{\mathbf{i}=1}^{\mathbf{k}}\sigma^\mathbf{2}{}_\mathbf{i}\mathbf{X_i}\mathcal{P}(\mathbf{E_i}|\mathbf{F_i})\mathbf{Y_i^\star}$, so, $\mathbf{ZZ}^\star = \mathbf{AZ}^\star$. Similarly, $\mathbf{Z}^\star\mathbf{Z} = \mathbf{Z}^\star\mathbf{A}$. By the choice of $\mathbf{E_i}$ and $\mathbf{F_i}$, $\mathbf{Z}$ is a maximal element of $\mathfrak{D}$. Thus, $\mathbf{S}_\star(\mathbf{A}|\mathcal{S}, \mathcal{T})$ exists and is given by $\mathbf{S}_\star(\mathbf{A}|\mathcal{S}, \mathcal{T}) = \mathbf{2}\Sigma_{\mathbf{i}=1}^{\mathbf{k}}\sigma_\mathbf{i}\mathbf{X_i}\mathcal{P}(\mathbf{E_i}|\mathbf{F_i})\mathbf{Y_i^\star}$. □

Remark 12.3.12. The expression $2\Sigma_{i=1}^{k}\sigma_i\mathbf{X_i}\mathcal{P}(\mathbf{E_i}|\mathbf{F_i})\mathbf{Y_i^\star}$ provides a singular value decomposition of $\mathbf{S}_\star(\mathbf{A}|\mathcal{S}, \mathcal{T})$, since $2\mathcal{P}(\mathbf{E_i}|\mathbf{F_i})$ is a projector onto $\mathcal{C}(\mathbf{E_i}) \cap \mathcal{C}(\mathbf{F_i})$ and we can write $2\mathcal{P}(\mathbf{E_i}|\mathbf{F_i}) = \mathbf{D_i D_i^\star}$ with $\mathbf{D_i^\star D} = \mathbf{I}$.

Remark 12.3.13. Since $\mathbf{C} <^\star \mathbf{A} \Rightarrow \mathbf{C} <^- \mathbf{A}$, it follows that $\mathbf{S}_\star(\mathbf{A}|\mathcal{S}, \mathcal{T}) <^- \mathbf{S}(\mathbf{A}|\mathcal{S}, \mathcal{T})$. It is natural to ask when $\mathbf{S}_\star(\mathbf{A}|\mathcal{S}, \mathcal{T}) = \mathbf{S}(\mathbf{A}|\mathcal{S}, \mathcal{T})$? Except for trivial cases (e.g. conditions under which the star order and the minus order are equivalent) the answer is not known to us.

Theorem 12.3.14. *Let* $\mathbf{A}$ *be an* $m \times n$ *matrix. Let* $\mathcal{S}$ *and* $\mathcal{T}$ *be subspaces of* $\mathbb{C}^n$ *and* $\mathbb{C}^m$ *respectively and* $\mathbf{S}_\star(\mathbf{A}|\mathcal{S}, \mathcal{T})$, *the shorted matrix of* $\mathbf{A}$ *relative to* $\mathcal{S}$ *and* $\mathcal{T}$. *Then the following hold:*

(i) $(\mathbf{S}_\star(\mathbf{A}|\mathcal{S}, \mathcal{T}))^\star = \mathbf{S}_\star(\mathbf{A}^\star|\mathcal{S}, \mathcal{T})$.

(ii) *If* $\mathbf{A}$ *is hermitian, then so is* $\mathbf{S}_\star(\mathbf{A}|\mathcal{S},\mathcal{T})$.
(iii) *If* $\mathbf{A}$ *is nnd, then* $\mathbf{S}_\star(\mathbf{A}|\mathcal{S},\mathcal{T})$ *is also nnd.*
(iv) $(\mathbf{S}_\star(\mathbf{A}|\mathcal{S},\mathcal{T}))^\dagger = \mathbf{S}_\star(\mathbf{A}^\dagger|\mathcal{S},\mathcal{T})$ *and*
(v) *If* $\mathcal{S}',\mathcal{T}'$ *are subspaces of* $\mathbb{C}^n$ *and* $\mathbb{C}^m$ *respectively, then* $\mathbf{S}_\star(\mathbf{S}_\star(\mathbf{A}|\mathcal{S},\mathcal{T})|\mathcal{S}',\mathcal{T}') = \mathbf{S}_\star(\mathbf{A}|\mathcal{S}\cap\mathcal{S}',\mathcal{T}\cap\mathcal{T}')$.

Proof. (i) In notations of Theorem12.3.11 the result follows from (a) $\mathbf{A}$ and $\mathbf{A}^\star$ have same singular values and (b) $\mathcal{P}(\mathbf{E_i}|\mathbf{F_i}) = \mathcal{P}(\mathbf{F_i}|\mathbf{E_i})$.
(ii) If $\mathbf{A}$ is hermitian, then σ_i are real and $\mathbf{X} = \mathbf{Y}$. The proof now follows from (i).
(iii) is trivial.
(iv) If $\mathbf{A}$ has singular value decomposition given as $\mathbf{X}^\star\mathbf{AY} = \begin{pmatrix}\mathbf{D} & \mathbf{0}\\ \mathbf{0} & \mathbf{0}\end{pmatrix}$, where $\mathbf{D}$ is a positive definite diagonal matrix, then $\mathbf{A}^\dagger = \mathbf{Y}\begin{pmatrix}\mathbf{D} & \mathbf{0}\\ \mathbf{0} & \mathbf{0}\end{pmatrix}\mathbf{X}^\star$ is a singular value decomposition for $\mathbf{A}^\dagger$. Also, $\mathbf{C} <^\star \mathbf{A} \Leftrightarrow \mathbf{C}^\dagger <^\star \mathbf{A}^\dagger$. Thus, $\mathbf{S}_\star(\mathbf{A}^\dagger|\mathcal{T},\mathcal{S}) = 2\Sigma_{\mathrm{i=1}}^{\mathrm{k}}\sigma^{-1}{}_{\mathrm{i}}\mathbf{Y_i}\mathcal{P}(\mathbf{F_i}|\mathbf{E_i})\mathbf{X_i^\star} = 2\Sigma_{\mathrm{i=1}}^{\mathrm{k}}\sigma^{-1}{}_{\mathrm{i}}\mathbf{Y_i}\mathcal{P}(\mathbf{E_i}|\mathbf{F_i})\mathbf{X_i^\star}$. Now by direct verification we have $\mathbf{S}_\star(\mathbf{A}^\dagger|\mathcal{T},\mathcal{S})$ is a Moore-Penrose inverse of $\mathbf{S}_\star(\mathbf{A}|\mathcal{S},\mathcal{T})$.
(v) First note that each side has same column space $\mathcal{S}\cap\mathcal{S}'\cap\mathcal{C}(\mathbf{A})$ and same row space $\mathcal{T}\cap\mathcal{T}'\cap\mathcal{C}(\mathbf{A}^\star)$. So,

$$\rho(\mathbf{S}_\star(\mathbf{S}_\star(\mathbf{A}|\mathcal{S},\mathcal{T})|\mathcal{S}',\mathcal{T}')) = \rho(\mathbf{S}_\star(\mathbf{A}|\mathcal{S}\cap\mathcal{S}',\mathcal{T}\cap\mathcal{T}')).$$

Moreover,
$\mathbf{S}_\star(\mathbf{S}_\star(\mathbf{A}|\mathcal{S},\mathcal{T})|\mathcal{S}',\mathcal{T}') = \max\{\mathfrak{T}\}$, and $\mathbf{S}_\star(\mathbf{A}|\mathcal{S}\cap\mathcal{S}',\mathcal{T}\cap\mathcal{T}') = \max\{\mathfrak{E}\}$ where $\mathfrak{T} = \{\mathbf{C}\in\mathbb{C}^{m\times n} : \mathbf{C} <^\star \mathbf{S}_\star(\mathbf{A}|\mathcal{S},\mathcal{T}), \mathcal{C}(\mathbf{C}) \subseteq \mathcal{S}',\ \mathcal{C}(\mathbf{C}^\star) \subseteq \mathcal{T}'\}$ and $\{\mathfrak{E}\} = \{\mathbf{C}\in\mathbb{C}^{\mathbf{m}\times\mathbf{n}} : \mathbf{C} <^\star \mathbf{A},\ \mathcal{C}(\mathbf{C}) \subseteq \mathcal{S}\cap\mathcal{S}',\ \mathcal{C}(\mathbf{C}^\star) \subseteq \mathcal{T}\cap\mathcal{T}'\}$. Let $\mathbf{C}$ be such that $\mathbf{C} <^\star \mathbf{S}_\star(\mathbf{A}|\mathcal{S},\mathcal{T})$, $\mathcal{C}(\mathbf{C}) \subseteq \mathcal{S}'$ and $\mathcal{C}(\mathbf{C}^\star) \subseteq \mathcal{T}'$. Since $\mathbf{S}_\star(\mathbf{A}|\mathcal{S},\mathcal{T}) <^\star \mathbf{A}$, we have $\mathbf{C} <^\star \mathbf{A}$. Also, $\mathcal{C}(\mathbf{C}) \subseteq \mathcal{C}(\mathbf{A})\cap\mathcal{S}$, so, $\mathcal{C}(\mathbf{C}) \subseteq \mathcal{C}(\mathbf{A})\cap\mathcal{S}\cap\mathcal{S}'$. Similarly, $\mathcal{C}(\mathbf{C}^\star) \subseteq \mathcal{C}(\mathbf{A})\cap\mathcal{T}\cap\mathcal{T}'$. Therefore $\mathbf{C} \in \{\mathbf{C}\in\mathbb{C}^{\mathbf{m}\times\mathbf{n}} : \mathbf{C} <^\star \mathbf{A},\ \mathcal{C}(\mathbf{C}) \subseteq \mathcal{S}\cap\mathcal{S}',\ \mathcal{C}(\mathbf{C}^\star) \subseteq \mathcal{T}\cap\mathcal{T}'\}$ and hence, $\mathbf{C} <^\star \mathbf{S}_\star(\mathbf{A}|\mathcal{S}\cap\mathcal{S}',\mathcal{T}\cap\mathcal{T}')$. This further gives $\mathbf{S}_\star(\mathbf{S}_\star(\mathbf{A}|\mathcal{S},\mathcal{T})|\mathcal{S}',\mathcal{T}') <^\star \mathbf{S}_\star(\mathbf{A}|\mathcal{S}\cap\mathcal{S}',\mathcal{T}\cap\mathcal{T}')$. Thus,

$$\mathbf{S}_\star(\mathbf{S}_\star(\mathbf{A}|\mathcal{S},\mathcal{T})|\mathcal{S}',\mathcal{T}') = \mathbf{S}_\star(\mathbf{A}|\mathcal{S}\cap\mathcal{S}',\mathcal{T}\cap\mathcal{T}').$$ □

12.4 Infimum under the sharp order

In this section we discuss the existence of $\mathbf{A}\wedge\mathbf{B}$ and $\mathbf{A}\vee\mathbf{B}$ for matrices $\mathbf{A}$, $\mathbf{B}$ under the sharp order. In general neither the supremum $\mathbf{A}\vee\mathbf{B}$ nor the infimum $\mathbf{A} \wedge \mathbf{B}$ may exist. Supremum may fail to exist for obvious reason and the following example shows that the infimum may also fail to exist in general:

Example 12.4.1. Let $\mathbf{A} = \begin{pmatrix} 1 & 1 & 1 & -2 \\ 1 & 2 & 1 & -2 \\ 1 & 1 & 2 & -2 \\ 1 & 1 & 1 & -1 \end{pmatrix}$ and $\mathbf{B} = \begin{pmatrix} 0 & 0 & 0 & 0 \\ 0 & 1 & 0 & 0 \\ 0 & 0 & 1 & 0 \\ 0 & 0 & 0 & 1 \end{pmatrix} \in \mathbb{C}^{4\times 4}$.

Note that $\mathbf{A} = \mathbf{B} + \mathbf{K}$, where $\mathbf{K} = \begin{pmatrix} 1 \\ 1 \\ 1 \\ 1 \end{pmatrix} (\mathbf{1},\ \mathbf{1},\ \mathbf{1},\ -\mathbf{2})$. If $\mathbf{C}$ is a matrix dominated by $\mathbf{B}$ under the minus order, then $\mathbf{C}$ must be an idempotent matrix of the form $\begin{pmatrix} \mathbf{0} & \mathbf{0} \\ \mathbf{0} & \mathbf{H} \end{pmatrix}$, where $\mathbf{H}$ is an idempotent of order 3×3. Now, if $\mathbf{C} <^{\#} \mathbf{A}$ and $\mathbf{C} <^{\#} \mathbf{B}$, then $\mathbf{KC} = \mathbf{0}$ and $\mathbf{CK} = \mathbf{0}$. So, $(1,\ 1,\ 1,\ -2)\mathbf{C} = \mathbf{0}$ and $\mathbf{C} \begin{pmatrix} 1 \\ 1 \\ 1 \\ 1 \end{pmatrix} = \mathbf{0}$. This implies that $(1,\ 1,\ -2)\mathbf{H} = \mathbf{0}$ and $\mathbf{H} \begin{pmatrix} 1 \\ 1 \\ 1 \end{pmatrix} = \mathbf{0}$. Thus, columns of $\mathbf{H}$ are orthogonal to $(1,\ 1,\ -2)^{\star}$ and rows of $\mathbf{H}$ are orthogonal to $\begin{pmatrix} 1 \\ 1 \\ 1 \end{pmatrix}$. So, $\mathcal{C}(\mathbf{H}) \subseteq \mathcal{C}(\mathbf{L})$ and $\mathcal{C}(\mathbf{H}^{\star}) \subseteq \mathcal{C}(\mathbf{M}^{\star})$, where $\mathbf{L} = \begin{pmatrix} 1 & 1 \\ 1 & -1 \\ 1 & 0 \end{pmatrix}$ and $\mathbf{M}^{\star} = \begin{pmatrix} 1 & 1 \\ 1 & -1 \\ 2 & 0 \end{pmatrix}$, and therefore, $\mathbf{H} = \mathbf{LZM}$ for some matrix $\mathbf{Z}$. Since $\mathbf{H}$ is idempotent and $\mathbf{L}$ and $\mathbf{M}$ are full rank matrices, so, $\rho(\mathbf{H}) \leq \rho(\mathbf{ML}) = \mathbf{1}$ and $\mathbf{Z} = \mathbf{ZMLZ}$. To choose $\mathbf{C}$ of maximum rank in $\mathfrak{C}$, we must choose $\mathbf{Z} \in \{(\mathbf{ML})_{\mathbf{r}}^{-}\}$. A general solution to $\mathbf{Z}$ is given by $\mathbf{Z} = \begin{pmatrix} 2ab & b \\ a & \frac{1}{2} \end{pmatrix}$, where a, b are arbitrary complex numbers. Thus, there are several choices for matrix $\mathbf{C}$ and so, $\mathbf{A} \wedge \mathbf{B}$ does not exist.

Remark 12.4.2. In Example 12.4.1, $\mathbf{B} <^{-} \mathbf{A}$, so the minus infimum of $\mathbf{A}$, $\mathbf{B}$ exists and is equal to $\mathbf{B}$. Thus, infimum under different partial orders may be different. So one may ask: When do infimum and supremum under various partial orders coincide?

We now give a necessary and sufficient conditions under which the infimum under sharp order exists.

Theorem 12.4.3. *Let* $\mathbf{A}$, $\mathbf{B}$ *be square matrices having the same order and of index* ≤ 1, *over an arbitrary field.Let* $\rho(\mathbf{A}) = m$. *Let us first assume that* $\mathbf{A} = \begin{pmatrix} \mathbf{I_m} & \mathbf{0} \\ \mathbf{0} & \mathbf{0} \end{pmatrix}$. *Let* $\mathbf{U}$ *and* $\mathbf{V}$ *be matrices of order* $m \times r$ *and* $m \times s$ *respectively such that*

$$\mathcal{N}(\mathbf{B}-\mathbf{A}) \cap \mathcal{C}(\mathbf{A}) = \mathcal{C}\begin{pmatrix} \mathbf{U} \\ \mathbf{0} \end{pmatrix} \text{ and } \mathcal{N}(\mathbf{B^t}-\mathbf{A^t}) \cap \mathcal{C}(\mathbf{A^t}) = \mathcal{C}\begin{pmatrix} \mathbf{V} \\ \mathbf{0} \end{pmatrix},$$

where $r = \rho(\mathcal{N}(\mathbf{B}-\mathbf{A}) \cap \mathcal{C}(\mathbf{A}))$ *and* $s = \rho(\mathcal{N}(\mathbf{B^t}-\mathbf{A^t}) \cap \mathcal{C}(\mathbf{A^t}))$. *Then* $\mathbf{A} \wedge \mathbf{B}$ *exists if and only if* $r = s$ *and* $\mathbf{V^tU}$ *is invertible. When these conditions are satisfied,* $\mathbf{A} \wedge \mathbf{B} = \begin{pmatrix} \mathbf{U(V^tU)^{-1}V^t} & \mathbf{0} \\ \mathbf{0} & \mathbf{0} \end{pmatrix}$.

Proof. Let $\mathbf{C}$ be a matrix of index 1. Then
$\mathbf{C} <^{\#} \mathbf{A}$ and $\mathbf{C} <^{\#} \mathbf{B}$

$$\Leftrightarrow (\mathbf{A}-\mathbf{C})\mathbf{C} = \mathbf{C}(\mathbf{A}-\mathbf{C}) = \mathbf{0},\ (\mathbf{B}-\mathbf{C})\mathbf{C} = \mathbf{C}(\mathbf{B}-\mathbf{C}) = \mathbf{0}$$

$$\Leftrightarrow (\mathbf{A}-\mathbf{C})\mathbf{C} = \mathbf{C}(\mathbf{A}-\mathbf{C}) = \mathbf{0},\ (\mathbf{B}-\mathbf{A})\mathbf{C} = \mathbf{C}(\mathbf{B}-\mathbf{A}) = \mathbf{0}$$

$$\Rightarrow \mathcal{C}(\mathbf{C}) \subseteq \mathcal{N}(\mathbf{B}-\mathbf{A}) \cap \mathcal{C}(\mathbf{A}) \text{ and } \mathcal{C}(\mathbf{C^t}) \subseteq \mathcal{N}(\mathbf{B^t}-\mathbf{A^t}) \cap \mathcal{C}(\mathbf{A^t})$$

$$\Rightarrow \mathcal{C}(\mathbf{C}) \subseteq \mathcal{C}\begin{pmatrix} \mathbf{U} \\ \mathbf{0} \end{pmatrix} \text{ and } \mathcal{C}(\mathbf{C^t}) \subseteq \mathcal{C}\begin{pmatrix} \mathbf{V} \\ \mathbf{0} \end{pmatrix}$$

$$\Rightarrow \mathbf{C} = \begin{pmatrix} \mathbf{U\Delta V^t} & \mathbf{0} \\ \mathbf{0} & \mathbf{0} \end{pmatrix}.$$

Also, $\mathbf{C^2} = \mathbf{AC} \Rightarrow \mathbf{\Delta V^tU\Delta} = \mathbf{\Delta}$. So, $\mathbf{\Delta} = \mathbf{P(QV^tUP)_r^-Q}$ for some $\mathbf{P}$ and $\mathbf{Q}$. If $r = s$ and $\mathbf{V^tU}$ is invertible, then we have $\mathbf{UP(QV^tUP)_r^-QV^t} <^{\#} \mathbf{U(V^tU)^{-1}V^t}$. Thus, $\begin{pmatrix} \mathbf{U\Delta V^t} & \mathbf{0} \\ \mathbf{0} & \mathbf{0} \end{pmatrix}$ has maximal rank when $\mathbf{\Delta} = (\mathbf{V^tU})^{-1}$ and therefore, $\mathbf{A} \wedge \mathbf{B} = \begin{pmatrix} \mathbf{U(V^tU)^{-1}V^t} & \mathbf{0} \\ \mathbf{0} & \mathbf{0} \end{pmatrix}$.

Conversely, if either $r = s$ and $\rho(\mathbf{V^tU}) < r$ or $r \neq s$, then for each choice of matrices $\mathbf{P}$ and $\mathbf{Q}$, $\rho(\mathbf{UP(QV^tUP)_r^-QV^t}) \leq \rho(\mathbf{V^tU})$, so, a unique matrix $\mathbf{C}$ can not be found and therefore, $\mathbf{A} \wedge \mathbf{B}$ does not exist. □

We now consider the case when $\mathbf{A}$ does not have special form of Theorem 12.4.3.

Consider the setup in Theorem 12.4.3 except that $\mathbf{A}$ is an arbitrary square matrix. Let $(\mathbf{L}, \mathbf{R^t})$ be a rank factorization of $\mathbf{A}$. As $\mathbf{A}$ is of index 1, $\mathbf{R^tL}$ is invertible with $\rho(\mathbf{R^tL}) = \rho(\mathbf{A})$. Let the columns of $\mathbf{LU}$ form a basis of $\mathcal{N}(\mathbf{B} - \mathbf{A}) \cap \mathcal{C}(\mathbf{A})$ and rows of $\mathbf{RV}$ form a basis of $\mathcal{N}(\mathbf{B^t} - \mathbf{A^t}) \cap \mathcal{C}(\mathbf{A^t})$. Further, let $\mathbf{U_0}$ be a matrix columns of which form a basis of $\mathcal{C}(\mathbf{U})$ and $\mathbf{V_0}$ be a matrix columns of which form a basis of $\mathcal{C}(\mathbf{V})$. Then $\mathbf{U_0}$ is maximal being invariant under $\mathbf{R^tL}$ and $\mathbf{V_0}$ is maximal being invariant under $\mathbf{L^tR}$. Under this setup we have

Theorem 12.4.4. $\mathbf{A} \wedge \mathbf{B}$ *exists if and only if* $\mathbf{V_0^tU_0}$ *is invertible and when this condition is satisfied* $\mathbf{A} \wedge \mathbf{B} = \mathbf{LU_0(V_0^tU_0)^{-1}V_0^tR^t}$.

Proof. Let $\mathbf{C}$ be a matrix of index 1. Then $\mathbf{C} <^{\#} \mathbf{A}$ and $\mathbf{C} <^{\#} \mathbf{B} \Rightarrow \mathbf{C} = \mathbf{LHR^t}$ for some idempotent matrix $\mathbf{H}$. Also,

$$\begin{aligned}\mathbf{C} <^{\#} \mathbf{A} &\Leftrightarrow (\mathbf{A} - \mathbf{C})\mathbf{C} = \mathbf{C}(\mathbf{A} - \mathbf{C}) = \mathbf{0}\\ &\Leftrightarrow (\mathbf{I} - \mathbf{H})\mathbf{R^tLH} = \mathbf{HR^tL}(\mathbf{I} - \mathbf{H}) = \mathbf{0}.\end{aligned}$$

Thus, for each $\mathbf{Hy}$, $\mathbf{R^tLHy} = \mathbf{HR^tLHy} \in \mathcal{C}(\mathbf{H})$. So, $\mathcal{C}(\mathbf{H})$ is invariant under $\mathbf{R^tL}$. Similarly, $\mathcal{C}(\mathbf{H^t})$ is invariant under $\mathbf{L^tR}$. Now with $\mathbf{H}$ playing the role of $\mathbf{C}$, $\mathbf{U_0}$ of $\mathbf{U}$ and $\mathbf{V_0}$ of $\mathbf{V}$, $\mathbf{A_0} = \mathbf{I_m} = \mathbf{L}_\ell^{-1}\mathbf{A}(\mathbf{R^t})_r^{-1}$ and $\mathbf{B_0} = \mathbf{L}_\ell^{-1}\mathbf{B}(\mathbf{R^t})_r^{-1}$, where $\mathbf{L}_\ell^{-1}$ is a left inverse of $\mathbf{L}$ and $(\mathbf{R^t})_r^{-1}$ is a right inverse of $\mathbf{R^t}$ and we have the setup of Theorem 12.4.3. Therefore, $\mathbf{A_0} \wedge \mathbf{B_0}$ exists if and only if $\mathbf{V_0^tU_0}$ is invertible. When $\mathbf{V_0^tU_0}$ is invertible, $\mathbf{A_0} \wedge \mathbf{B_0} = \mathbf{U_0(V_0^tU_0)^{-1}V_0^t}$. Also, if $\mathbf{C_0} <^{\#} \mathbf{A_0}$, then $\mathbf{LC_0R^t} <^{\#} \mathbf{LAR^t}$. It is now easy to see that $\mathbf{A} \wedge \mathbf{B} = \mathbf{LU_0(V_0^tU_0)^{-1}V_0^tR^t}$. □

12.5 Exercises

(1) Let $\mathbf{A}$ and $\mathbf{B}$ be matrices of the same order such that $\mathbf{AB}^\star$ and $\mathbf{A}^\star\mathbf{B}$ are hermitian. Show that

$$\mathbf{A} \wedge \mathbf{B} = \mathbf{A}(\mathbf{I} - (\mathbf{I} - \mathbf{A}^\dagger\mathbf{B})(\mathbf{I} - \mathbf{A}^\dagger\mathbf{B})^\dagger) = (\mathbf{I} - (\mathbf{I} - \mathbf{B}\mathbf{A}^\dagger)^\dagger(\mathbf{I} - \mathbf{B}\mathbf{A}^\dagger))\mathbf{A}.$$

(2) Let $\mathbf{A}$ and $\mathbf{B}$ be matrices of the same order such that $\mathbf{A}$ and $\mathbf{B}$ commute and $\mathbf{A} - \mathbf{B}$ is range-hermitian. Show that

$$\mathbf{A} \wedge \mathbf{B} = \mathbf{A}(\mathbf{I} - (\mathbf{I} - \mathbf{A}^\dagger\mathbf{B})(\mathbf{I} - \mathbf{A}^\dagger\mathbf{B})^\dagger).$$

(3) Let $\mathbf{A}$ and $\mathbf{B}$ be unitary. Show that

$$\mathbf{A} \wedge \mathbf{B} = \mathbf{A}(\mathbf{I} - (\mathbf{I} - \mathbf{A}^\dagger\mathbf{B})(\mathbf{I} - \mathbf{A}^\dagger\mathbf{B})^\dagger) = (\mathbf{I} - (\mathbf{I} - \mathbf{B}\mathbf{A}^\dagger)^\dagger(\mathbf{I} - \mathbf{B}\mathbf{A}^\dagger))\mathbf{A}.$$

(4) Let $\mathbf{E}$ and $\mathbf{F}$ be orthogonal projectors. Show that $\mathbf{E} \vee \mathbf{F} = (\mathbf{E} + \mathbf{F})(\mathbf{E} + \mathbf{F})^\dagger = (\mathbf{E} + \mathbf{F})^\dagger(\mathbf{E} + \mathbf{F})$ and $\mathbf{E} \wedge \mathbf{F} = \mathbf{2}\mathcal{P}(\mathbf{E}, \mathbf{F})$.

(5) Prove or disprove the following statement: Let $\mathbf{A} = \mathbf{A_0} \oplus \mathbf{A_1}$ and $\mathbf{B} = \mathbf{A_0} \oplus \mathbf{kA_1}$, $k \neq 1$. Then $\mathbf{A} \vee \mathbf{B} = \mathbf{A}$.

(6) Let '$<^a$' denote either the left star order or the right star order. Prove that $\mathbf{A} \vee \mathbf{B}$ exists under '$<^a$'. Mimicking the star order define shorted matrix using '$<^a$' and also show that under each case this shorted matrix exists.

Chapter 13

Partial Orders of Modified Matrices

13.1 Introduction

The importance of studying partial orders and shorted operators of modified matrices lies in its potential for applications to Statistics and Networking. For example when there are model or data changes in a linear model, both partitioned matrices and rank 1 modification of a matrix play a big role in obtaining the revised inferences. Also, in electrical network theory when a new port is added or an existing port is deleted in an n-port network, the modified matrices play an important role in studying its physical properties. So, it is of interest to study the behavior of the partial orders of modified matrices. More specifically: Let $\mathbf{X}, \mathbf{Y}$ be a pair of matrices and $\mathbf{X} < \mathbf{Y}$, where '$<$' is some partial order on matrices. Suppose we append/delete a column/row to each of $\mathbf{X}, \mathbf{Y}$ and denote the new pair by $\mathbf{X_1}, \mathbf{Y_1}$ in either situation. We are interested in knowing if $\mathbf{X_1} < \mathbf{Y_1}$ and if not, what conditions will ensure that it will be so. Since the modifications obtained by appending/deleting a row can be treated by taking the transposes of the corresponding modifications by appending/deleting columns, we confine only to the case in which the matrices under consideration are modified by appending/deleting a column. We also study the same problem when we add a rank 1 matrix to both of $\mathbf{X}$ and $\mathbf{Y}$. In this chapter we study the partial orders of the modified matrices. The shorted operators of the modified matrices are dealt in Section 6 of Chapter 15.

In Section 13.2, we study the space pre-order for both types of modifications in concerned matrices. We first obtain necessary and sufficient conditions when matrices are modified by adding/deleting a column. We next take up the case when matrices are modified by adding a rank 1 matrix. This case appears to be highly computational though not difficult to understand. In Section 13.3, we give a detailed account of happenings in relation to the minus order. In Section 13.4, the same problem for the sharp order is taken up. As before in all the three sections 13.2-13.4, we consider matrices over general fields. In Sections 13.5 and 13.6, we consider matrices over the field of complex numbers where we study the behavior of the star order and the Löwner order for both types of modifications.

Enroute the study of various matrix order relations under both types of modifications, we see that all the conclusions are obtained after laborious computations, which are little more manageable in case of modifications obtained by appending a column than in case of modifications by adding rank 1 matrices. Keeping in view the length constraint of the monograph, we give proofs of some of basic results leading to all the other results and omit the proofs of others.

13.2 Space pre-order

Let $\mathbf{A}$ and $\mathbf{B}$ be matrices of order $m \times n$ over a field F (not necessarily field of complex numbers). Let $\mathbf{a}$ and $\mathbf{b}$ be m-column vectors over F. Let us write

$$\mathbf{A}_1 = (\mathbf{A} : \mathbf{a}) \text{ and } \mathbf{B}_1 = (\mathbf{B} : \mathbf{b}). \tag{13.2.1}$$

In this section we explore the conditions under which $\mathbf{A}_1 <^{\mathbf{s}} \mathbf{B}_1$, given $\mathbf{A} <^{\mathbf{s}} \mathbf{B}$ and vice versa. We start with

Lemma 13.2.1. *Consider the setup (13.2.1). Let $\mathbf{A}_1 <^{\mathbf{s}} \mathbf{B}_1$ and $\mathbf{b} \notin \mathcal{C}(\mathbf{B})$. Let $\begin{pmatrix} \mathbf{G} \\ \mathbf{c}^{\mathbf{t}} \end{pmatrix}$ and $\begin{pmatrix} \mathbf{H} \\ \mathbf{d}^{\mathbf{t}} \end{pmatrix}$ be two g-inverses of $\mathbf{B}_1$. Then $\mathbf{c}^{\mathbf{t}}\mathbf{A} = \mathbf{0}$ if and only if $\mathcal{C}(\mathbf{A}) \subseteq \mathcal{C}(\mathbf{B})$. If $\mathbf{c}^{\mathbf{t}}\mathbf{A} = \mathbf{0}$, then $\mathbf{d}^{\mathbf{t}}\mathbf{A} = \mathbf{0}$.*

Proof. Since $\begin{pmatrix} \mathbf{G} \\ \mathbf{c}^{\mathbf{t}} \end{pmatrix}$ and $\begin{pmatrix} \mathbf{H} \\ \mathbf{d}^{\mathbf{t}} \end{pmatrix}$ are g-inverses of $\mathbf{B}_1$ and $\mathbf{b} \notin \mathcal{C}(\mathbf{B})$, by Theorem 2.6.8, it follows that $\mathbf{G}$ and $\mathbf{H}$ are g-inverses of $\mathbf{B}$. Also, $\mathbf{c}^{\mathbf{t}}\mathbf{B} = \mathbf{0}$, $\mathbf{d}^{\mathbf{t}}\mathbf{B} = \mathbf{0}$, and $\mathbf{c}^{\mathbf{t}}\mathbf{b} = \mathbf{1}\mathbf{d}^{\mathbf{t}}\mathbf{b}$. Since $\mathbf{A}_1 <^{\mathbf{s}} \mathbf{B}_1$, we have $(\mathbf{B} : \mathbf{b}) \begin{pmatrix} \mathbf{G} \\ \mathbf{c}^{\mathbf{t}} \end{pmatrix} (\mathbf{A} : \mathbf{a}) = (\mathbf{A} : \mathbf{a})$ or equivalently $\mathbf{BGA} + \mathbf{bc}^{\mathbf{t}}\mathbf{A} = \mathbf{A}$ and $\mathbf{BGa} + \mathbf{bc}^{\mathbf{t}}\mathbf{a} = \mathbf{a}$. If $\mathbf{c}^{\mathbf{t}}\mathbf{A} = \mathbf{0}$, we have $\mathbf{BGA} = \mathbf{A}$, or equivalently $\mathcal{C}(\mathbf{A}) \subseteq \mathcal{C}(\mathbf{B})$. Since $\mathbf{d}^{\mathbf{t}}\mathbf{B} = \mathbf{0}$, it follows that $\mathbf{d}^{\mathbf{t}}\mathbf{A} = \mathbf{d}^{\mathbf{t}}\mathbf{BGA} = \mathbf{0}$.

On the other hand, if $\mathcal{C}(\mathbf{A}) \subseteq \mathcal{C}(\mathbf{B})$, then $\mathbf{bc}^{\mathbf{t}}\mathbf{A} = \mathbf{0}$. Since $\mathbf{b} \neq \mathbf{0}$, it follows that $\mathbf{c}^{\mathbf{t}}\mathbf{A} = \mathbf{0}$. □

Theorem 13.2.2. *Consider the setup (13.2.1). Then the following hold:*

(i) *Let* $\mathbf{b} \in \mathcal{C}(\mathbf{B})$. *Then* $\mathbf{A_1} <^s \mathbf{B_1}$ *if and only if* $\mathbf{A} <^s \mathbf{B}$ *and* $\mathbf{a} = \mathbf{AB^-b}$.

(ii) *Let* $\mathbf{b} \notin \mathcal{C}(\mathbf{B})$. *Then* $\mathbf{A_1} <^s \mathbf{B_1}$ *and* $\mathbf{c^tA} = \mathbf{0}$ *for some g-inverse* $\begin{pmatrix}\mathbf{G}\\ \mathbf{c^t}\end{pmatrix}$ *of* $\mathbf{B_1}$ *if and only if* $\mathbf{A} <^s \mathbf{B}$ *and* $\mathbf{a} \in \mathcal{C}(\mathbf{B} : \mathbf{b})$.

Proof. (i) Since $\mathbf{b} \in \mathcal{C}(\mathbf{B})$, $\mathbf{H} = \begin{pmatrix}\mathbf{B^-}\\ \mathbf{0}\end{pmatrix}$ is a g-inverse of $\mathbf{B_1}$, where $\mathbf{B^-}$ be any g-inverse of $\mathbf{B}$. Now,

$$\mathbf{A_1HB_1} = \mathbf{A_1} \Leftrightarrow \mathbf{AB^-B} = \mathbf{A} \text{ and } \mathbf{AB^-b} = \mathbf{a}.$$

Further,

$$\mathbf{B_1HA_1} = \mathbf{A_1} \Leftrightarrow \mathbf{BB^-a} = \mathbf{a}.$$

Hence,

$$\mathbf{A_1} <^s \mathbf{B_1} \Leftrightarrow \mathbf{A} <^s \mathbf{B},\ \mathbf{AB^-b} = \mathbf{a} \text{ and } \mathbf{BB^-a} = \mathbf{a} \Leftrightarrow \mathbf{A} <^s \mathbf{B},\ \mathbf{AB^-b} = \mathbf{a}.$$

(ii) Let $\begin{pmatrix}\mathbf{G}\\ \mathbf{c^t}\end{pmatrix}$ be a g-inverse of $\mathbf{B_1}$. Clearly, $\mathbf{c^tB} = \mathbf{0}$ and $\mathbf{c^tb} = \mathbf{1}$.

'Only if' part

$\mathbf{A_1} <^s \mathbf{B_1}$ and $\mathbf{c^tA} = \mathbf{0} \Rightarrow \mathcal{C}(\mathbf{A}) \subseteq \mathcal{C}(\mathbf{B})$, by Lemma 13.2.1. Further, $\mathbf{A_1} <^s \mathbf{B_1} \Rightarrow (\mathbf{B} : \mathbf{b})\begin{pmatrix}\mathbf{G}\\ \mathbf{c^t}\end{pmatrix}\mathbf{a} = \mathbf{a}$. Hence, $\mathbf{a} \in \mathcal{C}(\mathbf{B} : \mathbf{b})$.

Also, $\mathbf{A_1}\begin{pmatrix}\mathbf{G}\\ \mathbf{c^t}\end{pmatrix}\mathbf{B_1} = \mathbf{A_1} \Rightarrow (\mathbf{AG} + \mathbf{ac^t})\mathbf{B} = \mathbf{A} \Rightarrow \mathbf{AGB} = \mathbf{A}$, as $\mathbf{c^tB} = \mathbf{0}$.

Thus, $\mathbf{A_1} <^s \mathbf{B_1} \Rightarrow \mathbf{A} <^s \mathbf{B}$ and $\mathbf{a} \in \mathcal{C}(\mathbf{B} : \mathbf{b})$.

'If' part

Let $\mathbf{B^-}$ be a g-inverse of $\mathbf{B}$ and let $\mathbf{u} = \mathbf{B^-b}$. Then it is easy to check that $\begin{pmatrix}\mathbf{B^-} - \mathbf{uc^t}\\ \mathbf{c^t}\end{pmatrix}$ is a g-inverse of $\mathbf{B_1}$. Since $\mathbf{c^tB} = \mathbf{0}$ and $\mathcal{C}(\mathbf{A}) \subseteq \mathcal{C}(\mathbf{B})$, we have $\mathbf{c^tA} = \mathbf{0}$. It now follows that $\mathcal{C}(\mathbf{A} : \mathbf{a}) \subseteq \mathcal{C}(\mathbf{B} : \mathbf{b})$. Since $\mathcal{C}(\mathbf{A^t}) \subseteq \mathcal{C}(\mathbf{B^t})$, a simple verification shows that $\mathbf{A_1}\begin{pmatrix}\mathbf{B^-} - \mathbf{uc^t}\\ \mathbf{c^t}\end{pmatrix}\mathbf{B_1} = \mathbf{A_1}$. Further, $\mathbf{B_1}\begin{pmatrix}\mathbf{B^-} - \mathbf{uc^t}\\ \mathbf{c^t}\end{pmatrix}\mathbf{A_1} = \mathbf{A_1}$ also holds.

Thus, $\mathbf{A} <^s \mathbf{B}$ and $\mathbf{a} \in \mathcal{C}(\mathbf{B} : \mathbf{b}) \Rightarrow \mathbf{A_1} <^s \mathbf{B_1}$ and $\mathbf{c^tA} = \mathbf{0}$. □

Remark 13.2.3. Let $\mathbf{b} \notin \mathcal{C}(\mathbf{B})$. Let the 'if' part of Theorem 13.2.2 hold for some g-inverse $\begin{pmatrix}\mathbf{G}\\ \mathbf{c^t}\end{pmatrix}$ of $\mathbf{B_1}$. Then, by Lemma 13.2.1 it holds for each g-inverse of $\mathbf{B_1}$.

We now study the space pre-order for matrices which are modified by adding a rank 1 matrix.(Recall that each rank 1 matrix can be expressed as $\mathbf{xy^t}$ where $\mathbf{x}$ and $\mathbf{y}$ are non-null column vectors, not necessarily of the same order.)

Let $\mathbf{A}$ be an $m \times n$ matrix, $\mathbf{a}$ an m-column vector and $\mathbf{b}$ an n-column vector. Consider the following conditions:

(i) $\mathbf{a} \in \mathcal{C}(\mathbf{A})$ (13.2.2)

(ii) $\mathbf{b} \in \mathcal{C}(\mathbf{A^t})$ (13.2.3)

(iii) $\mathbf{b^t A^- a} + \mathbf{1} \neq \mathbf{0}$ for some g-inverse $\mathbf{A}^-$ of $\mathbf{A}$. (13.2.4)

Lemma 13.2.4. *Let* $\mathbf{A}$ *be an* $m \times n$ *matrix,* $\mathbf{a}$ *an* m*-column vector and* $\mathbf{b}$ *be an* n*-column vector. Then* $\rho(\mathbf{A} + \mathbf{ab^t}) = \rho\begin{pmatrix} \mathbf{A} & \mathbf{a} \\ -\mathbf{b^t} & -1 \end{pmatrix} - 1$. *Consequently,*

$$\rho(\mathbf{A} + \mathbf{ab^t}) = \begin{cases} \rho(\mathbf{A}) - 1 & \Leftrightarrow (13.2.2) - (13.2.4) \text{ hold} \\ \rho(\mathbf{A}) & \Leftrightarrow \text{excatly two of } (13.2.2) - (13.2.4) \text{ hold} \\ \rho(\mathbf{A}) + 1 & \Leftrightarrow \text{at most one of } (13.2.2) - (13.2.4) \text{ holds.} \end{cases}$$

Proof. Notice that we can write

$$\begin{pmatrix} \mathbf{A} + \mathbf{ab^t} & \mathbf{0} \\ \mathbf{0} & 1 \end{pmatrix} = \begin{pmatrix} \mathbf{I_m} & -\mathbf{a} \\ \mathbf{0} & 1 \end{pmatrix} \begin{pmatrix} \mathbf{A} & \mathbf{a} \\ -\mathbf{b^t} & 1 \end{pmatrix} \begin{pmatrix} \mathbf{I_n} & \mathbf{0} \\ \mathbf{b^t} & 1 \end{pmatrix}$$

Proof now follows from Theorem 2.6.14 and Theorem 13.2.2. □

Let $\mathbf{A}, \mathbf{B}$ be matrices of order $m \times n$, $\mathbf{a}$, $\mathbf{c}$ be m-column vectors and $\mathbf{b}, \mathbf{d}$ be n-column vectors. Let $\mathbf{A_2} = \mathbf{A} + \mathbf{ab^t}$, $\mathbf{B_2} = \mathbf{B} + \mathbf{cd^t}$; $\mathbf{E} = \mathbf{I} - \mathbf{BG}$, $\mathbf{F} = \mathbf{I} - \mathbf{GB}$ and $\beta = 1 + \mathbf{d^t Gc}$, where $\mathbf{G}$ is a g-inverse of $\mathbf{B}$. Let $\mathbf{A} <^{\mathbf{s}} \mathbf{B}$. We now explore the conditions under which $\mathbf{A_2} <^{\mathbf{s}} \mathbf{B_2}$ and vice versa. Notice that, in general, β depends on choice of g-inverse $\mathbf{G}$ of $\mathbf{B}$. However, if $\mathbf{c} \in \mathcal{C}(\mathbf{B})$ and $\mathbf{d} \in \mathcal{C}(\mathbf{B^t})$, then β is independent of the choice of g-inverse $\mathbf{G}$ of $\mathbf{B}$.

Theorem 13.2.5. *Let* $\mathbf{A}, \mathbf{B}$ *be matrices of order* $m \times n, \mathbf{a}, \mathbf{c}$ *be non-null* m*-column vectors and* $\mathbf{b}, \mathbf{d}$ *be non-null* n*-column vectors. Let* $\mathbf{c} \in \mathcal{C}(\mathbf{B})$, $\mathbf{d} \in \mathcal{C}(\mathbf{B^t})$ *and* $\beta = 0$. *Then any two of the following imply the third:*

(i) $\mathbf{A} <^{\mathbf{s}} \mathbf{B}$

(ii) $\mathbf{A_2} <^{\mathbf{s}} \mathbf{B_2}$

(iii) *There exists a g-inverse* $\mathbf{G}$ *of* $\mathbf{B}$ *satisfying*

(a) $\mathbf{A_2Gc} = \mathbf{0}$, $\mathbf{d^tGA_2} = \mathbf{0}$, *when* $\mathbf{a} \in \mathcal{C}(\mathbf{A})$, $\mathbf{b} \in \mathcal{C}(\mathbf{A^t})$.

(b) $\mathbf{A_2Gc} = \mathbf{0}$, $\mathbf{d^tGA} = \mathbf{0}$ *and* $\mathbf{b} \in \mathcal{C}(\mathbf{B^t})$, *when* $\mathbf{a} \in \mathcal{C}(\mathbf{A})$ *and* $\mathbf{b} \notin \mathcal{C}(\mathbf{A^t})$.

(c) $\mathbf{a} \in \mathcal{C}(\mathbf{B})$, $\mathbf{AGc} = \mathbf{0}$, *and* $\mathbf{d^tGA_2} = \mathbf{0}$ *when* $\mathbf{a} \notin \mathcal{C}(\mathbf{A})$ *and* $\mathbf{b} \in \mathcal{C}(\mathbf{A^t})$

(d) $\mathbf{a} \in \mathcal{C}(\mathbf{B})$, $\mathbf{b} \in \mathcal{C}(\mathbf{B^t})$, $\mathbf{b^tGc} = \mathbf{0}$, $\mathbf{AGc} = \mathbf{0}$, $\mathbf{d^tGa} = \mathbf{0}$ and $\mathbf{d^tGA} = \mathbf{0}$, *when* $\mathbf{a} \notin \mathcal{C}(\mathbf{A})$, $\mathbf{b} \notin \mathcal{C}(\mathbf{A^t})$.

Proof. (i) and (ii) $\Rightarrow$ (iii)
Since (i) holds, for each g-inverse $\mathbf{G}$ of $\mathbf{B}$ we have $\mathbf{A} = \mathbf{AGB} = \mathbf{BGA}$. Also, since (ii) holds $\mathbf{A_2} = \mathbf{A_2G_2B_2} = \mathbf{B_2G_2A_2}$ for each g-inverse $\mathbf{G_2}$ of $\mathbf{B_2}$. By Theorem 2.6.22(v), $\mathbf{G}$ is also a g-inverse of $\mathbf{B_2}$. So, $\mathbf{A_2} = \mathbf{A_2GB_2} = \mathbf{B_2GA_2}$. Consider first $\mathbf{A_2} = \mathbf{A_2GB_2}$. Now

$$\begin{aligned}\mathbf{A_2} &= \mathbf{A_2GB_2}\\ &\Rightarrow \mathbf{A} + \mathbf{ab^t} = (\mathbf{A} + \mathbf{ab^t})\mathbf{G}(\mathbf{B} + \mathbf{cd^t})\\ &= \mathbf{AGB} + \mathbf{AGcd^t} + \mathbf{ab^tGB} + \mathbf{ab^tGcd^t}. \qquad (13.2.5)\\ \mathbf{A_2} &= \mathbf{B_2GA_2}\\ &\Rightarrow \mathbf{A} + \mathbf{ab^t} = (\mathbf{B} + \mathbf{cd^t})\mathbf{G}(\mathbf{A} + \mathbf{ab^t})\\ &= \mathbf{BGA} + \mathbf{cd^tGA} + \mathbf{BGab^t} + \mathbf{cd^tGab^t}. \qquad (13.2.6)\end{aligned}$$

We consider the four cases according as '$\mathbf{a} \in \mathcal{C}(\mathbf{A})$ and $\mathbf{b} \in \mathcal{C}(\mathbf{A^t})$' or '$\mathbf{a} \in \mathcal{C}(\mathbf{A})$ and $\mathbf{b} \notin \mathcal{C}(\mathbf{A^t})$' or '$\mathbf{a} \notin \mathcal{C}(\mathbf{A})$ and $\mathbf{b} \in \mathcal{C}(\mathbf{A^t})$' or '$\mathbf{a} \notin \mathcal{C}(\mathbf{A})$ and $\mathbf{b} \notin \mathcal{C}(\mathbf{A^t})$.'

Let $\mathbf{a} \in \mathcal{C}(\mathbf{A})$ and $\mathbf{b} \in \mathcal{C}(\mathbf{A^t})$.
Since $\mathbf{b} \in \mathcal{C}(\mathbf{A^t}) \subseteq \mathcal{C}(\mathbf{B^t})$, (13.2.5) gives $\mathbf{A} + \mathbf{ab^t} = \mathbf{A} + \mathbf{AGcd^t} + \mathbf{ab^t} + \mathbf{ab^tGcd^t}$. So, $\mathbf{AGcd^t} + \mathbf{ab^tGcd^t} = \mathbf{0}$ or $\mathbf{A_2Gcd^t} = \mathbf{0}$. As $\mathbf{d}$ is non-null, we have $\mathbf{A_2Gc} = \mathbf{0}$.
Similarly, (13.2.6) $\Rightarrow$ $\mathbf{cd^tGA_2} = \mathbf{0}$, and since $\mathbf{c}$ is non-null, we get $\mathbf{d^tGA_2} = \mathbf{0}$.

Next, let $\mathbf{a} \in \mathcal{C}(\mathbf{A})$ and $\mathbf{b} \notin \mathcal{C}(\mathbf{A^t})$.
From (13.2.5), we have

$$\begin{aligned}\mathbf{A} + \mathbf{ab^t} &= \mathbf{AGB} + \mathbf{AGcd^t} + \mathbf{ab^tGB} + \mathbf{ab^tGcd^t}\\ &\Rightarrow \mathbf{ab^t} = \mathbf{AGcd^t} + \mathbf{ab^tGB} + \mathbf{ab^tGcd^t}\\ &\Rightarrow \mathbf{ab^t}(\mathbf{I} - \mathbf{GB}) = \mathbf{A_2Gcd^t}.\end{aligned}$$

If $\mathbf{b} \notin \mathcal{C}(\mathbf{B^t})$, then $\mathbf{b^t}(\mathbf{I} - \mathbf{GB}) \neq \mathbf{0}$.
So, $\mathcal{C}(\mathbf{ab^t}(\mathbf{I} - \mathbf{GB})) \subseteq \mathcal{C}(\mathbf{B^t})$ but $\mathcal{C}(\mathbf{A_2Gcd^t}) \subseteq \mathcal{C}(\mathbf{B^t})$. Therefore, $\mathbf{b} \in \mathcal{C}(\mathbf{B^t})$ and $\mathbf{A_2Gcd^t} = \mathbf{0}$. Since, $\mathbf{d}$ is non-null, we have $\mathbf{A_2Gc} = \mathbf{0}$.
Now, (13.2.6) $\Rightarrow$ $\mathbf{A} + \mathbf{ab^t} = \mathbf{A} + \mathbf{cd^tGA} + \mathbf{ab^t} + \mathbf{cd^tGab^t}$, since $\mathbf{a} \in \mathcal{C}(\mathbf{A}) \subseteq \mathcal{C}(\mathbf{B})$. So, $\mathbf{cd^tGA} + \mathbf{cd^tGab^t} = \mathbf{0}$. Since $\mathbf{c}$ is non-null, $\mathbf{d^tGA} + \mathbf{d^tGab^t} = \mathbf{0}$. Notice that $\mathbf{d^tGA} \in \mathcal{C}(\mathbf{A^t})$ and $\mathbf{d^tGab^t}$ does not belong to $\mathcal{C}(\mathbf{A^t})$, so, $\mathbf{d^tGA} = \mathbf{0}$ and $\mathbf{d^tGab^t} = \mathbf{0}$. As $\mathbf{b}$ is non-null, then $\mathbf{d^tGa} = \mathbf{0}$. But then $\mathbf{a} \in \mathcal{C}(\mathbf{A})$ and $\mathbf{d^tGA} = \mathbf{0}$, we have $\mathbf{d^tGa} = \mathbf{0}$.

The case when $\mathbf{a} \notin \mathcal{C}(\mathbf{A})$ and $\mathbf{b} \in \mathcal{C}(\mathbf{A}^t)$ is similar to the case when $\mathbf{a} \in \mathcal{C}(\mathbf{A})$, $\mathbf{b} \notin \mathcal{C}(\mathbf{A}^t)$.

Now, let $\mathbf{a} \notin \mathcal{C}(\mathbf{A})$ and $\mathbf{b} \notin \mathcal{C}(\mathbf{A}^t)$.
From (13.2.5), we have

$$\mathbf{A} + \mathbf{ab}^t = \mathbf{AGB} + \mathbf{AGcd}^t + \mathbf{ab}^t\mathbf{GB} + \mathbf{ab}^t\mathbf{Gcd}^t.$$

Since $\mathbf{AGB} = \mathbf{A}$, we have $\mathbf{ab}^t - \mathbf{ab}^t\mathbf{GB} - \mathbf{ab}^t\mathbf{Gcd}^t = \mathbf{AGcd}^t$. It follows that $\mathbf{AGc} = \mathbf{0}$, $\mathbf{b}^t = \mathbf{b}^t\mathbf{GB} + \mathbf{b}^t\mathbf{Gcd}^t \in \mathcal{C}(\mathbf{B}^t)$ and $\mathbf{b}^t\mathbf{Gc} = \mathbf{0}$.
Similarly, (13.2.6) yields $\mathbf{d}^t\mathbf{GA} = \mathbf{0}, \mathbf{a} \in \mathcal{C}(\mathbf{B})$ and $\mathbf{d}^t\mathbf{Ga} = \mathbf{0}$.

(i) and (iii) $\Rightarrow$ (ii)
Let (i) and (iii) hold. By (iii) there exists a g-inverse $\mathbf{G}$ of $\mathbf{B}$ such that
(1) $\mathbf{A_2Gc} = \mathbf{0}$, $\mathbf{d}^t\mathbf{GA_2} = \mathbf{0}$, in case $\mathbf{a} \in \mathcal{C}(\mathbf{A})$, $\mathbf{b} \in \mathcal{C}(\mathbf{A}^t)$.
(2) $\mathbf{A_2Gc} = \mathbf{0}$, $\mathbf{d}^t\mathbf{GA} = \mathbf{0}$ and $\mathbf{b} \in \mathcal{C}(\mathbf{B}^t)$, when $\mathbf{a} \in \mathcal{C}(\mathbf{A})$, and $\mathbf{b} \notin \mathcal{C}(\mathbf{A}^t)$
(3) $\mathbf{d}^t\mathbf{GA_2} = \mathbf{0}$, $\mathbf{AGc} = \mathbf{0}$, and $\mathbf{a} \in \mathcal{C}(\mathbf{B})$ when $\mathbf{a} \notin \mathcal{C}(\mathbf{A})$, $\mathbf{b} \in \mathcal{C}(\mathbf{A}^t)$ and
(4) $\mathbf{a} \in \mathcal{C}(\mathbf{B})$, $\mathbf{b} \in \mathcal{C}(\mathbf{B}^t), \mathbf{A_2Gc} = \mathbf{0}$, and $\mathbf{d}^t\mathbf{GA_2} = \mathbf{0}$, when $\mathbf{a} \notin \mathcal{C}(\mathbf{A})$, and $\mathbf{b} \notin \mathcal{C}(\mathbf{A}^t)$.

By Theorem 2.6.22(v), $\mathbf{G}$ is a g-inverse of $\mathbf{B}$. Let $\mathbf{a} \in \mathcal{C}(\mathbf{A})$ and $\mathbf{b} \in \mathcal{C}(\mathbf{A}^t)$. Notice that $A_2\mathbf{G_2B_2} = \mathbf{A} + \mathbf{A_2Gcd}^t + \mathbf{ab}^t\mathbf{GB}$. Since $\mathbf{A_2Gc} = \mathbf{0}$ and $\mathbf{b} \in \mathcal{C}(\mathbf{A}^t) \subseteq \mathcal{C}(\mathbf{B}^t)$ we have $\mathbf{A_2G_2B_2} = \mathbf{A_2}$. Again, since $\mathbf{a} \in \mathcal{C}(\mathbf{A}) \subseteq \mathcal{C}(\mathbf{B})$ and $\mathbf{d}^t\mathbf{GA_2} = \mathbf{0}$, we have $\mathbf{B_2GA_2} = \mathbf{A} + \mathbf{cd}^t\mathbf{GA_2} + \mathbf{BGab}^t = \mathbf{A_2}$. The Proof in the other cases is easy and is left to the reader. Proof of (ii) and (iii) $\Rightarrow$ (i) is straightforward. □

Remark 13.2.6. In (i) and (ii) $\Rightarrow$ (iii), the result holds for every g-inverse $\mathbf{G}$ of $\mathbf{B}$.

Theorem 13.2.7. *Let* $\mathbf{A}$ *and* $\mathbf{B}$ *be matrices of order* $m \times n$, $\mathbf{a}, \mathbf{c}$ *be non-null m-column vectors and* $\mathbf{b}, \mathbf{d}$ *be non-null n-column vectors. Let* $\mathbf{c} \in \mathcal{C}(\mathbf{B})$, $\mathbf{d} \in \mathcal{C}(\mathbf{B}^t)$ *and* $\beta \neq 0$. *Then any two of the following three implies the third:*

(i) $\mathbf{A} <^s \mathbf{B}$
(ii) $\mathbf{A_2} <^s \mathbf{B_2}$ *and*
(iii) $\mathbf{a} \in \mathcal{C}(\mathbf{B})$ *and* $\mathbf{b} \in \mathcal{C}(\mathbf{B}^t)$

Proof. (i) and (ii) $\Rightarrow$ (iii)
First note that $\rho(\mathbf{B}) = \rho(\mathbf{B_2})$, by Lemma 13.2.4. Further, $\mathcal{C}(\mathbf{B_2}) \subseteq \mathcal{C}(\mathbf{B})$ and $\mathcal{C}(\mathbf{B}_2^t) \subseteq \mathcal{C}(\mathbf{B}^t)$, since $\mathbf{c} \in \mathcal{C}(\mathbf{B})$ and $\mathbf{d} \in \mathcal{C}(\mathbf{B}^t)$. Hence, $\mathcal{C}(\mathbf{B_2}) = \mathcal{C}(\mathbf{B})$ and $\mathcal{C}(\mathbf{B}_2^t) = \mathcal{C}(\mathbf{B}^t)$.

Since $\mathbf{A_2} <^s \mathbf{B_2}$, in view of the above there exist matrices $\mathbf{R}$ and $\mathbf{S}$ such that $\mathbf{A} + \mathbf{ab}^t = \mathbf{BR} = \mathbf{SB}$. Also, $\mathbf{A} <^s \mathbf{B}$, it follows that $\mathbf{a} \in \mathcal{C}(\mathbf{B})$ and

$\mathbf{b} \in \mathcal{C}(\mathbf{B}^t)$.
(i) and (iii) $\Rightarrow$ (ii)
Since $\mathbf{a} \in \mathcal{C}(\mathbf{B})$ and $\mathbf{A} <^s \mathbf{B}$, we have $\mathcal{C}(\mathbf{A}+\mathbf{ab}^t) \subseteq \mathcal{C}(\mathbf{B_2})$. Again, since $\mathbf{b} \in \mathcal{C}(\mathbf{B}^t)$ and $\mathbf{A} <^s \mathbf{B}$, we have $\mathcal{C}(\mathbf{A}+\mathbf{ab}^t)^t \subseteq \mathcal{C}(\mathbf{B}^t) = \mathcal{C}(\mathbf{B_2})^t$. Hence, $\mathbf{A_2} <^s \mathbf{B_2}$.
(ii) and (iii) $\Rightarrow$ (i)
Let $\mathbf{G}$ be a g-inverse of $\mathbf{B}$. By Theorem 2.6.22, $\mathbf{H} = \mathbf{G} - \dfrac{\mathbf{Gcd}^t\mathbf{G}}{\beta}$ is a g-inverse of $\mathbf{B_2}$. Since $\mathbf{A_2} <^s \mathbf{B_2}$, we have $\mathbf{A_2HB_2} = \mathbf{B_2HA_2} = \mathbf{A_2}$.
$\mathbf{A_2} = \mathbf{A_2HB_2} = \mathbf{AGB} - \mathbf{A}\dfrac{\mathbf{Gcd}^t\mathbf{G}}{\beta}\mathbf{B} + \mathbf{AGcd}^t - \mathbf{A}\dfrac{\mathbf{Gcd}^t\mathbf{G}}{\beta}\mathbf{cd}^t + \mathbf{ab}^t\mathbf{GB} - \mathbf{ab}^t\dfrac{\mathbf{Gcd}^t\mathbf{G}}{\beta}\mathbf{B} + \mathbf{ab}^t\mathbf{Gcd}^t - \mathbf{ab}^t\dfrac{\mathbf{Gcd}^t\mathbf{G}}{\beta}\mathbf{cd}^t = \mathbf{AGB} - \mathbf{A}\dfrac{\mathbf{Gcd}^t}{\beta} + \mathbf{AGcd}^t - (\beta-1)\mathbf{A}\dfrac{\mathbf{Gcd}^t}{\beta} + \mathbf{ab}^t - \dfrac{\mathbf{ab}^t\mathbf{Gcd}^t}{\beta} + \mathbf{ab}^t\mathbf{Gcd}^t - (\beta-1)\dfrac{\mathbf{ab}^t\mathbf{Gcd}^t}{\beta} = \mathbf{AGB} + \mathbf{ab}^t$.
Hence, $\mathbf{A} = \mathbf{AGB}$.
Similarly, $\mathbf{A_2} = \mathbf{B_2HA_2} \Rightarrow \mathbf{A} = \mathbf{BGA}$. Thus, $\mathbf{A} <^s \mathbf{B}$. □

Corollary 13.2.8. *Consider the setup of Theorem 13.2.7. Further, let* $\mathbf{a} \in \mathcal{C}(\mathbf{A})$ *and* $\mathbf{b} \in \mathcal{C}(\mathbf{A}^t)$. *Then* $\mathbf{A} <^s \mathbf{B}$ *if and only if* $\mathbf{A_2} <^s \mathbf{B_2}$.

Assume that $\mathbf{c} \in \mathcal{C}(\mathbf{B})$ and $\mathbf{d} \notin \mathcal{C}(\mathbf{B}^t)$. We now explore the conditions for $\mathbf{A_2} <^s \mathbf{B_2}$ to hold when $\mathbf{A} <^s \mathbf{B}$. For this we first prove the following lemma:

Lemma 13.2.9. *Let* $\mathbf{X}$ *be an* $m \times n$ *matrix, the vector* $\mathbf{x} \in \mathcal{C}(\mathbf{X})$ *and* $\mathbf{y} \notin \mathcal{C}(\mathbf{X}^t)$. *Let* $\mathbf{X}^-$ *be a g-inverse of* $\mathbf{X}$ *such that* $\mathbf{y}^t\mathbf{X}^-\mathbf{x} = \lambda \neq \mathbf{0}$. *Then there exists a g-inverse* $\mathbf{G}$ *of* $\mathbf{X}$ *such that* $\mathbf{y}^t\mathbf{Gx} = \mathbf{0}$.

Proof. Note that $\mathbf{y}^t\mathbf{X}^-\mathbf{x} \neq \mathbf{0}$. So, both $\mathbf{x}$ and $\mathbf{y}$ are non-null. By Lemma 2.6.1, there exists a vector $\mathbf{z}$ such that $\mathbf{z}^t\mathbf{X}^t = \mathbf{0}$ and $\mathbf{z}^t\mathbf{y} = \mathbf{1}$. Moreover, $\mathbf{x} \neq \mathbf{0} \Rightarrow$ there exits a vector $\mathbf{w}$ such that $\mathbf{w}^t\mathbf{x} = 1$. Let $\mathbf{G} = \mathbf{X}^- - \lambda\mathbf{zw}^t$. Then $\mathbf{G}$ is a g-inverse of $\mathbf{X}$ and $\mathbf{y}^t\mathbf{Gx} = \mathbf{0}$. □

Corollary 13.2.10. *Let* $\mathbf{X}$ *be an* $m \times n$ *matrix, the vector* $\mathbf{y} \in \mathcal{C}(\mathbf{X}^t)$ *and* $\mathbf{x} \notin \mathcal{C}(\mathbf{X})$. *Let* $\mathbf{X}^-$ *be a g-inverse of* $\mathbf{X}$ *such that* $\mathbf{y}^t\mathbf{X}^-\mathbf{x} \neq \mathbf{0}$. *Then there exists a g-inverse* $\mathbf{G}$ *of* $\mathbf{X}$ *such that* $\mathbf{y}^t\mathbf{Gx} = \mathbf{0}$.

We now prove the following:

Theorem 13.2.11. *Let* $\mathbf{A}$ and $\mathbf{B}$ *be matrices of order* $m \times n$, $\mathbf{a}, \mathbf{c}$ *be non-null* m*-column vectors and* $\mathbf{b}, \mathbf{d}$ *be non-null* n*-column vectors. Let* $\mathbf{c} \in \mathcal{C}(\mathbf{B})$,

$\mathbf{d} \notin \mathcal{C}(\mathbf{B}^t)$ *and* $\mathbf{G}$ *be a g-inverse of* $\mathbf{B}$ *such that* $\mathbf{d}^t\mathbf{G}\mathbf{c} = \mathbf{0}$. *Then any two of the following three implies the third:*

(i) $\mathbf{A} <^s \mathbf{B}$

(ii) $\mathbf{A_2} <^s \mathbf{B_2}$ *and*

(iii) $\mathbf{a} \in \mathcal{C}(\mathbf{B})$ *and (either '*$\mathbf{A_2Gc} = \mathbf{0}$, *and* $\mathbf{b} \in \mathcal{C}(\mathbf{B}^t)$*') or* $\mathbf{b} \notin \mathcal{C}(\mathbf{B}^t)$, $\mathbf{d} = \mu\mathbf{b} + \mathbf{B}^t\mathbf{v}$ *and* $\mathbf{A_2Gc} = \dfrac{\mathbf{a}}{\mu}$ *for some vector* $\mathbf{v}$ *and some scalar* $\mu \neq 0$.

Proof. (i) and (ii) $\Rightarrow$ (iii)
Since (i) holds, $\mathbf{A} = \mathbf{AGB} = \mathbf{BGA}$. Also, $\mathbf{d}^t\mathbf{Gc} = \mathbf{0}, \beta = \mathbf{1}$. Hence, $\mathbf{H} = \mathbf{G} - \mathbf{Gcd}^t\mathbf{G}$ is a g-inverse of $\mathbf{B_2}$. Since (ii) holds, $\mathbf{A_2} = \mathbf{A_2HB_2} = \mathbf{B_2HA_2}$.
$\mathbf{A_2} = \mathbf{B_2HA_2}$
$\Rightarrow \mathbf{A}+\mathbf{ab}^t = \mathbf{A}-\mathbf{BGcd}^t\mathbf{GA}+\mathbf{BGab}^t-\mathbf{BGcd}^t\mathbf{Gab}^t+\mathbf{cd}^t\mathbf{GA}+\mathbf{cd}^t\mathbf{Gab}^t$
$\Rightarrow \mathbf{ab}^t = \mathbf{BGab}^t$, since $\mathbf{d}^t\mathbf{Gc} = \mathbf{0}$, $\mathbf{A} = \mathbf{BGA}$ and $\mathbf{BGc} = \mathbf{c}$. It follows that $\mathbf{a} = \mathbf{BGa}$ or $\mathbf{a} \in \mathcal{C}(\mathbf{B})$, as $\mathbf{b}$ is non-null.
Now, $\mathbf{A_2} = \mathbf{A_2HB_2} \Rightarrow \mathbf{A} + \mathbf{ab}^t = \mathbf{AGB} + \mathbf{ab}^t\mathbf{GB} - \mathbf{AGcd}^t\mathbf{GB} - \mathbf{ab}^t\mathbf{Gcd}^t\mathbf{GB} + \mathbf{AGcd}^t + \mathbf{ab}^t\mathbf{Gcd}^t$, since $\mathbf{d}^t\mathbf{Gc} = \mathbf{0}$
$\Rightarrow (\mathbf{ab}^t - \mathbf{AGcd}^t - \mathbf{ab}^t\mathbf{Gcd}^t)(\mathbf{I} - \mathbf{GB}) = \mathbf{0}$.
If $\mathbf{b} \in \mathcal{C}(\mathbf{B}^t)$, then $\mathbf{A_2Gcd}^t = \mathbf{0}$, since $\mathbf{d} \notin \mathcal{C}(\mathbf{B}^t)$. So, $\mathbf{A_2Gc} = \mathbf{0}$, as $\mathbf{d}$ is non-null.
If $\mathbf{b} \notin \mathcal{C}(\mathbf{B}^t)$, then $\mathbf{A_2Gc} \neq \mathbf{0}$ and $\mathbf{ab}^t + \mathbf{A_2Gcd}^t = \mathbf{TB}$ for some matrix $\mathbf{T}$. Hence, $\mathbf{A_2Gcd}^t = \mathbf{TB} - \mathbf{ab}^t$. Since $\mathbf{A_2Gc} \neq \mathbf{0}$, there exists a vector $\mathbf{u}$ such that $\mathbf{u}^t\mathbf{A_2Gc} = 1$. So, $\mathbf{d}^t = \mathbf{u}^t\mathbf{TB} - \mathbf{u}^t\mathbf{ab}^t$ or $\mathbf{d} = \mathbf{B}^t\mathbf{v} + \mu\mathbf{b}$ with $\mu \neq 0$. (As $\mu = 0$ gives $\mathbf{d} \in \mathcal{C}(\mathbf{B}^t)$.) notice that μ is unique, since $\mathbf{b} \notin \mathcal{C}(\mathbf{B}^t)$. Further, $(\mathbf{ab}^t - \mathbf{A_2Gcd}^t)(\mathbf{I} - \mathbf{GB}) = \mathbf{0}$. So, $(\mathbf{ab}^t - \mathbf{A_2Gc}(\mu\mathbf{b}^t + \mathbf{vB}))(\mathbf{I} - \mathbf{GB}) = \mathbf{0}$ or $(\mathbf{a} - \mu\mathbf{A_2Gc})\mathbf{b}^t(\mathbf{I} - \mathbf{GB}) = \mathbf{0}$.
Since $\mathbf{b} \notin \mathcal{C}(\mathbf{B}^t)$, $\mathbf{b}^t(\mathbf{I} - \mathbf{GB}) \neq \mathbf{0}$. So, $\mathbf{A_2Gc} = \dfrac{1}{\mu}\mathbf{a}$.
(i) and (iii) $\Rightarrow$ (ii)
Let (i) and (iii) hold. As (i) holds, there exists a g-inverse $\mathbf{G}$ of $\mathbf{B}$ such that $\mathbf{A} = \mathbf{AGB} = \mathbf{BGA}$.

Let $\mathbf{a} \in \mathcal{C}(\mathbf{B})$, $\mathbf{b} \in \mathcal{C}(\mathbf{B}^t)$.

Since (iii) holds, $\mathbf{A_2Gc} = \mathbf{0}$. Also, $\mathbf{G} - \mathbf{Gcd}^t\mathbf{G}$ is a g-inverse of $\mathbf{B_2}$. Now, $\mathbf{A_2G_2B_2} = \mathbf{A}+\mathbf{ab}^t\mathbf{GB}-\mathbf{AGcd}^t-\mathbf{ab}^t\mathbf{Gcd}^t+\mathbf{AGcd}^t+\mathbf{ab}^t\mathbf{Gcd}^t = \mathbf{A} + \mathbf{ab}^t - \mathbf{A_2Gcd}^t\mathbf{GB} + \mathbf{A_2Gcd}^t = \mathbf{A} + \mathbf{ab}^t = \mathbf{A_2}$, since $\mathbf{b} \in \mathcal{C}(\mathbf{B}^t)$.

Similarly, $\mathbf{B_2G_2A_2} = \mathbf{A_2}$ holds.

Now, let $\mathbf{a} \in \mathcal{C}(\mathbf{B})$ and $\mathbf{b} \notin \mathcal{C}(\mathbf{B}^t)$.

Since (i) holds, there exists a g-inverse $\mathbf{G}$ of $\mathbf{B}$ such that $\mathbf{A} = \mathbf{AGB} = \mathbf{BGA}$. Also, (iii) holds, so, $\mathbf{d} = \mu\mathbf{b} + \mathbf{B}^t\mathbf{v}$ for some scalar

$\mu \neq 0$ and $\mathbf{A_2Gc} = \dfrac{\mathbf{a}}{\mu}$.

It is easy to check that (ii) holds.

Proof of (ii) and (iii) $\Rightarrow$ (i) is similar. □

Theorem 13.2.12. *et* $\mathbf{A}, \mathbf{B}$ *be matrices of order* $m \times n$, $\mathbf{a}$, $\mathbf{c}$ *be* m*-column vectors and* $\mathbf{b}, \mathbf{d}$ *be* n*-column vectors such that* $\mathbf{A} <^{\mathbf{s}} \mathbf{B}$. *Let* $\mathbf{c} \notin \mathcal{C}(\mathbf{B})$ *and* $\mathbf{d^t} \in \mathcal{C}(\mathbf{B^t})$ *and* $\mathbf{G}$ *be a g-inverse of* $\mathbf{B}$ *such that* $\mathbf{d^tGc} = \mathbf{0}$. *Then any two of the following imply the third:*

(i) $\mathbf{A} <^{\mathbf{s}} \mathbf{B}$

(ii) $\mathbf{A_2} <^{\mathbf{s}} \mathbf{B_2}$ *and*

(iii) $\mathbf{b} \in \mathcal{C}(\mathbf{B^t})$ *and* *either* $\{\mathbf{d^tGA_2} = \mathbf{0}$ *and* $\mathbf{a} \in \mathcal{C}(\mathbf{B})\}$ *or* $\{\mathbf{a} \notin \mathcal{C}(\mathbf{B})$, $\mathbf{c} = \mathbf{Bu} + \eta\mathbf{a}$ *for some vector* $\mathbf{u}$ *and some scalar* $\eta \neq 0$ *and* $\mathbf{d^tGA_2} = \dfrac{\mathbf{b^t}}{\eta}.\}$

Proof is similar to the proof of Theorem 13.2.9.

To see when $\mathbf{A_2} <^{\mathbf{s}} \mathbf{B_2}$ in case $\mathbf{c} \notin \mathcal{C}(\mathbf{B})$ and $\mathbf{d} \notin \mathcal{C}(\mathbf{B^t})$, we need some more preparation. We first prove the following lemmas:

Lemma 13.2.13. *Let* $\mathbf{B}$ *be an* $m \times n$ *matrix and* $\mathbf{B_2} = \mathbf{B} + \mathbf{cd^t}$ *where* $\mathbf{c}$ *and* $\mathbf{d}$ *are respectively* m *and* n*-column vectors. Let* $\mathbf{c} \notin \mathcal{C}(\mathbf{B})$ *and* $\mathbf{d} \notin \mathcal{C}(\mathbf{B^t})$. *then there exists a matrix* $\mathbf{G}$ *that is a g-inverse of both* $\mathbf{B}$ *and* $\mathbf{B_2}$.

Proof. By Lemma 13.2.4, $\rho(\mathbf{B_2}) = \rho(\mathbf{B}) + 1$. Theorem 2.3.19 guarantees the existence of such a g-inverse. □

Lemma 13.2.14. *Under the setup of Lemma 13.2.13, let* $\mathbf{G}$ *be a common g-inverse of* $\mathbf{B}$ *and* $\mathbf{B_2}$. *Then* $\mathbf{BGc} = \mathbf{0}$, $\mathbf{d^tGB} = \mathbf{0}$ *and* $\mathbf{d^tGc} = \mathbf{1}$.

Proof. Since $\mathbf{G}$ is g-inverse of $\mathbf{B_2}$, so, $\mathbf{B_2GB_2} = \mathbf{B_2}$. Now,

$$\begin{aligned}
&\mathbf{B_2GB_2} = \mathbf{B_2}\\
&\qquad \Rightarrow \mathbf{BGcd^t} + \mathbf{cd^tGB} + \mathbf{cd^tGcd^t} = \mathbf{cd^t}\\
&\qquad \text{or } \mathbf{cd^tGB} = (\mathbf{I} - \mathbf{BG} - \mathbf{cd^tG})\mathbf{cd^t}\\
&\qquad \Rightarrow \mathbf{cd^tGB} = \mathbf{0} \text{ and } (\mathbf{I} - \mathbf{BG} - \mathbf{cd^tG})\mathbf{cd^t} = \mathbf{0},\\
&\qquad \text{since } \mathcal{C}(\mathbf{cd^tGB}) \subseteq \mathcal{C}(\mathbf{B^t}) \text{and } \mathcal{C}((\mathbf{I} - \mathbf{BG} - \mathbf{cd^tG})\mathbf{cd^t}) \nsubseteq \mathcal{C}(\mathbf{B^t}).
\end{aligned}$$

So, $\mathbf{d^tGB} = \mathbf{0}$ and $(\mathbf{I} - \mathbf{BG} - \mathbf{cd^tG})\mathbf{c} = \mathbf{0}$. However, $(\mathbf{I} - \mathbf{BG} - \mathbf{cd^tG})\mathbf{c} = \mathbf{0} \Rightarrow \mathbf{BGc} = \mathbf{0}$ and $\mathbf{d^tGt} = \mathbf{1}$, as $\mathbf{BGc} \in \mathcal{C}(\mathbf{B})$ and $\mathbf{c} \notin \mathcal{C}(\mathbf{B})$. □

Theorem 13.2.15. *Let* $\mathbf{A}$ and $\mathbf{B}$ *be matrices of order* $m \times n$, $\mathbf{a}$, $\mathbf{c}$ *be non-null* m*-vectors and and* $\mathbf{b}$, $\mathbf{d}$ *be non-null* n*-column vectors such that* $\mathbf{c} \notin \mathcal{C}(\mathbf{B})$ *and* $\mathbf{d} \notin \mathcal{C}(\mathbf{B}^t)$. *Then any two of the following imply the third:*

(i) $\mathbf{A} <^s \mathbf{B}$

(ii) $\mathbf{A_2} <^s \mathbf{B_2}$ *and*

(iii) $\mathbf{a} = \mathbf{Bu} + \alpha\mathbf{c}$, *where* $\mathbf{u}$ *and* α *are arbitrary subject to* $\mathbf{a} \neq \mathbf{0}$ *and* $\mathbf{b} = \mathbf{v}^t\mathbf{B} + \theta\mathbf{d}$, *where* $\mathbf{v}$ *and* θ *are arbitrary subject to* $\mathbf{b} \neq \mathbf{0}$.

Proof. By Lemmas 13.2.13 and 13.2.14, there exists a common g-inverse $\mathbf{G}$ of $\mathbf{B}$ and $\mathbf{B_2}$ such that $\mathbf{BGc} = \mathbf{0}$, $\mathbf{d}^t\mathbf{GB} = \mathbf{0}$ and $\mathbf{d}^t\mathbf{Gc} = 1$. Now,
$\mathbf{A_2} = \mathbf{A_2GB_2}$
$\Leftrightarrow \mathbf{A} + \mathbf{ab}^t = (\mathbf{A} + \mathbf{ab}^t)\mathbf{G}(\mathbf{B} + \mathbf{cd}^t)$ and
$\mathbf{A_2} = \mathbf{B_2GA_2}$
$\Leftrightarrow \mathbf{A} + \mathbf{ab}^t = \mathbf{BGA} + \mathbf{BGab}^t + \mathbf{cd}^t\mathbf{GA} + \mathbf{cd}^t\mathbf{Gab}^t$
(i) and (ii) $\Rightarrow$ (iii)
As $\mathbf{BGc} = \mathbf{0}$, $\mathbf{d}^t\mathbf{GB} = \mathbf{0}$ and $\mathbf{A} <^s \mathbf{B}$, we have $\mathbf{AGc} = \mathbf{0}$ and $\mathbf{d}^t\mathbf{GA} = \mathbf{0}$. So, (i) and (ii) $\Rightarrow \mathbf{ab}^t(\mathbf{I} - \mathbf{GB}) = \mathbf{ab}^t\mathbf{Gcd}^t$ and $(\mathbf{I} - \mathbf{BG})\mathbf{ab}^t = \mathbf{cd}^t\mathbf{Gab}^t$ $\Rightarrow \mathbf{b}^t(\mathbf{I} - \mathbf{GB}) = \mathbf{b}^t\mathbf{Gcd}^t$ and $(\mathbf{I} - \mathbf{BG})\mathbf{a} = \mathbf{cd}^t\mathbf{Ga}$, since $\mathbf{a}$, $\mathbf{b}$ are non-null
$\Rightarrow \mathbf{b} = \mathbf{v}^t\mathbf{B} + \theta\mathbf{d}$ for some $\mathbf{v}$ and θ; and $\mathbf{a} = \mathbf{Bu} + \alpha\mathbf{c}$, for some $\mathbf{u}$ and α. As $\mathbf{a}$, $\mathbf{b}$ are non-null $\mathbf{v}^t\mathbf{B} + \theta\mathbf{d} \neq \mathbf{0}$ and $\mathbf{Bu} + \alpha\mathbf{c} \neq \mathbf{0}$.
(i) and (iii) $\Rightarrow$ (ii)
$\mathbf{BGc} = \mathbf{0}$ and $\mathbf{d}^t\mathbf{GB} = \mathbf{0}$, $\mathbf{b} = \mathbf{v}^t\mathbf{B} + \theta\mathbf{d}$ and $\mathbf{a} = \mathbf{Bu} + \alpha\mathbf{c} \Rightarrow \theta = \mathbf{b}^t\mathbf{Gc}$ and $\alpha = \mathbf{d}^t\mathbf{Ga}$. So, $\mathbf{ab}^t(\mathbf{I} - \mathbf{GB}) = \mathbf{ab}^t\mathbf{Gcd}^t$ and $(\mathbf{I} - \mathbf{BG})\mathbf{ab}^t = \mathbf{cd}^t\mathbf{Gab}^t$. Also, (i) $\Rightarrow \mathbf{AGc} = \mathbf{0}$ and $\mathbf{d}^t\mathbf{GA} = \mathbf{0}$. Now (ii) follows.
(ii) and (iii) $\Rightarrow$ (i)
As already seen (iii) $\Rightarrow \mathbf{ab}^t(\mathbf{I} - \mathbf{GB}) = \mathbf{ab}^t\mathbf{Gcd}^t$ and $(\mathbf{I} - \mathbf{BG})\mathbf{ab}^t = \mathbf{cd}^t\mathbf{Gab}^t$. So, (ii) and (iii) $\Rightarrow \mathbf{A} = \mathbf{AGB} + \mathbf{AGcd}^t = \mathbf{BGA} + \mathbf{cd}^t\mathbf{GA}$. Now,
$\mathbf{A} = \mathbf{AGB} + \mathbf{AGcd}^t \Rightarrow \mathbf{A} - \mathbf{AGB} = \mathbf{AGcd}^t = \mathbf{0}$ and
$\mathbf{A} = \mathbf{BGA} + \mathbf{cd}^t\mathbf{GA} \Rightarrow \mathbf{A} - \mathbf{BGA} = \mathbf{cd}^t\mathbf{GA} = \mathbf{0}$.
Since $\mathbf{c} \notin \mathcal{C}(\mathbf{B})$ and $\mathbf{d} \notin \mathcal{C}(\mathbf{B}^t)$, we have $\mathbf{A} <^s \mathbf{B}$. □

13.3 Minus order

In this section, we study the effect of modifications on matrices related to each other under minus order. Given matrices $\mathbf{A}$ and $\mathbf{B}$ of the same order such that $\mathbf{A} <^- \mathbf{B}$, we first give necessary and sufficient conditions under which the new matrices $\mathbf{A_1}$ and $\mathbf{B_1}$ obtained by appending/deleting a column also satisfy $\mathbf{A_1} <^- \mathbf{B_1}$. We also study the converse problem namely, if $\mathbf{A_1} <^- \mathbf{B_1}$, when does $\mathbf{A} <^- \mathbf{B}$ hold? Let $\mathbf{A}$ and $\mathbf{B}$ be matrices of the same order such that $\mathbf{A} <^- \mathbf{B}$. If the matrices $\mathbf{A}$ and $\mathbf{B}$ are modified by addition of rank 1 matrices, the conditions under which new matrices are similarly related and vice versa have also been explored. For matrices $\mathbf{A}$ and $\mathbf{B}$ of the same order $m \times n$ and m-column vectors $\mathbf{a}$, $\mathbf{b}$ we write, as in Section 2, $\mathbf{A_1} = (\mathbf{A} : \mathbf{a})$ and $\mathbf{B_1} = (\mathbf{B} : \mathbf{b})$. We start with the following:

Theorem 13.3.1. *Let* $\mathbf{A}$ *and* $\mathbf{B}$ *be matrices of order* $m \times n$ *and* $\mathbf{a}$ *and* $\mathbf{b}$ *be* m*-column vectors. Then any two of the following imply the third:*

(i) $\mathbf{A} <^- \mathbf{B}$
(ii) $\mathbf{A_1} <^- \mathbf{B_1}$ *and*
(iii) *Either* '$\mathbf{a} \in \mathcal{C}(\mathbf{A})$, *and* $\mathbf{b} - \mathbf{a} \in \mathcal{C}(\mathbf{B} - \mathbf{A})$,' *or* ($\mathbf{b} \notin \mathcal{C}(\mathbf{B})$ and exactly one of '$\mathbf{a} \in \mathcal{C}(\mathbf{A})$ and $\mathbf{b} - \mathbf{a} \in \mathcal{C}(\mathbf{B} - \mathbf{A})$)' *holds.*

Proof. (i) and (ii) $\Rightarrow$ (iii)
Since (i) and (ii) hold, by Theorem 3.3.5 and Remark 3.3.9,

$$\rho(\mathbf{B}) = \rho(\mathbf{A}) + \rho(\mathbf{B} - \mathbf{A}) \tag{13.3.1}$$

and

$$\rho(\mathbf{B_1}) = \rho(\mathbf{A_1}) + \rho(\mathbf{B_1} - \mathbf{A_1}). \tag{13.3.2}$$

Moreover, $\rho(\mathbf{B}) \leq \rho(\mathbf{B_1}) \leq \rho(\mathbf{B_1}) + \mathbf{1}$ and similar inequalities hold for '$\mathbf{A_1}$ and $\mathbf{A}$' and for '$\mathbf{B_1} - \mathbf{A_1}$ and $\mathbf{B} - \mathbf{A}$'.

Let $\mathbf{b} \in \mathcal{C}(\mathbf{B})$. Then $\rho(\mathbf{B_1}) = \rho(\mathbf{B})$. Using (13.3.1) and (13.3.2), we have $\rho(\mathbf{A}) \leq \rho(\mathbf{A_1}) + \rho(\mathbf{B_1} - \mathbf{A_1}) = \rho(\mathbf{B_1}) = \rho(\mathbf{B}) = \rho(\mathbf{A}) + \rho(\mathbf{B} - \mathbf{A})$. Thus, $\rho(\mathbf{B_1} - \mathbf{A_1}) \leq \rho(\mathbf{B} - \mathbf{A})$. However, $\rho(\mathbf{B} - \mathbf{A}) \leq \rho(\mathbf{B_1} - \mathbf{A_1})$ always holds, so, $\rho(\mathbf{B_1} - \mathbf{A_1}) = \rho(\mathbf{B} - \mathbf{A})$. Therefore, $\rho(\mathbf{A_1}) = \rho(\mathbf{A})$ and consequently $\mathbf{a} \in \mathcal{C}(\mathbf{A})$ and $\mathbf{b} - \mathbf{a} \in \mathcal{C}(\mathbf{B} - \mathbf{A})$.

Let $\mathbf{b} \notin \mathcal{C}(\mathbf{B})$. Then $\rho(\mathbf{B_1}) = \rho(\mathbf{B}) + \mathbf{1}$. Then by (13.3.1), we have $\rho(\mathbf{B_1}) = \rho(\mathbf{B}) + \mathbf{1} = \rho(\mathbf{A}) + \rho(\mathbf{B} - \mathbf{A}) + \mathbf{1}$. Using this in (13.3.2), we have $\rho(\mathbf{A_1}) + \rho(\mathbf{B_1} - \mathbf{A_1}) = \rho(\mathbf{B_1}) = \rho(\mathbf{B}) + \mathbf{1} = \rho(\mathbf{A}) + \rho(\mathbf{B} - \mathbf{A}) + \mathbf{1}$. So, exactly one of '$\rho(\mathbf{A_1}) = \rho(\mathbf{A}) + \mathbf{1}$ and $\rho(\mathbf{B_1} - \mathbf{A_1}) = \rho(\mathbf{B} - \mathbf{A})$' and '$\rho(\mathbf{B_1} - \mathbf{A_1}) = \rho(\mathbf{B} - \mathbf{A}) + \mathbf{1}$ and $\rho(\mathbf{A_1}) = \rho(\mathbf{A})$' can hold. Hence, exactly one of '$\mathbf{a} \in \mathcal{C}(\mathbf{A})$, and $\mathbf{b} - \mathbf{a} \in \mathcal{C}(\mathbf{B} - \mathbf{A})$' holds.

(i) and (iii) $\Rightarrow$ (ii)
Let (i) and (iii) hold. In case, $\mathbf{a} \in \mathcal{C}(\mathbf{A})$, and $\mathbf{b} - \mathbf{a} \in \mathcal{C}(\mathbf{B} - \mathbf{A})$, we have

$\rho(\mathbf{A_1}) = \rho(\mathbf{A})$ and $\rho(\mathbf{B_1} - \mathbf{A_1}) = \rho(\mathbf{B} - \mathbf{A})$. Since (i) holds, we have $\rho(\mathbf{B}) = \rho(\mathbf{A}) + \rho(\mathbf{B} - \mathbf{A})$. Also $\mathbf{b} \in \mathcal{C}(\mathbf{B})$. So,

$$\rho(\mathbf{A_1}) + \rho(\mathbf{B_1} - \mathbf{A_1}) = \rho(\mathbf{A}) + \rho(\mathbf{B} - \mathbf{A}) = \rho(\mathbf{B}) = \rho(\mathbf{B_1}).$$

Hence, $\mathbf{b} \in \mathcal{C}(\mathbf{B})$ and (ii) holds.
Let $\mathbf{b} \notin \mathcal{C}(\mathbf{B})$. If $\mathbf{a} \in \mathcal{C}(\mathbf{A})$, then we have $\rho(\mathbf{A_1}) = \rho(\mathbf{A})$, and $\rho(\mathbf{B_1} - \mathbf{A_1}) = \rho(\mathbf{B} - \mathbf{A}) + \mathbf{1}$. Therefore,

$$\rho(\mathbf{A_1}) + \rho(\mathbf{B_1} - \mathbf{A_1}) = \rho(\mathbf{A}) + \rho(\mathbf{B} - \mathbf{A}) + \mathbf{1} = \rho(\mathbf{B}) + \mathbf{1} = \rho(\mathbf{B_1}),$$

so, (ii) holds.
Similarly, if $\mathbf{b} - \mathbf{a} \in \mathcal{C}(\mathbf{B} - \mathbf{A})$ and $\mathbf{a} \in \mathcal{C}(\mathbf{A})$, we have $\rho(\mathbf{B_1} - \mathbf{A_1}) = \rho(\mathbf{B} - \mathbf{A})$ and $\rho(\mathbf{A_1}) = \rho(\mathbf{A}) + \mathbf{1}$. So, $\rho(\mathbf{A_1}) + \rho(\mathbf{B_1} - \mathbf{A_1}) = \rho(\mathbf{B_1})$. Thus, (ii) holds.

For (ii) and (iii) $\Rightarrow$ (i), the proof is similar to the above. □

We shall now consider the minus order for matrices that have been modified by adding rank 1 matrices. Let the notations and terminology be the same as those introduced before Theorem 13.2.5. We shall explore necessary and sufficient conditions for $\mathbf{A_2} <^- \mathbf{B_2}$ to hold when we know that $\mathbf{A} <^- \mathbf{B}$ holds.

Since the minus order implies the space pre-order, so, in view of Theorems 13.2.5, 13.2.7, 13.2.11 and 13.2.15; we just need to obtain the additional conditions in each of the cases when $\mathbf{AGA} = \mathbf{A}$ or/and $\mathbf{A_2G_2A_2} = \mathbf{A_2}$ hold. No proofs are included, as they are highly computational but straightforward.

Theorem 13.3.2. *Let* $\mathbf{A}, \mathbf{B}$ *be matrices of order* $m \times n$, $\mathbf{a}, \mathbf{c}$ *be* m*-column vectors and* $\mathbf{b}, \mathbf{d}$ *be* n*-column vectors. Let* $\mathbf{c} \in \mathcal{C}(\mathbf{B})$, $\mathbf{d} \in \mathcal{C}(\mathbf{B^t})$ *and* $\beta = 0$.

(a) *Let* $\mathbf{a} \in \mathcal{C}(\mathbf{A})$ *and* $\mathbf{b} \in \mathcal{C}(\mathbf{A^t})$. *Let* $\mathbf{G}$ *be a g-inverse of* $\mathbf{B}$ *such that* $\mathbf{A_2Gc} = \mathbf{0}$ *and* $\mathbf{d^tGA_2} = \mathbf{0}$. *Then any two of the following imply the third:*

   (i) $\mathbf{A} <^- \mathbf{B}$
   (ii) $\mathbf{A_2} <^- \mathbf{B_2}$
   (iii) $1 + \mathbf{b^tGa} = \mathbf{0}$.

(b) *Let* $\mathbf{a} \in \mathcal{C}(\mathbf{A})$ *and* $\mathbf{b} \notin \mathcal{C}(\mathbf{A^t})$. *Let* $\mathbf{G}$ *be a g-inverse of* $\mathbf{B}$ *such that* $\mathbf{A_2Gc} = \mathbf{0}$ *and* $\mathbf{d^tGA} = \mathbf{0}$. *Then any two of the following imply the third:*

   (i) $\mathbf{A} <^- \mathbf{B}$
   (ii) $\mathbf{A_2} <^- \mathbf{B_2}$ *and* $\mathbf{b^tGA} = \mathbf{0}$

(iii) $\mathbf{b} \in \mathcal{C}(\mathbf{B} - \mathbf{A})^t$.

(c) *Let* $\mathbf{a} \notin \mathcal{C}(\mathbf{A})$ *and* $\mathbf{b} \in \mathcal{C}(\mathbf{A}^t)$. *Let* $\mathbf{G}$ *be a g-inverse of* $\mathbf{B}$ *such that* $\mathbf{AGc} = \mathbf{0}$ *and* $\mathbf{d}^t\mathbf{GA_2} = \mathbf{0}$. *Then any two of the following imply the third:*

(i) $\mathbf{A} <^- \mathbf{B}$
(ii) $\mathbf{A_2} <^- \mathbf{B_2}$ *and* $\mathbf{AGa} = \mathbf{0}$
(iii) $\mathbf{a} \in \mathcal{C}(\mathbf{B} - \mathbf{A})$.

(d) *Let* $\mathbf{a} \notin \mathcal{C}(\mathbf{A})$ *and* $\mathbf{b} \notin \mathcal{C}(\mathbf{A}^t)$. *Let* $\mathbf{G}$ *be a g-inverse of* $\mathbf{B}$ *such that* $\begin{pmatrix} \mathbf{A} \\ \cdots \\ \mathbf{b}^t \end{pmatrix} \mathbf{Gc} = \mathbf{0}$ *and* $\mathbf{d}^t G \left(\mathbf{A} : \mathbf{a}\right) = \mathbf{0}$. *Then any two of the following imply the third:*

(i) $\mathbf{A} <^- \mathbf{B}$
(ii) $\mathbf{A_2} <^- \mathbf{B_2}$, $\mathbf{AGa} = \mathbf{0}$ *and* $\mathbf{b}^t\mathbf{GA} = \mathbf{0}$,
(iii) $\mathbf{b}^t\mathbf{Ga} = 1$.

Theorem 13.3.3. *Let* $\mathbf{A}, \mathbf{B}$ *be matrices of the same order* $m \times n, \mathbf{a}, \mathbf{c}$ *be* m*-column vectors and* $\mathbf{b}, \mathbf{d}$ *be* n*-column vectors. Let* $\mathbf{c} \in \mathcal{C}(\mathbf{B})$, $\mathbf{d} \in \mathcal{C}(\mathbf{B}^t)$ *and* $\beta \neq 0$.

(a) *Let* $\mathbf{a} \in \mathcal{C}(\mathbf{A})$ *and* $\mathbf{b} \in \mathcal{C}(\mathbf{A}^t)$. *Then any two of the following imply the third:*

(i) $\mathbf{A} <^- \mathbf{B}$
(ii) $\mathbf{A_2} <^- \mathbf{B_2}$, $\mathbf{AGa} = \mathbf{a}$ *and* $\mathbf{b}^t\mathbf{GA} = \mathbf{b}^t$
(iii) *If* $1 + \mathbf{b}^t\mathbf{Ga} = \mathbf{0}$, *then at least one of* $\mathbf{A_2Gc}$ *and* $\mathbf{d}^t\mathbf{GA_2}$ *is null. If* $1 + \mathbf{b}^t\mathbf{Ga} \neq \mathbf{0}$, *then there exists a non-null scalar* α *such that* $\mathbf{A_2Gc} = \alpha\mathbf{a}$ **and** $\mathbf{d}^t\mathbf{GA_2} = \frac{\beta}{\alpha}(1 + \mathbf{b}^t\mathbf{Ga})\mathbf{b}^t$.

(b) *Let* $\mathbf{a} \in \mathcal{C}(\mathbf{A})$ *and* $\mathbf{b} \notin \mathcal{C}(\mathbf{A}^t)$. *Then any two of the following imply the third:*

(i) $\mathbf{A} <^- \mathbf{B}$
(ii) $\mathbf{A_2} <^- B_2$ *and* $\mathbf{AGa} = \mathbf{a}$,
(iii) *either '*$\mathbf{b} \in \mathcal{C}(\mathbf{B} - \mathbf{A})^t$ *and at least one of the* $\mathbf{A_2Gc}$ *and* $\mathbf{d}^t\mathbf{GA_2}$ *is null' or '*$\mathbf{A_2Gc}$ *and* $\mathbf{d}^t\mathbf{GA_2}$ *are non-null,* $\mathbf{a} = \theta\mathbf{A_2Gc}$, *where* $\theta \neq 0$ *is arbitrary and* $\mathbf{b} = (\mathbf{B} - \mathbf{A})^t\mathbf{u} + \frac{1}{\beta\theta}\mathbf{d}$, *where* $\mathbf{u}$ *is arbitrary subject to the condition that* $\mathbf{b} \notin \mathcal{C}(\mathbf{A}^t)$.*'*

(c) *Let* $\mathbf{a} \notin \mathcal{C}(\mathbf{A})$ *and* $\mathbf{b} \in \mathcal{C}(\mathbf{A}^t)$. *Then any two of the following imply the third:*

(i) $\mathbf{A} <^- \mathbf{B}$

(ii) $\mathbf{A_2} <^- \mathbf{B_2}$ *and* $\mathbf{b^tGA} = \mathbf{b^t}$,

(iii) *either* '$\mathbf{a} \in \mathcal{C}(\mathbf{B}-\mathbf{A})$ *and at least one of* $\mathbf{A_2Gc}$ *and* $\mathbf{d^tGA_2}$ *is null'* or '$\mathbf{A_2Gc}$ *and* $\mathbf{d^tGA_2}$ *are non-null,* $\mathbf{b} = \delta\mathbf{A_2^tG^td}$ *where* $\delta \neq 0$ *is arbitrary and* $\mathbf{a} = (\mathbf{B}-\mathbf{A})\mathbf{v} + \frac{1}{\beta\delta}\mathbf{c}$, *where* $\mathbf{v}$ *is arbitrary subject to the condition that* $\mathbf{a} \notin \mathcal{C}(\mathbf{A})$.'

(d) *Let* $\mathbf{a} \notin \mathcal{C}(\mathbf{A})$ *and* $\mathbf{b} \notin \mathcal{C}(\mathbf{A^t})$. *Then any two of the following imply the third:*

(i) $\mathbf{A} <^- \mathbf{B}$

(ii) $\mathbf{A_2} <^- \mathbf{B_2}$

(iii) $\mathbf{b^tGa} - \frac{\mathbf{b^tGcd^tGa}}{\beta} = 1$ *and at least one of the following holds:*

(ν_1) : - $\mathbf{AG}(\mathbf{a}:\mathbf{c}) = \mathbf{0}$, $\mathbf{b} - \frac{\mathbf{b^tGc}}{\beta}\mathbf{d} \in \mathcal{C}(\mathbf{B}-\mathbf{A})^\mathbf{t}$ *or*

(ν_2) : - $(\mathbf{d^t}:\mathbf{b^t})\,\mathbf{GA} = \mathbf{0}$, *and* $\mathbf{a} - \frac{\mathbf{d^tGa}}{\beta}\mathbf{c} \in \mathcal{C}(\mathbf{B}-\mathbf{A})$.

Theorem 13.3.4. *Let* $\mathbf{A}, \mathbf{B}$ *be matrices of order* $m \times n$, $\mathbf{a}$, $\mathbf{c}$ *be* m*-column vectors and* $\mathbf{b}$, $\mathbf{d}$ *be* n*-column vectors. Let* $\mathbf{c} \in \mathcal{C}(\mathbf{B})$ *and* $\mathbf{d} \notin \mathcal{C}(\mathbf{B^t})$. *Let* $\mathbf{a} \in \mathcal{C}(\mathbf{B})$, $\mathbf{b} \in \mathcal{C}(\mathbf{B^t})$ *and* $\mathbf{G}$ *be a g-inverse of* $\mathbf{B}$ *such that* $\mathbf{d^tGc} = \mathbf{0}$. *Then any two of the following imply the third:*

(i) $\mathbf{A} <^- \mathbf{B}$

(ii) $\mathbf{A_2} <^- \mathbf{B_2}$

(iii) (a) $\mathbf{A_2Gc} = \mathbf{0}$ *and* $1 + \mathbf{b^tGa} = \mathbf{0}$, *when* $\mathbf{a} \in \mathcal{C}(\mathbf{A})$ *and* $\mathbf{b} \in \mathcal{C}(\mathbf{A^t})$.

(b) $\mathbf{A_2Gc} = \mathbf{0}$ *and* $\mathbf{b^tGA} = \mathbf{0}$, *when* $\mathbf{a} \in \mathcal{C}(\mathbf{A})$ *and* $\mathbf{b} \notin \mathcal{C}(\mathbf{A^t})$.

(c) $\mathbf{d^tGA_2} = \mathbf{0}$ *and* $\mathbf{AGa} = \mathbf{0}$ *when* $\mathbf{a} \notin \mathcal{C}(\mathbf{A})$ *and* $\mathbf{b} \in \mathcal{C}(\mathbf{A^t})$.

(d) $\mathbf{A_2Gc} = \mathbf{0}$, $\mathbf{d^tGA_2} = \mathbf{0}$, $\mathbf{b^tGa} = 1$, $\mathbf{AGa} = \mathbf{0}$ *and* $\mathbf{b^tGA} = \mathbf{0}$, *when* $\mathbf{a} \notin \mathcal{C}(\mathbf{A})$ *and* $\mathbf{b} \notin \mathcal{C}(\mathbf{A^t})$.

Theorem 13.3.5. *Let* $\mathbf{A}, \mathbf{B}$ *be matrices of order* $m \times n, \mathbf{a}, \mathbf{c}$ *be* m*-column vectors and* $\mathbf{b}, \mathbf{d}$ *be* n*-column vectors. Let* $\mathbf{c} \in \mathcal{C}(\mathbf{B})$ *and* $\mathbf{d} \notin \mathcal{C}(\mathbf{B^t})$. *Let* $\mathbf{a} \in \mathcal{C}(\mathbf{B})$, $\mathbf{b} \notin \mathcal{C}(\mathbf{B^t})$ *and* $\mathbf{G}$ *be a g-inverse of* $\mathbf{B}$ *such that* $\mathbf{d^tGc} = \mathbf{0}$. *Then any two of the following imply the third:*

(i) $\mathbf{A} <^- \mathbf{B}$

(ii) $\mathbf{A_2} <^- \mathbf{B_2}$

(iii) (a) $\mathbf{d} = \mu\mathbf{b} + (\mathbf{B}-\mathbf{A})^\mathbf{t}\mathbf{u}$ *for some scalar* $\mu \neq 0$, *for some vector* u *and* $\mathbf{A_2Gc} = \dfrac{\mathbf{a}}{\mu}$, *when* $\mathbf{a} \in \mathcal{C}(\mathbf{A})$.

(b) $\mathbf{d} = \mu\mathbf{b} + (\mathbf{B}-\mathbf{A})^\mathbf{t}\mathbf{z}$, $\mathbf{z^ta} = -\mu$, $\mathbf{z^tGc} + \mathbf{1} = \mathbf{0}$, $\mathbf{b^tGc}\mu = \mathbf{1}$ *for some scalar* μ *and some vector* z *and* $\mathbf{a}, \mathbf{c} \in \mathcal{C}(\mathbf{B}-\mathbf{A})$, *when* $\mathbf{a} \notin \mathcal{C}(\mathbf{A})$.

Theorem 13.3.6. *Let* $\mathbf{A}, \mathbf{B}$ *be matrices of order* $m \times n$, $\mathbf{a}, \mathbf{c}$ *be* m*-column vectors and* $\mathbf{b}, \mathbf{d}$ *be* n*-column vectors. Let* $\mathbf{c} \notin \mathcal{C}(\mathbf{B})$ *and* $\mathbf{d} \notin \mathcal{C}(\mathbf{B^t})$. *Further, let* $\mathbf{a} \in \mathcal{C}(\mathbf{B})$ *and* $\mathbf{b} \in \mathcal{C}(\mathbf{B^t})$ *and* $\mathbf{G}$ *be a common g-inverse of* $\mathbf{B}$ *and* $\mathbf{B_2}$. *Then any two of the following imply the third:*

(i) $\mathbf{A} <^- \mathbf{B}$
(ii) $\mathbf{A_2} <^- \mathbf{B_2}$
(iii) (a) $1 + \mathbf{b^t G a} = \mathbf{0}$, *when* $\mathbf{a} \in \mathcal{C}(\mathbf{A})$ *and* $\mathbf{b} \in \mathcal{C}(\mathbf{A^t})$.
(b) $\mathbf{b^t G A} = \mathbf{0}$, *when* $\mathbf{a} \in \mathcal{C}(\mathbf{A})$ *and* $\mathbf{b} \notin \mathcal{C}(\mathbf{A^t})$.
(c) $\mathbf{AGa} = \mathbf{0}$, *when* $\mathbf{a} \notin \mathcal{C}(\mathbf{A})$ *and* $\mathbf{b} \in \mathcal{C}(\mathbf{A^t})$.
(d) $\mathbf{AGa} = \mathbf{0}$, $\mathbf{b^t G A} = \mathbf{0}$ *and* $\mathbf{b^t G a} = 1$, *when* $\mathbf{a} \notin \mathcal{C}(\mathbf{A})$ *and* $\mathbf{b} \notin \mathcal{C}(\mathbf{A^t})$.

Theorem 13.3.7. *Let* $\mathbf{A}, \mathbf{B}$ *be matrices of order* $m \times n$, $\mathbf{a}, \mathbf{c}$ *be* m*-column vectors and* $\mathbf{b}, \mathbf{d}$ *be* n*-column vectors. Let* $\mathbf{c} \notin \mathcal{C}(\mathbf{B})$ *and* $\mathbf{d} \notin \mathcal{C}(\mathbf{B^t})$. *Further, let* $\mathbf{a} \in \mathcal{C}(\mathbf{B})$ *and* $\mathbf{b} \notin \mathcal{C}(\mathbf{B^t})$ *and* $\mathbf{G}$ *be a common g-inverse of* $\mathbf{B}$ *and* $\mathbf{B_2}$. *Then any two of the following imply the third:*

(i) $\mathbf{A} <^- \mathbf{B}$
(ii) $\mathbf{A_2} <^- \mathbf{B_2}$
(iii) $\mathbf{A_2} <^s \mathbf{B_2}$, $\mathbf{a} \in \mathcal{C}(\mathbf{A})$ *and* $\mathbf{b^t G A} = \mathbf{0}$.

Theorem 13.3.8. *Let* $\mathbf{A}, \mathbf{B}$ *be matrices of order* $m \times n$, $\mathbf{a}, \mathbf{c}$ *be* m*-column vectors and* $\mathbf{b}, \mathbf{d}$ *be* n*-column vectors. Let* $\mathbf{c} \notin \mathcal{C}(\mathbf{B})$ *and* $\mathbf{d} \notin \mathcal{C}(\mathbf{B^t})$. *Further, let '*$\mathbf{a} \notin \mathcal{C}(\mathbf{B})$ *and* $\mathbf{b} \notin \mathcal{C}(\mathbf{B^t})$*' and* $\mathbf{G}$ *be a common g-inverse of* $\mathbf{B}$ *and* $\mathbf{B_2}$. *Then any two of the following imply the third:*

(i) $\mathbf{A} <^- \mathbf{B}$
(ii) $\mathbf{A_2} <^- \mathbf{B_2}$
(iii) $\mathbf{b^t G c} \neq 0$, $\mathbf{d^t G a} \neq 0$, $\mathbf{c} = \dfrac{(\mathbf{I} - \mathbf{BG})\mathbf{a}}{\mathbf{d^t G a}}$ *and* $\mathbf{d} = \dfrac{(\mathbf{I} - (\mathbf{GB^t}))\mathbf{b}}{\mathbf{b^t G c}}$.
Further, $\mathbf{a} = (\mathbf{I} - \mathbf{AG})\xi$ *and* $\mathbf{b} = (\mathbf{I} - \mathbf{GA})^t \eta$ *for some* ξ *and* η.

13.4 Sharp order

Let $\mathbf{A}, \mathbf{B}$ be matrices of order $n \times n$ and of index ≤ 1. $\mathbf{a}$, $\mathbf{b}$, $\mathbf{c}$ and $\mathbf{d}$ be n-vectors and $\mathbf{A_2} = \begin{pmatrix} \mathbf{A} & \mathbf{a} \\ \mathbf{b^t} & \alpha \end{pmatrix}$ and $\mathbf{B_2} = \begin{pmatrix} \mathbf{B} & \mathbf{c} \\ \mathbf{d^t} & \beta \end{pmatrix}$. Given $\mathbf{A} <^{\#} \mathbf{B}$, we wish to explore the necessary and sufficient conditions under which (a) the matrix $\mathbf{A_2}$ and $\mathbf{B}$ are of index ≤ 1 and (b) $\mathbf{A_2} <^{\#} \mathbf{B_2}$. We also study when $\mathbf{A_2} <^{\#} \mathbf{B_2}$ implies $\mathbf{A} <^{\#} \mathbf{B}$. Also, given matrices $\mathbf{A}, \mathbf{B}$ of the same order and of index ≤ 1, vectors $\mathbf{a}$, $\mathbf{b}$ and scalar α, we wish to determine all the vectors $\mathbf{c}$, $\mathbf{d}$ and the scalar β such that whenever $\mathbf{A} <^{\#} \mathbf{B}$, we have $\mathbf{A_2} <^{\#} \mathbf{B_2}$. We first prove the following theorem which determines when the matrix $\mathbf{A_2}$ is of index ≤ 1.

Theorem 13.4.1. *Let* $\mathbf{A}$ *be a matrix of order* $n \times n$ *and index* ≤ 1. *Let* $\mathbf{G}$ *be the group inverse of* $\mathbf{A}$. *Then* $\mathbf{A_2}$ *is of index* ≤ 1 *if and only if*

either $\mathbf{a} \in \mathcal{C}(\mathbf{A}),\ \mathbf{b} \in \mathcal{C}(\mathbf{A^t}),\ \alpha = \mathbf{b^tGa}$ *and* $q = \mathbf{b^t(G^2)a + 1 \neq 0}$
in which case $\rho(\mathbf{A_2}) = \rho(\mathbf{A})$

or $\alpha \neq \mathbf{b^tGa}$ *and at least one of* $\mathbf{a} \in \mathcal{C}(\mathbf{A}),\ \mathbf{b} \in \mathcal{C}(\mathbf{A^t})$ *holds*
in which case $\rho(\mathbf{A_2}) = \rho(\mathbf{A}) + 1$

or $\mathbf{b^t(I - AG)a \neq 0}$ *in which case* $\rho(\mathbf{A_2}) = \rho(\mathbf{A}) + \mathbf{2}$.

Proof. 'If part'
Let $\mathbf{a} \in \mathcal{C}(\mathbf{A}),\ \mathbf{b} \in \mathcal{C}(\mathbf{A^t}),\ \alpha = \mathbf{b^tGa}$ and $\mathbf{b^t(G^2)a + 1 \neq 0}$. Consider the matrix $\mathbf{G_2} = \dfrac{\mathbf{1}}{\mathbf{r}}\begin{pmatrix} \mathbf{X} & \mathbf{XGa} \\ \mathbf{b^tGX} & \mathbf{b^tGXGa} \end{pmatrix}$, where

$$\mathbf{X} = \mathbf{rG - G^2ab^tG - Gab^tG^2 + kGab^tG} \text{ and } \mathrm{k} = \frac{\mathbf{b^tG^3a}}{\mathrm{r}}.$$

We can check that $\mathbf{A_2^2G_2} = \mathbf{A_2}$. So, $\rho(\mathbf{A_2}) \leq \rho(\mathbf{A_2^2}) \leq \rho(\mathbf{A_2})$. Thus, $\rho(\mathbf{A_2}) = \rho(\mathbf{A})$, equivalently $\mathbf{A_2}$ is of index ≤ 1.
In case $\alpha \neq \mathbf{b^tGa}$ and at least one of $\mathbf{a} \in \mathcal{C}(\mathbf{A}),\ \mathbf{b} \in \mathcal{C}(\mathbf{A^t})$ holds, take

$$\mathbf{G_2} = \frac{\mathbf{1}}{\mathbf{q}}\begin{pmatrix} \mathbf{X} & k\mathbf{(I - AG)a - Ga} \\ k\mathbf{b^t(I - AG) - b^tG} & 1 \end{pmatrix}$$

where $\mathbf{X} = \mathbf{qG + Gab^tG - (G + kI)ab^t(I - AG) - (I - AG)ab^tG(G + kI)}$ and $k = \dfrac{r}{q}$. Then $\mathbf{A_2^2G_2} = \mathbf{A_2}$. Thus index of $\mathbf{A_2}$ is ≤ 1.

In case $\mathbf{b^t(I - AG)a \neq 0}$, take $\mathbf{G_2} = \dfrac{\mathbf{1}}{\mathbf{p}}\begin{pmatrix} \mathbf{X} & \mathbf{(I - AG)a} \\ \mathbf{b^t(I - AG)} & \mathbf{0} \end{pmatrix}$, where $\mathbf{X} = \mathbf{pG - Gab^t(I - AG) - (I - AG)ab^tG - k(I - AG)ab^t(I - AG)}$ and $k = \dfrac{q}{p}$ and $\mathbf{G_2}$ satisfies $\mathbf{{A_2}^2G_2} = \mathbf{A_2}$, so that $\mathbf{A_2}$ is of index ≤ 1.
'Only if' part
Let $\mathbf{A_2}$ be of index ≤ 1. In case $\rho(\mathbf{A_2}) = \rho(\mathbf{A})$, then from the proof of Lemma 13.2.4, it follows that $\mathbf{a} \in \mathcal{C}(\mathbf{A}),\ \mathbf{b} \in \mathcal{C}(\mathbf{A^t}),\ \alpha = \mathbf{b^tGa}$. If possible, let $\mathbf{b^tG^2a + 1 = 0}$. Then consider the vector $\mathbf{w} = \begin{pmatrix} \mathbf{G^2a} \\ \mathbf{0} \end{pmatrix}$. Then

$$\mathbf{A_2w} = \begin{pmatrix} \mathbf{AG^2a} \\ \mathbf{b^tG^2a} \end{pmatrix} = \begin{pmatrix} \mathbf{Ga} \\ -1 \end{pmatrix} \text{ and} \mathbf{{A^2}_2w = 0}.$$

Since $\mathbf{A_2}$ is of index 1, $\mathbf{A_2 w} \neq \mathbf{0}$. Thus, $\mathbf{w} \in \mathcal{N}(\mathbf{A^2}_2)$, but $\mathbf{w} \in \mathcal{N}(\mathbf{A_2})$ and this can not happen. So, $\mathbf{b^t G^2 a} + 1 \neq \mathbf{0}$.

Similarly, in case $\rho(\mathbf{A_2}) = \rho(\mathbf{A}) + \mathbf{1}$ take $\mathbf{w} = \begin{pmatrix} \mathbf{G^2 a} \\ 1 \end{pmatrix}$ to show $\alpha \neq \mathbf{b^t G a}$.

In case $\rho(\mathbf{A_2}) = \rho(\mathbf{A}) + \mathbf{2}$ take $\mathbf{w} = \begin{pmatrix} \mathbf{Ga} + \mathbf{z} \\ 1 \end{pmatrix}$, where $\mathbf{z}$ is a solution of $\begin{pmatrix} \mathbf{A} \\ \mathbf{b^t} \end{pmatrix} = \begin{pmatrix} \mathbf{0} \\ \mathbf{b^t Ga} - \alpha \end{pmatrix}$ and show that $\mathbf{b^t(I - AG)a} \neq \mathbf{0}$. □

Lemma 13.4.2. *Let* $\mathbf{A}$, $\mathbf{B}$ *be matrices of order* $n \times n$ *and index* ≤ 1. *Let* $\mathbf{A} <^{\#} \mathbf{B}$. *Then* $(\mathbf{B} - \mathbf{A})\mathbf{x} = \mathbf{0}$ *if and only if* $\mathbf{x} \in \mathcal{C}(\mathbf{A}) \oplus \mathcal{N}(\mathbf{B})$ *for any* $\mathbf{x} \in \mathbf{F^n}$.

Proof. Since $\mathbf{B}$ is of index ≤ 1, $F^n = \mathcal{C}(\mathbf{B}) \oplus \mathcal{N}(\mathbf{B})$. Also, $\mathbf{A} <^{\#} \mathbf{B}$, so, $\mathbf{A} <^{-} \mathbf{B}$. This further gives $\mathcal{C}(\mathbf{B}) = \mathcal{C}(\mathbf{A}) \oplus \mathcal{C}(\mathbf{B} - \mathbf{A})$. It follows that $F^n = \mathcal{C}(\mathbf{A}) \oplus \mathcal{C}(\mathbf{B} - \mathbf{A}) \oplus \mathcal{N}(\mathbf{B})$. Now, each $\mathbf{x} \in \mathbf{F^n}$ can be written as $\mathbf{x} = \mathbf{x_1} + \mathbf{x_2} + \mathbf{x_3}$, where $\mathbf{x_1} \in \mathcal{C}(\mathbf{A}), \mathbf{x_2} \in \mathcal{C}(\mathbf{B} - \mathbf{A})$ and $\mathbf{x_3} \in \mathcal{N}(\mathbf{B})$. Therefore,

$$(\mathbf{B} - \mathbf{A})\mathbf{x} = (\mathbf{B} - \mathbf{A})\mathbf{x_1} + (\mathbf{B} - \mathbf{A})\mathbf{x_2} + (\mathbf{B} - \mathbf{A})\mathbf{x_3}.$$

Since, $\mathbf{A} <^{\#} \mathbf{B} \Rightarrow \mathbf{A^2} = \mathbf{BA} = \mathbf{AB}$, we have $(\mathbf{B} - \mathbf{A})\mathbf{x_1} = \mathbf{0}$. Since, $\mathbf{x_3} \in \mathcal{N}(\mathbf{B})$, we have $(\mathbf{B} - \mathbf{A})\mathbf{x_3} = \mathbf{0}$. Therefore, $(\mathbf{B} - \mathbf{A})\mathbf{x} = (\mathbf{B} - \mathbf{A})\mathbf{x_2}$. So, $(\mathbf{B} - \mathbf{A})\mathbf{x} = \mathbf{0} \Leftrightarrow (\mathbf{B} - \mathbf{A})\mathbf{x_2} = \mathbf{0} \Leftrightarrow \mathbf{x} \in \mathcal{C}(\mathbf{A}) \oplus \mathcal{N}(\mathbf{B})$. □

Remark 13.4.3. Notice that we only need $\mathbf{A} < \# \mathbf{B}$ for Lemma 13.4.2 to hold.

Corollary 13.4.4. *Let* $\mathbf{A}$, $\mathbf{B}$ *be matrices of order* $n \times n$ *and index* ≤ 1. *Let* $\mathbf{A} <^{\#} \mathbf{B}$. $\mathbf{y^t}(\mathbf{B} - \mathbf{A}) = \mathbf{0}$ *if and only if* $\mathbf{y} \in \mathcal{C}(\mathbf{A^t}) \oplus \mathcal{N}(\mathbf{B^t})$.

Let $\mathbf{A}$, $\mathbf{B}$ be matrices of the same order $n \times n$ and index ≤ 1. Let us assume that $\mathbf{A_2}$ is of index ≤ 1 in next few theorems. Also, note that if $\mathbf{A} <^{\#} \mathbf{B}$ and $\mathbf{G}$ denotes the group inverse of $\mathbf{B}$, then $\mathbf{G}$ is a commuting g-inverse of $\mathbf{A}$. We record the following equations for our use in this section:

$\mathbf{A_2^2} = \mathbf{A_2 B_2} = \mathbf{B_2 A_2}$ if and only if

$$\mathbf{A^2} + \mathbf{ab^t} = \mathbf{AB} + \mathbf{ad^t} = \mathbf{BA} + \mathbf{cb^t} \tag{13.4.1}$$

$$\mathbf{Aa} + \alpha\mathbf{a} = \mathbf{b^t} = \mathbf{Ac} + \beta\mathbf{a} = \mathbf{Ba} + \alpha\mathbf{c} \tag{13.4.2}$$

$$\mathbf{b^t A} + \alpha\mathbf{b^t} = \mathbf{b^t B} + \alpha\mathbf{d^t} = \mathbf{d^t A} + \beta\mathbf{b^t} \tag{13.4.3}$$

$$\mathbf{b^t a} + \alpha^{\mathbf{2}} = b^t\mathbf{c} + \alpha\beta = \mathbf{d^t a} + \beta\alpha. \tag{13.4.4}$$

Theorem 13.4.5. *Let* $\mathbf{A}$, $\mathbf{B}$ *be matrices of order* $n \times n$ *and index* ≤ 1. *Let* $\mathbf{A_2}$ *be of index* ≤ 1 *and the vectors* $\mathbf{a}$, $\mathbf{b}$ *be non-null. Let* $\mathbf{G}$ *denote the group inverse of* $\mathbf{B}$. *Then any two of the following three statements implies the third:*

(i) $\mathbf{A} <^{\#} \mathbf{B}$
(ii) $\mathbf{A_2} <^{\#} \mathbf{B_2}$ *and*
(iii) $\mathbf{a} \in \mathcal{C}(\mathbf{A}) \oplus \mathcal{N}(\mathbf{B})$, $\mathbf{b} \in \mathcal{C}(\mathbf{A^t}) \oplus \mathcal{N}(\mathbf{B^t})$, $\mathbf{a} = \mathbf{c}, \mathbf{b} = \mathbf{d}$ *and* $\alpha = \beta$.

Remark 13.4.6. In the setup Theorem 13.4.5,
(a) the condition (ii) and the fact that $\mathbf{a} = \mathbf{c}, \mathbf{b} = \mathbf{d}$ and $\alpha = \beta$ together imply $\mathbf{a} \in \mathcal{C}(\mathbf{A}) \oplus \mathcal{N}(\mathbf{B})$ and $\mathbf{b} \in \mathcal{C}(\mathbf{A^t}) \oplus \mathcal{N}(\mathbf{B^t})$ and
(b) if (i) holds and either $\mathbf{a} \notin \mathcal{C}(\mathbf{A}) \oplus \mathcal{N}(\mathbf{B})$ or $\mathbf{b} \notin \mathcal{C}(\mathbf{A^t}) \oplus \mathcal{N}(\mathbf{B^t})$, then there can not exist a matrix $\mathbf{B_2}$ of index ≤ 1 such that (ii) holds.

Theorem 13.4.7. *Let* $\mathbf{A}$, $\mathbf{B}$ *be matrices of order* $n \times n$ *and index* ≤ 1. *Let* $\mathbf{A_2}$ *be of index* ≤ 1 *and* $\mathbf{a} = \mathbf{0}$, $\mathbf{b} \neq \mathbf{0}$. *Let* $\mathbf{G}$ *denote the group inverse of* $\mathbf{B}$. *Then any two of the following three statements implies the third:*

(i) $\mathbf{A} <^{\#} \mathbf{B}$
(ii) $\mathbf{A_2} <^{\#} \mathbf{B_2}$ *and*
(iii) $\mathbf{c} = \mathbf{0}$, $\mathbf{b^t} \in \mathcal{C}(\mathbf{A^t})$, *either* $\beta = 0$ *and* $\mathbf{d} = \mathbf{b} + (\mathbf{B} - \mathbf{A})^{\mathbf{t}}\mathbf{y}$ *for vector* $\mathbf{y}$ *or* $\beta \neq 0$ *is arbitrary and* $\mathbf{d} = \mathbf{b} - \beta\mathbf{G^t}\mathbf{b} + (\mathbf{I} - \mathbf{AG})^{\mathbf{t}}\eta$, η *arbitrary.*

Corollary 13.4.8. *Let* $\mathbf{A}$, $\mathbf{B}$ *be matrices of order* $n \times n$ *and index* ≤ 1. *Let* $\mathbf{A_2}$ *be of index* ≤ 1 *and* $\mathbf{a} \neq \mathbf{0}$, $\mathbf{b} = \mathbf{0}$ *and* $\alpha = 0$. *Let* $\mathbf{G}$ *denote the group inverse of* $\mathbf{B}$. *Then any two of the following three statements implies the third:*

(i) $\mathbf{A} <^{\#} \mathbf{B}$
(ii) $\mathbf{A_2} <^{\#} \mathbf{B_2}$ *and*
(iii) $\mathbf{d} = \mathbf{0}$ *and either* $\beta = 0$ *and* $\mathbf{c} = \mathbf{a} + (\mathbf{B} - \mathbf{A})\mathbf{w}$, *for some vector* $\mathbf{w}$ *or* $\beta \neq 0$ *is arbitrary,* $\mathbf{a} \in \mathcal{C}(\mathbf{A})$ *and* $\mathbf{c} = \mathbf{a} - \beta\mathbf{G}\mathbf{a} + (\mathbf{I} - \mathbf{AG})\xi$ *for arbitrary* ξ.

Theorem 13.4.9. *Let* $\mathbf{A}$, $\mathbf{B}$ *be matrices of order* $n \times n$ *and index* ≤ 1. *Let* $\mathbf{A_2}$ *be of index* ≤ 1 *and* $\mathbf{a} = \mathbf{0}$, $\mathbf{b} \neq \mathbf{0}$ *and* $\alpha \neq 0$. *Let* $\mathbf{G}$ *denote the group inverse of* $\mathbf{B}$. *Then any two of the following three statements implies the third:*

(i) $\mathbf{A} <^{\#} \mathbf{B}$
(ii) $\mathbf{A_2} <^{\#} \mathbf{B_2}$ *and*

(iii) $\mathbf{c} = \mathbf{0}, \beta = \alpha$, *and* $\mathbf{d} = \mathbf{b} - \frac{1}{\alpha}(\mathbf{B} - \mathbf{A})^{\mathbf{t}}\mathbf{b}$.

Corollary 13.4.10. *Let* $\mathbf{A}$, $\mathbf{B}$ *be matrices of order* $n \times n$ *and index* ≤ 1. *Let* $\mathbf{A_2}$ *be of index* ≤ 1 *and* $\mathbf{a} \neq \mathbf{0}$, $\mathbf{b} = \mathbf{0}$ *and* $\alpha \neq 0$. *Let* $\mathbf{G}$ *denote the group inverse of* $\mathbf{B}$. *Then any two of the following three statements implies the third:*

(i) $\mathbf{A} <^{\#} \mathbf{B}$
(ii) $\mathbf{A_2} <^{\#} \mathbf{B_2}$ *and*
(iii) $\mathbf{d} = \mathbf{0}$, $\beta = \alpha$, *and* $\mathbf{c} = \mathbf{a} - \frac{1}{\alpha}(\mathbf{B} - \mathbf{A})\mathbf{a}$.

Theorem 13.4.11. *Let* $\mathbf{A}$, $\mathbf{B}$ *be matrices of order* $n \times n$ *and index* ≤ 1. *Let* $\mathbf{B_2}$ *be of index* ≤ 1 *and* $\mathbf{a} = \mathbf{0}$, $\mathbf{b} = \mathbf{0}$ *and* $\alpha \neq 0$. *Let* $\mathbf{G}$ *denote the group inverse of* $\mathbf{B}$. *Then any two of the following three statements implies the third:*

(i) $\mathbf{A} <^{\#} \mathbf{B}$
(ii) '$\mathbf{B_2}$ *is of index* ≤ 1*and* $\mathbf{A_2} <^{\#} \mathbf{B_2}$' *and*
(iii) $\mathbf{c} = \mathbf{d} = \mathbf{0}$, *and* $\beta = \alpha$.

Note that in this case $\mathbf{A_2}$ is of index ≤ 1.

Theorem 13.4.12. *Let* $\mathbf{A}$, $\mathbf{B}$ *be matrices of order* $n \times n$ *and index* ≤ 1. *Let* $\mathbf{B_2}$ *be of index* ≤ 1 *and* $\mathbf{a} = \mathbf{0}$, $\mathbf{b} = \mathbf{0}$ *and* $\alpha = 0$. *Let* $\mathbf{G}$ *denote the group inverse of* $\mathbf{B}$. *Then any two of the following three statements implies the third:*

(i) $\mathbf{A} <^{\#} \mathbf{B}$
(ii) '$\mathbf{B_2}$ *is of index* ≤ 1 *and* $\mathbf{A_2} <^{\#} \mathbf{B_2}$' *and*
(iii) $\mathbf{c} = (\mathbf{I} - \mathbf{AG})\xi$, $\mathbf{d} = (\mathbf{I} - \mathbf{AG})\eta$ *and either at least one of* $\mathbf{c} \in \mathcal{C}(\mathbf{B})$ *and* $\mathbf{d} \in \mathcal{C}(\mathbf{B^t})$ *holds and* $\beta \neq \eta^{\mathbf{t}}\mathbf{Gc}$ *or* $\mathbf{c} \in \mathcal{C}(\mathbf{B})$ *and* $\mathbf{d} \in \mathcal{C}(\mathbf{B^t})$ *and* $\beta = \eta^{\mathbf{t}}\mathbf{Gc}$, $\eta^{\mathbf{t}}\mathbf{G^2c} \neq \mathbf{0}$ *or* $\mathbf{c} \notin \mathcal{C}(\mathbf{B})$ *and* $\mathbf{d} \notin \mathcal{C}(\mathbf{B^t})$ *and* $\eta^{\mathbf{t}}(\mathbf{I} - \mathbf{BGc}) \neq \mathbf{0}$.

Let $\mathbf{A}$, $\mathbf{B}$ be matrices of order $n \times n$ and index ≤ 1. Assume now $\mathbf{B_2}$ to be of index ≤ 1. We shall now obtain the class of all index ≤ 1 matrices $\mathbf{A_2}$ such that $\mathbf{A} <^{\#} \mathbf{B}$ and $\mathbf{A_2} <^{\#} \mathbf{B_2}$ are equivalent.

Theorem 13.4.13. *Let* $\mathbf{A}$, $\mathbf{B}$ *be matrices of order* $n \times n$ *and index* ≤ 1. *Let* $\mathbf{B_2}$ *be of index* ≤ 1 *and* $\mathbf{c} \neq \mathbf{0}, \mathbf{d} \neq \mathbf{0}$. *Let* $\mathbf{G}$ *be the group inverse of* $\mathbf{B}$. *Then any two of the following imply the third:*

(i) $\mathbf{A} <^{\#} \mathbf{B}$

(ii) '$\mathbf{A_2}$ *is of index* ≤ 1 *and* $\mathbf{A_2} <^{\#} \mathbf{B_2}$' *and*

(iii) $\mathbf{a} = \mathbf{c} \in \mathcal{C}(\mathbf{A}) \oplus \mathcal{N}(\mathbf{B}), \mathbf{b} = \mathbf{d} \in \mathcal{C}(\mathbf{A^t}) \oplus \mathcal{N}(\mathbf{B^t})$ *and* $\beta = \alpha$.

Theorem 13.4.14. *Let* $\mathbf{A}$, $\mathbf{B}$ *be matrices of order* $n \times n$ *and index* ≤ 1. *Let* $\mathbf{B_2}$ *be of index* ≤ 1 *and* $\mathbf{c} = \mathbf{0}, \mathbf{d} \neq \mathbf{0}$ *and* $\beta = 0$. *Let* $\mathbf{G}$ *be the group inverse of* $\mathbf{B}$. *Then any two of the following imply the third:*

(i) $\mathbf{A} <^{\#} \mathbf{B}$

(ii) '$\mathbf{A_2}$ *is of index* ≤ 1 *and* $\mathbf{A_2} <^{\#} \mathbf{B_2}$' *and*

(iii) $\mathbf{a} = \mathbf{0}$, $\alpha = \mathbf{0}$, $\mathbf{d} \in \mathcal{C}(\mathbf{B^t})$ *and* $\mathbf{b}$ *is a projection of* $\mathbf{d}$ *in* $\mathcal{C}(\mathbf{A^t})$ *along* $\mathcal{C}(\mathbf{B^t} - \mathbf{A^t}) \oplus \mathcal{N}(\mathbf{B^t})$.

Corollary 13.4.15. *Let* $\mathbf{A}$, $\mathbf{B}$ *be matrices of order* $n \times n$ *and index* ≤ 1. *Let* $\mathbf{B_2}$ *be of index* ≤ 1 *and* $\mathbf{c} \neq \mathbf{0}, \mathbf{d} = \mathbf{0}$ *and* $\beta = 0$. *Let* $\mathbf{G}$ *be the group inverse of* $\mathbf{B}$. *Then any two of the following imply the third:*

(i) $\mathbf{A} <^{\#} \mathbf{B}$

(ii) '$\mathbf{A_2}$ *is of index* ≤ 1 *and* $\mathbf{A_2} <^{\#} \mathbf{B_2}$' *and*

(iii) $\mathbf{b} = \mathbf{0}$, $\alpha = \mathbf{0}$, $\mathbf{c} \in \mathcal{C}(\mathbf{B})$ *and* $\mathbf{a}$ *is the projection in* $\mathcal{C}(\mathbf{A})$ *along* $\mathcal{C}(\mathbf{B} - \mathbf{A}) \oplus \mathcal{N}(\mathbf{B})$.

Theorem 13.4.16. *Let* $\mathbf{A}$, $\mathbf{B}$ *be matrices of order* $n \times n$ *and index* ≤ 1. *Let* $\mathbf{B_2}$ *be of index* ≤ 1 *and* $\mathbf{c} = \mathbf{0}, \mathbf{d} \neq \mathbf{0}$ *and* $\beta \neq 0$. *Let* $\mathbf{G}$ *be a group inverse of* $\mathbf{B}$. *Then any two of the following imply the third:*

(i) $\mathbf{A} <^{\#} \mathbf{B}$

(ii) '$\mathbf{A_2}$ *is of index* 1 *and* $\mathbf{A_2} <^{\#} \mathbf{B_2}$' *and*

(iii) $\mathbf{a} = \mathbf{0}$ *and either* $\alpha = 0$, $\mathbf{b} = \mathbf{A^t}(\mathbf{S^-}\mathbf{A^t}\mathbf{d} + (\mathbf{I} - \mathbf{S^-}\mathbf{S}\xi))$, *where* ξ *is arbitrary,* $\mathbf{A^t}\mathbf{d} \in \mathcal{C}(\mathbf{S})$ *and* $\mathbf{S} = (\mathbf{B} - \beta\mathbf{I})^{\mathbf{t}}$ *or* $\alpha = \beta$, *and* $\mathbf{b} = \beta\mathbf{T^-}\mathbf{d} + (\mathbf{I} - \mathbf{T^-}\mathbf{T})\eta$, *where* η *is arbitrary and* $\mathbf{T} = (\mathbf{A} - \mathbf{B} + \beta\mathbf{I})^{\mathbf{t}}$.

Theorem 13.4.17. *Let* $\mathbf{A}$, $\mathbf{B}$ *be matrices of order* $n \times n$ *and index* ≤ 1. *Let* $\mathbf{B_2}$ *be of index* ≤ 1 *and* $\mathbf{c} = \mathbf{0}, \mathbf{d} = \mathbf{0}$ *and* $\beta \neq 0$. *Let* $\mathbf{G}$ *be a group inverse of* $\mathbf{B}$. *Then any two of the following imply the third:*

(i) $\mathbf{A} <^{\#} \mathbf{B}$

(ii) '$\mathbf{A_2}$ *is of index* ≤ 1 *and* $\mathbf{A_2} <^{\#} \mathbf{B_2}$' *and*

(iii) *either* '$\alpha = 0$, $\mathbf{a} = \mathbf{0}, \mathbf{b} = \mathbf{0}$' *or* β *is an eigen-value of* $\mathbf{A}$ *and* $\mathbf{b}$ *is an eigen vector of* $\mathbf{A^t}$ *corresponding to eigen-value* β *or* '$\alpha = \beta$, $\mathbf{a} = \mathbf{0}, \mathbf{b} = \mathbf{0}$' *or* β *is an eigen-value of* $\mathbf{B} - \mathbf{A}$ *and* $\mathbf{b}$ *is an eigen vector of* $(\mathbf{B} - \mathbf{A})^{\mathbf{t}}$ *corresponding to eigen-value* β *or* $\alpha =$

β, $\mathbf{a} = \mathbf{0}, \mathbf{b} = \mathbf{0}$' or β is an eigen-value of $(\mathbf{B} - \mathbf{A})^{\mathbf{t}}$ and $\mathbf{a}$ is an eigen vector of $(\mathbf{B} - \mathbf{A})$ corresponding to eigen-value β.

Theorem 13.4.18. *Let $\mathbf{A}$, $\mathbf{B}$ be matrices of order $n \times n$ and index ≤ 1. Let $\mathbf{B_2}$ be of index ≤ 1 and $\mathbf{c} = \mathbf{0}, \mathbf{d} = \mathbf{0}$ and $\beta = 0$. Let $\mathbf{G}$ be a group inverse of $\mathbf{B}$. Then any two of the following imply the third:*

(i) $\mathbf{A} <^{\#} \mathbf{B}$
(ii) '$\mathbf{A_2}$ *is of index* ≤ 1 *and* $\mathbf{A_2} <^{\#} \mathbf{B_2}$' *and*
(iii) $\mathbf{a} = \mathbf{0}$, $\mathbf{b} = \mathbf{0}$ *and* $\alpha = 0$.

Let $\mathbf{A}, \mathbf{B}$ be matrices of order $n \times n$ and of index ≤ 1. Let $\mathbf{a}$, $\mathbf{b}$, $\mathbf{c}$ and $\mathbf{d}$ be n-vectors and $\mathbf{A_3} = \mathbf{A} + \mathbf{a}\mathbf{b}^{\mathbf{t}}$, $\mathbf{B_3} = \mathbf{B} + \mathbf{c}\mathbf{d}^{\mathbf{t}}$. Further, let $\mathbf{A_3}$ and $\mathbf{B_3}$ be of index ≤ 1. To determine necessary and sufficient conditions such that $\mathbf{A} <^{\#} \mathbf{B}$ and $\mathbf{A_3} <^{\#} \mathbf{B_3}$ are equivalent, we give the following theorems:

Theorem 13.4.19. *Let $\mathbf{A}$ be a matrix of order $n \times n$ and index ≤ 1. Let $\mathbf{G}$ be a commuting g-inverse of $\mathbf{A}$. Then $\mathbf{A_3}$ is of index ≤ 1 if and only if either $\mathbf{a} \in \mathcal{C}(\mathbf{A})$, $\mathbf{b} \in \mathcal{C}(\mathbf{A}^{\mathbf{t}})$, $\alpha = -\mathbf{1} - \mathbf{b}^{\mathbf{t}}\mathbf{G}\mathbf{a} = \mathbf{0}$ and $r = \mathbf{b}^{\mathbf{t}}\mathbf{G}^{\mathbf{2}}\mathbf{a} \neq \mathbf{0}$ (in which case $\rho(\mathbf{A_3}) = \rho(\mathbf{A}) - \mathbf{1}$) or $\alpha = -1 - \mathbf{b}^{\mathbf{t}}\mathbf{G}\mathbf{a} \neq \mathbf{0}$ and at least one of $\mathbf{a} \in \mathcal{C}(\mathbf{A})$, $\mathbf{b} \in \mathcal{C}(\mathbf{A}^{\mathbf{t}})$ holds (in which case $\rho(\mathbf{A_3}) = \rho(\mathbf{A})$) or $\mathbf{b}^{\mathbf{t}}(\mathbf{I} - \mathbf{G}\mathbf{A})\mathbf{a} \neq \mathbf{0}$ holds (in which case $\rho(\mathbf{A_3}) = \rho(\mathbf{A}) + \mathbf{1}$.)*

Theorem 13.4.20. *Let $\mathbf{A}, \mathbf{B}$ be matrices of order $n \times n$ and same index ≤ 1 such that $\mathbf{A} <^{\#} \mathbf{B}$ and $\mathbf{G}$ be the group inverse of $\mathbf{B}$. (Notice that $\mathbf{G}$ in such a case is a commuting g-inverse of $\mathbf{A}$.). Let $\mathbf{A_3}$ and $\mathbf{B_3}$ be of index ≤ 1 and $\rho(\mathbf{A_3}) = \rho(\mathbf{A}) - \mathbf{1}$. Then $\mathbf{A_3} <^{\#} \mathbf{B_3}$ if and only if exactly one of the following holds:*

(i) $\mathbf{A_3a} = \mathbf{0}$, $\mathbf{A_3c} = \mathbf{0}$, *in which case either* $\mathbf{b}^{\mathbf{t}}\mathbf{A_3} = \mathbf{0}$ *and* $\mathbf{c} \in \mathcal{N}(\mathbf{A})$ *or* $\mathbf{b}^{\mathbf{t}}\mathbf{A_3} = \mathbf{0}$ *and* $\mathbf{d} \in \mathcal{N}(\mathbf{A_3})^{\mathbf{t}}$ *or* $\mathbf{d} = (\frac{\mathbf{1}}{\chi^{\mathbf{t}}\mathbf{b}})(\mathbf{b}\mathbf{a}^{\mathbf{t}}\mathbf{b} + (\mathbf{I} - (\mathbf{A_3}\mathbf{G})^{\mathbf{t}})\mathbf{w})$, *where* χ *is a vector such that* $\mathbf{c} = (\mathbf{I} - \mathbf{G}\mathbf{A_3})\chi$.
(ii) $\mathbf{A_3a} = \mathbf{0}$, $\mathbf{d}^{\mathbf{t}}\mathbf{G}\mathbf{A} = \mathbf{0}$ *and* $\mathbf{A}\mathbf{G}\mathbf{c} = \mathbf{0}$ *in which case* $\mathbf{b}^{\mathbf{t}}$ *is a left eigen value of* $\mathbf{A}$ *corresponding to eigen value* $\mathbf{b}^{\mathbf{t}}\mathbf{a} \neq \mathbf{0}$.
(iii) $\mathbf{A_3a} \neq \mathbf{0}$, $\mathbf{A_3c} \neq \mathbf{0}$, *and* $\mathbf{d}^{\mathbf{t}}\mathbf{G}\mathbf{A} \neq \mathbf{0}$, *in which case*

$$\mathbf{c} = \alpha\mathbf{a} + (\mathbf{I} - \mathbf{A}\mathbf{G})\zeta \text{ and } \mathbf{d} = \frac{\mathbf{b}}{\alpha} + (\mathbf{I} - (\mathbf{A}\mathbf{G})^{\mathbf{t}})\eta$$

for arbitrary vectors ζ *and* η.

Theorem 13.4.21. *Let $\mathbf{A}, \mathbf{B}$ be matrices of order $n \times n$ and of index ≤ 1 such that $\mathbf{A} <^{\#} \mathbf{B}$ and $\mathbf{G}$ be the group inverse of $\mathbf{B}$. (Notice that $\mathbf{G}$ in* *such a case is a commuting g-inverse of* $\mathbf{A}$.*). Let* $\mathbf{A_3}$ *and* $\mathbf{B_3}$ *be of index* ≤ 1 *and* $\rho(\mathbf{A_3}) = \rho(\mathbf{A})$. *Then* $\mathbf{A_3} <^{\#} \mathbf{B_3}$ *if and only if exactly one of the following holds:*

(i) $\mathbf{b^t A_3} = \mathbf{0}$ *and* $\mathbf{AGc} = \mathbf{0}$, *in which case* $\mathbf{c} \in \mathcal{N}(\mathbf{A})$, $\mathbf{d}$ *is arbitrary and either* $\mathbf{A_3 a} = \mathbf{0}$ *or* $\mathbf{b} \in \mathcal{N}(\mathbf{A^t})$.

(ii) $\mathbf{b^t A_3} = \mathbf{0}$ *and* $\mathbf{d^t A_3} = \mathbf{0}$ *in which case, if* $\mathbf{b^t Ga} = \mathbf{0}$, *then* $\mathbf{c}$ *is arbitrary otherwise either* $\mathbf{a}$, $\mathbf{c} \in \mathcal{N}(\mathbf{A_3})$ *or* $\mathbf{a} \in \mathcal{N}(\mathbf{A_3})$ *and* $\mathbf{c}$ *is arbitrary or* $\mathbf{c} = \dfrac{\mathbf{b^t Ga}}{\mathbf{d^t Ga}}\mathbf{a} \in \mathcal{N}(\mathbf{A_3})$.

(iii) $\mathbf{b^t A_3} \neq \mathbf{0}$ *and* $\mathbf{AGc} \neq \mathbf{0}$, *and* $\mathbf{d^t A_3} \neq \mathbf{0}$, *in which case* $\mathbf{c} - \alpha\mathbf{a} \in \mathcal{N}(\mathbf{A})$ *and* $\mathbf{d} - \dfrac{\mathbf{1}}{\alpha}\mathbf{b^t} \in \mathcal{N}(\mathbf{A_3})$.

Theorem 13.4.22. *Let* $\mathbf{A}, \mathbf{B}$ *be matrices of order* $n \times n$ *and of index* ≤ 1 *such that* $\mathbf{A} <^{\#} \mathbf{B}$ *and* $\mathbf{G}$ *be the group inverse of* $\mathbf{B}$. *(Notice that* $\mathbf{G}$ *in such a case is a commuting g-inverse of* $\mathbf{A}$.*). Let* $\mathbf{A_3}$ *and* $\mathbf{B_3}$ *be of index* ≤ 1, $\mathbf{u} = (\mathbf{I} - \mathbf{AG})\mathbf{a}$, $\mathbf{v^t} = \mathbf{b^t}(\mathbf{I} - \mathbf{AG})$ *and* $\rho(\mathbf{A_3}) = \rho(\mathbf{A}) + \mathbf{1}$. *Then* $\mathbf{A_3} <^{\#} \mathbf{B_3}$ *if and only if exactly one of the following holds:*

(i) $\mathbf{Aa} = \mathbf{0}$ *and* $\mathbf{d^t u} = \mathbf{0}$, *in which case* $\mathbf{a} \in \mathcal{N}(\mathbf{A})$, $\mathbf{c}$ *is arbitrary and* $\mathbf{d}$ *is such that* $\mathbf{d^t a} = \mathbf{0}$.

(ii) $\mathbf{Aa} = \mathbf{0}$ *and* $\mathbf{Ac} = \mathbf{0}$, *in which case* $\mathbf{a}, \mathbf{c} \in \mathcal{N}(\mathbf{A})$, *and* $\mathbf{d}$ *is arbitrary.*

(iii) $\mathbf{Aa} \neq \mathbf{0}$, $\mathbf{Ac} \neq \mathbf{0}$ *and* $\mathbf{d^t u} \neq \mathbf{0}$, *in which case we choose* $\mathbf{d}$ *such that* $\mathbf{d^t u} \neq \mathbf{0}$ *and* $\mathbf{c} = \dfrac{\mathbf{p}}{\mathbf{d^t u}}\mathbf{a} + (\mathbf{I} - \mathbf{AG})\theta$, *where* θ *is arbitrary.*

13.5 Star order

Let $\mathbf{A}$ and $\mathbf{B}$ be complex $m \times n$ matrices. In this section, we first obtain necessary and sufficient conditions for $\mathbf{A_1} <^{\star} \mathbf{B_1}$ to hold when $\mathbf{A} <^{\star} \mathbf{B}$ holds and vice versa. The conditions we obtain for the star are much simpler than the ones obtained for the minus order and for the sharp order. Recall $\mathbf{A_1} = (\mathbf{A} : \mathbf{a})$ and $\mathbf{B_1} = (\mathbf{B} : \mathbf{b})$, where $\mathbf{a}$ and $\mathbf{b}$ are m-column vectors. We reiterate the vectors and matrices in this and the next section are over $\mathbb{C}$, the field of complex numbers.

Theorem 13.5.1. *Let* $\mathbf{A}$ and B *be matrices of order* $m \times n$ *and* $\mathbf{a}, \mathbf{b}$ *be* m*-column vectors. Then any two of the following three implies the third:*

(i) $\mathbf{A} <^{\star} \mathbf{B}$

(ii) $\mathbf{A_1} <^{\star} \mathbf{B_1}$

(iii) *either* '$\mathbf{a} = \mathbf{0}$ *and* $\mathbf{b} \in \mathcal{C}(\mathbf{A}^{\perp})$' *or* '$\mathbf{a} \neq \mathbf{0}$, $\mathbf{b} = \mathbf{a}$ *and* $\mathbf{b} \in \mathcal{C}(\mathbf{B} - \mathbf{A})^{\perp}$.'

Proof. (i) and (ii) $\Rightarrow$ (iii)
Since (i) and (ii) hold, we have

$$\mathbf{A}^{\star}(\mathbf{B} - \mathbf{A}) = \mathbf{0}, \quad (\mathbf{B} - \mathbf{A})\mathbf{A}^{\star} = \mathbf{0} \tag{13.5.1}$$

$$\mathbf{A}_1^{\star}(\mathbf{B}_1 - \mathbf{A}_1) = \mathbf{0}, \quad (\mathbf{B}_1 - \mathbf{A}_1)\mathbf{A}_1^{\star} = \mathbf{0}. \tag{13.5.2}$$

From (13.5.2), we have

$$\begin{pmatrix} \mathbf{A}^{\star} \\ \mathbf{a}^{\star} \end{pmatrix} (\mathbf{B} - \mathbf{A} : \mathbf{b} - \mathbf{a}) = \mathbf{0}$$

and

$$(\mathbf{B} - \mathbf{A} : \mathbf{b} - \mathbf{a}) \begin{pmatrix} \mathbf{A}^{\star} \\ \mathbf{a}^{\star} \end{pmatrix} = \mathbf{0};$$

equivalently, $\mathbf{A}^{\star}(\mathbf{B} - \mathbf{A}) = \mathbf{0}$, $\mathbf{A}^{\star}(\mathbf{b} - \mathbf{a}) = \mathbf{0}$, $\mathbf{a}^{\star}(\mathbf{B} - \mathbf{A}) = \mathbf{0}$ and $(\mathbf{B} - \mathbf{A})\mathbf{a}^{\star} + (\mathbf{b} - \mathbf{a})\mathbf{a}^{\star} = \mathbf{0}$. Using (13.5.1) in $(\mathbf{B} - \mathbf{A})\mathbf{a}^{\star} + (\mathbf{b} - \mathbf{a})\mathbf{a}^{\star} = \mathbf{0}$, we have $(\mathbf{b} - \mathbf{a})\mathbf{a}^{\star} = \mathbf{0} \Rightarrow \mathbf{a} = \mathbf{0}$ or $\mathbf{b} = \mathbf{a}$.
In case $\mathbf{a} = \mathbf{0}, \mathbf{A}^{\star}(\mathbf{b} - \mathbf{a}) = \mathbf{0} \Rightarrow \mathbf{A}^{\star}(\mathbf{b}) = \mathbf{0}$. So, $\mathbf{b} \in \mathcal{C}(\mathbf{A}^{\perp})$.
In case $\mathbf{b} = \mathbf{a}$, $\mathbf{a}^{\star}(\mathbf{B} - \mathbf{A}) = \mathbf{0} \Rightarrow \mathbf{b} \in \mathcal{C}(\mathbf{B} - \mathbf{A})^{\perp}$.
Thus, (iii) holds.
(i) and (iii) $\Rightarrow$ (ii)
Consider $\mathbf{A}_1^{\star}(\mathbf{B}_1 - \mathbf{A}_1) = \begin{pmatrix} \mathbf{A}^{\star}(\mathbf{B} - \mathbf{A}) & \mathbf{A}^{\star}(\mathbf{b} - \mathbf{a}) \\ \mathbf{a}^{\star}(\mathbf{B} - \mathbf{A}) & \mathbf{a}^{\star}(\mathbf{b} - \mathbf{a}) \end{pmatrix}$.
Let $\mathbf{a} = \mathbf{0}$ and $\mathbf{b} \in \mathcal{C}(\mathbf{A}^{\perp})$. Then $\mathbf{A}^{\star}\mathbf{b} = \mathbf{0}$. Using $\mathbf{A}^{\star}(\mathbf{B} - \mathbf{A})$, we have $\mathbf{A}_1^{\star}(\mathbf{B}_1 - \mathbf{A}_1) = \mathbf{0}$.
Next, let $\mathbf{a} \neq \mathbf{0}$, $\mathbf{b} = \mathbf{a}$ and $\mathbf{b} \in \mathcal{C}(\mathbf{B} - \mathbf{A})^{\perp}$. Clearly, $\mathbf{A}_1^{\star}(\mathbf{B}_1 - \mathbf{A}_1) = \mathbf{0}$. Similarly, $(\mathbf{B}_1 - \mathbf{A}_1)\mathbf{A}_1^{\star} = \mathbf{0}$.
The proof for (ii) and (iii) $\Rightarrow$ (ii) is equally easy. □

We now study the star order for matrices $\mathbf{A_2} = \mathbf{A} + \mathbf{ab}^{\star}$, $\mathbf{B_2} = \mathbf{B} + \mathbf{cd}^{\star}$ obtained from matrices $\mathbf{A}$, $\mathbf{B}$ by adding to them rank 1 matrices $\mathbf{ab}^{\star}$ and $\mathbf{cd}^{\star}$ respectively. Two methods to deal with the problem, first of which is similar to the methods adopted for the minus order and the sharp order, are presented. The first method is similar to the method adopted for minus and sharp orders. The second method is simple, intuitive and reveals very clearly as to what happens when matrices are subjected to such modifications.

Theorem 13.5.2. *Let* $\mathbf{A}, \mathbf{B}$ *be complex matrices of order* $m \times n$. *Let* $\mathbf{a}, \mathbf{c} \in \mathbb{C}^{\mathbf{m}}$ *and* $\mathbf{b}, \mathbf{d} \in \mathbb{C}^{\mathbf{n}}$, $\mathbf{A_2} = \mathbf{A} + \mathbf{ab}^{\star}$ *and* $\mathbf{B_2} = \mathbf{B} + \mathbf{cd}^{\star}$. *Then any two of the following three imply the third:*

(i) $\mathbf{A} <^\star \mathbf{B}$
(ii) $\mathbf{A_1} <^\star \mathbf{B_1}$
(iii) (a) $(\mathbf{ab}^\star - \mathbf{cd}^\star)^\star \mathbf{A_2} = \mathbf{0}$ *and* $\mathbf{A_2}(\mathbf{ab}^\star - \mathbf{cd}^\star)^\star = \mathbf{0}$, *when* $\mathbf{a} \in \mathcal{C}(\mathbf{A})$ *and* $\mathbf{b} \in \mathcal{C}(\mathbf{A}^\star)$.
(b) $(\mathbf{ab}^\star - \mathbf{cd}^\star)^\star \mathbf{A_2} = \mathbf{0}$ *and* $\mathbf{A_2}(\mathbf{ab}^\star - \mathbf{cd}^\star)^\star + \mathbf{ab}^\star(\mathbf{B} - \mathbf{A})^\star = \mathbf{0}$, *when* $\mathbf{a} \in \mathcal{C}(\mathbf{A})$ *and* $\mathbf{b} \notin \mathcal{C}(\mathbf{A}^\star)$.
(c) $(\mathbf{ab}^\star - \mathbf{cd}^\star)^\star \mathbf{A_2} + (\mathbf{B} - \mathbf{A})^\star \mathbf{ab}^\star = \mathbf{0}$ *and* $\mathbf{A_2}(\mathbf{ab}^\star - \mathbf{cd}^\star)^\star = \mathbf{0}$, *when* $\mathbf{a} \notin \mathcal{C}(\mathbf{A})$ *and* $\mathbf{b} \in \mathcal{C}(\mathbf{A}^\star)$.
(d) $(\mathbf{ab}^\star - \mathbf{cd}^\star)^\star \mathbf{A_2} + (\mathbf{B} - \mathbf{A})^\star \mathbf{ab}^\star = \mathbf{0}$ *and* $\mathbf{A_2}(\mathbf{ab}^\star - \mathbf{cd}^\star)^\star + \mathbf{ab}^\star(\mathbf{B} - \mathbf{A})^\star = \mathbf{0}$, *when* $\mathbf{a} \notin \mathcal{C}(\mathbf{A})$ *and* $\mathbf{b} \notin \mathcal{C}(\mathbf{A}^\star)$.

Proof is straightforward.

As an alternative to Theorem 13.5.2, notice that to obtain necessary and sufficient conditions for $\mathbf{A} <^\star \mathbf{B} \Leftrightarrow \mathbf{A_2} <^\star \mathbf{B_2}$ is equivalent to finding first a necessary and sufficient conditions for $\mathbf{A} <^\star \mathbf{B} \Leftrightarrow \mathbf{A} + \mathbf{xy}^\star <^\star \mathbf{B} + \mathbf{xy}^\star$ and then for $\mathbf{A} <^\star \mathbf{B} \Leftrightarrow \mathbf{A} + \mathbf{wz}^\star <^\star \mathbf{B} - \mathbf{wz}^\star$ for suitable column vectors $\mathbf{x}$, $\mathbf{y}$, $\mathbf{z}$ and $\mathbf{w}$. In the following theorems we consider these special cases, which are interesting in their own right owing to the simple structure of necessary and sufficient conditions.

Theorem 13.5.3. *Let* $\mathbf{A}, \mathbf{B}$ *be complex matrices of order* $m \times n$. *Let* $\mathbf{a} \in \mathbb{C}^m$, $\mathbf{b} \in \mathbb{C}^n$, $\mathbf{X} = \mathbf{A} + \mathbf{ab}^\star$ *and* $\mathbf{Y} = \mathbf{B} + \mathbf{ab}^\star$. *Then any two of the following three statements imply the third:*

(i) $\mathbf{A} <^\star \mathbf{B}$
(ii) $\mathbf{X} <^\star \mathbf{Y}$
(iii) $\mathbf{a} \in \mathcal{C}(\mathbf{B} - \mathbf{A}^\perp)$ *and* $\mathbf{b} \in \mathcal{C}((\mathbf{B} - \mathbf{A})^\star)^\perp$.

Proof. Notice that (i) is equivalent to

$$\mathbf{A}^\star(\mathbf{B} - \mathbf{A}) = \mathbf{0}, \quad (\mathbf{B} - \mathbf{A})\mathbf{A}^\star = \mathbf{0}$$

and (ii) is equivalent to

$$\mathbf{A}^\star(\mathbf{B} - \mathbf{A}) + (\mathbf{ab}^\star)^\star(\mathbf{B} - \mathbf{A}) = \mathbf{0}, \quad (\mathbf{B} - \mathbf{A})\mathbf{A}^\star + (\mathbf{B} - \mathbf{A})(\mathbf{ab}^\star)^\star = \mathbf{0}.$$

Now the proof is easy. □

Remark 13.5.4. Theorem holds with suitable modifications even when we add an arbitrary $m \times n$ matrix to $\mathbf{A}$ and $\mathbf{B}$.

Theorem 13.5.5. *Let* $\mathbf{A}, \mathbf{B}$ *be complex matrices of the same order* $m \times n$, $\mathbf{a} \in \mathbb{C}^m$ *and* $\mathbf{b} \in \mathbb{C}^n$. *Let* $\mathbf{X} = \mathbf{A} + \mathbf{ab}^\star$ *and* $\mathbf{Y} = \mathbf{B} - \mathbf{ab}^\star$. *Then any two of the following three statements imply the third:*

(i) $\mathbf{A} <^{\star} \mathbf{B}$

(ii) $\mathbf{X} <^{\star} \mathbf{Y}$

(iii) $\mathbf{a}$ *and* $\mathbf{b}$ *are singular vectors of* $\mathbf{A}$ *corresponding to the same singular value and* $\mathbf{b} = -\dfrac{-\mathbf{A}^{\star}\mathbf{a}}{\mathbf{a}^{\star}\mathbf{a}}$ *according as* $\mathbf{a} \in \mathcal{C}(\mathbf{A})$ *or* $\mathbf{b} \in \mathcal{C}(\mathbf{A}^{\star})$.

Proof. (i) and (ii) $\Rightarrow$ (iii)
Since (i) and (ii) hold, we have

$$\mathbf{A}^{\star}(\mathbf{B}-\mathbf{A}) = \mathbf{0}, \quad (\mathbf{B}-\mathbf{A})\mathbf{A}^{\star} = \mathbf{0} \tag{13.5.3}$$

and

$$\begin{aligned} \mathbf{A}^{\star}(\mathbf{B}-\mathbf{A}) + (\mathbf{ab}^{\star})^{\star}(\mathbf{B}-\mathbf{A}) - 2\mathbf{A}^{\star}\mathbf{ab}^{\star} - 2(\mathbf{ab}^{\star})^{\star}\mathbf{ab}^{\star} &= \mathbf{0}, \\ (\mathbf{B}-\mathbf{A})\mathbf{A}^{\star} + (\mathbf{B}-\mathbf{A})(\mathbf{ab}^{\star})^{\star} - 2(\mathbf{ab}^{\star})^{\star}\mathbf{A}^{\star} - 2\mathbf{ab}^{\star}(\mathbf{ab}^{\star})^{\star} &= \mathbf{0}. \end{aligned} \tag{13.5.4}$$

Using (13.5.3) in (13.5.4), we have

$$(\mathbf{ab}^{\star})^{\star}(\mathbf{B}-\mathbf{A}) - 2\mathbf{A}^{\star}\mathbf{ab}^{\star} - 2(\mathbf{ab}^{\star})^{\star}\mathbf{ab}^{\star} = \mathbf{0} \tag{13.5.5}$$

and

$$(\mathbf{B}-\mathbf{A})(\mathbf{ab}^{\star})^{\star} - 2(\mathbf{ab}^{\star})^{\star}\mathbf{A}^{\star} - 2\mathbf{ab}^{\star}(\mathbf{ab}^{\star})^{\star} = \mathbf{0}. \tag{13.5.6}$$

Let $\mathbf{a} \in \mathcal{C}(\mathbf{A})$. Therefore, from (13.5.3) we have $\mathbf{a}^{\star}(\mathbf{B}-\mathbf{A}) = \mathbf{0}$. Substituting in (13.5.5) and noting that $\mathbf{a}$ and $\mathbf{b}$ are non-null vectors, we have $\mathbf{A}^{\star}\mathbf{a} + \mathbf{ba}^{\star}\mathbf{a} = \mathbf{0} \Rightarrow \mathbf{b} = \dfrac{-\mathbf{A}^{\star}\mathbf{a}}{\mathbf{a}^{\star}\mathbf{a}}$. Substituting in (13.5.6) and simplifying we have $\mathbf{AA}^{\star}\mathbf{a}^{\star} = \dfrac{\mathbf{a}^{\star}\mathbf{AA}^{\star}\mathbf{a}}{\mathbf{a}^{\star}\mathbf{a}}\mathbf{a}$. Thus, $\dfrac{\mathbf{a}^{\star}\mathbf{AA}^{\star}\mathbf{a}}{\mathbf{a}^{\star}\mathbf{a}}$ is a singular value of $\mathbf{A}$ corresponding to singular vector $\mathbf{a}$.

Let $\mathbf{b} \in \mathcal{C}(\mathbf{A}^{\star})$. Then a similar computation shows that $\dfrac{\mathbf{a}^{\star}\mathbf{AA}^{\star}\mathbf{a}}{\mathbf{a}^{\star}\mathbf{a}}$ is a singular value of $\mathbf{A}$ corresponding to singular vector $\mathbf{b}$. □

13.6 Löwner order

Once again in this section too, all matrices are over the field $\mathbb{C}$ of complex numbers. Let $\mathbf{A}$ and $\mathbf{B}$ be two hermitian matrices of the same order and $\mathbf{A_2} = \begin{pmatrix} \mathbf{A} & \mathbf{u} \\ \mathbf{u}^{\star} & \alpha \end{pmatrix}$ and $\mathbf{B_2} = \begin{pmatrix} \mathbf{B} & \mathbf{x} \\ \mathbf{x}^{\star} & \beta \end{pmatrix}$. Given $\mathbf{A} <^{\mathbf{L}} \mathbf{B}$, we explore the conditions under which $\mathbf{A_2} <^{\mathbf{L}} \mathbf{B_2}$ holds and vice versa. We have the following:

Theorem 13.6.1. *Let* $\mathbf{A}$ *and* $\mathbf{B}$ *be hermitian matrices of order* $n \times n$, $\mathbf{u}, \mathbf{x} \in \mathbb{C}^{\mathbf{n}}$ *and* α, β *complex scalars. Let* $\mathbf{A} <^{\mathbf{L}} \mathbf{B}$. *Then* $\mathbf{A_2} <^{\mathbf{L}} \mathbf{B_2}$ *if and only if* $\mathbf{x} = \mathbf{u} + (\mathbf{B}-\mathbf{A})\mathbf{w}$ *for some* $\mathbf{w}$ *and* $\beta \geq \alpha + \mathbf{w}^{\star}(\mathbf{B}-\mathbf{A})\mathbf{w}$.

Proof. Let $\mathbf{A} <^{\mathbf{L}} \mathbf{B}$. Then $\mathbf{B} - \mathbf{A}$ is *nnd*. Now, $\mathbf{B_2} - \mathbf{A_2}$ is *nnd* if and only if $\mathbf{x} - \mathbf{u} \in (\mathbf{B} - \mathbf{A})$ and $\beta - \alpha - (\mathbf{x} - \mathbf{u})^\star(\mathbf{B} - \mathbf{A})^-(\mathbf{x} - \mathbf{u}) \geq \mathbf{0}$. Let $\mathbf{x} - \mathbf{u} = (\mathbf{B} - \mathbf{A})\mathbf{w}$. Then $\mathbf{x} = \mathbf{u} + (\mathbf{B} - \mathbf{A})\mathbf{w}$ and $(\mathbf{x} - \mathbf{u})^\star(\mathbf{B} - \mathbf{A})^-(\mathbf{x} - \mathbf{u}) = \mathbf{w}^\star(\mathbf{B} - \mathbf{A})\mathbf{w}$. The result now follows. □

Corollary 13.6.2. *Let* $\mathbf{A}$ *and* $\mathbf{B}$ *be nnd matrices of order* $n \times n$ *such that* $\mathbf{A} <^{\mathbf{L}} \mathbf{B}$. *Let* $\mathbf{A_2}$ *be nnd. Then* $\mathbf{B_2}$ *is nnd and* $\mathbf{A_2} <^{\mathbf{L}} \mathbf{B_2}$ *if and only if* $\mathbf{x} = \mathbf{u} + (\mathbf{B} - \mathbf{A})\mathbf{w}$ *for some* $\mathbf{w}$ *and* $\beta \geq \alpha + \mathbf{w}^\star(\mathbf{B} - \mathbf{A})\mathbf{w}$.

We now study the following problem:
Given $\mathbf{A}$ and $\mathbf{B}$ be *nnd* matrices of the same order $n \times n$ such that $\mathbf{A} <^{\mathbf{L}} \mathbf{B}$. Let $\mathbf{A_3} = \mathbf{A} + \mathbf{ab}^\star$ and $\mathbf{B_3} = \mathbf{B} + \mathbf{cd}^\star$. When does $\mathbf{A_3} <^{\mathbf{L}} \mathbf{B_3}$ hold? To answer this, we investigate when the following are true:

(a) $\mathbf{A} + \mathbf{aa}^\star <^{\mathbf{L}} \mathbf{B} + \mathbf{bb}^\star$,
(b) $\mathbf{A} + \mathbf{aa}^\star <^{\mathbf{L}} \mathbf{B} - \mathbf{bb}^\star$, and
(c) $\mathbf{A} - \mathbf{aa}^\star <^{\mathbf{L}} \mathbf{B} - \mathbf{bb}^\star$.

Notice that whenever $\mathbf{A} - \mathbf{aa}^\star$ is *nnd*, $\mathbf{A} - \mathbf{aa}^\star <^{\mathbf{L}} \mathbf{B} + \mathbf{bb}^\star$ is always true.

Theorem 13.6.3. *Let* $\mathbf{A}$ *and* $\mathbf{B}$ *be hermitian matrices of the same order* $n \times n$ *such that* $\mathbf{A} <^{\mathbf{L}} \mathbf{B}$ *and* $\mathbf{a}, \mathbf{b}$ *n-column vectors. Then*

$$\mathbf{A} + \mathbf{aa}^\star <^{\mathbf{L}} \mathbf{B} + \mathbf{bb}^\star \text{ if and only if } \mathbf{a} = ((\mathbf{B} - \mathbf{A}) + \mathbf{bb}^\star)\mathbf{w},$$

where $\mathbf{w}$ *is arbitrary vector subject to* $\mathbf{w}^\star(\mathbf{B} - \mathbf{A})\mathbf{w} \leq 1$.

Proof. $(\mathbf{B} + \mathbf{bb}^\star) - (\mathbf{A} + \mathbf{aa}^\star)$ is *nnd* $\Leftrightarrow \mathbf{T} = (\mathbf{B} - \mathbf{A}) + \mathbf{bb}^\star - \mathbf{aa}^\star$ is *nnd*. Now, $\mathbf{T}$ is *nnd* if and only if $\mathbf{S} = \begin{pmatrix} (\mathbf{B} - \mathbf{A}) + \mathbf{bb}^\star & \mathbf{a} \\ \mathbf{a}^\star & 1 \end{pmatrix}$ is *nnd*. The matrix $\mathbf{S}$ is *nnd* $\Leftrightarrow \mathbf{a} \in \mathcal{C}((\mathbf{B} - \mathbf{A}) + \mathbf{bb}^\star)$ and $1 \geq \mathbf{a}^\star(\mathbf{B} - \mathbf{A} + \mathbf{bb}^\star)^-\mathbf{a} \Leftrightarrow \mathbf{a} = ((\mathbf{B} - \mathbf{A}) + \mathbf{bb}^\star)\mathbf{w}$ for some $\mathbf{w}$ and $1 \geq \mathbf{w}^\star(\mathbf{B} - \mathbf{A})\mathbf{w}$. □

Theorem 13.6.4. *Let* $\mathbf{A}$ *and* $\mathbf{B}$ *be nnd matrices of order* $n \times n$ *such that* $\mathbf{A} <^{\mathbf{L}} \mathbf{B}$ *and* $\mathbf{a}, \mathbf{b}$ *n-column vectors. Then* $\mathbf{A} + \mathbf{aa}^\star <^{\mathbf{L}} \mathbf{B} - \mathbf{bb}^\star$ *if and only if* $(\mathbf{b} : \mathbf{a}) = (\mathbf{B} - \mathbf{A})\mathbf{T}$ *for some matrix* $\mathbf{T}$ *and the* 2×2 *matrix* $\mathbf{I} - \mathbf{T}^\star(\mathbf{B} - \mathbf{A})\mathbf{T}$ *is nnd.*

Proof. $\mathbf{A} + \mathbf{aa}^\star <^{\mathbf{L}} \mathbf{B} - \mathbf{bb}^\star \Leftrightarrow$ the matrix $\mathbf{Z} = (\mathbf{B} - \mathbf{A}) - (\mathbf{bb}^\star + \mathbf{aa}^\star)$ is *nnd*. As the 2×2 matrix $\mathbf{I}$ is always *nnd*, the matrix $\mathbf{Z}$ is *nnd* $\Leftrightarrow$ the matrix $\mathbf{R} = \begin{pmatrix} \mathbf{B} - \mathbf{A} & (\mathbf{b} : \mathbf{a}) \\ (\mathbf{b} : \mathbf{a})^\star & \mathbf{I} \end{pmatrix}$ is *nnd* $\Leftrightarrow$ $(\mathbf{b} : \mathbf{a}) = (\mathbf{B} - \mathbf{A})\mathbf{T}$ for some matrix $\mathbf{T}$ and the 2×2 matrix $\mathbf{I} - \mathbf{T}^\star(\mathbf{B} - \mathbf{A})\mathbf{T}$ is *nnd*.

The remaining case can be proved similarly. □

Chapter 14

Equivalence Relations on Generalized and Outer Inverses

14.1 Introduction

The classes of generalized inverses and outer inverses of a matrix are big, so big that they can be best described as a forest. At the first glance, exploration of a forest may appear to be an arduous task. However, if the forest is divided into zones based on certain well defined criteria, then the exploration may be more tractable or may even become quite easy. If one pursues, one may find a pattern in the madness of the forest.

The class of all generalized inverses and the class of all outer inverses of a matrix $\mathbf{A}$ form our forest. We define certain equivalence relations on these classes. The equivalence classes are the zones into which this forest gets divided and the hierarchy of the g-inverses is what we find. These lead to nice diagrammatic representation of all the g-inverses and outer inverses of a matrix revealing a well structured hierarchy. In Section 14.2, we define an equivalence relation on the class of all g-inverses of a matrix using the minus order. We explore the properties of the equivalence classes including some interesting characterizations. In Section 14.3, we study the equivalence relations based on special types of partial orders such as the star order and the sharp order. In Section 14.4, we define an equivalence relation on outer inverses and obtain characterizations of the hierarchy of outer inverses. Section 14.5 develops a scheme for diagrammatic representation of g-inverses and outer inverses depicting the hierarchy based on the results obtained in the earlier sections. Finally, we construct a ladder of g-inverses and outer inverses of a matrix and their reflexive g-inverses in Section 14.6.

14.2 Equivalence relation on g-inverses of a matrix

Let $\mathbf{A}$ be a matrix of order $m \times n$ over a field F. Let $\{\mathbf{A}^-\}$ be the class of all g-inverses of $\mathbf{A}$. We begin by defining a relation "$\backsim$" on $\{\mathbf{A}^-\}$ as follows:

Definition 14.2.1. Let $\mathbf{G_1}$ and $\mathbf{G_2}$ belong to $\{\mathbf{A}^-\}$. We say $\mathbf{G_1} \backsim \mathbf{G_2}$ if $\mathbf{AG_1} = \mathbf{AG_2}$ and $\mathbf{G_1A} = \mathbf{G_2A}$ (or equivalently, $\mathbf{G_1AG_2} = \mathbf{G_2AG_1}$).

It is easy to verify that the relation $\backsim$ is an equivalence relation on $\{\mathbf{A}^-\}$ and therefore partitions the set $\{\mathbf{A}^-\}$ into a disjoint union of equivalence classes. If $\mathbf{G} \in \{\mathbf{A}^-\}$, we denote the equivalence class that contains $\mathbf{G}$ by $\overline{\mathrm{Eq}}(\mathbf{G}|\mathbf{A})$. We next show that each $\overline{\mathrm{Eq}}(\mathbf{G}|\mathbf{A})$ contains a unique reflexive g-inverse of $\mathbf{A}$.

Theorem 14.2.2. *Let $\mathbf{A}$ be a matrix of order $m \times n$. Then each equivalence class under relation "$\backsim$" on $\{\mathbf{A}^-\}$, as in Definition 14.2.1, contains a unique reflexive g-inverse of $\mathbf{A}$.*

Proof. Consider any equivalence class $\mathfrak{T}$ and let $\mathbf{G} \in \mathfrak{T}$. Let $\mathbf{G_0} = \mathbf{GAG}$. Clearly, $\mathbf{G_0}$ is a reflexive g-inverse of $\mathbf{A}$. Further, $\mathbf{AG_0} = \mathbf{AGAG} = \mathbf{AG}$ and $\mathbf{G_0A} = \mathbf{GAGA} = \mathbf{GA}$. Hence, $\mathbf{G_0} \backsim \mathbf{G}$. Thus, $\mathbf{G_0} \in \mathfrak{T}$.

Moreover, if $\mathbf{G_1}$ and $\mathbf{G_2}$ are reflexive g-inverses of $\mathbf{A}$ such that $\mathbf{G_1} \backsim \mathbf{G_2}$. Then $\mathbf{G_1} = \mathbf{G_1AG_1} = \mathbf{G_2AG_1} = \mathbf{G_2AG_2} = \mathbf{G_2}$. Thus, each equivalence class contains a unique reflexive g-inverse of $\mathbf{A}$. □

Remark 14.2.3. Let $\mathbf{A}$ be a matrix of order $m \times n$ and of rank r, where $0 < r < min\{m, n\}$. Let $\mathbf{G}$ be a g-inverse of $\mathbf{A}$ with rank s $(s \geq r)$. Then by Theorem 2.3.18, it follows that there exist non- singular matrices $\mathbf{P}$ and $\mathbf{Q}$ such that $\mathbf{A} = \mathbf{P}\mathbf{diag}\left(\mathbf{I_r}\ , \mathbf{0}\right)\mathbf{Q}$ and $\mathbf{G} = \mathbf{Q}^{-1}\mathbf{diag}\left(\mathbf{I_s}\ , \mathbf{0}\right)\mathbf{P}^{-1}$. It is easy to see that the reflexive g-inverse of $\mathbf{A}$ belonging to $\overline{\mathrm{Eq}}(\mathbf{G}|\mathbf{A})$ is given by $\mathbf{G_0} = \mathbf{Q}^{-1}\mathbf{diag}\left(\mathbf{I_r}\ , \mathbf{0}\right)\mathbf{P}^{-1}$.

Theorem 14.2.4. *Let $\mathbf{A}$ be a matrix of order $m \times n$. Let $\mathbf{G_0} \in \{\mathbf{A}_r^-\}$ and $\mathbf{G_1} \in \overline{\mathrm{Eq}}(\mathbf{G_0}|\mathbf{A})$. Then $\mathbf{G_0} <^- \mathbf{G_1}$.*

Proof. Clearly, $\mathbf{AG_0} = \mathbf{AG_1}$ and $\mathbf{G_0A} = \mathbf{G_1A}$, as $\mathbf{A}$ is a g-inverse of $\mathbf{G_0}$. Hence, $\mathbf{G_0} <^- \mathbf{G_1}$. □

Remark 14.2.5. Let $\mathbf{A}$ and $\mathbf{G_0}$ be as in Theorem 14.2.4. Then

$$\mathbf{G_0} = \inf\{\mathbf{G} : \mathbf{G} \in \overline{\mathrm{Eq}}(\mathbf{G_0}|\mathbf{A})\}.$$

Let $\mathbf{G_0}$ be a given reflexive g-inverse of $\mathbf{A}$. We now determine the class all g-inverses $\mathbf{G}$ (of a specified rank) of $\mathbf{A}$ such that $\mathbf{G} \in \overline{\mathrm{Eq}}(\mathbf{G_0}|\mathbf{A})$.

Theorem 14.2.6. *Let* $\mathbf{A}$ *be a matrix of order* $m \times n$ *of rank* r *such that* $0 < r < min\{m, n\}$. *Let* $\mathbf{A} = \mathbf{P}\mathbf{diag}\left(\mathbf{I_r}\ , \mathbf{0}\right)\mathbf{Q}$ *be a normal form of* $\mathbf{A}$, *where* $\mathbf{P}$ *and* $\mathbf{Q}$ *are non-singular matrices. Let* $\mathbf{G_0} \in \{\mathbf{A_r^-}\}$ *such that* $\mathbf{G_0} = \mathbf{Q}^{-1}\begin{pmatrix}\mathbf{I_r} & \mathbf{L}\\ \mathbf{M} & \mathbf{ML}\end{pmatrix}\mathbf{P}^{-1}$ *for some matrices* $\mathbf{L}$ *and* $\mathbf{M}$. *Then an* $n \times m$ *matrix* $\mathbf{G}$ *of rank* s *is in the equivalence class* $\overline{\mathrm{Eq}}(\mathbf{G_0}|\mathbf{A})$ *if and only if* $\mathbf{G} = \mathbf{Q}^{-1}\begin{pmatrix}\mathbf{I_r} & \mathbf{L}\\ \mathbf{M} & \mathbf{ML}+\mathbf{S}\end{pmatrix}\mathbf{P}^{-1}$, *where* $\mathbf{S}$ *is arbitrary matrix of rank* $s - r$.

Proof. Notice that $\mathbf{G} \in \overline{\mathrm{Eq}}(\mathbf{G_0}|\mathbf{A})$ if and only if $\mathbf{G_0} = \mathbf{GAG}$. Every g-inverse $\mathbf{G}$ of $\mathbf{A}$ is of the form $\mathbf{G} = \mathbf{Q}^{-1}\begin{pmatrix}\mathbf{I_r} & \mathbf{T_1}\\ \mathbf{T_2} & \mathbf{T_3}\end{pmatrix}\mathbf{P}^{-1}$ for some matrices $\mathbf{T_1}, \mathbf{T_2}$ and $\mathbf{T_3}$. Now, $\mathbf{G_0} = \mathbf{GAG} \Leftrightarrow \mathbf{T_2} = \mathbf{M}$ and $\mathbf{T_1} = \mathbf{L}$. Thus, $\mathbf{G} = \mathbf{Q}^{-1}\begin{pmatrix}\mathbf{I_r} & \mathbf{L}\\ \mathbf{M} & \mathbf{T_3}\end{pmatrix}\mathbf{P}^{-1}$. Further $\rho(\mathbf{G}) = s \Leftrightarrow \rho(\mathbf{T_3} - \mathbf{ML}) = s - r$. Take $\mathbf{S} = \mathbf{T_3} - \mathbf{ML}$. □

Corollary 14.2.7. *Let* $\mathbf{A}$ *be a matrix of order* $m \times n$ *and of rank* r. *Let* $\mathbf{G_0}$ *be a reflexive g-inverse of* $\mathbf{A}$. *Then the class of all g-inverses* $\mathbf{G}$ *of* $\mathbf{A}$ *having rank* s $(s > r)$ *such that* $\mathbf{G} \in \overline{\mathrm{Eq}}(\mathbf{G_0}|\mathbf{A})$ *is given by* $\mathbf{Q}^{-1}\mathbf{diag}\left(\mathbf{I_r}\ , \mathbf{N}\right)\mathbf{P}^{-1}$, *where* $\mathbf{N}$ *is an arbitrary matrix of rank* $s - r$ *and the non- singular matrices* $\mathbf{P}$ *and* $\mathbf{Q}$ *are such that* $\mathbf{A} = \mathbf{P}\mathbf{diag}\left(\mathbf{I_r}\ , \mathbf{0}\right)\mathbf{Q}$ *and* $\mathbf{G_0} = \mathbf{Q}^{-1}\mathbf{diag}\left(\mathbf{I_r}\ , \mathbf{0}\right)\mathbf{P}^{-1}$.

Corollary 14.2.8. *Let* $\mathbf{A}$ *be a matrix of order* $m \times n$ *of rank* r. *Let* $\mathbf{G_0}$ *be a reflexive g-inverse of* $\mathbf{A}$. *Then an* $m \times n$ *matrix* $\mathbf{G}$ *of rank* s *such that* $\mathbf{G} \in \overline{\mathrm{Eq}}(\mathbf{G_0}|\mathbf{A})$ *if and only if there exists non- singular matrices* $\mathbf{R}$ *and* $\mathbf{S}$ *such that* $\mathbf{A} = \mathbf{R}\mathbf{diag}\left(\mathbf{I_r}\ , \mathbf{0}\right)\mathbf{S}$ *and* $\mathbf{G_0} = \mathbf{S}^{-1}\mathbf{diag}\left(\mathbf{I_r}\ , \mathbf{0}\right)\mathbf{R}^{-1}$ *and* $\mathbf{G} = \mathbf{S}^{-1}\mathbf{diag}\left(\mathbf{I_s}\ , \mathbf{0}\right)\mathbf{R}^{-1}$.

The following theorem is yet another characterization of g-inverses belonging to an equivalence class.

Theorem 14.2.9. *Let* $\mathbf{A}$ *be a matrix of order* $m \times n$ *and of rank* r. *Let* $\mathbf{G_0}$ *be a reflexive g-inverse of* $\mathbf{A}$. *Then the class of all* $\mathbf{G} \in \overline{\mathrm{Eq}}(\mathbf{G_0}|\mathbf{A})$ *is given by* $\mathbf{G} = \mathbf{G_0} + (\mathbf{I} - \mathbf{G_0A})\mathbf{U}(\mathbf{I} - \mathbf{AG_0})$ *where* $\mathbf{U}$ *is arbitrary.*

Proof. Notice that $\mathbf{A}(\mathbf{G} - \mathbf{G_0}) = \mathbf{0} \Leftrightarrow \mathcal{C}(\mathbf{G} - \mathbf{G_0}) \subseteq \mathcal{N}(\mathbf{A})$ and $(\mathbf{G} - \mathbf{G_0})\mathbf{A} = \mathbf{0} \Leftrightarrow \mathcal{C}(\mathbf{G} - \mathbf{G_0})^{\mathbf{t}} \subseteq \mathcal{N}(\mathbf{A^t})$. Now the result follows easily. □

Notice that in Theorem 14.2.4, we have proved that if $\mathbf{G} \in \overline{\text{Eq}}(\mathbf{G_0}|\mathbf{A})$. Then $\mathbf{G_0} <^{-} \mathbf{G}$ where $\mathbf{G_0}$ is a reflexive g-inverse of $\mathbf{A}$. We now show that every matrix $\mathbf{Y}$ satisfying $\mathbf{G_0} <^{-} \mathbf{Y} <^{-} \mathbf{G}$ also belongs to $\overline{\text{Eq}}(\mathbf{G_0}|\mathbf{A})$.

Theorem 14.2.10. *Let* $\mathbf{A}$ *be an* $m \times n$ *matrix and let* $\mathbf{G_0} \in \{\mathbf{A_r^-}\}$. *Let* $\mathbf{G} \in \overline{\text{Eq}}(\mathbf{G_0}|\mathbf{A})$. *Then each matrix* $\mathbf{Y}$ *satisfying* $\mathbf{G_0} <^{-} \mathbf{Y} <^{-} \mathbf{G}$ *belongs to* $\overline{\text{Eq}}(\mathbf{G_0}|\mathbf{A})$.

Proof. Let $\mathbf{Y}$ be of rank s. Then by Theorem 14.2.9, there exists non-singular matrices $\mathbf{R}$ and $\mathbf{S}$ such that $\mathbf{A} = \mathbf{R}\mathbf{diag}\left(\mathbf{I_r}, \mathbf{0_{s-r}}, \mathbf{0}\right)\mathbf{S}$, $\mathbf{G_0} = \mathbf{S^{-1}}\mathbf{diag}\left(\mathbf{I_r}, \mathbf{0_{s-r}}, \mathbf{0}\right)\mathbf{R^{-1}}$ and $\mathbf{G} = \mathbf{S^{-1}}\mathbf{diag}\left(\mathbf{I_0}, \mathbf{I_{s-r}}, \mathbf{0}\right)\mathbf{R^{-1}}$. Since $\mathbf{Y} <^{-} \mathbf{G}$ it follows that $\mathbf{Y} = \mathbf{S^{-1}}\mathbf{diag}\left(\mathbf{T_{s\times s}}, \mathbf{0}\right)\mathbf{R^{-1}}$, where $\mathbf{T}$ is idempotent. Since $\mathbf{G_0} <^{-} \mathbf{Y} <^{-} \mathbf{G}$, we have $\mathbf{diag}\left(\mathbf{I_r}, \mathbf{0}\right) <^{-} \mathbf{T} <^{-} \mathbf{I_s}$. As each of the matrices $\mathbf{diag}\left(\mathbf{I_r}, \mathbf{0}\right)$, $\mathbf{T}$ and $\mathbf{I_s}$ are idempotent, we have by Theorem 3.6.4, $\mathbf{T} = \mathbf{diag}\left(\mathbf{I_r}, \mathbf{0}\right) + \mathbf{diag}\left(\mathbf{0}, \mathbf{M}\right)$, where $\mathbf{M}$ is idempotent. Clearly, $\mathbf{AG_0} = \mathbf{AY}$ and $\mathbf{G_0A} = \mathbf{YA}$. Hence, $\mathbf{Y} \in \overline{\text{Eq}}(\mathbf{G_0}|\mathbf{A})$. □

Corollary 14.2.11. *Let* $\mathbf{A}$ *be an* $m \times n$ *matrix and let* $\mathbf{G_0} \in \{\mathbf{A_r^-}\}$. *Let* $\mathbf{G_1}, \mathbf{G_2} \in \overline{\text{Eq}}(\mathbf{G_0}|\mathbf{A})$ *such that* $\mathbf{G_1} <^{-} \mathbf{G_2}$. *Let* $\mathbf{Y}$ *be a matrix such that* $\mathbf{G_1} <^{-} \mathbf{Y} <^{-} \mathbf{G_2}$. *Then* $\mathbf{Y} \in \overline{\text{Eq}}(\mathbf{G_0}|\mathbf{A})$.

Proof. By Theorem 14.2.4, $\mathbf{G_0} <^{-} \mathbf{G_1}$. Since $\mathbf{G_1} <^{-} \mathbf{Y}$, we have $\mathbf{G_0} <^{-} \mathbf{Y}$. The corollary now follows from Theorem 14.2.10. □

Let $\mathbf{A}$ be an $m \times n$ matrix and let $\mathbf{G_0} \in \{\mathbf{A_r^-}\}$. Let $\mathbf{G_{s-r}}$ be a matrix of rank s such that $\mathbf{G_{s-r}} \in \overline{\text{Eq}}(\mathbf{G_0}|\mathbf{A})$. Does there exist a matrix $\mathbf{A_{s-r}} \in \{\mathbf{G_{s-r}^-}\}$ such that $\mathbf{A_{s-r}} \in \overline{\text{Eq}}(\mathbf{A}|\mathbf{G_0})$? If so, what is the class of all such $\mathbf{A_{s-r}}$? Determination of all such $\mathbf{A_{s-r}}$ will be helpful in the diagrammatic representation in Section 14.5. We prove

Theorem 14.2.12. *Let* $\mathbf{A}$ *be an* $m \times n$ *matrix and let* $\mathbf{G_0} \in \{\mathbf{A_r^-}\}$. *Let* $\mathbf{G_{s-r}}$ *be a matrix of rank* $s(> r)$ *such that* $\mathbf{G_{s-r}} \in \overline{\text{Eq}}(\mathbf{G_0}|\mathbf{A})$. *As in Corollary 14.2.8, let* $\mathbf{A} = \mathbf{R}\mathbf{diag}\left(\mathbf{I_r}, \mathbf{0}\right)\mathbf{S}$ *and* $\mathbf{G_0} = \mathbf{S^{-1}}\mathbf{diag}\left(\mathbf{I_r}, \mathbf{0}\right)\mathbf{R^{-1}}$ *and* $\mathbf{G_{s-r}} = \mathbf{S^{-1}}\mathbf{diag}\left(\mathbf{I_s}, \mathbf{0}\right)\mathbf{R^{-1}}$, *where* $\mathbf{R}$ *and* $\mathbf{S}$ *are non-singular matrices. The class of all matrices* $\mathbf{A_{s-r}}$ *such that* $\mathbf{A_{s-r}} \in \overline{\text{Eq}}(\mathbf{A}|\mathbf{G_0}) \cap \{\mathbf{G_{s-r}^-}\}$ *is given by* $\mathbf{A} + \mathbf{R}\begin{pmatrix}\mathbf{0}\\ \mathbf{I_{s-r}}\\ \mathbf{M}\end{pmatrix}\left(\mathbf{0}\ \mathbf{I_{s-r}}\ \mathbf{L}\right)\mathbf{S}$ *where* $\mathbf{L}$ *and* $\mathbf{M}$ *are arbitrary.*

Proof. $\mathbf{A_{s-r}} \in \{\mathbf{G^-_{s-r}}\}$ if and only if

$$\mathbf{A_{s-r}} = \mathbf{R}\begin{pmatrix} \mathbf{I_r} & \mathbf{0} & \mathbf{J_1} \\ \mathbf{0} & \mathbf{I_{s-r}} & \mathbf{L} \\ \mathbf{J_2} & \mathbf{M} & \mathbf{ML} + \mathbf{J_1J_2} \end{pmatrix}\mathbf{S}$$

for some matrices $\mathbf{J_1}$, $\mathbf{J_2}$, $\mathbf{L}$ and $\mathbf{M}$. Now, $\mathbf{G_0A_0} = \mathbf{G_0A_{s-r}}$ if and only if $\mathbf{J_1} = \mathbf{0}$. Further, $\mathbf{A_0G_0} = \mathbf{A_{s-r}G_0}$ if and only if $\mathbf{J_2} = \mathbf{0}$. Thus,

$$\mathbf{A_{s-r}} \in \overline{\mathrm{Eq}}(\mathbf{A}|\mathbf{G_{s-r}}) \cap \mathbf{A_{s-r}} \in \{\mathbf{G^-_{s-r}}\} \Leftrightarrow \mathbf{A_{s-r}} = \mathbf{R}\begin{pmatrix} \mathbf{I_r} & \mathbf{0} & \mathbf{0} \\ \mathbf{0} & \mathbf{I_{s-r}} & \mathbf{L} \\ \mathbf{0} & \mathbf{M} & \mathbf{ML} \end{pmatrix}\mathbf{S}$$

for some matrices $\mathbf{L}$ and $\mathbf{M}$ or equivalently, $\mathbf{A_{s-r}} = \mathbf{A} + \mathbf{B}$, where $\mathbf{B} = \mathbf{R}\begin{pmatrix} \mathbf{0} \\ \mathbf{I_{s-r}} \\ \mathbf{M} \end{pmatrix}(\mathbf{0}\ \mathbf{I_{s-r}}\ \mathbf{L})\,\mathbf{S}$ for some matrices $\mathbf{L}$ and $\mathbf{M}$. □

The matrix $\mathbf{A_{s-r}}$ of Theorem 14.2.12 admits an alternative expression, which we give in our next theorem.

Theorem 14.2.13. *Consider the same setup as in Theorem 14.2.12. The class of all matrices* $\mathbf{A_{s-r}} \in \overline{\mathrm{Eq}}(\mathbf{A}|\mathbf{G_0}) \cap \{\mathbf{G^-_{s-r}}\}$ *is given by*

$$\mathbf{A_{s-r}} = \mathbf{A} + (\mathbf{I} - \mathbf{AG_0})(\mathbf{G_{s-r}} - \mathbf{G_0})^-{}_{\mathbf{r}}(\mathbf{I} - \mathbf{G_0A}).$$

Proof. Let $\mathbf{R}, \mathbf{S}, \mathbf{L}, \mathbf{M}$ and $\mathbf{B}$ be as in the proof of Theorem 14.2.12. Notice that $\mathbf{G_0R}\begin{pmatrix} 0 \\ \mathbf{I_{s-r}} \\ \mathbf{M} \end{pmatrix} = \mathbf{0}$ and $(\mathbf{0}\ \mathbf{I_{s-r}}\ \mathbf{L})\,\mathbf{SG_0} = \mathbf{0}$ for all matrices $\mathbf{L}$ and $\mathbf{M}$. Hence, $\mathcal{C}(\mathbf{B}) \subseteq \mathcal{C}(\mathbf{I} - \mathbf{AG_0})$ and $\mathcal{C}(\mathbf{B^t}) \subseteq \mathcal{C}(\mathbf{I} - \mathbf{G_0A})$. The rest is computational. □

Several important remarks are in order. Their importance lies in the fact that they will be instrumental in designing the diagrammatic representation of the generalized inverses of a given matrix.

Remark 14.2.14. Notice that $\rho(\mathbf{A_{s-r}}) = \rho(\mathbf{A}) + \rho(\mathbf{B})$.

Remark 14.2.15. In Theorem 14.2.13, if we let

$$\begin{pmatrix} \mathbf{0} \\ \mathbf{I_{s-r}} \\ \mathbf{M} \end{pmatrix} = (\mathbf{u_1} : \ldots : \mathbf{u_{s-r}}) \text{ and } (\mathbf{0},\ \mathbf{I_{s-r}},\ \mathbf{L}) = (\mathbf{v_1^t} : \ldots : \mathbf{v^t_{s-r}}),$$

then $\mathbf{B} = \mathbf{Ru_1v_1^tS} \oplus \ldots \oplus \mathbf{Ru_{s-r}v^t_{s-r}S}$.

Remark 14.2.16. Let $\mathbf{G_i} = \mathbf{S}^{-1}\mathbf{diag}\,(\mathbf{I_{r+i}}\ ,\ \mathbf{0})\mathbf{R}^{-1}$ and $\mathbf{A_i} = \mathbf{A} + \sum_{j=1}^{i} \mathbf{R u_j v_j^t S}$. Then $\mathbf{A_i} \in \overline{\text{Eq}}(\mathbf{A}|\mathbf{G_0}) \cap \{(\mathbf{G_i^-})_r\}$, $i = 1, \ldots, s-r$.

Remark 14.2.17. Write $\mathbf{A_0} = \mathbf{A}$. Then for each $i = 0, 1, \ldots, s-r-1$, we have $\overline{\text{Eq}}(\mathbf{G_{i+1}}|\mathbf{A_{i+1}}) \subseteq \overline{\text{Eq}}(\mathbf{G_i}|\mathbf{A_i})$, where $\mathbf{A_i}$ and $\mathbf{G_i}$ are defined in Remark 14.2.16.

Remark 14.2.18. Notice that s can take values from $r+1$ to $min\{m, n\}$.

Remark 14.2.19. We also have $\mathbf{G_i} <^- \mathbf{G_{i+1}}$ and $\mathbf{A_i} <^- \mathbf{A_{i+1}}$ for each $i = 0, 1, \ldots, s-r-1$.

Consider an $m \times n$ matrix $\mathbf{A}$. Each equivalence class of g-inverses under the equivalence relation as in Definition 14.2.1 is triggered by a reflexive g-inverse of $\mathbf{A}$. Let us ask the following question:
Given a pair of equivalence classes, is it possible for some member of one to dominate a member of the other?

The answer to this question is in affirmative. In fact, given a g-inverse $\mathbf{G}$, we shall determine the set all equivalence classes such that each one of them contains a g-inverse $\mathbf{G_1}$ of $\mathbf{A}$ such that $\mathbf{G} <^- \mathbf{G_1}$ and also characterize all such g-inverses $\mathbf{G_1}$.

Theorem 14.2.20. *Let* $\mathbf{A}$ *be an* $m \times n$ *matrix of rank* r. *Let* $\mathbf{G_0}$ *be a reflexive g-inverse of* $\mathbf{A}$ *and* $\mathbf{G_1}$ *be a g-inverse of* $\mathbf{A}$ *of rank* s $(r < s < min\{m, n\})$ *such that* $\mathbf{G_1} \in \overline{\text{Eq}}(\mathbf{G_0}|\mathbf{A})$. *Let* $\mathbf{P}$ *and* $\mathbf{Q}$ *be non-singular matrices such that* $\mathbf{A} = \mathbf{P}\mathbf{diag}\,(\mathbf{I_r}\ ,\ \mathbf{0}\ ,\ \mathbf{0})$, $\mathbf{G_0} = \mathbf{Q}^{-1}\mathbf{diag}\,(\mathbf{I_r}\ ,\ \mathbf{0}\ ,\ \mathbf{0})\,\mathbf{P}^{-1}$ *and* $\mathbf{G_1} = \mathbf{Q}^{-1}\mathbf{diag}\,(\mathbf{I_r}\ ,\ \mathbf{I_{s-r}}\ ,\ \mathbf{0})\,\mathbf{P}^{-1}$. *(This is guaranteed by Remark 14.2.3). Let* $\mathbf{G_2} = \mathbf{Q}^{-1}\begin{pmatrix}\mathbf{I_r} & \mathbf{L_1} & \mathbf{L_2}\\ \mathbf{M_1} & \mathbf{M_1L_1} & \mathbf{M_1L_2}\\ \mathbf{M_2} & \mathbf{M_2L_1} & \mathbf{M_2L_2}\end{pmatrix}\mathbf{P}^{-1}$ *be reflexive g-inverse of* $\mathbf{A}$, *where* $\mathbf{L_1}$ *is of order* $r \times (s-r)$, $\mathbf{M_1}$ *is of order* $(s-r) \times r$ *and the partition is determined accordingly. Then there exists a* $\mathbf{G_3} \in \overline{\text{Eq}}(\mathbf{G_2}|\mathbf{A})$ *such that* $\mathbf{G_1} <^- \mathbf{G_3}$ *if and only if* $\mathcal{C}(\mathbf{L_1}) \subseteq \mathcal{C}(\mathbf{L_2})$, $\mathcal{C}(\mathbf{M_1^t}) \subseteq \mathcal{C}(\mathbf{M_2^t})$ *and* $\rho(\mathbf{L_2}) + \rho(\mathbf{M_2}) \leq min\{m-s, n-s\}$. *(Notice that every reflexive g-inverse of* $\mathbf{A}$ *is of the form given above.)*

Proof. By Theorem 14.2.8, $\mathbf{G_3} \in \overline{\text{Eq}}(\mathbf{G_2}|\mathbf{A})$ if and only if for some matrix $\mathbf{S} = \begin{pmatrix} \mathbf{S_{11}} & \mathbf{S_{12}} \\ \mathbf{S_{21}} & \mathbf{S_{22}} \end{pmatrix}$,

$$\mathbf{G_3} = \mathbf{Q}^{-1} \begin{pmatrix} \mathbf{I_r} & \mathbf{L_1} & \mathbf{L_2} \\ \mathbf{M_1} & \mathbf{M_1L_1} + \mathbf{S_{11}} & \mathbf{M_1L_2} + \mathbf{S_{12}} \\ \mathbf{M_2} & \mathbf{M_2L_1} + \mathbf{S_{21}} & \mathbf{M_2L_2} + \mathbf{S_{22}} \end{pmatrix} \mathbf{P}^{-1}.$$

Now, $\mathbf{G_1} <^- \mathbf{G_3}$ if and only if there exists a reflexive g-inverse $\mathbf{H}$ of $\mathbf{G_1}$ such that $\mathbf{G_1H} = \mathbf{G_3H}$ and $\mathbf{HG_1} = \mathbf{HG_3}$. Notice that every reflexive g-inverse of $\mathbf{G_1}$ is of the form $\mathbf{H} = \mathbf{P} \begin{pmatrix} \mathbf{I_r} & \mathbf{0} & \mathbf{E_1} \\ \mathbf{0} & \mathbf{I_{s-r}} & \mathbf{E_2} \\ \mathbf{F_1} & \mathbf{F_2} & \mathbf{F_1E_1} + \mathbf{F_2E_2} \end{pmatrix} \mathbf{Q}$ for some matrices $\mathbf{E_1}, \mathbf{E_2}, \mathbf{F_1}$ and $\mathbf{F_2}$ of appropriate orders. It is easy to check that

$$\mathbf{G_1H} = \mathbf{G_3H} \Leftrightarrow \begin{cases} \mathbf{L_2F_1} = \mathbf{0} & (14.2.1) \\ \mathbf{L_2} = -\mathbf{L_2F_2} & (14.2.2) \\ \mathbf{M_1} = -\mathbf{S_{12}F_1} & (14.2.3) \\ \mathbf{S_{11}} + \mathbf{S_{12}F_2} = \mathbf{I} & (14.2.4) \\ \mathbf{M_2} = -\mathbf{S_{22}F_1} & (14.2.5) \\ \mathbf{S_{21}} = -\mathbf{S_{22}F_2} & (14.2.6) \end{cases}$$

$$\mathbf{HG_1} = \mathbf{HG_3} \Leftrightarrow \begin{cases} \mathbf{E_1M_2} = \mathbf{0} & (14.2.7) \\ \mathbf{L_1} = -\mathbf{E_1S_{21}} & (14.2.8) \\ \mathbf{L_2} = -\mathbf{E_1S_{22}} & (14.2.9) \\ \mathbf{M_1} = -\mathbf{E_2M_2} & (14.2.10) \\ \mathbf{S_{11}} + \mathbf{E_2S_{21}} = \mathbf{I} & (14.2.11) \\ \mathbf{S_{12}} = -\mathbf{E_2S_{22}}. & (14.2.12) \end{cases}$$

'Only if' part

(14.2.2) and (14.2.10) imply $\mathcal{C}(\mathbf{L_1}) \subseteq \mathcal{C}(\mathbf{L_2})$, $\mathcal{C}(\mathbf{M_1}^t) \subseteq \mathcal{C}(\mathbf{M_2^t})$ respectively. Since $\mathbf{L_2F_1} = \mathbf{0}$, it follows that $\rho(\mathbf{F_1}) \leq n - s - \rho(\mathbf{L_2})$. Since $\mathbf{M_2} = -\mathbf{S_{22}F_1}$, it follows that $\rho(\mathbf{M_2}) \leq \rho(\mathbf{F_1}) \leq n - s - \rho(\mathbf{L_2})$. Similarly, from (14.2.7) and (14.2.9) it follows that $\rho(\mathbf{L_2}) \leq m - s - \rho(\mathbf{M_2})$. Thus, $\rho(\mathbf{L_2}) + \rho(\mathbf{M_2}) \leq min\{m - s, n - s\}$.

'If' part

We first exhibit $\mathbf{S_{22}}, \mathbf{F_1}$ and $\mathbf{E_1}$ satisfying equations (14.2.5), (14.2.9), (14.2.1) and (14.2.7). In view of the equations (14.2.5) and (14.2.9), the equations (14.2.1) and (14.2.7) are the same. Let $\rho(\mathbf{L_2}) = a$ and $\rho(\mathbf{M_2}) = b$. We know that $a + b \leq min\{m - s, n - s\}$. Let $\mathbf{M_2} = \mathbf{H_1}\mathbf{diag}\left(\mathbf{I_a}\ ,\ \mathbf{0}\right)\mathbf{H_2}$ and $\mathbf{L_2} = \mathbf{J_1}\mathbf{diag}\left(\mathbf{I_b}\ ,\ \mathbf{0}\right)\mathbf{J_2}$ be normal forms of $\mathbf{M_2}$ and $\mathbf{L_2}$ respectively, where $\mathbf{H_1}, \mathbf{H_2}, \mathbf{J_1}$ and $\mathbf{J_2}$ are non-singular matrices of appropriate orders.

$\mathbf{M_2}$ is of order $(n-s)\times r$ and $\mathbf{L_2}$ is of order $r\times(n-s)$. Consider

$$\mathbf{S_{22}} = \mathbf{H_1}\begin{pmatrix} \mathbf{0} & \mathbf{I_a} & \mathbf{0} \\ \mathbf{I_b} & \mathbf{0} & \mathbf{0} \\ \mathbf{0} & \mathbf{0} & \mathbf{0} \end{pmatrix}\mathbf{J_2}, \mathbf{F_1} = \mathbf{J_2}^{-1}\begin{pmatrix} \mathbf{0} & \mathbf{0} & \mathbf{0} \\ -\mathbf{I_a} & \mathbf{0} & \mathbf{0} \\ \mathbf{0} & \mathbf{0} & \mathbf{0} \end{pmatrix}\mathbf{H_2} \text{ and}$$

$$\mathbf{E_1} = \mathbf{J_1}\begin{pmatrix} \mathbf{0} & -\mathbf{I_b} & \mathbf{0} \\ \mathbf{0} & \mathbf{0} & \mathbf{0} \\ \mathbf{0} & \mathbf{0} & \mathbf{0} \end{pmatrix}\mathbf{H_1}^{-1}.$$

Notice that $\mathbf{S_{22}}$ is of order $(n-s)\times(m-s)$. Since

$$a+b\leq min\{m-s,n-s\},$$

the above construction of $\mathbf{S_{22}}$ is possible. It is easy to verify that $\mathbf{S_{22}}, \mathbf{F_1}$ and $\mathbf{E_1}$ as constructed above satisfy the equations (14.2.1), (14.2.5), (14.2.7) and (14.2.9). Since $\mathcal{C}(\mathbf{L_1})\subseteq\mathcal{C}(\mathbf{L_2})$, there exists an $\mathbf{F_2}$ satisfying (14.2.7). Construct $\mathbf{S_{21}} = -\mathbf{S_{22}F_2}$. So, (14.2.6) holds. Also, $-\mathbf{E_1S_{21}} = \mathbf{E_1S_{22}F_2} = -\mathbf{L_2F_2} = \mathbf{L_1}$, thus, (14.2.8) is also satisfied. Since $\mathcal{C}(\mathbf{M_1^t})\subseteq\mathcal{C}(\mathbf{M_2^t})$, there exists a matrix $\mathbf{E_2}$ satisfying (14.2.10). Let $\mathbf{S_{12}} = -\mathbf{E_2S_{22}}$. Then (14.2.12) is satisfied. By post-multiplying (14.2.12) by F_1 we have, $\mathbf{S_{12}F_1} = -\mathbf{E_2S_{22}F_1} = -\mathbf{E_2M_2} = \mathbf{M_1}$. Thus, (14.2.3) holds. Now, $\mathbf{E_2S_{21}} = -\mathbf{E_2S_{22}F_2} = \mathbf{S_{12}F_2}$, we let $\mathbf{S_{11}} = \mathbf{I}-\mathbf{S_{12}F_2} = \mathbf{I}-\mathbf{E_2S_{21}}$. Then both (14.2.4) and (14.2.11) are satisfied. Thus, we have exhibited $\mathbf{S_{22}}, \mathbf{F_1}$ and $\mathbf{F_2}$, where $\mathbf{F_1}$ and $\mathbf{E_2}$ are such that the equations (14.2.1)- (14.2.12) are all satisfied. Hence, $\mathbf{G_1} <^- \mathbf{G_3}$. In the process, we have also exhibited a reflexive g-inverse $\mathbf{H}$ such that $\mathbf{G_1H} = \mathbf{G_3H}$ and $\mathbf{HG_1} = \mathbf{HG_3}$. □

In the setup of Theorem 14.2.20, let $\mathcal{C}(\mathbf{L_1})\subseteq\mathcal{C}(\mathbf{L_2})$, $\mathcal{C}(\mathbf{M_1}^t)\subseteq\mathcal{C}(\mathbf{M_2}^t)$ and $\rho(\mathbf{L_2})+\rho(\mathbf{M_2})\leq min\{m-s,n-s\}$. We now obtain the class of all g-inverses $\mathbf{G_3}\in\overline{\mathrm{Eq}}(\mathbf{G_2}|\mathbf{A})$ such that $\mathbf{G_1} <^- \mathbf{G_3}$.

Theorem 14.2.21. *Consider the setup of Theorem 14.2.20. Let* $\mathcal{C}(\mathbf{L_1})\subseteq\mathcal{C}(\mathbf{L_2})$, $\mathcal{C}(\mathbf{M_1^t})\subseteq\mathcal{C}(\mathbf{M_2^t})$ *and* $\rho(\mathbf{L_2})+\rho(\mathbf{M_2})\leq min\{m-s,n-s\}$. *Let* $\mathbf{M_2} = \mathbf{H_1}\mathbf{diag}(\mathbf{I_a}, \mathbf{0})\mathbf{H_2}$ *and* $\mathbf{L_2} = \mathbf{J_1}\mathbf{diag}(\mathbf{I_b}, \mathbf{0})\mathbf{J_2}$ *be normal forms of* $\mathbf{M_2}$ *and* $\mathbf{L_2}$ *respectively, where* $\mathbf{H_1}, \mathbf{H_2}, \mathbf{J_1}$ *and* $\mathbf{J_2}$ *are non-singular matrices of appropriate orders. Then the class of all* $\mathbf{G_3}\in\overline{\mathrm{Eq}}(\mathbf{G_2}|\mathbf{A})$ *such that* $\mathbf{G_1} <^- \mathbf{G_3}$ *is given by* $\mathbf{G_3} = \mathbf{G_2}+\mathbf{Q}^{-1}\begin{pmatrix} \mathbf{0} & \mathbf{0} & \mathbf{0} \\ \mathbf{0} & \mathbf{S_{11}} & \mathbf{S_{12}} \\ \mathbf{0} & \mathbf{S_{21}} & \mathbf{S_{22}} \end{pmatrix}\mathbf{P}^{-1}$,

where (i) $\mathbf{S_{22}} = \mathbf{H_1}\begin{pmatrix} \mathbf{B_{11}} & \mathbf{B_{12}} \\ \mathbf{B_{21}} & \mathbf{B_{22}} \end{pmatrix}\mathbf{H_2}$ *with* $\mathbf{B_{11}}, \mathbf{B_{12}}$ *and* $\mathbf{B_{21}}$ *arbitrary matrices of order* $a\times b, a\times(m-s-b), (n-s-a)\times b$ *respectively and*

$\mathbf{B_{22}} = (\mathbf{I} - \mathbf{Y_2^-Y_2})\mathbf{D}(\mathbf{I} - \mathbf{W_2W_2^-}), \mathbf{D}, \mathbf{Y_2}$ *and* $\mathbf{W_2}$ *arbitrary solutions to* $\mathbf{Y_2B_{21}} = \begin{pmatrix} \mathbf{I_b} \\ \mathbf{0} \end{pmatrix}$ *and* $\mathbf{B_{21}W_2} = \begin{pmatrix} \mathbf{I_a} & \mathbf{0} \end{pmatrix}$.

(ii) $\mathbf{S_{21}} = \mathbf{S_{22}}(\mathbf{L_2^-L_1} + (\mathbf{I} - \mathbf{L_2^-L_2})\mathbf{Z})$, *where* $\mathbf{L_2^-}$ *is some g-inverse of* $\mathbf{L_2}$ *and* $\mathbf{Z}$ *is arbitrary.*

(iii) $\mathbf{S_{12}} = (\mathbf{M_1M_2^-} + \mathbf{T}(\mathbf{I} - \mathbf{M_2M_2^-})\mathbf{Z})\mathbf{S_{22}}$, *where* $\mathbf{M_2^-}$ *is some g-inverse of* $\mathbf{M_2}$ *and* $\mathbf{T}$ *is arbitrary.*

(iv) $\mathbf{S_{11}} = \mathbf{I} + (\mathbf{M_1M_2^-} + \mathbf{T}(\mathbf{I} - \mathbf{M_2M_2^-})\mathbf{Z})\mathbf{S_{22}}(\mathbf{L_2^-L_1} + (\mathbf{I} - \mathbf{L_2^-L_2})\mathbf{Z})$.

Proof. We first show that if $\rho(\mathbf{L_2}) + \rho(\mathbf{M_2}) \leq min\{m-s, n-s\}$, then there exist matrices $\mathbf{S_{22}}, \mathbf{E_1}$ and $\mathbf{F_1}$ satisfying the equations:

$$\mathbf{M_2} = -\mathbf{S_{22}F_1},\ \mathbf{L_2} = -\mathbf{E_1S_{22}} \text{ and } \mathbf{E_1S_{22}F_1} = \mathbf{0}.$$

Let $\rho(\mathbf{M_2}) = a$ and $\rho(\mathbf{M_2}) = b$. We know that $a + b \leq min\{m-s, n-s\}$. Let $\mathbf{M_2} = \mathbf{H_1}\mathbf{diag}\left(\mathbf{I_a}, \mathbf{0}\right)\mathbf{H_2}$ and $\mathbf{L_2} = \mathbf{J_1}\mathbf{diag}\left(\mathbf{I_b}, \mathbf{0}\right)\mathbf{J_2}$ be normal forms of $\mathbf{M_2}, \mathbf{L_2}$ respectively, where $\mathbf{H_1}, \mathbf{H_2}, \mathbf{J_1}$ and $\mathbf{J_2}$ are non-singular matrices of appropriate orders. Write $\mathbf{S_{22}} = \mathbf{H_1}\begin{pmatrix} \mathbf{B_{11}} & \mathbf{B_{12}} \\ \mathbf{B_{21}} & \mathbf{B_{22}} \end{pmatrix}\mathbf{J_2}$ and $\mathbf{F_1} = \mathbf{J_2}^{-1}\begin{pmatrix} \mathbf{W_{11}} & \mathbf{W_{12}} \\ \mathbf{W_{21}} & \mathbf{W_{22}} \end{pmatrix}\mathbf{H_2}$ and $\mathbf{E_1} = \mathbf{J_1}\begin{pmatrix} \mathbf{Y_{11}} & \mathbf{Y_{12}} \\ \mathbf{Y_{21}} & \mathbf{Y_{22}} \end{pmatrix}\mathbf{H_1}^{-1}$. Thus, by equation (14.2.1), $\mathbf{L_2F_1} = \mathbf{0} \Leftrightarrow \mathbf{J_1}\begin{pmatrix} \mathbf{I_b} & \mathbf{0} \\ \mathbf{0} & \mathbf{0} \end{pmatrix}\mathbf{J_2J_2^{-1}}\begin{pmatrix} \mathbf{W_{11}} & \mathbf{W_{12}} \\ \mathbf{W_{21}} & \mathbf{W_{22}} \end{pmatrix}\mathbf{H_2} = \mathbf{0}$ or $\mathbf{W_{11}} = \mathbf{0}$ and $\mathbf{W_{12}} = \mathbf{0}$. Similarly, equation (14.2.7) yields $\mathbf{Y_{11}} = \mathbf{0}$ and $\mathbf{Y_{21}} = \mathbf{0}$. Now, $\mathbf{M_2} = -\mathbf{S_{22}F_1} \Leftrightarrow \mathbf{B_{12}W_{21}} = \mathbf{I_a}, \mathbf{B_{12}W_{22}} = \mathbf{0}$ and $\mathbf{B_{22}}\begin{pmatrix} \mathbf{W_{21}} & \mathbf{W_{22}} \end{pmatrix} = \mathbf{0}$. Clearly, $\mathbf{B_{12}}$ is an $a \times ((n-s)-b)$ matrix. Since $a < (n-s)-b$, we can find a matrix $\mathbf{B_{12}}$ of rank a. Let $\mathbf{B_{12}}$ be any arbitrary matrix of order $a \times ((n-s)-b)$ such that $\rho(\mathbf{B_{12}}) = a$. Choose and fix a $\mathbf{B_{12}}$. The matrix $\mathbf{W_{21}}$ is of order $(n-s-b) \times a$ and $\mathbf{W_{22}}$ is of order $(n-s-b) \times (r-a)$. Let $\mathbf{W_{21}}$ be an arbitrary right inverse of $\mathbf{B_{12}}$ and $\mathbf{W_{22}} = (\mathbf{I} - \mathbf{W_{21}B_{12}})\mathbf{Z}$, where $\mathbf{Z}$ is an arbitrary matrix of order $(n-s-b) \times (r-a)$. $\mathbf{B_{22}}$ is an arbitrary matrix such that $\mathcal{C}(\mathbf{B_{22}^t}) \subseteq \mathcal{N}(\mathbf{W_{21}W_{22}})^{\mathbf{t}}$. Also, $\mathbf{L_2} = -\mathbf{E_1S_{22}} \Leftrightarrow \mathbf{Y_{12}B_{21}} = -\mathbf{I_b}, \mathbf{Y_{12}B_{22}} = \mathbf{0}, \mathbf{Y_{22}B_{21}} = \mathbf{0}$ and $\mathbf{Y_{22}B_{22}} = \mathbf{0}$. Notice that $\mathbf{B_{21}}$ is a matrix of order $(m-s-a) \times b$. Since $b \leq m-s-a$, there exists a matrix $\mathbf{B_{21}}$ of order $(m-s-a) \times b$ with rank b. Let $\mathbf{B_{21}}$ be an arbitrary matrix of order $(m-s-a) \times b$ with rank b. Choose and fix $\mathbf{B_{21}}$. Then $\mathbf{Y_{12}}$ is an arbitrary left inverse of $-\mathbf{B_{21}}$ and $\mathbf{Y_{22}} = \mathbf{R}(\mathbf{I} - \mathbf{B_{21}Y_{12}})$, where $\mathbf{R}$ is an arbitrary matrix of appropriate order. Then $\mathbf{B_{22}}$ must satisfy $\begin{pmatrix} \mathbf{Y}_{12} \\ \mathbf{Y}_{22} \end{pmatrix} \mathbf{B}_{22} = \mathbf{0}$. Thus, $\mathbf{B}_{22}$ is an arbitrary solution of the matrix equation $\begin{pmatrix} \mathbf{Y}_{12} \\ \mathbf{Y}_{22} \end{pmatrix} \mathbf{B}_{22} = \mathbf{0}$ and $\mathbf{B}_{22}(\mathbf{W}_{21}\mathbf{W}_{22}) = \mathbf{0}$, which is given by $\mathbf{B}_{22} = (\mathbf{I} - \begin{pmatrix} \mathbf{Y}_{12} \\ \mathbf{Y}_{22} \end{pmatrix}^{-} \begin{pmatrix} \mathbf{Y}_{12} \\ \mathbf{Y}_{22} \end{pmatrix})\mathbf{D}(\mathbf{I} - (\mathbf{W}_{21}\ , \mathbf{W}_{22})(\mathbf{W}_{21}\ , \mathbf{W}_{22})^{-})$, where $\mathbf{D}$ is arbitrary and "$\begin{pmatrix} \mathbf{Y}_{12} \\ \mathbf{Y}_{22} \end{pmatrix}^{-}$ and $(\mathbf{W}_{21}\ , \mathbf{W}_{22})^{-}$" are arbitrary g-inverses of $\begin{pmatrix} \mathbf{Y}_{12} \\ \mathbf{Y}_{22} \end{pmatrix}$ and $(\mathbf{W}_{21}\ , \mathbf{W}_{22})$ respectively. The matrix $\mathbf{B}_{11}$ is arbitrary. The rest of the proof is now easy. □

Consider an $m \times n$ matrix $\mathbf{A}$ and an arbitrary reflexive g-inverse $\mathbf{G}_0$ of $\mathbf{A}$. Does the equivalence class $\overline{\mathrm{Eq}}(\mathbf{G}_0|\mathbf{A})$ determine $\mathbf{A}$ in the sense that if $\overline{\mathrm{Eq}}(\mathbf{G}_0|\mathbf{A}) = \overline{\mathrm{Eq}}(\mathbf{G}_0|\mathbf{B})$ for some matrix $\mathbf{B}$ having $\mathbf{G}_0$ as a g-inverse, should we have $\mathbf{A} = \mathbf{B}$? This indeed is so, as shown by the following theorem:

Theorem 14.2.22. *Let* $\mathbf{A}$ *and* $\mathbf{B}$ *be matrices of the same order and same rank* r. *Let* $\mathbf{A}$ *and* $\mathbf{B}$ *have a common reflexive g-inverse* $\mathbf{G}_0$ *such that* $\overline{\mathrm{Eq}}(\mathbf{G}_0|\mathbf{A}) = \overline{\mathrm{Eq}}(\mathbf{G}_0|\mathbf{B})$. *Then* $\mathbf{A} = \mathbf{B}$.

Proof. By Theorem 2.3.18, there exist non-singular matrices $\mathbf{P}$ and $\mathbf{Q}$ such that $\mathbf{A} = \mathbf{P}\mathbf{diag}\left(\mathbf{I_r}\ , \mathbf{0}\right)\mathbf{Q}$ and $\mathbf{G}_0 = \mathbf{Q}^{-1}\mathbf{diag}\left(\mathbf{I_r}\ , \ \mathbf{0}\right)\mathbf{P}^{-1}$. By Corollary 14.2.7, $\mathbf{G} \in \overline{\mathrm{Eq}}(\mathbf{G}_0|\mathbf{A})$ if and only if $\mathbf{G} = \mathbf{Q}^{-1}\mathbf{diag}\left(\mathbf{I_r}\ , \ \mathbf{S}\right)\mathbf{P}^{-1}$ for some matrix $\mathbf{S}$. Since $\mathbf{G}_0$ is reflexive g-inverse of $\mathbf{B}$, we can write $\mathbf{B} = \mathbf{P}\begin{pmatrix} \mathbf{I_r} & \mathbf{T}_1 \\ \mathbf{T}_2 & \mathbf{T}_2\mathbf{T}_1 \end{pmatrix}\mathbf{Q}$ for some matrices $\mathbf{T}_1$ and $\mathbf{T}_2$. Since $\mathbf{G} \in \overline{\mathrm{Eq}}(\mathbf{G}_0|\mathbf{B})$, we have $\mathbf{B}\mathbf{G}_0 = \mathbf{B}\mathbf{G}$ and $\mathbf{G}_0\mathbf{B} = \mathbf{G}\mathbf{B}$. However, $\mathbf{B}\mathbf{G}_0 = \mathbf{P}\begin{pmatrix} \mathbf{I_r} & \mathbf{0} \\ \mathbf{T}_2 & \mathbf{0} \end{pmatrix}\mathbf{P}^{-1}$ and $\mathbf{B}\mathbf{G} = \mathbf{P}\begin{pmatrix} \mathbf{I_r} & \mathbf{T}_1\mathbf{S} \\ \mathbf{T}_2 & \mathbf{T}_2\mathbf{T}_1\mathbf{S} \end{pmatrix}\mathbf{P}^{-1}$, so, $\mathbf{B}\mathbf{G}_0 = \mathbf{B}\mathbf{G}$ for all $\mathbf{G} \in \overline{\mathrm{Eq}}(\mathbf{G}_0|\mathbf{A})$. So, $\mathbf{T}_1\mathbf{S} = \mathbf{0}$ for all $\mathbf{S}$ and therefore, $\mathbf{T}_1 = \mathbf{0}$. Similarly, $\mathbf{G}_0\mathbf{B} = \mathbf{G}\mathbf{B}$ for all $\mathbf{G} \in \overline{\mathrm{Eq}}(\mathbf{G}_0|\mathbf{A}) \Rightarrow \mathbf{T}_2 = \mathbf{0}$. Hence, $\mathbf{A} = \mathbf{B}$. □

14.3 Equivalence relations on subclasses of g-inverses

We begin this section by identifying the equivalence classes under the equivalence relation "$\backsim$" that contain two of the special reflexive g-inverses namely the Moore-Penrose inverse and the group inverse.

Theorem 14.3.1. (a) *Let* $\mathbf{A} \in \mathbb{C}^{\mathbf{m}\times\mathbf{n}}$ *such that* $\rho(\mathbf{A}) = r$ $(0 < r < min\{m,n\})$. *Then an* $n \times m$ *matrix* $\mathbf{G} \in \overline{\mathrm{Eq}}(\mathbf{A}^{\dagger}|\mathbf{A})$ *if and only if* $\mathbf{G} \in \{\mathbf{A}^{-}_{\ell\mathbf{m}}\}$.
(b) *Let* $\mathbf{A}$ *be an* $n \times n$ *matrix of index* ≤ 1 *with rank* r $(0 < r < n)$. *Then an* $n \times n$ *matrix* $\mathbf{G} \in \overline{\mathrm{Eq}}(\mathbf{A}^{\#}|\mathbf{A})$ *if and only if* $\mathbf{G} \in \{\mathbf{A}^{-}_{\mathbf{com}}\}$.

Proof follows easily from Theorem 14.2.6.

Thus, the equivalence class of $\mathbf{A}^{\dagger}$, the Moore-Penrose inverse of a matrix $\mathbf{A}$ consists of the class of all minimum norm least squares g-inverses of $\mathbf{A}$. Similarly the equivalence class of $\mathbf{A}^{\#}$, the group inverse of $\mathbf{A}$ consists of all commuting g-inverses of $\mathbf{A}$. It is therefore natural for us to think about other subclasses of g-inverses like $\{\mathbf{A}^{-}_{\ell}\}$, the class of all least squares g-inverses, $\{\mathbf{A}^{-}_{\mathbf{m}}\}$, the class of minimum norm g-inverses, $\{\mathbf{A}^{-}_{\ell\mathbf{m}}\}$, the class of minimum norm least squares g-inverses, $\{\mathbf{A}^{-}_{\mathbf{c}}\}$, $\{\mathbf{A}^{-}_{\mathbf{a}}\}$ and $\{\mathbf{A}^{-}_{\mathbf{com}}\}$, the commuting g-inverses and their status in relation to the equivalence relation "$\backsim$". If we consider any of these classes of g-inverses and define relations on them in a manner analogous to the Definition 14.2.1, what are the inter-relationships among the equivalence classes of these relations? It turns out that the relations defined are each an equivalence relation and several results concerning them are similar to those of the relation "$\backsim$."

For the study of equivalence relations on the classes of all least square g-inverses, the class of minimum norm g-inverses, and the class of minimum norm least square g-inverses, we consider the matrices over the field of complex numbers $\mathbb{C}$. For the study of equivalence relations on the other subclasses, we consider square matrices of index ≤ 1 over a general field.

We start with the study of the equivalence relations on the classes of the least squares g-inverses and minimum norm g-inverses. We present the results corresponding to these relations side by side because of the duality between these two classes (see Theorem 2.5.15).

Definition 14.3.2. Let $\mathbf{A} \in \mathbb{C}^{\mathbf{m}\times\mathbf{n}}$ and $\mathbf{G_1}$ and $\mathbf{G_2} \in \{\mathbf{A}^{-}_{\ell}\}$. The relation "$\backsim_{\ell}$" on $\{\mathbf{A}^{-}_{\ell}\}$ defined as $\mathbf{G_1} \backsim_{\ell} \mathbf{G_2}$ if $\mathbf{AG_1} = \mathbf{AG_2}$ and $\mathbf{G_1A} = \mathbf{G_2A}$ (or equivalently if $\mathbf{G_1AG_1} = \mathbf{G_2AG_2}$).

Similarly the relation "$\backsim_m$" on $\{\mathbf{A}^{-}_{\mathbf{m}}\}$ is defined.

Clearly, the relations "$\backsim_{\ell}$" and "$\backsim_m$" are equivalence relations on $\{\mathbf{A}^{-}_{\ell}\}$ and $\{\mathbf{A}^{-}_{\mathbf{m}}\}$ respectively. Further there is a unique reflexive g-inverse in each equivalence class corresponding to each of these relations. We denote the equivalence class containing the least squares reflexive g-inverse $\mathbf{G_0}$ by

$\overline{\mathrm{Eq}_\ell}(\mathbf{G}|\mathbf{A})$. $\overline{\mathrm{Eq}_m}(\mathbf{G}|\mathbf{A})$ is similarly defined. Notice that both $\overline{\mathrm{Eq}_\ell}(\mathbf{G}|\mathbf{A})$ and $\overline{\mathrm{Eq}_m}(\mathbf{G}|\mathbf{A})$ are subsets of the equivalence class $\overline{\mathrm{Eq}}(\mathbf{G}|\mathbf{A})$. Before we elaborate on properties of these relations we note the following:

Theorem 14.3.3. *Let $\mathbf{A} \in \mathbb{C}^{\mathbf{m}\times\mathbf{n}}$ and $\mathfrak{T}$ be any equivalence class of $\{\mathbf{A}^-\}$ under the relation "$\backsim$". If $\mathfrak{T}$ contains a least squares g-inverse $\mathbf{G}$, then it contains a reflexive least squares g-inverse $\mathbf{G_1}$. Consequently, $\mathfrak{T} = \overline{\mathrm{Eq}_\ell}(\mathbf{G}|\mathbf{A})$. Similarly, if $\mathfrak{T}$ contains a minimum norm g-inverse $\mathbf{G}$, then it contains a reflexive minimum norm g-inverse $\mathbf{G_2}$. Consequently, $\mathfrak{T} = \overline{\mathrm{Eq}_\mathrm{m}}(\mathbf{G}|\mathbf{A})$.*

Proof. Let $\mathbf{G_1} = \mathbf{GAG}$. Then $\mathbf{G_1}$ is a reflexive g-inverse such that $\mathbf{G_1}\backsim_\ell\mathbf{G}$ and $\mathbf{AG_1} = \mathbf{AG}$. Since $\mathbf{AG}$ is hermitian, so is $\mathbf{AG_1}$. So, $\mathbf{G_1}$ is the reflexive least squares g-inverse belonging to $\mathfrak{T}$. We already know that $\overline{\mathrm{Eq}_\ell}(\mathbf{G_1}|\mathbf{A}) = \overline{\mathrm{Eq}_\ell}(\mathbf{G}|\mathbf{A}) \subseteq \mathfrak{T}$. So, let $\mathbf{H} \in \mathfrak{T}$. As $\mathbf{H}\backsim_\ell\mathbf{G}$, we have $\mathbf{AH} = \mathbf{AG}$. Thus, $\mathbf{AH}$ is hermitian giving $\mathbf{H}$ as a least squares g-inverse of $\mathbf{A}$. Also, $\mathbf{H}\backsim_\ell\mathbf{G}\backsim_\ell\mathbf{G_1}$. Hence, $\mathbf{H} \in \overline{\mathrm{Eq}_\ell}(\mathbf{G_1}|\mathbf{A}) = \overline{\mathrm{Eq}_\ell}(\mathbf{G}|\mathbf{A})$, proving $\mathfrak{T} = \overline{\mathrm{Eq}_\ell}(\mathbf{G}|\mathbf{A})$. □

Theorem 14.3.4. *Let $\mathbf{A} \in \mathbb{C}^{\mathbf{m}\times\mathbf{n}}$. If $\mathbf{G_0} \in \{\mathbf{A}^-_{\ell\mathbf{r}}\}$ and $\mathbf{G_1} \in \overline{\mathrm{Eq}_\ell}(\mathbf{G_0}|\mathbf{A})$, then $\mathbf{G_0} < \star\mathbf{G_1}$. If $\mathbf{G_0} \in \{\mathbf{A}^-_{\mathbf{mr}}\}$, and $\mathbf{G_1} \in \overline{\mathrm{Eq}_\mathrm{m}}(\mathbf{G_0}|\mathbf{A})$, then $\mathbf{G_0}\star < \mathbf{G_1}$. Thus, $\mathbf{G_0} = \inf\{\mathbf{G} : \mathbf{G} \in \overline{\mathrm{Eq}_\ell}(\mathbf{G_0}|\mathbf{A})\}$, if $\mathbf{G_0} \in \{\mathbf{A}^-_{\ell\mathbf{r}}\}$ and $\mathbf{G_0} = \inf\{\mathbf{G} : \mathbf{G} \in \overline{\mathrm{Eq}_\mathrm{m}}(\mathbf{G_0}|\mathbf{A})\}$, if $\mathbf{G_0} \in \{\mathbf{A}^-_{\mathbf{mr}}\}$.*

We now give some characterizations of g-inverses in an equivalence class under each of the relations "$\backsim_\ell$" and "$\backsim_m$".

Theorem 14.3.5. *Let $\mathbf{A} \in \mathbb{C}^{\mathbf{m}\times\mathbf{n}}$ be a matrix of rank r. Let $\mathbf{G_0} \in \{\mathbf{A}^-_{\ell\mathbf{r}}\}$. Let $\mathbf{G}$ be an $n \times m$ matrix of rank s. Then $\mathbf{G} \in \overline{\mathrm{Eq}_\ell}(\mathbf{G_0}|\mathbf{A})$ if and only if there exist a unitary matrix $\mathbf{U}$ and a non-singular matrix $\mathbf{V}$ such that $\mathbf{A} = \mathbf{U}\mathbf{diag}\left(\mathbf{I_r},\ \mathbf{0}\right)\mathbf{V}$, $\mathbf{G_0} = \mathbf{V}^{-1}\mathbf{diag}\left(\mathbf{I_r},\ \mathbf{0}\right)\mathbf{U}^\star$ and $\mathbf{G} = \mathbf{V}^{-1}\mathbf{diag}\left(\mathbf{I_r}\ ,\ \mathbf{I_{r-s}}\ ,\ \mathbf{0}\right)\mathbf{U}^\star$.*

Similarly, let $\mathbf{G_0} \in \{\mathbf{A}^-_{\mathbf{mr}}\}$. Then an $n \times m$ matrix $\mathbf{G}$ of rank s such that $\mathbf{G} \in \overline{\mathrm{Eq}_\mathrm{m}}(\mathbf{G_0}|\mathbf{A})$ if and only if there exists a unitary matrix $\mathbf{U}$ and a non-singular matrix $\mathbf{V}$ such that $\mathbf{A} = \mathbf{U}\mathbf{diag}\left(\mathbf{I_r},\ \mathbf{0}\right)\mathbf{V}$, $\mathbf{G_0} = \mathbf{V}^{-1}\mathbf{diag}\left(\mathbf{I_r},\ \mathbf{0}\right)\mathbf{U}^\star$ and $\mathbf{G} = \mathbf{V}^{-1}\mathbf{diag}\left(\mathbf{I_r}\ ,\ \mathbf{I_{r-s}}\ ,\ \mathbf{0}\right)\ \mathbf{U}^\star$.

Theorem 14.3.6. *Let $\mathbf{A} \in \mathbb{C}^{\mathbf{m}\times\mathbf{n}}$ be matrix of rank r. Let $\mathbf{U}$ and $\mathbf{V}$ are unitary matrices such that $\mathbf{A} = \mathbf{U}\mathbf{diag}\left(\triangle\ ,\ \mathbf{0}\right)\mathbf{V}^\star$ be a singular value decomposition of $\mathbf{A}$, where $\triangle$ is a positive definite diagonal matrix. Let $\mathbf{G_0} \in \{\mathbf{A}^-_{\ell\mathbf{r}}\}$ such that $\mathbf{G_0} = \mathbf{V}\begin{pmatrix}\triangle^{-1} & \mathbf{0}\\ \mathbf{M} & \mathbf{0}\end{pmatrix}\mathbf{U}^\star$ for some matrix $\mathbf{M}$. Then* *an $n \times m$ matrix* $\mathbf{G}$ *of rank* s *such that* $\mathbf{G} \in \overline{\mathrm{Eq}_\ell}(\mathbf{G_0}|\mathbf{A})$ *if and only if* $\mathbf{G} = \mathbf{V}\begin{pmatrix}\triangle^{-1} & \mathbf{0}\\ \mathbf{M} & \mathbf{N}\end{pmatrix}\mathbf{U}^\star$, *where* $\mathbf{N}$ *is an arbitrary matrix of rank* $s - r$.

We can define a relation "$\sim_{\ell m}$" on $\{\mathbf{A}^-_{\ell\mathbf{m}}\}$ in a manner similar to that in Definition 14.3.4. It is interesting to see that not only this relation is an equivalence relation on $\{\mathbf{A}^-_{\ell\mathbf{m}}\}$, but the full set $\{\mathbf{A}^-_{\ell\mathbf{m}}\}$ forms the only equivalence class of this set under the relation and the unique reflexive g-inverse in it is $\mathbf{A}^\dagger$, the Moore-Penrose inverse of a matrix $\mathbf{A}$. Note that in Theorem 14.3.1, we have shown that this is the equivalence class $\overline{\mathrm{Eq}}(\mathbf{A}^\dagger|\mathbf{A})$. Thus, $\{\mathbf{A}^-_{\ell\mathbf{m}}\} = \overline{\mathrm{Eq}}(\mathbf{A}^\dagger|\mathbf{A})$.

We now consider square matrices of index not greater than 1 over a general field.

Definition 14.3.7. Let $\mathbf{A}$ be a square matrix of index ≤ 1. Let $\mathbf{G_1}, \mathbf{G_1} \in \{\mathbf{A}^-_{\mathbf{c}}\}$. The relation "$\sim_c$" on $\{\mathbf{A}^-_{\mathbf{c}}\}$ is defined as: $\mathbf{G_1} \sim_{\mathbf{c}} \mathbf{G_1}$ if $\mathbf{AG_1} = \mathbf{AG_2}$ and $\mathbf{G_1A} = \mathbf{G_2A}$.
Similarly, Let $\mathbf{G_1}, \mathbf{G_1} \in \{\mathbf{A}^-_{\mathbf{a}}\}$. The relation "$\sim_a$" on $\{\mathbf{A}^-_{\mathbf{a}}\}$ is defined as: $\mathbf{G_1} \sim_{\mathbf{a}} \mathbf{G_1}$ if $\mathbf{AG_1} = \mathbf{AG_2}$ and $\mathbf{G_1A} = \mathbf{G_2A}$.

The relations "$\sim_c$" and "$\sim_a$" are equivalence relations. We denote the equivalence classes under "$\sim_c$" and "$\sim_a$" as $\overline{\mathrm{Eq}_c}(\mathbf{G}|\mathbf{A})$ and $\overline{\mathrm{Eq}_a}(\mathbf{G}|\mathbf{A})$ respectively. The following theorem is easy to prove.

Theorem 14.3.8. *Let* $\mathbf{A}$ *be a square matrix of index* ≤ 1. *Then the following hold:*

(i) *There exists a unique* χ*-inverse* $\mathbf{G_0}$ *in each equivalence class under* "$\sim_c$".

(ii) *Let* $\mathbf{A} = \mathbf{P}\,\mathbf{diag}\,(\mathbf{T}\ ,\ \mathbf{0})\ \mathbf{P}^{-1}$ *and* $\mathbf{G_0} = \mathbf{P}\begin{pmatrix}\mathbf{T}^{-1} & \mathbf{L}\\ \mathbf{0} & \mathbf{0}\end{pmatrix}\mathbf{P}^{-1}$, *where* $\mathbf{P}$ *and* $\mathbf{T}$ *are non-singular and* $\mathbf{L}$ *is a fixed matrix. Then* $\mathbf{G} \in \overline{\mathrm{Eq}_c}(\mathbf{G_0}|\mathbf{A})$ *if and only if* $\mathbf{G} = \mathbf{P}\begin{pmatrix}\mathbf{T}^{-1} & \mathbf{L}\\ \mathbf{0} & \mathbf{N}\end{pmatrix}\mathbf{P}^{-1}$, *for some matrix* $\mathbf{N}$. *Also,* $\mathbf{G_0} < \#\mathbf{G},\ \forall\ \mathbf{G} \in \overline{\mathrm{Eq}_c}(\mathbf{G_0}|\mathbf{A})$.

(iii) *There exists a unique* ρ*-inverse* $\mathbf{G_0}$ *in each equivalence class under* "$\sim_a$".

(iv) *Let* $\mathbf{A} = \mathbf{P}\,\mathbf{diag}\,(\mathbf{T}\ ,\ \mathbf{0})$ *and* $\mathbf{G_0} = \mathbf{P}\begin{pmatrix}\mathbf{T}^{-1} & \mathbf{0}\\ \mathbf{M} & \mathbf{0}\end{pmatrix}\mathbf{P}^{-1}$, *where* $\mathbf{P}$ *and* $\mathbf{T}$ *are non-singular and* $\mathbf{M}$ *is a fixed matrix. Then an* $n \times n$ *matrix*

$\mathbf{G} \in \overline{\mathrm{Eq_a}}(\mathbf{G_0}|\mathbf{A})$ *if and only if* $\mathbf{G} = \mathbf{P}\begin{pmatrix}\mathbf{T}^{-1} & \mathbf{0}\\ \mathbf{M} & \mathbf{N}\end{pmatrix}\mathbf{P}^{-1}$, *for some matrix* $\mathbf{N}$. *Also,* $\mathbf{G_0} \# < \mathbf{G}$, *for all* $\mathbf{G} \in \overline{\mathrm{Eq_c}}(\mathbf{G_0}|\mathbf{A})$.

Definition 14.3.9. Let $\mathbf{A}$ be a square matrix of index ≤ 1. Let $\mathbf{G_1},\ \mathbf{G_2} \in \{\mathbf{A}^-_{\mathbf{com}}\}$. The relation "$\sim_{com}$" on $\{\mathbf{A}^-_{\mathbf{com}}\}$ is defined as: $\mathbf{G_1}\sim_{\mathbf{com}}\mathbf{G_2}$ if $\mathbf{AG_1} = \mathbf{AG_2}$.

It is easy to check that the relation "$\sim_{com}$" on $\{\mathbf{A}^-_{\mathbf{com}}\}$ is an equivalence relation and $\{\mathbf{A}^-_{\mathbf{com}}\}$ is the only equivalence class of $\{\mathbf{A}^-_{\mathbf{com}}\}$ under this relation. Note that the reflexive g-inverse in this equivalence class is $\mathbf{A}^{\#}$. By Theorem 14.3.1, $\{\mathbf{A}^-_{\mathbf{com}}\} = \overline{\mathrm{Eq}}(\mathbf{A}^{\#}|\mathbf{A})$.

We have seen in earlier chapters that if $\mathbf{A}$ is a range hermitian matrix, we have $\{\mathbf{A}^-_{\mathbf{c}}\} = \{\mathbf{A}^-_{\mathbf{m}}\}$ and $\{\mathbf{A}^-_{\mathbf{a}}\} = \{\mathbf{A}^-_{\ell}\}$. Therefore for a range hermitian matrix the relations "$\sim_c$" and "$\sim_m$", "$\sim_a$" and "$\sim_\ell$" coincide.

14.4 Equivalence relation on the outer inverses of a matrix

Let $\mathbf{A}$ be an $m \times n$ matrix. Recall that an $n \times m$ matrix $\mathbf{X}$ is called an outer inverse of $\mathbf{A}$, if $\mathbf{A}$ is a g-inverse of $\mathbf{X}$. The class of all outer inverses of $\mathbf{A}$ is denoted by $\{\mathbf{A}_-\}$. We start this section by obtaining a characterization of the outer inverses of a matrix.

Theorem 14.4.1. *Let* $\mathbf{A}$ *be an* $m \times n$ *matrix of rank* r (> 0). *Let* $\mathbf{A} = \mathbf{P}\mathrm{diag}\left(\mathbf{I_r}\ ,\ \mathbf{0}\right)\mathbf{Q}$ *be a normal form of* $\mathbf{A}$, *where* $\mathbf{P}$ *and* $\mathbf{Q}$ *are non-singular. Then the class of all outer inverses of* $\mathbf{A}$ *is given by*

$$\mathbf{X} = \mathbf{Q}^{-1}\begin{pmatrix}\mathbf{L} & \mathbf{M}\\ \mathbf{N} & \mathbf{NM}\end{pmatrix}\mathbf{P}^{-1},$$

where $\mathbf{L}$ *is an arbitrary idempotent matrix of order* $r \times r$ *and* $\mathbf{M}, \mathbf{N}$ *are arbitrary matrices such that* $\mathcal{C}(\mathbf{M}) \subseteq \mathcal{C}(\mathbf{L})$ *and* $\mathcal{C}(\mathbf{N^t}) \subseteq \mathcal{C}(\mathbf{L^t})$.

Proof. Direct verification. □

Remark 14.4.2. Let $\mathbf{A}$, $\mathbf{X}$ and $\mathbf{L}$ be as in Theorem 14.4.1. Then $\mathbf{X}$ is an outer inverse of $\mathbf{A}$ with $\rho(\mathbf{X}) = s$ $(0 \leq s \leq r)$ if and only if $\rho(\mathbf{L}) = s$.

Here is another characterization of the outer inverses of a matrix that can be easily deduced from Theorem 2.3.18.

Theorem 14.4.3. *Let* $\mathbf{A}$ *and* $\mathbf{X}$ *be matrices of rank* r *and* s $(s \leq r)$ *respectively. Then* $\mathbf{X}$ *is an outer inverse of* $\mathbf{A}$, *if and only if there exist non-singular matrices* $\mathbf{R}$ *and* $\mathbf{S}$ *such that* $\mathbf{A} = \mathbf{R}\mathbf{diag}\left(\mathbf{I_r}\ ,\ \mathbf{0}\right)\mathbf{S}$ *and* $\mathbf{X} = \mathbf{S}^{-1}\begin{pmatrix}\mathbf{I_s} & \mathbf{0}\\ \mathbf{0} & \mathbf{0}\end{pmatrix}\mathbf{R}^{-1}$.

The following result on outer inverses is analogous to Theorem 14.2.10 and Corollary 14.2.11 for g-inverses.

Theorem 14.4.4. *Let* $\mathbf{A}$ *be an* $m \times n$ *matrix and* $\mathbf{G}$ *a reflexive g-inverse of* $\mathbf{A}$. *If* $\mathbf{X}$ *is an* $n \times m$ *matrix such that* $\mathbf{X} <^- \mathbf{G}$, *then* $\mathbf{X}$ *and* $\mathbf{G} - \mathbf{X}$ *are outer inverses of* $\mathbf{A}$.

Proof. Since $\mathbf{X} <^- \mathbf{G}$, there exists a matrix $\mathbf{Y}$ such that $\mathbf{G} = \mathbf{X} \oplus \mathbf{Y}$. Since $\mathbf{GAG} = \mathbf{G}$, we have $\mathbf{X}+\mathbf{Y} = (\mathbf{X}+\mathbf{Y})\mathbf{A}(\mathbf{X}+\mathbf{Y})$. Rewriting we have $\mathbf{X} - (\mathbf{X}+\mathbf{Y})\mathbf{AX} = (\mathbf{X}+\mathbf{Y})\mathbf{AY} - \mathbf{Y}$. Now, $\mathcal{C}(\mathbf{X^t}) \cap \mathcal{C}(\mathbf{Y^t}) = \mathbf{0}$, so, we have $\mathbf{X} - (\mathbf{X}+\mathbf{Y})\mathbf{AX} = (\mathbf{X}+\mathbf{Y})\mathbf{AY} - \mathbf{Y} = \mathbf{0}$. Thus, $\mathbf{X} - \mathbf{XAX} = \mathbf{YAX}$ and $\mathbf{Y} - \mathbf{YAY} = \mathbf{XAY}$. Again, since $\mathcal{C}(\mathbf{X}) \cap \mathcal{C}(\mathbf{Y}) = \mathbf{0}$, it follows that both $\mathbf{X}$ and $\mathbf{Y}$ are outer inverses of $\mathbf{A}$. Further $\mathbf{XAY} = \mathbf{0}$ and $\mathbf{YAX} = \mathbf{0}$. □

Remark 14.4.5. Let $\mathbf{A}$ and $\mathbf{X}$ be as in Theorem 14.4.4. If $\mathbf{X} <^- \mathbf{G}$, then $\mathbf{X}$ is an outer inverse of every reflexive g-inverse of $\mathbf{G}$.

Let $\mathbf{X}$ be an outer inverse of $\mathbf{A}$. We now identify an important reflexive g-inverse of $\mathbf{X}$.

Theorem 14.4.6. *Let* $\mathbf{X}$ *is an outer inverse of* $\mathbf{A}$. *Then* $\mathbf{AXA}$ *is a reflexive g-inverse of* $\mathbf{X}$ *and* $\mathbf{A} \in \overline{\mathrm{Eq}}(\mathbf{AXA}|\mathbf{X})$.

Proof. Since $\mathbf{X}$ is an outer inverse of $\mathbf{A}$, $\mathbf{A}$ is an g-inverse of $\mathbf{X}$. Hence, $\mathbf{AXA} \in \{\mathbf{X_r^-}\}$. Moreover, since $\mathbf{AX} = \mathbf{AXAX}$ and $\mathbf{XA} = \mathbf{XAXA}$ it follows that $\mathbf{A} \in \overline{\mathrm{Eq}}(\mathbf{AXA}|\mathbf{X})$. □

We now define an equivalence relation on the outer inverses of a matrix.

Definition 14.4.7. Let $\mathbf{A}$ be an $m \times n$ matrix. Let $\mathbf{X_1}, \mathbf{X_2} \in \{\mathbf{A_-}\}$. Define $\mathbf{X_1} \cong \mathbf{X_2}$ if $\mathbf{AX_1A} = \mathbf{AX_2A}$.

Theorem 14.4.8. *The relation "*$\cong$*" is an equivalence relation on* $\{\mathbf{A_-}\}$.

Proof is trivial.

Thus, the relation "$\cong$" partitions $\{\mathbf{A_-}\}$, the set of outer inverses into mutually disjoint equivalence classes. We denote the equivalence class containing an $\mathbf{X} \in \{\mathbf{A_-}\}$ by $\underline{\mathrm{Eq}}(\mathbf{X}|\mathbf{A})$.

Theorem 14.4.9. *Let* $\mathbf{A}$ *be an* $m \times n$ *matrix and* $\mathbf{X} \in \{\mathbf{A}_-\}$. *Then for each* $\mathbf{Y} \in \underline{\mathrm{Eq}}(\mathbf{X}|\mathbf{A})$, $\rho(\mathbf{Y}) = \rho(\mathbf{X})$.

Let $\mathbf{A}$ be an $m \times n$ matrix of rank r. Let $\mathbf{G_0}$ be a reflexive g-inverse of $\mathbf{A}$. Then $\overline{\mathrm{Eq}}(\mathbf{G_0}|\mathbf{A})$ contains g-inverses of $\mathbf{A}$ of all ranks with $r \leq s \leq min\{m, n\}$. However, if $\mathbf{X}$ is an outer inverse, then $\underline{\mathrm{Eq}}(\mathbf{X}|\mathbf{A})$ contains all outer inverses of same rank as $\mathbf{X}$.

We now identify all the matrices $\mathbf{Y}$ such that $\mathbf{Y} \in \underline{\mathrm{Eq}}(\mathbf{X}|\mathbf{A})$.

Theorem 14.4.10. *Let* $\mathbf{A}$ *be an* $m \times n$ *matrix of rank* r. *Let* $\mathbf{X}$ *be an outer inverse of* $\mathbf{A}$. *Following Theorem 14.3.1, let* $\mathbf{A} = \mathbf{P}\mathbf{diag}\,(\mathbf{I_r}\ , \mathbf{0})\,\mathbf{Q}$ *be a normal form of* $\mathbf{A}$ *and let* $\mathbf{X} = \mathbf{Q}^{-1}\begin{pmatrix} \mathbf{L} & \mathbf{M} \\ \mathbf{N} & \mathbf{NM} \end{pmatrix}\mathbf{P}^{-1}$, *where* $\mathbf{L}$ *is an idempotent matrix of order* $r \times r$ *and* $\mathbf{LM} = \mathbf{M}$ *and* $\mathbf{NL} = \mathbf{N}$. *Then* $\mathbf{Y} \in \underline{\mathrm{Eq}}(\mathbf{X}|\mathbf{A})$ *if and only if* $\mathbf{Y} = \mathbf{Q}^{-1}\begin{pmatrix} \mathbf{L} & \mathbf{T_1} \\ \mathbf{T_2} & \mathbf{T_2T_1} \end{pmatrix}\mathbf{P}^{-1}$, *where* $\mathbf{T_1}$ *and* $\mathbf{T_2}$ *are arbitrary matrices such that* $\mathbf{LT_1} = \mathbf{T_1}$ *and* $\mathbf{T_2L} = \mathbf{T_2}$.

Proof. We first note that a matrix $\mathbf{Y} \in \underline{\mathrm{Eq}}(\mathbf{X}|\mathbf{A})$ if and only if $\mathbf{Y} = \mathbf{Q}^{-1}\begin{pmatrix} \mathbf{T_3} & \mathbf{T_1} \\ \mathbf{T_2} & \mathbf{T_2T_1} \end{pmatrix}\mathbf{P}^{-1}$, where $\mathbf{T_3}$ is idempotent, $\mathbf{T_3T_1} = \mathbf{T_1}$ and $\mathbf{T_2T_3} = \mathbf{T_2}$. Now, $\mathbf{AXA} = \mathbf{AYA}$ if and only if $\mathbf{T_3} = \mathbf{L}$. In fact, $\mathbf{AXA} = \mathbf{P}\begin{pmatrix} \mathbf{L} & \mathbf{0} \\ \mathbf{0} & \mathbf{0} \end{pmatrix}\mathbf{Q}$ and $\mathbf{AYA} = \mathbf{P}\begin{pmatrix} \mathbf{T_3} & \mathbf{0} \\ \mathbf{0} & \mathbf{0} \end{pmatrix}\mathbf{Q}$. □

Remark 14.4.11. For any $m \times n$ matrix $\mathbf{A}$ of rank r, there is a one-one correspondence between the idempotent matrices of order $r \times r$ and the equivalence classes of $\{\mathbf{A}_-\}$ under "$\cong$." In particular, the class of all reflexive g-inverses of $\mathbf{A}$ forms an equivalence class of $\{\mathbf{A}_-\}$ under "$\cong$." The equivalence class that contains the null matrix has only null matrix in it.

Let $\mathbf{A}$ be an $m \times n$ matrix and let $\mathbf{X}$ be an outer inverse of $\mathbf{A}$. Notice that $\mathbf{A} = \mathbf{AXA} \oplus (\mathbf{A} - \mathbf{AXA})$. Following [Goller (1986)], $\mathbf{AXA}$ is a rank decomposition matrix of $\mathbf{A}$. (In general, if $\mathbf{A} = \mathbf{B} \oplus \mathbf{C}$, then $\mathbf{B}$ (as also $\mathbf{C}$) is called a rank decomposition matrix of $\mathbf{A}$.) Thus, each equivalence class of $\{\mathbf{A}_-\}$ under "$\cong$" corresponds to a unique rank decomposition matrix. As seen in Remark 14.4.2, each equivalence class of $\{\mathbf{A}_-\}$ under "$\cong$" is determined by a unique idempotent matrix, so, each rank decomposition matrix is determined by a unique idempotent. In fact, we can make a stronger statement leading to another characterization of the equivalence classes of $\{\mathbf{A}_-\}$ under "$\cong$."

Theorem 14.4.12. *Let* $\mathbf{A}$ *be an* $m \times n$ *matrix. Then every rank decomposition matrix* $\mathbf{D_A}$ *determines an equivalence class of* $\{\mathbf{A}_-\}$ *under "*$\cong$*" given by* $\{\mathbf{A^-D_AA^-} : \mathbf{A^-} \in \{\mathbf{A^-}\}\}$.

Proof. First notice that if $\mathbf{D_A}$ is a rank decomposition matrix of $\mathbf{A}$, then $\mathbf{AA^-D_A} = \mathbf{D_AA^-A} = \mathbf{D_AA^-D_A} = \mathbf{D_A}$ for all $\mathbf{A^-}$. Clearly, $\mathbf{A^-D_AA^-}$ is an outer inverse of $\mathbf{A}$, since for all $\mathbf{A^-}$,

$$\mathbf{A^-D_AA^-AA^-D_AA^-} = \mathbf{A^-D_AA^-D_AA^-} = \mathbf{A^-D_AA^-}.$$

Further, if $\mathbf{A_1^-}$ and $\mathbf{A_2^-}$ are two g-inverses of $\mathbf{A}$, then

$$\mathbf{AA_1^-D_AA_1^-A} = \mathbf{D_A} = \mathbf{AA_2^-D_AA_2^-A}.$$

Hence, for all $\mathbf{A^-} \in \{\mathbf{A^-}\}$, the outer inverses $\mathbf{A^-D_AA^-}$ are related to each other under "$\cong$."

Let $\mathbf{A} = \mathbf{P}\mathrm{\mathbf{diag}}(\mathbf{I_r}, \mathbf{0})\mathbf{Q}$. The matrix $\mathbf{X} = \mathbf{GD_AG}$ is an outer inverse for some g-inverse $\mathbf{G}$ of $\mathbf{A}$. By Theorem 14.4.10, $\mathbf{X} = \mathbf{Q}^{-1}\begin{pmatrix}\mathbf{L} & \mathbf{M}\\ \mathbf{N} & \mathbf{NM}\end{pmatrix}\mathbf{P}^{-1}$, where $\mathbf{L}$ is an idempotent and matrices $\mathbf{M}$ and $\mathbf{N}$ are such that $\mathbf{LM} = \mathbf{M}$ and $\mathbf{NL} = \mathbf{N}$. Let $\mathfrak{T}$ be the equivalence class of $\mathbf{X}$. Now, $\mathbf{Z} = \mathbf{AXA}$ is a rank decomposition matrix and by discussion in the last para, for each g-inverse $\mathbf{A^-}$ of $\mathbf{A}, \{\mathbf{A^-ZA^-}\} \subseteq \mathfrak{T}$.

Let $\mathbf{Y} \in \mathfrak{T}$. Since $\mathbf{Y}$ an outer inverse of $\mathbf{A}$ and $\mathbf{Y} \cong \mathbf{X}$, we have $\mathbf{Y} = \mathbf{Q}^{-1}\begin{pmatrix}\mathbf{L} & \mathbf{T_1}\\ \mathbf{T_2} & \mathbf{T_2T_1}\end{pmatrix}\mathbf{P}^{-1}$, where $\mathbf{T_1}$ and $\mathbf{T_2}$ are some matrices such that $\mathbf{LT_1} = \mathbf{T_1}$, $\mathbf{T_2L} = \mathbf{T_2}$. Consider $\mathbf{H} = \mathbf{Q}^{-1}\begin{pmatrix}\mathbf{I_r} & \mathbf{T_1}\\ \mathbf{T_2} & \mathbf{J}\end{pmatrix}\mathbf{P}^{-1}$. It is easy to see that $\mathbf{H}$ is a g-inverse of $\mathbf{A}$ and $\mathbf{Y} = \mathbf{HZH}$. Thus,

$$\mathfrak{T} = \{\mathbf{A^-D_AA^-} : \mathbf{A^-} \in \{\mathbf{A^-}\}\}. \qquad \square$$

We have seen in the proof of Theorem 14.4.12 that for a given rank decomposition matrix $\mathbf{D_A}$ of $\mathbf{A}$, there are possibly several g-inverses $\mathbf{H}$ of $\mathbf{A}$ such that $\mathbf{HD_AH} = \mathbf{Y}$ for some outer inverse $\mathbf{Y}$ of $\mathbf{A}$. It may be interesting to ask the following:

Given an outer inverse $\mathbf{Y}$ of $\mathbf{A}$ in an equivalence class of a rank decomposition matrix $\mathbf{D_A}$ of $\mathbf{A}$, what is the class of all g-inverses $\mathbf{H}$ of $\mathbf{A}$ such that $\mathbf{HD_AH} = \mathbf{Y}$?

The answer gives a nice link to the equivalence classes of g-inverses of $\mathbf{D_A}$.

Theorem 14.4.13. *Let* $\mathbf{A}$ *be an* $m \times n$ *matrix and let* $\mathbf{D_A}$ *be a rank decomposition matrix of* $\mathbf{A}$. *Let* $\mathbf{H_1}, \mathbf{H_2} \in \{\mathbf{A^-}\}$. *Then* $\mathbf{H_1D_AH_2} = \mathbf{H_2D_AH_1}$ *if and only if* $\mathbf{H_1}, \mathbf{H_2}$ *belong to the same equivalence of g-inverses of* $\mathbf{D_A}$.

Proof. Since $\mathbf{D_A}$ is a rank decomposition matrix of $\mathbf{A}$, each g-inverse of $\mathbf{A}$ is also a g-inverse of $\mathbf{D_A}$. It is clear that $\mathbf{H_1} \in \overline{\mathrm{Eq}}(\mathbf{H_1D_AH_1}|\mathbf{D_A})$. By Definition 14.2.1, $\mathbf{H_2} \in \overline{\mathrm{Eq}}(\mathbf{H_1D_AH_1}|\mathbf{D_A})$ if and only if $\mathbf{H_1D_AH_2} = \mathbf{H_2D_AH_1}$. □

Let $\mathbf{B}$ be an $m \times n$ matrix and $\mathbf{X}$ be an outer inverse of $\mathbf{B}$. Does there exist a reflexive g-inverse $\mathbf{B_r^-}$ of $\mathbf{B}$ such that $\mathbf{X} <^- \mathbf{B_r^-}$? If so, what is the class of all $\mathbf{B_r^-}$ that dominate $\mathbf{X}$ under minus. The answer is contained in the following:

Theorem 14.4.14. *Let* $\mathbf{B}$ *be an* $m \times n$ *matrix and* $\mathbf{X}$ *be an outer inverse of* $\mathbf{B}$*. Then there exists a reflexive g-inverse* $\mathbf{B_r^-}$ *of* $\mathbf{B}$ *such that* $\mathbf{X} <^- \mathbf{B_r^-}$*. The class of all such* $\mathbf{B_r^-}$ *is given by* $\mathbf{X} + (\mathbf{I} - \mathbf{XB})(\mathbf{B} - \mathbf{BXB})_r^-(\mathbf{I} - \mathbf{BX})$.

Proof. First notice the striking similarity between this and Theorem 14.2.9. In fact, we show that the statement of the present theorem is almost a restatement of Theorem 14.2.9. If $\mathbf{X}$ is an outer inverse of $\mathbf{B}$, then $\mathbf{B}$ is a g-inverse of $\mathbf{X} \in \overline{\mathrm{Eq}}(\mathbf{BXB}|\mathbf{X})$ under the relation "$\backsim$". Further, if $\mathbf{X} <^- \mathbf{B_r^-}$, then $(\mathbf{B_r^-} - \mathbf{X})\mathbf{BX} = \mathbf{XB}(\mathbf{B_r^-} - \mathbf{X}) = \mathbf{0}$, since $\mathbf{B}$ is a g-inverse of $\mathbf{B_r^-}$. Thus, $\mathbf{B_r^-BXB} = \mathbf{XBXB} = \mathbf{XB}$. Similarly, $\mathbf{BXBB_r^-} = \mathbf{BX}$. Hence, $\mathbf{B_r^-} \in \overline{\mathrm{Eq}}(\mathbf{X}|\mathbf{BXB})$. Now, take $\mathbf{A} = \mathbf{X}$ and $\mathbf{G_{s-r}} = \mathbf{B}$ and apply Theorem 14.2.8, we get the class of all $\mathbf{B_r^-}$ such that $\mathbf{X} <^- \mathbf{B_r^-}$ as $\mathbf{X} + (\mathbf{I} - \mathbf{XBXB})(\mathbf{B} - \mathbf{BXB})_r^-(\mathbf{I} - \mathbf{BXBX})$ where $(\mathbf{B} - \mathbf{BXB})_r^-$ is an arbitrary g-inverse of $\mathbf{B} - \mathbf{BXB}$. Since $\mathbf{XBX} = \mathbf{X}$, the result follows. □

Given an outer inverse $\mathbf{X}$ of $\mathbf{B}$, what is the the class of all reflexive g-inverse $\mathbf{B_r^-}$ of $\mathbf{B}$ such that $\mathbf{X} <^- \mathbf{B_r^-}$? We obtained a characterization in Theorem 14.4.14 analogous to Theorem 14.2.9. We now obtain a characterization similar to Theorem 14.2.13.

Theorem 14.4.15. *Let* $\mathbf{B}$ *be an* $m \times n$ *matrix of rank* r *and let* $\mathbf{X}$ *be an outer inverse of* $\mathbf{B}$ *with* $\rho(\mathbf{X}) = s(< r)$*. Let* $\mathbf{B} = \mathbf{R}\,\mathbf{diag}\left(\mathbf{I_r}\,,\,\mathbf{0}\right)\mathbf{S}$ *and* $\mathbf{X} = \mathbf{S}^{-1}\mathbf{diag}\left(\mathbf{I_s}\,,\,\mathbf{0}\right)\mathbf{R}^{-1}$*. Then the class of all reflexive g-inverses* $\mathbf{B_r^-}$ *of* $\mathbf{B}$ *such that* $\mathbf{X} <^- \mathbf{B_r^-}$ *is given by* $\mathbf{B_r^-} = \mathbf{S}^{-1}\begin{pmatrix} \mathbf{I_s} & \mathbf{0} & \mathbf{0} \\ \mathbf{0} & \mathbf{I_{r-s}} & \mathbf{L} \\ \mathbf{0} & \mathbf{M} & \mathbf{ML} \end{pmatrix}\mathbf{R}^{-1}$*, where* $\mathbf{L}$, $\mathbf{M}$ *are arbitrary.*

Proof is easy.

Remark 14.4.16. Consider an equivalence class under the relation "$\backsim$." It contains g-inverses of $\mathbf{A}$ of all ranks s such that $\rho(\mathbf{A}) \leq s \leq min\{m, n\}$. Let

$s < min\{m, n\}$. Also, there exist distinct g-inverses of $\mathbf{A}$ namely $\mathbf{G_1}$ and $\mathbf{G_2}$ such that $\mathbf{G_1} <^- \mathbf{G_2}$. On the other hand, all the outer inverses of $\mathbf{A}$ in the same equivalence class under the relation "$\cong$" have same rank. Further, if $\mathbf{X_1}$ and $\mathbf{X_2}$ are distinct outer inverses belonging to the same equivalence class under the relation "$\cong$", then neither $\mathbf{X_1} <^- \mathbf{X_2}$ nor $\mathbf{X_2} <^- \mathbf{X_1}$.

The Drazin inverse $\mathbf{A^D}$ of a square matrix $\mathbf{A}$ is a well known outer inverse $\mathbf{A}$. We now identify the equivalence class $\underline{\text{Eq}}(\mathbf{A^D}|\mathbf{A})$ of $\mathbf{A^D}$.

Theorem 14.4.17. *Let* $\mathbf{A} = \mathbf{P}\,\mathbf{diag}\left(\mathbf{T}\ ,\ \mathbf{N}\right)\mathbf{P}^{-1}$ *be the core-nilpotent decomposition of a square matrix* $\mathbf{A}$, *where* $\mathbf{P}$ *and* $\mathbf{T}$ *are non-singular matrices and* $\mathbf{N}$ *is a nilpotent matrix. Then* $\mathbf{G} \in \underline{\text{Eq}}(\mathbf{A^D}|\mathbf{A})$ *if and only if* $\mathbf{G} = \mathbf{P}\begin{pmatrix}\mathbf{T}^{-1} & \mathbf{R_1}\\ \mathbf{R_2} & \mathbf{R_2TR_1}\end{pmatrix}\mathbf{P}^{-1}$, *where* $\mathbf{R_1} = \mathbf{W}(\mathbf{I} - \mathbf{NN}^-)$ *and* $\mathbf{R_2} = (\mathbf{I} - \mathbf{N}^-\mathbf{N})\mathbf{Z}$ *for some matrices* $\mathbf{W}$ *and* $\mathbf{Z}$ *of appropriate orders and for some g-inverse* $\mathbf{N}^-$ *of* $\mathbf{N}$.

Proof. By Theorem 2.4.26, we have $\mathbf{A^D} = \mathbf{Pdiag}\left(\mathbf{T}^{-1}\ ,\mathbf{0}\right)\mathbf{P}^{-1}$. So, $\mathbf{AA^DA} = \mathbf{Pdiag}\left(\mathbf{T}\ ,\mathbf{0}\right)\mathbf{P}^{-1}$. Let $\mathbf{G} = \mathbf{P}\begin{pmatrix}\mathbf{L} & \mathbf{R_1}\\ \mathbf{R_2} & \mathbf{R_3}\end{pmatrix}\mathbf{P}^{-1}$. Now, $\mathbf{G} \in \underline{\mathbf{Eq}}(\mathbf{A^D}|\mathbf{A})$ if and only if $\mathbf{AGA} = \mathbf{AA^DA}$ and $\mathbf{GAG} = \mathbf{G}$. Moreover, $\mathbf{AGA} = \mathbf{AA^DA} \Leftrightarrow \mathbf{L} = \mathbf{T}^{-1}, \mathbf{R_1N} = \mathbf{0}, \mathbf{NR_2} = \mathbf{0}$ and $\mathbf{NR_3N} = \mathbf{0}$. Also, $\mathbf{GAG} = \mathbf{G} \Leftrightarrow \mathbf{R_2TR_1} + \mathbf{R_3NR_3} = \mathbf{R_3} \Leftrightarrow \mathbf{R_3N} = \mathbf{0}$ and $\mathbf{R_2TR_1} = \mathbf{R_3}$. So, the result follows. □

Remark 14.4.18. Let $\mathbf{A}$ be a square matrix and $\mathbf{A^D}$ be its Drazin inverse. Then $\mathbf{G} \in \underline{\mathbf{Eq}}(\mathbf{A^D}|\mathbf{A})$ if and only if there exists a matrix $\mathbf{S}$ such that (i) $\mathbf{G} = \mathbf{A^D} + \mathbf{S}$ (ii) $\rho(\mathbf{G}) = \rho(\mathbf{A^D})$, (iii) $\mathbf{A^DSA^D} = \mathbf{0}$; (iv) $(\mathbf{A} - (\mathbf{A^D})^\#)\mathbf{S} = \mathbf{0}$ and (v) $\mathbf{S}(\mathbf{A} - (\mathbf{A^D})^\#) = \mathbf{0}$. Recall that $(\mathbf{A} - (\mathbf{A^D})^\#)$ is the nilpotent part of $\mathbf{A}$.

Let $\mathbf{A}$ be an $m\times n$ matrix of rank $r(> 1)$ over $\mathbb{C}$. An $n\times m$ matrix $\mathbf{G}$ such that $\mathbf{GAG} = \mathbf{G}$, $\mathbf{GA}$ and $\mathbf{AG}$ are hermitian is called an $\mathbf{A}_{-\ell\mathbf{m}}$ outer inverse of $\mathbf{A}$. We shall now identify the equivalence class $\underline{\text{Eq}}(\mathbf{A}_{-\ell\mathbf{m}}|\mathbf{A})$ of $\mathbf{A}_{-\ell\mathbf{m}}$. We also note that if $\mathbf{A} = \mathbf{Udiag}\left(\triangle\ ,\mathbf{0}\right)\mathbf{V}^\star$ is a singular value decomposition of $\mathbf{A}$, where $\mathbf{U}$ and $\mathbf{V}$ are unitary matrices and $\triangle$ is a positive definite diagonal matrix, then $\mathbf{G}$ is an $\mathbf{A}_{-\ell\mathbf{m}}$ g-inverse of rank $s(0 < s < r)$ of $\mathbf{A}$ if and only if $\mathbf{G} = \mathbf{Vdiag}\left(\mathbf{R}\ ,\mathbf{0}\right)\mathbf{U}^\star$, where $\mathbf{R}$ is a $\triangle_{-\ell\mathbf{m}}$. Moreover, $\mathbf{G}$ is an $\mathbf{A}_{-\ell\mathbf{m}}$ if and only if $\mathbf{A}$ is an $\mathbf{G}^-_{\ell\mathbf{m}}$.

Theorem 14.4.19. *Let* $\mathbf{A}$ *and* $\mathbf{G}$ *be as in the preceding para and* $\mathbf{H}$ *be an* $n \times m$ *matrix. Then* $\mathbf{H} \in \underline{\mathbf{Eq}}(\mathbf{G}|\mathbf{A})$ *if and only if* $\mathbf{H} = \mathbf{V}\begin{pmatrix} \mathbf{R} & \mathbf{S_1} \\ \mathbf{S_2} & \mathbf{S_2} \Delta \mathbf{S_1} \end{pmatrix}\mathbf{U}^\star$ *for some matrices* $\mathbf{S_1}$ *and* $\mathbf{S_2}$ *such that* $\mathcal{C}(\mathbf{S_1}) \subseteq \mathcal{C}(\mathbf{R})$ *and* $\mathcal{C}(\mathbf{S_1}^t) \subseteq \mathcal{C}(\mathbf{R}^t)$.

Proof is by straightforward verification.

Remark 14.4.20. Let $\mathbf{G}$ be an $\mathbf{A}_{-\ell m}$. Then $\mathbf{G}$ is the unique $\mathbf{A}_{-\ell m}$ such that $\mathbf{G} \in \underline{\mathbf{Eq}}(\mathbf{G}|\mathbf{A})$.

In Theorem 14.2.22, we have seen that an equivalence class $\overline{\mathrm{Eq}}(\mathbf{G}|\mathbf{A})$ under the equivalence relation "$\backsim$" determines the matrix $\mathbf{A}$. However, a similar statement does not hold for outer inverses as the the following example shows.

Example 14.4.21. Let

$$\mathbf{A} = \begin{pmatrix} 1 & 0 & 0 \\ 0 & 1 & 0 \\ 0 & 0 & 0 \end{pmatrix}, \mathbf{B} = \begin{pmatrix} 1 & 0 & 0 \\ 0 & 2 & 0 \\ 0 & 0 & 0 \end{pmatrix} \text{ and } \mathbf{G} = \begin{pmatrix} 1 & 0 & 0 \\ 0 & 0 & 0 \\ 0 & 0 & 0 \end{pmatrix}.$$

Then $\mathbf{G}$ is an outer inverse of both $\mathbf{A}$ and $\mathbf{B}$ and $\underline{\mathrm{Eq}}(\mathbf{G}|\mathbf{A}) = \underline{\mathbf{Eq}}(\mathbf{G}|\mathbf{B})$. However, $\mathbf{A} \neq \mathbf{B}$.

14.5 Diagrammatic representation of the g-inverses and outer inverses

In this section we provide a diagrammatic representation of the inverses (g-inverses and outer inverses) of a matrix depicting the hierarchy of these inverses under minus order. The equivalence relations developed in the previous three sections play an important role in obtaining a neat diagrammatic representation of these generalized inverses. In fact, we show below the hierarchical representation of generalized inverses belonging to any one equivalence class under the equivalence relation "$\backsim$" yields the hierarchical representation of generalized inverses belonging to all other equivalence classes under the same equivalence relation. In the process of developing the diagram, we have been able to demonstrated the power and importance of some of the other results obtained in the earlier sections. We illustrate the diagram using a 4×4 matrix over $\mathbf{GF(2)}$.

We start with a series of theorems to justify the statement: 'the hierarchical representation of generalized inverses belonging to any one equivalence class' under the equivalence relation "$\backsim$" yields the hierarchical representation of generalized inverses belonging to all other equivalence classes under the same equivalence relation'. Notice that Theorem 14.2.6 establishes a one-one correspondence of g-inverses in the distinct equivalence classes. More precisely, let $\mathbf{A} = \mathbf{P}\mathbf{diag}\,(\mathbf{I_r}\ ,\mathbf{0})\,\mathbf{Q}$ be an $m \times n$ matrix of rank r, where $\mathbf{P}$ and $\mathbf{Q}$ are non-singular. Let $\mathbf{G_i} = \mathbf{Q}^{-1}\begin{pmatrix}\mathbf{I_r} & \mathbf{L_i}\\ \mathbf{M_i} & \mathbf{M_iL_i}\end{pmatrix}\mathbf{P}^{-1}$, $i = 1,2$ be two reflexive g-inverses of $\mathbf{A}$. If $\mathbf{G} \in \overline{\mathrm{Eq}}(\mathbf{G_1}|\mathbf{A})$, then $\mathbf{G} = \mathbf{Q}^{-1}\begin{pmatrix}\mathbf{I_r} & \mathbf{L_1}\\ \mathbf{M_1} & \mathbf{M_1L_1}+\mathbf{S}\end{pmatrix}\mathbf{P}^{-1}$ for some matrix $\mathbf{S}$. The matrix $\mathbf{H} = \mathbf{Q}^{-1}\begin{pmatrix}\mathbf{I_r} & \mathbf{L_2}\\ \mathbf{M_2} & \mathbf{M_2L_2}+\mathbf{S}\end{pmatrix}\mathbf{P}^{-1}$ is a g-inverse in $\overline{\mathrm{Eq}}(\mathbf{G_2}|\mathbf{A})$. Similarly, if we take an $\mathbf{H} \in \overline{\mathrm{Eq}}(\mathbf{G_2}|\mathbf{A})$, where $\mathbf{H} = \mathbf{Q}^{-1}\begin{pmatrix}\mathbf{I_r} & \mathbf{L_2}\\ \mathbf{M_2} & \mathbf{M_2L_2}+\mathbf{T}\end{pmatrix}\mathbf{P}^{-1}$ for some matrix $\mathbf{T}$, then we see $\mathbf{G} = \mathbf{Q}^{-1}\begin{pmatrix}\mathbf{I_r} & \mathbf{L_1}\\ \mathbf{M_1} & \mathbf{M_1L_1}+\mathbf{T}\end{pmatrix}\mathbf{P}^{-1} \in \overline{\mathrm{Eq}}(\mathbf{G_1}|\mathbf{A})$. This gives a one-one correspondence between the two equivalence classes $\overline{\mathrm{Eq}}(\mathbf{G_1}|\mathbf{A})$ and $\overline{\mathrm{Eq}}(\mathbf{G_2}|\mathbf{A})$. We begin with the following:

Theorem 14.5.1. *Let* $\mathbf{A} = \mathbf{P}\mathbf{diag}\,(\mathbf{I_r}\ ,\mathbf{0})\,\mathbf{Q}$ *be an* $m \times n$ *matrix of rank* r, *where* $\mathbf{P}$ *and* $\mathbf{Q}$ *are non-singular and let* $\mathbf{G_i} = \mathbf{Q}^{-1}\begin{pmatrix}\mathbf{I_r} & \mathbf{L_i}\\ \mathbf{M_i} & \mathbf{M_iL_i}\end{pmatrix}\mathbf{P}^{-1}$, $i = 1,2$ *be two reflexive g-inverses of* $\mathbf{A}$. *Further, for* $i = 1,2,\ j = 1,2$, *let* $\mathbf{G_{ij}} = \mathbf{Q}^{-1}\begin{pmatrix}\mathbf{I_r} & \mathbf{L_i}\\ \mathbf{M_i} & \mathbf{M_iL_i}+\mathbf{S_j}\end{pmatrix}\mathbf{P}^{-1}$. *(Notice that* $\mathbf{G_{ij}} \in \overline{\mathrm{Eq}}(\mathbf{G_i}|\mathbf{A})$ *for* $j = 1,2$.*) Then the following are equivalent:*

(i) $\mathbf{G_{11}} <^- \mathbf{G_{12}}$
(ii) $\mathbf{S_1} <^- \mathbf{S_2}$ *and*
(iii) $\mathbf{G_{21}} <^- \mathbf{G_{22}}$.

Proof. We show (i) $\Leftrightarrow$ (ii). Now,

$$\begin{aligned}\mathbf{G_{11}} <^- \mathbf{G_{12}} &\Leftrightarrow \begin{pmatrix}\mathbf{I_r} & \mathbf{L_1}\\ \mathbf{M_1} & \mathbf{M_1L_1}+\mathbf{S_1}\end{pmatrix} <^- \begin{pmatrix}\mathbf{I_r} & \mathbf{L_1}\\ \mathbf{M_1} & \mathbf{M_1L_1}+\mathbf{S_2}\end{pmatrix}\\ &\Leftrightarrow r + \rho(\mathbf{S_1}) + \rho(\mathbf{S_2}-\mathbf{S_1}) = r + \rho(\mathbf{S_2})\\ &\Leftrightarrow \mathbf{S_1} <^- \mathbf{S_2}.\end{aligned}$$

Proof of (ii) $\Leftrightarrow$ (iii) is similar. □

Thus, the one-one correspondence carries over to the hierarchical relationship also.

Theorem 14.5.2. *In the setup of Theorem 14.5.1,*

$$\mathbf{Q}^{-1}\begin{pmatrix}\mathbf{I_r} & \mathbf{L_2}\\ \mathbf{M_2} & \mathbf{M_2L_2+S}\end{pmatrix}\mathbf{P}^{-1} <^{-} \mathbf{Q}^{-1}\begin{pmatrix}\mathbf{I_r} & \mathbf{L_1}\\ \mathbf{M_1} & \mathbf{M_1L_1+T}\end{pmatrix}\mathbf{P}^{-1}$$
$$\Leftrightarrow \mathbf{Q}^{-1}\begin{pmatrix}\mathbf{I_r} & \mathbf{L_1}\\ \mathbf{M_1} & \mathbf{M_1L_1+S}\end{pmatrix}\mathbf{P}^{-1} <^{-} \mathbf{Q}^{-1}\begin{pmatrix}\mathbf{I_r} & \mathbf{L_2}\\ \mathbf{M_2} & \mathbf{M_2L_2+T}\end{pmatrix}\mathbf{P}^{-1}.$$

Proof. Clearly, it is enough to show

$$\begin{pmatrix}\mathbf{I_r} & \mathbf{L_2}\\ \mathbf{M_2} & \mathbf{M_2L_2+S}\end{pmatrix} <^{-} \begin{pmatrix}\mathbf{I_r} & \mathbf{L_1}\\ \mathbf{M_1} & \mathbf{M_1L_1+T}\end{pmatrix}$$
$$\Leftrightarrow \begin{pmatrix}\mathbf{I_r} & \mathbf{L_1}\\ \mathbf{M_1} & \mathbf{M_1L_1+S}\end{pmatrix} <^{-} \begin{pmatrix}\mathbf{I_r} & \mathbf{L_2}\\ \mathbf{M_2} & \mathbf{M_2L_2+T}\end{pmatrix}.$$

We can rewrite

$$\begin{pmatrix}\mathbf{I_r} & \mathbf{L_1}\\ \mathbf{M_1} & \mathbf{M_1L_1+S}\end{pmatrix} = \begin{pmatrix}\mathbf{I_r} & \mathbf{0}\\ \mathbf{M_1} & \mathbf{I}\end{pmatrix}\begin{pmatrix}\mathbf{I_r} & \mathbf{0}\\ \mathbf{0} & \mathbf{S}\end{pmatrix}\begin{pmatrix}\mathbf{I_r} & \mathbf{L_1}\\ \mathbf{0} & \mathbf{I}\end{pmatrix} = \mathbf{X}\begin{pmatrix}\mathbf{I_r} & \mathbf{0}\\ \mathbf{0} & \mathbf{S}\end{pmatrix}\mathbf{Y},$$

where $\mathbf{X} = \begin{pmatrix}\mathbf{I_r} & \mathbf{0}\\ \mathbf{M_1} & \mathbf{I}\end{pmatrix}$ and $\mathbf{Y} = \begin{pmatrix}\mathbf{I_r} & \mathbf{L_1}\\ \mathbf{0} & \mathbf{I}\end{pmatrix}$ are non-singular matrices. Now,

$$\mathbf{X}\begin{pmatrix}\mathbf{I_r} & \mathbf{0}\\ \mathbf{0} & \mathbf{S}\end{pmatrix}\mathbf{Y} <^{-} \begin{pmatrix}\mathbf{I_r} & \mathbf{L_2}\\ \mathbf{M_2} & \mathbf{M_2L_2+T}\end{pmatrix}$$
$$\Leftrightarrow \begin{pmatrix}\mathbf{I_r} & \mathbf{0}\\ \mathbf{0} & \mathbf{S}\end{pmatrix} <^{-} \mathbf{X}^{-1}\begin{pmatrix}\mathbf{I_r} & \mathbf{L_2}\\ \mathbf{M_2} & \mathbf{M_2L_2+T}\end{pmatrix}\mathbf{Y}^{-1}.$$

However,

$$\mathbf{X}^{-1}\begin{pmatrix}\mathbf{I_r} & \mathbf{L_2}\\ \mathbf{M_2} & \mathbf{M_2L_2+T}\end{pmatrix}\mathbf{Y}^{-1} = \begin{pmatrix}\mathbf{I_r} & \mathbf{L_2-L_1}\\ \mathbf{M_2-M_1} & \mathbf{(L_2-L_1)(M_2-M_1)+T}\end{pmatrix}.$$

Write $\mathbf{L} = \mathbf{L_2} - \mathbf{L_1}$ and $\mathbf{M} = \mathbf{M_2} - \mathbf{M_1}$. So,

$$\begin{pmatrix}\mathbf{I_r} & \mathbf{0}\\ \mathbf{0} & \mathbf{S}\end{pmatrix} <^{-} \begin{pmatrix}\mathbf{I_r} & \mathbf{L}\\ \mathbf{M} & \mathbf{ML+T}\end{pmatrix}$$
$$\Leftrightarrow r + \rho(\mathbf{S}) + \rho\begin{pmatrix}\mathbf{0} & \mathbf{L}\\ \mathbf{M} & \mathbf{ML+T-S}\end{pmatrix} = \mathbf{r} + \rho(\mathbf{T})$$
$$\Leftrightarrow \rho(\mathbf{S}) + \rho\begin{pmatrix}\mathbf{0} & \mathbf{L}\\ \mathbf{M} & \mathbf{ML+T-S}\end{pmatrix} = \rho(\mathbf{T})$$
$$\Leftrightarrow \rho\begin{pmatrix}\mathbf{0} & \mathbf{L_2-L_1}\\ \mathbf{M_2-M_1} & \mathbf{(M_2-M_1)(L_2-L_1)+T-S}\end{pmatrix} = \rho(\mathbf{T}) - \rho(\mathbf{S})$$
$$\Leftrightarrow \rho\begin{pmatrix}\mathbf{0} & \mathbf{L_1-L_2}\\ \mathbf{M_1-M_2} & \mathbf{(M_1-M_2)(L_1-L_2)+T-S}\end{pmatrix} = \rho(\mathbf{T}) - \rho(\mathbf{S})$$
$$\Leftrightarrow \begin{pmatrix}\mathbf{I_r} & \mathbf{0}\\ \mathbf{0} & \mathbf{S}\end{pmatrix} <^{-} \begin{pmatrix}\mathbf{0} & \mathbf{L_1-L_2}\\ \mathbf{M_1-M_2} & \mathbf{(M_1-M_2)(L_1-L_2)+T-S}\end{pmatrix}$$
$$\Leftrightarrow \mathbf{Q}^{-1}\begin{pmatrix}\mathbf{I_r} & \mathbf{L_1}\\ \mathbf{M_1} & \mathbf{M_1L_1+S}\end{pmatrix}\mathbf{P}^{-1} <^{-} \mathbf{Q}^{-1}\begin{pmatrix}\mathbf{I_r} & \mathbf{L_2}\\ \mathbf{M_2} & \mathbf{M_2L_2+T}\end{pmatrix}\mathbf{P}^{-1}.$$

□

Corollary 14.5.3. *In the setup of Theorem 14.5.2,*

$$\mathbf{Q}^{-1}\begin{pmatrix}\mathbf{I_r} & \mathbf{L_1}\\ \mathbf{M_1} & \mathbf{M_1L_1}+\mathbf{S}\end{pmatrix}\mathbf{P}^{-1} <^- \mathbf{Q}^{-1}\begin{pmatrix}\mathbf{I_r} & \mathbf{L_2}\\ \mathbf{M_2} & \mathbf{M_2L_2}+\mathbf{T}\end{pmatrix}\mathbf{P}^{-1}$$

if and only if

(i) $\mathbf{S} <^- \mathbf{T}$
(ii) $\mathcal{C}(\mathbf{M_1}-\mathbf{M_2}) \subseteq \mathcal{C}(\mathbf{T}-\mathbf{S})$
(iii) $\mathcal{C}(\mathbf{L_1}-\mathbf{L_2})^t \subseteq \mathcal{C}(\mathbf{T}-\mathbf{S})^t$ *and*
(iv) $(\mathbf{L_1}-\mathbf{L_2})(\mathbf{T}-\mathbf{S})^-(\mathbf{M_1}-\mathbf{M_2}) = \mathbf{0}$.

Proof. From the proof of Theorem 14.5.2

$$\mathbf{Q}^{-1}\begin{pmatrix}\mathbf{I_r} & \mathbf{L_1}\\ \mathbf{M_1} & \mathbf{M_1L_1}+\mathbf{S}\end{pmatrix}\mathbf{P}^{-1} <^- \mathbf{Q}^{-1}\begin{pmatrix}\mathbf{I_r} & \mathbf{L_2}\\ \mathbf{M_2} & \mathbf{M_2L_2}+\mathbf{T}\end{pmatrix}\mathbf{P}^{-1}$$

$$\Leftrightarrow \rho\begin{pmatrix}\mathbf{0} & \mathbf{L_2}-\mathbf{L_1}\\ \mathbf{M_2}-\mathbf{M_1} & (\mathbf{M_2}-\mathbf{M_1})(\mathbf{L_2}-\mathbf{L_1})+\mathbf{T}-\mathbf{S}\end{pmatrix} = \rho(\mathbf{T})-\rho(\mathbf{S}).$$

However,

$$\rho\begin{pmatrix}\mathbf{0} & \mathbf{L_2}-\mathbf{L_1}\\ \mathbf{M_2}-\mathbf{M_1} & (\mathbf{M_2}-\mathbf{M_1})(\mathbf{L_2}-\mathbf{L_1})+\mathbf{T}-\mathbf{S}\end{pmatrix} \geq \rho(\mathbf{T}-\mathbf{S}).$$

So, $\rho(\mathbf{T})-\rho(\mathbf{S}) \geq \rho(\mathbf{T}-\mathbf{S})$. Further, equality sign holds

$$\Leftrightarrow \rho(\mathbf{T})-\rho(\mathbf{S}) = \rho(\mathbf{T}-\mathbf{S})$$

$$\Leftrightarrow \rho\begin{pmatrix}\mathbf{0} & \mathbf{L_1}-\mathbf{L_2}\\ \mathbf{M_1}-\mathbf{M_2} & (\mathbf{M_1}-\mathbf{M_2})(\mathbf{L_1}-\mathbf{L_2})+\mathbf{T}-\mathbf{S}\end{pmatrix} = \rho(\mathbf{T}-\mathbf{S}).$$

Now the result follows. □

Let $\mathbf{A} = \mathbf{P}\mathbf{diag}\,(\mathbf{I_r}\ ,\ \mathbf{0})\,\mathbf{Q}$ be an $m\times n$ matrix of rank r, where $\mathbf{P}$ and $\mathbf{Q}$ are non-singular. In view of Theorem 14.4.10 and Remark 14.4.2, each equivalence class under the relation " $\cong$" is determined by some outer inverse $\mathbf{X_L} = \mathbf{Q}^{-1}\begin{pmatrix}\mathbf{L} & \mathbf{0}\\ \mathbf{0} & \mathbf{0}\end{pmatrix}\mathbf{P}^{-1}$ of $\mathbf{A}$, where $\mathbf{L}$ is idempotent. Distinct idempotent matrices $\mathbf{L}$ of order $r\times r$ lead to distinct equivalence classes. In fact,

$$\underline{\mathrm{Eq}}(\mathbf{X_L}|\mathbf{A}) = \left\{\mathbf{Q}^{-1}\begin{pmatrix}\mathbf{L} & \mathbf{T_1}\\ \mathbf{T_2} & \mathbf{T_2T_1}\end{pmatrix}\mathbf{P}^{-1}\right\},$$

where $\mathbf{T_1}$ and $\mathbf{T_2}$ are arbitrary subject to $\mathbf{LT_1} = \mathbf{T_1}$ and $\mathbf{T_2L} = \mathbf{T_2}$.

We now show that every reflexive g-inverse of $\mathbf{A}$ is above a unique outer inverse in each equivalence class under the relation " $\cong$".

Theorem 14.5.4. *Let* $\mathbf{A} = \mathbf{P}\mathbf{diag}\,(\mathbf{I_r}\ ,\ \mathbf{0})\,\mathbf{Q}$ *be an* $m\times n$ *matrix of rank* r *and* $\mathbf{G} = \mathbf{Q}^{-1}\begin{pmatrix}\mathbf{I_r} & \mathbf{N}\\ \mathbf{M} & \mathbf{MN}\end{pmatrix}\mathbf{P}^{-1}$ *be a reflexive g-inverse of* $\mathbf{A}$, *where* $\mathbf{P}$ *and*

$\mathbf{Q}$ *are non-singular. Let* $\mathbf{X_L} = \mathbf{Q}^{-1}\begin{pmatrix}\mathbf{L} & \mathbf{0}\\ \mathbf{0} & \mathbf{0}\end{pmatrix}\mathbf{P}^{-1}$, *where* $\mathbf{L}$ *is an idempotent matrix of order* $r\times r$. *Then there is a unique outer inverse of* $\mathbf{X} \in \underline{\mathbf{Eq}}(\mathbf{X_L}|\mathbf{A})$ *such that* $\mathbf{X} <^- \mathbf{G}$ *and is given by* $\mathbf{X} = \mathbf{Q}^{-1}\begin{pmatrix}\mathbf{L} & \mathbf{LN}\\ \mathbf{ML} & \mathbf{MLN}\end{pmatrix}\mathbf{P}^{-1}$.

Proof. Notice that $\mathbf{X} \in \underline{\mathbf{Eq}}(\mathbf{X_L}|\mathbf{A}) \Leftrightarrow \mathbf{Q}^{-1}\begin{pmatrix}\mathbf{L} & \mathbf{S_1}\\ \mathbf{S_2} & \mathbf{S_2S_1}\end{pmatrix}\mathbf{P}^{-1}$, where $\mathbf{LS_1} = \mathbf{S_1}$ and $\mathbf{S_2L} = \mathbf{S_2}$. Now,

$$\mathbf{X} <^- \mathbf{G} \Leftrightarrow \begin{pmatrix}\mathbf{L} & \mathbf{S_1}\\ \mathbf{S_2} & \mathbf{S_2S_1}\end{pmatrix} <^- \begin{pmatrix}\mathbf{I_r} & \mathbf{N}\\ \mathbf{M} & \mathbf{MN}\end{pmatrix}$$

$$\Leftrightarrow \begin{pmatrix}\mathbf{I_r} & \mathbf{0}\\ \mathbf{0} & \mathbf{0}\end{pmatrix} <^- \begin{pmatrix}\mathbf{I_r} & \mathbf{N}-\mathbf{S_1}\\ \mathbf{M}-\mathbf{S_2} & (\mathbf{M}-\mathbf{S_2})(\mathbf{N}-\mathbf{S_1})\end{pmatrix}$$

However,

$$\rho\begin{pmatrix}\mathbf{I_r} & \mathbf{N}-\mathbf{S_1}\\ \mathbf{M}-\mathbf{S_2} & (\mathbf{M}-\mathbf{S_2})(\mathbf{N}-\mathbf{S_1})\end{pmatrix} = r.$$

So,

$$\begin{pmatrix}\mathbf{I_r} & \mathbf{0}\\ \mathbf{0} & \mathbf{0}\end{pmatrix} <^- \begin{pmatrix}\mathbf{I_r} & \mathbf{N}-\mathbf{S_1}\\ \mathbf{M}-\mathbf{S_2} & (\mathbf{M}-\mathbf{S_2})(\mathbf{N}-\mathbf{S_1})\end{pmatrix}$$
$$\Leftrightarrow \rho\begin{pmatrix}\mathbf{I_r}-\mathbf{L} & \mathbf{N}-\mathbf{S_1}\\ \mathbf{M}-\mathbf{S_2} & (\mathbf{M}-\mathbf{S_2})(\mathbf{N}-\mathbf{S_1})\end{pmatrix}$$
$$= r - \rho(\mathbf{L}) = \rho(\mathbf{I_r}-\mathbf{L})$$
$$\Leftrightarrow \mathcal{C}(\mathbf{N}-\mathbf{S_1}) \subseteq \mathcal{C}(\mathbf{I_r}-\mathbf{L}) \text{ and } \mathcal{C}((\mathbf{M}-\mathbf{S_2})^{\mathbf{t}}) \subseteq \mathcal{C}((\mathbf{I_r}-\mathbf{L})^{\mathbf{t}}).$$

Let $\mathcal{C}(\mathbf{N}-\mathbf{S_1}) \subseteq \mathcal{C}(\mathbf{I_r}-\mathbf{L})$ and $\mathcal{C}((\mathbf{M}-\mathbf{S_2})^{\mathbf{t}}) \subseteq \mathcal{C}((\mathbf{I_r}-\mathbf{L})^{\mathbf{t}})$. Then $\mathbf{N}-\mathbf{S_1} = (\mathbf{I_r}-\mathbf{L})\mathbf{U}$ and $\mathbf{M}-\mathbf{S_2} = \mathbf{V}(\mathbf{I_r}-\mathbf{L})$ for some matrices $\mathbf{U}$ and $\mathbf{V}$. So, $\mathbf{L}(\mathbf{N}-\mathbf{S_1}) = \mathbf{0}$ and $(\mathbf{M}-\mathbf{S_2})\mathbf{L} = \mathbf{0}$ or $\mathbf{LN} = \mathbf{LS_1} = \mathbf{S_1}$ and $\mathbf{ML} - \mathbf{S_2L} = \mathbf{S_2}$. Hence, $\mathbf{S_1} = \mathbf{LN}$ and $\mathbf{S_2} = \mathbf{ML}$.

If $\mathbf{S_1} = \mathbf{LN}$ and $\mathbf{S_2} = \mathbf{ML}$, then $\mathbf{N}-\mathbf{S_1} = \mathbf{N}-\mathbf{LN} = (\mathbf{I}-\mathbf{L})\mathbf{N}$ and $\mathbf{M}-\mathbf{S_2} = \mathbf{M}(\mathbf{I}-\mathbf{L})$. Thus, $\mathcal{C}(\mathbf{N}-\mathbf{S_1}) \subseteq \mathcal{C}(\mathbf{I_r}-\mathbf{L})$ and $\mathcal{C}((\mathbf{M}-\mathbf{S_2})^{\mathbf{t}}) \subseteq \mathcal{C}((\mathbf{I_r}-\mathbf{L})^{\mathbf{t}}) \Leftrightarrow \mathbf{S_1} = \mathbf{LN}$ and $\mathbf{S_2} = \mathbf{ML}$. □

Corollary 14.5.5. *Let* $\mathbf{A} = \mathbf{P}\mathbf{diag}\left(\mathbf{I_r}\ ,\mathbf{0}\right)\mathbf{Q}$ *be an* $m \times n$ *matrix of rank* r, *where* $\mathbf{P}$ *and* $\mathbf{Q}$ *are non-singular. Let* $\mathbf{X_{L_i}} = \mathbf{Q}^{-1}\begin{pmatrix}\mathbf{L_i} & \mathbf{0}\\ \mathbf{0} & \mathbf{0}\end{pmatrix}\mathbf{P}^{-1}$ *and* $\mathbf{G_i} = \mathbf{Q}^{-1}\begin{pmatrix}\mathbf{L_i} & \mathbf{N_i}\\ \mathbf{M_i} & \mathbf{M_iN_i}\end{pmatrix}\mathbf{P}^{-1} \in \underline{\mathbf{Eq}}(\mathbf{X_{L_i}}|\mathbf{A})$ *where* $i = 1,\ 2$, *and each* $\mathbf{L_i}$ *is idempotent. Then* $\mathbf{G_1} <^- \mathbf{G_2}$ *if and only if* $\mathbf{L_1L_2} = \mathbf{L_2L_1} = \mathbf{L_1}$, *equivalently* $\mathbf{L_1} <^- \mathbf{L_2}, \mathbf{N_1} = \mathbf{L_1N_2}$ *and* $\mathbf{M_1} = \mathbf{M_2L_1}$.

We are now ready to describe the scheme of the diagram. Let $\mathbf{A} = \mathbf{P}\mathbf{diag}\left(\mathbf{I_r}\ , \mathbf{0}\right)\mathbf{Q}$ be a given matrix, where $\mathbf{P}$ and $\mathbf{Q}$ are non-singular. By Theorem 14.2.6, the equivalence class of $\mathbf{G_0} = \mathbf{Q}^{-1}\mathbf{diag}\left(\mathbf{I_r}\ , \mathbf{0}\right)\mathbf{P}^{-1}$ is given by $\overline{\mathrm{Eq}}(\mathbf{G_0}|\mathbf{A}) = \{\mathbf{Q}^{-1}\mathbf{diag}\left(\mathbf{I_r}\ , \mathbf{S}\right)\mathbf{P}^{-1} : \mathbf{S} \text{ is arbitrary}\}$.

In the diagram, every g-inverse in an equivalence class is denoted by a node and we identify a node with the g-inverse it represents, thus making no distinction between the two. Two nodes differing by rank 1 are connected by a line if the node of lower rank is below the node with higher rank under the minus order. Let $\mathcal{S}_j = \{\mathbf{Q}^{-1}\mathbf{diag}\left(\mathbf{I_r}\ , \mathbf{S}\right)\mathbf{P}^{-1}\}$, where the matrix $\mathbf{S}$ is an arbitrary matrix with $\rho(\mathbf{S}) = j$. Then $\mathbf{G_0}$ is connected to every g-inverse in $\mathcal{S}_j$, by Theorem 14.2.4. A g-inverse $\mathbf{Q}^{-1}\mathbf{diag}\left(\mathbf{I_r}\ , \ \mathbf{S}\right)\mathbf{P}^{-1} \in \mathcal{S}_\mathbf{j}$ is connected to a g-inverse $\mathbf{Q}^{-1}\mathbf{diag}\left(\mathbf{I_r}\ , \ \mathbf{T}\right)\mathbf{P}^{-1} \in \mathcal{S}_{\mathbf{j+1}} \Leftrightarrow \mathbf{S} <^- \mathbf{T}$, by Theorem 14.5.1. This completes the diagram for the g-inverses in $\overline{\mathrm{Eq}}(\mathbf{G_0}|\mathbf{A})$. In view of Theorems 14.2.4 and 14.5.1, the diagram for g-inverses of $\overline{\mathrm{Eq}}(\mathbf{G}|\mathbf{A})$ where $\mathbf{G} = \mathbf{Q}^{-1}\begin{pmatrix}\mathbf{I_r} & \mathbf{N}\\ \mathbf{M} & \mathbf{MN}\end{pmatrix}\mathbf{P}^{-1}$ is any other reflexive g-inverse is just a replica of the diagram for the g-inverses where the correspondence is given by $\mathbf{Q}^{-1}\mathbf{diag}\left(\mathbf{I_r}\ , \mathbf{S}\right)\mathbf{P}^{-1} \longleftrightarrow \mathbf{Q}^{-1}\begin{pmatrix}\mathbf{I_r} & \mathbf{N}\\ \mathbf{M} & \mathbf{MN}+\mathbf{S}\end{pmatrix}\mathbf{P}^{-1}$.

Let $\{\mathbf{L}_\alpha, \alpha \in \Omega\}$ be the class of all idempotent matrices of order $r \times r$. For each $\alpha \in \Omega$, write $\mathbf{X}_{\mathbf{L}_\alpha} = \mathbf{Q}^{-1}\mathbf{diag}\left(\mathbf{L}_\alpha\ , \mathbf{0}\right)\mathbf{P}^{-1}$. Then the equivalence classes under the relation "$\cong$" are given by $\{\underline{\mathrm{Eq}}(\mathbf{X}_{\mathbf{L}_\alpha}|\mathbf{A})\}_{\alpha\in\Omega}$, where $\underline{\mathrm{Eq}}(\mathbf{X}_{\mathbf{L}_\alpha}|\mathbf{A}) = \left\{\mathbf{Q}^{-1}\begin{pmatrix}\mathbf{L}_\alpha & \mathbf{N}\\ \mathbf{M} & \mathbf{MN}\end{pmatrix}\mathbf{P}^{-1} : \mathbf{L}_\alpha\mathbf{N} = \mathbf{N}, \mathbf{M}\mathbf{L}_\alpha = \mathbf{M}\right\}$. Consider two idempotent matrices $\mathbf{L_1}$ and $\mathbf{L_2}$ of order $r \times r$ and of ranks s and $s+1$ respectively, $0 \leq s \leq r-1$. Consider

$$\mathbf{X_2} = \mathbf{Q}^{-1}\begin{pmatrix}\mathbf{L_2} & \mathbf{N_2}\\ \mathbf{M_2} & \mathbf{M_2N_2}\end{pmatrix}\mathbf{P}^{-1} \in \underline{\mathrm{Eq}}(\mathbf{Q}^{-1}\mathbf{diag}\left(\mathbf{L_2}\ , \mathbf{0}\right)\mathbf{P}^{-1}|\mathbf{A}),$$

where $\mathbf{M_2}$ and $\mathbf{N_2}$ satisfy $\mathbf{L_2N_2} = \mathbf{N_2}$ and $\mathbf{M_2L_2} = \mathbf{M_2}$. Then by Corollary 14.5.5, there exists an outer inverse in the equivalence class $\underline{\mathrm{Eq}}(\mathbf{Q}^{-1}\mathbf{diag}\left(\mathbf{L_1}\ , \mathbf{0}\right)\mathbf{P}^{-1}|\mathbf{A})$ if and only if $\mathbf{L_1} <^- \mathbf{L_2}$. In such a case the unique outer inverse $\mathbf{X_1} \in \underline{\mathrm{Eq}}(\mathbf{Q}^{-1}\mathbf{diag}\left(\mathbf{L_1}\ , \mathbf{0}\right)\mathbf{P}^{-1}|\mathbf{A})$ such that $\mathbf{X_1} <^- \mathbf{X_2}$ is given by $\mathbf{X_1} = \mathbf{Q}^{-1}\begin{pmatrix}\mathbf{L_1} & \mathbf{L_1N_2}\\ \mathbf{M_2L_1} & \mathbf{M_2L_1N_2}\end{pmatrix}\mathbf{P}^{-1}$.

From the preceding discussion, it is clear that if we have the complete diagram with respect to one equivalence class, then we can construct the diagram for the complete class of g-inverses. In the case of outer inverses same can be achieved by using the Corollary 14.5.5 and the above discussion. Thus, a complete diagram can be constructed.

Example 14.5.6. For this example, let $\mathbf{F} = \mathbf{GF}(2)$. Let $\mathbf{A} = \begin{pmatrix} \mathbf{B} & \mathbf{0} \\ \mathbf{0} & \mathbf{0} \end{pmatrix}$ be a 4×4 matrix, where $\mathbf{B} = \begin{pmatrix} 1 & 1 \\ 0 & 1 \end{pmatrix}$. It is easy to see that that $\mathbf{A} = \mathbf{P}\begin{pmatrix} \mathbf{I_2} & \mathbf{0} \\ \mathbf{0} & \mathbf{0} \end{pmatrix}$ is a normal form of $\mathbf{A}$, where $\mathbf{P} = \begin{pmatrix} \mathbf{B} & \mathbf{0} \\ \mathbf{0} & \mathbf{I} \end{pmatrix}$. Notice that $\mathbf{B}^{-1} = \mathbf{B}$ and therefore each g-inverse of $\mathbf{A}$ is of the form $\begin{pmatrix} \mathbf{I_2} & \mathbf{N} \\ \mathbf{M} & \mathbf{K} \end{pmatrix}\mathbf{P}$, where $\mathbf{N}, \mathbf{M}, \mathbf{K}$ are 2×2 matrices. Thus, there are $4096 = 2^{12}$ g-inverses of $\mathbf{A}$. Since every reflexive g-inverse of $\mathbf{A}$ is of the form $\begin{pmatrix} \mathbf{I_2} & \mathbf{N} \\ \mathbf{M} & \mathbf{MN} \end{pmatrix}\mathbf{P}$, there are $256 = 2^8$ reflexive g-inverse of $\mathbf{A}$. So, there are there are $16 = 2^4$ matrices of order 2×2 that play important role in enumerating all the g-inverses of $\mathbf{A}$ in a particular equivalence class. We first list out these 16 matrices:

Rank 0: $\mathbf{S_1} = \begin{pmatrix} 0 & 0 \\ 0 & 0 \end{pmatrix}$

Rank 1: $\mathbf{S_2} = \begin{pmatrix} 1 & 0 \\ 0 & 0 \end{pmatrix}$, $\mathbf{S_3} = \begin{pmatrix} 0 & 1 \\ 0 & 0 \end{pmatrix}$, $\mathbf{S_4} = \begin{pmatrix} 1 & 1 \\ 0 & 0 \end{pmatrix}$, $\mathbf{S_5} = \begin{pmatrix} 0 & 0 \\ 1 & 0 \end{pmatrix}$,

$\mathbf{S_6} = \begin{pmatrix} 0 & 0 \\ 0 & 1 \end{pmatrix}$, $\mathbf{S_7} = \begin{pmatrix} 0 & 0 \\ 1 & 1 \end{pmatrix}$, $\mathbf{S_8} = \begin{pmatrix} 1 & 0 \\ 1 & 0 \end{pmatrix}$, $\mathbf{S_9} = \begin{pmatrix} 0 & 1 \\ 0 & 1 \end{pmatrix}$, $\mathbf{S_{10}} = \begin{pmatrix} 1 & 1 \\ 1 & 1 \end{pmatrix}$

Rank 2: $\mathbf{S_{11}} = \begin{pmatrix} 1 & 0 \\ 0 & 1 \end{pmatrix}$, $\mathbf{S_{12}} = \begin{pmatrix} 1 & 0 \\ 1 & 1 \end{pmatrix}$, $\mathbf{S_{13}} = \begin{pmatrix} 1 & 1 \\ 0 & 1 \end{pmatrix}$,

$\mathbf{S_{14}} = \begin{pmatrix} 0 & 0 \\ 0 & 0 \end{pmatrix}$, $\mathbf{S_{15}} = \begin{pmatrix} 0 & 1 \\ 1 & 1 \end{pmatrix}$, $\mathbf{S_{16}} = \begin{pmatrix} 1 & 1 \\ 1 & 0 \end{pmatrix}$.

Let $\mathbf{G_1} = \mathbf{diag}\left(\mathbf{I_2}\ ,\ \mathbf{0}\right)\mathbf{P} = \mathbf{diag}\left(\mathbf{I_2}\ ,\mathbf{0}\right)\mathbf{diag}\left(\mathbf{B}\ ,\ \mathbf{I}\right) = \mathbf{A}$. Consider the equivalence class $\overline{\mathrm{Eq}}(\mathbf{G_1}|\mathbf{A})$. Notice that $\mathbf{A}$ is of index 1 and $\mathbf{G_1}$ is the group inverse of $\mathbf{A}$. The g-inverses in $\overline{\mathrm{Eq}}(\mathbf{G_1}|\mathbf{A})$ are given by $\mathbf{G_i} = \mathbf{G_1} + \mathbf{diag}\left(\mathbf{0}\ ,\ \mathbf{S_i}\right)\mathbf{P}, i = 1, \ldots, 16$. Moreover, if $\mathbf{H_1} = \begin{pmatrix} \mathbf{I_2} & \mathbf{N} \\ \mathbf{M} & \mathbf{MN} \end{pmatrix}\mathbf{P}$ is another reflexive g-inverse of $\mathbf{A}$, then $\overline{\mathrm{Eq}}(\mathbf{H_1}|\mathbf{A}) = \{\mathbf{H_i} : i = 1, \ldots, 16.\}$, where $\mathbf{H_i} = \mathbf{H_1} + \mathbf{diag}\left(\mathbf{0}\ ,\ \mathbf{S_i}\right)\mathbf{P}$. Notice the one-one correspondence between the g-inverses of $\overline{\mathrm{Eq}}(\mathbf{G_1}|\mathbf{A})$ and those in $\overline{\mathrm{Eq}}(\mathbf{H_1}|\mathbf{A})$. Henceforth we concentrate on the equivalence class $\overline{\mathrm{Eq}}(\mathbf{G_1}|\mathbf{A})$. Note that $\rho(\mathbf{G_1}) = \mathbf{2}$, $\rho(\mathbf{G_i}) = \mathbf{3}$ for $i = 2, \ldots, 10$ and $\rho(\mathbf{G_i}) = \mathbf{4}$ for $i = 11, \ldots, 16$. In view of Theorem 14.2.5, $\mathbf{G_1} <^- \mathbf{G_i}, i = 2, \ldots, 10$. In fact $\mathbf{G_1} <^- \mathbf{G_i}, i = 2, \ldots, 16$. In order to determine the g-inverses among $\mathbf{G_{11}}$ to $\mathbf{G_{16}}$ which are above $\mathbf{G_2}$ under he minus order, we need to determine which matrices amongst $\mathbf{S_{11}}$ to $\mathbf{S_{16}}$ lie above $\mathbf{S_2}$. It is easy to see that $\mathbf{S_2} <^- \mathbf{S_{1i}}, i = 1, 2, 3, 5$. By

Theorem 14.5.3, it follows that $\mathbf{G_2} <^- \mathbf{G_{1i}}, i = 1, 2, 3, 5$. Similar checking with respect to $\mathbf{S_3}$ to $\mathbf{S_{10}}$ yields the following table, which enumerates for each of $\mathbf{G_2}$ to $\mathbf{G_{10}}$, the g-inverses of rank 4 belonging to $\overline{\text{Eq}}(\mathbf{G_1}|\mathbf{A})$ that are above g-inverses of rank 3. For example, in Table 14.1, we notice that $\mathbf{G_8}$ is below $\mathbf{G_{11}}, \mathbf{G_{12}}, \mathbf{G_{14}}$ and $\mathbf{G_{16}}$ under the minus order. Consider $\overline{\text{Eq}}(\mathbf{H_1}|\mathbf{A})$. In view of Theorem 14.5.1, if we replace $\mathbf{G}$ by $\mathbf{H}$ in the table, we get for each of g-inverse of rank 3, the g-inverse of rank 4 which are above $\mathbf{H_1}$ under the minus order. Thus for example, $\mathbf{H_5}$ is below $\mathbf{H_{13}}, \mathbf{H_{14}}, \mathbf{H_{15}}$ and $\mathbf{H_{16}}$.

Table 14.1 Rank 3 g-inverses below the rank 4 g-inverses of $\overline{\text{Eq}}(\mathbf{G_1}|\mathbf{A})$

g-inverse of rank 3	**g-inverse of rank 4**
g-inverse of rank 3	g-inverse of rank 4
G_2	$G_{11}, G_{12}, G_{13}, G_{15}$
G_3	$G_{12}, G_{14}, G_{15}, G_{16}$
G_4	$G_{11}, G_{13}, G_{14}, G_{16}$
G_5	$G_{13}, G_{14}, G_{15}, G_{16}$
G_6	$G_{11}, G_{12}, G_{13}, G_{16}$
G_7	$G_{11}, G_{12}, G_{14}, G_{15}$
G_8	$G_{11}, G_{12}, G_{14}, G_{16}$
G_9	$G_{11}, G_{13}, G_{14}, G_{15}$
G_{10}	$G_{12}, G_{13}, G_{15}, G_{16}$

Let us now do the same for $\mathbf{H_1} = \begin{pmatrix} \mathbf{I_2} & \mathbf{N} \\ \mathbf{0} & \mathbf{0} \end{pmatrix} \mathbf{P}$, where $\mathbf{N} = \begin{pmatrix} 0 & 0 \\ 0 & 1 \end{pmatrix}$. We shall now examine if any of the $\mathbf{G}$'s is below any of the $\mathbf{H}$'s. We first check that whether $\mathbf{G_1}$ is below any of $\mathbf{H_2}$ to $\mathbf{H_{10}}$. Recall that $\mathbf{G_1} = \begin{pmatrix} \mathbf{I_2} & \mathbf{0} \\ \mathbf{0} & \mathbf{0} \end{pmatrix} \mathbf{P}$ and $\mathbf{H_i} = \begin{pmatrix} \mathbf{I_2} & \mathbf{N} \\ \mathbf{0} & \mathbf{S_i} \end{pmatrix} \mathbf{P}$, $i = 2, \ldots, 10$.

Table 14.2 g-inverses in $\overline{\text{Eq}}(\mathbf{G_1}|\mathbf{A})$ below g-inverses in $\overline{\text{Eq}}(\mathbf{H_1}|\mathbf{A})$

g-inverse of rank 3	**g-inverse of rank 4**
G_2	H_{11}, H_{13}
G_4	H_{11}, H_{13}
G_5	H_{14}, H_{15}
G_7	H_{14}, H_{15}
G_8	H_{12}, H_{16}
G_{10}	H_{12}, H_{16}

By Corollary 14.5.3, $\mathbf{G_1} <^- \mathbf{H_i} \Leftrightarrow \mathcal{C}(\mathbf{N^t}) \subseteq \mathcal{C}(\mathbf{S_i})$. ($\mathbf{0}(\mathbf{S_i}-\mathbf{0})^-\mathbf{N} = \mathbf{0}$, $\mathbf{0} <^- \mathbf{S_i}$, and $\mathcal{C}(\mathbf{0}) \subseteq \mathcal{C}(\mathbf{S_i}-\mathbf{0})$ hold automatically.) Hence, $\mathbf{G_1} <^- \mathbf{H_3}$, $\mathbf{G_1} <^- \mathbf{H_6}$ and $\mathbf{G_1} <^- \mathbf{H_9}$. Again by using Corollary 14.5.3, we arrive at the following table which enumerates for each of $\mathbf{G_2}$ to $\mathbf{G_{10}}$, the matrices $\mathbf{H_{11}}$ to $\mathbf{H_{16}}$ which are above it under the minus order.

Thus, $\mathbf{G_3}, \mathbf{G_6}$ and $\mathbf{G_9}$ are not below any of $\mathbf{H_{11}}$ to $\mathrm{H_{16}}$. From Theorem 14.5.2, it is clear that the g-inverses of rank 3 in the equivalence class $\overline{\mathrm{Eq}}(\mathbf{H_1}|\mathbf{A})$ which are below the inverses from among $\mathbf{G_{11}} - \mathbf{H_{16}}$ in $\overline{\mathrm{Eq}}(\mathbf{G_1}|\mathbf{A})$ are obtained by interchanging $\mathbf{G}$ and $\mathbf{H}$ in the Table 14.2.

We shall now turn our attention to the equivalence classes of the outer inverses of $\mathbf{A}$. As mentioned previously, these equivalence classes are linked to idempotent matrices of order 2×2. We now enumerate them according to their rank.

Rank 2: $\mathbf{L_1} = \mathbf{I_2}$

Rank 1: $\mathbf{L_2} = \begin{pmatrix}1 & 0\\0 & 0\end{pmatrix}$, $\mathbf{L_3} = \begin{pmatrix}1 & 1\\0 & 0\end{pmatrix}$, $\mathbf{L_4} = \begin{pmatrix}0 & 0\\1 & 1\end{pmatrix}$, $\mathbf{L_5} = \begin{pmatrix}1 & 0\\1 & 0\end{pmatrix}$,

$\mathbf{L_6} = \begin{pmatrix}0 & 1\\0 & 1\end{pmatrix}$, $\mathbf{L_7} = \begin{pmatrix}0 & 0\\0 & 1\end{pmatrix}$

Rank 0: $\mathbf{L_8} = \begin{pmatrix}0 & 0\\0 & 0\end{pmatrix}$.

Thus, there are 8 equivalence classes of the outer inverses of $\mathbf{A}$ under the relation "$\cong$". The equivalence class $\mathcal{O}_1 = \underline{\mathrm{Eq}}(\begin{pmatrix}\mathbf{L_1} & \mathbf{0}\\ \mathbf{0} & \mathbf{0}\end{pmatrix}\mathbf{P}|\mathbf{A})$ contains all the reflexive g-inverses of $\mathbf{A}$. There are six equivalence classes of the outer inverses of rank 1, namely, $\mathcal{O}_i = \underline{\mathrm{Eq}}(\begin{pmatrix}\mathbf{L_i} & \mathbf{0}\\ \mathbf{0} & \mathbf{0}\end{pmatrix}\mathbf{P}|\mathbf{A}), \mathbf{i} = \mathbf{2}, \ldots \mathbf{7}$. Finally there is one equivalence class corresponding to $\mathbf{L_8} = \mathbf{0}$ and is $\underline{\mathrm{Eq}}(\mathbf{0}|\mathbf{A}) = \mathbf{0}$. Using Theorem 14.2.12, it can be shown that there are exactly 16 matrices in each of $\mathcal{O}_i$ $i = 2, \ldots 7$. Using the same one can enumerate the 16 matrices in $\mathcal{O}_1$ as:

1. $\begin{pmatrix}1&0&0&0\\0&0&0&0\\0&0&0&0\\0&0&0&0\end{pmatrix}\mathbf{P} = \begin{pmatrix}1&1&0&0\\0&0&0&0\\0&0&0&0\\0&0&0&0\end{pmatrix}$, **2.** $\begin{pmatrix}1&0&0&0\\0&0&0&0\\1&0&0&0\\0&0&0&0\end{pmatrix}\mathbf{P} = \begin{pmatrix}1&1&0&0\\0&0&0&0\\1&1&0&0\\0&0&0&0\end{pmatrix}$,

3. $\begin{pmatrix}1&0&0&0\\0&0&0&0\\0&0&0&0\\1&0&0&0\end{pmatrix}\mathbf{P} = \begin{pmatrix}1&1&0&0\\0&0&0&0\\0&0&0&0\\1&1&0&0\end{pmatrix}$, **4.** $\begin{pmatrix}1&0&0&0\\0&0&0&0\\1&0&0&0\\1&0&0&0\end{pmatrix}\mathbf{P} = \begin{pmatrix}1&1&0&0\\0&0&0&0\\1&1&0&0\\1&1&0&0\end{pmatrix}$,

5. $\begin{pmatrix}1&0&1&0\\0&0&0&0\\0&0&0&0\\0&0&0&0\end{pmatrix}\mathbf{P}=\begin{pmatrix}1&1&1&0\\0&0&0&0\\0&0&0&0\\0&0&0&0\end{pmatrix}$, **6.** $\begin{pmatrix}1&0&1&0\\0&0&0&0\\1&0&0&0\\0&0&0&0\end{pmatrix}\mathbf{P}=\begin{pmatrix}1&1&1&0\\0&0&0&0\\1&1&1&0\\0&0&0&0\end{pmatrix}$,

7. $\begin{pmatrix}1&0&1&0\\0&0&0&0\\0&0&0&0\\1&0&1&0\end{pmatrix}\mathbf{P}=\begin{pmatrix}1&1&0&0\\0&0&0&0\\0&0&0&0\\1&1&1&0\end{pmatrix}$, **8.** $\begin{pmatrix}1&0&1&0\\0&0&0&0\\1&1&0&0\\1&0&1&0\end{pmatrix}\mathbf{P}=\begin{pmatrix}1&1&1&0\\0&0&0&0\\1&1&1&0\\1&1&1&0\end{pmatrix}$,

9. $\begin{pmatrix}1&0&0&1\\0&0&0&0\\0&0&0&0\\0&0&0&0\end{pmatrix}\mathbf{P}=\begin{pmatrix}1&1&0&1\\0&0&0&0\\0&0&0&0\\0&0&0&0\end{pmatrix}$, **10.** $\begin{pmatrix}1&0&0&1\\0&0&0&0\\1&0&0&1\\0&0&0&0\end{pmatrix}\mathbf{P}=\begin{pmatrix}1&1&0&1\\0&0&0&0\\1&1&0&1\\0&0&0&0\end{pmatrix}$,

11. $\begin{pmatrix}1&0&0&1\\0&0&0&0\\0&0&0&0\\1&0&0&1\end{pmatrix}\mathbf{P}=\begin{pmatrix}1&1&0&1\\0&0&0&0\\0&0&0&0\\1&0&0&1\end{pmatrix}$, **12.** $\begin{pmatrix}1&0&0&1\\0&0&0&0\\1&0&0&1\\1&0&0&1\end{pmatrix}\mathbf{P}=\begin{pmatrix}1&1&0&1\\0&0&0&0\\1&1&0&1\\1&1&0&1\end{pmatrix}$,

13. $\begin{pmatrix}1&0&1&1\\0&0&0&0\\0&0&0&0\\0&0&0&0\end{pmatrix}\mathbf{P}=\begin{pmatrix}1&1&1&1\\0&0&0&0\\0&0&0&0\\0&0&0&0\end{pmatrix}$, **14.** $\begin{pmatrix}1&0&1&1\\0&0&0&0\\1&0&1&1\\0&0&0&0\end{pmatrix}\mathbf{P}=\begin{pmatrix}1&1&1&1\\0&0&0&0\\1&1&1&1\\0&0&0&0\end{pmatrix}$,

15. $\begin{pmatrix}1&0&1&1\\0&0&0&0\\0&0&0&0\\1&0&1&1\end{pmatrix}\mathbf{P}=\begin{pmatrix}1&1&1&1\\0&0&0&0\\0&0&0&0\\1&1&1&1\end{pmatrix}$, **16.** $\begin{pmatrix}1&0&1&1\\0&0&0&0\\1&0&1&1\\1&0&1&1\end{pmatrix}\mathbf{P}=\begin{pmatrix}1&1&1&1\\0&0&0&0\\1&1&1&1\\1&1&1&1\end{pmatrix}$.

It is easy to see by Theorem 14.5.4, that the outer inverses of rank 1 below $\mathbf{A}$ in each of the equivalence classes $\mathcal{O}_i$ $i = 2, \ldots 7$ are given by

$$\begin{pmatrix}1&1&0&0\\0&0&0&0\\0&0&0&0\\0&0&0&0\end{pmatrix}\text{ in }\mathcal{O}_2,\quad \begin{pmatrix}1&0&0&0\\0&0&0&0\\0&0&0&0\\0&0&0&0\end{pmatrix}\text{ in }\mathcal{O}_3,$$

$$\begin{pmatrix}0&0&0&0\\1&0&0&0\\0&0&0&0\\0&0&0&0\end{pmatrix}\text{ in }\mathcal{O}_4,\quad \begin{pmatrix}1&1&0&0\\1&1&0&0\\0&0&0&0\\0&0&0&0\end{pmatrix}\text{ in }\mathcal{O}_5,$$

$$\begin{pmatrix}0&1&0&0\\0&1&0&0\\0&0&0&0\\0&0&0&0\end{pmatrix}\text{ in }\mathcal{O}_6,\quad \begin{pmatrix}0&0&0&0\\0&1&0&0\\0&0&0&0\\0&0&0&0\end{pmatrix}\text{ in }\mathcal{O}_7.$$

These details are diagrammatically represented in the following figure:

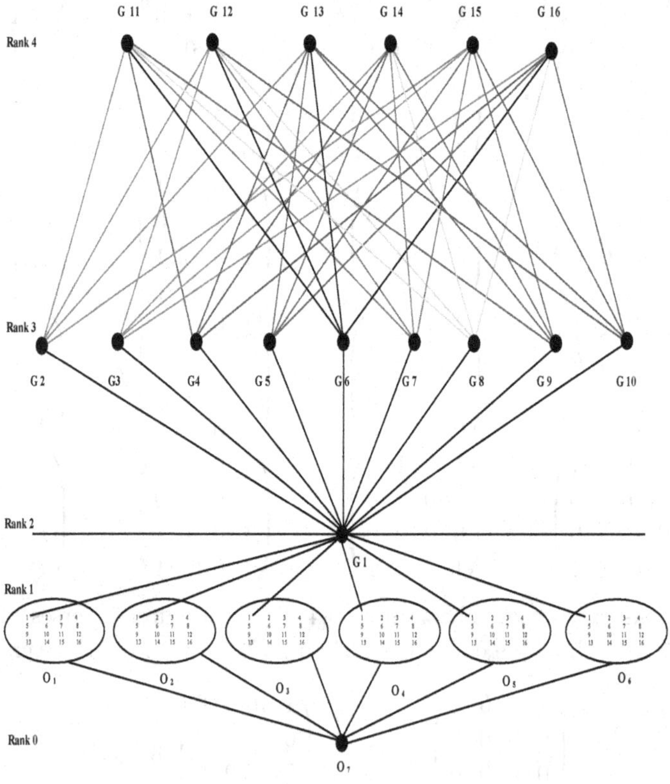

Fig. 14.1 *An example of a network.*

14.6 The Ladder

In this section we construct a ladder of matrices and their reflexive g-inverses (on either side of each step) which are in hierarchy above and below a given matrix **A** and a reflexive g-inverse of **A** using Theorems 14.2.12 and 14.2.11. We illustrate the ladder construction using the same 4×4 matrix of the previous section.

From the Figure 14.1, we notice that there is a path connecting the outer inverses and the g-inverses of the matrix **A**. One such path is

$$\mathbf{O_8} \to \mathbf{O_{11}} \to \mathbf{G_1} \to \mathbf{G_2} \to \mathbf{G_{11}}.$$

Equivalently,

$$\mathbf{O_8} <^- \mathbf{O_{11}} <^- \mathbf{G_1} <^- \mathbf{G_2} <^- \mathbf{G_{11}}.$$

Notice that $\mathbf{O_8}$ is the outer inverse of **A** with rank 0 and $\mathbf{G_{11}}$ is a g-inverse of of **A** with rank 4. From the same figure, it is also clear that such a path beginning from $\mathbf{O_8}$ is not unique, since $\mathbf{O_8} \to \mathbf{O_{21}} \to \mathbf{G_1} \to \mathbf{G_5} \to \mathbf{G_{16}}$ is yet another path. In fact, as you see in the figure, there are several paths that connect $\mathbf{O_8}$ to $\mathbf{G_1}$ and end in a g-inverse of **A** with rank 4.

Write $\mathbf{A_{-2}} = \mathbf{O_8} = \mathbf{0}, \mathbf{A_{-1}} = \mathbf{O_{11}}, \mathbf{A_0} = \mathbf{A}, \mathbf{A_1} = \mathbf{G_2}$ and $\mathbf{A_2} = \mathbf{G_{11}}$. Then $\mathbf{A_{-2}} \to \mathbf{A_{-1}} \to \mathbf{A_0} \to \mathbf{A_1} \to \mathbf{A_2}$ is also a path in the sense that $\mathbf{A_{-2}} <^- \mathbf{A_{-1}} <^- \mathbf{A_0} <^- \mathbf{A_1} <^- \mathbf{A_2}$. Further, $\mathbf{A_{-2}}, \mathbf{A_{-1}}, \mathbf{A_0}, \mathbf{A_1}$, and $\mathbf{A_2}$ are the reflexive g-inverses of $\mathbf{O_8}, \mathbf{O_{11}}, \mathbf{G_1}, \mathbf{G_2}$ and $\mathbf{G_{11}}$ respectively.

We now consider the ordered pairs $(\mathbf{A_{-2}}, \mathbf{O_8}), (\mathbf{A_{-1}}, \mathbf{O_{11}}), (\mathbf{A_0}, \mathbf{G_1})$, $(\mathbf{A_1}, \mathbf{G_2})$, $(\mathbf{A_2}, \mathbf{G_{11}})$. These pairs have the following properties:

(i) Within each pair, the coordinate matrices are reflexive g-inverses of each other.

(ii) Rank of the coordinate matrices of each successive pair beginning with $(\mathbf{A_{-2}}, \mathbf{O_8})$ is one more than the rank of the coordinate matrices of its predecessor.

(iii) Consider any two consecutive pairs. The first(the second) coordinate matrix of the former pair is below the first (respectively the second) coordinate matrix of the later pair under the minus order.

Thus, we can think of the these various pairs forming a ladder in which each pair resembles a step of the ladder. We formalize all this discussion into a definition of the ladder as follows:

Let **A** be an $m \times n$ matrix of rank r. Let $\alpha = \min\{m, n\}$. Write $\mathbf{A_0} = \mathbf{A}$. Let $\mathbf{G_0}$ be a reflexive g-inverse of **A**. Let $\mathbf{G_j}$, $j = -r, \ldots, -1$, outer inverses of **A**, $\mathbf{G_j}$, $j = 1, \ldots, r$, the g-inverses of **A** and $\mathbf{A_j}$, $j = -r, \ldots, \alpha - r$ be matrices satisfy the following conditions:

(i) $\rho(\mathbf{G_j}) = r + j$, $j = -r, \ldots, \alpha - r$,

(ii) $\mathbf{G_j} <^- \mathbf{G_{j+1}}$, $j = -r, \ldots, \alpha - r - 1$;
(iii) $\mathbf{A_j}$ is a reflexive g-inverse of $\mathbf{G_j}$, $j = -r, \ldots, \alpha - r$ and
(iv) $\mathbf{A_j} <^- \mathbf{A_{j+1}}$, $j = -r, \ldots, \alpha - r - 1$.

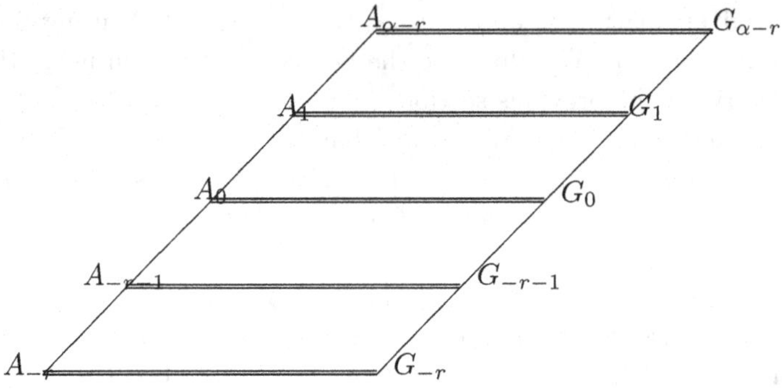

Fig. 14.2 *Ladder.*

We denote the step $(\mathbf{A_0}, \mathbf{G_0})$ as the ground level, $(\mathbf{A_{-r}}, \mathbf{G_{-r}}) = (\mathbf{0}, \mathbf{0})$ as the basement and $(\mathbf{A_{\alpha-r}}, \mathbf{G_{\alpha-r}})$ as the first floor. Thus, the ladder connects the basement to the first floor via a landing in the ground level. A ladder has the following features:

(i) Each step is characterized by a matrix and a reflexive g-inverse of the matrix.
(ii) Consider two consecutive steps $(\mathbf{A_j}, \mathbf{G_j})$ and $(\mathbf{A_{j+1}}, \mathbf{G_{j+1}})$. Then $\rho(\mathbf{A_{j+1}}) = \rho(\mathbf{A_j}) + \mathbf{1}$, $\rho(\mathbf{G_{j+1}}) = \rho(\mathbf{G_j}) + \mathbf{1}$, $\mathbf{A_j} <^- \mathbf{A_{j+1}}$, and $\mathbf{G_j} <^- \mathbf{G_{j+1}}$. Thus, it makes sense to call the step $(\mathbf{A_{j+1}}, \mathbf{G_{j+1}})$ as the step just above the step $(\mathbf{A_j}, \mathbf{G_j})$.

The ladder defined above is very flexible in the sense, that the landing (the ground level) can be adjusted anywhere between the levels $-r$ to $\alpha - r$. For example 14.5.6, we have already constructed a ladder. Does a ladder exist for every matrix $\mathbf{A}$ and a reflexive g-inverse $\mathbf{G}$? What is the class of all such ladders, in case one exists? We shall explore all this in the rest of the section.

Let $\mathbf{A}$ be an $m \times n$ matrix of rank r. Let $\mathbf{G_0}$ be reflexive g-inverse of $\mathbf{A}$. We can take $\mathbf{A} = \mathbf{P}\mathbf{diag}\left(\mathbf{I_r}\ , \mathbf{0}\right)\mathbf{Q}$ and $\mathbf{G_0} = \mathbf{Q}^{-1}\mathbf{diag}\left(\mathbf{I_r}\ , \mathbf{0}\right)\mathbf{P}^{-1}$ for some non-singular matrices $\mathbf{P}$ and $\mathbf{Q}$. Write $\mathbf{A_0} = \mathbf{A}$, and define for each $j = -r, \ldots, \alpha - r$, $\alpha = \min\{m, n\}$

$$\mathbf{A_j} = \mathbf{P}\mathbf{diag}\left(\mathbf{I_{r+j}}\ , \mathbf{0}\right)\mathbf{Q},\ \mathbf{G_j} = \mathbf{Q}^{-1}\mathbf{diag}\left(\mathbf{I_{r+j}}\ , \mathbf{0}\right)\mathbf{P}^{-1}.$$

Then clearly,

(i) $\rho(\mathbf{G_j}) = r + j,\ j = -r, \ldots, \alpha - r$
(ii) $\mathbf{G_j} <^- \mathbf{G_{j+1}},\ j = -r, \ldots, \alpha - r - 1$
(iii) $\mathbf{A_j}$ is a reflexive g-inverse of $\mathbf{G_j},\ j = -r, \ldots, \alpha - r$ and
(iv) $\mathbf{A_j} <^- \mathbf{A_{j+1}},\ j = -r, \ldots, \alpha - r - 1.$

Thus, we have constructed a ladder $(\mathbf{A_j}, \mathbf{G_j}),\ j = -r, \ldots, \alpha - r$. We shall now explore the possibilities of obtaining all possible ladders passing through $(\mathbf{A}, \mathbf{G_0})$, where $\mathbf{G_0}$ is a reflexive g-inverse of $\mathbf{A}$. The following theorems are useful towards this end:

Theorem 14.6.1. *Let* $\mathbf{A}$ *be an* $m \times n$ *matrix of rank* r *and* $\mathbf{G_0}$ *be reflexive g-inverse of* $\mathbf{A}$. *Write* $\mathbf{A_0} = \mathbf{A}$. *Let* $\mathbf{G_{-i}}$ *be an outer inverse of* $\mathbf{A}$ *with rank* $r - i, i = 1, \ldots r$ *such that* $\mathbf{G_{-i}} <^- \mathbf{G_{-i+1}},\ i = 1, \ldots, r$. *Then*

(i) *There exists non-singular matrices* $\mathbf{P}$ *and* $\mathbf{Q}$ *such that* $\mathbf{A} = \mathbf{P}\mathbf{diag}\left(\mathbf{I_r}\ ,\ \mathbf{0}\right)\mathbf{Q}$ *and* $\mathbf{G_{-i}} = \mathbf{Q}^{-1}\mathbf{diag}\left(\mathbf{I_{r-i}}\ ,\ \mathbf{0}\right)\mathbf{P}^{-1},\ i = 0, 1, \ldots r.$
(ii) *There exist matrices* $\mathbf{A_{-i}},\ i = 1, \ldots r$ *such that*
   (a) $\mathbf{A_{-i}}$ *is a reflexive g-inverse of* $\mathbf{G_{-i}},\ i = 0, 1, \ldots, r$ *and*
   (b) $\mathbf{A_{-i}} <^- \mathbf{A_{-i+1}},\ i = 1, \ldots, r.$
(iii) *General form of* $\mathbf{A_{-i}}$ *satisfying the condition in (b) is given by*

$$\mathbf{A_{-i}} = \mathbf{Q}^{-1}\begin{pmatrix}\mathbf{M} & \mathbf{0}\\ \mathbf{0} & \mathbf{0}\end{pmatrix}\mathbf{P}^{-1}, \text{ where } \mathbf{M} = \begin{pmatrix}\mathbf{I_{r-i}} & \mathbf{u_i} & \cdot & \mathbf{u_1}\\ \mathbf{m_i}^{\mathbf{t}} & \mathbf{0} & \cdot & \mathbf{0}\\ \cdot & \cdot & \cdot & \cdot\\ \cdot & \cdot & \cdot & \cdot\\ \mathbf{m_1}^{\mathbf{t}} & \mathbf{0} & \cdot & \mathbf{0}\end{pmatrix},$$

where at least one of $\mathbf{u_j}$ *and* $\mathbf{m_j}$ *is null for* $j = 1, \ldots, i$.

Proof is by induction on i.

Let $\mathbf{A}$ be an $m \times n$ matrix of rank r. Given an outer inverse $\mathbf{G_{-i}}$ of $\mathbf{A}$ with rank $r - i$, what is the class of all outer inverses $\mathbf{G_{-i+1}}$ of $\mathbf{A}$ such that $\mathbf{G_{-i}} <^- \mathbf{G_{-i+1}}$? The answer is contained in the following:

Theorem 14.6.2. *Let* $\mathbf{A}$ *be an* $m \times n$ *matrix of rank* r. *Let* $\mathbf{G_{-i}}$ *be an outer inverse of* $\mathbf{A}$. *Let* $\mathbf{A} = \mathbf{R}\mathbf{diag}\left(\mathbf{I_r}\ ,\ \mathbf{0}\right)\mathbf{S}$ *and* $\mathbf{G_{-i}} = \mathbf{S}^{-1}\mathbf{diag}\left(\mathbf{I_{r-i}}\ ,\ \mathbf{0}\right)\mathbf{R}^{-1}$ *Then the class of all outer inverses* $\mathbf{G_{-i+1}}$ *of* $\mathbf{A}$ *of rank* $r - i + 1$ *such that* $\mathbf{G_{-i}} <^- \mathbf{G_{-i+1}}$ *is given by*

$$\mathbf{G_{-i+1}} = \mathbf{S}^{-1}\begin{pmatrix}\mathbf{I_{r-i}} & \mathbf{0} & \mathbf{0}\\ \mathbf{0} & 1 & \mathbf{u^t}\\ \mathbf{0} & \mathbf{m} & \mathbf{mu^t}\end{pmatrix}\mathbf{R}^{-1}, \text{ where } u \text{ and } m \text{ are arbitrary.}$$

Proof is analogous to Theorem 14.4.16.

We now turn our attention to g-inverses of $\mathbf{A}$. We first prove

Theorem 14.6.3. *Let* $\mathbf{A}$ *be an* $m \times n$ *matrix of rank* r. *Let* $\mathbf{G_i}$ *be a g-inverse of rank* $r+i$ *of* $\mathbf{A}$. *Write* $\mathbf{A} = \mathbf{P}\mathbf{diag}\left(\mathbf{I_r}\ ,\mathbf{0}\right)\mathbf{Q}$ *and* $\mathbf{G_i} = \mathbf{Q}^{-1}\begin{pmatrix}\mathbf{I_r} & \mathbf{L_1}\\ \mathbf{M_1} & \mathbf{M_1L_1}+\mathbf{S}\end{pmatrix}\mathbf{P}^{-1}$ *for some matrices* $\mathbf{L_1}, \mathbf{M_1}$ *and for some matrix* $\mathbf{S}$ *of rank* i. *Then the class of all g-inverses* $\mathbf{G_{i+1}}$ *of rank* $r+i+1$ *such that* $\mathbf{G_i} <^- \mathbf{G_{i+1}}$ *is given by*

$$\mathbf{G_{i+1}} = \mathbf{Q}^{-1}\begin{pmatrix}\mathbf{I_r} & \mathbf{L_2}\\ \mathbf{M_2} & \mathbf{M_2L_2}+\mathbf{S}+\mathbf{uv^t}\end{pmatrix}\mathbf{P}^{-1},$$

where $\mathbf{u}, \mathbf{v}$ *are arbitrary subject to* $\rho(\mathbf{S}+\mathbf{uv^t}) = i+1$, $\mathbf{L_2} = \mathbf{L_1}+\mathbf{z^tuv^t}$ *and* $\mathbf{M_2} = \mathbf{M_1}+\mathbf{uv^tw}$ *with* $\mathbf{z}$, *and* $\mathbf{w}$ *are arbitrary subject to either* $\mathbf{z^tu} = \mathbf{0}$ *or* $\mathbf{v^tw} = \mathbf{0}$.

Proof follows from Theorem 14.5.3.

Let $\mathbf{A}$ be an $m \times n$ matrix of rank r and $\mathbf{G_0}$ be reflexive g-inverse of $\mathbf{A}$. We shall now restrict our attention to g-inverses in the equivalence class $\overline{\mathrm{Eq}}(\mathbf{G_0}|\mathbf{A})$. Given g-inverses $\mathbf{G_1}, \ldots, \mathbf{G_{\alpha-r}}$ in $\overline{\mathrm{Eq}}(\mathbf{G_1}|\mathbf{A})$ such that $\mathbf{G_i} <^- \mathbf{G_{i+1}}$, $i = 0, 1, \ldots, \alpha - r = 1$, we find the class of all matrices $\mathbf{A_1}, \ldots, \mathbf{A_{\alpha-r}}$ such that $(\mathbf{A_i}, \mathbf{G_i})$ is a step of a ladder.

Theorem 14.6.4. *Let* $\mathbf{A}$ *be an* $m \times n$ *matrix of rank* r *and* $\alpha = \min\{m, n\}$. *Let* $\mathbf{G_0}$ *be reflexive g-inverse of* $\mathbf{A}$. *Let* $\mathbf{G_i}, i = 0, 1, \ldots, r$ *be g-inverses of* $\mathbf{A}$ *satisfying the following conditions:*

(i) $\mathbf{G_i} \in \overline{\mathrm{Eq}}(\mathbf{G_0}|\mathbf{A})$ *for all* i
(ii) $\rho(\mathbf{G_i}) = r + i$ *for all* i *and*
(iii) $\mathbf{G_i} <^- \mathbf{G_{i+1}}$, $i = 0, 1, \ldots, \alpha - r - 1$.

Then the following hold:

(a) *There exist non-singular matrices* $\mathbf{P}$ *and* $\mathbf{Q}$ *such that*

$$\mathbf{A} = \mathbf{P}\mathbf{diag}\left(\mathbf{I_r}\ ,\ \mathbf{0}\right)\mathbf{Q} \text{ and } \mathbf{G_i} = \mathbf{Q}^{-1}\mathbf{diag}\left(\mathbf{I_{r+i}}\ ,\ \mathbf{0}\right)\mathbf{P}^{-1},$$

$i = 0, 1, \ldots, \alpha - r$.
(b) *There exist matrices* $\mathbf{A_i}$, $i = 1, \ldots, \alpha - r$ *such that* (i) $\mathbf{A_i}$ is a reflexive g-inverse of $\mathbf{G_i}$, $i = 0, 1, \ldots, \alpha - r - 1$, where $\mathbf{A_0} = \mathbf{A}$ and (ii) $\mathbf{A_i} <^- \mathbf{A_{i+1}}$, $i = 0, 1, \ldots, \alpha - r - 1$.
(c) General form of $\mathbf{A_i}$ is given by Theorem 14.2.4.

Proof is by induction on i.

Let **A** be an $m \times n$ matrix of rank r. Theorems 14.6.1 and 14.6.2 completely characterize all possibilities for the part of the ladder from basement to ground level. Theorem 14.6.3 completely characterize all possibilities for the g-inverse part of each step in a ladder and Theorem 14.6.4 characterizes all possible steps of a ladder if we choose to restrict to an equivalence class. The characterizations of $\mathbf{A}_i$'s if $\mathbf{G}_i$'s cut across the equivalence classes appears to be cumbersome. It would be nice if a neat characterization of the same can be obtained.

Chapter 15

Applications

15.1 Introduction

In the preceding chapter, we established hierarchies in various classes of generalized inverses of a matrix using the matrix partial orders developed earlier in this monograph. In this chapter, we give some applications of matrix partial orders and shorted operators in statistics and electrical networks. In Section 15.2, we recall a few basic results on point estimation in linear models. In Section 15.3 we make a systematic comparison of a pair of linear models and interpret statistically the implication of the design matrix of one model being below the one of the other model under various orders. Section 15.4 starts with interpreting the BLUE and dispersion of BLUE in terms of a shorted operator. We then give applications of shorted operator to the recovery of inter-block information in incomplete block designs. In Section 15.5, we give applications of the parallel sum and the shorted operator to testing in a linear model. Not surprisingly, the shorted operator proves to be of superior use here. In Section 15.6, we consider modifications of shorting mechanisms in n-port reciprocal resistive networks endowed with a shorting mechanism and obtain the modified shorted operators. This may prove useful in sensitivity analysis of shorting mechanism when there is a scope of choice. We also consider adding one more port to the network and obtain the modified shorted operator.

15.2 Point estimation in a general linear model

In this section, we gather a few results on the point estimation of linear parametric functions in a general linear model that are needed in the following sections. For proofs of these results and an excellent exposition of linear models in general, we refer the reader to [Sengupta and Jammalamadaka (2003)].

Consider the linear model $\mathbf{y} = \mathbf{X}\beta + \varepsilon$, where (i) $\mathbf{X}$ is a known $n \times m$ matrix, known as the model matrix and β is an unknown non-stochastic vector of parameters in $\mathbb{R}^m$, (ii) ε is a random vector (unobservable) in $\mathbb{R}^n$ such that its mean, $E(\varepsilon) \neq \mathbf{0}$ and its dispersion matrix, $D(\varepsilon) \neq \sigma^2\mathbf{V}$, where σ is an unspecified positive constant and $\mathbf{V}$ is a known $n \times n$ non-negative definite matrix. (Thus, $E(\mathbf{y}) = \mathbf{X}\beta$ and $D(\mathbf{y}) = \sigma^2\mathbf{V}$.) The vector $\mathbf{y}$ is referred to as the random vector of observations.

Henceforth, we denote this linear model by the triple $(\mathbf{y}, \mathbf{X}\beta, \sigma^2\mathbf{V})$.

Lemma 15.2.1. *The observation vector* $\mathbf{y} \in \mathcal{C}(\mathbf{X} : \mathbf{V})$ *with probability 1.*

Proof. Consider the vector $\zeta = (\mathbf{I} - \mathbf{P_V})(\mathbf{y} - \mathbf{X}\beta) = (\mathbf{I} - \mathbf{P_V})\varepsilon$. It is easy to see that $E(\zeta) = \mathbf{0}$ and $D(\zeta) = \mathbf{0}$. Hence, for each i, ζ_i, the ith component of ζ is zero with probability 1. Thus, $\mathbf{y} - \mathbf{X}\beta \in \mathcal{C}(\mathbf{V})$ with probability 1 or equivalently $\mathbf{y} = \mathbf{X}\beta + \mathbf{V}\mathbf{u}$ for some vector $\mathbf{u}$. It follows that $\mathbf{y} \in \mathcal{C}(\mathbf{X} : \mathbf{V})$ with probability 1. □

Lemma 15.2.2. *(Covariance Adjustment) Let* $\mathbf{u}$ *and* $\mathbf{v}$ *be random vectors with finite first and second moments and* $E(\mathbf{v}) = \mathbf{0}$. *Then the vector* $\mathbf{v}$ *is uncorrelated with the vector* $\mathbf{u} - \mathbf{B}\mathbf{v}$ *if and only if* $\mathbf{B}\mathbf{v} = Cov(\mathbf{u}, \mathbf{v})(\mathbf{D}(\mathbf{v}))^{-}\mathbf{v}$ *with probability 1, where* $Cov(\mathbf{u}, \mathbf{v})$ *denotes the covariance matrix between vectors* $\mathbf{u}$ *and* $\mathbf{v}$.

See Proposition 3.1.2 of [Sengupta and Jammalamadaka (2003)] for a proof.

Definition 15.2.3. Let $(\mathbf{y}, \mathbf{X}\beta, \sigma^2\mathbf{V})$ be a linear model and $\mathbf{A}$ be a matrix with m columns. We say $\mathbf{A}\beta$ is estimable, if there exists a matrix $\mathbf{C}$ such that $E(\mathbf{C}\mathbf{y}) = \mathbf{A}\beta$ for all $\beta \in \mathbb{R}^{\mathbf{m}}$.

Theorem 15.2.4. *Let* $(\mathbf{y}, \mathbf{X}\beta, \sigma^2\mathbf{V})$ *be a linear model and* $\mathbf{A}$ *be a matrix with* m *columns. Then* $\mathbf{A}\beta$ *is estimable if and only if* $\mathcal{C}(\mathbf{A}^{\mathbf{t}}) \subseteq \mathcal{C}(\mathbf{X}^{\mathbf{t}})$.

See Proposition 7.2.4 [Sengupta and Jammalamadaka (2003)].

Definition 15.2.5. Let $(\mathbf{y}, \mathbf{X}\beta, \sigma^2\mathbf{V})$ be a linear model. A linear function $\mathbf{m}^{\mathbf{t}}\mathbf{y}$ is said to be a linear zero function if $E(\mathbf{m}^{\mathbf{t}}\mathbf{y}) = \mathbf{0}$ for all $\beta \in \mathbb{R}^{\mathbf{m}}$.

Remark 15.2.6. The class of all linear zero functions in the linear model $(\mathbf{y}, \mathbf{X}\beta, \sigma^2\mathbf{V})$ is given by $\{\mathbf{m}^{\mathbf{t}}\mathbf{y},\ \mathbf{m} \in \mathcal{C}(\mathbf{I} - \mathbf{P_X})\}$.

Definition 15.2.7. Let $(\mathbf{y}, \mathbf{X}\beta, \sigma^2\mathbf{V})$ be a linear model, $\mathbf{A}$ be a matrix with m columns and $\beta \in \mathbb{R}^{\mathbf{m}}$. Let $\mathbf{A}\beta$ be estimable. Then $\mathbf{Ly}$ is said to be BLUE of $\mathbf{A}\beta$ if (i) $E(\mathbf{Ly}) = \mathbf{A}\beta$ for all $\beta \in \mathbb{R}^m$ and (ii) $D(\mathbf{Ly}) <^{\mathbf{L}} \mathbf{D}(\mathbf{My})$ for or all $\beta \in \mathbb{R}^m$ and all $\mathbf{My}$ satisfying $E(\mathbf{My}) = \mathbf{A}\beta$.

Thus, BLUE of $\mathbf{A}\beta$ has the smallest dispersion matrix amongst the class of linear unbiased estimators of $\mathbf{A}\beta$. Expressed in terms of the Löwner order, $D(\mathbf{Ly}) = \inf_{\mathbf{E(My)=A}\beta}\{\mathbf{D}(\mathbf{My})\}$. We also note that Blue of an estimable function of β is unique.

Theorem 15.2.8. *Let $(\mathbf{y}, \mathbf{X}\beta, \sigma^2\mathbf{V})$ be a linear model. Then $\mathbf{Ly}$ is the BLUE of its expectation if and only if it is uncorrelated with each linear zero function.*

See Proposition 4.3.2 [Sengupta and Jammalamadaka (2003)].

Theorem 15.2.9. *Let $(\mathbf{y}, \mathbf{X}\beta, \sigma^2\mathbf{V})$ be a linear model. Then the following hold:*

(i) *The BLUE of $\mathbf{X}\beta$, known as the vector of fitted values $\hat{\mathbf{y}}$ of $\mathbf{y}$ is given by $\hat{\mathbf{y}} = (\mathbf{I} - \mathbf{V}(\mathbf{I} - \mathbf{P_X})((\mathbf{I} - \mathbf{P_X})\mathbf{V}(\mathbf{I} - \mathbf{P_X}))^-(\mathbf{I} - \mathbf{P_X}))\mathbf{y}$.*

(ii) $\mathbf{e} = \mathbf{y} - \hat{\mathbf{y}}$, *the vector of residuals is given by*

$$\mathbf{V}(\mathbf{I} - \mathbf{P_X})((\mathbf{I} - \mathbf{P_X})\mathbf{V}(\mathbf{I} - \mathbf{P_X}))^-(\mathbf{I} - \mathbf{P_X})\mathbf{y}.$$

(iii) $D(\hat{\mathbf{y}}) = \sigma^2(\mathbf{V} - \mathbf{V}(\mathbf{I} - \mathbf{P_X})((\mathbf{I} - \mathbf{P_X})\mathbf{V}(\mathbf{I} - \mathbf{P_X}))^-(\mathbf{I} - \mathbf{P_X})\mathbf{V})$.

(iv) $D(\mathbf{e}) = \sigma^2(\mathbf{V}(\mathbf{I} - \mathbf{P_X})((\mathbf{I} - \mathbf{P_X})\mathbf{V}(\mathbf{I} - \mathbf{P_X}))^-(\mathbf{I} - \mathbf{P_X})\mathbf{V})$.

Proof. Since $E(\mathbf{y}) = \mathbf{X}\beta$ and every linear zero function is a linear function of $(\mathbf{I} - \mathbf{P_X})\mathbf{y}$, it follows that every linear unbiased estimator of $\mathbf{X}\beta$ is of the form $\mathbf{y} - \mathbf{B}(\mathbf{I} - \mathbf{P_X})\mathbf{y}$ for some matrix $\mathbf{B}$. By Theorem 15.2.8, $\mathbf{y} - \mathbf{B}(\mathbf{I} - \mathbf{P_X})\mathbf{y}$ is the BLUE of $\mathbf{X}\beta$ if and only if it is uncorrelated with $(\mathbf{I} - \mathbf{P_X})\mathbf{y}$. So, by Lemma 15.2.2, $\mathbf{y} - \mathbf{B}(\mathbf{I} - \mathbf{P_X})\mathbf{y}$ is the BLUE if and only if

$$\begin{aligned}\mathbf{B}(\mathbf{I} - \mathbf{P_X})\mathbf{y} &= Cov(\mathbf{y}, (\mathbf{I} - \mathbf{P_X})\mathbf{y}(\mathbf{D}((\mathbf{I} - \mathbf{P_X})\mathbf{y}))^-(\mathbf{I} - \mathbf{P_X})\mathbf{y}) \\ &= \mathbf{V}(\mathbf{I} - \mathbf{P_X})((\mathbf{I} - \mathbf{P_X})\mathbf{V}(\mathbf{I} - \mathbf{P_X}))^-(\mathbf{I} - \mathbf{P_X})\mathbf{y}.\end{aligned}$$

Thus, $\hat{\mathbf{y}} = (\mathbf{I} - \mathbf{V}(\mathbf{I} - \mathbf{P_X})((\mathbf{I} - \mathbf{P_X})\mathbf{V}(\mathbf{I} - \mathbf{P_X}))^-(\mathbf{I} - \mathbf{P_X}))\mathbf{y}$.
(ii),(iii) and (iv) are simple consequences of (i). □

Remark 15.2.10. In the setup of Theorem 15.2.9, if $\mathbf{A}\beta$ is estimable, then there exists a matrix $\mathbf{L}$ such that $\mathbf{A} = \mathbf{LX}$. The BLUE of $\mathbf{A}\beta$ is $\mathbf{L}\hat{\mathbf{y}}$.

Theorem 15.2.11. *Let* $(\mathbf{y}, \mathbf{X}\beta, \sigma^2\mathbf{V})$ *be a linear model with* $\mathcal{C}(\mathbf{X}) \subseteq \mathcal{C}(\mathbf{V})$. *Then*

(i) $\hat{\mathbf{y}} = \widehat{\mathbf{X}\beta} = \mathbf{X}(\mathbf{X}^t\mathbf{V}^-\mathbf{X})^-\mathbf{X}^t\mathbf{V}^-\mathbf{y}$ *and*
(ii) $D(\hat{\mathbf{y}}) = \sigma^2\mathbf{X}(\mathbf{X}^t\mathbf{V}^-\mathbf{X})^-\mathbf{X}^t$, *where* $\mathbf{V}^-$ *is an any g-inverses of* $\mathbf{V}$.

Proof. First note that $\mathbf{X}^t\mathbf{V}^-\mathbf{X}$ is invariant under choices of g-inverse of $\mathbf{V}$, since $\mathcal{C}(\mathbf{X}) \subseteq \mathcal{C}(\mathbf{V})$. Clearly, $\rho(\mathbf{X}) = \rho(\mathbf{X}^t\mathbf{V}^-\mathbf{X})$ for each $\mathbf{V}^-$, for we can choose a positive definite g-inverse $\mathbf{V}^-$ of $\mathbf{V}$. So, $(\mathbf{X}^t\mathbf{V}^-\mathbf{X})^-\mathbf{X}^t\mathbf{V}^-$ is a g-inverse of $\mathbf{X}$. Thus, $E(\mathbf{X}(\mathbf{X}^t\mathbf{V}^-\mathbf{X})^-\mathbf{X}^t\mathbf{V}^-\mathbf{y}) = \mathbf{X}\beta$ for all β. Further,

$$\begin{aligned}&Cov(\mathbf{X}(\mathbf{X}^t\mathbf{V}^-\mathbf{X})^-\mathbf{X}^t\mathbf{V}^-\mathbf{y}, (\mathbf{I}-\mathbf{P_X})\mathbf{y})\\ &= \mathbf{X}(\mathbf{X}^t\mathbf{V}^-\mathbf{X})^-\mathbf{X}^t\mathbf{V}^-\sigma^2\mathbf{V}(\mathbf{I}-\mathbf{P_X}) = \mathbf{0},\end{aligned}$$

since, $\mathbf{X}^t\mathbf{V}^-\mathbf{V} = \mathbf{X}^t$. Therefore (i) follows.
For (ii),

$$\begin{aligned}D(\hat{\mathbf{y}}) = Cov(\hat{\mathbf{y}}, \hat{\mathbf{y}}) &= \sigma^2\mathbf{X}(\mathbf{X}^t\mathbf{V}^-\mathbf{X})^-\mathbf{X}^t\mathbf{V}^-\mathbf{V}(\mathbf{X}(\mathbf{X}^t\mathbf{V}^-\mathbf{X})^-\mathbf{X}^t\mathbf{V}^-)^t\\ &= \sigma^2\mathbf{X}(\mathbf{X}^t\mathbf{V}^-\mathbf{X})^-\mathbf{X}^t(\mathbf{X}(\mathbf{X}^t\mathbf{V}^-\mathbf{X})^-\mathbf{X}^t\mathbf{V}^-)^t\\ &= \sigma^2\mathbf{X}(\mathbf{X}^t\mathbf{V}^-\mathbf{X})^-\mathbf{X}^t\mathbf{V}^-(\mathbf{X}(\mathbf{X}^t\mathbf{V}^-\mathbf{X})^-\mathbf{X}^t)^t\\ &= \sigma^2\mathbf{X}(\mathbf{X}^t\mathbf{V}^-\mathbf{X})^-\mathbf{X}^t\mathbf{V}^-\mathbf{X}(\mathbf{X}^t\mathbf{V}^-\mathbf{X})^-\mathbf{X}^t\\ &= \sigma^2\mathbf{X}(\mathbf{X}^t\mathbf{V}^-\mathbf{X})^-\mathbf{X}^t\end{aligned}$$

and hence the result. □

Remark 15.2.12. If $\mathbf{V} = \mathbf{I}$, then $\hat{\mathbf{y}} = \mathbf{P_X}\mathbf{y}$.

Remark 15.2.13. For the linear model $(\mathbf{y}, \mathbf{X}\beta, \sigma^2\mathbf{V})$, $\ell^t\mathbf{y}$ is the BLUE of its expectation if and only if $\mathbf{V}\ell \in \mathcal{C}(\mathbf{X})$. Further, if $\mathbf{V} = \mathbf{I}$, then $\ell^t\mathbf{y}$ is the BLUE of its expectation if and only if $\ell \in \mathcal{C}(\mathbf{X})$.

We now consider linear constraints $\mathbf{A}\beta = \mathbf{0}$ on the parametric vector β, where $\mathbf{A}\beta$ is estimable. We have

Theorem 15.2.14. *Let* $(\mathbf{y}, \mathbf{X}\beta, \sigma^2\mathbf{V})$ *be a linear model where* β *is subject to the constraint* $\mathbf{A}\beta = \mathbf{0}$. *Let* $\mathbf{A}\beta$ *be estimable. Then the constrained model is equivalent to the unconstrained model* $(\mathbf{y}, \mathbf{X}(\mathbf{I} - \mathbf{A}^-\mathbf{A})\theta, \sigma^2\mathbf{V})$. *Further,* $\rho(\mathbf{X}) = \rho(\mathbf{X}(\mathbf{I} - \mathbf{A}^-\mathbf{A})) + \rho(\mathbf{A})$.

Proof. We first note that $\mathbf{A}\beta = \mathbf{0} \Leftrightarrow \beta = (\mathbf{I} - \mathbf{A}^-\mathbf{A})\theta$, for some θ. Thus, the linear model $(\mathbf{y}, \mathbf{X}\beta, \sigma^2\mathbf{V})$ subject to the constraint $\mathbf{A}\beta = \mathbf{0}$ is equivalent to the linear model $(\mathbf{y}, \mathbf{X}(\mathbf{I} - \mathbf{A}^-\mathbf{A})\theta, \sigma^2\mathbf{V})$.

To prove the second statement, we write $\mathbf{X} = \mathbf{X}\mathbf{A}^-\mathbf{A} + \mathbf{X}(\mathbf{I} - \mathbf{A}^-\mathbf{A})$. Since $\mathbf{A}\beta$ is estimable, by Theorem 15.2.4, we have $\mathcal{C}(\mathbf{A}^t) \subseteq \mathcal{C}(\mathbf{X}^t)$. We now show that $\mathcal{C}(\mathbf{X}\mathbf{A}^-\mathbf{A}) \cap \mathcal{C}(\mathbf{X}(\mathbf{I} - \mathbf{A}^-\mathbf{A})) = \{\mathbf{0}\}$. Let $\mathbf{u}$ and $\mathbf{v}$ be vectors such that $\mathbf{X}\mathbf{A}^-\mathbf{A}\mathbf{u} = (\mathbf{X}(\mathbf{I} - \mathbf{A}^-\mathbf{A}))\mathbf{v}$. Pre-multiplying by

$\mathbf{AX^-}$, we have $\mathbf{AX^-XA^-Au} = \mathbf{AX^-(X(I - A^-A))v}$ or equivalently, $\mathbf{Au} = \mathbf{0}$, since $\mathbf{AX^-X} = \mathbf{A}$. Thus, $\mathbf{XA^-Au} = \mathbf{0} = (\mathbf{X}(\mathbf{I} - \mathbf{A^-A}))\mathbf{v}$. Similarly, we can show that $\mathcal{C}(\mathbf{XA^-A})^t \cap \mathcal{C}(\mathbf{X}(\mathbf{I} - \mathbf{A^-A}))^t = \{\mathbf{0}\}$. Thus, $\rho(\mathbf{X}) = \rho(\mathbf{XA^-A}) + \rho(\mathbf{X}(\mathbf{I} - \mathbf{A^-A}))$. However,

$$\begin{aligned} \rho(\mathbf{A}) &= \rho(\mathbf{AX^-XA^-A}) \leq \rho(\mathbf{XA^-A}) \leq \rho(\mathbf{A}) \\ &\Rightarrow \rho(\mathbf{A}) = \rho(\mathbf{XA^-A}). \end{aligned}$$

Hence, $\rho(\mathbf{X}) = \rho(\mathbf{X}(\mathbf{I} - \mathbf{A^-A})) + \rho(\mathbf{A})$. □

Consider the linear model $M=(\mathbf{y}, \mathbf{X}\beta, \sigma^2\mathbf{V})$. Let $\mathbf{X}$ and β be partitioned conformably as $\mathbf{X} = (\mathbf{X_1} : \mathbf{X_2})$ and $\beta = \begin{pmatrix}\beta_1 \\ \beta_2\end{pmatrix}$ respectively such that $\mathbf{X}\beta = \mathbf{X_1}\beta_1 + \mathbf{X_2}\beta_2$. Suppose we are interested in estimable linear functions of β_1 only. Let M^* denote the reduced model $((\mathbf{I} - \mathbf{P_{X_2}})\mathbf{y}, (\mathbf{I} - \mathbf{P_{X_2}})\mathbf{X_1}\beta_1, \sigma^2(\mathbf{I} - \mathbf{P_{X_2}})\mathbf{V}(\mathbf{I} - \mathbf{P_{X_2}}))$. The following theorem describes the relationship of the linear models M^* and M:

Theorem 15.2.15. *Let the linear models M and M^* be defined as above. Then*

(i) *The sets of linear zero functions of M and M^* coincide.*
(ii) $\mathbf{p}^t\beta_1$ *is estimable in M^* if and only if it is estimable in M.*
(iii) *Let $\mathbf{p}^t\beta_1$ be estimable under M^*. The the BLUE of $\mathbf{p}^t\beta_1$ under M and under M^* coincide.*
(iv) *The dispersion matrices of BLUE's of $(\mathbf{I} - \mathbf{P_{X_2}})\mathbf{X_1}\beta_1$ under M and under M^* coincide.*
(v) *The residual sums of squares under M and under M^* are identical.*

See Proposition 7.10.1 [Sengupta and Jammalamadaka (2003)].

15.3 Comparison of models when model matrices are related under matrix partial orders

In this section, we consider the linear models when their model matrices are related to each other under different order relations on matrices such as: the space pre-order, the minus order, the left/right star order or star order and study the consequences with respect to the estimability and finding BLUE's.

We choose and fix two arbitrary linear models $M_1=(\mathbf{y}, \mathbf{X}_1\beta, \sigma^2\mathbf{V})$ and $M_2=(\mathbf{y}, \mathbf{X}_2\beta, \sigma^2\mathbf{V})$ and refer to the model matrices of M_1 and M_2 for proving various theorems of this section. We start with the space pre-order.

Theorem 15.3.1. *Let M_1 and M_2 be linear models for which $\mathcal{C}(\mathbf{X}_1) \subseteq \mathcal{C}(\mathbf{V})$. Then $\mathbf{X}_1 <^s \mathbf{X}_2$ if and only if* (i) *every estimable linear parametric function under M_1 is also estimable under M_2 and* (ii) *for each ℓ such that $\ell^t\mathbf{y}$ is the BLUE of its expectation under M_1 is also the BLUE of its expectation under M_2.*

Proof. We show that (i) is equivalent to $\mathcal{C}(\mathbf{X}_1^t) \subseteq \mathcal{C}(\mathbf{X}_2^t)$ and (ii) is equivalent to $\mathcal{C}(\mathbf{X}_1) \subseteq \mathcal{C}(\mathbf{X}_2)$. Thus, $\mathbf{X}_1 <^s \mathbf{X}_2$ if and only if (i) and (ii) hold.

Let (i) hold and $\mathbf{p} \in \mathcal{C}(\mathbf{X}_1^t)$. Then $\mathbf{p}^t\beta$ is estimable under M_1. Since (i) holds $\mathbf{p}^t\beta$ is estimable under M_2, so, $\mathbf{p} \in \mathcal{C}(\mathbf{X}_2^t)$. Thus, $\mathcal{C}(\mathbf{X}_1^t) \subseteq \mathcal{C}(\mathbf{X}_2^t)$. Now, let $\mathcal{C}(\mathbf{X}_1^t) \subseteq \mathcal{C}(\mathbf{X}_2^t)$ and $\mathbf{p}^t\beta$ be estimable under M_1. By Theorem 15.2.4, $\mathbf{p} \in \mathcal{C}(\mathbf{X}_1^t)$ and so, $\mathbf{p} \in \mathcal{C}(\mathbf{X}_2^t)$. Thus, $\mathbf{p}^t\beta$ is estimable under M_2.

Let (ii) hold and $\mathbf{m} \in \mathcal{C}(\mathbf{X}_1)$. Since $\mathcal{C}(\mathbf{X}_1) \subseteq \mathcal{C}(\mathbf{V})$, we have $\mathbf{m} \in \mathcal{C}(\mathbf{V})$. So, $\mathbf{m} = \mathbf{V}\mathbf{a}$ for some vector $\mathbf{a}$. By Remark 15.2.13, $\mathbf{m}^t\mathbf{y}$ is the BLUE of its expectation under M_1 and by (ii) $\mathbf{m}^t\mathbf{y}$ is the BLUE of its expectation under M_2. Thus, $\mathbf{m} = \mathbf{V}\mathbf{a} \in \mathcal{C}(\mathbf{X}_2)$ and so, $\mathcal{C}(\mathbf{X}_1) \subseteq \mathcal{C}(\mathbf{X}_2)$.

Let $\mathcal{C}(\mathbf{X}_1) \subseteq \mathcal{C}(\mathbf{X}_2)$ and $\mathbf{p}^t\mathbf{y}$ be the BLUE of its expectation under M_1. Then $\mathbf{V}\mathbf{p} \in \mathcal{C}(\mathbf{X}_1)$. Since $\mathcal{C}(\mathbf{X}_1) \subseteq \mathcal{C}(\mathbf{X}_2)$, we have $\mathbf{V}\mathbf{p} \in \mathcal{C}(\mathbf{X}_2)$. Therefore, by Remark 15.2.13, $\mathbf{p}^t\mathbf{y}$ is the BLUE of its expectation under M_2. □

Corollary 15.3.2. *Let M_1 and M_2 be linear models and let $\mathbf{V}$ be positive definite. Then $\mathbf{X}_1 <^s \mathbf{X}_2$ if and only if* (i) *and* (ii) *of Theorem 15.3.1 hold.*

The condition $\mathcal{C}(\mathbf{X}_1) \subseteq \mathcal{C}(\mathbf{V})$ can not be dispensed with from Theorem 15.3.1. We give the following example:

Example 15.3.3. Let

$$\mathbf{X}_1 = \begin{pmatrix} 1 & 1 \\ 0 & 1 \\ 0 & 0 \end{pmatrix} \begin{pmatrix} 1 & 1 & 1 \\ 0 & 0 & 1 \end{pmatrix}, \ \mathbf{X}_2 = \begin{pmatrix} 1 & 1 \\ 0 & 0 \\ 0 & 1 \end{pmatrix} \begin{pmatrix} 1 & 1 & 1 \\ 0 & 0 & 1 \end{pmatrix} \text{ and } \mathbf{V} = \begin{pmatrix} 1 & 0 & 0 \\ 0 & 0 & 0 \\ 0 & 0 & 0 \end{pmatrix}.$$

Then for each $\mathbf{m}$, $\mathbf{V}\mathbf{m} \in \mathcal{C}(\mathbf{X}_1)$ and $\mathbf{V}\mathbf{m} \in \mathcal{C}(\mathbf{X}_2)$ and $\mathcal{C}(\mathbf{X}_1) \nsubseteq \mathcal{C}(\mathbf{X}_2)$.

Remark 15.3.4. Consider the set up of Theorem 15.3.1. Let $\mathcal{C}(\mathbf{X}_1) \subseteq \mathcal{C}(\mathbf{V})$ and $\mathbf{X}_1 <^s \mathbf{X}_2$. It is clear that the BLUE of $\mathbf{X}_1\beta$ under M_1 continues to be its BLUE under M_2. However, the dispersion matrices of the BLUE of $\mathbf{X_1}\beta$ under M_1 and M_2 are not comparable.

Before we prove a theorem similar to Theorem 15.3.1 for the minus order, we show that if $\mathcal{C}(\mathbf{X_1}) \subseteq \mathcal{C}(\mathbf{V})$ and $\mathbf{X_1} <^- \mathbf{X_2}$, then the dispersion matrix of the BLUE of $\mathbf{X_1}\beta$ under the linear model M_1 is below the the dispersion matrix of the BLUE of $\mathbf{X_1}\beta$ under the linear model M_2 in the Löwner order.

Theorem 15.3.5. *Let M_1 and M_2 be linear models for which $\mathcal{C}(\mathbf{X_1}) \subseteq \mathcal{C}(\mathbf{V})$ and $\mathbf{X_1} <^- \mathbf{X_2}$. Then the following hold:*

(i) *The linear model M_1 is equivalent to $M_3 = (\mathbf{y}, \mathbf{L_1}\theta_1, \sigma^2\mathbf{V})$ and the linear model M_2 is equivalent to $M_4 = (\mathbf{y}, \mathbf{L_1}\theta_1 + \mathbf{L_2}\theta_2, \sigma^2\mathbf{V})$, where $(\mathbf{L_1} : \mathbf{L_2})$ is a full column rank matrix such that $\rho(\mathbf{L_1}) = \rho(\mathbf{X_1})$ and $\rho(\mathbf{L_1} : \mathbf{L_2}) = \rho(\mathbf{X_2})$*

(ii) *θ_1 is estimable under both M_3 and M_4.*

(iii) *The dispersion matrix of the BLUE of θ_1 under M_3 is below the dispersion matrix of the BLUE of θ_1 under M_4 under the Löwner order.*

(iv) *There is no estimable linear function of β common to the linear models M_1 and $M_5 = (\mathbf{y}, (\mathbf{X_2} - \mathbf{X_1})\beta, \sigma^2\mathbf{V})$.*

Proof. (i) Since $\mathbf{X_1} <^- \mathbf{X_2}$ there exist matrices $\mathbf{L_1}$, $\mathbf{L_2}$, $\mathbf{N_1}$ and $\mathbf{N_2}$ such that $(\mathbf{L_1}, \mathbf{N_1})$ and $\left((\mathbf{L_1} : \mathbf{L_2}), \begin{pmatrix}\mathbf{N_1}\\ \mathbf{N_2}\end{pmatrix}\right)$ are rank factorizations of $\mathbf{X_1}$ and $\mathbf{X_2}$ respectively. Write $\theta_1 = \mathbf{N_1}\beta$ and $\theta_2 = \mathbf{N_2}\beta$. Clearly (i) holds.
(ii) and (iv) are easy to prove.
(iii) Since $\mathcal{C}(\mathbf{X_1}) \subseteq \mathcal{C}(\mathbf{V})$, we have $\mathcal{C}((\mathbf{I}-\mathbf{P_{L_2}})\mathbf{L_1}) \subseteq \mathcal{C}((\mathbf{I}-\mathbf{P_{L_2}})\mathbf{V}(\mathbf{I}-\mathbf{P_{L_2}}))$. Also, $\mathcal{C}(\mathbf{L_1}) = \mathcal{C}(\mathbf{X_1}) \subseteq \mathcal{C}(\mathbf{V})$, so, $\mathbf{L_1^t}\mathbf{V}^-\mathbf{L_1}$ is invariant under choices of g-inverses of $\mathbf{V}$. Further, $\rho(\mathbf{L_1^t}\mathbf{V}^-\mathbf{L_1}) = \rho(\mathbf{L_1})$. Hence $\mathbf{L_1^t}\mathbf{V}^-\mathbf{L_1}$ is non-singular. Now,

$$\begin{aligned}\rho((\mathbf{I}-\mathbf{P_{L_2}})\mathbf{L_1}) &= \rho(\mathbf{L_1}) - \mathbf{d}(\mathcal{C}(\mathbf{L_1}) \cap \mathcal{N}(\mathbf{I}-\mathbf{P_{L_2}}))\\ &= \rho(\mathbf{L_1}) - \mathbf{d}(\mathcal{C}(\mathbf{L_1}) \cap \mathcal{C}(\mathbf{L_2})) = \rho(\mathbf{L_1}).\end{aligned}$$

Hence, $\mathbf{L_1^t}(\mathbf{I}-\mathbf{P_{L_2}})((\mathbf{I}-\mathbf{P_{L_2}})\mathbf{V}^-(\mathbf{I}-\mathbf{P_{L_2}}))^-(\mathbf{I}-\mathbf{P_{L_2}})\mathbf{L_1}$ is non-singular. Let $\hat{\theta_1}$ and $\widehat{\hat{\theta_1}}$ be the BLUEs of θ_1 under M_3 and M_4. Using Theorems 15.2.10 and 15.2.14, we have $D(\hat{\theta_1}) = \sigma^2(\mathbf{L_1^t}\mathbf{V}^-\mathbf{L_1})^{-1}$ and $D(\widehat{\hat{\theta_1}}) = \sigma^2(\mathbf{L_1^t}(\mathbf{I}-\mathbf{P_{L_2}})((\mathbf{I}-\mathbf{P_{L_2}})\mathbf{V}^-(\mathbf{I}-\mathbf{P_{L_2}}))^-(\mathbf{I}-\mathbf{P_{L_2}})\mathbf{L_1})^{-1}$. Now, $\mathbf{L_1^t}\mathbf{V}^-\mathbf{L_1} = \mathbf{L_1^t}\mathbf{V}^-\mathbf{V}\mathbf{V}^-\mathbf{L_1}$. Also, the matrix

$$\mathbf{V} - \mathbf{V}(\mathbf{I}-\mathbf{P_{L_2}})((\mathbf{I}-\mathbf{P_{L_2}})\mathbf{V}^-(\mathbf{I}-\mathbf{P_{L_2}}))^-(\mathbf{I}-\mathbf{P_{L_2}})\mathbf{V}$$

is *nnd*. Hence,

$$\mathbf{L}_1^t\mathbf{V}^-\mathbf{V}\mathbf{V}^-\mathbf{L}_1 - \mathbf{L}_1^t\mathbf{V}^-\mathbf{V}(\mathbf{I}-\mathbf{P}_{\mathbf{L}_2})((\mathbf{I}-\mathbf{P}_{\mathbf{L}_2})\mathbf{V}^-(\mathbf{I}-\mathbf{P}_{\mathbf{L}_2}))^-(\mathbf{I}-\mathbf{P}_{\mathbf{L}_2})\mathbf{V}\mathbf{V}^-\mathbf{L}_1$$

is *nnd*. So, $D(\hat{\theta_1}) <^L D(\widehat{\theta_1})$. □

We are now ready to prove the following:

Theorem 15.3.6. *Let* $M_1=(\mathbf{y}, \mathbf{X}_1\theta, \sigma^2\mathbf{V})$ *and* $M_2=(\mathbf{y}, \mathbf{X}_2\beta, \sigma^2\mathbf{V})$ *be any two linear models. Then* $\mathbf{X}_1 <^- \mathbf{X}_2$ *if and only if there exists a matrix* $\mathbf{A}$ *such that* $\mathcal{C}(\mathbf{A}^t) \subseteq \mathcal{C}(\mathbf{X}_2^t)$ *and* M_1 *is the model* M_2 *constrained by* $\mathbf{A}\beta = \mathbf{0}$.

Proof. 'If' part follows by Theorem 15.2.14.
'Only if' part
Let $\mathbf{A} = \mathbf{X}_2 - \mathbf{X}_1$. Since $\mathbf{X}_1 <^- \mathbf{X}_2$, we have $\mathbf{A} <^- \mathbf{X}_2$. So, there exists a g-inverse $\mathbf{A}^-$ of $\mathbf{A}$ such that $\mathbf{A}^-\mathbf{A} = \mathbf{A}^-\mathbf{X}_2$ and $\mathbf{A}\mathbf{A}^- = \mathbf{X}_2\mathbf{A}^-$. It is clear that $\mathcal{C}(\mathbf{A})^t \subseteq \mathcal{C}(\mathbf{X}_2^t)$ and therefore, $\mathbf{A}\beta$ is estimable. Further, $\mathbf{X}_1 = \mathbf{X}_2(\mathbf{I}-\mathbf{A}^-\mathbf{A})$. Let M_2 be constrained under $\mathbf{A}\beta = \mathbf{0}$. Then $\beta = (\mathbf{I}-\mathbf{A}^-\mathbf{A})\theta$ and therefore, M_1 is the model M_2 constrained by $\mathbf{A}\beta = \mathbf{0}$.

□

We now investigate the interpretations of the left star order in the linear models.

Theorem 15.3.7. *Let* $M_1^*=(\mathbf{y}, \mathbf{X}_1\beta, \sigma^2\mathbf{I})$ *and* $M_2^*=(\mathbf{y}, \mathbf{X}_2\beta, \sigma^2\mathbf{I})$ *be any two linear models. Then* $\mathbf{X}_1 \star < \mathbf{X}_2$ *if and only if*

(i) *The linear models* M_1^* *and* $M=(\mathbf{y}, (\mathbf{X}_2 - \mathbf{X}_1)\beta, \sigma^2\mathbf{I})$ *have no common estimable linear functions of* β.
(ii) $\mathbf{X}_1\beta$ *is estimable under the model* M_2^* *and*
(iii) *The BLUE of* $\mathbf{X}_1\beta$ *under the model* M_1^* *is also its BLUE under* M_2^* *and the dispersion matrix of* $\mathbf{X}_1\beta$ *under the model* M_1^* *is same as under model* M_2^*.

Proof. 'If' part
Let the rows of $\mathbf{L}_1$ and $\mathbf{L}_2$ form the basis of the row subspaces of $\mathbf{X}_1$ and $\mathbf{X}_2 - \mathbf{X}_1$ respectively. Since (i) holds, $\mathcal{C}(\mathbf{L}_1^t) \cap \mathcal{C}(\mathbf{L}_2^t) = \{\mathbf{0}\}$. Let $\mathbf{X}_1 = \mathbf{T}_1\mathbf{L}_1$ and $\mathbf{X}_2 - \mathbf{X}_1 = \mathbf{T}_2\mathbf{L}_2$ for some matrices $\mathbf{T}_1$ and $\mathbf{T}_2$. Clearly, $(\mathbf{T}_1, \mathbf{L}_1)$ and $(\mathbf{T}_2, \mathbf{L}_2)$ are rank factorizations of $\mathbf{X}_1$ and $\mathbf{X}_2 - \mathbf{X}_1$ respectively. Write $\mathbf{L}_i\beta = \theta_i$, for $i = 1, 2$. Then $M_1^*=(\mathbf{y}, \mathbf{T}_1\theta_1, \sigma^2\mathbf{I})$ and $M_2^*=(\mathbf{y}, \mathbf{T}_1\theta_1 + \mathbf{T}_2\theta_2, \sigma^2\mathbf{I})$ respectively. Since $\mathbf{X}_1\beta$ is estimable

under both the models, therefore θ_1 is estimable under both the models. Let $\hat{\theta}_1$ and $\hat{\hat{\theta}}_1$ denote the BLUEs of θ_1 under $(\mathbf{y}, \mathbf{T_1}\theta_1, \sigma^2\mathbf{I})$ and $(\mathbf{y}, \mathbf{T_1}\theta_1 + \mathbf{T_2}\theta_2, \sigma^2\mathbf{I})$ respectively. Now, $D(\hat{\theta}_1) = \sigma^2(\mathbf{T}_1^t\mathbf{T}_1)^{-1}$ and $D(\hat{\hat{\theta}}_1) = \sigma^2((\mathbf{T}_1^t\mathbf{T}_1)^{-1} + \mathbf{T}_1^t\mathbf{T}_2(\mathbf{T}_2^t\mathbf{T}_2)^{-1}\mathbf{T}_2^t\mathbf{T}_1)$. As $D(\hat{\theta}_1) = \mathbf{D}(\hat{\hat{\theta}}_1)$, we have $\mathbf{T}_1^t\mathbf{T}_2 = \mathbf{0}$. Hence, $\mathbf{X}_1^t(\mathbf{X_2} - \mathbf{X_1}) = \mathbf{0}$, implying $\mathbf{X_1} \star < \mathbf{X_2}$.
'Only if' part follows, since

$$\mathbf{X_1} \star < \mathbf{X_2} \Rightarrow \mathbf{X_1} = \mathbf{U}\mathbf{diag}\begin{pmatrix}\mathbf{I} & \mathbf{0} & \mathbf{0}\end{pmatrix}\mathbf{Q} \text{ and } \mathbf{X_2} = \mathbf{U}\mathbf{diag}\begin{pmatrix}\mathbf{I} & \mathbf{I} & \mathbf{0}\end{pmatrix}\mathbf{Q}$$

for some orthogonal matrix $\mathbf{U}$ and non-singular matrix $\mathbf{Q}$. □

Corollary 15.3.8. *Let* $M_1^* = (\mathbf{y}, \mathbf{X_1}\beta, \sigma^2\mathbf{I})$ *and* $M_2^* = (\mathbf{y}, \mathbf{X_2}\beta, \sigma^2\mathbf{I})$ *be any two linear models such that* $\mathbf{X_1} \star < \mathbf{X_2}$. *Then the BLUE of every estimable linear function of the parameters under* M_1^* *is a linear zero function under* $M = (\mathbf{y}, (\mathbf{X_2} - \mathbf{X_1})\beta, \sigma^2\mathbf{I})$ *and vice versa.*

Our next theorem is an application of Corollary 8.2.10 on Löwner order for estimation under linear models.

Theorem 15.3.9. *Let* $(\mathbf{y}, \mathbf{X}\beta, \sigma^2\mathbf{V})$ *be a linear model such that* $\mathbf{V}$ *is nnd. Let* $\mathbf{A}\beta$ *be estimable with* $\widehat{\mathbf{A}\beta}$ *as its BLUE. If* $\mathbf{Ly}$ *is an unbiased estimator of* $\mathbf{A}\beta$, *then* (i) $tr(D(\widehat{\mathbf{A}\beta})) \leq tr(D(\mathbf{Ly}))$ *and* (ii) $det(D(\widehat{\mathbf{A}\beta})) \leq det(D(\mathbf{Ly}))$.

Proof. Since $\mathbf{Ly}$ is an unbiased estimator of $\mathbf{A}\beta$, and $\widehat{\mathbf{A}\beta}$ is its BLUE, $D(\widehat{\mathbf{A}\beta}) <^L D(\mathbf{Ly})$. The result follows from Corollary 8.2.10. □

15.4 Shorted operators - Applications

In this section we shall give some interpretations and applications of shorted operators in Statistics. We first give the interpretation of BLUE in a linear model.

Theorem 15.4.1. *Let* $(\mathbf{y}, \mathbf{X}\beta, \sigma^2\mathbf{V})$ *be a linear model in which both the matrices* $\mathbf{X}$ *and* $\mathbf{V}$ *can be rank deficient. Let* $\mathbf{y} = \mathbf{X}\beta + \mathbf{V}\mathbf{u}$ *(Lemma 15.2.1). Then the following hold:*

(i) *The BLUE of* $\mathbf{X}\beta$ *is* $\widehat{\mathbf{y}} = S(\mathbf{V}|\mathcal{C}(\mathbf{X}))\mathbf{u} + \mathbf{X}\beta$.
(ii) $D(\widehat{\mathbf{y}}) = \sigma^2 S(\mathbf{V}|\mathcal{C}(\mathbf{X}))$.
(iii) $D(\mathbf{e}) = \sigma^2(\mathbf{V} - \mathbf{V}|\mathcal{C}(\mathbf{X}))$.

Proof follows from Theorem 15.2.9 and the definition of a shorted operator.

In Corollary 10.3.6, we showed that if $\mathbf{A}$ and $\mathbf{B}$ are two *nnd* matrices of the same order $n \times n$ such that $\mathbf{A} <^{\mathbf{L}} \mathbf{B}$ and a is subspace $\mathcal{S}$ of $\mathbb{C}^n$, then $\mathbf{S}(\mathbf{A}|\mathcal{S}) <^{\mathbf{L}} \mathbf{S}(\mathbf{B}|\mathcal{S})$. We give a statistical interpretation and a statistical proof of this theorem due to Sengupta, [Sengupta (2009)]. The above result is equivalent to the following:

Theorem 15.4.2. *Let* $\mathbf{A}$ *and* $\mathbf{B}$ *be two nnd matrices of the same order such that* $\mathbf{A} <^{\mathbf{L}} \mathbf{B}$. *Write* $\mathbf{A} = \begin{pmatrix} \mathbf{A}_{11} & \mathbf{A}_{12} \\ \mathbf{A}_{21} & \mathbf{A}_{22} \end{pmatrix}$ *and* $\mathbf{B} = \begin{pmatrix} \mathbf{B}_{11} & \mathbf{B}_{12} \\ \mathbf{B}_{21} & \mathbf{B}_{22} \end{pmatrix}$, *where* $\mathbf{A}_{ij}$ *and* $\mathbf{B}_{ij}$ *are of the same order for all* $i,\ j = 1, 2$ *and* $\mathbf{A}_{11}$ *is of order* $r \times r$, $(r < n)$. *Then* $\mathbf{A}_{11} - \mathbf{A}_{12}\mathbf{A}_{22}^{-}\mathbf{A}_{21} <^{\mathbf{L}} \mathbf{B}_{11} - \mathbf{B}_{12}\mathbf{B}_{22}^{-}\mathbf{B}_{21}$.

Proof. *(Statistical).* Let

$$\begin{pmatrix} \mathbf{Y}_1 \\ \mathbf{Y}_2 \end{pmatrix} \sim \mathbf{N}_n \left(0, \begin{pmatrix} \mathbf{B}_{11} & \mathbf{B}_{12} \\ \mathbf{B}_{21} & \mathbf{B}_{22} \end{pmatrix}\right)$$

and

$$\begin{pmatrix} \mathbf{Y}_1 \\ \mathbf{Y}_2 \\ \mathbf{Z} \end{pmatrix} \sim \mathbf{N}_{n+s} \left(0, \begin{pmatrix} \mathbf{B} & \mathbf{C} \\ \mathbf{C}^t & \mathbf{I} \end{pmatrix}\right),$$

where $s = \rho(\mathbf{B} - \mathbf{A})$ and $(\mathbf{C}, \mathbf{C}^t)$ is a rank factorization of $\mathbf{B} - \mathbf{A}$. Then $D(\mathbf{Y}_1|\mathbf{Y}_2) = \mathbf{B}_{11} - \mathbf{B}_{12}\mathbf{B}_{22}^{-}\mathbf{B}_{21}$ and $D(\mathbf{Y}_1|\mathbf{Y}_2, \mathbf{Z}) = \mathbf{A}_{11} - \mathbf{A}_{12}\mathbf{A}_{22}^{-}\mathbf{A}_{21}$. Since $D(\mathbf{Y}_1|\mathbf{Y}_2, \mathbf{Z}) <^{\mathbf{L}} \mathbf{D}(\mathbf{Y}_1|\mathbf{Y}_2)$, the result follows. □

Consider a tribal population on which several anthropometric measurements are made. Let $\mathbf{X}_1$ be the vector of measurements on the face and $\mathbf{X}_2$ be the vector of measurements on the remaining part of the body. Let $\begin{pmatrix} \mathbf{X}_1 \\ \mathbf{X}_2 \end{pmatrix}$ have distributions $\mathbf{N}_n(\mu, \mathbf{\Sigma})$ in population 1 and $\mathbf{N}_\mathbf{n}(\nu, \mathbf{\Lambda})$ in population 2. Suppose the random vector $\begin{pmatrix} \mathbf{X}_1 \\ \mathbf{X}_2 \end{pmatrix}$ has smaller dispersion in population 1 than in population 2, i.e. $\mathbf{\Sigma} <^{\mathbf{L}} \mathbf{\Lambda}$. then by Theorem 15.4.2, we have the following: the conditional dispersion of facial measurements given the measurements of the rest of the body, namely $D(\mathbf{X}_1|\mathbf{X}_2)$ is also smaller in population 1 than in population 2.

We now give another application of the shorted matrix in the recovery of inter-block information in the incomplete block design experiments.

Let

$$\mathbf{Ax} = \mathbf{a} \tag{15.4.1}$$

Let

and

$$\mathbf{Bx} = \mathbf{b} \tag{15.4.2}$$

be the normal equations for deriving intra-block and inter-block estimators respectively and

$$(\mathbf{A} + \mathbf{B})\mathbf{x} = \mathbf{a} + \mathbf{b} \tag{15.4.3}$$

be the normal equations for deriving combined intra-inter-block estimators. Clearly, each of the equations (15.4.1)-(15.4.3) is consistent. Also, $\mathbf{A}$, $\mathbf{B}$ and $\mathbf{A}+\mathbf{B}$ are *nnd* matrices. The linear function $\mathbf{p^t x}$ is unique for all solutions of (15.4.1) if and only if $\mathbf{p} \in \mathcal{C}(\mathbf{A^t})$. In the recovery of inter-block information it is of interest to identify those linear functions $\mathbf{p^t x}$ with $\mathbf{p} \in \mathcal{C}(\mathbf{A^t})$ for which the substitution of a solution of (15.4.1) or (15.4.3) leads to identical answers. We exhibit a solution to this problem whenever $S(\mathbf{A}|\mathbf{B}) = \mathbf{C}$ exists.

Theorem 15.4.3. *Let* $\mathbf{A}$ *and* $\mathbf{B}$ *be nnd matrices of the same order such that the shorted matrix* $S(\mathbf{A}|\mathbf{B}) = \mathbf{C}$ *exists. Let* $\mathbf{p} \in \mathcal{C}(\mathbf{A})$. *Then* $\mathbf{p^t}(\mathbf{A}+\mathbf{B})^-(\mathbf{a}+\mathbf{b}) = \mathbf{p^t}(\mathbf{A})^-\mathbf{a}$ *for all* $\mathbf{a} \in \mathcal{C}(\mathbf{A})$ *and* $\mathbf{b} \in \mathcal{C}(\mathbf{B})$ *if and only if* $\mathbf{p} \in \mathcal{C}(\mathbf{A}-\mathbf{C})$.

Proof. 'If' part
First note that $\mathcal{C}(\mathbf{A}-\mathbf{C}) \subseteq \mathcal{C}(\mathbf{A})$. Let $\mathbf{x_0}$ satisfy (15.4.1). Since $\rho(\mathbf{A}+\mathbf{B}) = \rho(\mathbf{A}-\mathbf{C}) + \rho(\mathbf{B}+\mathbf{C})$, (see proof of Theorem 11.2.11) and $\mathbf{b}-\mathbf{Bx_0} \in \mathcal{C}(\mathbf{B}) = \mathcal{C}(\mathbf{B}+\mathbf{C})$, we have $(\mathbf{A}-\mathbf{C})(\mathbf{A}+\mathbf{B})^-(\mathbf{b}-\mathbf{Bx_0}) = \mathbf{0}$. Therefore,

$$\begin{aligned}
&(\mathbf{A}-\mathbf{C})(\mathbf{A}+\mathbf{B})^-(\mathbf{a}+\mathbf{b})\\
&\quad = (\mathbf{A}-\mathbf{C})(\mathbf{A}-\mathbf{C}+\mathbf{C}+\mathbf{B})^-((\mathbf{A}+\mathbf{B})\mathbf{x_0}+\mathbf{b}-\mathbf{Bx_0})\\
&\quad = (\mathbf{A}-\mathbf{C})(\mathbf{B}+\mathbf{C}+\mathbf{A}-\mathbf{C})^-(\mathbf{A}+\mathbf{B})\mathbf{x_0}\\
&\quad = (\mathbf{A}-\mathbf{C})(\mathbf{B}+\mathbf{A})^-(\mathbf{A}+\mathbf{B})\mathbf{x_0}, \textbf{ since } \mathcal{C}(\mathbf{A}-\mathbf{C}) \subseteq \mathcal{C}(\mathbf{A}+\mathbf{B})\\
&\quad = (\mathbf{A}-\mathbf{C})\mathbf{x_0}\\
&\quad = (\mathbf{A}-\mathbf{C})\mathbf{A}^-\mathbf{a}.
\end{aligned}$$

Hence, $\mathbf{p^t}(\mathbf{A}+\mathbf{B})^-(\mathbf{a}+\mathbf{b}) = \mathbf{p^t}(\mathbf{A})^-\mathbf{a}$ for all $\mathbf{p} \in \mathcal{C}(\mathbf{A}-\mathbf{C})$.
'Only if' part
Now,
$\mathbf{p^t}(\mathbf{A}+\mathbf{B})^-(\mathbf{a}+\mathbf{b}) = \mathbf{p^t A^- a}$ for all $\mathbf{a} \in \mathcal{C}(\mathbf{A})$ and for all $\mathbf{b} \in \mathcal{C}(\mathbf{B})$
$\Rightarrow \mathbf{p^t}(\mathbf{A}+\mathbf{B})^-\mathbf{b} = \mathbf{0}$ for all $\mathbf{b} \in \mathcal{C}(\mathbf{B})$
$\Rightarrow \mathbf{p^t}(\mathbf{A}+\mathbf{B})^-\mathbf{B} = \mathbf{0}$.

Since $\mathbf{p} \in \mathcal{C}(\mathbf{A})$, so, $\mathbf{p} = \mathbf{Ax}$ for some $\mathbf{x}$. Now, $\mathbf{p^t}(\mathbf{A}+\mathbf{B})^-\mathbf{B} = \mathbf{0} \Rightarrow \mathbf{x^t}\mathcal{P}(\mathbf{A}|\mathbf{B}) = \mathbf{0}$. However, as $\mathcal{C}(\mathcal{P}(\mathbf{A}|\mathbf{B})) = \mathcal{C}(\mathbf{S}(\mathbf{A}|\mathbf{B})) = \mathcal{C}(\mathbf{C})$ and $\mathbf{A}^- \in \{\mathbf{C}^-\}$, we have $\mathbf{x^t C} = \mathbf{0}$. This implies $\mathbf{x^t} = \theta^{\mathbf{t}}(\mathbf{I} - \mathbf{CA}^-)$ for some θ. Hence $\mathbf{p} = \mathbf{Ax} = \mathbf{A}(\mathbf{I} - \mathbf{A}^-\mathbf{C})\theta = (\mathbf{A} - \mathbf{C})\theta$ for some θ. □

15.5 Application of parallel sum and shorted operator to testing in linear models

Consider the linear model $(\mathbf{y}, \mathbf{X}\beta, \sigma^2\mathbf{I})$ and let $\mathbf{y} \sim \mathbf{N_n}(\mathbf{X}\beta, \sigma^2\mathbf{I})$. Consider the linear hypothesis $H_0 : \mathbf{A}\beta = \xi$, that is assumed to be consistent i.e., $\xi \in \mathcal{C}(\mathbf{A})$. As shown in the Proposition 5.3.6 of [Sengupta and Jammalamadaka (2003)], the testable part of the hypothesis is $H_{01} : \mathbf{TA}\beta = \mathbf{T}\xi$, where $\mathbf{T}$ is an arbitrary matrix such that $\mathcal{C}(\mathbf{A^tT^t}) = \mathcal{C}(\mathbf{A^t}) \cap \mathcal{C}(\mathbf{X^t})$. Recall that $\mathcal{C}(\mathcal{P}(\mathbf{X^tX}|\mathbf{A^tA})) = \mathcal{C}(\mathbf{A^t}) \cap \mathcal{C}(\mathbf{X^t})$. Also, for the shorted operator $\mathbf{S}(\mathbf{X^tX}|\mathcal{C}(\mathbf{A^t}))$, we have $\mathcal{C}(\mathbf{S}(\mathbf{X^tX}|\mathcal{C}(\mathbf{A^t}))) = \mathcal{C}(\mathbf{A^t}) \cap \mathcal{C}(\mathbf{X^t})$. Choose $\mathbf{T} = (\mathcal{P}(\mathbf{X^tX}|\mathbf{A^tA}))\mathbf{A}^-$ or $(S(\mathbf{X^tX}|\mathcal{C}(\mathbf{A^t})))\mathbf{A}^-$, where $\mathbf{A}^-$ is any g-inverse of $\mathbf{A}$, then $\mathcal{C}(\mathbf{A^tT^t}) = \mathcal{C}(\mathbf{A^t}) \cap \mathcal{C}(\mathbf{X^t})$. If $\mathbf{T} = (\mathcal{P}(\mathbf{X^tX}|\mathbf{A^tA}))\mathbf{A}^-$, then $\mathbf{TA} = (\mathcal{P}(\mathbf{X^tX}|\mathbf{A^tA}))\mathbf{A}^-\mathbf{A} = \mathcal{P}(\mathbf{X^tX}|\mathbf{A^tA})$ and if $\mathbf{T} = (\mathbf{S}(\mathbf{X^tX}|\mathcal{C}(\mathbf{A^t})))\mathbf{A}^-$, then $\mathbf{TA} = \mathbf{S}(\mathbf{X^tX}|\mathcal{C}(\mathbf{A^t}))$. Note that $\mathbf{TA}$ is symmetric in both the cases. Either way $D(\mathbf{TA}\hat{\beta}) = \sigma^2\mathbf{TA}(\mathbf{X^tX})^-$. Further, if $\mathbf{TA} = \mathbf{S}(\mathbf{X^tX}|\mathcal{C}(\mathbf{A^t}))$, then $D(\mathbf{TA}\hat{\beta})\mathbf{TA} = \sigma^2\mathbf{TA}$. Henceforth, we assume $\mathbf{TA} = \mathbf{S}(\mathbf{X^tX}|\mathcal{C}(\mathbf{A^t}))$. Clearly, $\dfrac{(\mathbf{X^tX})^-}{\sigma^2}$ is a g-inverse of $D(\mathbf{TA}\hat{\beta})$. Further, if $\mathbf{u} = \mathbf{TA}\hat{\beta} - \mathbf{g}$, where $\mathbf{g} = \mathbf{Th}$, then the variance ratio F-Test for testing the hypothesis H_{01} is $\mathbf{u^t}\dfrac{(\mathbf{X^tX})^-}{\sigma^2}\mathbf{u}$, which follows χ^2_v, where $\mathbf{v} = \rho(\mathbf{TA}) = \mathbf{tr}((\mathbf{X^tX})^-)\mathbf{TA}$.

15.6 Shorted operator adjustment for modification of network or mechanism

Let us consider a reciprocal resistive electrical network with n ports with impedance matrix $\mathbf{A}$. Let $(i_1,\ i_2,\ \ldots\ i_n)$ be a permutation of $(1,\ 2, \ldots\ n)$. Suppose that $i^{th}_{r+1}, \ldots,\ i^{th}_n$ ports of the network are shorted. If $\mathcal{S}$ denotes the subspace spanned by $e_{i_1}, \ldots, e_{i_r}$, where e_{i_j} is the i^{th}_j column of the identity matrix. Then $\mathbf{S}(\mathbf{A}|\mathcal{S})$ is the shorted operator corresponding to this shorting mechanism. We wish to study the shorted operator when the shorting mechanism undergoes some modifications.

In this section we consider the following modifications to the shorting mechanism:

(a) In addition to the ports shorted earlier, one more port is shorted.
(b) Amongst the ports shorted earlier, we choose one and undo the shorting operation on this port.

We also consider a modification of the network by adding a new port. We obtain the modified shorted operator corresponding to a suitably chosen shorting mechanism. In fact, we obtain slightly more general results from which the results for all the above modifications follow easily.

Let $\mathbf{A}$ be an *nnd* matrix of order $n \times n$. Let $\mathcal{S}$ be a subspace of $\mathbb{C}^n$ and the shorted matrix $\mathbf{S}(\mathbf{A}|\mathcal{S})$ be available. Let $\mathbf{x}$ and $\mathbf{y}$ be vectors such that $\mathbf{x} \notin \mathcal{S}$ and $\mathbf{y} \in \mathcal{S}$. Let $\mathcal{L}_u$ denote the subspace spanned by the vector $\mathbf{u}$. We obtain $\mathbf{S}(\mathbf{A}|\mathcal{S} + \mathcal{L}_x)$ and $\mathbf{S}(\mathbf{A}|\mathcal{S} \cap \mathcal{L}_y^{\perp})$. Again, let $\mathbf{B} = \begin{pmatrix} \mathbf{A} & \mathbf{x} \\ \mathbf{x}^t & c \end{pmatrix}$, where $\mathbf{B}$ is *nnd*. Let $\mathbf{u_1}, \mathbf{u_2}, \ldots \mathbf{u_r}$ form an ortho-normal basis of $\mathcal{S}$. Let $d_1, d_2, \ldots d_r$ be given real numbers and $\mathcal{S}_*$ be the subspace of $\mathbb{C}^{n+1}$ generated by $\begin{pmatrix} \mathbf{u_1} \\ d_1 \end{pmatrix}, \begin{pmatrix} \mathbf{u_2} \\ d_2 \end{pmatrix} \ldots \begin{pmatrix} \mathbf{u_r} \\ d_r \end{pmatrix}$. We obtain $\mathbf{S}(\mathbf{B}|\mathcal{S}_*)$. Before we can embark upon doing this we need the following:

Lemma 15.6.1. *Let* $\mathbf{B} = \begin{pmatrix} \mathbf{A} & \mathbf{x} \\ \mathbf{x}^\star & c \end{pmatrix}$ *be a hermitian matrix and* $\mathbf{G} = \begin{pmatrix} \mathbf{H} & \mathbf{y} \\ \mathbf{y}^\star & d \end{pmatrix}$ *be a hermitian g-inverse of* $\mathbf{B}$. *Then the following hold:*

(i) $\mathbf{x} \notin \mathcal{C}(\mathbf{A})$ *if and only if* $\mathbf{A}\mathbf{y} = \mathbf{0}$, $\mathbf{y}^\star\mathbf{x} = \mathbf{1}$ *and* $d = 0$.
(ii) $\mathbf{H}$ *is a g-inverse of* $\mathbf{A}$ *if* $\mathbf{x} \notin \mathcal{C}(\mathbf{A})$.
(iii) *Let* $\theta = 1 - \mathbf{y}^\star\mathbf{x} - cd \neq \mathbf{0}$. *then*

$$\mathbf{T} = \mathbf{H} + \frac{1}{\theta}(\mathbf{H}\mathbf{x} + c\mathbf{y})\mathbf{y}^\star + \frac{1}{\theta}\mathbf{y}(\mathbf{x}^\star\mathbf{H} + c\mathbf{y}^\star) + \frac{d}{\theta^2}(\mathbf{H}\mathbf{x} + c\mathbf{y})(\mathbf{H}\mathbf{x} + c\mathbf{y})^\star$$

is g-inverse of $\mathbf{A}$.
(iv) *Let* $\mathbf{y}^\star\mathbf{x} + cd = 1$, $d \neq \mathbf{0}$ *and* $\mathbf{A}\mathbf{y} + d\mathbf{x} = \mathbf{0}$. *Then* $\mathbf{H} - \frac{1}{d}\mathbf{y}\mathbf{y}^\star$ *is g-inverse of* $\mathbf{A}$.
(v) *Let* $\mathbf{y}^\star\mathbf{x} + cd = 1$, $d \neq \mathbf{0}$ *and* $t = \|\mathbf{A}\mathbf{y} + d\mathbf{x}\|^2 \neq \mathbf{0}$. *Write* $\xi = \frac{1}{t}(\mathbf{I} - \mathbf{H}\mathbf{A} - \mathbf{y}\mathbf{x}^\star)(\mathbf{A}\mathbf{y} + d\mathbf{x})$. *Then* $\mathbf{R} = \mathbf{H} + \xi\mathbf{y}^\star + \mathbf{y}\xi^\star + d\xi\xi^\star$ *is a g-inverse of* $\mathbf{A}$.
(vi) *Let* $\mathbf{y}^\star\mathbf{x} = 1$, $d = \mathbf{0}$ *and* $\mathbf{A}\mathbf{y} \neq \mathbf{0}$. *Write* $t = \|\mathbf{A}\mathbf{y}\|^2$ *and* $\xi = \frac{1}{t}(\mathbf{I} - \mathbf{H}\mathbf{A} - \mathbf{y}\mathbf{x}^\star)\mathbf{A}\mathbf{y}$. *Then* $\mathbf{R} = \mathbf{H} + \xi\mathbf{y}^\star + \mathbf{y}\xi^\star$, *is a g-inverse*

of $\mathbf{A}$.

For a proof see [Bhimasankaram (1988a)].

Before we can actually start giving the shorted operators, we fix a setup to be used subsequently.

$\mathbf{S}^{\maltese}$: Let $\mathbf{A}$ be an *nnd* matrix of order $n \times n$. Let $\mathcal{S}$ be a subspace of $\mathbb{C}^n$ and $\mathbf{q_1}, \mathbf{q_2}, \ldots, \mathbf{q_{n-r}}$ form an orthonormal basis of $\mathcal{S}^{\perp}$. Write $\mathbf{Q} = (\mathbf{q_1}, \mathbf{q_2}, \ldots, \mathbf{q_{n-r}})$. Let $\mathbf{Q}$, $\mathbf{AQ}$; $\mathbf{Q^{\star}AQ}$ and $(\mathbf{Q^{\star}AQ})^{-}$ be available. Clearly, $\mathbf{S}(\mathbf{A}|\mathcal{S}) = \mathbf{A} - \mathbf{AQ}(\mathbf{Q^{\star}AQ})^{-}\mathbf{Q^{\star}A}$.

Let $\mathbf{x} \notin \mathcal{S}$. To obtain $\mathbf{S}(\mathbf{A}|\mathcal{S} + \mathcal{L}_x)$, we proceed as follows:

Algorithm 15.6.2.
Step 1: Compute $\mathbf{u} = \mathbf{P}_{\mathcal{S}^{\perp}}\mathbf{x} = \mathbf{QQ^{\star}x}$.
Step 2: Extend $\mathbf{u}$ to an orthonormal basis $\mathbf{u}, \mathbf{p_1}, \mathbf{p_2}, \ldots, \mathbf{p_{n-r-1}}$ of $\mathcal{S}^{\perp}$. (This can be achieved by performing Gram Schmidt orthogonalization process on $\mathbf{u}, \mathbf{q_1}, \mathbf{q_2}, \ldots, \mathbf{q_{n-r}}$.) Let $\mathbf{P_1} = (\mathbf{p_1}, \mathbf{p_2}, \ldots, \mathbf{p_{n-r-1}})$ and $\mathbf{P} = (\mathbf{P_1}, \mathbf{u})$.

Note that $\mathbf{P} = \mathbf{QQ^{\star}P}$ and $\mathbf{Q^{\star}P}$ is unitary. Moreover,

$$\mathbf{S}(\mathbf{A}|\mathcal{S}) = \mathbf{A} - \mathbf{AP}(\mathbf{P^{\star}AP})^{-}\mathbf{P^{\star}A}.$$

Step 3: Compute $(\mathbf{P^{\star}AP})^{-} = \mathbf{P^{\star}Q}(\mathbf{Q^{\star}AQ})^{-}\mathbf{Q^{\star}P}$.
Step 4: Compute $\mathbf{AP_1} = \mathbf{AQQ^{\star}P_1}$ and $(\mathbf{P_1^{\star}AP_1})^{-}$ using Lemma 15.6.1.
Step 5: Compute $\mathbf{S}(\mathbf{A}|\mathcal{S} + \mathcal{L}_x) = \mathbf{A} - \mathbf{AP_1}(\mathbf{P_1^{\star}AP_1})^{-}\mathbf{P_1^{\star}A}$.

Remark 15.6.3. Notice that $\mathcal{C}(\mathbf{P_1}) = (\mathcal{S} + \mathcal{L}_{\mathbf{x}})^{\perp}$. For, $\mathbf{P_1y} = \mathbf{0}$, whenever $\mathbf{y} \in \mathcal{S}$ as $\mathcal{C}(\mathbf{P_1}) \subseteq \mathcal{S}^{\perp}$. Further, $\mathbf{P_1x} = \mathbf{P_1}(\mathbf{P}_{\mathcal{S}}\mathbf{x} + \mathbf{u})$. Again $\mathbf{P_1P}_{\mathcal{S}}\mathbf{x} = \mathbf{0}$, since $\mathbf{P}_{\mathcal{S}}\mathbf{x} \in \mathcal{S}$ and $\mathbf{P_1u} = \mathbf{0}$ by construction. Hence, $\mathbf{P_1x} = \mathbf{0}$. Thus, $\mathcal{C}(\mathbf{P_1}) \subseteq (\mathcal{S} + \mathcal{L}_{\mathbf{x}})^{\perp}$. Now $d(\mathcal{S} + \mathcal{L}_x) = r + 1$ and $\rho(\mathbf{P_1}) = n - r - 1$. So, the result follows.

Let us consider the setup $\mathbf{S}^{\maltese}$ fixed before Algorithm 15.6.2. Let $\mathbf{y} \in \mathcal{S}$. Then it is quite easy to obtain $\mathbf{S}(\mathbf{A}|\mathcal{S} \cap \mathcal{L}_y^{\perp})$. So, we have

Theorem 15.6.4. *If* $\mathbf{T} = \mathbf{S}(\mathbf{A}|\mathcal{S})$, *then* $\mathbf{S}(\mathbf{A}|\mathcal{S} \cap \mathcal{L}_y^{\perp}) = \mathbf{T} - \dfrac{1}{\mathbf{y^{\star}Ty}}\mathbf{Tyy^{\star}T}$.

Proof follows from the fact that $\mathbf{S}(\mathbf{A}|\mathcal{S} \cap \mathcal{T}) = \mathbf{S}(\mathcal{S}(\mathbf{A}|\mathcal{S})|\mathcal{T})$.

Consider a reciprocal resistive electrical network with n ports having impedance matrix $\mathbf{A}$. Let $i_1^{th}, i_2^{th}, \ldots, i_{n-r}^{th}$ ports of the network be shorted. Let $\mathcal{S}$ be space generated by the vectors $\mathbf{e_{i_{n-r+1}}}, \ldots, \mathbf{e_{i_n}}$, where $(i_1, i_2, \ldots i_n)$ is a permutation of $(1, 2, \ldots n)$ and $\mathbf{e_j}$ denotes the jth column of the identity matrix. Let $\mathbf{S}(\mathbf{A}|\mathcal{S})$ be the corresponding shorted operator. Suppose we want to unshort the port $\mathbf{e_{i_j}}$, $1 \leq j \leq n-r$. Then the resultant shorted operator is given by $\mathbf{S}(\mathbf{A}|\mathcal{S}+\mathcal{L}_{e_{i_j}})$ and can be computed using the Algorithm 15.6.2. (In this case the Steps 1 and 2 are not necessary since $\mathbf{e_{i_j}} \in \mathcal{S}^{\perp}$ and is part of the orthonormal basis $\mathbf{e_{i_1}}, \ldots, \mathbf{e_{i_{n-r}}}$.)

Consider the same setup as before. Suppose we want to short one more port say the i_j^{th}, $n-r < j \leq n$ in addition to the already shorted ports $i_1, i_2, \ldots, i_{n-r}$. The resulting shorted operator is given by $\mathbf{S}(\mathbf{A}|\mathcal{S} \cap \mathcal{L}_{e_{i_j}}^{\perp})$ and can be easily obtained by using Theorem 15.6.4.

Once again let us consider the setup $\mathbf{S}^{\maltese}$ fixed before Algorithm 15.6.2. Let $\mathbf{B} = \begin{pmatrix} \mathbf{A} & \mathbf{x} \\ \mathbf{x^t} & c \end{pmatrix}$, where $\mathbf{x} \in \mathcal{C}(\mathbf{A})$ and $c - \mathbf{x^t A^- x} \geq \mathbf{0}$, so that $\mathbf{B}$ is *nnd*. Let $\mathbf{w_1}, \mathbf{w_2}, \ldots \mathbf{w_r}$ form an orthonormal basis of $\mathcal{S}$. Let $d_1, d_2, \ldots d_r$ be given complex numbers and $\mathcal{S}_*$ be the subspace of $\mathbb{C}^{n+1}$ generated by $\begin{pmatrix} \mathbf{w_1} \\ d_1 \end{pmatrix}, \begin{pmatrix} \mathbf{w_2} \\ d_2 \end{pmatrix} \cdots \begin{pmatrix} \mathbf{w_r} \\ d_r \end{pmatrix}$. In order to obtain $\mathbf{S}(\mathbf{B}|\mathcal{S}_*)$ we proceed as follows:

Algorithm 15.6.5.

Step 1: Compute $\mathbf{v} = -\frac{\sum_{\mathrm{i=1}}^{\mathrm{r}} d_{\mathrm{i}} \mathbf{w_i}}{\sum_{\mathrm{i=1}}^{\mathrm{r}} d_{\mathrm{i}}^2}$, if $\sum d_i^2 \neq 0$ and set $\mathbf{v} = \mathbf{0}$, if $\sum d_i^2 = 0$.

Form $\mathbf{T} = \begin{pmatrix} \mathbf{Q} & \mathbf{v} \\ \mathbf{0} & 1 \end{pmatrix}$ and note that $\mathbf{T^\star T} = \mathbf{I}$ and $\mathcal{C}(\mathbf{T}) = \mathcal{S}_*^{\perp}$.

Step 2: Compute $\mathbf{T^\star B T} = \begin{pmatrix} \mathbf{Q^\star A Q} & \mathbf{Q^\star y} \\ \mathbf{y^\star Q} & \mathbf{v^\star y + x^\star v + c} \end{pmatrix}$, where $\mathbf{y} = \mathbf{Av} + \mathbf{x}$.

Step 3: Compute

$$(\mathbf{T^\star B T})^- = \begin{pmatrix} (\mathbf{Q^\star A Q})^- & \mathbf{0} \\ \mathbf{0} & 0 \end{pmatrix} + \bar{\mathbf{h}} \begin{pmatrix} (\mathbf{Q^\star A Q})^- \mathbf{Q^\star y} \\ -1 \end{pmatrix} (\mathbf{y^\star Q}(\mathbf{Q^\star A Q})^-, \ -\mathbf{1}),$$

$$\text{where } \bar{h} = \begin{cases} 0, & \text{if } h = \mathbf{v^\star y + x^\star v + c - y^\star Q (Q^\star A Q)^- Q^\star y = 0} \\ \frac{1}{h}, & \text{if } h \neq 0 \end{cases}$$

Step 4: Compute $\mathbf{S}(\mathbf{B}|\mathcal{S}_*) = \mathbf{B} - \mathbf{BT}(\mathbf{T^\star BT})^- \mathbf{T^\star B}$.

Remark 15.6.6. If $h = 0$, then $\mathbf{S}(\mathbf{B}|\mathcal{S}_*) = \begin{pmatrix} \mathbf{S}(\mathbf{A}|\mathcal{S}) & \mathbf{S}(\mathbf{A}|\mathcal{S})\xi \\ \xi^\star \mathbf{S}(\mathbf{A}|\mathcal{S}) & c - \xi^\star \mathbf{S}(\mathbf{A}|\mathcal{S})\xi \end{pmatrix}$, where $\mathbf{x} = \mathbf{A}\xi$. Since $\mathbf{x} \in \mathcal{C}(\mathbf{A})$, there exists a ξ such that $\mathbf{x} = \mathbf{A}\xi$.

Remark 15.6.7. Notice that $h = 0$ if and only if $\rho(\mathbf{T}^\star\mathbf{B}\mathbf{T}) = \rho(\mathbf{Q}^\star\mathbf{A}\mathbf{Q})$.

Remark 15.6.8. If $\mathbf{y} = \mathbf{A}\mathbf{v} + \mathbf{x} \in \mathcal{S}$, then $\mathbf{Q}^\star\mathbf{y} = \mathbf{0}$ and $\mathbf{S}(\mathbf{B}|\mathcal{S}_*) = \begin{pmatrix} \mathbf{S}(\mathbf{A}|\mathcal{S}) & \mathbf{S}(\mathbf{A}|\mathcal{S})\xi \\ \xi^\star\mathbf{S}(\mathbf{A}|\mathcal{S}) & c - \bar{h}(\mathbf{x}^\star\mathbf{v} + \mathbf{c})^2 \end{pmatrix}$, where ξ is as in Remark 15.6.6.

Consider the setup $\mathbf{S}^{\maltese}$ as described before Algorithm 15.6.2. Suppose one more port is added to the network and let the new impedance matrix be $\mathbf{B} = \begin{pmatrix} \mathbf{A} & \mathbf{x} \\ \mathbf{x}^\star & c \end{pmatrix}$. Then the shorted operator corresponding to the shorting of $i_1^{th}, \ldots, i_r^{th}$ ports is given by $\mathbf{S}(\mathbf{B}|\mathcal{S}_*)$, where $\mathcal{S}_*$ is the space generated by the vectors $e_{i_{r+1}}, \ldots, e_{i_n}$. The vectors $e_{i_1}, \ldots, e_{i_r}, e_{n+1}$ form an orthonormal basis of $\mathcal{S}_*^\perp$. (Here $d_1, \ldots, d_r$ are all zero.) The rest follows immediately.

Chapter 16

Some Open Problems

The open problems in any area of research can always be found with some effort but when they are recorded in a readily available format in some accessible source along with the progress made in attempting a solution they become an asset. In this chapter, we enlist some problems that we could not resolve while writing this monograph. We also provide some background information relating to the problems:- how these problems arose and what little is known to us about them.

16.1 Simultaneous diagonalization

Simultaneous diagonalization of various types played a major role in our exposition on the partial orders. [Eckart and Young (1939)] showed that, given two matrices $\mathbf{A_1}$ and $\mathbf{A_2}$ of order $m \times n$ over the complex field $\mathbb{C}$, there exist unitary matrices $\mathbf{U}$ and $\mathbf{V}$ of orders $m \times m$ and $n \times n$ respectively such that $\mathbf{A_i} = \mathbf{UD_iV^\star}$ for $i = 1, 2$, where $\mathbf{D_1}$ and $\mathbf{D_2}$ are real diagonal matrices if and only if $\mathbf{A_1A_2^\star}$ and $\mathbf{A_1^\star A_2}$ are hermitian matrices.

In Chapter 2, we extended this result for a finite number (more than 2) of matrices and provided necessary and sufficient conditions for simultaneous singular value decomposition of several matrices in Theorem 2.7.19. This lead us to the following question:

Problem 16.1.1. *Let $\mathbf{A_1}$ and $\mathbf{A_2}$ be matrices of order $m \times n$ over the complex field $\mathbb{C}$. Obtain necessary and sufficient conditions for the existence of unitary matrices $\mathbf{U}$ and $\mathbf{V}$ of orders $m \times m$ and $n \times n$ respectively such that $\mathbf{A_i} = \mathbf{UD_iV^\star}$ for $i = 1, 2$, where $\mathbf{D_1}$ and $\mathbf{D_2}$ are diagonal matrices, not necessarily real.*

16.2 Matrices below a given matrix under sharp order

Let $\mathbf{A}$ and $\mathbf{B}$ be square matrices of index ≤ 1 over an algebraically closed field F. Further, let $\mathbf{B} = \mathbf{P}\mathbf{diag}(\mathbf{J_1}, \ldots, \mathbf{J_r}, \mathbf{0})\mathbf{P^{-1}}$ be the Jordan form of $\mathbf{B}$, where $\mathbf{J_i}$ for $i = 1, \ldots, r$ are non-singular Jordan blocks and $\mathbf{A} = \mathbf{P}\mathbf{diag}(\mathbf{D_1}, \ldots, \mathbf{D_r}, \mathbf{0})\mathbf{P^{-1}}$, where $\mathbf{P}$ is a non-singular matrix, $\mathbf{D_{i_j}} = \mathbf{J_{i_j}}$, $j = 1, \ldots, s$ for some sub-permutation $(i_1, \ldots, i_s)$ of $\{1, \ldots, r\}$ and $\mathbf{D_t} = \mathbf{0}$ for $t \in \{1, \ldots, r\} \cap \{i_1, \ldots, i_s\}^c$. In Chapter 4, we observed that $\mathbf{A} <^{\#} \mathbf{B}$. We proved the following converse in special case in Theorem 4.3.13:

Let $\mathbf{A}$ and $\mathbf{B}$ are square matrices of index ≤ 1 over an algebraically closed field F and $\mathbf{B}$ has the Jordan decomposition mentioned above with the further condition that the geometric multiplicity of each of its non-null eigen-values is 1. Then $\mathbf{A} <^{\#} \mathbf{B}$ implies that $\mathbf{A}$ must be of the form mentioned above.

Problem 16.2.1. *Let* $\mathbf{B}$ *be a square matrix of index* ≤ 1 *over an algebraically closed field* F*. Characterize the class of all matrices* $\mathbf{A}$ *of index* ≤ 1 *such that* $\mathbf{A} <^{\#} \mathbf{B}$ *in terms of Jordan form of the matrix* $\mathbf{B}$.

16.3 Partial order combining the minus and sharp orders

Let $\mathbf{A}$ and $\mathbf{B}$ be square matrices of the same order. Let $\mathbf{A} = \mathbf{A_1} + \mathbf{A_2}$ and $\mathbf{B} = \mathbf{B_1} + \mathbf{B_2}$ be the core-nilpotent decompositions of $\mathbf{A}$ and $\mathbf{B}$ respectively, where $\mathbf{A_1}$ is core part of $\mathbf{A}$, $\mathbf{B_1}$ is core part of $\mathbf{B}$, $\mathbf{A_2}$ is nilpotent part of $\mathbf{A}$ and $\mathbf{B_1}$ is nilpotent part of $\mathbf{B}$. In chapter 4, using the sharp order on the core parts and the minus order on the nilpotent parts, we defined $(\#, -)$ order, written as $\mathbf{A} <^{\#,-} \mathbf{B}$ if $\mathbf{A_1} <^{\#} \mathbf{B_1}$ and $\mathbf{A_2} <^{-} \mathbf{B_2}$. This order was introduced by Mitra and Hartwig [Mitra and Hartwig (1992a)] under the name C-N order. We had shown that this defines a partial order that implies the minus order (Corollary 4.4.19 and Theorem 4.4.20). Now, suppose $\mathbf{A} <^{-} \mathbf{B}$ and $\mathbf{A_1} <^{\#} \mathbf{B_1}$. Clearly, this too defines a partial order. When does this relation imply the $(\#, -)$ order? In other words when $\mathbf{A} <^{-} \mathbf{B}$ and $\mathbf{A_1} <^{\#} \mathbf{B_1}$ implies $\mathbf{A} <^{\#,-} \mathbf{B}$? This leads to the following:

Problem 16.3.1. *Let* $\mathbf{A}$ *and* $\mathbf{B}$ *be square matrices of the same order. Let* $\mathbf{A} = \mathbf{A_1} + \mathbf{A_2}$ *and* $\mathbf{B} = \mathbf{B_1} + \mathbf{B_2}$ *be the core-nilpotent decompositions of* $\mathbf{A}$ *and* $\mathbf{B}$ *respectively. What are necessary and sufficient conditions under which* $\mathbf{A} <^{-} \mathbf{B}$ *and* $\mathbf{A_1} <^{\#} \mathbf{B_1}$ *implies* $\mathbf{A} <^{\#,-} \mathbf{B}$?

Let $\mathfrak{C}$ denote the class of square matrices whose core part is range-hermitian. Let the setup be that of Problem 16.3.1. Define $\mathbf{A} <^{\#,\star} \mathbf{B}$ if $\mathbf{A_1} <^{\#} \mathbf{B_1}$ and $\mathbf{A_2} <^{\star} \mathbf{B_2}$. It is easy to see that this is a partial order on the class $\mathfrak{C}$ which implies the star order. Now, let $\mathbf{A}$ and $\mathbf{B} \in \mathfrak{C}$ such that $\mathbf{A} <^{\star} \mathbf{B}$ and $\mathbf{A_1} <^{\#} \mathbf{B_1}$. It is clear that this too defines a partial order on $\mathfrak{C}$. When does $\mathbf{A} <^{\star} \mathbf{B}$ and $\mathbf{A_1} <^{\#} \mathbf{B_1}$ together imply $\mathbf{A} <^{\#,\star} \mathbf{B}$. So, we have

Problem 16.3.2. *Let* $\mathbf{A}$ *and* $\mathbf{B}$ *be square matrices of the same order. Let* $\mathbf{A} = \mathbf{A_1} + \mathbf{A_2}$ *and* $\mathbf{B} = \mathbf{B_1} + \mathbf{B_2}$ *be the core-nilpotent decompositions of* $\mathbf{A}$ *and* $\mathbf{B}$ *respectively. What are necessary and sufficient conditions for under which* $\mathbf{A} <^{\star} \mathbf{B}$ *and* $\mathbf{A_1} <^{\#} \mathbf{B_1}$ *implies* $\mathbf{A} <^{\#,\star} \mathbf{B}$?

16.4 When is a $\mathcal{G}$-based order relation a partial order?

In Chapter 7, we obtained some sufficient conditions for a $\mathcal{G}$-based order to be a partial order on its support (Theorem 7.2.13). We also obtained a necessary and sufficient condition in a special case (Corollary 7.2.33). The question that still remains to be answered is 'When is a $\mathcal{G}$-based based order a partial order?' This prompts us to include the following:

Problem 16.4.1. *Obtain a necessary and sufficient condition such that a* $\mathcal{G}$*-base order relation is a partial order on its support.*

Problem 16.4.2. *Obtain a necessary and sufficient condition such that a* $\mathcal{G}$*-base order relation defines a partial order on the class of all matrices possibly rectangular over a field* F.

16.5 Parallel sum and g-inverses

Let $\mathbf{A}$ and $\mathbf{B}$ be parallel summable matrices over $\mathbb{C}$, the field of complex numbers. We have seen that Theorem 9.2.20 gives a nice expression for the Moore-Penrose inverse of the parallel sum of $\mathbf{A}$ and $\mathbf{B}$ in terms of g-inverses of $\mathbf{A}$ and $\mathbf{B}$ and the orthogonal projectors onto the column and row spaces of the parallel sum. Even when the matrices $\mathbf{A}$ and $\mathbf{B}$ are over a general field, one can construct a reflexive g-inverse of the parallel sum in terms of g-inverses of $\mathbf{A}$ and $\mathbf{B}$ and oblique projectors onto the row and column spaces of the parallel sum (Remark 9.2.21). Can we generate all reflexive g-inverses of the parallel sum by varying over the g-inverses and projectors involved in the above construction?

Problem 16.5.1. *Let* $\mathbf{A}$ *and* $\mathbf{B}$ *be parallel summable. Let* $\mathbf{G}$ *be a g-inverse of* $\mathcal{P}(\mathbf{A}|\mathbf{B})$. *Write* $\mathbf{P_G} = \mathcal{P}(\mathbf{A}|\mathbf{B})\mathbf{G}$ *and* $\mathbf{Q_G} = \mathbf{G}\mathcal{P}(\mathbf{A}|\mathbf{B})$. *Does the class of matrices*

$$\{\mathbf{Q_G}(\mathbf{A}^- + \mathbf{B}^-)\mathbf{P_G} : \mathbf{G} \in \{\mathcal{P}(\mathbf{A}|\mathbf{B})^-\}, \mathbf{A}^- \in \{\mathbf{A}^-\} \text{ and } \mathbf{B}^- \in \{\mathbf{B}^-\}\}$$

exhaust the class of all reflexive g-inverses of $\mathcal{P}(\mathbf{A}|\mathbf{B})$?

Suppose $\mathbf{A}$ and $\mathbf{B}$ be are parallel summable. What about the parallel summability of their g-inverses? This leads us to the following

Problem 16.5.2. *Let* $\mathbf{A}$ *and* $\mathbf{B}$ *be parallel summable.*

(a) *Do there exist g-inverses* $\mathbf{A}^-$ *and* $\mathbf{B}^-$ *which are parallel summable?*
(b) *Obtain necessary and sufficient conditions under which the Moore-Penrose inverses of* $\mathbf{A}$ *and* $\mathbf{B}$ *are parallel summable, assuming that* $\mathbf{A}$ *and* $\mathbf{B}$ *are over the field of complex numbers.*
(c) *If* $\mathbf{A}$ *and* $\mathbf{B}$ *are square matrices of index not exceeding 1, is* $\mathcal{P}(\mathbf{A}|\mathbf{B})$, *their parallel sum of index not exceeding* 1 *too?*
(d) *Let* $\mathbf{A}$ *and* $\mathbf{B}$ *be square matrices and of index not exceeding* 1. *Obtain necessary and sufficient conditions under which the group inverses of* $\mathbf{A}$ *and* $\mathbf{B}$ *are parallel summable.*

16.6 Shorted operator and a maximization problem

Problem 16.6.1. *Let* $\mathbf{A}$ *be an nnd matrix of order* $n \times n$ *and* $\mathcal{S}$ *be a subspace of* $\mathbb{C}^n$. *We have seen in Theorem 10.3.11 that the shorted operator* $\mathbf{S}(\mathbf{A}|\mathcal{S})$ *is the maximal element of* $\{\mathbf{D} : \mathbf{D} <^- \mathbf{A} \text{ and } \mathcal{C}(\mathbf{D}) \subseteq \mathcal{S}\}$. *What happens if we maximize over all matrices* $\mathbf{D}$ *such that* $\mathbf{D} <^\star \mathbf{A}$?
This leads us to the following:

Problem 16.6.2. *Let* $\mathfrak{T} = \{\mathbf{D} : \mathbf{D} <^\star \mathbf{A} \text{ and } \mathcal{C}(\mathbf{D}) \subseteq \mathcal{S}\}$. *We know* $\max\{\mathfrak{T}\}$ *exists. When does this maximum coincide with* $\mathbf{S}(\mathbf{A}|\mathcal{S})$? *(Notice that if* $\mathbf{A}$ *and* $\mathbf{D}$ *are idempotent, it is certainly the case.) More generally, let* $\mathbf{A}$ *be an* $m \times n$ *matrix over* $\mathbb{C}$, *the field of complex numbers. When* $\mathbf{S}_\star(\mathbf{A}|\mathcal{S}, \mathcal{T}) = \mathbf{S}(\mathbf{A}|\mathcal{S}, \mathcal{T})$?

16.7 The ladder problem

We had seen in Chapter 14 that Theorems 14.6.4 and 14.2.4 characterize all possible steps of a ladder, if we choose to restrict to an equivalence class. What if the g-inverses cut across the equivalence classes? So, we have

Problem 16.7.1. *Consider the setup of Theorem 14.6.4 except that the g-inverses* $\mathbf{G_i}$ *can cut across the equivalence classes. Obtain a characterization of all* $\mathbf{A_i}$ *such that* $(\mathbf{A_i},\ \mathbf{G_i})$*, varying over* i *form the part of a ladder from the ground level to the first floor.*

Appendix A

Relations and Partial Orders

A.1 Introduction

This monograph presumes basic knowledge in relations and orders on a set. The object of this appendix is to provide the definitions and simple results on these topics aided by several illustrative examples. We hope that this will help the reader who is familiar with these concepts but needs a quick brush up. Readers wishing to go for more details can consult standard text books, such as [Vagner (1952)] and Classical Algebra by P.M. Cohn.

A.2 Relations

The relations we consider in Mathematics are an abstraction of the relations we see every day in our life. If the objects are represented by x and y, we can write an ordered pair as (x, y) or as (y, x). In general the ordered pairs (x, y) and (y, x) are different. In fact if (x, y) and (w, z) are two ordered pairs, then

$$(x, y) = (w, z) \text{ if and only if } x = w \text{ and } y = z.$$

Definition A.2.1. Let $\mathfrak{A}$ and $\mathfrak{B}$ be two sets. Then the set $\mathfrak{A} \times \mathfrak{B}$, known as Cartesian product of $\mathfrak{A}$ and $\mathfrak{B}$ is the set

$$\{(x, y) : x \in \mathfrak{A} \text{ and } y \in \mathfrak{B}\}.$$

Definition A.2.2. A relation from a set $\mathfrak{A}$ to a set $\mathfrak{B}$ is a subset of $\mathfrak{A} \times \mathfrak{B}$.

Let $\mathbf{R}$ be relation from a set $\mathfrak{A}$ to a set $\mathfrak{B}$. If an ordered pair $(x, y) \in \mathbf{R}$, then we say x is related to y under the relation $\mathbf{R}$ and denote it as $x\mathbf{R}y$.

A relation from a set $\mathfrak{A}$ to itself is called a relation in the set $\mathfrak{A}$ or on the set $\mathfrak{A}$.

Definition A.2.3. A relation $\mathbf{R}$ in a set $\mathfrak{A}$ is called

(i) **Reflexive** if $x\mathbf{R}x$ (or $(x,x) \in \mathbf{R}$), for each $x \in \mathfrak{A}$.
(ii) **Anti-symmetric** if for $x, y \in \mathfrak{A}$, $x\mathbf{R}y$ and $y\mathbf{R}x$, then $x = y$.
(iii) **Symmetric** if for $x, y \in \mathfrak{A}$, $x\mathbf{R}y$, then $y\mathbf{R}x$.
(iv) **Transitive** if for x, y and $z \in \mathfrak{A}$, $x\mathbf{R}y$ and $y\mathbf{R}z$, then $x\mathbf{R}z$.

Example A.2.4. Let $\mathfrak{A} = \{x_1, x_2, \ldots, x_n\}$ be any finite set. Then

(a) $\mathbf{R} = \{(x_1, x_1), (x_2, x_2) \ldots, (x_n, x_n)\}$ is a reflexive relation on $\mathfrak{A}$. What are the other properties $\mathbf{R}$ enjoys?
(b) $\mathbf{R} = \{(x_1, x_1), (x_1, x_2), (x_2, x_2)\}$ is an anti-symmetric relation.
(c) $\mathbf{R} = \{(x_1, x_1), (x_1, x_2), (x_2, x_1), (x_2, x_2)\}$ is a symmetric relation.

Example A.2.5. Let $\mathfrak{A} = \mathbb{N}$, the set of natural numbers. Let $\mathbf{R}$ be the usual 'less than' relation on $\mathfrak{A}$ Clearly, this is an anti-symmetric as well as a transitive relation. What about reflexivity?

Definition A.2.6. A relation $\mathbf{R}$ on a set $\mathfrak{A}$ is called an **equivalence relation** if it is a reflexive, symmetric and transitive relation.

Example A.2.7. The relation $\mathfrak{A} \times \mathfrak{A}$ is an equivalence relation on $\mathfrak{A}$. The relation in Example A.2.4(a) is also an equivalence relation. What about $\emptyset$, the null relation?

Definition A.2.8. A relation $\mathbf{R}$ on a set $\mathfrak{A}$ is called a **partial order** if it is a reflexive, anti-symmetric and transitive relation.

Example A.2.9. The well known relation '$A \subseteq B$' on the set $\wp(U)$, of all subsets of the set U is clearly a partial order on the set U.

Example A.2.10. Let M denote the set of all finite subsets of the real numbers having the same number of elements arranged in (say) decreasing order. We define a relation on M as follows:

Let $X = \{x_1, x_2, \ldots, x_n\}$ and $Y = \{y_1, y_2, \ldots, y_n\}$ be any two elements of M. We say Y majorizes X, if $\sum_{i=1}^{j} x_i = \sum_{i=1}^{j} y_i$ for each $j = 1, 2, \ldots (n-1)$ and $\sum_{i=1}^{n} x_i = \sum_{i=1}^{n} y_i$.

When Y majorizes X, we write $X \preceq Y$.

Clearly, '$\preceq$' is a partial order on M.

A set equipped with a partial order is also referred to as a **Poset**. Whenever a relation $\mathbf{R}$ is a partial order on a set $\mathfrak{A}$, it is a standard practice to write **R** as '$\prec$' and we also adopt it. Thus, $(\mathfrak{A}, \prec)$ is a set $\mathfrak{A}$ with a partial order '$\prec$' on it.

Definition A.2.11. A relation **R** on a set $\mathfrak{A}$ is called an **pre-order** if it is a reflexive and transitive relation.

Thus, every partial order and also every equivalence relation is a pre-order.

Remark A.2.12. A subset $\mathfrak{T}$ of a relation **R** on a set $\mathfrak{A}$ may or may not inherit a property that **R** possesses. For any subset $\mathfrak{B}$ of a partially ordered set $\mathfrak{A}$, if the partial order on $\mathfrak{A}$ is also a partial order on $\mathfrak{B}$, then it shall be referred to as partial order on $\mathfrak{B}$.

Definition A.2.13. Let $(\mathfrak{A}, \prec)$ be a Poset. An element $a \in \mathfrak{A}$ is called a maximal element of $\mathfrak{A}$ with respect to the partial order $\prec$ if there exists no element $x_0 \in \mathfrak{A}$ other than a such that $a \prec x_0$, i.e. if there exists an element $x_0 \in \mathfrak{A}$ such that $a \prec x_0$, then $x_0 = a$.

An element $a \in \mathfrak{A}$ is called greatest element of $\mathfrak{A}$ if for each $x \in \mathfrak{A}$, $x \prec a$.

Similarly, an element $a \in \mathfrak{A}$ is called a minimal element of $\mathfrak{A}$ with respect to the partial order $\prec$ if there exists no element $x_0 \in \mathfrak{A}$ other than a such that $x_0 \prec a$.

An element $a \in \mathfrak{A}$ is called least element of $\mathfrak{A}$ if for each $x \in \mathfrak{A}$, $a \prec x$.

Definition A.2.14. Let $\mathfrak{B}$ be a subset of a poset $(\mathfrak{A}, \prec)$. An element $a \in \mathfrak{A}$ is called an upper bound of $\mathfrak{B}$ if for all $x \in \mathfrak{B}$, $x \prec a$. The least element amongst all the upper bounds of $\mathfrak{B}$ is called Supremum of $\mathfrak{B}$. Whenever, Supremum of a subset $\mathfrak{B}$ exists we write $\sup\{\mathfrak{B}\}$. If $a,\ b \in \mathfrak{A}$, we denote the the Supremum of $a,\ b$ as $a \vee b$ and read it as 'a join b'.

One can similarly define Infimum of a subset in analogous manner. For $a,\ b \in \mathfrak{A}$, we denote the the Infimum of $a,\ b$ as $a \wedge b$ and read it as 'a meet b'.

Definition A.2.15. A Poset $(\mathfrak{A}, \prec)$ is said to form a lattice if for each pair of elements $a,\ b \in \mathfrak{A}$, Supremum of $a,\ b$ and Infimum of $a,\ b$ exist in $\mathfrak{A}$.

A Poset $(\mathfrak{A}, \prec)$ is called join semi-lattice if Supremum of $a,\ b$ exists for each pair of elements $a,\ b \in \mathfrak{A}$. Analogously we have a meet semi-lattice.

Definition A.2.16. Let the set $\mathfrak{A}$ be equipped with two partial orders $\prec_1$ and $\prec_2$. Then "$\prec_1$" is said to be finer than "$\prec_2$" if for $a, b \in \mathfrak{A}$, $a \prec_1 b \Longrightarrow a \prec_2 b$.

A.3 Semi-groups and groups

We briefly introduce the concepts of semi-group and group.

Definition A.3.1. A **binary operation** $*$ on a set $\mathfrak{S}$ is a function with domain $\mathfrak{S} \times \mathfrak{S}$ and codomain $\mathfrak{S}$.

Example A.3.2. Consider the set $\mathbb{Z}$ of integers. The operation of addition as well as multiplication are binary operations on $\mathbb{Z}$, but the operation of division is not a binary operation on $\mathbb{Z}$.

Definition A.3.3. Let $\mathfrak{S}$ be a set with at least one element equipped with a binary operation $*$ is called a semi-group if
(S1.) '$*$' is associative, i.e. $\forall\ a, b, c \in \mathfrak{S}$, $a * (b * c) = (a * b) * c$.

We usually denote the binary operation multiplicatively.

If there exists an element $e \in \mathfrak{S}$ such that $\forall\ a \in \mathfrak{S}$, $ae = ea = a$, then $\mathfrak{S}$ is said to have an **identity** e. Whenever a semi-group has an identity, it is unique and is called the identity of $\mathfrak{S}$. An element 0 satisfying $a0 = 0a = 0, \forall\ a \in \mathfrak{S}$ is called a **Zero** element of $\mathfrak{S}$. If a semi-group has a Zero element, it is unique and is called as the **Zero** of $\mathfrak{S}$.

We generally write 1 for the identity of a semi-group and 0 for the zero of the semi-group, if it exists.

Definition A.3.4. An element a of a semi-group $\mathfrak{S}$ is called an idempotent element (or simply an idempotent) if $a^2 = a$ and denote by $E(\mathfrak{S})$, the set of all idempotent elements of $\mathfrak{S}$.

Definition A.3.5. In a semi-group $\mathfrak{S}$ with zero, 0, a non-zero element $a \in \mathfrak{S}$ is called a nilpotent if there is a positive integer $n \in \mathbb{N}$ such that $a^n = 0$ and the smallest such integer is called the index of nilpotency of a. It is clear that the index of nilpotency of any element is ≥ 2.

Definition A.3.6. Let $\mathfrak{S}$ be a semi-group. An element $a \in \mathfrak{S}$ is called regular if $a = axa$ for some $x \in \mathfrak{S}$.

The element x in Linear Algebra terminology is called g-inverse of the element a. Moreover, if $a \in \mathfrak{S}$ is regular with $a = axa$ for some $x \in \mathfrak{S}$, then the two elements ax and xa are idempotent elements of $\mathfrak{S}$. If we let $y = xax$, then y satisfies $aya = a$ and $yay = y$. Thus y is also regular.

Definition A.3.7. Let $\mathfrak{S}$ be a semi-group. An element $b \in \mathfrak{S}$ is called r-inverse of an element a if $a = aba, b = bab$

Thus, every regular element in a semi-group has an r-inverse. Further if b is an r-inverse of a, then a is an r-inverse of b. The notion 'r-inverse' of an element in Linear Algebra terminology is the same as reflexive g-inverse of the element.

Definition A.3.8. A semi-group $\mathfrak{S}$ is called a regular semi-group, if each of its elements is regular.

Definition A.3.9. A semi-group $\mathfrak{S}$ is called an inverse semi-group if each of its elements has a unique r-inverse.

Definition A.3.10. A semi-group $\mathfrak{S}$ with identity 1 is called unit regular, if $\forall\ a \in \mathfrak{S}$, there exists an invertible element $u \in \mathfrak{S}$ such that $aua = a$.

Theorem A.3.11. (i) A semi-group $\mathfrak{S}$ is an inverse semi-group if and only if $\mathfrak{S}$ is regular and for each $a, b \in E(\mathfrak{S}), ab = ba$ if and only if $\mathfrak{S}$ is regular and $E(\mathfrak{S})$ is a sub semi-group.

(ii) For each pair of elements a, b of an inverse semi-group $\mathfrak{S}$, let g, h and k respectively denote the r-inverses of a, b and ab, the $k = hg$.

Definition A.3.12. A semi-group G is called a group if

G2. there exists an element $e \in G$ such that for each $a \in G, ae = a$.
G3. For each $a \in G$, there exists $b \in G$ such that $ab = e$.

Equivalently,

G2$'$. If there exists an element $f \in G$ such that for each $a \in G, fa = a$.
G3$'$. For each $a \in G$, there exists a $b \in G$ such that $ba = f$.

Equivalently,

G2$''$. If there exists an element $e \in G$ such that for each $a \in G, ea = a = ae$.
G3$''$. For each $a \in G$, there exists a $b \in G$ such that $ba = e = ab$.

A.4 Semi-groups and partial orders

Definition A.4.1. A partial order on a semi-group $\mathfrak{S}$ is called a natural partial order if it is defined by means of multiplication of $\mathfrak{S}$.

For any semi-group $\mathfrak{S}$, the relation on $E(\mathfrak{S})$ defined by

$$e \leq f \text{ if } e = ef = fe, \ \forall e, \ f \in E(\mathfrak{S}) \tag{A.1}$$

is a partial order. This partial order on $E(\mathfrak{S})$ is known as the natural (or canonical) ordering of idempotent elements of a semi-group $\mathfrak{S}$.

The relation '$\leq$' defined on any inverse semi-group $\mathfrak{S}$ as

$$a \leq b \text{ if } a = eb \text{ for some } e \in E(\mathfrak{S}) \text{ and } a, \ b \in \mathfrak{S} \tag{A.2}$$

is also an example of a natural partial order. We call this partial order the Vagner order.

Theorem A.4.2. *Let $\mathfrak{S}$ be an inverse semi-group equipped with Vagner order [Vagner (1952)]. Suppose $a, b \in \mathfrak{S}$. Let g and h be the r-inverses of a and b respectively. Then the following are equivalent:*

(i) $a \leq b$
(ii) $a \in \{be : e \in E(\mathfrak{S})\} = bE(\mathfrak{S})$
(iii) $ag = ah$
(iv) $ga = ha$ *and*
(v) $a = aha$.

Moreover, if $a \leq b$ and $c \in \mathfrak{S}$, then $ac \leq bc$, $ca \leq cb$ and $g \leq h$.

Another example of a natural partial order on a regular semi-group defined in 1980 independently by Hartwig and Nambooripad, is the relation

$$a \leq b \text{ if and only if } a = eb = bf \text{ for some } e, \ f \in E(\mathfrak{S}). \tag{A.3}$$

This relation coincides with relation (A.1) on $E(\mathfrak{S})$. For an inverse semi-group the relation (A.3) coincides with the partial order (A.2).

Theorem A.4.3. *For a regular semi-group $\mathfrak{S}$, the following are equivalent:*

(i) $a = eb = bf$ *for some* $e, f \in E(\mathfrak{S})$
(ii) $a = aa'b = ba''a$ *for some* $a', a'' \in V(a = \{x \in \mathfrak{S} : a = axa, x = xax\}$
(iii) $a = aa'b = ba'a$ *for some* $a'' \in V(a)$
(iv) $a'a = a'b, aa' = ba'$ *for some* $a' \in V(a)$
(v) $a = ab^{\star}b = bb^{\star}a = ab^{\star}a$ *for some* $b^{\star} \in V(b)$
(vi) $a = axb = bxa, a = axa, b = bxb$ *for some* $x \in \mathfrak{S}$
(vii) $a = eb$ *for some idempotent* $e \in a\mathfrak{S}$ *and* $a\mathfrak{S} \subseteq b\mathfrak{S}$
(viii) *For each idempotent* $f \in b\mathfrak{S}$, *there exists an idempotent* $e \in a\mathfrak{S}$ *such that* $e \leq f$ *and* $a = eb$
(ix) $a = ab'a$ *for some* $b' \in V(b)$, $a\mathfrak{S} \subseteq b\mathfrak{S}$
(x) $a = xb = by, xa = a$ *for some* $x, y \in \mathfrak{S}$ *and*
(xi) $a = eb = bx$ *for some* $e \in E(\mathfrak{S})$, $x \in \mathfrak{S}$.

A.5 Involution

Definition A.5.1. An involution 'star' denoted as $\star$ on a semi-group $\mathfrak{S}$ is a map $a \to a^\star$ of $\mathfrak{S}$ into itself, satisfying

(i) $a^{\star\star} = a$ and
(ii) $(ab)^\star = a^\star b^\star$.

Obviously an involution is a bijective map on $\mathfrak{S}$.

Definition A.5.2. An involution $\star$ on a semi-group $\mathfrak{S}$ is called proper if $a^\star a = a^\star b = b^\star a = b^\star b \Rightarrow a = b$.

Definition A.5.3. A proper $\star$-semi-group is a semi-group with $\star$ as a specified proper involution on it.

Notice that if the semi-group has the zero element '0', then an involution $\star$- is proper if and only if $a^\star a = 0 \Rightarrow a = 0$. (star cancelation law)

Example A.5.4. Any inverse semi-group with $\star$-defined as $a^\star = g$, the unique pseudo inverse of a, ia a proper $\star$-semi-group.

Example A.5.5. The set $\mathbb{C}^{m \times m}$ of all the $m \times m$ matrices with involution $\star$- as the conjugate transpose i.e. $A^\star = (A^-)'$ is a proper $\star$-semi-group. More generally, the ring $\mathfrak{B}(H)$ of all bounded linear operators on the complex Hilbert space H with with involution $\star$ as $T^\star$ = adjoint of T, is a proper $\star$-semi-group.

Definition A.5.6. In a $\star$-semi-group $\mathfrak{S}$, an element $a \in \mathfrak{S}$ is called a $\star$-regular if $aa^\star$ and $aa^\star$ are both regular. Equivalently, $a \in \mathfrak{S}$ is $\star$-regular if and only if there exists a solution (that is necessarily unique) to the equations: $axa = a, xax = x, (ax)^\star = ax, (xa)^\star = xa$.

A.6 Compatibility of partial orders with algebraic operations

Definition A.6.1. Let $(\mathfrak{S}, \leq)$ be a partially ordered set. If $\mathfrak{S}$ is a semi-group, the we say the order relation '$\leq$' is compatible from right or left with semi-group structure of $\mathfrak{S}$, if for any $a, b, c \in \mathfrak{S}$, $a \leq b \Rightarrow ac \leq bc$ or $ca \leq cb$ respectively.

Example A.6.2. Vagner's natural partial order is compatible on both sides with multiplication.

When a semi-group has a partial order compatible with its semi-group structure, we call the semi-group a partially ordered semi-group.

A.7 Partial orders induced by convex cones

Definition A.7.1. Let $\mathcal{C}$ be a non-empty subset of a real vector space V (e.g. $\mathbb{R}^n$). $\mathcal{C}$ is called a convex cone if for each $x, y \in \mathcal{C}$, $\lambda_1, \lambda_2 \geq 0$, $\lambda_1 x + \lambda_2 y \in \mathcal{C}$.

It is clear that $0 \in \mathcal{C}$.

Definition A.7.2. The cone ordering on any subset W of a real vector space V induced by $\mathcal{C}$ is a relation on W defined as follows: For $x, y \in W$, $x \leq y \Longleftrightarrow y - x \in \mathcal{C}$.

Notice that the cone ordering defines a pre-order on W. Moreover, if W is itself a convex cone, then the relation $\leq$ also satisfies the following additional relations: If $x \leq y$, then for all $z \in W$, $x + z \leq y + z$ and for all $\lambda \geq 0$, $\lambda x \leq \lambda y$. Conversely, any pre-order on a subset W of V that satisfies both the above properties is a cone ordering induced by the cone $\mathcal{C} = \{x : x \geq 0\}$.

A.8 Creating new partial orders from old partial orders

Given a set, it may have several partial orders on it. We can use these to induce new partial orders on the set. We include here two techniques.

Definition A.8.1. If $<_k$, $k \in I$ are partial orders on a set S, then the relation $a \ll b$ if $a <_k b$, for each $k \in I$; where I is an index set, defines a partial order on S.

This partial order is called as the intersection of $<_k$, $k \in I$.

Definition A.8.2. Let S be a set and $<_1$, $<_2$ be two partial orders on S. Suppose $<_2$ is finer than $<_1$ and P is a property satisfied by elements of S. Further, if $a, b \in S$ are such that $a <_2 b$ and the element a does not satisfy

P, then so does not b. We can define a new relation $\ll$ on S as follows: For $a, b \in S$, $a \ll b$ if $a <_1 b$ if the element a satisfies P and $a <_2 b$ otherwise.

Clearly, $<$ is a partial order on S. This partial order is called a combination of two partial orders $<_1$, and $<_2$.

Bibliography

Albert A. (1969) Conditions for positive and nonnegative definiteness in terms of pseudo inverses, SIAM J. Appl. Math. 17: 434-440.

Ander W.N. (Jr.) Duffin R.J. (1963) Series and parallel addition of matrices. J. Math. Anal. Appl. 11: 576-574.

Anderson W.N.(Jr.) (1971) Shorted operators I, SIAM J. Appl. Math. 20:520-525.

Anderson, W.N.(Jr.) and Trapp, G.E. (1975) Shorted operators. II, SIAM J. Appl. Math. 28:60-71, (this concept first introduced by Krein [880]).

Ando T. (1979) Generalized Schur complements. Linear Algebra and its Applications 27: 173-186.

Baksalary J.K. (1986) A relationship between the star and minus orderings. Linear Algebra and its applications 127:157-169.

Baksalary J.K. and Hauke J. (1987) Partial orderings of matrices referring to singular values or eigen-values. Linear Algebra and its applications 96:17-26.

Baksalary J.K. and Hauke J. (1990) A Further Algebraic version of Cochran's Theorem and Matrix Partial Orderings. Linear Algebra and its applications 127:157-169.

Baksalary J.K., Hauke J., Liu Xiaoji and Liu Sanyang (2004) Relationships between partial orders of matrices and their powers. Linear Algebra and its applications 379: 277-287.

Baksalary J.K., Kala, R. and Klaczynski, K. (1983). The Matrix inequality $M \geq B^*MM$. Linear Algebra and its applications 54:77-86.

Baksalary J.K. and Mitra S.K. (1989) The Left-star and right-Star Partial Orderings. Technical Report A 220, Department of Mathematical Sciences, University of Tampere, Tampere, Finland.

Baksalary J.K., Nordström Kenneth and Styan George P.H. (1990) Löwner Ordering Antitonicity of generalized Inverses of Hermitian matrices. Linear Algebra and its applications 127:171-182.

Baksalary J.K., Pukelshiem F. and Styan G.P.H. (1989) Some properties of the matrix partial orderings, Linear Algebra and its Applications 96:17-26.

Baksalary J.K. and Pukelshiem F. (1991) On Löwner, minus, and star order of nonnegative definite matrices and their squares, Linear Algebra and its applications 151:135-141.

Ben-Israel A (1969) On matrices of index zero or one. SIAM J. Appl. Math. 17(6):1118-1121.

Ben-Israel A and Greville T.N.E. (2001) *Generalized inverses: Theory and applications*, Springer.

Bhagwat K.V and Subramanian R. Inequalities between means of positive operations, math. Proc. Cambridge Philos. Soc. 393-401.

Bhimasankaram P. (1971a) Some contributions to the theory, applications and computation of generalized inverses of matrices, Doctoral dissertation, Indian Statistical Institute.

Bhimasankaram P. (1971b) On generalized inverses of partitioned matrices, Sankhyã, Series A, 33, 311-314.

Bhimasankaram P. (1988a) On Genralized Inverses of a Block in a Partitioned matrix, Linear Algebra and its applications 109, 131-143.

Bhimasankaram P. (1988b) Rank factorization of a matrix and its applications, Math. Sci. 13, no.1, 4-14.

Bhimasankaram P. and Malik Saroj (2007) Shorted Operators of Partitioned matrices and Applications, Linear Algebra and its applications 425, 150-161.

Bhimasankaram P. and Mathew Thomas. (1993) On ordering properties of Generalized inverse of Nonnegative Definite matrices, Linear Algebra and its applications 183:131-146.

Boullion T.L. and Odell P.L. (1971) *Generalized Inverse Matrices*, John Wiley & Sons, New York.

Butler C.A. and Morley T.D. (1988) A note on the shorted operator, SIAM J. Matrix Anal. Appl. 9, no. 2, 147-155.

Campbell S.L. (1977) Drazin generalized inverses, Linear Algebra and Appl. 18: no. 1, 53-57.

Campbell S.L. (1979/80) Continuity of the Drazin inverse. Linear and Multilinear Algebra 8 no. 3, 265-268.

Campbell S.L. and Meyer C.D.(Jr) (1978) Weak Drazin inverses, Linear Algebra and its Applications 20 no. 2:167-178.

Campbell S.L. and Meyer C.D.(Jr) (1991) *Generalized Inverses of Linear Transformations*, Pitman (Advanced Publishing Program), Boston, Mass., 1979 (reprinted by Dover, 1991).

Carlson D. (1975) Matrix decompositions involving the Schur complement SIAM J. Appl. Math. 28: 577-587.

Carlson D. (1986) What are Schur complements, anyway? Linear Algebra and its Applications 74: 257-275.

Carlson D. and Haynsworth Emilie V. (1983) Complementable and Almost Definite Matrices, Linear Algebra and its Applications 52/53: 157-176.

Drazin M.P. (1958) Pseudo inverse associative rings and semi-groups. Amer. Math. Monthly 65: 506-514.

Drazin M.P. (1978) Natural structure on semi-groups with involution. Bull. Amer. Math. Soc. 84: 139-141.

Eckart C. and Young G. (1939) A principal axis transformation for non Hermitian matrices, Bull. Amer. Math. Soc. 45: 118-121.

Englefield M.H. (1966) The commuting inverses of a square matrix, Proc. Cambridge Philos. Soc. 62: 667-671.

Erdelyi I. (1967) On the matrix equation $\mathbf{Ax} = \lambda\mathbf{Bx}$ J. Math. Anal. Appl. 17:117-232.

Goller H. (1986) Shorted operators and rank decomposition matrices. Linear Algebra and its applications 81: 207-236.

G.H Golub and L.Van Loan *Matrix Computations* (3rd Ed.)(1996), John Hopkin University studies in mathematical sciences.

González N. Castro, Koliha J.J. and Yimin Wei (2000) Perturbation of the Drazin inverse for matrices with equal eigen projections at zero. Linear Algebra and its applications 312:181-189.

Gonzákez N. Castro and Koliha J.J. and Straškraba, (2001) Perturbation of Drazin inverse. Soochow J. Math 27(20): 201-211.

Größ J (1987) A note on a partial ordering in the set of Hermitian matrices, SIAM J. Matrix Anal. Appl. 18, no. 4, 887-892.

Größ J (1997) Some Remarks on partialorderings of Hermitian matrices, Linear and Multilinear Algebra 42, 53-60.

Größ J (2006) Remarks on the sharp partial order and the ordering of squares of matrices, Linear Algebra and its applications 417: 87-93.

Hartwig R.E. (1975) 1-2 inverses and invariance of $\mathbf{BA}^{\dagger}\mathbf{C}$, Linear Algebra and its applications 9: 271-275.

Hartwig R.E. (1976) Block generalized inverses, Arch. Rational Mech. Anal. 61: 197-251.

Hartwig R.E. (1978) A note on partial ordering of Positive Semi-definite Matrices, Linear and Multilinear algebra 6: 223-226.

Hartwig R.E. (1979) Pseudo Lattice properties of the star-orthogonal partial ordering for star-regular rings, Proceedings of the American mathematical society 77(3).

Hartwig R.E. (1980) How to order regular elements. Math Japonica 25: 1-13.

Hartwig R.E. (1981) A note on rank-additivity, Linear and Multilinear algebra. 10: 50-61.

Hartwig R.E. and Drazin M.P. (1982) Lattice properties of the star order for matrices. J. Math. Anal. Appl. 86: 145-161.

Hartwig R.E. and Raphael L. (1992) Maximal elementa under the Three Partial Orders. Linear Algebra and its applications 175: 39-61.

Hartwig R.E. and Spindleböck Klaus (1983). Some Closed Form Formulae for the intersection of two special matrices under the star order, 13:323-331.

Hartwig R.E. and Spindleböck Klaus (1984) Matrices for which $\mathbf{A}^*$ and $\mathbf{A}^{\dagger}$ commute. Linear and Multilinear Algebra 14: 241-256.

Hartwig R.E. and Styan G.P.H. (1986) On some characterizations of the "star" partial ordering and rank subtractivity. Linear Algebra and its applications 82: 145-161.

Hartwig R.E. and Styan G.P.H. (1987) Partially ordered idempotent matrices, in Proceedings of the Second International Tampere Conference in Statistics (Pukkila T. and Puntanen S, Eds.), Department of Mathematical Sciences, Univ. of Tampere, 361-383.

Hauke Jan and Markiewicz (1995) On partial orderings of Rectangular matrices. Linear Algebra and its applications 219: 187-193.

Hestenes M.R. (1961) Relative Hermitian matrices, Pacific J. Math. 11, 225-245.

Horn R.A. and Johnson C.R. (1985) *Matrix analysis*, Cambridge Press.

Horn R.A. and Johnson C.R. (1999, Reprint) *Topics in Matrix analysis*, Cambridge Press.

Jain S.K., Mitra S.K. and Werner H.-J. (1996) Extensions of G-based matrix partial orders, SIAM J. Matrix Anal. Appl. 17 no. 4, 834-850.

Jain S.K., Srivastava Ashish K., Blackwood B. and Prasad K.M.(2009) Shorted Operators Relative to a Partial Order in Regular rings, Communications in Algebra, to appear.

Khatri C.G. (1968) Some results for the singular multivariate regression models, Series A, 30, 267-280.

Lewis T.O. and Newman T.G. (1968) Pseudo inverses of positive semi-definite matrices. SIAM J. Appl. Math. 16: 701-703.

Löwner K. (1934) Über monotone Matrixfunktionen, Math. Z. 38: 177-216.

Marshall Albert W. and Olkin Ingram (1979) *Inequalities: The Theory of Majorization and Its applications*, Academic Press, Orlando, Florida.

Mitch H. (1986) A natural partial order for semi-groups, Proc. Amer. Math. Soc. 97: 384-388.

Mitra S.K. (1968) A new class of g-inverses of square matrices. Sankhyã Ser. A 30: 323-330.

Mitra Sujit Kumar (1982a) Simultaneous Diagonalization of rectangular matrices, Linear Algebra and its applications 47: 139-150.

Mitra Sujit Kumar (1982b) Properties of the Fundamental bordered matrix used in linear estimation, Statistics and Probability: Essays in Honor of C.R. Rao © North Holland Publishing Company 505-509.

Mitra S.K. (1986b) The Minus Partial Order and the Shorted Matrix. Linear Algebra and its applications 83: 1-27.

Mitra S.K. (1987) On group inverses and the sharp order. Linear Algebra and its applications 92: 17-37.

Mitra S.K. (1988) Infimum of a pair of matrices. Linear Algebra and its applications 105: 163-182.

Mitra S.K. (1989) Block Independence in Generalized inverse: A coordinate free look, Statistical Data Analysis and Inference 429-443.

Mitra S.K. (1990) Shorted matrices in star and related orderings, circuits Systems signal Process 9: 197-212.

Mitra S.K. (1991) Matrix Partial Orders through Generalized inverses: Unified Theory. Linear Algebra and its applications 148: 237-263.

Mitra S.K. (1992b) On $\mathcal{G}$-based extensions of the sharp order. Linear and Multilinear algebra 31: 147-151.

Mitra S.K. (1994) Separation Theorems. Linear Algebra and its applications 208/209:239-256.

Mitra S.K. (1999) Diagrammatic presentation of inner and outer inverses: S-diagrams, Linear Algebra and its Applications 287: no. 1-3, 271-288.

Mitra Sujit Kumar and Bhimasankaram P. (1971) Generalized Inverses of Partitioned Matrices and recalculation of least squares estimates for data or model changes, Sankhyã, Series A, 33(4), 395-410.

Mitra S.K. and Hartwig R.E. (1992a) Partial orders based on outer inverses, Linear Algebra and its Applications 176: 3-20.

Mitra S.K. and Puri M.L. (1973) On parallel Sum and Difference of Matrices. Journal of Mathematical Analysis and Applications 44: 92-97.

Mitra S.K. and Puri M.L. (1979) Shorted operators and generalized inverses of matrices, Linear Algebra and its Applications 25: 45-56.

Mitra S.K. and Puri M.L. (1982c) Shorted matrices An extended concept and some applications. Linear Algebra and its applications 42: 57-79.

Mitra S.K. and Puri M.L. (1983) The fundamental Bordered matrix of Linear estimation and Duffin-Morley General Linear Electromechanical systems, Applicable Analysis 14: 214-258.

Mitra S.K., Puntanen S. and Styan G.P.H. (1994) shorted matrices in Linear Statistical Models: A Review Report A 287, Department of Mathematical Sciences, University of Tampere, Tampere, Finland.

Mitra S.K. and Odell P.L. (1986a) On parallel summability of matrices. Linear Algebra and its applications 74: 239-255.

Morley T.M. and William W.L. (1990a) Parallel sums of operators, Proc. Symposia in Pure Mathematics 51: 129-133.

Morley T.M. and William W.L. (1990b) Parallel sums and norm convergence, Circuits, Systems and Signal Processing 9: 213-222.

Nambooripad K.S.S. (1980) The natural partial order on a regular semi-group. Proc. Edinburgh Math. Soc. 23: 249-260.

Pringle R.M. and Ranyer A.A. (1971) *Generalized inverse Matrices and the applications to Statistics*, Griffin, London.

Rao A.R. and Bhimasankaram P. (2000) *Linear Algebra*, Hindustan Book Agency, Delhi, India.

Rao C.R. and Mitra S.K. (1971) *Generalized Inverses of Matrices and Its applications*, John Wiley, New York.

Rao C.R., Mitra S.K. and Bhimasankaram P. (1972) Determination of a matrix by its subclasses of generalized inverses, Sankhya Ser. A 34: 5-8.

Robert P. (1968) On the group inverse of a linear transformation, J. Math. Anal. Appl. 22: 658-669.

Sumbamurthy P. (1987) Characterizations of a matrix by its subclass of g-inverse. Sankhya Ser. A 49: 412-414.

Sengupta Debasis (2009), Personal communication.

Sengupta Debasis and Jammalamadaka Sreenivasa Rao (2003) *Linear Models An Integrated Approach*, World Scientific, Singapore.

Seshu S. and Reed M.B. (1961) *Linear Graphs and Electrical Networks*, Addison-Wesley, Reading Massachusetts.

Vagner V. (1952) Generalized groups, Dokl. Akad. Nauk SSSR, 84: 1119-1122 (Russian).

Index